Studies in Computational Intelligence

Volume 1056

Series Editor

Janusz Kacprzyk, Polish Academy of Sciences, Warsaw, Poland

The series “Studies in Computational Intelligence” (SCI) publishes new developments and advances in the various areas of computational intelligence—quickly and with a high quality. The intent is to cover the theory, applications, and design methods of computational intelligence, as embedded in the fields of engineering, computer science, physics and life sciences, as well as the methodologies behind them. The series contains monographs, lecture notes and edited volumes in computational intelligence spanning the areas of neural networks, connectionist systems, genetic algorithms, evolutionary computation, artificial intelligence, cellular automata, self-organizing systems, soft computing, fuzzy systems, and hybrid intelligent systems. Of particular value to both the contributors and the readership are the short publication timeframe and the world-wide distribution, which enable both wide and rapid dissemination of research output.

Indexed by SCOPUS, DBLP, WTI Frankfurt eG, zbMATH, SCImago.

All books published in the series are submitted for consideration in Web of Science.

Muhammad Alshurideh · Barween Hikmat Al Kurdi · Ra'ed Masa'deh · Haitham M. Alzoubi · Said Salloum
Editors

The Effect of Information Technology on Business and Marketing Intelligence Systems

Volume 2

Springer

Editors
Muhammad Alshurideh
Department of Management, College of Business Administration
University of Sharjah
Sharjah, United Arab Emirates

Department of Marketing, School of Business
The University of Jordan
Amman, Jordan

Barween Hikmat Al Kurdi
Department of Marketing, Faculty of Economics and Administrative Sciences
The Hashemite University
Zarqa, Jordan

Haitham M. Alzoubi
Skyline University College
Sharjah, United Arab Emirates

Ra'ed Masa'deh
Management Information Systems Department, School of Business
University of Jordan
Aqaba, Jordan

Said Salloum
School of Comupting, Science and Engineering
University of Salford
Salford, England

ISSN 1860-949X ISSN 1860-9503 (electronic)
Studies in Computational Intelligence
ISBN 978-3-031-12381-8 ISBN 978-3-031-12382-5 (eBook)
https://doi.org/10.1007/978-3-031-12382-5

This Springer imprint is published by the registered company Springer Nature Switzerland AG
The registered company address is: Gewerbestrasse 11, 6330 Cham, Switzerland

Contents

Business and Data Analytics

Innovation, Entrepreneurship and Leadership

Knowledge Management

Corporate Governance and Performance

The Impact of the Quality of Medical Information Systems on Job Performance in Private Hospitals in Jordan

Nida'a Al-Husban, Sulieman Ibraheem Shelash Al-Hawary, Doa'a Ahmad Odeh Al-Husban, Riad Ahmad Mohammed Abazeed, Bayan Anwar Al-Azzam, Ibrahim Rashed Soliaman AlTaweel, Mohammad Fathi Almaaitah, Ayat Mohammad, and Muhammad Turki Alshurideh

Abstract The study aims to identify the impact the quality of medical information systems on the job performance in Private Hospitals in Jordan. The study population consisted of all employees with job titles (doctor, nurse, pharmacist, laboratory technician) in private hospitals in Jordan. Data were primarily gathered through self-reported questionnaires creating by Google Forms which were distributed to a purposive sample via email. A self-reported questionnaire that consists of two main sections along with a section regarding control variables was used as the measurement instrument. Statistical program AMOSv24 was used to test the study hypotheses. The results showed that the highest impact was for ease of use, followed by ease of learning, then flexibility, then reliability, and finally the lowest impact was for responsiveness. Based on the results of the study, researchers recommend managers

N. Al-Husban · S. I. S. Al-Hawary (✉) · B. A. Al-Azzam
Faculty of Economics and Administrative Sciences, Department of Business Administration, Al Al-Bayt University, P.O. Box 130040, Mafraq 25113, Jordan
e-mail: dr_sliman73@aabu.edu.jo

D. A. O. Al-Husban
Faculty of Alia College, Department of Human Fundamental Sciences, Al-Balqa Applied University, Amman, Jordan
e-mail: D_husban@bau.edu.jo

R. A. M. Abazeed
Faculty of Finance and Business Administration, Business Management and Public Administration, Department of Business Administration, Al Al-Bayt University, P.O. Box 130040, Mafraq 25113, Jordan

I. R. S. AlTaweel
Faculty of Business School Al Russ City, Department of Business Administration, Qassim University, P.O. Box 6502, Al Russ 51452, Saudi Arabia
e-mail: toiel@qu.edu.sa

M. F. Almaaitah
Faculty of Economic and Administration Sciences, Department of Business Administration & Public Administration, Al Al-Bayt University Jordan, P.O. Box 130040, Mafraq 25113, Jordan
e-mail: m.maaitah@aabu.edu.jo

M. Alshurideh et al. (eds.), *The Effect of Information Technology on Business and Marketing Intelligence Systems*, Studies in Computational Intelligence 1056,
https://doi.org/10.1007/978-3-031-12382-5_45

and decision-makers in private hospitals in Jordan to hold training courses in the field of using medical information systems for all employees in private hospitals, and to focus on maintaining these courses, and raising their levels in line with continuous technological developments and changes.

Keywords Quality · Medical information systems · Job performance Private Hospitals · Jordan

1 Introduction

The phenomenon of globalization has emerged, which has entered the pyramid of the new world system, which calls for a shift from the local market to the global market, and from the concept of time to the concept of synchronization, where the person in his place has become running his business and marketing his product in another place (Aldaihani & Ali, 2018; Allahow et al., 2018; Mohammad et al., 2013). Hence, commercial markets and administrative work have shifted from multiple markets to a single market, and from the unified management to the system of multiple advanced technological managements that work as one body to carry out the knowledge and cultural exchange of information under an important concept (business), this concept plays a major role in the success of any project whatsoever (Al-Hawary & Hadad, 2016; Al-Hawary et al., 2010).

The business system faces many challenges, problems and obstacles that forced it to have a clear strategic approach in light of its managerial work towards achieving its desired goals and interests, and then ensuring that this approach remains safe and away from the complexity, entanglement and challenges (Aldaihani et al., 2020; Al-Hawary & Al-Hamwan, 2017; Al-Hawary & Ismael, 2010). Hence, information systems have an important role, as this system represents (the heart in relation to the body) if it is disturbed, the whole body is disturbed, and if it is safe, it helps in the success of the managerial system (Altamony et al., 2012; Alzoubi et al., 2021a, 2021b; Svoboda et al., 2021).

Modern organizations focus their efforts to continuously improve the level of their performance and the performance of their employees and allocate a lot of their budgets to find ways to raise performance and achieve high levels of productivity (Alshura et al., 2016). The concept of performance is related to the behavior of the

A. Mohammad
Business and Finance Faculty, The World Islamic Science and Education University (WISE), P.O. Box 1101, Amman 11947, Jordan

M. T. Alshurideh
Department of Marketing, School of Business, The University of Jordan, Amman 11942, Jordan
e-mail: malshurideh@sharjah.ac.ae; m.alshurideh@ju.edu.jo

Department of Management, College of Business, University of Sharjah, 27272 Sharjah, United Arab Emirates

individual and the organization, and it occupies a special place within any organization as the final product of the outcome of all its activities (Al-Hawary & Shdefat, 2016; Al-Lozi et al., 2017; Al-Lozi et al., 2018). Human behavior is the determinant of job performance, which is the outcome of the interaction between the nature of the individual, his upbringing and the situation in which he is located, and performance does not appear as a result of forces or pressures stemming from within the individual himself, but as a result of the process of interaction and compatibility between the internal forces of the individual and the external forces surrounding him (Al-Nady et al., 2013; Altarifi et al., 2015; Al-Hawary & Mohammed, 2017; Alshurideh et al., 2022).

Information systems and cognitive technology have developed in recent times, and it has become necessary to have a contemporary and advanced medical information system to compete with the challenges and problems it faces (Alhalalmeh et al., 2020; Al-Hawary & AlDafiri, 2017; Aldaihani & Ali, 2018; Aldaihani et al., 2020; Ghazal et al., 2021). Hence the importance of this research by providing a broad understanding of the issue of the quality of health information systems, as the quality of information systems has become evolving and has practical importance in business management; the health information system is a central force to achieve the desired goal of work (Aburayya et al., 2020b, c; Al-Maroof et al., 2021). Thus, this study comes to fill the literary gap in this field, and the concept of employee performance and its implications on the performance of the organization and the performance of the economy as a whole, which increased the importance of this study, so this study came to clarify the impact of the quality of information systems on the job performance in private hospitals.

2 Theoretical Framework and Hypotheses Development

2.1 Medical Information Systems

The information and communications revolution and its advanced and renewable applications in the third decade of the third millennium, of the twenty-first century, brought about a managerial reality different from what it was in previous decades, and the manifestations of the new reality are manifested in the decline of many intellectual concepts and systems on which previous managerial studies relied, which making modern administrative methods almost irrelevant to the reality of managerial thought and its applications in the past decades (Aldaihani & Ali, 2018; Aldaihani et al., 2020; Taryam et al., 2020).

The information revolution is the tool of globalization for the new global system. These technologies are undergoing radical and profound transformations that have increased the speed of the communication process, so that technical obstacles, political considerations and geographical boundaries are no longer an obstacle to the developmental tide of this service (Al-Lozi et al., 2018; Aburayya et al., 2020a; Al-Hawary & Al-Syasneh, 2020). The information revolution, with its enormous power and ability, has become the main nerve for all possible changes in various aspects of life in this era (Al- Quran et al., 2020). Information systems are considered one of the most important requirements of this era in all its fields, and its various theoretical and practical aspects, and at all levels and activities related to the performance of organizations, and a major resource of the organization and an important source of its success (Ahmed et al., 2020; Alshurideh, 2022; Tariq et al., 2022).

The rapid development in the medical field and medical services in recent decades has increased the demand for appropriate medical information to make medical and managerial decisions, as health managers and policy-makers need appropriate and accurate information to measure the program's effectiveness and follow-up progress to achieve the desired goals, so the investment in medical information systems will be justified because it helps in improving performance and making decisions regarding the detection and control of medical problems (Abu Qaaud et al., 2011; Al-Hawary & Al-Namlan, 2018; Alolayyan et al., 2018; Mohammad et al., 2014, 2020).

The World Health Organization defines medical information systems as "integrated efforts to collect and process medical data and transform it into information and knowledge for use in decision-making and policy implementation at all levels of medical services to improve its efficiency and effectiveness". Lippeveld (2000) Defined medical information systems as "a set of elements, components and procedures for an organization that aims to obtain information" that supports managerial and medical decisions at all levels of the medical system. Medical information systems are defined as "a group of special computer systems, which are used to provide medical information" Patients and auditors in particular, and the hospital in general, regards to a set of functions that enable investors to enter, maintain and review information, and issue statistics and reports that help in making medical, therapeutic and managerial decisions.

The joint commission on accreditation of hospitals (JCAH) defines it as the degree of adherence to generally recognized contemporary standards of good practice and expected outcomes for a specific service, diagnosis, or medical problem. Good medical services are those that follow the standards and foundations that were followed and taught by the founding leaders of the medical profession in society. Quality of medical care may be defined; as an accessible, safe, responsible and fair medical care. This means that the service provider provides the right care for the right condition at the right time in the right way. Patients easily get care at the right time in an accurate manner, and they have to have full information about the risks in order to avoid unsafe procedures and full information about the benefits of the care they receive, respect the rights of both patients and providers (Al-Hawary & Abu-Laimon, 2013; Al-Hawary et al., 2011). The quality of the medical service is measured by the availability of the five dimensions that Berry et al. reached in the medical service provided by the hospital, and these dimensions are:

Quality of medical information systems are those systems that aim to provide medical information for patients, auditors and the hospital itself, which helps to make, develop and improve the level of effective medical, treatment and managerial decisions. The quality of medical information systems is the systems with good and specific specifications and characteristics, easy to use and at a high level of accuracy. Good medical information systems are the systems that are able to raise the level of the organization's performance and achieve its goals with the least effort and time (Al-Hawary, 2012; Alzoubi et al., 2020; Zu'bi et al.,2012).

Ease of use: The ability of systems to be used easily in order to achieve a specific goal, and usability also means how efficient those systems, clarity in design, and human interaction with them (Al-Hawary & Al-Smeran, 2017; Alshurideh et al., 2020).

Response time: The time it takes for a system or functional unit to respond to a specific input or is the total time it takes to respond to a service request, which is the sum of the service time and waiting time, which is the time it takes to do the work you requested.

Flexibility: It expresses the extent to which the systems are able and responsive to changes, or it is the ability of the system to absorb disturbances, reorganize, and continue to perform functions with the same efficiency as before.

Ease of learning: The extent to which systems are able to understand, use, deal with, and use them in the correct ways (Al-Hawary & Hussien, 2017).

Reliability: The ability of the systems to perform the required tasks (completely free from technical errors); It is a measure of the period during which the systems can perform their agreed function without interruption. The term reliability can also be used to indicate the extent to which a particular process or function is likely to achieve the outputs required (Al-Hawary & Metabis, 2012).

2.2 *Job Performance*

There is an increasing interest in the issue of job performance, and the effectiveness and efficiency of organizations in carrying out their functions and achieving the goals for which they were established, whether these are profit or service organizations (AlShehhi et al., 2020; Alsuwaidi et al., 2020). Hence the focus on human resources management and the exploitation of available opportunities and available energies, in order to provide services quickly and with high quality, so that organizations can respond to the needs and expectations of customers, in a way that guarantees them reaching their goals by raising the level of performance to ensure their continuity, growth and development (Zwick, 2006; AlHamad et al., 2022; Shamout et al., 2022).

Job performance is the main axis around which the efforts of organizations focus as it constitutes their most important goals. Organizations expect human resources to perform the functions assigned to them efficiently and effectively. Performance refers to the degree of achievement and completion of the tasks that constitute the job of the working individual, and it reflects how the individual achieves the requirements of the job, and often there is confusion and overlap between performance and effort, effort refers to the energy expended, while performance is measured on the basis of the results achieved by the individual (Al-Nady et al., 2013; Mohammad, 2020; Al Kurdi et al., 2020). Huang et al. (2015) defines job performance as the achievements of an organization through the goals set and includes the outputs that are achieved or achieved through the contribution of individuals or the team in achieving the strategic goals of the organization.

Willcocks and Smith (1995) defined Job performance as "a continuous process of defining, measuring and developing performance in an organization by linking each individual's job performance and goals with the organization's tasks and objectives in full." We believe that performance is the levels of business achievement, which expresses the extent to which individuals or organizations carry out their responsibilities and tasks using the available capabilities and resources in the best possible way. In other words, it is the outputs and goals that the organization seeks to achieve through its employees, and therefore it is a concept that reflects both the goals and the means necessary to achieve them, that is, it is a concept that links the aspects of activity and the goals that organizations seek to achieve through the tasks and duties performed by the workers in those organizations. They are embodied in the form of useful products and services.

2.3 *Medical Information Systems and Job Performance*

Al-Gharabawi study (2014) found a positive relationship between computerized health information systems and job performance in UNRWA health centers in the Gaza Strip. Abugabah et al. found an impact of computerized information systems on user performance depends on 4 influencing factors: system quality, information quality, technological components, and human characteristics. This study presented most of the previous evaluation models with an explanation of the similarities and differences between them and the model under development, where the researchers considered that there is a need to develop this model to be a comprehensive model that facilitates the process of understanding the impact of the use of information systems on user performance. Based on the above, the hypothesis of the study can be formulated. as follows:

There is a statistically significant effect of the quality of health information systems on job performance in private hospitals in Jordan.

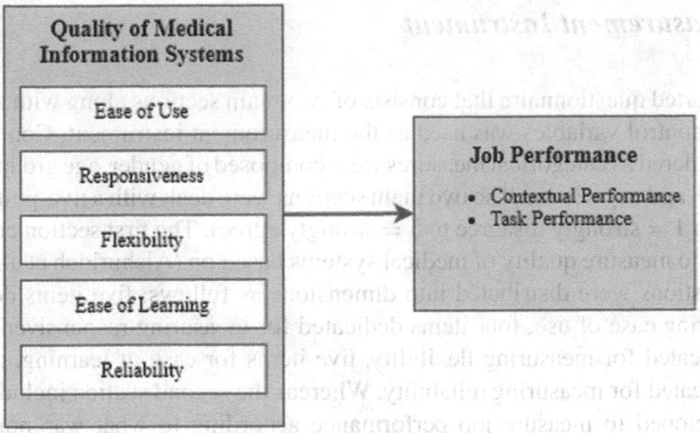

Fig. 1 Conceptual model

3 Study Model

See Fig. 1.

4 Methodology

4.1 Population and Sample Selection

A qualitative method based on a questionnaire was used in this study for data collection and sample selection. The major aim of the study was to examine the impact of quality of medical systems on job performance. The study population consisted of all employees with job titles (doctor, nurse, pharmacist, laboratory technician) in private hospitals in Jordan. Data were primarily gathered through self-reported questionnaires creating by Google Forms which were distributed to a purposive sample via email. In total, (205) responses were received including (14) invalid to statistical analysis due to uncompleted or inaccurate. Hence, the final sample contained (191) responses suitable to analysis requirements, where it proved to be sufficient to the extent that was predictable and allowed for a presumption of data saturation (Sekaran & Bougie, 2016).

4.2 *Measurement Instrument*

A self-reported questionnaire that consists of two main sections along with a section regarding control variables was used as the measurement instrument. Control variables considered as categorical measures were composed of gender, age group, educational level, and experience. The two main sections were dealt with a five-point Likert scale (from 1 = strongly disagree to 5 = strongly agree). The first section contained (23) items to measure quality of medical systems based on (Alshurideh et al., 2017). These questions were distributed into dimensions as follows: five items dedicated for measuring ease of use, four items dedicated for measuring responsiveness, five items dedicated for measuring flexibility, five items for ease of learning, and four items dedicated for measuring reliability. Whereas the second section included eight items developed to measure job performance according to what was pointed by (Al-Hawary & Metabis, 2012).

5 Findings

5.1 *Measurement Model Evaluation*

This study was conducted structural equation modeling (SEM) to test hypotheses, which represents a contemporary statistical technique for testing and estimating the relationship between factors and variables (Wang & Rhemtulla, 2021). Accordingly, the reliability and validity of the constructs were tested using confirmatory factor analysis (CFA) through the statistical program AMOSv24. Table 1 summarizes the results of convergent and discriminant validity, as well the indicators of reliability.

Table 1 shows that the standard loading values for the individual items were within the domain (0.615–0.866), these values greater than the minimum retention of the elements based on their standard loads (Al-Lozi et al., 2018; Sung et al., 2019). Average variance extracted (AVE) is a summary indicator of the convergent validity of constructs that must be above 0.50 (Howard, 2018). The results indicate that the AVE values were greater than 0.50 for all constructs, thus the used measurement model has an appropriate convergent validity. Rimkeviciene et al. (2017) suggested the comparison approach as a way to deal with discriminant validity assessment in covariance-based SEM. This approach is based on comparing the values of maximum shared variance (MSV) with the values of AVE, as well as comparing the values of square root of AVE ($\sqrt{AVE}$) with the correlation between the rest of the structures. The results show that the values of MSV were smaller than the values of AVE, and that the values of $\sqrt{AVE}$ were higher than the correlation values among the rest of the constructs. Therefore, the measurement model used is characterized by discriminative validity. The internal consistency measured through Cronbach's Alpha coefficient (α) and compound reliability by McDonald's Omega coefficient (ω) was conducted as indicators to evaluate measurement model. The results listed

Table 1 Results of validity and reliability tests

Constructs	1	2	3	4	5	6	7
1. EU	**0.758**						
2. RP	0.497	**0.725**					
3. FL	0.533	0.538	**0.741**				
4. EL	0.567	0.477	0.443	**0.755**			
5. RL	0.482	0.509	0.581	0.592	**0.765**		
6. CP	0.711	0.694	0.722	0.693	0.708	**0.737**	
7. TP	0.702	0.715	0.712	0.699	0.695	0.645	**0.753**
VIF	2.384	1.885	2.642	2.539	1.286	–	–
Loadings range	0.615–0.866	0.692–0.758	0.653–0.791	0.681–0.806	0.724–0.813	0.652–0.816	0.705–0.824
AVE	0.575	0.526	0.549	0.571	0.585	0.543	0.566
MSV	0.506	0.468	0.513	0.498	0.508	0.482	0.510
Internal consistency	0.865	0.815	0.856	0.868	0.845	0.823	0.837
Composite reliability	0.869	0.816	0.858	0.869	0.849	0.825	0.839

Note EU: ease of use, RP: responsiveness, FL: flexibility, EL: ease of learn, RL: reliability, CP: contextual performance, TP: task performance, Bold fonts indicate to square root of average variance extracted

in Table 1 demonstrated that both values of Cronbach's Alpha coefficient and McDonald's Omega coefficient were greater than 0.70, which is the lowest limit for judging on measurement reliability (De Leeuw et al., 2019).

5.2 *Structural Model*

The structural model illustrated no multicollinearity issue among predictor constructs because variance inflation factor (VIF) values are below the threshold of 5, as shown in Table 1 (Hair et al., 2017). This result is supported by the values of model fit indices shown in Figs. 1 and 2.

The results in Fig. 1 indicated that the chi-square to degrees of freedom (CMIN/DF) was 1.975, which is less than 3 the upper limit of this indicator. The values of the goodness of fit index (GFI), the comparative fit index (CFI), and the Tucker-Lewis index (TLI) were upper than the minimum accepted threshold of 0.90. Moreover, the result of root mean square error of approximation (RMSEA) indicated to value 0.046, this value is a reasonable error of approximation because it is less than the higher limit of 0.08. Consequently, the structural model used in this study was recognized as a fit model for predicting the DEP and generalization of its result (Ahmad et al., 2016; Shi et al., 2019). To verify the results of testing the study

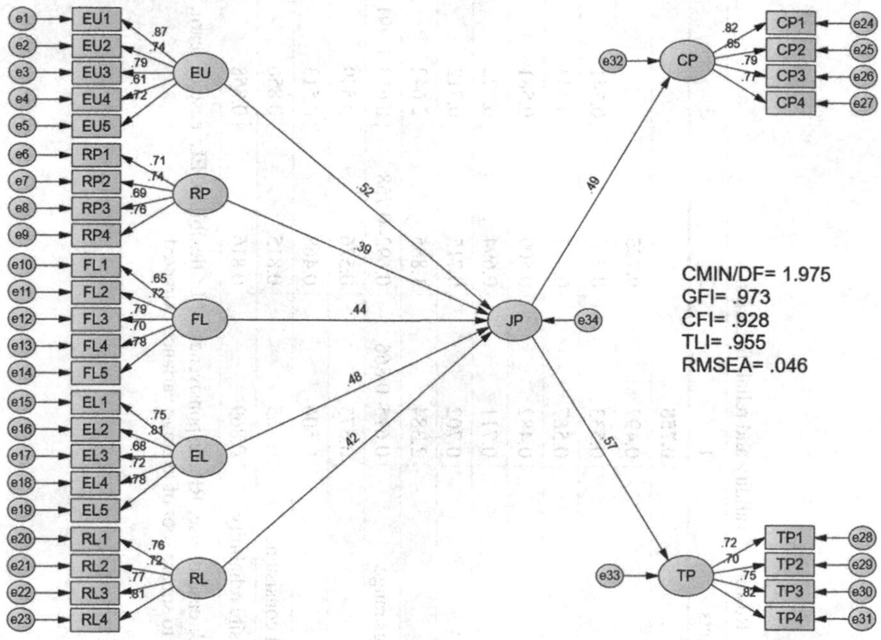

Fig. 2 SEM results of the quality of medical systems effect on job performance

Table 2 Hypothesis testing

Relation	Standard Beta	*t* value	*p* value
EU → JP	0.521	22.68***	0.000
RP → JP	0.394	15.06*	0.03
FL → JP	0.438	18.29**	0.004
EL → JP	0.480	21.74***	0.000
RL → JP	0.423	17.03**	0.002

Note EU: ease of use, RP: responsiveness, FL: flexibility, EL: ease of learn, RL: reliability, JP: job performance, * $p < 0.05$, ** $p < 0.01$, *** $p < 0.001$

hypotheses, structural equation modeling (SEM) was used, the results of which are listed in Table 2.

The results demonstrated in Table 2 show that the highest impact was for ease of use ($\beta = 0.521$, $t = 22.68$, $p = 0.000$), followed by ease of learn ($\beta = 0.480$, $t = 21.74$, $p = 0.000$), then flexibility ($\beta = 0.438$, $t = 18.29$, $p = 0.004$), then reliability($\beta = 0.423$, $t = 17.03$, $p = 0.002$), and finally the lowest impact was for responsiveness ($\beta = 0.394$, $t = 15.06$, $p = 0.03$). Thus, all research hypotheses were supported based on these results.

6 Discussion

The results showed that the level of quality of medical information systems; Medium, and this indicates that private hospitals are keen to have modern medical information systems with high efficiency, due to their awareness of the importance and effectiveness of the quality of medical information systems. This has a significant and positive impact on the quality of services provided as a whole within the hospital. It was also found that the level of job performance in private hospitals in Jordan is average, attributing this to the stability and job security that hospitals provide to their employees, in addition to the high quality of working life compared to other hospitals working in the same field in the public sector, which increased the level of performance.

The results show that there is an impact of the quality of medical information systems on the job performance in private hospitals in Jordan, and this result can be justified by the fact that most of the work carried out by hospital employees depends mainly on the various applications of the medical information system, as the more varied the information systems used and the higher their efficiency in the implementation of the work, this led to the development of performance and speed of completion. Meaning that the higher the efficiency of the quality of medical information systems, the higher the level of functionality.

The results show that there is an impact of ease of use on the job performance in private hospitals in Jordan. It is evident from the above that a high percentage of employees consider themselves qualified to deal with the modern system used in the hospital, and this indicates that they have high capabilities in dealing with information, which increases work efficiency and improves job level. It was also found that there is an impact of response time on the job performance in private hospitals in Jordan, and this result is attributed to the use of the various applications of the medical information system inside the hospital, and there is a comprehensive system that computerizes and links all departments and specializations, and it is relied on daily treatment as a form for the sick. The relevance and modernity of the software used serves to meet the user's information needs in a timely manner, which has a significant impact on the completion of tasks and functions and thus improves job performance.

The results show that there is an impact of flexibility on the job performance in private hospitals in Jordan, and the researchers attribute this result to the fact that the information provided by the medical system is commensurate with the needs of the job, and this indicates the follow-up of the senior management and the specialized technical team to the workflow based on the use of the medical system and with periodic feedback in order to develop the system, increase its effectiveness, and make it more compatible with work requirements, thus, job performance is improved. It was also found that there is an effect of ease of learning on the job performance in private hospitals in Jordan, and the researchers attribute this result to the conditions of employment followed by private hospitals, such as subjecting the employee to an exam to measure his computer skills and taking this into account when deciding to hire. Employees possess good levels of computer use as a result of the employment policy inside the hospital that requires tests to measure computer skills, in addition to the good relations linking employees with the specialized technical department, which facilitates understanding the users' needs of the system and addressing the problems they face during their use.

The results show that there is an impact of reliability on the job performance in private hospitals in Jordan, and this result is attributed to the good level of information provided by the medical system, which is commensurate with the needs of the job, in addition to the follow-up of the higher management of the workflow based on the use of the medical system and its effective management, and its implementation and management. More compatible with the work requirements and the flexibility of the hospital's organizational structure, which allows the flow of information easily, and all this has led to an improvement in the job performance.

7 Recommendations

Based on the results of the study, researchers recommend managers and decision-makers in private hospitals in Jordan to hold training courses in the field of using medical information systems for all employees in private hospitals, and to focus on maintaining these courses, and raising their levels in line with continuous technological developments and changes. Implementation of awareness campaigns targeting employees in the medical fields on the importance of medical information systems in raising the level of job performance and the performance of hospitals in general. Helping the hospital to improve the quality of services provided to patients by increasing staff productivity, speeding up the completion of administrative and medical transactions, facilitating procedures, improving decision-making and reducing errors.

References

Abu Qaaud, F., Al-Shoura, M., & Al-Hawary, S. I. (2011). The impact of the service marketing mix in the service quality of health services from the viewpoint of patients in government hospitals in Amman "A field study." *Abhath Al-Yarmouk, 27*(1B), 417–441.

Aburayya, A., Alshurideh, M., Al Marzouqi, A., Al Diabat, O., Alfarsi, A., Suson, R., …, & Salloum, S. A. (2020c). An empirical examination of the effect of TQM practices on hospital service quality: An assessment study in UAE hospitals. *Systematic Reviews in Pharmacy*, 11(9), 347–362.

Aburayya, A., Alshurideh, M., Al Marzouqi, A., Al Diabat, O., Alfarsi, A., Suson, R., ... & Alzarouni, A. (2020b). Critical success factors affecting the implementation of tqm in public hospitals: A case study in UAE Hospitals. *Systematic Reviews in Pharmacy*, 11(10), 230–242.

Aburayya, A., Alshurideh, M., Albqaeen, A., Alawadhi, D., & Ayadeh, I. (2020c). An investigation of factors affecting patients waiting time in primary health care centers: An assessment study in Dubai. *Management Science Letters, 10*(6), 1265–1276.

Ahmad, S., Zulkurnain, N., & Khairushalimi, F. (2016). Assessing the validity and reliability of a measurement model in structural equation modeling (SEM). *British Journal of Mathematics & Computer Science, 15*(3), 1–8. https://doi.org/10.9734/BJMCS/2016/25183

Ahmed, A., Alshurideh, M., Al Kurdi, B., & Salloum, S. A. (2020, October). Digital transformation and organizational operational decision making: A systematic review. In International Conference on Advanced Intelligent Systems and Informatics (pp. 708–719). Springer, Cham.

Al- Quran, A. Z., Alhalalmeh, M. I., Eldahamsheh, M. M., Mohammad, A. A., Hijjawi, G. S., Almomani, H. M., & Al-Hawary, S. I. (2020). Determinants of the green purchase intention in Jordan: The moderating effect of environmental concern. *International Journal of Supply Chain Management (IJSCM), 9*(5), 366–371.

Aldaihani, F. M. F., & Ali, N. A. B. (2018). Impact of Electronic Customer Relationship Management on customers satisfaction of the five stars hotels in Kuwait. *Global Journal of Management and Business Research.*

Aldaihani, F. M. F., Ali, N. A. B., Hashim, H. B., & Kamal, N. (2020). Impact of social customer relationship management on customer retention of Islamic banks in Kuwait: The mediating role of customer empowerment. *International Journal of Supply Chain Management (IJSCM), 9*(1), 330.

Alhalalmeh, M. I., Almomani, H. M., Altarifi, S., Al-Quran, A. Z., Mohammad, A. A., & Al-Hawary, S. I. (2020). The nexus between corporate social responsibility and organizational performance in Jordan: The mediating role of organizational commitment and organizational citizenship behavior. *Test Engineering and Management, 83*(July), 6391–6410.

AlHamad, A., Alshurideh, M., Alomari, K., Kurdi, B., Alzoubi, H., Hamouche, S., & Al-Hawary, S. (2022). The effect of electronic human resources management on organizational health of telecommunications companies in Jordan. *International Journal of Data and Network Science*, 6(2), 429–438. *International Journal of Data and Network Science* 6 (2), 429–438.

Al-Hawary, S. I. (2012). Health care services quality at private hospitals, from patient's perspective: A comparative study between Jordan and Saudi Arabia. *African Journal of Business Management, 6*(22), 6516–6529.

Al-Hawary, S. I. S., & AlDafiri, M. F. S. (2017). Effect of the components of information technology adoption on employees performance of interior Ministry of Kuwait State. *International Journal of Academic Research in Economics and Management Sciences, 6*(2), 149–169.

Al-Hawary, S. I. S., & Mohammed, A. K. (2017). Impact of team work traits on organizational citizenship behavior from the viewpoint of the employees in the education directorates in North Region of Jordan. *Global Journal of Management and Business*, 17(2-A), 23–40.

Al-Hawary, S. I., & Abu-Laimon, A. A. (2013). The impact of TQM practices on service quality in cellular communication companies in Jordan. *International Journal of Productivity and Quality Management, 11*(4), 446–474.

Al-Hawary, S. I., & Al-Hamwan, A. (2017). Environmental analysis and its impact on the competitive capabilities of the commercial banks operating in Jordan. *International Journal of Academic Research in Accounting, Finance and Management Sciences, 7*(1), 277–290.

Al-Hawary, S. I., & Al-Namlan, A. (2018). Impact of electronic human resources management on the organizational learning at the private hospitals in the state of Qatar. *Global Journal of Management and Business Research: A Administration and Management, 18*(7), 1–11.

Al-Hawary, S. I., & Al-Smeran, W. (2017). Impact of electronic service quality on customers satisfaction of Islamic Banks in Jordan. *International Journal of Academic Research in Accounting, Finance and Management Sciences, 7*(1), 170–188.

Al-Hawary, S. I., & Batayneh, A. M. (2010). The effect of marketing communication tools on non-Jordanian students' choice of Jordanian Public Universities: A field study. *International Management Review, 6*(2), 90–99.

Al-Hawary, S. I., & Hadad, T. F. (2016). The effect of strategic thinking styles on the enhancement competitive capabilities of commercial banks in Jordan. *International Journal of Business and Social Science, 7*(10), 133–144.

Al-Hawary, S. I., & Hussien, A. J. (2017). The impact of electronic banking services on the customers loyalty of commercial banks in Jordan. *International Journal of Academic Research in Accounting, Finance and Management Sciences, 7*(1), 50–63.

Al-Hawary, S. I., & Ismael, M. (2010). The effect of using information technology in achieving competitive advantage strategies: A field study on the Jordanian Pharmaceutical Companies. *Al Manara for Research and Studies, 16*(4), 196–203.

Al-Hawary, S. I., & Metabis, A. (2012). Service quality at Jordanian commercial banks: What do their customers say? *International Journal of Productivity and Quality Management, 10*(3), 307–334.

Al-Hawary, S. I., & Shdefat, F. (2016). Impact of human resources management practices on employees' satisfaction a field study on the Rajhi Cement Factory. *International Journal of Academic Research in Accounting, Finance and Management Sciences, 6*(4), 274–286.

Al-Hawary, S. I., Alghanim, S., & Mohammad, A. (2011). Quality level of health care service provided by king Abdullah educational hospital from patient's viewpoint. *Interdisciplinary Journal of Contemporary Research in Business, 2*(11), 552–572.

Allahow, T. J. A. A., Al-Hawary, S. I. S., & Aldaihani, F. M. F. (2018). Information Technology and Administrative Innovation of the Central Agency for Information Technology in Kuwait. *Global Journal of Management and Business Research.*

Al-Lozi, M., Almomani, R. Z., & Al-Hawary, S. I. (2017). Impact of talent management on achieving organizational excellence in Arab Potash Company in Jordan. *Global Journal of Management and Business Research: A Administration and Management, 17*(7), 15–25.

Al-Lozi, M., Almomani, R. Z., & Al-Hawary, S. I. (2018). Talent Management strategies as a critical success factor for effectiveness of Human Resources Information Systems in commercial banks working in Jordan. *Global Journal of Management and Business Research: A Administration and Management, 18*(1), 30–43.

Al-Maroof, R., Ayoubi, K., Alhumaid, K., Aburayya, A., Alshurideh, M., Alfaisal, R., & Salloum, S. (2021). The acceptance of social media video for knowledge acquisition, sharing and application: A comparative study among YouTube users and TikTok users' for medical purposes. *International Journal of Data and Network Science, 5*(3), 197–214.

Al-Nady, B. A., Al-Hawary, S. I., & Alolayyan, M. (2013). Strategic management as a key for superior competitive advantage of sanitary ware suppliers in kingdom of Saudi Arabia. *International Journal of Management and Information Technology, 7*(2), 1042–1058.

Alolayyan, M., Al-Hawary, S. I., Mohammad, A. A. & Al-Nady, B. A. (2018). Banking service quality provided by commercial banks and customer satisfaction. A structural equation modelling approaches. *International Journal of Productivity and Quality Management, 24*(4), 543–565.

AlShehhi, H., Alshurideh, M., Al Kurdi, B., & Salloum, S. A. (2020, October). The impact of ethical leadership on employees performance: A systematic review. In *International Conference on Advanced Intelligent Systems and Informatics* (pp. 417–426). Springer, Cham.

Alshura, M. S. K., Nusair, W. K. I., & Aldaihani, F. M. F. (2016). Impact of internal marketing practices on the organizational commitment of the employees of the insurance companies in Jordan. *International Journal of Academic Research in Economics and Management Sciences, 5*(4), 168–187.

Alshurideh, M. (2022). Does electronic customer relationship management (E-CRM) affect service quality at private hospitals in Jordan? *Uncertain Supply Chain Management, 10*(2), 1–8.

Alshurideh, M. T., Al Kurdi, B., Alzoubi, H. M., Ghazal, T. M., Said, R. A., AlHamad, A. Q., ... & Al-kassem, A. H. (2022). Fuzzy assisted human resource management for supply chain management issues. *Annals of Operations Research*, 1–19.

Alshurideh, M., Al Kurdi, B., Salloum, S. A., Arpaci, I., & Al-Emran, M. (2020). Predicting the actual use of m-learning systems: A comparative approach using PLS-SEM and machine learning algorithms. *Interactive Learning Environments*, 1–15.

Alshurideh, M., Al-Hawary, S. I., Batayneh, A. M., Mohammad, A., & Al-Kurdi, B. (2017). The impact of Islamic Banks' service quality perception on Jordanian customers loyalty. *Journal of Management Research, 9*(2), 139–159.

Alsuwaidi, M., Alshurideh, M., Al Kurdi, B., & Salloum, S. A. (2020, October). Performance appraisal on employees' motivation: A comprehensive analysis. In *International Conference on Advanced Intelligent Systems and Informatics* (pp. 681–693). Springer, Cham.

Altamony, H., Masa'deh, R., Alshurideh, M., Obeidat, B. (2012) Information systems for competitive advantage: Implementation of an organisational strategic management process. *Innovation and Sustainable Competitive Advantage: From Regional Development to World Economies*. 583–592.

Altarifi, S., Al-Hawary, S. I. S., & Al Sakkal, M. E. E. (2015). Determinants of E-shopping and its effect on consumer purchasing decision in Jordan. *International Journal of Business and Social Science, 6*(1), 81–92.

Alzoubi, H. M., Alshurideh, M., Al Kurdi, B., & Inairat, M. (2020). Do perceived service value, quality, price fairness and service recovery shape customer satisfaction and delight? A practical study in the service telecommunication context. *Uncertain Supply Chain Management, 8*(3), 579–588.

Alzoubi, H., Alshurideh, M., Akour, I., Al Shraah, A., & Ahmed, G. (2021a) Impact of information systems capabilities and total quality management on the cost of quality. *Journal of Legal, Ethical and Regulatory Issues*, 24(Special Issue 6), 1–11.

Alzoubi, H. M., Alshurideh, M., & Ghazal, T. M. (2021b). Integrating BLE Beacon technology with Intelligent Information Systems IIS for operations' performance: A managerial perspective. In *The International Conference on Artificial Intelligence and Computer Vision* (pp. 527–538). Springer, Cham.

de Leeuw, E., Hox, J., Silber, H., Struminskaya, B., & Vis, C. (2019). Development of an international survey attitude scale: Measurement equivalence, reliability, and predictive validity. *Measurement Instruments for the Social Sciences, 1*(1), 9. https://doi.org/10.1186/s42409-019-0012-x

Ghazal, T. M., Alshurideh, M. T., & Alzoubi, H. M. (2021, June). Blockchain-Enabled Internet of Things (IoT) platforms for pharmaceutical and biomedical research. In *The International Conference on Artificial Intelligence and Computer Vision* (pp. 589–600). Springer, Cham.

Hair, J. F., Babin, B. J., & Krey, N. (2017). Covariance-based structural equation modeling in the journal of advertising: Review and recommendations. *Journal of Advertising, 46*(1), 163–177. https://doi.org/10.1080/00913367.2017.1281777

Howard, M. C. (2018). The convergent validity and nomological net of two methods to measure retroactive influences. *Psychology of Consciousness: Theory, Research, and Practice, 5*(3), 324–337. https://doi.org/10.1037/cns0000149

Huang, S., Lee, C.-H., & Yen, C. (2015). How business process reengineering affects information technology investment and employee performance under different performance measurement. *Information Systems Frontier, 17*, 1133–1144.

Mohammad, A. A. (2020). The effect of customer empowerment and customer engagement on marketing performance: The mediating effect of brand community membership. *Business: Theory and Practice, 21*(1), 30–38.

Mohammad, A. A. S., Saleem Khlif Alshura, M., Al-Hawary, S. I. S., Al-Syasneh, M. S., & Alhajri, T. M. S. (2020) The influence of internal marketing practices on the employees' intention to leave: A study of the private hospitals in Jordan. *International Journal of Advanced Science and Technology, 29*(5), 1174-1189

Mohammad, A. A., Alkayed, W., Irtameh, H., & Alafi, K. (2013). Unexploited business opportunities attracting Islamic tourists to Australia. *European Journal of Business and Management, 5*(30), 133–143.

Mohammad, K. I., Alafi, K. K., Mohammad, A. A., Gamble, J., & Creedy, D. (2014). Jordanian women's dissatisfaction with childbirth care: Dissatisfaction with childbirth care. *International Nursing Review, 61*(2), 278–284.

Rimkeviciene, J., Hawgood, J., O'Gorman, J., & De Leo, D. (2017). Construct validity of the acquired capability for suicide scale: Factor structure, convergent and discriminant validity. *Journal of Psychopathology and Behavioral Assessment, 39*(2), 291–302. https://doi.org/10.1007/s10862-016-9576-4

Sekaran, U., & Bougie, R. (2016). *Research methods for business: A skill-building approach* (7th edn). Wiley.

Shamout, M., Elayan, M., Rawashdeh, A., Kurdi, B., & Alshurideh, M. (2022). E-HRM practices and sustainable competitive advantage from HR practitioner's perspective: A mediated moderation analysis. *International Journal of Data and Network Science, 6*(1), 165–178.

Shi, D., Lee, T., & Maydeu-Olivares, A. (2019). Understanding the model size effect on SEM fit indices. *Educational and Psychological Measurement, 79*(2), 310–334. https://doi.org/10.1177/0013164418783530

Sung, K.-S., Yi, Y. G., & Shin, H.-I. (2019). Reliability and validity of knee extensor strength measurements using a portable dynamometer anchoring system in a supine position. *BMC Musculoskeletal Disorders, 20*(1), 1–8. https://doi.org/10.1186/s12891-019-2703-0

Svoboda, P., Ghazal, T. M., Afifi, M. A., Kalra, D., Alshurideh, M. T., & Alzoubi, H. M. (2021, June). Information systems integration to enhance operational customer relationship management in the pharmaceutical industry. In *The International Conference on Artificial Intelligence and Computer Vision* (pp. 553–572). Springer, Cham.

Tariq, E., Alshurideh, M., Akour, I., & Al-Hawary, S. (2022). The effect of digital marketing capabilities on organizational ambidexterity of the information technology sector. *International Journal of Data and Network Science, 6*(2), 401–408.

Taryam, M., Alawadhi, D., Aburayya, A., Albaqa'een, A., Alfarsi, A., Makki, I., … & Salloum, S. A. (2020). Effectiveness of not quarantining passengers after having a negative COVID-19 PCR test at arrival to Dubai airports. *Systematic Reviews in Pharmacy*, 11(11), 1384–1395.

Wang, Y. A., & Rhemtulla, M. (2021). Power Analysis for parameter estimation in structural equation modeling: A discussion and tutorial. *Advances in Methods and Practices in Psychological Science, 4*(1), 1–17. https://doi.org/10.1177/2515245920918253

Willcocks, L., & Smith, G. (1995). IT-enabled business process reengineering: Organizational and human resource dimensions. *The Journal of Strategic Information Systems, 4*(3), 279–301.
Zu'bi, Z., Al-Lozi, M., Dahiyat, S., Alshurideh, M., & Al Majali, A. (2012). Examining the effects of quality management practices on product variety. *European Journal of Economics, Finance and Administrative Sciences*, 51(1), 123–139.
Zwick, T. (2006). The impact of training intensity on establishments productivity. *Labor Economics, 11*, 715–740.

Impact of Human Resources Management Strategies on Organizational Learning of Islamic Banks in Jordan

Ibrahim Rashed Soliaman AlTaweel, Riad Ahmad Mohammed Abazeed, Mohammad Fathi Almaaitah, Dheifallah Ibrahim Mohammad, Doa'a Ahmad Odeh Al-Husban, Sulieman Ibraheem Shelash Al-Hawary, Faraj Mazyed Faraj Aldaihani, Anber Abraheem Shlash Mohammad, and Ayat Mohammad

Abstract The major aim of the study was to examine the impact of Human Resources Management strategies on Organizational Learning. Therefore, it focused on Islamic banks operating in Jordan. Data were primarily gathered through self-reported questionnaires creating by Google Forms which were distributed to a purposive sample of managers via email. The statistical program AMOSv24 was used to test the study hypotheses and achieve objectives. The study results demonstrated that Human Resources Management strategies have a positive impact relationship on Organizational Learning of Islamic banks operating in Jordan. Based on the study

I. R. S. AlTaweel
Faculty of Business School, Al Russ City, Department of Business Administration, Qussim University, P.O. Box 6502, Al Russ 51452, Saudi Arabia
e-mail: toiel@qu.edu.sa

R. A. M. Abazeed
Faculty of Finance and Business Administration, Department of Business Administration, Al Al-Bayt University, P.O. Box 130040, Mafraq 25113, Jordan

M. F. Almaaitah
Faculty of Economic and Administration Sciences, Department of Business Administration& Public Administration, Al Al-Bayt University Jordan, P.O. Box 130040, Mafraq 25113, Jordan
e-mail: m.maaitah@aabu.edu.jo

D. I. Mohammad
Ministry of Education, Amman, Jordan

D. A. O. Al-Husban
Faculty of Alia College, Department of Human Fundamental Sciences, Al-Balqa Applied University, Amman, Jordan
e-mail: D_husban@bau.edu.jo

S. I. S. Al-Hawary (✉)
Faculty of Economics and Administrative Sciences, Department of Business Administration, Al Al-Bayt University, P.O. Box 130040, Mafraq 25113, Jordan
e-mail: dr_sliman73@aabu.edu.jo

M. Alshurideh et al. (eds.), *The Effect of Information Technology on Business and Marketing Intelligence Systems*, Studies in Computational Intelligence 1056,
https://doi.org/10.1007/978-3-031-12382-5_46

results, researchers recommend managers of Islamic banks in Jordan to develop mechanisms that reward employees for learning new things, in a way that enhances adaptive and generative learning processes, and enhances confidence building among employees which allowing organizational learning processes of both adaptive and generative types.

Keywords Human Resources Management strategies · Organizational Learning · Islamic banks · Jordan

1 Introduction

In light of the great technological advances witnessed by the business world in recent decades, organizations are operating in an environment of intense competition and uncertainty conditions, which in turn led these organizations to research more in ways to deal with these conditions, and to identify the competitors' capabilities and the need of markets (Al-Hawary & Al-Hamwan, 2017; Al-Hawary & Hadad, 2016; Al-Hawary & Ismael, 2010). Organizational learning has become a prerequisite for business organizations to keep themselves updated with technological and competitive developments, Organizational learning involves the ability of an organization to adapt to emerging circumstances in its business environment and its ability to keep up with or compete with its competitors (Al-Hawary & Al-Namlan, 2018; Al-Hawary & Al-Syasneh, 2020). It is necessary to have a system that serves as an early warning for these organizations to predict future changes in their business environment, making the functions, activities and functions of human resources management in business organizations more complex and sophisticated, and need a large amount of data and information that help organizations to make decisions in a manner that supports and achieve its objectives through the management of their resources in an optimal way.

Organizations need to increase the pace of organizational learning to be able to cope with the dramatic changes in the contemporary business environment, and to achieve this; they need to be more open and receive new ideas and concepts (Al-Hawary et al., 2020). They need to innovate products and modern methods, train human resources and encourage innovation and creativity while doing business (Al-Hawary et al., 2018; Alameeri et al., 2021; Nuseir et al., 2021; Alzoubi et al., 2022). Organizations are treating human resources as intellectual energy, a source of information and knowledge, and innovations. They have also increased attention to

F. M. F. Aldaihani
Kuwait Civil Aviation, Ishbiliyah Bloch 1, Street 122, Home 1, Kuwait

A. A. S. Mohammad
Faculty of Administrative and Financial Sciences, Marketing Department, Petra University, P.O. Box 961343, Amman 11196, Jordan

A. Mohammad
The World Islamic Science and Education University (WISE), P.O. Box 1101, Amman 11947, Jordan

these resources and created a healthy environment (Al-Hawary & Aldaihani, 2016; Al-Hawary & Alwan, 2016; AlHamad et al., 2022; Alshurideh et al., 2022). It has also increased attention to these resources and created a healthy environment work (Alameeri et al., 2020; Alshurideh et al., 2019). Despite the increasing importance of the study of organizational learning, and its role in achieving the goals of the organization, the concept of organizational learning has become one of the basic concepts that managers, researchers and practitioners have been interested in enhancing job satisfaction.

Human resources management is one of the most important functions of the organizations, where it performs many tasks, activities and functions that will provide efficient and qualified human resources suitable for the current and future needs of the organization, which should be able to contribute to the achievement of its objectives efficiently and effectively (Al Kurdi et al., 2021; Al-Hawary, 2011, 2015; Shamout et al., 2022). Past and current studies in strategic human resources management have indicated that human resources management already has an impact on the organizational outputs, and several studies have indicated that there is a relationship between human resources management and organizational performance (Huselid, 1995; Arthur 1994 Youndt. et al., 1996; Alkalha et al., 2012). Human resources management practice considered as an important source of competitive advantage (Wright et al., 2001).

2 Theoretical Framework and Hypotheses Development

2.1 *Strategic Human Resources Management*

The concept of strategic human resources management has played an important and leading role in management research for the past three decades (Boxall & Purcell, 2011). Continuous analysis in the field of strategic human resources management shows how this field has achieved added strategic value and contributed to organizational success. Human resources represent the critical and necessary resources in the performance of the organizations. The integration of human resources management with the organization's strategy results in more organizational effectiveness improves organizational performance and enhances business continuity, improvement and efficiency (Holbeche, 2001; Metabis & Al-Hawary, 2013; Zu'bi et al., 2012).

HRM is defined as a strategic and coherent approach to the management of individuals contributing to the achievement of organizational goals. This term was introduced as an alternative to the concept of people management. Human resources management has taken into consideration the focus on the strategic dimension of human resources management in the organizations, and in order to focus more on the concept of human resources management, the strategic dimension is added to refer to human resources management as a strategic function not only for building and activating organizations but also for seeking and creating sustainable competitive advantage (Holbeche, 2001; Obeidat et al., 2021), which means that the strategic dimension of human resources management requires to set long-term goals for these organizations, and take plans to allocate their resources for the achievement of long-term goals (Altamony et al., 2012).

Human resource management practices represent an intermediary factor between HRM strategies and HRM outputs (Al-Hawary & Nusair, 2017; Al-Hawary & Shdefat, 2016). Sheppeck and Militello (2000) divided the HRM strategy into four divisions: business skills and policies, supportive work environment, measurement and promotion of performance, and market organization. Guest (1997) divided them into three categories: differentiation through innovation, focus on quality, and decreasing cost. Stavrou-Costea (2005) concluded that effective human resource management may be a necessary factor in the success of the organization, and Lee and Lee's (2007) referred that HRM practices including: Training, teams, compensation and incentives, human resources planning, performance appraisal, and job stability, improve the organizational performance.

Strategic HRM represents a great insight into the integration of HRM functions with both strategic planning and operations in organizations, especially in circumstances where consumers and employees are constantly interacting and communicating. Strategic human resources management can be described as a strategy of an integrated organizational strategy that includes an effective environmental survey that works to achieve the organization's objectives (Nankervis et al., 2005; Schuler & Jackson, 1999). The strategic approach to human resources management requires from the manager of human resources to be responsive to the changes that may occur in the organization and the environment in which the organization operates (Devanna et al., 1981). Wilton (2006) found that large organizations adopt higher levels of HR practices than small organizations.

Strategic human resource management focuses on individuals as means to develop competitive advantage (Barber et al., 1999; Becker & Huselid, 1999). The human resources manager plays an important and critical role in the development of business plans, and strategies to serve the objectives of the Organization (Artis et al., 1999). Highly effective communication and information-sharing between managers and their subordinates reinforces them to engage in strategy formulation and implementation (Barber et al., 1999; Becker & Huselid, 2006). Strategic HRM requires integration of HR practices, and address of these practices as a whole rather than focusing on individual HRM practices (Al-Hawary & Metabis, 2012; Becker & Huselid, 2006; Combs et al., 2006). Integration of human resource practices leads the organization to achieve sustainable competitive advantage, because they are complex, unique, specific, accurate and difficult to imitate by other competitors (Al-Hawary et al., 2013; Pfeffer, 2005; Wright et al., 2003). This study adopts five strategies, which represent three basic functions of human resources management. These strategies are as follows:

Strategy for human resources formation: This strategy is concerned with ensuring timely access to human resources, and retention of the workforce to achieve synergies between performance requirements and organizational conditions of the environment. This strategy is designed in the light of the general human resources strategy with the aim of acquiring a workforce that meets the specifications of personal characteristics, skills and knowledge as well as possessing the ability of career advancement and future professional development. The HR formation strategy consists of a number of sub-activities that are interrelated in an integrated that enables organization to meet the human resources needs according to specific characteristics and traits to fill specific functions, the most important of these functions is the recruitment of human resources and then the selection and appointment of human resources, and these functions are one of the main activities in this strategy. Armstrong (2006: 154) noted that the human resource formation strategy works to provide the Organization with qualified human resource in the right quantity and quality, while sustaining these human resources and in harmony with the culture of the Organization.

Strategy of training and development: Training and development is a cornerstone in the advancement and success of organizations, so training is one of the supporting activities in the organization (Al-Hawary & Abu-Laimon, 2013; Alshraideh et al., 2017). The training activity is responsible for upgrading cognitive skills; this is in addition to the need to pay attention to the external environment and the need to diversify training programs in line with the environment in which the organization operates (Al-Hawary & Al-Kumait, 2017). Training and development department must have a strategic role in the organization, and in the design of the training process to achieve efficiency and efficiency in the completion of the work carried out by the organization in its field. The development and training function represents the next job of the recruitment function, and aims to teach the selected workers during the recruitment process the skills and knowledge necessary to perform the jobs. Training is known as planned efforts. Through development, skills and knowledge are expanded, and in order to achieve the goal of training and development. The skills and knowledge acquired during training must be transferred to practical application while performing tasks and duties (Mondy et al., 2012). The subject of training is one of the most important topics, as the training aims to give the trainees knowledge, skills and attitudes in order to develop their performance and professional development.

Work Quality Strategy: The quality of work life is represented by increasing the participation of employees in the decisions made by managers and reflected in the form of improving productivity through the optimal use of individuals more than money, in addition to improving security and job health, return and job satisfaction and reduce the turnover rate (Moroccan, 2007; Al-Hawajreh et al., 2011). Al-Salem believes that the quality of work life is represented in the positive aspects of the work environment from the perspective of employees. Whenever the management of the organization has overcome the difficulties in achieving the work of its employees, and provides an appropriate organizational environment and climate and respect of the employees, the result of this is the enhancement of the conviction of the employees and the spirit of loyalty and belonging to the organization, which enhances job satisfaction, and reduces the work turnover in the organization.

Performance appraisal Strategy: This is a function that measures and evaluates the results of employees' performance, which identifies their opportunities for promotion and career path (Al-Hawary& Alajmi, 2017; Al-Lozi et al., 2018a, b; AlShehhi et al., 2020; Mohammad et al., 2020). This function is an incentive for further learning and development to keep pace with requirements and provides feedback on how employees benefit from the training and development they have received. And their need for other training and development programs (Byars, 2006; Alshurideh, 2019; Al Kurdi et al., 2020; Alsuwaidi et al., 2020; Kurdi et al., 2020).

Incentive and Remuneration Strategy: This strategy includes all the financial payments, bonuses and non-financial benefits provided by the organization to the employees for carrying out their work and activities, in order to attract qualified human resources, and maintain what is available in the organization, compensation is one of the most important factors that affect the motivation of individuals to grow, develop and continue Learning. Motivation is a critcal tool for greater productivity and improves the performance of the entire organization and achieves its goals (Al-Lozi et al., 2017; Casio, 2013).

2.2 Organizational Learning

Thomas and Allen define organizational learning as "the process that leads to continuous learning of the organization." Sun and Scott points out that organizational learning expresses "the learning process used in the organization, dealing with the question of how individuals learn within the organization." Hodgkinson argues that organizational learning occurs as a result of individuals' continuous interaction with each other during the learning process, resulting in their acquisition of experience. Farago refers to organizational learning as "all systems, mechanisms and processes used to continuously improve the capabilities of individuals, to achieve specific objectives concerning individuals and the organization." From the above definitions, it is noted that organizational learning reflects a process with specific components and mechanisms. This process is ongoing, as well as the need for interaction to improve the capacity of all members of the organization.

Finger and Brand argues that organizational learning is an activity and the process by which an organization can reach the stage of learning. This process generates new and useful ways of thinking to teach individuals how to work together. As a result of organizational learning, the organization reaches out to what is known as the Learning Organization. Organizational learning may be defined as: the continuous life cycle of the organization and the shared and planned vision to remain within the organizational community through a systematic effort and growing awareness stemming from the knowledge and experience of the leadership of the organization, and its competitive culture and strategy to bring about continuous change through monitoring information constantly, and to benefit from its expertise and experience and solve its problems to bring it and its personnel to the level that ensures the achievement of its policy, and management goals with the highest degree of efficiency and competition, and take sound decisions and improve Organizational performance.

The real development of the organizational learning concept in the early 1990s by Senge, who introduced the idea of adaptive learning and generative learning (Al-Hawary & Obiadat, 2021). Adaptive learning focuses on adapting to what happens in an organizational environment, this type of learning helps the organization to survive, but does not provide organizational learning. Sun and Scott notes that this type of learning does not require much time and cost. Generative learning builds new capabilities and deliberately discards previous working methods. Generative learning is necessary in the operational aspect of an organization operating in a highly changing environment. This type of learning has high costs. Wijnhoven points out that what distinguishes adaptive learning from generative learning is that adaptive learning relates to simple change that relates to the first stage of the learning process, whereas generative learning relates to the advanced and complementary phase of adaptive learning, which improves the organization's ability to discover abilities to modify behavior and create new knowledge and experience. Organizational learning consists of two dimensions:

Adaptive Learning: This relates to the subordinates' learning of the skills needed to accomplish new business, and the business processes that lead to gradual development in order to ensure that these companies survive and continue.

Generative Learning: It relates to learning in which subordinates are interested in trying to imagine the future of the work that concerns them and try to design, and requires that subordinates have a high degree of autonomy to try good methods.

3 Study Model

See Fig. 1.

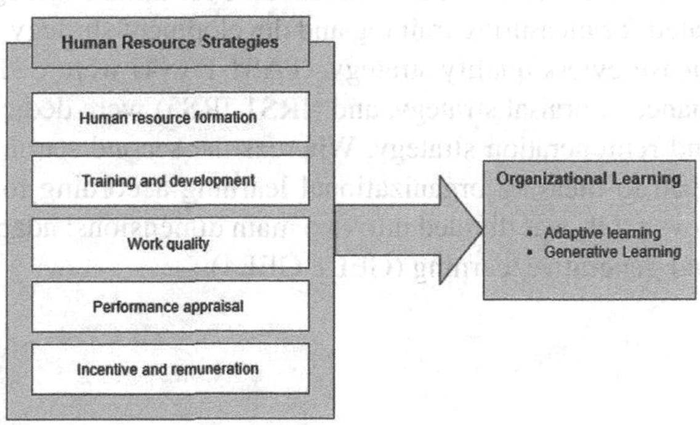

Fig. 1 Research model

H1: There is a statistically significant impact of Human Resources Management strategies on Organizational Learning of Islamic banks in Jordan.

4 Methodology

4.1 Population and Sample Selection

A qualitative method based on a questionnaire was used in this study for data collection and sample selection. The major aim of the study was to examine the impact of HRMS on OL. Therefore, it focused on Islamic banks operating in Jordan. Data were primarily gathered through self-reported questionnaires creating by Google Forms which were distributed to a purposive sample of managers via email. In total, (284) responses were received including (9) invalid to statistical analysis due to uncompleted or inaccurate. Hence, the final sample contained (275) responses suitable to analysis requirements, where it proved to be sufficient to the extent that was predictable and allowed for a presumption of data saturation (Sekaran & Bougie, 2016).

4.2 Measurement Instrument

A self-reported questionnaire that consists of two main sections along with a section regarding control variables was used as the measurement instrument. Control variables considered as categorical measures were composed of gender, age group, educational level, and experience. The two main sections were dealt with a five-point Likert scale (from 1 = strongly disagree to 5 = strongly agree). The first section contained (23) items to measure human resource management strategies based on (Al-Hawary & Metabis, 2012; Becker & Huselid, 2006; Combs et al., 2006). These items were distributed into dimensions as follows: items (HRF1-HRF5) were used to measure human resource formation strategy, (TDS1-TDS5) were dedicated for measuring training and development strategy, (WQS1WQS4) were used to measurework quality strategy, (PAS1-PAS4) were dedicated for measuring performance appraisal strategy, and (IRS1-IRS5) were dedicated for measuring incentive and remuneration strategy. Whereas the second section included eightitems developed to measure organizational learning according to what was pointed by Wijnhoven. OL was divided into two main dimensions: adaptive learning (ADL1-ADL4) and generative learning (GEL1-GEL4).

5 Findings

5.1 Measurement Model Evaluation

This study was conducted structural equation modeling (SEM) to test hypotheses, which represents a contemporary statistical technique for testing and estimating the relationship between factors and variables (Wang & Rhemtulla, 2021). Accordingly, the reliability and validity of the constructs were tested using confirmatory factor analysis (CFA) through the statistical program AMOSv24. Table 1 summarizes the results of convergent and discriminant validity, as well the indicators of reliability.

Table 1 shows that the standard loading values for the individual items were within the domain (0.592–0.881), these values greater than the minimum retention of the elements based on their standard loads (Al-Lozi et al., 2018a, b; Sung et al., 2019). Average variance extracted (AVE) is a summary indicator of the convergent validity of constructs that must be above 0.50 (Howard, 2018). The results indicate that the AVE values were greater than 0.50 for all constructs, thus the used measurement model has an appropriate convergent validity. Rimkeviciene et al. (2017) suggested the comparison approach as a way to deal with discriminant validity assessment in covariance-based SEM. This approach is based on comparing the values of maximum shared variance (MSV) with the values of AVE, as well as comparing the values of square root of AVE ($\sqrt{AVE}$) with the correlation between the rest of the structures. The results show that the values of MSV were smaller than the values of AVE, and that the values of $\sqrt{AVE}$ were higher than the correlation values among the rest of the constructs. Therefore, the measurement model used is characterized by discriminative validity. The internal consistency measured through Cronbach's Alpha coefficient (α) and compound reliability by McDonald's Omega coefficient (ω) was conducted as indicators to evaluate measurement model. The results listed in Table 1 demonstrated that both values of Cronbach's Alpha coefficient and McDonald's Omega coefficient were greater than 0.70, which is the lowest limit for judging on measurement reliability (de Leeuw et al., 2019).

5.2 Structural Model

The structural model illustrated no multicollinearity issue among predictor constructs because variance inflation factor (VIF) values are below the threshold of 5, as shown in Table 1 (Hair et al., 2017).This result is supported by the values of model fit indices shown in Fig. 1 (Fig. 2).

The results in Fig. 1 indicated that the chi-square to degrees of freedom (CMIN/DF) was 1.454, which is less than 3 the upper limit of this indicator. The values of the goodness of fit index (GFI), the comparative fit index (CFI), and the Tucker-Lewis index (TLI) were upper than the minimum accepted threshold of 0.90.

Table 1 Results of validity and reliability tests

Constructs	1	2	3	4	5	6	7
1. HRF	**0.743**						
2. TDS	0.458	**0.759**					
3. WQS	0.385	0.496	**0.745**				
4. PAS	0.492	0.538	0.513	**0.737**			
5. IRS	0.538	0.506	0.497	0.528	**0.726**		
6. ADL	0.671	0.695	0.711	0.662	0.682	**0.765**	
7. GEL	0.702	0.655	0.680	0.635	0.675	0.698	**0.781**
VIF	3.591	2.663	2.067	3.359	1.938	–	–
LR	0.627–0.833	0.682–0.821	0.607–0.861	0.592–0.837	0.674–0.761	0.702–0.830	0.706–0.881
AVE	0.552	0.577	0.555	0.544	0.527	0.585	0.610
MSV	0.468	0.503	0.394	0.510	0.437	0.462	0.507
α	0.856	0.868	0.829	0.822	0.845	0.848	0.858
ω	0.859	0.871	0.831	0.824	0.847	0.849	0.861

Note VIF: variance inflation factor, LR: loading ranges, AVE: average variance extracted, MSV: maximum shared variance, α: Cronbach's alpha, ω: MacDonald's omega, Bold font refer to $\sqrt{AVE}$

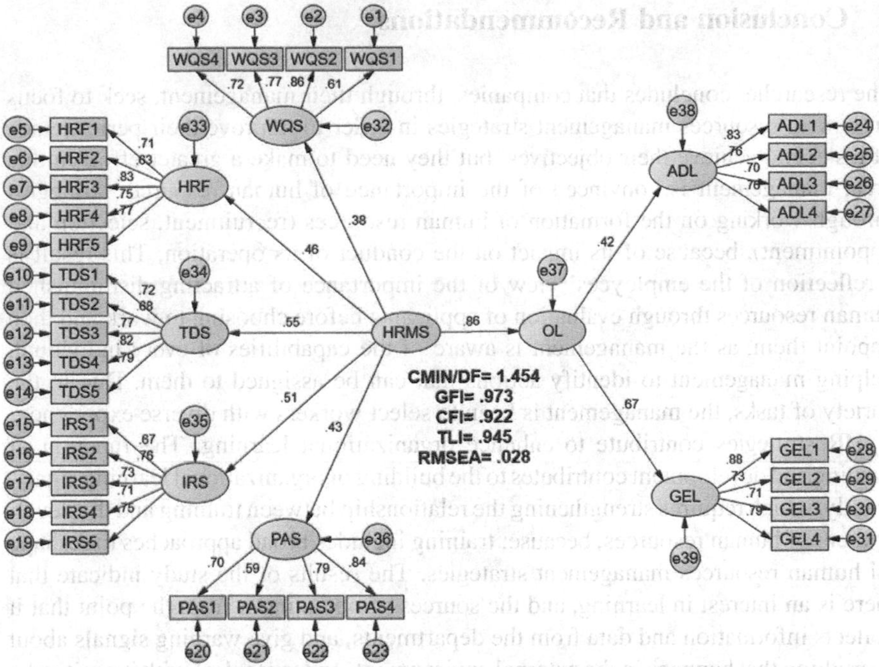

Fig. 2 SEM results of the HRMS effect on OL

Moreover, the result of root mean square error of approximation (RMSEA) indicated to value 0.028, this value is a reasonable error of approximation because it is less than the higher limit of 0.08. Consequently, the structural model used in this study was recognized as a fit model for predicting the DEP and generalization of its result (Ahmad et al., 2016; Shi et al., 2019).To verify the results of testing the study hypotheses, structural equation modeling (SEM) was used, the results of which are listed in Table 2.These results demonstrated that HRMS has a positive impact relationship on OL ($\beta = 0.858$, $t = 41.86$, $p = 0.000$), which justifies support for the study's major hypothesis.

Table 2 Hypothesis testing

Relation	Standard Beta	*t* value	*p* value
HRMS → OL	0.858	41.86***	0.000

Note HRMS: human resource management strategies, OL: organizational learning, * p < 0.05, ** p < 0.01, *** p < 0.001

6 Conclusion and Recommendations

The researcher concludes that companies, through their management, seek to focus on human resources management strategies in order to improve their performance of tasks and achieve their objectives, but they need to make a greater effort in this area, management is convinced of the importance of human resources strategies through working on the formation of human resources (recruitment, selection and appointment), because of its impact on the conduct of its operation, This result is a reflection of the employees' view of the importance of attracting distinguished human resources through evaluation of applicants before choosing to work and then appoint them, as the management is aware of the capabilities of every individual, helping management to identify actions that can be assigned to them. Due to the variety of tasks, the management is keen to select workers with diverse experience.

HR strategies contribute to enhance organizational learning. The function of training and development contributes to the building of organizational learning significantly, which requires strengthening the relationship between training and the development of human resources, because; training includes broad approaches to reshape all human resources management strategies. The results of the study indicate that there is an interest in learning, and the sources of obtaining it from the point that it collects information and data from the departments, and give warning signals about everything that happens in the external environment, and try to deal with it positively, and works to achieve the principle of participation among employees in the expertise and experience they possess, these operations are based on the formal structure owned by organization, and enhance the flow of knowledge through it, in addition to the application, and activation of work teams in most of the operations.

The results of the study showed an impact of human resources strategies (training and development, quality of work life, incentives and rewards) on organizational learning. The results show that there is an impact of training and development strategies, quality of work life, incentives and rewards and no impact of human resources formation and performance evaluation strategies on organizational learning. This result indicates that organizational learning processes need a long time and to be reinforced through incentives and rewards. It also needs a good organizational climate mechanism that leads to the consolidation and validation of information through training and knowledge programs. The result of the study agreed with Al-Kasasbeh et al., Momeni and Pourdsaid and Kumar and Idris (2006).

Based on the study results, researchers recommend managers of Islamic banks in Jordan to develop mechanisms that reward employees for learning new things, in a way that enhances adaptive and generative learning processes, and enhances confidence building among employees which allowing organizational learning processes of both adaptive and generative types. In addition; emphasizing the need to give workers the opportunity to freely choose when carrying out work duties, in a way that achieves competence and flexibility in carrying out work and enhances their satisfaction. Also departments management have to collect information and ideas from their practical experience in a way that enhances the enrichment of adaptive and generative organizational learning processes, and transfers them to other workers in various departments.

References

Ahmad, S., Zulkurnain, N., & Khairushalimi, F. (2016). Assessing the validity and reliability of a measurement model in structural equation modeling (SEM). *British Journal of Mathematics & Computer Science, 15*(3), 1–8. https://doi.org/10.9734/BJMCS/2016/25183

Al Kurdi, B., & Alshurideh, M., Al afaishat, T. (2020). Employee retention and organizational performance: Evidence from banking industry. *Management Science Letters,* 10(16), 3981–3990.

Al Kurdi, B., Elrehail, H., Alzoubi, H., Alshurideh, M., & Al-Adaila, R. (2021). The interplay among HRM practices, job satisfaction and intention to leave: An empirical investigation. *Journal of Legal, Ethical and Regulatory, 24*(1), 1–14.

Alameeri, K. A., Alshurideh, M. T., & Al Kurdi, B. (2021). The effect of Covid-19 pandemic on business systems' innovation and entrepreneurship and how to cope with it: a theatrical view. *The Effect of Coronavirus Disease (COVID-19) on Business Intelligence*, 334, 275–288.

Alameeri, K., Alshurideh, M., Al Kurdi, B., & Salloum, S. A. (2020, October). The effect of work environment happiness on employee leadership. In *International Conference on Advanced Intelligent Systems and Informatics* (pp. 668–680). Springer, Cham.

AlHamad, A., Alshurideh, M., Alomari, K., Kurdi, B., Alzoubi, H., Hamouche, S., & Al-Hawary, S. (2022). The effect of electronic human resources management on organizational health of telecommunications companies in Jordan. *International Journal of Data and Network Science,* 6(2), 429–438. *International Journal of Data and Network Science* 6 (2), 429–438.

Al-Hawajreh, K., AL-Zeaud, H., Al-Hawary, S. I., & Mohammad, A. A. (2011). The influence of top management support and commitment on total quality management indicators from managers and heads of departments viewpoint: A case study of Sahab Industrial City. *Jordan Journal of Business Administration, 7*(4), 557–576.

Al-Hawary, S. I. (2011). Human resource management practices in ZAIN cellular communications company operating in Jordan. *Perspectives of Innovations, Economics and Business, 8*(2), 26–34.

Al-Hawary, S. I. (2015). Human resource management practices as a success factor of knowledge management implementation at health care sector in Jordan. *International Journal of Business and Social Science, 6*(11/1), 83–98.

Al-Hawary, S. I., & Abu-Laimon, A. A. (2013). The impact of TQM practices on service quality in cellular communication companies in Jordan. *International Journal of Productivity and Quality Management, 11*(4), 446–474.

Al-Hawary, S. I., & Alajmi, H. M. (2017). Organizational commitment of the employees of the ports security affairs of the State of Kuwait: The impact of human recourses management practices. *International Journal of Academic Research in Economics and Management Sciences, 6*(1), 52–78.

Al-Hawary, S. I., & Aldaihani, F. M. (2016). Customer relationship management and innovation capabilities of Kuwait airways. *International Journal of Academic Research in Economics and Management Sciences, 5*(4), 201–226.

Al-Hawary, S. I., & Al-Hamwan, A. (2017). Environmental analysis and its impact on the competitive capabilities of the commercial banks operating in Jordan. *International Journal of Academic Research in Accounting, Finance and Management Sciences, 7*(1), 277–290.

Al-Hawary, S. I., & Al-Kumait, Z. (2017). Training Programs and their effect on the employees performance at King Hussain Bin Talal Development Area at Al—Mafraq Governate in Jordan. *International Journal of Academic Research in Economics and Management Sciences, 6*(1), 258–274.

Al-Hawary, S. I., & Al-Namlan, A. (2018). Impact of electronic human resources management on the organizational learning at the private hospitals in the state of Qatar. *Global Journal of Management and Business Research: A Administration and Management, 18*(7), 1–11.

Al-Hawary, S. I., & Hadad, T. F. (2016). The effect of strategic thinking styles on the enhancement competitive capabilities of commercial banks in Jordan. *International Journal of Business and Social Science, 7*(10), 133–144.

Al-Hawary, S. I., & Ismael, M. (2010). The effect of using information technology in achieving competitive advantage strategies: A field study on the Jordanian pharmaceutical companies. *Al Manara for Research and Studies, 16*(4), 196–203.

Al-Hawary, S. I., & Metabis, A. (2012). Implementation of internal marketing in Jordan Banks. *International Journal of Data Analysis and Information, 4*(1), 37–53.

Al-Hawary, S. I., & Nusair, W. (2017). Impact of human resource strategies on perceived organizational support at Jordanian Public Universities. *Global Journal of Management and Business Research: A Administration and Management, 17*(1), 68–82.

Al-Hawary, S. I., & Shdefat, F. (2016). Impact of human resources management practices on employees' satisfaction a field study on the Rajhi Cement Factory. *International Journal of Academic Research in Accounting, Finance and Management Sciences, 6*(4), 274–286.

Al-Hawary, S. I., AL-Awawdeh, W., & Abden, M. A. (2012). The impact of the leadership style on organizational commitment: A field study on Kuwaiti Telecommunications Companies. *ALEDARI*, (130), 53–102.

Al-Hawary, S. I., Al-Qudah, K., Abutayeh, P., Abutayeh, S., & Al-Zyadat, D. (2013). The Impact of internal marketing on employee's job satisfaction of commercial banks in Jordan. *Interdisciplinary Journal of Contemporary Research in Business, 4*(9), 811–826.

Al-Hawary, S. I. S., & Alwan, A. M. (2016). Knowledge management and its effect on strategic decisions of Jordanian Public Universities. *Journal of Accounting-Business & Management, 23*(2), 24–44.

Al-Hawary, S. I. S., & Obiadat, A. A. (2021). Does mobile marketing affect customer loyalty in Jordan? *International Journal of Business Excellence, 23*(2), 226–250.

Al-Hawary, S. I. S., Abdul Aziz Allahow, T. J., & Aldaihani, F. M. F. (2018). Information technology and administrative innovation of the central agency for information technology in Kuwait. *Global Journal of Management and Business*, 18(11-A), 1–16.

Al-Hawary, S. I. S., Mohammad, A. S., Al-Syasneh, M. S., Qandah, M. S. F., & Alhajri, T. M. S. (2020). Organisational learning capabilities of the commercial banks in Jordan: Do electronic human resources management practices matter? *International Journal of Learning and Intellectual Capital, 17*(3), 242–266.

Al-Hawary, S. I., & Al-Syasneh, M. S. (2020). Impact of dynamic strategic capabilities on strategic entrepreneurship in presence of outsourcing of five stars hotels in Jordan. *Business: Theory and Practice, 21*(2), 578–587.

Alkalha, Z., Al-Zu'bi, Z., Al-Dmour, H., Alshurideh, M., & Masa'deh, R. (2012). Investigating the effects of human resource policies on organizational performance: An empirical study on commercial banks operating in Jordan. *European Journal of Economics, Finance and Administrative Sciences*, 51(1), 44–64.

Al-Lozi, M., Almomani, R. Z., & Al-Hawary, S. I. (2017). Impact of talent management on achieving organizational excellence in Arab Potash Company in Jordan. *Global Journal of Management and Business Research: A Administration and Management, 17*(7), 15–25.

Al-Lozi, M., Almomani, R. Z., & Al-Hawary, S. I. (2018a). Talent Management strategies as a critical success factor for effectiveness of Human Resources Information Systems in commercial banks working in Jordan. *Global Journal of Management and Business Research: A Administration and Management, 18*(1), 30–43.

AlShehhi, H., Alshurideh, M., Al Kurdi, B., & Salloum, S. A. (2020, October). The impact of ethical leadership on employees performance: A systematic review. In *International Conference on Advanced Intelligent Systems and Informatics* (pp. 417–426). Springer, Cham.

Alshraideh, A. T. R., Al-Lozi, M., & Alshurideh, M. T. (2017). The impact of training strategy on organizational loyalty via the mediating variables of organizational satisfaction and organizational performance: An empirical study on Jordanian agricultural credit corporation staff. *Journal of Social Sciences (COES&RJ-JSS)*, 6(2), 383–394.
Alshurideh, D. M. (2019). Do electronic loyalty programs still drive customer choice and repeat purchase behaviour? *International Journal of Electronic Customer Relationship Management, 12*(1), 40–57.
Alshurideh, M. T., Al Kurdi, B., Alzoubi, H. M., Ghazal, T. M., Said, R. A., AlHamad, A. Q., ... & Al-kassem, A. H. (2022). Fuzzy assisted human resource management for supply chain management issues. *Annals of Operations Research*, 1–19.
Alshurideh, M., Kurdi, B. A., Shaltoni, A. M., & Ghuff, S. S. (2019). Determinants of pro-environmental behaviour in the context of emerging economies. *International Journal of Sustainable Society, 11*(4), 257–277.
Alsuwaidi, M., Alshurideh, M., Al Kurdi, B., & Salloum, S. A. (2020, October). Performance appraisal on employees' motivation: a comprehensive analysis. In *International Conference on Advanced Intelligent Systems and Informatics* (pp. 681–693). Springer, Cham.
Altamony, H., Masa'deh, R., Alshurideh, M., Obeidat, B. (2012) Information systems for competitive advantage: Implementation of an organisational strategic management process. *Innovation and Sustainable Competitive Advantage: From Regional Development to World Economies*, 583–592.
Alzoubi, H., Alshurideh, M., Kurdi, B., Akour, I., & Aziz, R. (2022). Does BLE technology contribute towards improving marketing strategies, customers' satisfaction and loyalty? The role of open innovation. *International Journal of Data and Network Science, 6*(2), 449–460.
Armstrong, M. (2006). *A hand book of human resource management practice* (10th ed.). Kogan Page Limited.
Armstrong, M. (2012). *Armstrong's hand book of human resource management practice*. London, Philadelphia and New Delhi: Kogan Page Limited.
Arthur, D. (2001). Electronic Recruitment. In *The employee recruitment and retention handbook*. U.S.A: Amacon, 001.
Artis, C. R., Becker, B., & Huselid, M. A. (1999). Strategic human resource management at Lucent. *Human Resource Management, 38*(4), 309–313.
Barber, D., Huselid, M. A., & Becker, B. (1999). Strategic human resource management at Quantum. *Human Resource Management, 38*(4), 321–328.
Becker, B., & Huselid, M. A. (1999). Overview: Strategic human resource management in five leading firms. *Human Resource Management, 38*(4), 287–301.
Becker, B., & Huselid, M. A. (2006). Strategic human resource management: Where do we go from here. *Journal of Management, 32*(6), 898–925.
Boxall, P., & Purcell, J. (2011). *Strategy and human resource management* (3rd ed.). Palgrave McMillan.
Byars, L. L., & Rue, L. W. (2006). *Human resource management* (2nd ed.). McGraw Hill Irwin.
Combs, J., Liu, Y. M., Hall, A., & Ketchen, D. (2006). How much do high performance work practices matter? A meta-analysis of their effects on organizational performance. *Personnel Psychology, 59*, 501–528.
de Leeuw, E., Hox, J., Silber, H., Struminskaya, B., & Vis, C. (2019). Development of an international survey attitude scale: Measurement equivalence, reliability, and predictive validity. *Measurement Instruments for the Social Sciences, 1*(1), 9. https://doi.org/10.1186/s42409-019-0012-x
Devanna, M. A., Fombrun, C., & Tichy, N. (1981). Human resource management: A strategic perspective. *Organizational Dynamics, 9*(3), 51–67.
Guest, D. (1997). Human resource management and performance: A review and research agenda. *International Journal of Human Resource Management, 8*(3), 263–276.

Hair, J. F., Babin, B. J., & Krey, N. (2017). Covariance-based structural equation modeling in the journal of advertising: Review and recommendations. *Journal of Advertising, 46*(1), 163–177. https://doi.org/10.1080/00913367.2017.1281777

Holbeche, L. (2001). *Aligning human resources and business strategy*. Butterworth-Heinemann.

Howard, M. C. (2018). The convergent validity and nomological net of two methods to measure retroactive influences. *Psychology of Consciousness: Theory, Research, and Practice, 5*(3), 324–337. https://doi.org/10.1037/cns0000149

Huselid, M. A. (1995). The impact of human resource management practices on turnover, productivity, and corporate financial performance. *Academy of Management Journal, 38*, 635–672.

Kurdi, B., Alshurideh, M., & Alnaser, A. (2020). The impact of employee satisfaction on customer satisfaction: Theoretical and empirical underpinning. *Management Science Letters, 10*(15), 3561–3570.

Lee, F. & Lee, F. (2007). The relationships between HRM practices, leadership style, competitive strategy and business performance in Taiwanese steel industry. Paper presented at the 13th Asia Pacific Management Conference, Melbourne, Australia.

Metabis, A., & Al-Hawary, S. I. (2013). The impact of internal marketing practices on services quality of commercial banks in Jordan. *International Journal of Services and Operations Management, 15*(3), 313–337.

Mohammad, A. A., Alshura, M. S., Al-Hawary, S. I. S., Al-Syasneh, M. S., & Alhajri, T. M. (2020). The influence of Internal Marketing Practices on the employees' intention to leave: A study of the private hospitals in Jordan. *International Journal of Advanced Science and Technology, 29*(5), 1174–1189.

Mondy, Wayne R., Noe, Robert M. & Mondy, Judy bandy (2012). Human resource management, (2nd ed.). Boston: Person Education Limited

Nankervis, A., Compton, R., & Baird, M. (2005). *Human resource management: strategies and processes* (5th ed.). Thomson Learning Australia.

Nuseir, M. T., Aljumah, A., & Alshurideh, M. T. (2021). How the business intelligence in the new startup performance in UAE during covid-19: the mediating role of innovativeness. The Effect of Coronavirus Disease (COVID-19) on Business Intelligence, 334, 63–79.

Obeidat, U., Obeidat, B., Alrowwad, A., Alshurideh, M., Masadeh, R., & Abuhashesh, M. (2021). The effect of intellectual capital on competitive advantage: The mediating role of innovation. *Management Science Letters, 11*(4), 1331–1344.

Pfeffer, J. (2005). Producing sustainable competitive advantage through the effective management of people. *Academy of Management Executive, 19*(4), 95–106.

Rimkeviciene, J., Hawgood, J., O'Gorman, J., & De Leo, D. (2017). Construct validity of the acquired capability for suicide scale: factor structure, convergent and discriminant validity. *Journal of Psychopathology and Behavioral Assessment, 39*(2), 291–302. https://doi.org/10.1007/s10862-016-9576-4

Schuler, R. S., & Jackson, S. E. (Eds.). (1999). *Strategic human resource management*. Blackwell.

Sekaran, U., & Bougie, R. (2016). *Research methods for business: A skill-building approach* (Seventh edition). Wiley.

Shamout, M., Elayan, M., Rawashdeh, A., Kurdi, B., & Alshurideh, M. (2022). E-HRM practices and sustainable competitive advantage from HR practitioner's perspective: A mediated moderation analysis. *International Journal of Data and Network Science, 6*(1), 165–178.

Sheppeck, M., & Militello, J. (2000). Strategic HR configurations and organizational performance. *Human Resource Management, 39*(1), 5–16.

Shi, D., Lee, T., & Maydeu-Olivares, A. (2019). Understanding the model size effect on SEM Fit Indices. *Educational and Psychological Measurement, 79*(2), 310–334. https://doi.org/10.1177/0013164418783530

Stavrou-Costea, E. (2005). The challenges of human resource management towards organizational effectiveness. *Journal of European Industrial Training, 29*(2), 112–134.

Sung, K.-S., Yi, Y. G., & Shin, H.-I. (2019). Reliability and validity of knee extensor strength measurements using a portable dynamometer anchoring system in a supine position. *BMC Musculoskeletal Disorders, 20*(1), 1–8. https://doi.org/10.1186/s12891-019-2703-0

Wang, Y. A., & Rhemtulla, M. (2021). Power Analysis for Parameter Estimation in Structural Equation Modeling: A Discussion and Tutorial. *Advances in Methods and Practices in Psychological Science, 4*(1), 1–17. https://doi.org/10.1177/2515245920918253

Wilton, N. (2006). Strategic choice and organisational context in HRM in the UK hotel sector. *The Service Industries Journal, 26*(8), 903–919.

Wright, P. M., Dunford, B. B., & Snell, S. A. (2001). Human resources and resource based view of the firm. *Journal of Management, 27*(6), 701–721.

Wright, P. M., Gardner, T. M., & Moynihan, L. M. (2003). The impact of HR practices on the performance of business units. *Human Resource Management Journal, 13*(3), 21–36.

Youndt, M., Snell, S., Dean, J., & Lepak, D. (1996). Human resource management, manufacturing strategy, and firm performance. *The Academy of Management Journal, 39*(4), 836–866.

Zu'bi, Z., Al-Lozi, M., Dahiyat, S., Alshurideh, M., & Al Majali, A. (2012). Examining the effects of quality management practices on product variety. *European Journal of Economics, Finance and Administrative Sciences*, 51(1), 123–139.

The Impact of Organizational Structure Characteristic on Administrative Communication Efficiency: Evidence from Telecommunication Companies in Jordan

Mohammad Fathi Almaaitah, Doa'a Ahmad Odeh Al-Husban, Riad Ahmad Mohammed Abazeed, Ibrahim Rashed Soliaman AlTaweel, Nida'a Al-Husban, Muhammad Turki Alshurideh, Sulieman Ibraheem Shelash Al-Hawary, Ayat Mohammad, and Anber Abraheem Shlash Mohammad

Abstract The major aim of the study was to examine the impact of organizational structure characteristic on administrative communication efficiency. Therefore, it focused on telecommunication companies operating in Jordan. Data were primarily gathered through self-reported questionnaires creating by Google Forms which were distributed to a purposive sample of managers at different levels via email.

M. F. Almaaitah
Faculty of Economic and Administration Sciences, Department of Business Administration & Public Administration, Al Al-Bayt University Jordan, P.O. Box 130040, Mafraq 25113, Jordan
e-mail: m.maaitah@aabu.edu.jo

D. A. O. Al-Husban
Faculty of Alia College, Department of Human Fundamental Sciences, Al-Balqa Applied University, Amman, Jordan
e-mail: D_husban@bau.edu.jo

R. A. M. Abazeed
Faculty of Finance and Business Administration, Department of Business Administration, Al Al-Bayt University, P.O. BOX 130040, Mafraq 25113, Jordan

I. R. S. AlTaweel
Faculty of Business School, Al Russ City, Department of Business Administration, Qussim University, P.O. BOX 6502, Al Russ 51452, Saudi Arabia
e-mail: toiel@qu.edu.sa

N. Al-Husban
Faculty of Economics and Administrative Sciences, Department of Business Administration, Al Al-Bayt University, P.O. BOX 130040, Mafraq 25113, Jordan

M. T. Alshurideh
Department of Marketing, School of Business, The University of Jordan, Amman 11942, Jordan
e-mail: malshurideh@sharjah.ac.ae; m.alshurideh@ju.edu.jo

Department of Management, College of Business, University of Sharjah, 27272 Sharjah, United Arab Emirates

M. Alshurideh et al. (eds.), *The Effect of Information Technology on Business and Marketing Intelligence Systems*, Studies in Computational Intelligence 1056,
https://doi.org/10.1007/978-3-031-12382-5_47

Structural equation modeling (SEM) was used to test hypotheses. The study results showed that organizational structure characteristic has a positive impact on administrative communication efficiency. Based on the results of the study, the researchers recommend managers and decision makers to pay more attention to the dimensions related to the characteristics of the organizational structure (specialization and flexibility) that will affect the enhancement of the level of effectiveness of administrative communication.

Keywords Organizational Structure Characteristic · Administrative Communication Efficiency · Telecommunication companies · Jordan

1 Introduction

Human organization has existed and developed since the dawn of history. Man, by nature, cannot live in isolation from others, so human groups emerged, and these groups did not find a way to organize except by choosing a leader for this group who addresses its problems, settles its affairs and achieves harmony among its members. With the passage of time, organizations appeared that soon played a major role in our daily lives, whether at the level of individual life or on the level of society as a whole or the scope of the economy (Al-Hawary, 2011; Al-Hawary & Al-Hamwan, 2017; Al-Nady et al., 2016). It is difficult to imagine modern societies without these organizations that have spread rapidly to provide goods and services to the public, which calls for studying these organizations, and their structural construction to provide the appropriate conditions for their survival, growth and continuity and creativity, and providing goods and services to the beneficiaries (Al-Hawary, 2015; Al-Hawary & Aldaihani, 2016; Alsharari & Alshurideh, 2020; Bebba et al., 2017). Also, good communication means the blood that runs in the veins of organizations, and lack of effective communication is often the problem of organizations and one of their important obstacles (Aljumah et al., 2021; Alshurideh et al., 2014; Alyammahi et al., 2020). Both sides of the communication process have different desires, trends and needs, and this represents an important obstacle to the success of this process (Alshurideh et al., 2019; Alameeri et al., 2020; Alshurideh, 2022).

S. I. S. Al-Hawary (✉)
Faculty of Economics and Administrative Sciences, Department of Business Administration, Al Al-Bayt University, P.O. BOX 130040, Mafraq 25113, Jordan
e-mail: dr_sliman73@aabu.edu.jo

A. Mohammad
Business and Finance Faculty, The World Islamic Science and Education University (WISE), P.O. Box 1101, Amman 11947, Jordan

A. A. S. Mohammad
Faculty of Administrative and Financial Sciences, Marketing Department, Petra University, P.O. Box 961343, Amman 11196, Jordan

Communication can be defined as a social process by which an individual communicates with others to achieve common interests or benefits through cooperation, or conflicts and incompatibility in the event of competition and conflict as a method of dealing (Al-Hawary & Batayneh, 2010). It can also be said that the process of communication, as defined by the Oxford Dictionary, is "the transfer of ideas and information and their communication or exchange by words, writing or sign." Information or ideas are exchanged between a sender and a receiver or a sender and two receivers. When we speak, we want someone to hear us. When we write, we want someone to read us, when we use gestures and smiles, we want someone to receive and respond to them with similar gestures. The importance of the organizational structure in the life and continuity of organizations, whether governmental or private, highlights its interest in dividing and distributing work among individuals and its role in unifying efforts in order to achieve the goals of the organization, preventing overlapping of powers and reducing conflicts over competencies, and its role is also in the optimal use of human competencies in terms of distributing job roles, defining activities and facilitating the process of managerial communication between managerial levels (Al-Hawary et al., 2011; Nuseir et al., 2021; AlHamad et al., 2022; Alshurideh et al., 2022; Shamout et al., 2022). The current study comes to identify the impact of the characteristics of the organizational structure on the effectiveness of administrative communication.

2 Theoretical Framework and Hypotheses Development

2.1 Organizational Structure Characteristics

The issue of the organizational structure has occupied and continues to be an important space in administrative and organizational thought, and it is still receiving increasing attention by thinkers, consultants and managers; this is because it is a vital way to help organizations accomplish their work and achieve their goals efficiently and effectively (Al-Hawary, 2009; Al-Hawary et al., 2012; Agha et al., 2021). Because it is a major variable that affects many variables and organizational aspects of any organization. The attention of thinkers and researchers focused on determining the dimensions of the organizational structure and their relationship to the organizational performance, its efficiency, effectiveness, flexibility, adaptability, and other criteria for the success of organizations that were mentioned in many studies related to measuring the performance and success of organizations (Alhalalmeh et al., 2020; Al-Hawary & Mohammed, 2017; Mohammad et al., 2020; Al-Hawary & AlDafiri, 2017; Kabrilyants et al., 2021). The managerial literature indicated that there is no general agreement on one comprehensive definition of the organizational structure, as the views and concepts they presented varied in terms of comprehensiveness and depth, but they see that the organizational structure is only a means and a tool to achieve the goals of the organization (Al-Hawajreh et al., 2011; Aburayya et al., 2020).

Robbins defines the organizational structure as "the way in which tasks are organized and the main roles of employees are defined, and the information exchange system is identified and the coordination mechanisms and interaction patterns necessary between the different departments and their employees are identified." Hall believes that the idea of the organizational structure can be similar to the structure of the building that has pillars, walls, ceilings and cabins, and it is this structure that determines the movements and activities of the individuals who are inside it. Daft (1986) defines the organizational structure as a reflection of the organizational map, which is a clear representation of all activities and operations in organizations.

Organizational structure is a complex concept and it cannot be considered as a single element. Some writers link the organizational structure to the chain of authority, the scope of supervision, the division of labor, and so on (Ahmad et al., 2021; Alameeri et al., 2021). As is the difference in defining the definition of the organizational structure, there is a difference in defining its characteristics. Robbins refers to that difference that the organizational structure consists of complex, formal, and centralization. As for the team of researchers from Aston School, they see that the organizational structure consists of formality, specialization and centralization, formality and profiling. In view of the lack of consensus on the main characteristics of the organizational structure, and the opinion of most researchers about the main characteristics of the organizational structure revolved around: (complexity, formality, centralization, flexibility, and specialization), the study will adopt these characteristics as dimensions of the organizational structure, which are as follows.

Complexity: It refers to "the amount and volume of work that reflects lines of authority, verticality, geographical dimension of organizational units, and difficulty of coordination". The organizational structure also represents the number of activities, jobs and departments within the organization, the degree of specialization, division of work, managerial levels, and geographical locations (Al-Hawary & Al-Syasneh, 2020; Mohammad, 2020). The more these elements increase, the more complex the organizational structure. The importance of complexity arises in that when the degree of complexity of the organizational structure increases, the organization's need for communication, coordination and control increases. This places an increasing burden on management towards more communication and better control (Al Kurdi et al., 2020; Alshura et al., 2016).

Formalization: The term formality refers to the extent to which the organization relies on laws, regulations, rules, and directives to pressure the behavior of individuals during work, and the degree of formality may change significantly between organizations and within a single organization, as there are certain businesses that do not need high formality. Formality varies according to the standard of craftsmanship, or the degree of skill, and it also varies according to the managerial levels (Mohammad et al., 2020; Al-Hawary & Shdefat, 2016; Mohammad et al., 2014; Allahow et al., 2018). The nature of work in the management owned by the manager increases as he rises to the top in the organizational structure of the organization.

Centralization: It refers to the degree of distribution of decision-making power in the organization at any level, and the more the decision-making right is limited to the higher levels of the organization, the more this indicates a high degree of centralization in the organization and vice versa (Ahmed et al., 2020; Al-Tarawneh et al, 2012; Harahsheh et al., 2021).

Specialization: It means "the degree of dividing the organization's tasks into separate sub-works from each other. If the specialization is accurate, we will find that each of the workers performs limited tasks, but if the specialization is inaccurate, we find each worker performs a wide range of tasks."

Flexibility: It refers to the ability of the organizational structure to self-change, and it depends on the degree of response in the structure to the processes of change resulting from the internal and external developments and influences imposed by its self-capacity for expansion and absorption (Al-Lozi et al., 2017; Al-Hawary & Al-Syasneh, 2020; Alkitbi et al., 2020).

2.2 Administrative Communication

Studies and research indicate that communication in public organizations represents the lifeblood of these organizations, and the backbone on which they depend to provide their services to all beneficiaries. In this regard, it should be noted that public organizations, in their essence and orientations, are institutions that work for the common good, and not for the benefit of a particular group or party at the expense of others. Therefore, the basis for its survival depends on its ability to communicate with the broad segments of people who find in the services of these institutions a safe and sure haven away from the cases of attrition and waste that result from the behavior of some business organizations (Al-Hawary et al., 2020; Al-Nady et al., 2013; Metabis & Al-Hawary, 2013). Also, public organizations is a term that refers to all institutions that provide goods, services and ideas and provide advice and guidance to citizens based on their orientation to the citizen and their adoption of methods that would advance the society and its members. On this basis, the organizations' orientations are often societal and ethical.

Daft defined communication as: "the use of words, letters or any similar means to share information about a topic or an event." Effective communication means "the occurrence of a perfect match between the meaning sent by the sender and the meaning understood by the receiver. Whereas, efficient communication means that communication takes place at the lowest possible cost, and communication in order to be successful must be efficient and effective at the same time." (Al-Hawary & Metabis, 2012; Al-Hawary & Nusair, 2017; Al-Hawary & Obiadat, 2021). The efficiency of the communication process depends on the extent to which the two parties to the communication (the sender and receiver) understand each other. Some management scholars, in defining the effectiveness of communication, tend to determine the return on behavior that is issued by an individual after the communication, without paying attention to the individual from whom that behavior originates. Therefore, effectiveness is defined as "a characteristic of the behavior in which you expend energy and have a financial impact."

2.3 *The Organizational Structure and the Effectiveness of Administrative Communication*

The organizational structure is a mechanism to achieve the goals that organizations seek to achieve, as the organization has been found to achieve certain goals. A good organizational structure helps to focus efforts to link activities with the goals to be achieved, clarify how work flows, and avoid duplication of work and conflict in tasks. The organizational structure serves the organization by enabling it to make optimal use of the available resources, whether financial or human, which leads to an increase in the organization's ability to implement its goals easily and smoothly, facilitating and avoiding overlap and duplication between the various administrative processes and activities, defining the roles of individuals working in the organization, and reducing pressures Work, and contributes to increasing the ability of organizations to deal with internal and external variables that the organization may be exposed to, and helps avoid work differences.

The success of the organization in building a successful organizational structure depends on its ability to work to find an appropriate work environment and its ability to achieve a high degree of adaptation, congruence and harmonization between its organizational structure and its objectives, as well as the efficiency of the human element, and its material resources, this shows the importance of the organizational structure of the organization. Without a good appropriate organizational structure, the organizational processes go in a chaotic manner that is not based on a solid practical basis, as the organization flounders and deviates from its path and its goals, becoming useless and heading for decline, in addition to wasting human and material resources. An inappropriate organizational structure has negative effects in terms of low Morale and motivation of employees, improper decision-making, increased organizational and job conflicts, decreased ambition and creativity, and increased expenditures. Based on the above, the hypothesis of the study can be formulated as follows.

There is an effect of the characteristics of the organizational structure on the effectiveness of administrative communication.

3 Study Model

See Fig. 1.

Fig. 1 Research model

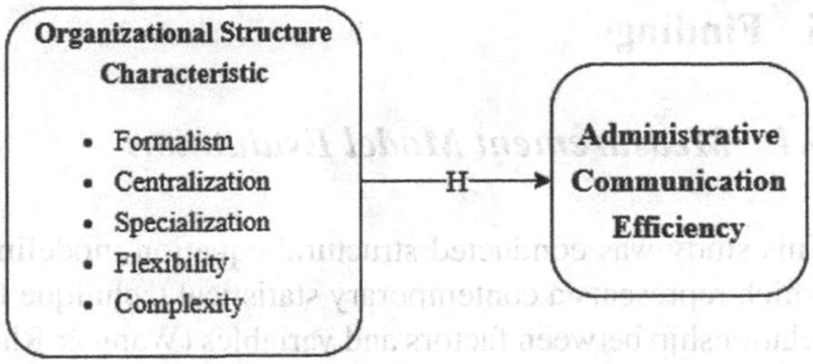

4 Methodology

4.1 Population and Sample Selection

A qualitative method based on a questionnaire was used in this study for data collection and sample selection. The major aim of the study was to examine the impact of organizational structure characteristic on administrative communication efficiency. Therefore, it focused on telecommunication companies operating in Jordan. Data were primarily gathered through self-reported questionnaires creating by Google Forms which were distributed to a purposive sample of managers at different levels via email. In total, (173) responses were received including (18) invalid to statistical analysis due to uncompleted or inaccurate. Hence, the final sample contained (155) responses suitable to analysis requirements, where it proved to be sufficient to the extent that was predictable and allowed for a presumption of data saturation (Sekaran & Bougie, 2016).

4.2 Measurement Instrument

A self-reported questionnaire that consists of two main sections along with a section regarding control variables was used as the measurement instrument. Control variables considered as categorical measures were composed of gender, age group, educational level, and experience. The two main sections were dealt with a five-point Likert scale (from 1 = strongly disagree to 5 = strongly agree). The first section contained (21) items to measure organizational structure based on (Daft, 1986). These items were distributed into dimensions as follows: four items dedicated for measuring formalism (FO1-FO4), four items dedicated for measuring centralization (CE1-CE4), five items dedicated for measuring specialization (SP1-SP5), four items dedicated for measuring flexibility (FL1-FL4), and four items dedicated for measuring complexity (CO1-CO4). Whereas the second section included six items developed to measure administrative communication efficiency (ACE1-ACE6) according to what was pointed by (Al-Hawary & Batayneh, 2010).

5 Findings

5.1 Measurement Model Evaluation

This study was conducted structural equation modeling (SEM) to test hypotheses, which represents a contemporary statistical technique for testing and estimating the relationship between factors and variables (Wang & Rhemtulla, 2021). Accordingly, the reliability and validity of the constructs were tested using confirmatory factor analysis (CFA) through the statistical program AMOSv24. Table 1 summarizes the results of convergent and discriminant validity, as well the indicators of reliability.

Table 1 shows that the standard loading values for the individual items were within the domain (0.641–0.864), these values greater than the minimum retention of the elements based on their standard loads (Al-Lozi et al., 2018; Sung et al., 2019). Average variance extracted (AVE) is a summary indicator of the convergent validity of constructs that must be above 0.50 (Howard, 2018). The results indicate that the AVE values were greater than 0.50 for all constructs, thus the used measurement model has an appropriate convergent validity. Rimkeviciene et al. (2017) suggested the comparison approach as a way to deal with discriminant validity assessment in covariance-based SEM. This approach is based on comparing the values of maximum shared variance (MSV) with the values of AVE, as well as comparing the values of square root of AVE ($\sqrt{AVE}$) with the correlation between the rest of the structures. The results show that the values of MSV were smaller than the values of AVE, and that the values of $\sqrt{AVE}$ were higher than the correlation values among the

Table 1 Results of validity and reliability tests

Constructs	1	2	3	4	5	6
1. FO	**0.769**					
2. CE	0.514	**0.779**				
3. SP	0.442	0.439	**0.756**			
4. FL	0.527	0.495	0.552	**0.774**		
5. CO	0.439	0.537	0.506	0.486	**0.776**	
6. ACE	0.698	0.715	0.708	0.725	0.702	**0.754**
VIF	1.647	1.254	2.778	2.366	2.013	–
Loadings range	0.691–0.884	0.653–0.846	0.642–0.822	0.711–0.824	0.675–0.836	0.641–0.864
AVE	0.591	0.607	0.571	0.598	0.602	0.568
MSV	0.517	0.531	0.492	0.503	0.524	0.484
Internal consistency	0.848	0.854	0.866	0.855	0.854	0.883
Composite reliability	0.851	0.860	0.868	0.856	0.857	0.886

Note Bold fonts refer to square root of average variance extracted

rest of the constructs. Therefore, the measurement model used is characterized by discriminative validity. The internal consistency measured through Cronbach's Alpha coefficient (α) and compound reliability by McDonald's Omega coefficient (ω) was conducted as indicators to evaluate measurement model. The results listed in Table 1 demonstrated that both values of Cronbach's Alpha coefficient and McDonald's Omega coefficient were greater than 0.70, which is the lowest limit for judging on measurement reliability (De Leeuw et al., 2019).

5.2 *Structural Model*

The structural model illustrated no multicollinearity issue among predictor constructs because variance inflation factor (VIF) values are below the threshold of 5, as shown in Table 1 (Hair et al., 2017).This result is supported by the values of model fit indices shown in Fig. 2.

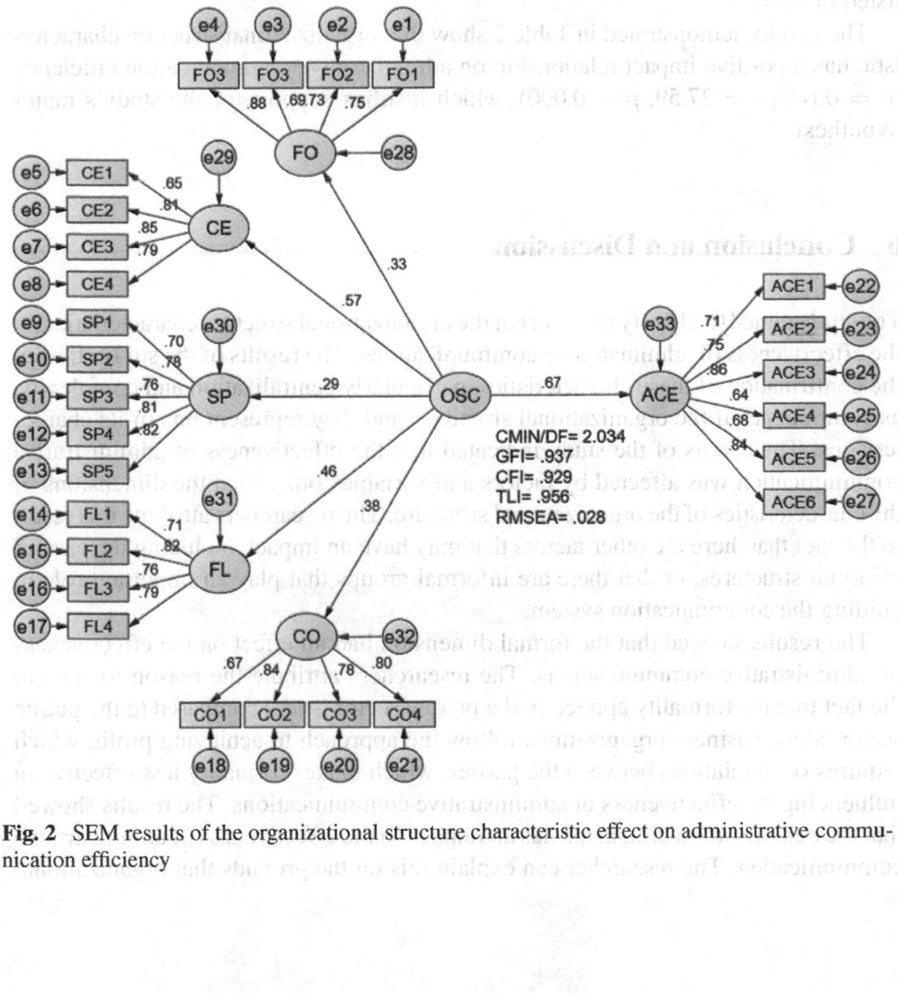

Fig. 2 SEM results of the organizational structure characteristic effect on administrative communication efficiency

Table 2 Hypothesis testing

Relation	Standard Beta	*t* value	*p* value
OSC → ACE	0.675	27.59***	0.000

Note OSC: organizational structure characteristic, ACE: administrative communication efficiency,* $p < 0.05$, ** $p < 0.01$, *** $p < 0.001$

The results in Fig. 2 indicated that the chi-square to degrees of freedom (CMIN/DF) was 2.034, which is less than 3 the upper limit of this indicator. The values of the goodness of fit index (GFI), the comparative fit index (CFI), and the Tucker-Lewis index (TLI) were upper than the minimum accepted threshold of 0.90. Moreover, the result of root mean square error of approximation (RMSEA) indicated to value 0.028, this value is a reasonable error of approximation because it is less than the higher limit of 0.08. Consequently, the structural model used in this study was recognized as a fit model for predicting the DEP and generalization of its result (Ahmad et al., 2016; Shi et al., 2019).To verify the results of testing the study hypotheses, structural equation modeling (SEM) was used, the results of which are listed in Table 2.

The results demonstrated in Table 2 show that organizational structure characteristic has a positive impact relationship on administrative communication efficiency ($\beta = 0.675$, $t = 27.59$, $p = 0.000$), which justifies support for the study's major hypothesis.

6 Conclusion and Discussion

The study aimed to identify the effect of the organizational structure characteristics on the effectiveness of administrative communications. The results of the study showed the confirmation of these characteristics, particularly centralization and complexity, as components of the organizational structure, and they represent one of its characteristics. The results of the study indicated that the effectiveness of administrative communication was affected by factors and variables other than the dimensions of the characteristics of the organizational structure. The researchers attribute this result to the fact that there are other factors that may have an impact, including the case of informal structures, or that there are informal groups that play an important role in guiding the communication system.

The results showed that the formal dimension had an effect on the effectiveness of administrative communications. The researchers attribute the reason for this to the fact that the formality applied in the private sector is less compared to the public sector, since business organizations follow the approach to achieving profit, which requires open relations between the parties, which makes formality less effective in influencing the effectiveness of administrative communications. The results showed that the centralization dimension has an impact on the effectiveness of administrative communication. The researcher can explain this on the grounds that organizational structures in business organizations follow a horizontal communication pattern that allows a kind of delegation to act in situations that do not require the intervention of senior management. This appears through the activities and processes carried out by the middle and lower departments of reporting and defining the duties entrusted to managers. It is clear from the results that the dimension of specialization has an impact on the effectiveness of administrative communication. The researchers attribute this result to the fragmentation of business that characterizes the organizational structure, in addition to the decrease in the number of working departments, in order to reduce the costs involved, as the private sector announces the principle of efficiency and effectiveness and not the principle of social goals.

The results indicated that the flexibility dimension has an effect on the effectiveness of administrative communication. From the researcher's point of view, this result is due to the fact that business organizations operate in an unstable environment, which makes the response to the environment highly dynamic, in addition to keeping pace with technological developments. The results indicated that the dimension of formality has an impact on the effectiveness of administrative communications, which means that the lines of communication between the various departments and departments affiliated to it are somewhat unclear to a high degree due to the interference of stakeholders in many activities and operations. It is clear that the complexity dimension has an impact on the effectiveness of administrative communication. This indicates that the organizational structure does not have a large number of administrative levels, which means that the vertical differentiation in the organizational structure does not exist, as well as that the horizontal departments are not a large number, which means that the horizontal differentiation is one of the components of the organizational structure.

7 Recommendations

Based on the results of the study, the researchers recommend managers and decision makers to pay more attention to the dimensions related to the characteristics of the organizational structure (specialization and flexibility) that will affect the enhancement of the level of effectiveness of administrative communication, and to try to pay attention to the dimensions related to the effectiveness of administrative communication and study them at the individual level and at the organizational level as a tool to measure the extent to which the administrative communication system has been verified and its effectiveness to achieve the set goals, In addition to raising awareness and the participation of managers in decision-making, especially with regard to the administrative communication system to increase its effectiveness through the selection of the communication system, lines of communication and effective means of communication, and preparing managers to improve the communication system as an administrative approach. And finally, designing the organizational structure in a way that achieves the effectiveness of administrative communication and facilitates the process of administrative communication between the president and subordinates in all directions.

References

Aburayya, A., Alshurideh, M., Al Marzouqi, A., Al Diabat, O., Alfarsi, A., Suson, R., ... & Alzarouni, A. (2020). Critical success factors affecting the implementation of tqm in public hospitals: A case study in UAE Hospitals. *Systematic Reviews in Pharmacy*, 11(10), 230–242.

Agha, K., Alzoubi, H. M., & Alshurideh, M. T. (2021, June). Measuring reliability and validity instruments of technologically driven cognitive intrusion towards work-life balance. In *The International Conference on Artificial Intelligence and Computer Vision* (pp. 601–614). Springer, Cham.

Ahmad, A., Alshurideh, M. T., Al Kurdi, B. H., & Salloum, S. A. (2021). Factors impacts organization digital transformation and organization decision making during Covid19 Pandemic. In *The effect of coronavirus disease (COVID-19) on business intelligence* (pp. 95–106). Springer, Cham.

Ahmad, S., Zulkurnain, N., & Khairushalimi, F. (2016). Assessing the validity and reliability of a measurement model in structural equation modeling (SEM). *British Journal of Mathematics & Computer Science, 15*(3), 1–8. https://doi.org/10.9734/BJMCS/2016/25183.

Ahmed, A., Alshurideh, M., Al Kurdi, B., & Salloum, S. A. (2020, October). Digital transformation and organizational operational decision making: a systematic review. In *International Conference on Advanced Intelligent Systems and Informatics* (pp. 708–719). Springer, Cham.

Alameeri, K. A., Alshurideh, M. T., & Al Kurdi, B. (2021). The effect of Covid-19 pandemic on business systems' innovation and entrepreneurship and how to cope with it: a theatrical view. *The Effect of Coronavirus Disease (COVID-19) on Business Intelligence*, 334, 275–288.

Alameeri, K., Alshurideh, M., Al Kurdi, B., & Salloum, S. A. (2020, October). The effect of work environment happiness on employee leadership. In *International Conference on Advanced Intelligent Systems and Informatics* (pp. 668–680). Springer, Cham.

Alhalalmeh, M. I., Almomani, H. M., Altarifi, S., Al- Quran, A. Z., Mohammad, A. A., & Al-Hawary, S. I. (2020). The nexus between corporate social responsibility and organizational performance in Jordan: The mediating role of organizational commitment and organizational citizenship behavior. *Test Engineering and Management, 83*(July), 6391–6410.

AlHamad, A., Alshurideh, M., Alomari, K., Kurdi, B., Alzoubi, H., Hamouche, S., & Al-Hawary, S. (2022). The effect of electronic human resources management on organizational health of telecommunications companies in Jordan. *International Journal of Data and Network Science*, 6(2), 429–438. *International Journal of Data and Network Science* 6 (2), 429–438.

Al-Hawajreh, K., AL-Zeaud, H., Al-Hawary, S. I., & Mohammad, A. A. (2011). The influence of top management support and commitment on total quality management indicators from managers and heads of departments viewpoint: A case study of Sahab Industrial City. *Jordan Journal of Business Administration, 7*(4), 557–576.

Al-Hawary, S. I. (2009). The effect of the leadership style on the effectiveness of the organization: A field study at Zarqa Private University. *The Egyptian Journal for Commercial Studies, 33*(1), 361–393. http://library.mans.edu.eg/eulc_v5/Libraries/start.aspx?fn=DigitalLibraryViewIssues&ScopeID=1.1.&item_id=370428.108.

Al-Hawary, S. I. (2011). Human resource management practices in ZAIN cellular communications company operating in Jordan. *Perspectives of Innovations, Economics and Business, 8*(2), 26–34.

Al-Hawary, S. I. (2015). Human resource management practices as a success factor of knowledge management implementation at health care sector in Jordan. *International Journal of Business and Social Science, 6*(11/1), 83–98.

Al-Hawary, S. I. S., & AlDafiri, M. F. S. (2017). Effect of the Components of information technology adoption on employees performance of interior ministry of Kuwait State. *International Journal of Academic Research in Economics and Management Sciences, 6*(2), 149–169.

Al-Hawary, S. I. S., & Mohammed, A. K. (2017). Impact of team work traits on organizational citizenship behavior from the viewpoint of the employees in the education directorates in North Region of Jordan. *Global Journal of Management and Business*, 17(2-A), 23–40.

Al-Hawary, S. I. S., & Obiadat, A. A. (2021). Does mobile marketing affect customer loyalty in Jordan? *International Journal of Business Excellence, 23*(2), 226–250.

Al-Hawary, S. I. S., Mohammad, A. S., Al-Syasneh, M. S., Qandah, M. S. F., & Alhajri, T. M. S. (2020). Organisational learning capabilities of the commercial banks in Jordan: Do electronic human resources management practices matter? *International Journal of Learning and Intellectual Capital, 17*(3), 242–266.

Al-Hawary, S. I., & Aldaihani, F. M. (2016). Customer relationship management and innovation capabilities of Kuwait airways. *International Journal of Academic Research in Economics and Management Sciences, 5*(4), 201–226.

Al-Hawary, S. I., & Al-Hamwan, A. (2017). Environmental analysis and its impact on the competitive capabilities of the commercial banks operating in Jordan. *International Journal of Academic Research in Accounting, Finance and Management Sciences, 7*(1), 277–290.

Al-Hawary, S. I., & Al-Syasneh, M. S. (2020). Impact of dynamic strategic capabilities on strategic entrepreneurship in presence of outsourcing of five stars hotels in Jordan. *Business: Theory and Practice, 21*(2), 578–587.

Al-Hawary, S. I., & Batayneh, A. M. (2010). The effect of marketing communication tools on Non-Jordanian students' choice of Jordanian Public Universities: A Field Study. *International Management Review, 6*(2), 90–99. https://www.questia.com/library/journal/1P3-2143106641/the-effect-of-marketing-communication-tools-on-non-jordanian.

Al-Hawary, S. I., & Metabis, A. (2012). Implementation of Internal Marketing in Jordan Banks. *International Journal of Data Analysis and Information, 4*(1), 37–53.

Al-Hawary, S. I., & Nusair, W. (2017). Impact of human resource strategies on perceived organizational support at Jordanian Public Universities. *Global Journal of Management and Business Research: A Administration and Management, 17*(1), 68–82.

Al-Hawary, S. I., & Shdefat, F. (2016). Impact of human resources management practices on employees' satisfaction a field study on the Rajhi Cement Factory. *International Journal of Academic Research in Accounting, Finance and Management Sciences, 6*(4), 274–286.

Al-Hawary, S. I., AL-Awawdeh, W., & Abden, M. A. (2012). The impact of the leadership style on organizational commitment: A field study on Kuwaiti telecommunications companies. *ALEDARI*, (130), 53–102. http://ipa.gov.om/index.php/nums/index/20.

Al-Hawary, S. I., AL-Zeaud, H., & Batayneh, A. M. (2011). The relationship between transformational leadership and employee's satisfaction at Jordanian Private Hospitals. *Business and Economic Horizons, 5*(2), 35–46. https://doi.org/10.15208/beh.2011.13.

Aljumah, A., Nuseir, M. T., & Alshurideh, M. T. (2021). The impact of social media marketing communications on consumer response during the COVID-19: Does the brand equity of a University Matter. *The effect of coronavirus disease (COVID-19) on business intelligence*, 367–384.

Alkitbi, S. S., Alshurideh, M., Al Kurdi, B., & Salloum, S. A. (2020, October). Factors affect customer retention: A systematic review. In International Conference on Advanced Intelligent Systems and Informatics (pp. 656–667). Springer, Cham.

Al Kurdi, B., & Alshurideh, M., Al afaishat, T. (2020). Employee retention and organizational performance: Evidence from banking industry. *Management Science Letters*, 10(16), 3981–3990.

Allahow, T. J. A. A., Al-Hawary, S. I. S., & Aldaihani, F. M. F. (2018). Information technology and administrative innovation of the central agency for information technology in Kuwait. *Global Journal of Management and Business*, 18(11-A), 1–16.

Al-Lozi, M., Almomani, R. Z., & Al-Hawary, S. I. (2017). Impact of talent management on achieving organizational excellence in Arab Potash Company in Jordan. *Global Journal of Management and Business Research: A Administration and Management, 17*(7), 15–25.
Al-Lozi, M. S., Almomani, R. Z. Q., & Al-Hawary, S. I. S. (2018). Talent Management strategies as a critical success factor for effectiveness of Human Resources Information Systems in commercial banks working in Jordan. *Global Journal of Management and Business Research: A Administration and Management, 18*(1), 30–43.
Al-Nady, B. A., Al-Hawary, S. I., & Alolayyan, M. (2013). Strategic management as a key for superior competitive advantage of sanitary ware suppliers in Kingdom of Saudi Arabia. *International Journal of Management and Information Technology, 7*(2), 1042–1058.
Al-Nady, B. A., Al-Hawary, S. I., & Alolayyan, M. (2016). The role of time, communication, and cost management on project management success: An empirical study on sample of construction projects customers in Makkah City, Kingdom of Saudi Arabia. *International Journal of Services and Operations Management, 23*(1), 76–112.
Alsharari, N. M., & Alshurideh, M. T. (2020). Student retention in higher education: the role of creativity, emotional intelligence and learner autonomy. *International Journal of Educational Management. International Journal of Educational Management*, 35 (1), 233–247.
Alshura, M. S. K., Nusair, W. K. I., & Aldaihani, F. M. F. (2016). Impact of internal marketing practices on the organizational commitment of the employees of the Insurance Companies in Jordan. *International Journal of Academic Research in Economics and Management Sciences, 5*(4), 168–187.
Alshurideh, M. (2022). Does electronic customer relationship management (E-CRM) affect service quality at private hospitals in Jordan? *Uncertain Supply Chain Management, 10*(2), 325–332.
Alshurideh, M. T., Shaltoni, A., & Hijawi, D. (2014). Marketing communications role in shaping consumer awareness of cause-related marketing campaigns. *International Journal of Marketing Studies, 6*(2), 163–168.
Alshurideh, M. T., Al Kurdi, B., Alzoubi, H. M., Ghazal, T. M., Said, R. A., AlHamad, A. Q., … & Al-kassem, A. H. (2022). Fuzzy assisted human resource management for supply chain management issues. *Annals of Operations Research*, 1–19.
Alshurideh, M., Salloum, S. A., Al Kurdi, B., & Al-Emran, M. (2019, February). Factors affecting the social networks acceptance: an empirical study using PLS-SEM approach. In *Proceedings of the 2019 8th International Conference on Software and Computer Applications* (pp. 414–418).
Al-Tarawneh, K. A., Mohammad Alhamadani, S. Y., & Mohammad, A. A. (2012). Transformational leadership and marketing effectiveness in commercial banks in Jordan. *European Journal of Economics, Finance and Administrative Sciences, 46*, 71–87.
Alyammahi, A., Alshurideh, M., Al Kurdi, B., & Salloum, S. A. (2020, October). The impacts of communication ethics on workplace decision making and productivity. In *International Conference on Advanced Intelligent Systems and Informatics* (pp. 488–500). Springer, Cham.
Bebba, I., Bentafat, A., & Al-Hawary, S. I. S. (2017). An evaluation of the performance of higher educational institutions using data envelopment analysis: An empirical study on Algerian higher educational institutions. *Global Journal of Human Social Science Research, 17*(8), 21–30.
Daft, Richard. (1986). Organization theory and design (West publishing Company), 2 edition.
de Leeuw, E., Hox, J., Silber, H., Struminskaya, B., & Vis, C. (2019). Development of an international survey attitude scale: Measurement equivalence, reliability, and predictive validity. *Measurement Instruments for the Social Sciences, 1*(1), 9. https://doi.org/10.1186/s42409-019-0012-x.
Hair, J. F., Babin, B. J., & Krey, N. (2017). Covariance-based structural equation modeling in the journal of advertising: Review and recommendations. *Journal of Advertising, 46*(1), 163–177. https://doi.org/10.1080/00913367.2017.1281777.
Harahsheh, A., Houssien, A., Alshurideh, M. & AlMontaser, M. (2021). The effect of transformational leadership on achieving effective decisions in the presence of psychological capital as an intermediate variable in private Jordanian universities in light of the corona pandemic. *The Effect of Coronavirus Disease (COVID-19) on Business Intelligence*, 334, 221–243.

Howard, M. C. (2018). The convergent validity and nomological net of two methods to measure retroactive influences. *Psychology of Consciousness: Theory, Research, and Practice, 5*(3), 324–337. https://doi.org/10.1037/cns0000149.

Kabrilyants, R., Obeidat, B., Alshurideh, M., & Masadeh, R. (2021). The role of organizational capabilities on e-business successful implementation. *International Journal of Data and Network Science, 5*(3), 417–432.

Metabis, A., & Al-Hawary, S. I. (2013). The Impact of internal marketing practices on services quality of commercial banks in Jordan. *International Journal of Services and Operations Management, 15*(3), 313–337.

Mohammad, A. A. (2020). The effect of customer empowerment and customer engagement on marketing performance: the mediating effect of brand community membership. *Business: Theory and Practice, 21*(1), 30–38.

Mohammad, A. A., Altarifi, S. M., & Alafi, K. (2014). The Impact of corporate social responsibility toward employees on company performance: A Jordanian study. *Interdisciplinary Journal of Contemporary Research in Business, 6*(5), 255–270.

Mohammad, A. A., Alshura, M. S., Al-Hawary, S. I. S., Al-Syasneh, M. S., & Alhajri, T. M. (2020). The influence of Internal Marketing Practices on the employees' intention to leave: A study of the private hospitals in Jordan. *International Journal of Advanced Science and Technology, 29*(5), 1174–1189.

Nuseir, M. T., Al Kurdi, B. H., Alshurideh, M. T., & Alzoubi, H. M. (2021, June). Gender discrimination at workplace: Do artificial intelligence (AI) and machine learning (ML) have opinions about it. In *The International Conference on Artificial Intelligence and Computer Vision* (pp. 301–316). Springer, Cham.

Rimkeviciene, J., Hawgood, J., O'Gorman, J., & De Leo, D. (2017). Construct validity of the acquired capability for suicide scale: factor structure, convergent and discriminant validity. *Journal of Psychopathology and Behavioral Assessment, 39*(2), 291–302. https://doi.org/10.1007/s10862-016-9576-4.

Sekaran, U., & Bougie, R. (2016). *Research methods for business: A skill-building approach* (7th edn). Wiley.

Shamout, M., Elayan, M., Rawashdeh, A., Kurdi, B., & Alshurideh, M. (2022). E-HRM practices and sustainable competitive advantage from HR practitioner's perspective: A mediated moderation analysis. *International Journal of Data and Network Science, 6*(1), 165–178.

Shi, D., Lee, T., & Maydeu-Olivares, A. (2019). Understanding the model size effect on SEM fit indices. *Educational and Psychological Measurement, 79*(2), 310–334. https://doi.org/10.1177/0013164418783530.

Sung, K.-S., Yi, Y. G., & Shin, H.-I. (2019). Reliability and validity of knee extensor strength measurements using a portable dynamometer anchoring system in a supine position. *BMC Musculoskeletal Disorders, 20*(1), 1–8. https://doi.org/10.1186/s12891-019-2703-0.

Wang, Y. A., & Rhemtulla, M. (2021). Power analysis for parameter estimation in structural equation modeling: A discussion and tutorial. *Advances in Methods and Practices in Psychological Science, 4*(1), 1–17. https://doi.org/10.1177/2515245920918253.

Develop a Causal Model for the Impact of Critical Success Factors of the Strategic Information System in Promoting Human Resources Management Strategies in the Social Security Corporation

Kamel Mohammad Al-hawajreh, Muhammad Bajes Al-Majali, Menahi Mosallam Alqahtani, Basem Yousef Ahmad Barqawi, Sulieman Ibraheem Shelash Al-Hawary, Enas Ahmad Alshuqairat, Ayat Mohammad, Muhammad Turki Alshurideh, and Anber Abraheem Shlash Mohammad

Abstract This study aimed to analyze the impact of critical success factors of strategic information system on the strengthening human resources management strategies in the Social Security Corporation. The study population was made up of 120 members of the directors and chiefs of sections of the Social Security Corporation. Structural equation modeling (SEM) was conducted to test hypotheses. The study found a statistically significant impact of the critical factors of the strategic information system on the human resources management strategies of the Social Security Corporation. The study recommended that attention should be given to

K. M. Al-hawajreh · M. B. Al-Majali
Business Faculty, Mu'tah University, Karak, Jordan

M. M. Alqahtani
Administration Department, Community College of Qatar, Doha, Qatar
e-mail: mena7i@icloud.com

B. Y. A. Barqawi
Faculty of Administration & Financial Sciences Petra University, Business Administration Department, Amman, Jordan

S. I. S. Al-Hawary (✉)
Department of Business Administration, School of Business, Al Al-Bayt University, P.O. Box 130040, Mafraq 25113, Jordan
e-mail: dr_sliman73@aabu.edu.jo

E. A. Alshuqairat
Faculty of Money and Management, Management Department, The World Islamic Science University, P.O. Box 1101, Amman 11947, Jordan

A. Mohammad
Business and Finance Faculty, The World Islamic Science and Education University (WISE), Postal Code 11947, P.O. Box 1101, Amman, Jordan

M. Alshurideh et al. (eds.), *The Effect of Information Technology on Business and Marketing Intelligence Systems*, Studies in Computational Intelligence 1056,
https://doi.org/10.1007/978-3-031-12382-5_48

strengthening top management support for its important role in strengthening human resources management strategies and the strategic information system.

Keywords Strategic Information Systems · Human Resources Management Strategies · Social Security Corporation · Jordan

1 Introduction

Information is a basic pillar for the effectiveness of decisions in the organization, and for the success of the strategic information system, it is necessary to identify the critical success factors for the strategic information system, which is one of the important and necessary means for the development and distinction of organizations, which in turn constitutes a basic process for the success or failure of the strategic information system, which must be adhered to and taken into consideration (AlTaweel & Al-Hawary, 2021; Al-Hawary & Al-Syasneh, 2020; Al-Quran et al., 2020; Al-Hawary et al., 2020; Al-Hawary & Obiadat, 2021; Allahow et al., 2018; Al-Hawary & Alhajri, 2020; Altarifi et al., 2015; Al-Hawary & Hussien, 2017; Al-Hawary & Al-Smeran, 2017; Al-Hawary & Al-Menhaly, 2016; Al-Hawary & Ismael, 2010).

Since the human resource is considered one of the main and necessary resources for any organization, and this resource is the first responsible for the implementation of all functions and activities related to the organization, therefore, this resource is the main nerve for organizations (Alhalalmeh et al., 2020; Hijjawi & Mohammad, 2019; Al-Hawary, 2015; Al-Lozi et al., 2017; Al-Hawary & Mohammed, 2017; Al-Hawary et al., 2012; Al-Hawary, 2009), and in order for these human resources to accomplish their various activities, they urgently need to find strategies to manage human resources and support them through a strategic information system capable of processing data and information and supporting the decision maker in achieving and completing various activities in organizations in general and the Social Security Corporation especially (Abu Zayyad et al., 2021; Alaali et al., 2021). Therefore, there is a link between the completion of the tasks and activities of the management of the organization and its various activities, especially the strategy and the strategic information system in the organization, where the information system and human resources are an integral part of the strategies that organizations must care about and interact with to achieve the vision and objectives of the organization, it plays an

M. T. Alshurideh
Department of Marketing, School of Business, The University of Jordan, Amman 11942, Jordan
e-mail: m.alshurideh@ju.edu.jo; malshurideh@sharjah.ac.ae

Department of Management, College of Business, University of Sharjah, 27272 Sharjah, United Arab Emirates

A. A. S. Mohammad
Faculty of Administrative and Financial Sciences, Marketing Department, Petra University, B.O. Box 961343, Amman 11196, Jordan

important and essential role in achieving the success and distinction of organizations (Al-Dmour et al., 2021; Alzoubi et al., 2022).

Human resource management strategies are indispensable in all business organizations, as they need to provide comprehensive strategic information, and this information is a capital resource for organizations, which in turn leads to an increase in the effectiveness of human resource management outputs, to achieve this, an efficient and effective strategic information system must be harnessed through commitment by the organization to identify the critical success factors for this system (Alameeri et al., 2021; Al-Hawary & Al-Namlan, 2018; Alshurideh et al., 2019a, b). Human resources are considered a very important main element and they play an important and essential role in achieving the success and distinction of organizations, and the effectiveness of human resources management strategies greatly contribute to this, so it is necessary to pay attention to these strategies by the organization to ensure the achievement of excellence and provide high levels of performance (Alkalha et al., 2012; AlHamad et al., 2022; Alshurideh et al., 2022).

The effectiveness of human resource management strategies is determined by understanding the organization's internal and external environment, its requirements and basic variables (Al Kurdi et al., 2021; Shamout et al., 2022). To achieve this, a strategic information system must be harnessed and work to identify the critical success factors for this system. Given the social and economic importance of the Social Security Corporation, this importance has raised in the mind of the researcher the importance of having an effective system for human resource management as it represents the cornerstone in all the activities carried out by this national institution, and from this point of view, and to determine the effectiveness of the management strategies Human resources applied in the Social Security Corporation. The researchers conducted a number of interviews with a number of department heads at the various administrative levels in the Social Security Corporation and it was found that there is a lack of awareness among managers of the concept and application of a number of human resource management strategies, in addition to a lack of awareness of the importance of critical success factors for the strategic information system applied in the institution. This study seeks to identify the role of critical success factors for the strategic information system in enhancing human resource management strategies in the Social Security Corporation (Al-Dmour et al., 2021; Ashal et al., 2021; Ben-Abdallah et al., 2022).

2 Theoretical Framework and Hypotheses Development

2.1 *Critical Success Factors of a Strategic Information System*

The critical success factors are among the necessary topics in management literature in general, and information systems literature in particular. With regard to applications in the field of information systems, it should be noted that strategic information systems are in themselves one of the most important critical success factors in the life of organizations (Al-Taie and Al-Khafaji, 2009; Altamony et al., 2012).

Critical success factors play an important role in the failure or success of the information system and the organization, as they contribute significantly to helping business organizations identify information systems that need to be developed, and they also serve as a planning tool for information systems and help direct the efforts of business organizations towards developing and using strategic plans In identifying the critical issues accompanying the implementation of plans and directing the resources that are under the authority of managers towards important areas (Al-Zoubi, 2005; Alzoubi et al., 2021a).

Information system refers to a system for converting inputs or data from internal and external sources into information and communicating this information to managers at all organizational levels, and at all functions to enable and support managers in making successful and effective decisions in a timely manner to plan, direct and control the activities for which they are responsible (Esmeray, 2016; Svoboda et al., 2021).

Researchers and those interested in the field of strategic information systems agreed that this system represents the most important and main source for improving business competitiveness in an era when information and knowledge has become the most important, basic and decisive economic resource to ensure growth, survival and prosperity in the business environment, and understanding the competitive environment with its variables, dimensions and movement is the main premise and the basis for building, developing, enhancing and sustaining competitive advantage.

The strategic information system makes a substantial contribution to achieving this goal only when this system achieves strategic harmony and integration with other business resources, specifically the organization's strategy and infrastructure (Alzoubi et al., 2021b; Tariq et al., 2022). Obviously, the importance of any information system is determined by its output level, meaning the level of use, benefits, results, or achievements that result from, from this point of view, the importance of strategic information systems stems from the role of its outputs (strategic information) that the organization's strategic management needs to carry out its activities and activities efficiently and effectively, this information is of a special nature as that information necessary for making strategic decisions at the level of the organization (Masrek et al., 2009).

Researchers differed in defining a specific and agreed concept of critical success factors, as a result of differing opinions and viewpoints in addition to the difference in the cognitive orientations of researchers, as (Syed et al., 2018) indicated that organizations that are interested in identifying critical success factors, this gives them the ability to develop its business, improving its competitive position, and achieving its planned organizational goals, since these factors represent important areas of focus, in addition to the fact that understanding these factors helps organizations to effectively confront risk situations, and thus enhances their ability to invest available opportunities in the future, and focus on identifying strengths and weaknesses in job.

Olszak and Ziemba (2012) believes that the critical success factors represent the processes, systems, and policies that organizations meticulously pursue to achieve a high and distinctive competitive position in the market and to adopt a certain type of mechanisms and methods that achieve success, growth and development for the organization as a competitive strategy and strategic position. Ranong and Phuenngam (2009) defined the critical success factors as the sum of results and the elements that help the organization to obtain effective and successful competitive performance, as there is a certain sequence of steps that must be followed to achieve the effectiveness and success of the organization, if the organization reaches unsatisfactory results, this indicates that there is a failure in the efforts made by the organization.

One of the main objectives achieved by the critical success factors is to adequately ensure the effectiveness and success of any organization, and to maintain a distinguished level of performance to implement its operations. Al-Abadi and Al-Atabi (2014) pointed out that the critical success factors are those vital factors and elements identified by the senior management for the success of the organization's strategic plan and the achievement of its goals, mission and distinction in the organization's internal and external environment.

Among the most prominent dimensions of the critical success factors for a strategic information system: **The commitment of senior management to support the strategic information system**, it has been described as representing the level of importance that management places on the successful completion of the information system, and in general, the commitment of senior management to support the strategic information system refers to the emotional or psychological commitment shown by senior management towards the information system. **Interaction between the beneficiaries of the strategic information system and its specialists**, It is represented in the level of coordination, communication, partnership and teamwork between the beneficiaries and specialists in the information system in terms of the amount and quality of communications and the exchange of ideas between them and the nature of the activities carried out by them. The beneficiaries and specialists must focus on the main objective of having a sound and effective strategic information system in order to interact appropriately. **The level of readiness of workers in the system**, it refers to the organization's ability through its employees to achieve successful adoption, use and benefit from information technology and systems (Fathian et al., 2018). **Users understand system requirements**, it is represented by the extent to which users understand the objectives, tasks and outputs of operations related to the information system, i.e. knowledge and comprehensive and integrated understanding in the field of information systems.

2.2 Human Resource Management Strategies

The idea of human resource management strategies in nomenclature and content was formed at the beginning of the nineties of the last century, basing its existence on the ideas that came from the Human Resources Management School; it did not cancel the previous ideas, but tried to add to it what could enhance its work at the present time. The directions of this new label for human resource management can be understood in the human resource management strategy by asking the following question: "How can coordination between human resource management activities and the organization's strategy?" (Al-Hawary & Alajmi, 2017).

Human resource management strategies are one of the most important elements of the success of the organization's strategies. It is not possible to find a strategy for an organization without including the human resource strategy, as goals can only be achieved through the human element as leaders or as subordinates in the organization (Jleida & Samir, 2018; Al-Hawary & Al-Hamwan, 2017; Al-Nady et al., 2013, 2016; Al-Hawary & Hadad, 2016). The strategy is defined as a set of procedures or behaviors that show how the organization will move from the activity in which it is currently working to the activity in which it hopes to work in the future, bearing in mind external opportunities and threats, and internal strengths and weaknesses (Dessler, 2015). From another point of view, the strategy is defined as a long-term plan taken by the organization as a base for making decisions based on its identification of its current and future mission. It is based on defining the range of products and markets it deals with, the uses of the resources available, the competitive advantage it enjoys, and the impact of compatibility between its various administrative functions and business activities, in order to achieve the organization's internal cohesion and enable it to move freely and adapt to what links it to its external environment and to reach its basic goals and objectives in a balanced manner (Alameeri et al., 2020; Alshurideh et al., 2019a, b; Schrager & Madansky, 2013).

The strategies of human resource management gain great importance through its contribution to the preparation and implementation of the organization's strategy, such as attracting experienced and competent workers, training and developing employees, motivating them and mobilizing their energies towards making more efforts and creativity to achieve the organization's goals and increasing its ability to gain opportunities and avoid environmental threats, as well as its ability to make optimal use of the organization's financial and material resources in line with the general strategic directions of the organization (Matei, 2013; Shah et al., 2020, 2021). The main objective of human resource management strategies is to attract creative human talents with experience and efficiency and try to take care of these talents and preserve them through their development, training, development, motivation and protection of their interests in order to achieve the ambitions and goals of the organization efficiently and effectively.

The researchers' viewpoints varied regarding the number of strategies that make up the human resource management strategy, as well as their nomenclature. Accordingly, the researcher decided to address the following strategies, which he looked at in terms of their importance in the organization in which the study was applied and were as follows: **The recruitment strategy** is considered the most important among the human resource management strategies, as the construction and implementation of this strategy and the quality of the human resources employed result in the rest of the organization's functions, and thus the quality of the organization's performance as a whole (Al-Hawary & Nusair, 2017; Lee et al., 2022a, b). **Training and development strategy** refers to the administrative and organizational efforts related to the state of continuity aimed at making a modification in skills, behavior and knowledge within the current or future characteristics of the employee in order to be able to meet the requirements of his work or to better develop his practical performance in order to achieve the goals of the organization (Al-Hawary, 2011; Alshraideh et al., 2017). The **performance evaluation strategy** is defined as the organized description of the strengths and weaknesses associated with the job, whether collectively or individually, and serves two main purposes in the organization: providing managers and workers with the information necessary for decision-making, and improving and developing the performance of employees, which is positively reflected on the performance of the organization (Al-Hawary & Abu-Laimon, 2013; Al-Hawary et al., 2013; Lim et al, 2017). **Compensation strategy** is a system related to wages, rewards, incentives, and benefits that the organization provides to employees within certain bases in return for the employees' completion of their job tasks so that the compensation system is characterized by justice and equity and aims to attract human competencies and preserve human talents within the organization (Metabis & Al-Hawary, 2013; Mohammad et al., 2020).

2.3 *Strategic Information Systems and Human Resource Management Strategies*

Hammadi (2020) concluded that there is a relationship between human resource management strategies and strategic success in organizations, and that the information technology component of strategic information systems has a significant impact on competitive advantage. Holding training courses for new employees to qualify them in the field of practices related to the application of software because of their role in improving their efficiency and effectiveness. Al-Tayyar and Al-Nama (2018) emphasized that the inputs of strategic information systems and the sources of obtaining them, whether internal or external, are considered the main source of information used in making important decisions in the organization, as well as trying to reduce the difficulties facing the decision maker by using the strategic information system to provide the appropriate information to the decision maker when making strategic decisions, as well as using software to support the strategic decision (Youssef & Yahia, 2018).

That senior management adopts effective policies and vision in highlighting the role of critical success factors for information systems to achieve sustainable competitive advantage with the presence of environmental sensing. Al-Samarrai believes that there is a need for the company's management to invest in the strong positive relationship between the commitment of senior management to support the strategic information system and the effectiveness of the strategic decision to achieve more of this effectiveness and to make it more competitive. There is strong evidence that strategic information systems support and enhance the company's ability and flexibility to find competitive strategies in response to external environmental changes and that the effects of strategic information systems on the company's performance vary based on the competitive strategy. Accordingly, the study hypothesis can be formulated as follows.

There is a statistically significant effect of the critical success factors of the strategic information system in enhancing the strategies of human resources management in the Social Security Corporation.

3 Study Model

See Fig. 1.

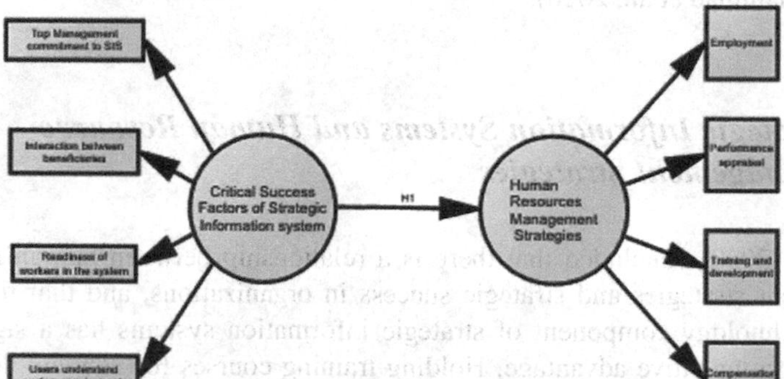

Fig. 1 Research model

4 Methodology

4.1 Population and Sample Selection

A quantitative method based on a questionnaire was used in this study for data collection and sample selection. The major aim of the study was to examine the impact of critical success factors of strategic information system on reinforcement of human resources management strategies in the Social Security Corporation. Data were primarily gathered through the questionnaires created by researchers which were distributed to a random sample of (165) respondent. In total, (145) responses were received including (25) invalid to statistical analysis due to uncompleted or inaccurate. Hence, the final sample contained (120) responses suitable to analysis requirements that were formed a response rate of (82.76%), where it proved to be sufficient to the extent that was predictable and allowed for a presumption of data saturation (Sekaran & Bougie, 2016).

4.2 Measurement Instrument

A questionnaire that consists of two main sections along with a section regarding control variables was used as the measurement instrument. Control variables considered as categorical measures were composed of gender, age, educational level, and experience. The two main sections were dealt with a five-point Likert scale (from 1 = strongly disagree to 5 = strongly agree). The first section contained (22) items to critical success factors of the strategic information system based on. These questions were distributed into dimensions as follows: five items dedicated for measuring Senior management commitment to support the strategic information system, six items dedicated for measuring Interaction between beneficiaries, five items dedicated for measuring readiness of workers in the system, and six items dedicated for measuring users understand system requirements. Whereas the second section included (20) items developed to measure human resource management strategies according to what was pointed by (Lim et al., 2017). This variable was divided into four dimensions: employment that was measured through five items, performance appraisal which measured by five items, training and development was measured using five items, and compensation systems that was measured by five items.

5 Findings

5.1 Measurement Model Evaluation

This study was conducted structural equation modeling (SEM) to test hypotheses, which represents a contemporary statistical technique for testing and estimating the relationship between factors and variables (Wang & Rhemtulla, 2021). Accordingly, the reliability and validity of the constructs were tested using confirmatory factor analysis (CFA) through the statistical program AMOSv26. Table 1 summarizes the results of convergent and discriminant validity, as well the indicators of reliability.

Table 1 shows that the standard loading values for the individual items were within the domain (0.573–0.654), these values greater than the minimum retention of the elements based on their standard loads (Sung et al., 2019). Average variance extracted (AVE) is a summary indicator of the convergent validity of constructs that must be above 0.50 (Howard, 2018). The results indicate that the AVE values were greater than 0.50 for all constructs, thus the used measurement model has an appropriate convergent validity. Rimkeviciene et al. (2017) suggested the comparison approach as a way to deal with discriminant validity assessment in covariance-based SEM. This approach is based on comparing the values of maximum shared variance (MSV) with the values of AVE, as well as comparing the values of square root of AVE ($\sqrt{AVE}$) with the correlation between the rest of the structures. The results show that the values of MSV were smaller than the values of AVE, and that the values of $\sqrt{AVE}$ were higher than the correlation values among the rest of the constructs. Therefore, the measurement model used is characterized by discriminative validity. The internal consistency measured through Cronbach's Alpha coefficient (α) and compound reliability by McDonald's Omega coefficient (ω) was conducted as indicators to evaluate measurement model. The results listed in Table 1 demonstrated that both values of Cronbach's Alpha coefficient and McDonald's Omega coefficient were greater than 0.70, which is the lowest limit for judging on measurement reliability (de Leeuw et al., 2019).

5.2 Structural Model

The structural model illustrated no multicollinearity issue among predictor constructs because variance inflation factor (VIF) values are below the threshold of 5, as shown in Table 1 (Hair et al., 2017). This result is supported by the values of model fit indices shown in Fig. 1 (Fig. 2).

The results in Fig. 1 indicated that the chi-square to degrees of freedom (CMIN/DF) was 2.133, which is less than 3 the upper limit of this indicator. The values of the goodness of fit index (GFI), the comparative fit index (CFI), and the Tucker-Lewis index (TLI) were upper than the minimum accepted threshold of 0.90.

Table 1 Results of validity and reliability tests

Constructs	1	2	3	4	5	6	7	8
1. TMC	**0.702**							
2. IBB	0.557	**0.719**						
3. RWS	0.582	0.588	**0.752**					
4. UUSR	0.533	0.561	0552	**0.754**				
5. E	0.754	0.622	0.576	0.693	**0.788**			
6. TD	0.719	0.678	0.695	0.627	0.799	**0.711**		
7. PA	0.622	0.679	0.684	0.744	0.637	0.641	**0.694**	
8. CS	0.655	0.621	0.608	0.639	0.595	0.622	0.671	**0.799**
VIF	2.727	2.906	2.171	3.374	–	–	–	–
Loadings range	0.573–0.654	0.701–0.792	0.751–0.836	0.796–0.832	0.716–0.826	0.796–0.872	0.796–0.869	0.810–0.899
AVE	0.863	0.847	0.838	0.898	0.886	0.826	0.0.858	0.835
MSV	0.544	0.502	0.513	0.511	0.516	0.513	0.488	0.522
Internal consistency	0.847	0.832	0.875	0.895	0.844	0.911	0.864	0.873
Composite reliability	0.887	0.891	0.893	0.915	0.892	0.902	0.879	0.856

Note DPRO: digital promotion, DPRI: digital pricing, ATTR: attraction, MADB: marketing database, PERC: perception, INFL: influence, MOTI: motivation, DESI: desire, bold fonts in the table indicate to root square of AVE

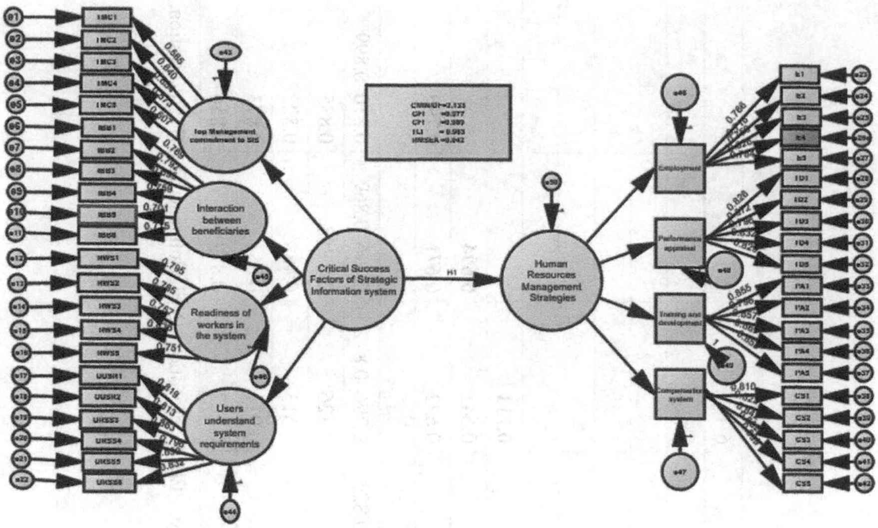

Fig. 2 SEM results of the digital marketing effect on mental image

Moreover, the result of root mean square error of approximation (RMSEA) indicated to value 0.049, this value is a reasonable error of approximation because it is less than the higher limit of 0.08. Consequently, the structural model used in this study was recognized as a fit model for predicting the DEP and generalization of its result (Ahmad et al., 2016; Shi et al., 2019). To verify the results of testing the study hypotheses, structural equation modeling (SEM) was used, the results of which are listed in Table 2.

The results demonstrated in Table 2 show that all critical success factors of the strategic information system dimensions have a positive impact relationship on human resources management strategies except Top management commitment to

Table 2 Hypothesis testing

Hypothesis	Relation	Standard Beta	*t* value	*p* value
H1	Top management commitment to support the strategic information system → human resources management strategies	0.084	1.041	0.300
H2	Interaction between the beneficiaries of the strategic information system → human resources management strategies	0.246	3.260	0.001
H3	The level of readiness of workers in the system → human resources management strategies	0.272	3.411	0.001
H4	Users understand system requirements → human resources management strategies	0.352	3.921	0.000

Note * $p < 0.05$, ** $p < 0.01$, *** $p < 0.001$

support the strategic information system which has no impact on human resources management strategies ($\beta = 0.084$, $t = 1.041$, $p = 0.300$). However, the results indicated that the highest impact was for combination ($\beta = 0.352$, $t = 3.921$, $p = 0.000$), followed by socialization ($\beta = 0.272$, $t = 3.411$, $p = 0.001$), and finally the lowest impact was for internalization ($\beta = 0.246$, $t = 3.260$, $p = 0.001$).

6 Discussion

The results concluded that the Social Security Corporation adopts the critical success factors of the strategic information system, which are the interaction between the beneficiaries of the strategic information system and its specialists, the level of readiness of workers in the system, and the users' understanding of the system's requirements, it enhances the organization's management's interest in these factors because of their impact on the management of the organization's business from the strategic side, especially in the aspects of interaction between users and users' understanding of the aspects and operations of the information system, as well as the level of readiness of workers in the system to work. The critical success factors of the strategic information system enable the Social Security Corporation to practice recruitment activities (recruitment and appointment) in the various departments of the Human Resources Department and in a manner that enhances the possibility of creating databases that serve these activities and in a manner that enhances the possibility of obtaining the highest competencies and in order to achieve the strategic objectives of the Social Security Corporation.

The results confirmed that the management of the Social Security Corporation considers the interaction activities between the beneficiaries of the strategic information system and its specialists, and the level of readiness of the workers in the system from the normal routine activities in the institution and not from the strategic activities at the level of the Social Security Corporation. In addition, the Social Security Corporation exercises its internal tasks and operations with the aspects related to the performance evaluation strategy and related to the keenness of these leaders to give the process of evaluating the performance of workers great importance as it contributes to spreading the spirit of competition among workers and uses the results of performance evaluation to determine the training needs in the Social Security Corporation and to determine promotions And bonuses, and follow-up on the performance of employees by comparing the previous performance with the current performance at the level of departments and sections in the Social Security Corporation.

The results showed that the Social Security Corporation exercises its internal functions and operations in the aspects related to the compensation systems strategy in a tight manner through information systems and linking it to the performance evaluation and training strategy related to the keenness of these leaders to give the compensation systems strategy great importance as it contributes to formulating and determining the strategy of wages, rewards and compensation that achieves satisfaction, and the use of this strategy in determining the levels of workers' achievement of their tasks in a way that qualifies them to determine the type of reward, in addition to the Social Security Corporation's reliance on the information provided by the performance appraisal system to determine the rewards and incentives that outstanding workers deserve, in addition to the Social Security Corporation conducting continuous wage surveys And salaries for the purpose of linking them to the cost of living.

7 Recommendations

Based on the previous results, the researchers recommend the need for the management of the General Organization for Social Security to pay attention to the commitment of the senior management to support the strategic information system, as it represents one of the critical success factors for the strategic information system. The study recommends that leaders in the organization strengthen the performance evaluation strategy because of its impact on enhancing many activities related to human resource management and linking them to strategic information systems to benefit from them in activating electronically and increasing the speed in making strategic decisions, and the need for the institution to pay attention to the recruitment strategy because of its reflection in attracting and employing competencies from human resources and working to link them electronically in terms of recruitment and models, leading to appointment in vacant positions in the institution.

References

Jleida, A., & Samir, S. (2018). *The impact of human resource management strategies on the performance of workers in Libyan telecom companies, unpublished master's thesis*. Middle East University.

Abu Zayyad, H. M., Obeidat, Z. M., Alshurideh, M. T., Abuhashesh, M., Maqableh, M., & Masa'deh, R. E. (2021). Corporate social responsibility and patronage intentions: the mediating effect of brand credibility. *Journal of Marketing Communications, 27*(5), 533–510.

Ahmad, S., Zulkurnain, N., & Khairushalimi, F. (2016). Assessing the validity and reliability of a measurement model in structural equation modeling (SEM). *British Journal of Mathematics & Computer Science, 15*(3), 1–8. https://doi.org/10.9734/BJMCS/2016/25183.

Al Kurdi, B., Elrehail, H., Alzoubi, H., Alshurideh, M., & Al-Adaila, R. (2021). The interplay among HRM practices, job satisfaction and intention to leave: An empirical investigation. *Journal of Legal, Ethical and Regulatory, 24*(1), 1–14.

Al-Quran, A. Z., Alhalalmeh, M. I., Eldahamsheh, M. M., Mohammad, A. A., Hijjawi, G. S., Almomani, H. M., & Al-Hawary, S. I. (2020). Determinants of the green purchase intention in Jordan: The moderating effect of environmental concern. *International Journal of Supply Chain Management, 9*(5), 366–371.

Alaali, N., Al Marzouqi, A., Albaqaeen, A., Dahabreh, F., Alshurideh, M., Alrwashdh, S., Iyadeh, I., Salloum, S., Aburayya, A. (2021). The impact of adopting corporate governance strategic performance in the tourism sector: A case study in the Kingdom of Bahrain. *Journal of Legal, Ethical and Regulatory*, 24 (Special Issue 1), 1–18.

Al-Abadi, S.-R., & Al-Atabi, T. N. (2014). The impact of critical success elements in achieving sustainable competitive advantage: a field research for the views of a sample of managers in Al-Mu'tasim general company for construction contracting. *Journal of Economic and Administrative Sciences, 20*(80), 204–235.

Alameeri, K. A., Alshurideh, M. T., & Al Kurdi, B. (2021). The effect of Covid-19 pandemic on business systems' innovation and entrepreneurship and how to cope with it: A theatrical view. *The Effect of Coronavirus Disease (COVID-19) on Business Intelligence*, 334, 275–288.

Alameeri, K., Alshurideh, M., Al Kurdi, B., & Salloum, S. A. (2020, October). The effect of work environment happiness on employee leadership. In International Conference on Advanced Intelligent Systems and Informatics (pp. 668–680). Springer, Cham.

Al-Dmour, R., AlShaar, F., Al-Dmour, H., Masa'deh, R., & Alshurideh, M. T. (2021). The effect of service recovery justices strategies on online customer engagement via the role of "Customer Satisfaction" during the Covid-19 pandemic: An empirical study. *The Effect of Coronavirus Disease (COVID-19) on Business Intelligence*, 334, 325–346.

Alhalalmeh, M. I., Almomani, H. M., Altarifi, S., Al- Quran, A. Z., Mohammad, A. A., & Al-Hawary, S. I. (2020). The nexus between corporate social responsibility and organizational performance in Jordan: The mediating role of organizational commitment and organizational citizenship behavior. *Test Engineering and Management, 83*(July), 6391–6410.

AlHamad, A., Alshurideh, M., Alomari, K., Kurdi, B., Alzoubi, H., Hamouche, S., & Al-Hawary, S. (2022). The effect of electronic human resources management on organizational health of telecommunications companies in Jordan. *International Journal of Data and Network Science, 6*(2), 429–438. *International Journal of Data and Network Science*, 6 (2), 429–438.

Al-Hawary, S. I. (2015). Human resource management practices as a success factor of knowledge management implementation at health care sector in Jordan. *International Journal of Business and Social Science, 6*(11/1), 83–98.

Al-Hawary, S. I. (2009). The effect of the leadership style on the effectiveness of the organization: A field study at Zarqa Private University. *The Egyptian Journal for Commercial Studies, 33*(1), 361–393.

Al-Hawary, S. I. (2011). Human resource management practices in Zain cellular communications company operating in Jordan. *Perspectives of Innovations, Economics and Business, 8*(2), 26–34.

Al-Hawary, S. I., & Alajmi, H. M. (2017). Organizational commitment of the employees of the ports security affairs of the state of Kuwait: The impact of human recourses management practices. *International Journal of Academic Research in Economics and Management Sciences, 6*(1), 52–78.

Al-Hawary, S. I., & Al-Hamwan, A. (2017). Environmental analysis and its impact on the competitive capabilities of the commercial banks operating in Jordan. *International Journal of Academic Research in Accounting, Finance and Management Sciences, 7*(1), 277–290.

Al-Hawary, S. I., & Al-Namlan, A. (2018). Impact of electronic human resources management on the organizational learning at the private hospitals in the state of Qatar. *Global Journal of Management and Business Research: A Administration and Management, 18*(7), 1–11.

Al-Hawary, S. I., & Hadad, T. F. (2016). The effect of strategic thinking styles on the enhancement competitive capabilities of commercial banks in Jordan. *International Journal of Business and Social Science, 7*(10), 133–144.

Al-Hawary, S. I., & Ismael, M. (2010). The effect of using information technology in achieving competitive advantage strategies: A field study on the Jordanian Pharmaceutical Companies. *Al Manara for Research and Studies, 16*(4), 196–203.

Al-Hawary, S. I., & Nusair, W. (2017). Impact of human resource strategies on perceived organizational support at Jordanian Public Universities. *Global Journal of Management and Business Research: A Administration and Management, 17*(1), 68–82.

Al-Hawary, S. I., AL-Awawdeh, W., & Abden, M. A. (2012). The impact of the leadership style on organizational commitment: A field study on Kuwaiti Telecommunications Companies. *ALEDARI*, (130), 53–102.

Al-Hawary, S. I., Al-Qudah, K., Abutayeh, P., Abutayeh, S., & Al-Zyadat, D. (2013). The impact of internal marketing on employee's Job satisfaction of commercial banks in Jordan. *Interdisciplinary Journal of Contemporary Research in Business, 4*(9), 811–826.

Al-Hawary, S. I. S., & Alhajri, T. M. S. (2020). Effect of electronic customer relationship management on customers' electronic satisfaction of communication companies in Kuwait. *Calitatea, 21*(175), 97–102.

Al-Hawary, S. I. S., & Mohammed, A. K. (2017). Impact of team work traits on organizational citizenship behavior from the viewpoint of the employees in the education directorates in North Region of Jordan. *Global Journal of Management and Business*, 17(2-A), 23–40.

Al-Hawary, S. I. S., & Obiadat, A. A. (2021). Does mobile marketing affect customer loyalty in Jordan? *International Journal of Business Excellence, 23*(2), 226–250.

Al-Hawary, S. I. S., Mohammad, A. S., Al-Syasneh, M. S., Qandah, M. S. F., & Alhajri, T. M. S. (2020). Organisational learning capabilities of the commercial banks in Jordan: Do electronic human resources management practices matter? *International Journal of Learning and Intellectual Capital, 17*(3), 242–266.

Al-Hawary, S. I., & Abu-Laimon, A. A. (2013). The impact of TQM practices on service quality in cellular communication companies in Jordan. *International Journal of Productivity and Quality Management, 11*(4), 446–474.

Al-Hawary, S. I., & Al-Menhaly, S. (2016). The quality of E-government services and its role on achieving beneficiaries satisfaction. *Global Journal of Management and Business Research: A Administration and Management, 16*(11), 1–11.

Al-Hawary, S. I., & Al-Smeran, W. (2017). Impact of electronic service quality on customers satisfaction of Islamic Banks in Jordan. *International Journal of Academic Research in Accounting, Finance and Management Sciences, 7*(1), 170–188.

Al-Hawary, S. I., & Al-Syasneh, M. S. (2020). Impact of dynamic strategic capabilities on strategic entrepreneurship in presence of outsourcing of five stars hotels in Jordan. *Business: Theory and Practice*, 21(2), 578–587.

Al-Hawary, S. I., & Hussien, A. J. (2017). The impact of electronic banking services on the customers loyalty of commercial banks in Jordan. *International Journal of Academic Research in Accounting, Finance and Management Sciences, 7*(1), 50–63.

Alkalha, Z., Al-Zu'bi, Z., Al-Dmour, H., Alshurideh, M., & Masa'deh, R. (2012). Investigating the effects of human resource policies on organizational performance: An empirical study on commercial banks operating in Jordan. *European Journal of Economics, Finance and Administrative Sciences*, 51(1), 44–64.

Allahow, T. J. A. A., Al-Hawary, S. I. S., & Aldaihani, F. M. F. (2018). Information technology and administrative innovation of the central agency for information technology in Kuwait. *Global Journal of Management and Business*, 18(11-A), 1–16.

Al-Lozi, M., Almomani, R. Z., & Al-Hawary, S. I.(2017). Impact of talent management on achieving organizational excellence in Arab Potash Company in Jordan. *Global Journal of Management and Business Research: A Administration and Management*, 17(7), 15–25.

Al-Nady, B. A., Al-Hawary, S. I., & Alolayyan, M. (2013). Strategic management as a key for superior competitive advantage of sanitary ware suppliers in Kingdom of Saudi Arabia. *International Journal of Management and Information Technology, 7*(2), 1042–1058.

Al-Nady, B. A., Al-Hawary, S. I., & Alolayyan, M. (2016). The role of time, communication, and cost management on project management success: An empirical study on sample of construction projects customers in Makkah City, Kingdom of Saudi Arabia. *International Journal of Services and Operations Management, 23*(1), 76–112.

Alshraideh, A. T. R., Al-Lozi, M., & Alshurideh, M. T. (2017). The impact of training strategy on organizational loyalty via the mediating variables of organizational satisfaction and organizational performance: An empirical study on Jordanian agricultural credit corporation staff. *Journal of Social Sciences (COES&RJ-JSS)*, 6(2), 383–394.

Alshurideh, M. T., Al Kurdi, B., Alzoubi, H. M., Ghazal, T. M., Said, R. A., AlHamad, A. Q., ... & Al-kassem, A. H. (2022). Fuzzy assisted human resource management for supply chain management issues. *Annals of Operations Research*, 1–19.

Alshurideh, M., Al Kurdi, B., & Salloum, S. A. (2019a). Examining the main mobile learning system drivers' effects: A mix empirical examination of both the Expectation-Confirmation Model (ECM) and the Technology Acceptance Model (TAM). In *International Conference on Advanced Intelligent Systems and Informatics* (pp. 406–417). Springer, Cham.

Alshurideh, M., Kurdi, B. A., Shaltoni, A. M., & Ghuff, S. S. (2019b). Determinants of pro-environmental behaviour in the context of emerging economies. *International Journal of Sustainable Society, 11*(4), 257–277.

Al-Taie, Muhammad Abd Hussein, and Al-Khafaji, Nima Abbas. (2009). *Strategic information systems: a perspective of strategic advantage* (1st Edition), Amman. Jordan, House of Culture for Publishing and Distribution.

Altamony, H., Masa'deh, R., Alshurideh, M., Obeidat, B. (2012). Information systems for competitive advantage: Implementation of an organisational strategic management process. In Innovation and Sustainable Competitive Advantage: From Regional Development to World Economies (pp. 583–592).

Altarifi, S., Al-Hawary, S. I. S., & Al Sakkal, M. E. E. (2015). Determinants of E-shopping and its effect on consumer purchasing decision in Jordan. *International Journal of Business and Social Science, 6*(1), 81–92.

AlTaweel, I. R., & Al-Hawary, S. I. (2021). The mediating role of innovation capability on the relationship between strategic agility and organizational performance. *Sustainability, 13*(14), 7564.

Al-Tayyar, M. R., Al-Nama'a, A. Z. (2018). The role of strategic information systems in supporting the production-on-demand strategy: An exploratory study in the General Company for the manufacture of ready-made garments in Mosul. *Al-Rafidain Development Journal*, 37(120), 32–46.

Alzoubi, H. M., Alshurideh, M., & Ghazal, T. M. (2021a). Integrating BLE Beacon technology with intelligent information systems IIS for operations' performance: A managerial perspective. In The International Conference on Artificial Intelligence and Computer Vision (pp. 527–538). Springer, Cham.

Alzoubi, H., Alshurideh, M., Akour, I., Al Shraah, A., & Ahmed, G. (2021b). Impact of information systems capabilities and total quality management on the cost of quality. *Journal of Legal, Ethical and Regulatory Issues*, 24(Special Issue 6), 1–11.

Alzoubi, H., Alshurideh, M., Kurdi, B., Akour, I., & Aziz, R. (2022). Does BLE technology contribute towards improving marketing strategies, customers' satisfaction and loyalty? The role of open innovation. *International Journal of Data and Network Science, 6*(2), 449–460.

Al-Zoubi, H. A. (2005). *Strategic information systems: A strategic introduction* (1st ed.). Jordan, Dar Wael for publishing and distribution.

Ashal, N., Alshurideh, M., Obeidat, B., Masa'deh, R. (2021) The impact of strategic orientation on organizational performance: Examining the mediating role of learning culture in Jordanian telecommunication companies. *Academy of Strategic Management Journal*, 21(Special Issue 6), 1–29.

Ben-Abdallah, R., Shamout, M., & Alshurideh, M. (2022) Business development strategy model using EFE, IFE and IE analysis in a high-tech company: An empirical study. *Academy of Strategic Management Journal*, 21(Special Issue 2), 1–9.

de Leeuw, E., Hox, J., Silber, H., Struminskaya, B., & Vis, C. (2019). Development of an international survey attitude scale: Measurement equivalence, reliability, and predictive validity. *Measurement Instruments for the Social Sciences, 1*(1), 9. https://doi.org/10.1186/s42409-019-0012-x.

Dessler, G. (2015). Human Resource Management (14"hed). England, Edinburgh, Harlow: Pearson Education Limited.
Esmeray, A. (2016). The impact of accounting information systems (AIS) on firm performance: Empirical evidence in Turkish small and medium sized enterprises. *International Review of Management and Marketing, 6*(2), 233–236.
Fathian, M., Akhavan, P., & Hoorali, M. (2018). E-readiness assessment of non-profit ICT SMEs in a developing country: The case of Iran. *Technovation, 28*(9), 578–590.
Hair, J. F., Babin, B. J., & Krey, N. (2017). Covariance-based structural equation modeling in the journal of advertising: Review and recommendations. *Journal of Advertising, 46*(1), 163–177. https://doi.org/10.1080/00913367.2017.1281777.
Hammadi, A. A. (2020). The role of human resources management strategies in achieving strategic success: An exploratory study of the opinions of managers in the office of the Iraqi Ministry of Electricity. *Anbar University Journal of Economic and Administrative Sciences, 12*(29), 373–400.
Hijjawi, G. S., & Mohammad, A. (2019). Impact of organizational ambidexterity on organizational conflict of Zain telecommunication company in Jordan. *Indian Journal of Science and Technology, 12*, 26.
Howard, M. C. (2018). The convergent validity and nomological net of two methods to measure retroactive influences. *Psychology of Consciousness: Theory, Research, and Practice, 5*(3), 324–337. https://doi.org/10.1037/cns0000149.
Lee, K., Azmi, N., Hanaysha, J., Alshurideh, M., & Alzoubi, H. (2022a). The effect of digital supply chain on organizational performance: An empirical study in Malaysia manufacturing industry. *Uncertain Supply Chain Management, 10*(2), 1–16.
Lee, K., Ramiz, P., Hanaysha, J., Alzoubi, H., & Alshurideh, M. (2022b). Investigating the impact of benefits and challenges of IOT adoption on supply chain performance and organizational performance: An empirical study in Malaysia. *Uncertain Supply Chain Management, 10*(2), 1–14.
Lim, S., Wang, T, K., & Lee, S. Y. (2017). Shedding new light on strategic human resource management: The impact of human resource management practices and human resources on the perception of federal agency mission accomplishment. *Public Personnel Management,* 46(2), 91–117.
Masrek, M. N., Jamaludin, A., & Hashim, D. M. (2009). Determinants of strategic utilization of information systems: A conceptual framework. *Journal of Software, 4*(6), 591–598.
Matei, S. (2013). Conceptual clarification of strategic human resource management and the Way it is implemented. *SEA: Practical Application of Science,* 1(1), 182–191.
Metabis, A., & Al-Hawary, S. I. (2013). The Impact of internal marketing practices on services quality of commercial banks in Jordan. *International Journal of Services and Operations Management, 15*(3), 313–337.
Mohammad, A. A., Alshura, M. S., Al-Hawary, S. I. S., Al-Syasneh, M. S., & Alhajri, T. M. (2020). The influence of Internal Marketing Practices on the employees' intention to leave: A study of the private hospitals in Jordan. *International Journal of Advanced Science and Technology, 29*(5), 1174–1189.
Olszak, C. M., & Ziemba, E. (2012). Critical success factors for implementing business intelligence systems in small and medium enterprises on the example of upper Silesia, Poland. *Interdisciplinary Journal of Information, Knowledge, and Management, 7*(2), 129–150.
Ranong, P., & Phuenngam, W. (2009). Critical success factors for effective risk management procedures in financial industries. Unpublished Master's Thesis, Umea University, Umea, Sweden.
Rimkeviciene, J., Hawgood, J., O'Gorman, J., & De Leo, D. (2017). Construct validity of the acquired capability for suicide scale: factor structure, convergent and discriminant validity. *Journal of Psychopathology and Behavioral Assessment, 39*(2), 291–302. https://doi.org/10.1007/s10862-016-9576-4.
Schrager, J. E., & Madansky, A. (2013). Behavioral strategy: A foundational view. *Journal of Strategy and Management, 6*(1), 81–95.

Sekaran, U., & Bougie, R. (2016). *Research methods for business: A skill-building approach* (Seventh edition). Wiley.

Shah, S. F., Alshurideh, M. T., Al-Dmour, A., & Al-Dmour, R. (2021). Understanding the influences of cognitive biases on financial decision making during normal and COVID-19 pandemic situation in the United Arab Emirates. *The Effect of Coronavirus Disease (COVID-19) on Business Intelligence,* 334, 257.

Shah, S. F., Alshurideh, M., Al Kurdi, B., & Salloum, S. A. (2020, October). The Impact of the behavioral factors on investment decision-making: a systemic review on financial institutions. In International Conference on Advanced Intelligent Systems and Informatics (pp. 100–112). Springer, Cham.

Shamout, M., Elayan, M., Rawashdeh, A., Kurdi, B., & Alshurideh, M. (2022). E-HRM practices and sustainable competitive advantage from HR practitioner's perspective: A mediated moderation analysis. *International Journal of Data and Network Science, 6*(1), 165–178.

Shi, D., Lee, T., & Maydeu-Olivares, A. (2019). Understanding the model size effect on SEM fit indices. *Educational and Psychological Measurement, 79*(2), 310–334. https://doi.org/10.1177/0013164418783530.

Sung, K.-S., Yi, Y. G., & Shin, H.-I. (2019). Reliability and validity of knee extensor strength measurements using a portable dynamometer anchoring system in a supine position. *BMC Musculoskeletal Disorders, 20*(1), 1–8. https://doi.org/10.1186/s12891-019-2703-0.

Svoboda, P., Ghazal, T. M., Afifi, M. A., Kalra, D., Alshurideh, M. T., & Alzoubi, H. M. (2021, June). Information systems integration to enhance operational customer relationship management in the pharmaceutical industry. In The International Conference on Artificial Intelligence and Computer Vision (pp. 553–572). Springer, Cham.

Syed, R., Bandara, W., French, E., & Stewart, G. (2018). Getting it right! Critical success factors of BPM in the public sector: A systematic literature review. *Australasian Journal of Information Systems, 22,* 1–39.

Tariq, E., Alshurideh, M., Akour, I., & Al-Hawary, S. (2022). The effect of digital marketing capabilities on organizational ambidexterity of the information technology sector. *International Journal of Data and Network Science, 6*(2), 401–408.

Wang, Y. A., & Rhemtulla, M. (2021). Power analysis for parameter estimation in structural equation modeling: A discussion and tutorial. *Advances in Methods and Practices in Psychological Science, 4*(1), 1–17. https://doi.org/10.1177/2515245920918253.

Youssef, B. A. R., & Yehia, N. Z. (2018). Implications of the outputs of the strategic information system in promoting entrepreneurship: a case study in private hospitals in Erbil/Kurdistan Region, *Tikrit Journal of Administrative and Economic Sciences, 3* (43), 137–160.

Factors Affecting Local Employees Sectorial Choice (Public vs Private), the Case of Abu Dhabi, UAE

Mohammad Mousa Eldahamsheh, Main Naser Alolayyan, Hanan Mohammad Almomani, Ali Zakariya Al-Quran, Fuad N. Al-Shaikh, Mohammed Saleem Khlif Alshura, Menahi Mosallam Alqahtani, Sulieman Ibraheem Shelash Al-Hawary, and Anber Abraheem Shlash Mohammad

Abstract The purpose of this research study was to explore the factors that affect emirates employee sectorial choice (public vs private). And, to examine if there is a difference between these factors related to local employees working in the public sector and the private sector. A quantitative approach was deployed in this study to determine the most important factors that affect local employees' career choice

M. M. Eldahamsheh
Strategic management, Aqaba, Jordan

M. N. Alolayyan
Health Management and Policy Department, Faculty of Medicine, Jordan University of Science and Technology, Ar-Ramtha, Jordan
e-mail: mnalolayyan@just.edu.jo

H. M. Almomani · A. Z. Al-Quran · S. I. S. Al-Hawary (✉)
Department of Business Administration, School of Business, Al al-Bayt University, P.O.BOX 130040, Mafraq 25113, Jordan
e-mail: dr_sliman73@aabu.edu.jo

A. Z. Al-Quran
e-mail: ali.z.al-quran@aabu.edu.jo

F. N. Al-Shaikh
Department of Business Administration, Faculty of Economics and Administrative Sciences, Yarmouk University, 21163, P.O Box 566, Irbid, Jordan
e-mail: eco_fshaikh@yu.edu.jo

M. S. K. Alshura
Management Department, Faculty of Money and Management, The World Islamic Science University, P.O.Box1101, Amman 11947, Jordan

M. M. Alqahtani
Administration Department, Community College of Qatar, Doha, Qatar
e-mail: mena7i@icloud.com

A. A. S. Mohammad
Marketing, Marketing Department, Faculty of Administrative and Financial Sciences, Petra University, P.O.Box: 961343, Amman 11196, Jordan

M. Alshurideh et al. (eds.), *The Effect of Information Technology on Business and Marketing Intelligence Systems*, Studies in Computational Intelligence 1056,
https://doi.org/10.1007/978-3-031-12382-5_49

toward the private sector. The population of this study was the work force in the state of Abu Dhabi, those whom are employed in the public sector and in the private sector. Simple random sampling technique used to choose appropriate sample from this population. From each sector a random sample has been chosen from different industry that falls under the public sector and the same procedure has been followed for the private sector. The statistical tool used for the purpose of this study was Chi-square. Chi-square analysis is used to check for differences in the categories between the two samples. This study reveals the difference between public sector employees and private sector employees and their job characteristics. The public sector employees experience more autonomy and feedback than their colleague in the private sector. The study provided practical and theoretical recommendations.

Keywords Local employees · Sectorial choice · Abu Dhabi · UAE

1 Introduction

Career choice is considered the most important decision job seekers face. Choosing a career is not an easy decision, as this decision will affect their satisfaction, motivation, turnover, earnings, and productivity (Gagné & Deci, 2005; Alameeri et al., 2020; Allozi et al., 2022). In contrast, choosing the right career will increase productivity, and reduce turnover, and job burnout (Al Kurdi et al., 2020a, b). In this regard many theories were established to explain how people make their career choice. For example, Holland theory of vocational types, Super's developmental self-concept theory, Roe's personality theory of career choice (Osipow & Fitzgerald, 1996), social cognitive career theory (Brown, 2002), and the theory of planned behavior (Ajzen, 1991) provide an explanations of how people chose their career from different perspectives (sociological, personality, self-concept…etc.).

Switching between careers is sometimes considered an easy decision to make, but sometimes it is a difficult decision, especially when the opportunity to find another job is limited and competitive (Al Kurdi et al., 2021; Hansen, 2012; Hayajneh et al., 2021). In the Middle East area in general, and in the Gulf region specifically, job seekers prefer to work for the government (Al-Waqfi & Forstenlechner, 2012). United Arab Emirates is one such country in the Gulf region with employment rates that reveal high rates of people who prefer to work for government. According to Aldhaheri et al. (2017) in the United Arab Emiratis the employment rate in the government for locals is 89%, while only 11% are working in the private sector. Many factors affect employee's career choice. A number of studies have shown that many factors affect employee preferences for career and sector (private or public), factors such as job motivations, job characteristics, social influence (Albugamy, 2014; AlShehhi et al., 2020) and employer branding (Moroko & Uncles, 2008).

The private sector in the United Arab Emirate is currently suffering a shortage of Emiratis employees (Al-Waqfi & Forstenlechner, 2012). The majority of Emiratis target public sector as their preferred employment destination (Aldhaheriet al., 2017).

Local job seekers in this country prefer to stay in unemployment pool for years to get a job in the public sector rather than starting immediate job in the private sector. The future policy of the country is to minimize dependence on oil as the main source of income, and finding other sources of income, such as depend on the service and production sectors.

This study is considered the first quantitative large-scale testing research for Emirates employment sectorial choice and the factors affecting their career choice preferences toward the private sector versus the public sector. Al-Waqfi and Forstenlechner (2012) in their research paper highlighted the need for quantitative research on local's job seekers in United Arab Emirates. Research also needed to explore external factors for job seekers career choice (Aldhaheri et al., 2017). The research added to the body of knowledge in career choice by examining job motivations, job characteristics, employer branding, and social influence. In addition, this study fills the gap in the literature that largely ignored the differences between the public sector employees and the private sector employees regarding their sectorial choice.

2 Theoretical Framework and Hypotheses Development

2.1 *Career Choice*

Career choice is considered a very critical and complex decision for each human being (Alshurideh, 2016; Alshurideh et al., 2015a). The stage in person's life in which he/she will transfer from dependent to independent person, where people think about their future and what they want to be (Alshurideh, 2019; Alshurideh et al., 2015b). A number of studies have shown that many factors affect employee preferences for career and sector (private or public), factors such as job motivations, job characteristics, social influence (Albugamy, 2014; Almazrouei et al., 2020; Alshurideh et al., 2019) and employer branding (Moroko & Uncles, 2008).

2.2 *Factors Affect Employee Preferences for Career and Sector*

Job Motivations: Motivations are considered an important factor for employees; when employee gets motivated, he/she will be able to do his job in more efficient and in effective manner. Job motivation (Al-bawaia et al., 2022; Alshurideh, 2022; Ammari et al., 2017). According to George and Jones (2012, p. 157) "can be defined as the psychological forces within a person that determine the direction of that person's behavior in an organization, effort level, and persistence in the face of obstacles". The differences in motivation between public and private sector employees have been discussed in many research articles (Al Shebli et al., 2021; Al-Hawary, 2015;

Al-Hawary & Abu-Laimon, 2013; Al-Hawary et al., 2020). Many of these studies focused on two different types of motivations, intrinsic and extrinsic motivations (Al-Hawary & Al-Namlan, 2018; Al-Hawary et al., 2013; Bettayeb et al., 2020; Metabis & Al-Hawary, 2013; Rashid & Rashid, 2012). According to Rayn and Deci (2000, p. 55) " *intrinsic motivations*, which refers to doing something because it is inherently interesting and enjoyable, and *extrinsic motivations,* which refers to doing something because it leads to a separable outcome". Thomas (2009) in his book indicated four types of intrinsic rewards or motivation; the four main intrinsic rewards are: sense of meaningfulness, a sense of choice, a sense of competence, and a sense of progress. Other studies indicated different types of extrinsic motivations, such as financial rewards (Al-Hawary, 2011; Al-Hawary & Nusair, 2017; Ng et al., 2016) prestige (Ng et al., 2016), pay (Lewis & Frank, 2002; Al-Lozi et al., 2018; Al-Hawary & Alajmi, 2017; Al-Hawary et al., 2020; Mohammad et al., 2020). Similar to Ng et al. (2016) and Dur and Zoutenbier (2015) found out that intrinsic motivations for public sector employees are stronger than private sector employees. However, Dur and Zoutenbier (2015) stated that altruism and laziness are more prevalent in the behavior of public sector employees. Extrinsic and intrinsic motivations were considered very important for both employees (public and private sectors); still, public employees' value extrinsic motivations more than private sector employees (Maidani, 1991).

DeSantis and Durst (1996) and Alsuwaidi et al. (2020) found out that pay is considered very important factor for choosing between public and private sector, especially between young employees, which prefer the private sector. Dan (2015) found out that there are no differences between employee benefits and the sectorial choice. Public sector employees were less concerned about financial rewards than private sector employees (Alshurideh et al., 2014; Rashid & Rashid, 2012). In contrast, Lewis and Frank (2002) argued that employees prefer the public sector because of the high security. Based on the above literature, the study hypotheses can be formulated as:

H1: There is a difference between extrinsic motivations and employee sectorial choice (Public vs Private).

H2: There is a difference between intrinsic motivations and employee sectorial choice (Public vs Private).

Job Characteristics: A job characteristic is considered a very important factor as it has many effects on work related outcomes. The job characteristics theory (JCT) developed by Hackman and Oldham (Faturochman, 2016) is widely used as a framework to study how particular job characteristics impact job outcomes, including job satisfaction. According to (Faturochman, 2016, p. 9) "job characteristics theory describes the relationship between job characteristics and individual response to work". The theory states that there is five core job characteristics: (a) skill variety, (b) task identity, (c) task significance, (d) autonomy, and (e) job feedback (Faturochman, 2016). Many studies linked job characteristic with organization performance, but few studies focused on the relationship between job characteristics and the employment sector.

DeSantis and Durst (1996) found out that employees in both sectors are concerned with a pleasant work atmosphere, friendly work environment, and job that are important for them and for the society. Solomon (1986) argued that some job characteristics that promote efficiency were more prevalent for the private sector employees. Lewis and Frank (2002) found out that job security considered very important characteristics, and employees who value job security will be more likely to work for the public sector. Based on the above literature, the study hypotheses can be formulated as:

H3: There is a difference between job characteristic and employee sectorial choice (Public vs Private).

Social Influence on Career Choice: Social influence plays a major role for job seekers in their decision for the career path. Studies indicted many social factors that affect job seekers career chose, factors such as, family, friends, and peers (AlHamad et al., 2022; Alkalha et al., 2012; Alshurideh et al., 2022; Alzoubi et al., 2020). Kulkarni and Nithyanand (2013) examined why social influence is considered a key factors in job choice decision for young job seekers. Kulkarni and Nithyanand indicated that there are two reasons on how social influence affects the decision of job seekers. The First reason was the pressure from their family. The second reason was to show their worth for their peer group, they want to be seen equal to their peers at work not want to be looked down upon.

In addition, Mishkin et al. (2016) studied career choice among women from the perspective of the theory of planned behavior. Based upon the theory of planned behavior, the researchers studied the effect of attitudes, subjective norms, and perceived behavioral control and their impact on career choice. Mishkin et al. stated that subjective norms; which are considered a social factor that explain how a person changes his/her behavior based upon the social pressure, was the most significant factor that affect women's career choice.

Further, Mishkin et al. (2016) investigated the role of role model on student's career choice. Both of these factors (subjective norms and role model) were positively affected students career choice (Mishkin et al., 2016). In similar study, (Korkmaz, 2015) investigated the social factors affecting student's career choice in science and technology. Korkmaz indicated that there was strong impact for father and mother on student career choice. In the same manner, Law and Arthur (2003) investigated the factors affecting school student career choice toward nursing as profession. Law and Arthur found out that parents, social career masters, and friends significantly affected their career choice. Based on the above literature, the study hypotheses can be formulated as:

H4: There is a difference between social influence and employee sectorial choice (Public vs Private).

Employer Branding and Job Seekers Career Choice: Private organizations conduct their business in a competitive environment. Competition between organizations has many forms. There is a competition on resources, technology, suppliers, human capital…etc. Attracting the most talent human capital is primary concern of the employer. This will enhance the fit between organization and people, which will contribute to the strategic goals of organizations (Wallace et al., 2014). According to Wallace et al. (2014) employer branding can be defined as "the package of functional, economic, and psychological benefits provided by employment, and identified with the employing company" (p. 21). Moroko and Uncles (2008) attempt to provide characteristics of successful employer brand. There are two characteristics of successful employer brand. First, attractiveness and accuracy, which mean, that employee can assure an employer to be attractive if he supports awareness, differentiation, and relevance. Secondly, employees consider the employer brand successful if he supports consistency; in which there are a consistency between employer brand and employer experience, and organization culture and value (Moroko & Uncles, 2008).

Looking to employer brand from the perspective of potential employee, Jain and Bhatt (2015) pointed out the importance of understanding potential employees' perspective about employer brand. Jain and Bhatt declared that potential employees are willing to work for the employer if they perceived his brand to support company stability, work-life balance, and job security. Based on the above literature, the study hypotheses can be formulated as:

H5: There is a difference between employer branding and employee sectorial choice (Public vs Private).

3 Study Model

See Fig. 1.

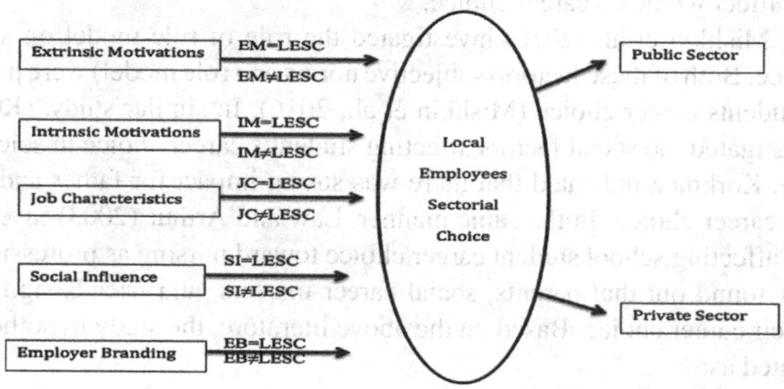

Fig. 1 Career choice conceptual model

4 Research Methodology

This is an exploratory study in which, it investigated and explored the factors that affect local employees career choice. A quantitative approach was deployed in this study to determine the most important factors that affect local employees career choice toward the private sector. The statistical tool used for the purpose of this study was Chi-square. Chi-square analysis is used to check for differences in the categories between the two samples.

4.1 Data Collection

The population of this study were the work force in the state of Abu Dhabi, those whom are employed in the public sector and in the private sector. Simple random sampling technique used to choose appropriate sample from this population. From each sector a random sample has been chosen from different industry that falls under the public sector and the same procedure has been followed for the private sector. Two samples were selected from both sectors (public and private), and a comparison between these two samples were conducted. The researcher deployed questionnaire to collect data from selected respondents, and five-point Likert Scale has been used to measure the variables.

4.2 Population

The population that concerns this study was the local employees working in United Arab Emirates, in specific, the state of Abu Dhabi. Employees working in the public and the private sector in the state of Abu Dhabi were considered the population of this study. Information about this population was gathered from Statistics Centre—Abu Dhabi (SCAD) and the Ministry of Human Resource and Emiratization (MHRE).

4.3 Sampling Frame and Sample

The method of simple random sampling was used to develop the sample of the research under discussion. This type of sampling is preferred when we are looking for a representative sample, it also allows for using statistical tests in data analysis (Neuman, 2006). Randomization in the sample helps the researcher in generalizing the results on the population, as it is free from bias and prejudice and it is a representative of the population. The researcher selects one sample from each sector (Public

Table 1 Questionnaire items obtained from the review of related literature

Construct	Author/s
Motivations (intrinsic and extrinsic)	Rashid and Rashid (2012), Dan (2015), Park and Word (2012), Lee and Wilkins (2011), Ko and Jun (2015)
Job characteristics	Johari and Yahya (2016), Faturochman (2016), Kim (2016)
Social influence	Pavel (2015), Eesley and Wang (2017), Al subait et al., (2017)
Employer branding	Jain and Bhatt (2015), Saini et al., (2015)

and Private), from different organization. Mostly, the industry under the investigation was from the service industry. The overall sample size was 344 respondents, 350 respondents from the public sector and 150 from the private sector.

4.4 *Instrumentation*

The questionnaire was developed upon reviewing the literature related to this dissertation. Table 1 represents the research constructs and the questionnaire items obtained from review of related literature by the researcher.

4.5 *Validity*

To assure the content validity the questions to be asked for the respondents supposed to cover all parts of the construct under investigation. Content validity should measure all elements of the construct and it involves three steps; identify of the content from the construct definition, from all the definitions chose a sample, and developing indicators that hit all of the parts of the construct definition (Neuman, 2006).

4.6 *Reliability*

For intrinsic and extrinsic work motivation, this part the researcher borrowed some questions from the work extrinsic and intrinsic motivation scale (WEIMS) developed by Deci and Ryan (Tremblay et al., 2009). The part of job characteristics was borrowed from Job characteristics inventory, developed by Hackman and Oldman (Faturochman, 2016). For employer branding the researcher borrowed some questions from the work of Tanwar and Prasad (Tanwar & Prasad, 2017). Finally, for social influence scale the researcher borrowed some questions from the career influence inventory (Pavel, 2015). After conducting the pilot study, the Cronbach Alpha scores was 0.848 which indicates a strong reliability of the instrument.

5 Data Analysis

The aim of this research was to explore the differences between the public and the private sector employees in selecting their career sector. The statistical techniques employed to achieve this aim was Chi Square using SPSS (Statistical Package for Social Sciences). According to (Cooper & Schindler, 2014, p. 653) "chi-square (χ2) test is a test of significance used for nominal and ordinal measurements". Chi-square (χ 2) goodness-of-fit test to determine how closely observed frequencies or probabilities match expected frequencies or probabilities. It can be calculated for nominal, ordinal, interval, or ratio data (Leedy & Ormrod, 2013). This test used to determine whether a relationship observed in a contingency table is statistically significant (Christensen et al., 2011).

Hypothesis (1) was tested by five items (salary, working in prestigious organization, opportunity for career advancement, benefits, and job security). As it can be shown from Table 2, the table shows that all P-values for Chi-Square test are 0.001, and that is less than the established significant 0.05. Also, Table 1 shows that all Spearman Correlation are negative which indicate that the private sector employees are less extrinsically motivated than the public sector employees.

Overall, total extrinsic motivations combined have P-values for Chi-Square test is 0.001 and that is less than the established significant 0.05. Therefore, the hypothesis (H1) is supported. Correlation value equal (−0.608) this indicates strong negative relationship. We conclude that There is a significant difference between the public and the private sector employees regarding extrinsic motivation in favor to the public sector.

Table 2 Chi-square and Spearman correlation tests for extrinsic motivations

#	Item	Chi-square value	P-value	Spearman correlation
1	Salary	208.437	0.001	−0.753
2	Work in a prestigious organization	173.322	0.001	−0.640
3	Opportunity for career advancement (promotion)	32.916	0.001	−0.212
4	Benefits (health insurance, car, housing, bonus, leaves, etc.)	130.225	0.001	−0.593
5	Job security	218.865	0.001	−0.723
Total	Extrinsic motivations	654.404	0.001	−0.608

Table 3 Chi-square and Spearman correlation tests for intrinsic motivations

#	Item	Chi-square value	P-value	Spearman correlation
1	Interesting and challenging work	60.028	0.001	−0.368
2	Opportunity for personal growth and development (training, skills, languages, etc.)	30.296	0.001	−0.243
3	Sense of achievement	47.995	0.001	−0.349
4	Opportunity to serve the society	47.573	0.001	−0.337
5	I want to be a useful to the society	52.318	0.001	−0.344
Total	Intrinsic motivations	224.622	0.001	−0.308

This hypothesis (2) was tested by five items (interesting and challenging work, opportunity for personal growth and development, sense of achievement, opportunity to serve the society, useful to the society). Table 3 shows that all P-values for Chi-Square test are 0.001, and that is less than the established significant 0.05. Also, Table 3 shows that all Spearman Correlation are negative which indicate that the private sector employees are less intrinsically motivated than the public sector employees. Overall, total intrinsic motivations combined have P-values for Chi-Square test is 0.001 and that is less than the established significant 0.05. Therefore, the hypothesis (H2) is supported. Correlation value equal (−0.308) this indicates weak negative relationship. We conclude that there is a significant difference between the public and the private sector employees regarding extrinsic motivation in favor to the public sector.

Hypothesis (3) was tested by five items (the job requires a lot of cooperative work with other people, the job denies me any chance to use my personal initiative or judgment in carrying out the work, the job gives me considerable opportunity for independence and freedom in how I do the work, the supervisor and co-workers on this job almost never give me any feedback about how well I am doing in the work, supervisor often let me know how well they think I am performing the job, the job itself provides very few clues about whether or not I am performing well). Table 4 shows that all P-values for Chi-Square test are less than 0.05, and that is less than the established significant 0.05. Also, Table 4 shows that all Spearman Correlation except for item 2 are negative. The positive correlation for item 2 means that the private sector employees have more independent in performing their jobs.

Overall, total job characteristics combined have P-values for Chi-Square test is 0.001 and that is less than the established significant 0.05. Therefore, the hypothesis (H3) is supported. Correlation value equal (−0.037) this indicates very weak negative relationship. We conclude that there is a significant difference between the public and the private sector employees regarding job characteristics.

Hypothesis 4 was tested by five items (family, friends, and peers have a great impact on my decision, I believe on myself and he know what I want). As it can be

Table 4 Chi-square and Spearman correlation tests for job characteristics

#	Item	Chi-square value	P-value	Spearman correlation
1	The job requires a lot of cooperative work with other people	10.624	0.031	−0.162
2	The job denies me any chance to use my personal initiative or judgment in carrying out the work	54.135	0.001	0.220
3	The job gives me considerable opportunity for independence and freedom in how I do the work	104.587	0.001	−0.471
4	The supervisor and co-workers on this job almost never give me any feedback about how well I am doing in my work	20.383	0.001	−0.223
5	The job itself provides very few clues about whether or not I am performing well	15.804	0.001	−0.039
Total	Job characteristics	11.937	0.018	−0.037

shown from Table 5, the table shows that all P-values for Chi-Squair test are less than 0.05, and that is less than the established significant 0.05. Also, Table 5 shows that all Spearman Correlation are negative which indicate that the private sector employees are less affected by social influence than the public sector employees.

Overall, total social influence combined have P-values for Chi-Square test is 0.001 and that is less than the established significant 0.05 (see Appendices4). Therefore,

Table 5 Chi-square and Spearman correlation tests for social influence

#	Item	Chi-square value	P-value	Spearman correlation
1	My family has great impact on my decision (parents, spouse, siblings, children)	22.774	0.001	−0.243
2	My friends have great impact on my decision	18.212	0.001	−0.132
3	Peers have a great impact on my decision	12.489	0.014	−0.179
4	Professors and teachers in my university and school have great impact on my decision	23.989	0.001	−0.240
5	I believe on myself and I know what I want	13.840	0.008	−0.179
Total	Social Influence	66.052	0.00 1	−0.140

Table 6 Chi-square and Spearman correlation tests for employer branding

#	Item	Chi-square value	P-value	Spearman correlation
1	An employer reputation in the market for looking after and valuing employees	99.786	0.001	−0.450
2	Understanding the important of family or life outside the work	11.528	0.009	−0.181
3	Definitive and strong company values	56.045	0.001	−0.349
4	Flexible work hours	29.114	0.001	−0.187
5	Ethical practices	61.627	0.001	−0.405
Total	Employer brand	182.562	0.001	−0.309

the hypothesis (H4) is supported. Correlation value equal (−0.140) this indicates very week negative relationship. We conclude that there is a significant difference between the public and the private sector employees regarding social influence in favor to the public sector.

Hypothesis 5 was tested by five items (an employer reputation in the market for looking after and valuing employees, understanding the important of family or life outside the work, definitive and strong company values, flexible work hours, ethical practice). As it can be shown from Table 6, the table shows that all P-values for Chi-Square test are 0.001, and that is less than the established significant 0.05. Also, Table 6 shows that all Spearman Correlation are negative which indicate that the private sector employees have less employer brand practices than the public sector employees. Overall, total employer brand combined have P-values for Chi-Square test is 0.001 and that is less than the established significant 0.05. Therefore, the hypothesis (H5) is supported. Correlation value equal (−0.309) this indicates weak negative relationship. We conclude that There is a significant difference between the public and the private sector employees regarding employer branding in favor to the public sector.

6 Conclusions

The result of the study shows that there is a difference between public and private employees and sectorial choice (public vs private) for all hypothesis. The first variable in this study was extrinsic motivations. The first item measure extrinsic motivations was salary. In terms of salaries the public sector employees were very motivated by salary and they considered salary as the reason they choose this sector, in contrast private sector employees were not motivated by salaries and they did not considered salary as the reason they choose this sector. This result contradicts with previous studies (Rexhaj, 2011). The reason behind this contradiction is the salary gap between the public sector and the private sector, the gap can reach 5 times between the two sectors in favor for the public sector. As for work in prestigious organization the public sector employees see their organization as prestigious ones, and it was the reason they choose this sector. While, the employees in the private sector did not think this is the reason they choose this sector. This is due to the cultural perspective which connect salaries with prestige. The third item for measuring extrinsic motivation is opportunity for career advancement (promotion). The public sector employees were considering this item very important for them and affect their decision. In contrast, the private sector employees do not consider this item the reason behind their sectorial choice. This finding is not consistent with previous studies (Lee & Wilkins, 2011). The reason behind this result is that the private sector employees consider their job as a temporary one, and this will give them the experience to find a job in the public sector. Added to that, finally, the competition between locals on public jobs, especially with the increasing numbers of locals whom hold bachelor's degrees.

As for the benefits (health insurance, car, housing, bonus, leaves, etc.) and job security, both were considered very important for the public sector. The public sector employees considered these factors very important and affect their sectorial choice. In the contrary, the private sector employees did not consider these factors in their interest when it comes to sectorial choice. These finding contradicted with previous studies (Lee & Wilkins, 2011). The interpretation of why private sector employees feels unsecure and did not have benefits as their counterparts in the public sector is the hiring and firing policies practices followed by the private sector, and the absences of benefits in the private sector and the threat of expatriates. As stated by (Daleure, 2016), the researcher found out that job seekers prefer working in the public sector in UAE because they believe that the public sector had high salaries, better benefits and working conditions.

The second variable measured the difference between employee's sectorial choice was intrinsic motivations. For interesting and challenging work, public sector employees considered their work interesting and challenging while the private sector employees did not agree that their work is challenging and interesting. The reason behind this result is that the hiring polices imposed on the private sector and the employees themselves. Opportunity for personal growth and development shows a difference between employees in the public sector and employees in the private sector. This item was very important for public sector employees. In contrast the private sector employees did not consider this factor as a reason to work in the private sector. As discussed in the previous section private sector employees considered working in this sector as a temporary job, as soon as they get a job in the public sector they will leave immediately.

The third variable that measures sectorial choice between local employees in the public sector and employees in the private sector was job characteristics. For the purpose of this study, this variable measure job autonomy and job feedback, and it shows differences between employees' sectorial choice. The public sector employees have more job autonomy and receive feedback about their performance. In the contrary, the private sector employees have less autonomy in their job and receive lees feedback about their performance. These finding is supported with previous study (Kim, 2016).

The fourth variable that measures sectorial choice between employees in the public sector and the private sector is social influence. All the factors that measure social influence are family, friends, peers, professors, and believe on oneself. For the public employees; family, friends, and believing in themselves, have strong impact on their decision toward the public sector. The same can be said for the private sector employees but with less impact. These finding confirmed with previous research (Mishkin et al., 2016, Workman, 2015; Ng, Burke, & Fiksenbaum, 2008). The finding can be explained due to the type of relationship between people and the culture of the Arab countries. Arab countries are considered a collectivistic societies base upon Hofstede cultural dimensions, individuals are embedded in groups with strong ties and loyalties and are expected to place collective interests over personal interests (Klasing, 2013).

The last variable that measures sectorial choice between the public and the private sector employees was employer brand. The results indicate differences between the two sectors regarding this variable. Public sector organizations are characterized by looking after and valuing their employees more than the private sector organizations. Also, organizations in both sectors appears to have an understanding of the important of life outside the work in favor to the public sector companies. Moreover, the public sector organizations have stronger company values than the private sector organizations (Daleure, 2016).

7 Implication

This study is considered the first point, this study is considered the first quantitative large-scale testing research for Emirates employment sectorial choice and the factors affecting their career choice preferences toward the private sector versus the public sector. The research added to the body of knowledge in career choice by examining job motivations, job characteristics, employer branding, and social influence. The second point of this study was to fill the gap in the literature that largely ignored the differences between the public sector employees and the private sector employees regarding their sectorial choice.

Two types of motivation were discussed in this study, extrinsic motivations and intrinsic motivations. The finding and conclusions of this study approved that there is a difference in the type of motivations that affect local employees decisions regarding sectorial choice. In general, prior studies showed that public sector employees are intrinsically motivated while private sector employees are extrinsically motivated.

The finding of this study somewhat surprising with prior studies, public sector employees considered both types of motives were the reason behind their sectorial choice decision. In contrast, private sector employees did not consider these factors as important as their counterpart in the public sector regarding their sectorial choice decision. This finding has important impactions for increasing the number of locals in the private sector, by understanding their motives toward the private sector. Also, a review for the national program to encourage the employment of UAE nationals is needed. This program facilitates the employment of UAE nationals within specified occupations across the private sector.

In addition, this study reveals the difference between public sector employees and private sector employees and their job characteristics. The public sector employees experience more autonomy and feedback than their colleague in the private sector. An implication of this result is the possibility of the role of the human resources practice in these sectors. Furthermore, this study showed the impact of family and friends on employees sectorial choice decision. This finding may help us in understand the role played by the family and friends on the decision process. It also explains the weak role played by the educational system to deploy more locals in the private sector. Finally, one of the issues that emerged in this study was the difference between the two sectors in regard to employer branding. This finding has an important implication related to the private companies. These companies must shed the light on their image, and how to enhance this image to be more attractive for the locals as their desire sectorial choice.

8 Limitations

This study achieved its aim and its objective; however, it is not without limitations. It is difficult to generalize the finding of this study due to the industry that were willing to participate in this study (service industry only), and only three companies were participated in this study. Added to that, the study only covers one state from seven states in UAE (the state of Abu Dhabi).

Another limitation was the bureaucracy procedures to get the approval for collecting data from the government organizations. Even if the researcher inquired into general information that is supposed to be available to the public, the procedures were time consuming and the researcher's request was often rejected. Finally, as English is the second language for the researcher the process of developing this study was challenging for the researcher. In addition, the researcher did not have any idea about the academic writing style, it was challenging for the researcher.

9 Recommendations

This study provided practical and theoretical recommendations, the practical recommendations concern the leaders and decision makers about how to address this problem based on the results of this study. The theoretical recommendations concern the researchers and what further studies can be conducted based on the results of this study. Leaders and decision makers must focus on job security and the benefits that the private sector offer to the locals working in this sector. The study reveals a fear from the locals working in the private sector regarding the continuity of their jobs and the fair benefits they deserve. Locals face a competition from expatriates, expatriates are willing to work more for less. Leaders and Decision makers must start to develop a pension and retirement plan that guarantee the rights of locals, to make the local feels more secure and protected from any prejudiced action from their companies. Leaders and decision makers must focus on the human resource practice and performance assessment in the private sector. The result of the study disclose that local employees have little autonomy and did not receive significant a feedback from their supervisors. As mentioned in the finding section, private companies must hire a specific number of locals in their companies.

References

Ajzen, I. (1991). The theory of planned behavior. *Organizational Behavior and Human Decision Processes, 50*(2), 179–211.

Al Kurdi, B., Alshurideh, M., & Al afaishat, T. (2020a). Employee retention and organizational performance: Evidence from banking industry. *Management Science Letters, 10*(16), 3981–3990.

Al Kurdi, B., Alshurideh, M., & Alnaser, A. (2020b). The impact of employee satisfaction on customer satisfaction: Theoretical and empirical underpinning. *Management Science Letters, 10*(15), 3561–3570.

Al Kurdi, B., Elrehail, H., Alzoubi, H., Alshurideh, M., & Al-Adaila, R. (2021). The interplay among HRM practices, job satisfaction and intention to leave: An empirical investigation. *Journal of Legal, Ethical and Regulatory, 24*(1), 1–14.

Al Shebli, K., Said, R. A., Taleb, N., Ghazal, T. M., Alshurideh, M. T., & Alzoubi, H. M. (2021, June). RTA's employees' perceptions toward the efficiency of artificial intelligence and big data utilization in providing smart services to the residents of Dubai. In *The International Conference on Artificial Intelligence and Computer Vision* (pp. 573–585). Springer.

Al Subait, A., Ali, A., Andijani, A. I., Altuwaijry, M. A., Algarni, S. M., Alduhaimi, T. S., ... & El metwally, A. (2017). Factors influencing the career choice among medical university students of King Saud bin Abdul-Aziz University, Riyadh Saudi Arabia; a cross-sectional study design. *The Saudi Journal for Dental Research, 8*, 73–78.

Alameeri, K., Alshurideh, M., Al Kurdi, B., & Salloum, S. A. (2020, October). The effect of work environment happiness on employee leadership. In *International Conference on Advanced Intelligent Systems and Informatics* (pp. 668–680). Springer.

Al-bawaia, E., Alshurideh, M., Obeidat, B., Masa'deh, R. (2022) The impact of corporate culture and employee motivation on organization effectiveness in Jordanian banking sector. *Academy of Strategic Management Journal, 21*(Special Issue 2), 1–18.

Albugamy, R. T. (2014). *Institutional and personal influences on career choice: A study on MBA students in Saudi Arabia.* Doctoral thesis, Brunel University, London. http://bura.brunel.ac.uk/handle/2438/10388.

AlDhaheri, R., Jabeen, F., Hussain, M., & Abu-Rahma, A. (2017). Career choice of females in the private sector: Empirical evidence from the United Arab Emirates. *Higher Education, Skills and Work-Based Learning, 7*(2), 179–197.

AlHamad, A., Alshurideh, M., Alomari, K., Kurdi, B., Alzoubi, H., Hamouche, S., & Al-Hawary, S. (2022). The effect of electronic human resources management on organizational health of telecommunications companies in Jordan. *International Journal of Data and Network Science, 6*(2), 429–438.

Al-Hawary, S. I. (2011). Human resource management practices in ZAIN cellular communications company operating in Jordan. *Perspectives of Innovations, Economics and Business, 8*(2), 26–34.

Al-Hawary, S. I. (2015). Human resource management practices as a success factor of knowledge management implementation at health care sector in Jordan. *International Journal of Business and Social Science, 6*(11/1), 83–98.

Al-Hawary, S. I., & Abu-Laimon, A. A. (2013). The impact of TQM practices on service quality in cellular communication companies in Jordan. *International Journal of Productivity and Quality Management, 11*(4), 446–474.

Al-Hawary, S. I., & Alajmi, H. M. (2017). Organizational commitment of the employees of the ports security affairs of the state of Kuwait: The impact of human recourses management practices. *International Journal of Academic Research in Economics and Management Sciences, 6*(1), 52–78.

Al-Hawary, S. I., & Al-Namlan, A. (2018). Impact of electronic human resources management on the organizational learning at the private hospitals in the state of Qatar. *Global Journal of Management and Business Research: A Administration and Management, 18*(7), 1–11.

Al-Hawary, S. I., & Nusair, W. (2017). Impact of human resource strategies on perceived organizational support at Jordanian public universities. *Global Journal of Management and Business Research: A Administration and Management, 17*(1), 68–82.

Al-Hawary, S. I., Al-Qudah, K., Abutayeh, P., Abutayeh, S., & Al-Zyadat, D. (2013). The impact of internal marketing on employee's job satisfaction of commercial banks in Jordan. *Interdisciplinary Journal of Contemporary Research in Business, 4*(9), 811–826.

Al-Hawary, S. I. S., Mohammad, A. S., Al-Syasneh, M. S., Qandah, M. S. F., & Alhajri, T. M. S. (2020). Organisational learning capabilities of the commercial banks in Jordan: Do electronic human resources management practices matter? *International Journal of Learning and Intellectual Capital, 17*(3), 242–266.

Alkalha, Z., Al-Zu'bi, Z., Al-Dmour, H., Alshurideh, M., & Masa'deh, R. (2012). Investigating the effects of human resource policies on organizational performance: An empirical study on commercial banks operating in Jordan. *European Journal of Economics, Finance and Administrative Sciences, 51*(1), 44–64.

Allozi, A., Alshurideh, M., AlHamad, A., & Al Kurdi, B. (2022). Impact of transformational leadership on the job satisfaction with the moderating role of organizational commitment: Case of UAE and Jordan manufacturing companies. *Academy of Strategic Management Journal, 21*, 1–13.

Al-Lozi, M., Almomani, R. Z., & Al-Hawary, S. I. (2018). Talent Management strategies as a critical success factor for effectiveness of Human Resources Information Systems in commercial banks working in Jordan. *Global Journal of Management and Business Research: A Administration and Management, 18*(1), 30–43.

Almazrouei, F. A., Alshurideh, M., Kurdi, B. A., & Salloum, S. A. (2020, October). Social media impact on business: a systematic review. In *International Conference on Advanced Intelligent Systems and Informatics* (pp. 697–707). Springer.

AlShehhi, H., Alshurideh, M., Al Kurdi, B., & Salloum, S. A. (2020, October). The impact of ethical leadership on employees performance: A systematic review. In *International Conference on Advanced Intelligent Systems and Informatics* (pp. 417–426). Springer.

Alshurideh, D. M. (2019). Do electronic loyalty programs still drive customer choice and repeat purchase behaviour? *International Journal of Electronic Customer Relationship Management, 12*(1), 40–57.

Alshurideh, M. (2022). Does electronic customer relationship management (E-CRM) affect service quality at private hospitals in Jordan? *Uncertain Supply Chain Management, 10*(2), 325–332.

Alshurideh, M. T. (2016). Exploring the main factors affecting consumer choice of mobile phone service provider contracts. *International Journal of Communications, Network and System Sciences, 9*(12), 563–581.

Alshurideh, M. T., Shaltoni, A., & Hijawi, D. (2014). Marketing communications role in shaping consumer awareness of cause-related marketing campaigns. *International Journal of Marketing Studies, 6*(2), 163–168.

Alshurideh, M. T., Al Kurdi, B., Alzoubi, H. M., Ghazal, T. M., Said, R. A., AlHamad, A. Q., ... & Al-kassem, A. H. (2022). Fuzzy assisted human resource management for supply chain management issues. *Annals of Operations Research*, 1–19.

Alshurideh, M., Alhadid, A. Y., & Barween, A. (2015a). The effect of internal marketing on organizational citizenship behavior an applicable study on the University of Jordan employees. *International Journal of MaRketing Studies, 7*(1), 138–145.

Alshurideh, M., Bataineh, A., Alkurdi, B., & Alasmr, N. (2015b). Factors affect mobile phone brand choices–Studying the case of Jordan universities students. *International Business Research, 8*(3), 141–155.

Alshurideh, M., Salloum, S. A., Al Kurdi, B., & Al-Emran, M. (2019, February). Factors affecting the social networks acceptance: an empirical study using PLS-SEM approach. In *Proceedings of the 2019 8th International Conference on Software and Computer Applications* (pp. 414–418).

Alsuwaidi, M., Alshurideh, M., Al Kurdi, B., & Salloum, S. A. (2020, October). Performance appraisal on employees' motivation: a comprehensive analysis. In *International Conference on Advanced Intelligent Systems and Informatics* (pp. 681–693). Springer.

Al-Waqfi, M. A., & Forstenlechner, I. (2012). Of private sector fear and prejudice: The case of young citizens in an oil-rich Arabian gulf economy. *Personnel Review, 41*(5), 609–629.

Alzoubi, H. M., Alshurideh, M., Al Kurdi, B., & Inairat, M. (2020). Do perceived service value, quality, price fairness and service recovery shape customer satisfaction and delight? A practical study in the service telecommunication context. *Uncertain Supply Chain Management, 8*(3), 579–588.

Ammari, G., Alkurdi, B., Alshurideh, A., & Alrowwad, A. (2017). Investigating the impact of communication satisfaction on organizational commitment: A practical approach to increase employees' loyalty. *International Journal of Marketing Studies, 9*(2), 113–133.

Bettayeb, H., Alshurideh, M. T., & Al Kurdi, B. (2020). The effectiveness of mobile learning in UAE universities: A systematic review of motivation, self-efficacy, usability and usefulness. *International Journal of Control and Automation, 13*(2), 1558–1579.

Brown, D. (2002). *Career choice and development* (4th ed.). http://www.borbelytiborbors.extra.hu/ZSKF/CareerDevelopment.pdf.

Christensen, L. B., Johnson, R. B., & Turner, L. A. (2011). *Research methods, design, and analysis* (11th ed.). Allyn and Bacon.

Cooper, D. R., & Schindler, P. S. (2014). *Business research methods* (12th ed.). McGraw-Hill Irwin.

Daleure, G. M. (2016). Holistic sustainability as key to Emiratization: Links between job satisfaction in the private sector and young Emirati adult unemployment. *FIRE: Forum for International Research in Education, 3*(2), 5–24.

Dan, Z. (2015). Differences in work motivation between public and private sector organizations. *American Journal of Business, Economics and Management, 3*(2), 86–91.

DeSantis, V. S., & Durst, S. L. (1996). Comparing job satisfaction among public- and private-sector employees. *American Review of Public Administration, 26*(3), 327–343.

Dur, R., & Zoutenbier, R. (2015). Intrinsic motivations of public sector employees: evidence for Germany, German. *Economic Review, 16*(3), 343–366.

Eesley, C., & Wang, Y. (2017). Social influence in career choice: Evidence from a randomized field experiment on entrepreneurial mentorship. *Research Policy, 46*(3), 636–650.

Faturochman. (2016). The job characteristics theory: A review. *Bulletin Psikologi, 5*(2), 1–13. https://doi.org/10.22146/bpsi.13552.

Gagné, M., & Deci, E. L. (2005). Self-determination theory and work motivation. *Journal of Organizational Behavior, 26*(4), 331–362.

George, J. M., & Jones, G. R. (2012). *Understanding and managing organizational behavior* (6th ed.). Prentice Hall.

Hansen, J. R. (2012). From Public to Private Sector: Motives and explanations for sector switching. *Public Management Review, 16*(4), 590–607. https://doi.org/10.1080/14719037.2012.743575

Hayajneh, N., Suifan, T., Obeidat, B., Abuhashesh, M., Alshurideh, M., & Masa'deh, R. (2021). The relationship between organizational changes and job satisfaction through the mediating role of job stress in the Jordanian telecommunication sector. *Management Science Letters, 11*(1), 315–326.

Jain, N., & Bhatt, P. (2015). Employment preferences of job applicants: Unfolding employer branding determinants. *Journal of Management Development, 34*(6), 634–652.

Johari, J., & Yahya, K. K. (2016). Job characteristics, work involvement, and job performance of public servants. *European Journal of Training and Development, 40*(7), 554–575.

Kim, S. (2016). Job characteristics, public service motivation, and work performance in Korea. *Gestion Et Management Public, 5*(1), 7–24.

Klasing, M. J. (2013). Cultural dimensions, collective values and their importance for institutions. *Journal of Comparative Economics, 41*(2), 447–467.

Ko, K., & Jun, K. (2015). A comparative analysis of job motivation and career preference of Asian undergraduate students. *Public Personnel Management, 44*(2), 192–213.

Korkmaz, H. (2015). Factors influencing students' career chooses in science and technology: Implications for high school science curricula. *Procedia—Social and Behavioral Sciences, 197*(25), 966–972.

Kulkarni, M., & Nithyanand, S. (2013). Social influence and job choice decisions. *Employee Relations, 35*(2), 139–156.

Law, W., & Arthur, D. (2003). What factors influence Hong Kong school students in their choice of a career in nursing? *International Journal of Nursing Studies, 40*(1), 23–32.

Lee, Y., & Wilkins, V. M. (2011). More similarities or more differences? Comparing public and nonprofit managers' job motivations. *Public Administration Review*, 45–56.

Leedy, P. D., & Ormrod, J. E. (2013). *Practical research: Planning and design* (10th ed.). Pearson.

Lewis, G. B., & Frank, S. A. (2002). Who wants to work for the government? *Public Administration Review, 62*(4), 395–404.

Maidani, E. A. (1991). Comparative study of Herzberg's two-factor theory of job satisfaction among public and private sectors. *Public Personnel Management, 20*(4), 441–448.

Metabis, A., & Al-Hawary, S. I. (2013). The impact of internal marketing practices on services quality of commercial banks in Jordan. *International Journal of Services and Operations Management, 15*(3), 313–337.

Mishkin, H., Wangrowicz, N., Dori, D., & Dori, Y. J. (2016). Career choice of undergraduate engineering students. *Procedia Social and Behavioral Sciences, 228*(1), 222–228. https://doi.org/10.1016/j.sbspro.2016.07.033

Mohammad, A. A., Alshura, M. S., Al-Hawary, S. I. S., Al-Syasneh, M. S., & Alhajri, T. M. (2020). The influence of Internal Marketing Practices on the employees' intention to leave: A study of the private hospitals in Jordan. *International Journal of Advanced Science and Technology, 29*(5), 1174–1189.

Moroko, L., & Uncles, M. D. (2008). Characteristics of successful employer brands. *Journal of Brand Management, 16*(3), 160–175.

Neuman, W. L. (2006). *Social research methods: Qualitative and quantitative approaches* (7th ed.). Allyn and Bacon.

Ng, E. S., Gossett, C. W., & Chinyoka, S., & Obasi, I. (2016). Public vs private sector employment: An exploratory study of career choice among graduate management students in Botswana. *Personnel Review, 45*(6), 1367–1385.

Osipow, S. H., & Fitzgerald, L. F. (1996). Theories of career development. A comparison of the theories. *Theories of career development* (4th ed.). Allyn Ans Bacon.

Park, S. M., & Word, J. (2012). Driven to service: Intrinsic and extrinsic motivation for public and nonprofit managers. *Public Personnel Management, 41*(4), 705–734.

Pavel, L. (2015). The role of social influence in career choice. *Journal of Innovation in Psychology, Education and Didactis, 19*(2), 247–254.

Rashid, S., & Rashid, U. (2012). Work motivation differences between public and private sector. *American International Journal of Social Science, 1*(2), 24–33.

Rayn, R. M., & Deci, E. L. (2000). Intrinsic and extrinsic motivations: Classic definitions and new directions. *Contemporary Educational Psychology, 25*, 54–67.

Rexhaj, B. (2011). *Motivation to work in for-profit and not-for-profit organizations*. Doctoral dissertation. http://lnu.diva-

Saini, G. K., Gopal, A., & Kumari, N. (2015). Employer brand and job application decisions: Insights from the best employers. *Management and Labour Studies, 40*(1–2), 34–51.

Solomon, E. E. (1986). Private and public sector managers: An empirical investigation of job characteristics and organizational climate. *Journal of Applied Psychology, 71*(2), 247–259.

Tanwar, K., & Prasad, A. (2017). Employer brand scale development and validation: A second-order factor approach. *Personnel Review, 46*(2), 389–409.

Thomas, K. W. (2009). *Intrinsic motivation at work: What really drives employee engagement* (2nd ed.). Berrett-Koehler.

Tremblay, M. A., Blanchard, C. M., Taylor, S., Pelletier, L. G., & Villeneuve, M. (2009). Work extrinsic and intrinsic motivation scale: Its value for organizational psychology research. *Canadian Psychological Association, 41*(4), 213–226.

Wallace, M., Lings, I., Cameron, R., & Sheldon, N. (2014). Attracting and retaining staff: The role of branding and industry image. *Workforce development* (pp. 19–36). Springer.

Workman, J. L. (2015). Parental influence on exploratory students' college choice, major, and career decision making. *College Student Journal, 49*(1), 23–30.

The Impact of Job Insecurity on Employees Job Performance Among Employees Working at Save the Children in Jordan

Yahia Salim Melhem, Isra Ali Hamad BanyHani, Fatima Lahcen Yachou Aityassine, Abdullah Matar Al-Adamat, Main Naser Alolayyan, Sulieman Ibraheem Shelash Al-Hawary, Ayat Mohammad, Menahi Mosallam Alqahtani, and Muhammad Turki Alshurideh

Abstract This study aims to investigate the impact of job insecurity on job performance among employees of Save the Children Jordan. The population of this study is 254 employees of which 153 employees are the recommended sample size for Save the Children Jordan. In total (254) questionnaires were distributed to the target sample, of which (248) questionnaires were retrieved. The final number of valid questionnaires for analysis was (160). This study used a quantitative, descriptive statistics and other tools to test the study model. SEM analysis using PLS considering that present study has a sample of a small size relatively, therefore, it's vital to validate the garnered results using such SEM type. The findings in the study illustrated that job insecurity is negatively related to job performance. Basc on the study results; Save the Children management is advised to provide more attention to Job insecurity issue among its employees and to find suitable means and apply adequate interventions to cope with this phenomenon and its consequences that expected to last for relatively a long period of time.

Y. S. Melhem
Business Management, Business Administration Department, School of Business, Yarmouk University, Irbid, Jordan
e-mail: ymelhem@yu.edu.jo

I. A. H. BanyHani
Department of Business Administration, School of Business, Yarmouk University, Irbid, Jordan

F. L. Y. Aityassine
Department of Financial and Administrative Sciences, Irbid University College, Al-Balqa' Applied University, As-Salt, Jordan
e-mail: Fatima.yassin@bau.edu.jo

A. M. Al-Adamat
School of Business, Department of Business Administration & Public Administration, Al Al-Bayt University Jordan, P.O. Box 130040, Mafraq 25113, Jordan
e-mail: aaladamat@aabu.edu.jo

M. Alshurideh et al. (eds.), *The Effect of Information Technology on Business and Marketing Intelligence Systems*, Studies in Computational Intelligence 1056,
https://doi.org/10.1007/978-3-031-12382-5_50

Keywords Job insecurity · Job performance · Save the children · Jordan

1 Introduction

In response to the devastating Syrian crisis and refugee's emergency situation, humanitarian work and relief has been booming recently in Jordan and the region where too many local and international Non-Governmental Organizations (NGOs) were established (Musa et al., 2014). One major focus of these NGOs is how to prepare their staff to respond to refugee needs with high quality services. However, since these NGOs are operationally based on donors' funding of certain projects, most of their staff gets recruited on project base, which means that by the end of the fund and the project duration the employees' contract is ended, and this creates a feeling of job insecurity which might negatively impact job performance.

Save the Children Jordan is one of those big NGOs operating in Jordan, especially when a merger took place between Save the Children Jordan office and international office to generate a big member with more than 300 staff and 600 volunteers. The merger includes a physical relocation, asset transfer, organizational restructuring and strategy refinement. These huge organizational changes expected to impact employees job insecurity level and thus their job performance at many local and international NGOs and specifically the case at Save the Children in Jordan.

Recently, job insecurity has become a very hot topic for researchers, being one of the most important stressors in working life (De Cuyper et al., 2008; Al Kurdi et al., 2020; Allozi et al., 2022) especially in the last 15 years, job insecurity researches has grown significantly (Lee et al., 2018). Due to the rivalry and market globalization, many organizations tend to achieve more adaptation to the market changes therefore

M. N. Alolayyan
Faculty of Medicine, Health Management and Policy Department, Jordan University of Science and Technology, Ar-Ramtha, Jordan
e-mail: mnalolayyan@just.edu.jo

S. I. S. Al-Hawary (✉)
Business Management, Faculty of Finance and Business Administration, Department of Business Administration, Al Al-Bayt University, P.O. Box 130040, Mafraq 25113, Jordan
e-mail: dr_sliman73@aabu.edu.jo

A. Mohammad
Business and Finance Faculty, The World Islamic Science and Education University (WISE), 11947, P.O. Box 1101, Amman, Jordan

M. M. Alqahtani
Administration Department, Community College of Qatar, Doha, Qatar
e-mail: mena7i@icloud.com

M. T. Alshurideh
Department of Marketing, School of Business, The University of Jordan, Amman 11942, Jordan
e-mail: m.alshurideh@ju.edu.jo; malshurideh@sharjah.ac.ae

Department of Management, College of Business, University of Sharjah, Sharjah 27272, UAE

more mergers, acquisitions and other organizational changes appear (AlTaweel & Al-Hawary, 2021, Al-Hawary & Al-Syasneh, 2020; Al-Hawary & Al-Hamwan, 2017; Hirsch & De Soucey, 2006), which are expected to put those organization's employees in doubt about their future careers. Job insecurity emerged as one of the most critical and alarming issues in contemporary organization, and has been studied among scholars and researchers (Sverke & Hellgren, 2002; Sverke et al., 2002).

Organizations try to improve the employees' performance adopting different approaches and measures (Allahow et al., 2018; Al-Nady et al., 2013; Mohammad ct al., 2020; Alhalalmeh et al., 2020; Alolayyan et al., 2018; Alshurideh et al., 2017; Al-Hawary & Al-Kumait, 2017; Al-Hawary et al., 2011, 2013). However, few organizations including NGOs pay enough attention to the feeling of job insecurity among their employees and its impact on organizational outcomes, such as employees' satisfaction, turnover, productivity and overall performance. This might create subtle unwanted consequences including low levels of employee performance where employees' performance is vital and critical to the organizations' success and overall performance (Al Kurdi et al., 2021; Alameeri et al., 2020; Hayajneh et al., 2021). Managers and researchers have not fully realized the impact between these two important constructs. The present study aims to contribute to the understanding of the job insecurity–job performance relationship by contributing with additional perspective to the researchers' debate in the literature that claims different views for the relationship (AlShehhi et al., 2020; Alsuwaidi et al., 2020). Hence, most research on the effect of job insecurity on performance has so far reported mixed results (Al Shebliet al., 2021). This research contributes to support one view over the other. Since several studies covered the relationship between job insecurity and other variables, the researcher couldn't find big focus on the Job performance as a major dependent variable in the literature and it has received relatively little empirical attention in the literature as a consequence of job insecurity construct (Rosen et al., 2010; Al-bawaia et al., 2022). Even recent researchers, where tackling job performance among other outcomes with no focus on the direct Job Insecurity-Job performance relationship (Jiang & Lavayss, 2018). This can be considered as a second major contribution for this research. Therefore this study aims to investigate the impact of job insecurity on job performance.

2 Theoretical Framework and Hypotheses Development

2.1 Job Insecurity

Job security has been defined in the literature as the probability of an employee to keep his/her job. On the contrary, job insecurity is defined as the perception of a potential threat to continuity in his or her current job (De Witte, 2000; Probst, 2003). In growing societies, where companies need to keep up with the technological changes and the continuous demand, it is common to experience job security. In periods of economic growth or expansion, organizations experience more demands and this in turn increases the amount of jobs available, and the probabilities of maintaining the actual employment thus job security will be prevalence. In contrast, when organizations experience loss of demand, they are forced to downsize thus job insecurity will be prevalence (Jimenez & Didona, 2017). On counterpart, some researchers argue that job insecurity is a common organizational phenomenon that does not necessarily comprise organizational changes, such as restructuring or downsizing, they reported in their studies that even in the same environment, employees could experience different levels of job insecurities (Huang et al., 2017).

Other researchers agreed that the global definition is best fit and has been most applied for organizations passed thru a crisis or change (De Witte, 1999). Whereas scholars who treat job insecurity as a multidimensional construct instead address not just job loss but also the loss of desired job features, certain dimensions or privileges in the job itself; such as opportunities for promotion, pay or other areas (Kurdi et al., 2020; Lee et al., 2006; Mauno et al., 2005). Researchers who adopt this multi-dimensional definition argue that job insecurity refers not only to the degree of uncertainty, but also to the components of job insecurity including: the severity of the threat concerning job continuity, the importance of the job feature to the individual, the perceived threat of the occurrence of a total negative effect on the job situation, the total importance of the changes and powerlessness and inability of the individuals to control the above mentioned factors.

2.2 *Job Performance*

Employees' performance defined as the quantity, quality, and timeliness to deliver an output with efficiency and effectiveness according to a pre-defined measure set by the supervisor or the organization (Al-Hawary & AlDafiri, 2017; Mathis & Jackson, 2009). Another definition Aguinis (2009) described performance as an employee perception on his behavior and to what extent it contributes to the organizational success rather than the real outcomes of their work. Quality and quantity of performance was defined as "a measure of how well the job was done" and "ratings of the quantity or volume of work produced," respectively (Viswesvaran et al., 1996). Primary studies often reported performance information from multiple sources, including objective assessment, supervisor ratings, and self-ratings.

2.3 *Job Insecurity and Job Performance*

As job insecurity may disturb performance at work, it becomes important for organizations to know the factors that might buffer the negative effects of job insecurity (Probst et al., 2018), being a potential moderator in the job insecurity-performance relationship (Sverke et al., 2006; Alshurideh, 2022). And since several studies covered the relationship between job insecurity and other variables, the researcher couldn't find any focus on the Job performance as the major dependent variable in the literature (Jiang & Lavaysse, 2018).

The performance of employee will also get affected to the extent of feeling job insecurity (Alshraideh et al., 2017; Al-Dhuhouri et al., 2020; Ashal et al., 2021). Literature shows that there is an association between job insecurity and self-related performance (AlMehrzi et al., 2020). Moreover, studies also reveal that employees being afraid of job insecurity perform more to become valuable, in order to confirm their stay with organization. Other studies show negative relationship between job insecurity and well-being of employees, their attitudes, job involvement and trust over management (Alaali et al., 2021; Shoss, 2017; Sverke, et al., 2002).

It was clear in the literature that numerous studies have found negative relationship between job insecurity and job satisfaction, describing that an increase of job insecurity is associated to a decrease in job satisfaction (Reisel et al., 2010; Jiang & Probst, 2019). What is less clear in the literature is the relationship between job insecurity and performance, there is an evidence that job insecurity in some instances may leads to either increased or decreased levels of productivity or performance. Thus, a debate between researchers about job insecurity and job performance relationship raged. Although some researchers claimed that job insecurity negatively impact the job attitudes and then the performance, with the majority of studies considered that job insecure employees perform worse than job secure employees (Cheng & Chan, 2008). However, the counterpart researchers argued that job insecure employees are motivated to do necessary actions and extra effort to alter the risk of losing their jobs, as employees think that higher performing individuals are unlikely to be laid off, which implies the positive relationship between job insecurity and performance (Hartley et al., 1990). To investigate whether job insecurity might be a double effect factor or not, a recent research by (Lam et al., 2015) studied the relationship between job insecurity and organizational citizenship behavior (OCB). In his study he finds that as job insecurity increases from low to medium, OCB decreases and people deteriorate their social exchanges, and once job insecurity continues to increase, the employees become interested in OCB to impress their management thus enhance their chances of survival. Similar effects between job insecurity and self-reported job performance have also been reported by Selenko et al. Hence, these recent studies also suggest the need to continue examining potential moderators that might contribute to job insecurity and its associated performance. Based on the above literature the study hypotheses may be formulated as".

H1: Job insecurity will be a negative predictor for job performance among employees of Save the Children Jordan.

3 Methodology

This study used a quantitative, descriptive statistics and other tools to test a model that examined the effect of job insecurity on job performance among employees working at Save the Children in Jordan. Data were collected using self-administered questionnaire in regards to two main variables; job insecurity (the independent variable), job performance (the dependent variable) from employees working at Save the Children Jordan.

3.1 *Population and Sampling*

The population of this study is 254 employees of which 153 employees are the recommended sample size for Save the Children Jordan according to Sekaran (2003). However, due to the small population size it is preferable in this study to consider total population sampling to examine all Save the Children employees and avoid sample selection. This ensures that all concerns of this study will be better captured by the respondents. According to the above, 254 questioners is distributed to survey Job Insecurity especially after the unification, targeting different levels, positions, experiences, ages, genders, educational levels, salaries, etc. and focusing on certain geographical working areas where save the Children operating in (e.i. Zaatari camp, Azraq camp; and host communities in Amman, Irbid, Mafraq, Karak and Maan).

In total (254) questionnaires were distributed to the target sample, of which (248) questionnaires were retrieved. Hence, the response rate was (97%). Following that, validity of the questionnaire was checked for statistical analysis purposes, (88) questionnaires were deemed impropriate for further analysis due to inconsistent answers or consistent pattern through the entire questionnaire (either a low or high value on Likert scale). A number of reverse questions were added to enhance the robustness of the analysis and the findings. The exclusion rate of a 35% is perhaps due to participants' reluctance to provide factual data in judging the management behavior and decisions regarding downsizing and job security. Hence, the final number of valid questionnaires for analysis were (160). SPSS v.25 was used for data coding, using frequencies and means. The sample characteristics is summarized as:

Half of the sample were females (n = 79, 49.4%), this indicates good diversity in the sample gender. Majority of the sample were Jordanians (n = 148, 92.5%). The majority of the sample were relatively young and had aged less than 40 years (n = 119, 74%), with bachelor or higher education qualification (n = 114, 71%) among the sample, which is expected as Jordan classified with having one of the highest ratios of literacy among Middle East countries. Majority of the sample, had more than 5 years working in the humanitarian sector (n = 96, 60%), this is expected to allow for better understanding for the questionnaire items. Majority of the sample worked less than 5 years in Save the Children Jordan (n = 145, 69%) a sound of job inscurity is evident with a possibility of a high employees' turnover rate, thus the renewed contracts seems to be conditioned with full commitment and high performance which fits the study design and concerns. In terms of the basic salary, about 40% of the workers in this organization are getting paid above 1,000 JOD which reflects the very good salary scale and the good packages this organization offer compared to other local organizations, especially during the difficult Jordanian economic situation. Majority of the sample were from Programs department (n = 92, 57.5%) which is in alignment with the organizational human capital distribution ratio (35% Support Departments; 65% Programs Department) and aligned with the Non-Governmental Organizations (NGOs) human capital normal distribution as these organizations scope of work is mainly based on programs implementation in the areas of deprivation. Hence, Job location reported that the majority of the sample were field based (n = 125, 78.1%) as some support jobs like: HR, Safety and Security, IT and logistics are also based in the to provide the programs with the necessary support.

3.2 *Questionnaire Design*

A survey questionnaire is used as an instrument with certain statistical tools to measure two variables, those designated to be used in this study. Through an extensive review of studies that used these variables, the researchers assessed reliability, validity, number of items, and the feasibility of each variable. These variables are:

1. **Job insecurity:** Job insecurity was measured in its global concept using De Witte (2000) scale which is a four items scale (2 items for cognitive job insecurity and 2 items for affective job insecurity). This scale was fruitfully employed in previous cross-cultural studies, exhibiting excellent validity and reliability of which it was the most used scale for job insecurity since 2003 (Lee et al., 2018). "Chances are, I will soon lose my job", "I am sure I can keep my job" (reverse coded), "I feel insecure about the future of my job", and "I think I might lose my job in the near future". Respondents were asked to rate these items on a 5-point Likert type scale, ranging from 1 ("strongly disagree") to 5 ("strongly agree"). It's worth mentioning that all previously mentioned different conceptualization of job insecurity in the literature have generated various measurement instruments for job insecurity (Huang et al., 2010). In a review for many empirical studies more than 10 different measures were adopted for job insecurity. Some studies use a single-item measure, asking if respondents perceive their jobs as insecure (e.g. Debus et al., 2012; Jiang & Probst, 2019); other studies rely on self-developed scales. Still the most frequently adopted measures are De Witte's (2000) four item scale, which appears in 46 studies conducted in Europe and Asia (Chiu et al., 2015). In the case of Save the Children Jordan, De Witte's (2000) Job Insecurity four item scale will be adopted as it is frequently used in studies on job insecurity and has a reasonably high reliability across different studies.
2. **Job Performance:** Job performance was measured using 29 items and as per a recent scale of which this particular one was developed in order to be useful in any

job-related context with a high probability of applicability worldwide (Carlos & Rodrigues, 2016). In Carlos & Rodrigues article they developed a job performance measure that was applicable across jobs and cultures after an extensive literature review and based on studies that were developed in different cultural and job-related contexts to end up with two main dimensions (task performance and contextual performance) and eight sub dimensions: job knowledge, organizational skills, efficiency, persistent effort, cooperation, organizational consciousness, personal characteristics and interpersonal and relational skills. Confirmatory factorial analysis was used in order to test their relevance. The dimensions 'personal characteristics' and 'persistent effort' were merged. The resulting 29 item scale presents appropriate psychometric properties. This scale has been selected for this research for its appropriateness and comprehensiveness for Job performance measurement. Both task performance and contextual performance has been measured as the follows: For Task Performance, 3 main items have been employed: 4 items in Job Knowledge, 5 items in Organizational skills, and 3 items under the efficiency. For Contextual Performance, 4 main items have been used: 5 items for Persistent effort, 3 items for the Cooperation, 5 items for Organizational conscientiousness and 4 items for Interpersonal and relational skills.

3.3 Data Normality

Skewness and Kurtosis values were examined to confirm normal distribution of the data. Kurtosis measure the extent to which data-tail is heavy or light in compare to normal distribution, and Skewness measure the extent to which data distribution differ from the normal distribution as suggested by (Sposito et al., 1983) who also suggested that Skewness and Kurtosis values should be within (+2) and (−2) range to confirm normality for SEM analysis purposes. Table 1 confirmed normality for study variables as Skewness recorded (−1.377) (−0.726) for job insecurity and job performance respectively, whereas Kurtosis recorded (1.719) and (−1.096) for job insecurity and job performance respectively.

Table 1 Skewness and kurtosis values for study variables (N = 160)

Construct	Skewness	Kurtosis
Job insecurity	−1.377	1.719
Job performance	−0.726	−1.096

Table 2 Summary of model fit indices

Indices	Cutoff criteria	Estimate	Interpretation
Chi-Square	–	387.818	–
NFI	>0.90	0.904	**Confirmed**
SRMR	<0.08	0.036	**Confirmed**

4 SEM Analysis Using SMART PLS

Recently, SMART PLS has emerged as a method of choice in different social sciences arena due its high capabilities in assessing complex models using regression-based ordinary least squares (OLS) estimation method with the goal of explaining the latent constructs variance by minimizing the error terms and maximizing the R^2 values of the endogenous constructs particularly with samples of small size (Hair et al., 2017), therefore, highlights SEM analysis using PLS considering that present study has a sample of a small size relatively, therefore, its vital to validate the garnered results using such SEM type.

4.1 SEM Analysis

Anderson and Gerbing (1988) suggested the use of two steps to complete SEM analysis, testing the measurement model using CFA and structural model testing.

Step 1: Measurement model & CFA:

The initial running of CFA reported poor model fit as (Hair et al., 2017) recommended values for model fit indices in variance structure analysis including Normed Fit Index (NFI) and Standardized Root Mean Residual (SRMR) to achieve good model fit. Therefore, number of items with low loading deleted from the model, as (Hair et al., 2017) recommended that items loading should exceed (0.70) to achieve good model fit. Following that (14) items were deleted from job performance construct and (1) item from job insecurity, and a sound of good model fit was achieved as summarized in Table 2. Table 3 summarizes deleted items.

After identifying final measurement model, statistical validity and reliability of the model need to be confirmed before proceeding with testing structural model.

4.2 Model Reliability

Reliability of the model was confirmed through Composite Reliability [CR] and Cronbach's α as suggested by Sekaran and Bougie (2016) and Hair et al. (2017). The recommended values for both criteria need to exceed (0.70) which was confirmed for present study variables as presented in Table 4.

Table 3 Deleted items from measurement model

Variable	Item
Job performance	Task_Perf_Knowledge_2
	Task_Perf_Knowledge_3
	Task_Perf_Skills_1
	Task_Perf_Skills_5
	Task_Perf_Efficiency_1
	Task_Perf_Efficiency_3
	Contex_Perf_1
	Contex_Perf_5
	Task_Perf_Cooperation_1
	Task_Perf_Cooperation_3
	Task_Perf_Conscient_2
	Task_Perf_Conscient_4
	Task_Perf_Interpersonal_3
	Task_Perf_Interpersonal_4
Job insecurity	Job_Insecurity_3

Table 4 Values of Cronbach α and Composite Reliability

Variable	Cronbach's α	CR
Job insecurity	0.805	0.807
Job performance	0.974	0.974

4.3 *Model Validity*

1. **Convergent validity:**

The majority of the model items load exceeded the recommended level (0.70) and significant at (0.05) level as (T) statistics of significance exceeded (1.96) as depicted in previous figure, therefore convergent validity was confirmed. Table 5 summarize final measurement model items loading and significance.

The second criterion to confirm convergent validity is the Average Variance Extracted (AVE), as Hair et al. (2017) suggested that AVE values should exceed (0.50) to confirm convergent validity of the model, which was confirmed in present study as summarizes in Table 6.

2. **Discriminant validity:**

PLS provides Fornell–Larcker criterion to confirm discriminant validity, which was confirmed for present study as presented in Table 7.

The final measurement model was confirmed, therefore its suitable to proceed with testing structural model.

Table 5 Loading and significance for final measurement model items

Variable	Item	Loading	(T)
Job insecurity	Job_Insecurity_1	0.863	20.452
	Job_Insecurity_2	0.867	15.019
	Job_Insecurity_4	0.813	13.349
Job performance	Task_Perf_Knowledge_1	0.893	65.212
	Task_Perf_Knowledge_4	0.854	40.488
	Task_Perf_Skills_2	0.866	49.486
	Task_Perf_Skills_3	0.883	50.365
	Task_Perf_Skills_4	0.835	32.420
	Task_Perf_Efficiency_2	0.852	32.869
	Contex_Perf_2	0.866	44.477
	Contex_Perf_3	0.847	32.926
	Contex_Perf_4	0.868	40.473
	Task_Perf_Cooperation_2	0.831	31.507
	Task_Perf_Conscient_1	0.782	20.009
	Task_Perf_Conscient_3	0.859	37.944
	Task_Perf_Conscient_5	0.832	27.231
	Task_Perf_Interpersonal_1	0.899	66.059
	Task_Perf_Interpersonal_2	0.859	38.323

Table 6 Values of AVE

Variable	AVE
Job insecurity	0.584
Job performance	0.713

Table 7 Fornell–Larcker discriminant validity

Variable	Job insecurity	Job performance	Locus of control
Job insecurity	**0.764**		
Job performance	−0.325	**0.844**	

Step 2: Structural model:

To test the proposed hypotheses, structural model needs to be tested. Testing structural model requires path testing through examining coefficient of determination (R^2) which represent the amount of variance that independent variable explain in the dependent variable, and path coefficient (β) that represent the influence of the variable and finally the significance of each path coefficient at (0.05) level through (T) statistics as suggested by Hair et al. (2017). Structural model that examine the influence of job insecurity on job performance was tested, and result reported that job insecurity explained ($R^2 = 11\%$) of the variance in job performance, which indicate low variance, moreover, the influence was seen to be negative as ($\beta = -0.331$) and significant as ($T = 4.851$) which is evidently exceeding (1.96), therefore confirming the first hypothesis. See Figs. 1 and 2.

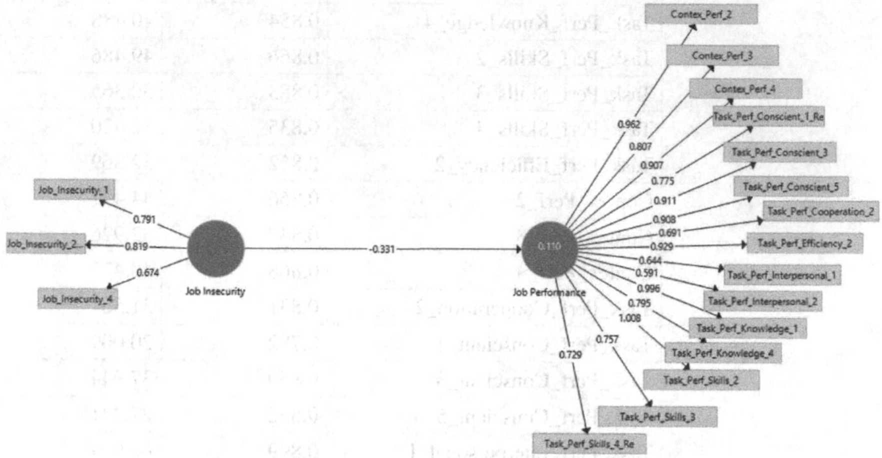

Fig. 1 Path testing for job insecurity as a predictor for job performance

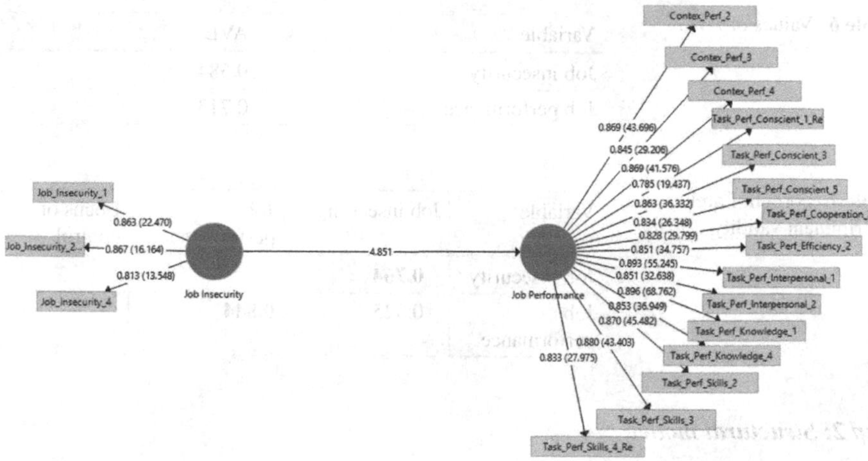

Fig. 2 Path significance for job insecurity as a predictor for job performance

5 Discussion

The researchers intended to understand the relationship and the prediction power of job insecurity on job performance. The findings in the study illustrated that job insecurity is negatively related to job performance. According to the literature, most studies on job insecurity and job performance agreed to have a negative relationship between Job insecurity and Job Performance indicate that this finding is logical and expected. The conclusion of these studies is that individuals who experience high levels of job insecurity generally perform worse when compared with individuals who experience low levels of job insecurity (Cheng & Chan, 2008; Schreurs et al., 2012; James, 2012; Awan & Salam, 2014; Wang et al., 2015). Moreover, an interesting weekly diary study conducted in 2012 on the buffering role of social support in the relationship between Job Insecurity and Employee Performance showed that during weeks in which individuals felt insecure about their job, they performed worse in terms of meeting organizational objectives and fulfilling the requirements of their job. However, they found that this negative relationship is less pronounced when employees feel supported by their supervisor (moderator) (Schreurset al., 2012). Moreover, a recent study conducted in the Sultanate of Oman, concluded that there is a significant negative effect of job insecurity on employee performance (Sanyal et al., 2018). In such instance, all the above studies and findings support the view that job insecurity has a negative relationship with performance which is the case in this study.

However, the result in this study contradicts with some researchers who argued there is no significant negative association observed between job insecurity and performance. Berth stressed in his study that employees do not increase or decrease their discretionary efforts when experiencing job insecurity and he claimed that employees' motivation was a main reason to be unaffected by the external threat of organizational restructuring. Moreover, some other studies showed a positive and significant relationship between job insecurity and job performance or a positive but not significant relationship (Probst et al., 2007; Staufenbiel & König, 2010). Moreover, some other studies showed a positive and significant relationship between job insecurity and job performance or a positive but not significant relationship (Probst et al., 2007; Staufenbiel & König, 2010). According to the above, this study revealed the negative relationship between Job insecurity and Job Performance to contribute with one more vote to the side of the researchers stressed on the negative relationship between job insecurity and job performance among all the literature's argument.

6 Limitations of the Study and Direction for Future Research

Using self-administered survey which may allow for subjectivity although it is the most appropriate method that matches the study design. The study conducted for Save the Children Jordan office, thus considered as a case study with limitation for the generalizability of the findings and the results. Therefore Future research is needed to expand the population and to consider other NGOs in Jordan, for further validation, expanding acceptability, generalizability and applicability of this model. The conceptualization of job insecurity might be considered as a limitation. As we assessed only the core perception of job insecurity as the probability of losing the current job. Although using a reliable scale is an advantage however, some job insecurity measures are more differentiated and include additional aspects like the probability of losing valued job features, so it would be interesting to see whether these aspects have a similar correlational pattern or not. Future research is needed to investigate which individual factors (i.e. other personality traits) could have either a stronger or weaker moderation effect compared to LOC on the relationship between job insecurity and job performance in the presence of high job insecurity.

This study focuses on individual aspects in job insecurity and job performance, and did not include organizational aspects. However, organizational aspects such as organizational culture might influence the relationship between job insecurity and performance. Future research could also explore the generalizability to other cultures and countries. Such intercultural research should also try to explain the effect that studies from English speaking countries found relationships between job insecurity and performance that are double as high as in non-English speaking countries this could refer to the laws and regulations in every country as in some countries firing people could be legally difficult such as Germany. Future research could study how organizational factors such as culture, justice, commitment, etc. could affect the relationship between job insecurity and job performance.

7 Managerial Implications

Despite that this study revealed that Save the Children employees have a high level of performance, this high performance is expected to be a short term status as those employees are expected to leave the organization whenever they secure a better stable job which could impact the organization negatively with high turnover rate, loss of institutional memory, losing competent staff, high cost of recruitment and training, beside other implications. According to the literature, Save the Children management is advised to provide more attention to Job insecurity issue among its employees and to find suitable means and apply adequate interventions to cope with this phenomenon and its consequences that expected to last for relatively a long period of time. Management needs to listen to employees' concerns, increase their participation in change-related decision making, and spend more time clearly explaining the aim of the changes using accurate and unbiased procedures. More focus on employees' long term contracts instead of the short term contracts to improve staff stability and job security. Human Resources Management in Save the Children need to ensure that new comers orientation program matches the skills needed in the workplace and may need to expand to areas that teach about moderating stressors at work environment such as job insecurity and having the abilities to achieve work goals, too.

References

Aguinis, H. (2009). *Performance management*. Pearson/Prentice Hall.

Al Kurdi, B., Alshurideh, M., & Al afaishat, T. (2020). Employee retention and organizational performance: Evidence from banking industry. *Management Science Letters, 10*(16), 3981–3990.

Al Kurdi, B., Elrehail, H., Alzoubi, H., Alshurideh, M., & Al-Adaila, R. (2021). The interplay among HRM practices, job satisfaction and intention to leave: An empirical investigation. *Journal of Legal, Ethical and Regulatory, 24*(1), 1–14.

Al Shebli, K., Said, R. A., Taleb, N., Ghazal, T. M., Alshurideh, M. T., & Alzoubi, H. M. (2021, June). RTA's employees' perceptions toward the efficiency of artificial intelligence and big data utilization in providing smart services to the residents of Dubai. In *The International Conference on Artificial Intelligence and Computer Vision* (pp. 573–585). Springer.

Alaali, N., Al Marzouqi, A., Albaqaeen, A., Dahabreh, F., Alshurideh, M., Alrwashdh, S., Iyadeh, I., Salloum, S., & Aburayya, A. (2021) The impact of adopting corporate governance strategic performance in the tourism sector: A case study in the Kingdom of Bahrain. *Journal of Legal, Ethical and Regulatory, 24*(Special Issue 1), 1–18.

Alameeri, K., Alshurideh, M., Al Kurdi, B., & Salloum, S. A. (2020, October). The effect of work environment happiness on employee leadership. In *International Conference on Advanced Intelligent Systems and Informatics* (pp. 668–680). Springer.

Alhalalmeh, M. I., Almomani, H. M., Altarifi, S., Al-Quran, A. Z., Mohammad, A. A., & Al-Hawary, S. I. (2020). The nexus between corporate social responsibility and organizational performance in Jordan: The mediating role of organizational commitment and organizational citizenship behavior. *Test Engineering and Management, 83*, 6391–6410.

Al-bawaia, E., Alshurideh, M., Obeidat, B., & Masa'deh, R. (2022) The impact of corporate culture and employee motivation on organization effectiveness in Jordanian banking sector. *Academy of Strategic Management Journal, 21*(Special Issue 2), 1–18.

Al-Dhuhouri, F. S., Alshurideh, M., Al Kurdi, B., & Salloum, S. A. (2020, October). Enhancing our understanding of the relationship between leadership, team characteristics, emotional intelligence and their effect on team performance: A critical review. In *International Conference on Advanced Intelligent Systems and Informatics* (pp. 644–655). Springer.

Al-Hawary, S. I., & Al-Hamwan, A. (2017). Environmental analysis and its impact on the competitive capabilities of the commercial banks operating in Jordan. *International Journal of Academic Research in Accounting, Finance and Management Sciences, 7*(1), 277–290.

Al-Hawary, S. I., & Al-Kumait, Z. (2017). Training programs and their effect on the employees performance at King Hussain Bin Talal development area at Al - Mafraq governate in Jordan. *International Journal of Academic Research in Economics and Management Sciences, 6*(1), 258–274.

Al-Hawary, S. I., Al-Qudah, K., Abutayeh, P., Abutayeh, S., & Al-Zyadat, D. (2013). The impact of internal marketing on employee's job satisfaction of commercial banks in Jordan. *Interdisciplinary Journal of Contemporary Research in Business, 4*(9), 811–826.

Al-Hawary, S. I., & Al-Syasneh, M. S. (2020). Impact of dynamic strategic capabilities on strategic entrepreneurship in presence of outsourcing of five stars hotels in Jordan. *Business: Theory and Practice, 21*(2), 578–587.

Al-Hawary, S. I., Al-Zeaud, H., & Batayneh, A. M. (2011). The relationship between transformational leadership and employee's satisfaction at Jordanian Private Hospitals. *Business and Economic Horizons, 5*(2), 35–46.

Al-Hawary, S. I. S., & AlDafiri, M. F. S. (2017). Effect of the components of information technology adoption on employees performance of interior ministry of Kuwait State. *International Journal of Academic Research in Economics and Management Sciences, 6*(2), 149–169.

Allahow, T. J. A. A., Al-Hawary, S. I. S., & Aldaihani, F. M. F. (2018). Information technology and administrative innovation of the central agency for information technology in Kuwait. *Global Journal of Management and Business, 18*(11-A), 1–16.

Allozi, A., Alshurideh, M., AlHamad, A., & Al Kurdi, B. (2022). Impact of transformational leadership on the job satisfaction with the moderating role of organizational commitment: Case of UAE and Jordan manufacturing companies. *Academy of Strategic Management Journal, 21*, 1–13.

AlMehrzi, A., Alshurideh, M., & Al Kurdi, B. (2020). Investigation of the key internal factors influencing knowledge management, employment, and organisational performance: A qualitative study of the UAE hospitality sector. *International Journal of Innovation, Creativity and Change, 14*(1), 1369–1394.

Al-Nady, B. A., Al-Hawary, S. I., & Alolayyan, M. (2013). Strategic management as a key for superior competitive advantage of sanitary ware suppliers in Kingdom of Saudi Arabia. *International Journal of Management and Information Technology, 7*(2), 1042–1058.

Alolayyan, M., Al-Hawary, S. I., Mohammad, A. A., & Al-Nady, B. A. (2018). Banking service quality provided by commercial banks and customer satisfaction. A structural equation modelling approaches. *International Journal of Productivity and Quality Management, 24*(4), 543–565.

AlShehhi, H., Alshurideh, M., Al Kurdi, B., & Salloum, S. A. (2020, October). The impact of ethical leadership on employees performance: A systematic review. In *International Conference on Advanced Intelligent Systems and Informatics* (pp. 417–426). Springer.

Alshraideh, A. T. R., Al-Lozi, M., & Alshurideh, M. T. (2017). The impact of training strategy on organizational loyalty via the mediating variables of organizational satisfaction and organizational performance: An empirical study on Jordanian agricultural credit corporation staff. *Journal of Social Sciences (COES&RJ-JSS), 6*(2), 383–394.

Alshurideh, M. (2022). Does electronic customer relationship management (E-CRM) affect service quality at private hospitals in Jordan? *Uncertain Supply Chain Management, 10*(2), 325–332.

Alshurideh, M., Al-Hawary, S. I., Batayneh, A. M., Mohammad, A., & Al-Kurdi, B. (2017). The impact of Islamic Banks' service quality perception on Jordanian customers loyalty. *Journal of Management Research, 9*(2), 139–159.

Alsuwaidi, M., Alshurideh, M., Al Kurdi, B., & Salloum, S. A. (2020, October). Performance appraisal on employees' motivation: A comprehensive analysis. In *International Conference on Advanced Intelligent Systems and Informatics* (pp. 681–693). Springer.

AlTaweel, I. R., & Al-Hawary, S. I. (2021). The mediating role of innovation capability on the relationship between strategic agility and organizational performance. *Sustainability, 13*(14), 7564.

Anderson, J. C., & Gerbing, D. W. (1988). Structural equation modeling in practice: A review and recommended two-step approach. *Psychological Bulletin, 103*(3), 411.

Ashal, N., Alshurideh, M., Obeidat, B., & Masa'deh, R. (2021) The impact of strategic orientation on organizational performance: Examining the mediating role of learning culture in Jordanian telecommunication companies. *Academy of Strategic Management Journal, 21*(Special Issue 6), 1–29.

Awan, W. A., & Salam, A. (2014). Identifying the relationship between job insecurity and employee performance–An evidence from private colleges in Larkana, Pakistan.

Carlos, V. S., & Rodrigues, R. G. (2016). Development and validation of a self-reported measure of job performance. *Social Indicators Research, 126*(1), 279–307.

Cheng, G. H. L., & Chan, D. K. S. (2008). Who suffers more from job insecurity? A meta-analytic review. *Applied Psychology, 57*(2), 272–303.

Chiu, S. F., Lin, S. T., & Han, T. S. (2015). Employment status and employee service-oriented organizational citizenship behaviour. *Career Development International., 20*(2), 133–146.

Cuyper, N. D., Bernhard-Oettel, C., Berntson, E., Witte, H. D., & Alarco, B. (2008). Employability and employees' well-being: Mediation by job insecurity 1. *Applied Psychology, 57*(3), 488–509.

De Witte, H. (2000). Work ethic and job insecurity: Assessment and consequences for well-being, satisfaction and performance at work. *From Group to Community*, 325–350.

Debus, M. E., Probst, T. M., König, C. J., & Kleinmann, M. (2012). Catch me if I fall! Enacted uncertainty avoidance and the social safety net as country-level moderators in the job insecurity–job attitudes link. *Journal of Applied Psychology, 97*(3), 690–698.

Hair, J. F., Jr., Sarstedt, M., Ringle, C. M., & Gudergan, S. P. (2017). *Advanced issues in partial least squares structural equation modeling*. SAGE Publications.

Hartley, J., Jacobson, D., Klandermans, B., & Van Vuuren, T. (1990). *Job insecurity: Coping with jobs at risk*. Sage Publications Ltd.

Hayajneh, N., Suifan, T., Obeidat, B., Abuhashesh, M., Alshurideh, M., & Masa'deh, R. (2021). The relationship between organizational changes and job satisfaction through the mediating role of job stress in the Jordanian telecommunication sector. *Management Science Letters, 11*(1), 315–326.

Hirsch, P. M., & Soucey, M. D. (2006). Organizational restructuring and its consequences: Rhetorical and structural. *Annual Review of Sociology, 32*, 171–189.

Huang, G. H., Lee, C., Ashford, S., Chen, Z., & Ren, X. (2010). Affective job insecurity: A mediator of cognitive job insecurity and employee outcomes relationships. *International Studies of Management & Organization, 40*(1), 20–39.

Huang, G. H., Wellman, N., Ashford, S. J., Lee, C., & Wang, L. (2017). Deviance and exit: The organizational costs of job insecurity and moral disengagement. *Journal of Applied Psychology, 102*(1), 26–42.

James, G. (2012). How to achieve true job security. www.inc.com/geoffrey-james/how-to-achieve-true-job-security.html.

Jiang, L., & Lavaysse, L. M. (2018). Cognitive and affective job insecurity: A meta-analysis and a primary study. *Journal of Management, 44*(6), 2307–2342.

Jiang, L., & Probst, T. M. (2019). The moderating effect of trust in management on consequences of job insecurity. *Economic and Industrial Democracy, 40*(2), 409–433.

Jimenez, H., & Didona, T. (2017). Perceived job security and its effects on job performance: Unionized VS. non unionized organizations. The *International Journal of Social Sciences and Humanities Invention, 4*(7).

Kurdi, B., Alshurideh, M., & Alnaser, A. (2020). The impact of employee satisfaction on customer satisfaction: Theoretical and empirical underpinning. *Management Science Letters, 10*(15), 3561–3570.

Lam, C. F., Liang, J., Ashford, S. J., & Lee, C. (2015). Job insecurity and organizational citizenship behavior: Exploring curvilinear and moderated relationships. *Journal of Applied Psychology, 100*(2), 499–510.

Lee, C., Bobko, P., & Chen, Z. X. (2006). Investigation of the multidimensional model of job insecurity in China and the USA. *Applied Psychology, 55*(4), 512–540.

Lee, C., Huang, G. H., & Ashford, S. J. (2018). Job insecurity and the changing workplace: Recent developments and the future trends in job insecurity research. *Annual Review of Organizational Psychology and Organizational Behavior, 5*, 335–359.

Mathis, R. L., & Jackson, J. H. (2009). *Human Resource Management*. South-Western Cengage Learning.

Mauno, S., Kinnunen, U., Mäkikangas, A., & Nätti, J. (2005). Psychological consequences of fixed-term employment and perceived job insecurity among health care staff. *European Journal of Work and Organizational Psychology, 14*(3), 209–237.

Mohammad, A. A., Alshura, M. S., Al-Hawary, S. I. S., Al-Syasneh, M. S., & Alhajri, T. M. (2020). The influence of Internal Marketing Practices on the employees' intention to leave: A study of the private hospitals in Jordan. *International Journal of Advanced Science and Technology, 29*(5), 1174–1189.

Musa S., Jonathan W., and Christina K. (2014). *A review of the response to Syrian refugees in Jordan: Coping with the crisis.* Center for Strategic Studies- UNHCR.

Probst, T. M. (2003). Exploring employee outcomes of organizational restructuring: A Solomon four-group study. *Group & Organization Management, 28*(3), 416–439.

Probst, T. M., Jiang, L., & Benson, W. (2018). Job insecurity and anticipated job loss: A primer and exploration of possible. *The Oxford handbook of job loss and job search.*

Probst, T. M., Stewart, S. M., Gruys, M. L., & Tierney, B. W. (2007). Productivity, counter productivity and creativity: The ups and downs of job insecurity. *Journal of Occupational and Organizational Psychology, 80*(3), 479–497.

Reisel, W. D., Probst, T. M., Chia, S. L., Maloles, C. M., & König, C. J. (2010). The effects of job insecurity on job satisfaction, organizational citizenship behavior, deviant behavior, and negative emotions of employees. *International Studies of Management & Organization, 40*(1), 74–91.

Rosen, C. C., Chang, C. H., Djurdjevic, E., & Eatough, E. (2010). Occupational stressors and job performance: An updated review and recommendations. In *New developments in theoretical and conceptual approaches to job stress.* Emerald Group Publishing Limited.

Sanyal, S., Hisam, M., & BaOmar, Z. (2018). Loss of job security and its impact on employee performance—A study in Sultanate of Oman. *International Journal of Innovative Research & Growth., 7*(6), 74–91.

Schreurs, B. H., Hetty van Emmerik, I. J., Günter, H., & Germeys, F. (2012). A weekly diary study on the buffering role of social support in the relationship between job insecurity and employee performance. *Human Resource Management, 51*(2), 259–279.

Sekaran, U. (2003). *Research methods for business* (4th ed.). Wiley.

Sekaran, U., & Bougie, R. (2016). *Research methods for business: A skill building approach.* Wiley.

Shoss, M. K. (2017). Job insecurity: An integrative review and agenda for future research. *Journal of Management, 43*(6), 1911–1939.

Sposito, V. A., Hand, M. L., & Skarpness, B. (1983). On the efficiency of using the sample kurtosis in selecting optimal LP estimators. *Communications in Statistics-Simulation and Computation, 12*(3), 265–272.

Staufenbiel, T., & König, C. J. (2010). A model for the effects of job insecurity on performance, turnover intention, and absenteeism. *Journal of Occupational and Organizational Psychology, 83*(1), 101–117.

Sverke, M., & Hellgren, J. (2002). The nature of job insecurity: Understanding employment uncertainty on the brink of a new millennium. *Applied Psychology, 51*(1), 23–42.

Sverke, M., Hellgren, J., & Näswall, K. (2002). No security: A meta-analysis and review of job insecurity and its consequences. *Journal of Occupational Health Psychology, 7*(3), 242–265.

Sverke, M., Hellgren, J., & Näswall, K. (2006). *Job insecurity: A literature review.* Arbetslivsinstitutet.

Viswesvaran, C., Ones, D. S., & Schmidt, F. L. (1996). Comparative analysis of the reliability of job performance ratings. *Journal of Applied Psychology, 81*(5), 557–574.

Wang, H. J., Lu, C. Q., & Siu, O. L. (2015). Job insecurity and job performance: The moderating role of organizational justice and the mediating role of work engagement. *Journal of Applied Psychology, 100*(4), 1249–1269.

Witte, H. D. (1999). Job insecurity and psychological well-being: Review of the literature and exploration of some unresolved issues. *European Journal of Work and Organizational Psychology, 8*(2), 155–177.

The Impact of Strategic Thinking on Performance of Non-Governmental Organizations in Jordan

Fuad N. Al-Shaikh, Yahia Salim Melhem, Ola Mashriqi, Ziad Mohd Ali Smadi, Mohammed Saleem Khlif Alshura, Ali Zakariya Al-Quran, Hanan Mohammad Almomani, Sulieman Ibraheem Shelash Al-Hawary, and Ayat Mohammad

Abstract This study aimed to address the issue of strategic thinking at non-governmental organizations in Jordan & investigate whether strategic thinking has any impact on the performance of these organizations. It also aimed at finding out whether the top manager's demographic characteristics have moderating effects on the relationship between strategic thinking and organization performance. The study sample consisted of (262) top managers of non-governmental organizations in Jordan. The primary data of the study collected by questionnaire. SPSS, AMOS, and Smart PLS software utilized to conduct the statistical analysis process. The results show a positive impact of strategic thinking on the on performance. Furthermore, the

F. N. Al-Shaikh · O. Mashriqi
Department of Business Administration, Faculty of Economics and Administrative Sciences, Yarmouk University, P.O. Box 566, Irbid 21163, Jordan
e-mail: eco_fshaikh@yu.edu.jo

Y. S. Melhem
Business Management, Business Administration Department, School of Business, Yarmouk University, Irbid, Jordan
e-mail: ymelhem@yu.edu.jo

Z. M. A. Smadi · A. Z. Al-Quran · H. M. Almomani · S. I. S. Al-Hawary (✉)
Department of Business Administration, School of Business Al Al, Bayt University, P.O. BOX 130040, Mafraq 25113, Jordan
e-mail: dr_sliman73@aabu.edu.jo

Z. M. A. Smadi
e-mail: ziad38in@aabu.edu.jo

A. Z. Al-Quran
e-mail: ali.z.al-quran@aabu.edu.jo

M. S. K. Alshura
Management Department, Faculty of Money and Management, the World Islamic Science University, P.O. Box 1101, Amman 11947, Jordan

A. Mohammad
Business and Finance Faculty, The World Islamic Science and Education University, P.O. Box 1101, Amman 11947, Jordan

M. Alshurideh et al. (eds.), *The Effect of Information Technology on Business and Marketing Intelligence Systems*, Studies in Computational Intelligence 1056,
https://doi.org/10.1007/978-3-031-12382-5_51

study will hopefully be beneficial for policymakers & all organizations interested in boosting the role of NGOs. This research recommends the organizations to promote and support the adoption of strategic thinking factors for NGO managers and related institutions to improve the level of performance and quality of services provided such as developing training programs specialized in strategic thinking topics and organizing training workshops to discuss the real cases and companies' experiences inside and outside Jordan.

Keywords Strategic thinking · Organization performance · Non-governmental organization

1 Introduction

Strategic thinking (ST) is a critical component in the strategic management literature (Altamony et al., 2012; Al-Hawary & Hadad, 2016). A combination of careful blending of alternative strategic approaches and the impact of different decisions on an organization's value-creating process is essential. ST entails maintaining perspective on an organization's overall approach and coordinating the multiple dimensions of a combination of strategy (Pisapia, 2009). Most organizations operate under extreme pressure to attain efficiency and effectiveness (AlTaweel & Al-Hawary, 2021; Al-Hawary & Al-Syasneh, 2020; Allahow et al., 2018; Al-Nady et al., 2013). Hence, strategic thinking is needed to show direction and adapt to the ever-evolving environmental challenges (Alshraideh et al., 2017; Alaali et al., 2021). Managers with strategic thinking acumen who can scan the salient ecological trends and events with maximum predictability must be of paramount importance to the successful change process that positively impacts organizational performance (Harzing & Ruysseveldt, 2004; Al-Dmour et al., 2021; Lee et al., 2022a, b).

Strategic management and strategic thinking involve a series of decisions that companies could make strategically in the process of developing an effective and long-lasting business strategy to succeed and compete well at the internal levels (Alhalalmeh et al., 2020; Al-Hawary et al., 2020; Alyammahi et al., 2020). A company often adopts strategic planning process through business documentation, team management, time management and effective utilization of HR resources to achieve their desired levels of productivity and business growth in comparison to their competitors (Al-Hawary & Al-Namlan, 2018; Al-Hawary et al., 2013a, ¾b; Al-Lozi et al., 2017, 2018). Companies strive hard to create a strategic thinking environment within the internal business culture to make creativity and innovation through strategic decision making in year around processes (Hamadneh et al., 2021; Odeh et al., 2021). Strategic thinking enhances the ability of the companies to think strategically and differently than routine business processes and to get competitive edge over their competitors by identifying emerging business opportunities (Kitonga et al., 2016).

Non-governmental organizations (NGOs) have grown enormously to the point where they are now involved with every aspect of humanitarian work and relief. One primary focus of NGOs is how to prepare their staff to respond to the needs of the beneficiaries from their services. According to AlNasser (2016), Jordan now has many NGOs that act on an international scale through communication, coordination, training, fund-raising, and project implementation. They work with different local and international organizations such as environmental communities, community centers, women's groups, cultural foundations, and IT circles, all in the common purpose of empowering women and youth as well as improving their chances in the future.

The research work on the importance of strategic thinking in the process of strategic planning through which companies can achieve desired business growth and success. It provides necessary direction and guidelines to the management of a company to think strategically about limited business resources and time and people who are important for a success of a business. Strategic thinking enhances the ability of a business to effectively utilize available resources and advance the company towards its business goals and objectives. The research indicates that strategic thinking has become essential for companies to adapt their business processes and operations according to changing business environment and market conditions. With the lack of research related to NGOs in Jordan and growing interest in strategic thinking, it is necessary to shed more light on this area. So, the present study aims to address the issue of strategic thinking at NGOs in Jordan and investigate whether strategic thinking has any impact on the performance of these organizations.

2 Theoretical Framework and Hypotheses Development

2.1 Strategic Thinking

Strategic thinking is seen as the generation and application of unique business ideas intended to create a competitive advantage for a firm or business (Hamel & Prahalad, 1994). According to Mahdavian et al. (2014), ST is a way of thinking that has an essential place in the modern world and plays an essential role in significant issues at different organizational levels. Several researchers of strategic management emphasize that ST is one of the capabilities of high- performing leaders and enables thinkers to discover opportunities & understanding of the past, present, and future of the organization in order to survive, grow and achieve competitive advantage (Obeidat et al., 2021; Shamout et al., 2022; Shirvani & Shojaie, 2011). Bonn (2005) emphasizes that ST provides thinking and future vision that can change the competing rules of organizations and stresses that strategic thinking has been identified by a panel of experts as one of the ten most critical areas for future management research. Leadership and strategy theorists also noted that strategic thinking is required at multiple organizational levels; the need for thinking skills moves deeper at all levels of organizations (Ahmad et al., 2021; Alzoubi et al., 2022; Goldman & Casey, 2010).

Strategic thinking has many concepts. Porter (1991) states that ST is the glue that holds together the many systems and initiatives within the company, Mintzberg notes many practitioners and theorists have wrongly assumed that strategic planning, ST, and strategy making are all synonymous, at least in best practice. It is a matter of common sense to realize the existence and importance of strategic thinking as a concept. According to Horne and Wooton (2000), ST involves gathering information, formulating ideas, and planning action. Keelin and Arnold (2002) consider that the critical ability to be a strategic thinker is to have a strategic perspective and the ability to create clarity out of complex and disconnected details. Bonn (2005) defines ST as a way of solving strategic problems that combines a rational and convergent approach with creative and divergent thought processes. He argued that ST at the individual level consists of three main elements: systems thinking, creativity, and future vision. It can be seen from the above discussion that strategic thinking is conceptualized in different ways by different researchers. The present study adopts Bonn's (2005) definition, which includes three dimensions: systems thinking, creativity, and future vision.

Different models were proposed by strategic management thinkers to conceptualize ST. Bonn (2005) scrutinizes strategic thinking in three levels individual, group, and organization. On an individual level identifies three dimensions of ST: systems thinking, creativity, and future vision. He also concludes that competent decision-makers in strategic thinking have obtained more diverse conceptual systems than those who lack this capacity. Bonn believes that the role of strategic thinking is to try to innovate and envision a different future for the organization, which may lead to the redefinition of the company's primary or industrial strategies. Mintzberg (1998) argues that strategic thinking is a combination of looking backward, outlook, looking around, and completing it because he thinks that a real strategic thinker is the one who can connect all perspectives and visualize their cooperation in his mind. Kaufman et al. (2003) views Strategic thinking as a response to everyday problems. He argues that strategic thinking focuses on creating a better future by making the future and also on increasing value goals in society by obtaining profitable results. Kaufman et al. (2003) outlines six critical factors as success factors that chart strategic thinking. These factors are changing existing models and defining new broad lines of thinking, planning, activity, evaluation, and continuous improvement.

Many scholars seem to agree that the main dimensions of strategic thinking are Systems thinking, creativity, and having a future vision (Morrisey, 2004; Mintzberg, 1998; Kaufman et al., 2003; Bonn, 2005; Ashal et al., 2021). The present research also uses the same dimensions for strategic thinking. A brief description of each of these dimensions is provided in the following sections.

Systems thinking: Systems thinking is a way to see holistically. In other words, system thinking creates a conceptual framework to represent clearer models and makes clear how they are changed effectively, and it is formed based on a system viewpoint. A strategic thinker should always have a complete model of value creation factors and understand their collaboration well (Alameeri et al., 2020; Liedtka, 1998;

Marquardt, 1996). System thinking sees a problem or an opportunity as a part of the whole situation or system (Senge, 1990; Liedtka, 1998; Bonn, 2005; Alzoubi et al., 2021).

Creativity: Creativity is another essential element of ST. In a rapidly changing competitive business world, strategy and creativity are not random phenomena. Many organizations are looking for ways to find new perspectives and think about products, markets, challenges, and opponents. It is necessary to combine creativity with a strategic thinking process that allows employees and organizations to utilize more effectively and leverage the power of their minds in the best way (Alsharari & Alshurideh, 2020; Bonn, 2005; Mintzberg, 1998). According to Amabile (2013) creativity disciplined work style and skills in generating ideas. These cognitive processes include the ability to use wide, flexible categories for synthesizing information and the ability to break out of perceptual and performance "scripts." The personality processes include self-discipline and a tolerance for ambiguity.

Future vision: The third element of ST has a future vision. The latest theories in the world of leadership emphasize the roles of leaders in arranging view choosing its content (Bonn, 2005). The view of an organization defines its way and direction, which helps the organization heightens its success. Some believe that the view is created while others believe that it exists but should be discovered. In any case, the organizational view consists of the process of creation (Collins & Porras, 1989; Nuseir et al., 2021). ST must be accompanied by a strong understanding of the ultimate goal and a positive vision of the future of the organization.

2.2 *Organizational Performance*

According to Zheng et al. (2010) organizational performance stimulation has always been a priority in all sectors since it is directly associated with the value creation of the entity. Organizations are continually striving for better results, influence, and competitive advantage. Evaluating the performance of the organization has always been of interest to management teams and researchers (Al Kurdi et al., 2020; Al-Hawary, 2011, 2015; Al-Hawary & Abu-Laimon, 2013; Al-Hawary et al., 2013a, 2013b). In this regard, some researchers focused on determining definitions and how to measure organizational performance. The term performance is used extensively in the field of business management, and it varies according to organization objective. According to Hamon (2003), performance is a set of appropriate standards that enable to provide judgment on activities results. Borman and Motowidlo (1997) identified two types of employee behavior that are necessary for organizational effectiveness: task performance and contextual performance. Task performance refers to behaviors that are directly involved in producing goods or services or activities that provide indirect support for the organization's core technical processes. On the other hand, contextual performance is defined as individual efforts that are not directly related to their primary task functions. However, these behaviors are essential because they formulate the organizational, social, and psychological contexts (Wang et al., 2015; AlShehhi et al., 2020).

In the first decade of the twenty-first century, according to Robbins (2000) and Mouzas (2006), each of effectiveness and efficiency have their distinct meaning. Effectiveness focuses on achieving mission, goals, and vision, whereas efficiency relates to the optimal use of resources to achieve the desired output. Al-Hawary and Al-Syasneh (2020), argues that organizational performance should focus on the capability and ability of an organization to efficiently utilize the available resources to achieve accomplishments consistent with the set objectives of the company. Effectiveness oriented companies are concerned with output, quality, creation of value-added, innovation, and cost reduction. It measures the degree to which a business achieves its goals or the way outputs interact with the economic and social environment. Usually, effectiveness determines the policy objectives of the organization or the degree to which an organization realizes its own goals (Zheng et al., 2010; Al-Hawary, 2009; Al-Hawary & Mohammed, 2017; Al-Hawary & Nusair, 2017). Meyer and Herscovitch (2001) analyzed organizational effectiveness through organizational commitment. The workplace may take various forms, such as the relationship between leaders and staff, employee's identification with the organization, involvement in the decision-making process, psychological attachment felt by an individual.

According to Al-Hawary (2009), organizational effectiveness helps to assess the progress towards mission fulfillment and goal achievement. To improve organizational effectiveness, management should strive for better communication, interaction, leadership, direction, adaptability, and a positive environment. Pinprayong & Siengthai (2012) distinguishes between business efficiency & organizational efficiency. Business efficiency reveals the performance of input and output ratio, while organizational efficiency reflects the improvement of internal processes of the organization, such as organizational structure, culture, and community. Excellent organizational efficiency could improve entities' performance in terms of management, productivity, quality, and profitability. Therefore, management must ensure success in both areas.

In summary, effectiveness and efficiency are two essential dimensions of organizational performance. According to the researcher's best knowledge and based on the nature of the study population (non-profit organizations), effectiveness and efficiency are more suitable for the present research to measure organizational performance. In the other hand, strategic thinking associated with creativity and innovation to improve business performance and competitiveness of business; ST identified and revealed different strategic management approaches such as open-minded idea creation by mangers and enhanced learning within internal culture of the organizations. So, the organizations could success in developing a learning business environment within their business structure to succeed in the processes of strategic thinking to enhance innovation and creativity and uniqueness of ideas within the organization.

2.3 Strategic Thinking and Performance

Bonn (2005) showed that the development of strategies at the individual, community, and organization levels helps to develop the strategic thinking of organizations, which in turn helps to achieve organizational objectives. Jelenc and Swiercz (2010) explored the relationship between leader cognitive models of ST and the performance of Croatian firms. The results showed a positive relationship between leader cognitive models of ST and firm performance. Al-Qatamin and Esam (2018) addressed the effect of ST on the achievement of a competitive advantage. The Results showed that ST skills have a significant effect on competitive advantage. Moghadam et al. (2018) studied the impact of ST on innovation performance in Sistan and Baluchistan province customs administrations in Afghanistan. Results showed that all ST dimensions significantly influence innovation performance. Based on the literature review above studies, the study hypotheses can be formulated as:

There is a positive impact of strategic thinking on NGOs performance in Jordan

3 Research Model

As can be seen from the research model in Fig. 1, strategic thinking is the independent variable, and its dimensions are systems thinking, creativity, and future vision. These dimensions were adopted by well-known researchers in the field of strategic management (Morrisey, 2004; Mintzberg, 1998; Kaufman et al., 2003; Bonn, 2005). These dimensions/variables are hypothesized to affect the dependent variable (i.e., efficiency and effectiveness).

The model below represents the predicted association and relationships between strategic thinking represented by system thinking, creativity and future vision. Hence, strategic thinking is expected to impact organizational performance. Organization performance can be measured by two significant variables: efficiency and effectiveness.

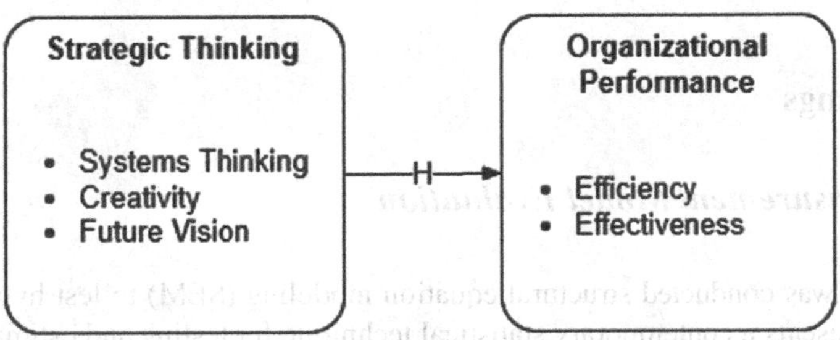

Fig. 1 Research model

4 Methodology

4.1 Population and Sample Selection

A qualitative method based on a questionnaire was used in this study for data collection and sample selection. The major aim of the study was to examine the impact of strategic thinking on organizational performance. Therefore, it focused on non-governmental organizations (NGOs) operating in Jordan. Data were primarily gathered through self-reported questionnaires creating by Google Forms which were distributed to (275) senior managers via email. In total, (263) responses were received including (6) invalid to statistical analysis due to uncompleted or inaccurate. Hence, the final sample contained (257) responses suitable to analysis requirements that were formed a response rate of (93.45%), where it proved to be sufficient to the extent that was predictable and allowed for a presumption of data saturation (Sekaran & Bougie, 2016).

4.2 Measurement Instrument

A self-reported questionnaire that consists of two main sections along with a section regarding control variables was used as the measurement instrument. Control variables considered as categorical measures were composed of gender, age group, educational level, and experience. The two main sections were dealt with a five-point Likert scale (from 1 = strongly disagree to 5 = strongly agree). The first section contained (14) items to measure strategic thinking based on (Alomari, 2020; Dixit et al., 2021). These questions were distributed into dimensions as follows: four items dedicated for measuring systems thinking, five items dedicated for measuring creativity, and five items dedicated for measuring future vision. Whereas the second section included nine items developed to measure organizational performance according to what was pointed by (Jami Pour & Asarian, 2019; Ling, 2019). These items divided into two dimensions: five items for measuring efficiency and four items for effectiveness.

5 Findings

5.1 Measurement Model Evaluation

This study was conducted structural equation modeling (SEM) to test hypotheses, which represents a contemporary statistical technique for testing and estimating the relationship between factors and variables (Wang & Rhemtulla, 2021). Accordingly, the reliability and validity of the constructs were tested using confirmatory factor

Table 1 Results of validity and reliability tests

Constructs	1	2	3	4	5
1. Systems thinking	**0.760**				
2. Creativity	0.428	**0.775**			
3. Future vision	0.337	0.384	**0.756**		
4. Efficiency	0.674	0.682	0.471	**0.779**	
5. Effectiveness	0.584	0.665	0.387	0.628	**0.745**
VIF	1.054	1.784	1.674	--	--
Loadings range	0.688–0.847	0.658–0.839	0.674–0.822	0.652–0.894	0.681–0.852
AVE	0.578	0.600	0.571	0.606	0.555
MSV	0.497	0.502	0.488	0.436	0.510
Internal consistency	0.842	0.880	0.865	0.854	0.858
Composite reliability	0.845	0.882	0.869	0.859	0.861

Note Bold fonts demonstrate the values of $\sqrt{AVE}$

analysis (CFA) through the statistical program AMOSv24. Table 1 summarizes the results of convergent and discriminant validity, as well the indicators of reliability.

Table 1 shows that the standard loading values for the individual items were within the domain (0.652–0.894), these values greater than the minimum retention of the elements based on their standard loads (Al-Lozi et al., 2018; Sung et al., 2019). Average variance extracted (AVE) is a summary indicator of the convergent validity of constructs that must be above 0.50 (Howard, 2018). The results indicate that the AVE values were greater than 0.50 for all constructs, thus the used measurement model has an appropriate convergent validity. Rimkeviciene et al. (2017) suggested the comparison approach as a way to deal with discriminant validity assessment in covariance-based SEM. This approach is based on comparing the values of maximum shared variance (MSV) with the values of AVE, as well as comparing the values of square root of AVE ($\sqrt{AVE}$) with the correlation between the rest of the structures. The results show that the values of MSV were smaller than the values of AVE, and that the values of $\sqrt{AVE}$ were higher than the correlation values among the rest of the constructs. Therefore, the measurement model used is characterized by discriminative validity. The internal consistency measured through Cronbach's Alpha coefficient (α) and compound reliability by McDonald's Omega coefficient (ω) was conducted as indicators to evaluate measurement model. The results listed in Table 1 demonstrated that both values of Cronbach's Alpha coefficient and McDonald's Omega coefficient were greater than 0.70, which is the lowest limit for judging on measurement reliability (De Leeuw et al., 2019).

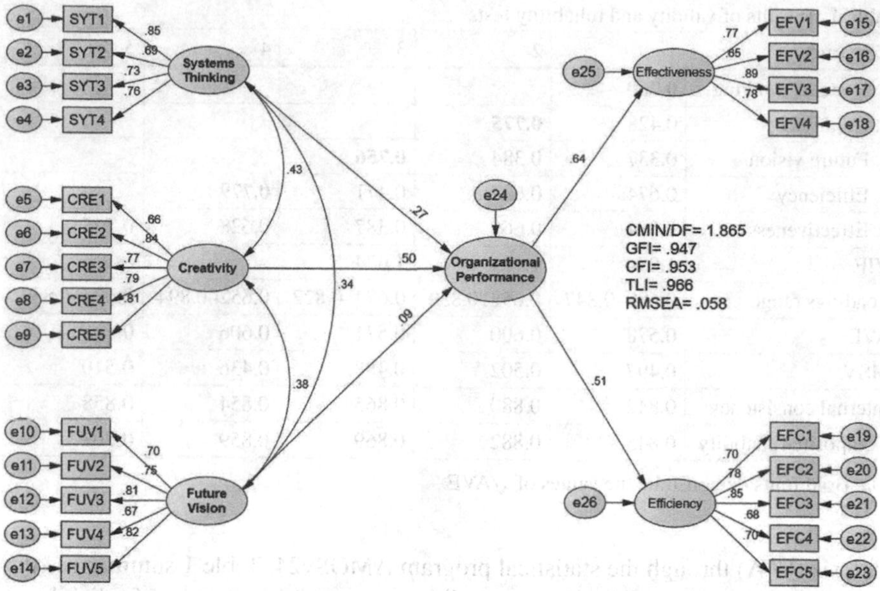

Fig. 2 SEM results of the strategic thinking effect on organizational performance

5.2 *Structural Model*

The structural model illustrated no multicollinearity issue among predictor constructs because variance inflation factor (VIF) values are below the threshold of 5, as shown in Table 1 (Hair et al., 2017). This result is supported by the values of model fit indices shown in Fig. 2.

The results in Fig. 2 indicated that the chi-square to degrees of freedom (CMIN/DF) was 1.865, which is less than 3 the upper limit of this indicator. The values of the goodness of fit index (GFI), the comparative fit index (CFI), and the Tucker-Lewis index (TLI) were upper than the minimum accepted threshold of 0.90. Moreover, the result of root mean square error of approximation (RMSEA) indicated to value 0.058, this value is a reasonable error of approximation because it is less than the higher limit of 0.08. Consequently, the structural model used in this study was recognized as a fit model for predicting the DEP and generalization of its result (Ahmad et al., 2016; Shi et al., 2019). To verify the results of testing the study hypotheses, structural equation modeling (SEM) was used, the results of which are listed in Table 2.

The results demonstrated in Table 2 show that the highest impact was for creativity ($\beta = 0.504$, $t = 3.489$, $p = 0.001$) followed by systems thinking ($\beta = 0.275$, $t = 2.974$, $p = 0.003$), while future vision ($\beta = 0.088$, $t = 0.978$, $p = 0.849$) had not impact on organizational performance.

Table 2 Hypothesis testing

Hypothesis	Relation	Standard Beta	*t* value	*p* value
H1	Systems Thinking → Organizational Performance	0.275	2.974**	0.003
H2	Creativity → Organizational Performance	0.504	3.489**	0.001
H3	Future Vision →; Organizational Performance	0.088	0.978	0.849

Note $^*p < 0.05$, $^{**}p < 0.01$, $^{***}\ p < 0.001$

6 Discussion

Understanding the linkage between thinking strategically and organizations performance sets the stage for exploring how managers exploit managerial activities and value creation. Based on the findings and hypotheses presented in this research, the result of the first hypothesis showed that the level of strategic thinking in NGOs in Jordan is moderate, with a mean of (3.49).

There is a positive impact of strategic thinking on performance of NGOs in Jordan; structural testing reported that strategic thinking through its three dimensions namely, systems thinking, creativity and future vision achieved moderate variance in organization performance in term of Effectiveness (R2 = 38%) and achieved low variance in Organization Performance in terms of Efficiency as (R2 = 7%). ST at NGOs in Jordan had a little impact on performance especially in efficiency side. These findings are Inconsistent with previous studies (Jelenc & Swiercz, 2010; Al-Qatamin & Esam, 2018; Moghadam et al.,;2018). At the same time the results showed a high level of organizations efficiency (4.35), this indicates the elements of strategic thinking are not directed towards achieving the goal and there are other factors that affect on performance efficiency. So, the researcher recommends for future studies to know the influencing factors that have achieved the success of the organizations and greatly affect in their efficiency and continuity.

7 Managerial Implications

The researcher suggests some recommendation to promote and support the adoption of strategic thinking factors for NGO managers to build a model of value creation, understand their collaboration well and sees a problem or an opportunity as a part of the whole situation or system and combine creativity with a strategic thinking process that allows employees and organizations to utilize more effectively and leverage the power of their minds in the best way which corresponds with (Bonn, 2005; Liedtka, 1998; Mintzberg, 1998; Senge, 1990).

Strategic thinking is a necessity to address the problems facing organizations in creative ways that allow them to succeed and to provide what is new in their field through the exploitation of resources, knowledge, and skills. So, the organizations need to develop managerial awareness of the importance of strategic thinking and be more flexible to meet unexpected circumstances and adapt quickly to sudden changes to achieve organization objectives and developing their internal resources that in turn enable them to address future problems. The award scheme is an appropriate way to encourage managers and employees to think strategically because this process can motivate them to collaborate. Such an award plan can be presented Influencing members, which leads to important results in management decisions and strategies. It is necessary to encourage initiative and innovation among employees and managers to create a regulatory environment in which everyone motivates innovation in line with Strategic thinking.

8 Limitations and Direction for Future Work

The limitation of this study is based on self-reported data, causal relationships among the independent and dependent variables may be affected by other factors that cannot be controlled. A longitudinal study is needed in order to build a solid ground for the causal direction of relationships. This study using a quantitative approach, so, in order to gain a more in-depth knowledge of the impact of strategic thinking on NGOs' performance, there is a need for a more qualitative approach to research needs to take place. It would be essential for future research to focus on a smaller population to enable a qualitative approach to be used. The researcher believes that this will enable more critical questions to be asked in an interview process.

References

Ahmad, A., Alshurideh, M. T., Al Kurdi, B. H., & Alzoubi, H. M. (2021, June). Digital strategies: A systematic literature review. In *The International Conference on Artificial Intelligence and Computer Vision* (pp. 807–822). Springer.

Ahmad, S., Zulkurnain, N., & Khairushalimi, F. (2016). Assessing the validity and reliability of a measurement model in Structural Equation Modeling (SEM). *British Journal of Mathematics & Computer Science, 15*(3), 1–8. https://doi.org/10.9734/BJMCS/2016/25183

Al Kurdi, B., Alshurideh, M., & Al afaishat, T. (2020). Employee retention and organizational performance: Evidence from banking industry. *Management Science Letters, 10*(16), 3981–3990.

Alaali, N., Al Marzouqi, A., Albaqaeen, A., Dahabreh, F., Alshurideh, M., Alrwashdh, S., Iyadeh, I., Salloum, S., & Aburayya, A. (2021) The impact of adopting corporate governance strategic performance in the tourism sector: A case study in the Kingdom of Bahrain. *Journal of Legal, Ethical and Regulatory, 24*(Special Issue 1), 1–18.

Alameeri, K., Alshurideh, M., Al Kurdi, B., & Salloum, S. A. (2020, October). The effect of work environment happiness on employee leadership. In *International Conference on Advanced Intelligent Systems and Informatics* (pp. 668–680). Springer.

Al-Dmour, R., AlShaar, F., Al-Dmour, H., Masa'deh, R., & Alshurideh, M. T. (2021). The effect of service recovery justices strategies on online customer engagement via the role of "customer satisfaction" during the covid-19 pandemic: An empirical study. *The Effect of Coronavirus Disease (COVID-19) on Business Intelligence, 334*, 325–346.

Alhalalmeh, M. I., Almomani, H. M., Altarifi, S., Al-Quran, A. Z., Mohammad, A. A., & Al-Hawary, S. I. (2020). The nexus between corporate social responsibility and organizational performance in Jordan: The mediating role of organizational commitment and organizational citizenship behavior. *Test Engineering and Management, 83*(July), 6391–6410.

Al-Hawary, S. I. (2009). The effect of the leadership style on the effectiveness of the organization: A field study at Zarqa Private University. *The Egyptian Journal for Commercial Studies, 33*(1), 361–393.

Al-Hawary, S. I. (2011). The Effect of banks governance on banking performance of the Jordanian commercial banks: Tobin's Q model "An Applied Study." *International Research Journal of Finance and Economics, 71*, 34–47.

Al-Hawary, S. I. (2015). Human resource management practices as a success factor of knowledge management implementation at health care sector in Jordan. *International Journal of Business and Social Science, 6*(11/1), 83–98.

Al-Hawary, S. I., & Abu-Laimon, A. A. (2013). The impact of TQM practices on service quality in cellular communication companies in Jordan. *International Journal of Productivity and Quality Management, 11*(4), 446–474.

Al-Hawary, S. I., & Al-Namlan, A. (2018). Impact of electronic human resources management on the organizational learning at the private hospitals in the State of Qatar. *Global Journal of Management and Business Research: A Administration and Management, 18*(7), 1–11.

Al-Hawary, S. I., & Al-Syasneh, M. S. (2020). Impact of dynamic strategic capabilities on strategic entrepreneurship in presence of outsourcing of five stars hotels in Jordan. *Business: Theory and Practice, 21*(2), 578–587.

Al-Hawary, S. I., & Hadad, T. F. (2016). The effect of strategic thinking styles on the enhancement competitive capabilities of commercial banks in Jordan. *International Journal of Business and Social Science, 7*(10), 133–144.

Al-Hawary, S. I., & Nusair, W. (2017). Impact of human resource strategies on perceived organizational support at Jordanian Public Universities. *Global Journal of Management and Business Research: A Administration and Management, 17*(1), 68–82.

Al-Hawary, S. I., Al-Hawajreh, K., Al-Zeaud, H., & Mohammad, A. (2013a). The impact of market orientation strategy on performance of commercial banks in Jordan. *International Journal of Business Information Systems, 14*(3), 261–279.

Al-Hawary, S. I., Al-Qudah, K., Abutayeh, P., Abutayeh, S., & Al-Zyadat, D. (2013b). The impact of internal marketing on employee's job satisfaction of commercial banks in Jordan. *Interdisciplinary Journal of Contemporary Research in Business, 4*(9), 811–826.

Al-Hawary, S. I. S., & Mohammed, A. K. (2017). Impact of team work traits on organizational citizenship behavior from the viewpoint of the employees in the education directorates in North Region of Jordan. *Global Journal of Management and Business, 17*(2-A), 23–40.

Al-Hawary, S. I. S., Mohammad, A. S., Al-Syasneh, M. S., Qandah, M. S. F., & Alhajri, T. M. S. (2020). Organisational learning capabilities of the commercial banks in Jordan: Do electronic human resources management practices matter? *International Journal of Learning and Intellectual Capital, 17*(3), 242–266.

Allahow, T. J. A. A., Al-Hawary, S. I. S., & Aldaihani, F. M. F. (2018). Information technology and administrative innovation of the central agency for information technology in Kuwait. *Global Journal of Management and Business, 18*(11-A), 1–16.

Al-Lozi, M., Almomani, R. Z., & Al-Hawary, S. I. (2017). Impact of talent management on achieving organizational excellence in Arab Potash Company in Jordan. *Global Journal of Management and Business Research: A Administration and Management, 17*(7), 15–25.

Al-Lozi, M., Almomani, R. Z., & Al-Hawary, S. I. (2018). Talent Management strategies as a critical success factor for effectiveness of Human Resources Information Systems in commercial banks

working in Jordan. *Global Journal of Management and Business Research: A Administration and Management, 18*(1), 30–43.
Al-Nady, B. A., Al-Hawary, S. I., & Alolayyan, M. (2013). Strategic management as a key for superior competitive advantage of sanitary ware suppliers in Kingdom of Saudi Arabia. *International Journal of Management and Information Technology, 7*(2), 1042–1058.
AlNasser, H. W. (2016). *New social enterprises in Jordan: Redefining the meaning of civil society.*
Al-Qatamin, A. A., & Esam, A. M. (2018). Effect of strategic thinking skills on dimensions of competitive advantage: empirical evidence from Jordan. Economics, Management And Marketing (MAC-EMM 2018), 8.
Alomari, Z. S. (2020). Does human capital moderate the relationship between strategic thinking and strategic human resource management? *Management Science Letters, 10*(2020), 565–574. https://doi.org/10.5267/j.msl.2019.9.024
Alsharari, N. M., & Alshurideh, M. T. (2020). Student retention in higher education: the role of creativity, emotional intelligence and learner autonomy. *International Journal of Educational Management, 35*(1), 233–247.
AlShehhi, H., Alshurideh, M., Al Kurdi, B., & Salloum, S. A. (2020, October). The impact of ethical leadership on employees performance: A systematic review. In *International Conference on Advanced Intelligent Systems and Informatics* (pp. 417–426). Springer.
Alshraideh, A. T. R., Al-Lozi, M., & Alshurideh, M. T. (2017). The impact of training strategy on organizational loyalty via the mediating variables of organizational satisfaction and organizational performance: An empirical study on Jordanian agricultural credit corporation staff. *Journal of Social Sciences (COES&RJ-JSS), 6*(2), 383–394.
Altamony, H., Masa'deh, R., Alshurideh, M., Obeidat, B. (2012) Information systems for competitive advantage: Implementation of an organisational strategic management process. In *Innovation and Sustainable Competitive Advantage: From Regional Development to World Economies* (pp. 583–592).
AlTaweel, I. R., & Al-Hawary, S. I. (2021). The mediating role of innovation capability on the relationship between strategic agility and organizational performance. *Sustainability, 13*(14), 7564.
Alyammahi, A., Alshurideh, M., Al Kurdi, B., & Salloum, S. A. (2020, October). The impacts of communication ethics on workplace decision making and productivity. In International Conference on Advanced Intelligent Systems and Informatics (pp. 488–500). Springer.
Alzoubi, H., Alshurideh, M., Kurdi, B., Akour, I., & Aziz, R. (2022). Does BLE technology contribute towards improving marketing strategies, customers' satisfaction and loyalty? The role of open innovation. *International Journal of Data and Network Science, 6*(2), 449–460.
Alzoubi, H. M., Alshurideh, M., & Ghazal, T. M. (2021, June). Integrating BLE Beacon technology with intelligent information systems IIS for operations' performance: A managerial perspective. In *The International Conference on Artificial Intelligence and Computer Vision* (pp. 527–538). Springer.
Amabile, T. (2013). *Componential theory of creativity* (Teresa M. Amabile. Harvard Business School: Encyclopedia of Management Theory [E. H. Kessler (Ed.)]). Sage Publications.
Ashal, N., Alshurideh, M., Obeidat, B., & Masa'deh, R. (2021) The impact of strategic orientation on organizational performance: Examining the mediating role of learning culture in Jordanian telecommunication companies. *Academy of Strategic Management Journal, 21*(Special Issue 6), 1–29.
Bonn, I. (2005). Improving strategic thinking: A multilevel approach. *Leadership & Organization Development Journal., 26*(5), 336–354.
Borman, W. C., & Motowidlo, S. J. (1997). Task performance and contextual performance: The meaning for personnel selection research. *Human Performance, 10*(2), 99–109.
Collins, J. C., & Porras, J. I. (1989). Making impossible dreams come true. *Stanford BusinessSchool Magazine, 57*(1), 12–19.
De Leeuw, E., Hox, J., Silber, H., Struminskaya, B., & Vis, C. (2019). Development of an international survey attitude scale: Measurement equivalence, reliability, and predictive validity.

Measurement Instruments for the Social Sciences, 1(1), 9. https://doi.org/10.1186/s42409-019-0012-x

Dixit, S., Singh, S., Dhir, S., & Dhir, S. (2021). Antecedents of strategic thinking and its impact on competitive advantage. *Journal of Indian Business Research, 13*(4), 437–458. https://doi.org/10.1108/JIBR-08-2020-0262

Goldman, E. F., & Casey, A. (2010). Building a culture that encourages strategic thinking. *Journal of Leadership & Organizational Studies, 17*(2), 119–128.

Hair, J. F., Babin, B. J., & Krey, N. (2017). Covariance-based structural equation modeling in the journal of advertising: Review and recommendations. *Journal of Advertising, 46*(1), 163–177. https://doi.org/10.1080/00913367.2017.1281777

Hamadneh, S., Hassan, J., Alshurideh, M., Al Kurdi, B., & Aburayya, A. (2021). The effect of brand personality on consumer self-identity: The moderation effect of cultural orientations among British and Chinese consumers. *Journal of Legal, Ethical and Regulatory Issues, 24*, 1–14.

Hamel, G., & Prahalad, C. K. (1994). *Competing for the future*. Harvard Business School Press.

Hamon, T. T. (2003). "Organizational effectiveness as explained by the social structure in a

Harzing, A., & Ruysseveldt, J. (2004). *International human resource management*. Sage.

Horne, T. & Wooton, S. (2000) *Strategic Thinking: A Step-by-Step Approach to Strategy* (2nd ed.).

Howard, M. C. (2018). The convergent validity and nomological net of two methods to measure retroactive influences. *Psychology of Consciousness: Theory, Research, and Practice, 5*(3), 324–337. https://doi.org/10.1037/cns0000149

Jami Pour, M., & Asarian, M. (2019). Strategic orientations, knowledge management (KM) and business performance: An exploratory study in SMEs using clustering analysis. *Kybernetes, 48*(9), 1942–1964. https://doi.org/10.1108/K-05-2018-0277

Jelenc, L., & Swiercz, P. M. (2010). Cognitive models of strategic thinking and firm performance: The croatian experience. In *ICSB World Conference Proceedings* (p. 1). International Council for Small Business (ICSB).

Kaufman, R., Browne, H., Watkins, R., & Leigh, D. (2003). *Strategic Planning for Success*.

Keelin, T., & Arnold, R. (2002). Five habits of highly strategic thinkers. *Journal of Business Strategy, 23*(5), 38–42.

Kitonga, D. M., Bichanga, W. O., & Meuma, B. K. (2016). Stra tegic leadership and organizational performance in not-for-profit organizations in Nairobi County in Kenya. *International Journal of Scientific & Technology Research, 5*(5), 17–27.

Lee, K., Azmi, N., Hanaysha, J., Alshurideh, M., & Alzoubi, H. (2022a). The effect of digital supply chain on organizational performance: An empirical study in Malaysia manufacturing industry. *Uncertain Supply Chain Management, 10*(2), 1–16.

Lee, K., Ramiz, P., Hanaysha, J., Alzoubi, H., & Alshurideh, M. (2022b). Investigating the impact of benefits and challenges of IOT adoption on supply chain performance and organizational performance: An empirical study in Malaysia. *Uncertain Supply Chain Management, 10*(2), 1–14.

Liedtka, J. M. (1998). Strategic thinking: Can it be taught? *Long Range Planning, 31*(1), 120–129.

Ling, Y.-H. (2019). Influence of corporate social responsibility on organizational performance: Knowledge management as moderator. *VINE Journal of Information and Knowledge Management Systems*, VJIKMS-11-2018-0096. https://doi.org/10.1108/VJIKMS-11-2018-0096.

Mahdavian, M., Mirabi, V., & Haghshenas, F. (2014). A study of the impact of strategic thinking on the performance of Mashhad municipal managers. *Management Science Letters, 4*(4), 679–690.

Marquardt, M. (1996). *Building The Learning Organization*. McGraw hill.

Meyer, J. P., & Herscovitch, L. (2001). Commitment in the workplace: Toward a general model. *Human Recource Managment Review, 11*(3), 299–326.

Mintzberg, H. (1998). Strategic thinking as seeing, In B. Garratt (Ed.), *Developing Strategic Thought: Reinventing the Art of Direction*. McGraw-Hill.

Moghadam, H., Haddadi, E., & Kikha, A. (2018). Studying the effect of strategic thinking on innovation performance (Case study: Sistan and Baluchestan Customs Administration). *Revista Publicando, 5*(15), 1123–1135.

Morrisey, G. L. (2004). *A guide to strategic thinking: Building your planning foundation*. Recording for the Blind & Dyslexic.
Mouzas, S. (2006). Efficiency versus effectiveness in business networks. *Journal of Business Research, 59*(10–11), 1124–2113.
Nuseir, M. T., Aljumah, A., & Alshurideh, M. T. (2021). How the business intelligence in the new startup performance in UAE during COVID-19: The mediating role of innovativeness. *The Effect of Coronavirus Disease (COVID-19) on Business Intelligence, 334*, 63–79.
Obeidat, U., Obeidat, B., Alrowwad, A., Alshurideh, M., Masadeh, R., & Abuhashesh, M. (2021). The effect of intellectual capital on competitive advantage: The mediating role of innovation. *Management Science Letters, 11*(4), 1331–1344.
Odeh, R. B. M., Obeidat, B. Y., Jaradat, M. O., & Alshurideh, M. T. (2021). The transformational leadership role in achieving organizational resilience through adaptive cultures: the case of Dubai service sector. *International Journal of Productivity and Performance Management*, Vol. ahead-of-print No. ahead-of-print. https://doi.org/10.1108/IJPPM-02-2021-0093.
Pinprayong, B., & Siengtai, S. (2012). Restructuring for organizational efficiency in the banking sector in thailand: A case study of SIAM commercial bank. *Far East Journal of Psychology and Business, 8*(2), 29–42.
Pisapia, J. (2009). *The Strategic Leader: New Tactic for a Globalizing World*. IAP.
Porter, M. E. (1991). Towards a dynamic theory of strategy. *Strategic Management Journal, 12*(S2), 95–117.
Rimkeviciene, J., Hawgood, J., O'Gorman, J., & De Leo, D. (2017). Construct validity of the acquired capability for suicide scale: Factor structure, convergent and discriminant validity. *Journal of Psychopathology and Behavioral Assessment, 39*(2), 291–302. https://doi.org/10.1007/s10862-016-9576-4
Robbins, S. P. (2000). *Organizational Theory: Structure, Design, and Application*. Prentice-Hall.
Sekaran, U., & Bougie, R. (2016). *Research Methods for Business: A Skill-Building Approach* (Seventh edition). Wiley.
Senge, P. M. (1990). *The Fifth Discipline: The Art and Practice of the Learning Organization*. Currency Doubleday.
Shamout, M., Elayan, M., Rawashdeh, A., Kurdi, B., & Alshurideh, M. (2022). E-HRM practices and sustainable competitive advantage from HR practitioner's perspective: A mediated moderation analysis. *International Journal of Data and Network Science, 6*(1), 165–178.
Shi, D., Lee, T., & Maydeu-Olivares, A. (2019). Understanding the model size effect on SEM fit indices. *Educational and Psychological Measurement, 79*(2), 310–334. https://doi.org/10.1177/0013164418783530
Shirvani, A., & Shojaie, S. (2011). A review on the leader's role in creating a culture that encourages strategic thinking. *Procedia-Social and Behavioral Sciences, 30*, 2074–2078.
Singh, S. K., Burgess, T. F., Heap, J., Almatrooshi, B., & Farouk, S. (2016). Determinants of organizational performance: a proposed framework. *International Journal of Productivity and Performance Management*.
Sung, K.-S., Yi, Y. G., & Shin, H.-I. (2019). Reliability and validity of knee extensor strength measurements using a portable dynamometer anchoring system in a supine position. *BMC Musculoskeletal Disorders, 20*(1), 1–8. https://doi.org/10.1186/s12891-019-2703-0
Wang, H. J., Lu, C. Q., & Siu, O. L. (2015). When does job security affect job performance? *I/O at Work*.
Wang, Y. A., & Rhemtulla, M. (2021). Power analysis for parameter estimation in structural equation modeling: A discussion and tutorial. *Advances in Methods and Practices in Psychological Science, 4*(1), 1–17. https://doi.org/10.1177/2515245920918253
Zheng, W., Yang, B., & McLean, G. N. (2010). Linking organizational culture, structure, strategy, and organizational effectiveness: Mediating role of knowledge management. *Journal of Business Research, 63*(7), 763–771.

The Influence of Electronic Human Resource Management on Intention to Leave: An Empirical Study of International NGOs in Jordan

Menahi Mosallam Alqahtani, Hanan Mohammad Almomani, Sulieman Ibraheem Shelash Al-Hawary, Kamel Mohammad Al-Hawajreh, Ayat Mohammad, Mohammad Issa Ghafel Alkhawaldeh, Yahia Salim Melhem, Muhammad Turki Alshurideh, and Shoroq Haidar Al-Qudah

Abstract The aim of the study was to examine the impact of electronic human resource management on intention to leave. Therefore, it focused on international non-governmental organizations (NGOs) operating in the humanitarian sector. Data were primarily gathered through self-reported questionnaires created by Google

---

M. M. Alqahtani
Administration Department, Community College of Qatar, Doha, Qatar
e-mail: mena7i@icloud.com

H. M. Almomani · S. I. S. Al-Hawary (✉) · S. H. Al-Qudah
Department of Business Administration, School of Business, Al Al-Bayt University, P.O. BOX 130040, Mafraq 25113, Jordan
e-mail: dr_sliman73@aabu.edu.jo

S. H. Al-Qudah
e-mail: malshurideh@sharjah.ac.ae

K. M. Al-Hawajreh
Business Faculty, Mu'tah University, Karak, Jordan

A. Mohammad
Business and Finance Faculty, The World Islamic Science and Education University (WISE), P.O Box 1101, Amman 11947, Jordan

M. I. G. Alkhawaldeh
Building and Land Tax, Zarqa Municipality, Ministry of Local Administration, Zarqa, Jordan

Y. S. Melhem
Business Management, Business Administration Department, School of Business, Yarmouk University, Irbid, Jordan
e-mail: ymelhem@yu.edu.jo

M. T. Alshurideh
Department of Marketing, School of Business, The University of Jordan, Amman 11942, Jordan
e-mail: m.alshurideh@ju.edu.jo

Department of Management, College of Business, University of Sharjah, Sharjah 27272, UAE

M. Alshurideh et al. (eds.), *The Effect of Information Technology on Business and Marketing Intelligence Systems*, Studies in Computational Intelligence 1056,
https://doi.org/10.1007/978-3-031-12382-5_52

Forms which were distributed to a random sample of (620) employees. Structural equation modeling (SEM) was conducted to test hypotheses. The results showed that electronic human resource management had a negative impact on intention to leave. Moreover, the results indicated that the highest impact was for e-performance evaluation. Based on the study results, the researcher recommends organizations to involve employees in databases that enable them to follow the electronic human resources management practices, which is to follow up on the performance appraisal and how it works.

Keywords Electronic human resource management · Intention to leave · International NGOs · Jordan

1 Introduction

Technological development and electronic tools have contributed to the various sciences development such as management, where today employees have a greater awareness than employees in the past, and new technologies have created a new employee's generation and a change in the organizational structure (Al-Hawary & Alhajri, 2020; Al-Hawary & Al-Menhaly, 2016; Al-Hawary & Al-Smeran, 2017; Al-Hawary & Hussien, 2017). Where the changes are organized in such a way that the organizations absence on the global communication network means the loss of huge capital resources, business organizations are nowadays facing many challenges such as globalization, competitive value chain and technological changes (Al-Hawary & Ismael, 2010; Al-Nady et al., 2013; Altarifi et al., 2015; Allahow et al., 2018; Al-Hawary & Al-Syasneh, 2020; Al-Hawary & Obiadat, 2021; AlTaweel & Al-Hawary, 2021). The rapid information and communication technology and its ability to apply it in various organizational fields led to the information technology penetration into human resource processes and systems, which led to a new approach to human resource management called "electronic human resource management" (Al-Hawary et al., 2020; Al-Hawary & Al-Namlan, 2018).Where electronic human resources management is one of the latest topics in the human resources management science as it aims to improve procedures to speed up the human resources functions operations, reduce costs and free managers from administrative restrictions to perform strategic roles (Alkalha et al., 2012; AlHamad et al., 2022; Alshurideh et al., 2022). In addition, increasing efficiency and benefiting from the different dimensions of technology, the rapid development of systems that provide electronic human resources services to employees and managers has allowed more information in a more convenient way that enables them to benefit from this information in the organization resources, so it automates human resources tasks and practices and transforms them from traditional and paper-based activities to electronic activities, allowing organizations to anticipate and transform environmental changes to create the necessary competitive advantage (Al Kurdi et al., 2021; Almaazmi et al., 2020; Harahsheh et al., 2021; Shamout et al., 2022).

What concerns us about the leave work intention is monitoring changes taking place in the workforce size, and there is no doubt that the organization's management is striving to reduce this phenomenon, because the high intention to leave work means that there is a large exit and entry situation to and from the organization (Alhalalmeh et al., 2020; Al-Hawary et al., 2013a, 2013b; Mohammad et al., 2020). This means that the efficiency degree of its workforce will be relatively low due to modernity and lack of experience, which makes the organization face many negative situations (Ahmed et al., 2020; Ahmad et al., 2021a, b; Odeh et al., 2021). As aspiring organizations today seek to achieve excellence and competitiveness, stay in the business world, retain their employees, and reduce the leave work intention by increasing job loyalty levels and achieving high job satisfaction levels, which helps to achieve goals (Abu Qaaud et al., 2011; Metabis & Al-Hawary, 2013; Al-Hawary et al., 2013a, 2013b; Al-Hawary & Harahsheh, 2014; Al-Nady et al., 2016; Al-Lozi et al., 2017; Alshurideh et al., 2017; Alolayyan et al., 2018; Al-Hawary & Al-Khazaleh, 2020). This study provides a scientific explanation to the impact of electronic human resources management on the intention to leave work. Hence the study idea came to identify the impact and importance of the effectiveness of electronic human resources management practices in the intention to leave work, through which it contributes to putting the organization on the right track.

2 Literature Review and Hypotheses Development

2.1 *Electronic HR Management Practices*

The human resource management department in many organizations focuses on human capital development and advocates human resource planning, performance management, recruitment and employee selection, as human resource management (HRM) scholars have sought to understand the ways in which the efficiency of human resource (HR) practices relates to their employees' attitude (Al-Hawary, 2015; Al-Hawary & Abu-Laimon, 2013). The human resource department is responsible for increasing human resources effectiveness in the organization in order to achieve the individual and society goals (Al-Hawary, 2011; Al-Hawary & Alajmi, 2017). There are many views of human resource management practices concept, which are defined as "a set of functions, activities and programs related to the human resources management affairs in the organization, and aiming to achieve the individuals, organization and society goals (Al-Hawary & Nusair, 2017; Al-Hawary & Mohammed, 2017). These functions, activities and programs include developing human resources strategy, analyzing organization jobs, planning human resources, evaluating employees performance, training and developing them, determining their salaries, wages and additional benefits, motivating them, activating them and addressing their problems to coordinate their goals and needs with the organization needs in which they work, all within the framework of specific organizational and societal context" (al-Bari Dora and Al-Sabbagh, 2010). Al-Hawary and Nusair (2017) explained that it is a group of administrative activities and efforts that seek to obtain the organization's needs of human resources, develop, motivate and maintain them in order to achieve organizational goals with the highest efficiency and effectiveness level. Akinyemi (2011) defined that it is the organization formal structure which responsible for all decisions, strategies, principles, processes and practices in addition to the total functions, activities and methods related to the management of employees. The researchers dealt with human resource practices differed about its practices, depending on the study community or its nature. Muhammad et al. sees that human resource management practices are represented by (motivation, selection and appointment, training), while Al-Hawary (2015) sees that the practices are represented by (Recruitment and selection, human resource planning, training and development, compensation), while) expanded the practices related to human resource management, which were represented by (human resource planning, recruitment and selection, training and development, job analysis and design, motivation, performance evaluation, Participation in decision-making).One of the technology various dimensions is improving efficiency as it has enabled human resource managers to hire fewer employees who play a more important role in the organization. There are also many viewpoints of the concept of e-HRM practices as) explained as an umbrella term describing the integration of information technology and HR services to automate and facilitate service delivery. The researchers addressed electronic human resource management practices from multiple points of view, but they agreed on the mechanism by which the practices are applied, as the human resource management practices represented by: recruitment, selection, training, performance evaluation and compensation, aim to ensure that employees possess the necessary attributes for effective organizational performance, and to provide a behavioral indicator and appropriate reinforcement to guide and motivate desired behaviors. This can be done via using the main electronic human resource management practices which are (e-recruitment, e-selection, e-training) (Al-Hawary & Nusair, 2017; AlMehrzi et al., 2020; Alshurideh et al., 2021).) believe that the practices are (e-learning, electronic recruitment, selection and appointment, e-training). While Swaroop (2012) sees that electronic human resources management practices are (electronic profile, electronic recruitment, electronic selection, e-learning, e-training, e-compensation, e-leave).When reviewing the literature related to electronic human resources management practices, which were addressed by researchers from different cultural destinations and in different environmental conditions, the researcher believes that electronic human resources management practices represented in (electronic recruitment, electronic learning, electronic performance evaluation, electronic compensation) can be addressed in the current study and may be appropriate for the study community. **Electronic Recruitment**, the mechanism of electronic recruitment works by posting vacancies on the organizations' websites or an online recruitment site. Recruitment refers to the overall process of attracting, selecting, and appointing suitable candidates for jobs. The recruitment process includes analyzing job requirements, attracting employees to work, screening and selecting applicants, and integrating the new employee into the organization-learning: **E-training** is seen as the preferred learning channel because it can be accessed from different parts of the world with a mouse click on the Internet, and e-training can happen anywhere. The researchers also revealed in the literature on training that the main objective of e-training is to enhance job performance and the extent of satisfaction that the trainee feels. **Electronic performance appraisal**: Macwan and Sajja (2013) define performance appraisal as a formal management procedure that provides an assessment of an individual's performance quality in an organization. As Swaroop (2012) stated that electronic performance appraisal means the company's website use to conduct an online assessment of employees' skills, knowledge and performance over the Internet. **Electronic compensation** is an effective reward and incentives system to enable organizations to respond quickly to changes in the environment, as the rewards and incentives system has become a tool for organizational effectiveness, as the electronic compensation system allows all organization employees to apply electronically in terms of reducing human resources management burden (Al-Khayyal et al., 2021; Alshurideh, 2019, 2022).

2.2 *Intention to Leave Work*

Human capital is one of the most important organizational assets and it is the most important pillars for maintaining organizational stability, growth stability and market control. Without the human element presence, organizations would not continue to work, and this is observed in major organizations and the rewards, training and loans they provide to their employees (Alshraideh et al., 2017; Al Suwaidi et al., 2020; Nuseir et al., 2021). Where the organizations owners always strive to strengthen the employees' loyalty and spirit, which contributes to the increasing productivity process and continuing to work (Alsuwaidi et al., 2020; Al Shebli et al., 2021). Thus intent to leave is a behavior directed toward leaving the organization, such as looking for a new job and quitting. In this regard, it is important to meet employee's needs and demands towards organizational support (Ammari et al., 2017; Al-bawaia et al., 2022). It is known that the intention to leave work has a negative impact on organizational efficiency and job satisfaction, as this phenomenon causes the organization to bear additional costs as a result of replacing an employee with another. If the employee is not satisfied with his work, this will reflect on his performance, and therefore the quality of service provided by this organization will decrease (Al-Dhuhouri et al., 2020; Al Mehrez et al., 2020). When the employee feels that organization does not fulfill his desires and needs; he will have the idea of leaving work and looking for the nearest presented opportunity (Alameeri et al., 2020; AlShehhi et al., 2020). In this case, the organization will have significant costs, whether direct costs (replacement, recruitment and selection, temporary staff) or indirect costs (morale, pressure on remaining staff, new staff learning) which are perhaps more important than direct costs. Intention to leave the job is the employee's plan to forget the current job and look for another in the near future. Amora (2014) defined the intention to leave work as the general framework for planning the behavior resulting from the worker's intention and the worker's determination to leave work. Mohammad et al. also defined it as the employee's intention to voluntarily end the work relationship that binds him to the institution in which he works. There are many things that lead to the intention of leaving work, including the imbalance between work and life that employee may suffers, whether managers or employees, which may be represented by several situations, including that there is extra work, or a lack of compatibility with the organization's schedule, as these situations and work pressure can lead to conflicts between work and family (Al Kurdi et al., 2020a, b; Allozi et al., 2022). Thus, the stress that the employee feels without an increase in wages, promotions or vacations, this stress hinders the employee's performance and he may think with the intention of leaving work.

2.3 Electronic HR Management Practices and Intent to Leave

The researchers concluded that there are different relationships between electronic human resources management practices in the intention to leave work. Ali and Khalifa (2013) indicated that there is no direct statistically significant relationship between human resources management practices represented in performance evaluation and the intention to leave work, while the incentives were linked to the performance evaluation results in order to give the employee an incentive to improve his performance. When human resources department managers in organizations undertake the selecting workers process, weights and criteria are set for this process in order to ensure the human elements provision that tend to voluntary practices from the beginning of their integration into employment. Al-Hawary and Alajmi (2017) also indicated that employee's moral factors have an impact on the intention to leave work and the importance of maintaining the positive aspects of these factors in order to avoid reaching negative results, which may eventually lead to the employee leaving his work. While Issam indicated that there is a statistically significant effect of human resources strategies on employees' intention to leave work, the results also showed a statistically significant effect of the perceived organizational support on the employees' intention to leave work While Al-Hawary et al., (2013a, 2013b) indicated that there is a statistically significant effect of human resources strategies on employees' intention to leave work, the results also showed a statistically significant effect of the perceived organizational support on the employees' intention to leave work. Mohammad et al. (2020) indicated that employees job performance is greatly affected by modern human resource management process, and the greatest impact was for job analysis and design, followed by the workers participation in decision-making and performance evaluation, and that all practices positively affect workers performance, the better the application of these practices, the better the performance of workers. Based on above literature, the study hypotheses may be formulated as:

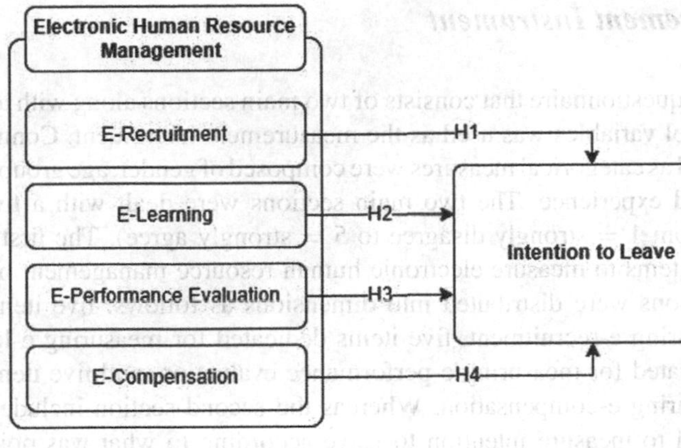

Fig. 1 Research model

There is a significant effect of human resource management practices on intention to leave.

3 Study Model

See Fig. 1.

4 Methodology

4.1 Population and Sample Selection

A qualitative method based on a questionnaire was used in this study for data collection and sample selection. The major aim of the study was to examine the impact of electronic human resource management on intention to leave. Therefore, it focused on international Non-Governmental Organizations (NGOs) operating in the humanitarian sector. Data were primarily gathered through self-reported questionnaires created by Google Forms which were distributed to a random sample of (620) employees via email. In total, (568) responses were received including (16) invalid to statistical analysis due to uncompleted or inaccurate. Hence, the final sample contained (552) responses suitable to analysis requirements that were formed a response rate of (89.03%), where it proved to be sufficient to the extent that was predictable and allowed for a presumption of data saturation (Sekaran & Bougie, 2016).

4.2 Measurement Instrument

A self-reported questionnaire that consists of two main sections along with a section regarding control variables was used as the measurement instrument. Control variables considered as categorical measures were composed of gender, age group, educational level, and experience. The two main sections were dealt with a five-point Likert scale (from 1 = strongly disagree to 5 = strongly agree). The first section contained (20) items to measure electronic human resource management based on (). These questions were distributed into dimensions as follows: five items dedicated for measuring e-recruitment, five items dedicated for measuring e-learning, five items dedicated for measuring e-performance evaluation, and five items dedicated for measuring e-compensation. Whereas the second section included seven items developed to measure intention to leave according to what was pointed by (Mohammad et al., 2020).

5 Findings

5.1 Measurement Model Evaluation

This study was conducted structural equation modeling (SEM) to test hypotheses, which represents a contemporary statistical technique for testing and estimating the relationship between factors and variables (Wang & Rhemtulla, 2021). Accordingly, the reliability and validity of the constructs were tested using confirmatory factor analysis (CFA) through the statistical program AMOSv24. Table 1 summarizes the results of convergent and discriminant validity, as well the indicators of reliability.

Table 1 shows that the standard loading values for the individual items were within the domain (0.531–0.864), these values greater than the minimum retention of the elements based on their standard loads (Al-Lozi et al., 2018; Sung et al., 2019). Average variance extracted (AVE) is a summary indicator of the convergent validity of constructs that must be above 0.50 (Howard, 2018). The results indicate that the AVE values were greater than 0.50 for all constructs, thus the used measurement model has an appropriate convergent validity. Rimkeviciene et al. (2017) suggested the comparison approach as a way to deal with discriminant validity assessment in covariance-based SEM. This approaches based on comparing the values of maximum shared variance (MSV) with the values of AVE, as well as comparing the values of square root of AVE ($\sqrt{AVE}$) with the correlation between the rest of the structures. The results show that the values of MSV were smaller than the values of AVE, and that the values of $\sqrt{AVE}$ were higher than the correlation values among the rest of the constructs. Therefore, the measurement model used is characterized by discriminative validity. The internal consistency measured through Cronbach's Alpha coefficient (α) and compound reliability by McDonald's Omega coefficient (ω) was conducted as indicators to evaluate measurement model. The results listed in Table

Table 1 Results of validity and reliability tests

Constructs	1	2	3	4	5
1. E-Recruitment	**0.740**				
2. E-Learning	0.425	**0.753**			
3. E-Performance Evaluation	0.364	0.348	**0.761**		
4. E-Compensation	0.333	0.452	0.425	**0.766**	
5. Intention to Leave	0.625	0.604	0.633	0.597	**0.761**
VIF	2.064	1.778	1.649	1.947	- -
Loadings range	0.642–0.811	0.681–0.834	0.703–0.812	0.655–0.864	0.531–0.860
AVE	0.548	0.566	0.579	0.587	0.580
MSV	0.371	0.438	0.505	0.425	0.511
Internal consistency	0.854	0.866	0.870	0.874	0.900
Composite reliability	0.857	0.867	0.873	0.876	0.904

Note Bold fonts in the table indicate to square root of average variance extracted

1 demonstrated that both values of Cronbach's Alpha coefficient and McDonald's Omega coefficient were greater than 0.70, which is the lowest limit for judging on measurement reliability (De Leeuw et al., 2019).

5.2 Structural Model

The structural model illustrated no multicollinearity issue among predictor constructs because variance inflation factor (VIF) values are below the threshold of 5, as shown in Table 1 (Hair et al., 2017). This result is supported by the values of model fit indices shown in Fig. 2.

The results in Fig. 2 indicated that the chi-square to degrees of freedom (CMIN/DF) was 1.644, which is less than 3 the upper limit of this indicator. The values of the goodness of fit index (GFI), the comparative fit index (CFI), and the Tucker-Lewis index (TLI) were upper than the minimum accepted threshold of 0.90. Moreover, the result of root mean square error of approximation (RMSEA) indicated to value 0.026, this value is a reasonable error of approximation because it is less than the higher limit of 0.08. Consequently, the structural model used in this study was recognized as a fit model for predicting the intention to leave and generalization of its result (Ahmad et al., 2016; Shi et al., 2019). To verify the results of testing the study hypotheses, structural equation modeling (SEM) was used, the results of which are listed in Table 2.

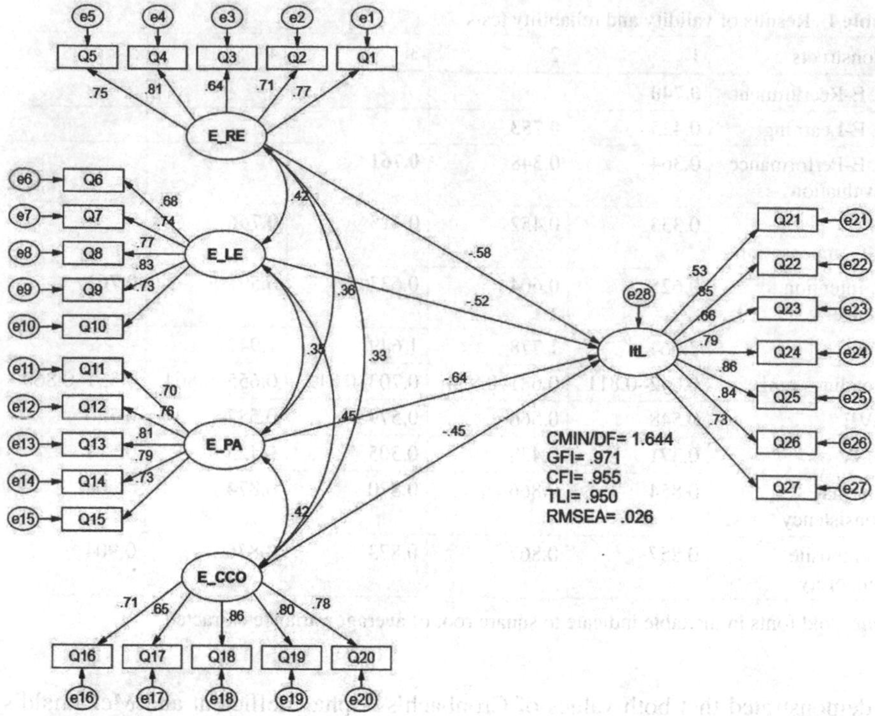

Fig. 2 SEM results of the E-HRM effect on intention to leave

Table 2 Hypothesis testing

Hypothesis	Relation	Standard beta	*t* value	*p* value
H1	E-Recruitment → Intention to Leave	−0.584***	18.611	0.000
H2	E-Learning → Intention to Leave	−0.522***	16.302	0.000
H3	E-Performance Evaluation → Intention to Leave	−0.641***	22.597	0.000
H4	E-Compensation → Intention to Leave	−0.451**	12.095	0.002

Note * $p < 0.05$, ** $p < 0.01$, *** $p < 0.001$

The results demonstrated in Table 2 show that electronic human resource management had a negative impact on intention to leave. Moreover, the results indicated that the highest impact was for e-performance evaluation (β = 0.641, t = 22.597, p = 0.000), followed by e-recruitment (β = 0.584, t = 18.611, p = 0.000), then e-learning (β = 0.522, t = 16.597, p = 0.000), and finally the lowest impact was for e-compensation (β = 0.451, t = 12.095, p = 0.002).

6 Discussions

The results showed the presence of a statistically insignificant effect of the electronic recruitment variable on the intention to leave work. Perhaps the reason behind this is that the electronic recruitment process is considered modern and not applied in all organizations, even if it is applied, the process of sorting applications and matching them with standards is not completely electronic, which in turn affects the employee and makes him think leave work. The results also showed a statistically significant effect of the e-learning variable in the intention to leave work, and the reason is that the process of teaching new employees the skills required to perform work tasks is based on the use of electronic systems in an average manner, as well as the case for determining training needs and evaluating the training process results. There was also a statistically significant impact of the electronic performance evaluation variable in the intention to leave work, which is a logical result in light of the partial reliance on employee performance evaluation through the use of the Internet and the presence of fixed standards based on evaluating employees performance by electronic tools that limit human intervention. In addition to the decisions related to employee's promotion, are still below the required level, as the performance evaluation results are obtained in a traditional way, not completely electronically. As it gives the employee the opportunity to get rid of negative reasons such as unfairness in performance appraisal. Finally, the results showed that there is an impact of the electronic compensation variable on the intention to leave work, and this means that informing employees of incentives through electronic methods is still below the required level, and that the use of electronic systems is partially applied in determining rewards through electronic programs, and the adoption of these programs is still weak in determining the relative jobs value.

7 Recommendations

Based on the study results, the researcher recommends organizations to involve employees in databases that enables them to follow the electronic human resources management practices, which is to follow up on the performance appraisal and how it works, which is done electronically, as well as identifying granting incentives mechanism and following up on training programs provided by the organization through modern technology tools. The researcher also recommends involving employees in paid training courses to raise their satisfaction and efficiency with work, in addition to providing an appropriate work environment for employees to reduce leaving work idea. The study also recommends that the electronic performance evaluation criteria should be clear to all employees and that they be familiar with their performance evaluation results, as this increases the employee's belief in the electronic evaluation fairness and integrity.

References

Abu Qaaud, F., Al-Shoura, M., & Al-Hawary, S. I. (2011). The impact of the service marketing mix in the service quality of health services from the viewpoint of patients in government hospitals in Amman " a field study ". *Abhath Al-Yarmouk, 27*(1B), 417–441.

Ahmad, A., Alshurideh, M. T., Al Kurdi, B. H., & Salloum, S. A. (2021b). Factors impacts organization digital transformation and organization decision making during covid19 pandemic. In *The Effect of Coronavirus Disease (COVID-19) on Business Intelligence* (pp. 95–106). Springer.

Ahmad, A., Alshurideh, M., Al Kurdi, B., Aburayya, A., & Hamadneh, S. (2021b). Digital transformation metrics: A conceptual view. *Journal of Management Information & Decision Sciences, 24*(7), 1–18.

Ahmad, S., Zulkurnain, N., & Khairushalimi, F. (2016). Assessing the validity and reliability of a measurement model in Structural Equation Modeling (SEM). *British Journal of Mathematics & Computer Science, 15*(3), 1–8. https://doi.org/10.9734/BJMCS/2016/25183

Ahmed, A., Alshurideh, M., Al Kurdi, B., & Salloum, S. A. (2020, October). Digital transformation and organizational operational decision making: a systematic review. In *International Conference on Advanced Intelligent Systems and Informatics* (pp. 708–719). Springer.

Akinyemi, B. (2011). An assessment of human resource development climate in Rwanda private sector organizations. *International Bulletin of Business Administration, 12*(1), 12–32.

Al Kurdi, B., Alshurideh, M., & Al afaishat, T. (2020a). Employee retention and organizational performance: Evidence from banking industry. *Management Science Letters, 10*(16), 3981–3990.

Al Kurdi, B., Alshurideh, M., & Alnaser, A. (2020b). The impact of employee satisfaction on customer satisfaction: Theoretical and empirical underpinning. *Management Science Letters, 10*(15), 3561–3570.

Al Kurdi, B., Elrehail, H., Alzoubi, H., Alshurideh, M., & Al-Adaila, R. (2021). The interplay among HRM practices, job satisfaction and intention to leave: An empirical investigation. *Journal of Legal, Ethical and Regulatory, 24*(1), 1–14.

Al Mehrez, A. A., Alshurideh, M., Al Kurdi, B., & Salloum, S. A. (2020, October). Internal factors affect knowledge management and firm performance: a systematic review. In *International Conference on Advanced Intelligent Systems and Informatics* (pp. 632–643). Springer.

Al Shebli, K., Said, R. A., Taleb, N., Ghazal, T. M., Alshurideh, M. T., & Alzoubi, H. M. (2021, June). RTA's employees' perceptions toward the efficiency of artificial intelligence and big data utilization in providing smart services to the residents of Dubai. In *The International Conference on Artificial Intelligence and Computer Vision* (pp. 573–585). Springer.

Al Suwaidi, F., Alshurideh, M., Al Kurdi, B., & Salloum, S. A. (2020, October). The impact of innovation management in SMEs performance: a systematic review. In *International Conference on Advanced Intelligent Systems and Informatics* (pp. 720–730). Springer.

Alameeri, K., Alshurideh, M., Al Kurdi, B., & Salloum, S. A. (2020, October). The effect of work environment happiness on employee leadership. In *International Conference on Advanced Intelligent Systems and Informatics* (pp. 668–680). Springer.

al-Bari Dora, A., & al-Sabbagh, Z. (2010). *Human Resources Management in the Twenty-First Century, a Nazmi Approach* (2nd ed.). Dar Wael for Publishing and Distribution.

Al-bawaia, E., Alshurideh, M., Obeidat, B., Masa'deh, R. (2022) The impact of corporate culture and employee motivation on organization effectiveness in Jordanian banking sector. *Academy of Strategic Management Journal, 21*(Special Issue 2), 1–18.

Al-Dhuhouri, F. S., Alshurideh, M., Al Kurdi, B., & Salloum, S. A. (2020, October). Enhancing our understanding of the relationship between leadership, team characteristics, emotional intelligence and their effect on team performance: A Critical Review. In *International Conference on Advanced Intelligent Systems and Informatics* (pp. 644–655). Springer.

Alhalalmeh, M. I., Almomani, H. M., Altarifi, S., Al- Quran, A. Z., Mohammad, A. A., & Al-Hawary, S. I. (2020). The nexus between corporate social responsibility and organizational performance in Jordan: The mediating role of organizational commitment and organizational citizenship behavior. *Test Engineering and Management, 83*(July), 6391–6410.

AlHamad, A., Alshurideh, M., Alomari, K., Kurdi, B., Alzoubi, H., Hamouche, S., & Al-Hawary, S. (2022). The effect of electronic human resources management on organizational health of telecommunications companies in Jordan. *International Journal of Data and Network Science, 6*(2), 429–438.
Al-Hawary, S. I. (2011). Human resource management practices in ZAIN cellular communications company operating in Jordan. *Perspectives of Innovations, Economics and Business, 8*(2), 26–34.
Al-Hawary, S. I. (2015). Human resource management practices as a success factor of knowledge management implementation at health care sector in Jordan. *International Journal of Business and Social Science, 6*(11/1), 83–98.
Al-Hawary, S. I., & Abu-Laimon, A. A. (2013). The Impact of TQM practices on service quality in cellular communication companies in Jordan. *International Journal of Productivity and Quality Management, 11*(4), 446–474.
Al-Hawary, S. I., & Alajmi, H. M. (2017). Organizational commitment of the employees of the ports security affairs of the state of Kuwait: The impact of human recourses management practices. *International Journal of Academic Research in Economics and Management Sciences, 6*(1), 52–78.
Al-Hawary, S. I., & Al-Khazaleh A, M. (2020). The mediating role of corporate image on the relationship between corporate social responsibility and customer retention. *Test Engineering and Management, 83*(516), 29976–29993.
Al-Hawary, S. I., & Al-Menhaly, S. (2016). The quality of E-government services and its role on achieving beneficiaries satisfaction. *Global Journal of Management and Business Research: A Administration and Management, 16*(11), 1–11.
Al-Hawary, S. I., & Al-Namlan, A. (2018). Impact of electronic human resources management on the organizational learning at the private hospitals in the state of Qatar. *Global Journal of Management and Business Research: A Administration and Management, 18*(7), 1–11.
Al-Hawary, S. I., & Al-Smeran, W. (2017). Impact of electronic service quality on customers satisfaction of islamic banks in Jordan. *International Journal of Academic Research in Accounting, Finance and Management Sciences, 7*(1), 170–188.
Al-Hawary, S. I., & Al-Syasneh, M. S. (2020). Impact of dynamic strategic capabilities on strategic entrepreneurship in presence of outsourcing of five stars hotels in Jordan. *Business: Theory and Practice, 21*(2), 578–587.
Al-Hawary, S. I., & Harahsheh, S. (2014). Factors affecting jordanian consumer loyalty toward cellular phone brand. *International Journal of Economics and Business Research, 7*(3), 349–375.
Al-Hawary, S. I., & Hussien, A. J. (2017). The impact of electronic banking services on the customers loyalty of commercial banks in Jordan. *International Journal of Academic Research in Accounting, Finance and Management Sciences, 7*(1), 50–63.
Al-Hawary, S. I., & Ismael, M. (2010). The effect of using information technology in achieving competitive advantage strategies: A field study on the Jordanian pharmaceutical companies. *Al Manara for Research and Studies, 16*(4), 196–203.
Al-Hawary, S. I., & Nusair, W. (2017). Impact of human resource strategies on perceived organizational support at Jordanian public universities. *Global Journal of Management and Business Research: A Administration and Management, 17*(1), 68–82.
Al-Hawary, S. I., Al-Hawajreh, K., AL-Zeaud, H., & Mohammad, A. (2013a). The impact of market orientation strategy on performance of commercial banks in Jordan. *International Journal of Business Information Systems, 14*(3), 261–279.
Al-Hawary, S. I., Al-Qudah, K., Abutayeh, P., Abutayeh, S., & Al-Zyadat, D. (2013b). The impact of internal marketing on employee's job satisfaction of commercial banks in Jordan. *Interdisciplinary Journal of Contemporary Research in Business, 4*(9), 811–826.
Al-Hawary, S. I. S., & Alhajri, T. M. S. (2020). Effect of electronic customer relationship management on customers' electronic satisfaction of communication companies in Kuwait. *Calitatea, 21*(175), 97–102.

Al-Hawary, S. I. S., & Mohammed, A. K. (2017). Impact of team work traits on organizational citizenship behavior from the viewpoint of the employees in the education directorates in north region of Jordan. *Global Journal of Management and Business, 17*(2-A), 23–40.

Al-Hawary, S. I. S., & Obiadat, A. A. (2021). Does mobile marketing affect customer loyalty in Jordan? *International Journal of Business Excellence, 23*(2), 226–250.

Al-Hawary, S. I. S., Mohammad, A. S., Al-Syasneh, M. S., Qandah, M. S. F., & Alhajri, T. M. S. (2020). Organisational learning capabilities of the commercial banks in Jordan: Do electronic human resources management practices matter? *International Journal of Learning and Intellectual Capital, 17*(3), 242–266.

Ali, M., & Khalifa, A. (2013), The impact of human resource management practices on the intention to leave work, the role of organizational citizenship behavior as a mediating variable, a field study applied to Sudanese commercial banks. *Journal of Economic Sciences*, Sudan University of Science and Technology, College of Business Studies.

Alkalha, Z., Al-Zu'bi, Z., Al-Dmour, H., Alshurideh, M., & Masa'deh, R. (2012). Investigating the effects of human resource policies on organizational performance: An empirical study on commercial banks operating in Jordan. *European Journal of Economics, Finance and Administrative Sciences, 51*(1), 44–64.

Al-Khayyal, A., Alshurideh, M., Al Kurdi, B., & Salloum, S. A. (2021). Factors influencing electronic service quality on electronic loyalty in online shopping context: data analysis approach. In *Enabling AI Applications in Data Science* (pp. 367–378). Springer.

Allahow, T. J. A. A., Al-Hawary, S. I. S., & Aldaihani, F. M. F. (2018). Information technology and administrative innovation of the central agency for information technology in Kuwait. *Global Journal of Management and Business, 18*(11-A), 1–16.

Allozi, A., Alshurideh, M., AlHamad, A., & Al Kurdi, B. (2022). Impact of transformational leadership on the job satisfaction with the moderating role of organizational commitment: Case of UAE and Jordan manufacturing companies. *Academy of Strategic Management Journal, 21*, 1–13.

Al-Lozi, M., Almomani, R. Z., & Al-Hawary, S. I. (2017). Impact of talent management on achieving organizational excellence in Arab Potash company in Jordan. *Global Journal of Management and Business Research: A Administration and Management, 17*(7), 15–25.

Al-Lozi, M., Almomani, R. Z., & Al-Hawary, S. I. (2018). Talent Management strategies as a critical success factor for effectiveness of Human Resources Information Systems in commercial banks working in Jordan. *Global Journal of Management and Business Research: A Administration and Management, 18*(1), 30–43.

Almaazmi, J., Alshurideh, M., Al Kurdi, B., & Salloum, S. A. (2020, October). The effect of digital transformation on product innovation: a critical review. In *International Conference on Advanced Intelligent Systems and Informatics* (pp. 731–741). Springer.

AlMehrzi, A., Alshurideh, M., & Al Kurdi, B. (2020). Investigation of the key internal factors influencing knowledge management, employment, and organisational performance: A qualitative study of the UAE hospitality sector. *Int. J. Innov. Creat. Chang, 14*(1), 1369–1394.

Al-Nady, B. A., Al-Hawary, S. I., & Alolayyan, M. (2013). Strategic management as a key for superior competitive advantage of sanitary ware suppliers in Kingdom of Saudi Arabia. *International Journal of Management and Information Technology, 7*(2), 1042–1058.

Al-Nady, B. A., Al-Hawary, S. I., & Alolayyan, M. (2016). The role of time, communication, and cost management on project management success: An empirical study on sample of construction projects customers in Makkah City, Kingdom of Saudi Arabia. *International Journal of Services and Operations Management, 23*(1), 76–112.

Alolayyan, M., Al-Hawary, S. I., Mohammad, A. A., & Al-Nady, B. A. (2018). Banking service quality provided by commercial banks and customer satisfaction. A structural equation modelling approaches. *International Journal of Productivity and Quality Management, 24*(4), 543–565.

AlShehhi, H., Alshurideh, M., Al Kurdi, B., & Salloum, S. A. (2020, October). The impact of ethical leadership on employees performance: A systematic review. In *International Conference on Advanced Intelligent Systems and Informatics* (pp. 417–426). Springer.

Alshraideh, A. T. R., Al-Lozi, M., & Alshurideh, M. T. (2017). The impact of training strategy on organizational loyalty via the mediating variables of organizational satisfaction and organizational performance: An empirical study on Jordanian agricultural credit corporation staff. Journal of Social Sciences (COES&RJ-JSS), 6(2), 383–394.

Alshurideh, D. M. (2019). Do electronic loyalty programs still drive customer choice and repeat purchase behaviour? *International Journal of Electronic Customer Relationship Management, 12*(1), 40–57.

Alshurideh, M. (2022). Does electronic customer relationship management (E-CRM) affect service quality at private hospitals in Jordan? *Uncertain Supply Chain Management, 10*(2), 325–332.

Alshurideh, M. T., Al Kurdi, B., & Salloum, S. A. (2021). The moderation effect of gender on accepting electronic payment technology: A study on United Arab Emirates consumers. *Review of International Business and Strategy, 31*(3), 375–396.

Alshurideh, M. T., Al Kurdi, B., Alzoubi, H. M., Ghazal, T. M., Said, R. A., AlHamad, A. Q., ... & Al-kassem, A. H. (2022). Fuzzy assisted human resource management for supply chain management issues. *Annals of Operations Research*, 1–19.

Alshurideh, M., Al-Hawary, S. I., Batayneh, A. M., Mohammad, A., & Al-Kurdi, B. (2017). The impact of Islamic Banks' service quality perception on Jordanian customers loyalty. *Journal of Management Research, 9*(2), 139–159.

Alsuwaidi, M., Alshurideh, M., Al Kurdi, B., & Salloum, S. A. (2020, October). Performance appraisal on employees' motivation: a comprehensive analysis. In *International Conference on Advanced Intelligent Systems and Informatics* (pp. 681–693). Springer.

Altarifi, S., Al-Hawary, S. I. S., & Al Sakkal, M. E. E. (2015). Determinants of E-shopping and its effect on consumer purchasing decision in Jordan. *International Journal of Business and Social Science, 6*(1), 81–92.

AlTaweel, I. R., & Al-Hawary, S. I. (2021). The mediating role of innovation capability on the relationship between strategic agility and organizational performance. *Sustainability, 13*(14), 7564.

Ammari, G., Alkurdi, B., Alshurideh, A., & Alrowwad, A. (2017). Investigating the impact of communication satisfaction on organizational commitment: A practical approach to increase employees' loyalty. *International Journal of Marketing Studies, 9*(2), 113–133.

Amora, Reem (2014), The impact of employees' moral factors on intentions to leave work, "A case study on the Syrian telecom company, Syriatel", Damascus University.

de Leeuw, E., Hox, J., Silber, H., Struminskaya, B., & Vis, C. (2019). Development of an international survey attitude scale: Measurement equivalence, reliability, and predictive validity. *Measurement Instruments for the Social Sciences, 1*(1), 9. https://doi.org/10.1186/s42409-019-0012-x

Hair, J. F., Babin, B. J., & Krey, N. (2017). Covariance-based structural equation modeling in the journal of advertising: Review and recommendations. *Journal of Advertising, 46*(1), 163–177. https://doi.org/10.1080/00913367.2017.1281777

Harahsheh, A., Houssien, A., Alshurideh, M. & AlMontaser, M. (2021). The effect of transformational leadership on achieving effective decisions in the presence of psychological capital as an intermediate variable in private Jordanian Universities in light of the corona pandemic. *The Effect of Coronavirus Disease (COVID-19) on Business Intelligence, 334*, 221–243.

Howard, M. C. (2018). The convergent validity and nomological net of two methods to measure retroactive influences. *Psychology of Consciousness: Theory, Research, and Practice, 5*(3), 324–337. https://doi.org/10.1037/cns0000149

Macwan, N., & Sajja, P. S. (2013, March). Modeling performance appraisal using soft computing techniques: Designing neuro-fuzzy application. In *2013 international conference on intelligent systems and signal processing (ISSP)* (pp. 403–407).

Metabis, A., & Al-Hawary, S. I. (2013). The impact of internal marketing practices on services quality of commercial banks in Jordan. *International Journal of Services and Operations Management, 15*(3), 313–337.

Mohammad, A. A., Alshura, M. S., Al-Hawary, S. I. S., Al-Syasneh, M. S., & Alhajri, T. M. (2020). The influence of Internal Marketing Practices on the employees' intention to leave: A study of the private hospitals in Jordan. *International Journal of Advanced Science and Technology, 29*(5), 1174–1189.

Nuseir, M. T., Aljumah, A., & Alshurideh, M. T. (2021). How the business intelligence in the new startup performance in UAE during COVID-19: The mediating role of innovativeness. *The Effect of Coronavirus Disease (COVID-19) on Business Intelligence, 334*, 63–79.

Odeh, R. B. M., Obeidat, B. Y., Jaradat, M. O., & Alshurideh, M. T. (2021). The transformational leadership role in achieving organizational resilience through adaptive cultures: the case of Dubai service sector. *International Journal of Productivity and Performance Management*, Vol. ahead-of-print No. ahead-of-print. https://doi.org/10.1108/IJPPM-02-2021-0093.

Rimkeviciene, J., Hawgood, J., O'Gorman, J., & De Leo, D. (2017). Construct validity of the acquired capability for suicide scale: Factor structure, convergent and discriminant validity. *Journal of Psychopathology and Behavioral Assessment, 39*(2), 291–302. https://doi.org/10.1007/s10862-016-9576-4

Sekaran, U., & Bougie, R. (2016). *Research methods for business: A skill-building approach* (7th ed.). Wiley.

Shamout, M., Elayan, M., Rawashdeh, A., Kurdi, B., & Alshurideh, M. (2022). E-HRM practices and sustainable competitive advantage from HR practitioner's perspective: A mediated moderation analysis. *International Journal of Data and Network Science, 6*(1), 165–178.

Shi, D., Lee, T., & Maydeu-Olivares, A. (2019). Understanding the model size effect on SEM fit indices. *Educational and Psychological Measurement, 79*(2), 310–334. https://doi.org/10.1177/0013164418783530

Sung, K.-S., Yi, Y. G., & Shin, H.-I. (2019). Reliability and validity of knee extensor strength measurements using a portable dynamometer anchoring system in a supine position. *BMC Musculoskeletal Disorders, 20*(1), 1–8. https://doi.org/10.1186/s12891-019-2703-0

Swaroop, K. R. (2012). E-HRM and how it will reduce the Cost in Organization. *Asia Pacific Journal of Marketing & Management Review., 1*(4), 133–139.

Wang, Y. A., & Rhemtulla, M. (2021). Power analysis for parameter estimation in structural equation modeling: A discussion and tutorial. *Advances in Methods and Practices in Psychological Science, 4*(1), 1–17. https://doi.org/10.1177/2515245920918253

The Mediating Effect of Organizational Commitment on the Relationship Between Work Life Balance and Intention to Leave

Hanan Mohammad Almomani, Hasan Aleassa, Kamel Mohammad Al-Hawajreh, Fatima Lahcen Yachou Aityassine, Raed Ismael Ababneh, Sulieman Ibraheem Shelash Al-Hawary, Muhammad Turki Alshurideh, Ayat Mohammad, and Anber Abraheem Shlash Mohammad

Abstract This study aimed to measure the mediating effect of organizational commitment on the relationship between work life balance and intention to leave. The study population is consisted of (800) working women in king Abdullah university

H. M. Almomani
Department of Business Administration, School of Business, Al Al-Bayt University, P.O. BOX 130040, Mafraq 25113, Jordan

H. Aleassa
Business Management, Business Administration Department, School of Business, Yarmouk University, Irbid, Jordan

K. M. Al-Hawajreh
Business Faculty, Mu'tah University, Karak, Jordan
e-mail: hawajreh2005@gmail.com

F. L. Y. Aityassine
Department of Financial and Administrative Sciences, Irbid University College, Al-Balqa' Applied University, Irbid, Jordan
e-mail: Fatima.yassin@bau.edu.jo

R. I. Ababneh
Policy, Planning, and Development Program, Department of International Affairs, College of Arts and Sciences, Qatar University, 2713 Doha, Qatar

S. I. S. Al-Hawary (✉)
Business Management, Department of Business Administration, Faculty of Finance and Business Administration, Al Al-Bayt University, P.O. BOX 130040, Mafraq 25113, Jordan
e-mail: dr_sliman73@aabu.edu.jo

M. T. Alshurideh
Department of Management, College of Business, University of Sharjah, Sharjah 27272, UAE
e-mail: malshurideh@sharjah.ac.ae

A. Mohammad
Business and Finance Faculty, The World Islamic Science and Education University (WISE), P.O. Box 1101, Amman 11947, Jordan

M. Alshurideh et al. (eds.), *The Effect of Information Technology on Business and Marketing Intelligence Systems*, Studies in Computational Intelligence 1056,
https://doi.org/10.1007/978-3-031-12382-5_53

hospital. To collect the primary data questionnaire survey was distributed to (200) working women and (144) questionnaires were returned. The study uses ordinary least square (OLS) technique to run simple regression and test hypothesis where the work life balance is used as independent variable and intention to leave as dependent variable, for mediating effect the study used the organization commitment variable. After analysis, study found that affective commitment has fully mediated the relationship between work life balance and intention to leave. While normative commitment has partially mediated the relationship between work life balance and intention to leave among working women in King Abdullah University Hospital. In light of previous findings, the researchers recommends adoption of work life balance program in Jordanian hospitals, to improve the motivation and commitment of women employees for better organizational performance.

Keywords Work life balance · Organizational commitment · Intention to leave · King Abdullah university hospital · Jordan

1 Introduction

Today Arab women occupy an important position in modern societies and contribute effectively in advancing and developing their countries economy. Women have become a key partner in success of their organization (Aburayya et al., 2020; Kabrilyants et al., 2021). However, they face extra demand due of their roles in work and home; they should do their best at work and take care their families when baking home. The main factor that leads to the success of women in their work is the ability of achieving balance between family obligations and work demands. Therefore, the failure to achieve this balance will negatively influence their duties towards home and work, that leading to the occurrence of physical and psychological problems and consequently, reducing her life satisfaction (Al Khayyal et al., 2020; Nuseir et al., 2021).

Women working in the Middle East are not an exception to the rule, the perfect balance between work and personal life remains the most important challenge faced by women in Middle East as they are first responsible for the welfare of their families and care for their children (Abu et al., 2012). And while, she made strides to have a more flexible treatment in the workplace in terms of working hours and the establishment of units for the care of children in the workplace and work-sharing arrangements; she still face social pressure to comply with standards that force them to be ideal mother and ideal employee and so on a daily basis.

The conflict between work and family is considered as an important issue for organizations and individuals due to its association with negative results (Lyness &

A. A. S. Mohammad
Marketing Department, Faculty of Administrative and Financial Sciences, Petra University, P.O. Box 961343, Amman 11196, Jordan
e-mail: mohammad197119@yahoo.com

Judiesch, 2008). For example, the conflict between work role and family life role is associated with increased absenteeism and labor force turnover, low performance in the work place, and poor physical and mental health (Agha et al., 2021; Mohammad et al., 2020). One of negative consequences of work life balance is intention to leave, which is costly for organization. The cost of Replacing one hospital nurse has been estimated to be around $42,000 in Canada (Strachotaet al, 2003). Current studies in the Health Industry are paying attention to how to address the high cost of labor turnover and staff shortage (Chaaban, 2006). Although the significant of work life balance, little attention is paid to this issue in Arab word. The best of researcher knowledge no studies have been conducted in Arab countries (include Jordan) to investigate the effect of work life balance on intention to leave among working women. Therefore, the current study is carried out to fill the gap in literature by investigate the mediating effect of organizational commitment on the relationship between work life balance and intention to leave.

2 Theoretical Framework and Hypotheses Development

2.1 Work Life Balance

The work life balance of employees has come to focus during second half of the twentieth century (Naithani, 2010). The research on working mother has started during 1960 (Lewis et al., 2007). However, before 1970, the work life and family life were treated separately (Blunsdo et al., 2006). After 1970, various research articles were published regarding this issue. In 1977, when some research was conducted on interdependence between work life and family life (Pleck, 1977).

There are only quite general and no homogenous definitions of the term work-life balance. There has been a wealth of earlier revisions trying to deliver a theoretical definition of the concept of work-life balance. For example, Fleetwood (1977) defined work-life balance as “the relationship between cultural and institutional spaces and times of work and non-work in civilizations where income is primarily created and distributed through labor markets”. Greenblatt (2002) characterizes work-life balance by "the absence of unacceptable levels of conflict between work and non-work demands". And Voydanoff (2005) defines work-family-balance as “a global assessment that work and family resources are sufficient to meet work and family demands such that participation is effective in both domains”. This definition serves as a basic concept for this thesis as it shed lights elements that entail work family-balance and thus are in the center of interest. Another Definition of the work-life balance was introduced by the Rigby & O’Brien (2010) that work-life balance is a split of time between the work and personal life based on the priorities. In this definition, there is clear indication between the time management for both lives, one is personal life and other is official life. Priorities of the people may be different based on their perceptions, but the main thing is the consideration of both lives rather just focusing on the single one. Because if the focus is shifted to the single side of the life only, then it is likely that balance would go out and effectiveness of the life may be compromised (Irene, 2014).

2.2 Organizational Commitment

The theoretical concept of the organizational commitment is elaborated along with the definitions of the organization commitment. The concept of the organizational commitment has taken a considerable importance over the years and become an essential feature of the researches that are connected with the employees and the organizations (Alhalalmeh et al., 2020; Al-Hawary & Alajmi, 2017; Al-Hawary et al., 2020; Alketbi et al., 2020; Metabis & Al-Hawary, 2013) It has been defined as a psychological state that combine an employee to an organization, that lead to reduce the turnover of employee (Allen, 2000). There are several definitions being introduced by the researchers about the organizational commitment (Al-Hawary & Abu-Laimon, 2013; Al-Hawary & Al-Namlan, 2018; Al-Hawary et al., 2013; Al-Lozi et al., 2018). One definition of the organizational commitment is, "it is a multidimensional approach in nature which intends to analyze the attitude of the employee for the achievement of the organizational goals in addition to the membership willingness" (Al-Hawary, 2015; Al-Lozi et al., 2017; Meyer & Herscovitch, 2001). This definition is clearly based on two points where one is targeting the intention of the employee to be part of the organization and the other is merely based on the working attitude of the employee because both these things indicate the commitment of the employee with the organization. The definition is, "Organizational commitment is basically a person's attachment with the organization with the intention to stay with it along with the identification of the organizational goals so that extra effort can also be put in the consideration of achieving this" (Meyer et al, 2002). As per the analysis of all these definitions presented above, it shows that there are several things which are common among these definitions and these are termed to be part of the organizational commitment. One important factor, which is common in most of the definitions, is the commitment of the employee to stay in the organization for a longer period of time by providing a quality work with continuity (Al Kurdi et al., 2020a, b; Alameeri et al., 2020). The other thing, which is being focused on the definitions, is the goal of the organization and goal congruence. Organizational goals are simple that these are the aims of the organizations but the goal congruence is basically the unity of goals, goal of the employee and of the organization (Al-Hawary & Al-Syasneh, 2020; Allozi et al., 2022). This means that employees act in such a way that their goals are also attached with the goals of the organizations.

There are basically three models of the organizational commitment which are also known as commitment models. Meyer et al. (1993) explain that an employee's commitment reflected a need, desire and obligation to maintain membership in an organization. So the models of organization commitment can be outlined as follows: ***Continuance Commitment*** occurs when an employee remains in organization because there are largely out of need, as lack of alternatives or costs associated with leaving, such as lost income, or an inability to transfer skills and education to another organization are the primary antecedents of continuance commitment (Meyer et al, 2002). ***Affective Commitment*** has been linked to positive work-related behaviors like organizational citizenship behavior (Meyer et al., 2002). Affective commitment model is linked with the wish of the employee to stay in the organization due to multiple reasons perceived by the employee himself. it is associated positively with organizational citizenship behaviors and negatively with turnover cognitions. ***Normative Commitment*** is more emotional and sense of liability-based model as compared to the other than two models discussed above. The basis of this model is different from the other two models and results in a more positive way for the organization. The employee feeling of moral obligation to stay in the organization because he feels that the organization has spent a lot of the resources in the grooming and the teachings of the employee.

2.3 Intention to Leave

Intention to leave is one of the measurements of performance (Abu Qaaud et al., 2011; Al-Nady et al., 2013; Al-Nady et al., 2016; Alshurideh et al., 2017; Alolayyan et al., 2018; Allahow et al., 2018; Al-Quran et al., 2020; AlTaweel & Al-Hawary, 2021; Altarifi et al., 2015; Al-Hawary & Alhajri, 2020; Al-Hawary & Obiadat, 2021). Intention to leave is defined as the planning of an employee to leave the current job and search for another job in near future time (AlShehhi et al., 2020; Alsuwaidi et al., 2020; Alshurideh, 2022). It is the willingness of the employees to leave the organization. There are several definitions being introduced of the intention to leave. According to the definition given by the Chaaban (2006) it is being defined as the intention of turnover whereas the definition given by Barak and Nissly (2001) it means that a serious consideration of leaving the job at present. However, the definition given by the (Chan & Morrison, 2000) is a little bit changed from the above two definitions. They said that intention to leave is about having a deliberate desire to leave the current employment. Aiken et al. (2002) defined this as a prediction of the future behavior because the intent is linked with the future action, which is based on the behavior, so it is very much integrated definition. It is clear that there have been similarities in the definitions being proposed by the plenty of writers. This "Employee Retention" is one of the major problems that are faced by present organizations (Ammari et al., 2017; Al-bawaia et al., 2022; Al Shebli et al., 2021). To maintain their workforce and to reduce the rate of erosion the organizations deliver visions into staff retention strategies, techniques and measures. In order to retain valuable employees, every employee should be given a proper attention.

2.4 *Organizational Commitment, Work Life Balance and Intention to Leave*

The research aims to investigate the work life policies and their effect on organizational commitment and intention to leave the organization of female employees in Jordan. Arab Muslim female has to perform more responsibilities in family than women in developed countries do. Hence, a work life balance is more difficult for them, which may lead to leave the organization. Effective work-life balance policies can solve this issue, which is the focus of this research. Kim (2014) concluded that there is a direct relationship between work-life balance with affective commitment but not with in-role performance. However, affective commitment can affect the in-role performance and hence, there is an indirect relationship between work-life balance and in-role performance. Tummers et al. (2013), concluded that development and career opportunities as well as work atmosphere are most influential factor that affect the turnover intention. Lee et al. (2013) found that individual related features such as marital status, education level etc. and work related features such as Work unit and teaching hospital are correlated with nurses' intention to leave their organization. Seven indicators were used to measure the quality of work life. Among them four (supportive milieu with job security and professional recognition, work arrangement and workload, work or home life balance, and nursing staffing and patient care) are correlated with nurses' intention to leave their organization. Nwagbara and Akanji (2012) found that women face more work life imbalances than men do. Women have more responsibilities, which pressurize themselves and reduce their motivation and commitment. Sakthivel and Jayakrishnan (2012) thought that working load interfere with their family greatly, but their family do not interfere with working significantly. The study also found that there is a positive relationship between work life balance and organizational commitment. Noor (2011) found that there is a negative correlation between perceived work life balance satisfaction and intentions to leave the organizations. The result of the study also indicated that job satisfaction and organizational commitment can act as partial mediator to define the relationship between work life balance and intentions to leave the organizations. Malone (2010) identified some factors, which causes dissatisfaction of the employees such as attitudes of the owner and/or managers, fairness, equality etc. These issues cannot be solved by work life balance policies as they are related to personal attitude. Kamel (2013) concluded that Affective Commitment can fully mediate the relationship between Quality of Work Life and Intention to Leave. Work life balance, organization commitment and intention to leave the organization: all are important factors for better human resource management and overall performance of the organization. From above studies, it is clear that these three are related to each other. Based on the literature above, the study hypotheses may be formulated as:

Organizational commitment mediates the relationship between work life balance and intention to leave among working women in King Abdullah University Hospital.

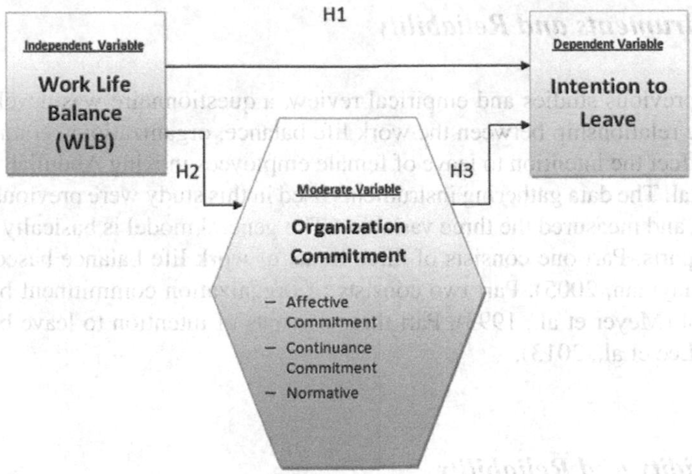

Fig. 1 The research model based on theoretical framework

3 Model and Variable of the Study

Based on previous literature the researcher prepared the following model in Fig. 1 that contains independent variable (work life balance) that measure their effect on dependent variable (intention to leave) through the mediating variable (organization commitment).

4 Research Methodology

4.1 Population and Sample

The female employees in king Abdullah University Hospital were the target population of the study with approximately (800) female employees. The sample frame comprised of working women in King Abdullah University Hospital at all degree levels, single and married with at least one year of experience of being employed with KAUH, the sample of the current study consist of 200 employees, which represent 25 percent of the study population. A sample population consisting of working women in King Abdullah University Hospital to achieve the goal of this study.

4.2 *Instruments and Reliability*

Based on previous studies and empirical review, a questionnaire was developed to explore the relationship between the work life balance, organizational commitment and how affect the intention to leave of female employees in King Abdullah University Hospital. The data gathering instruments used in this study were previously used instrument and measured the three variables. The general model is basically divided into three parts. Part one consists of three items of work life balance based on the work of (Hayman, 2005). Part two consists of organization commitment based on the work of (Meyer et al., 1993), Part three consists of intention to leave based on the work (Lee et al., 2013).

4.3 *Validity and Reliability*

Validity refers to the instrument should contain items related to the study variables, and it measures them accurately and clearly. To examine the validity of the survey and to ensure no biases with multiple interpretations or inappropriate and unclear wording, the tool of this study questionnaire was in English, but it was translated from English to Arabic by the researcher, then evaluated, the questionnaire had been sent to a number of specialist in business administration field.Their comments and recommendations were taken into consideration to improve some of the questionnaire items in a manner that fits the study variables measurements.

Reliability an indication of stability and internal consistency. It was established by calculating Cronbachs' Alpha coefficient which is an evidence indicating how well the questions are positively correlated to one another. Obtaining (Alpha $\geq$ 0.60) considered in practice for management science and Humanities is acceptable in general (Sekaran, 2003). Reliability coefficient for the main content of the questionnaire values ranged between (0.69) and (0.87), Hence it is acceptance values for the purposes of our analysis and test hypothesis, and the following Table 1 shows the results of Cronbach s Alpha for each dimension.

5 Analysis

H1.1: Affective Commitment mediates the relationship between work life balance and intention to leave among working women in King Abdullah University Hospital. The following model illustrates the nature of relationship between dependent, independent and mediator variables (Fig. 2).

To test this hypothesis, we used path analysis using regression technique results, the results are presented in Table 2, as illustrated in the below table, we noticed B and t turn out to be insignificant; this means that Affective Commitment fully

Table 1 Scales reliability

Variable measured	Dimensions	Number of items	Cronbach's Alpha
Work Life Balance	Work interference with personal life (WIPL)	7	0.87
	Personal life interference with work (PLIW)	4	0.71
Organization commitment	Affective commitment	6	0.80
	Continuance commitment	6	0.78
	Normative commitment	6	0.79
Intention to leave		6	0.69

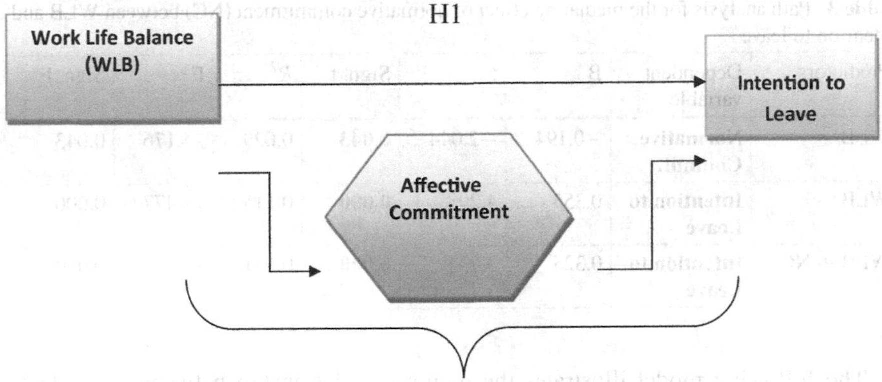

Fig. 2 First mediate model

mediates relationship between Work Life Balance (WLB) and Intention to leave. So, Hypothesis 1 is supported (Accepted). This explanation supported by previous research like (Kamel, 2013).

H1.2: Normative Commitment mediates the relationship between work life balance and intention to leave among working women in King Abdullah University Hospital.

Table 2 Path analysis for the mediating effect of affective commitment between WLB and intention to leave

Predictors	dependent variable	B	t	Sign. t	R^2	F	Sign. F
WLB	**Affective Commit.**	−0.705	−8.468	**0.000**	0.336	71.714	0.000
WLB	**Intention to Leave**	0.355	4.298	**0.000**	0.115	18.477	0.000
WLB + AC	**Intention to Leave**	0.179	1.82	**0.071**	0.171	14.553	0.000

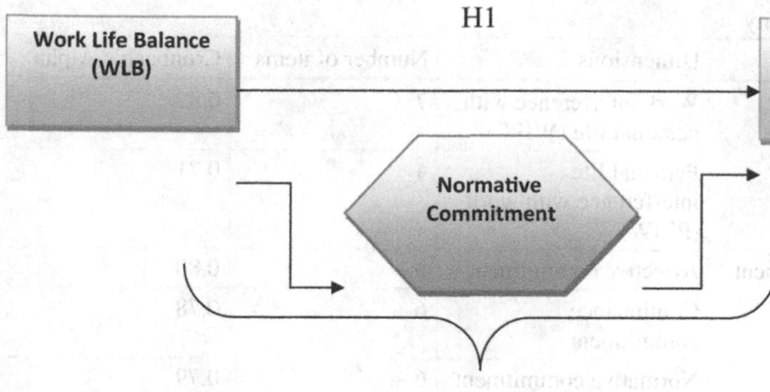

Fig. 3 Second mediate model

Table 3 Path analysis for the mediating effect of normative commitment (NC) between WLB and intention to leave

Predictors	Dependent variable	B	t	Sign. t	R^2	F	Sign. F
WLB	**Normative Commit.**	−0.194	−2.044	**0.043**	0.029	4.176	0.043
WLB	**Intention to Leave**	0.355	4.298	**0.000**	0.115	18.477	0.000
WLB + NC	**Intention to Leave**	0.325	3.934	**0.000**	0.141	11.62	0.000

The following model illustrates the nature of relationship between dependent, independent and moderator variables (Fig. 3).

To test this hypothesis, we used path analysis using regression technique results, the results are presented in Table 3, as illustrated in the below table, we noticed B and t remained significant; this means that Normative Commitment (NC) partially mediates relationship between Work Life Balance (WLB) and Intention to leave. So, Hypothesis 1.2 is supported (Accepted).

6 Conclusion and Discussion

The balance between the work and life is essential for a healthy land happy life because extra time spends on the work leads to work stress and extra time spends at home leads to the unsuccessful life. A high turnover is not considered a good sign for a corporate organization. If too many employees leave a particular organization, other experts or professional will think that there is a problem in the organization for which the employees left. Therefore, they will not be willing to join the organization (Davis & Kalleberg, 2006). Thus, intention to leave of the employees creates adverse effect on organizations. However, women in organization are increasing, they have dual roles at work and at home, this put extra pressure on women, they struggle than men to balance between demand of work and life. Organizational commitment is further being explained as the employee's commitment to continue to stay in the organization and the integration of the self-goals with the goals of the organization because it is the way to pay for the affiliation with organization.

The aim of the study is to examine the mediate effect of the Organizational commitment on the relationship between work life balance and intention to leave among working women in King Abdullah University Hospital. The study found that affective commitment has fully mediated the relationship between work life balance and intention to leave. While normative commitment has partially mediated the relationship between work life balance and intention to leave among working women in King Abdullah University, the results of the study are consistent with Kim (2014), he concluded that there is a direct relationship between work-life balances with affective commitment, and Nwagbara and Akanji (2012) found that women face more work life imbalances than men do. Women have more responsibilities, which pressurize themselves and reduce their motivation and commitment. And Kamel (2013) concluded that Affective Commitment can fully mediate the relationship between Quality of Work Life and Intention to Leave. Work life balance, organization commitment and intention to leave the organization. The research conducted by Clutterbuck (2003) showed that the employees having been provided with work life opportunities believed more in commitment to the work and to the organization.

So, there must be some flexibilities for women during their working hours in order to have an ability to create the balance between the work and life such as compressed hours, short time and flexible timings, those are very effective policies for women. The theories of the work life balance are applied differently to women because their perception and the attitude are different from those of men. For more illustration, let's take this example: it is about a mother, who has plenty tasks to perform and these tasks cannot be done by a man like feeding the child, taking care of the child at home and also handling the routine tasks like house cleaning and cooking (Clutterbuck, 2003). These factors clearly show that women require more time to spend at home, but this would create the lack of balance between the work and life at home; so basically, if a little flexibility is given to women, they will easily maintain the balance. So, these are basically the major differences between men and women, these differences have caused more emphasizing on the work life balance for women contradictory to the less importance given to men.

The employees of the organizations with work life opportunities are more satisfied with the organization and have more willingness to work for the organization. This helps organization in retention of staff. (Booth & Jan, 2008). Work pressure is one of the causes of the high turnover. This is a state where the employee finds himself stuck in the work and has the specified deadline coming which puts an immense pressure on the tissues that it becomes harder for him to concentrate. Employee facing such situation may suffer from physiological and physical problems, so they may think about leaving organization (Obeidat et al., 2019; Kristensen, 2000; Alshurideh et al., 2022).

7 Recommendations

Based on the study results, the researchers recommend managers and decision makers to adopt work life balance program, to improve the motivation and commitment of women employees for better organizational performance. The recommended programs as safety, reward, promotion, flexible hour, training, payment, childcare support, parental leave, and to provide work life balance program is necessary for woman in Jordanian hospitals to increase job satisfaction that lead to enhance commitment for organization and reduce intention to leave. The organizations must enhance the participation of employee in decision making and locus of control to improved organization commitment. Finally, organization should give special attention to commitment its vision and goals in such a way that is consistent with the employee values to enhance affective commitment.

References

Abu Qaaud, F., Al-Shoura, M., & Al-Hawary, S. I. (2011). The impact of the service marketing mix in the service quality of health services from the viewpoint of patients in government hospitals in Amman "a field study." *Abhath Al-Yarmouk, 27*(1B), 417–441.

Aburayya, A., Alshurideh, M., Al Marzouqi, A., Al Diabat, O., Alfarsi, A., Suson, R., ... & Alzarouni, A. (2020). Critical success factors affecting the implementation of TQM in public hospitals: A case study in UAE Hospitals. *Systematic Reviews in Pharmacy, 11*(10), 230–242.

Agha, K., Alzoubi, H. M., & Alshurideh, M. T. (2021, June). Measuring reliability and validity instruments of technologically driven cognitive intrusion towards work-life balance. In *The International Conference on Artificial Intelligence and Computer Vision* (pp. 601–614). Springer.

Aiken, L. H., Clarke, S. P., & Sloane, D. M. (2002). Hospital staffing, organization, and quality of care: Cross-national findings. *International Journal for Quality in Health Care, 14*(1), 5–14.

Alameeri, K., Alshurideh, M., Al Kurdi, B., & Salloum, S. A. (2020, October). The effect of work environment happiness on employee leadership. In *International Conference on Advanced Intelligent Systems and Informatics* (pp. 668–680). Springer.

Alhalalmeh, M. I., Almomani, H. M., Altarifi, S., Al-Quran, A. Z., Mohammad, A. A., & Al-Hawary, S. I. (2020). The nexus between corporate social responsibilty and organizational performance in Jordan: The mediating role of organizational commitment and organizational citizenship behavior. *Test Engineering and Management, 83*(July), 6391–6410.

Al Khayyal, A. O., Alshurideh, M., Al Kurdi, B., & Salloum, S. A. (2020, October). Women empowerment in UAE: A systematic review. In *International Conference on Advanced Intelligent Systems and Informatics* (pp. 742–755). Springer.

Al Kurdi, B., Alshurideh, M., & Alafaishat, T. (2020a). Employee retention and organizational performance: Evidence from banking industry. *Management Science Letters, 10*(16), 3981–3990.

Al Kurdi, B., Alshurideh, M., & Alnaser, A. (2020b). The impact of employee satisfaction on customer satisfaction: Theoretical and empirical underpinning. *Management Science Letters, 10*(15), 3561–3570.

Al Shebli, K., Said, R. A., Taleb, N., Ghazal, T. M., Alshurideh, M. T., & Alzoubi, H. M. (2021, June). RTA's employees' perceptions toward the efficiency of artificial intelligence and big data utilization in providing smart services to the residents of Dubai. In *The International Conference on Artificial Intelligence and Computer Vision* (pp. 573–585). Springer.

Al-bawaia, E., Alshurideh, M., Obeidat, B., Masa'deh, R. (2022) The impact of corporate culture and employee motivation on organization effectiveness in Jordanian banking sector. *Academy of Strategic Management Journal, 21*(Special Issue 2), 1–18.
Al-Hawary, S. I. (2015). Human resource management practices as a success factor of knowledge management implementation at health care sector in Jordan. *International Journal of Business and Social Science, 6*(11/1), 83–98.
Al-Hawary, S. I., & Abu-Laimon, A. A. (2013). The impact of TQM practices on service quality in cellular communication companies in Jordan. *International Journal of Productivity and Quality Management, 11*(4), 446–474.
Al-Hawary, S. I., & Alajmi, H. M. (2017). Organizational commitment of the employees of the ports security affairs of the State of Kuwait: The impact of human recourses management practices. *International Journal of Academic Research in Economics and Management Sciences, 6*(1), 52–78.
Al-Hawary, S. I., & Al-Namlan, A. (2018). Impact of electronic human resources management on the organizational learning at the private hospitals in the State of Qatar. *Global Journal of Management and Business Research: A Administration and Management, 18*(7), 1–11.
Al-Hawary, S. I., & Al-Syasneh, M. S. (2020). Impact of dynamic strategic capabilities on strategic entrepreneurship in presence of outsourcing of five stars hotels in Jordan. *Business: Theory and Practice, 21*(2), 578–587.
Al-Hawary, S. I., Al-Qudah, K., Abutayeh, P., Abutayeh, S., & Al-Zyadat, D. (2013). The impact of internal marketing on employee's job satisfaction of commercial banks in Jordan. *Interdisciplinary Journal of Contemporary Research in Business, 4*(9), 811–826.
Al-Hawary, S. I. S., & Alhajri, T. M. S. (2020). Effect of electronic customer relationship management on customers' electronic satisfaction of communication companies in Kuwait. *Calitatea, 21*(175), 97–102.
Al-Hawary, S. I. S., & Obiadat, A. A. (2021). Does mobile marketing affect customer loyalty in Jordan? *International Journal of Business Excellence, 23*(2), 226–250.
Al-Hawary, S. I. S., Mohammad, A. S., Al-Syasneh, M. S., Qandah, M. S. F., & Alhajri, T. M. S. (2020). Organisational learning capabilities of the commercial banks in Jordan: Do electronic human resources management practices matter? *International Journal of Learning and Intellectual Capital, 17*(3), 242–266.
Al-Quran, A. Z., Alhalalmeh, M. I., Eldahamsheh, M. M., Mohammad, A. A., Hijjawi, G. S., Almomani, H. M., & Al-Hawary, S. I. (2020). Determinants of the green purchase intention in Jordan: The moderating effect of environmental concern. *International Journal of Supply Chain Management, 9*(5), 366–371.
Alketbi, S., Alshurideh, M., & Al Kurdi, B. (2020). The Influence of service quality on customers' retention and loyalty in the UAE hotel sector with respect to the impact of customer' satisfaction, trust, and commitment: A qualitative study. *International Journal of Innovation, Creativity and Change, 14*(7), 734–754.
Allahow, T. J. A. A., Al-Hawary, S. I. S., & Aldaihani, F. M. F. (2018). Information technology and administrative innovation of the central agency for information technology in Kuwait. *Global Journal of Management and Business, 18*(11-A), 1–16.
Allen, T. D. (2000). Family-supportive work environments: The role of organization perceptions. *Journal of Vocational Behavior., 58*, 414–435.
Allozi, A., Alshurideh, M., AlHamad, A., & Al Kurdi, B. (2022). Impact of transformational leadership on the job satisfaction with the moderating role of organizational commitment: Case of UAE and Jordan manufacturing companies. *Academy of Strategic Management Journal, 21*, 1–13.
Al-Lozi, M., Almomani, R. Z., & Al-Hawary, S. I. (2017). Impact of talent management on achieving organizational excellence in Arab Potash Company in Jordan. *Global Journal of Management and Business Research: A Administration and Management, 17*(7), 15–25.
Al-Lozi, M., Almomani, R. Z., & Al-Hawary, S. I. (2018). Talent Management strategies as a critical success factor for effectiveness of Human Resources Information Systems in commercial banks

working in Jordan. *Global Journal of Management and Business Research: A Administration and Management, 18*(1), 30–43.
Al-Nady, B. A., Al-Hawary, S. I., & Alolayyan, M. (2013). Strategic management as a key for superior competitive advantage of sanitary ware suppliers in Kingdom of Saudi Arabia. *International Journal of Management and Information Technology, 7*(2), 1042–1058.
Al-Nady, B. A., Al-Hawary, S. I., & Alolayyan, M. (2016). The role of time, communication, and cost management on project management success: an empirical study on sample of construction projects customers in Makkah City, Kingdom of Saudi Arabia. *International Journal of Services and Operations Management, 23*(1), 76–112.
Alolayyan, M., Al-Hawary, S. I., Mohammad, A. A., & Al-Nady, B. A. (2018). Banking service quality provided by commercial banks and customer satisfaction. A structural equation modelling approaches. *International Journal of Productivity and Quality Management, 24*(4), 543–565.
AlShehhi, H., Alshurideh, M., Al Kurdi, B., & Salloum, S. A. (2020, October). The impact of ethical leadership on employees performance: A systematic review. In *International Conference on Advanced Intelligent Systems and Informatics* (pp. 417–426). Springer.
Alshurideh, M. (2022). Does electronic customer relationship management (E-CRM) affect service quality at private hospitals in Jordan? *Uncertain Supply Chain Management, 10*(2), 325–332.
Alshurideh, M., Al-Hawary, S. I., Batayneh, A. M., Mohammad, A., & Al-Kurdi, B. (2017). The impact of Islamic banks' service quality perception on Jordanian customers loyalty. *Journal of Management Research, 9*(2), 139–159.
Alshurideh, M. T., Al Kurdi, B., Alzoubi, H. M., Ghazal, T. M., Said, R. A., AlHamad, A. Q., ... & Al-kassem, A. H. (2022). Fuzzy assisted human resource management for supply chain management issues. *Annals of Operations Research*, 1–19.
Alsuwaidi, M., Alshurideh, M., Al Kurdi, B., & Salloum, S. A. (2020, October). Performance appraisal on employees' motivation: a comprehensive analysis. In *International Conference on Advanced Intelligent Systems and Informatics* (pp. 681–693). Springer.
Altarifi, S., Al-Hawary, S. I. S., & Al Sakkal, M. E. E. (2015). Determinants of E-shopping and its effect on consumer purchasing decision in Jordan. *International Journal of Business and Social Science, 6*(1), 81–92.
AlTaweel, I. R., & Al-Hawary, S. I. (2021). The mediating role of innovation capability on the relationship between strategic agility and organizational performance. *Sustainability, 13*(14), 7564.
Ammari, G., Alkurdi, B., Alshurideh, A., & Alrowwad, A. (2017). Investigating the impact of communication satisfaction on organizational commitment: A practical approach to increase employees' loyalty. *International Journal of Marketing Studies, 9*(2), 113–133.
Barak, M. E.M., Nissly, J. A. (2001). Antecedents to retention and turnover among child welfare, social work, and other human service employees: What can we learn from past research? A review and meta analysis. *Social Service Review,75*(4), 625–661.
Booth, A. L., & Jan, C. (2008). Job satisfaction and family happiness: The part-time work puzzle. *The Economic Journal, 118*(526), 77–99.
Chaaban, H. A. (2006). Job satisfaction, organizational commitment and turnover intent among nurse anesthetists in Michigan. *Dissertation Abstracts International, 67*(01), UMI No. 3206380.
Chan, E., & Morrison, P. (2000). Factors influencing the retention and turnover intentions of registered nurses in a Singapore hospital. *Nursing and Health Sciences., 2*, 113–121.
Clutterbuck, D. (2003). *Managing work-life balance: A guide for HR in achieving organizational and individual change*. Chartered Institute of Personnel and Development.
Davis, A. E., & Kalleberg, A. L. (2006). Family-friendly organizations? Work and family programs in the 1990s. *Work and Occupations., 33*(2), 191–223.
Greenblatt, E. (2002). Work/life balance: Wisdom or whining. *Organizational Dynamics, 31*, 177–194.
Hayman, J. (2005). Psychometric assessment of an instrument designed to measure work life balance. *Research and Practice in Human Resource Management, 13*(1), 85–91.

Irene, O.-P. (2014). *Work-life balance and health of women: A qualitative study of a mining company in Ghana*. University of Bergen.

Kabrilyants, R., Obeidat, B., Alshurideh, M., & Masadeh, R. (2021). The role of organizational capabilities on e-business successful implementation. *International Journal of Data and Network Science, 5*(3), 417–432.

Kamel, M. M. (2013). The mediating role of affective commitment in the relationship between quality of work life and intention to leave. *Life Science Journal, 10*(4), 1062–1067.

Kim, H. K. (2014). Work-life balance and employees' performance: The mediating role of affective commitment. *Global Business and Management Research: An International Journal, 6*(1), 37–51.

Kristensen, T. S. (2000). *A new tool for assessing psychosocial factors at work: The Copenhagen psychosocial questionnaire*. National Institute of Health.

Lee, Y.-W., Dai, Y.-T., Park, C.-G., & McCreary, L. L. (2013). Predicting quality of work life on nurses' intention to leave. *Journal of Nursing Scholarship, 45*(2), 160–168.

Lyness, K. S., & Judiesch, M. K. (2008). Can a manager have a life and a career? International and multisource perspectives on work-life balance and career advancement potential. *Journal of Applied Psychology, 93*(4), 789–805.

Malone, E. K. (2010). *Work-life balance and organizational commitment of women in construction in the United States*. Doctoral thesis, University of Florida, United States.

Metabis, A., & Al-Hawary, S. I. (2013). The impact of internal marketing practices on services quality of commercial banks in Jordan. *International Journal of Services and Operations Management, 15*(3), 313–337.

Meyer, J. P., Allen, N. J., & Smith, C. A. (1993). Extension and test of a three-component conceptualization. *Journal of Applied Psychology, 78*(4), 538–551.

Meyer, J. P., & Herscovitch, L. (2001). Commitment in the workplace: Toward a general model. *Human Resource Management Review, 11*(3), 299–326.

Mohammad, A. A., Alshura, M. S., Al-Hawary, S. I. S., Al-Syasneh, M. S., & Alhajri, T. M. (2020). The influence of Internal Marketing Practices on the employees' intention to leave: A study of the private hospitals in Jordan. *International Journal of Advanced Science and Technology, 29*(5), 1174–1189.

Meyer, J. P., Stanley, D. J., Herscovitch, L., & Topolnytsky, L. (2002). Affective, continuance, and normative commitment to the organization: A meta-analysis of antecedents, correlates, and consequences. *Journal of Vocational Behavior, 61*, 20–52.

Naithani, P. (2010). Overview of work-life balance discourse and its relevance in current economic scenario. *Asian Social Science, 6*(6), 148–155.

Noor, K. M. (2011). Work-life balance and intention to leave among academics in Malaysian Public Higher Education Institutions. *International Journal of Business and Social Science, 2*(11), 240–248.

Nuseir, M. T., Al Kurdi, B. H., Alshurideh, M. T., & Alzoubi, H. M. (2021, June). Gender discrimination at workplace: Do Artificial Intelligence (AI) and Machine Learning (ML) have opinions about it. In *The International Conference on Artificial Intelligence and Computer Vision* (pp. 301–316). Springer.

Nwagbara, U., &Akanji, B. O. (2012, March). The impact of work-life balance on the commitment and motivation of Nigerian women employees. *International Journal of Academic Research in Business and Social Sciences, 2*(3). www.hrmars.com.

Obeidat, Z. M., Alshurideh, M. T., Al Dweeri, R., & Masa'deh, R. (2019). The influence of online revenge acts on consumers psychological and emotional states: Does revenge taste sweet. In *Proceedings of the 33rd International Business Information Management Association Conference, IBIMA* (pp. 4797–4815).

Pleck, J. H. (1977). The work-family role system. *Social Problems, 24*(4), 417–427.

Raddaha, A. H. A., Alasad, J., Albikawi, Z. F., Batarseh, K. S., Realat, E. A., Saleh, A. A., & Froelicher, E. S. (2012). Jordanian nurses' job satisfaction and intention to quit. *Leadership in Health Services Journal, 25*(3), 216–231.

Rigby, M., & O'Brien, S. (2010). Trade Union interventions in the work life balance work. *Employment Society, 24*(2), 1–18.
Sakthivel, D., & Jayakrishnan, J. (2012). Work life balance and organizational commitment for nurses. *Asian Journal of Business and Management Sciences, 2*(5), 1–6.
Sekaran, U. (2003). *Research Methods for Business*. Wiley.
Strachota, E., Normandin, P., O'Brien, N., Clary, M., & Krukow, B. (2003). Reason registered nurses leave or change employment status. *Journal of Nursing Administration, 33*(2), 111–117.
Tummers, L. G., Groeneveld, S. M., & Lankhaar, M. (2013). Why do nurses intend to leave their organization? A large-scale analysis in long-term care. *Journal of Advanced Nursing, 69*(12), 2826–2838.
Voydanoff, P. (2005). Toward a conceptualization of perceived work-family fit and balance: A demands and resources approach. *Journal of Marriage and Family, 67*(4), 822–836.

Electronic HR Practices as a Critical Factor of Employee Satisfaction in Private Hospitals in Jordan

Main Naser Alolayyan, Reham Zuhier Qasim Almomani, Shoroq Haidar Al-Qudah, Sulieman Ibraheem Shelash Al-Hawary, Anber Abraheem Shlash Mohammad, Kamel Mohammad Al-hawajreh, Raed Ismael Ababneh, Muhammad Turki Alshurideh, and Abdullah Ibrahim Mohammad

Abstract The aim of the study was to examine the impact of e-human resource management on employee satisfaction. Therefore, it focused on the private hospitals in Jordan. Data were primarily gathered through self-reported questionnaircs creating by Google Forms which were distributed to a purposive sample of (320) physicians.

M. N. Alolayyan
Health Management and Policy Department, Faculty of Medicine, Jordan University of Science and Technology, Ar-Ramtha, Jordan
e-mail: mnalolayyan@just.edu.jo

R. Z. Q. Almomani
Business Administration, Amman, Jordan

S. H. Al-Qudah · S. I. S. Al-Hawary (✉)
Department of Business Administration, School of Business, Al Al-Bayt University, P.O.BOX 130040, Mafraq 25113, Jordan
e-mail: dr_sliman73@aabu.edu.jo

A. A. S. Mohammad
Marketing Department, Faculty of Administrative and Financial Sciences, Petra University, P.O.Box: 961343, Amman 11196, Jordan

K. M. Al-hawajreh
Business Faculty, Mu'tah University, Karak, Jordan

R. I. Ababneh
Policy, Planning, and Development Program, Department of International Affairs, College of Arts and Sciences, Qatar University, 2713 Doha, Qatar

M. T. Alshurideh
Department of Management, College of Business, University of Sharjah, 27272 Sharjah, United Arab Emirates
e-mail: malshurideh@sharjah.ac.ae; m.alshurideh@ju.edu.jo

Department of Marketing, School of Business, The University of Jordan, Amman 11942, Jordan

A. I. Mohammad
Department of Basic Scientific Sciences, Al-Huson University College, Al-Balqa Applied University, Irbid, Jordan

M. Alshurideh et al. (eds.), *The Effect of Information Technology on Business and Marketing Intelligence Systems*, Studies in Computational Intelligence 1056,
https://doi.org/10.1007/978-3-031-12382-5_54

In total, (285) responses were received including (11) invalid to statistical analysis due to uncompleted or inaccurate. Hence, the final sample contained (274) responses. Structural equation modeling (SEM) was conducted to test hypotheses. The results showed that E-human resource practices had a positive impact on employee satisfaction except e-compensation. They also indicated that the highest impact was for e-development. According to the study's findings, managers and decision-makers should follow the contract's implementation of all employee rights, as well as the employee's commitment to carry out all of the duties entrusted to him, ensure the employee's job stability, and work to ensure that an organization offers advanced technologies based on employee performance evaluation.

Keywords Electronic HR practices · Employee satisfaction · Private Hospitals · Jordan

1 Introduction

In today's information-based economy, an electronic human resources management system is a must to deal with the challenges of human resources in the twenty-first century (Al-Hawary et al., 2020; AlHamad et al., 2022). Web-based human resource management solutions, often known as an electronic human resource management system, are becoming increasingly popular in businesses (Al-Hawary & Al-Namlan, 2018; Swaroop, 2012). As surveys of human resources consultants' show, the number of organizations adopting the electronic human resource management system (E-HRM) and the depth of applications within organizations is constantly increasing, the rapid development of the Internet over the last decade has led to an improvement in the implementation and application of electronic human resource management. Academic interest in electronic human resources management has risen as a result of the publication of numerous issues of journals devoted to human resources and the existence of a group of empirical studies in the field (E-HRM). HRIS is known as E-HRM, according to Stone and Dulebohn, because organizations have enabled online HR transactions. Because it automates human resources tasks and practices and transforms them from traditional and paper-based activities to electronic activities, the rapid development of systems for providing electronic human resources services to employees and managers helps them obtain additional information in a more convenient way so that they can benefit from the information in organizational resources (Alkalha et al., 2012; Alshurideh et al., 2022).

Satisfaction is a term that refers to a mix of psychological, functional, and environmental variables that contribute to an employee's happiness at work (Al-Hawary & Obiadat, 2021; Alhalalmeh et al., 2020; Al-Hawary & Alhajri, 2020; Al-Hawary & Al-Khazaleh, 2020; Alolayyan et al., 2018; Alshurideh et al., 2017; Al-Hawary & Al-Menhaly, 2016; Al-Nady et al., 2016; Al-Hawary & Al-Smeran, 2017; Al-Hawary et al., 2017; Al-Hawary & Harahsheh, 2014; Altarifi et al., 2015; Al-Hawary & Hussien, 2017; Al-Hawary, 2013; Al-Hawary et al., 2013a, b). Furthermore, it is an understanding of the difference between what a person expects from his job and what he actually gets (Al Kurdi et al., 2021). Positive feedback on the extent to which the job satisfies the employee's aims and objectives is included in job satisfaction. Employees typically attempt to attain their goals and fulfill their wishes by giving the appropriate performance, as well as the employee's knowledge of specific changes and conditions that may affect him while at work (Alsuwaidi et al., 2020; Al-bawaia et al., 2022). Organizations are interested in the topic of job satisfaction because a higher degree of job satisfaction correlates with a lower percentage of employee absenteeism, implying that there is a link between job satisfaction and workplace productivity (Alameeri et al., 2020; Kurdi et al., 2020). Employee satisfaction is one of the issues that must be examined from time to time owing to variances in human nature as a result of current improvements in our world, and it is still a research topic (Al Kurdi et al., 2020; AlShehhi et al., 2020). The purpose of the study was to see how electronic human resource management affected employee satisfaction.

2 Theoretical Framework

2.1 Electronic HR Management Practices

Because the human resources department focuses on the human aspect, it is one of the most crucial administrative roles on which any corporation is built (Al-Hawary, 2009; Al-Hawary et al., 2012; Al-Hawary & Mohammed, 2017). Organizational success or failure is determined by the efficiency and effectiveness of human resource management techniques. Human resource management is defined as the department in charge of increasing the effectiveness of human resources in the organization in order to achieve individual and societal goals, and it is thus the department in charge of enabling the organization to build, maintain, and develop its strategic advantages (Al-Hawary & Nusair, 2017).

Human resource management, according to Al-Hawary (2011), is the total of functions within an organization, including hiring the right personnel and managing labors relations, as well as both the strategic and operational perspective of the firm's needs. They are identified procedures and activities that support common institutional goals by combining the demands of the organization and the needs of the individuals who work for it, according to Keshway (2008). Electronic human resources management (E-HRM) is described by Parry and Tyson as the use of technology in human resource management functions and communication via technology directed through networks between the organization and its personnel.

Electronic human resource management is defined by Schramm (2006) as a method of implementing human resource management strategies, processes, and policies in an organization through direct and conscious directed support based on various online technologies. According to AlTaweel and Al-Hawary (2021), it is a technology that allows human resources functions to establish new areas of contribution to company performance. Electronic human resource management strives to provide a comprehensive, thorough, and continuous information system for employees and occupations at a fair cost, as well as support future planning and policy formulation. Automate personnel information and make human resource demand and supply discrepancies easier to track.

Researchers measured a set of techniques for electronic human resource management, and their opinions on the procedures vary. For example, Mohammad et al. (2020) see electronic human resource management practices in (work analysis and design, recruitment, selection, compensation and benefits), whereas) see electronic human resource management practices in (work analysis and design, recruitment, selection, compensation and benefits) (electronic polarization, electronic selection, electronic training, electronic compensation, electronic performance evaluation, and electronic communication). The researcher believes that the practices of electronic human resources management represented in (electronic recruitment and selection, electronic development, electronic performance evaluation, electronic compensation, and electronic communication) can be addressed after reviewing studies related to the practices of electronic human resources management that were addressed by researchers from various cultural destinations.

It may be appropriate for the current study's population. **Electronic recruitment:** E-recruitment, according to Swaroop (2012), is the use of a company's website to attract prospects and receive electronic applications. E-recruitment has a number of advantages, including 24/7 access to online opportunities, the comfort of job examination without the physical stress of an interview, and the ability for applicants to obtain a thorough grasp of the business and its culture before joining it. **Electronic development:** Organizations have resorted to developing training and integrating it with technology in order to effectively trains their employees. As a result, organizations have begun to use a new term called e-training, whose main goal is to improve job performance and the extent of satisfaction that the trainee feels, as well as to create a productive workforce. E-training, according to Al-Hawary and Alajmi (2017), is "skills training using modern computer technologies." **Electronic performance appraisal**: This practice is based on evaluating human resource performance via the Internet, which means that managers can provide performance information directly to the human resources department via electronic forms, reducing the use of paper for supervisor and control, as well as the time and cost of the strike for resource management. Managers can also use human self-service tools to enter performance appraisal management outcomes in real time. **Electronic Compensation:** Electronic compensation is one way to a compensation toolkit that allows a company to gather, store, analyze, and use compensation data and information. The use of a company's website to arrange employee remuneration is defined by Swaroop (2012) as an organization's pay strategy should be able to attract the right kind of personnel, keep them, and ensure that they are treated fairly. **Electronic communication**: The ability to communicate solely through electronic means, removing the need for physical contact and allowing for geographical dispersion of organization members, as well as online collaboration via e-mail, discussion boards, and chats, as well as telephone and fax communications, promises to eliminate time and space constraints (Lee & Kim, 2009).

2.2 Employee Satisfaction

Job satisfaction is one of the most important factors in determining the efficiency and effectiveness of business organizations (Alshurideh, 2022; Hayajneh et al., 2021). Employees are an important part of the process of achieving the company's mission and vision, so they must meet the organization's performance standards to ensure the quality of their work (Alzoubi et al., 2022). We considered a number of negative consequences of job unhappiness, including disloyalty, greater authoritarianism, and a rise in the number of accidents. Employees require a working environment that allows them to operate freely without being hindered from achieving their full potential (Alshraideh et al., 2017; Alzoubi & Inairat, 2020). Job satisfaction is a complicated and varied notion that can mean different things to different people (Aburayya et al., 2020; Alshurideh et al., 2012). For example, what satisfies one employee may not satisfy another. For all employees, there is no single measure of job satisfaction. As a result of exposure to a group of psychological, social, professional, and material elements, it describes the psychological or emotional state that the employee achieves at the degree of certain satisfaction.

Job satisfaction and motivation are linked, although the basis of this correlation is unclear. Job satisfaction, according to Al-Lozi et al. (2017), is described as a feeling of inner loyalty and pride experienced when performing a specific task. Employees' typical feelings and feelings about their work are referred to as job satisfaction (Al-Dmour et al., 2021; Allozi et al., 2022). Positive attitudes toward the workplace indicate job satisfaction, whereas negative attitudes toward the job indicate job discontent (Al-Khayyal et al., 2020; Armstrong, 2006). According to George and Jones (2008), job satisfaction is a set of feelings and beliefs that people have about their current job. Job satisfaction levels can range from extreme satisfaction to complete dissatisfaction. In addition to attitudes about their jobs as a whole, people can have attitudes about different aspects of their careers such as the type of work they do, coworkers, supervisors, or developers, and their pay.

Job satisfaction, according to Hoppock is "any combination of psychological, biological, and environmental variables that leads a person to honestly state that I am content with my job". Job satisfaction is the sum of a worker's positive and negative feelings about their work. Job circumstances were discovered by Metabis and Al-Hawary (2013). It is a significant factor for job satisfaction for workers who work in challenging working conditions; as a result, employees in these situations are dissatisfied with their jobs, and it was required for management to improve working conditions in order to promote employee happiness. As a result, their overall work performance will improve (Alketbi et al., 2020; Sultan et al., 2021).

2.3 E-HR Management Practices and Employee Satisfaction

Employee satisfaction and computerized human resource management approaches have varied correlations, according to the researchers. According to Wadi (2018), there is a favorable link between the benefits of an electronic human resources management (E-HRM) in terms of streamlining work procedures, decreasing the load and efficacy of communications, and transparency between satisfaction and satisfaction. Employees gave feedback on how to utilize the system, the ease with which it could be used, its suitability for work, and the level to which technical capabilities were available. In addition to Al-Shakhanbeh (2015), who referred to the identification of the role of electronic human resources management in the quality of job performance, the study's findings revealed that there is a statistically significant effect of human cadres in the quality of job performance, as well as the presence of a statistically significant effect of human cadres and electronic human resources management software in continuous improvement in Orange Company.

Through the mediating function of employee satisfaction, Khan et al. (2019) investigated the impact of human resource management methods (selection and appointment, training and development, performance evaluation, and compensation) on employee performance in six Pakistani public institutions. And it came up with a set of conclusions, the most important of which are: Human resource practices have a significant impact on faculty members' job performance, with job satisfaction serving as a mediating factor; in addition, human resource management practices have a positive impact on job satisfaction; and job satisfaction has a positive impact on improving faculty members' performance in these universities.

Madanat and Khasawneh (2018) investigated the effectiveness of human resource management practices (manpower planning, selection and appointment, training and development, compensation, and performance evaluation) and their impact on employee satisfaction in 15 commercial and Islamic banks. It came to a set of conclusions, the most important of which is that the banking sector uses human resource management practices at a high level, with the exception of compensation, which is at a medium level, employee satisfaction is at a medium level, and there is a strong and positive statistical relationship between human resource management practices and job satisfaction. Based on above literature review, the study hypotheses may be formulated as:

E-HR management practices have a significant impact on employee satisfaction

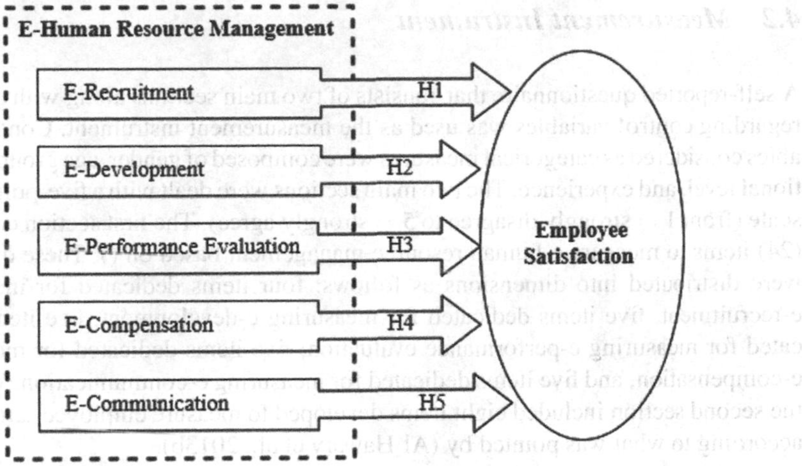

Fig. 1 Research model

3 Study Model

See Fig. 1.

4 Methodology

4.1 Population and Sample Selection

A qualitative method based on a questionnaire was used in this study for data collection and sample selection. The major aim of the study was to examine the impact of e-human resource management on employee satisfaction. Therefore, it focused on the private hospitals in Jordan. Data were primarily gathered through self-reported questionnaires creating by Google Forms which were distributed to a purposive sample of (320) physicians via email. In total, (285) responses were received including (11) invalid to statistical analysis due to uncompleted or inaccurate. Hence, the final sample contained (274) responses suitable to analysis requirements that were formed a response rate of (85.6%), where it proved to be sufficient to the extent that was predictable and allowed for a presumption of data saturation (Sekaran & Bougie, 2016).

4.2 Measurement Instrument

A self-reported questionnaire that consists of two main sections along with a section regarding control variables was used as the measurement instrument. Control variables considered as categorical measures were composed of gender, age group, educational level, and experience. The two main sections were dealt with a five-point Likert scale (from 1 = strongly disagree to 5 = strongly agree). The first section contained (24) items to measure e-human resource management based on (). These questions were distributed into dimensions as follows: four items dedicated for measuring e-recruitment, five items dedicated for measuring e-development, five items dedicated for measuring e-performance evaluation, five items dedicated for measuring e-compensation, and five items dedicated for measuring e-communication. Whereas the second section included eight items developed to measure employee satisfaction according to what was pointed by (Al-Hawary et al., 2013b).

5 Findings

5.1 Measurement Model Evaluation

This study was conducted structural equation modeling (SEM) to test hypotheses, which represents a contemporary statistical technique for testing and estimating the relationship between factors and variables (Wang & Rhemtulla, 2021; Al-Hawary & Al-Syasneh, 2020; Al-Adamat et al., 2020; Al-Gasawneh & Al-Adamat, 2020). Accordingly, the reliability and validity of the constructs were tested using confirmatory factor analysis (CFA) through the statistical program AMOSv24. Table 1 summarizes the results of convergent and discriminant validity, as well the indicators of reliability.

Table 1 shows that the standard loading values for the individual items were within the domain (0.572–0.911), these values greater than the minimum retention of the elements based on their standard loads (Al-Lozi et al., 2018; Sung et al., 2019). Average variance extracted (AVE) is a summary indicator of the convergent validity of constructs that must be above 0.50 (Howard, 2018). The results indicate that the AVE values were greater than 0.50 for all constructs, thus the used measurement model has an appropriate convergent validity. Rimkeviciene et al. (2017) suggested the comparison approach as a way to deal with discriminant validity assessment in covariance-based SEM. This approach is based on comparing the values of maximum shared variance (MSV) with the values of AVE, as well as comparing the values of square root of AVE ($\sqrt{AVE}$) with the correlation between the rest of the structures. The results show that the values of MSV were smaller than the values of AVE, and that the values of $\sqrt{AVE}$ were higher than the correlation values among the rest of the constructs. Therefore, the measurement model used is characterized by discriminative validity. The internal consistency measured through Cronbach's Alpha

Table 1 Results of validity and reliability tests

Constructs	1	2	3	4	5	6
1. e-recruitment (ERC)	**0.757**					
2. e-development (EDV)	0.351	**0.737**				
3. e-performance evaluation (EPE)	0.375	0.355	**0.764**			
4. e-compensation (ECP)	0.418	0.468	0.458	**0.769**		
5. e-communication (ECM)	0.436	0.455	0.528	0.392	**0.774**	
6. employee satisfaction (ESA)	0.628	0.602	0.671	0.652	0.660	**0.771**
VIF	2.544	1.825	1.671	2.237	2.060	–
Loadings range	0.572–0.855	0.694–0.782	0.703–0.821	0.743–0.815	0.675–0.831	0.643–0.911
AVE	0.572	0.543	0.583	0.592	0.599	0.594
MSV	0.458	0.338	0.420	0.497	0.503	0.514
Internal consistency	0.836	0.854	0.871	0.877	0.879	0.918
Composite reliability	0.840	0.856	0.875	0.879	0.881	0.921

coefficient (α) and compound reliability by McDonald's Omega coefficient (ω) was conducted as indicators to evaluate measurement model. The results listed in Table 1 demonstrated that both values of Cronbach's Alpha coefficient and McDonald's Omega coefficient were greater than 0.70, which is the lowest limit for judging on measurement reliability (De Leeuw et al., 2019).

5.2 Structural Model

The structural model illustrated no multicollinearity issue among predictor constructs because variance inflation factor (VIF) values are below the threshold of 5, as shown in Table 1 (Hair et al., 2017). This result is supported by the values of model fit indices shown in Fig. 2.

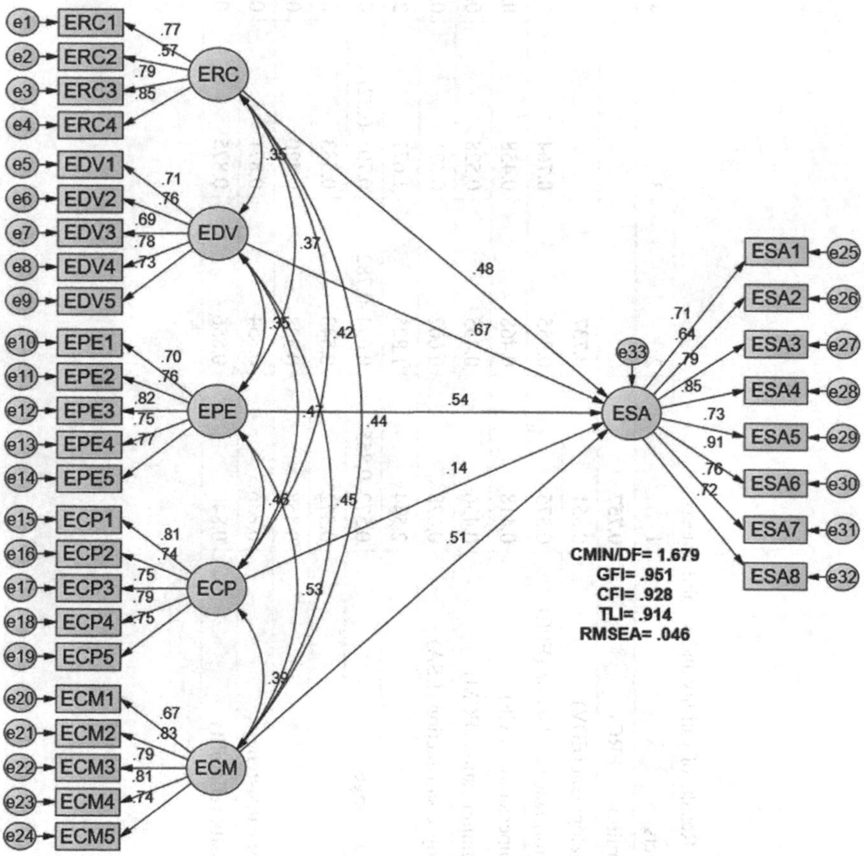

Fig. 2 SEM results of the e-human resource management effect on employee satisfaction

Table 2 Hypothesis testing

Hyp	Relation	Standard beta	t value	p value
H1	e-recruitment→employee satisfaction	0.485*	8.847	0.03
H2	e-development→employee satisfaction	0.671***	14.873	0.000
H3	e-performance evaluation→employee satisfaction	0.542**	10.597	0.002
H4	e-compensation→employee satisfaction	0.137	0.897	0.188
H5	e-communication→employee satisfaction	0.615***	13.042	0.000

Note * $p < 0.05$, ** $p < 0.01$, *** $p < 0.001$

The results in Fig. 2 indicated that the chi-square to degrees of freedom (CMIN/DF) was 1.679, which is less than 3 the upper limit of this indicator. The values of the goodness of fit index (GFI), the comparative fit index (CFI), and the Tucker-Lewis index (TLI) were upper than the minimum accepted threshold of 0.90. Moreover, the result of root mean square error of approximation (RMSEA) indicated to value 0.046, this value is a reasonable error of approximation because it is less than the higher limit of 0.08. Consequently, the structural model used in this study was recognized as a fit model for predicting the employee satisfaction and generalization of its result (Ahmad et al., 2016; Shi et al., 2019). To verify the results of testing the study hypotheses, structural equation modeling (SEM) was used, the results of which are listed in Table 2.

The results demonstrated in Table 2 show that e-human resource management dimensions had a positive impact on employee satisfaction except e-compensation ($\beta = 0.137, t = 0.897, p = 0.188$). They also indicated that the highest impact was for e-development ($\beta = 0.671$, $t = 14.873$, $p = 0.000$), followed by e-communication ($\beta = 0.615$, $t = 13.042$, $p = 0.000$), then e-performance evaluation ($\beta = 0.542$, $t = 10.597$, $p = 0.002$), and finally the lowest impact was for e-recruitment ($\beta = 0.485$, $t = 8.847$, $p = 0.03$).

6 Results Discussion

The study concluded that electronic human resource management practices (selection and appointment, development, compensation and incentives, performance evaluation, communication) have a statistically significant impact on achieving employee satisfaction, which agreed with the findings of Khan et al. (2019), who found an impact of human resource management practices (selection and appointment, training and development, performance evaluation, communication) on achieving employee satisfaction. Employee satisfaction was also found to have a statistically significant effect on the electronic human resources management variable in the study. The reason for this could be that the systems that underpin electronic human resource management, such as electronic performance appraisal, protect the worker's right to be free from manipulation by others. This finding is in line with Atyani and Abu

Salmi's research (2014). In addition, the variables of polarization and computerized selection have a statistically significant effect on employee satisfaction. Electronic systems are utilized in the sorting of job applications in this activity, and advanced applications are sifted and applications that meet the standards and conditions are picked using electronically operated systems, because these systems are so vital in increasing employee satisfaction. Finally, the electronic training and development variable has a statistically significant effect on employee satisfaction. Electronic systems are used to teach new employees the skills they need to accomplish their jobs and to provide them with appropriate training based on their needs being closely monitored. Employee satisfaction is increased as the results of training programs are evaluated utilizing specific technological technologies. This finding is consistent with Ruël et al. (2004), who found that using computerized human resource management in the training process can help finish the training and development process at the lowest feasible cost.

The electronic performance evaluation has a statistically significant effect on employee satisfaction. Employee performance is reviewed using the Internet, and data on employee performance is collected using computerized systems in order to make judgments about promotions or determine training needs. This practice has the most impact compared to others. Because of the transparency that electronic performance appraisal systems provide. The benefit of electronic human resources management, according to Stone et al. (2006), contributes to the dissemination of information about employees and access by managers, the human resources department, and employees themselves, as well as the identification of job opportunities and performance feedback (Al-Dhuhouri et al., 2020; Odeh et al., 2021).

The electronic compensation variable has a statistically significant effect on employee satisfaction. Electronic programs determine the relative and compensating value of occupations, and employees are alerted about incentives electronically, demonstrating transparency among employees and so leading to employee satisfaction. Employee satisfaction is influenced by the electronic communication variable because each employee receives a personalized email from the company, which is intended to keep employees informed about changes to the company's rules and procedures. Employee happiness is a result of working hours that are determined by a prior agreement between them.

7 Recommendations

According to the study's findings, managers and decision-makers should follow the contract's implementation of all employee rights, as well as the employee's commitment to carry out all of the duties entrusted to him, ensure the employee's job stability, and work to ensure that an organization offers advanced technologies based on employee performance evaluation, by making the most significant investment in contemporary technologies imaginable. Finally, electronic compensation mechanisms must be improved. Pay close attention to electronic human resource procedures, and place a strong emphasis on employee satisfaction, since this will help them work better.

References

Aburayya, A., Alshurideh, M., Alawadhi, D., Alfarsi, A., Taryam, M., & Mubarak, S. (2020). An investigation of the effect of lean six sigma practices on healthcare service quality and patient satisfaction: Testing the mediating role of service quality in Dubai primary healthcare sector. *Journal of Advanced Research in Dynamical and Control Systems, 12*(8), 56–72.

Ahmad, S., Zulkurnain, N., & Khairushalimi, F. (2016). Assessing the validity and reliability of a measurement model in Structural Equation Modeling (SEM). *British Journal of Mathematics & Computer Science, 15*(3), 1–8. https://doi.org/10.9734/BJMCS/2016/25183

Al-Adamat, A., Al-Gasawneh, J., & Al-Adamat, O. (2020). The impact of moral intelligence on green purchase intention. *Management Science Letters, 10*(9), 2063–2070.

Alameeri, K., Alshurideh, M., Al Kurdi, B., & Salloum, S. A. (2020, October). The effect of work environment happiness on employee leadership. In *International Conference on Advanced Intelligent Systems and Informatics* (pp. 668–680). Springer.

Al-bawaia, E., Alshurideh, M., Obeidat, B., Masa'deh, R. (2022) The impact of corporate culture and employee motivation on organization effectiveness in Jordanian banking sector. *Academy of Strategic Management Journal, 21*(Special Issue 2), 1–18.

Al-Dhuhouri, F. S., Alshurideh, M., Al Kurdi, B., & Salloum, S. A. (2020, October). Enhancing our understanding of the relationship between leadership, team characteristics, emotional intelligence and their effect on team performance: A critical review. In *International Conference on Advanced Intelligent Systems and Informatics* (pp. 644–655). Springer.

Al-Dmour, R., AlShaar, F., Al-Dmour, H., Masa'deh, R., & Alshurideh, M. T. (2021). The effect of service recovery justices strategies on online customer engagement via the role of "customer satisfaction" during the covid-19 pandemic: An empirical study. *The Effect of Coronavirus Disease (COVID-19) on Business Intelligence, 334*, 325–346.

Al-Gasawneh, J. A., & Al-Adamat, A. M. (2020). The relationship between perceived destination image, social media interaction and travel intentions relating to Neom city. *Academy of Strategic Management Journal, 19*(2), 1–12.

Alhalalmeh, M. I., Almomani, H. M., Altarifi, S., Al- Quran, A. Z., Mohammad, A. A., & Al-Hawary, S. I. (2020). The nexus between corporate social responsibility and organizational performance in Jordan: The mediating role of organizational commitment and organizational citizenship behavior. *Test Engineering and Management, 83*, 6391–6410.

AlHamad, A., Alshurideh, M., Alomari, K., Kurdi, B., Alzoubi, H., Hamouche, S., & Al-Hawary, S. (2022). The effect of electronic human resources management on organizational health of telecommunications companies in Jordan. *International Journal of Data and Network Science, 6*(2), 429–438.

Al-Hawary, S. I. (2009). The effect of the leadership style on the effectiveness of the organization: A field Study at Zarqa Private University. *The Egyptian Journal for Commercial Studies, 33*(1), 361–393.

Al-Hawary, S. I. (2011). Human resource management practices in ZAIN cellular communications company operating in Jordan. *Perspectives of Innovations, Economics and Business, 8*(2), 26–34.

Al-Hawary, S. I. (2013). The role of perceived quality and satisfaction in explaining customer brand loyalty: Mobile phone service in Jordan. *International Journal of Business Innovation and Research, 7*(4), 393–413.

Al-Hawary, S. I., & Alajmi, H. M. (2017). Organizational commitment of the employees of the ports security affairs of the State of Kuwait: The impact of human recourses management practices. *International Journal of Academic Research in Economics and Management Sciences, 6*(1), 52–78.

Al-Hawary, S. I., & Al-Khazaleh A, M. (2020). The mediating role of corporate image on the relationship between corporate social responsibility and customer retention. *Test Engineering and Management, 83*(516), 29976–29993.

Al-Hawary, S. I., & Al-Menhaly, S. (2016). The quality of E-government services and its role on achieving beneficiaries satisfaction. *Global Journal of Management and Business Research: A Administration and Management, 16*(11), 1–11.

Al-Hawary, S. I., & Al-Namlan, A. (2018). Impact of electronic human resources management on the organizational learning at the private hospitals in the State of Qatar. *Global Journal of Management and Business Research: A Administration and Management, 18*(7), 1–11.

Al-Hawary, S. I., & Al-Smeran, W. (2017). Impact of electronic service quality on customers satisfaction of Islamic Banks in Jordan. *International Journal of Academic Research in Accounting, Finance and Management Sciences, 7*(1), 170–188.

Al-Hawary, S. I., & Al-Syasneh, M. S. (2020). Impact of dynamic strategic capabilities on strategic entrepreneurship in presence of outsourcing of five stars hotels in Jordan. *Business: Theory and Practice, 21*(2), 578–587.

Al-Hawary, S. I., & Harahsheh, S. (2014). Factors affecting jordanian consumer loyalty toward cellular phone brand. *International Journal of Economics and Business Research, 7*(3), 349–375.

Al-Hawary, S. I., & Hussien, A. J. (2017). The Impact of electronic banking services on the customers loyalty of commercial banks in Jordan. *International Journal of Academic Research in Accounting, Finance and Management Sciences, 7*(1), 50–63.

Al-Hawary, S. I., & Nusair, W. (2017). Impact of human resource strategies on perceived organizational support at Jordanian Public Universities. *Global Journal of Management and Business Research: A Administration and Management, 17*(1), 68–82.

Al-Hawary, S. I., Al-Awawdeh, W., & Abden, M. A. (2012). The impact of the leadership style on organizational commitment: A field study on Kuwaiti telecommunications companies. *ALEDARI, 130*, 53–102.

Al-Hawary, S. I., Al-Hawajreh, K., Al-Zeaud, H., & Mohammad, A. (2013a). The impact of market orientation strategy on performance of commercial banks in Jordan. *International Journal of Business Information Systems, 14*(3), 261–279.

Al-Hawary, S. I., Al-Qudah, K., Abutayeh, P., Abutayeh, S., & Al-Zyadat, D. (2013b). The impact of internal marketing on employee's job satisfaction of commercial banks in Jordan. *Interdisciplinary Journal of Contemporary Research in Business, 4*(9), 811–826.

Al-Hawary, S. I., Batayneh, A. M., Mohammad, A. A., & Alsarahni, A. H. (2017). Supply chain flexibility aspects and their impact on customers satisfaction of pharmaceutical industry in Jordan. *International Journal of Business Performance and Supply Chain Modelling, 9*(4), 326–343.

Al-Hawary, S. I. S., & Alhajri, T. M. S. (2020). Effect of electronic customer relationship management on customers' electronic satisfaction of communication companies in Kuwait. *Calitatea, 21*(175), 97–102.

Al-Hawary, S. I. S., & Mohammed, A. K. (2017). Impact of team work traits on organizational citizenship behavior from the viewpoint of the employees in the education directorates in North Region of Jordan. *Global Journal of Management and Business, 17*(2-A), 23–40.

Al-Hawary, S. I. S., & Obiadat, A. A. (2021). Does mobile marketing affect customer loyalty in Jordan? *International Journal of Business Excellence, 23*(2), 226–250.

Al-Hawary, S. I. S., Mohammad, A. S., Al-Syasneh, M. S., Qandah, M. S. F., & Alhajri, T. M. S. (2020). Organisational learning capabilities of the commercial banks in Jordan: Do electronic human resources management practices matter? *International Journal of Learning and Intellectual Capital, 17*(3), 242–266.

Alkalha, Z., Al-Zu'bi, Z., Al-Dmour, H., Alshurideh, M., & Masa'deh, R. (2012). Investigating the effects of human resource policies on organizational performance: An empirical study on commercial banks operating in Jordan. *European Journal of Economics, Finance and Administrative Sciences, 51*(1), 44–64.

Alketbi, S., Alshurideh, M., & Al Kurdi, B. (2020). The influence of service quality on customers' retention and loyalty in the UAE hotel sector with respect to the impact of customer' satisfaction, trust, and commitment: A qualitative study. *International Journal of Innovation, Creativity and Change, 14*(7), 734–754.

Al-Khayyal, A., Alshurideh, M., Al Kurdi, B., & Aburayya, A. (2020). The impact of electronic service quality dimensions on customers' E-shopping and E-loyalty via the impact of E-satisfaction and E-trust: A qualitative approach. *International Journal of Innovation, Creativity and Change, 14*(9), 257–281.

Al Kurdi, B., Alshurideh, M., & Al afaishat, T. (2020). Employee retention and organizational performance: Evidence from banking industry. *Management Science Letters, 10*(16), 3981–3990.

Al Kurdi, B., Elrehail, H., Alzoubi, H., Alshurideh, M., & Al-Adaila, R. (2021). The interplay among HRM practices, job satisfaction and intention to leave: An empirical investigation. *Journal of Legal, Ethical and Regulatory, 24*(1), 1–14.

Allozi, A., Alshurideh, M., AlHamad, A., & Al Kurdi, B. (2022). Impact of transformational leadership on the job satisfaction with the moderating role of organizational commitment: Case of UAE and Jordan manufacturing companies. *Academy of Strategic Management Journal, 21*, 1–13.

Al-Lozi, M. S., Almomani, R. Z. Q., & Al-Hawary, S. I. S. (2018). Talent management strategies as a critical success factor for effectiveness of human resources information systems in commercial banks working in Jordan. *Global Journal of Management and Business Research: A Administration and Management, 18*(1), 30–43.

Al-Lozi, M., Almomani, R. Z., & Al-Hawary, S. I. (2017). Impact of talent management on achieving organizational excellence in Arab Potash Company in Jordan. *Global Journal of Management and Business Research: A Administration and Management, 17*(7), 15–25.

Al-Nady, B. A., Al-Hawary, S. I., & Alolayyan, M. (2016). The role of time, communication, and cost management on project management success: An empirical study on sample of construction projects customers in Makkah City, Kingdom of Saudi Arabia. *International Journal of Services and Operations Management, 23*(1), 76–112.

Alolayyan, M., Al-Hawary, S. I., Mohammad, A. A., & Al-Nady, B. A. (2018). Banking service quality provided by commercial banks and customer satisfaction. A structural equation modelling approaches. *International Journal of Productivity and Quality Management*, *24*(4), 543–565.

Al-Shakhanbeh, M.-Q. (2015). *The quality of electronic human resources management systems and their impact on the quality of job performance: A case study on Orange Telecom/Jordan*. Middle East University.

AlShehhi, H., Alshurideh, M., Al Kurdi, B., & Salloum, S. A. (2020, October). The impact of ethical leadership on employees performance: A systematic review. In *International Conference on Advanced Intelligent Systems and Informatics* (pp. 417–426). Springer.

Alshraideh, A. T. R., Al-Lozi, M., & Alshurideh, M. T. (2017). The impact of training strategy on organizational loyalty via the mediating variables of organizational satisfaction and organizational performance: An empirical study on Jordanian agricultural credit corporation staff. *Journal of Social Sciences (COES&RJ-JSS), 6*(2), 383–394.

Alshurideh, M. (2022). Does electronic customer relationship management (E-CRM) affect service quality at private hospitals in Jordan? *Uncertain Supply Chain Management, 10*(2), 1–8.

Alshurideh, M. T., Al Kurdi, B., Alzoubi, H. M., Ghazal, T. M., Said, R. A., AlHamad, A. Q., ... & Al-kassem, A. H. (2022). Fuzzy assisted human resource management for supply chain management issues. *Annals of Operations Research*, 1–19.

Alshurideh, M., Al-Hawary, S. I., Batayneh, A. M., Mohammad, A., & Al-Kurdi, B. (2017). The impact of Islamic Banks' service quality perception on Jordanian customers loyalty. *Journal of Management Research, 9*(2), 139–159.

Alshurideh, M., Masa'deh, R. M. D. T., & Alkurdi, B. (2012). The effect of customer satisfaction upon customer retention in the Jordanian mobile market: An empirical investigation. *European Journal of Economics, Finance and Administrative Sciences, 47*(12), 69–78.

Alsuwaidi, M., Alshurideh, M., Al Kurdi, B., & Salloum, S. A. (2020, October). Performance appraisal on employees' motivation: a comprehensive analysis. In *International Conference on Advanced Intelligent Systems and Informatics* (pp. 681–693). Springer.

Altarifi, S., Al-Hawary, S. I. S., & Al Sakkal, M. E. E. (2015). Determinants of E-shopping and its effect on consumer purchasing decision in Jordan. *International Journal of Business and Social Science, 6*(1), 81–92.

AlTaweel, I. R., & Al-Hawary, S. I. (2021). The mediating role of innovation capability on the relationship between strategic agility and organizational performance. *Sustainability, 13*(14), 7564.

Al-Wahshi, A. H. M. (2020). The impact of human resource management practices on employee satisfaction in public universities: A case study of the United Arab Emirates University.

Alzoubi, H. M., & Inairat, M. (2020). Do perceived service value, quality, price fairness and service recovery shape customer satisfaction and delight? A practical study in the service telecommunication context. *Uncertain Supply Chain Management, 8*(3), 579–588.

Alzoubi, H., Alshurideh, M., Kurdi, B., Akour, I., & Aziz, R. (2022). Does BLE technology contribute towards improving marketing strategies, customers' satisfaction and loyalty? The role of open innovation. *International Journal of Data and Network Science, 6*(2), 449–460.

Armstrong, M. (2006). *A Handbook of Human Resource Management Practice* (10th ed.). Kogan Page Publishing.

Atyani, M., & Abu Salma, A. (2014). the impact of the practice of human resources departments work ethics on achieving employee satisfaction (a field study in cellular communications companies in Jordan). *Studies—Administrative Sciences, 41*(2), 388–401.

De Leeuw, E., Hox, J., Silber, H., Struminskaya, B., & Vis, C. (2019). Development of an international survey attitude scale: Measurement equivalence, reliability, and predictive validity. *Measurement Instruments for the Social Sciences, 1*(1), 9. https://doi.org/10.1186/s42409-019-0012-x

George, J. M., & Jones, G.R. (2008). *Understanding and Managing Organizational Behavior* (5th ed., p. 78). Pearson/Prentice Hall.

Hair, J. F., Babin, B. J., & Krey, N. (2017). Covariance-based structural equation modeling in the journal of advertising: Review and recommendations. *Journal of Advertising, 46*(1), 163–177. https://doi.org/10.1080/00913367.2017.1281777

Hayajneh, N., Suifan, T., Obeidat, B., Abuhashesh, M., Alshurideh, M., & Masa'deh, R. (2021). The relationship between organizational changes and job satisfaction through the mediating role of job stress in the Jordanian telecommunication sector. *Management Science Letters, 11*(1), 315–326.

Howard, M. C. (2018). The convergent validity and nomological net of two methods to measure retroactive influences. *Psychology of Consciousness: Theory, Research, and Practice, 5*(3), 324–337. https://doi.org/10.1037/cns0000149

Keshway, B. (2008). *Human Resources Management* (translator). Dar Al-Farouk Publishing.

Khan, M., Yusoff, R., Hussain, A., & Ismail, F. (2019). The mediating effect of job satisfaction on the relationship of HR practices and employee job performance: Empirical evidence from higher education sector. *International Journal of Organizational Leadership, 8*, 78–94.

Kurdi, B., Alshurideh, M., & Alnaser, A. (2020). The impact of employee satisfaction on customer satisfaction: Theoretical and empirical underpinning. *Management Science Letters, 10*(15), 3561–3570.

Lee, J. J., & Kim, B. B. (2009). *Encyclopedia of Information Science and Technology* (2nd ed., p. 7).

Madanat, H., & Khasawneh, A. (2018). Level of effectiveness of human resources management practices and its impact on employee's satisfaction in the banking sector of Jordan. *Journal of Organizational Culture, Communications and Conflict, 22*(1), 1–19.

Metabis, A., & Al-Hawary, S. I. (2013). The impact of internal marketing practices on services quality of commercial banks in Jordan. *International Journal of Services and Operations Management, 15*(3), 313–337.

Mohammad, A. A., Alshura, M. S., Al-Hawary, S. I. S., Al-Syasneh, M. S., & Alhajri, T. M. (2020). The influence of internal marketing practices on the employees' intention to leave: A study of the private hospitals in Jordan. *International Journal of Advanced Science and Technology, 29*(5), 1174–1189.

Odeh, R. B. M., Obeidat, B. Y., Jaradat, M. O., & Alshurideh, M. T. (2021). The transformational leadership role in achieving organizational resilience through adaptive cultures: the case of Dubai service sector. *International Journal of Productivity and Performance Management*, Vol. ahead-of-print No. ahead-of-print. https://doi.org/10.1108/IJPPM-02-2021-0093.

Rimkeviciene, J., Hawgood, J., O'Gorman, J., & De Leo, D. (2017). Construct validity of the acquired capability for suicide scale: Factor structure, convergent and discriminant validity. *Journal of Psychopathology and Behavioral Assessment, 39*(2), 291–302. https://doi.org/10.1007/s10862-016-9576-4

Ruël, H., Bondarouk, T., & Looise, J. K. (2004). E-HRM: Innovation or irritation: An explorative empirical study in five large companies on webbased HRM. *Management Review, 15*(3), 364–380.

Schramm, J. (2006) HR technology competencies (New Roles for HR Professionals). *HR Magazine, 51*(4), 3.

Sekaran, U., & Bougie, R. (2016). *Research Methods for Business: A Skill-Building Approach* (7th ed.). Wiley.

Shi, D., Lee, T., & Maydeu-Olivares, A. (2019). Understanding the model size effect on SEM fit indices. *Educational and Psychological Measurement, 79*(2), 310–334. https://doi.org/10.1177/0013164418783530

Stone, D. L., Stone-Romero, E. F., & Lukaszewski, K. (2006). Factors affecting the acceptance and effectiveness of electronic human resource systems. *Human Resource Management Review, 16*(2), 229–244.

Sultan, R. A., Alqallaf, A. K., Alzarooni, S. A., Alrahma, N. H., AlAli, M. A., & Alshurideh, M. T. (2021, June). How students influence faculty satisfaction with online courses and do the age of faculty matter. In *The International Conference on Artificial Intelligence and Computer Vision* (pp. 823–837). Springer.

Sung, K.-S., Yi, Y. G., & Shin, H.-I. (2019). Reliability and validity of knee extensor strength measurements using a portable dynamometer anchoring system in a supine position. *BMC Musculoskeletal Disorders, 20*(1), 1–8. https://doi.org/10.1186/s12891-019-2703-0

Swaroop, K. R. (2012). E-HRM and how it will reduce the cost in organization. *Asia Pacific Journal of Marketing & Management Review., 1*(4), 133–139.

Wadi, R. A. L. (2018). The advantages of applying the electronic human resources management system (HRM) (E- in the UNRWA institutions in the Gaza Strip and the extent of employee satisfaction with it), *26*(2).

Wang, Y. A., & Rhemtulla, M. (2021). Power analysis for parameter estimation in structural equation modeling: A discussion and tutorial. *Advances in Methods and Practices in Psychological Science, 4*(1), 1–17. https://doi.org/10.1177/2515245920918253

Employee Empowerment and Intention to Quit: The Mediating Role of Work Engagement: Evidence from the Information Technology Sector in Jordan

Yahia Salim Melhem, Fuad N. Al-Shaikh, Shatha Mamoun Fayez, Sulieman Ibraheem Shelash Al-Hawary, Muhammad Turki Alshurideh, Ayat Mohammad, Kamel Mohammad Al-hawajreh, Anber Abraheem Shlash Mohammad, and Barween H. Al Kurdi

Abstract The objective of this study to understands the role of work engagement as a mediator in the relationship between employee empowerment and intention to quit. A quantitative approach was adopted in order to achieve this study aim: 318 responses were collected via an e-questionnaire from the respondents belonging to IT sector employees in Jordan, one of the most significant Jordanian sectors. The Statistical Analysis Package for Social Science (SPSS) was used for data analysis, and the Sobel test was further used to assess the mediating impact of work engagement on the relationship between employee empowerment and employees' intentions to quit. Findings indicated that that work engagement has a mediating effect between employee empowerment and employees' intention to quit. In light of the

Y. S. Melhem · S. M. Fayez
Business Administration Department, School of Business, Yarmouk University, Irbid, Jordan
e-mail: ymelhem@yu.edu.jo

F. N. Al-Shaikh
Department of Business Administration, Faculty of Economics and Administrative Sciences, Yarmouk University, P.O. Box 566, Irbid 21163, Jordan
e-mail: eco_fshaikh@yu.edu.jo

S. I. S. Al-Hawary (✉)
Department of Business Administration, School of Business, Al al-Bayt University, P.O. Box 130040, Mafraq 25113, Jordan
e-mail: dr_sliman73@aabu.edu.jo

M. T. Alshurideh
Department of Marketing, School of Business, The University of Jordan, Amman 11942, Jordan
e-mail: m.alshurideh@ju.edu.jo

A. Mohammad
Business and Finance Faculty, The World Islamic Science and Education University (WISE), 11947, P.O. Box 1101, Amman, Jordan

K. M. Al-hawajreh
Business Faculty, Mu'tah University, Karak, Jordan

M. Alshurideh et al. (eds.), *The Effect of Information Technology on Business and Marketing Intelligence Systems*, Studies in Computational Intelligence 1056,
https://doi.org/10.1007/978-3-031-12382-5_55

findings of this research, the most important recommendations include that information technology companies should focus on enhancing employee empowerment within organizations so that the employee becomes more affiliated and engaged in the organization in which they work. More future studies on empowerment, intention to quit and work engagement could be conducted in other organizations.

Keywords Work engagement · Intention to quit · Employee empowerment · Information technology · Jordan

1 Introduction

Empowering employees is one of the modern administrative trends in the growth of human resources, which is responsible for enhancing capability and competitiveness in organizations and enhancing quality and sustainability (Alshurideh et al., 2019; Al Khayyal et al., 2020). Empowered employees are in better position to make appropriate choices to solve particular problems on their own (Alhalalmeh et al., 2020; Al-Hawary et al., 2020; Andrew & Sofian, 2012; Mohammad et al., 2020). So, individuals who feel more efficient in their ability to perform the work successfully would be more satisfied in their work (Al-Lozi et al., 2018; Saks, 2006), be more committed and have less desire to leave their job (Al-Hawary & Alajmi, 2017; Shuck et al., 2011)., and show positive performance in their work, more than those individuals who have lower levels of empowerment (Al-Lozi et al., 2017; Meyerson & Kline, 2008).

Work engagement has become a very common concept and a topic of significant concern in management and psychology over the last 20 years. Getting an engaged workforce is a strategic advantage for organizations, as it is strongly associated with desirable organizational outcomes (Han, 2015; Yin, 2018). Organizations with high levels of work engagement increased earnings per share by 3.9 times more than those with low levels of work engagement (Alameeri et al., 2020; Gallup, 2010). Disengaged employees, on the other hand, cost organizations because they are less productive (Alyammahi et al., 2020; Richman, 2006). As a result, having engaged employees is regarded as the fourth most important management challenge facing business organizations (Al-Hawary & Al-Namlan, 2018; Aselstine & Alletson, 2006).

In the modern era, organizations strive to maintain talent and explore potential ways to enhance work engagement with their employees and teams (Al-Hawary,

A. A. S. Mohammad
Marketing Department, Faculty of Administrative and Financial Sciences, Petra University, P.O. Box 961343, Amman 11196, Jordan

B. H. Al Kurdi
Department of Marketing, Faculty of Economics and Administrative Sciences, The Hashemite University, Zarqa, Jordan
e-mail: barween@hu.edu.jo

2015; Al-Hawary & Abu-Laimon, 2013). Hence, workforce stability is a powerful competitive strategy that is expected to become increasingly important in the future, and employees' intention to quit continues to be a topic of interest among management researchers (Al-Hawary et al., 2013; Metabis & Al-Hawary, 2013; Nuseir et al., 2021). Alfes et al. (2013) indicated that engaged employees tend to have less intention to leave their organizations.

"Employees' intention to quit," according to Fugate et al. (2012), is a very costly issue, especially when it is accompanied by cognitive withdrawal. The term "occupational stability" contrasts with the term "intention to quit" and means the survival and continuation of individuals in the same job for a long period (Agha et al., 2021; Whitebook & Sakai, 2003). Studying, in the Jordanian context, the impact of employee empowerment on employees' intention to quit is considered a necessary input to the process of promoting sustainable job opportunities, while reports received from individuals and organizations indicate that the rate of employees' intention to quit has a significant impact on organizational performance (AlShehhi et al., 2020; Kurdi et al., 2020). The rate of job turnover is a challenge to the workflow, given the time and effort needed by human resources employees to maintain employment levels and productivity in organizations (Abu Jadayil, 2011; Al Kurdi et al., 2020). From the review above, it can be concluded that, in spite of the volume of research on employee empowerment, intentions to quit, and work engagement, most such effort is still conceptual at best, confined to a limited, scope or lacks empirical testing (Alsuwaidi et al., 2020; Al Shebli et al., 2021; Al-bawaia et al., 2022). Hence, this research study aims to explore the impact of employee empowerment on intentions to quit, and to identify the role of work engagement as a mediator, in the rather unexplored context and.

2 Literature Review and Hypotheses Development

2.1 Employee Empowerment

Employee empowerment, according to Selvi and Maheswari (2020), is a process in which an organization's employees are given authority, power, responsibility, resources and liberty to make decisions and complete job responsibilities. It also grants employees control and allows for the distribution of responsibilities and control at all levels (Ammari et al., 2017). There is no doubt that attention to the concept of empowering employees constitutes an essential and decisive element for organizations in Jordan, especially in light of the trend towards adopting and applying modern administrative concepts, such as total quality management, re-engineering management processes and comprehensive performance planning (Alshurideh et al., 2015, 2020) Empowerment leads the individual to a sense of belonging to the organization, which in turn leaves positive effects on the employee's psyche and sense of importance and appreciation in the organization. Empowerment is seen as a way to involve employees in the decision-making process, enrich their skills and experience, and establish a sense of their importance at work (Al-Hawary & Al-Syasneh, 2020; Moye & Henkin, 2006).

Employee empowerment contributes to changing the individual's perception and attitude towards their organization or their job (Ghaniyoun et al., 2017). Empowerment has been considered a vital subject in management recently. There is common support to grant employees adequate scope of freedom in their work meaning and responsibilities to enable them to apply this scope towards achieving the overall aims of the organization (Al-Dhuhouri et al., 2020; AlTaweel & Al-Hawary, 2021).

Due to the many dimensions of empowerment, and because of a dynamic process in a dynamic environment, and many elements involved in various steps of this process, empowerment is very complex to define (Al Mehrez et al., 2020; Al-Hawary & Alwan, 2016; Robbins et al., 2002). Conger and Kanungo (1988) defined empowerment as the motivating principle of self-efficacy, According to Khalili et al. (2016), empowerment is an administrative approach in which managers and other organizing members participate in influencing decision-making (i.e., decision-making cooperation). This definition is in line with that of Baird and Wang (2010, p. 577), who defined employee empowerment as "the delegation of power and responsibility from higher levels in the organizational hierarchy to lower level employees, particularly decision-making power."

Previous studies present many dimensions that can be used to measure empowerment within organizations. To achieve the purpose of our study, competence, impact, self-determination and meaning were used as dimensions to measure employee empowerment. These four perceptions together reflect a positive attitude towards work, and from that, empowerment is a motivating component that supports individuals and qualifies them to possess independent capabilities that enable them to make decisions commensurate with the goals of the organization (Al Suwaidi et al., 2020; Alaali et al., 2021). These dimensions are; **Meaning**, is related to the individual's sense of the importance of his/her work on a personal level and within the work environment (Zhang & Bartol, 2010). **Competence**, it is the individual's ability to perform tasks successfully and this refers to self-efficacy. **Self-Determination,** it has also been described as a feeling of control, freedom of choice, and self-determination to initiate action (Salman, 2015). **Impact**, it represents the degree to which the individual sees that his/her behaviour at work makes a difference in results (Zhang & Bartol, 2010).

2.2 *Intention to Quit*

Individual employees' plans to leave their jobs can be referred to as intentions to quit (Al-Hawary & Aldaihani, 2016; Al-Hawary & Nusair, 2017; Al-Hawary & Mohammed, 2017; Williams & Hazer, 1986); such intentions to quit can be either voluntary or involuntary. The biggest predictor of actual turnover is employees' intention to quit (Joseph et al., 2007) and the study of Williams and Hazer (1986) indicates that there is a strong correlation between the employee's intention to quit work and the actual behaviour of quitting. We conclude from the foregoing that employees' intention to quit is defined as the individual's unwillingness to continue working, and decision to think carefully and seriously about leaving the job and trying to search for new job opportunities. intention to quit is one of the phenomena that have a high cost to the organization, and it crystallizes in two directions; the first is the decision to leave the work, which incurs large losses to the organization, and the second is the loss of morale among workers who decide to stay in the organization in the presence of these intentions (Odeh et al., 2021; Lee et al., 2022a, b). Based on the above, the study of intention to quit becomes one of the constants that the organization must study in order to avoid many material and moral costs.

2.3 *Work Engagement*

Successful organizations consider engagement to be important, and it is one of the hot topics highlighted in human resource management insights today. Some researchers refer to it as employee engagement, while others refer to it as work engagement. Even so, there is no difference between the two terms when it comes to explaining employee engagement. Employee engagement is a solution that managers can recommend when dealing with employees' motivation issues (Oehler & Adair, 2019; Alshurideh, 2019, 2022). Engagement was interpreted by Kwon and Park (2019) as an individual's mental, emotional, and cognitive condition that leads to the achievement of organizational goals. Hence, in occupational health psychology, work engagement is one of the most common outcomes (Lesener et al., 2020). Therefore, the issue of work engagement has become a topic that has captured the interest of many researchers and practitioners in the field of management and organizational behaviour, as it is a basic and decisive element in the success of business organizations (Garg & Singh, 2019).

Recent research showed high levels of disengagement from work, a situation that continues to cause problems for leaders around the world. Massive employee disengagement has recently been reported (Hewitt, 2013). According to Baron (2013), employee engagement is beneficial; engaged employees work harder, perform better, and provide better customer service, believe in the values of the organization, search and explore opportunities to improve organizational performance, can be relied upon and perform job requirements to the fullest, all of which led to improvement in the bottom line. Work engagement is described as involvement, commitment, enthusiasm, optimism, and energy, which stem from each employee's attitudes and behaviours in carrying out his or her responsibilities (Na-Nan et al., 2020).

Most researchers have dealt with work engagement as a three-dimensional construction, comprising: first, **vigour**, characterized by high levels of energy, mental resilience at work, investment in the performance of the tasks of the job, and perseverance in the face of difficulties; second, **dedication,** including the ability to solve work problems and awareness of work events and challenges (Chirkowska-Smolak,

2012); third, **absorption**, which refers to being fully focused, taking pleasure in being preoccupied with work and feeling the passage of time quickly (Cole et al., 2012).

2.4 Work Engagement as a Mediator

Work engagement has long been a source of interest for managers and researchers, who have discovered that it has a strong link to a variety of positive consequences, e.g., job satisfaction, financial returns, better productivity, lower intention to quit and employee innovation and creativity (Anderson, 2017; Polo-Vargas et al., 2019). Organizations depend heavily on the abilities and skills of their employees in a rapidly changing business environment. Where the competitive business environment is very high, engaged employees are essential for successful organizations that want to remain competitive; this requires individuals with dedication, absorption, energies and ambition; employees who are engaged have a plethora of "resources" to invest in their jobs (Karatepe et al., 2020). They are passionate about their job, completely immersed in their work activities, and tenacious when faced with obstacles. Work engagement is a crucial indicator of job effectiveness and organizational performance, according to several reports that distilled the average impact observed in hundreds of studies. Furthermore, recent research has shown that employee engagement contributes to key organizational outcomes, such as creativity and innovation, customer satisfaction, improved financial performance, and decreased turnover rate (Bakker & Leiter, 2017).

Work engagement and psychological empowerment are critical in coping with work-related deviations and improving organizational effectiveness (Ugwu et al., 2014). Stander and Rothmann (2010) analyzed the relationship between psychological empowerment, work engagement and job insecurity among employees working in a government and manufacturing organization. The findings show that job security, psychological empowerment, and work engagement all have significant connections between them. On the other hand, the study also revealed four dimensions of psychological empowerment (self-determination, competence, meaning, impact) that predict engagement (vigour, dedication and absorption) of the employee.

In the information technology (IT) sector, organizations get to be more powerful and stronger by increasing the employees' experience. Turnover, on the other hand, poses a threat to successful organizations since it results in the loss of skilled employees. It is also an additional expense for the organizations. As a result, the managers attempt to keep turnover to a minimum (Alshurideh, 2019; Alshurideh et al., 2012). However, the turnover intention must be handled in order to keep turnover under control. This is a difficult undertaking because there are so many factors that affect turnover intentions (Özkan, 2021). Based on the above literature review, the study hypotheses may be formulated as:

Work engagement of IT companies in Jordan significantly mediates the relationship between employee empowerment and employees' intention to quit.

3 Research Model

The systematic treatment of the research problem in light of its theoretical framework and field implications requires designing a hypothetical scheme that indicates the logical relationship between the research variables. First, the model assumes that a four-variable construct, consisting of meaning, impact, competence, and self-determination, represents employee empowerment. This model further asserts a directional relationship between the empowerment of workers and this construct. Second, the model assumes that work engagement is a compound of Vigour, dedication, and absorption of three variable constructs. In addition, this model states a directional correlation between work engagement and this construct. Third, the model assumes that (Employee empowerment) and work engagement have a directional relationship. In Fig. 1, the hypothesized ten variables are described in this model.

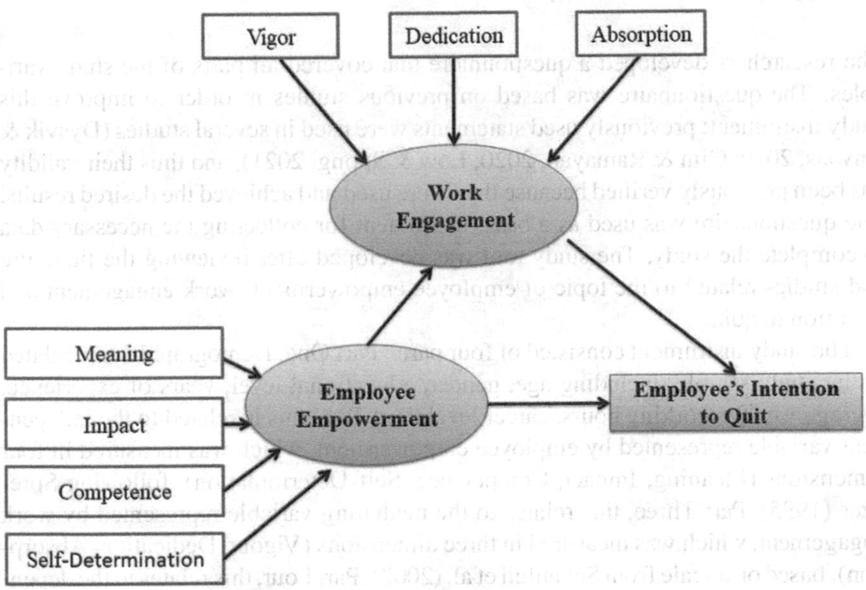

Fig. 1 Research model

4 Methodology

4.1 Population and Sample

The population in this study consisted of all IT employees working in IT companies that participated in this study, in Amman and the North of Jordan, with a total number of 1568. For the study sample, convenience sampling was used, which is a sort of non-probability sampling. Convenience sampling is a way of collecting research data from a conveniently available number of respondents who are easily accessible. An e-questionnaire was distributed, to which the response rate was 20%, with 318 questionnaires returned, while 312 questionnaires were valid for analysis.

4.2 Questionnaire Design

The researchers developed a questionnaire that covered all parts of the study variables. The questionnaire was based on previous studies in order to improve this study instrument; previously used statements were used in several studies (Dysvik & Kuvaas, 2010; Gim & Ramayah, 2020; Low & Spong, 2021), and thus their validity has been previously verified because they were used and achieved the desired results. The questionnaire was used as a basic instrument for collecting the necessary data to complete the study. The study tool was developed after reviewing the literature and studies related to the topic of employee empowerment, work engagement and intention to quit.

The study instrument consisted of four parts: Part One, Demographic data related to the study sample, including age, gender, educational level, years of experience, average weekly working hours, career level. Part Two, this is related to the independent variable represented by employee empowerment, which was measured in four dimensions (Meaning, Impact, Competence, Self-Determination), following Spreitzer (1995). Part Three, this relates to the mediating variable represented by work engagement, which was measured in three dimensions (Vigour, Dedication, Absorption), based on a scale from Schaufeli et al. (2002). Part Four, this relates to the dependent variable represented by the intention to quit (ITQ), which was measured based on a scale from Begley and Czajka (1993). A Likert scale method was employed from 1 (Strongly agree) to 5 (Strongly disagree) to measure all variables, except demographic variables.

Table 1 Cronbach's α values

Factors	Cronbach's alpha	Number of items
Employee empowerment	0.904	18
Competence	0.669	4
Impact	0.640	3
Meaning	0.715	4
Self determination	0.761	7
Work engagement	0.949	13
Absorption	0.818	4
Dedication	0.912	4
Vigour	0.879	5
Employees' intention to quit	0.839	9

4.3 Validity and Reliability

Content validity

The researcher also tested the validity of the questionnaire in order to identify the links to the questions and their relevance to the variables of the study, and to ensure the consistency of the answers of the sample members, in achieving the objectives of the study and answering its questions. The questionnaire was also reviewed by academic experts in the field from Yarmouk University; it took two weeks to complete the validity phase. After completing the face and content validity phase with academics, the researcher took corrections based on the recommendations of these scholars. The questionnaire after the content validity check. The questionnaire was then adjusted and revised according to their comments.

Instrument reliability

The reliability of the study tool was verified using the Cronbach's alpha reliability coefficient. The recommended reliability coefficient exceeds 0.70 according to Sekaran and Bougie (2016), while other researchers accept 0.60 as a cut-off point (Sekaran & Bougie, 2013; Al-Adamat et al., 2020; Al-Gasawneh & Al-Adamat, 2020). Cronbach's alpha was employed in this study to determine the research tool's reliability, as well as the consistency of the items with each other and with all items in general. This is illustrated in Table 1.

5 Analysis

The researcher employed the Sobel Test to examine whether work engagement mediates the relation between empowerment and intention to quit. Figure 2 indicates that

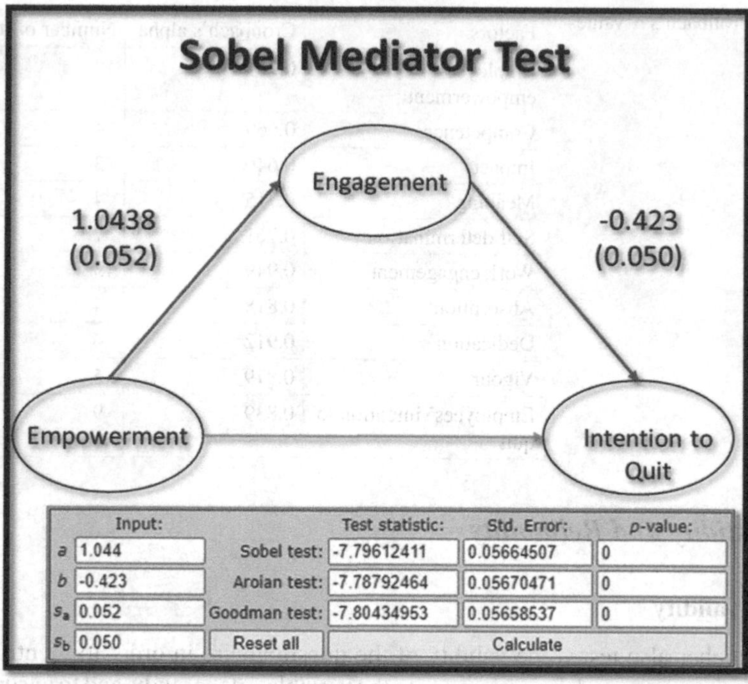

Fig. 2 Sobel mediator test

the indirect effect of empowerment on employees' intention to quit after work engagement is included in the model estimate is: Test statistic −7.79612411 and Std. Error 0.05664507; this result is significant, since the *p*-value is 0.000, which is less than 0.05. Thus, we accept the hypothesis that work engagement has a significant impact on mediating the relationship between empowerment and intention to quit at $\alpha \leq 0.05$.

6 Discussion

The purpose of this study is to determine the mediating role of work engagement on the relationship between the independent variable and the dependent variable. The information technology sector contributes significantly to national economies and employment (Thomas & Lucas, 2019). The researcher concludes that all participants have responded positively to all dimensions of the employee empowerment scale. This can be explained by highlighting the employees' awareness of the importance of their work and their confidence in performing their job tasks efficiently, as they possess all the skills and capabilities that qualify them to do so. Nevertheless, they do not feel a high level of self-determination in the job they do in their surrounding environment. The researcher employed the Sobel Test to examine whether work engagement mediates the relation between empowerment and intention to quit. The study results showed the indirect effect of empowerment on employees' intention to quit after work engagement is included in the model estimate is Test statistic −7.79612411 and Std. Error 0.05664507; this result is significant, since the *p*-value is 0.000, which is less than 0.05. To test the hypothesis that engagement mediates the relationship between empowerment and intention to quit, the Sobel test was used. The results show this hypothesis is validated and work engagement has a mediating effect between employee empowerment and intention to quit according to the Sobel test.

7 Recommendations and Direction for Future Research

In light of the results, the researcher makes some recommendations that she believes may contribute to enhancing and improving employee empowerment and work engagement in organizations in general and in information technology companies in particular: it is required from the Information technology companies to focus on enhancing the employee empowerment within organizations so that the employee becomes more affiliated and engaged in the organization in which he/she works, through working to enhance the dimensions of empowerment among employees, because of the positive impact of empowerment on improving work engagement, and to notifying the employee of the importance of the work they do and providing incentives and material and moral rewards to them in appreciation of their efforts, because of the importance of finding meaning in one's work and its positive impact on improving work engagement. Finally, to focus on developing the competence of employees by designing training programs that increase their skills and ability to do work, because of the importance of the dimension of competence and its positive impact on work engagement.

The current study examined the relationship between empowerment and the intention to quit work through the mediating variable of work engagement, and thus paves the way towards conducting more studies on empowerment and its relationship with other variables, such as creativity, employee well-being and work stress. And examining the relationship between empowerment and the intention to quit in other organizational contexts, such as the industrial sector and the health sector. Future studies can examine the relationship between work engagement and other variables, such as leadership, good citizenship behaviours, performance, and organizational justice. The researcher recommends conducting studies on different sectors for the same dimensions of this study and using different quantitative and qualitative methodologies to come up with results that can be generalized with greater confidence.

References

AbuJadayil, W. (2011). Main factors causing workers turnover in Jordan industrial sector. *Jordan Journal of Mechanical and Industrial Engineering, 5*(2), 161–166.

Agha, K., Alzoubi, H. M., & Alshurideh, M. T. (2021, June). Measuring reliability and validity instruments of technologically driven cognitive intrusion towards work-life balance. In *The International Conference on Artificial Intelligence and Computer Vision* (pp. 601–614). Cham: Springer.

Al Khayyal, A. O., Alshurideh, M., Al Kurdi, B., & Salloum, S. A. (2020, October). Women empowerment in UAE: A systematic review. In *International Conference on Advanced Intelligent Systems and Informatics* (pp. 742–755). Cham: Springer.

Al Suwaidi, F., Alshurideh, M., Al Kurdi, B., & Salloum, S. A. (2020, October). The impact of innovation management in SMEs performance: A systematic review. In *International Conference on Advanced Intelligent Systems and Informatics* (pp. 720–730). Cham: Springer.

Al Mehrez, A. A., Alshurideh, M., Al Kurdi, B., & Salloum, S. A. (2020, October). Internal factors affect knowledge management and firm performance: A systematic review. In *International Conference on Advanced Intelligent Systems and Informatics* (pp. 632–643). Cham: Springer.

Al Shebli, K., Said, R. A., Taleb, N., Ghazal, T. M., Alshurideh, M. T., & Alzoubi, H. M. (2021, June). RTA's employees' perceptions toward the efficiency of artificial intelligence and big data utilization in providing smart services to the residents of Dubai. In *The International Conference on Artificial Intelligence and Computer Vision* (pp. 573–585). Cham: Springer.

Al Kurdi, B., Alshurideh, M., & Al Afaishat, T. (2020). Employee retention and organizational performance: Evidence from banking industry. *Management Science Letters, 10*(16), 3981–3990.

Alaali, N., Al Marzouqi, A., Albaqaeen, A., Dahabreh, F., Alshurideh, M., Alrwashdh, S., Iyadeh, I., Salloum, S., & Aburayya, A. (2021). The impact of adopting corporate governance strategic performance in the tourism sector: A case study in the Kingdom of Bahrain. *Journal of Legal, Ethical and Regulatory, 24*(Special Issue 1), 1–18.

Al-Adamat, A., Al-Gasawneh, J., & Al-Adamat, O. (2020). The impact of moral intelligence on green purchase intention. *Management Science Letters, 10*(9), 2063–2070.

Alameeri, K., Alshurideh, M., Al Kurdi, B., & Salloum, S. A. (2020, October). The effect of work environment happiness on employee leadership. In *International Conference on Advanced Intelligent Systems and Informatics* (pp. 668–680). Cham: Springer.

Al-bawaia, E., Alshurideh, M., Obeidat, B., & Masa'deh, R. (2022). The impact of corporate culture and employee motivation on organization effectiveness in Jordanian banking sector. *Academy of Strategic Management Journal, 21*(Special Issue 2), 1–18.

Al-Dhuhouri, F. S., Alshurideh, M., Al Kurdi, B., & Salloum, S. A. (2020, October). Enhancing our understanding of the relationship between leadership, team characteristics, emotional intelligence and their effect on team performance: A critical review. In *International Conference on Advanced Intelligent Systems and Informatics* (pp. 644–655). Cham: Springer.

Alfes, K., Shantz, A., Truss, C., & Soane, E. (2013). The link between perceived human resource management practices, engagement and employee behavior: A moderated mediation model. *The International Journal of Human Resource Management, 24*(2), 330–351.

Al-Gasawneh, J. A., & Al-Adamat, A. M. (2020). The relationship between perceived destination image, social media interaction and travel intentions relating to Neom city. *Academy of Strategic Management Journal, 19*(2), 1–12.

Alhalalmeh, M. I., Almomani, H. M., Altarifi, S., Al- Quran, A. Z., Mohammad, A. A., & Al-Hawary, S. I. (2020). The nexus between corporate social responsibility and organizational performance in Jordan: The mediating role of organizational commitment and organizational citizenship behavior. *Test Engineering and Management, 83*, 6391–6410.

Al-Hawary, S. I. S., & Mohammed, A. K. (2017). Impact of team work traits on organizational citizenship behavior from the viewpoint of the employees in the education directorates in north region of Jordan. *Global Journal of Management and Business, 17*(2-A), 23–40.

Al-Hawary, S. I., & Al-Syasneh, M. S. (2020). Impact of dynamic strategic capabilities on strategic entrepreneurship in presence of outsourcing of five stars hotels in Jordan. *Business: Theory and Practice, 21*(2), 578–587.

Al-Hawary, S. I. (2015). Human resource management practices as a success factor of knowledge management implementation at health care sector in Jordan. *International Journal of Business and Social Science, 6*(11/1), 83–98.

Al-Hawary, S. I., & Abu-Laimon, A. A. (2013). The impact of TQM practices on service quality in cellular communication companies in Jordan. *International Journal of Productivity and Quality Management, 11*(4), 446–474.

Al-Hawary, S. I., & Alajmi, H. M. (2017). Organizational commitment of the employees of the ports security affairs of the state of Kuwait: The impact of human recourses management practices. *International Journal of Academic Research in Economics and Management Sciences, 6*(1), 52–78.

Al-Hawary, S. I., & Aldaihani, F. M. (2016). Customer relationship management and innovation capabilities of Kuwait Airways. *International Journal of Academic Research in Economics and Management Sciences, 5*(4), 201–226.

Al-Hawary, S. I., & Al-Namlan, A. (2018). Impact of electronic human resources management on the organizational learning at the private hospitals in the state of Qatar. *Global Journal of Management and Business Research: A Administration and Management, 18*(7), 1–11.

Al-Hawary, S. I., Al-Qudah, K., Abutayeh, P., Abutayeh, S., & Al-Zyadat, D. (2013). The impact of internal marketing on employee's job satisfaction of commercial banks in Jordan. *Interdisciplinary Journal of Contemporary Research in Business, 4*(9), 811–826.

Al-Hawary, S. I. S., & Alwan, A. M. (2016). Knowledge management and its effect on strategic decisions of Jordanian Public Universities. *Journal of Accounting-Business & Management, 23*(2), 24–44.

Al-Hawary, S. I. S., Mohammad, A. S., Al-Syasneh, M. S., Qandah, M. S. F., & Alhajri, T. M. S. (2020). Organisational learning capabilities of the commercial banks in Jordan: Do electronic human resources management practices matter? *International Journal of Learning and Intellectual Capital, 17*(3), 242–266.

Al-Hawary, S. I., & Nusair, W. (2017). Impact of human resource strategies on perceived organizational support at Jordanian Public Universities. *Global Journal of Management and Business Research: A Administration and Management, 17*(1), 68–82.

Al-Lozi, M., Almomani, R. Z., & Al-Hawary, S. I. (2017). Impact of talent management on achieving organizational excellence in Arab Potash Company in Jordan. *Global Journal of Management and Business Research: A Administration and Management, 17*(7), 15–25.

Al-Lozi, M., Almomani, R. Z., & Al-Hawary, S. I. (2018). Talent management strategies as a critical success factor for effectiveness of human resources information systems in commercial banks working in Jordan. *Global Journal of Management and Business Research: A Administration and Management, 18*(1), 30–43.

AlShehhi, H., Alshurideh, M., Al Kurdi, B., & Salloum, S. A. (2020, October). The impact of ethical leadership on employees performance: A systematic review. In *International Conference on Advanced Intelligent Systems and Informatics* (pp. 417–426). Cham: Springer.

Alshurideh, M., Salloum, S. A., Al Kurdi, B., & Al-Emran, M. (2019, February). Factors affecting the social networks acceptance: an empirical study using PLS-SEM approach. In *Proceedings of the 2019 8th International Conference on Software and Computer Applications* (pp. 414–418).

Alshurideh, M. (2019). Do electronic loyalty programs still drive customer choice and repeat purchase behaviour? *International Journal of Electronic Customer Relationship Management, 12*(1), 40–57.

Alshurideh, M. (2022). Does electronic customer relationship management (E-CRM) affect service quality at private hospitals in Jordan? *Uncertain Supply Chain Management, 10*(2), 325–332.

Alshurideh, M., Alhadid, A. Y., & Barween, A. (2015). The effect of internal marketing on organizational citizenship behavior an applicable study on the University of Jordan employees. *International Journal of Marketing Studies, 7*(1), 138–145.

Alshurideh, M., Gasaymeh, A., Ahmed, G., Alzoubi, H., & Kurd, B. (2020). Loyalty program effectiveness: Theoretical reviews and practical proofs. *Uncertain Supply Chain Management, 8*(3), 599–612.

Alshurideh, M., Nicholson, M., & Xiao, S. (2012). The effect of previous experience on mobile subscribers' repeat purchase behaviour. *European Journal of Social Sciences, 30*(3), 366–376.

Alsuwaidi, M., Alshurideh, M., Al Kurdi, B., & Salloum, S. A. (2020, October). Performance appraisal on employees' motivation: a comprehensive analysis. In *International Conference on Advanced Intelligent Systems and Informatics* (pp. 681–693). Cham: Springer.

AlTaweel, I. R., & Al-Hawary, S. I. (2021). The mediating role of innovation capability on the relationship between strategic agility and organizational performance. *Sustainability, 13*(14), 7564.

Alyammahi, A., Alshurideh, M., Al Kurdi, B., & Salloum, S. A. (2020, October). The impacts of communication ethics on workplace decision making and productivity. In *International Conference on Advanced Intelligent Systems and Informatics* (pp. 488–500). Cham: Springer.

Ammari, G., Alkurdi, B., Alshurideh, A., & Alrowwad, A. (2017). Investigating the impact of communication satisfaction on organizational commitment: A practical approach to increase employees' loyalty. *International Journal of Marketing Studies, 9*(2), 113–133.

Anderson, L. A. (2017). Employee engagement as a key strategy for change. Retrieved September 5, 2021, from https://blog.beingfirst.com/employee-engagement-as-a-key-strategy-for-change

Andrew, O., & Sofian, S. (2012). Individual factors and work outcomes of employee engagement. *Procedia-Social and Behavioral Sciences, 40*, 498–508.

Aselstine, K., & Alletson, K. (2006). A new deal for the 21st century workplace. *Ivey Business Journal, 70*(4), 1–9.

Baird, K., & Wang, H. (2010). Employee empowerment: Extent of adoption and influential factors. *Personnel Review*.

Bakker, A. B., & Leiter, M. (2017). Strategic and proactive approaches to work engagement. *Organizational Dynamics, 46*(2), 67–75.

Baron, A. (2013). What do engagement measures really mean? *Strategic HR Review*.

Begley, T. M., & Czajka, J. M. (1993). Panel analysis of the moderating effects of commitment on job satisfaction, intent to quit, and health following organizational change. *Journal of Applied Psychology, 78*(4), 552.

Chirkowska-Smolak, T. (2012). Does work engagement burn out? The person-job fit and levels of burnout and engagement in work. *Polish Psychological Bulletin, 43*(2), 76–85.

Cole, M. S., Walter, F., Bedeian, A. G., & O'Boyle, E. H. (2012). Job burnout and employee engagement: A meta-analytic examination of construct proliferation. *Journal of Management, 38*(5), 1550–1581.

Conger, J. A., & Kanungo, R. N. (1988). The empowerment process: Integrating theory and practice. *Academy of Management Review, 13*(3), 471–482.

Dysvik, A., & Kuvaas, B. (2010). Exploring the relative and combined influence of mastery-approach goals and work intrinsic motivation on employee turnover intention. *Personnel Review, 39*(5), 622–638.

Fugate, M., Prussia, G. E., & Kinicki, A. J. (2012). Managing employee withdrawal during organizational change: The role of threat appraisal. *Journal of Management, 38*(3), 890–914.

Gallup, Inc. (2010). Employee engagement: What's your engagement ratio? Washington, DC.

Garg, N., & Singh, P. (2019). Work engagement as a mediator between subjective well-being and work-and-health outcomes. *Management Research Review, 43*(6), 735–752.

Ghaniyoun, A., Shakeri, K., & Heidari, M. (2017). The association of psychological empowerment and job burnout in operational staff of Tehran emergency center. *Indian Journal of Critical Care Medicine: Peer-Reviewed, Official Publication of Indian Society of Critical Care Medicine, 21*(9), 563–567.

Gim, G. C., & Ramayah, T. (2020). Predicting turnover intention among auditors: Is WIPL a mediator? *The Service Industries Journal, 40*(9–10), 726–752.

Han, Y. (2015). *A study on employee engagement program in full service hotel*. A professional paper submitted in Partial Fulfillment of the Requirements for The Master of Science Hotel Administration. University of Nevada, Las Vegas.

Hewitt, A. (2013). Trends in Global Employee Engagement.

Joseph, D., Ng, K. Y., Koh, C. & Ang, S. (2007). Turnover of information technology professionals: A narrative review, meta-analytic structural equation modeling, and model development. *MIS Quarterly, 31*(3), 547–577.

Karatepe, O. M., Rezapouraghdam, H., & Hassannia, R. (2020). Job insecurity, work engagement and their effects on hotel employees' non-green and nonattendance behaviors. *International Journal of Hospitality Management, 87*, 102472.

Khalili, H., Sameti, A., & Sheybani, H. (2016). A study on the effect of empowerment on customer orientation of employees. *Global Business Review, 17*(1), 38–50.

Kurdi, B., Alshurideh, M., & Alnaser, A. (2020). The impact of employee satisfaction on customer satisfaction: Theoretical and empirical underpinning. *Management Science Letters, 10*(15), 3561–3570.

Kwon, K., & Park, J. (2019). The life cycle of employee engagement theory in HRD research. *Advances in Developing Human Resources, 21*(3), 352–370.

Lee, K., Azmi, N., Hanaysha, J., Alshurideh, M., & Alzoubi, H. (2022a). The effect of digital supply chain on organizational performance: An empirical study in Malaysia manufacturing industry. *Uncertain Supply Chain Management, 10*(2), 1–16.

Lee, K., Ramiz, P., Hanaysha, J., Alzoubi, H., & Alshurideh, M. (2022b). Investigating the impact of benefits and challenges of IOT adoption on supply chain performance and organizational performance: An empirical study in Malaysia. *Uncertain Supply Chain Management, 10*(2), 1–14.

Lesener, T., Gusy, B., Jochmann, A., & Wolter, C. (2020). The drivers of work engagement: A meta analytic review of longitudinal evidence. *Work and Stress, 34*(3), 259–278.

Low, M. P., & Spong, H. (2021). Predicting employee engagement with micro-level corporate social responsibility (CSR) practices in the public accounting firms. *Social Responsibility Journal*.

Metabis, A., & Al-Hawary, S. I. (2013). The impact of internal marketing practices on services quality of commercial banks in Jordan. *International Journal of Services and Operations Management, 15*(3), 313–337.

Meyerson, S. L., & Kline, T. J. B. (2008). Psychological and environmental empowerment: Antecedents and consequences. *Leadership and Organization Development Journal, 29*(5), 444–460.

Mohammad, A. A., Alshura, M. S., Al-Hawary, S. I. S., Al-Syasneh, M. S., & Alhajri, T. M. (2020). The influence of internal marketing practices on the employees' intention to leave: A study of the private hospitals in Jordan. *International Journal of Advanced Science and Technology, 29*(5), 1174–1189.

Moye, M. J., & Henkin, A. B. (2006). Exploring associations between employee empowerment and interpersonal trust in managers. *Journal of management development*.

Na-Nan, K., Pukkeeree, P., & Chaiprasit, K. (2020). Employee engagement in small and medium-sized enterprises in Thailand: The construction and validation of a scale to measure employees. *International Journal of Quality & Reliability Management, 37*(9/10), 1325–1343.

Nuseir, M. T., Al Kurdi, B. H., Alshurideh, M. T., & Alzoubi, H. M. (2021, June). Gender discrimination at workplace: Do artificial intelligence (AI) and machine learning (ML) have opinions about it. In *The International Conference on Artificial Intelligence and Computer Vision* (pp. 301–316). Cham: Springer.

Odeh, R. B. M., Obeidat, B. Y., Jaradat, M. O., & Alshurideh, M. T. (2021). The transformational leadership role in achieving organizational resilience through adaptive cultures: The case of Dubai service sector. *International Journal of Productivity and Performance Management*. https://doi.org/10.1108/IJPPM-02-2021-0093

Oehler, K., & Adair, C. (2019). *Trends in Global Employee Engagement*. AON Hewitt, Illinois.

Özkan, A. H. (2021). A meta-analysis of the variables related to turnover intention among IT personnel. *Kybernetes, 50*(3), 1584–1600.

Polo-Vargas, J. D., Fernandez-Ríos, M., Bargsted, M., Ramírez-Vielma, R., Mebarak, M., Zambrano-Curcio, M., et al. (2019). Relationships between work design, engagement, and life satisfaction. *Psicología Desde El Caribe, 35*(4), 98–108. https://doi.org/10.14482/psdc.35.4.158.74

Richman, A. (2006). Everyone wants an engaged workforce. How can you create it? *Workspan, 49*(1), 36–39.

Robbins, T. L., Crino, M. D., & Fredendall, L. D. (2002). An integrative model of the empowerment process. *Human Resource Management Review, 12*(3), 419–443.

Saks, A. M. (2006). Antecedents and consequences of employee engagement. *Journal of Managerial Psychology, 21*(7), 600–619.

Salman, H. M. (2015). The impact of human resources' psychological empowerment on organizational commitment in banking sector local commercial banks in Gaza Strip. Master of Business Administration, The Islamic University of Gaza.

Schaufeli, W. B., Salanova, M., González-Romá, V., & Bakker, A. B. (2002). The measurement of engagement and burnout: A Two sample confirmatory factor analytic approach. *Journal of Happiness Studies, 3*, 71–92.

Sekaran, U., & Bougie, R. (2013). Research methods for business: A skill-building approach (6th ed.). West Sussex, UK: John Wiley & Sons Ltd.

Sekaran, U., & Bougie, R. (2016). *Research methods for business: A skill-building approach* (7th ed.). Wiley.

Selvi, M. S., & Maheswari, G. S. (2020). Effects of Employee empowerment on organizational success. *Journal of Xi'an University of Architecture & Technology, 12*(3), 2018–2025.

Shuck, B., Reio, T. G., Jr., & Rocco, T. S. (2011). Employee engagement: An examination of antecedent and outcome variables. *Human Resource Development International, 14*(4), 427–445.

Spreitzer, G. M. (1995). Psychological empowerment in the workplace: Dimensions, measurement, and validation. *Academy of Management Journal, 38*(5), 1442–1465.

Stander, M. W., & Rothmann, S. (2010). Psychological empowerment, job insecurity and employee engagement. *SA Journal of Industrial Psychology, 36*(1), 1–8.

Thomas, B., & Lucas, K. (2019). Development and validation of the workplace dignity scale. *Group and Organization Management, 44*(1), 72–111.

Ugwu, F. O., Onyishi, I. E., & Rodríguez-Sánchez, A. M. (2014). Linking organizational trust with employee engagement: The role of psychological empowerment. *Personnel Review, 43*(3), 377–400.

Whitebook, M., & Sakai, L. (2003). Turnover begets turnover: An examination of job and occupational instability among child care center staff. *Early Childhood Research Quarterly, 18*(3), 273–293.

Williams, L. J., & Hazer, J. T. (1986). Antecedents and consequences of satisfaction and commitment in turnover models: A reanalysis using latent variable structural equation methods. *Journal of Applied Psychology, 71*(2), 219–231.

Yin, N. (2018). The influencing outcomes of job engagement: An interpretation from the social exchange theory. *International Journal of Productivity and Performance Management, 67*(5), 873–889.

Zhang, X., & Bartol, K. (2010). Linking empowering leadership and employee creativity: The influence of psychological empowerment, intrinsic motivation, and creative process engagement. *Academy of Management Journal*.

Impact of Manufacturing Flexibility on Response to Customer Requirements of Manufacturing Companies in King Abdullah II Ibn Al Hussein Industrial City in Jordan

Ziad Mohd Ali Smadi, Eyass Ahmad AL-Qaisi, Main Naser Alolayyan, Ali Zakariya Al-Quran, Abdullah Matar Al-Adamat, Anber Abraheem Shlash Mohammad, Muhammad Turki Alshurideh, Sulieman Ibraheem Shelash Al-Hawary, and D. Barween Al Kurdi

Abstract This study aims to examine the impact of manufacturing flexibility on the response to customer requirements which is the dependent variable. This study was applied on 210 random samples of 458 manufacturing companies in King Abdullah II Ibn Al Hussein Industrial City. The researcher used the questionnaire tool to collect the primary data from the units of observation. The data was analyzed by using SPSS program. The study findings showed the impact of manufacturing flexibility

Z. M. A. Smadi · E. A. AL-Qaisi · A. Z. Al-Quran · S. I. S. Al-Hawary (✉)
Department of Business Administration, School of Business, Al al-Bayt University, P.O. Box 130040, Mafraq 25113, Jordan
e-mail: dr_sliman73@aabu.edu.jo

A. Z. Al-Quran
e-mail: ali.z.al-quran@aabu.edu.jo

M. N. Alolayyan
Health Management and Policy Department, Faculty of Medicine, Jordan University of Science and Technology, Ar-Ramtha, Jordan
e-mail: mnalolayyan@just.edu.jo

A. M. Al-Adamat
School of Business, Department of Business Administration & Public Administration, Al al-Bayt University, P.O. Box 130040, Mafraq 25113, Jordan
e-mail: aaladamat@aabu.edu.jo

A. A. S. Mohammad
Marketing Department, Faculty of Administrative and Financial Sciences, Petra University, P.O. Box 961343, Amman 11196, Jordan

M. T. Alshurideh
Department of Marketing, School of Business, The University of Jordan, Amman 11942, Jordan
e-mail: m.alshurideh@ju.edu.jo; malshurideh@sharjah.ac.ae

Department of Management, College of Business, University of Sharjah, 27272 Sharjah, United Arab Emirates

M. Alshurideh et al. (eds.), *The Effect of Information Technology on Business and Marketing Intelligence Systems*, Studies in Computational Intelligence 1056,
https://doi.org/10.1007/978-3-031-12382-5_56

on response to customer requirements. The study recommends increasing of job rotations of employees among company's activities to amend labor flexibility.

Keywords Manufacturing flexibility · Response to customer requirements · King Abdullah II Ibn Al Hussein · Industrial city · Jordan

1 Introduction

In light of the rapid development that the world is witnessing in all life aspects and the increasing in the products and services diversity that did not exist in the past, the pressure on business organizations has increased in their race to meet this diversity as much as possible in a way that reflects positively on the business organization and preserves it from decline (Al-Hawary & Al-Syasneh, 2020; Al-Hawary & Obiadat, 2021; Al-Nady et al., 2013; AlTaweel & Al-Hawary, 2021). Therefore, many business organizations are making rapid and continuous changes in order to respond to developing or changing products and diversifying them to adapt to the world new situation and to enhance their competitive position, this leads organizations to activating a competitive weapon that depends on their ability to deal with variables, which is manufacturing flexibility (Al-Nady et al., 2016; Allahow et al., 2018; Al Kurdi et al., 2020a, b).

The most important thing for business organizations today is their ability to meet customer's requirements, and one of the means to achieve this is by providing high product diversity in the least amount of time while maintaining or improving quality and reducing costs (Al-Hawary, 2013; Al-Hawary & Abu-Laimon, 2013; Al-Hawary & Al-Menhaly, 2016; Al-Hawary & Al-Smeran, 2017; Al-Hawary & Harahsheh, 2014; Al-Hawary & Hussien, 2017; Al-Hawary et al., 2013, 2017; Alolayyan et al., 2018; Alshurideh et al., 2017; Metabis & Al-Hawary, 2013). Therefore, manufacturing flexibility represented by (machines flexibility, product mix flexibility, new product flexibility, labor flexibility, operations flexibility, process flexibility) is a weapon with which organizations face changes in customer requirements. As for Jordan, which has adopted the economic openness option, Jordanian business organizations today find themselves forced to keep pace with the development they have made in order to continue in the market and maintain their market position.

The study importance comes from the importance of responding to customers' requirements by industrial companies because of their essential role in developing, improving and preserving the company from failure if it cannot meet these requirements in a timely manner (Ahmed et al., 2020; Ahmad et al., 2021a, b; Odeh et al., 2021). In addition to manufacturing flexibility importance in meeting the change in products according to circumstances, times and places, especially with the increase

D. B. Al Kurdi
Department of Marketing, Faculty of Economics and Administrative Sciences, The Hashemite University, Zarqa, Jordan
e-mail: barween@hu.edu.jo

of requirements changes in recent times, the most recent of which was during the Corona Virus (COVID-19) crisis, which impact has become clear on all life aspects, including industrial companies (Alameeri et al., 2021; Nuseir et al., 2021; Taryam et al., 2020). It should be noted here that some companies responded quickly to the changes accompanying this crisis, which reflected positively on their position and market share (Alshurideh et al., 2021a, b, c; Khasawneh et al., 2021a, b; Leo et al., 2021).

The application of this study to the companies of King Abdullah II Ibn Al-Hussein Industrial City is of special importance, as it is considered the oldest industrial city in Jordan, as it was built in 1984 and is considered the largest industrial gathering in the Kingdom, where companies number reached 458 industrial companies, according to the Industrial Cities Company Jordanian website, this reflects its influential role in developing the Jordan economy. It was found by the study literature reviewing that there is a lack of studies that dealt with the relationship between manufacturing flexibility and its variables and its impact on the speed, flexibility and reliability of responding to customer requirements. And after the spread of Corona virus globally and the subsequent changes in consumer behavior and the sudden change in the humans' requirements in general, this is a strong test of the extent to which the manufacturing flexibility in companies affects the speed response to the sudden change in demands.

Responding to customers requirements and needs is a familiar slogan for many companies and one of the most important things that companies do by following them in order to continue their development and success in their work (Al-Hawary & Alhajri, 2020; Altarifi et al., 2015; Mohammad et al., 2020; Williamson, 1991). Manufacturing flexibility is becoming an increasingly important ability to design and operate manufacturing systems, as these systems operate in highly variable and unpredictable environments (Ghannajeh et al., 2015; Al-Dhuhouri et al., 2020; Alsuwaidi et al., 2020). When looking at previous studies on the issue of manufacturing flexibility, we find that most of them dealt with the impact of manufacturing flexibility on production, cost or performance, as in the study of Alsada and Allawi (2017) and (Kumar et al., 2017), and did not link manufacturing flexibility with responding to customers' requirements (Alkitbi et al., 2020; Alshamsi et al., 2020; Sweiss et al., 2021). This study came to find the effect of manufacturing flexibility with responding to customer's requirements.

2 Literature Review and Hypotheses Development

2.1 Manufacturing Flexibility

Continuous developments in manufacturing systems technology have made operations more capable, efficient and flexible to meet changing requirements and this has been particularly evident in the automotive industry. Operations are not longer limited to the production of a single product but can continually adapt to a range of products and demands. Flexibility of the production process has become an inevitable necessity imposed by the competition that characterizes the contemporary business environment, as a prerequisite for achieving business strategic goals of survival and growth. In light of the modern technology use, this aims to reduce unused energy by using automation, as the use of flexibility in manufacturing works to exploit idle energy. As for manufacturing flexibility concept, it was defined by Zhang et al. (2003) as the organization's ability to manage production resources and cases of meeting different customer demands uncertainty.

Daoud and Jawad (2016), defined manufacturing flexibility as the organization's ability to respond or adapt to the environment changes, and it can enhance the company's competitive performance by providing a variety of products that meet customers different tastes quickly while maintaining the production system performance. Ojstersek and Buchmeister (2020) defined it as the ability of a manufacturing system to respond effectively and efficiently to uncertain surrounding changes (manufacturing system and global demand). Accordingly, manufacturing flexibility can be defined as the organization's ability to interact with external or internal changes that occur suddenly or can be predicted, through the use of available resources quickly and using the least effort and expenses possible to meet customers demand (Al Mehrez et al., 2020; Alshurideh, 2019, 2022; Alzoubi et al., 2022; Tariq et al., 2022a, b).

Several researchers dealt with different manufacturing flexibility dimensions, where Gregory et al. (2019) took eight dimensions of manufacturing flexibility, while Daoud and Jawad (2016) addressed four dimensions of manufacturing flexibility, and Upton (1994) identified fifteen dimensions of manufacturing flexibility, which are: Production path flexibility, product flexibility, mix flexibility, activity flexibility, status flexibility, scale flexibility, program flexibility, long-run and short-run flexibility, stretch flexibility, machine flexibility, labor flexibility, design change flexibility, operational flexibility, and process flexibility. As for Benjaafar and Ramakrishman (1996), they differentiate between the different types of flexibility as they are divided into product flexibility and process flexibility; the first refers to a variety of factory options for a particular product. While process flexibility, is defined as one of the characteristics of an industrial process to work under various dynamic operating conditions. The researcher has chosen six dimensions of manufacturing flexibility from the dimensions identified by (Upton), which are the most appropriate to the researched companies' nature, and they are as follows: **Machine Flexibility**: Machine flexibility is one of the most fundamental and important dimensions of manufacturing flexibility that many researchers have discussed. It has been defined as the ability to make changes to the machine system with minimal interference in production (Gregory et al., 2019). **Labor Flexibility** which is the ability to multitask a labor, operator, or human resource within the manufacturing system without affecting efficiency (Kumar et al., 2017). **New Product Flexibility**: which is the company's ability to bring new parts and products to market in response to a changing business environment (Alamro, 2014). **Product Mix Flexibility**: This means that the mix of items delivered to the market can be changed quickly while maintaining cost-effectiveness, which achieved when product and process flexibility achieved together.

Process Flexibility: which is the ability to change the steps required to complete a required task, allowing the system to complete many tasks (Gregory et al., 2019). Process flexibility in a manufacturing system is related to the range of parts types that the system can produce without major preparation processes. **Operation Flexibility**: is a property of the part, and it means that the part can be produced with alternative process plans, where the process plan means a series of operations required to produce the part.

2.2 *Responding to Customer Requirements*

A customer is that person who buys something, whether goods or services, and is thus a customer of a company that sells products for making a profit purpose. As for customer requirements according to the researcher, is the characteristics or specifications that must be available in the product or service in order to receive approval from customers (Alshurideh et al., 2012; Alzoubi et al., 2020; Al-Dmour et al., 2021b; Abuhashesh et al., 2021). It is divided into two types: Service requirements: These are the intangible aspects of the purchased product that the customer expects to occur, such as delivery on time, good dealing with the seller when purchasing, ease of payment, and so on (Al-Dmour et al., 2021a; Alshurideh et al., 2021a, b, c). Final good requirements: which are the tangible characteristics, features, or specifications that the consumer expects to find in the product or service.

As for responding to customer requirements, it is often seen as a flexible response, but the response must be characterized by reliability and speed. Responsiveness is defined as encompassing the full range of values related to timely product development and delivery, as well as reliable scheduling and flexible performance (Heizer et al., 2017). Flexible responsiveness may be seen as the ability to adapt to changes in a market where design innovations and volumes fluctuate wildly. Heizer et al. (2017) defined response speed as speed in product development, speed in production, and speed in delivery. speed response to the customers' requirements and needs can be raised by reducing the time to deliver products to the market, reducing the time of the product manufacturing cycle, reducing the customer cycle time, and reducing the time of operations conversion or change.

2.3 *Manufacturing Flexibility and Responding to Customer Requirements*

Many managers see flexibility not as the end but as a means to many ends. As managers view flexibility as a tool for improving manufacturing as it is an essential component of the production process, it must be developed until the best production is reached. Daoud and Jawad (2016) studied the impact of manufacturing flexibility in achieving competitive advantage and found that the level of manufacturing flexibility positively affects the competitive advantage, and the company's abilities to achieve competitive advantage represented by increasing interest and developing manufacturing flexibility. Zhang et al. (2003) explain Strong, positive, and direct relationships between flexible manufacturing efficiency and volume flexibility, and flexible manufacturing efficiency and mix elasticity. Size Flexibility and product mix flexibility have strong, positive, and direct relationships with customer satisfaction. As customers want a quick response and a variety of updated products and competitors achieve higher performance levels than were considered possible a few years ago, manufacturing flexibility, above all other measures of manufacturing performance, is cited as a solution. More flexibility in manufacturing processes means more ability to move with customer needs, respond to competitive pressures and move closer to the market (Slack, 2005). Accordingly, the research hypothesis can be formulated as follows:

There is a statistically significant impact of manufacturing flexibility on responding to customers' requirements in industrial companies in King Abdullah II Ibn Al Hussein Industrial City.

3 Study Model

See Fig. 1.

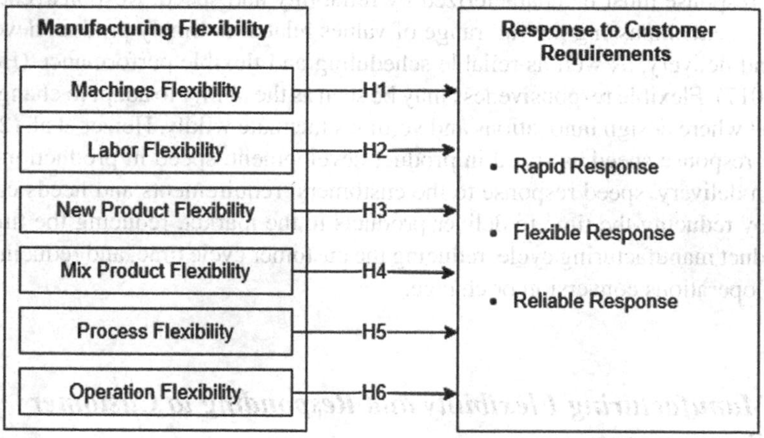

Fig. 1 Research model

4 Methodology

4.1 Population and Sample Selection

A qualitative method based on a questionnaire was used in this study for data collection and sample selection. The major aim of the study was to examine the impact of manufacturing flexibility on response to customer requirements. Therefore, it focused on manufacturing companies in King Abdullah II Ibn Al Hussein industrial city in Amman. Data were primarily gathered through self-reported questionnaires creating by Google Forms which were distributed to a random sample of (250) respondents via email. In total, (210) responses were received including (22) invalid to statistical analysis due to uncompleted or inaccurate. Hence, the final sample contained (188) responses suitable to analysis requirements that were formed a response rate of (89.5%), where it proved to be sufficient to the extent that was predictable and allowed for a presumption of data saturation (Sekaran & Bougie, 2016).

4.2 Measurement Instrument

A self-reported questionnaire that consists of two main sections along with a section regarding control variables was used as the measurement instrument. Control variables considered as categorical measures were composed of gender, age group, educational level, and experience. The two main sections were dealt with a five-point Likert scale (from 1 = strongly disagree to 5 = strongly agree). The first section contained (31) items to measure manufacturing flexibility based on (Upton, 1994). These questions were distributed into dimensions as follows: five items dedicated for measuring machines flexibility, six items dedicated for measuring labor flexibility, five items dedicated for measuring new product flexibility, six items dedicated for measuring mix product flexibility, four items dedicated for measuring process flexibility, and five items dedicated for measuring operation flexibility. Whereas the second section included (12) items developed to measure response to customer requirements according to what was pointed by (Al-Hawary & Hussien, 2017). This variable was a second-order construct divided into three first-order constructs. Rapid response measured by four items, flexible response measured through four items, and reliable response measured using four items.

5 Findings

5.1 Measurement Model Evaluation

This study was conducted structural equation modeling (SEM) to test hypotheses, which represents a contemporary statistical technique for testing and estimating the relationship between factors and variables (Wang & Rhemtulla, 2021). Accordingly, the reliability and validity of the constructs were tested using confirmatory factor analysis (CFA) through the statistical program AMOSv24. Table 1 summarizes the results of convergent and discriminant validity, as well the indicators of reliability.

Table 1 shows that the standard loading values for the individual items were within the domain (0.664–0.8871), these values greater than the minimum retention of the elements based on their standard loads (Al-Lozi et al., 2018; Sung et al., 2019). Average variance extracted (AVE) is a summary indicator of the convergent validity of constructs that must be above 0.50 (Howard, 2018). The results indicate that the AVE values were greater than 0.50 for all constructs, thus the used measurement model has an appropriate convergent validity. Rimkeviciene et al. (2017) suggested the comparison approach as a way to deal with discriminant validity assessment in covariance-based SEM. This approach is based on comparing the values of maximum shared variance (MSV) with the values of AVE, as well as comparing the values of square root of AVE ($\sqrt{}$AVE) with the correlation between the rest of the structures. The results show that the values of MSV were smaller than the values of AVE, and that the values of $\sqrt{}$AVE were higher than the correlation values among the rest of the constructs. Therefore, the measurement model used is characterized by discriminative validity. The internal consistency measured through Cronbach's Alpha coefficient (α) and compound reliability by McDonald's Omega coefficient (ω) was conducted as indicators to evaluate measurement model. The results listed in Table 1 demonstrated that both values of Cronbach's Alpha coefficient and McDonald's Omega coefficient were greater than 0.70, which is the lowest limit for judging on measurement reliability (De Leeuw et al., 2019).

5.2 Structural Model

The structural model illustrated no multicollinearity issue among predictor constructs because variance inflation factor (VIF) values are below the threshold of 5, as shown in Table 1 (Hair et al., 2017). This result is supported by the values of model fit indices shown in Fig. 1.

The results in Fig. 2 indicated that the chi-square to degrees of freedom (CMIN/DF) was 2.716, which is less than 3 the upper limit of this indicator. The values of the goodness of fit index (GFI), the comparative fit index (CFI), and the Tucker–Lewis index (TLI) were upper than the minimum accepted threshold of 0.90. Moreover, the result of root mean square error of approximation (RMSEA) indicated

Table 1 Results of validity and reliability tests

Constructs	1	2	3	4	5	6	7	8	9
1. MAF	**0.749**								
2. LAF	0.312	**0.774**							
3. NPF	0.297	0.318	**0.759**						
4. MPF	0.336	0.361	0.310	**0.761**					
5. PRF	0.374	0.338	0.375	0.394	**0.768**				
6. OPF	0.402	0.394	0.413	0.305	0.341	**0.757**			
7. RAR	0.512	0.487	0.526	0.425	0.624	0.533	**0.765**		
8. FLR	0.556	0.602	0.597	0.554	0.571	0.509	0.625	**0.775**	
9. RER	0.571	0.597	0.555	0.498	0.592	0.579	0.614	0.622	**0.767**
VIF	2.014	1.664	1.974	1.535	2.487	2.645	–	–	–
Loadings range	0.681–0.801	0.731–0.825	0.694–0.849	0.674–0.871	0.715–0.855	0.715–0.834	0.664–0.825	0.731–0.854	0.715–0.831
AVE	0.561	0.599	0.576	0.579	0.590	0.573	0.586	0.601	0.588
MSV	0.418	0.513	0.467	0.435	0.406	0.502	0.458	0.511	0.429
I.C	0.862	0.897	0.869	0.890	0.848	0.867	0.846	0.855	0.849
C.R	0.864	0.900	0.871	0.891	0.851	0.870	0.849	0.857	0.851

Note MAF: machines flexibility, LAF: labor flexibility, NPF: new product flexibility, MPF: mix product flexibility, PRF: process flexibility, OPF: operation flexibility, RAR: rapid response, FLR: flexible response, RER: reliable response, I.C: internal consistency, C.R: composite reliability

to value 0.056, this value is a reasonable error of approximation because it is less than the higher limit of 0.08. Consequently, the structural model used in this study was recognized as a fit model for predicting the response to customer requirements and generalization of its result (Ahmad et al., 2016; Shi et al., 2019). To verify the results of testing the study hypotheses, structural equation modeling (SEM) was used, the results of which are listed in Table 2.

The results demonstrated in Table 2 show that manufacturing flexibility dimensions, which were new product flexibility ($\beta = 0.368, t = 4.305, p = 0.000$), machines flexibility ($\beta = 0.202$, $t = 2.469$, $p = 0.014$), and operation flexibility ($\beta = 0.177$, $t = 2.146$, $p = 0.033$) had impact on response to customer requirements. Despite, the results indicated that labor flexibility ($\beta = 0.088, t = 1.240, p = 0.216$), mix product flexibility ($\beta = 0.047$, $t = 0.530$, $p = 0.597$), and process flexibility ($\beta = 0.006$, $t = 0.064$, $p = 0.949$) had no impact on response to customer requirements.

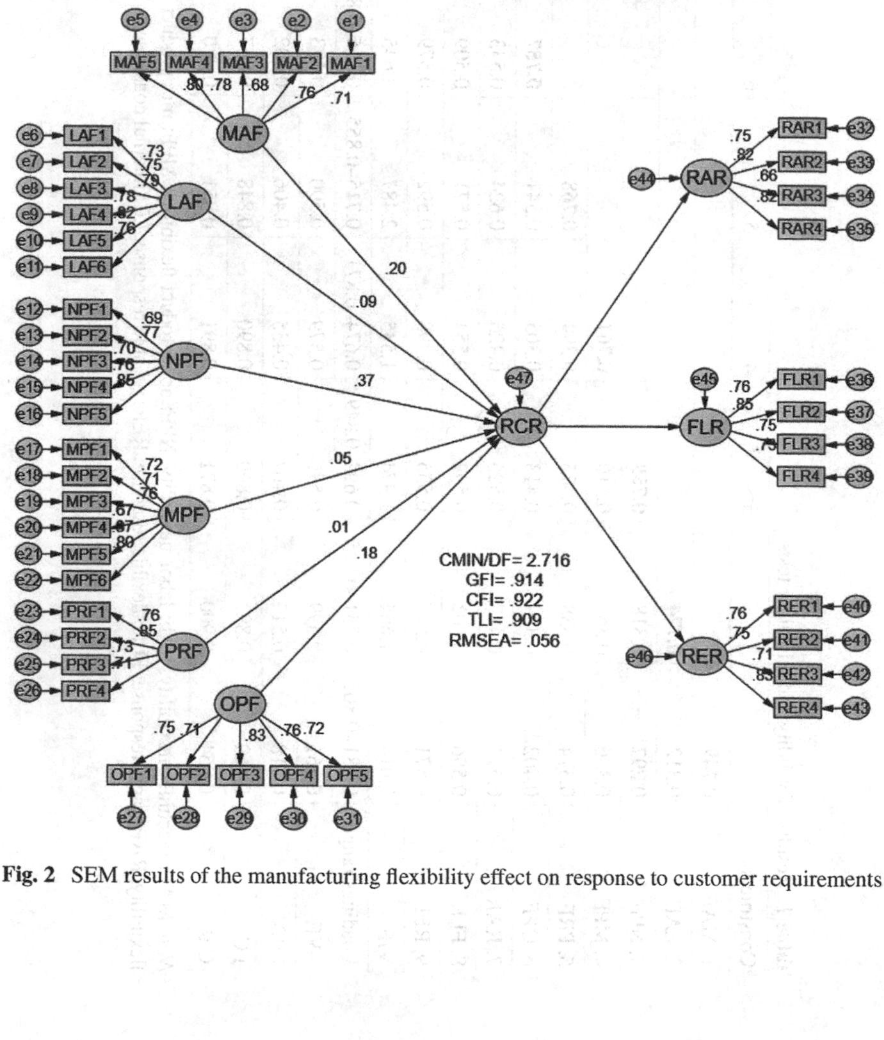

Fig. 2 SEM results of the manufacturing flexibility effect on response to customer requirements

Table 2 Hypothesis testing

Hypothesis	Relation	Standard beta	*t* value	*p* value
H1	MAF → RCR	0.202	2.469*	0.014
H2	LAF → RCR	0.088	1.240	0.216
H3	NPF → RCR	0.368	4.305***	0.000
H4	MPF → RCR	0.047	0.530	0.597
H5	PRF → RCR	0.006	0.064	0.949
H6	OPF → RCR	0.177	2.146*	0.033

Note MAF: machines flexibility, LAF: labor flexibility, NPF: new product flexibility, MPF: mix product flexibility, PRF: process flexibility, OPF: operation flexibility, RCR: response to customer requirements, * $p < 0.05$, ** $p < 0.01$, *** $p < 0.001$

6 Discussion

The study results showed that the nature of the used machines in the researched companies is characterized by ease of preparation and configuration as a result of the extensive use of advanced technology in manufacturing, which in turn helps in the goods production in appropriate quantities that correspond to the customers' demand quantity. Also, the machines used in manufacturing can be operated in certain mechanisms, which contribute to increasing production, while reducing the time needed to modify the process on the machine, and this indicates those machines flexibility and their ability to produce well in addition to saving worker time and effort.

Study results revealed the multiplicity of skills and experiences possessed by workers in King Abdullah II Ibn Al Hussein Industrial City, their experience levels, and their obtaining the appropriate academic qualifications that qualify them to perform various tasks at low costs and within the company's capabilities. The company's various needs also require its production lines to move employees from one site to another according to the company's production need, production type, or required expertise type, to remain and compete in the market, and to provide employees with diverse skills and experiences. This is in line with Bernardes and Hanna (2008) study which showed that systems allow change within predetermined parameters and response usually refers to system behavior that involves purposeful change in time.

Study results show the companies' desire to go in line with the market demand, and to cover that demand in light of the necessary machinery and equipment availability, as well as the participation of their employees and their concerted efforts in order to reach good production levels, and diversification in products to achieve success in introducing different products types to the market, and satisfy consumers' desires and needs as well. Results also indicated the companies' desire to maintain production and competition with products and the desire to improve performance, and this result could also be due to the availability of many experiences and workers scientific qualifications in those companies, which can determine the ability to perform many operations on the same production line, as well as the desire through the workers in these companies to interact and achieve higher production levels, and all these factors help in increasing the production levels.

The study results also show the workers ability to maintain the production processes level, and this is also due to the nature of the machines and equipment used, and their accuracy in work and production, and also for drawing plans in production processes an important role in maintaining the products' types, and maintaining the cost as well. The study also revealed the companies' desire for competition and continuity in the markets, and to work on the development and modernization of products issued by industrial companies operating in King Abdullah II Ibn Al Hussein Industrial City. Also, companies must take into account all customers' requirements in their main objectives. The reason for the high response level among companies may be attributed to raising the competition level they have with other companies, and raising the trust level between companies and customers, which is beneficial to companies' sustainability and continuity, maintaining their level, and employing the largest possible number of working employees, who are the ones who help provide market and customer requirements.

Study results were clear, showing the existence of an impact of manufacturing flexibility application in its dimensions (machine flexibility, new product flexibility, and operating flexibility) in responding to customer requirements in industrial companies in King Abdullah II Ibn Al Hussein Industrial City. The reason for this may be the companies' high professionalism work and their follow-up to the development and used equipment and machines modernization, and also due to the companies providing new products to the markets on a permanent basis, which covers the changed consumer's need, This is also due to enabling companies to produce many multiple products at the same time to cover all consumers tastes and desires, and the reason for the effect may also be due to carrying out many operations on the same production line, and the possibility of changing the steps and processes necessary to produce a particular product easily, and within the specified time With the provision of efforts in light of providing many workers capabilities and experiences, as well as the clear impact of operating flexibility by setting manufacturing plans and goals, and intensively ensuring coverage of market requirements with all its elements, as traders and consumers, this results agreed with the results of Daoud and Jawad (2016), study which confirmed the existence of an impact of manufacturing flexibility in achieving competitive advantage.

The study also showed that there is no impact on the dimensions of manufacturing flexibility (labor flexibility, product mix flexibility, and process flexibility) in responding to customers' requirements. It is worth noting that it is noted that the relationship between the flexibility of the product mix and the response to customer requirements, although it did not reach the statistical significance level, but it is inverse, and this may be due to the fact that some of the measures taken in one of the dimensions of manufacturing flexibility may negatively affect other dimensions.

7 Recommendations

In light of the results and their discussion, the study recommends reducing the time spent when switching from one process to another on the same machines. And the increase in the employee's turnover rate within the company's various activities, which ensures an increase in the employment flexibility, and work to organize and speed up time, to introduce new products to the market. The study also recommends training workers to raise their skills to move from producing a specific product to producing another product quickly, and training workers to change the process on the same machine quickly and in a short time, taking care of developing several operations plans for each manufacturing process, so that the course of operations can be changed to produce a product when a machine malfunctions or is unable to operate it for any reason.

References

Abuhashesh, M. Y., Alshurideh, M. T., & Sumadi, M. (2021). The effect of culture on customers' attitudes toward Facebook advertising: The moderating role of gender. *Review of International Business and Strategy, 31*(3), 416–437.

Ahmad, A., Alshurideh, M. T., Al Kurdi, B. H., & Salloum, S. A. (2021a). Factors impacts organization digital transformation and organization decision making during Covid-19 pandemic. In *The Effect of Coronavirus Disease (COVID-19) on Business Intelligence* (pp. 95–106). Cham: Springer.

Ahmad, A., Alshurideh, M., Al Kurdi, B., Aburayya, A., & Hamadneh, S. (2021b). Digital transformation metrics: A conceptual view. *Journal of Management Information & Decision Sciences, 24*(7), 1–18.

Ahmad, S., Zulkurnain, N., & Khairushalimi, F. (2016). Assessing the validity and reliability of a measurement model in structural equation modeling (SEM). *British Journal of Mathematics & Computer Science, 15*(3), 1–8. https://doi.org/10.9734/BJMCS/2016/25183

Ahmed, A., Alshurideh, M., Al Kurdi, B., & Salloum, S. A. (2020, October). Digital transformation and organizational operational decision making: a systematic review. In *International Conference on Advanced Intelligent Systems and Informatics* (pp. 708–719). Cham: Springer.

Al Mehrez, A. A., Alshurideh, M., Al Kurdi, B., & Salloum, S. A. (2020, October). Internal factors affect knowledge management and firm performance: A systematic review. In *International Conference on Advanced Intelligent Systems and Informatics* (pp. 632–643). Cham: Springer.

Al Kurdi, B., Alshurideh, M., & Al Afaishat, T. (2020a). Employee retention and organizational performance: Evidence from banking industry. *Management Science Letters, 10*(16), 3981–3990.

Al Kurdi, B., Alshurideh, M., & Alnaser, A. (2020b). The impact of employee satisfaction on customer satisfaction: Theoretical and empirical underpinning. *Management Science Letters, 10*(15), 3561–3570.

Alameeri, K. A., Alshurideh, M. T., & Al Kurdi, B. (2021). The effect of Covid-19 pandemic on business systems' innovation and entrepreneurship and how to cope with it: A theatrical view. In *The Effect of Coronavirus Disease (COVID-19) on Business Intelligence* (Vol. 334), pp. 275–288.

Alamro, A. S. (2014). The impact of new product flexibility (NPF) on operational performance: Evidence from Jordanian manufacturing companies. Semantic Scholar (Corpus ID: 202632411).

Al-Dhuhouri, F. S., Alshurideh, M., Al Kurdi, B., & Salloum, S. A. (2020, October). Enhancing our understanding of the relationship between leadership, team characteristics, emotional intelligence and their effect on team performance: A critical review. In *International Conference on Advanced Intelligent Systems and Informatics* (pp. 644–655). Cham: Springer.

Al-Dmour, A., Al-Dmour, H., Al-Barghuthi, R., Al-Dmour, R., & Alshurideh, M. T. (2021a). Factors influencing the adoption of E-payment during pandemic outbreak (COVID-19): Empirical evidence. In *The Effect of Coronavirus Disease (COVID-19) on Business Intelligence* (Vol. 334), pp. 133–154.

Al-Dmour, R., AlShaar, F., Al-Dmour, H., Masa'deh, R., & Alshurideh, M. T. (2021b). The effect of service recovery justices strategies on online customer engagement via the role of "Customer Satisfaction" during the Covid-19 pandemic: An empirical study. In *The Effect of Coronavirus Disease (COVID-19) on Business Intelligence* (Vol. 334), pp. 325–346.

Al-Hawary, S. I., & Al-Syasneh, M. S. (2020). Impact of dynamic strategic capabilities on strategic entrepreneurship in presence of outsourcing of five stars hotels in Jordan. *Business: Theory and Practice, 21*(2), 578–587.

Al-Hawary, S. I. (2013). The role of perceived quality and satisfaction in explaining customer brand loyalty: Mobile phone service in Jordan. *International Journal of Business Innovation and Research, 7*(4), 393–413.

Al-Hawary, S. I., & Abu-Laimon, A. A. (2013). The impact of TQM practices on service quality in cellular communication companies in Jordan. *International Journal of Productivity and Quality Management, 11*(4), 446–474.

Al-Hawary, S. I. S., & Alhajri, T. M. S. (2020). Effect of electronic customer relationship management on customers' electronic satisfaction of communication companies in Kuwait. *Calitatea, 21*(175), 97–102.

Al-Hawary, S. I., Al-Hawajreh, K., Al-Zeaud, H., & Mohammad, A. (2013). The impact of market orientation strategy on performance of commercial banks in Jordan. *International Journal of Business Information Systems, 14*(3), 261–279.

Al-Hawary, S. I., & Al-Menhaly, S. (2016). The quality of E-government services and its role on achieving beneficiaries satisfaction. *Global Journal of Management and Business Research: A Administration and Management, 16*(11), 1–11.

Al-Hawary, S. I., & Al-Smeran, W. (2017). Impact of electronic service quality on customers satisfaction of Islamic Banks in Jordan. *International Journal of Academic Research in Accounting, Finance and Management Sciences, 7*(1), 170–188.

Al-Hawary, S. I., Batayneh, A. M., Mohammad, A. A., & Alsarahni, A. H. (2017). Supply chain flexibility aspects and their impact on customers satisfaction of pharmaceutical industry in Jordan. *International Journal of Business Performance and Supply Chain Modelling, 9*(4), 326–343.

Al-Hawary, S. I., & Harahsheh, S. (2014). Factors affecting Jordanian consumer loyalty toward cellular phone brand. *International Journal of Economics and Business Research, 7*(3), 349–375.

Al-Hawary, S. I., & Hussien, A. J. (2017). The impact of electronic banking services on the customers loyalty of commercial banks in Jordan. *International Journal of Academic Research in Accounting, Finance and Management Sciences, 7*(1), 50–63.

Al-Hawary, S. I. S., & Obiadat, A. A. (2021). Does mobile marketing affect customer loyalty in Jordan? *International Journal of Business Excellence, 23*(2), 226–250.

Alkitbi, S. S., Alshurideh, M., Al Kurdi, B., & Salloum, S. A. (2020, October). Factors affect customer retention: A systematic review. In *International Conference on Advanced Intelligent Systems and Informatics* (pp. 656–667). Cham: Springer.

Allahow, T. J. A. A., Al-Hawary, S. I. S., & Aldaihani, F. M. F. (2018). Information technology and administrative innovation of the central agency for information technology in Kuwait. *Global Journal of Management and Business, 18*(11-A), 1–16.

Al-Lozi, M. S., Almomani, R. Z. Q., & Al-Hawary, S. I. S. (2018). Talent management strategies as a critical success factor for effectiveness of human resources information systems in commercial banks working in Jordan. *Global Journal of Management and Business Research: A Administration and Management, 18*(1), 30–43.

Al-Nady, B. A., Al-Hawary, S. I., & Alolayyan, M. (2013). Strategic management as a key for superior competitive advantage of sanitary ware suppliers in Kingdom of Saudi Arabia. *International Journal of Management and Information Technology, 7*(2), 1042–1058.

Al-Nady, B. A., Al-Hawary, S. I., & Alolayyan, M. (2016). The role of time, communication, and cost management on project management success: An empirical study on sample of construction projects customers in Makkah City, Kingdom of Saudi Arabia. *International Journal of Services and Operations Management, 23*(1), 76–112.

Alolayyan, M., Al-Hawary, S. I., Mohammad, A. A., & Al-Nady, B. A. (2018). Banking service quality provided by commercial banks and customer satisfaction. A structural Equation Modelling Approaches. *International Journal of Productivity and Quality Management, 24*(4), 543–565.

Alsada, R. A. A., & Allawi, J. S. (2017). Production flexibility and its impact on reducing costs, a field study in public companies for leather industries in Baghdad. *Journal of Administration and Economy,* (117), 2018.

Alshamsi, A., Alshurideh, M., Al Kurdi, B., & Salloum, S. A. (2020, October). The influence of service quality on customer retention: A systematic review in the higher education. In *International Conference on Advanced Intelligent Systems and Informatics* (pp. 404–416). Cham: Springer.

Alshurideh, M. T., Al Kurdi, B., & Salloum, S. A. (2021a). The moderation effect of gender on accepting electronic payment technology: A study on United Arab Emirates consumers. *Review of International Business and Strategy, 31*(3), 375–396.

Alshurideh, M. T., Hassanien, A. E., & Masa'deh, R. (2021b). *The Effect of Coronavirus Disease (COVID-19) on Business Intelligence*. Cham: Springer.

Alshurideh, M. T., Kurdi, B. A., AlHamad, A. Q., Salloum, S. A., Alkurdi, S., Dehghan, A., et al. (2021c). Factors affecting the use of smart mobile examination platforms by universities' postgraduate students during the COVID 19 pandemic: An empirical study. *In Informatics, 8*(2), 1–21. Multidisciplinary Digital Publishing Institute.

Alshurideh, D. M. (2019). Do electronic loyalty programs still drive customer choice and repeat purchase behaviour? *International Journal of Electronic Customer Relationship Management, 12*(1), 40–57.

Alshurideh, M. (2022). Does electronic customer relationship management (E-CRM) affect service quality at private hospitals in Jordan? *Uncertain Supply Chain Management, 10*(2), 325–332.

Alshurideh, M., Al-Hawary, S. I., Batayneh, A. M., Mohammad, A., & Al-Kurdi, B. (2017). The impact of Islamic banks' service quality perception on Jordanian customers loyalty. *Journal of Management Research, 9*(2), 139–159.

Alshurideh, M., Masa'deh, R. M. D. T., & Alkurdi, B. (2012). The effect of customer satisfaction upon customer retention in the Jordanian mobile market: An empirical investigation. *European Journal of Economics, Finance and Administrative Sciences, 47*(12), 69–78.

Alsuwaidi, M., Alshurideh, M., Al Kurdi, B., & Salloum, S. A. (2020, October). Performance appraisal on employees' motivation: A comprehensive analysis. In *International Conference on Advanced Intelligent Systems and Informatics* (pp. 681–693). Cham: Springer.

Altarifi, S., Al-Hawary, S. I. S., & Al Sakkal, M. E. E. (2015). Determinants of E-shopping and its effect on consumer purchasing decision in Jordan. *International Journal of Business and Social Science, 6*(1), 81–92.

AlTaweel, I. R., & Al-Hawary, S. I. (2021). The mediating role of innovation capability on the relationship between strategic agility and organizational performance. *Sustainability, 13*(14), 7564.

Alzoubi, H. M., Alshurideh, M., Al Kurdi, B., & Inairat, M. (2020). Do perceived service value, quality, price fairness and service recovery shape customer satisfaction and delight? A practical study in the service telecommunication context. *Uncertain Supply Chain Management, 8*(3), 579–588.

Alzoubi, H., Alshurideh, M., Kurdi, B., Akour, I., & Aziz, R. (2022). Does BLE technology contribute towards improving marketing strategies, customers' satisfaction and loyalty? The role of open innovation. *International Journal of Data and Network Science, 6*(2), 449–460.

Benjaafar, S., & Ramakrishman, R. (1996). Modeling, measurement and evaluation of sequencing flexibility in manufacturing systems. *International Journal of Production Research, 34*, 1195–1220.

Bernardes, E. S., & Hanna, M. D. (2008). A theoretical review of flexibility, agility and responsiveness in the operations management literature toward a conceptual definition of customer responsiveness. *International Journal of Operations & Production Management, 29*(1), 30–53.

Daoud, G. Q., & Jawad, K. A. (2016). The impact of manufacturing flexibility on achieving competitive advantage, an applied study in a selected sample of companies of the Iraqi Ministry of Industry and Minerals. *Journal of Administration and Economy,* (106).

de Leeuw, E., Hox, J., Silber, H., Struminskaya, B., & Vis, C. (2019). Development of an international survey attitude scale: Measurement equivalence, reliability, and predictive validity. *Measurement Instruments for the Social Sciences, 1*(1), 9. https://doi.org/10.1186/s42409-019-0012-x

Ghannajeh, A. M., AlShurideh, M., Zu'bi, M. F., Abuhamad, A., Rumman, G. A., Suifan, T., & Akhorshaideh, A. H. O. (2015). A qualitative analysis of product innovation in Jordan's pharmaceutical sector. *European Scientific Journal, 11*(4).

Gregory, S., Bastias, A., Molenaar, K. R., Gregory, S., Bastias, A., Molenaar, K. R., Potter, L., & Kremer, G. (2019). Assessing the influence of manufacturing flexibility on facility construction costs. In *IISE Annual Conference and Expo, Orlando*. Abstract ID: 584464.

Hair, J. F., Babin, B. J., & Krey, N. (2017). Covariance-based structural equation modeling in the journal of advertising: Review and recommendations. *Journal of Advertising, 46*(1), 163–177. https://doi.org/10.1080/00913367.2017.1281777

Heizer, J., Render, B., &Munson, C. (2017). *Operations Management Sustainability and Supply Chain Management* (12th ed.).

Howard, M. C. (2018). The convergent validity and nomological net of two methods to measure retroactive influences. *Psychology of Consciousness: Theory, Research, and Practice, 5*(3), 324–337. https://doi.org/10.1037/cns0000149

Khasawneh, M. A., Abuhashesh, M., Ahmad, A., Alshurideh, M. T., & Masa'deh, R. (2021a). Determinants of e-word of mouth on social media during COVID-19 outbreaks: An empirical study. In *The Effect of Coronavirus Disease (COVID-19) on Business Intelligence* (pp. 347–366). Cham: Springer.

Khasawneh, M. A., Abuhashesh, M., Ahmad, A., Masa'deh, R., & Alshurideh, M. T. (2021b). Customers online engagement with social media influencers' content related to COVID-19. In *The Effect of Coronavirus Disease (COVID-19) on Business Intelligence* (pp. 385–404). Cham: Springer.

Kumar, S., Goyal, A., & Singhal, S. (2017). Manufacturing flexibility and its effect on system performance. *11*(2), 105–112.

Leo, S., Alsharari, N. M., Abbas, J., & Alshurideh, M. T. (2021). From offline to online learning: A qualitative study of challenges and opportunities as a response to the COVID-19 pandemic in the UAE higher education context. In *The Effect of Coronavirus Disease (COVID-19) on Business Intelligence* (pp. 203–217). Cham: Springer.

Metabis, A., & Al-Hawary, S. I. (2013). The impact of internal marketing practices on services quality of commercial banks in Jordan. *International Journal of Services and Operations Management, 15*(3), 313–337.

Mohammad, A. A., Alshura, M. S., Al-Hawary, S. I. S., Al-Syasneh, M. S., & Alhajri, T. M. (2020). The influence of internal marketing practices on the employees' intention to leave: A study of the private hospitals in Jordan. *International Journal of Advanced Science and Technology, 29*(5), 1174–1189.

Nuseir, M. T., Aljumah, A., & Alshurideh, M. T. (2021). How the business intelligence in the new startup performance in UAE during COVID-19: The mediating role of innovativeness. In *The Effect of Coronavirus Disease (COVID-19) on Business Intelligence* (Vol. 334), pp. 63–79.

Odeh, R. B. M., Obeidat, B. Y., Jaradat, M. O., & Alshurideh, M. T. (2021). The transformational leadership role in achieving organizational resilience through adaptive cultures: the case of Dubai service sector. *International Journal of Productivity and Performance Management*. https://doi.org/10.1108/IJPPM-02-2021-0093

Ojstersek, R., & Buchmeister, B. (2020). The impact of manufacturing flexibility and multicriteria optimization on the sustainability of manufacturing systems. *Symmetry, 12*, 157.

Rimkeviciene, J., Hawgood, J., O'Gorman, J., & De Leo, D. (2017). Construct validity of the acquired capability for suicide scale: Factor structure, convergent and discriminant validity. *Journal of Psychopathology and Behavioral Assessment, 39*(2), 291–302. https://doi.org/10.1007/s10862-016-9576-4

Sekaran, U., & Bougie, R. (2016). *Research Methods for Business: A Skill-Building Approach* (7th ed.). Wiley.

Shi, D., Lee, T., & Maydeu-Olivares, A. (2019). Understanding the model size effect on SEM fit indices. *Educational and Psychological Measurement, 79*(2), 310–334. https://doi.org/10.1177/0013164418783530

Slack, N. (2005). The flexibility of manufacturing systems. *International Journal of Operations & Production Management, 25*(12), 1190–1200.

Sung, K.-S., Yi, Y. G., & Shin, H.-I. (2019). Reliability and validity of knee extensor strength measurements using a portable dynamometer anchoring system in a supine position. *BMC Musculoskeletal Disorders, 20*(1), 1–8. https://doi.org/10.1186/s12891-019-2703-0

Sweiss, N., Obeidat, Z. M., Al-Dweeri, R. M., Mohammad Khalaf Ahmad, A., Obeidat, A. M., & Alshurideh, M. (2021). The moderating role of perceived company effort in mitigating customer misconduct within Online Brand Communities (OBC). *Journal of Marketing Communications*, 1–24.

Tariq, E., Alshurideh, M., Akour, E., Al-Hawaryd, S., & Al Kurdi, B. (2022a). The role of digital marketing, CSR policy and green marketing in brand development at UK. *International Journal of Data and Network Science, 6*(3), 1–10.

Tariq, E., Alshurideh, M., Akour, I., & Al-Hawary, S. (2022b). The effect of digital marketing capabilities on organizational ambidexterity of the information technology sector. *International Journal of Data and Network Science, 6*(2), 401–408.

Taryam, M., Alawadhi, D., Aburayya, A., Albaqa'een, A., Alfarsi, A., Makki, I., et al. (2020). Effectiveness of not quarantining passengers after having a negative COVID-19 PCR test at arrival to Dubai airports. *Systematic Reviews in Pharmacy, 11*(11), 1384–1395.

Upton, D. M. (1994). The management of manufacturing flexibility. *California Management Review, 36*(2).

Wang, Y. A., & Rhemtulla, M. (2021). Power analysis for parameter estimation in structural equation modeling: A discussion and tutorial. *Advances in Methods and Practices in Psychological Science, 4*(1), 1–17. https://doi.org/10.1177/2515245920918253

Williamson, P. J. (1991). Supplier strategy and customer responsiveness: Managing the links. *Business Strategy Review*, 75–90.

Zhang, Q., Vonderembse, M., & Lim, J. (2003). Manufacturing flexibility: Defining and analyzing relationships among competence, capability, and customer satisfaction. *Journal of Operations Management, 21*(2), 173–191.

Impact of Strategic Vigilance on Competitive Capabilities in Jordanian Insurance Companies

Refd Safi Jamil Al-Khasswneh, Ayat Mohammad, Fuad N. Al-Shaikh, Yahia Salim Melhem, Majed Kamel Ali Al-Azzam, Main Naser Alolayyan, Abdullah Matar Al-Adamat, and Sulieman Ibraheem Shelash Al-Hawary

Abstract The study aimed to identify the effect of strategic vigilance on competitive capabilities in Jordanian insurance companies. The study population included (288) individuals, who are the managers of the upper and middle level at the Jordanian insurance companies, which are (24) companies. The study sample is chosen from study population and this sample was chosen by equal stratified random method, which is estimated at (167) respondent according to the sample selection schedule. The questionnaire was used as a tool for the study, the final sample was consisted of

R. S. J. Al-Khasswneh · A. Mohammad
Business and Finance Faculty, The World Islamic Science and Education University (WISE), P.O. Box 1101, Amman 11947, Jordan

F. N. Al-Shaikh · M. K. A. Al-Azzam
Department of Business Administration, Faculty of Economics and Administrative Sciences, Yarmouk University, P.O. Box 566, Irbid 21163, Jordan
e-mail: eco_fshaikh@yu.edu.jo

M. K. A. Al-Azzam
e-mail: Majedaz@yu.edu.jo

Y. S. Melhem
Business Administration Department, School of Business, Yarmouk University, Irbid, Jordan
e-mail: ymelhem@yu.edu.jo

M. N. Alolayyan
Health Management and Policy Department, Faculty of Medicine, Jordan University of Science and Technology, Ar-Ramtha, Jordan
e-mail: mnalolayyan@just.edu.jo

A. M. Al-Adamat
School of Business, Department of Business Administration & Public Administration, Al al-Bayt University, P.O. Box 130040, Mafraq 25113, Jordan
e-mail: aaladamat@aabu.edu.jo

S. I. S. Al-Hawary (✉)
Department of Business Administration, Faculty of Finance and Business Administration, Al al-Bayt University, P.O. Box 130040, Mafraq 25113, Jordan
e-mail: dr_sliman73@aabu.edu.jo

M. Alshurideh et al. (eds.), *The Effect of Information Technology on Business and Marketing Intelligence Systems*, Studies in Computational Intelligence 1056,
https://doi.org/10.1007/978-3-031-12382-5_57

(138) individual. Two statistical analysis programs (IBM SPSS 24) (SMARTPLS, v. 3.2.6) to analyze the study data and test its hypotheses. The study showed a statistically significant impact of the strategic vigilance on competitive capabilities in Jordan insurance companies. According to the results, the study recommended the need to enhance interest strategic vigilance information in Jordanian insurance companies through the establishment of units and departments for the strategic vigilance in them, as they are important as strategic information resource.

Keywords Strategic vigilance · Competitive capabilities · Insurance companies · Jordan

1 Introduction

Business organizations are currently witnessing large and rapid developments, because they operate within their external environment, as they are affected and affected by them, and as a result, organizations must keep pace with these developments on an ongoing basis to ensure their continuity and survival (Al-Nady et al., 2016; Allahow et al., 2018; Al-Hawary & Obiadat, 2021; AlTaweel & Al-Hawary, 2021). The concept of strategic vigilance with its various dimensions (environmental, technological, competitive, and marketing) has played a major role in this field because it provides the organization with a strategic ability to ensure its survival within the successful business system. And because organizations are naturally working to achieve profits and enhance their marketing and competitive capabilities, this has had a significant impact by linking the strategic vigilance process to creating a state of enhancing their competitive capabilities, which in turn achieves organization overall goals (Alaali et al., 2021; Al-Hawary & Al-Hamwan, 2017; Al-Hawary & Al-Syasneh, 2020). It enhances its competitive capabilities in its various dimensions, it has become imperative for organizations to create and adopt competitive capabilities that are unique from their competing organizations (Altamony et al., 2012; Ahmad et al., 2021; Shamout et al., 2022; Tariq et al., 2022), and these capabilities and mechanisms for exploiting and applying may differ from one organization to another, and because this era is the information age, many organizations have adopted the idea of information searching proactively to develop their competitive capabilities, and because Information is the main source that enables organizations to stay in the field of competition and thus continuity (Al-Nady et al., 2013; Al-Hawary & Alwan, 2016; Alzoubi et al., 2021; Kabrilyants et al., 2021). Business organizations are looking for information in their internal and external environments to reach a high level of competitiveness and create added value for their products and services (Alshurideh, 2022; Alzoubi et al., 2022; Svoboda et al., 2021). The study importance appears in the fact that it deals with two important variables of management, strategic vigilance plays the role of a stimulus to ensure the organization survival in conducting business and competition, while competitive capabilities may contribute to organization market share raising by exploiting the available market opportunities and avoiding threats and thus raising profitability levels.

The studied insurance companies in Jordan face intense competition, as each company seeks to attract the customer, meet his needs, and achieve his desires and aspirations, for this reason, Jordanian insurance companies must strengthen attention to strategic vigilance in order to develop competitive capabilities, considering that it is one of the mainstays to establish the competition rules, thus giving it a competitive position that makes it able to overcome the competitors threats. Through this study, the researcher tries to identify competitive capabilities and highlight their important role in creating an added value for the organization, which would guarantee it an advanced position among business organizations (Al Kurdi et al., 2020a, b; Alshurideh et al., 2020; Alzoubi et al., 2020; Al-Dmour et al., 2021; Ghazal et al., 2021; Alshurideh et al., 2022).

2 Literature Review and Hypotheses Development

2.1 Strategic Vigilance

The idea of building and integrating vigilance with strategy came as an extension of several previous concepts such as strategy first and then strategic planning to strategic management and finally the idea of smart organizations. Where strategic vigilance is one of the means that enables organizations to reach the strategic intelligence, organizations with the great developments taking place in the technical and technological sector have become interested in reaching a high stage of administrative and strategic intelligence, as strategic intelligence depends largely on the strategic vigilance process out puts. It is represented in monitoring the environment and collecting proactive information about it in order to use it to achieve the organization strategic objectives of the organization (Al-Hawary & Ismael, 2010; Mouloud, 2014; Al-Hawary & Hadad, 2016). Thus, the strategic vigilance concept came to enhance these axes in a smarter way by not being satisfied with information that is easy to notice, but rather searching for information in a proactive and conscious manner to remove ambiguity from the organization's work environment and, in turn, build a safe future for its continued performance in an effective manner (Zu'bi et al., 2012; Al-Hawary et al., 2013; Morabiti & Ben Tafat, 2016; Aburayya et al., 2020a, b).

Based on the foregoing, strategic vigilance is defined as the organizations race to obtain information through which organizations can listen to their external environment in order to open the appropriate opportunities to exploit them, and to monitor and avoid risks to reduce their effects, this process includes the collection, analysis and dissemination of strategic information with the aim of feeding strategic decisions, which, in turn, contributes to making strategic decisions for the organization (Sahnoun, 2018), On the other hand, Hajar (2018, 27) defined it as an organization that seeks to know the business environment and anticipate changes. It is an informational process through which organizations listen to their environment in order to be able to make decisions. This process is classified within the information systems group that allow managers to lead in difficult times, and the strategic vigilance process can be considered as a system consisting of sub-systems affected by information flows coming from its overall environment. Decision makers must keep pace with changes and adapt to them, and this can only be done through several activities, including strategic vigilance. Competitiveness is no longer based only on reducing costs, but on the speed of response and adaptation to the large and rapid changes taking place in the dynamic work environment of organizations, in addition to the fact that the proper and proactive analysis of environmental signals avoids the organization from any upcoming danger and leads it towards exploiting the opportunities coming from its external environment (Alhalalmeh et al., 2020; Al-Lozi et al., 2017, 2018). Through vigilance concept, many different types and concepts of vigilance emerge. Researchers have differed about the classifications of strategic vigilance according to the purpose for which this strategic system is to be applied (Zhou et al., 2019; Alshurideh et al., 2019; Alameeri et al., 2020).

Marketing vigilance: The concept of marketing (commercial) vigilance is considered one of the modern and advanced concepts in the field of strategic management and in the field of marketing alike. Rapid technological developments and environmental challenges facing business organizations, organizations must be keen to realize, follow up, monitor and confront these changes and conflicts occurring in their environment. Marketing, as organizations must create a database of the marketing environment in order to take the necessary measures and take precautions and caution.

Competitive vigilance: Also known as competitive inquiry, it is the process through which the organization identifies its current and potential competitors, as it studies the environment in which competing organizations live by collecting and analyzing information and then extracting the results and applying them in making decisions. It also allows the organization to build a solid information base that enables it to study competitors deeply.

Technological vigilance: Technological vigilance has been defined as the organized effort by the organization to monitor, receive, analyze, re-publish and retrieve accurate and comprehensive information about certain events, as it contributes to the appropriate and accurate distribution of information within the organization for the purpose of using it in decision-making. As it helps to face rapid technological changes and monitor technological change in a proactive and orderly manner, which contributes to the organization's survival on the safe side of any rapid technological threat that may threaten its survival.

Environmental vigilance: Environmental vigilance is concerned with the remaining factors of the organization external environment that were not addressed by the previous types of vigilance, and of course they are no less important than the previous types, and from the types of environmental vigilance (social vigilance, economic vigilance, and political and legislative vigilance). Environmental vigilance has been defined as monitoring all developments that could affect the organization's activity and that were not taken into account in the previous types.

2.2 *Competitive Capabilities*

Competitive advantage is one of the organizational performance measurements (Al-Hawary, 2011, 2013; Al-Hawary & Abu-Laimon, 2013; Al-Hawary & Al-Namlan, 2018). It is the organization's ability to provide products and services more efficiently and effectively compared to other competing organizations, thus achieving what is known as competitive advantage. Therefore, the advantage is not achieved for organizations without access to competitive capabilities, because the competitive advantage depends in essence on internal and external factors. The external factors are political, technological and cultural factors, while the internal factors are the organization's ability to possess resources and capabilities that are not available to its competitors (Metabis & Al-Hawary, 2013; Alshurideh et al., 2017; Alolayyan et al., 2018; Al-Hawary et al., 2020; Al-Quran et al., 2020; Mohammad et al., 2020; AlHamad et al., 2022). The importance of competitive capabilities lies in providing the opportunity for organizations to overcome the narrowness of the local market, as it helps to provide a competitive environment that ensures organizations efficiency in production. If the organization competitive ability is available, this will help the organization to reach an effective method and approach to ensure economic efficiency and promote economic growth in addition to developing creativity and innovation skills in the organization (Al-Hawary & Hadad, 2016; Alsharari & Alshurideh, 2020; AlShehhi et al., 2020; Alsuwaidi et al., 2020).

Cost leadership capabilities: Cost leadership capabilities require the organization to have the ability to offer products with low costs and an acceptable quality, by seeking to reduce costs in various ways, and that the organization pays high attention to cost control and employee performance monitoring to ensure optimal utilization of resources. As the industry becomes more mature and costs decrease, this will increase organization market share and thus increase sales amount, which means an increase in profits.

Flexibility capabilities: Flexibility is defined as the organizations ability to effect change from one product to another, from one market to another, or from one customer to another, at the lowest possible costs and time, in addition to quickly adapting to changes, disturbances and complexities in the markets and in production volume, where flexibility is one of the most prominent trends with critical importance of the organization's competitive and strategic capabilities, as it is considered a mainstay in the organizations success (Al-Hawary & Al-Menhaly, 2016; Al-Hawary & Al-Smeran, 2017; Al-Hawary & Hussien, 2017; Al-Hawary et al., 2017; Alolayyan et al., 2018).

Speed response capabilities: This priority refers to the organization's ability to provide products and services according to the schedule agreed upon between them and consumers, as many international companies such as FedEx have adopted this competitiveness as they considered it one of their most important competitive priorities within their framework.

2.3 Strategic Vigilance and Competitive Capabilities

A study conducted by Haroun and Hamo (2019) showed that there is a statistically significant relationship between strategic vigilance and competitive capabilities, and that the use of strategic vigilance functions enhances the organization competitive capabilities. Mesbah and Boukhamkham (2019) found that the organization relies in formulating its competitive strategies on the information that vigilance provides about competitors and everything that would affect its competitive position. Whereas, Sanaz and Yaghub (2019) study, which was conducted with the aim of identifying the factors affecting competitive advantages, showed that the organization needs to be strategically agile to reach competitiveness and achieve competitive advantages, this was agreed by the study (Arokodare et al., 2019), which found that strategic agility enhances the company performance, thus achieving competitive capabilities and access to global markets. From the foregoing, we conclude that strategic vigilance increases the organization competitiveness by providing information that contributes to making decisions related to improving competitive capabilities, which the researcher previously concluded is directly related to customers. And the researcher defined it as the organization ability to create and discover the aspirations, desires and needs of customers and analyze them in order to adapt their production capabilities to simulate and apply those aspirations in a proactive manner that guarantees the organization's singularity with this product or service. It works to ensure the organizations competitive advantages continuity and sustainability. The research hypothesis can be formulated as follows.

There is a statistically significant effect of strategic vigilance on competitive capabilities in Jordanian insurance companies.

3 Study Model

See Fig. 1.

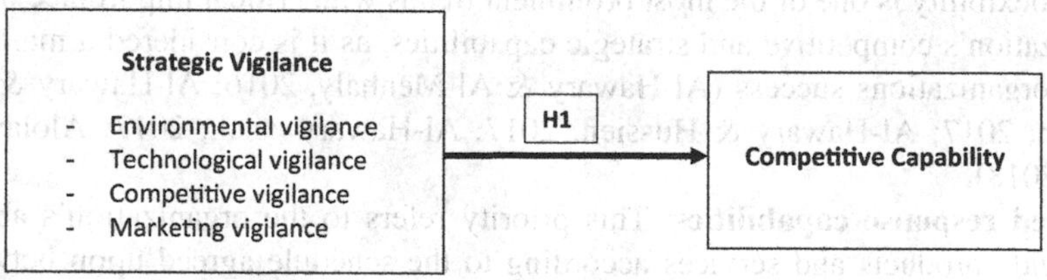

Fig. 1 Research model

4 Methodology

4.1 Study Population

The study population, which is the Jordanian insurance companies, operates under the umbrella of the Jordanian Federation of Insurance Companies, a union that was established in (1989) and includes in its membership all the 24 Jordanian insurance companies (www.joif.org), where insurance companies work is based on several Pivotal aspects include life insurance, car insurance, fire insurance, and marine insurance. Since the union establishment, the insurance sector has witnessed a relative increase in the intensity of competition, especially in the past five years. The researcher found the need to study this sector, especially after the increase in employees' numbers in it in the past decade, where employees' numbers in insurance companies reached (2017)—the last approved statistic—(3124) employees, including employees of the union, in addition to employees of insurance companies (except for sales people) and the insurance administration. The study population consisted of workers in the upper and middle management of these companies, where the managers was (288) from various specializations.

4.2 Study Sample

The study relied on an equal stratified random sample consisting of (167) individuals from the upper and middle management level in the Jordanian insurance companies. The appropriate sample size is (167) individuals. The reason for choosing (167) individuals as a sample for the study is due to the fact that the study population consists of (288) individuals, and according to (Sekaran & Bougie, 2016, 295), the appropriate sample size is (167).

4.3 Study Tool

The study tool includes 32 item to measure the study variables, 20 items were to measure the independent variable, strategic vigilance with its dimensions (environmental vigilance, technological vigilance, competitive vigilance, marketing vigilance) based on the following previous studies: (Haroun & Hamo, 2019), (Zhou et al., 2019) as follows: Items (5-1) measure environmental vigilance, items (10-6) measure technological vigilance, items (15-11) measure competitive vigilance, and items (10-6) measure competitive vigilance. (16-16) marketing vigilance. (12) items were to measure the dependent variable (competitive capabilities) based on the following previous studies. As follows: items (24-21) measure cost leadership capabilities, items (28-25) measure flexibility capabilities, and items (32-29) measure responsiveness capabilities.

4.4 Measurement Model Analysis

SmartPLS 3.2.9 software was used to test the model with latent variables and to perform measurement model tests. First, this study found no outliers that exceeded the permissible limit therefore the outliers number remain as it is, after that the measurement model was examined to examine the reliability and validity of the instruments (Hair et al., 2011; Al-Adamat et al., 2020; Al-Gasawneh & Al-Adamat, 2020). Sekaran and Bougie (2016) pronounced reliability as the way an instrument measures the concept it is meant to measure consistently, while validity was defined as the ability of the instrument to measure the concept it is intended to measure.

The PLS measurement model analysis comprised reliability and validity (internal consistency and convergent and discriminate validity). The measurement model measures are as follows: all the item loadings should be more than 0.7 or 0.6 (Hair et al., 2017; Tan et al., 2017). The composite reliability (CR) value ought to be 0.7 or greater; and the average variance extracted (AVE) should be at least 0.5 (Hair et al., 2010, 2017). Further. for validity testing, the convergent validity reflects whether a specific item estimates a latent variable that it is expected to measure (Tan et al., 2017; Urbach & Ahlemann, 2010), while the AVE assesses the measure of change that a build catches from its contrasting markers and the sum because of the estimation mistake (Fornell & Larcker, 1981; Ringle et al., 2015; Tan et al., 2017). Table 1 presents the latest reliable as well as valid measurement model conducted using Smart PLS version 3.2.9.

Table 1 shows the results of the reliability and validity assessments of the measurement model. As Table 1 which shows that all items were above 0.6. The results also depict that the CR of all constructs exceeded the threshold value of 0.7, ranging from 0.839 to 0.906. This illustrates that the measurement model of this study is reliable (Hair et al., 2017). Additionally, the validity was examined based on its convergent and discriminate validity for the measurement model. The convergent validity of the measures was accepted, with AVE values greater than the recommended level of 0.5, ranging from 0.592 to 0.763 (Hair et al., 2017).

Discriminate validity is confirmed when item loading should load more highly on own construct than on other constructs. Hence, researchers assessed discriminate validity by differentiating the square root of the AVE value of each construct with all the other constructs to be larger than the correlations between two factors (Barclay et al., 1995; Fornell & Larcker, 1981). Thus, Table 2 reveals that the correlation scores between each construct with itself is greater than all other constructs. This indicates that the discriminant validity was met by all constructs and it is acceptable. Therefore, it can be concluded that the measurement model is reliable and valid. Based on this, the data' reliability and validity of the instrument were confirmed.

Table 1 Results of reflective measurements model—a summary

Constructs	Indicators	Internal consistency reliability			
		Loading	(AVE)	Composite reliability	Crombach's alpha
		>0.60	>0.50	0.70–0.90	0.60–0.90
Strategic vigilance					
Environmental vigilance	Q1	0.700	0.67	0.92	0.90
	Q2	0.886			
	Q3	0.879			
	Q4	0.813			
	Q5	0.751			
Technological vigilance	Q6	0.731	0.64	0.92	0.85
	Q7	0.814			
	Q8	0.814			
	Q9	0.788			
	Q10	0.806			
Competitive vigilance	Q11	0.852	0.66	0.94	0.87
	Q12	0.822			
	Q13	0.799			
	Q14	0.787			
	Q15	0.789			
Marketing vigilance	Q16	0.808	0.60	0.93	0.89
	Q17	0.734			
	Q18	0.855			
	Q19	0.862			
	Q20	0.801			
Competitive capabilities					
	Q21	0.765	0.60	0.95	0.95
	Q22	0.736			
	Q23	0.755			
	Q24	0.773			
	Q25	0.769			
	Q26	0.809			
	Q27	0.763			
	Q28	0.808			
	Q29	0.757			
	Q30	0.743			
	Q31	0.773			
	Q32	0.853			

Table 2 Discriminate validity of constructs

Construct	EV	TV	CV	MV
Environmental vigilance	1.00			
Technological vigilance	0.656	1.00		
Competitive vigilance	0.511	0.618	1.00	
Marketing vigilance	0.420	0.583	0.589	1.00

4.5 Structural Model Analysis

The structural model consists of the constructs or latent variables and the paths that connect them with each other (Hair et al., 2017). Figure 2 is the schematic diagram of the structural model, which starts with EV, TV, CV, and MV. The arrows linking the constructs are determined by the direction of the five hypotheses proposed in the paper. The standardized estimate for the structural model of this study shows the direct relationship between Strategic Vigilance dimensions and Competitive Capability. To clarify more, the coefficient of path Values ranges from −1 to +1, which expresses the depth of the relationship between any two constructs (Hair et al., 2017).

Ramayah et al. (2016) specified the critical values for significance in 1-tailed tests as follows: $p < 10\%$ (1.64), $p < 5\%$ (1.96), and $p < 1$ (2.58); also, as usual, the researchers in marketing utilized the significance level of $p < 5\%$. Hair et al. (2017) stated that the bias-corrected bootstrap confidence intervals (lower limit, upper limit) allow testing whether a path coefficient is significantly different from zero. If the confidence interval for an estimated path coefficient does not include zero, this means there is a significant effect of this path (Hair et al., 2017). Therefore, the

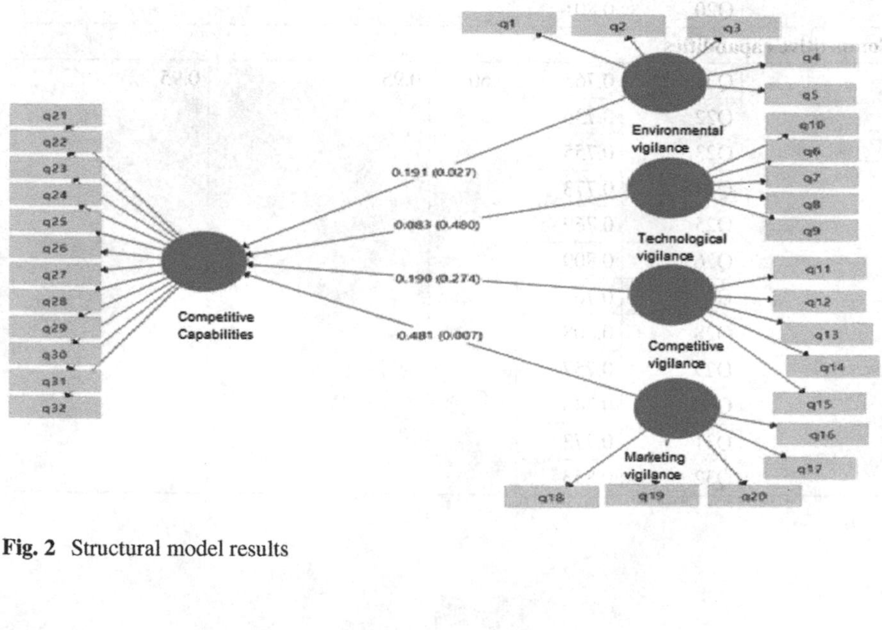

Fig. 2 Structural model results

Table 3 Summary of the structural model results

Hypothesis	Relationship (direct)	Direct effect (β)	*t*-statistic	*p* values	Decision
H1.1	Direct relationship EV → CC	0.191	2.211	0.027	Supported**
H1.2	Direct relationship TV → CC	0.083	0.708	0.480	Not supported
H1.3	Direct relationship MV → CC	0.190	1.095	0.274	Not supported
H1.4	Direct relationship CV → CC	0.481	2.710	0.007	Supported*
H1	Direct relationship SV → CC	0.836	23.087	0.000	Supported**

Note Significance level at ** $p < 0.01$, * $p < 0.05$ (one-tailed). LL, lower limit at 5%; UL, upper limit at 95% confidence interval; BC, bias corrected

researcher set 5000 subsamples with a replacement from the bootstrap cases equal to the original set of data which is 138 and the results of the structural model for this study are shown in Table 3.

As can be observed in Table 3, the path coefficients of EV were found to have a significant relationship with CC; the result indicated that ($\beta = 0.191$; *t*-statistic of 2.211), which supported H1.1. Otherwise, the path coefficients of TV were found to have insignificant relationship with CC; where the result indicated that ($\beta = 0.083$; *t*-statistic of 0.708), which that mean that H1.2 is not supported. Also, the path coefficients of MV were found to have insignificant relationship with CC; where the result indicated that ($\beta = 0.190$; *t*-statistic of 1.095), which means that H1.3 is not supported. As for the path coefficients of CV were found to have a significant relationship with CC; the result indicated that ($\beta = 0.481$; *t*-statistic of 2.710), which supported H1.4. Furthermore, the path coefficients of SV were found to have a significant relationship with CC; the result indicated that ($\beta = 0.836$; *t*-statistic of 23.087), Moreover (R^2) = (0.698 which supported H1.

5 Discussions

study results showed that the strategic vigilance system adoption helps to control all components of its environment through the four basic types of adoption of strategic vigilance: commercial vigilance, technological vigilance, competitive vigilance, and environmental vigilance, as the use of strategic vigilance functions enhances the organization competitiveness and allows it to reach high net sales levels and achieve significant income and profit margins. According to these results and studies, this is evidence of the insurance companies' awareness of the need to follow up and monitor all information within their work environment (insurance), and reinforces the insurance companies' awareness that information represents a strong point for the company if it is strategically monitored to avoid its risks and exploit its opportunities. It also indicates that Jordanian insurance companies' ability to understand and apply strategic vigilance in the company contributes vitally to improving and enhancing competitive capabilities, and leads the company towards differentiation, which gives it an advantage over its competitors in the market.

6 Recommendations

In light of the study findings, the researcher reached a number of recommendations that can be taken by Jordanian insurance companies' managers and decision makers in order to advance their current situation. Researchers recommended the necessity of establishing units for strategic vigilance within insurance companies because of their significant role in raising competitive capabilities levels within these companies. The researcher also recommended the need to form special work teams equipped with the minimum capabilities to enhance interest in the monitoring and collecting information process.

References

Aburayya, A., Alshurideh, M., Al Marzouqi, A., Al Diabat, O., Alfarsi, A., Suson, R., et al. (2020a). Critical success factors affecting the implementation of TQM in public hospitals: A case study in UAE hospitals. *Systematic Reviews in Pharmacy, 11*(10), 230–242.

Aburayya, A., Alshurideh, M., Alawadhi, D., Alfarsi, A., Taryam, M., & Mubarak, S. (2020b). An investigation of the effect of lean six sigma practices on healthcare service quality and patient satisfaction: Testing the mediating role of service quality in Dubai primary healthcare sector. *Journal of Advanced Research in Dynamical and Control Systems, 12*(8), 56–72.

Ahmad, A., Alshurideh, M. T., Al Kurdi, B. H., & Alzoubi, H. M. (2021, June). Digital strategies: A systematic literature review. In *The International Conference on Artificial Intelligence and Computer Vision* (pp. 807–822). Cham: Springer.

Al Kurdi, B., Alshurideh, M., & Al Afaishat, T. (2020a). Employee retention and organizational performance: Evidence from banking industry. *Management Science Letters, 10*(16), 3981–3990.

Al Kurdi, B., Alshurideh, M., & Alnaser, A. (2020b). The impact of employee satisfaction on customer satisfaction: Theoretical and empirical underpinning. *Management Science Letters, 10*(15), 3561–3570.

Alaali, N., Al Marzouqi, A., Albaqaeen, A., Dahabreh, F., Alshurideh, M., Alrwashdh, S., Iyadeh, I., Salloum, S., & Aburayya, A. (2021). The impact of adopting corporate governance strategic performance in the tourism sector: A case study in the Kingdom of Bahrain. *Journal of Legal, Ethical and Regulatory, 24*(Special Issue 1), 1–18.

Al-Adamat, A., Al-Gasawneh, J., & Al-Adamat, O. (2020). The impact of moral intelligence on green purchase intention. *Management Science Letters, 10*(9), 2063–2070.

Alameeri, K., Alshurideh, M., Al Kurdi, B., & Salloum, S. A. (2020, October). The effect of work environment happiness on employee leadership. In *International Conference on Advanced Intelligent Systems and Informatics* (pp. 668–680). Cham: Springer.

Al-Dmour, R., AlShaar, F., Al-Dmour, H., Masa'deh, R., & Alshurideh, M. T. (2021). The effect of service recovery justices strategies on online customer engagement via the role of "Customer Satisfaction" during the Covid-19 pandemic: An empirical study. In *The Effect of Coronavirus Disease (COVID-19) on Business Intelligence* (Vol. 334), pp. 325–346.

Al-Gasawneh, J. A., & Al-Adamat, A. M. (2020). The relationship between perceived destination image, social media interaction and travel intentions relating to Neom city. *Academy of Strategic Management Journal, 19*(2), 1–12.

Alhalalmeh, M. I., Almomani, H. M., Altarifi, S., Al-Quran, A. Z., Mohammad, A. A., & Al-Hawary, S. I. (2020). The nexus between corporate social responsibility and organizational performance in Jordan: The mediating role of organizational commitment and organizational citizenship behavior. *Test Engineering and Management, 83*, 6391–6410.

AlHamad, A., Alshurideh, M., Alomari, K., Kurdi, B., Alzoubi, H., Hamouche, S., & Al-Hawary, S. (2022). The effect of electronic human resources management on organizational health of telecommunications companies in Jordan. *International Journal of Data and Network Science, 6*(2), 429–438.

Al-Hawary, S. I., & Al-Syasneh, M. S. (2020). Impact of dynamic strategic capabilities on strategic entrepreneurship in presence of outsourcing of five stars hotels in Jordan. *Business: Theory and Practice, 21*(2), 578–587.

Al-Hawary, S. I. (2011). The effect of banks governance on banking performance of the Jordanian commercial banks: Tobin's Q model an applied study. *International Research Journal of Finance and Economics, 71*, 34–47.

Al-Hawary, S. I. (2013). The role of perceived quality and satisfaction in explaining customer brand loyalty: Mobile phone service in Jordan. *International Journal of Business Innovation and Research, 7*(4), 393–413.

Al-Hawary, S. I., & Abu-Laimon, A. A. (2013). The impact of TQM practices on service quality in cellular communication companies in Jordan. *International Journal of Productivity and Quality Management, 11*(4), 446–474.

Al-Hawary, S. I., & Al-Hamwan, A. (2017). Environmental analysis and its impact on the competitive capabilities of the commercial banks operating in Jordan. *International Journal of Academic Research in Accounting, Finance and Management Sciences, 7*(1), 277–290.

Al-Hawary, S. I., Al-Hawajreh, K., Al-Zeaud, H., & Mohammad, A. (2013). The impact of market orientation strategy on performance of commercial banks in Jordan. *International Journal of Business Information Systems, 14*(3), 261–279.

Al-Hawary, S. I., & Al-Menhaly, S. (2016). The quality of e-government services and its role on achieving beneficiaries satisfaction. *Global Journal of Management and Business Research: A Administration and Management, 16*(11), 1–11.

Al-Hawary, S. I., & Al-Namlan, A. (2018). Impact of electronic human resources management on the organizational learning at the private hospitals in the state of Qatar. *Global Journal of Management and Business Research: A Administration and Management, 18*(7), 1–11.

Al-Hawary, S. I., & Al-Smeran, W. (2017). Impact of electronic service quality on customers satisfaction of Islamic Banks in Jordan. *International Journal of Academic Research in Accounting, Finance and Management Sciences, 7*(1), 170–188.

Al-Hawary, S. I. S., & Alwan, A. M. (2016). Knowledge management and its effect on strategic decisions of Jordanian Public Universities. *Journal of Accounting-Business & Management, 23*(2), 24–44.

Al-Hawary, S. I., Batayneh, A. M., Mohammad, A. A., & Alsarahni, A. H. (2017). Supply chain flexibility aspects and their impact on customers satisfaction of pharmaceutical industry in Jordan. *International Journal of Business Performance and Supply Chain Modelling, 9*(4), 326–343.

Al-Hawary, S. I., & Hadad, T. F. (2016). The effect of strategic thinking styles on the enhancement competitive capabilities of commercial banks in Jordan. *International Journal of Business and Social Science, 7*(10), 133–144.

Al-Hawary, S. I., & Hussien, A. J. (2017). The impact of electronic banking services on the customers loyalty of commercial banks in Jordan. *International Journal of Academic Research in Accounting, Finance and Management Sciences, 7*(1), 50–63.
Al-Hawary, S. I., & Ismael, M. (2010). The effect of using information technology in achieving competitive advantage strategies: A field study on the Jordanian pharmaceutical companies. *Al Manara for Research and Studies, 16*(4), 196–203.
Al-Hawary, S. I. S., Mohammad, A. S., Al-Syasneh, M. S., Qandah, M. S. F., & Alhajri, T. M. S. (2020). Organisational learning capabilities of the commercial banks in Jordan: Do electronic human resources management practices matter? *International Journal of Learning and Intellectual Capital, 17*(3), 242–266.
Al-Hawary, S. I. S., & Obiadat, A. A. (2021). Does mobile marketing affect customer loyalty in Jordan? *International Journal of Business Excellence, 23*(2), 226–250.
Allahow, T. J. A. A., Al-Hawary, S. I. S., & Aldaihani, F. M. F. (2018). Information technology and administrative innovation of the central agency for information technology in Kuwait. *Global Journal of Management and Business, 18*(11-A), 1–16.
Al-Lozi, M., Almomani, R. Z., & Al-Hawary, S. I. (2017). Impact of talent management on achieving organizational excellence in Arab Potash Company in Jordan. *Global Journal of Management and Business Research: A Administration and Management, 17*(7), 15–25.
Al-Lozi, M., Almomani, R. Z., & Al-Hawary, S. I. (2018). Talent management strategies as a critical success factor for effectiveness of human resources information systems in commercial banks working in Jordan. *Global Journal of Management and Business Research: A Administration and Management, 18*(1), 30–43.
Al-Nady, B. A., Al-Hawary, S. I., & Alolayyan, M. (2013). Strategic management as a key for superior competitive advantage of sanitary ware suppliers in Kingdom of Saudi Arabia. *International Journal of Management and Information Technology, 7*(2), 1042–1058.
Al-Nady, B. A., Al-Hawary, S. I., & Alolayyan, M. (2016). The role of time, communication, and cost management on project management success: An empirical study on sample of construction projects customers in Makkah City, Kingdom of Saudi Arabia. *International Journal of Services and Operations Management, 23*(1), 76–112.
Alolayyan, M., Al-Hawary, S. I., Mohammad, A. A., & Al-Nady, B. A. (2018). Banking service quality provided by commercial banks and customer satisfaction. A structural equation modelling approaches. *International Journal of Productivity and Quality Management, 24*(4), 543–565.
Al-Quran, A. Z., Alhalalmeh, M. I., Eldahamsheh, M. M., Mohammad, A. A., Hijjawi, G. S., Almomani, H. M., & Al-Hawary, S. I. (2020). Determinants of the green purchase intention in Jordan: The moderating effect of environmental concern. *International Journal of Supply Chain Management, 9*(5), 366–371.
Alsharari, N. M., & Alshurideh, M. T. (2020). Student retention in higher education: The role of creativity, emotional intelligence and learner autonomy. *International Journal of Educational Management, 35*(1), 233–247.
AlShehhi, H., Alshurideh, M., Al Kurdi, B., & Salloum, S. A. (2020, October). The impact of ethical leadership on employees performance: A systematic review. In *International Conference on Advanced Intelligent Systems and Informatics* (pp. 417–426). Cham: Springer.
Alshurideh, M. T., Al Kurdi, B., Alzoubi, H. M., Ghazal, T. M., Said, R. A., AlHamad, A. Q., et al. (2022). Fuzzy assisted human resource management for supply chain management issues. *Annals of Operations Research*, 1–19.
Alshurideh, M. (2022). Does electronic customer relationship management (E-CRM) affect service quality at private hospitals in Jordan? *Uncertain Supply Chain Management, 10*(2), 325–332.
Alshurideh, M., Al-Hawary, S. I., Batayneh, A. M., Mohammad, A., & Al-Kurdi, B. (2017). The impact of Islamic banks' service quality perception on Jordanian customers loyalty. *Journal of Management Research, 9*(2), 139–159.
Alshurideh, M., Gasaymeh, A., Ahmed, G., Alzoubi, H., & Kurd, B. (2020). Loyalty program effectiveness: Theoretical reviews and practical proofs. *Uncertain Supply Chain Management, 8*(3), 599–612.

Alshurideh, M., Kurdi, B. A., Shaltoni, A. M., & Ghuff, S. S. (2019). Determinants of pro-environmental behaviour in the context of emerging economies. *International Journal of Sustainable Society, 11*(4), 257–277.

Alsuwaidi, M., Alshurideh, M., Al Kurdi, B., & Salloum, S. A. (2020, October). Performance appraisal on employees' motivation: A comprehensive analysis. In *International Conference on Advanced Intelligent Systems and Informatics* (pp. 681–693). Cham: Springer.

Altamony, H., Masa'deh, R., Alshurideh, M., & Obeidat, B. (2012). Information systems for competitive advantage: Implementation of an organisational strategic management process. In *Innovation and Sustainable Competitive Advantage: From Regional Development to World Economies*, pp. 583–592.

AlTaweel, I. R., & Al-Hawary, S. I. (2021). The mediating role of innovation capability on the relationship between strategic agility and organizational performance. *Sustainability, 13*(14), 7564.

Alzoubi, H. M., Alshurideh, M., & Ghazal, T. M. (2021, June). Integrating BLE beacon technology with intelligent information systems IIS for operations' performance: A managerial perspective. In *The International Conference on Artificial Intelligence and Computer Vision* (pp. 527–538). Cham: Springer.

Alzoubi, H. M., Alshurideh, M., Al Kurdi, B., & Inairat, M. (2020). Do perceived service value, quality, price fairness and service recovery shape customer satisfaction and delight? A practical study in the service telecommunication context. *Uncertain Supply Chain Management, 8*(3), 579–588.

Alzoubi, H., Alshurideh, M., Kurdi, B., Akour, I., & Aziz, R. (2022). Does BLE technology contribute towards improving marketing strategies, customers' satisfaction and loyalty? The role of open innovation. *International Journal of Data and Network Science, 6*(2), 449–460.

Arokodare, M. A., Asikhia, O. U., & Makinde, G. O. (2019). Strategic agility and firm performance: The moderating role of organizational culture. *Business Management Dynamics, 9*(3), 1–12.

Barclay, M. N. I., McPherson, A., & Dixon, J. (1995). Selenium content of a range of foods. *Journal of Food Composition Analysis, 8*, 307–318.

Fornell, C., & Larcker, D. (1981). Evaluating structural equation models with unobservable variables and measurement error. *Journal of Marketing Research, 18*(1), 39–50.

Hair, J. F., Black, B., Babin, B., & Anderson, R. E. (2010). *Multivariate Data Analysis* (7th ed.). Pearson Prentice Hall.

Hair, J., Hollingsworth, C. L., Randolph, A. B., & Chong, A. Y. L. (2017). An updated and expanded assessment of PLS-SEM in information systems research. *Industrial Management and Data Systems, 117*(3), 442–458. https://doi.org/10.1108/IMDS-04-2016-0130

Hair, J. F., Ringle, C. M., & Sarstedt, M. (2011). PLS-SEM: Indeed, a silver bullet. *Journal of Marketing Theory and Practice, 19*(2), 139–152.

Hajar, B. (2018). *The role of strategic vigilance in enterprise development: A case study of the Algeria Telecom Corporation—Mostaganem.* Unpublished Master's thesis, Faculty of Economics, Commercial and Management Sciences, Abdelhamid Ibn Badis University, Mostaganem, People's Democratic Republic of Algeria.

Haroun, D., & Hamo, N. (2019). Strategic vigilance and its contribution to enhancing the competitiveness of Algerian economic institutions: A case study of the Condor Foundation. *Journal of the Economic Researcher: University of August 20, 1955 Skikda, 12*(7), 381–403.

Kabrilyants, R., Obeidat, B., Alshurideh, M., & Masadeh, R. (2021). The role of organizational capabilities on e-business successful implementation. *International Journal of Data and Network Science, 5*(3), 417–432.

Mesbah, A., & Boukhamkham, A. F. (2019). The role of strategic vigilance in developing the competitive advantage of the economic enterprise: A case study in the Eastern Regional Directorate for the mobile operator Ooredoo. *Journal of Economic Studies: Abdelhamid Mehri University—Constantine 2—Faculty of Economic, Commercial and Management Sciences, 6*, 23–46.

Metabis, A., & Al-Hawary, S. I. (2013). The impact of internal marketing practices on services quality of commercial banks in Jordan. *International Journal of Services and Operations Management, 15*(3), 313–337.

Mohammad, A. A., Alshura, M. S., Al-Hawary, S. I. S., Al-Syasneh, M. S., & Alhajri, T. M. (2020). The influence of internal marketing practices on the employees' intention to leave: A study of the private hospitals in Jordan. *International Journal of Advanced Science and Technology, 29*(5), 1174–1189.

Morabiti, Y., & Ben Tafat, A. (2016). The role of strategic vigilance in preventing marketing crises for service institutions: A case study of the Mobilis Foundation, Ouargla Unit. A magister message that is not published. Kasdi Merbah University, Ouargla, Ouargla.

Mouloud, Z. (2014). *The reality of economic intelligence in Algerian economic institutions: A case study of the Anou Foundation, the Mostaganem branch*. Unpublished Master's thesis, Faculty of Economics, Commercial and Management Sciences, Abdelhamid Ben Badis University, Mostaganem, People's Democratic Republic of Algeria.

Ramayah, T., Cheah, J., Chuah, F., Ting, H., & Memon, M. A. (2016). Partial least squares structural equation modeling (PLS-SEM) using SmartPLS 3.0: An updated and practical guide to statistical analysis.

Ringle, C. M., Wende, S., & Becker, J. M. (2015). *Smart PLS*. SmartPLS GmbH. Retrieved from www.smartpls.com

Sahnoun, H. (2018). The impact of strategic vigilance in supporting creativity in Algerian organizations: A field study at the Foundation for Fatty Materials "Sibos—Label Annaba." *Journal of Economic and Administrative Sciences, 2*(43), 135–145.

Sanaz, D., & Yaghub, R. (2019). Strategic agility in telecom industry: The effective factors on competitive advantages. *Middle East Journal of Management, 6*(1), 1–20.

Sekaran, U., & Bougie, R. (2016). Research Methods for Business: A Skill Building Approach. John Wiley & Sons.

Shamout, M., Elayan, M., Rawashdeh, A., Kurdi, B., & Alshurideh, M. (2022). E-HRM practices and sustainable competitive advantage from HR practitioner's perspective: A mediated moderation analysis. *International Journal of Data and Network Science, 6*(1), 165–178.

Svoboda, P., Ghazal, T. M., Afifi, M. A., Kalra, D., Alshurideh, M. T., & Alzoubi, H. M. (2021, June). Information systems integration to enhance operational customer relationship management in the pharmaceutical industry. In *The International Conference on Artificial Intelligence and Computer Vision* (pp. 553–572). Cham: Springer.

Tan, P., He, L., Cui, J., Qian, C., Cao, X., Lin, M., & Wang, R. F. (2017). Assembly of the WHIP-TRIM14-PPP6C mitochondrial complex promotes RIG-I-mediated antiviral signaling. *Molecular Cell, 68*(2), 293–307.

Tariq, E., Alshurideh, M., Akour, I., & Al-Hawary, S. (2022). The effect of digital marketing capabilities on organizational ambidexterity of the information technology sector. *International Journal of Data and Network Science, 6*(2), 401–408.

Urbach, N., & Ahlemann, F. (2010). Structural equation modeling in information systems research using partial least squares. *Journal of Information Technology Theory and Application, 11*(2), 5–40.

Zhou, J., Movando, F., & Saunders, S. (2019). The relationship between marketing agility and financial performance under different levels of market turbulence. *Industrial Marketing Management Journal, 83*, 31–41.

Zu'bi, Z., Al-Lozi, M., Dahiyat, S., Alshurideh, M., & Al Majali, A. (2012). Examining the effects of quality management practices on product variety. *European Journal of Economics, Finance and Administrative Sciences, 51*(1), 123–139.

The Impact of Emotional Intelligence (EI) on Teamwork Performance in Information Technology Sector in Jordan

Majed Kamel Ali Al-Azzam, Marah Jameel Albash, Ziad Mohd Ali Smadi, Reham Zuhier Qasim Almomani, Ali Zakariya Al-Quran, Sulieman Ibraheem Shelash Al-Hawary, Mohammad Mousa Eldahamsheh, Anber Abraheem Shlash Mohammad, and Abdullah Ibrahim Mohammad

Abstract The study aimed to investigate the impact of emotional intelligence on teamwork performance. The study population consisted of all employees working in the IT private sector in Jordan. From all IT companies working in this sector, four IT firms, at King Hussein Business Park, were selected with total number of (80) IT employees. A questionnaire was developed as the study instrument, which was composed of ten dimensions with (65) items. Multiple regression analysis was used for testing the study hypotheses. The study results found that emotional intelligence

M. K. A. Al-Azzam · M. J. Albash
Department of Business Administration, Faculty of Economics and Administrative Sciences, Yarmouk University, P.O. Box 566, Irbid 21163, Jordan
e-mail: Majedaz@yu.edu.jo

Z. M. A. Smadi · A. Z. Al-Quran · S. I. S. Al-Hawary (✉)
Department of Business Administration, School of Business, Al al-Bayt University, P.O. BOX 130040, Mafraq 25113, Jordan
e-mail: dr_sliman73@aabu.edu.jo

Z. M. A. Smadi
e-mail: ziad38in@aabu.edu.jo

A. Z. Al-Quran
e-mail: ali.z.al-quran@aabu.edu.jo

R. Z. Q. Almomani
Business Administration, Amman, Jordan

M. M. Eldahamsheh
Strategic Management, Amman, Jordan

A. A. S. Mohammad
Marketing Department, Faculty of Administrative and Financial Sciences, Petra University, P.O. Box 961343, Amman 11196, Jordan

A. I. Mohammad
Department of Basic Scientific Sciences, Al-Huson University College, Al-Balqa Applied University, Irbid, Jordan

M. Alshurideh et al. (eds.), *The Effect of Information Technology on Business and Marketing Intelligence Systems*, Studies in Computational Intelligence 1056,
https://doi.org/10.1007/978-3-031-12382-5_58

has a significant impact on teamwork performance. Based on the study results, the researchers recommend managers and decision makers to build good relationships with each other to improve the communication of information and ideas among the team members, and to have enough ability to overwhelm disputes occurring during the practice of work activities, which will make the work be achieved in smoother and faster mode.

Keywords Emotional intelligence · Teamwork performance · Information technology sector · Jordan

1 Introduction

Information technology (IT) sector is one of the most advanced sectors in Jordanian economy, where its growth rates are the highest in Jordan at a rate of 50% per year. Its future development is considered as a gateway for modernizing economic and social life in the kingdom (Batool 2013). Previous research has proven that intellectual quotient (IQ) is not the main factor that can predict the success and brilliance of employees at work. Emotional intelligence (EI) is an important element to determine employee's skills and competencies in communication and information technology, where the person with high (EI) is more able to build relationships and manage conflicts with others, because of their elevated capacity to understand the needs of the others they engage with (Lopes et al., 2005). EI which is considered as a set of skills and competencies, mainly based on knowledge, understanding, management and leadership of personal feelings and emotions between the individual and others, as well as internally between the individual and himself (Marzuki et al., 2015; Alameeri et al., 2020; Odeh et al. 2021). Newman and Joseph (2010) stated that in an organization, emotion regulation relates to job performance. It is well known that the good use and management of emotions leads to a positive influence on motivation (which in turn affects productivity and performance).

Moreover, increased emotional intelligence leads to improved personal and professional human interactions, which leads to a higher level of job performance of the employees in return (Batool 2013; Alsharari & Alshurideh, 2020). Emotional intelligence is an important predictor of key organizational outcomes including job satisfaction (Daus & Ashkanasy 2005; Al-Dhuhouri et al. 2020). According to the theory of emotional intelligence, a person who can understand his feelings and emotions and has full awareness of them as well as his ability to control stress can certainly have better relationships with colleagues and supervisors, hence, better job performance (Kafetsios & Zampetakis 2008; Obeidat et al. 2019). In addition, most of the organizations seek for an employee, who has the ability to cope with change and respond. Accordingly, Moghadam, Tehrani, and Amin (2011) proved in their study that the employee who has high (EI) affects positively in coping with situations and doing tasks in effective ways, especially in teamwork.

Information technology (IT) employees were selected for this study for the importance of IT sector, which is a highly value-added sector contributing significantly to gross domestic product (GDP) in Jordan which makes IT sector an important study sector to enhance employee performance individually or in teamwork. In addition, IT employees' job requires consistent concentration when they are coding programs, coping with stress. Also, they need to recognize their own as well as others' emotions in view of the nature of their work that follows teamwork.

According to several studies (e.g: Stys & Brown 2004; Singh, 2006), EI integrative components, such as consistent interactions and communications, cooperation with others, adaptability to stresses and conflicts, and self-motivation are very critical factors that increase the levels of the individual's productivity leading to the organizational performance. Therefore, EI factors have been examined against teamwork performance. According to Extremera and Rey (2016), EI factors have a significant role in the employee's efficiency and performance. He added that self-motivation, as an EI variable, is the inner force that drives individuals to accomplish personal and organizational goals, but regarding to social skills (such as, communication and cooperation) by virtue of the nature of IT staff's job they need to communicate with their teams as well as customers who communicate either through email or face to face to solve their problems or to convey their ideas to them, so for these reasons this study finds it is very important to reveal the levels of common EI social skills that are represented in (Communication, Cooperation, Adaptability to situation, Stress and Conflict Management, and motivation) as indicated by EI models, such Goleman Model of IE (1998), Bar-on Model, Singh model (2006), at employees of Jordanian IT sector, and investigate their impact on their teamwork performance.

2 Theoretical Background and Hypotheses Development

2.1 Intelligence (EI)

For long time, Emotional intelligence (EI) concept has developed gradually according to its increasing importance at all scientific fields so that the term of (EI) has become an essential at the level of individuals, teams, and even organizations that aspire to improve and develop its performance. Emotional intelligence (EI) is a defined as a conceptualization that has been discussed progressively by the literature of social psychology. Recently, emotional intelligence has been given much importance in research especially in psychological research. The concept of emotional intelligence was early introduced by Salovey and Mayer (1990). Who defined emotional intelligence as the individual's capability to manage his own emotions and the emotions of other individuals, to distinguish among them and to utilize information for facilitating one's reasoning, activities, and thinking (Mayer & Salovey, 1997).

Furthermore, Emotional intelligence as specified by Singh is the personal ability to respond appropriately and successfully to a wide variety of emotional stimuli that is drawn from within the person and his surrounding environment. Emotional intelligence is also defined as the consciousness in using feelings and emotions and employs them by standards and methods of personal knowledge to cope with current situations and problems (Stys & Brown 2004). What's more, emotional intelligence is described as the ability of an individual to adapt to, choose, and change a situation through an emotional process (Gignac, 2010).

Another definition of Emotional intelligence is that it composed of a set of significant parts of individual's internal and external dealings, mindset, personality, stress management abilities, temperament adjustment and all these have significant impact on the individual and teamwork Performance (Fallahzadeh, 2011; Alshurideh et al. 2019; Alzoubi et al., 2021). Lastly, Bradberry and Greaves (2009) defined emotional intelligence as the individual's capability, aptitude, recognition assignment, accurate assessment and control of his senses against other individuals and groups. Depending on all above definitions of Emotional intelligence, the researcher can define the concept as "Knowing the self's emotions and the emotions of others and employing this knowledge intelligently to choose the best method that ensures the successful management of different life situations".

Individuals who have higher emotional intelligence are more expected to regulate, understand, and control emotions excellently in themselves as well as in the other individuals (Wijekoon et al. 2017; AlShehhi et al. 2020). There is substantial evidence that emotional intelligence is a better predictor of success than intelligence quotient. On the basis of intelligence quotient, one may get into a field but the ultimate success in that field depends upon emotional intelligence (Johar et al., 2019).

People who have a high level of emotional intelligence know their own feelings and manage them well, beside they understand and treat others' feelings properly. Accordingly, they are distinguished in all areas of life, more satisfied with themselves, efficient in their lives, and able to control their mental structure pushing their production forward (Huang & Lee, 2019). On the other hand, people with low level of emotional intelligence do not have the ability that can enable them to cope with stressful events, and thus they are characterized by depression, frustration and low level of production (Allozi et al., 2022; Zirak & Ahmadian, 2015).

Emotional intelligence helps individuals resolve emotional problems in their daily life, enhance their perception of positive emotions, and reduce negative emotions, eventually bringing life satisfaction (Chew et al.2013; Agha et al., 2021). Generally, it is believed that individuals who perceive more positive and fewer negative emotions in their daily life have higher subjective well-being (Fallahzadeh, 2011; Nuseir et al. 2021). Overall, people with a high level of Emotional intelligence will still demonstrate better social adjustment ability and emotional regulation skills than will do those low in Emotional intelligence and their Emotional intelligence will have a positive connection with life satisfaction (Mayer et al. 2000). In addition, Ciarrochi et al. (2000) found that this positive effect continued to exist even after general intellectual and character factors were controlled.

2.2 *Team's Job Performance*

Job performance is one of the most important factors for the organizations 'success. Today, many organizations seek to improve their performance through several methods, such as: adopting the team's approach, engaging employees in decision-making and policy, and motivating employees financially and morally (Al-Hawary et al., 2013; Al-Nady et al., 2013; Metabis & Al-Hawary, 2013; Al-Hawary, 2015; Al-Nady et al., 2016; Al-Lozi et al., 2017; Al-Hawary & Alajmi, 2017; Al-Hawary & Al-Namlan, 2018; Al-Lozi et al. 2018; Mohammad et al., 2020; Allahow et al. 2018; Alhalalmeh et al., 2020; Al-Hawary et al., 2020; Al-Hawary & Al-Syasneh, 2020; AlTaweel & Al-Hawary, 2021). Job performance is a fundamental concept for all organizations, because it is a phenomenon that is described as comprehensive and fundamental to all fields of administrative knowledge, as well as the dimension of the organization's survival (Saeed & Shakeel, 2013; Amarneh et al. 2021; Shamout et al., 2022).

There are several definitions of job performance; Wahyudi (2013) defined Job performance as "the output of an employee doing business". According to Zhang et al. (2014), job performance is "The glossary of administrative science terms defined performance as: "achieving a set of duties and responsibilities based on the performance rate of the trained efficient worker". Taylor and Beh (2013) defined job performance as "the process of monitoring and analyzing the employees' work and behavior, judging the degree of their competence in the work they are currently doing, and judging the prospects of future progress in their work and the extent to which they can have more responsibilities".

According to Scott et al. (2015), job performance is closely related to the nature of the work done by the employee and his understanding of the operations he performs in various types until the job tasks are accomplished. Therefore, the performance is the outcome of human behavior in the framework of the techniques and procedures that guide the work towards achieving the desired goals (Al Kurdi et al., 2021; Scott et al., 2015).

According to Strom et al. (2014), there are two main factors making up the job performance of the working staff: Task Performance (Role Behavior) and Contextual Performance. Task Performance is traditionally defined as the ability of the employee to perform his or her duties and responsibilities according to the task description, whereas Contextual Performance is the individual effort that is indirectly related to the basic tasks and activities of the job but related to stimulating tasks and processes that constitute the organizational, social and psychological environment. In other words, while performing the functional task means fulfilling job requirements successfully, contextual performance is related to the quality of social relations with the different community members: young people, the elderly, and clients, a factor that is not always linked to the job (Strom et al., 2014).

Scott et al. (2015) divided contextual performance into two sub-dimensions: facilitation of personal relationships and job dedication. Facilitation of personal relationships consists of well-planned behaviors, tends to teamwork, and support of peer performance, whereas job dedication consists of disciplined internal behaviors such as work-related initiative, overtime work, and following of workplace rules.

2.3 The Relationship Between EI and Team' Job Performance

The extensive body of research advocates that capabilities of emotional intelligence contribute to excellent performance (Brouzos et al., 2014; Parker et al., 2004). Emotional intelligence's abilities are considered four times more significant than intelligence quotient (IQ) in deciding professional success and prestige. Emotional intelligence is a substantial predictor of performance in different work fields. Emotional skills, abilities and knowledge contribute to the enhancement of teamwork learning, their professional development for the success of higher levels of accomplishment, career success, personal wellbeing, and leadership (Chew et al., 2013).

Bar-On indicates that emotionally intelligent individuals are committed, excited and eager in their life, therefore they are usually found at the highest level of performance in their work team. Goleman (1995) expresses that emotional intelligence represents 80% of all learning while cognitive capabilities represent about 20%. Although it is a strong claim, but yet it needs to be adequately investigated that to what degree emotional intelligence really correlates to work performance, at individuals and teams.

Mayer et al. (2000) proposed that emotional intelligence improves with the increasing of age. They also reported that as emotional intelligence expanded, work-based success also improved and likewise, capabilities of communicating inspirational thoughts were also stimulated. Alternatively, as emotional intelligence diminished, oppositional behavior increased. Excellent emotional intelligence aptitudes are related to outstanding ability and capability to deal with one's own feelings and the feelings of other individuals (Brouzos et al., 2014). Based on the above literature the hypotheses study can be formulated as:

There is a significant impact of emotional intelligence (EI) on Teamwork Performance (TWP)

3 The Study Model

Below is a model that illustrates both the independent and dependent variables examined in this research study. EI's variables and their items were adopted from Bar-On, Bradberry and Greaves (2009), Extremera and Rey (2016), where TWP's variables

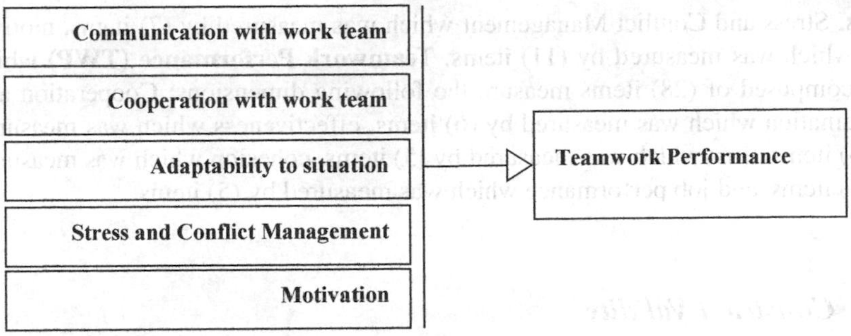

Fig. 1 Theoretical model

and their items were adopted from Froebel and Marchington, Finn et al., Higgs. All the variables' items were adapted to fit the study's environment (Fig. 1).

4 Methodology

4.1 Population and Sample of the Study

The study population consisted of all employees working in the IT private sector in Jordan. From all IT companies working in IT private sector in Jordan, four IT firms, at King Hussein Business Park (MICROSOFT, ORACLE, CISCO, IATA) were selected with a total number of (80) IT employees. Therefore, (80) questionnaires were provided to the respondents. After two weeks, (78) questionnaires were retrieved with percent of (97.5 %) of the total number of distributed questionnaires. Then, six questionnaires were excepted because they were improperly filled. Therefore, the final questionnaire's number was (72), with percent of (90%) of the total sample's number, which is very accepted for data analysis.

4.2 The Study Instrument

The questionnaire was developed as an instrument for this study, which was developed based on the literature relating to the current study topic and its concepts. **First section**: it includes the questions relating to the demographic data of the sample (Gender, Age, Scientific Qualification, and Experience). **Second section**: it includes the two main variables: **Emotional Intelligence (EI),** which was composed of (37) items measuring the following dimensions: Communication with the work team which was measured by (6) items, cooperation with the work team which was measured by (6) items, adaptability to situation which was measured by (7) items, Stress and Conflict Management which was measured by (7) items, motivation which was measured by (11) items. **Teamwork Performance (TWP)** which was composed of (28) items measure the following dimensions: Cooperation and coordination which was measured by (6) items, effectiveness which was measured by (6) items, trust which was measured by (5) items, cohesion which was measured by (6) items, and job performance which was measured by (5) items.

4.3 Construct Validity

To test the construct validity of the study, pilot test was carried out, in which (30) questionnaires were provided to (30) IT employees who were from the main study sample, (27) Questionnaires were retrieved and were valid for testing the construct validity and reliability of the study instrument. Construct validity measures the extent to which the objectives of the instrument have been achieved, and whether the instrument is able to measure its content for which it was designed. Correlation coefficient (Pearson Correlation) was found to determine the degree of item's correlation with the total score of its dimension. Items with negative or less than 0.25 degree of correlation are considered low and preferably deleted. (Tables 1 and 2) show the results of the construct validity.

Table 1 shows that correlations' coefficients between each item of EI with its main dimension ranges between (0.465–0.869) exceeding the cut-off level (0.25), and they are significant at ($\alpha \leq 0.05$), which confirms that all EI scale items are valid to measure EI variable.

Table 2 shows that correlations' coefficients between each item of TWP with its main dimension range between (0.560–0.883) exceeding the cut-off level (0.25), and they are significant at ($\alpha \leq 0.01$), which confirms that all TWP scale items are valid to measure TWP variable.

4.4 Testing Reliability

The Reliability of the study instrument was measured by calculating the value of Cronbach's Alpha coefficient, where the result is statistically acceptable if the Cronbach's Alpha coefficient is greater than (0.60), and the more this value is close to 1 (i.e. 100 %), the more the reliability of the study instrument is acceptable.

Table 3 shows that Cronbach Alpha for EI's dimensions and TWP Dimensions ranged between (0.838–0.941) and the total Cronbach Alpha for EI value is (0.917), which confirms that EI and TWP scale can be described by reliability and all data collected for measuring EI and TWP are reliable to measure all dimensions of EI and TWP.

Table 1 Correlation coefficients between each EI's item and its dimension

EI dimensions									
Communication with work team		Cooperation with work team		Adaptability to situation		Stress and conflict management		Motivation	
Item num	Correlation	Item num	Correlation	Item num	Correlation	Item num	Correlation	Item num	Correlation
1	0.713**	1	0.692**	1	0.856**	1	0.761**	1	0.710**
2	0.814**	2	0.817**	2	0.663**	2	0.768**	2	0.626**
3	0.637**	3	0.803**	3	0.627**	3	0.680**	3	0.465*
4	0.485*	4	0.788**	4	**0.403**	4	**0.341**	4	0.792**
5	0.590**	5	0.623**	5	0.794**	5	0.810**	5	0.748**
6	0.757**	6	0.645**	6	0.869**	6	0.757**	6	0.867**
7	**0.434**			7	0.843**	7	0.710**	7	0.739**

** Correlation is significant at the 0.01 level (2-tailed)
* Correlation is significant at the 0.05 level (2-tailed)

Table 2 Correlation coefficients between each TWP's item and its dimension (TWP) dimensions

Cooperation and coordination		Effectiveness		Trust		Cohesion		Job performance	
Item	Pearson correlation	Item	Pearson correlation	Item	Pearson correlation	Item	Pearson correlation	Item	Pearson correlation
1	**0.326**	1	0.628**	1	0.866**	1	0.881**	1	0.842**
2	0.883**	2	0.740**	2	0.714**	2	0.801**	2	0.851**
3	0.560**	3	0.670**	3	0.759**	3	0.800**	3	0.840**
4	**0.115**	4	0.879**	4	0.648**	4	0.804**	4	0.830**
5	0.874**	5	0.800**	5	0.584**	5	0.650**	5	0.820**
6	0.727**	6	0.819**			6	0.787**		

** Correlation is significant at the 0.01 level (2-tailed)
* Correlation is significant at the 0.05 level (2-tailed)

Table 3 Cronbach Alpha values for "EI" dimensions and TWP dimensions

Variables	(Cronbach Alpha)	Num. of items
Communication with work team	0.910	6
Cooperation with work team	0.909	6
Adaptability to situation	0.898	6
Stress and conflict management	0.917	6
motivation	0.902	7
Cooperation and coordination	0.862	4
Effectiveness	0.838	6
Trust	0.858	5
Cohesion	0.836	6
Job performance	0.941	5

4.5 Testing Normality (Kolmogorov–Smirnov & Shapiro Tests)

Kolmogorov-Smirnov test and Shapiro test were used for testing normality of data distribution, which is used in case the number of cases is ≥ 50. In these tests, data are described as normally distributed in case the significance is > (0.05).

Based on the tests' data shown in Table 4, the data of the two main variables (EI and TWP) was normally distributed, where the significance values for all dimensions were greater than 5%.

4.6 The Study Model Fit

According to the nature of data in this study, parametric methods are considered the most appropriate methods for measuring the impact of variables. Parametric methods include the diagnosis of "Multicollinearity" between independent variables and "Autocorrelation", where the ideal situation in multiple regression requires that the independent variables should be strongly associated with the dependent variable, at the same time should not be correlated with each other, where the strong correlation between the independent variables lead to a decrease in the value of (R), because the independent variables share the same variance of the dependent variable creating difficulty of determining the relative importance of each independent variable.

Table 4 Normal distribution test of data based on (K-S) values

EI variables	Mean	SD	Kolmogorov–Smirnov (Sig.)	Shapiro–Wilk (Sig.)
Communication with work team	4.0447	0.55113	0.200	0.327
Cooperation with work team	3.9127	0.61151	0.089	0.207
Adaptability to situation	3.9365	0.63108	0.096	0.263
Stress and Conflict Management	3.8095	0.65361	0.122	0.234
Motivation	3.9437	0.54884	0.113	0.166
TWP variables	*Mean*	*SD*	*Kolmogorov–***Smirnov** (**Sig.)**	*Shapiro–Wilk (Sig.)*
Cooperation and coordination	3.8571	0.57563	0.102	0.192
Effectiveness	3.9968	0.72318	0.123	0.077
Trust	3.9238	0.63396	0.132	0.200
Cohesion	3.9762	0.62425	0.235	0.317
Job performance	3.2762	0.7402	0.200	0.252

5 Testing Hypotheses

Two main tests were used for confirming that the study is devoid of multicollinearity; variance inflation factor (VIF) which should be (≤10), and tolerance which should be (≥1), and less 5 should VIF between ≥ 1 and ≤ 5. Table 5 shows that the values of VIF and Tolerance have met these two conditions, which indicates that there is not "Multicollinearity" between the independent variables, and thus they are suitable for multiple linear regression analysis relating to the main hypothesis.

Multiple regression analysis was employed to test the study hypotheses which states that: "There is a significant impact of emotional intelligence (EI) on Teamwork Performance (TWP)". In this model, emotional intelligence acts as the dependent variable and Teamwork Performance, as the independent variables. From the result as shown in Table 5, the regression model was statistically significant ($F = 357.106$; $R2 = 0.86$; $P = 0.000$). The adjR2 is 0.86, which means that 86% of the variation in Teamwork Performance can be explained by emotional intelligence. The proposed model was adequate as the F-statistic = 357.106 were significant at the 5% level ($p < 0.05$). This indicates that the overall model was reasonable fit and there was a statistically significant association between emotional intelligence and Teamwork Performance.

Table 5 also shows that Communication with work team (ß = 0.707, $p < 0.05$), Cooperation with work team (ß = 0.811; $p < 0.05$), Adaptability to situation (ß = 0.154, $p < 0.05$), Stress and conflict management (ß = 0.419; $p < 0.05$), and

Table 5 Regression summary of emotional intelligence and teamwork performance

Model	Standardized coefficients ß	T	SIG	Collinearity statistics	
				Tolerance	VIF
Communication with work team	0.707	3.500	0.002	0.459	2.180
Cooperation with work team	0.811	6.293	0.000	0.264	3.787
Adaptability to situation	0.154	2.336	0.034	0.287	3.481
Stress and conflict management	0.419	3.101	0.007	0.337	2.963
Motivation	0.703	3.116	0.003	0.367	2.728

Notes Adj. R 2 = 0.86; Sig. F = 0.000; F-value = 357.106; dependent variable, Teamwork Performance
$p < 0.05$

Motivation ($ß = 0.703; p < 0.05$) had a significant effect on Teamwork Performance. Cooperation with work team had the highest effect on Teamwork Performance. This provides evidence to support H1a, H1b, H1c, H1d, and H1e.

6 Discussion

The results showed that dimension of emotional intelligence (as a whole) has come within the high level, from the study sample individuals' perspective, where communication with work team obtained the highest degree, while stress and conflict management got the lowest degree. The follows explain detailed discussion of emotional intelligence and its dimensions.

The study results found that Communication with work team has been classified in the first rank, which means that the relationships among all teamwork members are based on respect and appreciation feelings even there is difference in opinions and ideas, which is regarded very important for teamwork performance. Also, the results confirmed that teamwork members are mostly having a dialogue mode when facing problems or variance in suggestions, which means that IT private sector companies reinforce their employees' interaction and integration by providing training and development programs that grows their personal and social skills.

The study results found that motivation was found at a high level with the fifth rank in IT industry in Jordan. This result proves that IT employees in Jordan are practicing their work with motivating emotions that support them to do work actively. This intrapersonal emotional motivation may be attributed to external motivation provided by the organization's management to increase their employees' job satisfaction, and thus increases their performance. This result indicates that IT employees may have stability motives, which are (motivation and emotional stability), that are considered integrated properties drawn from intrapersonal emotional intelligence.

The study results found that Adaptability to Situation was found at a high level with the third rank in IT private sector in Jordan. This result explains that teamwork employees in IT sector in Jordan behave emotionally adapting their thinking according to the situation conditions. This means they have ability to be flexible to accept new changes in their companies. The results proved that employee of IT sector in Jordan has a high level of emotional intelligence by understanding emotions and feelings of his team members. By this respond to the external environment represented in other employees in the organization but outside the teamwork, and IT sector employee has typically the concept of emotional intelligence defined by Goleman (1995) that emotional intelligence is the individual's ability to understand his emotions, the emotions of others, their knowledge and directions, control and deal with them positively, motivate the self, manage emotions and relationships with others effectively.

The study results found that Cooperation with the Work Team was found at a high level with the second rank in IT private sector in Jordan. This explains that teamwork members have good understanding of the team goals and objectives, and thus try to work in a cooperative manner to achieve those objectives effectively. The results indicated the teamwork member has willingness to help each other by providing suitable information to his colleagues for purpose of helping them to accomplish their job's tasks. This cooperative behavior goes beyond this advantage when the teamwork member tries to facilitate the tasks required from the colleagues as initiative to work team even if the colleagues have not asked the help. But this advantage must be activated in case of absence of some members, which means that some job' s positions are related to specific employees but not integrated to others.

The study results found that Conflict and Stress Management was found at a high level with the fourth rank in IT private sector in Jordan. Teamwork has many stressful situations that need a person who deals with the stresses and conflicts flexibility and wisely by control of emotions. IT sector mostly depends on cooperation and interaction among team members, which entails stresses tolerance and flexible ability to manage teamwork-based conflicts. This result confirms that IT employees in Jordan have ability to manage stressful situations and deal with conflicts intelligently complying to the concept of Conflict management included in Goleman model (1995) and Singh model (2006) that expressed conflict management as The individual's ability to negotiate, resolve disputes, deal with difficult characters and long-term problems in a diplomatic and tactical method, in addition to the encouragement of open debate, as well as focus on latent and invisible conflict points.

The study results found that the Teamwork Performance has come in the high level at IT sector in Jordan. This refers to that IT private sector in Jordan depends on teamwork as a cooperative and collaborative work. Such this work enhances the investment of the organizational capabilities and core competences in a synergy mode. This high level of teamwork performance in IT sector is attained as outcome of multi-diversified skills' combination at the individuals of work team. This interpretation is supported by Frykman et al. (2014), where they asserted that any professional skills, such as communicative skills, positive interaction, conflicts solving, have become very critical for an organization aspiring for competitive position. Such these highly required skills can be achieved just by teamwork, therefore managers and employers are looking for individuals who can practice their work in very collaborative environment (Frykman et al., 2014).

The study found that emotional intelligence at employees working in IT sectors has a significant impact on teamwork performance. This result explains that emotional intelligence at individuals belonging to teamwork supports and motivates the employees' ability and persistence to achieve their work activities effectively increasing their teamwork performance. This result supports Bar-On opinion who mentioned that emotionally intelligent individuals are committed, excited and eager in their life, therefore they are usually found at the highest level of performance in their work team. Also, the results support opinion of Goleman (1995), who expressed that emotional intelligence represents 80% of all learning while cognitive capabilities represent about 20%, which means that when IT teamwork's' employees are found intelligently emotional, they will be more eligible to learning. Moreover, this result can be interpreted by Chew et al. (2013), by conforming those Emotional skills, abilities and knowledge enhance effectively team work learning, their professional work achievement, career success, and leadership. Also, supported by Qualter et al. found that Emotional intelligence is a substantial predictor of performance in different work fields. The result is consistent with the study of Higgs, which confirmed that Emotional intelligence is strongly correlated to the performance of employees' staff. The result is consistent with the study of Law et al. (2007), which found that Emotional intelligence is a true predictive tool about job performance, and it comes consistently with the study of Lee et al. (2018), which showed Positive correlation between emotional intelligence and teamwork skills post-workshop. Also, the result comes consistently with the study of Johar et al. (2019), which found emotional intelligence with its dimensions (emotional awareness, emotional management, and social emotional awareness) positively correlated with academic performance.

7 Recommendations

According to the results of the study, the researcher recommends managers and decision makers to build good relationships among employees to facilitate the communication of information and ideas among the team members. This flexible information communication between the team members should motivate the innovative and creative thinking at the IT employees effectively. Teamwork members should activate the process of knowledge sharing and application. Knowledge sharing among the team members will make the new concept easily shared and valuable at all the employees, and according to which they can develop their work performance. IT employees should have enough ability to overwhelm disputes occurring during the practice of work activities, which will make the work be achieved in smoother and faster mode. This positive behavior of IT employee may be supported and be more enhanced when IT employee has an initiative to give the most priority to the work duties more than personal stubbornness. IT employees should try to get an opportunity that may lead to their work improvement and their skills development, whether inside or outside the organization. Therefore, they should try to be included in training programs that may develop their options' capabilities and improve their functional skills so that they can be able to adapt to new business processes and new IT development. Finally, Leaders of companies in Jordanian IT should build proper mechanism that can enable the teamwork members to work in an integrative manner. Which mean that an employee must be distributed in his team position based on his skills and capabilities. Such this mechanism of employees' distribution should lead the teamwork employees to practice their work activities interactively and consistently overcoming some functional gaps penetrating teamwork performance.

References

Agha, K., Alzoubi, H. M., & Alshurideh, M. T. (2021, June). Measuring reliability and validity instruments of technologically driven cognitive intrusion towards work-life balance. In *The International Conference on Artificial Intelligence and Computer Vision* (pp. 601–614). Springer.

Al Kurdi, B., Elrehail, H., Alzoubi, H., Alshurideh, M., & Al-Adaila, R. (2021). The interplay among HRM practices, job satisfaction and intention to leave: An empirical investigation. *Journal of Legal, Ethical and Regulatory, 24*(1), 1–14.

Alameeri, K., Alshurideh, M., Al Kurdi, B., & Salloum, S. A. (2020, October). The effect of work environment happiness on employee leadership. In *International Conference on Advanced Intelligent Systems and Informatics* (pp. 668–680). Springer.

Al-Dhuhouri, F. S., Alshurideh, M., Al Kurdi, B., & Salloum, S. A. (2020, October). Enhancing our understanding of the relationship between leadership, team characteristics, emotional intelligence and their effect on team performance: A critical review. In *International Conference on Advanced Intelligent Systems and Informatics* (pp. 644–655). Springer.

Alhalalmeh, M. I., Almomani, H. M., Altarifi, S., Al-Quran, A. Z., Mohammad, A. A., & Al-Hawary, S. I. (2020). The nexus between corporate social responsibilty and organizational performance in Jordan: The mediating role of organizational commitment and organizational citizenship behavior. *Test Engineering and Management, 83*, 6391–6410.

Al-Hawary, S. I. (2015). Human resource management practices as a success factor of knowledge management implementation at health care sector in Jordan. *International Journal of Business and Social Science, 6*(11/1), 83–98.

Al-Hawary, S. I., & Alajmi, H. M. (2017). Organizational commitment of the employees of the ports security affairs of the State of Kuwait: The impact of human recourses management practices. *International Journal of Academic Research in Economics and Management Sciences, 6*(1), 52–78.

Al-Hawary, S. I., & Al-Namlan, A. (2018). Impact of electronic human resources management on the organizational learning at the private hospitals in the State of Qatar. *Global Journal of Management and Business Research: A Administration and Management, 18*(7), 1–11.

Al-Hawary, S. I., & Al-Syasneh, M. S. (2020). Impact of dynamic strategic capabilities on strategic entrepreneurship in presence of outsourcing of five stars hotels in Jordan. *Business: Theory and Practice, 21*(2), 578–587.

Al-Hawary, S. I., Al-Qudah, K., Abutayeh, P., Abutayeh, S., & Al-Zyadat, D. (2013). The impact of internal marketing on employee's job satisfaction of commercial banks in Jordan. *Interdisciplinary Journal of Contemporary Research in Business, 4*(9), 811–826.

Al-Hawary, S. I. S., Mohammad, A. S., Al-Syasneh, M. S., Qandah, M. S. F., & Alhajri, T. M. S. (2020). Organisational learning capabilities of the commercial banks in Jordan: Do electronic human resources management practices matter? *International Journal of Learning and Intellectual Capital, 17*(3), 242–266.

Allahow, T. J. A. A., Al-Hawary, S. I. S., & Aldaihani, F. M. F. (2018). Information technology and administrative innovation of the Central Agency for Information Technology in Kuwait. *Global Journal of Management and Business, 18*(11-A), 1–16.

Allozi, A., Alshurideh, M., AlHamad, A., & Al Kurdi, B. (2022). Impact of transformational leadership on the job satisfaction with the moderating role of organizational commitment: Case of UAE and Jordan manufacturing companies. *Academy of Strategic Management Journal, 21*, 1–13.

Al-Lozi, M., Almomani, R. Z., & Al-Hawary, S. I. (2017). Impact of talent management on achieving organizational excellence in Arab Potash Company in Jordan. *Global Journal of Management and Business Research: A Administration and Management, 17*(7), 15–25.

Al-Nady, B. A., Al-Hawary, S. I., & Alolayyan, M. (2013). Strategic management as a key for superior competitive advantage of sanitary ware suppliers in Kingdom of Saudi Arabia. *International Journal of Management and Information Technology, 7*(2), 1042–1058.

Al-Nady, B. A., Al-Hawary, S. I., & Alolayyan, M. (2016). The role of time, communication, and cost management on project management success: An empirical study on sample of construction projects customers in Makkah City, Kingdom of Saudi Arabia. *International Journal of Services and Operations Management, 23*(1), 76–112.

Alsharari, N. M., & Alshurideh, M. T. (2020). Student retention in higher education: the role of creativity, emotional intelligence and learner autonomy. *International Journal of Educational Management, 35*(1), 233–247.

AlShehhi, H., Alshurideh, M., Al Kurdi, B., & Salloum, S. A. (2020, October). The impact of ethical leadership on employees performance: A systematic review. In *International Conference on Advanced Intelligent Systems and Informatics* (pp. 417–426). Springer.

Alshurideh, M., Salloum, S. A., Al Kurdi, B., & Al-Emran, M. (2019, February). Factors affecting the social networks acceptance: an empirical study using PLS-SEM approach. In *Proceedings of the 2019 8th International Conference on Software and Computer Applications* (pp. 414–418).

AlTaweel, I. R., & Al-Hawary, S. I. (2021). The mediating role of innovation capability on the relationship between strategic agility and organizational performance. *Sustainability, 13*(14), 7564.

Alzoubi, H., Alshurideh, M., Akour, I., Shishan, F., Aziz, R., & Al Kurdi, B. (2021) Adaptive intelligence and emotional intelligence as the new determinant of success in organizations. An empirical study in Dubai's real estate. *Journal of Legal, Ethical and Regulatory Issues, 24*(Special Issue 6), 1–15.

Amarneh, B. M., Alshurideh, M. T., Al Kurdi, B. H., & Obeidat, Z. (2021, June). The impact of COVID-19 on E-learning: Advantages and challenges. In *The International Conference on Artificial Intelligence and Computer Vision* (pp. 75–89). Cham: Springer.

Batool, B. F. (2013). Emotional intelligence and effective leadership. *Journal of Business Studies Quarterly, 5*, 84–94.

Bradberry, T., & Greaves, J. (2009). *Emotional Intelligence*. Publisher Group West.

Brouzos, A., Misailidi, P., & Hadjimattheou, A. (2014). Associations between emotional intelligence, socio-emotional adjustment, and academic success in childhood the influence of age. *Canadian Journal of School Psychology, 29*(2), 83–99.

Chew, B. H., Zain, A. M., & Hassan, F. (2013). Emotional intelligence and academic performance in first and final year medical students: A cross-sectional study. *BMC Medical Education, 13*(44), 1–10.

Ciarrochi, J. V., Chan, A. Y., & Caputi, P. (2000). A critical evaluation of the emotional intelligence construct. *Personality and Individual Differences, 28*, 539–561.

Daus, C. S., & Ashkanasy, N. M. (2005). The case for the ability-based model of emotional intelligence in organizational behaviour. *Journal of Organizational Behavior, 6*(4), 453–466.

Extremera, N., & Rey, L. (2016). Ability emotional intelligence and life satisfaction: Positive and negative affect as mediators. *Personality and Individual Differences, 102*, 98–101.

Fallahzadeh, H. (2011). The Relationship between Emotional Intelligence and Academic Success in medical science students in Iran. *Procedia-Social and Behavioral Sciences*, 1461–1466.

Frykman, M., Hasson, H., Macfarlane, F., Muntlin, A., & Schwarz, U. (2014). Functions of behavior change interventions when implementing multi-professional teamwork at an emergency department: A comparative case study. *BMC Health Services Research, 14*, 212–228.

Gignac, G. E. (2010). On a nomenclature for emotional intelligence research. *Industrial and Organizational Psychology, 3*, 131–135.

Goleman, D. (1995). *Emotional Intelligence*. Bantam.

Goleman, D. (1998). *Working with Emotional Intelligence*. Bantom Books.

Huang, N., & Lee, H. (2019). Ability emotional intelligence and life satisfaction: Humor style as a mediator. *Social Behavior and Personality: An International Journal, 47*(5), e7805.

Johar, N., Ehsan, N., & Khan, M. (2019). Association of emotional intelligence with academic performance of medical students. *Pak Armed Forces Med, 69*(3), 455–459.

Kafetsios, K., & Zampetakis, L. A. (2008). Emotional intelligence and job satisfaction: Testing the mediatory role of positive and negative affect at work. *Personality and Individual Differences, 44*(3), 710–720.

Law, K., Wong, C., Huang, G., & Lim, X. (2007). The effects of emotional intelligence on job performance and life satisfaction for the research and development scientists in China. *Asia Pacific Journal of Management, 25*(1), 51–69.

Lee, C., Bristow, M., & Wong, J. (2018). Emotional intelligence and teamwork skills among undergraduate engineering and nursing students: A Pilot Study. *Journal of Research in Interprofessional Practice and Education, 8*(1).

Lopes, P. N., Salovey, P., Cote, S., & Beers, M. (2005). Emotion regulation ability and the quality of social interaction. *Emotion, 5*(1), 113–118.

Marzuki, N., Saad, Z., & Mustaffa, C. (2015). Emotional intelligence: Its relations to communication and information technology skills. *Asian Social Science, 11*(115), 267–273.

Mayer, J., & Salovey, P. (1997). What is emotional intelligence? In P. Salovey & D. J. Sluyter (Eds.), *Emotional Development and Emotional Intelligence: Educational Implications* (pp. 3–34). Harper Collins.

Mayer, J. D., Salovey, P., & Caruso, D. R. (2000). Models of emotional intelligence. In R. J. Sternberg (Ed.), *Handbook of Emotional Intelligence* (pp. 396–420). Cambridge University Press.

Metabis, A., & Al-Hawary, S. I. (2013). The impact of internal marketing practices on services quality of commercial banks in Jordan. *International Journal of Services and Operations Management, 15*(3), 313–337.

Moghadam, A. H., Tehrani, M., & Amin, F. (2011). Study of the relationship between emotional intelligence (ei) and management decision making styles. *World Applied Sciences Journal*, 1017–1025.

Mohammad, A. A., Alshura, M. S., Al-Hawary, S. I. S., Al-Syasneh, M. S., & Alhajri, T. M. (2020). The influence of internal marketing practices on the employees' intention to leave: A study of the private hospitals in Jordan. *International Journal of Advanced Science and Technology, 29*(5), 1174–1189.

Newman, D., & Joseph, D. (2010). Emotional intelligence and job performance: The importance of emotion regulation and emotional labor context. *Industrial and Organizational Psychology, 3*, 159–164.

Nuseir, M. T., Aljumah, A., & Alshurideh, M. T. (2021). How the business intelligence in the new startup performance in UAE during COVID-19: The mediating role of innovativeness. *The Effect of Coronavirus Disease (COVID-19) on Business Intelligence, 334*, 63–79.

Obeidat, Z. M., Alshurideh, M. T., Al Dweeri, R., & Masa'deh, R. (2019). The influence of online revenge acts on consumers psychological and emotional states: Does revenge taste sweet. In *Proceedings of the 33rd International Business Information Management Association Conference, IBIMA* (pp. 4797–4815).

Odeh, R. B. M., Obeidat, B. Y., Jaradat, M. O., & Alshurideh, M. T. (2021). The transformational leadership role in achieving organizational resilience through adaptive cultures: the case of Dubai service sector. *International Journal of Productivity and Performance Management, Vol. ahead-of-print No. ahead-of-print*. https://doi.org/10.1108/IJPPM-02-2021-0093.

Parker, J. D., Summerfeldt, L. J., Hogan, M. J., & Majeski, S. (2004). Emotional intelligence and academic success: Examining the transition from high school to university. *Personality and Individual Differences, 36*, 163–172.

Saeed, R., & Shakeel, M. (2013). Ethical behavior and employees job performance in education sector of Pakistan. *Middle-East Journal of Scientific Research, 18*(4), 524–529.

Salovey, P., & Mayer, J. (1990). Emotional intelligence. *Imagination, Cognition, and Personality, 9*, 185–211. https://doi.org/10.2190/DUGG-P24E-52WK-6CDG

Scott, B. A., Garza, A. S., Conlon, D. E., & Kim, Y. J. (2015). Why do managers act fairly in the first place? A daily investigation of "hot" and "cold" motives and discretion. *Academy of Management Journal, 5*(1), 37–57.

Shamout, M., Elayan, M., Rawashdeh, A., Kurdi, B., & Alshurideh, M. (2022). E-HRM practices and sustainable competitive advantage from HR practitioner's perspective: A mediated moderation analysis. *International Journal of Data and Network Science, 6*(1), 165–178.

Singh, D. (2006). *Emotional Intelligence at Work* (3rd ed.). Response Books.

Strom, D. L., Sears, K. L., & Kelly, K. M. (2014). Work engagement: The roles of organizational justice and leadership style in predicting engagement among employees. *Journal of Leadership & Organizational Studies, 21*(1), 71–82.

Stys, Y. & Brown, S. (2004). *Review of the emotional intelligence literature and implications for corrections*. Research Branch, Correctional Service of Canada, 340 Laurier Ave., West, Ottawa, Ontario, K1A 0P9.

Taylor, J., & Beh, L. (2013). The impact of pay-for-performance schemes on the performance of Australian and Malaysian government employees. *Public Management Review, 15*(8), 1090–1115.

Wahyudi, A. (2013). The Impact of work ethics on performance using job satisfaction and affective commitment as mediating variables: Evidences from lecturers in Central Java. *Issues in Social and Environmental Accounting, 7*(3), 165–184.

Wijekoon, C. N., Amaratunge, H., De Silva, Y., Senanayake, S., Jayawardane, P., & Senarath, U. (2017). Emotional intelligence and academic performance of medical undergraduates: A cross-sectional study in a selected university in Sri Lanka. *BMC Medical Education, 17*, 176.

Zhang, Y., Lepine, J., Buckman, B., & Wei, F. (2014). It's not fair … or is it?: The role of justice and leadership in explaining work stressor–job performance relationships. *Academy of Management Journal, 57*(3), 675–697.

Zirak, M., & Ahmadian, E. (2015). Relationship between emotional intelligence & academic success emphasizing on creative thinking. *Mediterranean Journal of Social Sciences, 6*(5S2), 561–570.

The Impact of Functional Withdrawal on Organizational Commitment as Perceived by Nurses Working in Public Hospitals in Jordan

Raed Ismael Ababneh, Bashaier Hatem Khasawneh, Reham Zuhier Qasim Almomani, Main Naser Alolayyan, Ziad Mohd Ali Smadi, Hanan Mohammad Almomani, Fatima Lahcen Yachou Aityassine, Sulieman Ibraheem Shelash Al-Hawary, and Dheifallah Ibrahim Mohammad

Abstract The major aim of the study was to examine the impact of functional withdrawal on organizational commitment. Therefore, it focused on Jordanian public hospitals operating in the north region. Data were primarily gathered through self-reported questionnaires created by Google Forms which were distributed to a random

R. I. Ababneh
Policy, Planning, and Development Program, Department of International Affairs, College of Arts and Sciences, Qatar University, 2713 Doha, Qatar

B. H. Khasawneh
Department of Public Administration, Faculty of Economics and Administrative Sciences, Yarmouk University, P.O. Box 566, Irbid 21163, Jordan

R. Z. Q. Almomani
Business Administration, Amman, Jordan

M. N. Alolayyan
Health Management and Policy Department, Faculty of Medicine, Jordan University of Science and Technology, Ar-Ramtha, Jordan
e-mail: mnalolayyan@just.edu.jo

Z. M. A. Smadi · H. M. Almomani · S. I. S. Al-Hawary (✉)
Department of Business Administration, School of Business, Al al-Bayt University, P.O. Box 130040, Mafraq 25113, Jordan
e-mail: dr_sliman73@aabu.edu.jo

Z. M. A. Smadi
e-mail: ziad38in@aabu.edu.jo

F. L. Y. Aityassine
Department of Financial and Administrative Sciences, Irbid University College, Al-Balqa' Applied University, As-Salt, Jordan
e-mail: Fatima.yassin@bau.edu.jo

D. I. Mohammad
Ministry of Education, Amman, Jordan

M. Alshurideh et al. (eds.), *The Effect of Information Technology on Business and Marketing Intelligence Systems*, Studies in Computational Intelligence 1056,
https://doi.org/10.1007/978-3-031-12382-5_59

sample of (306) nurses via email. Structural equation modeling (SEM) was conducted to test hypotheses. The results demonstrated in Table 2 show that functional withdrawal has a negative impact on organizational commitment. In light of the findings of the study, managers and decision makers in hospital management should provide support, material and moral incentives, and educational, development and promotion opportunities for employees, especially nurses, with the need to focus on knowing the needs of nurses to increase their connection with their organization and reduce their tendency to search for new opportunities and leave work.

Keywords Functional withdrawal · Organizational commitment public hospitals · Jordan

1 Introduction

Organizations seek to find and provide an appropriate environmental and practical environment in which the employee feels his importance and his role in achieving the organization's goals (Al-Hawary & Al-Hamwan, 2017; AlTaweel & Al-Hawary, 2021). Despite the efforts made by these organizations to satisfy the workers and improve their performance (Al-Nady et al., 2013; Allahow et al., 2018; Al-Hawary & Al-Syasneh, 2020; Mohammad et al., 2020), the workers must be exposed to pressures and circumstances that push them to carry out practices and behaviors that have a negative impact on the individual and the organization, which may make the individual fall under the practice and behavior of the so-called career withdrawal, which is one of the most behaviors (Sweissi & Mahjar, 2017). Job withdrawal is one of the most important administrative problems faced by organizations, which troubles working managers and researchers in organizational affairs because of its negative effects and the material and moral loss it causes to the organization and to individuals, especially if the organization spends a lot of its financial resources on training and developing the employee to become an expert and efficient (Al-Hawary, 2009; Al-Hawary & Nusair, 2017; Al-Hawary et al., 2012; Sweissi & Mahjar, 2017).

Job withdrawal occurs when the employee is not satisfied with his work, which negatively affects job performance and production, as well as commitment to the job and the organization (Al Kurdi et al., 2021; Allozi et al., 2022). As for psychological withdrawal, it is a set of actions and behaviors that an individual performs to escape and stay away from performing the tasks and duties assigned to him, where the individual is physically present and mentally and psychologically absent from work, among these practices include daydreaming, building informal social relationships, pretending to be busy, surfing the Internet and doing other work during the official working hours that have nothing to do with the job duties he performs (Sweissi & Mahjar, 2017). Job withdrawal, both physical and psychological, plays an important role in the decrease in organizational loyalty, which increases the losses of organizations and job instability, which in turn will negatively affect the level of overall performance (Al Kurdi et al., 2020a, b; Alameeri et al., 2020).

The health sector is one of the important sectors in which the medical staff has a decisive role in providing treatment services to patients in an efficient and effective manner (Aburayya et al., 2020a, b). However, the presence of job withdrawal behaviors, especially for nurses, will increase the degree of their instability and loyalty to the hospital in which they work, and thus lower the level of health services provided to citizens. It was noticed that the percentage of employees late to come to work on time and the increase in spending of official working hours outside the departments in which they work through social conversations, in addition to the observation of nurses' preoccupation with using modern means of communication, smart phones and social media in abundance during working hours with no attention to achievement and complete the tasks required (Aburayya et al., 2020c, d; Alshurideh, 2014).

The mentioned practices fall under the organizational term, which is job withdrawal, as it is considered one of the negative and unhealthy phenomena in organizations because of its undesirable effects at the individual, group and organizational levels (AlShehhi et al., 2020; Hayajneh et al., 2021). Withdrawal is a way for an individual to escape from the reality of work, whose negative effects are reflected on the organization, which requires striving and working hard to avoid them by controlling these behaviors and standing at the level of their practice and finding appropriate solutions to them (Alsuwaidi et al., 2020; Al-bawaia et al., 2022). The interest of researchers and business organizations has increased recently in issues related to organizational loyalty, as it is one of the most important goals that organizations seek to gain from their employees to increase the strength of the employee's attachment and his love to work because of its impact on job performance. Therefore, this study seeks to find out the effect of practicing job withdrawal behaviors on the organizational commitment of workers in the Jordanian health sector (Alshurideh, 2019, 2022; Alshurideh et al., 2020; Alzoubi et al., 2022).

2 Theoretical Framework and Hypotheses Development

2.1 *Functional Withdrawals*

Functional withdrawal is defined as a set of behaviors and practices carried out by an individual that express separation from his organization and his unwillingness to work in it (Redmond, 2010). Functional withdrawal is also defined as the employee's interruption or cessation of work in his organization with his desire, choice, and transfer to work in another organization (Al-Ghanim, 2003). Kaplan et al. (2009) defines functional withdrawal as a set of behaviors and practices that the employee performs when he decides not to stay in his work for reasons beyond his control so that he is less accomplished, performing and participating in work. There are two types of functional withdrawal, namely physical withdrawal and psychological withdrawal, and that both types It is not separated from the other and that they lead to the result of each other, but this does not mean that the employee practices all behaviors at the same time. The functional withdrawal includes two types.

Psychological Withdrawal: Psychological withdrawal behaviors represent the individual's presence physically, but his absence mentally and psychologically (Sweissi & Mahjar, 2017). Lim et al. (2002) defined psychological withdrawal as a set of behaviors that the employee performs, which is represented by his physical presence and his mental absence from his work environment, which is represented by several behaviors, namely daydreaming, electronic surfing, pretending to be busy, doing side jobs, and building informal social relationships. From an organizational point of view, daydreaming is defined as showing the employee his preoccupation with his work and camouflaging, but he is originally busy and thinking about other things that are not related to work, and it is a way to relieve work pressure and escape from reality without seeking to solve the problem, so that these dreams become a means of escaping and withdrawing from the organization (Sweissi & Mahjar, 2017). Building informal social relations is clearly a bad withdrawal behavior that negatively affects the functioning and safety of work on the one hand, and it has a positive impact by increasing cooperation and love of teamwork and cooperation between individuals at work. It is considered entertainment for the employee that distances him from the work atmosphere. As for doing side jobs (**Moon Lighting**), it is the individual's exploitation of work time and resources to complete other tasks outside the scope of work or for the benefit of a job (Lim et al., 2002).

In this case, the individual is physically present at his place of work, but all his energy and effort are concentrated in another personal work that is not related to his current work. **Looking busy**: It is for the employee to appear intentionally that he is busy with something even though he is not doing any task, and the reason for this may be in order to avoid his manager assigning him other tasks. By pretending to organize papers and files or sitting at a computer pretending to complete a work-related task when in reality there is no work to be done (Hanish & Hulin, 1990). As for **Cyber loafing** is the use of the Internet and social media during official working hours for personal purposes and not for work purposes, and this behavior is considered a way to combat negative working conditions (Chen & Lim, 2012).

Physical withdrawal, which is defined as a set of behaviors and actions that the individual practices to avoid performing the tasks and duties entrusted to him, and the behaviors of physical withdrawal are represented in several practices, including (**Tardiness**): It is the person's arrival late to work or leaving early before the end of the work. **Long breaks**: There are several reasons that drive the employee, the most important of which is the feeling of dissatisfaction with work, the lack of incentive to encourage working hard and boredom from the work routine. **Missing Meetings**, it is the employee's ignoring important meetings and work meetings that are part of his job. **Absenteeism**: It is the employee's failure to attend work, either with an excuse, because of illness or an emergency, or it may be without justification and is represented by the employee's absence for one or more days from work (Harrison & Price, 2003). And the last one is to leave work with the individual's desire or to resign (**Quitting**) due to circumstances related to the work itself or the employee, and it is the last solution for the employee and the highest degree of physical withdrawal.

2.2 Organizational Commitment

The concept of organizational commitment is one of the prevailing organizational concepts and one of the goals that all organizations strive to achieve (Alhalalmeh et al., 2020; Al-Hawary, 2013; Al-Hawary & Alajmi, 2017; Al-Hawary & Al-Smeran, 2017; Al-Hawary & Obiadat, 2021; Al-Hawary et al., 2020). Commitment, in the linguistic and idiomatic sense, expresses sincerity and fulfillment of the covenant, and it is an individual's feeling of love and devotion to something, his attachment, and his feeling of responsibility towards this thing (Al-Hawary et al., 2013; Al-Hawary & Abu-Laimon, 2013; Metabis & Al-Hawary, 2013; Al-Hawary, 2015; Al-Lozi et al., 2017; Al-Hawary & Al-Namlan, 2018; Al-Lozi et al., 2018; Al-Quran et al., 2020). Steers and Mowady (1974) defined commitment as: the strength of belonging and integration of the individual within the organization in which he works and his effective contribution to it, with a strong desire to continue working and belief in the organization's goals. It is the extent of the strong connection and integration between the individual and the organization, which leads him to accept the goals and values of the organization and work hard to achieve what the organization desires to achieve, and the desire to continue working in it and remain as a member (Al-Hawary, 2011; Al-Hawary & Aldaihani, 2016; Al-Hawary & Hadad, 2016; Al-Nady et al., 2016).

Judeh (2004) defined commitment as: a state of connection between an individual and his organization represented by his belief in the goals and values of this organization and through which he expresses his desire to work in it. Al-Fahdawi (2004) defined it as the level of the employee's positive feeling about his organization and commitment to its values and objectives, and the constant feeling of connection, pride in it, and his love for working, regardless of the financial return that he will derive from the work. Also, Al-Rawashdah (2007) defined organizational commitment as the unwillingness of the individual to leave his work in the organization in which he works for any reason, whether financial or other reasons, and his feeling that he is an integral part of the organization and his attachment to it and its values while striving to achieve its goals.

The dimensions of organizational commitment differ according to the entrance and axis through which it is studied, as the dimensions of organizational commitment varied. Organizational commitment consists of several components linked to each other and not separate, as these dimensions were classified in several models, including the model (Allen & Meyer, 1996), which is the model adopted by the study, where it classified the dimensions of organizational commitment as follows; **Affective Commitment**: It is the individual's feeling of pride in belonging to the organization in which he works and adopting its activities in a positive way, in addition to his feeling that he is part of the organization and that its problems belong to him, as the individual feels that he has strong relationships and friendships that work to keep him in the organization with a brotherly atmosphere that attracts and attracts him To continue working in it (Awaida, 2008). **Normative Commitment**: It expresses the individual's sense of obligation to remain in the organization and not to think about leaving it, and the feeling of guilt when leaving it because of a moral and ethical commitment that works to keep him in the organization. **Continuance Commitment**: It is the individual's assessment of the benefits that he will derive and the losses that will result from staying with the organization or leaving it. This evaluation is affected by several factors, including the abundance or lack of available alternatives and jobs, the number of years of work in the organization itself, and the difference in systems between organizations (Alshraideh et al., 2017; Al-Khayyal et al., 2020; Alketbi et al., 2020).

2.3 *Functional Withdrawal and Organizational Commitment*

Making the organization the perfect place for the employee, motivating him, strengthening relations between the members of the organization, and improving working conditions, which leads in one way or another to a feeling of commitment to the organizations. Many studies have shown that these practices differ from one person to another depending on personal factors and characteristics such as gender, age and other characteristics that enable a person to deal with the circumstances surrounding him (Turan, 2015). Studies have shown that the practice of these behaviors is associated with several organizational factors such as the level of job satisfaction and job and organizational commitment (Bennett & Naumann, 2004). The employee's withdrawal practices reduce the effort he exerts at work, while not paying attention to the interest of the organization, and his interest is in preserving the material benefit that he achieves from staying at work in it (Harrison et al., 2006).

Several studies have mentioned functional withdrawal behaviors within the list of behaviors that the employee commits by stealing and exploiting work time by practicing behaviors that do not serve the interest and workflow, but rather negatively reflect on the performance of the organization and achieve the individual interest of the person who practices these behaviors (Brock et al., 2013). The employee's practice of functional withdrawal behavior is also considered as a disregard for the work and its requirements in a negative way that is reflected on the quality of the services provided. The researcher believes that being late for work becomes a withdrawal behavior if it occurs frequently and almost daily and in all periods of work and without justification or an official excuse, as the employee does not feel guilty because of his practice of this behavior. Based on the foregoing theoretical literature, the study hypothesis can be formulated as follows.

There is a significant effect of functional withdrawal on organizational commitment.

3 Study Model

See Fig. 1.

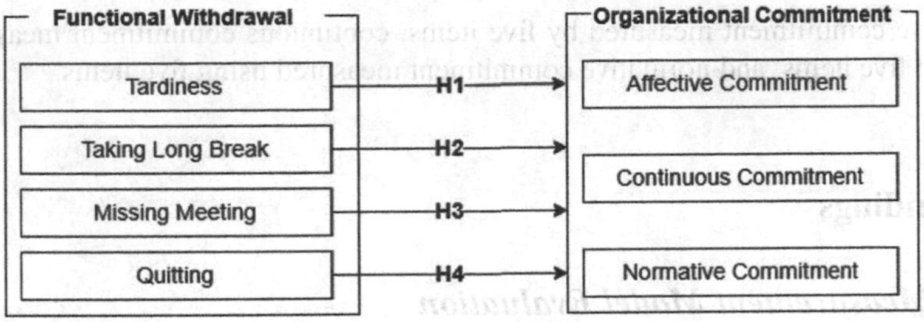

Fig. 1 Research model

4 Methodology

4.1 *Population and Sample Selection*

A qualitative method based on a questionnaire was used in this study for data collection and sample selection. The major aim of the study was to examine the impact of functional withdrawal on organizational commitment. Therefore, it focused on Jordanian public hospitals operating in the north region. Data were primarily gathered through self-reported questionnaires created by Google Forms which were distributed to a random sample of (306) nurses via email. In total, (285) responses were received including (18) invalid to statistical analysis due to uncompleted or inaccurate. Hence, the final sample contained (267) responses suitable to analysis requirements that were formed a response rate of (87.25%), where it proved to be sufficient to the extent that was predictable and allowed for a presumption of data saturation (Sekaran & Bougie, 2016).

4.2 *Measurement Instrument*

A self-reported questionnaire that consists of two main sections along with a section regarding control variables was used as the measurement instrument. Control variables considered as categorical measures were composed of gender, age group, educational level, and experience. The two main sections were dealt with a five-point Likert scale (from 1 = strongly disagree to 5 = strongly agree). The first section contained (17) items to measure functional withdrawal based on (Chen & Lim, 2012). These questions were distributed into dimensions as follows: four items dedicated for measuring tardiness, four items dedicated for measuring taking long break, four items dedicated for measuring missing meeting, and five items dedicated for measuring quitting. Whereas the second section included (15) items developed to measure organizational commitment according to what was pointed by (Allen & Meyer, 1996). This variable was a second-order construct divided into three first-order constructs.

Affective commitment measured by five items, continuous commitment measured through five items, and normative commitment measured using five items.

5 Findings

5.1 Measurement Model Evaluation

This study was conducted structural equation modeling (SEM) to test hypotheses, which represents a contemporary statistical technique for testing and estimating the relationship between factors and variables (Al-Adamat et al., 2020; Al-Gasawneh & Al-Adamat, 2020; Wang & Rhemtulla, 2021). Accordingly, the reliability and validity of the constructs were tested using confirmatory factor analysis (CFA) through the statistical program AMOSv24. Table 1 summarizes the results of convergent and discriminant validity, as well the indicators of reliability.

Table 1 shows that the standard loading values for the individual items were within the domain (0.543–0.913), these values greater than the minimum retention of the elements based on their standard loads (Al-Lozi et al., 2018; Sung et al., 2019). Average variance extracted (AVE) is a summary indicator of the convergent validity of constructs that must be above 0.50 (Howard, 2018). The results indicate that the AVE values were greater than 0.50 for all constructs, thus the used measurement model has an appropriate convergent validity. Rimkeviciene et al. (2017) suggested the comparison approach as a way to deal with discriminant validity assessment in covariance-based SEM. This approach is based on comparing the values of maximum shared variance (MSV) with the values of AVE, as well as comparing the values of square root of AVE ($\sqrt{AVE}$) with the correlation between the rest of the structures. The results show that the values of MSV were smaller than the values of AVE, and that the values of $\sqrt{AVE}$ were higher than the correlation values among the rest of the constructs. Therefore, the measurement model used is characterized by discriminative validity. The internal consistency measured through Cronbach's Alpha coefficient (α) and compound reliability by McDonald's Omega coefficient (ω) was conducted as indicators to evaluate measurement model. The results listed in Table 1 demonstrated that both values of Cronbach's Alpha coefficient and McDonald's Omega coefficient were greater than 0.70, which is the lowest limit for judging on measurement reliability (De Leeuw et al., 2019).

5.2 Structural Model

The structural model illustrated no multicollinearity issue among predictor constructs because variance inflation factor (VIF) values are below the threshold of 5, as shown

Table 1 Results of validity and reliability tests

Constructs	1	2	3	4	5	6	7
1. Tardiness	**0.762**						
2. Taking long break	0.301	**0.741**					
3. Missing meeting	0.394	0.341	**0.744**				
4. Quitting	0.461	0.332	0.406	**0.764**			
5. Affective commitment	0.553	0.503	0.624	0.637	**0.740**		
6. Continuous commitment	0.512	0.645	0.534	0.622	0.524	**0.756**	
7. Normative commitment	0.603	0.614	0.550	0.574	0.584	0.647	**0.775**
VIF	2.514	2.066	1.581	2.368	–	–	–
Loadings range	0.703–0.814	0.645–0.834	0.543–0.864	0.675–0.913	0.675–0.824	0.670–0.826	0.707–0.865
AVE	0.580	0.549	0.553	0.583	0.547	0.571	0.600
MSV	0.418	0.461	0.501	0.466	0.513	0.409	0.511
Internal consistency	0.845	0.826	0.827	0.872	0.857	0.866	0.880
Composite reliability	0.846	0.828	0.829	0.874	0.858	0.869	0.882

Note Bold fonts refers to average variance extracted

Table 2 Hypothesis testing

Hypothesis	Relation	Standard beta	*t* value	*p* value
H1	Tardiness → organizational commitment	−0.557**	−14.164	0.002
H2	Taking long break → organizational commitment	−0.486*	−11.754	0.03
H3	Missing meeting → organizational commitment	−0.602***	−16.297	0.000
H4	Quitting → organizational commitment	−0.521**	−13.065	0.008

Note * $p < 0.05$, ** $p < 0.01$, *** $p < 0.001$

in Table 1 (Hair et al., 2017). This result is supported by the values of model fit indices shown in Fig. 1.

The results in Fig. 2 indicated that the chi-square to degrees of freedom (CMIN/DF) was 2.715, which is less than 3 the upper limit of this indicator. The values of the goodness of fit index (GFI), the comparative fit index (CFI), and the Tucker–Lewis index (TLI) were upper than the minimum accepted threshold of 0.90. Moreover, the result of root mean square error of approximation (RMSEA) indicated to value 0.051, this value is a reasonable error of approximation because it is less than the higher limit of 0.08. Consequently, the structural model used in this study was recognized as a fit model for predicting the organizational commitment and generalization of its result (Ahmad et al., 2016; Shi et al., 2019). To verify the results of testing the study hypotheses, structural equation modeling (SEM) was used, the results of which are listed in Table 2.

The results demonstrated in Table 2 show that functional withdrawal has a negative impact on organizational commitment. Moreover, the results indicated that the highest impact was for missing meeting ($\beta = -0.602$, $t = -16.297$, $p = 0.000$), followed by tardiness ($\beta = -0.557$, $t = -14.164$, $p = 0.002$), then quitting ($\beta = -0.521$, $t = -13.065$, $p = 0.008$), and finally the lowest impact was for taking long break ($\beta = -0.486$, $t = -11.754$, $p = 0.03$).

6 Discussion

The results of the study revealed that there is insignificant effect of physical withdrawal on organizational commitment, which may be due to the employee's exposure to personal and social conditions that push him to practice these behaviors in a certain period of his work and does not mean his lack of love for work and the organization. The study results inconsistent with (Dahlke, 1996). It was also found that there was insignificant effect of physical withdrawal on emotional commitment. The employee's practice of one of the physical withdrawal behaviors does not mean the employee's lack of emotional commitment with his organization and place of work,

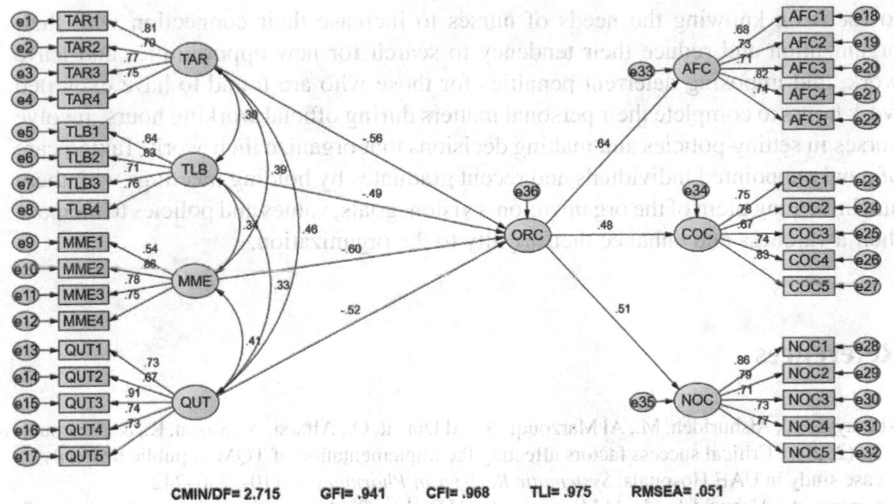

Fig. 2 SEM results of the functional withdrawal effect on organizational commitment

or his unhappiness with working in it, but the employee may be exposed to circumstances that push him to practice such behaviors such as the transportation crisis and family emergency circumstances, or obtaining work in a place close to his residence, which leads him to leave this Place and resignation and be more stable. The result of this study inconsistent with Somer (2009), Taris et al. (2004), and Somers and Binbaum (2000), which showed a negative relationship between physical withdrawal and organizational loyalty.

The study also found that there was insignificant effect of the remaining dimensions of physical withdrawal on continuous organizational commitment. The researchers attribute this result to the fact that the employee often seeks the best opportunities with the best financial return and the nature of the work that is more comfortable and closest to his place of residence and with better job privileges. He linked in his previous work is the payoff and the lack of better opportunities. It was also found that there is a negative, statistically significant effect of leaving work on organizational commitment, with no statistical significance for the remaining dimensions of physical withdrawal on organizational loyalty. In its conclusion, the study results agreed with Somers and Binbaum (2000) and Falkenburg and Schyns (2007).

7 Recommendations

In light of the findings of the study, managers and decision makers in hospital management should provide support, material and moral incentives, and educational, development and promotion opportunities for employees, especially nurses, with the need to focus on knowing the needs of nurses to increase their connection with their organization and reduce their tendency to search for new opportunities and leave work, and imposing deterrent penalties for those who are found to have exploited work hours to complete their personal matters during official working hours. Involve nurses in setting policies and making decisions that organize their work. Taking care of newly appointed individuals and recent graduates by holding meetings with them and informing them of the organization's vision, goals, values and policies to increase their awareness and enhance their loyalty to the organization.

References

Aburayya, A., Alshurideh, M., Al Marzouqi, A., Al Diabat, O., Alfarsi, A., Suson, R., & Alzarouni, A. (2020a). Critical success factors affecting the implementation of TQM in public hospitals: A case study in UAE Hospitals. *Systematic Reviews in Pharmacy, 11*(10), 230–242.

Aburayya, A., Alshurideh, M., Al Marzouqi, A., Al Diabat, O., Alfarsi, A., Suson, R., & Salloum, S. A. (2020b). An empirical examination of the effect of TQM practices on hospital service quality: An assessment study in UAE hospitals. *Systematic Reviews in Pharmacy, 11*(9), 347–362.

Aburayya, A., Alshurideh, M., Alawadhi, D., Alfarsi, A., Taryam, M., & Mubarak, S. (2020c). An investigation of the effect of lean six sigma practices on healthcare service quality and patient satisfaction: Testing the mediating role of service quality in Dubai primary healthcare sector. *Journal of Advanced Research in Dynamical and Control Systems, 12*(8), 56–72.

Aburayya, A., Alshurideh, M., Albqaeen, A., Alawadhi, D., & Ayadeh, I. (2020d). An investigation of factors affecting patients waiting time in primary health care centers: An assessment study in Dubai. *Management Science Letters, 10*(6), 1265–1276.

Ahmad, S., Zulkurnain, N., & Khairushalimi, F. (2016). Assessing the validity and reliability of a measurement model in structural equation modeling (SEM). *British Journal of Mathematics & Computer Science, 15*(3), 1–8. https://doi.org/10.9734/BJMCS/2016/25183

Al Kurdi, B., Alshurideh, M., & Al Afaishat, T. (2020a). Employee retention and organizational performance: Evidence from banking industry. *Management Science Letters, 10*(16), 3981–3990.

Al Kurdi, B., Alshurideh, M., & Alnaser, A. (2020b). The impact of employee satisfaction on customer satisfaction: Theoretical and empirical underpinning. *Management Science Letters, 10*(15), 3561–3570.

Al Kurdi, B., Elrehail, H., Alzoubi, H., Alshurideh, M., & Al-Adaila, R. (2021). The interplay among HRM practices, job satisfaction and intention to leave: An empirical investigation. *Journal of Legal, Ethical and Regulatory, 24*(1), 1–14.

Al-Adamat, A., Al-Gasawneh, J., & Al-Adamat, O. (2020). The impact of moral intelligence on green purchase intention. *Management Science Letters, 10*(9), 2063–2070.

Alameeri, K., Alshurideh, M., Al Kurdi, B., & Salloum, S. A. (2020, October). The effect of work environment happiness on employee leadership. In *International Conference on Advanced Intelligent Systems and Informatics* (pp. 668–680). Cham: Springer.

Al-bawaia, E., Alshurideh, M., Obeidat, B., Masa'deh, R. (2022). The impact of corporate culture and employee motivation on organization effectiveness in Jordanian banking sector. *Academy of Strategic Management Journal, 21*(Special Issue 2), 1–18.

Al-Fahdawi, K. (2004). The effects of organizational justice on organizational loyalty: A field study of the central departments in the southern governorates of Jordan. *The Arab Journal of Management, 2*, 23–47.

Al-Gasawneh, J. A., & Al-Adamat, A. M. (2020). The relationship between perceived destination image, social media interaction and travel intentions relating to Neom city. *Academy of Strategic Management Journal, 19*(2), 1–12.

Al-Ghanim, W. (2003). *Attitudes towards job dropout and its relationship to performance: An applied study on airport customs in the Kingdom of Saudi Arabia*. Master's thesis, Naif Arab University for Security Sciences, Riyadh, Saudi Arabia.
Alhalalmeh, M. I., Almomani, H. M., Altarifi, S., Al-Quran, A. Z., Mohammad, A. A., & Al-Hawary, S. I. (2020). The nexus between corporate social responsibility and organizational performance in Jordan: The mediating role of organizational commitment and organizational citizenship behavior. *Test Engineering and Management, 83*, 6391–6410.
Al-Hawary, S. I., & Al-Syasneh, M. S. (2020). Impact of dynamic strategic capabilities on strategic entrepreneurship in presence of outsourcing of five stars hotels in Jordan. *Business: Theory and Practice, 21*(2), 578–587.
Al-Hawary, S. I. (2009). The effect of the leadership style on the effectiveness of the organization: A field study at Zarqa Private University. *The Egyptian Journal for Commercial Studies, 33*(1), 361–393.
Al-Hawary, S. I. (2011). Human resource management practices in ZAIN cellular communications company operating in Jordan. *Perspectives of Innovations, Economics and Business, 8*(2), 26–34.
Al-Hawary, S. I. (2013). The role of perceived quality and satisfaction in explaining customer brand loyalty: Mobile phone service in Jordan. *International Journal of Business Innovation and Research, 7*(4), 393–413.
Al-Hawary, S. I. (2015). Human resource management practices as a success factor of knowledge management implementation at health care sector in Jordan. *International Journal of Business and Social Science, 6*(11/1), 83–98.
Al-Hawary, S. I., & Abu-Laimon, A. A. (2013). The impact of TQM practices on service quality in cellular communication companies in Jordan. *International Journal of Productivity and Quality Management, 11*(4), 446–474.
Al-Hawary, S. I., & Alajmi, H. M. (2017). Organizational commitment of the employees of the ports security affairs of the state of Kuwait: The impact of human recourses management practices. *International Journal of Academic Research in Economics and Management Sciences, 6*(1), 52–78.
Al-Hawary, S. I., Al-Awawdeh, W., & Abden, M. A. (2012). The impact of the leadership style on organizational commitment: A field study on Kuwaiti telecommunications companies. *ALEDARI, 130*, 53–102.
Al-Hawary, S. I., & Aldaihani, F. M. (2016). Customer relationship management and innovation capabilities of Kuwait airways. *International Journal of Academic Research in Economics and Management Sciences, 5*(4), 201–226.
Al-Hawary, S. I., & Al-Hamwan, A. (2017). Environmental analysis and its impact on the competitive capabilities of the commercial banks operating in Jordan. *International Journal of Academic Research in Accounting, Finance and Management Sciences, 7*(1), 277–290.
Al-Hawary, S. I., & Al-Namlan, A. (2018). Impact of electronic human resources management on the organizational learning at the private hospitals in the state of Qatar. *Global Journal of Management and Business Research: A Administration and Management, 18*(7), 1–11.
Al-Hawary, S. I., Al-Qudah, K., Abutayeh, P., Abutayeh, S., & Al-Zyadat, D. (2013). The impact of internal marketing on employee's job satisfaction of commercial banks in Jordan. *Interdisciplinary Journal of Contemporary Research in Business, 4*(9), 811–826.
Al-Hawary, S. I., & Al-Smeran, W. (2017). Impact of electronic service quality on customers satisfaction of Islamic Banks in Jordan. *International Journal of Academic Research in Accounting, Finance and Management Sciences, 7*(1), 170–188.
Al-Hawary, S. I., & Hadad, T. F. (2016). The effect of strategic thinking styles on the enhancement competitive capabilities of commercial banks in Jordan. *International Journal of Business and Social Science, 7*(10), 133–144.
Al-Hawary, S. I. S., Mohammad, A. S., Al-Syasneh, M. S., Qandah, M. S. F., & Alhajri, T. M. S. (2020). Organisational learning capabilities of the commercial banks in Jordan: Do electronic human resources management practices matter? *International Journal of Learning and Intellectual Capital, 17*(3), 242–266.

Al-Hawary, S. I., & Nusair, W. (2017). Impact of human resource strategies on perceived organizational support at Jordanian Public Universities. *Global Journal of Management and Business Research: A Administration and Management, 17*(1), 68–82.

Al-Hawary, S. I. S., & Obiadat, A. A. (2021). Does mobile marketing affect customer loyalty in Jordan? *International Journal of Business Excellence, 23*(2), 226–250.

Alketbi, S., Alshurideh, M., & Al Kurdi, B. (2020). The influence of service quality on customers' retention and loyalty in the UAE hotel sector with respect to the impact of customer' satisfaction, trust, and commitment: A qualitative study. *International Journal of Innovation, Creativity and Change, 14*(7), 734–754.

Al-Khayyal, A., Alshurideh, M., Al Kurdi, B., & Aburayya, A. (2020). The impact of electronic service quality dimensions on customers' e-shopping and e-loyalty via the impact of e-satisfaction and e-trust: A qualitative approach. *International Journal of Innovation, Creativity and Change, 14*(9), 257–281.

Allahow, T. J. A. A., Al-Hawary, S. I. S., & Aldaihani, F. M. F. (2018). Information technology and administrative innovation of the central agency for information technology in Kuwait. *Global Journal of Management and Business, 18*(11-A), 1–16.

Allen, N. J., & Meyer, J. P. (1996). "Affective, continuance, & normative commitment to the organization: An examination of construct validity", the University of Western Ontario. *Journal of Vocational Behavior, 49*(3), 76–252.

Allozi, A., Alshurideh, M., AlHamad, A., & Al Kurdi, B. (2022). Impact of transformational leadership on the job satisfaction with the moderating role of organizational commitment: Case of UAE and Jordan manufacturing companies. *Academy of Strategic Management Journal, 21*, 1–13.

Al-Lozi, M., Almomani, R. Z., & Al-Hawary, S. I. (2017). Impact of talent management on achieving organizational excellence in Arab Potash Company in Jordan. *Global Journal of Management and Business Research: A Administration and Management, 17*(7), 15–25.

Al-Lozi, M., Almomani, R. Z., & Al-Hawary, S. I. (2018). Talent management strategies as a critical success factor for effectiveness of human resources information systems in commercial banks working in Jordan. *Global Journal of Management and Business Research: A Administration and Management, 18*(1), 30–43.

Al-Nady, B. A., Al-Hawary, S. I., & Alolayyan, M. (2013). Strategic management as a key for superior competitive advantage of sanitary ware suppliers in Kingdom of Saudi Arabia. *International Journal of Management and Information Technology, 7*(2), 1042–1058.

Al-Nady, B. A., Al-Hawary, S. I., & Alolayyan, M. (2016). The role of time, communication, and cost management on project management success: An empirical study on sample of construction projects customers in Makkah City, Kingdom of Saudi Arabia. *International Journal of Services and Operations Management, 23*(1), 76–112.

Al-Quran, A. Z., Alhalalmeh, M. I., Eldahamsheh, M. M., Mohammad, A. A., Hijjawi, G. S., Almomani, H. M., & Al-Hawary, S. I. (2020). Determinants of the green purchase intention in Jordan: The moderating effect of environmental concern. *International Journal of Supply Chain Management, 9*(5), 366–371.

Al-Rawashdah, K. S. (2007). *School Decision-Making and a Sense of Security and Organizational Loyalty* (1st ed.). Dar Hamed for Publishing and Distribution.

AlShehhi, H., Alshurideh, M., Al Kurdi, B., & Salloum, S. A. (2020, October). The impact of ethical leadership on employees performance: A systematic review. In *International Conference on Advanced Intelligent Systems and Informatics* (pp. 417–426). Cham: Springer.

Alshraideh, A. T. R., Al-Lozi, M., & Alshurideh, M. T. (2017). The impact of training strategy on organizational loyalty via the mediating variables of organizational satisfaction and organizational performance: An empirical study on Jordanian agricultural credit corporation staff. *Journal of Social Sciences (COES&RJ-JSS), 6*(2), 383–394.

Alshurideh, D. M. (2019). Do electronic loyalty programs still drive customer choice and repeat purchase behaviour? *International Journal of Electronic Customer Relationship Management, 12*(1), 40–57.

Alshurideh, M. (2014). The factors predicting students' satisfaction with universities' healthcare clinics' services. *Dirasat: Administrative Sciences, 41*(2), 451–464.

Alshurideh, M. (2022). Does electronic customer relationship management (E-CRM) affect service quality at private hospitals in Jordan? *Uncertain Supply Chain Management, 10*(2), 325–332.

Alshurideh, M., Gasaymeh, A., Ahmed, G., Alzoubi, H., & Kurd, B. (2020). Loyalty program effectiveness: Theoretical reviews and practical proofs. *Uncertain Supply Chain Management, 8*(3), 599–612.

Alsuwaidi, M., Alshurideh, M., Al Kurdi, B., & Salloum, S. A. (2020, October). Performance appraisal on employees' motivation: A comprehensive analysis. In *International Conference on Advanced Intelligent Systems and Informatics* (pp. 681–693). Cham, Springer.

AlTaweel, I. R., & Al-Hawary, S. I. (2021). The mediating role of innovation capability on the relationship between strategic agility and organizational performance. *Sustainability, 13*(14), 7564.

Alzoubi, H., Alshurideh, M., Kurdi, B., Akour, I., & Aziz, R. (2022). Does BLE technology contribute towards improving marketing strategies, customers' satisfaction and loyalty? The role of open innovation. *International Journal of Data and Network Science, 6*(2), 449–460.

Awaida, E. (2008). *The impact of job satisfaction on organizational loyalty among workers in NGOs in Gaza Governorate*. Master's thesis, Islamic University—College of Commerce, Gaza, Palestine.

Bennett, N., & Naumann, E. S. (2004). Withholding effort at work: Understanding and preventing shrinking, job neglect, social loafing, and free riding. In *Management Organizational Deviance* (Ch. 5), pp. 113–126.

Brock, M., Martin, L., & Buckly, R. (2013). Time theft in organizations: The development of the time banditry questionnaire. *International Journal of Selection and Assessment, 21*(3), 309–321.

Chen, J., & Lim, V. G. (2012). Cyberloafing at workplace: Gain or drain on work? *Behaviour on Information Technology, 31*(4), 343–353.

Dahlke, G. (1996). Absenteeism & organizational commitment. *Nursing Management, 27*(10), 30–30.

De Leeuw, E., Hox, J., Silber, H., Struminskaya, B., & Vis, C. (2019). Development of an international survey attitude scale: Measurement equivalence, reliability, and predictive validity. *Measurement Instruments for the Social Sciences, 1*(1), 1–10. https://doi.org/10.1186/s42409-019-0012-x

Falkenburg, K., & Schyns, B. (2007). Work satisfaction, organizational commitment & withdrawal behaviors. *Management Research News, 30*(10), 708–723.

Hair, J. F., Babin, B. J., & Krey, N. (2017). Covariance-based structural equation modeling in the journal of advertising: Review and recommendations. *Journal of Advertising, 46*(1), 163–177. https://doi.org/10.1080/00913367.2017.1281777

Hanisch, K. A., & Hulin, C. L. (1990). General attitude & organizational withdrawal: An evaluation of causal model. *Journal of Vocational Behaviour*, 110–128.

Harrison, D. A., Newman, D. A., & Roth, P. L. (2006). How important are job attitudes? Meta-analytic comparisons for integrative behavioral outcomes and time sequence. *Academy of Management Journal, 49*(2), 305–325.

Harrison, D. A., & Price, K. H. (2003). Context and consistency in absenteeism: Studying social and dispositional influences across multiple settings. *Human Resource Management Review, 13*(2), 203–225.

Hayajneh, N., Suifan, T., Obeidat, B., Abuhashesh, M., Alshurideh, M., & Masa'deh, R. (2021). The relationship between organizational changes and job satisfaction through the mediating role of job stress in the Jordanian telecommunication sector. *Management Science Letters, 11*(1), 315–326.

Howard, M. C. (2018). The convergent validity and nomological net of two methods to measure retroactive influences. *Psychology of Consciousness: Theory, Research, and Practice, 5*(3), 324–337. https://doi.org/10.1037/cns0000149

Judeh, A. (2004). Methods of coping with stressful life events and their relationship to mental health among a sample of Al-Aqsa University students: Research presented to the First Educational Conference, Palestine, pp. 667–696.

Kaplan, S., Bradley, J. C., Lachman, J. N., & Hayness, D. (2009). On the role of positive and negative affectivity in job performance: A meta-analytic investigation. *Journal of Applied Psychology, 94*(1), 162–170.

Lim, V. K. G., Teo, T. S. H., & Loo, G. L. (2002). How do I loaf here? Let me count the way. *Communications of ACM, 45*, 66–70.

Metabis, A., & Al-Hawary, S. I. (2013). The impact of internal marketing practices on services quality of commercial banks in Jordan. *International Journal of Services and Operations Management, 15*(3), 313–337.

Mohammad, A. A., Alshura, M. S., Al-Hawary, S. I. S., Al-Syasneh, M. S., & Alhajri, T. M. (2020). The influence of internal marketing practices on the employees' intention to leave: A study of the private hospitals in Jordan. *International Journal of Advanced Science and Technology, 29*(5), 1174–1189.

Redmond, B. F. (2010). *Lateness, absenteeism, turnover, and burnout: Am I likely to miss work?* The Pennsylvania State University World Campus.

Rimkeviciene, J., Hawgood, J., O'Gorman, J., & De Leo, D. (2017). Construct validity of the acquired capability for suicide scale: Factor structure, convergent and discriminant validity. *Journal of Psychopathology and Behavioral Assessment, 39*(2), 291–302. https://doi.org/10.1007/s10862-016-9576-4

Sekaran, U., & Bougie, R. (2016). *Research Methods for Business: A Skill-Building Approach* (7th ed.). Wiley.

Shi, D., Lee, T., & Maydeu-Olivares, A. (2019). Understanding the model size effect on SEM fit indices. *Educational and Psychological Measurement, 79*(2), 310–334. https://doi.org/10.1177/0013164418783530

Somer, M. J. (2009). The combined influence of affective, normative & continuous commitment on employee withdrawal. *Journal of Vocational Behaviour, 7*(1), 75–81.

Somers, M., & Binbaum, D. (2000). Exploring the relationship between commitment profile and work attitudes, employee withdrawal, and job performance. *Public Personal Management, 29*(3), 353–366.

Steers, R. T., & Mowady, L. (1974). Organizational commitment, job satisfaction & turnover among psychiatric technician. *Journal of Applied Psychology, 59*(5), 603–609.

Sung, K.-S., Yi, Y. G., & Shin, H.-I. (2019). Reliability and validity of knee extensor strength measurements using a portable dynamometer anchoring system in a supine position. *BMC Musculoskeletal Disorders, 20*(1), 1–8. https://doi.org/10.1186/s12891-019-2703-0

Sweissi, D., & Mahjar, Y. (2017). The reality of psychological withdrawal from work as a form of organizational misconduct among employees: A field study on a number of employees of the public hospital institution in Djelfa. *Afaq Al-Ulum Magazine*, (9), 63–74.

Taris, T., Horn, J., Schaufili, W., & Schreurs, P. (2004). Inequity, burnout and psychological withdrawal among teachers: A dynamic exchange model. *Anxiety, Stress & Coping, 17*(1), 103–122.

Turan, A. (2015). Examining the impact of Machiavellianism on psychological withdrawal, physical withdrawal, & antagonistic behavior. *Global Management & Business Research: An International Research, 7*(3), 87–103.

Wang, Y. A., & Rhemtulla, M. (2021). Power analysis for parameter estimation in structural equation modeling: A discussion and tutorial. *Advances in Methods and Practices in Psychological Science, 4*(1), 1–17. https://doi.org/10.1177/2515245920918253

The Impact of Human Resources Agility on Job Performance in—Islamic Banks Operating in Jordan

Ibrahim Yousef Al-Armeti, Majed Kamel Ali Al-Azzam, Mohammad Issa Ghafel Alkhawaldeh, Ayat Mohammad, Yahia Salim Melhem, Raed Ismael Ababneh, Sulieman Ibraheem Shelash Al-Hawary, and Muhammad Turki Alshurideh

Abstract The study aimed to identify the impact of human resource agility on job performance in Islamic banks operating in Jordan. The study population consists of all managers in the upper management, middle management, and lower management in these banks. The study adopted the stratified random sample that is suitable for the managers in the upper, middle management and the lower management, who

I. Y. Al-Armeti · A. Mohammad
Business and Finance Faculty, the World Islamic Science and Education University (WISE), Postal Code 11947, P.O Box 1101, Amman, Jordan

M. K. A. Al-Azzam
Department of Business Administration-Faculty of Economics and Administrative Sciences, Yarmouk University, P.O Box 566-Zip Code 21163, Irbid, Jordan
e-mail: Majedaz@yu.edu.jo

M. I. G. Alkhawaldeh
Directorates of Building and Land Tax, Ministry of Local Administration, Zarqa municipality, Amman, Jordan

Y. S. Melhem
Business Administration Department, School of Business, Yarmouk University, Irbid, Jordan
e-mail: ymelhem@yu.edu.jo

R. I. Ababneh
Policy, Planning, and Development Program, Department of International Affairs, College of Arts and Sciences, Qatar University, 2713 Doha, Qatar

S. I. S. Al-Hawary (✉)
Department of Business Administration, School of Business, Al Al-Bayt University, P.O.BOX 130040, Mafraq 25113, Jordan
e-mail: dr_sliman73@aabu.edu.jo

M. T. Alshurideh
Department of Marketing, School of Business, The University of Jordan, Amman 11942, Jordan
e-mail: m.alshurideh@ju.edu.jo; malshurideh@sharjah.ac.ae

Department of Management, College of Business, University of Sharjah, 27272 Sharjah, United Arab Emirates

M. Alshurideh et al. (eds.), *The Effect of Information Technology on Business and Marketing Intelligence Systems*, Studies in Computational Intelligence 1056,
https://doi.org/10.1007/978-3-031-12382-5_60

are (148) respondents. To achieve the goals of the study and test its hypotheses, the researcher used the descriptive analytical approach, and collected data with a questionnaire used by the researcher as a main performance of collecting information. Smart PLS program was used to test study hypotheses. The study results showed a statistically significant effect of human resource agility on job performance in Islamic banks operating in Jordan. Based on the results of the study, the study recommends assisting decision makers in solving problems, among which was to enhance interest in practicing adaptive capacity, creativity in problem solving, professional flexibility, and learning skills.

Keywords Human resource agility · Job performance · Islamic banks · Jordan

1 Introduction

In the past, organizations dealt negative with individuals and did not provide them with the appropriate work environment, they considered individual like a machine and does not have any moral interests, and the organizations were only thinking of achieving their goal without looking at individuals' needs and desires. The organizations later began to recognize their employees' rights with the increasing control of the external environment factors, which eventually made them recognize the human element value and consider it the most important component, and one of the most important basic resources, in addition the added value that it can be added by individuals. Job performance concept appeared in the managerial literature as part of organizations overall performance, then the researchers' interest increased in line with their interest in the organization's overall performance, rather more researchers try to explore variables that affect job performance to improve and take their advantage (Al Kurdi et al., 2021; Allozi et al., 2022; Dhani & Sharma, 2017; Hayajneh et al., 2021; Kashyap, 2014).

Human resources agility is one of the most important topics that researchers and managers have paid great attention to, in response to the intense competition between business organizations in the markets, and the technological development (Al-Hawary, 2011, 2015; AlTaweel & Al-Hawary, 2021; Shakhour et al., 2021). Organizations has also contributed to the pursuit of using different methods from competitors that put them first in the local and global markets, and this will be not achieved without workers partnership.

Jordan banking sector contributes to economic development by providing job opportunities and services that promote well-being and increase people purchasing power, which calls for an attempt to advance this vital sector and to improve all the variables related to it for survival, growth and continuity. This puts on the shoulders of every bank to try to find a special advantage that makes it ahead of competition and owning a larger market share compared to other banks, and since the number of Islamic banks in the world is increasing, each bank has to provide its own services through which it can gain a larger number of customers (Al Kurdi et al., 2020a, b).

It is worth noting that this can be achieved by investing in human resources and raising their performance levels, as attention to the human resource agility results in distinctive, efficient and effective performance (the ability to adapt, creativity in solving problems, professional flexibility, learning work skills, bearing work pressures), therefore, each organization must provide an appropriate work environment for employees and pay attention to raising their skills and training them in a way that positively affects employees performance (performance of the context, performance of tasks).

The concept of human resource agility is attractive to researchers and scholars because it is one of the new topics that Arab literature lacks in, especially in its analytical form through studying the impact of human resource agility on job performance. Islamic banks operating in Jordan, like other business organizations, are exposed to high competition between them, and thus they must preserve their human resources and seek to improve the level of their work skills by using human resources agility that support job performance, as human resources agility must have a positive impact on employees job performance and those who contribute to raising the organizational performance levels, which may necessarily lead to organizational goals achievement and continuity (Alkalha et al., 2012; Al-bawaia et al., 2022).

2 Literature Review and Hypotheses Development

2.1 *Human Resources Agility*

Organizations operate in a fast-moving environment that requires rethinking of the human resources through planned education, helping to develop key competencies that enable individuals to maintain focus, and perform well in current and future jobs (Al-Quran et al., 2020; Al-Hawary & Abu-Laimon, 2013; Al-Hawary & Al-Namlan, 2018). Developing key competencies in an employee is essential to improving their performance and to empower them with capabilities in an agile context to improve performance (Al-Hawary et al., 2013; Al-Lozi et al., 2017, 2018).Since changing organizations have become more the rule than the exception, the human element agility has become as well (Jager et al., 2019), the effective use of human resource agility is often a critical factor in organizations in the long term to succeed over competitors, so organizations depend on agile human resources highly. Human resources, in their general sense, are known as job performance capabilities to enhance organizational competitive opportunities to survive in the business world, turning them into opportunities that can be exploited (Altamony et al., 2012; Alavi, 2016; Al-Hawary & Alajmi, 2017; Shamout et al., 2022). Agile organizations are able to meet the customers' needs and develop new, high-quality products with high competitiveness. Human element is an important element to develop his knowledge, ideas, judgment and cooperation quickly in line with the new working conditions called (human resource agility) or (individual agility) (Alhalalmeh et al, 2020;

Alshurideh et al., 2022). The aspect of HR agility is very important and has practical implications for the direct methods used in intervening for the technological, cultural and control changes needed for the organization (Khambayat & Affiliations, 2019). We can define HR agility as the human resources ability to respond quickly to market changes and flexibly deal with the unexpected to survive in the business environment. Human resources must be flexible and trainable to adapt quickly and easily to market conditions, also known as enhancing capabilities to survive in a globally volatile work environment (Muduli, 2013).

In the past decade, human resource agility training has been utilized in manufacturing and service operations. (Iravani & Krishnamurthy, 2006). Human resource agility has many dimensions, and the majority of researchers used the same human resource agility dimensions, which are (the ability to adapt, creativity in problem solving, professional flexibility, work stress tolerance, learning work skills) (Mubarak, 2012; Al-kasasbeh et al., 2016; AlHamad et al., 2022). Some researchers have used human resource agility dimensions as follows (flexibility, adaptation, motivation, training, participation, and empowerment) (Al-Hawary & Al-Syasneh, 2020). Other researchers dealt with the dimensions of workforce agility as (flexibility, creativity, quality) (Mubarak, 2012).

It is noted that there are similar and close dimensions when researchers study the subject of human resource agility, but a few have adopted the dimensions (proactive, adaptive, and flexible) (Al-Hawary et al., 2020). For achieving the study objectives, the dimensions were adopted as follows, **adaptability:** the employee's ability to change and adjust himself and his behavior better in the organization, adapting between people and cultures, learning new skills and responsibilities (Alavi et al., 2014), **problem solving innovation:** is to start activities that have a positive impact on the organization, and it is expected from them to change and start activities that lead to solving problems and improving work. (Alavi et al., 2014). **Professional flexibility:** It is the employee effective performance under pressure resulting from changing the environment and the applied strategies and solutions, which is reflected in a positive impact that leads to new ideas and technological techniques and dealing with the unexpected situation and pressures (Alavi et al., 2014), this may result in the employee's ability to tackle problems to achieve the organization's goals and move flexibly and effectively (Muduli, 2013). **Bearing work pressure:** It is the employee's ability to withstand the conditions that lead to stress and pressures in the situations that the individual is exposed to at his work, stress does not match the physical and psychological changes with individuals who cannot deal with work requirements (Mohammad et al., 2020). **Learning work skills:** It is learning new ways to perform jobs and teaching workers new skills through training programs provided by the organization, which is the mainstay of human resources that keeps pace with rapid developments and helps in training workers to prepare for a new job or profession in the organization (Allahow et al., 2018).

2.2 *Job Performance*

Performance has received great attention in human resource studies because of its importance to the individual (Al-Hawary & Alajmi, 2017). Organization and task performance are defined as the implementation of job burdens for employee responsibilities and duties (Mohammad, 2019). The success of performing the task depends mainly on the workoutcome, which has a strong relationship with the organization strategic objectives, customer satisfaction and economic development (Al-Hawary & Hadad, 2016; Al-Hawary & Obiadat, 2021; Al-Nady et al., 2013, 2016; Alolayyan et al., 2018; Alshurideh et al., 2017a, 2017b; Muhtasom et al., 2017). General concept of job performance can be developed as the extent to which the managerial processes that are accomplished in a specific period of time conform to the pre-established plans, identify weaknesses, and develop scientific and practical solutions. And give workers the chance to follow their needs, self-esteem, self-follow and respect for others (Al-Hawary & Hadad, 2016; AlShehhi et al., 2020; Alsuwaidi et al., 2020).

Employee performance is the main factor to achieve the organization goals that must be focused on, by achieving performance in light of work such as: job positions, job satisfaction, commitment to accomplish the task (Ghani et al., 2016; Al-Dhuhouri et al., 2020). It is also certain that job performance refers to the total value expected to organize the separate behaviour that the individual implements within a short period of time so that we obtain the expected quality and quantity from an employee in a particular job, and this is evidence that individual performance is the basis of motivation, willing and ability to do this task (Mohamad & Jais, 2016; Nuseir et al., 2021). Here we can define job performance in general as a set of processes that determine and measure the employee's performance and their rewards for the efforts they make to achieve what is required. Also, job performance may be defined as the degree of achievement and tasks completion that make up an individual's job that is reflected in job requirements fulfilment that often overlap between performance and effort, while effort refers to the energy expended, performance is measured on the results achieved by the individual (Al-Haddad, 2016; Al Mehrez et al., 2020). While some defined job performance as the employee's ability to complete the tasks assigned to him and contribute to the achievement of organization's goals. The employee's performance is inspired by a sense of job affiliation and thus to obtain efficiency and effectiveness in the workplace. (Adagbabiri & Okolie, 2019; Odeh et al., 2021).

There are several dimensions that were addressed in previous studies regarding job performance, but the focus of most researchers revolved around two of these dimensions, namely "context performance" and "task performance". In this study, we address these two dimensions; **Tasks performance:** which is related to the work plan construction and through which all planning and coordination skills are used to achieve the required performance. **Contextual performance:** called organizational citizenship behavior such as protecting, volunteering, and helping others to develop at work.

Although job performance dimensions are task performance and contextual performance, but it is necessary to separate them when processing, it is not a condition that they occur together. For example, the employee may be high-performance in performing the tasks, but is weak in providing a good environment to increase organization effectiveness or provide assistance to others. Task performance is the activities and plans that help transform raw materials into goods and services by obtaining individuals and using distributors to sell goods and services using all planning and supervision skills to ensure that the organization performs its functions efficiently and effectively. Contextual performance is what concerned with achieving the organization effectiveness through providing a good environment for performing tasks, including behaviours such as staying motivated and making extra effort, volunteering to carry out activities, helping and cooperating with others, a commitment to follow organizational rules and procedures, providing support to others and defending the organization goals.

2.3 Human Resource Agility and Job Performance

Taran (2019) asserts that healthcare organizations are subject to change and this may lead organizations to invest in developing workforce agility and resilience where social change in people is facilitated by creating societies that adapt in an optimistic way and find opportunities that benefit society even through negative changes. Workforce agility can be built by encouraging the capabilities of: learning, teamwork, problem solving, information seeking and decision-making, and failure to adapt to change is a major barrier to workforce agility, which necessitates companies to develop workforce agility dynamic ability (Onyait, 2019). According to Harikrishnan and Suresh (2018), age, health status, and work environment are the critical factors that influence workforce agility. Ogbo & Wilfred (2014) confirmed that there is an impact of the diversity of workforce agility on improving organizational performance, and that education has an impact on workforce agility, which often affects the increase in corporate profitability. Azuara (2015) presented an evidence supports the need for more future research, in order to deepen the understanding of the mechanisms related to human resource agility evolution and the main challenges that leaders face in implementing workforce agility. There are many studies that confirm the positive relationship between the individual components of workforce agility and organizational intelligence, and the positive relationship between workforce diversity and agility (Sohrabi, Asari & Hozoori, 2014). The workforce agility also has a significant impact on organizational memory (Al-faouri, Al-nsour & Al-kasasbeh, 2013). Today's business environment has put pressures on the organization to build human resources agility (workforce), improve the performance of its ability to adapt to the environment, its effectiveness and its implementation on time, agility has emerged as a new single solution to manage a dynamic environment and has become a vital factor for the success of contemporary organizations today (Al-faouri, Al-nsour & Al-kasasbeh, 2013; Lee et al., 2022a, b). Effective use of the workforce is a critical factor in the long-term for the organization to succeed and that agility depends on smart people rather than systems and human resource capabilities that affect job performance and enhance chances of organization to survive in the competitive environment (Alavi, 2016). Based on the literature review above, the study hypothesis can be formulated as follows:

There is no statistically significant effect of human resource agility on job performance in Islamic banks operating in Jordan.

3 Methodology

3.1 Population and Sample Selection

A qualitative method based on a questionnaire was used in this study for data collection and sample selection. The major aim of the study was to examine the impact of human resources agility on job performance. Therefore, it focused on Islamic banks operating in Jordan. Data were primarily gathered through self-reported questionnaires creating by Google Forms which were distributed to a random stratified sample of (242) managers via email. In total, (173) responses were received including (25) invalid to statistical analysis due to uncompleted or inaccurate. Hence, the final sample contained (148) responses suitable to analysis requirements that were formed a response rate of (61.16%), where it proved to be sufficient to the extent that was predictable and allowed for a presumption of data saturation (Sekaran & Bougie, 2016).

3.2 Measurement Instrument

A self-reported questionnaire that consists of two main sections along with a section regarding control variables was used as the measurement instrument. Control variables considered as categorical measures were composed of gender, age group, educational level, and experience. The two main sections were dealt with a five-point Likert scale (from 1 = strongly disagree to 5 = strongly agree). The first section contained (21) items to measure human resources agility based on (Hozoori, 2014). These questions were distributed into dimensions as follows: three items dedicated for measuring adaptation ability, four items dedicated for measuring problem solving creativity, five items dedicated for measuring professional flexibility, five items dedicated for measuring learning work skills, and four items dedicated for measuring bearing work pressure. Whereas the second section included (10) items developed to measure job performance according to what was pointed by (Arocas & Del Valle, 2021; Xie & Li, 2021). This variable was a second-order construct divided into two first-order constructs. Task performance measured by three items, and contextual performance measured using seven items.

4 Findings

4.1 Measurement Model Evaluation

This study was conducted structural equation modeling (SEM) to test hypotheses, which represents a contemporary statistical technique for testing and estimating the relationship between factors and variables (Al-Adamat et al., 2020; Al-Gasawneh et al., 2020; Wang & Rhemtulla, 2021). Accordingly, the reliability and validity of the constructs were tested using confirmatory factor analysis (CFA) through the statistical program AMOSv24. Table 1 summarizes the results of convergent and discriminant validity, as well the indicators of reliability.

Table 1 shows that the standard loading values for the individual items were within the domain (0.697–0.909), these values greater than the minimum retention of the elements based on their standard loads (Al-Lozi et al., 2018; Sung et al., 2019). Average variance extracted (AVE) is a summary indicator of the convergent validity of constructs that must be above 0.50 (Howard, 2018). The results indicate that the AVE values were greater than 0.50 for all constructs, thus the used measurement model has an appropriate convergent validity. Rimkeviciene et al. (2017) suggested the comparison approach as a way to deal with discriminant validity assessment in covariance-based SEM. This approaches based on comparing the values of maximum shared variance (MSV) with the values of AVE, as well as comparing the values of square root of AVE ($\sqrt{}$AVE) with the correlation between the rest of the structures. The results show that the values of MSV were smaller than the values of AVE, and that the values of $\sqrt{}$AVE were higher than the correlation values among the rest of the constructs. Therefore, the measurement model used is characterized by discriminative validity. The internal consistency measured through Cronbach's Alpha coefficient (α) and compound reliability by McDonald's Omega coefficient (ω) was conducted as indicators to evaluate measurement model. The results listed in Table 1 demonstrated that both values of Cronbach's Alpha coefficient and McDonald's Omega coefficient were greater than 0.70, which is the lowest limit for judging on measurement reliability (De Leeuw et al., 2019).

4.2 Structural Model

The structural model illustrated no multicollinearity issue among predictor constructs because variance inflation factor (VIF) values are below the threshold of 5, as shown

Table 1 Results of validity and reliability tests

Constructs	1	2	3	4	5	6	7
1. ADA	**0.765**						
2. PSC	0.384	**0.858**					
3. PRF	0.415	0.425	**0.798**				
4. LWS	0.388	0.391	0.411	**0.831**			
5. BWP	0.402	0.362	0.382	0.441	**0.845**		
6. TAP	0.594	0.612	0.564	0.637	0.581	**0.772**	
7. COP	0.541	0.574	0.577	0.625	0.557	0.639	**0.816**
VIF	2.047	1.985	1.564	2.294	2.554	–	–
Loadings range	0.697–0.855	0.814–0.889	0.743–0.869	0.721–0.905	0.701–0.909	0.721–0.834	0.710–0.901
AVE	0.585	0.737	0.637	0.691	0.714	0.596	0.666
MSV	0.412	0.512	0.425	0.502	0.487	0.461	0.506
Internal consistency	0.806	0.916	0.895	0.915	0.907	0.812	0.814
Composite reliability	0.808	0.918	0.897	0.918	0.908	0.815	0.816

Note ADA: adaptation ability, PSC: problem solving creativity, PRF: professional flexibility, LWS: learning work skills, BWP: bearing work pressure, TAP: task performance, COP: contextual performance, bold fonts refer to square root of AVE

Fig. 1 Research model

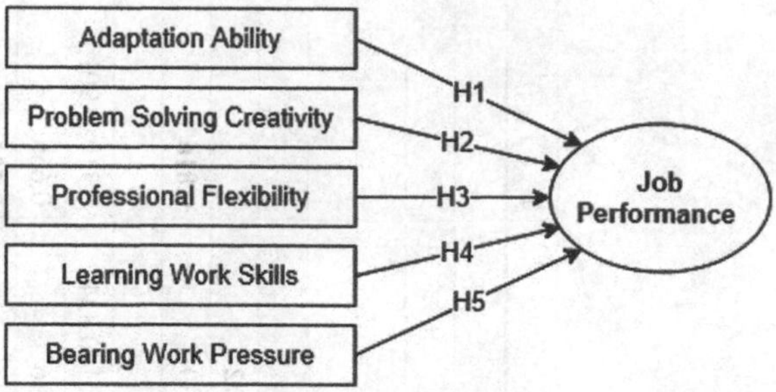

in Table 1 (Hair et al., 2017). This result is supported by the values of model fit indices shown in Fig. 1.

The results in Fig. 1 indicated that the chi-square to degrees of freedom (CMIN/DF) was 1.648, which is less than 3 the upper limit of this indicator (Fig. 2). The values of the goodness of fit index (GFI), the comparative fit index (CFI), and the Tucker-Lewis index (TLI) were upper than the minimum accepted threshold of 0.90. Moreover, the result of root mean square error of approximation (RMSEA) indicated to value 0.046, this value is a reasonable error of approximation because it is less than the higher limit of 0.08. Consequently, the structural model used in this study was recognized as a fit model for predicting the job performance and generalization of its result (Ahmad et al., 2016; Shi et al., 2019). To verify the results of testing the study hypotheses, structural equation modeling (SEM) was used, the results of which are listed in Table 2.

The results demonstrated in Table 2 show that most human resources agility dimensions had a positive impact on job performance except adaptation ability ($\beta = 0.107$, $t = 1.207$, $p = 0.228$) and problem-solving creativity ($\beta = 0.027$, $t = 0.508$, $p = 0.612$). However, the results indicated that the highest impact was for learning work skills ($\beta = 0.412$, $t = 8.905$, $p = 0.000$), followed by professional flexibility ($\beta = 0.352$, $t = 3.952$, $p = 0.000$), and finally the lowest impact was for bearing work pressure ($\beta = 0.268$, $t = 3.501$, $p = 0.000$).

5 Discussions

The results indicated that the bank employees deal with various situations and overcome the consequences, and this is what characterized employees with the ability to understand the complex ideas of effective adaptation to the environment, furthermore the employees are able to face sudden changes in the environment so that the bank employees can respond to unexpected changes in the internal and external environment. The researcher believes that the workers in Islamic banks have the responsibility towards performing their work so that the workers have creative solutions to solve work problems, and the workers present creative ideas to take decisions related

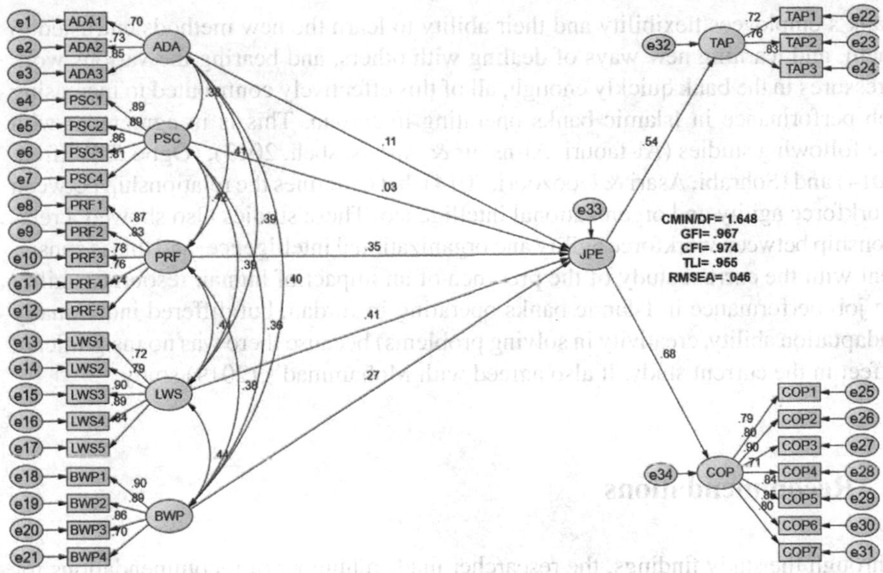

Fig. 2 SEM results of the human resources agility effect on job performance

Table 2 Hypothesis testing

Hypothesis	Relation	Standard Beta	*t* value	*p* value
H1	ADA → JPE	0.107	1.207	0.228
H2	PSC → JPE	0.027	0.508	0.612
H3	PRF → JPE	0.352	3.952***	0.000
H4	LWS → JPE	0.412	8.905***	0.000
H50	BWP → JPE	0.268	3.501***	0.000

Note ADA: adaptation ability, PSC: problem solving creativity, PRF: professional flexibility, LWS: learning work skills, BWP: bearing work pressure, JPE: job performance, * $p < 0.05$, ** $p < 0.01$, *** $p < 0.001$

to solving work problems so that teamwork and an effective communication system contribute to providing creative solutions to work problems, and that the work nature these banks Islamic need creativity in solving problems. It was found that employees of Islamic banks operating in Jordan are keen to participate in training courses and bear work pressures in order to achieve the investigated Islamic banks objectives.

Team members exchange knowledge and ideas, Staff actively participate in group discussions, work, and meetings in banks. Staff like to deal with additional responsibilities, and communicate effectively with colleagues to solve problems and make decisions in Islamic banks. The study found that dimensions of human resource agility availability affects Islamic banks performance, the bank's ability to adapt to environmental changes, and creatively solve problems facing the bank, as well as the bank's employees flexibility and their ability to learn the new methods entrusted to them, and learning new ways of dealing with others, and bearing the various work pressures in the bank quickly enough, all of this effectively contributed to increasing job performance in Islamic banks operating in Jordan. This is in agreement with the following studies (Al-faouri, Al-nsour & Al-kasasbeh, 2013), (Ogbo & Wilfred, 2014) and (Sohrabi, Asari & Hoozoori, 2014) that examines the relationship between workforce agility and organizational intelligence. These studies also showed a relationship between workforce agility and organizational intelligence, and this is consistent with the current study of the presence of an impact of human resources agility on job performance in Islamic banks operating in Jordan, but differed individually (adaptation ability, creativity in solving problems) because there was no insignificant effect in the current study. It also agreed with Mohammad's (2019) study.

6 Recommendations

Through the study findings, the researcher made a number of recommendations for managers and decision-makers in Islamic banks. The researcher aims behind these recommendations to enhance the desired benefits from human resource agility application, and among these recommendations: Strengthening the interest in practicing the ability to adapt, and the necessity of paying attention to the employees of Islamic banks operating in Jordan in order to face any sudden changes in the environment through training. And the promotion of interest in practicing creativity in solving problems and the need to develop an effective communication system in Islamic banks operating in Jordan in order to provide the best solutions to the problems faced by workers in the studied banks. Overcoming situations in order to achieve the highest level of performance among employees. Exploiting the creativity dimensions, the effective communication system, and the responsibility towards the employees' performance in solving the problems they have to achieve the highest levels of job performance.

References

Adagbabiri, M. M., & Okolie, U. C. (2019). Job design and employee performance in nigeria bottling company plc, benin city. Annals of SpriruHaret University, *19*(3), 10–11.

Ahmad, S., Zulkurnain, N., & Khairushalimi, F. (2016). Assessing the validity and reliability of a measurement model in Structural Equation Modeling (SEM). *British Journal of Mathematics & Computer Science, 15*(3), 1–8. https://doi.org/10.9734/BJMCS/2016/25183

Al Kurdi, B., Alshurideh, M. & Al afaishat, T. (2020a). Employee retention and organizational performance: Evidence from banking industry. Management Science Letters, *10*(16), 3981–3990.

Al Kurdi, B., Alshurideh, M., & Alnaser, A. (2020b). The impact of employee satisfaction on customer satisfaction: Theoretical and empirical underpinning. *Management Science Letters, 10*(15), 3561–3570.

Al Kurdi, B., Elrehail, H., Alzoubi, H., Alshurideh, M., & Al-Adaila, R. (2021). The interplay among HRM practices, job satisfaction and intention to leave: An empirical investigation. *Journal of Legal, Ethical and Regulatory, 24*(1), 1–14.
Al Mehrez, A. A., Alshurideh, M., Al Kurdi, B., & Salloum, S. A. (2020). Internal factors affect knowledge management and firm performance: A systematic review. In: International Conference on Advanced Intelligent Systems and Informatics (pp. 632–643). Springer, Cham.
Al-Quran, A. Z., Alhalalmeh, M. I., Eldahamsheh, M. M., Mohammad, A. A., Hijjawi, G. S., Almomani, H. M., & Al-Hawary, S. I. (2020). Determinants of the green purchase intention in Jordan: The moderating effect of environmental concern. *Int. J Sup. Chain. Mgt, 9*(5), 366–371.
Al-Adamat, A., Al-Gasawneh, J., & Al-Adamat, O. (2020). The impact of moral intelligence on green purchase intention. *Management Science Letters, 10*(9), 2063–2070.
Alavi, S., Abd. Wahab, D., Muhamad, N., & Arbab Shirani, B. (2014). Organic structure and organisational learning as the main antecedents of workforce agility. International Journal of Production Research, *52*(21), 6273–6295.
Alavi, S. (2016). The influence of workforce agility on external. manufacturing flexibility of Iranian SMEs. International Journal of Technological Learning Innovation and Development, *8*(1), 111–127. Doi:https://doi.org/10.1504/IJTLID.2016.075185.
Al-bawaia, E., Alshurideh, M., Obeidat, B., Masa'deh, R. (2022). The impact of corporate culture and employee motivation on organization effectiveness in Jordanian banking sector. Academy of Strategic Management Journal, *21*(Special Issue 2), 1–18.
Al-Dhuhouri, F. S., Alshurideh, M., Al Kurdi, B., & Salloum, S. A. (2020). Enhancing our understanding of the relationship between leadership, team characteristics, emotional intelligence and their effect on team performance: A Critical Review. In: International Conference on Advanced Intelligent Systems and Informatics (pp. 644–655). Springer, Cham.
Al-Faouri, A. H., Al-Nsour, M. M., & Al-Kasasbeh, M. M. (2014). The impact of workforce agility on organizational memory. *Knowledge Management Research & Practice, 12*(4), 432–442.
Al-Gasawneh, J. A., & Al-Adamat, A. M. (2020). The relationship between perceived destination image, social media interaction and travel intentions relating to Neom city. *Academy of Strategic Management Journal, 19*(2), 1–12.
Alhalalmeh, M. I., Almomani, H. M., Altarifi, S., Al- Quran, A. Z., Mohammad, A. A., & Al-Hawary, S. I. (2020). The nexus between corporate social responsibilty and organizational performance in Jordan: The mediating role of organizational commitment and organizational citizenship behavior. *Test Engineering and Management, 83*(July), 6391–6410.
AlHamad, A., Alshurideh, M., Alomari, K., Kurdi, B., Alzoubi, H., Hamouche, S., & Al-Hawary, S. (2022). The effect of electronic human resources management on organizational health of telecommunications companies in Jordan. International Journal of Data and Network Science, *6*(2), 429–438.
Al-Hawary, S. I. (2011). Human resource management practices in ZAIN cellular communications company operating in Jordan. *Perspectives of Innovations, Economics and Business, 8*(2), 26–34.
Al-Hawary, S. I. (2015). Human resource management practices as a success factor of knowledge management implementation at health care sector in Jordan. *International Journal of Business and Social Science, 6*(11/1), 83–98.
Al-Hawary, S. I. S., & Obiadat, A. A. (2021). Does mobile marketing affect customer loyalty in Jordan? *International Journal of Business Excellence, 23*(2), 226–250.
Al-Hawary, S. I. S., Mohammad, A. S., Al-Syasneh, M. S., Qandah, M. S. F., & Alhajri, T. M. S. (2020). Organisational learning capabilities of the commercial banks in Jordan: Do electronic human resources management practices matter? *International Journal of Learning and Intellectual Capital, 17*(3), 242–266.
Al-Hawary, S. I., & Abu-Laimon, A. A. (2013). The impact of TQM practices on service quality in cellular communication companies in Jordan. *International Journal of Productivity and Quality Management, 11*(4), 446–474.
Al-Hawary, S. I., & Alajmi, H. M. (2017). Organizational commitment of the employees of the ports security affairs of the State of Kuwait: The impact of human recourses management practices.

International Journal of Academic Research in Economics and Management Sciences, 6(1), 52–78.
Al-Hawary, S. I., & Al-Namlan, A. (2018). Impact of electronic human resources management on the organizational learning at the Private Hospitals in the State of Qatar. *Global Journal of Management and Business Research: A Administration and Management, 18*(7), 1–11.
Al-Hawary, S. I., & Al-Syasneh, M. S. (2020). Impact of dynamic strategic capabilities on strategic entrepreneurship in presence of outsourcing of five stars hotels in Jordan. Business: Theory and Practice, *21*(2), 578–587.
Al-Hawary, S. I., & Hadad, T. F. (2016). The effect of strategic thinking styles on the enhancement competitive capabilities of Commercial Banks in Jordan. *International Journal of Business and Social Science, 7*(10), 133–144.
Al-Hawary, S. I., Al-Qudah, K., Abutayeh, P., Abutayeh, S., & Al-Zyadat, D. (2013). The impact of internal marketing on employee's job satisfaction of Commercial Banks in Jordan. *Interdisciplinary Journal of Contemporary Research in Business, 4*(9), 811–826.
Alkalha, Z., Al-Zu'bi, Z., Al-Dmour, H., Alshurideh, M., & Masa'deh, R. (2012). Investigating the effects of human resource policies on organizational performance: An empirical study on commercial banks operating in Jordan. European Journal of Economics, Finance and Administrative Sciences, *51*(1), 44–64.
Al-kasasbeh. A., Abdul Halim., M., & Omar, K. (2016). E-HRM, workforce agility and organizational performance: A review paper toward theoretical framework. I J A B E R, *14*(15), 10671–10685.
Allahow, T. J. A. A., Al-Hawary, S. I. S., & Aldaihani, F. M. F. (2018). Information technology and administrative innovation of the central agency for information technology in Kuwait. Global Journal of Management and Business, *18*(11-A), 1–16.
Allozi, A., Alshurideh, M., AlHamad, A., & Al Kurdi, B. (2022). Impact of transformational leadership on the job satisfaction with the moderating role of organizational commitment: Case of UAE and Jordan manufacturing companies. *Academy of Strategic Management Journal, 21*, 1–13.
Al-Lozi, M. S., Almomani, R. Z. Q., & Al-Hawary, S. I. S. (2018). Talent management strategies as a critical success factor for effectiveness of human resources information systems in commercial banks working in Jordan. *Global Journal of Management and Business Research: A Administration and Management, 18*(1), 30–43.
Al-Lozi, M., Almomani, R. Z., & Al-Hawary, S. I. (2017). Impact of talent management on achieving organizational excellence in Arab Potash Company in Jordan. *Global Journal of Management and Business Research: A Administration and Management, 17*(7), 15–25.
Almahamid, S. (2018). Knowledge management processes and workforce agility: A theoretical perspective. *International Journal of Management and Applied Science, 4*(Issue7), 28–33.
Al-Nady, B. A., Al-Hawary, S. I., & Alolayyan, M. (2013). Strategic management as a key for superior competitive advantage of sanitary ware suppliers in Kingdom of Saudi Arabia. *International Journal of Management and Information Technology, 7*(2), 1042–1058.
Al-Nady, B. A., Al-Hawary, S. I., & Alolayyan, M. (2016). The role of time, communication, and cost management on project management success: An empirical study on sample of construction projects customers in Makkah City, Kingdom of Saudi Arabia. *International Journal of Services and Operations Management, 23*(1), 76–112.
Alolayyan, M., Al-Hawary, S. I., Mohammad, A. A., & Al-Nady, B. A. (2018). Banking service quality provided by commercial banks and customer satisfaction. A structural equation modelling approaches. International Journal of Productivity and Quality Management, *24*(4), 543–565.
AlShehhi, H., Alshurideh, M., Al Kurdi, B., & Salloum, S. A. (2020). The impact of ethical leadership on employees performance: A systematic review. In: International Conference on Advanced Intelligent Systems and Informatics (pp. 417–426). Springer, Cham.
Alshurideh, M. T., Al Kurdi, B., Alzoubi, H. M., Ghazal, T. M., Said, R. A., AlHamad, A. Q., & Alkassem, A. H. (2022). Fuzzy assisted human resource management for supply chain management issues. Annals of Operations Research, 1–19.

Alshurideh, M. T., Al-Hawary, S., Mohammad, A., Al-Hawary, A., & Al Kurdi, A. (2017a). The impact of Islamic bank's service quality perception on Jordanian customer's loyalty. *Journal of Management Research, 9*(2), 139–159.
Alshurideh, M., Al-Hawary, S. I., Batayneh, A. M., Mohammad, A., & Al-Kurdi, B.(2017b). The Impact of Islamic Banks' Service Quality Perception on Jordanian Customers Loyalty. Journal of Management Research, *9*(2), 139–159.
Alsuwaidi, M., Alshurideh, M., Al Kurdi, B., & Salloum, S. A. (2020). Performance appraisal on employees' motivation: a comprehensive analysis. In: International Conference on Advanced Intelligent Systems and Informatics (pp. 681–693). Springer, Cham.
Altamony, H., Masa'deh, R., Alshurideh, M., Obeidat, B. (2012). Information systems for competitive advantage: Implementation of an organisational strategic management process. Innovation and sustainable competitive advantage: From regional development to world economies (pp. 583–592).
AlTaweel, I. R., & Al-Hawary, S. I. (2021). The mediating role of innovation capability on the relationship between strategic agility and organizational performance. *Sustainability, 13*(14), 7564.
Arocas, R. L., & Del Valle, I. D. (2021). Does positive wellbeing predict job performance three months later? *Applied Research in Quality of Life, 16*(4), 1555–1569. https://doi.org/10.1007/s11482-020-09835-0
Azuara, A. V. (2015). A Human resource perspective on the development of workforce agility. Faculty of The George L. Graziadio School of Business and Management. Pepperdine University, East Eisenhower Parkway, United States.
de Leeuw, E., Hox, J., Silber, H., Struminskaya, B., & Vis, C. (2019). Development of an international survey attitude scale: Measurement equivalence, reliability, and predictive validity. *Measurement Instruments for the Social Sciences, 1*(1), 9. https://doi.org/10.1186/s42409-019-0012-x
Dhani, P., & Sharma, T. (2017). Effect of emotional intelligence on job performance of IT employees: A gender study. *Procedia Computer Science, 122*, 180–185.
Ghani, N. M. A., Yunus, N. S. N. M., & Bahry, N. S. (2016). Leader's personality traits and employees job performance in public sector, Putrajaya. *Procedia Economics and Finance, 37*, 46–51.
Hair, J. F., Babin, B. J., & Krey, N. (2017). Covariance-based structural equation modeling in the journal of advertising: Review and recommendations. *Journal of Advertising, 46*(1), 163–177. https://doi.org/10.1080/00913367.2017.1281777
Harikrishnan, R., & Suresh, M. (2018). An Integrative analysis of workforce agility of police office. International Journal of Engineering & Technology, *7*(2.33), 950–954.
Hayajneh, N., Suifan, T., Obeidat, B., Abuhashesh, M., Alshurideh, M., & Masa'deh, R. (2021). The relationship between organizational changes and job satisfaction through the mediating role of job stress in the Jordanian telecommunication sector. Management Science Letters, *11*(1), 315–326.
Howard, M. C. (2018). The convergent validity and nomological net of two methods to measure retroactive influences. *Psychology of Consciousness: Theory, Research, and Practice, 5*(3), 324–337. https://doi.org/10.1037/cns0000149
Jager, D., Vliet, V., Born, M., Molen, H. (2019). Using a portfolio-based process to develop agility among employees. Human Resource Development Quarterly, *2019*, 1–22. doi:https://doi.org/10.1002/hrdq.21337.
Kashyap, H. (2014). Managing workforce diversity in India. *Tactful Management Research Journal, 2*(Issue9), 1–4.
Lee, K., Azmi, N., Hanaysha, J., Alshurideh, M., & Alzoubi, H. (2022a). The effect of digital supply chain on organizational performance: An empirical study in Malaysia manufacturing industry. *Uncertain Supply Chain Management, 10*(2), 1–16.
Lee, K., Ramiz, P., Hanaysha, J., Alzoubi, H., & Alshurideh, M. (2022b). Investigating the impact of benefits and challenges of IOT adoption on supply chain performance and organizational

performance: An empirical study in Malaysia. *Uncertain Supply Chain Management, 10*(2), 1–14.

Metabis, A., & Al-Hawary, S. I. (2013). The impact of internal marketing practices on services quality of commercial banks in Jordan. *International Journal of Services and Operations Management, 15*(3), 313–337.

Mohamad, M., & Jais, J. (2016). Emotional intelligence and jop performance: A study among Malaysian teachers. *Procedia Economics and Finance, 35*(2016), 674–682. https://doi.org/10.1016/S2212-5671(16)00083-6

Mohammad, A. A., Alshura, M. S., Al-Hawary, S. I. S., Al-Syasneh, M. S., & Alhajri, T. M. (2020). The influence of internal marketing practices on the employees' intention to leave: A study of the private hospitals in Jordan. *International Journal of Advanced Science and Technology, 29*(5), 1174–1189.

Mohammad, A. M. (2019). Effect of workforce diversity onjob performance of hotels working in Jordan. International Journal of Business and Management, *14*(4), 87–85. doi:https://doi.org/10.5539/ijbm.v14n4p85.

Muduli, A. (2013). Workforce agility: A review of literature. *The IUP Journal of Management Research, XI, 1*(3), 56–65.

Muduli, A. (2016). Exploring the facilitators and mediators of workforce agility: An empirical study. *Management Research Review, 39*(12), 1567–1586. https://doi.org/10.1108/MRR-10-2015-0236

Muduli, A. (2017). Workforce agility: Examining the role of organizational practices and psychological empowerment. *Global Business and Organizational Excellence, 36*(5), 46–56. https://doi.org/10.1002/joe.21800

Muhtasom, A., Mus, H. A. R., Bijang, J., & Latief, B. (2017). Influence of servant leadership, organizational citizenship behaviour on organizational culture and employee performance at Star Hotel in Makassar. *Star, 486*(206), 71–88.

Nuseir, M. T., Aljumah, A., & Alshurideh, M. T. (2021). How the business intelligence in the new startup performance in UAE during COVID-19: The mediating role of innovativeness. The Effect of Coronavirus Disease (COVID-19) on Business Intelligence, *334*, 63–79.

Odeh, R. B. M., Obeidat, B. Y., Jaradat, M. O., & Alshurideh, M. T. (2021). The transformational leadership role in achieving organizational resilience through adaptive cultures: the case of Dubai service sector. International Journal of Productivity and Performance Management. Vol. ahead-of-print No. ahead-of-print. https://doi.org/10.1108/IJPPM-02-2021-0093

Ogbo, A. I., Anthony, K. A., & Ukpere, W. I. (2014). The effect of workforce diversity on organizational performance of selected firms in Nigeria. *Mediterranean Journal of Social Sciences, 5*(10), 231–236.

Naomi Kunda, O. (2019). *Achieving workforce agility in dynamic environments.* Faculty of business studies school of management. University of Vaasa, Vaasa, Finnish.

Rimkeviciene, J., Hawgood, J., O'Gorman, J., & De Leo, D. (2017). Construct validity of the acquired capability for suicide scale: Factor structure, convergent and discriminant validity. *Journal of Psychopathology and Behavioral Assessment, 39*(2), 291–302. https://doi.org/10.1007/s10862-016-9576-4

Sekaran, U., & Bougie, R. (2016). *Research methods for business: A skill-building approach* (Seventh edition). Wiley.

Shakhour, R., Obeidat, B., Jaradat, M., Alshurideh, M., Masa'deh, R. (2021). Agile-minded organizational excellence: Empirical investigation. Academy of Strategic Management Journal, *20*(Special Issue 6), 1–25.

Shamout, M., Elayan, M., Rawashdeh, A., Kurdi, B., & Alshurideh, M. (2022). E-HRM practices and sustainable competitive advantage from HR practitioner's perspective: A mediated moderation analysis. *International Journal of Data and Network Science, 6*(1), 165–178.

Shi, D., Lee, T., & Maydeu-Olivares, A. (2019). Understanding the model size effect on SEM fit indices. *Educational and Psychological Measurement, 79*(2), 310–334. https://doi.org/10.1177/0013164418783530

Sohrabi, R., Asari, M., & Hozoori, M. (2014). Relationship between workforce agility and organizational intelligence (Case Study: The Companies of "Iran High Council of Informatics"). *Asian Social Science, 10*(4), 279–287. doi:https://doi.org/10.5539/ass.v10n4p279.

Sung, K.-S., Yi, Y. G., & Shin, H.-I. (2019). Reliability and validity of knee extensor strength measurements using a portable dynamometer anchoring system in a supine position. *BMC Musculoskeletal Disorders, 20*(1), 1–8. https://doi.org/10.1186/s12891-019-2703-0

Taran, O. (2019). *Training program effectiveness in building workforce agility and resilience.* University of Walden, United States.

Wang, Y. A., & Rhemtulla, M. (2021). Power analysis for parameter estimation in structural equation modeling: A discussion and tutorial. *Advances in Methods and Practices in Psychological Science, 4*(1), 1–17. https://doi.org/10.1177/2515245920918253

Xie, B., & Li, M. (2021). Coworker Guanxi and job performance: Based on the mediating effect of interpersonal trust. *Technological Forecasting and Social Change, 171*, 120981. https://doi.org/10.1016/j.techfore.2021.120981

The Impact of Strategic Orientation on Organizational Ambidexterity at the Hotels Sector in Jordan

Sandy Fawzi Al-Barakat, Sana aNawaf Al-Nsour, Ziad Mohd Ali Smadi, Mohammad Mousa Eldahamsheh, Sulieman Ibraheem Shelash Al-Hawary, Fuad N. Al-Shaikh, and Muhammad Turki Alshurideh

Abstract The aim of the study was to examine the impact of strategic orientation on organizational ambidexterity. Therefore, it focused on five-star hotels in Jordan. Data were primarily gathered through self-reported questionnaires created by Google Forms which were distributed to a random-stratified sample of (215) managers. In total, (193) responses were received including (8) invalid to statistical analysis due to

S. F. Al-Barakat
Business and Finance Faculty, World Islamic Science and Education University (WISE), Amman, Jordan

S. Al-Nsour
Business and Finance Faculty, World Islamic Science and Education University (WISE), Postal Code 11947, P.O Box 1101, Amman, Jordan

Z. M. A. Smadi
Department of Business Administration, School of Business Al Al, Bayt University, P.O.BOX 130040, Mafraq 25113, Jordan
e-mail: ziad38in@aabu.edu.jo

M. M. Eldahamsheh
Strategic management, Mafraq, Jordan

S. I. S. Al-Hawary (✉)
Department of Business Administration, School of Business, Al Al-Bayt University, P.O.BOX 130040, Mafraq 25113, Jordan
e-mail: dr_sliman73@aabu.edu.jo

F. N. Al-Shaikh
Department of Business Administration-Faculty of Economics and Administrative Sciences, Yarmouk University, P.O Box 566-Zip Code 21163, Irbid, Jordan
e-mail: eco_fshaikh@yu.edu.jo

M. T. Alshurideh
Department of Marketing, School of Business, The University of Jordan, Amman 11942, Jordan
e-mail: m.alshurideh@ju.edu.jo; malshurideh@sharjah.ac.ae

Department of Management, College of Business, University of Sharjah, 27272 Sharjah, United Arab Emirates

M. Alshurideh et al. (eds.), *The Effect of Information Technology on Business and Marketing Intelligence Systems*, Studies in Computational Intelligence 1056,
https://doi.org/10.1007/978-3-031-12382-5_61

uncompleted or inaccurate. Hence, the final sample contained (185) responses suitable to analysis requirements. Structural equation modeling (SEM) was conducted to test hypotheses. The results showed that strategic orientation had a positive impact on organizational ambidexterity. Moreover, the results indicated that the highest impact was for technology orientation. Based on the study results, the researchers recommend hotel managers to develop strategic directions by updating their general strategies to contribute to increasing the market share compared to other competing hotels.

Keywords Strategic orientations · Organizational ambidexterity · Hotels' sector · Jordan

1 Introduction

Strategic directions represent the outlines of the organizations strategies when they little details of the strategy content and the strategy implementation, In order to make strategies and take decisions about organizations, there must be leaders with a vision that is compatible with creative and innovative thinking in drawing strategic directions, which is a tool for coordinating the organization efforts (Al-Hawary & Al-Syasneh, 2020; Al-Hawary & Hadad, 2016; Al-Nady et al., 2013; AlTaweel & Al-Hawary, 2021). The idea of organizational ambidexterity came as one of those trends through which organizations can face challenges and environmental changes and stand strong towards them for their survival and continuity in their work performance. Organizational ambidexterity topic has captured the attention of many researchers, as organizations must be exceptionally ambidextrous and simultaneous to strike a balance between exploratory and exploitative activities.

The tourism sector is based on a number of components, including the facilities availability, and tourist facilities that provide services that tourists need during their trips, and the most important of these establishments are hotels, as they are those facilities that provide places for housing and sleeping, in addition to many Of services it provides, so hotels presence in the countries has become indispensable, given the volume of services provided by such establishments, and among the most important hotels are the international five-star hotels. Through previous literature review, it was found that few studies employ the approach (strategic directions) in the study of organizational ambidexterity, especially in hotels in Jordan, so this research is a new contribution within the Jordanian environment, specifically the hotel sector, the research also provides evidence that can be used by hotel managers in Jordan to keep pace with the rapid developments by highlighting the importance of the impact of strategic directions in achieving organizational ambidexterity.

The hotels sector has become a major focus in the revitalizing tourism process in Jordan, as it is a supportive and enhancing sector for the tourism movement, and because of the great similarity in the services provided by hotels, and the rapid development and emergence of the globalization concept and its applications, the hotel sector has put the hotel sector in front of great challenges and competitors, especially the crises of (Corona crisis), emphasized to the senior and middle management in the hotel sector the importance of adaptation and the ability to move quickly towards exploring new opportunities for the purpose of survival in the current and future environment (Alaali et al., 2021; Aljumah et al., 2021; Al-Hamad et al., 2021; Nuseir et al., 2021). The challenges facing the hotel sector require adopting new methods that enable it to explore the opportunities surrounding it and at the same time working to exploit those opportunities according to structurally independent teams capable of changing their structures according to that environment requirements, and for the purpose of balancing these methods requires following what is called organizational ambidexterity. (O'reilly & Tushman, 2013). However, the development of (dynamic) capabilities is important for the hotel sector to balance the activities of exploring new opportunities and exploiting current opportunities, which represent the main challenge to reach the hotel sector level, which represents the transition from the current situation to the future change management, so it is necessary to adopt strategic directions through develop strategies characterized by high flexibility to move from one strategic option to another and according to the conditions imposed by the environment surrounding the hotel sector.

2 Literature Review and Hypotheses Development

2.1 Strategic Orientation

The performance concept always contributes to the strategic goals achievement, the performance concept has become included in strategic thinking and its directions, and the organizations' goal has now become to achieve high strategic performance levels in the field of their activities. It is necessary to improve performance and develop organizations, which is no longer optional but has become an essential condition to ensure survival and continuity (Wheelen & Hunger, 2008; Al-Hawary, 2011; Al-Hawary, 2013; Al-Hawary et al., 2013; Metabis & Al-Hawary, 2013; Al-Hawary & Abu-Laimon, 2013; Al-Nady et al., 2016: Al-Lozi et al., 2017; Alshurideh et al., 2017; Allahow et al., 2018; Alolayyan et al., 2018; Alhalalmeh et al., 2020; Al-Quran et al., 2020; Mohammad et al., 2020). Strategic directions are defined as a specific approach that the organization applies to reach superior and continuous performance in the course of its work, as strategic directions work to establish an approach linked to the goals of improving the organization's permanent performance, and reflecting managers' awareness of the surrounding environment and their reactions to environmental conditions (Jandab, 2013; Kabrilyants et al., 2021; Tariq et al., 2022). Strategic directions importance lies in the organization's compatibility of its resources with current opportunities, and it also helps the organization to be more compatible and adaptive to the surrounding environmental changes, and helps to explore new markets, work on developing radical innovations, and finding behaviors that help the organization achieve superior performance. (Al Suwaidi et al., 2020; Alameeri et al., 2021; Alshurideh et al., 2019; Alzoubi et al., 2022; Slater & Oslan, 2001).

There are dimensions of strategic directions, three of which have been selected because they represent a comprehensive set of strategic directions critical to the organization success. (Covin & Slevin, 1989; Narver & Slater, 1990) includes: **Market Orientation**: which is defined as the organization's culture that finds the most effective necessary behaviors to provide the greatest value to customers (Narver & Slater, 1990), market orientation is part of the organizations culture, which poses a challenge to any organization that has few technological capabilities available to it and which are scarce in these markets, and whose markets are characterized by stereotyping and must subject its requirements to continuous innovation to achieve more productivity at a lower cost. **Entrepreneurial orientation**: which is the organization ability that makes it possess the desire to innovate and to act independently and the tendency to take risks in order to meet the future customers' requirements in the market. It is also concerned with the organization strategic leadership aspects, and that today's business environment is more dynamic, so markets and organizations must achieve the competitive advantage that distinguishes them from other competitors. **Technology Orientation** which is defined as the ability to obtain modern technological knowledge and use it in the new products development, and it is also known as the use of advanced technologies in the development of new products, the new technology rapid integration, and the new product ideas proactive creation. (Gatignon & Xuereb, 1997, 78, 82).

2.2 *Strategic Ambidexterity*

Organizational ambidexterity has dealt with many controversial managerial definitions due to the different literary and scientific opinions that dealt with the concept of ambidexterity Organizational ambidexterity, or strategic ambidexterity, and all terms include the process of exploring and exploiting opportunities. Many researchers have addressed the organizational ambidexterity, most notably the organization's simultaneous pursuit of exploration and exploitation activities (Jansen et al., 2009). Ambidextrous organizations are superior organizations that enable innovation by exploiting their existing products and exploring new opportunities to promote more radical innovations (Andriopoulos & Lewis, 2010). Organizational ambidexterity is the organizations need to strike a balance between exploitation and exploration to achieve superior performance. The organizational ambidexterity importance is reflected in enabling organizations to achieve the goal of long-term survival in order to ensure continuity and future growth. (Yigit, 2013). Through the research and development processes in the organization, it enables the working managers to work through the information and various alternatives presence for the decisions in order to reduce the conflict between the exploration and the exploitation activities. (Prange & Schlegelmich, 2009).

There are two dimensions that the researcher identified for organizational ambidexterity as discussed by Bodwell and Chermack (2010) and Cao et al. (2009) which are: **Exploring Opportunities**: which represents the organization's ability to move quickly to explore new opportunities, prepare to adapt to volatile markets, leads to the new customers and markets emergence, new distribution channels creation, and the exploration of new opportunities requires new knowledge that is completely different and distinguished by research, difference, and flexibility. **Opportunity Exploitation**: which is the organization ability to improve activities to generate value in the short term, as it is designed to meet current customers' needs in the current markets, and seeks to expand the knowledge and skills available, and expand the current services and products while increasing the current distribution channels, and that the opportunities exploitation represents the activities of organizational learning, namely, refinement, efficiency, and implementation. And that exploitation is reflected in the organization's evaluation of opportunities in the environment that do not necessarily fall within the scope of its strategies, which is reflected in the identification of future opportunities (Brion & Mothe, 2010).

2.3 *Strategic Directions and Organizational Ambidexterity*

There are many organizations that are able to pursue innovative and proactive programs in a gradual or radical way, because of their full market orientation, and that linking exploration and exploitation activities leads to enabling new business units to practice market orientation behaviors towards new markets, and to respond quickly to them through its application, Therefore, managers must also arrange product markets in a proactive and rapid manner with market-oriented behaviors at the same time, by allocating the working managers in the organization to tangible and intangible resources (Li et al., 2008, p. 1022). The entrepreneurial orientation indicates that some proactive and innovative organizations adopt risk (Tuan, 2016), and the entrepreneurial orientation strikes a balance and link between exploration and exploitation as a dynamic task rather than as a static match (Westernan, Macfarlan & Iansitti, 2006, 230). There are few studies that dealt with the link between technology orientation and organizational ambidexterity, as pointed by Rotharrmal & Alexandre (2009). Managers in the organization must follow an approach based on modern technology use, through the development of a cost-oriented plan, and to promote high efficiencies and invest in existing markets for products (Kortmann, 2015, p. 3). And some creative organizations are discovering and entering new markets, and using new proactive technology, rather than focusing on existing capabilities and new product markets (Talke, Salomo, Kock, 2011; Almaazmi et al., 2020). Based on the above, study hypothesis can be formulated as follows:

There is a statistically significant impact of the strategic directions on organizational ambidexterity in the hotel sector in Jordan.

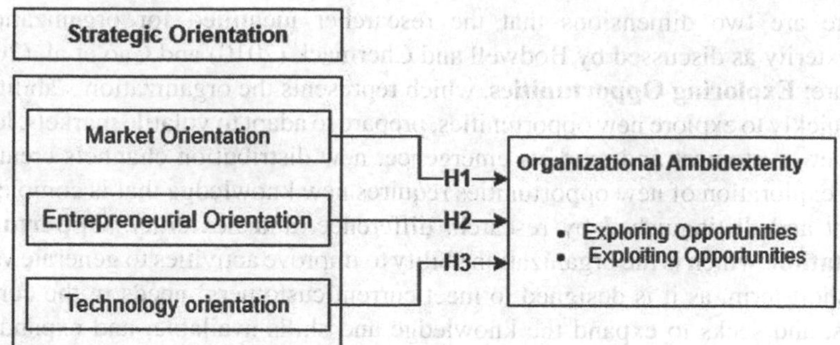

Fig. 1 Research model

3 Study Model

See Fig. 1.

4 Methodology

4.1 Population and Sample Selection

A qualitative method based on a questionnaire was used in this study for data collection and sample selection. The major aim of the study was to examine the impact of strategic orientation on organizational ambidexterity. Therefore, it focused on five-stars hotels in Jordan. Data were primarily gathered through self-reported questionnaires created by Google Forms which were distributed to a random-stratified sample of (215) managers via email. In total, (193) responses were received including (8) invalid to statistical analysis due to uncompleted or inaccurate. Hence, the final sample contained (185) responses suitable to analysis requirements that were formed a response rate of (86.04%), where it proved to be sufficient to the extent that was predictable and allowed for a presumption of data saturation (Sekaran & Bougie, 2016).

4.2 Measurement Instrument

A self-reported questionnaire that consists of two main sections along with a section regarding control variables was used as the measurement instrument. Control variables considered as categorical measures were composed of gender, age group, educational level, and experience. The two main sections were dealt with a five-point Likert scale (from 1 = strongly disagree to 5 = strongly agree). The first section contained (21) items to measure strategic orientation based on (Beliaeva et al., 2018; Tseng et al., 2019). These questions were distributed into dimensions as follows: seven items dedicated for measuring market orientation, seven items dedicated for measuring entrepreneurial orientation, and seven items dedicated for measuring technology orientation. Whereas the second section included (14) items developed to measure organizational ambidexterity according to what was pointed by (Peng et al., 2019). This variable was a second-order construct divided into two first-order constructs. Exploring opportunities measured by seven items and exploiting opportunities measured using seven items.

5 Findings

5.1 Measurement Model Evaluation

This study was conducted structural equation modeling (SEM) to test hypotheses, which represents a contemporary statistical technique for testing and estimating the relationship between factors and variables (Al-Hawary & Alhajri, 2020; Al-Hawary & Obiadat, 2021; Al-Hawary et al., 2020; Wang & Rhemtulla, 2021). Accordingly, the reliability and validity of the constructs were tested using confirmatory factor analysis (CFA) through the statistical program AMOSv24. Table 1 summarizes the results of convergent and discriminant validity, as well the indicators of reliability.

Table 1 shows that the standard loading values for the individual items were within the domain (0.637–0.864), these values greater than the minimum retention of the elements based on their standard loads (Al-Lozi et al., 2018; Sung et al., 2019). Average variance extracted (AVE) is a summary indicator of the convergent validity of constructs that must be above 0.50 (Howard, 2018). The results indicate that the AVE values were greater than 0.50 for all constructs, thus the used measurement model has an appropriate convergent validity. Rimkeviciene et al. (2017) suggested the comparison approach as a way to deal with discriminant validity assessment in covariance-based SEM. This approaches based on comparing the values of maximum shared variance (MSV) with the values of AVE, as well as comparing the values of square root of AVE ($\sqrt{AVE}$) with the correlation between the rest of the structures. The results show that the values of MSV were smaller than the values of AVE, and that the values of $\sqrt{AVE}$ were higher than the correlation values among the rest of the constructs. Therefore, the measurement model used is characterized by discriminative validity. The internal consistency measured through Cronbach's Alpha coefficient (α) and compound reliability by McDonald's Omega coefficient (ω) was conducted as indicators to evaluate measurement model. The results listed in Table 1 demonstrated that both values of Cronbach's Alpha coefficient and McDonald's

Table 1 Results of validity and reliability tests

Constructs	1	2	3	4	5
1. Market Orientation (MO)	**0.749**				
2. Entrepreneurial Orientation (EO)	0.294	**0.757**			
3. Technology Orientation (TO)	0.355	0.364	**0.744**		
4. Exploring Opportunities (ER)	0.415	0.605	0.597	**0.751**	
5. Exploiting Opportunities (ET)	0.524	0.578	0.622	0.615	**0.745**
VIF	2.641	2.078	2.235	– -	– -
Loadings range	0.637–0.824	0.673–0.854	0.656–0.802	0.694–0.833	0.688–0.864
AVE	0.561	0.574	0.554	0.563	0.555
MSV	0.415	0.495	0.503	0.425	0.510
Internal consistency	0.896	0.901	0.894	0.898	0.894
Composite reliability	0.899	0.903	0.896	0.900	0.897

Note Bold fonts refer to √AVE

Omega coefficient were greater than 0.70, which is the lowest limit for judging on measurement reliability (De Leeuw et al., 2019).

5.2 *Structural Model*

The structural model illustrated no multicollinearity issue among predictor constructs because variance inflation factor (VIF) values are below the threshold of 5, as shown in Table 1 (Hair et al., 2017). This result is supported by the values of model fit indices shown in Figs. 1 and 2.

The results in Fig. 1 indicated that the chi-square to degrees of freedom (CMIN/DF) was 1.739, which is less than 3 the upper limit of this indicator. The values of the goodness of fit index (GFI), the comparative fit index (CFI), and the Tucker-Lewis index (TLI) were upper than the minimum accepted threshold of 0.90. Moreover, the result of root mean square error of approximation (RMSEA) indicated to value 0.019, this value is a reasonable error of approximation because it is less than the higher limit of 0.08. Consequently, the structural model used in this study was recognized as a fit model for predicting the organizational ambidexterity and generalization of its result (Ahmad et al., 2016; Shi et al., 2019). To verify the results

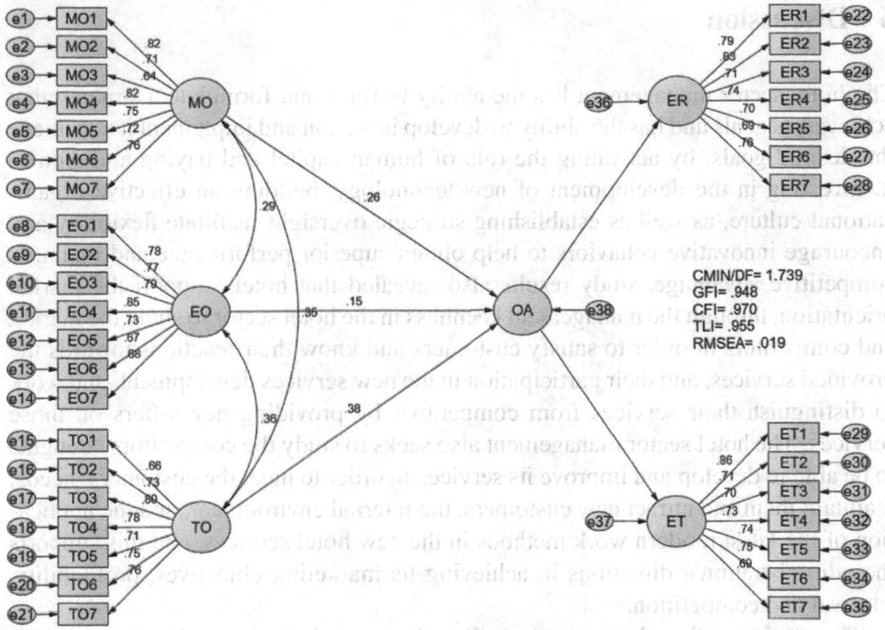

Fig. 2 SEM results of the strategic orientation effect on organizational ambidexterity

of testing the study hypotheses, structural equation modeling (SEM) was used, the results of which are listed in Table 2.

The results demonstrated in Table 2 show that strategic orientation dimensions had a positive impact on organizational ambidexterity. Moreover, the results indicated that the highest impact was for technology orientation (β = 0.380, t = 5.924, p = 0.000), followed by market orientation (β = 0.257, t = 4.560, p = 0.000), and finally the lowest impact was for entrepreneurial orientation (β = 0.151, t = 2.526, p = 0.012).

Table 2 Hypothesis testing

Hypothesis	Relation	Standard Beta	*t* value	*p* value
H1	Market Orientation → Organizational Ambidexterity	0.257***	4.560	0.000
H2	Entrepreneurial Orientation → Organizational Ambidexterity	0.151*	2.526	0.012
H3	Technology Orientation → Organizational Ambidexterity	0.380***	5.924	0.000

Note * p < 0.05, ** p < 0.01, *** p < 0.001

6 Discussion

The hotel sector management has the ability to form and formulate a strategy that achieves its goals and has the ability to develop its vision and implement it to achieve the desired goals, by activating the role of human capital and paying attention to it, investing in the development of new technology, building an effective organizational culture, as well as establishing strategic oversight facilitate flexibility and encourage innovative behaviors to help obtain superior performance and maintain competitive advantage, study results also revealed that hotels support the market orientation, through the management keenness in the hotel sector to study the market and competitors in order to satisfy customers and know their reactions towards the provided services, and their participation in the new services development, and work to distinguish their services from competitors by providing new offers on those services, The hotel sector management also seeks to study the competitors strengths to be able to develop and improve its services in order to meet the customer's needs, maintain them and attract new customers, the internal environment, and the application of the latest modern work methods in the new hotel services, and this supports the administration's directions in achieving its marketing objectives, profitability, survival and competition.

The results confirm the management's endeavor in the surveyed hotels to achieve leadership in work through the application of a number of practices that confirm the pioneering orientation, including their organizational structures review and development on a continuous basis to adapt to the surrounding developments, and encourage them to present new ideas with the intention of developing and simplifying work procedures and improving the provided service levels, as well as allocating sufficient budgets for the creating new services process, which contributes to achieving its planned goals in short and long-term strategic plans.

The results showed that the surveyed hotels management is taking advantage of the available opportunities to achieve its goals, and to confront environmental challenges in the hotel sector, which is most vulnerable to developments and environmental changes, especially since Jordan has suffered from political events in the Middle East region, as well as the health symptom that the whole world suffers from (Coronavirus). Which necessitates the surveyed hotels management to develop a risk strategy that includes procedures and practices to take advantage of the available opportunities with a view to improving the efficiency of the hotel services provided and investing in new opportunities. The results also confirm that the surveyed hotels management is able to explore the available opportunities and sense the environment in order to improve its services provided to customers and achieve its goals, and its ability to explore opportunities brings it more new customers and makes it able to develop and innovate new hotel services, and exploit the opportunities available by other hotels, as well as creating new outlets in providing its current services, and marketing hotel services that do not exist with competitors, or at least distinguished from competitors, which contributes to increasing the market share compared to competitors.

The results concluded that there is a statistically significant impact of the strategic orientation variable with its dimensions: (market orientation, pioneering orientation, and technology orientation) on organizational ambidexterity in the hotel sector in Jordan. Keeping pace with and using modern technology that fits with the internal environment, applying the latest modern work methods in new hotel services, and striving to achieve leadership in work, this positively affects its ability to achieve organizational ambidexterity in terms of exploring new opportunities and exploiting the current opportunity that is working towards achieving its goals. The application of the latest modern working methods in new hotel services, building new technological solutions to meet customers' needs, and taking the initiative in using advanced technologies in developing new services compared to competitors, supports the management's directions in achieving its marketing objectives, profitability, survival and competition, which means that technology orientation is considered an important factor in achieving organizational ambidexterity in terms of exploring new opportunities and exploiting the available opportunities that work towards achieving its goals (Alshurideh et al., 2012; Al Kurdi et al., 2020a, b; Alshamsi et al., 2020; Alzoubi et al., 2020; Alshurideh, 2022).

The results are agreed with the result of Rady and Qassem (2018), which showed the existence of a positive correlation and effect with significant between strategic directions and organizational ambidexterity. The results are also in agreement with the study of Hijjawi (2019), which showed that regulatory ambidexterity plays an important role in increasing the ability of the Jordanian telecommunications sector to face volatile environmental disruptions.

7 Recommendations

Based on the previous results, the study recommends hotel managers to develop strategic directions by updating their general strategies to contribute to increasing the market share compared to other competing hotels. Increasing interest in the market orientation through a better study of the market and competitors and following more advanced methods for collecting information that improve the quality of the hotel service. The study also recommends enhancing the technology orientation role, so that hotel management uses technology and keeps pace with it constantly, through its development and modernization, and the application of the latest modern work methods in new hotel services, encouraging and supporting new ideas presented by various managerial levels in the hotel sector with the aim of developing and simplifying work procedures and improving the service provided to customers.

References

Abdel-Wahab Ahmed Ali, J. (2013). The impact of creative and proactive strategic directions on new product development and marketing performance, an applied study on food manufacturing companies in Yemen. Unpublished master's thesis, Department of Business Administration, Middle East University, Jordan.

Ahmad, S., Zulkurnain, N., & Khairushalimi, F. (2016). Assessing the validity and reliability of a measurement model in structural equation modeling (SEM). *British Journal of Mathematics & Computer Science, 15*(3), 1–8. https://doi.org/10.9734/BJMCS/2016/25183

Alaali, N., Al Marzouqi, A., Albaqaeen, A., Dahabreh, F., Alshurideh, M., Alrwashdh, S., Iyadeh, I., Salloum, S., & Aburayya, A. (2021). The impact of adopting corporate governance strategic performance in the tourism sector: A case study in the Kingdom of Bahrain. Journal of Legal, Ethical and Regulatory, *24*(Special Issue 1), 1–18.

Al Kurdi, B., Alshurideh, M., & Al Afaishat, T. (2020a). Employee retention and organizational performance: Evidence from banking industry. *Management Science Letters, 10*(16), 3981–3990.

Al Kurdi, B., Alshurideh, M., & Alnaser, A. (2020b). The impact of employee satisfaction on customer satisfaction: Theoretical and empirical underpinning. *Management Science Letters, 10*(15), 3561–3570.

Al-Quran, A. Z., Alhalalmeh, M. I., Eldahamsheh, M. M., Mohammad, A. A., Hijjawi, G. S., Almomani, H. M., & Al-Hawary, S. I. (2020). Determinants of the green purchase intention in Jordan: The moderating effect of environmental concern. *International Journal of Supply Chain Management, 9*(5), 366–371.

Al Suwaidi, F., Alshurideh, M., Al Kurdi, B., & Salloum, S. A. (2020). The impact of innovation management in SMEs performance: A systematic review. In: International Conference on Advanced Intelligent Systems and Informatics (pp. 720–730). Cham: Springer.

Alameeri, K. A., Alshurideh, M. T., & Al Kurdi, B. (2021). The effect of Covid-19 pandemic on business systems' innovation and entrepreneurship and how to cope with it: A theatrical view. The Effect of Coronavirus Disease (COVID-19) on Business Intelligence, *334*, 275–288.

Alhalalmeh, M. I., Almomani, H. M., Altarifi, S., Al- Quran, A. Z., Mohammad, A. A., & Al-Hawary, S. I. (2020). The nexus between corporate social responsibilty and organizational performance in Jordan: The mediating role of organizational commitment and organizational citizenship behavior. *Test Engineering and Management, 83*(July), 6391–6410.

Al-Hamad, M., Mbaidin, H., AlHamad, A., Alshurideh, M., Kurdi, B., & Al-Hamad, N. (2021). Investigating students' behavioral intention to use mobile learning in higher education in UAE during Coronavirus-19 pandemic. *International Journal of Data and Network Science, 5*(3), 321–330.

Al-Hawary, S. I. (2011). The effect of banks governance on banking performance of the Jordanian Commercial Banks: Tobin's Q model "An applied study. *International Research Journal of Finance and Economics, 71*, 34–47.

Al-Hawary, S. I. (2013). The role of perceived quality and satisfaction in explaining customer brand loyalty: Mobile phone service in Jordan. *International Journal of Business Innovation and Research, 7*(4), 393–413.

Al-Hawary, S. I. S., & Alhajri, T. M. S. (2020). Effect of electronic customer relationship management on customers' electronic satisfaction of communication companies in Kuwait. *Calitatea, 21*(175), 97–102.

Al-Hawary, S. I. S., & Obiadat, A. A. (2021). Does mobile marketing affect customer loyalty in Jordan? *International Journal of Business Excellence, 23*(2), 226–250.

Al-Hawary, S. I. S., Mohammad, A. S., Al-Syasneh, M. S., Qandah, M. S. F., & Alhajri, T. M. S. (2020). Organisational learning capabilities of the commercial banks in Jordan: Do electronic human resources management practices matter? *International Journal of Learning and Intellectual Capital, 17*(3), 242–266.

Al-Hawary, S. I., & Abu-Laimon, A. A. (2013). The impact of TQM practices on service quality in cellular communication companies in Jordan. *International Journal of Productivity and Quality Management, 11*(4), 446–474.

Al-Hawary, S. I., & Al-Syasneh, M. S. (2020). Impact of dynamic strategic capabilities on strategic entrepreneurship in presence of outsourcing of five stars hotels in Jordan. Business: Theory and Practice, *21*(2), 578–587.

Al-Hawary, S. I., & Hadad, T. F. (2016). The effect of strategic thinking styles on the enhancement competitive capabilities of commercial Banks in Jordan. *International Journal of Business and Social Science, 7*(10), 133–144.

Al-Hawary, S. I., Al-Hawajreh, K., AL-Zeaud, H., & Mohammad, A. (2013). The impact of market orientation strategy on performance of Commercial Banks in Jordan. International Journal of Business Information Systems, *14*(3), 261–279.

Aljumah, A., Nuseir, M. T., & Alshurideh, M. T. (2021). The impact of social media marketing communications on consumer response during the COVID-19: Does the brand equity of a university matter. The Effect of Coronavirus Disease (COVID-19) on Business Intelligence, 367–384.

Allahow, T. J. A. A., Al-Hawary, S. I. S., & Aldaihani, F. M. F. (2018). Information technology and administrative innovation of the central agency for information technology in Kuwait. Global Journal of Management and Business, *18*(11-A), 1–16.

Al-Lozi, M. S., Almomani, R. Z. Q., & Al-Hawary, S. I. S. (2018). Talent Management strategies as a critical success factor for effectiveness of human resources information systems in commercial banks working in Jordan. *Global Journal of Management and Business Research: A Administration and Management, 18*(1), 30–43.

Al-Lozi, M., Almomani, R. Z., & Al-Hawary, S. I. (2017). Impact of talent management on achieving organizational excellence in Arab Potash Company in Jordan. *Global Journal of Management and Business Research: A Administration and Management, 17*(7), 15–25.

Almaazmi, J., Alshurideh, M., Al Kurdi, B., & Salloum, S. A. (2020). The effect of digital transformation on product innovation: a critical review. In International Conference on Advanced Intelligent Systems and Informatics (pp. 731–741). Cham: Springer.

Al-Nady, B. A., Al-Hawary, S. I., & Alolayyan, M. (2013). Strategic management as a key for superior competitive advantage of sanitary ware suppliers in Kingdom of Saudi Arabia. *International Journal of Management and Information Technology, 7*(2), 1042–1058.

Al-Nady, B. A., Al-Hawary, S. I., & Alolayyan, M. (2016). The role of time, communication, and cost management on project management success: An empirical study on sample of construction projects customers in Makkah City, Kingdom of Saudi Arabia. *International Journal of Services and Operations Management, 23*(1), 76–112.

Alolayyan, M., Al-Hawary, S. I., Mohammad, A. A., & Al-Nady, B. A. (2018). Banking service quality provided by commercial banks and customer satisfaction. A structural equation modelling approaches. International Journal of Productivity and Quality Management, *24*(4), 543–565.

Alshamsi, A., Alshurideh, M., Al Kurdi, B., & Salloum, S. A. (2020). The influence of service quality on customer retention: a systematic review in the higher education. In International Conference on Advanced Intelligent Systems and Informatics (pp. 404–416). Cham: Springer.

Alshurideh, M. (2022). Does electronic customer relationship management (E-CRM) affect service quality at private hospitals in Jordan? *Uncertain Supply Chain Management, 10*(2), 325–332.

Alshurideh, M., Al-Hawary, S. I., Batayneh, A. M., Mohammad, A., & Al-Kurdi, B. (2017). The impact of islamic banks' service quality perception on Jordanian customers loyalty. *Journal of Management Research, 9*(2), 139–159.

Alshurideh, M., Kurdi, B. A., Shaltoni, A. M., & Ghuff, S. S. (2019). Determinants of pro-environmental behaviour in the context of emerging economies. *International Journal of Sustainable Society, 11*(4), 257–277.

Alshurideh, M., Masa'deh, R. M. D. T., & Alkurdi, B. (2012). The effect of customer satisfaction upon customer retention in the Jordanian mobile market: An empirical investigation. *European Journal of Economics, Finance and Administrative Sciences, 47*(12), 69–78.

AlTaweel, I. R., & Al-Hawary, S. I. (2021). The mediating role of innovation capability on the relationship between strategic agility and organizational performance. *Sustainability, 13*(14), 7564.

Alzoubi, H. M., Alshurideh, M., Al Kurdi, B., & Inairat, M. (2020). Do perceived service value, quality, price fairness and service recovery shape customer satisfaction and delight? A practical study in the service telecommunication context. *Uncertain Supply Chain Management, 8*(3), 579–588.

Alzoubi, H., Alshurideh, M., Kurdi, B., Akour, I., & Aziz, R. (2022). Does BLE technology contribute towards improving marketing strategies, customers' satisfaction and loyalty? The role of open innovation. *International Journal of Data and Network Science, 6*(2), 449–460.

Andriopoulos, C., & Lewis, M. W. (2010). Managing innovation paradoxes: Ambidexterity lessons from leading product design companies. *Long Range Planning, 43*(1), 104–122.

Beliaeva, T., Shirokova, G., Wales, W., & Gafforova, E. (2018). Benefiting from economic crisis? Strategic orientation effects, trade-offs, and configurations with resource availability on SME performance. *International Entrepreneurship and Management Journal*. https://doi.org/10.1007/s11365-018-0499-2

Bodwell, W., & Chermack, T. (2010). Organizational ambidexterity: Integrating deliberate and emergent strategy with scenario planning. *Technological Forecasting & Social Change, 77*(2), 193–202.

Brion, S., Mothe, C., & Sabatier, M. (2010). The impact of organisational context and competences on innovation ambidexterity. *International Journal of Innovation Management, 14*(02), 151–178.

Cao, Q., Gedajlovic, E., & Zhang, H. (2009). Unpacking organizational ambidexterity: Dimensions, contingencies, and synergisticeffects. *Organization Science, 20*(4), 781–796.

Covin, J. G., & Slevin, D. P. (1989). Strategic management of small firms in hostile and benign environments. *Strategic Management Journal, 10*, 75–87.

De Leeuw, E., Hox, J., Silber, H., Struminskaya, B., & Vis, C. (2019). Development of an international survey attitude scale: Measurement equivalence, reliability, and predictive validity. *Measurement Instruments for the Social Sciences, 1*(1), 9. https://doi.org/10.1186/s42409-019-0012-x

Gatignon, H., & Xuereb, J. M. (1997). Strategic orientation of the firm and new product performance. *Journal of Marketing Research, 34*(1), 77–90.

Hair, J. F., Babin, B. J., & Krey, N. (2017). Covariance-based structural equation modeling in the journal of advertising: Review and recommendations. *Journal of Advertising, 46*(1), 163–177. https://doi.org/10.1080/00913367.2017.1281777

He, L. (2014). The perceived personal characteristics of entrepreneurial leaders. Unpublished Master's thesis, Edith Cown University.

Hijjawi, G. S., & Mohammad, A. (2019). Impact of organizational ambidexterity on organizational conflict of Zain Telecommunication Company in Jordan. *Indian Journal of Science and Technology, 12*, 26.

Howard, M. C. (2018). The convergent validity and nomological net of two methods to measure retroactive influences. *Psychology of Consciousness: Theory, Research, and Practice, 5*(3), 324–337. https://doi.org/10.1037/cns0000149

Jansen, J., Tempelaar, M., Bosch, F., & Volberda, H. (2009). Structural differentiation and ambidexterity: The mediating role of integration mechanisms. *Organization Science, 20*(4), 797–811.

Kabrilyants, R., Obeidat, B., Alshurideh, M., & Masadeh, R. (2021). The role of organizational capabilities on e-business successful implementation. *International Journal of Data and Network Science, 5*(3), 417–432.

Kortmann, S. (2015). The mediating role of strategic orientations on the relationship between ambidexterity-oriented decisions and innovative ambidexterity. *Journal of Product Innovation Management, 32*(5), 666–684.

Li, Y., Liu, Y., Duan, Y., & Li, M. (2008). Entrepreneurial orientation, strategic flexibilities and indigenous firm innovation in transitional China. *International Journal of Technology Management, 41*(1–2), 223–246.

Metabis, A., & Al-Hawary, S. I. (2013). The impact of internal marketing practices on services quality of Commercial Banks in Jordan. *International Journal of Services and Operations Management, 15*(3), 313–337.

Mohammad, A. A., Alshura, M. S., Al-Hawary, S. I. S., Al-Syasneh, M. S., & Alhajri, T. M. (2020). The influence of internal marketing practices on the employees' intention to leave: A study of the private hospitals in Jordan. *International Journal of Advanced Science and Technology, 29*(5), 1174–1189.

Narver, J. C., & Slater, S. F. (1990). The effect of a marketorientation on business profitability. *Journal of Marketing, 54*(3), 20–35.

Nuseir, M. T., Aljumah, A., & Alshurideh, M. T. (2021). How the business intelligence in the new startup performance in UAE during COVID-19: The mediating role of innovativeness. The Effect of Coronavirus Disease (COVID-19) on Business Intelligence, *334*, 63–79.

O'Reilly, C. A., III., & Tushman, M. L. (2013). Organizational ambidexterity: Past, present, and future. *Academy of Management Perspectives, 27*(4), 324–338.

Peng, M.Y.-P., Lin, K.-H., Peng, D. L., & Chen, P. (2019). Linking organizational ambidexterity and performance: The drivers of sustainability in high-tech firms. *Sustainability, 11*(14), 3931. https://doi.org/10.3390/su11143931

Prange, C., & Schlegelmilch, B. B. (2009). The role of ambidexterity in marketing strategy implementation: Resolving the exploration-exploitation dilemma. *Business Research, 2*(2), 215–240.

Radi, J. M., Qassem, & Naim, Z. (2018). Strategic orientation and its impact on achieving organizational ingenuity. Journal of Administration and Economics, (7.25), 139–105, Iraq.

Rimkeviciene, J., Hawgood, J., O'Gorman, J., & De Leo, D. (2017). Construct validity of the acquired capability for suicide scale: Factor structure, convergent and discriminant validity. *Journal of Psychopathology and Behavioral Assessment, 39*(2), 291–302. https://doi.org/10.1007/s10862-016-9576-4

Rothaermel, F. T., & Alexandre, M. T. (2009). Ambidexterity in technolo-gysourcing: The moderating role of absorptive capacity. *Organization Science, 20*(4), 759–780.

Sekaran, U., & Bougie, R. (2016). Research methods for business: A skill-building approach (Seventh edition). Wiley.

Shi, D., Lee, T., & Maydeu-Olivares, A. (2019). Understanding the model size effect on SEM fit indices. *Educational and Psychological Measurement, 79*(2), 310–334. https://doi.org/10.1177/0013164418783530

Slater, S. F., & Oslan E. M. (2001). Marketing's contribution to the implementation of business strategy: An empirical analysis. Strategic Management Journal, *22*(No. 11), 1055–1067.

Sung, K.-S., Yi, Y. G., & Shin, H.-I. (2019). Reliability and validity of knee extensor strength measurements using a portable dynamometer anchoring system in a supine position. *BMC Musculoskeletal Disorders, 20*(1), 1–8. https://doi.org/10.1186/s12891-019-2703-0

Talke, K., Salomo, S., & Kock, A. (2011). Top management team diversity and strategic innovation orientation: The relationship and consequences for innovativeness and performance. *Journal of Product Innovation Management, 28*(6), 819–832.

Tariq, E., Alshurideh, M., Akour, I., & Al-Hawary, S. (2022). The effect of digital marketing capabilities on organizational ambidexterity of the information technology sector. *International Journal of Data and Network Science, 6*(2), 401–408.

Tseng, C.-H., Chang, K.-H., & Chen, H.-W. (2019). Strategic orientation, environmental innovation capability, and environmental sustainability performance: The case of Taiwanese suppliers. *Sustainability, 11*(4), 1127. https://doi.org/10.3390/su11041127

Tuan, L. T. (2016). Organizational ambidexterity, entrepreneurial orientation, and I-deals: The moderating role of CSR. *Journal of Businessethics, 135*(1), 145–159.

Wang, Y. A., & Rhemtulla, M. (2021). Power analysis for parameter estimation in structural equation modeling: A discussion and tutorial. *Advances in Methods and Practices in Psychological Science, 4*(1), 1–17. https://doi.org/10.1177/2515245920918253

Westerman, G., McFarlan, F. W., & Iansiti, M. (2006). Organization design and effectiveness over the innovation life cycle. Organization Science, *17*(2), 230–238.

Wheelen, T. L. & Hunger, J. D. (2008). Strategic Management and Business Policy (11th edn). Upper Saddle River, New Jersey: Pearson Education Inc.

Yigit, M. (2013). Organizational ambidexterity: Balancing exploitation and exploration in organizations. Unpublished master dissertation, Blekinge Institute of Technology School of Management.

The Effect of Total Quality Management on the Organizational Reputation: The Moderating Role of Quality Standards in Jordanian Public Universities

Enas Ahmad Alshuqairat, Basem Yousef Ahmad Barqawi, Zaki Abdellateef Khalaf Khalaylah, Mohammed saleem khlif Alshura, Maali M. Al-mzary, Muhammad Turki Alshurideh, Sulieman Ibraheem Shelash Al-Hawary, Anber Abraheem Shlash Mohammad, and Ala Ahmed Hassan Odeibat

Abstract This study aimed to identify the modified role of quality standards in the impact of the application of total quality management on the organizational reputation of Jordanian public universities. The study population consisted of (5477) faculty members in five public universities. The study sample consisted of a proportional stratified sample of (390) individuals, and after the questionnaire was distributed with (410), 390 were recovered, valid for the purposes of statistical analysis. For the purposes of achieving the objectives of the study, multiple regression analysis and the

E. A. Alshuqairat · Z. A. K. Khalaylah · M. Alshura · A. A. H. Odeibat
Management Department, Faculty of Money and Management, The World Islamic Science University, P.O. Box 1101, Amman 11947, Jordan
e-mail: mhmadshura@yahoo.com

B. Y. A. Barqawi
Business Administration Department, Faculty of Administration & Financial Sciences, Petra University, Amman, Jordan

M. M. Al-mzary
Department of Applied Science, Irbid College, Al-Balqa Applied University, Mafraq, Jordan

M. T. Alshurideh
Department of Marketing, School of Business, The University of Jordan, Amman 11942, Jordan
e-mail: m.alshurideh@ju.edu.jo; malshurideh@sharjah.ac.ae

Department of Management, College of Business, University of Sharjah, 27272 Sharjah, United Arab Emirates

S. I. S. Al-Hawary (✉)
Department of Business Administration, School of Business, Al Al-Bayt University, P.O. BOX 130040, Mafraq 25113, Jordan
e-mail: dr_sliman73@aabu.edu.jo; dr_sliman@yahoo.com

A. A. S. Mohammad
Marketing Department, Faculty of Administrative and Financial Sciences, Petra University, P.O. Box: 961343, Amman 11196, Jordan

M. Alshurideh et al. (eds.), *The Effect of Information Technology on Business and Marketing Intelligence Systems*, Studies in Computational Intelligence 1056,
https://doi.org/10.1007/978-3-031-12382-5_62

multiple hierarchical regression analysis were used. The study results showed that quality standards play a statistically significant role on the impact of total quality management on organizational reputation. The researchers recommend the necessity for Jordanian public universities to pay attention to the standard because of its important role in the application of total quality management, as well as activating the role of accreditation and quality centers and supporting them with all the needs of qualified financial and human resources.

Keywords Total quality management · Quality standards · Organizational reputation · Jordanian Public University

1 Introduction

Total Quality Management, with its dimensions, forms an important interwoven chain in creating a significant competitive advantage and reputation, as evidenced by high-quality inputs and outputs (Aburayya et al., 2020a; Al-Hawary & Abu-Laimon, 2013). While performance indicators are the tools that are used to measure and control the main and subsidiary processes in the administrative systems in order to produce a high-quality end product that reflects a favorable image of the organizations outputs and activities (Al-Hawary & Al-Smeran, 2017; Aburayya et al., 2020b; Al-Quran et al., 2020; Al-Hawary & Obiadat, 2021; AlTaweel & Al-Hawary, 2021). Universities are the most important educational institutions that embrace students, and they are the most important inputs and outputs of the educational process, working to provide them with knowledge, experience, scientific and practical skills, that refine and develop their personality and enable them to provide service, product, and knowledge to their community, to assist it in achieving its goals (Al-Hawary & Alwan, 2016; Al-Hawary & Batayneh, 2010; Al-Hawary, 2010a, 2010b).

Universities are now working to attract and retain the largest number of students in the context of global competition through their reputation, and even help them achieve high performance in an era marked by the renaissance of knowledge and technology, as it no longer recognizes stereotypes, but has become characterized by the multiplicity and diversity of knowledge, and is considered an important indicator in ranking and reputation, and thus the pursuit of excellence in an era marked by the renaissance of knowledge and technology, as it no longer recognizes stereotype (Al-Hawary & Al-Khazaleh, 2020; Alshurideh et al., 2021). Researchers believe that the reputation of Jordanian public universities has been shattered in recent years, not only locally but also among Arab universities, because the modern experience of international university rankings is a modern and contemporary experience that universities in general use to raise their reputation levels. However, it remains constrained by a set of limited and narrow metrics, and because university education has grown dramatically in the last ten years, accompanied by a slew of barriers and challenges,, in addition to shortcomings in implementing the total quality management strategy and quality assurance standards as required, which raises standards and indicators of public university success and raises their reputation to wherever they want, which prompted them to seriously adopt the application of total quality management and academic accreditation in higher education institutions, which prompted them to seriously adopt the application of total quality management and academic accreditation in higher education institutions (Bakht, 2015; Shakhatra & Tarawneh, 2019; Al Kurdi et al., 2020a). This research was conducted with the goal of underlining our universities' involvement in the implementation and dedication to total quality management and assurance standards, as well as the fate of their academic and educational reputation on the Arab and worldwide levels.

2 Theoretical Framework and Hypotheses Development

2.1 *Total Quality Management*

Total Quality Management (TQM) is one of the most prominent modern concepts in management, and as a result of what the market witnessed of a great competitive image, to provide the best products and services to its customers and clients, this competition moved to a sector that was far from this competition due to the different services provided to the beneficiaries, to enter the education sector The circle of competition in providing its educational services in a better and better way than it was before (Al-Hawary, 2013; Al Nady et al., 2013; Al-Hawary& Al-Menhaly, 2016; Alshurideh et al., 2017; Al-Maroof et al., 2021). As a result, there has been a noticeable increase in the adoption of this concept in higher education institutions, and it has quickly spread from the industrial and production sectors to the education sector, where it has become urgent and necessary as a result of the rapid advancement of knowledge and technology, which has increased competition (Alhalalmeh et al., 2020; Al-Hawary & Al-Syasneh, 2020; Aljumah et al., 2021; Mohammad et al., 2020). And, given that the graduate is (the final service of the educational process), universities must meet and satisfy the needs of society by generating cadres capable of competition, progress, and improvement in an era of globalization and vast knowledge and technology.

The importance of quality in today's organizations is centered on working to keep up with the requirements and levels of quality, as well as the latest work technologies, and to meet the customer's requirements and desires (Abu Qaaud et al., 2011; Al-Hawary et al., 2017; Harahsheh et al., 2021; Hellman & Liu, 2013), as quality plays a major role in achieving the best levels of performance in all areas, and thus achieving expected satisfaction from customers (Alshurideh et al., 2012; Alzoubi et al., 2022; Hellman & Liu, 2013), as quality plays a major role in achieving the best levels of performance in all areas (Alshurideh et al., 2019; Paraschivescu & Caprioara, 2014). Total quality management, according to the researchers, is a philosophy that focuses on reaching the greatest and highest degree of operational quality and performance through the participation of working individuals, continuous improvement processes, and addressing the needs and wishes of beneficiaries (Krajewski et al., 2013, p.180; Alzoubi et al., 2020). While Hasham (2018) sees it as a management model based on the most efficient use of time, Hasham (2018) sees it as a strategic plan with clear goals that is characterized by constant evaluation, rectification, and successful review. This is due to the fact that quality management and time management work together to increase productivity by lowering mistake rates and preventing waste in the use of various resources, implying that quality management allows us to make better use of our time.

Today, total quality management has become one of the most important factors for achieving competitiveness in organizations, as the increasing demand for quality by customers makes the organization realize the critical importance in applying quality principles to meet this demand in a good way, and thus total quality management has become a major tool to achieve competitive advantage and performance higher (Al-Hawary & Alhajri, 2020; Obeidat et al., 2021), and its application has become one of the most important priorities in developed countries (Almansour, 2012; Alshamsi et al., 2020; Hassan et al., 2013).

2.2 *Organizational Reputation*

After the tremendous knowledge and technological revolution, our universities have recently experienced rapid and successive changes and challenges, prompting them to reconsider their level of performance by moving from local performance standards to global ones that raise their reputation, which they consider to be one of their intangible assets that they must possess and strive for, and it keeps it at a consistently good level, as it is one of the most important metrics (Al-Lozi et al., 2017; Alolayyan et al., 2018; Allahow et al., 2018). And, because the message of educational institutions is so important in the lives of people and nations at various stages of economic and social development, the goals are no longer traditional in terms of teaching, research, and knowledge, but have expanded to include all aspects of life, technical, scientific, and technological, forcing these institutions to interact with their society in order to understand its desires and needs, and because it is considered the basis for economic and social development, the goals have expanded to include all aspects of life, technical, scientific, and technological, forcing these institutions to interact (Ibrahim, 2011; Metabis & Al-Hawary, 2013; Al-Nady et al., 2016; Alshurideh, 2022).

Organizational reputation, according to Watson and Kitchen (2015), is a strategic asset that is directly tied to market prices. While another point of view suggests that it represents a kind of comprehensive collective representation of the audio-visual developments that the organization will build over time, as well as its organizational identity through performance and behavior, to later reflect the organization's many perspectives (Al-Hawary & Harahsheh, 2014; Al-Hawary & Hussien, 2017; Lee et al., 2022a, b). Othman and Mohammed (2015) divided the concept into four parts: an economic perspective, a characteristic, a behavior indication of competitive strength, and a behavioral indicator of competitive strength. From the standpoint of the market: It's what's referred to as a brand image in marketing research. The organization's point of view is as follows: It's what researchers observe since it's anchored in the experiences and accomplishments of the people who work there, embodies the organization's culture and identity, and demonstrates high performance and an appropriate working environment. It is concluded from these different definitions that the concept of organizational reputation is: the organization's ability to consolidate the relationship with all influential parties in society, by presenting strategies and plans capable of responding to the requirements and needs of its beneficiaries in the best way, in a way that ultimately leads to increasing and raising its value and competitive reputation among its peers.

The importance of organizational reputation is reflected in what he said (Qarfi & Sahrawi, 2016), that it is a social building that the organization obtains through its stakeholder relationships, and that its good reputation helps to consolidate its relationships with them and with influential parties in society. It is also regarded an essential component in decreasing the dangers that the company may face, as loyal customers bear a portion of the risk, as well as the fact that reputation adds value to the organization's products and services. In addition, the credibility and reliability of the organizational reputation emerges in times of crisis, because it increases organizational effectiveness, and increases the granting of confidence to the organization, by creating a distinct organizational climate that helps it improve and develop the required. The opinions of researchers varied and varied about determining the dimensions of organizational reputation. Walsh et al. (2009) believes that organizational reputation is represented in five important dimensions that are summarized as: orientation to the customer (the beneficiary), the good employer, the financial capacity of the organization, the quality of its products and services, the social and environmental responsibility of the organization. While Harrison indicated that there are eight important dimensions of reputation and they are complementary among them: leadership, financial strength, type of management, customer focus, social responsibility, reliability, and the nature and quality of communications.

2.3 *Quality Standards*

Quality is one of the most important topics that have become modern today in our society, after the interest in it in the higher education sector has increased significantly and remarkably, as it is considered one of the important means and modern methods for improving the quality and quality of education, and its elevation and advancement in our time, which is the era of excellence, development and quality. Quality has become an urgent necessity that requires educational institutions to catch up with it and adhere to it, for its continuity and better growth. Today, quality standards have become one of the important aspects in higher education, due to their direct impact on the process of improving the educational process (Nazir et al., 2022). The application of these standards is an urgent need to strive towards achieving interaction and communication with the variables that characterize our modern era, which bears the mark of knowledge, technology and intense competition (AlShehhi et al., 2020; Tariq et al., 2022).

The focus of attention in higher education institutions has now become the application of quality and its standards, as quality assurance depends on the application of methods and principles of total quality management on an ongoing basis, with the aim of development and improvement, and achieving the highest possible level in the practices, operations and outputs of educational institutions and the services provided by them. This is because the education system, like other systems, runs according to certain strategic plans and within a specific environment and conditions, taking into account all the circumstances and factors surrounding its environment in all its aspects.

Al-Sarayra (2008) and Al-Sayed (2012) define quality standards as a set of specifications that establish the requirements for quality systems in different higher education institutions, which it is necessary to put at the beginning of the application stages of the total quality management methodology, to help senior management in measuring the actual results on the basis of it. Without these specifications, the university will not be able to judge its performance, which are represented in admission policies, educational programs, which include (teaching methods, objectives, evaluation system, examinations, quality of faculty and administrative staff, quality of senior management, buildings and physical equipment), which It gives important outputs that meet the needs and requirements of the beneficiaries.

2.4 *Total Quality Management, Organizational Reputation and Quality Standards*

The researchers point out that quality-related decisions are strategic decisions that must be followed by senior management because they are one of the cornerstones of comprehensive quality management success. Top management should encourage internal and external clients to participate in decision-making, motivate them, and build and strengthen communication channels with them, as well as work on developing training and development methods to benefit from the capabilities and capabilities of its employees by empowering them and delegating powers to them, building a bridge of trust between internal and even external clients requires support. The senior management of quality policies, and their dissemination and inclusion of their strategies by making them an integral part of the organization's culture (Al Kurdi et al., 2020b; Abuhashesh et al., 2021; Odeh et al., 2021).

Al-Sarhan (2021) confirmed that total quality management in universities entails a set of processes through which the university aims to develop and improve the performance of its administrative and academic staff, work on developing study plans, and rehabilitate laboratories and libraries in order to excel in providing educational services, as evidenced by a Positive on the quality of its most important outputs (the graduates). According to Hiizer and Render, the concept of total quality management in higher education institutions encompasses all of the educational institution's functions and activities, including (teaching methods, curricula, infrastructure, scientific research, buildings, halls and laboratories, grants and missions, and government support), all of which work to improve the educational institution's evaluation. Total quality management in the educational institution: It is a systematic procedural examination of the institution and its academic programs to measure the methodology in it, in terms of planning goals and implementation, and measuring the compatibility between actual practice with work arrangements and planned results, as well as the institution's evaluation processes through learning, improvement, and self-evaluation.

Abu Al-Rub et al. (2010) and Al-Bilawi et al. (2015) pointed to the great importance of standards, if their application through the proper application of total quality management, would enable the administration to solve the problems it faces in scientific ways, and deal with it, through corrective and preventive measures to prevent its occurrence in the future, and there will be opportunities for optimal use of human and material resources, directing its capabilities towards continuous improvement in performance, and further improving the performance of employees through their participation and participation in the decision-making and decision-making process. Thus, we find that universities' interest in total quality and standards for its guarantee, which is reflected positively on their outputs, to provide the labor market with qualified graduates equipped with science and knowledge, and academic and educational experiences, which guarantee their ability, empowerment and involvement in the market locally, Arably and globally, in a way that reflects creativity and excellence. Accordingly, the study hypothesis can be formulated as follows:

There is a statistically significant role for quality standards as a moderate variable in improving the impact of total quality management on organizational reputation in Jordanian public universities.

3 Study Model

See Fig. 1.

4 Methodology

4.1 *Population and Sample Selection*

A qualitative method based on a questionnaire was used in this study for data collection and sample selection. The major aim of the study was to examine the impact of total quality management on organizational reputation through the mediating role of

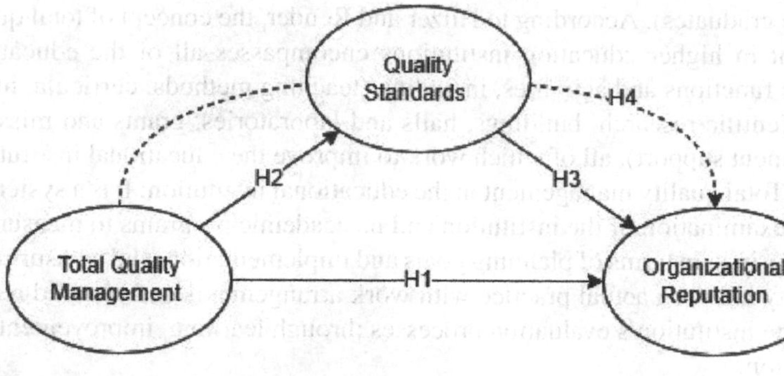

Fig. 1 Research model

quality standards. Therefore, it focused on Jordanian public universities. Data were primarily gathered through self-reported questionnaires creating by Google Forms which were distributed to a stratified sample of (410) faculty members via email. In total, (398) responses were received including nine invalids to statistical analysis due to uncompleted or inaccurate. Hence, the final sample contained (390) responses suitable to analysis requirements that were formed a response rate of (95.12%), where it proved to be sufficient to the extent that was predictable and allowed for a presumption of data saturation (Sekaran & Bougie, 2016).

4.2 Measurement Instrument

A self-reported questionnaire that consists of three main sections along with a section regarding control variables was used as the measurement instrument. Control variables considered as categorical measures were composed of gender, age group, educational level, and experience. The three main sections were dealt with a five-point Likert scale (from 1 = strongly disagree to 5 = strongly agree). The first section contained (20) items to measure total quality management based on (Al-Hawary & Abu-Laimon, 2013). These items were distributed into dimensions as follows: four items dedicated for measuring focus on beneficiary, four items dedicated for measuring senior management commitment, four items dedicated for measuring staff participation, four items dedicated for measuring continuous improvement, and four items dedicated for measuring training and development. The second section related to quality standards variable which contained six items developed according to. Whereas the third section included eight items developed to measure organizational reputation according to what was pointed by (Qarfi & Sahrawi, 2016).

5 Findings

5.1 Measurement Model Evaluation

This study was conducted structural equation modeling (SEM) to test hypotheses, which represents a contemporary statistical technique for testing and estimating the relationship between factors and variables (Wang & Rhemtulla, 2021). Accordingly, the reliability and validity of the constructs were tested using confirmatory factor analysis (CFA) through the statistical program AMOSv24. Table 1 summarizes the results of convergent and discriminant validity, as well the indicators of reliability.

Table 1 shows that the standard loading values for the individual items were within the domain (0.661–0.858), these values greater than the minimum retention of the elements based on their standard loads (Al-Lozi et al., 2018; Sung et al., 2019). Average variance extracted (AVE) is a summary indicator of the convergent validity of constructs that must be above 0.50 (Howard, 2018). The results indicate that the AVE values were greater than 0.50 for all constructs, thus the used measurement model has an appropriate convergent validity. Rimkeviciene et al. (2017) suggested the comparison approach as a way to deal with discriminant validity assessment in covariance-based SEM. This approach is based on comparing the values of maximum shared variance (MSV) with the values of AVE, as well as comparing the values of square root of AVE ($\sqrt{AVE}$) with the correlation between the rest of the structures. The results show that the values of MSV were smaller than the values of AVE, and that the values of $\sqrt{AVE}$ were higher than the correlation values among the rest of the constructs. Therefore, the measurement model used is characterized by discriminative validity. The internal consistency measured through Cronbach's Alpha coefficient (α) and compound reliability by McDonald's Omega coefficient (ω) was conducted as indicators to evaluate measurement model. The results listed in Table 1 demonstrated that both values of Cronbach's Alpha coefficient and McDonald's Omega coefficient were greater than 0.70, which is the lowest limit for judging on measurement reliability (De Leeuw et al., 2019).

5.2 Structural Model

The structural model illustrated no multicollinearity issue among predictor constructs because variance inflation factor (VIF) values are below the threshold of 5, as shown in Table 1 (Hair et al., 2017). This result is supported by the values of model fit indices shown in Figs. 1 and 2.

The results in Fig. 1 indicated that the chi-square to degrees of freedom (CMIN/DF) was 2.131, which is less than 3 the upper limit of this indicator. The values of the goodness of fit index (GFI), the comparative fit index (CFI), and the Tucker-Lewis index (TLI) were upper than the minimum accepted threshold of 0.90. Moreover, the result of root mean square error of approximation (RMSEA) indicated

Table 1 Results of validity and reliability tests

Constructs	1	2	3	4	5	6	7
1. FOB	**0.771**						
2. SMC	0.364	**0.741**					
3. SPA	0.445	0.425	**0.755**				
4. CIM	0.468	0.487	0.428	**0.776**			
5. TAD	0.382	0.394	0.468	0.395	**0.771**		
6. QSS	0.597	0.622	0.625	0.585	0.642	**0.761**	
7. ORE	0.625	0.635	0.645	0.671	0.665	0.605	**0.758**
VIF	2.157	1.885	2.364	1.154	1.647	–	–
Loadings range	0.711–0.834	0.685–0.804	0.674–0.858	0.733–0.815	0.661–0.814	0.681–0.825	0.683–0.821
AVE	0.594	0.549	0.570	0.602	0.594	0.580	0.575
MSV	0.464	0.512	0.384	0.336	0.491	0.438	0.449
Internal consistency	0.850	0.827	0.838	0.857	0.852	0.890	0.912
Composite reliability	0.853	0.829	0.840	0.858	0.854	0.892	0.915

Note FOB: focus on beneficiary, SMC: senior management commitment, SPA: staff participation, CIM: continuous improvement, TAD: training and development, QSS: quality standards, ORE: organizational reputation, Bold fonts in the table are the square root of average variance extracted

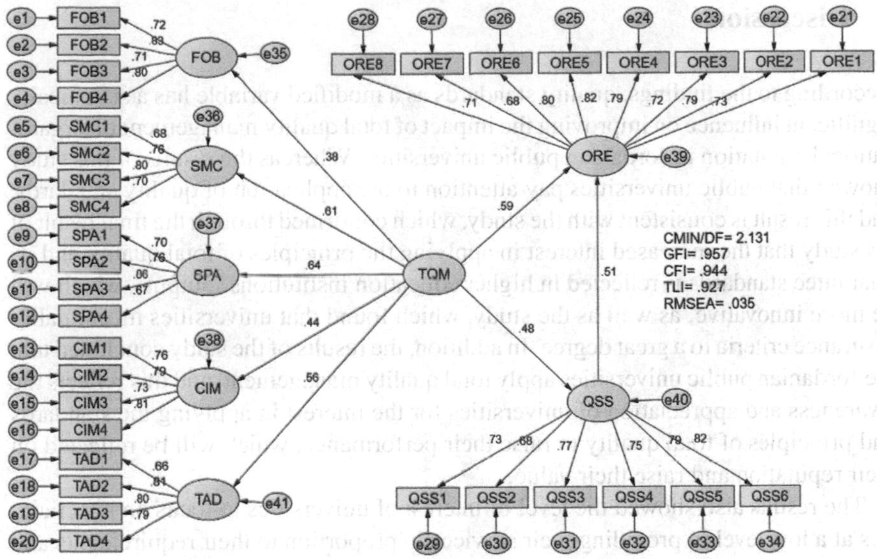

Fig. 2 SEM results of the TQM effect on OR through QS

to value 0.035, this value is a reasonable error of approximation because it is less than the higher limit of 0.08. Consequently, the structural model used in this study was recognized as a fit model for predicting the organizational reputation and generalization of its result (Ahmad et al., 2016; Shi et al., 2019). To verify the results of testing the study hypotheses, structural equation modeling (SEM) was used, the results of which are listed in Table 2.

The results demonstrated in Table 2 show that total quality management has a positive direct impact on organizational reputation ($\beta = 0.594$, $t = 14.568$, $p = 0.000$) and quality standards ($\beta = 0.483$, $t = 10.332$, $p = 0.002$). Moreover, it indicated that quality standards have a positive impact on organizational reputation ($\beta = 0.512$, $t = 13.625$, $p = 0.000$). Hence, quality standards have a partial mediation effect in the relationship between total quality management and organizational reputation, where the total effect was ($\beta = 0.841$, $t = 21.096$, $p = 0.000$) with indirect effect of ($\beta = 0.247$, $p = 0.000$).

Table 2 Hypothesis testing

Relation	Direct		Indirect		Total		
	β	p-value	β	p-value	β	t-value	p-value
TQM → ORE	0.594	0.000	–	–	0.594	14.568	0.000
TQM → QSS	0.483	0.002	–	–	0.482	10.332	0.002
QSS → OPE	0.512	0.000	–	–	0.512	13.625	0.000
TQM → QSS → ORE	0.594	0.000	0.247	0.000	0.841	21.096	0.000

6 Discussion

According to the findings, quality standards as a modified variable has a statistically significant influence on improving the impact of total quality management on organizational reputation in Jordanian public universities. Whereas the results of this study showed that public universities pay attention to the application of quality standards, and this result is consistent with the study, which confirmed through the final result of his study that the increased interest in applying the principles of total quality and its guarantee standards is reflected in higher education institutions' outputs, which will be more innovative, as well as the study, which found that universities meet quality assurance criteria to a great degree. In addition, the results of the study concluded that the Jordanian public universities apply total quality management, and this reflects the awareness and appreciation of universities for the interest in applying the standards and principles of total quality to raise their performance, which will be reflected on their reputation and raise their value.

The results also showed the level of interest of universities to focus on beneficiaries at a low level in providing their services in proportion to their requirements and desires, and this is far from the essence of the principle of focus on the beneficiary, which requires that the administration share with the beneficiaries of its services in presenting their ideas about the services they desire, as well as Take advantage of the feedback through them. Universities are also interested in raising their reputation, through the reputation of their distinguished faculty members, their interest in sending students to study abroad, and building good relations with different countries. The results indicated a high interest on the part of public universities (in academic programs), which is an important criterion for quality assurance in universities, as it is considered one of the factors that attract students.

7 Recommendations

Based on the findings of the study, researchers recommend providing university facilities and buildings, especially teaching halls, with devices, equipment and modern technology that support the educational process and achieve quality goals. Providing a model for measuring the quality of services provided in universities and providing feedback to this model on a continuous basis to identify strengths and weaknesses and work to improve and develop them to achieve quality goals in gaining the satisfaction of the beneficiaries of these services. The researchers recommend increasing the effectiveness of communication between university employees (faculty members and administrators) with the administration, by facilitating access to information, and enabling them to make the decision that enables them to achieve and accomplish their work and develop their level of performance, in addition to the adoption of modern methods by universities in providing their services and activities in a manner that ensures speed response in providing its services. Increasing the dissemination and marketing of the services of Jordanian public universities contributes to attracting more students and investors, and increases the focus of the public image of it in their minds; rather, it works to enhance the desire to highlight the university's brand.

References

Abu Qaaud, F., Al-Shoura, M., & Al-Hawary, S. I. (2011). The impact of the service marketing mix in the service quality of health services from the viewpoint of patients in Government Hospitals in Amman "A Field study." *Abhath Al-Yarmouk, 27*(1B), 417–441.

Abuhashesh, M. Y., Alshurideh, M. T., & Sumadi, M. (2021). The effect of culture on customers' attitudes toward Facebook advertising: The moderating role of gender. *Review of International Business and Strategy, 31*(3), 416–437.

Aburayya, A., Alshurideh, M., Al Marzouqi, A., Al Diabat, O., Alfarsi, A., Suson, R., & Alzarouni, A. (2020a). Critical success factors affecting the implementation of TQM in public hospitals: A case study in UAE Hospitals. *Systematic Reviews in Pharmacy, 11*(10), 230–242.

Aburayya, A., Alshurideh, M., Al Marzouqi, A., Al Diabat, O., Alfarsi, A., Suson, R., & Salloum, S. A. (2020b). An empirical examination of the effect of TQM practices on hospital service quality: An assessment study in UAE hospitals. *Syst. Rev. Pharm, 11*(9), 347–362.

Ahmad, S., Zulkurnain, N., & Khairushalimi, F. (2016). Assessing the validity and reliability of a measurement model in structural equation modeling (SEM). *British Journal of Mathematics & Computer Science, 15*(3), 1–8. https://doi.org/10.9734/BJMCS/2016/25183

Al Kurdi, B., Alshurideh, M., & Salloum, S. A. (2020a). Investigating a theoretical framework for e-learning technology acceptance. *International Journal of Electrical and Computer Engineering (IJECE), 10*(6), 6484–6496.

Al Kurdi, B., Alshurideh, M., Salloum, S., Obeidat, Z., & Al-dweeri, R. (2020b). An empirical investigation into examination of factors influencing university students' behavior towards elearning acceptance using SEM approach. *International Journal of Interactive Mobile Technologies, 14*(2), 19–24.

Al-Bilawi, H. H., Taima, R. A., Suleiman, S. A., Al-Qib, A. R., Saeed, Al-Mahdi, M., Al-Bandari, M. b. S., & Abdel-Baqi, M. A. (2015). In Taima, R. (Eds), *Total quality in education between indicators of excellence and accreditation standards: foundations and applications* (4th edition). Jordan: Amman: Dar Al Masirah for Publishing and Distribution.

Alhalalmeh, M. I., Almomani, H. M., Altarifi, S., Al- Quran, A. Z., Mohammad, A. A., & Al-Hawary, S. I. (2020). The nexus between corporate social responsibilty and organizational performance in Jordan: The mediating role of organizational commitment and organizational citizenship behavior. *Test Engineering and Management, 83*(July), 6391–6410.

Al-Hawary, S. I. (2010b). Marketing Public Higher Education: a Social Perspective. Al Manara for Research and Studies, *16*(4), 9–32. Retrieved from https://aabu.edu.jo/journal/manar/manarArt1648.html

Al-Hawary, S. I. (2010a). Factor underlying international students of Jordan Public Universities: Analytical study institutional factors. Al Manara for Research and Studies, *16*(1), 37–64. Retrieved from https://aabu.edu.jo/journal/manar/manarArt16110.html

Al-Hawary, S. I., & Al-Syasneh, M. S. (2020). Impact of dynamic strategic capabilities on strategic entrepreneurship in presence of outsourcing of five stars hotels in Jordan. Business: Theory and Practice, *21*(2), 578–587.

Al-Hawary, S. I., & Al-Khazaleh A, M. (2020b). The mediating role of corporate image on the relationship between corporate social responsibility and customer retention. Test Engineering and Management, *83*(516), 29976–29993.

Al-Hawary, S. I. (2013). The role of perceived quality and satisfaction in explaining customer brand loyalty: Mobile phone service in Jordan. *International Journal of Business Innovation and Research, 7*(4), 393–413.

Al-Hawary, S. I., & Abu-Laimon, A. A. (2013). The impact of TQM practices on service quality in cellular communication companies in Jordan. *International Journal of Productivity and Quality Management, 11*(4), 446–474.

Al-Hawary, S. I. S., & Alhajri, T. M. S. (2020). Effect of electronic customer relationship management on customers' electronic satisfaction of communication companies in Kuwait. *Calitatea, 21*(175), 97–102.

Al-Hawary, S. I., & Al-Menhaly, S. (2016). The quality of e-government services and its role on achieving beneficiaries satisfaction. *Global Journal of Management and Business Research: A Administration and Management, 16*(11), 1–11.

Al-Hawary, S. I., & Al-Smeran, W. (2017). Impact of electronic service quality on customers satisfaction of Islamic Banks in Jordan. *International Journal of Academic Research in Accounting, Finance and Management Sciences, 7*(1), 170–188.

Al-Hawary, S. I. S., & Alwan, A. M. (2016). Knowledge management and its effect on strategic decisions of Jordanian Public Universities. *Journal of Accounting-Business & Management, 23*(2), 24–44.

Al-Hawary, S. I., & Batayneh, A. M. (2010). The effect of marketing communication tools on Non-Jordanian students' choice of Jordanian Public Universities: A field study. *International Management Review, 6*(2), 90–99.

Al-Hawary, S. I., Batayneh, A. M., Mohammad, A. A., & Alsarahni, A. H. (2017). Supply chain flexibility aspects and their impact on customers satisfaction of pharmaceutical industry in Jordan. *International Journal of Business Performance and Supply Chain Modelling, 9*(4), 326–343.

Al-Hawary, S. I., & Harahsheh, S. (2014). Factors affecting Jordanian consumer loyalty toward cellular phone brand. *International Journal of Economics and Business Research, 7*(3), 349–375.

Al-Hawary, S. I., & Hussien, A. J. (2017). The impact of electronic banking services on the customers loyalty of Commercial Banks in Jordan. *International Journal of Academic Research in Accounting, Finance and Management Sciences, 7*(1), 50–63.

Al-Hawary, S. I. S., Mohammad, A. S., Al-Syasneh, M. S., Qandah, M. S. F., & Alhajri, T. M. S. (2020a). Organisational learning capabilities of the commercial banks in Jordan: Do electronic human resources management practices matter? *International Journal of Learning and Intellectual Capital, 17*(3), 242–266.

Al-Hawary, S. I. S., & Obiadat, A. A. (2021). Does mobile marketing affect customer loyalty in Jordan? *International Journal of Business Excellence, 23*(2), 226–250.

Aljumah, A., Nuseir, M. T., & Alshurideh, M. T. (2021). The impact of social media marketing communications on consumer response during the COVID-19: Does the brand equity of a university matter. The effect of coronavirus disease (COVID-19) on business intelligence, 367–384.

Allahow, T. J. A. A., Al-Hawary, S. I. S., & Aldaihani, F. M. F. (2018). Information technology and administrative innovation of the central agency for information technology in Kuwait. Global Journal of Management and Business, *18*(11-A), 1–16.

Al-Lozi, M., Almomani, R. Z., & Al-Hawary, S. I.(2017). Impact of Talent Management on Achieving Organizational Excellence in Arab Potash Company in Jordan. Global Journal of Management and Business Research: A Administration and Management, *17*(7), 15–25.

Al-Lozi, M. S., Almomani, R. Z. Q., & Al-Hawary, S. I. S. (2018). Talent Management strategies as a critical success factor for effectiveness of Human Resources Information Systems in commercial banks working in Jordan. *Global Journal of Management and Business Research: A Administration and Management, 18*(1), 30–43.

Almansour, Y. M. (2012). The impact of total quality management components on small & medium enterprises, financial performance in Jordan. *Journal of Arts, Science & Commerce, 3*(1), 87–91.

Al-Maroof, R. S., Alshurideh, M. T., Salloum, S. A., AlHamad, A. Q. M., & Gaber, T. (2021). Acceptance of Google Meet during the spread of Coronavirus by Arab university students. In Informatic, *8*(2), 1–17. Multidisciplinary Digital Publishing Institute.

Al-Nady, B. A., Al-Hawary, S. I., & Alolayyan, M. (2013). Strategic management as a key for superior competitive advantage of sanitary ware suppliers in Kingdom of Saudi Arabia. *International Journal of Management and Information Technology, 7*(2), 1042–1058.

Al-Nady, B. A., Al-Hawary, S. I., & Alolayyan, M. (2016). The role of time, communication, and cost management on project management success: An empirical study on sample of construction projects customers in Makkah City, Kingdom of Saudi Arabia. *International Journal of Services and Operations Management, 23*(1), 76–112.

Alolayyan, M., Al-Hawary, S. I., Mohammad, A. A., & Al-Nady, B. A. (2018). Banking service quality provided by commercial banks and customer satisfaction. A structural equation modelling approaches. International Journal of Productivity and Quality Management, *24*(4), 543–565.

Al-Quran, A. Z., Alhalalmeh, M. I., Eldahamsheh, M. M., Mohammad, A. A., Hijjawi, G. S., Almomani, H. M., & Al-Hawary, S. I. (2020). Determinants of the green purchase intention in Jordan: The moderating effect of environmental concern. *International Journal of Supply Chain Management, 9*(5), 366–371.

Abu Al-Rub, I., Qadada, I., Al-Wadi, M., & Al-Tai, R. (2010). Quality assurance in Higher Education Institutions: Research and studies. Jordan, Amman: Dar Safaa for Publishing and Distribution.

Al-Sarayrah, K., & Al-Assaf, L. (2008). Total quality management in higher education institutions between theory and practice. *The Arab Journal for Quality Assurance of University Education., 1*, 1–46.

Al-Sarhan, A. F. (2021). The impact of the applications of total quality standards and academic accreditation on improving individual and institutional performance of Jordanian Public Universities. *Journal of Islamic Management and Leadership, 6*(1), 14–38.

Al-Sayed, A. H. (2012). The priority of the quality of university education in Tindouf as a model. *The Arab Journal of Political Sciences, 33*, 4–5.

Alshamsi, A., Alshurideh, M., Al Kurdi, B., & Salloum, S. A. (2020). The influence of service quality on customer retention: a systematic review in the higher education. In International Conference on Advanced Intelligent Systems and Informatics (pp. 404–416). Cham: Springer.

AlShehhi, H., Alshurideh, M., Al Kurdi, B., & Salloum, S. A. (2020). The impact of ethical leadership on employees performance: A systematic review. In International Conference on Advanced Intelligent Systems and Informatics (pp. 417–426). Cham: Springer.

Alshurideh, M., Al-Hawary, S. I., Batayneh, A. M., Mohammad, A., & Al-Kurdi, B. (2017). The impact of Islamic Banks' service quality perception on Jordanian customers loyalty. Journal of Management Research, *9*(2), 139–159.

Alshurideh, M. T., Kurdi, B. A., AlHamad, A. Q., Salloum, S. A., Alkurdi, S., Dehghan, A., & Masa'deh, R. E. (2021). Factors affecting the use of smart mobile examination platforms by universities' postgraduate students during the COVID 19 pandemic: an empirical study. In Informatics, *8*(2), 1–21. Multidisciplinary Digital Publishing Institute.

Alshurideh, M. (2022). Does electronic customer relationship management (E-CRM) affect service quality at private hospitals in Jordan? *Uncertain Supply Chain Management, 10*(2), 325–332.

Alshurideh, M., Masa'deh, R. M. D. T., & Alkurdi, B. (2012). The effect of customer satisfaction upon customer retention in the Jordanian mobile market: An empirical investigation. *European Journal of Economics, Finance and Administrative Sciences, 47*(12), 69–78.

Alshurideh, M., Salloum, S. A., Al Kurdi, B., Monem, A. A., & Shaalan, K. (2019). Understanding the quality determinants that influence the intention to use the mobile learning platforms: A practical study. *International Journal of Interactive Mobile Technologies, 13*(11), 183–157.

AlTaweel, I. R., & Al-Hawary, S. I. (2021). The mediating role of innovation capability on the relationship between strategic agility and organizational performance. *Sustainability, 13*(14), 7564.

Alzoubi, H. M., Alshurideh, M., Al Kurdi, B., & Inairat, M. (2020). Do perceived service value, quality, price fairness and service recovery shape customer satisfaction and delight? A practical study in the service telecommunication context. *Uncertain Supply Chain Management, 8*(3), 579–588.

Alzoubi, H., Alshurideh, M., Kurdi, B., Akour, I., & Aziz, R. (2022). Does BLE technology contribute towards improving marketing strategies, customers' satisfaction and loyalty? The role of open innovation. *International Journal of Data and Network Science, 6*(2), 449–460.

Bakht, S. I. (2015). The importance of performance indicators in ranking universities at the global level: An analytical study of the attitudes of directors of sudanese governmental universities.

De Leeuw, E., Hox, J., Silber, H., Struminskaya, B., & Vis, C. (2019). Development of an international survey attitude scale: Measurement equivalence, reliability, and predictive validity. *Measurement Instruments for the Social Sciences, 1*(1), 1–10.

Hair, J. F., Babin, B. J., & Krey, N. (2017). Covariance-based structural equation modeling in the journal of advertising: Review and recommendations. *Journal of Advertising, 46*(1), 163–177. https://doi.org/10.1080/00913367.2017.1281777

Harahsheh, A., Houssien, A., Alshurideh, M. & AlMontaser, M. (2021). The effect of transformational leadership on achieving effective decisions in the presence of psychological capital as an intermediate variable in private Jordanian Universities in light of the corona pandemic. The effect of coronavirus disease (COVID-19) on business intelligence *334*, 221–243.

Hashem, S. E. (2018). Academic institutions are no different to any other: Total quality management does enhance performance. International Journal of Organizational Leadership, *7*, 348-373.

Hassan, M. U., Hassan, S., Shaukat, S., & Nawaz, M. S. (2013). Relationship between TQM elements and organizational performance: An empirical study of manufacturing sector of Pakistan. *Pakistan Journal of Commerce and Social Sciences, 7*(1), 1–18.

Hellman, P. & Liu, Y. (2013). Development of quality management systems: How have disruptive technological innovations in quality management affected organizations? Quality Innovation Prosperity, *17*(1), 104–119.

Howard, M. C. (2018). The convergent validity and nomological net of two methods to measure retroactive influences. *Psychology of Consciousness: Theory, Research, and Practice, 5*(3), 324–337. https://doi.org/10.1037/cns0000149

Ibrahim, I. (2011). Extent of practicing of the faculty members for their educational, research, and a comprehensive society services roles. Journal of the Educational & Psychological Researches, *1*(30), 193–220.

Krajewski, L. J., Ritzman, L. P., & Malhotra, M. K. (2013). *Operations Management Processes and Supply Chains*. Pearson Education Limited.

Lee, K., Azmi, N., Hanaysha, J., Alshurideh, M., & Alzoubi, H. (2022a). The effect of digital supply chain on organizational performance: An empirical study in Malaysia manufacturing industry. *Uncertain Supply Chain Management, 10*(2), 1–16.

Lee, K., Ramiz, P., Hanaysha, J., Alzoubi, H., & Alshurideh, M. (2022b). Investigating the impact of benefits and challenges of IOT adoption on supply chain performance and organizational performance: An empirical study in Malaysia. *Uncertain Supply Chain Management, 10*(2), 1–14.

Metabis, A., & Al-Hawary, S. I. (2013). The impact of internal marketing practices on services quality of Commercial Banks in Jordan. *International Journal of Services and Operations Management, 15*(3), 313–337.

Mohammad, A. A., Alshura, M. S., Al-Hawary, S. I. S., Al-Syasneh, M. S., & Alhajri, T. M. (2020). The influence of internal marketing practices on the employees' intention to leave: A study of the private hospitals in Jordan. *International Journal of Advanced Science and Technology, 29*(5), 1174–1189.

Nazir, J., Rahaman, S., Chunawala, S., Ahmed, G., Alzoubi, H., Alshurideh, M., AlHamad, A. (2022). Perceived factors affecting students academic performance. Academy of Strategic Management Journal, *21*(Special Issue 4), 1–15.

Obeidat, U., Obeidat, B., Alrowwad, A., Alshurideh, M., Masadeh, R., & Abuhashesh, M. (2021). The effect of intellectual capital on competitive advantage: The mediating role of innovation. *Management Science Letters, 11*(4), 1331–1344.

Odeh, R. B. M., Obeidat, B. Y., Jaradat, M. O., & Alshurideh, M. T. (2021). The transformational leadership role in achieving organizational resilience through adaptive cultures: the case of Dubai service sector. International Journal of Productivity and Performance Management. Vol. ahead-of-print No. ahead-of-print. https://doi.org/10.1108/IJPPM-02-2021-0093

Othman, A. H., & Mohamed, B. (2015). The impact of social responsibility in building and managing the reputation of organizations: An exploratory analytical study of the opinions of a sample of workers in Al-Rafidain and Al-Rasheed Banks. *Journal of Management and Economics, 103*, 128–114.

Paraschivescu, A., & Caprioara, F. (2014). Strategic Quality Management, Economy Transdisciplinarity. *Cognition, 17*(1), 19–27.

Qarfi, S., & Sahrawi, H. H. (2016). The role of social responsibility in supporting the reputation of the organization—an analytical study from the point of view of the administrations of the Rouiba Corporation for juices. *Al-Bahith Journal, No., 16*, 121–134.

Rimkeviciene, J., Hawgood, J., O'Gorman, J., & De Leo, D. (2017). Construct validity of the acquired capability for suicide scale: Factor structure, convergent and discriminant validity. *Journal of Psychopathology and Behavioral Assessment, 39*(2), 291–302. https://doi.org/10.1007/s10862-016-9576-4

Sekaran, U., & Bougie, R. (2016). Research methods for business: A skill-building approach (Seventh edition). Wiley.

Shakhatra, A. A., & Tarawneh, A. Y. (2019). The reality of administrative procedures practiced to raise the institutional reputation of Jordanian public universities from the point of view of academic administrations. *The Jordanian Educational Journal, 4*(4), 215–191.

Shi, D., Lee, T., & Maydeu-Olivares, A. (2019). Understanding the Model Size Effect on SEM Fit Indices. *Educational and Psychological Measurement, 79*(2), 310–334. https://doi.org/10.1177/0013164418783530

Sung, K.-S., Yi, Y. G., & Shin, H.-I. (2019). Reliability and validity of knee extensor strength measurements using a portable dynamometer anchoring system in a supine position. *BMC Musculoskeletal Disorders, 20*(1), 1–8. https://doi.org/10.1186/s12891-019-2703-0

Tariq, E., Alshurideh, M., Akour, I., & Al-Hawary, S. (2022). The effect of digital marketing capabilities on organizational ambidexterity of the information technology sector. *International Journal of Data and Network Science, 6*(2), 401–408.

Walsh, G., Mitchell, V. W., Jackson, P. R., & Beatty, S. E. (2009). Examining the antecedents and consequences of corporate reputation: A customer perspective. *British Journal of Management, 20*(2), 187–203.

Wang, Y. A., & Rhemtulla, M. (2021). Power analysis for parameter estimation in structural equation modeling: A discussion and tutorial. *Advances in Methods and Practices in Psychological Science, 4*(1), 1–17. https://doi.org/10.1177/2515245920918253

Watson, T., & Kitchen P. J. (2015). Reputation Management: Corporate Image and communication In Moutinho, L. & Southern, G. (Eds.) Strategic Marketing Management: A Process-Based Approach. Andover, Hampshire: Cengage Learning., Chapter 13. ISBN: 978-1-84480-000-1.

The Impact of Strategic Vigilance on Crisis Management in the Jordanian Dairy Companies: The Mediating Role of Organizational Learning

Zaki Abdellateef Khalaf Khalaylah, Mohammed Mufaddy AL-kasasbeh, Basem Yousef Ahmad Barqawi, Mohammed Saleem Khlif Alshura, Enas Ahmad Alshuqairat, Maali M. Al-mzary, Sulieman Ibraheem Shelash Al-Hawary, Muhammad Turki Alshurideh, and Barween Al Kurdi

Abstract The major aim of the study was to examine the impact of strategic vigilance on crisis management through the mediating role of organizational learning. Therefore, it focused on Jordanian dairy companies. Data were primarily gathered through self-reported questionnaires creating by Google Forms which were distributed to a sample of (285) managers via email. In total, (262) responses were

Z. A. K. Khalaylah · M. M. AL-kasasbeh · M. S. K. Alshura · E. A. Alshuqairat
Management Department, Faculty of Money and Management, The World Islamic Science University, P.O. Box1101, Amman 11947, Jordan

B. Y. A. Barqawi
Business Administration Department, Faculty of Administration & Financial Sciences, Petra University, Amman, Jordan

M. M. Al-mzary
Department of Applied Science, Al-Balqa Applied University, Irbid College, As-Salt, Jordan

S. I. S. Al-Hawary (✉)
Department of Business Administration, School of Business, Al Al-Bayt University, P.O. Box 130040, Mafraq 25113, Jordan
e-mail: dr_sliman73@aabu.edu.jo

M. T. Alshurideh
Department of Marketing, School of Business, The University of Jordan, Amman 11942, Jordan
e-mail: m.alshurideh@ju.edu.jo; malshurideh@sharjah.ac.ae

Department of Management, College of Business, University of Sharjah, 27272 Sharjah, United Arab Emirates

B. Al Kurdi
Department of Marketing, Faculty of Economics and Administrative Sciences, The Hashemite University, Zarqa, Jordan
e-mail: barween@hu.edu.jo

M. Alshurideh et al. (eds.), *The Effect of Information Technology on Business and Marketing Intelligence Systems*, Studies in Computational Intelligence 1056,
https://doi.org/10.1007/978-3-031-12382-5_63

received including (16) invalid to statistical analysis due to uncompleted or inaccurate. Hence, the final sample contained (246) responses suitable to analysis requirements. Structural equation modeling (SEM) was conducted to test hypotheses. The results showed that organizational learning plays a partial mediation effect on the relationship between strategic vigilance and crisis management. Based on the results, this study presents a set of recommendations to decision makers in order to better manage the crises facing their organizations.

Keywords Strategic vigilance · Crisis management · Organizational learning · Jordan

1 Introduction

Throughout history, human communities have experienced numerous cyclical or random crises. These crises have resulted in a slew of social, political, and economic ramifications (Coombs & Laufer, 2018). Simultaneously, it lowered the human development curve, which is critical for institutional growth and worldwide competitiveness in today's dynamic business environment (Zulkarnaini et al., 2019; Nuseir et al., 2021a, b). Internal environment variables, such as the bureaucratic regulatory environment, or external environment variables, such as elements of rapid technological progress and the move toward globalization, are commonly responsible for crises (Al-Hawary & Hadad, 2016; Allahow et al., 2018; Al-Hawary et al., 2020; Alhalalmeh et al., 2020; Al-Quran et al., 2020). These factors contribute to the creation of organizational instability, limiting employee innovation and management's capacity to make the appropriate decisions at the right time (Altarifi et al., 2015; Al-Hawary & Al-Syasneh, 2020; Al-Hawary & Obiadat, 2020; Coccia, 2020; Mohammad et al., 2020). With the great changes that the world has witnessed recently related to the dynamic business environment and the spread of the Corona pandemic (Ahmad et al., 2021; Alameeri et al., 2021; Al-Dmour et al., 2021a, b; Alshurideh et al., 2021; Shah et al., 2021), which showed the extent of the fragility of the global system in dealing with crises, organizations are suffering more in maintaining their survival and continuing to provide competitive products and services to their customers by focusing on reducing or avoiding the consequences of successive crises (AlTaweel & Al-Hawary, 2020; Alves et al., 2020; Al-Hawary & Alhajri, 2020).

The ability to analyze the organization's internal and external environments is critical in giving the knowledge needed to make the best decisions at the right time. This necessitated a rethinking of the administrative processes for assessing and forecasting changes in the corporate environment in order to capitalize on opportunities and avoid risks (Al-Hawary & Al-Hamwan, 2017; Al-Hawary & Hadad, 2016; Shalakah et al., 2019). As a result, strategic vigilance has arisen as a management trend that focuses on current information and invests it to forecast future changes in order to provide a prior grasp of the surrounding environment and anticipated adjustments (Alameeri et al., 2020; Jarallah, 2021). Furthermore, it enables the business to cope with large data, which aids in strengthening the firm's agility and enhancing its ability to attain a holistic strategic goal (Alzoubi et al., 2022; Bettahar & Aggoun, 2021). Strategic vigilance contributes to achieving organizational excellence (Jaaz & Jamal, 2021a, 2021b; Shakhour et al., 2021), improving creativity (Alaali et al., 2021; Ismaail, 2020), and increasing the ability to achieve competitive advantage (Altamony et al., 2012; Manhal & Hattab, 2018).

Furthermore, organizational learning is viewed as a strategic pattern and a driver for bringing about a fundamental shift in traditional thinking, as well as an important catalyst for implementing scientific collaboration and knowledge dissemination approaches (Al-Hawary & Aldaihani, 2016; Al-Hawary & Al-Namlan, 2018; Al-Hawary et al., 2020; Al-Lozi et al., 2017; Ashal et al., 2021; Park & Kim, 2018). Knowledge management methods enable learning organizations to be more flexible and adaptable to changing circumstances, as the accumulation of knowledge contributes to the development of employees' capabilities and skills, as well as motivating them to create products and services that meet the needs of customers (Al-Hawary, 2015; Al-Hawary & Alwan, 2016; Werlang & Rossetto, 2019). Moreover, organizational learning enables effective and efficient investment of various organization resources (Al-Hawary et al., 2020; Alshurideh et al., 2019b) and improved organizational performance (Al Kurdi et al., 2020a; Yuliansyah & Jermias, 2018).

The current study provides a framework for investing the potential of strategic vigilance in enhancing the organization's ability to monitor changes in the business environment and sense potential crises to deal with them proactively, in addition to the mediating role of organizational learning as one of the factors that limit the negative effects of crises. This research helps to throw light on Jordan's dairy industry, which is still reeling from a series of crises brought on by the country's quickly changing commercial environment and substantial technological advancements. Furthermore, it aids in the development of a dynamic strategic approach that enables the business to achieve long-term goals by depending on proactive methods for monitoring changes and reorganizing people and material resources to suit changing client demands. Therefore, this study came to examine the Impact of Organizational Learning On the relationship between Strategic Vigilance and Crisis Management in the Jordanian Dairy Companies.

2 Theoretical Framework and Hypotheses Development

2.1 Strategic Vigilance

By relying on contemporary technology and innovation activities, strategic vigilance plays a critical role in providing the firm with the essential knowledge on many areas of the business environment to discover available possibilities and avoid potential risks (Alshaer, 2020a, 2020b; Alsuwaidi et al., 2020). Strategic vigilance is defined by Drevon (2019) as a systematic, continuous, legal, and ethical process of collecting, analyzing, processing, and disseminating information, with the goal of assisting senior management, managers, or the entire organization in making better decisions and nurturing strategic thinking through a better understanding of the internal and external environment (Al-Nady et al., 2013, 2016; Hijjawi & Mohammad, 2019). According to Tawfik, early sensing and detection of changes in the business environment's trends is accomplished through the process of continuous monitoring and analysis of the influencing factors in order to extract accurate information and disseminate it throughout the organization's various departments. Furthermore, strategic vigilance was discussed as a sophisticated organizational process that tries to maintain the organization's sustainability and continuity of operations by producing reliable information that helps to reduce the uncertainty associated with decision-making (Alyammahi et al., 2020; Heintz et al., 2016).

The researchers identified the dimensions of strategic vigilance in six dimensions that include various aspects of the internal and external activities of the organization (Lesca, 2003; Drevon, 2019; Alshaer, 2020a, 2020b). **Technological vigilance**, which expresses measures and procedures aimed at revealing the latest technical developments in the business environment and seeking to obtain and employ them to serve the organization (Ahmed et al., 2020; Dawood & Abbas, 2018). **Competitive vigilance** is the organization's mechanism used to identify its current and potential competitors and determine their strengths and weaknesses to support management decisions in following and designing a marketing approach that enables the organization to lead its market sector (Boulifa, 2008; Amarneh et al., 2021). As for commercial vigilance, it focuses on analyzing customer needs and their continuous changes to design products and services that meet their aspirations (Alyasiri, 2019; Almaazmi et al., 2020). **Social vigilance** is the observation of changes in the demographic and cultural composition of community members, which allows the organization to invest its resources in creating products that meet these changes (Dawood & Abbas, 2018; Odeh et al., 2021). **Legal vigilance** is related to following up on the latest developments in the political environment in terms of regulations, laws and legislation governing the mechanism of work in the sectors related to the activities of the organization (Drevon, 2019; Tariq et al., 2022). Finally, **environmental vigilance**, which is concerned with identifying the latest developments, initiatives and laws aimed at controlling the investment of environmental resources and preserving the environment, in order to design strategies and organizational mechanisms that ensure enhancing the ability of subsequent generations to benefit from current resources (Alshurideh et al., 2019a; Lesca, 2003).

2.2 *Organizational Learning*

Organizational learning and creativity are two modern contemporary notions that have sparked significant interest in the previous two decades, as they were seen as a critical strategic necessity for companies' advancement and survival (Leo et al., 2021). Furthermore, studies have shown that organizational learning improves an organization's ability to respond to the uncertainty that comes with today's dynamic business environment (Nuseir et al., 2021a, b), creates an opportunity for long-term competitive advantage (Pratono et al., 2019; Obeidat et al., 2021), and transforms the organization into an integrated unit that strives to achieve a common vision (Yoon et al., 2018). As a result, the researchers used a variety of definitions to refer to the concept of organizational learning, which defined by Al-Amyan as "all continuous processes aimed at providing the individual with the necessary skills and knowledge based on prior experiences in order to bring about a relative and permanent change in his scientific inventory. "Organizational learning, according to (Arumugam & Munusamy, 2015), revolves around the organization's ability to improve its employees' skills and capabilities by motivating them to interact, accept positive criticism, and provide an organizational environment that promotes the exchange of knowledge and cumulative experiences. According to Al-Majali, organizational learning is a continuous activity in which an organization strives to organize and expand its employees' knowledge base and continually enhance it in order to reach common ideals that can be used to solve challenges. Some scholars have chosen to use the tripartite model to assess organizational learning, which encompasses the strategic, organizational, and cultural dimensions (Marquardt, 2002). Others such as Taher (2011) who used a four-dimensional model that included technology application, knowledge management, people empowerment, and learning dynamics to measure it.

2.3 *Crisis Management*

According to Karam (2018), crisis management is a set of actions intended at containing and minimizing the severity of a crisis so that it does not spiral out of control, resulting in a conflict, It is also a set of plans, strategies, methods and administrative activities appropriate to exceptional situations in order to control and contain problems and maintain the organization's balance. According to AlShobaki et al. (2017), disaster preparedness is a broad concept that encompasses planning and responding to a variety of disasters and emergencies. Crisis management is a method that allows an organization to respond in a systematic way to the impacts of a crisis, allowing the business to continue with its everyday activity of supplying products and services to consumers and making profits (Abu Hamadneh et al., 2021; Lee et al., 2022a; Zyead et al., 2017, b).

Furthermore, Sahina et al. (2015) suggested that crisis management is a continuous management process based on the organization's ability to foresee and perceive potential causes that cause crises so that they can be prevented or their effects on the organization's performance can be mitigated. Crisis management, on the other hand, entails making administrative decisions rapidly in crucial situations, as crises force decision-makers in the organization to take immediate action to mitigate the potential effects and risks associated with the occurrence of crises. Hence, crisis management can be described as the art of dealing with the exceptional circumstances facing the organization in the turbulent business environment and trying to mitigate its negative effects on the outcome of the organization's work.

2.4 Relationship of Strategic Vigilance in Crisis Management Through Organizational Learning

In today's corporate environment, strategic vigilance has become a necessity, as the organization's operations rely heavily on anticipating and analyzing information about the organization's many aspects and activities, as well as its surrounds, to aid in decision-making. Strategic vigilance, according to Alshaer (2020a, 2020b), is a preventive path that assists decision makers in the organization in anticipating adverse conditions and attempting to avoid them by restructuring organizational resources to face probable dangers. Furthermore, Al-Qutji & Al-Malahsen determined that crisis management is based on a set of organizational strategies to control the sources of the crisis and lessen the organization's negative consequences. As a result, good crisis management relies on correct information gathered in a timely way through strategic vigilance operations in monitoring and forecasting concurrent and possible changes in its analyses.

Furthermore, strategic vigilance contributes to the organization's knowledge store by giving information on many elements of its activities to multiple departments (Idris & Al-Rubaie, 2013). As a result, the organization's competitive advantages are enhanced, and its ability to tackle complex problems posed by the dynamic business environment is improved. It also helps employees increase their ability to adapt to the changing business environment by enhancing the creative components that result in innovative services and products that meet customers' expectations (Shalakah et al., 2019; Alshurideh, 2022).

Furthermore, knowledge generation and dissemination throughout the organization's numerous departments aids in enhancing the ability to detect and respond to change swiftly (Marquardt et al., 2002). Furthermore, understanding of the many facets of the organization's difficulties and crises helps to mitigate the bad effects on its performance while continually attempting to discover inventive and distinctive solutions. The study hypotheses can therefore be written as follows:

There is an impact of Organizational Learning On the relationship between Strategic Vigilance and Crisis Management in the Jordanian Dairy Companies.

3 Study Model

See Fig. 1.

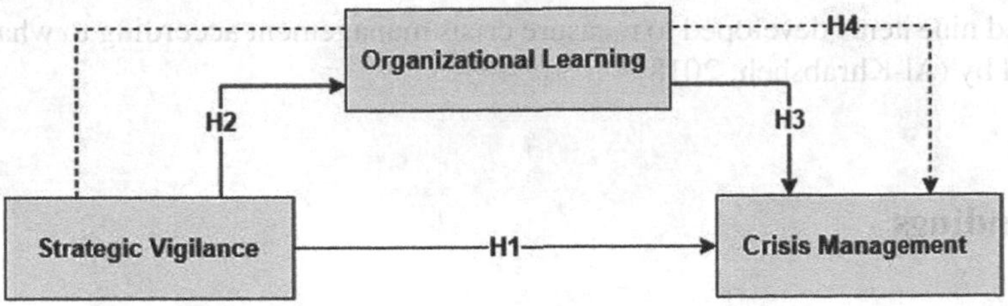

Fig. 1 Research model

4 Methodology

4.1 Population and Sample Selection

A qualitative method based on a questionnaire was used in this study for data collection and sample selection. The major aim of the study was to examine the impact of strategic vigilance on crisis management through the mediating role of organizational learning. Therefore, it focused on Jordanian dairy companies. Data were primarily gathered through self-reported questionnaires creating by Google Forms which were distributed to a sample of (285) managers via email. In total, (262) responses were received including (16) invalid to statistical analysis due to uncompleted or inaccurate. Hence, the final sample contained (246) responses suitable to analysis requirements that were formed a response rate of (86.31%), where it proved to be sufficient to the extent that was predictable and allowed for a presumption of data saturation (Sekaran & Bougie, 2016).

4.2 Measurement Instrument

A self-reported questionnaire that consists of three main sections along with a section regarding control variables was used as the measurement instrument. Control variables considered as categorical measures were composed of gender, age group, educational level, and experience. The three main sections were dealt with a five-point Likert scale (from 1 = strongly disagree to 5 = strongly agree). The first section contained (29) items to measure strategic vigilance based on Alshaer (2020a, 2020b) and Jaaz & Jamal (2021a, 2021b). These items were distributed into dimensions as follows: five items dedicated for measuring technological vigilance, five items dedicated for measuring business vigilance, five items dedicated for measuring social vigilance, five items dedicated for measuring competitive vigilance, four items dedicated for measuring environmental vigilance, and five items dedicated for measuring legal vigilance. The second section related to organizational learning which contained nine items developed according to (Al-Hawary et al., 2020). Whereas the third section included nine items developed to measure crisis management according to what was pointed by (Al-Khrabsheh, 2018).

5 Findings

5.1 Measurement Model Evaluation

This study was conducted structural equation modeling (SEM) to test hypotheses, which represents a contemporary statistical technique for testing and estimating the relationship between factors and variables (Wang & Rhemtulla, 2021). Accordingly, the reliability and validity of the constructs were tested using confirmatory factor analysis (CFA) through the statistical program AMOSv24. Table 1 summarizes the results of convergent and discriminant validity, as well the indicators of reliability.

Table 1 shows that the standard loading values for the individual items were within the domain (0.591–0.858), these values greater than the minimum retention of the elements based on their standard loads (Al-Lozi et al., 2018; Sung et al., 2019). Average variance extracted (AVE) is a summary indicator of the convergent validity of constructs that must be above 0.50 (Howard, 2018). The results indicate that the AVE values were greater than 0.50 for all constructs, thus the used measurement model has an appropriate convergent validity. Rimkeviciene et al. (2017) suggested the comparison approach as a way to deal with discriminant validity assessment in covariance-based SEM. This approaches based on comparing the values of maximum shared variance (MSV) with the values of AVE, as well as comparing the values of square root of AVE ($\sqrt{}$AVE) with the correlation between the rest of the structures. The results show that the values of MSV were smaller than the values of AVE, and that the values of $\sqrt{}$AVE were higher than the correlation values among the rest of the constructs. Therefore, the measurement model used is characterized by discriminative validity. The internal consistency measured through Cronbach's Alpha coefficient (α) and compound reliability by McDonald's Omega coefficient (ω) was conducted as indicators to evaluate measurement model. The results listed in Table 1 demonstrated that both values of Cronbach's Alpha coefficient and McDonald's Omega coefficient were greater than 0.70, which is the lowest limit for judging on measurement reliability (De Leeuw et al., 2019).

5.2 Structural Model

The structural model illustrated no multicollinearity issue among predictor constructs because variance inflation factor (VIF) values are below the threshold of 5, as shown in Table 1 (Hair et al., 2017). This result is supported by the values of model fit indices shown in Figs. 1 and 2.

Table 1 Results of validity and reliability tests

Constructs	1	2	3	4	5	6	7	8
1. TV	**0.753**							
2. BV	0.315	**0.762**						
3. SV	0.426	0.441	**0.759**					
4. CV	0.339	0.400	0.377	**0.757**				
5. EV	0.381	0.370	0.451	0.415	**0.764**			
6. LV	0.467	0.395	0.422	0.487	0.448	**0.749**		
7. OL	0.514	0.502	0.584	0.612	0.638	0.610	**0.738**	
8. CM	0.553	0.617	0.638	0.644	0.597	0.547	0.665	**0.742**
VIF	2.547	2.087	2.664	1.924	2.743	1.579	–	–
Loadings range	0.682–0.824	0.702–0.811	0.658–0.806	0.691–0.821	0.711–0.858	0.674–0.834	0.591–0.824	0.645–0.795
AVE	0.567	0.581	0.576	0.574	0.583	0.561	0.544	0.550
MSV	0.385	0.457	0.504	0.497	0.442	0.398	0.430	0.465
I.C	0.865	0.873	0.869	0.867	0.845	0.862	0.912	0.915
C.R	0.867	0.874	0.871	0.870	0.848	0.864	0.914	0.916

Note TV: Technological Vigilance, BV: Business Vigilance, SV: Social Vigilance, CV: Competitive Vigilance, EV: Environmental Vigilance, LV: Legal vigilance, OL: Organizational Learning, CM: Crisis Management, I.C: Internal Consistency, C.R: Composite Reliability

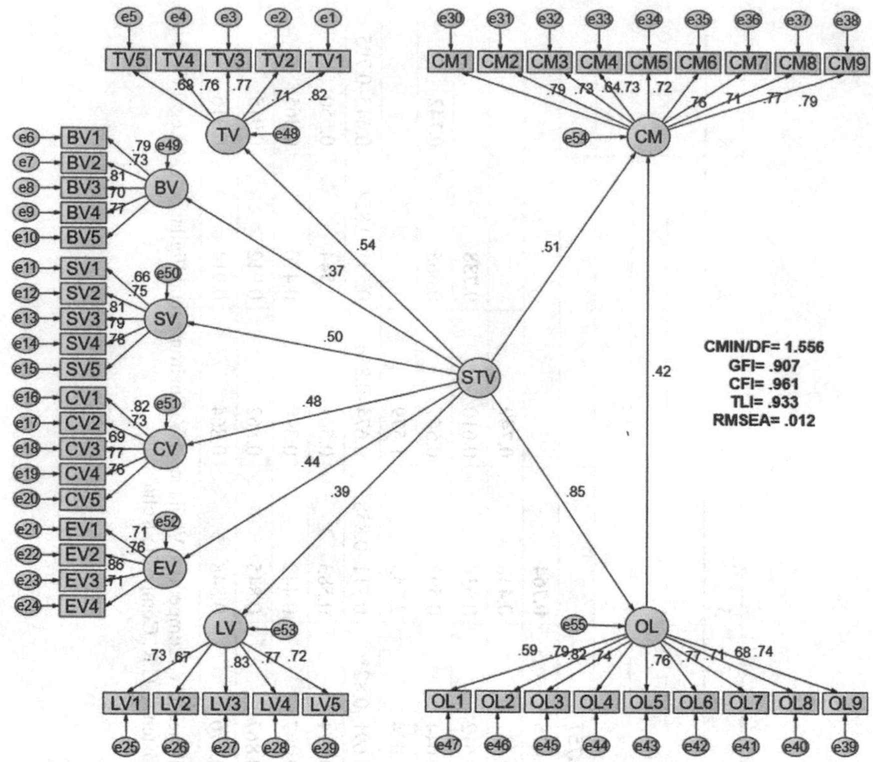

Fig. 2 SEM results of the strategic vigilance effect on crisis management through organizational learning

The results in Fig. 1 indicated that the chi-square to degrees of freedom (CMIN/DF) was 1.556, which is less than 3 the upper limit of this indicator. The values of the goodness of fit index (GFI), the comparative fit index (CFI), and the Tucker–Lewis index (TLI) were upper than the minimum accepted threshold of 0.90. Moreover, the result of root mean square error of approximation (RMSEA) indicated to value 0.012, this value is a reasonable error of approximation because it is less than the higher limit of 0.08. Consequently, the structural model used in this study was recognized as a fit model for predicting the crisis management and generalization of its result (Ahmad et al., 2016; Shi et al., 2019). To verify the results of testing the study hypotheses, structural equation modeling (SEM) was used, the results of which are listed in Table 2.

The results demonstrated in Table 2 show that strategic vigilance has a positive direct impact on crisis management ($\beta = 0.512$, $t = 7.086$, $p = 0.000$) and organizational learning ($\beta = 0.851$, $t = 15.045$, $p = 0.000$). Moreover, it indicated that organizational learning has a positive impact on crisis management ($\beta = 0.419$, $t = 7.778$, $p = 0.000$). Hence, organizational learning plays a partial mediation effect in

Table 2 Hypothesis testing

Relation	Direct Effect		Indirect Effect		Total Effect		
	β	p-value	β	p-value	β	t-value	p-value
Strategic Vigilance → Crisis Management	0.512	0.000			0.512	7.086	0.000
Strategic Vigilance → Organizational Learning	0.851	0.000			0.851	15.045	0.000
Organizational Learning → Crisis Management	0.419	0.000			0.419	7.778	0.000
Strategic Vigilance → Organizational Learning → Crisis Management	0.512	0.000	0.357	0.000	0.870	18.254	0.000

the relationship between strategic vigilance and crisis management, where the total effect was (β = 0.870, t = 18.254, p = 0.000) with indirect effect of (β = 0.3557, p = 0.000).

6 Discussion

Through the mediating role of organizational learning, the study attempted to examine the impact of strategic vigilance in crisis management. The findings revealed that strategic alertness has a direct impact on crisis management, which is consistent with previous research (Al-Tanayeeb, 2020). As a result, organizations' pursuit of a technological infrastructure, as well as training employees at various administrative levels to deal with the changing business environment, aids in the early detection of crises and the formulation of preventive plans to mitigate the negative effects of crises. Furthermore, the organization's constant concern and commitment, driven by the desire to achieve industrial leadership, drives it to monitor various aspects of business related to the nature of its activity, enhancing its ability to predict the conditions of the turbulent business environment and take proactive measures to achieve its strategic goals.

The study also found that strategic vigilance has a direct impact on organizational learning, which is consistent with previous findings (Dawood & Abbas, 2018). As a result, in addition to offering current technology to employees, permanent follow-up procedures for laws and legislative regulations contribute greatly to enhancing their ability to deal with difficulties and discover inventive solutions. Furthermore, incorporating environmentally friendly practices into different elements of the organization's operations raises employee awareness of the need to address environmental concerns and fosters a sense of social responsibility toward future generations.

Furthermore, the data revealed that organizational learning has an impact on crisis management, which is consistent with the study's conclusions (Gujarai, 2017). As a result, creating a work environment that encourages employees to share their previous experiences and skills, allowing them to participate in organizational decision-making, and continuously training them to deal with problems improves the organization's ability to detect flaws in the business environment early and take preventative measures to mitigate their effects and reduce the likelihood of them occurring. Moreover, the results showed that organizational learning plays a mediating role in the relationship between strategic vigilance and crisis management. Therefore, the organization's ability to use modern technology to follow up and control the variables of the business environment and the competitive environment enables it to build a stock of knowledge through which it can deal with the exceptional circumstances that organizations are going through. Moreover, this stock of knowledge, based mainly on accurate and objective information, contributes to supporting the decision-making process and reducing the uncertainty associated with it.

7 Recommendations

Based on the achieved results, this study presents a set of recommendations to decision makers in order to better manage the crises facing their organizations. First, focus on following a supportive management approach to the use of technology in the various activities of the organization because of its impact on enhancing its ability to sense changes in the business environment. Second, providing training programs for employees at various administrative levels on dealing with crises and scientific methods to identify the causes of problems and find appropriate solutions within the constraints of available material resources. Third, enhancing data integration through building an information management system and applying artificial intelligence methods because of their role in disseminating the acquired knowledge in various departments of the organization and enhancing the skills and capabilities of employees to create competitive products and services that take the organization to leadership positions. Finally, forming a crisis management team comprising employees from all administrative levels and training them to deal with fluctuations arising from the turbulent environment, in addition to assigning them to prepare studies and periodic reports on potential problems and the mechanism for dealing with them.

References

Ahmad, A., Alshurideh, M. T., Al Kurdi, B. H., & Salloum, S. A. (2021). Factors impacts organization digital transformation and organization decision making during Covid19 pandemic. In The effect of coronavirus disease (COVID-19) on business intelligence (pp. 95–106). Cham: Springer.

Ahmad, S., Zulkurnain, N., & Khairushalimi, F. (2016). Assessing the validity and reliability of a measurement model in structural equation modeling (SEM). *British Journal of Mathematics & Computer Science, 15*(3), 1–8. https://doi.org/10.9734/BJMCS/2016/25183.

Ahmed, A., Alshurideh, M., Al Kurdi, B., & Salloum, S. A. (2020). Digital transformation and organizational operational decision making: A systematic review. In International Conference on Advanced Intelligent Systems and Informatics (pp. 708–719). Cham: Springer.

Al Kurdi, B., Alshurideh, M., & Al afaishat, T. (2020a). Employee retention and organizational performance: Evidence from banking industry. *Management Science Letters, 10*(16), 3981–3990.

Al Kurdi, B., Alshurideh, M., & Alnaser, A. (2020b). The impact of employee satisfaction on customer satisfaction: Theoretical and empirical underpinning. *Management Science Letters, 10*(15), 3561–3570.

Al-Quran, A. Z., Alhalalmeh, M. I., Eldahamsheh, M. M., Mohammad, A. A., Hijjawi, G. S., Almomani, H. M., & Al-Hawary, S. I. (2020). Determinants of the green purchase intention in Jordan: The moderating effect of environmental concern. *Int. J Sup. Chain. Mgt, 9*(5), 366–371.

Alaali, N., Al Marzouqi, A., Albaqaeen, A., Dahabreh, F., Alshurideh, M., Alrwashdh, S., Iyadeh, I., Salloum, S., & Aburayya, A. (2021). The impact of adopting corporate governance strategic performance in the tourism sector: A case study in the Kingdom of Bahrain. Journal of Legal, Ethical and Regulatory, *24*(Special Issue 1), 1–18.

Alameeri, K., Alshurideh, M., Al Kurdi, B., & Salloum, S. A. (2020). The effect of work environment happiness on employee leadership. In International Conference on Advanced Intelligent Systems and Informatics (pp. 668–680). Cham: Springer.

Alameeri, K. A., Alshurideh, M. T., & Al Kurdi, B. (2021). The effect of Covid-19 pandemic on business systems' innovation and entrepreneurship and how to cope with it: A theatrical view. The effect of coronavirus disease (COVID-19) on business intelligence, *334*, 275–288.

Al-Dmour, R., AlShaar, F., Al-Dmour, H., Masa'deh, R., & Alshurideh, M. T. (2021b). The effect of service recovery justices strategies on online customer engagement via the role of "Customer Satisfaction" during the covid-19 pandemic: An empirical study. The effect of coronavirus disease (COVID-19) on business intelligence, *334*, 325–346.

Al-Dmour, A., Al-Dmour, H., Al-Barghuthi, R., Al-Dmour, R., & Alshurideh, M. T. (2021a). Factors influencing the adoption of e-payment during pandemic outbreak (COVID-19): Empirical evidence. The effect of coronavirus disease (COVID 19) on business intelligence, *334*, 133 154.

Alhalalmeh, M. I., Almomani, H. M., Altarifi, S., Al- Quran, A. Z., Mohammad, A. A., & Al-Hawary, S. I. (2020). The nexus between corporate social responsibilty and organizational performance in Jordan: The mediating role of organizational commitment and organizational citizenship behavior. *Test Engineering and Management, 83*(July), 6391–6410.

Al-Hawary, S. I., & Al-Syasneh, M. S. (2020). Impact of dynamic strategic capabilities on strategic entrepreneurship in presence of outsourcing of five stars hotels in Jordan. Business: Theory and Practice, *21*(2), 578–587.

Al-Hawary, S. I. (2015). Human resource management practices as a success factor of knowledge management implementation at health care sector in Jordan. *International Journal of Business and Social Science, 6*(11/1), 83–98.

Al-Hawary, S. I., & Aldaihani, F. M. (2016). Customer relationship management and innovation capabilities of Kuwait Airways. *International Journal of Academic Research in Economics and Management Sciences, 5*(4), 201–226.

Al-Hawary, S. I. S., & Alhajri, T. M. S. (2020). Effect of electronic customer relationship management on customers' electronic satisfaction of communication companies in Kuwait. *Calitatea, 21*(175), 97–102.

Al-Hawary, S. I., & Al-Hamwan, A. (2017). Environmental analysis and its impact on the competitive capabilities of the commercial banks operating in Jordan. *International Journal of Academic Research in Accounting, Finance and Management Sciences, 7*(1), 277–290.

Al-Hawary, S. I., & Al-Namlan, A. (2018). Impact of electronic human resources management on the organizational learning at the Private Hospitals in the State of Qatar. *Global Journal of Management and Business Research: A Administration and Management, 18*(7), 1–11.

Al-Hawary, S. I. S., & Alwan, A. M. (2016). Knowledge management and its effect on strategic decisions of Jordanian Public Universities. *Journal of Accounting-Business & Management, 23*(2), 24–44.

Al-Hawary, S. I., & Hadad, T. F. (2016). The effect of strategic thinking styles on the enhancement competitive capabilities of Commercial Banks in Jordan. *International Journal of Business and Social Science, 7*(10), 133–144.

Al-Hawary, S. I. S., Mohammad, A. S., Al-Syasneh, M. S., Qandah, M. S. F., & Alhajri, T. M. S. (2020). Organisational learning capabilities of the commercial banks in Jordan: Do electronic human resources management practices matter? *International Journal of Learning and Intellectual Capital, 17*(3), 242–266.

Al-Hawary, S. I. S., & Obiadat, A. A. (2021). Does mobile marketing affect customer loyalty in Jordan? *International Journal of Business Excellence, 23*(2), 226–250.

Al-Khrabsheh, A. A. (2018). Impact of strategic planning on crisis management in the profit and non-profit sector in Jordan. *Academy of Strategic Management Journal, 17*(5), 1–12.

Allahow, T. J. A. A., Al-Hawary, S. I. S., & Aldaihani, F. M. F. (2018). Information technology and administrative innovation of the central agency for information technology in Kuwait. Global Journal of Management and Business, *18*(11-A), 1–16.

Al-Lozi, M., Almomani, R. Z., & Al-Hawary, S. I. (2017). Impact of talent management on achieving organizational excellence in Arab Potash Company in Jordan. Global Journal of Management and Business Research: A Administration and Management, *17*(7), 15–25.

Al-Lozi, M. S., Almomani, R. Z. Q., & Al-Hawary, S. I. S. (2018). Talent Management strategies as a critical success factor for effectiveness of Human Resources Information Systems in commercial banks working in Jordan. *Global Journal of Management and Business Research: A Administration and Management, 18*(1), 30–43.

Almaazmi, J., Alshurideh, M., Al Kurdi, B., & Salloum, S. A. (2020). The effect of digital transformation on product innovation: a critical review. In International Conference on Advanced Intelligent Systems and Informatics (pp. 731–741). Cham: Springer.

Al-Nady, B. A., Al-Hawary, S. I., & Alolayyan, M. (2013). Strategic management as a key for superior competitive advantage of sanitary ware suppliers in Kingdom of Saudi Arabia. *International Journal of Management and Information Technology, 7*(2), 1042–1058.

Al-Nady, B. A., Al-Hawary, S. I., & Alolayyan, M. (2016). The role of time, communication, and cost management on project management success: An empirical study on sample of construction projects customers in Makkah City, Kingdom of Saudi Arabia. *International Journal of Services and Operations Management, 23*(1), 76–112.

Alshaer, S. A. (2020b). The effect of strategic vigilance on organizational ambidexterity in Jordanian Commercial Banks. Department of Management. The World Islamic Science & Education University, Jordan, Modern Applied Science, *14*(6), 82–89.

Alshaer, S. A. (2020a). The effect of strategic vigilance on organizational ambidexterity in Jordanian Commercial Banks. *Modern Applied Science, 14*(6), 82–89.

AlShobaki, M. J., Amuna, Y. M. A., & Naser, S. S. A. (2017). Strategic and operational planning as approach for crises management field study on UNRWA. International Journal of Information Technology and Electrical Engineering, *5*(6), 43–47.

Alshurideh, M. T., Hassanien, A. E., & Ra'ed Masa'deh. (2021). The effect of coronavirus disease (COVID-19) on business intelligence. Springer.

Alshurideh, M. (2022). Does electronic customer relationship management (E-CRM) affect service quality at private hospitals in Jordan? *Uncertain Supply Chain Management, 10*(2), 325–332.

Alshurideh, M., Kurdi, B. A., Shaltoni, A. M., & Ghuff, S. S. (2019a). Determinants of proenvironmental behaviour in the context of emerging economies. *International Journal of Sustainable Society, 11*(4), 257–277.

Alshurideh, M., Salloum, S. A., Al Kurdi, B., Monem, A. A., & Shaalan, K. (2019b). Understanding the quality determinants that influence the intention to use the mobile learning platforms: A practical study. *International Journal of Interactive Mobile Technologies, 13*(11), 183–157.

Alsuwaidi, M., Alshurideh, M., Al Kurdi, B., & Salloum, S. A. (2020). Performance appraisal on employees' motivation: a comprehensive analysis. In International Conference on Advanced Intelligent Systems and Informatics (pp. 681–693). Cham: Springer.

Altamony, H., Masa'deh, R., Alshurideh, M., Obeidat, B. (2012) Information systems for competitive advantage: Implementation of an organisational strategic management process. Innovation and Sustainable Competitive Advantage: From Regional Development to World Economies. 583–592.

Al-Tanayeeb, N. A. N. (2020). *The Impact of Strategic Vigilance in Crisis Management*. Zarqa University.

Altarifi, S., Al-Hawary, S. I. S., & Al Sakkal, M. E. E. (2015). Determinants of e-shopping and its effect on consumer purchasing decision in Jordan. *International Journal of Business and Social Science, 6*(1), 81–92.

AlTaweel, I. R., & Al-Hawary, S. I. (2021). The mediating role of innovation capability on the relationship between strategic agility and organizational performance. *Sustainability, 13*(14), 7564.

Alves, J. C., Tan, C. L., YuBo, L., & Wei, H. (2020). Crisis management for small business during the COVID-19 outbreak: Survival, resilience and renewal strategies of firms in Macau.

Alyammahi, A., Alshurideh, M., Al Kurdi, B., & Salloum, S. A. (2020). The impacts of communication ethics on workplace decision making and productivity. In International Conference on Advanced Intelligent Systems and Informatics (pp. 488–500). Cham: Springer.

Alyasiri, A., Mohammed, A., & Hussein, R. (2019). Measuring the impact of strategic vigilance in strategic intelligence. Analytical study of the views of a sample of workers in AL—FURAT general company for Chemical Industries. University of Kerbala. https://www.researchgate.net/publication/330182053_Measuring_The_Impact_Of_Strategic_Vigilance_in_Strategic_Intelligence.

Alzoubi, H., Alshurideh, M., Kurdi, B., Akour, I., & Aziz, R. (2022). Does BLE technology contribute towards improving marketing strategies, customers' satisfaction and loyalty? The role of open innovation. *International Journal of Data and Network Science, 6*(2), 449–460.

Amarneh, B. M., Alshurideh, M. T., Al Kurdi, B. H., & Obeidat, Z. (2021). The impact of COVID-19 on e-learning: Advantages and challenges. In The International Conference on Artificial Intelligence and Computer Vision (pp. 75–89). Cham: Springer.

Arumugam, T., Iis, K., & Munusamy, K. (2015). Conceptualizing organizational learning system model and innovativeness. *International Journal of Business and Social Science, 6*(3), 155–165.

Ashal, N., Alshurideh, M., Obeidat, B., Masa'deh, R. (2021) The impact of strategic orientation on organizational performance: Examining the mediating role of learning culture in Jordanian telecommunication companies. Academy of Strategic Management Journal, *21*(Special Issue 6), 1–29.

Bettahar, B., & Aggoun, A. (2021). The role of big data analysis and strategic vigilance in decision-making. In Big Data Analytics (pp. 107–120). Apple Academic Press.

Coccia, M. (2020). Critical decisions in crisis management: Rational strategies of decision making. *Journal of Economics Library, 7*(2), 81–96.

Coombs, W. T., & Laufer, D. (2018). Global crisis management–current research and future directions. *Journal of International Management, 24*(3), 199–203.

Dawood, F., & Abbas, A. (2018). The role of strategic vigilance in the operational performance of the banking sector: Field research in a sample of private banks. *European Journal of Business and Management, 10*(21), 1–18.

De Leeuw, E., Hox, J., Silber, H., Struminskaya, B., & Vis, C. (2019). Development of an international survey attitude scale: Measurement equivalence, reliability, and predictive validity. *Measurement Instruments for the Social Sciences, 1*(1), 9. https://doi.org/10.1186/s42409-019-0012-x.

Drevon, E., Maurel, D., & Dufour, C. (2019). Veillestratégique et prise de décision :Une revue de la literature. *Documentation Et Libraries, 64*(1), 28–34.

Gujari, R. (2017). *The Impact of Organizational Learning on Work Performance*. Osmania University College of Commerce and Business Management.

Hair, J. F., Babin, B. J., & Krey, N. (2017). Covariance-based structural equation modeling in the journal of advertising: Review and recommendations. *Journal of Advertising, 46*(1), 163–177. https://doi.org/10.1080/00913367.2017.1281777.

Hamadneh, S., Pedersen, O., & Al Kurdi, B. (2021). An investigation of the role of supply chain visibility into the Scottish blood supply chain. *Journal of Legal, Ethical and Regulatory Issues, 24*, 1–13.

Heintz, C., Karabegovic, M., & Molnar, A. (2016). The Co-evolution of honesty and strategic vigilance. *Frontiers in Psychology, 7 Article, 1503*, 1–13.

Hijjawi, G. S., & Mohammad, A. (2019). Impact of organizational ambidexterity on organizational conflict of Zain telecommunication company in Jordan. *Indian Journal of Science and Technology, 12*, 26.

Howard, M. C. (2018). The convergent validity and nomological net of two methods to measure retroactive influences. *Psychology of Consciousness: Theory, Research, and Practice, 5*(3), 324–337. https://doi.org/10.1037/cns0000149.

Idris, W. M. S., & Al-Rubaie, M. T. K. (2013). Examining the impact of strategic learning on strategic agility. *Journal of Management and Strategy, 4*(2), 70.

Inès, B. T. (2008). Identification des facteurs critique de succès pour la mise en place d'un dispositif de veillestratégique, thèse du doctorat. Tunis: ISG-Tunis.

Ismaail, A. H. (2020). Measuring the impact of strategic vigilance reflection on innovative marketing research edited by a sample of employees of the Iraqi Company for the manufacture and marketing of dates (Mixed Contribution). Enterpreneurship Journal for Finance and Business, *1*(2).

Jaaz, S. A., & Jamal, D. H. (2021a). The effect of the dimensions of strategic vigilance on organizational excellence) an applied study of premium Class Hotels In Baghdad. *Palarch's Journal of Archaeology of Egypt/egyptology, 18*(08), 2101–2137.

Jaaz, S. A., & Jamal, D. H. (2021b). The effect of the dimensions of strategic vigilance on organizational excellence: An applied study of premium class hotels in Baghdad. *Palarch's Journal of Archaeology of Egypt/egyptology, 18*(08), 2101–2137.

Jarallah, M. A. (2021). Strategic vigilance and its role in achieving the strategic direction an exploratory study of the opinions of a sample of senior administrative leaders at the Holy University of Karbala. *The Iraqi Magazinje for Managerial Sciences, 17*(70), 259–282.

Karam MG. (2018). The impact of strategic planning on crisis management styles in the 5-star hotels. Department of Hotel Studies, The Higher Institute of Tourism and Hotels in Alexandria, Egypt, Journal of Hotel & Business Management, 7(1), 1–9.

Lee, K., Azmi, N., Hanaysha, J., Alshurideh, M., & Alzoubi, H. (2022a). The effect of digital supply chain on organizational performance: An empirical study in Malaysia manufacturing industry. *Uncertain Supply Chain Management, 10*(2), 1–16.

Lee, K., Ramiz, P., Hanaysha, J., Alzoubi, H., & Alshurideh, M. (2022b). Investigating the impact of benefits and challenges of IOT adoption on supply chain performance and organizational performance: An empirical study in Malaysia. *Uncertain Supply Chain Management, 10*(2), 1–14.

Leo, S., Alsharari, N. M., Abbas, J., & Alshurideh, M. T. (2021). From offline to online learning: A qualitative study of challenges and opportunities as a response to the COVID-19 pandemic in the UAE higher education context. In The Effect of Coronavirus Disease (COVID-19) on Business Intelligence (pp. 203–217). Cham: Springer.

Lesca, H. (2003). Veillestratégique: la méthode L.E. Scanning, édition EMS, Management et Société, France: Cormelles-le-Royal.

Manhal, M. H., & Hattab, H. N. (2018). The effect of strategic vigilance on achieving sustainable competitive advantage: SWOT analysis as moderator. *Managerial Studies Journal, 10*(20), 302–332.

Marquardt, M. (2002). *Building the learning organization: Mastering the 5 elements for corporate learning (2nd edn).* Davies-Black Publishers.

Mohammad, A. A., Alshura, M. S., Al-Hawary, S. I. S., Al-Syasneh, M. S., & Alhajri, T. M. (2020). The influence of Internal Marketing Practices on the employees' intention to leave: A study of the private hospitals in Jordan. *International Journal of Advanced Science and Technology, 29*(5), 1174–1189.

Nuseir, M. T., Aljumah, A., & Alshurideh, M. T. (2021b). How the business intelligence in the new startup performance in UAE During COVID-19: The mediating role of innovativeness. The Effect of Coronavirus Disease (COVID-19) on Business Intelligence, *334*, 63–79.

Nuseir, M. T., Al Kurdi, B. H., Alshurideh, M. T., & Alzoubi, H. M. (2021a). Gender discrimination at workplace: Do artificial intelligence (AI) and machine learning (ML) have opinions about it. In The International Conference on Artificial Intelligence and Computer Vision (pp. 301–316). Cham: Springer.

Obeidat, U., Obeidat, B., Alrowwad, A., Alshurideh, M., Masadeh, R., & Abuhashesh, M. (2021). The effect of intellectual capital on competitive advantage: The mediating role of innovation. *Management Science Letters, 11*(4), 1331–1344.

Odeh, R. B. M., Obeidat, B. Y., Jaradat, M. O., & Alshurideh, M. T. (2021). The transformational leadership role in achieving organizational resilience through adaptive cultures: the case of Dubai service sector. International Journal of Productivity and Performance Management. Vol. ahead-of-print No. ahead-of-print. https://doi.org/10.1108/IJPPM-02-2021-0093.

Park, S., & Kim, E. J. (2018). Fostering organizational learning through leadership and knowledge sharing. Journal of Knowledge Management.

Pratono, A. H., Darmasetiawan, N. K., Yudiarso, A., &Jeong, B. G. (2019). Achieving sustainable competitive advantage through green entrepreneurial orientation and market orientation: The role of inter-organizational learning. The Bottom Line.

Rimkeviciene, J., Hawgood, J., O'Gorman, J., & De Leo, D. (2017). Construct validity of the acquired capability for suicide scale: Factor structure, convergent and discriminant validity. *Journal of Psychopathology and Behavioral Assessment, 39*(2), 291–302. https://doi.org/10.1007/s10862-016-9576-4.

Sahina, S., Ulubeylib, S., & Kazazaa, A. (2015). Innovative crisis management in construction: Approaches and the process. *Social and Behavioral Sciences, 195*, 2298–2305.

Sekaran, U., & Bougie, R. (2016). Research methods for business: A skill building approach (Seventh edition). Wiley.

Shah, S. F., Alshurideh, M. T., Al-Dmour, A., & Al-Dmour, R. (2021). Understanding the influences of cognitive biases on financial decision making during normal and COVID-19 pandemic situation in the United Arab Emirates. The Effect of Coronavirus Disease (COVID-19) on Business Intelligence, *334*, 257–274.

Shakhour, R., Obeidat, B., Jaradat, M., Alshurideh, M., Masa'deh, R. (2021). Agile-minded organizational excellence: Empirical investigation. Academy of Strategic Management Journal *20*(Special Issue 6), 1–25.

Shalakah, T. K., Hleehal, M. S., & Suliman, A. A. A. A. (2019). Strategic vigilance and its impact on the organization's vital capacities. Analytical Descriptive Study in UR Company for Engineering Industries/Dhi-Qar. Al Kut Journal of Economics and Administrative Sciences, *11*(33), 128–143.

Shi, D., Lee, T., & Maydeu-Olivares, A. (2019). Understanding the model size effect on SEM fit indices. *Educational and Psychological Measurement, 79*(2), 310–334. https://doi.org/10.1177/0013164418783530.

Sung, K.-S., Yi, Y. G., & Shin, H.-I. (2019). Reliability and validity of knee extensor strength measurements using a portable dynamometer anchoring system in a supine position. *BMC Musculoskeletal Disorders, 20*(1), 1–8. https://doi.org/10.1186/s12891-019-2703-0.

Tariq, E., Alshurideh, M., Akour, E., Al-Hawaryd, S., & Al Kurdi, B. (2022). The role of digital marketing, CSR policy and green marketing in brand development at UK. *International Journal of Data and Network Science, 6*(3), 1–10.

Wang, Y. A., & Rhemtulla, M. (2021). Power analysis for parameter estimation in structural equation modeling: A discussion and tutorial. *Advances in Methods and Practices in Psychological Science, 4*(1), 1–17. https://doi.org/10.1177/2515245920918253.

Werlang, N. B., & Rossetto, C. R. (2019). The effects of organizational learning and innovativeness on organizational performance in the service provision sector. Gestão&Produção, 26.

Yoon, D. Y., Han, S. H., Sung, M., & Cho, J. (2018). Informal learning, organizational commitment and self-efficacy: A study of a structural equation model exploring mediation. Journal of Workplace Learning.

Yuliansyah, Y., & Jermias, J. (2018). Strategic performance measurement system, organizational learning and service strategic alignment: Impact on performance. International Journal of Ethics and Systems.

Zulkarnaini, N. A. S., Shaari, R., & Sarip, A. (2019). Crisis management and human resource development: Towards research agenda. In International Conference on Applied Human Factors and Ergonomics (pp. 542–552). Cham: Springer.

Zyead, A., Abd El Mut'y, Z., & Sherifa Fouad, S. (2017). The impact of the implementation the concept of knowledge management on security crisis management. Review of Public Administration and Management, 5(2), 1–15.

The Impact of Total Quality Management on the Organizational Reputation

Enas Ahmad Alshuqairat, Nancy Abdullah Shamaileh, Mohammed Saleem Khlif Alshura, Zaki Abdellateef Khalaf Khalaylah, Maali M. Al-mzary, Basem Yousef Ahmad Barqawi, Sulieman Ibraheem Shelash Al-Hawary, Muhammad Turki Alshurideh, and Anber Abraheem Shlash Mohammad

Abstract The major aim of the study was to examine the impact of total quality management (TQM) on organizational reputation (OR). Therefore, it focused on the faculty members in five public universities, whose number is (5477) faculty members. As for the study sample, it consisted of a proportional stratified sample of (390) individuals, and after the questionnaire was distributed with (410), 390 were recovered, valid for the purposes of statistical analysis. Structural equation modeling (SEM) was conducted to test hypotheses. The results showed that all total quality management dimensions had a positive impact relationship on organizational reputation except focus on beneficiary. Based on the results of the study, researchers recommend managers and decision-makers in public universities to pay attention to

---

E. A. Alshuqairat · M. S. K. Alshura · Z. A. K. Khalaylah
Management Department, Faculty of Money and Management, The World Islamic Science University, P.O. Box1101, Amman 11947, Jordan

N. A. Shamaileh · S. I. S. Al-Hawary (✉)
Department of Business Administration, School of Business, Al Al-Bayt University, P.O.BOX 130040, Mafraq 25113, Jordan
e-mail: dr_sliman73@aabu.edu.jo; dr_sliman@yahoo.com

M. M. Al-mzary
Department of Applied Science, Irbid College, Al-Balqa Applied University, As-Salt, Jordan

B. Y. A. Barqawi
Business Administration Department, Faculty of Administration & Financial Sciences, Petra University, Amman, Jordan

M. T. Alshurideh
Department of Marketing, School of Business, The University of Jordan, Amman 11942, Jordan
e-mail: malshurideh@sharjah.ac.ae; mhmadshura@yahoo.com

Department of Management, College of Business, University of Sharjah, 27272 Sharjah, United Arab Emirates

A. A. S. Mohammad
Marketing Department, Faculty of Administrative and Financial Sciences, Petra University, B.O. Box: 961343, Amman 11196, Jordan

M. Alshurideh et al. (eds.), *The Effect of Information Technology on Business and Marketing Intelligence Systems*, Studies in Computational Intelligence 1056,
https://doi.org/10.1007/978-3-031-12382-5_64

the university management standard (higher leadership), which has the largest role in supporting and activating the application of total quality management.

Keywords Total quality management · Organizational reputation · Public universities · Jordan

1 Introduction

In today's higher education industry, the concept of total quality has gotten a lot of attention because it is seen as a competitive weapon that will lead to globalization. Universities are regarded as one of the most important tools for societal development and modernization, as well as an important tool for transitioning from their traditional role in activities and operations (educational, academic, and administrative) to a more developed and improving role in these processes to produce high-quality results and outputs regardless of the quantity produced (Alshamsi et al., 2020; Alsharari & Alshurideh, 2020), which prompted it to adopt the philosophy of total quality management in the educational process, and control it with assurance standards and specific indicators and standards, which many researchers considered the ideal solution to make a qualitative leap in the development of its services and outputs, which is reflected positively on its reputation and standing among its peers, this allows it to realize its strategic objectives in moving to global competitive positions among universities throughout the world (Al-Hamad et al., 2021; Leo et al., 2021; Sultan et al., 2021).

The reputation of universities today is considered one of the most prominent and most important standards, which constitute a major challenge that highlights their ability or not to respond quickly to the rapid changes in the surrounding environment, and even have the ability to resolve the vision of stakeholders and beneficiaries towards it, and reflects its ability to attract students and investors (Al Kurdi et al., 2020a; Nazi et al., 2022), which compel universities to join their communities and play a role in them by providing services to their various segments, linking their various specialties to their needs, and linking their research inputs to serve and solve their problems, as well as to create a strong interaction between their resources and the service and production sectors of society (Al-Ta'i & Al-Abadi, 2009; Al-Hawary, 2010a; Al-Hawary, 2010b; Al-Hawary & Batayneh, 2010). Universities today are witnessing efforts to improve their image and reputation by adopting the quality standards required by the most famous international university rankings today, because these rankings reflect a large part of the quality of higher education.

The significance of this research stems from the discussion of comprehensive quality management, including its dimensions and standards, which create an interwoven chain of relevance in creating a significant competitive advantage and reputation, as evidenced by high-quality inputs and outputs (Alshurideh et al., 2021; Amarneh et al., 2021). Given that we are discussing our public universities, the theoretical importance of emphasizing the role of our universities in their application and commitment to total quality management and assurance standards, as well as linking it to the fate of their academic and educational reputation on the Arab and international levels, is never less than the practical (Altamony et al., 2012; Batayneh et al., 2021).

Public universities today adopt a vision and message of excellence and a strong reputation through the provision of services (academic and pedagogical) and qualitative outputs of high quality and comprehensiveness in their plans and strategies. These are both strategic and operational plans that cover the foundations and concepts of comprehensive quality management in all of its dimensions, as well as the implementation of quality assurance standards in its programs and plans, as well as within specific indicators and standards. Researchers today believe that the reputation of Jordanian public universities has been shaken from before, not only locally, but also among Arab universities as well, because the modern experience of international university rankings is a modern and contemporary experience that universities in general use to raise the level of their reputation, but it is still within a framework limited to some limited and narrow indicators, and because university education has expanded significantly in the past ten years, and its expansion was accompanied by a number of obstacles and challenges (Alshurideh et al., 2019; Al Kurdi et al., 2020b; Alshurideh et al., 2020; AlHamad et al., 2021; Al-Maroof et al., 2021, Nuseir et al., 2021a, b), so this study came to examine the impact of total quality management with its dimensions (focus on the beneficiary, commitment of senior management, continuous improvement, participation of working individuals, training and development) on organizational reputation in Jordanian public universities.

2 Theoretical Framework and Hypotheses Development

2.1 Total Quality Management

Total Quality Management (TQM) is one of the most prominent modern management concepts, and as a result of what the market witnessed of a great competitive image, to provide the best products and services to its customers and clients, this competition moved to a sector that was far from this competition due to the different services provided to the beneficiaries, to enter the circle of competition of the education sector in providing its educational services in a cost-effective manner, after this concept was embodied in an integrated approach of modern management curricula, similar to other concepts that underwent renaissance, progress, and development, and after it was modified to keep pace with today's rapid changes in the environment, it began to focus its attention on achieving quality and excellence to gain customer satisfaction, and it evolved into a philosophy focused on achieving the level of high quality in the performance of operations, in addition to considering it a process that aims to create a distinct culture of performance through persistent pursuit between managers and employees in order to achieve the expectations of the beneficiaries, and to do all work as required from the first time with better quality, higher effectiveness and in the shortest time (Al-Hawary & Abu-Laimon, 2013; Al-Hawary et al., 2013; Ghoneim, 2018).

Total quality management, according to the researchers, is a philosophy that focuses on reaching the greatest and highest degree of quality and process performance through employee participation, continuous improvement procedures, and addressing the needs and aspirations of beneficiaries (Al-Hawary et al., 2017; Krajewski, et al., 2013; Metabis & Al-Hawary, 2013; Mohammad et al., 2020;). While Hasham (2021) sees it as a management model based on the most efficient use of time, Hasham (2021) sees it as a strategic plan with clear goals that is characterized by constant evaluation, rectification, and successful review. This is due to the consideration that quality management and time management go in one line towards achieving productivity, by reducing error rates, and avoiding waste in the exploitation of various resources, meaning that through quality management we can use time more effectively. We conclude that total quality management is the art of managing the entity of the organization, with all of its parts and components, for the purpose of achieving excellence, and it is a philosophy or set of principles that represent the foundation for any organization that follows a continuous improvement approach, to meet the requirements and needs of its customers in the long run, based on the multiple definitions of researchers about the concept in general.

2.2 *Organizational Reputation*

The concept of organizational reputation has attracted the attention of many researchers and academics, by noting its results on the work of commercial and service organizations, as the researchers indicated as an indicator for the future of organization (Abdel-Fattah, 2014). Researchers differed about the concept of organizational reputation, some of them saw it as a trait that comes from the advantages that are represented by the values that the beneficiaries realize (Walsh et al., 2004), whereas Watson and Kitchen (2015) see it as a strategic asset that is directly related to market values. While another point of view suggests that it represents a kind of comprehensive collective representation of the audio-visual developments that the organization will build over time, as well as its organizational identity through performance and behavior, to later reflect the organization's many perspectives. Mahdi and Kazem (2017) believe that organizational reputation is a set of relative values (originality, credibility, reliability, integrity, creativity, and social responsibility) that an individual feels about a company through the picture he has of it, and that it is one of its assets. As a result, it has a competitive advantage that helps it attract customers, earn their loyalty, and mitigate competitive threats and crises.

Watson and Kitchen (2015) indicate that the importance of organizational reputation is due to the fact that it is the thought of all stakeholders (internal and external) about the organization's reputation and internal identity, and it is a mixture and a time-spanned mixture of their evaluation (i.e. stakeholders) about the organization's overall performance, and its ability to face changes and adapt to it. Al-Sharifi (2019) stresses the importance of organizational reputation by highlighting its most important benefits, as it leads to: reducing costs, attracting customers, increasing profitability, setting differentiated prices, and setting competitive barriers to achieve relative stability of competitive future expectations.

According to Oncer & Yildiz (2012), the importance of organizational reputation is reflected in two important aspects: the aspect (external interest of the organization), which refers to the need to protect and improve the organization's reputation as a prerequisite for the continuation of its relationship with stakeholders and customers, which is reflected in its competitive performance; and the aspect (internal interest of the organization), which refers to the need to protect and improve the organization's reputation as a prerequisite for the continuation of While looking at the other side (internal interest), which portrays an acceptable and harmonious image of the organization's internal entity, which shapes its identity and works to improve its performance.

According to Walsh et al. (2009), organizational reputation is represented in five important dimensions that are summarized as: orientation to the customer (the beneficiary), the good employer, the organization's financial capacity, the quality of its products and services, and social responsibility. While Harrison identified eight important elements of reputation that are complementary among them, namely: leadership, financial strength, kind of management, customer focus, social responsibility, reliability, and the nature and quality of communications, adds to them. Market leadership, people quality, and ethical behavior are three more factors (Iwu-Egwuorwu, 2011; Al-Hawary et al., 2020a, 2020b; Al-Hawary & Al-Syasneh, 2020;).

This study will focus on five important dimensions of reputation as follows: **Innovation**: a process that results in unfamiliar ideas for a particular product or service (Al-Hawary, 2015; Al-Hawary & Aldaihani, 2016; AlTaweel & Al-Hawary, 2021; Fichman, 2001). It is the ability to present new ideas that, in turn, bring an organization a competitive advantage (Crossan & Apaydin, 2010; Almaazmi et al., 2020; Nuseir et al., 2021a, b). **Social responsibility**: Alhalalmeh et al. (2020) and Al-Hawary & Al-Khazaleh (2020) defined it as the commitment of business organizations to the community in which they operate, while it is seen as the customers' beliefs about the positive role of the organization toward society and the environment in general (Abu Zayyad et al., 2021; Al-Hawary, 2013; Al-Hawary & Obiadat, 2021; Alolayyan et al., 2018; Walsh et al., 2009). **The strength of the financial position**: It is what customers perceive about the organization in terms of its strength and profitability, and how they evaluate and use financial resources in a responsible manner (Shah et al., 2020, 2021; Walsh et al., 2009). **Product/service quality**: the quality of the service or product can help gain the competitive advantage of the organization, and to maintain a long-term relationship with customers and enhance their confidence in them (Abu Qaaud et al., 2011; Al-Hawary & Al-Menhaly, 2016; Al-Hawary & Al-Smeran, 2016; Alshurideh et al., 2017; Al-Hawary & Alhajri, 2020). **Attractiveness**: It means the ability of the organization to attract workers, investors and customers on an ongoing basis, and work to reduce costs and turnover of its employees. A reputable organization increases the desire of its employees to stay as well as customers (Al-Hawary & Harahsheh, 2014; Al-Hawary & Hussien, 2017; Ozbag, 2018).

2.3 *Total Quality Management and Organizational Reputation*

Universities are making attempts to improve their image and reputation by embracing the quality requirements mandated by the most well-known international university rankings, which reflect a significant component of higher education quality. Al-Taher concluded that the process of continuous improvement affects the achievement of competitive advantage, that the process of improvement and development is a shared responsibility among all university employees, that universities are comparing their overall performance with local and international universities, and that the university is working on task forces, conducting studies and analyses of competitive advantage to explore and benefit from strengths, weaknesses, opportunities, and threats, as well as benefiting from foreign expertise related to continuous improvement processes, and conducting studies and analyses of competitive advantage to explore and benefit from strengths, weaknesses, opportunities, and threats. According to Al-Drouqi & Al-Fitni (2021), total quality management is an integrated strategy that contributes significantly to the development of higher education institutions, that the administrative standard is the most important standard of total quality in education, and that the academic accreditation system is the most important mechanism for applying and ensuring quality.

Al-Sarhan (2021) found that public universities have a fundamental interest in two main elements: social responsibility and worker focus, in addition to a noticeable interest in the three TQM variables: organizational culture, focus on developing and improving the educational process, and adopting senior management. The term "total quality management" refers to the management of all aspects of a product's quality. By presenting results that the researcher summarized in the form of important points that dealt with how to pay attention and focus on aspects of total quality management with its principles, through its focus on the mechanisms and means that achieve the best results, (Abu Asba', 2020) confirmed that the five principles of total quality affect the achievement of a competitive advantage. Samir & Sobeih (2020) emphasized that any higher educational institution that wishes to invest and preserve its identity and brand (reputation) pays close attention to all aspects that affect its competitive reputation, such as advertising and promotion that best reflects its institutional image, strategic planning based on attracting components and elements that increase its reputation and raise its corporate brand competitively, and so on.

One of the most crucial variables influencing a student's choice of destination is the university's reputation (Ma, 2021; Al-Maroof et al., 2021). Dursun and Gumussoy

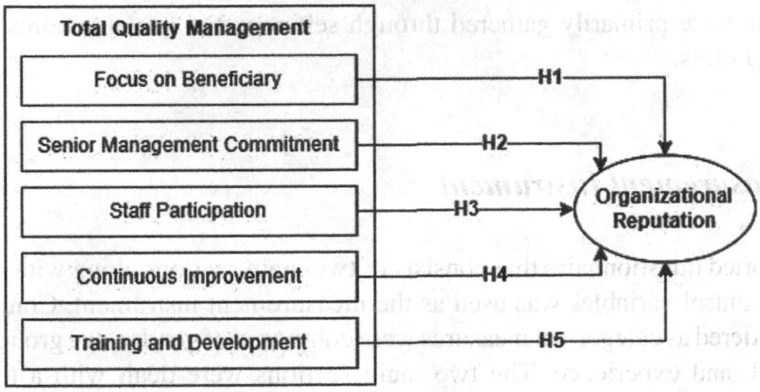

Fig. 1 Research model

(2021) underlined the impact of service quality and emotional dimension on university reputation, as well as demonstrating that personnel competency, academic leadership, and student orientation all have an impact on university reputation. Accordingly, the study hypothesis can be formulated as follows:

There is a statistically significant effect of total quality management with its dimensions (focus on the beneficiary, commitment of senior management, continuous improvement, staff participation, training and development) on the organizational reputation in Jordanian public universities.

3 Study Model

See Fig. 1.

4 Methodology

4.1 Population and Sample Selection

A qualitative method based on a questionnaire was used in this study for data collection and sample selection. The major aim of the study was to examine the impact of total quality management (TQM) on organizational reputation (OR). Therefore, it focused on the faculty members in five public universities, whose number is (5477) faculty members. As for the study sample, it consisted of a proportional stratified sample of (390) individuals, and after the questionnaire was distributed with (410), 390 were recovered, valid for the purposes of statistical analysis (Sekaran & Bougie,

2016). Data were primarily gathered through self-reported questionnaires creating by Google Forms.

4.2 Measurement Instrument

A self-reported questionnaire that consists of two main sections along with a section regarding control variables was used as the measurement instrument. Control variables considered as categorical measures were composed of gender, age group, educational level, and experience. The two main sections were dealt with a five-point Likert scale (from 1 = strongly disagree to 5 = strongly agree). The first section contained (20) items to measure total quality management based on (Al-Hawary & Abu-Laimon, 2013). These questions were distributed into dimensions as follows: four items dedicated for measuring focus on beneficiary, four items dedicated for measuring senior management commitment, four items dedicated for measuring staff participation, four items dedicated for measuring continuous improvement, and four items dedicated for measuring training and development. Whereas the second section included (12) items developed to measure organizational reputation according to what was pointed by (Abdel-Fattah, 2014).

5 Findings

5.1 Measurement Model Evaluation

This study was conducted structural equation modeling (SEM) to test hypotheses, which represents a contemporary statistical technique for testing and estimating the relationship between factors and variables (Wang & Rhemtulla, 2021). Accordingly, the reliability and validity of the constructs were tested using confirmatory factor analysis (CFA) through the statistical program AMOSv24. Table 1 summarizes the results of convergent and discriminant validity, as well the indicators of reliability.

Table 1 shows that the standard loading values for the individual items were within the domain (0.642–0.892), these values greater than the minimum retention of the elements based on their standard loads (Al-Lozi et al., 2018; Sung et al., 2019). Average variance extracted (AVE) is a summary indicator of the convergent validity of constructs that must be above 0.50 (Howard, 2018). The results indicate that the AVE values were greater than 0.50 for all constructs, thus the used measurement model has an appropriate convergent validity. Rimkeviciene et al. (2017) suggested the comparison approach as a way to deal with discriminant validity assessment in covariance-based SEM. This approaches based on comparing the values of maximum shared variance (MSV) with the values of AVE, as well as comparing the values of square root of AVE ($\sqrt{AVE}$) with the correlation between the rest of the structures.

Table 1 Results of validity and reliability tests

Constructs	1	2	3	4	5	6
1. FOB	**0.742**					
2. SMC	0.384	**0.741**				
3. SPA	0.332	0.392	**0.751**			
4. CIM	0.468	0.374	0.406	**0.759**		
5. TAD	0.501	0.425	0.481	0.443	**0.752**	
6. OR	0.637	0.587	0.662	0.681	0.597	**0.749**
VIF	1.814	1.662	1.876	1.036	1.574	–
Loadings range	0.642–0.793	0.673–0.824	0.705–0.811	0.735–0.771	0.681–0.831	0.662–0.892
AVE	0.551	0.549	0.564	0.576	0.566	0.562
MSV	0.415	0.394	0.497	0.502	0.445	0.381
Internal consistency	0.828	0.826	0.834	0.842	0.836	0.937
Composite reliability	0.830	0.828	0.837	0.844	0.838	0.939

Note FOB: focus on beneficiary, SMC: senior management commitment, SPA: staff participation, CIM: continuous improvement, TAD: training and development, OR: organizational reputation, Bold fonts in the table are the square root of average variance extracted

The results show that the values of MSV were smaller than the values of AVE, and that the values of $\sqrt{AVE}$ were higher than the correlation values among the rest of the constructs. Therefore, the measurement model used is characterized by discriminative validity. The internal consistency measured through Cronbach's Alpha coefficient (α) and compound reliability by McDonald's Omega coefficient (ω) was conducted as indicators to evaluate measurement model. The results listed in Table 1 demonstrated that both values of Cronbach's Alpha coefficient and McDonald's Omega coefficient were greater than 0.70, which is the lowest limit for judging on measurement reliability (De Leeuw et al., 2019).

5.2 *Structural Model*

The structural model illustrated no multicollinearity issue among predictor constructs because variance inflation factor (VIF) values are below the threshold of 5, as shown in Table 1 (Hair et al., 2017). This result is supported by the values of model fit indices shown in Figs. 1 and 2.

The results in Fig. 1 indicated that the chi-square to degrees of freedom (CMIN/DF) was 2.881, which is less than 3 the upper limit of this indicator. The values of the goodness of fit index (GFI), the comparative fit index (CFI), and the Tucker–Lewis index (TLI) were upper than the minimum accepted threshold of 0.90.

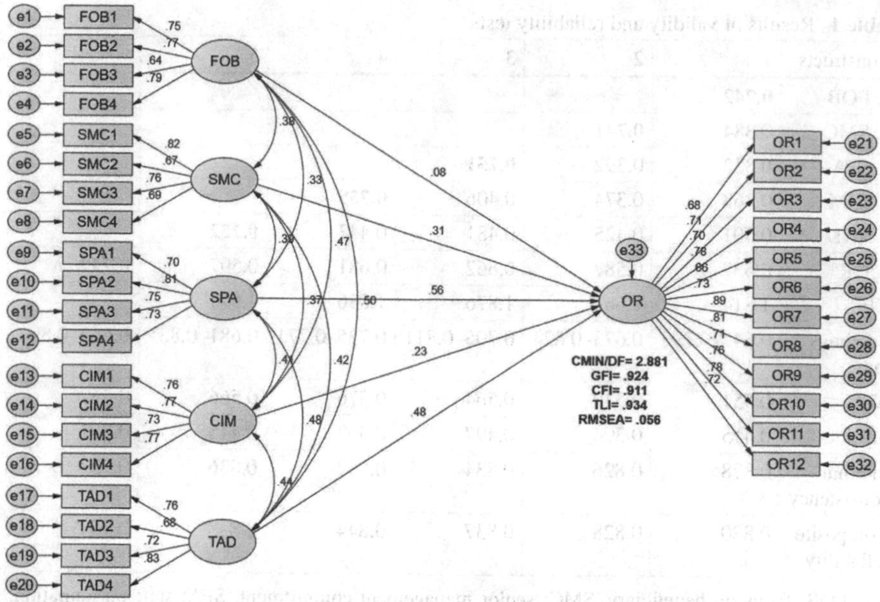

Fig. 2 SEM results of the TQM effect on OR

Moreover, the result of root mean square error of approximation (RMSEA) indicated to value 0.056, this value is a reasonable error of approximation because it is less than the higher limit of 0.08. Consequently, the structural model used in this study was recognized as a fit model for predicting the organizational reputation and generalization of its result (Ahmad et al., 2016; Shi et al., 2019). To verify the results of testing the study hypotheses, structural equation modeling (SEM) was used, the results of which are listed in Table 2.

The results demonstrated in Table 2 show that all total quality management dimensions had a positive impact relationship on organizational reputation except focus

Table 2 Hypothesis testing

Relation	Standard Beta	*t* value	*p* value
FOB → OR	0.083	1.750	0.081
SMC → OR	0.311*	3.297	0.017
SPA → OR	0.561***	9.364	0.000
CIM → OR	0.235*	2.226	0.046
TAC → OR	0.483***	7.529	0.000

Note FOB: focus on beneficiary, SMC: senior management commitment, SPA: staff participation, CIM: continuous improvement, TAD: training and development, OR: organizational reputation

$*p < 0.05$, $**p < 0.01$, $***p < 0.001$

on beneficiary ($\beta = 0.083$, $t = 1.750$, $p = 0.081$). Moreover, the results indicated that the highest impact was for staff participation ($\beta = 0.561$, $t = 9.364$, $p = 0.000$), followed by training and development ($\beta = 0.483$, $t = 7.529$, $p = 0.000$), then senior management commitment ($\beta = 0.311$, $t = 3.297$, $p = 0.017$), and finally the lowest impact was for continuous improvement ($\beta = 0.235$, $t = 2.226$, $p = 0.046$).

6 Discussion

The results showed that there is a positive and strong correlation between the TQM sub-variables and the organizational reputation variable, and that there is a statistically significant effect of TQM on the organizational reputation of Jordanian public universities, as it turns out that TQM was able to explain (25.1%) of the Variation in organizational reputation. This reflects universities' awareness and appreciation for the interest in applying the standards and practices of total quality to raise their performance, which will reflect on their reputation and raise their value. This is confirmed by a study (Obeid et al., 2021), which indicated the interest of the University of Jordan as one of the official universities in the application of total quality management, which contributed to raising the performance of its employees. As well as the study of Rabaa'a & Hamadat (2020), the result of which was that Jordanian private universities also implement total quality management at a high level.

The results showed that Jordanian public universities show great interest in their application of total quality management, but their interest is evident in some practices and not others. There is also a noticeable interest among Jordanian public universities with the participation of individuals in supporting the application of the practices of total quality, which is reflected positively on their overall performance. The findings also revealed that universities place a high value on subjecting faculty members to training courses that help them improve their academic performance, as well as providing financial support in the form of (financial rewards) to increase their research and scientific activity, which reflects the positive impact on faculty members' reputations at these institutions.

7 Recommendations

Based on the results of the study, researchers recommend managers and decision-makers in public universities to pay attention to the university management standard (higher leadership), which has the largest role in supporting and activating the application of total quality management, as the instability of administrative leaders and the inappropriateness of their scientific and administrative qualifications has an influential role on the level of management weakness. And activating the role of the accreditation and quality centers and supporting them with all the needs of qualified financial and human resources, which help them in achieving the effectiveness of the application of total quality management and conducting a periodic audit for the purposes of continuous improvement. The study also recommends activating the role of liaison officers in the centers, units and colleges of universities with the Accreditation and Quality Center more and supporting it from the technical and training aspects to make a periodic inspection for the purposes of showing weaknesses and shortcomings in the practices of total quality management.

References

Abdel-Fattah, I. S. H. (2014). Modeling the relationship between organizational reputation and competitiveness in the presence of the mental image as a modified variable: A field study on Port Said University. Journal of Financial and Business Research, 2, 551–618.

Asbaa, A., & Yahya, A. F. A. (2020). The role of total quality management in achieving competitive advantage in Yemeni private universities: An analytical study. *National University Journal, 14*, 203–237.

Abu Qaaud, F., Al-Shoura, M., & Al-Hawary, S. I. (2011). The impact of the service marketing mix in the service quality of health services from the viewpoint of patients in government hospitals in Amman "A Field study." *Abhath Al-Yarmouk, 27*(1B), 417–441.

Abu Zayyad, H. M., Obeidat, Z. M., Alshurideh, M. T., Abuhashesh, M., Maqableh, M., & Masa'deh, R. E. (2021). Corporate social responsibility and patronage intentions: the mediating effect of brand credibility. Journal of Marketing Communications, *27*(5), 510–533.

Ahmad, S., Zulkurnain, N., & Khairushalimi, F. (2016). Assessing the validity and reliability of a measurement model in structural equation modeling (SEM). *British Journal of Mathematics & Computer Science, 15*(3), 1–8. https://doi.org/10.9734/BJMCS/2016/25183

Al Kurdi, B., Alshurideh, M., & Salloum, S. A. (2020a). Investigating a theoretical framework for e-learning technology acceptance. *International Journal of Electrical and Computer Engineering (IJECE), 10*(6), 6484–6496.

Al Kurdi, B., Alshurideh, M., Salloum, S., Obeidat, Z., & Al-dweeri, R. (2020b). An empirical investigation into examination of factors influencing university students' behavior towards e-learning acceptance using SEM approach. *International Journal of Interactive Mobile Technologies, 14*(2), 19–24.

Al-Drouqi, S., Al-Toumi, B., & Al-Fatni, R. R. (2021). Developing the university education system in light of the comprehensive quality approach. Journal of Physical Education and Other Sciences, 7, 171–185.

Alhalalmeh, M. I., Almomani, H. M., Altarifi, S., Al- Quran, A. Z., Mohammad, A. A., & Al-Hawary, S. I. (2020). The nexus between corporate social responsibilty and organizational performance in Jordan: The mediating role of organizational commitment and organizational citizenship behavior. *Test Engineering and Management, 83*(July), 6391–6410.

AlHamad, M., Akour, I., Alshurideh, M., Al-Hamad, A., Kurdi, B., & Alzoubi, H. (2021). Predicting the intention to use google glass: A comparative approach using machine learning models and PLS-SEM. *International Journal of Data and Network Science, 5*(3), 311–320.

Al-Hamad, M., Mbaidin, H., AlHamad, A., Alshurideh, M., Kurdi, B., & Al-Hamad, N. (2021). Investigating students' behavioral intention to use mobile learning in higher education in UAE during Coronavirus-19 pandemic. *International Journal of Data and Network Science, 5*(3), 321–330.

Al-Hawary, S. I. (2010a). Factor underlying international students of Jordan Public Universities: Analytical study institutional factors. Al Manara for Research and Studies, *16*(1), 37–64. Retrieved from https://aabu.edu.jo/journal/manar/manarArt16110.html.

Al-Hawary, S. I. (2010b). Marketing public higher education: A social perspective. Al Manara for Research and Studies, *16*(4), 9–32. Retrieved from https://aabu.edu.jo/journal/manar/manarArt1648.html.

Al-Hawary, S. I. (2013). The role of perceived quality and satisfaction in explaining customer brand loyalty: Mobile phone service in Jordan. *International Journal of Business Innovation and Research, 7*(4), 393–413.

Al-Hawary, S. I. (2015). Human resource management practices as a success factor of knowledge management implementation at health care sector in Jordan. *International Journal of Business and Social Science, 6*(11/1), 83–98.

Al-Hawary, S. I. S., & Alhajri, T. M. S. (2020). Effect of electronic customer relationship management on customers' electronic satisfaction of communication companies in Kuwait. *Calitatea, 21*(175), 97–102.

Al-Hawary, S. I. S., & Obiadat, A. A. (2021). Does mobile marketing affect customer loyalty in Jordan? *International Journal of Business Excellence, 23*(2), 226–250.

Al-Hawary, S. I. S., Mohammad, A. S., Al-Syasneh, M. S., Qandah, M. S. F., & Alhajri, T. M. S. (2020a). Organisational learning capabilities of the commercial banks in Jordan: Do electronic human resources management practices matter? *International Journal of Learning and Intellectual Capital, 17*(3), 242–266.

Al-Hawary, S. I., & Abu-Laimon, A. A. (2013). The impact of TQM practices on service quality in cellular communication companies in Jordan. *International Journal of Productivity and Quality Management, 11*(4), 446–474.

Al-Hawary, S. I., & Aldaihani, F. M. (2016). Customer relationship management and innovation capabilities of Kuwait airways. *International Journal of Academic Research in Economics and Management Sciences, 5*(4), 201–226.

Al-Hawary, S. I., & Al-Khazaleh A, M. (2020b). The mediating role of corporate image on the relationship between corporate social responsibility and customer retention. Test Engineering and Management, *83*(516), 29976–29993.

Al-Hawary, S. I., & Al-Menhaly, S. (2016). The quality of e-government services and its role on achieving beneficiaries satisfaction. *Global Journal of Management and Business Research: A Administration and Management, 16*(11), 1–11.

Al-Hawary, S. I., & Al-Smeran, W. (2017). Impact of electronic service quality on customers satisfaction of Islamic Banks in Jordan. *International Journal of Academic Research in Accounting, Finance and Management Sciences, 7*(1), 170–188.

Al-Hawary, S. I., & Al-Syasneh, M. S. (2020). Impact of dynamic strategic capabilities on strategic entrepreneurship in presence of outsourcing of five stars hotels in Jordan. Business: Theory and Practice, *21*(2), 578–587.

Al-Hawary, S. I., & Batayneh, A. M. (2010). The effect of marketing communication tools on non-Jordanian students' choice of Jordanian Public Universities: A field study. *International Management Review, 6*(2), 90–99.

Al-Hawary, S. I., & Harahsheh, S. (2014). Factors affecting Jordanian consumer loyalty toward cellular phone brand. *International Journal of Economics and Business Research, 7*(3), 349–375.

Al-Hawary, S. I., & Hussien, A. J. (2017). The impact of electronic banking services on the customers loyalty of commercial banks in Jordan. *International Journal of Academic Research in Accounting, Finance and Management Sciences, 7*(1), 50–63.

Al-Hawary, S. I., Al-Qudah, K., Abutayeh, P., Abutayeh, S., & Al-Zyadat, D. (2013). The impact of internal marketing on employee's job satisfaction of commercial banks in Jordan. *Interdisciplinary Journal of Contemporary Research in Business, 4*(9), 811–826.

Al-Hawary, S. I., Batayneh, A. M., Mohammad, A. A., & Alsarahni, A. H. (2017). Supply chain flexibility aspects and their impact on customers satisfaction of pharmaceutical industry in Jordan. *International Journal of Business Performance and Supply Chain Modelling, 9*(4), 326–343.

Al-Lozi, M. S., Almomani, R. Z. Q., & Al-Hawary, S. I. S. (2018). Talent management strategies as a critical success factor for effectiveness of human resources information systems in

commercial banks working in Jordan. *Global Journal of Management and Business Research: A Administration and Management, 18*(1), 30–43.
Almaazmi, J., Alshurideh, M., Al Kurdi, B., & Salloum, S. A. (2020). The effect of digital transformation on product innovation: a critical review. In *International Conference on Advanced Intelligent Systems and Informatics* (pp. 731–741). Cham: Springer.
Al-Maroof, R. S., Alshurideh, M. T., Salloum, S. A., AlHamad, A. Q. M., & Gaber, T. (2021). Acceptance of Google Meet during the spread of Coronavirus by Arab university students. In Informatic, *8*(2), 1–17. Multidisciplinary Digital Publishing Institute.
Alolayyan, M., Al-Hawary, S. I., Mohammad, A. A., & Al-Nady, B. A. (2018). Banking service quality provided by commercial banks and customer satisfaction. A structural equation modelling approaches. International Journal of Productivity and Quality Management, *24*(4), 543–565.
Al-Sarhan, A. F. (2021). The impact of the applications of total quality standards and academic accreditation on improving individual and institutional performance of Jordanian public universities. *Journal of Islamic Management and Leadership, 6*(1), 14–38.
Alshamsi, A., Alshurideh, M., Al Kurdi, B., & Salloum, S. A. (2020). The influence of service quality on customer retention: a systematic review in the higher education. In *International Conference on Advanced Intelligent Systems and Informatics* (pp. 404–416). Cham: Springer.
Alsharari, N. M., & Alshurideh, M. T. (2020). Student retention in higher education: the role of creativity, emotional intelligence and learner autonomy. International Journal of Educational Management. International Journal of Educational Management *35*(1), 233–247.
Al-Sharifi, A. K. (2019). The role of strategic orientation in strategic reputation and its impact on competitive pressures: An exploratory study on mobile telecommunications companies in Iraq (Zain Iraq, Aseel, Umniah, Korek). *Economic Journal of Administrative Sciences, 25*(113), 191–220.
Alshurideh, M. T., Kurdi, B. A., AlHamad, A. Q., Salloum, S. A., Alkurdi, S., Dehghan, A., & Masa'deh, R. E. (2021). Factors affecting the use of smart mobile examination platforms by universities' postgraduate students during the COVID 19 pandemic: an empirical study. In Informatics, *8*(2), 1–21. Multidisciplinary Digital Publishing Institute.
Alshurideh, M., Al Kurdi, B., & Salloum, S. A. (2019). Examining the main mobile learning system drivers' effects: A mix empirical examination of both the expectation-confirmation model (ECM) and the technology acceptance model (TAM). In *International Conference on Advanced Intelligent Systems and Informatics* (pp. 406–417). Cham: Springer.
Alshurideh, M., Al Kurdi, B., Salloum, S. A., Arpaci, I., & Al-Emran, M. (2020). Predicting the actual use of m-learning systems: a comparative approach using PLS-SEM and machine learning algorithms. *Interactive Learning Environments*, 1–15.
Alshurideh, M., Al-Hawary, S. I., Batayneh, A. M., Mohammad, A., & Al-Kurdi, B. (2017). The impact of Islamic banks' service quality perception on Jordanian customers loyalty. *Journal of Management Research, 9*(2), 139–159.
Al-Tai, Y. J. S., & Al-Abadi, H. F. D. (2009). Customer relationship management (1st floor). Jordan, Amman: Al-Warraq Publishing and Distribution Corporation.
Altamony, H., Masa'deh, R., Alshurideh, M., Obeidat, B. (2012). Information systems for competitive advantage: Implementation of an organisational strategic management process. *Innovation and sustainable competitive advantage: From regional development to world economies* (pp. 583–592).
AlTaweel, I. R., & Al-Hawary, S. I. (2021). The mediating role of innovation capability on the relationship between strategic agility and organizational performance. *Sustainability, 13*(14), 7564.
Amarneh, B. M., Alshurideh, M. T., Al Kurdi, B. H., & Obeidat, Z. (2021). The impact of COVID-19 on e-learning: Advantages and challenges. In *The International Conference on Artificial Intelligence and Computer Vision* (pp. 75–89). Cham: Springer.
Batayneh, R. M. A., Taleb, N., Said, R. A., Alshurideh, M. T., Ghazal, T. M., & Alzoubi, H. M. (2021). IT governance framework and smart services integration for future development of Dubai infrastructure utilizing AI and big data, its reflection on the citizens standard of living. In *The*

International Conference on Artificial Intelligence and Computer Vision (pp. 235–247). Cham: Springer.

Crossan, M. M., & Apaydin, M. (2010). A multi-dimensional framework of organizational innovation: A systematic review of the literature. *Journal of Management Studies, 47*(6), 1154–1191.

de Leeuw, E., Hox, J., Silber, H., Struminskaya, B., & Vis, C. (2019). Development of an international survey attitude scale: Measurement equivalence, reliability, and predictive validity. *Measurement Instruments for the Social Sciences, 1*(1), 9. https://doi.org/10.1186/s42409-019-0012-x

Dursun, O., & Gumussoy, C. A. (2021). The effects of quality of services and emotional appeal on university reputation: Stakeholders' view. *Quality Assurance in Education., 29*(2/3), 166–182.

Fichman, R. G. (2001). The role of aggregation in the measurement of IT-related organizational innovation. MIS quarterly, 427–455. DOI: https://doi.org/10.2307/3250990.

Ghoneim, A. M. (2018). *Total Quality Management (I 2)*. Modern Library for Publishing and Distribution.

Hair, J. F., Babin, B. J., & Krey, N. (2017). Covariance-based structural equation modeling in the journal of advertising: Review and recommendations. *Journal of Advertising, 46*(1), 163–177. https://doi.org/10.1080/00913367.2017.1281777

Hashem, A. J. (2021). The prevalence of the culture of total quality among faculty members at the University of Kufa. *Rawafed Journal of Studies and Scientific Research in the Social Sciences and Humanities., 5*(1), 36–59.

Howard, M. C. (2018). The convergent validity and nomological net of two methods to measure retroactive influences. *Psychology of Consciousness: Theory, Research, and Practice, 5*(3), 324–337. https://doi.org/10.1037/cns0000149

Iwu-Egwuonwu, R. (2011). Corporate reputation and firm performance: Empirical literature evidence. *International Journal of Business & Management, 6*(4), 197–200.

Krajewski, L. J., Ritzman, L. P., & Malhotra, M. K. (2013). *Operations Management Processes and Supply Chains*. Pearson Education Limited.

Leo, S., Alsharari, N. M., Abbas, J., & Alshurideh, M. T. (2021). From offline to online learning: A qualitative study of challenges and opportunities as a response to the COVID-19 pandemic in the UAE higher education context. In *The Effect of Coronavirus Disease (COVID-19) on Business Intelligence* (pp. 203–217). Cham: Springer.

Ma, A. S. (2021). Assessing the effects of university reputation & city image on international student destination choice: Evidence from a flagship university in Taipei. *Education & Urban Society* (pp. 1–18).

Mahdi, A. M., & Kazem, S. M. (2017). Organizational commitment and its impact on building the organization's reputation—an analytical study of the opinions of a sample of workers in the Holy Karbala Health Department. *Karbala University Journal, 15*(2), 291–273.

Metabis, A., & Al-Hawary, S. I. (2013). The Impact of Internal Marketing Practices on Services Quality of Commercial Banks in Jordan. *International Journal of Services and Operations Management, 15*(3), 313–337.

Mohammad, A. A., Alshura, M. S., Al-Hawary, S. I. S., Al-Syasneh, M. S., & Alhajri, T. M. (2020). The influence of Internal Marketing Practices on the employees' intention to leave: A study of the private hospitals in Jordan. *International Journal of Advanced Science and Technology, 29*(5), 1174–1189.

Nazir, J., Rahaman, S., Chunawala, S., Ahmed, G., Alzoubi, H., Alshurideh, M., & AlHamad, A. (2022) Perceived factors affecting students academic performance. Academy of Strategic Management Journal, *21*(Special Issue 4), 1–15.

Nuseir, M. T., Al Kurdi, B. H., Alshurideh, M. T., & Alzoubi, H. M. (2021a). Gender discrimination at workplace: Do artificial intelligence (AI) and machine learning (ML) have opinions about it. In *The International Conference on Artificial Intelligence and Computer Vision* (pp. 301–316). Cham: Springer.

Nuseir, M. T., Aljumah, A., & Alshurideh, M. T. (2021b). How the business intelligence in the new startup performance in UAE during COVID-19: The mediating role of innovativeness. The Effect of Coronavirus Disease (COVID-19) on Business Intelligence, *334*, 63–79.

Obaid, H. S. A., Al-Mousawi, M. A., Ali, A.-A., & M. (2021). The impact of academic accreditation on the application of total quality at the University of Jordan. *Journal of Educational Studies and Research, 1*(1), 221–251.

Oncer, A. Z., & Ylldiz, M. L. (2012). The impact of ethical climate on relationship between corporate reputation and organizational identification. Social and Behavioral Sciences, *58*, 714–723.

Ozbag, G. K. (2018). The effects of the perceived reputation on counterproductive work behaviour. *Uluslararası Turizm, Ekonomi Ve İşletme Bilimleri Dergisi, 2*(2), 140–148.

Rabaa'a, O. A. R., & Hamadat, M. H. (2020). The degree of application of total quality management standards in Jordanian private universities from the viewpoint of faculty members. King Khalid University Journal of Educational Sciences, *31*(1), 40–63.

Rimkeviciene, J., Hawgood, J., O'Gorman, J., & De Leo, D. (2017). Construct validity of the acquired capability for suicide scale: Factor structure, convergent and discriminant validity. *Journal of Psychopathology and Behavioral Assessment, 39*(2), 291–302. https://doi.org/10.1007/s10862-016-9576-4

Samir, R., & Sobeih, Y. (2020). Institutional Brand Reputation Management within the Higher Education Institutes. (RAIS) Research Association for Interdisciplinary Studies. Conference Proceedings. 319–322. DOI: https://doi.org/10.5281/zenodo.3909991.

Sekaran, U., & Bougie, R. (2016). *Research methods for business: A skill-building approach* (7th ed.). Wiley.

Shah, S. F., Alshurideh, M. T., Al-Dmour, A., & Al-Dmour, R. (2021). Understanding the influences of cognitive biases on financial decision making during normal and COVID-19 pandemic situation in the United Arab Emirates. The Effect of Coronavirus Disease (COVID-19) on Business Intelligence, *334*, 257–274.

Shah, S. F., Alshurideh, M., Al Kurdi, B., & Salloum, S. A. (2020). The impact of the behavioral factors on investment decision-making: a systemic review on financial institutions. In *International Conference on Advanced Intelligent Systems and Informatics* (pp. 100–112). Cham: Springer.

Shi, D., Lee, T., & Maydeu-Olivares, A. (2019). Understanding the model size effect on SEM fit indices. *Educational and Psychological Measurement, 79*(2), 310–334. https://doi.org/10.1177/0013164418783530

Sultan, R. A., Alqallaf, A. K., Alzarooni, S. A., Alrahma, N. H., AlAli, M. A., & Alshurideh, M. T. (2021). How students influence faculty satisfaction with online courses and do the age of faculty matter. In *The International Conference on Artificial Intelligence and Computer Vision* (pp. 823–837). Cham: Springer.

Sung, K.-S., Yi, Y. G., & Shin, H.-I. (2019). Reliability and validity of knee extensor strength measurements using a portable dynamometer anchoring system in a supine position. *BMC Musculoskeletal Disorders, 20*(1), 1–8. https://doi.org/10.1186/s12891-019-2703-0

Walsh, G., & Wiedmann, K. P. (2004). A conceptualization of corporate reputation in Germany: An evaluation and extension of the RQ. *Corporate Reputation Review, 6*(4), 304–312.

Walsh, G., Mitchell, V. W., Jackson, P. R., & Beatty, S. E. (2009). Examining the antecedents and consequences of corporate reputation: A customer perspective. *British Journal of Management, 20*(2), 187–203.

Wang, Y. A., & Rhemtulla, M. (2021). Power analysis for parameter estimation in structural equation modeling: A discussion and tutorial. *Advances in Methods and Practices in Psychological Science, 4*(1), 1–17. https://doi.org/10.1177/2515245920918253

Watson, T, & Kitchen P. J. (2015). Reputation management: Corporate image and communication. In Moutinho, L., & Southern, G. (Eds.), *Strategic Marketing Management: A Process-Based Approach* (Chap. 13). Andover, Hampshire: Cengage Learning.

The Impact of Work-Life Balance on Organizational Commitment

Sulieman Ibraheem Shelash Al-Hawary, Maali M. Al-mzary, Ayat Mohammad, Nancy Abdullah Shamaileh, Anber Abraheem Shlash Mohammad, Muhammad Turki Alshurideh, Barween Al Kurdi, Kamel Mohammad Al-hawajreh, and Abdullah Ibrahim Mohammad

Abstract The study aimed to explore the effect of work-life balance on the dimensions of organizational commitment (affective, continuance, and normative commitment) by collecting data by means of a questionnaire distributed to a sample of workers in one of the big food companies. The results showed a positive significant

S. I. S. Al-Hawary (✉)
Department of Business Administration, School of Business, Al Al-Bayt University, P.O.BOX 130040, Mafraq 25113, Jordan
e-mail: dr_sliman73@aabu.edu.jo; dr_sliman@yahoo.com

M. M. Al-mzary
Department of Applied Science, Al-Balqa Applied University, Irbid college, As-Salt, Jordan

A. Mohammad
Business and Finance Faculty, the World Islamic Science and Education University (WISE), Postal Code 11947, P.O Box 1101, Amman, Jordan

N. A. Shamaileh
Department of Business Administration, School of Business, Al Al-Bayt University, P.O.BOX 130040, Mafraq 25113, Jordan

A. A. S. Mohammad
Marketing Department, Faculty of Administrative and Financial Sciences, Petra University, B.O.Box: 961343, Amman 11196, Jordan

M. T. Alshurideh
Department of Marketing, School of Business, The University of Jordan, Amman 11942, Jordan
e-mail: m.alshurideh@ju.edu.jo

Department of Management, College of Business, University of Sharjah, 27272 Sharjah, United Arab Emirates

B. Al Kurdi
Department of Marketing, Faculty of Economics and Administrative Sciences, The Hashemite University, Zarqa, Jordan
e-mail: barween@hu.edu.jo

K. M. Al-hawajreh
Business Faculty, Mu'tah University, Karak, Jordan

M. Alshurideh et al. (eds.), *The Effect of Information Technology on Business and Marketing Intelligence Systems*, Studies in Computational Intelligence 1056,
https://doi.org/10.1007/978-3-031-12382-5_65

effect of work-life balance on the affective and normative commitment, and a negative effect of work-life balance on continuance commitment. These results agreed with some previous studies and differed with others. The study recommended the need to pay more attention to the balance of work because it improves employee emotional attachment and obligation to the organization and reduces their assessment of their relationship with the organization through the costs they incur in the case of leaving the organization. Researchers are recommended to study work-life balance by focusing on its dimensions and not as a whole variable.

Keywords Work-life balance · Affective commitment · Continuance commitment · And normative commitment · Food industry

1 Introduction

Work-life balance (WLB) is of interest to researchers and organizations because of its impact on the level of organizational commitment of employees. It was found that this concept means providing a suitable work environment that balances the activities of work life and the activities of family life. One of the most important effects of WLB is that it leads to reducing employee intention to leave (Agha et al., 2021; Allen & Meyer, 1990) and improving organizational commitment (Alketbi et al., 2020; Wardana et al., 2020). In particular, a higher level of WLB improves employee emotional and normative adherence (Al Kurdi et al., 2020; Khalid & Ibrahim, 2018), and lowers the level of continuance commitment (Alameeri et al., 2020; Dávila, 2019; Shabir & Gani, 2020). Nevertheless, the results reached by the researchers are mixed due to the different samples or sectors that were covered or countries, so the current study was conducted to provide an applied addition and contribute to the generalization of the previous results.

2 Literature Review and Research Hypotheses

2.1 *Work-Life Balance*

WLB has been defined in several ways. First, WLB is a state of simultaneous satisfaction with work and life (Clark, 2000). Second, WLB is a situation of conflict absence between work and non-work requirements (Kim, 2014). Third, WLB is a case of resources adequacy by which an employee is able to participate in both work and family life in an effective manner (Voydanoff, 2005). Fourth, WLB refers to a

A. I. Mohammad
Department of Basic Scientific Sciences, Al-Huson University College, Al-Balqa Applied University, As-Salt, Jordan

situation in which time is split between work and personal life resting on priorities (Rigby & O'Brien-Smith, 2010). Fifth, WLB is a state between employee satisfaction and employee organizational commitment (AlShehhi et al., 2020; Peng, 2018). In fact, WLB is divided into three major types: time balance, content balance, and commitment balance. The first type refers to using time to the life and work equally. The second one means that refers to an individual pleasure in both work and life. Finally, commitment balance signifies showing psychological commitment to the roles at the same degree (Tayfun & Çatir, 2014). WLB can be explained using five models: the segmentation model, the spillover model, the compensation model, the instrumental model, and the conflict model. The segmentation model assumes that work life is different from non-work life since both of them represent life-separated aspects, while the spillover model advocates that one of these aspects have a positive or negative influence on the other one. The compensation model hypothesized that one may present what the other does not offer, whereas the instrumental model proposes that the activities of one lead to success in the other, and the conflict model assumes that these two domains conflict due to the high requirements for each of them (Alsuwaidi et al., 2020; Guest, 2002).

2.2 *Organizational Commitment*

From Allen and Meyer's (1990) view of point, OC is a psychological statethat associate an employee to an organization and therefore reduce his or her turnover intention. Hence, the concept was deemed as a multidimensional construct used to evaluate employee attitudes toward two key work-related aspects, which are employee membership willingness and employee inclination for organizational goals (Al-Hawary & Alajmi, 2017; Al-Hawary & Mohammed, 2017; Alhalalmeh et al., 2020; Mohammad et al., 2020; Meyer & Herscovitch, 2001). Based on such a definition, OC was introduced as an employee attachment with an organization as presented by employee intention to stay and employee extra effort to achieve organizational goals (Meyer et al., 2002; Al Shebli et al., 2021). According to Meyer and Allen, commitment to an organization reflects employee need, desire and obligation to maintain his or her membership in that organization. Therefore, OC has been conceptualized in terms of three dimensions: affective commitment, continuance commitment, and normative commitment (Allen & Meyer, 1990; Allozi et al., 2022). First, affective commitment is a state of positive work-related behaviors such as organizational citizenship behavior (Alshurideh et al., 2015; Meyer et al., 2002). It represents employee wish to stay with an organization due to personally perceived causes such as goal congruence, which a state in which an employee tries to link his personal goals with an organization's goals (Al-bawaia et al., 2022). For Dávila (2019), affective commitment represents employee emotional attachment, identification, and involvement in the organization. Such a dimension can be assessed using numerous statement s such as the one formulated by Marmol (2019), which was "I would be very happy to spend the rest of my career with this [organization]" (P. 115). Second, continuance commitment occurs when an employee remains with an organization due to a key reason, which is the need that tied the employee to the organization. In such a situation, the employee is aware of the costs of leaving the organization (Alshurideh, 2016; Dávila, 2019). In this regard, a need emerges based on the lack alternatives, leaving-associated costs, or in ability to transfer skills to a different organization (Meyer et al, 2002; Alkitbi et al., 2020). It is a result of a comparison between gains and losses in case of work leave (Alsharari & Alshurideh, 2020; Tayfun & Çatir, 2014). In other words, it is the cost incurred by the individual if he decides to leave the organization (Alshamsi et al., 2020; Koyuncu & Demirhan, 2021). An example of the items used to gauge continuance commitment is "It would be very hard for me to leave my [organization] right now, even if I wanted to" (Marmol, 2019, P. 115). Third, in Meyer and Allen's (1997) words, normative commitment refers to an employee obligation to continue employment. Likewise, Dávila (2019) defined normative commitment as employee feeling of obligation to carry on his or her work in the organization. It refers to the individual's sense of moral responsibility towards the organization (Alshurideh, 2022; Alshurideh et al., 2016; Koyuncu & Demirhan, 2021). Examples of the items that can be used to assess normative commitment include "One of the major reasons I continue to work for this [organization] is that I believe that loyalty is important and therefore feel a moral obligation to remain" (Marmol, 2019, P. 115). In short, it can be said that the emotional attachment to the organization is the reason behind the affective commitment, while the continuance commitment is related to the costs, and therefore its cause is the necessity that necessitates staying in the organization. The reason behind the normative commitment is the sense of responsibility towards the organization.

2.3 *WLB and Affective Commitment*

Many previous works on the effect of WLB on affective commitment revealed that WLB has a significant positive effect on affective commitment. Investigating such an effect using a sample encompassed of nurses, Tayfun and Çatir (2014) confirmed that WLB exerted a positive influence on affective commitment. Similarly, Shabir and Gani (2020) showed a significant positive relationship between WLB and affective commitment. Gathering data from Korean workers, Kim (2014) showed a positive association between WLB and affective commitment, which in turn affects employee in-role performance. On the other hand, the results of Al Momani (2017) indicated that WLB is negatively associated to affective commitment. Dávila (2019) studied the relationship between WLB practices and OC among academics at a Chilean university and established a positive bond between WLB in terms of family integration practices and affective commitment. Based on these results, the following hypothesis was presumed:

H1: There is a significant relationship between WLB and affective commitment.

2.4 WLB and Continuance Commitment

Generally, the effect of WLB on OC is well documented in the literature (Hariniet al., 2019; Malone, 2010; Sakthivel & Jayakrishnan, 2012; Wardana et al., 2020). However, prior works on the effect of WLB and continuance commitment showed different results. Khalid and Ibrahim (2018) tested the relationship between WLB and OC of radio journalists in Nigeria and concluded that WLB is a significant predictor of continuance commitment. Al Momani (2017) examined the mediating part of OC in the linkage between WLB and employee intention to leave using a sample consisted of women employees from a state hospital in Jordan and found that WLB had no significant effect on continuance commitment. Tayfun and Çatir (2014) examined the relationship between WLB and OC using a sample of nurses in Turkey and found that WLB had no significant effect on continuance commitment. Furthermore, Shabir and Gani (2020) investigated the relationship between WLB and OC among women employees in healthcare industry in India and found that WLB had a negative linkage with continuance commitment. For Dávila (2019), WBL, i.e., economic support and permissions for family responsibilities, is significantly related to continuance commitment. In order to explore the effect of WLB on continuance commitment using the current data, the following hypothesis was introduced:

H2: There is a significant relationship between WLB and continuance commitment.

2.5 WLB and Normative Commitment

For many scholars, WLB is positively related to normative commitment. Examples of previous studies in this regard include Shabir and Gani (2020) who investigated such a relationship in healthcare sector in India, as well as Tayfun and Çatir (2014) who used a sample of nurses working in a state hospital in Turkey. Explicitly, Dávila (2019) pointed out a significant positive relationship between WLB in terms of family integration and normative commitment. In their study on WLB and OC of radio journalists in Nigeria, Khalid and Ibrahim (2018) found a significant relationship between WLB and normative commitment. However, a study carried out by Al Momani (2017) on the mediating role of OC on the effect of WLB on employee intention to leave showed that WLB had a significant negative effect on normative commitment. Consequently, the following hypothesis was postulated:

H3: There is a significant relationship between WLB and normative commitment.

3 Research Methodology

3.1 Research Sample and Data Collection

The population of the study consisted of full-employment male and female employees in one large service company in food industry. Data were collected using a questionnaire administered to a randomly selected sample encompassed 250 employees, 173 questionnaires were returned, from which 11 were excluded due to incomplete responses. Therefore, the final number of the questionnaires used for data analysis purpose were 162 questionnaires.

3.2 Research Instrument

The questionnaire contains 20 items anchored using a five-pint Liker scale, in which 1 refers to strongly disagree and 5 signifies strongly agree. Employees were asked to report their agreement degree on items related to their assessment of WLB and OC statements. WLB was measured using 5 statements adopted from Marmol (2019, P. 115). Those statements were related to workload, work efficiency, family relationships, personal and self-care being, and personal wellness. OC dimensions are measured using 15 items distributed equally on affective commitment (5 items), continuance commitment (5 items), and normative commitment (5 items). These items were adapted from Dávila (2019, P. 18) who carried out a study on the relationship between WLB practices and OC among university academics, therefore, the item were reworded to suite the organization used in the current study.

3.3 Research Conceptual Model

The conceptual model of the research as shown in Fig. 1 displays the three suggested hypotheses of the research, in which WLB is assumed to exert a significant positive effect on affective commitment (AC) as stated in the first hypothesis (H1). WLB is hypothesized to show a significant positive effect on continuance commitment (CC) as indicated in the second hypothesis (H2). As well, WLB is presumed to display a significant positive effect on normative commitment (NC) as identified in the third hypothesis (H3).

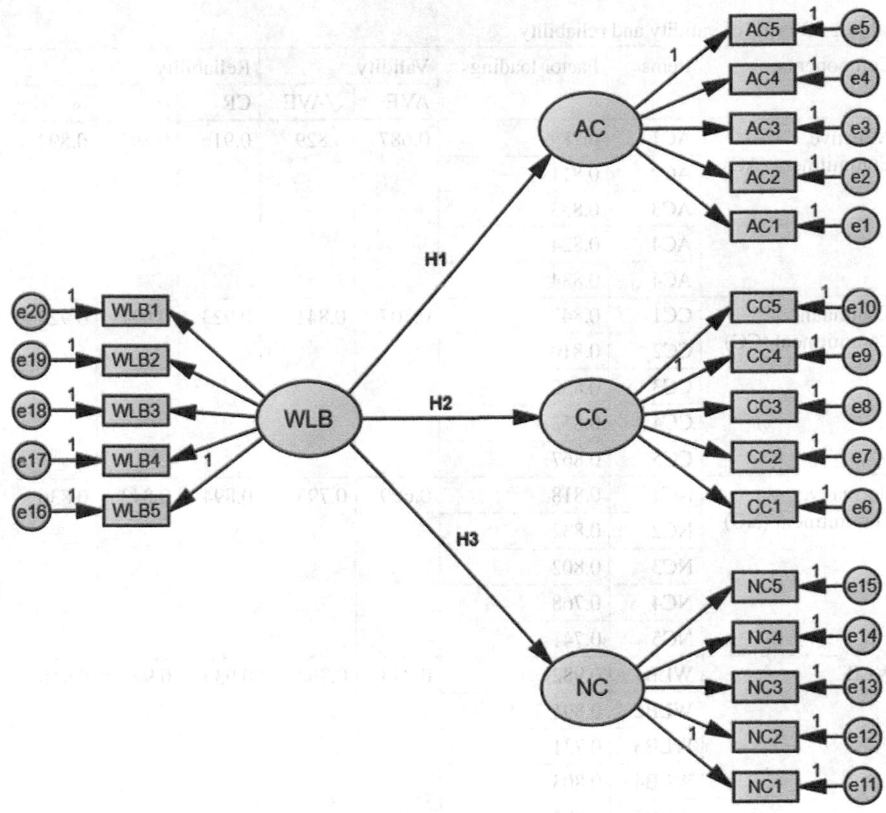

Fig. 1 Research conceptual model

3.4 *Multicollinearity, Validity and Reliability*

The results of multicollinearity statistics as assessed by variance inflation factor (VIF) and tolerance specify that the current data is free from multicollinearity since VIF values were less than 10and tolerance values were greater than 0.10 (Al-Hawary & Alhajri, 2020; Al-Nady et al., 2013; Altarifi et al., 2015).Exploratory factor analysis (EFA) was carried out first using IBM SPSS 24.0 in order to determine indicator loadings and to test validity in terms of convergent and discriminant values. Convergent validity was evaluated by the average variance extracted (AVE). Its value should be more than 0.50 (Al- Quran et al., 2020; Al-Hawary et al., 2020; AlTaweel & Al-Hawary, 2021), discriminant validity was evaluated using the square root of AVE values. Its value should be upper than the values of the correlation coefficients between any pair of the independent variables. The results in Table 1 show acceptable values of both convergent and discriminant validity. On the other hand, reliability was tested using Cronbach's alpha with a cut-off value of 0.70 (Mohammad et al., 2020), in addition to McDonald's Omega (ω) with a similar threshold value of 0.70

Table 1 Results of validity and reliability

Components	Items	Factor loadings	Validity		Reliability		
			AVE	√AVE	CR	α	ω
Affective Commitment (AC)	AC1	0.789	0.687	0.829	0.916	0.893	0.892
	AC2	0.811					
	AC3	0.833					
	AC4	0.824					
	AC4	0.884					
Continuance Commitment (CC)	CC1	0.842	0.707	0.841	0.923	0.931	0.929
	CC2	0.810					
	CC3	0.832					
	CC4	0.852					
	CC5	0.867					
Normative Commitment (NC)	NC1	0.818	0.629	0.793	0.894	0.843	0.836
	NC2	0.832					
	NC3	0.802					
	NC4	0.768					
	NC5	0.741					
WLB	WLB1	0.982	0.743	0.862	0.935	0.927	0.918
	WLB2	0.891					
	WLB3	0.731					
	WLB4	0.803					
	WLB5	0.883					

(Al-Hawary & Al-Syasneh, 2020)and composite reliability (CR) with a value greater than 0.70 (Mohammad, 2017; Al-Lozi et al., 2018; Al-Hawary & Obiadat, 2021). The results in Table 1 involve satisfactory values of reliability indexes. CR values ranged from 0.894 to 0.935, Cronbach's alpha coefficients ranged between 0.843 and 0.929, while McDonald's Omega values were from 0.836 to 0.929. Overall, both validity and reliability were assured.

3.5 *Descriptive Statistics and Correlation Matrix*

Means (M) and standard deviations (SD) were used to describe the degrees of research variables. It can be noted from Table 2 that the degrees of affective commitment, continuance commitment is moderate, and normative commitment were moderate (M = 3.56, SD = 0.841, M = 2.89, SD = 0.792, M = 2.95, SD = 0.852 respectively). The degree of WLB was also moderate (M = 3.48, SD = 0.951). In terms of Pearson correlation coefficients, the results in Table 2 illustrate that the research variables are

Table 2 Descriptive statistics and correlation matrix

Constructs	Mean	SD	(1)	(2)	(3)	(4)
(1)Affective commitment	3.56	0.841	1	–	–	–
(2)Continuance commitment	2.89	0.792	0.441**	1	–	–
(3)Normative commitment	2.95	0.852	0.452**	0.487**	1	–
(4)WLB	3.48	0.951	0.495**	0.396**	0.472**	1

** Correlation is significant at the 0.01 level (2-tailed)

Table 3 Results of model fit

Index	Measurement model	Structural model	Criteria	Result
CMIN/DF	2.381	2.381	<3.00	Accepted
GFI	0.924	0.924	>0.90	Accepted
CFI	0.964	0.964	>0.90	Accepted
RMSEA	0.072	0.072	<0.08	Accepted

significantly correlated, i.e., affective commitment had a significant correlation with both continuance commitment (r = 0.441) and normative commitment (r = 0.452). Affective commitment, continuance commitment, and normative commitment had significant correlations with WLB (r = 0.495, 0.487, 0.396 respectively).

3.6 Model Fit

Chi-square-to-degree of freedom ratio (CMIN/DF), the Goodness of Fit Index (GFI), the Comparative Fit Index (CFI), and the Root Mean Squared Approximation of Error (RMSEA) are used to test model fit. The results in Table 3 asserted that the fit indices of both measurement and structural models are accepted. CMIN/DF = 2.381, GFI = 0.924, CFI = 0.964 and RMSEA = 0.072. CMIN/DF is less than 3, GFI and CFI are greater than 0.90, and RMSEA is less than 0.08 (Al-Lozi et al., 2017; Al-Nady et al., 2016; Alolayyan et al., 2018).

3.7 Hypotheses Testing

Structural equation modeling (SEM) was applied using IBM AMOS 23.0 o test research hypotheses. Figure 2 shows the structural model of the research.

The detailed statistics as publicized in Table 4 indicate that WLB had a significant effect on all OC dimensions, i.e., affective commitment (β = 0.312, P = 0.000), continuance commitment (β = −0.171, P = 0.000), and normative commitment

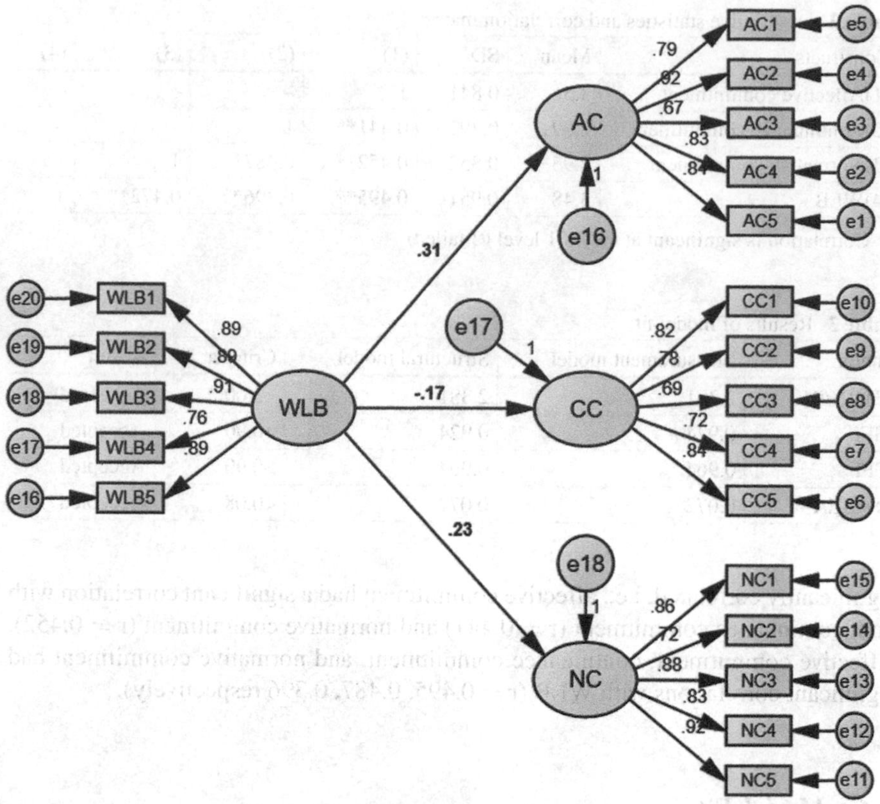

Fig. 2 Research structural model

(β = 0.233, P = 0.000). Based on these results, the three research hypotheses were supported, which means that WLB.

Table 4 Results of hypotheses testing

Hypotheses	Paths			β	P
H1	WLB	→	AC	0.312	0.000
H2	WLB	→	CC	−0.171	0.000
H5	WLB	→	NC	0.233	0.00

4 Results and Discussion

The aim of this study was to investigate the effects of work-balance life on organizational commitment dimensions (i.e., affective commitment, continuance commitment, and normative commitment. Hence, three hypotheses were suggested. The first hypothesis assumed that WLB has a significant effect on affective commitment (H1); the second hypothesis postulated that WLB has a significant effect on continuance commitment (H2), and the third one suggested that WLB exerted a significant effect on normative commitment (H3). All these hypotheses were supported by the current data. In terms of the directions of the effects, the results showed that WLB had positive effects on affective commitment and normative commitment and a negative effect on continuance commitment. In comparison with previous works, it was noted that the first result, i.e., the significant positive effect of WLB, was echoed by Tayfun and Çatir (2014). Moreover, the second result, i.e., the significant negative effect of WLB, was found by Shabir and Gani (2020). Finally, in line with the current third result, some previous studies (e.g., Dávila, 2019) pointed out that WLB had a significant positive effect on normative commitment. In fact, these results can be explained based on the nature of each dimension of OC construct. That is, the differences between these dimensions is the reason behind these results. Emotional encourages of the individual to continue in the organization because he/she feels happy. In the case of continuance commitment, the criterion for the employee to remain in the organization is the costs incurred by him/her if he/she leaves the organization. WLB also helps to improve normative commitment because the employee feels responsible for the organization.

5 Research Conclusion and Recommendations

The study concluded the importance of WLB in improving the level of organizational commitment of the employee, and therefore organizations must focus on the other side of the employee's life, which is the personal life and providing the appropriate organizational climate that does not make the employee obliged to stay in the organization because there is no other alternative.

References

Agha, K., Alzoubi, H. M., & Alshurideh, M. T. (2021). Measuring reliability and validity instruments of technologically driven cognitive intrusion towards work-life balance. In *The International Conference on Artificial Intelligence and Computer Vision* (pp. 601–614). Cham: Springer.

Al Kurdi, B., & Alshurideh, M., Al afaishat, T. (2020). Employee retention and organizational performance: Evidence from banking industry. *Management Science Letters, 10*(16), 3981–3990.

Al Momani, H. M. (2017). The mediating effect of organizational commitment on the relationship between work-life balance and intention to leave: Evidence from working women in Jordan. *International Business Research, 10*(6), 164–177.

Al- Quran, A. Z., Alhalalmeh, M. I., Eldahamsheh, M. M., Mohammad, A. A., Hijjawi, G. S., Almomani, H. M., & Al-Hawary, S. I. (2020). Determinants of the green purchase intention in Jordan: The moderating effect of environmental concern. *International Journal of Supply Chain Management, 9*(5), 366–371.

Al Shebli, K., Said, R. A., Taleb, N., Ghazal, T. M., Alshurideh, M. T., & Alzoubi, H. M. (2021). RTA's employees' perceptions toward the efficiency of artificial intelligence and big data utilization in providing smart services to the residents of Dubai. In *The International Conference on Artificial Intelligence and Computer Vision* (pp. 573–585). Cham: Springer.

Alameeri, K., Alshurideh, M., Al Kurdi, B., & Salloum, S. A. (2020). The effect of work environment happiness on employee leadership. In *International Conference on Advanced Intelligent Systems and Informatics* (pp. 668–680). Cham: Springer.

Al-bawaia, E., Alshurideh, M., Obeidat, B., Masa'deh, R. (2022). The impact of corporate culture and employee motivation on organization effectiveness in Jordanian banking sector. *Academy of Strategic Management Journal, 21*(Special Issue 2), 1–18.

Alhalalmeh, M. I., Almomani, H. M., Altarifi, S., Al-Quran, A. Z., Mohammad, A. A., & Al-Hawary, S. I. (2020). The nexus between corporate social responsibility and organizational performance in Jordan: The mediating role of organizational commitment and organizational citizenship behavior. *Test Engineering and Management, 83*(July), 6391–6410.

Al-Hawary, S. I., & Al-Syasneh, M. S. (2020). Impact of dynamic strategic capabilities on strategic entrepreneurship in presence of outsourcing of five stars hotels in Jordan. *Business: Theory and Practice, 21*(2), 578–587.

Al-Hawary, S. I. S., & Alhajri, T. M. S. (2020). Effect of electronic customer relationship management on customers' electronic satisfaction of communication companies in Kuwait. *Calitatea, 21*(175), 97–102.

Al-Hawary, S. I. S., & Mohammed, A. K. (2017). Impact of team work traits on organizational citizenship behavior from the viewpoint of the employees in the education directorates in North Region of Jordan. *Global Journal of Management and Business, 17*(2-A), 23–40.

Al-Hawary, S. I. S., & Obiadat, A. A. (2021). Does mobile marketing affect customer loyalty in Jordan? *International Journal of Business Excellence, 23*(2), 226–250.

Al-Hawary, S. I. S., Mohammad, A. S., Al-Syasneh, M. S., Qandah, M. S. F., & Alhajri, T. M. S. (2020). Organisational learning capabilities of the commercial banks in Jordan: Do electronic human resources management practices matter? *International Journal of Learning and Intellectual Capital, 17*(3), 242–266.

Al-Hawary, S. I., & Alajmi, H. M. (2017). Organizational commitment of the employees of the ports security affairs of the State of Kuwait: The impact of human recourses management practices. *International Journal of Academic Research in Economics and Management Sciences, 6*(1), 52–78.

Alketbi, S., Alshurideh, M., & Al Kurdi, B. (2020). The Influence of service quality on customers' retention and loyalty in the UAE hotel sector with respect to the impact of customer' satisfaction, trust, and commitment: A qualitative study. *International Journal of Innovation, Creativity and Change, 14*(7), 734–754.

Alkitbi, S. S., Alshurideh, M., Al Kurdi, B., & Salloum, S. A. (2020). Factors affect customer retention: A systematic review. In *International Conference on Advanced Intelligent Systems and Informatics* (pp. 656–667). Cham: Springer.

Allen, N., & Meyer, J. (1990). The measurement and antecedents of affective, continuance, and normative commitment to the organization. *Journal of Occupational Psychology, 63*, 1–18.

Allozi, A., Alshurideh, M., AlHamad, A., & Al Kurdi, B. (2022). Impact of transformational leadership on the job satisfaction with the moderating role of organizational commitment: Case of UAE and Jordan manufacturing companies. *Academy of Strategic Management Journal, 21*, 1–13.

Al-Lozi, M., Almomani, R. Z., & Al-Hawary, S. I. (2018). Talent Management strategies as a critical success factor for effectiveness of human resources information systems in commercial banks working in Jordan. *Global Journal of Management and Business Research: A Administration and Management, 18*(1), 30–43.

Al-Lozi, M., Almomani, R. Z., & Al-Hawary, S. I. (2017). Impact of talent management on achieving organizational excellence in Arab Potash Company in Jordan. *Global Journal of Management and Business Research: A Administration and Management, 17*(7), 15–25.

Al-Nady, B. A., Al-Hawary, S. I., & Alolayyan, M. (2013). Strategic management as a key for superior competitive advantage of sanitary ware suppliers in kingdom of Saudi Arabia. *International Journal of Management and Information Technology, 7*(2), 1042–1058.

Al-Nady, B. A., Al-Hawary, S. I., & Alolayyan, M. (2016). The role of time, communication, and cost management on project management success: An empirical study on sample of construction projects customers in Makkah City, Kingdom of Saudi Arabia. *International Journal of Services and Operations Management, 23*(1), 76–112.

Alolayyan, M. N., Al-Hawary, S. I. S., Mohammad, A. A. S., & Al-Nady, B. A. H. A. (2018). Banking service quality provided by commercial banks and customer satisfaction. A structural equation modelling approaches. *International Journal of Productivity and Quality Management, 24*(4), 543–565.

Alshamsi, A., Alshurideh, M., Al Kurdi, B., & Salloum, S. A. (2020, October). The influence of service quality on customer retention: a systematic review in the higher education. In *International Conference on Advanced Intelligent Systems and Informatics* (pp. 404–416). Cham: Springer.

Alsharari, N. M., & Alshurideh, M. T. (2020). Student retention in higher education: the role of creativity, emotional intelligence and learner autonomy. *International Journal of Educational Management. International Journal of Educational Management, 35*(1), 233–247.

AlShehhi, H., Alshurideh, M., Al Kurdi, B., & Salloum, S. A. (2020). The impact of ethical leadership on employees performance: A systematic review. In *International Conference on Advanced Intelligent Systems and Informatics* (pp. 417–426). Cham: Springer.

Alshurideh, M. (2016). Scope of customer retention problem in the mobile phone sector: A theoretical perspective. *Journal of Marketing and Consumer Research, 20*(2), 64–69.

Alshurideh, M. (2022). Does electronic customer relationship management (E-CRM) affect service quality at private hospitals in Jordan? *Uncertain Supply Chain Management, 10*(2), 325–332.

Alshurideh, M., Al Kurdi, B. H., Vij, A., Obiedat, Z., & Naser, A. (2016). Marketing ethics and relationship marketing-An empirical study that measure the effect of ethics practices application on maintaining relationships with customers. *International Business Research, 9*(9), 78–90.

Alshurideh, M., Alhadid, A. Y., & Barween, A. (2015). The effect of internal marketing on organizational citizenship behavior an applicable study on the University of Jordan employees. *International Journal of Marketing Studies, 7*(1), 138–145.

Alsuwaidi, M., Alshurideh, M., Al Kurdi, B., & Salloum, S. A. (2020). Performance appraisal on employees' motivation: a comprehensive analysis. In *International Conference on Advanced Intelligent Systems and Informatics* (pp. 681–693). Cham: Springer.

Altarifi, S., Al-Hawary, S. I. S., & Al Sakkal, M. E. E. (2015). Determinants of E-Shopping and its effect on consumer purchasing decision in Jordan. *International Journal of Business and Social Science, 6*(1), 81–92.

AlTaweel, I. R., & Al-Hawary, S. I. (2021). The Mediating role of innovation capability on the relationship between strategic agility and organizational performance. *Sustainability, 13*(14), 7564.

Clark, S. C. (2000). Work/family border theory: A new theory of work/family balance. *Human Relations, 53*(6), 747–770.

Dávila, M. (2019). Work-life balance practices and organizational commitment of academics at a Chilean University. *Revista Academia and Negocios, 4*(2), 13–22.

Guest, D. E. (2002). Perspectives on the study of work-life balance. *Social Science Information, 41*(2), 255–279.

Harini, S., Luddin, M. R., & Hamidah, H. (2019). Work life balance, job satisfaction, work engagement and organizational commitment among lecturers. *Journal of Engineering and Applied Sciences, 14*(7), 2195–2202.

Khalid, S. S., & Ibrahim, A. B. T. (2018). Relationship between work-life balance and employee commitment among professionals in Nigeria: A study on radio Journalists in Kano. *International Journal of Business and Technopreneurship, 8*(3), 257–268.

Kim, H. K. (2014). Work-life balance and employees' performance: The mediating role of affective commitment. *Global Business and Management Research, 6*(1), 37–51.

Koyuncu, D., & Demirhan, G. (2021). Quality of work life, organizational commitment, and organizational citizenship behaviour of teaching staff in higher education institutions. *Higher Education Governance and Policy, 2*(2), 98–109.

Malone, E. K. (2010). *Work-life balance and organizational commitment of women in construction in the United States (Doctoral Thesis)*. University of Florida.

Marmol, A. D. (2019). Dimensions of teachers' work-life balance and school commitment: Basis for policy review. *IOER International Multidisciplinary Research Journal, 1*(1), 110–120.

Meyer, J., & Allen, N. (1997). *Commitment in the workplace*. SAGE Publications.

Meyer, J. P., & Herscovitch, L. (2001). Commitment in the workplace: Toward a general model. *Human Resource Management Review., 11*(3), 299–326.

Meyer, J. P., Stanley, D. J., Herscovitch, L., & Topolnytsky, L. (2002). Affective, continuance and normative commitment to the organization: A meta-analysis of antecedents, correlates, and consequences. *Journal of Vocational Behavior, 61*, 20–52.

Mohammad, A. A. S. (2017). The impact of brand experiences, brand equity and corporate branding on brand loyalty: Evidence from Jordan. *International Journal of Academic Research in Accounting, Finance and Management Sciences, 7*(3), 58–69.

Mohammad, A. A. S., Alshura, M., S, K, Al-Hawary, S. I. S., Al-Syasneh, M. S., & Alhajri, T. M. S. (2020). The influence of internal marketing practices on the employees' intention to leave: A study of the private hospitals in Jordan. *International Journal of Advanced Science and Technology, 29*(5), 1174–1189.

Peng, F. (2018). Person-organization fit, work-family balance, and work attitude: The moderated mediating effect of supervisor support. *Social Behavior and Personality: An International Journal, 46*, 995–1010.

Rigby, M., & O'Brien-Smith, F. (2010). Trade union interventions in work-life balance. *Work, Employment and Society, 24*(2), 203–220.

Sakthivel, D., & Jayakrishnan, J. (2012). Work life balance and organizational commitment for nurses. *Asian Journal of Business and Management Sciences, 2*(5), 1–6.

Shabir, S., & Gani, A. (2020). Impact of work–life balance on organizational commitment of women health-care workers: Structural modeling approach. *International Journal of Organizational Analysis. International Journal of Organizational Analysis*, 1934–8835 (Emerald Publishing Limited). https://doi.org/10.1108/IJOA-07-2019-1820.

Tayfun, A., & Çatir, Ö. G. O. (2014). An empirical study into the relationship between work/life balance and organizational commitment. *ISGUC the Journal of Industrial Relations and Human Resources, 16*(1), 20–37.

Voydanoff, P. (2005). Toward a conceptualization of perceived work-family fit and balance: A demands and resources approach. *Journal of Marriage and Family, 67*(4), 822–836.

Wardana, M. C., Anindita, R., & Indrawati, R. (2020). Work life balance, turnover intention, and organizational commitment in nursing employees at X hospital, Tangerang, Indonesia. *Journal of Multidisciplinary Academic, 4*(4), 221–228.

The Influence of Enterprise Risk Management Framework Towards Company Performance at Conglomerate Group of Companies

Azman Makmor, Nurhanan Syafiah Abdul Razak, Musmuliadi Kamaluding, and Muhammad Alshurideh

Abstract This study explores the influence of Enterprise Risk Management (ERM) framework implementation on the company performance at a conglomerate group of companies, DRB-HICOM Berhad (the Group). This study intends to solve the problem in assessing the relationship between company performance and ERM implementation by focusing on the companies that standardized their ERM implementation towards all subsidiary companies with multi-type of businesses. In this theoretical analysis, quantitative research with correlational research design will be used, and sampling analysis will be done based on the data samples from 3 main business sectors (namely automotive, properties and services) in the Group and both primary and secondary data from Group subsidiary companies. The evaluation on the adoption of the Committee of Sponsoring Organization of the Treadway Commission (COSO) ERM framework within the subsidiary companies will be a key in assessing the ERM practices within the Group. At the same time, the company performance was measured from the company revenue achievement. Different from previous researches done by previous scholars where the studies were only analyzed on one

A. Makmor (✉) · N. S. A. Razak
MERITUS University, Kuala Lumpur, Malaysia
e-mail: azman.makmor@ctrm.com.my

Faculty of Business and Management, DRB, HICOM University of Automotive Malaysia, 26607 Pahang, Malaysia
e-mail: nurhanan@meritus.edu.my; hana.syafiah@gmail.com

M. Kamaluding
School of Civil Engineering, College of Engineering, University of Technology Mara, 40450 Shah Alam, Selangor, Malaysia
e-mail: musmuliadi@uitm.edu.my

M. Alshurideh
Marketing Department, School of Business, The University of Jordan, Amman, Jordan
e-mail: m.alshurideh@ju.edu.jo; malshurideh@sharjah.ac.ae

Department of Management, College of Business Administration, University of Sharjah, Sharjah, United Arab Emirates

M. Alshurideh et al. (eds.), *The Effect of Information Technology on Business and Marketing Intelligence Systems*, Studies in Computational Intelligence 1056,
https://doi.org/10.1007/978-3-031-12382-5_66

type of business (single way of ERM implementation only and without comparison) or were done on multi-type of businesses/companies (where too many ways of ERM implementation) which contributed to the failures in assessing the relationship between companies performance and ERM implementation, this study will provide a new solution on the above shortcomings by analyzing the companies (with different type businesses) that have an identical implementation of ERM framework across the Group. This may help future researchers and ERM practitioners to develop better standardization of ERM implementation based on the COSO framework. This paper also provides an insight into the best practices in the ERM to be implemented in other companies or organizations, which will impact the company in terms of financial and non-financial performances.

Keywords Enterprise Risk Management (ERM) · Company performance · Revenue · The Committee of Sponsoring Organization of the Treadway Commission (COSO)

1 Introduction

In the world of globalization in the 21st century has created a more complex, interdependent, and uncertain world than ever witnessed before. The opportunities and challenges have come together with significant risks to a business organization (Power, 2013). For instance, the well-known financial crisis in 1998 (Asian financial crisis) and 2007 (global financial crisis) led Lehman Brothers, the United States giant investment bank went bankrupt in the year 2008. The situation has continued and spread to other countries in the whole world.

As business opportunities and risks are constantly changing, organizations need to continuously identify, assess, manage, and monitor their business opportunities and threats (Protiviti, 2006). Effective risk management in the organization can minimize the impact on any crisis events and may create opportunities for the organization to succeed in today's business environment.

Enterprise Risk Management (ERM) is an integrated approach, structured and very systematic tool for management to comprehensively manage, evaluate and mitigate all the risks in the organization to successfully achieve the organization's objective (Dickinson, 2001). Exemplary ERM implementation will benefit the company in financial impact as company performance.

Many kinds of research tried to study, investigate, and examine the relationship between ERM implementation impacts the company's performance. However, due to a lack of reliable and standard methods, most of the studies failed to address the influence of ERM framework towards company's performance. The relation between company performances with ERM has still not been resolved and needs further study to be done in this field by scholars and researchers (Mikes & Kaplan, 2013; Fraser & Schoening-Thiessen, 2010).

In previous ERM researches, most of the studies are done on a specific type of business or organization, and no research has been done on the group of companies with diversified businesses. Generally, this study aims to envisage the influence of the ERM framework on company performance in a conglomerate group of companies with diversified businesses.

Diversifying the business will offering a lot of advantages and opportunities to the group of companies. It will offer a better choice to customers, build trust in the services or products offered, create a new potential range in an additional stream of revenue, and most important is it could spread the business risks among the companies in the group (Raymond, 2016). This also is well known as a conglomerate group of companies. Merriam-webster has defined 'conglomerate' as 'A widely diversified corporate' (Merriam-webster, 2020).

2 Background of the Study

This study is conducted in one of Malaysia's largest and most diverse conglomerates, DRB-HICOM Berhad ("DRB-HICOM" or the "Group"). As a conglomerate group of companies, DRB-HICOM involves many business sectors such as automotive, services, properties, hospitality, and education. Despite having a good business reputation in the country, running a diversified business requires the Group to cautiously manage their risks as an overextension of company resources, lack of expertise, high operation cost, and reduced innovation are the challenges DRB-HICOM must face in sustaining their businesses.

As a corporate practice in the Group, all the CEOs or Head of Companies will prepare their organization Annual Management Plan (AMP), containing company business objectives and strategic plans to sustain and grow the business. The AMP will be presented and to be approved by the Group General Managing Director (GMD). The performance of the AMPs will be monitored by various Board Committees and Senior Management members in the Group, namely Board Audit Committee (BAC), Board Nomination and Remuneration Committee (BNRC), and Board Risk and Sustainability Committee (BRSC).

The responsibilities in identifying principal risks and ensuring the implementation of an appropriate system to manage these risks are delegated to the Board Risk and Sustainability Committee (BRSC), supported by the Group Risk Management Committee and chaired by GMD (Group CG Report, 2019). To ensure the risk management is well-practiced within the Group, COSO ERM Framework and ISO 3100 are two primary standards that being references to the Group.

In Malaysia, Risk Management (RM) is governed through corporate governance, which set out from essential requirements which are coming from Bank Negara Corporate Governance (BNCG) policy, Malaysia Companies Act 2016, Malaysia Code of Corporate Governance (MCCG) 2017 by Securities Commission of Malaysia and Bursa Securities Main Market Listing Requirement (MMLR) by Bursa Malaysia.

Most of the companies in Malaysia are only complying with the requirements without gauging the effectiveness of the ERM implementation in their organizations. Even though they believe there is a good impact from the ERM implementation in their organization. Still, due to the non-existence of reliable method to determine the ERM effectiveness and performance, most organizations do not care about how they manage their risks as long as the implementation meets the requirements (Kraus & Lehner, 2012).

In response to this problem, the study envisage how the eight elements in the ERM framework established by COSO may affect company performance at the conglomerate group of companies.

3 Literature Review

In the past few decades, many researchers are being published by academic researchers and practitioners in conducting studies of characteristics and the implementation of ERM (Fraser & Simkins, 2010). In determining the focus and scope of this study, previous major studies on ERM have been studied to develop a group of study streams. Based on the previous 19 major studies on ERM, the studies stream has been categorized into four categories, as shown in Fig. 1.

Fig. 1 Table of ERM main study streams

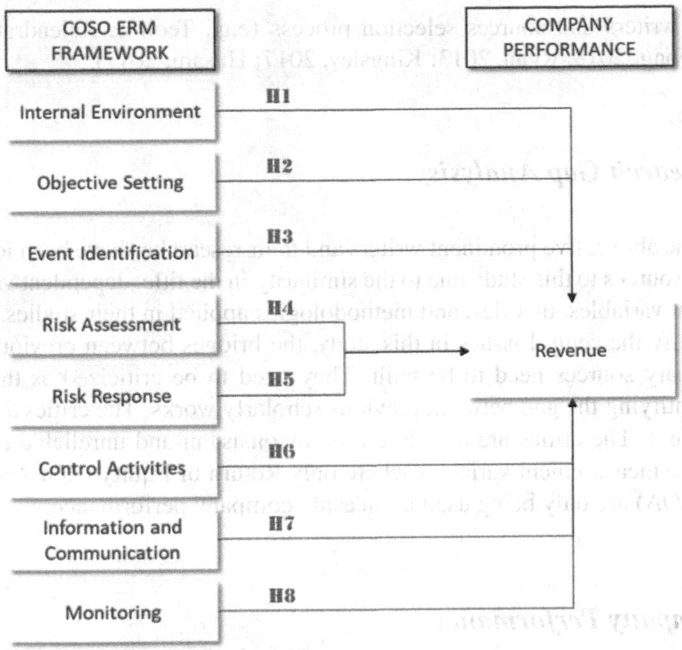

Fig. 2 Research conceptual framework

As shown in Fig. 1 above, the first category of ERM studies is the studies investigating ERM practice and characteristics (namely Colquit et al., 1999; Kleffner et al., 2003).

The second category is the studies making in-depth ERM case studies on specific business sectors (namely Aabo et al., 2005; Acharyya, 2009; Harrington et al., 2009; Mikes, 2009; Stroh, 2005).

The third category is the studies examining the relationship between ERM implementation and value creation (Gordon et al., 2009; McShane et al., 2011; Hoyt & Liebenberg, 2011; Pooser, 2012; Eckles et al., 2014; Grace et al., 2015).

Lastly, the fourth category is the studies that are analyzing the implementation of ERM determinants only (namely Beasley et al., 2005; Golshan & Rasid, 2012; Hoyt & Liebenberg, 2011; Liebenberg & Hoyt, 2003; Pagach & Warr, 2011; Razali et al., 2011).

Based on Fig. 1 above, the focus and scope of this study are under category three, which examining the relationship between ERM implementation and value creation (e.g., company performance). In-depth searching on the research that falls under this category has been done, and there are 34 kinds of research that have a similarity with this study's topics.

Further, post examining the research titles, theories, and methodologies, five kinds of research are found with close similarity with this study. Table 1 summarizes the prominent writers and sources selection process (e.g., Teoh & Rajendran, 2015; Kashif & Fong, 2018; Ryan, 2013; Kingsley, 2017; Husam, 2017).

3.1 Research Gap Analysis

Based on the above, five prominent writers and their researches have been identified as primary sources to this study due to the similarity in the title, dependent variables, independent variables, theories, and methodologies applied in their studies.

To identify the central issues in this study, the bridges between previous topics in the primary sources need to be built. They need to be criticized as this could help in identifying the gap between previous scholarly works. The critics are tabled as per Table 2. The critics are mainly on the inconsistent and unreliable company performance measurement variables where only Return of Equity (ROE) or Return of Asset (ROA) are only being used to measure company performance.

3.2 Company Performance

Company performance which is also defined as business performance is part of an organization's effectiveness. In this 21st century, company performance is focused on how organizations fully utilize their resources to improve their capabilities and abilities in achieving their objectives (Taouab & Issor, 2019; Alshurideh et al., 2020; Alshurideh et al., 2021; Alzoubi et al., 2021; Kabrilyants et al., 2021; Tariq et al., 2022).

As one of the essential management tools for improving business performance, ERM practices also help reduce different types of risk exposure to the business (Yang & Ishtiaq, 2018). A company with established ERM practices typically enjoys high operational performance and high revenue compared to a company with a lack of ERM practices (Callahan & Soileau, 2017). It is not doubted that there is a significant positive association between ERM practices and company business performance (Callahan & Soileau, 2017; Florio & Leoni, 2017; Zou & Hassan, 2017).

Based on the research gap analysis in 3.1, to have a good measurement of company performance, it is essential to measure both ***Financial Market Measurement*** (Market valuation-Tobin's Q) and ***Business Measurement*** (Revenue, Gross Profit, GP, Profit Before Tax, PBT, and growth rate) and accounting measurement (Return on Assets, ROA, and Return on Equity, ROE). The description of each element are described as below.

i. **Market Valuation (Tobin's Q):** (Q ratio or Tobin's Q) equals the market value of a company divided by its assets' replacement cost.
ii. **Revenue:** Units Sold x Sales Price.
iii. **Gross Profit (GP)**: Revenue—Cost of Goods Sold.

Table 1 Main writers and sources

ERM study stream category 3: studying the relationship between ERM framework and company performance (value)	Title	Theory	Methodology (Quantitative)
Sajjad and Engku (2017)		x	x
Kashif and Fong (2015)			x
Stephen et al. (2013)		x	x
Teoh and Rajendran (2015)	x	x	x
Kashif and Fong (2018)	x	x	x
Carolyn and Jared (2017)	x		x
Ryan (2013)	x	x	x
Kingsley (2017)	x	x	x
Tseng (2007)	x		x
Gordon et al. (2009)	x		x
Luis et al. (2020)	x		x
Rasid et al. (2012)	x		x
Musa et al. (2014)	x		x
Farah and Muneera (2017)	x	x	
Cristina and Giulia (2016)			x
Tahira and Faisal (2019)			x
Shoki (2019)	x	x	
Idris and Norlida (2016)	x		
Mohammad et al. (2019)	x		x
Yinka et al. (2018)	x		x
Tony et al. (2012)	x		
Alaa and Mukhtar (2017)	x		x
Pravaneh (2020)	x		x
Gullanut and Nopadol (2015)		x	
Izah and Ahmad (2011)	x		
Husam (2017)	x	x	x
Phillip and Nadine (2017)			x
Desender and Lafuente (2009)		x	
Kommunuri et al. (2009)	x		x
Ali et al. (2019)	x		x
Kuranaratne (2017)	x	x	x
Opiyo (2012)	x		
Annamalah et al. (2013)	x		x
Ramlee and Ahmad (2015)	x		x

Table 2 Critics on previous scholar works

Authors	Title	DV	IV	Scopes	Study results	Critics
Teoh and Rajendran (2015)	The impact of ERM of Firm Performance: Evidence from Malaysia	Company performance (Financial and Non-financial)	COSO ERM implementation Company size Company complexity BODs monitoring	Bursa Malaysia PLC Companies	Prove the significant relationship of ERM implementation and firm performance of PLCs in Malaysia	Low response rate (13.38%). Good DV with covering on Financial and Non-Financial
Kashif and Fong (2018)	ERM implementation and Firm Performance: Evidence from the Malaysia Oil and Gas Industry	Company performance (Return on Asset—ROA)	1.Internal environment 2.Objective setting 3.Event identification 4.Risk assessment 5.Risk response 6.Control activities 7.Information & communication 8.Monitoring	Oil and Gas listed companies in Bursa Malaysia	Found ERM implementation has a positive effect on the firm performance	Not reliable DV. Only ROA as DV

(continued)

Table 2 (continued)

Authors	Title	DV	IV	Scopes	Study results	Critics
Ryan (2013)	An empirical investigation into the association between ERM and Firm Financial Performance	Financial performance (KPI)	1.COSO ERM adoption 2.ERM maturity 3.Firm-specific variables	Listed companies	Found COSO and ERM implementation have a positive effect on the firm performance	Not reliable DV. Only ROA as DV
Kingsley (2018)	The impact of ERM and Firm Performance: Evidence from Sri Lankan banking and finance industry	Company performance (Return on Equity – ROE)	Same as above (8 components in COSO ERM Framework)	Sri Lanka finance and banking listed companies	Found COSO ERM framework adoption has an impact on the firm's performance	Not reliable DV. Only ROE as DV
Husam (2017)	COSO ERM implementation in Jordanian commercial banks and its impact on Financial Performance	Financial performance (ROA, ROE, D/E, CAR, NPL)	Same as above (8 components in COSO ERM Framework)	Jordan Commercial banks	Concluded the COSO ERM implementation is maintaining a high level of execution	Not reliable DV. Only ROA as DV

iv. **Profit Before Tax (PBT)**: Computed by getting the total sales revenue and then subtracting the cost of goods sold, operating expenses, and interest expense.
v. **Growth Rate**: (Current total revenue—Last year total revenue)/Last year total revenue
vi. **Return on Equity (ROE)**: Calculated by dividing net income by shareholders' equity. ROE is considered the return on net assets.
vii. **Return on Asset (ROA)**: Calculated by dividing net income by average total assets. ROA shows how much profit a company can generate from its assets.

This study is proposing for all these elements to be measured as company performance. Not all types of businesses are operating with the same business model, where the best measurements to measure their performances are different depending on their business model or type of business. Thus, incorporating all these elements like a company performance will produce higher accuracy in measuring company performance achievement.

However, due to the limitation of research time constraint, this study will only discuss the revenue as a measurement for the company performance, as in the DRB-HICOM Group, revenue is the most anticipated figure by the management as a company performance indicator in this diversified business of a conglomerate group of companies.

3.3 ERM Framework Components

One of the biggest problems and difficulties in implementing the ERM framework in an organization is that there are many definitions and practice standards in ERM (Aven, 2012). In this study, the COSO ERM framework is chosen as the independent variable. COSO ERM Framework is the most accepted and practiced ERM framework by organizations (Perera, 2019). There are eight components in the COSO ERM framework, which are explained as per below:

3.3.1 Internal Environment

The internal environment influences the establishment of company strategies and objectives (Al Mehrez et al., 2020; Alshurideh, 2019; ELSamen & Alshurideh, 2012). Consideration of the internal environment is essential in determining risk appetite and integrating ERM with related initiatives. ERM's supportive internal environment is a fundamental requirement for successfully implementing an effective ERM system in an organization. The internal environment represents the "tone of the top management" that encompasses the management attitude and awareness about the importance and the relevance of the ERM towards creating and preserving the company's value. The integrated framework's internal environment represents the organization's tone, including the risk management philosophy and risk appetite.

According to Kinyua et al. (2015), they recognize a significant association between internal control environment and financial performance. According to Liebenberg and Hoyt (2003), in one of their articles on "The determinants of enterprise risk management: evidence of the appointment of chief risk officers" states that this context, it is believed that ERM supportive internal environment facilities a culture of risk aligned decision making that would positively affect the company performance.

3.3.2 Objective Setting

Selects a strategy that is utilized throughout the enterprise and aligned with and linked to the strategy (Alaali et al., 2021; Ashal et al., 2021). The point when management sets its objectives is the point they set their risk. Management needs to develop their objectives by considering the corresponding risk of achieving them. According to Gates et al. (2012), most ERM frameworks assert that risk should be identified concerning the company objectives. As per the COSOs ERM integrated framework (2014), company objectives should be aligned within the company's risk appetite and risk tolerance levels. The company engaged in ERM should better understand the aggregated risk of different business activities, providing them with a more objective basis for resource allocation that will improve return on equity (ROE) and capital efficiency (Eikenhout, 2015). According to Hoyt and Liebenberg (2011), ERM strategy aims to reduce volatility by preventing aggregation of risk across different sources. In the DRB-HICOM Group, it's an essential responsibility for the CEO to establish company's long term objectives. When the objective setting and risk are aligned, this will help top management to consider any opportunities and threat in setting the objectives for the company. With different type of businesses in the Group will established a different type and nature of the risk for the companies and definitely the type of objectives also are determined by this factor. Company objective setting is always driven by the estimated revenue that has been set by the DRB-HICOM Group, and these expected revenue is always associated with the risk elements. To this effect, it is the time of the company's objective setting to set its desired risk appetite effectively.

3.3.3 Event Identification

Event identification takes into consideration both internal and external factors that affect event occurrence. By having a consideration on internal and external factors will help companies in the Group to predict any favorable or unfavorable situation that might become a obstacle for the companies to achieve their objectives. This component will stimulate companies to make an early identification and make a proper and suffice preparation to face any events which may impact negatively or positively to the company performance. As such, event identification minimizes the risk of facing business surprises will minimize volatility in return and improve company's financial performance. According to COSO's ERM integrated framework, it is very important for a companies to identify internal external events that will give an impact towards company in achieving their objectives and the impacts have to be distinguished between the threats and opportunities. The identified opportunities are channeled back into company's objective setting (COSO, 2004). By identifying the events and managing the threat and opportunities, will help companies to set a right strategies to ensure the company's long term objectives.

3.3.4 Risk Assessment

Allows an entity to consider how potential events might affect the achievement of objectives. Risk assessment encompasses assessing the likelihood and impact of events affecting the achievement of a company's objectives. It enables an organization to determine a more approach to address the risk factors emerging from the internal and external environment. Unlike the ISO 31000 standard on ERM (2009), which recognizes only the residual risk, ERM integrated framework assesses the risk on both inherent and residual basis. According to Solomon and Muntean (2012), a company's risk assessment based on leverage coefficients is required for the predicted behavior analysis for estimating future results. According to Deloitte and Touche LLP; Curtis and Currey (2012), Risk assessment is essential since it is how enterprises handle how significant each risk is to the achievement of their overall goals.

3.3.5 Risk Response

Management identifies risk response options and considers their effect on event likelihood and impact concerning risk tolerances and costs versus benefits and implements response options. Based on the risk assessment and in the light of the company's risk tolerance and appetite, management should decide upon the suitable response to each identified risk factor choosing amongst the risk avoidance, risk acceptance, risk-sharing, and risk reduction. An effective risk responding strategy is expected to have a positive impact on the company's performance. According to Vollmer (2015), a cost-effective and efficient risk response plan helps balance risk mitigation with the expected benefits of the strategic program.

3.3.6 Control Activities

Control activities ensure the completeness, accuracy, and validity of the various policies and procedures. Control activities are also designed to address the identified risk factors involves the management's policies and procedures to ensure that risk responses are effectively implemented. Some researchers assert that the effective implementation of control activities enhances operating efficiency and enhances company performance. According to Munene (2013), results established a significant relationship between an internal control system and financial performance.

Eniola and Akinselure (2016) state that effective internal controls will significantly improve financial performance by helping the organization to substantially reduce fraud perpetration. According to Beeler et al. (1999), internal controls provide an independent appraisal of the quality of managerial performance in carrying out assigned responsibilities for better revenue generation. Control activities usually strengthen the company's internal control functions, which enhances the efficiency and effectiveness of the operations affecting the company's performance positively.

3.3.7 Information and Communication

Information and communication are to process and refine large volumes of data into actionable information (Alshurideh et al., 2014; Alyammahi et al., 2020). Effective information communication channel is vital to achieving the intended benefits of an integrated risk management framework. Effective integration can only be achieved by ensuring effective communication among the related functions and people throughout the organization. This is a crucial feature that differentiates ERM from traditional silo-based risk management. The improvement in the organization's risk profile information is another potential source of value created by ERM (Eikenhout, 2015). According to Chaffey and Wood (2005), as cited by Olugbode et al. (2008), decisions tend to be more informed and effective when the communication is thorough and accurate. Enhanced communication of risk information lets the organizational managers make informed and risk-aligned decisions, leading to better performance.

3.3.8 Monitoring

The process of monitoring or assessing both the presence and functioning of the enterprise risk management components and the quality of the performance over time is a key factor in the process.

3.4 Relationship Between ERM Framework and Company Performance

Compared to other studies that have been done on studying the relationship between ERM framework and company performance, this research will provide better and reliable measurement on the study of the influence of ERM framework and company performance as this study will be tested not only on a single type of business but will be tested to many types of organization that exercising the same ERM implementation in their company as they being managed by a single risk management group at the holding company.

The hypothesis of this study is to propose a relationship between COSO ERM Framework (independent variables—control and change) and Company Performance (dependent variable—observes and measures). The description of the relationship between these two variables are constructed as below:

3.4.1 Internal Environment—Revenue

According to Kinyua et al. (2015), they recognize a significant association between internal control environment and financial performance. According to Liebenberg and Hoyt (2003), in one of their articles on "The determinants of enterprise risk management: evidence of the appointment of chief risk officers" states that this context, it is believed that ERM supportive internal environment facilities a culture of risk aligned decision making that would positively affect the company performance. Thus, this study derives its first hypothesis concerning ERM adoption as below:

H1: There is an impact from the internal environment in increasing the revenue at the conglomerate group of companies.

3.4.2 Objective Setting—Revenue

According to Gates et al. (2012), most ERM frameworks assert that risk should be identified concerning the company objectives. According to Hoyt and Liebenberg (2011), ERM strategy aims to reduce volatility by preventing aggregation of risk across different sources. The company engaged in ERM should better understand the aggregated risk of diverse business activities, providing them with a more objective basis for resource allocation that will improve return on equity (ROE) and capital efficiency (Eikenhout, 2015). Thus, this study derives its second hypothesis concerning ERM adoption as below:

H2: There is an impact from objective setting in increasing the revenue at the conglomerate group of companies.

3.4.3 Event Identification—Revenue

According to COSO's ERM integrated framework, the internal and external events affecting a firm's objectives must be identified, distinguishing between risks and opportunities. Opportunities are channeled back to management's strategy or objective-setting process (COSO, 2004). Thus, this study derives its third hypothesis concerning ERM adoption as below:

H3: There is an impact from event identification in increasing the revenue at the conglomerate group of companies.

3.4.4 Risk Assessment—Revenue

According to Solomon and Muntean (2012), a company's risk assessment based on leverage coefficients is required for the predicted behavior analysis for estimating future results. According to Deloitte and Touche LLP; Curtis and Currey (2012), Risk assessment is important since it is how enterprises get a handle on how significant each risk is to the achievement of their overall goals. Thus, this study derives its fourth hypothesis concerning ERM adoption as below:

H4: There is an impact from risk assessment in increasing the revenue at the conglomerate group of companies.

3.4.5 Risk Response—Revenue

According to Vollmer (2015), a cost-effective and efficient risk response plan helps balance risk mitigation with the expected benefits of the strategic program. Thus, this study derives its fifth hypothesis concerning ERM adoption as below:

H5: There is an impact from risk response in increasing the revenue at the conglomerate group of companies.

3.4.6 Control Activities—Revenue

According to Munene (2013), results established a significant relationship between an internal control system and financial performance. Eniola and Akinselure (2016) state that effective internal controls will significantly improve financial performance by helping the organization to significantly reduce fraud perpetration. According to Beeler et al. (1999), internal controls provide an independent appraisal of the quality of managerial performance in carrying out assigned responsibilities for better revenue generation. Thus, this study derives its sixth hypothesis concerning ERM adoption as below:

H6: There is an impact from control activities in increasing the revenue at the conglomerate group of companies.

3.4.7 Information and Communication—Revenue

The improvement in the organization's risk profile information is another potential source of value created by ERM (Eikenhout, 2015). According to Fisher and Kenny (2000), as cited by Olugbode et al. (2008), they suggest that organizations infuse information systems into their operations to enhance competitiveness and facilitate business growth and success. According to Chaffey and Wood (2005), as cited by Olugbode et al. (2008), decisions tend to be more informed and effective when the communication is thorough and accurate. Thus, this study derives its seventh hypothesis concerning ERM adoption as below:

H7: There is an impact from information and communication in increasing the revenue at the conglomerate group of companies.

3.4.8 Monitoring—Revenue

According to Wholey et al. (2010), government monitoring and evaluation are used to increase transparency, strengthen accountability, and improve performance. The monitoring function could be an ongoing process or timely evaluation aiming to decide whether further modifications are required for the company's ERM. Thus, this study derives its eighth hypothesis concerning ERM adoption as below:

H8: There is an impact from monitoring in increasing the revenue at the conglomerate group of companies.

4 Research Conceptual Framework

Research Underpinning Theory—The Theory of Business Performance Measurement

Business performance measurement could be measured from 3 functional analyses, such as an accounting perspective, marketing perspective, and operations perspective (Neely, 2002). These three perspectives are derived from the theories and concepts underpinning the field of business performance measurement. The theory of 'Risk in performance measurement' from Andrew Likierman has highlighted that risk management is an essential tool that needs to be integrated into business management (Likierman, 2007).

The research framework for this study is developed based on the gap found in previous literature. As explained in 2.6.2, there is a need to further the analysis in determining the influence of the ERM framework on company performance by incorporating reliable indicators to measure the relationship between the ERM framework and company performance.

From this study, it is expected to find the key determinants in the ERM framework that will impact company performance. A conceptual framework below illustrating the expectations to find through this research and also mapping out the relations between ERM framework determinants (Independent Variables) and company performance (Dependent Variables):

5 Research Methodology

The issues in this study will be studied and analyzed by adapting the quantitative research strategy. The research questions also will be addressed by using the quantitative research strategy. By using this research strategy, all the data are gathered and compiled from a survey questionnaire in analyzing the implementation and execution of the ERM framework in the Group subsidiaries.

This research target population for the survey is the whole population of a group of companies under DRB-HICOM Berhad (subsidiary companies). This is to allow to understand the current of ERM implementation ERM in the Group. Samples are clustered based on the type of business (business sectors) of the subsidiary companies. Cluster sampling design is used, and no stratification involves in population sampling.

The sampling unit for this study research is the number of subsidiary companies selected and available as a study sample. The sampling size for this study is the number of companies in DRB-HICOM Berhad (Group) that implemented ERM and practicing the same standard as per Group guidelines, which are around 40 to 45 companies. Due to this study is aiming to seek a practical ERM component to be used in performance measurement and recognize the current state of risk management in a conglomerate group of companies, it is utmost essential to obtain a survey feedback from an expert or person who is managing and governing the process of risk management within a subsidiary companies in the Group. The surveys are disseminate to all CEOs, Chief Risk Officer (CRO), Chief Financial Officer (CFO), or Risk Manager, Risk Management Committee members across the subsidiary companies in the Group.

On the statistical analysis method, descriptive statistics, cluster analysis, and factor analysis are the analysis methods that will be used in this study. Descriptive statistics are used in this research in describing the basic features of the collected data. This method provides a summary of the sample and the measures. Using this method helps simplify large amounts of data sensibly as each descriptive statistic reduces a lot of data into a more straightforward summary (Trochim, 2007). Cluster analysis is used in this study to analyze the influence of ERM implementation from different business sectors in the Group. Principal Component Analysis (PCA) is used to reduce the number of variables to components and use the principal component regression score. In this study, Partial Least Square (PLS) Regression is used to extracting latent factors when there remains factor variation in modeling, which is sometimes called a projection to latent structure that is still unobserved in structural equation modeling (Rannar et al., 1994).

In this research, Data Processing, the SPSS program has been identified to analyze the collected data and information. The original name for this software was Statistical Package for the Social Science (Quintero, 2012). For the current version (post-2015), the version name was changed to Statistical Product and Service Solutions (Hejase and Hejase, 2013). For Cluster Analysis, SPSS offers three methods, namely K-Mean Cluster, Hierarchical Cluster, and Two-Step Cluster. SPSS also be used for both Factor Analysis, i.e., Principal Component Analysis (PCA) and Partial Least Square (PLS) Regression.

Based on the regression model's data, it will be used to determine the relationship and influence of the ERM framework towards company performance.

6 Study Conclusion

By implementing risk management in an organization, the impact from crisis event to organization may be mitigated or reduced. The implementation also may create potential opportunities for organizations to achieve their business objectives and strategies. As a management tool, Enterprise Risk Management (ERM) is being used by an organization in managing, evaluating, and mitigating the risks that might become an obstacle to an organization in achieving its objectives. Well implementation of ERM will have a good impact on company financial performance. However, most of the previous studies could not resolve the problem in assessing the relationship between ERM implementation and a company's performance due to a lack of reliable and standard methods. Thus, the relationship between company performances with ERM has still not been resolved and needs further research in this area by academics.

Based on the literature study, there are none of the previous studies have been done on the conglomerate group of companies with having a standard ERM being implemented in all their subsidiaries companies. Many studies have been done to resolve the problem in assessing the relationship between ERM implementation and a company's performance, but due to a lack of reliable methods, they are unable to resolve the issue. As this study focuses on examining the influence of the ERM framework on company performance in a conglomerate group of companies, we could envisage how the COSO ERM framework will quantify the ERM implementation in the conglomerate group of companies and fill the literature gap.

As the number of implementation of ERM in organizations increases, the pressure in implementing the ERM creates a thought to researchers, risk practitioners, business owners, and organizations on whether ERM could generate a value and give a good impact as a return on the implementations.

Most of the scholars, law custodians, regulators, consultants, and corporate governance practitioners suggest that the implementation of ERM can positively impact the company's performance. This is viewed by many companies currently adopting the ERM framework in their organizations based on international standards.

This thesis provides an insight to organizations on the ERM implementation that will helps in determining factors that will influence the company performance. The benefits of this research are divided into several beneficial parties as follow:

Organization—Board of Directors, Chief Executive Officer (CEO), and senior management of the organization will prepare and develop much resilience and better strategies in achieving organizational objectives. Well implemented ERM framework in an organization will provide reliable insight for management in preparing the company's Annual Management Plan (AMP), company strategic initiatives, the annual budget for Capital Expenditure (CAPEX), and Operational Expenditure (OPEX) and cash flow projection.

Holding company—Having a well-implemented ERM framework in all subsidiary companies will help the holding company examine the ERM implementation gap among the subsidiary companies. Companies with a low level of ERM performance must be assessed on the ERM implementation in their organization to increase the effectiveness of the ERM framework in respective companies. By understanding the relationship between ERM performances and company performance, a holding company might determine any threat in advance or any potential opportunities for the subsidiaries. This will help the holding company provide a good and accurate insight for reporting to the shareholders or for public information to increase the interest of potential investors to invest in the group of companies.

Researchers—Lack of effective determinants for ERM is the main obstacle of good ERM implementation measurement (Beasley et al., 2008; Gordon et al., 2009; McShane et al., 2011; Pagach & Warr, 2010). This study provides an approach to determine the most influential factor in the ERM framework that will impact company performance and gauge the effectiveness of ERM framework implementation in the organization. Researchers may adopt this reliable approach to conduct further studies in ERM framework implementation in various fields of business.

Regulators—In Malaysia, most of the regulators/custodians that governing the risk management didn't put any obligation to recognize listed companies that comply with any ERM framework. However, the regulators are actively supporting in maintaining effective risk management in all organizations in Malaysia through various channels such as Malaysia Companies Act 2016 (Sect. 246), Bursa's Listing Requirement (Paragraph 15.12, 15.23, 15.26), Malaysia Code of Corporate Governance 2017 (Chapter 9.1, 9.2) and Bank Negara Risk Governance 2013. This study may guide the best practice of ERM framework implementation for regulators to prepare a proper guideline or policy in the future.

References

Aabo, T., Fraser, J. R. S., & Simkins, B. J. (2005). The Rise and Evolution of the chief risk officer: Enterprise risk management at hydro one. *Journal of Applied Corporate Finance, 17*(3), 62–75.

Acharyya, M. (2009). The influence of Enterprise Risk Management in insurer's stock market performance—an event analysis. In *2009 ERM Symposium.* Chicago, IL. http://www.societyofactuaries.org.

Alaali, N., Al Marzouqi, A., Albaqaeen, A., Dahabreh, F., Alshurideh, M., Alrwashdh, S., Iyadeh, I., Salloum, S., & Aburayya, A. (2021). the impact of adopting corporate governance strategic performance in the tourism sector: A case study in the Kingdom of Bahrain. *Journal of Legal, Ethical and Regulatory, 24*(Special Issue 1), 1–18.

Al Mehrez, A. A., Alshurideh, M., Al Kurdi, B., & Salloum, S. A. (2020, October). Internal factors affect knowledge management and firm performance: a systematic review. In *International Conference on Advanced Intelligent Systems and Informatics* (pp. 632–643). Cham: Springer.

Alshurideh, M. T., Shaltoni, A., & Hijawi, D. (2014). Marketing communications role in shaping consumer awareness of cause-related marketing campaigns. *International Journal of Marketing Studies, 6*(2), 163–168.

Alshurideh, D. M. (2019). Do electronic loyalty programs still drive customer choice and repeat purchase behaviour? *International Journal of Electronic Customer Relationship Management, 12*(1), 40–57.

Alshurideh, M., Gasaymeh, A., Ahmed, G., Alzoubi, H., & Kurd, B. (2020). Loyalty program effectiveness: Theoretical reviews and practical proofs. *Uncertain Supply Chain Management, 8*(3), 599–612.

Alshurideh, M. T., Hassanien, A. E., & Masa'deh, R. (2021). The effect of coronavirus disease (COVID-19) on business intelligence. Springer.

Alyammahi, A., Alshurideh, M., Al Kurdi, B., & Salloum, S. A. (2020). The impacts of communication ethics on workplace decision making and productivity. In *International Conference on Advanced Intelligent Systems and Informatics* (pp. 488–500). Cham: Springer

Alzoubi, H., Alshurideh, M., Akour, I., Al Shraah, A., & Ahmed, G. (2021) Impact of information systems capabilities and total quality management on the cost of quality. *Journal of Legal, Ethical and Regulatory Issues, 24*(Special Issue 6), 1–11.

Ashal, N., Alshurideh, M., Obeidat, B., Masa'deh, R. (2021) The impact of strategic orientation on organizational performance: Examining the mediating role of learning culture in Jordanian telecommunication companies. *Academy of Strategic Management Journal, 21*(Special Issue 6), 1–29.

Aven, T. (2012). Foundational issues in risk assessment and risk management. *Risk Analysis, 32*(10), 1647–1656.

Beasley, M., Clune, R., & Hermanson, D. (2005). Enterprise risk management: An empirical analysis of factors associated with the extent of implementation. *Journal of Accounting and Public Policy, 24*(6), 521–531.

Beasley, M., Pagach, D., & Warr, R. (2008). Information conveyed in hiring announcements of senior executives overseeing enterprise-wide risk management processes. *Journal of Accounting, Auditing & Finance, 23*(3), 311–333.

Beeler, J. D., Hunton, J. E., & Wier, B. (1999). Promotion performance of internal auditors: A survival analysis. *Internal Auditing, 14*(4), 3–14.

Callahan, C., & Soileau, J. (2017). Does enterprise risk management enhance operating performance? *Advances in Accounting, 37*, 122–139.

Chaffey, D., & Wood, S. (2005). *Business information management: Improving performance using information.* Pearson Education Ltd.

Colquitt, L. L., Hoyt, R. E., & Lee, R. B. (1999). Integrated risk management and the role of the risk manager. *Risk Management and Insurance Review, 2*(3), 43–61.

COSO. (2004). *Enterprise Risk Management—Integrated Framework Executive Summary.* https://www.coso.org/Documents/COSO-ERM-Executive-Summary.pdf.

Deloitte & Touche, L. L. P., Curtis, P., & Carey, M. (2012). *Risk assessment in practice, Committee of sponsoring organization of the Tread way Commission.*

Dickinson, G. (2001). *Enterprise risk management: Its origins and conceptual foundation. Geneva Papers on Risk and Insurance. Issues and Practice,* 360–366.

DRB-HICOM. (2020). *Discover us.* https://www.drb-hicom.com/about-us/discover-us/.

Eckles, D. L., Hoyt, R. E., & Miller, S. M. (2014). The impact of enterprise risk management on the marginal cost of reducing risk: Evidence from the insurance industry. *Journal of Banking and Finance, 43*, 247–261.

Eikenhout, L. (2015). *Risk Management and Performance in Insurance Companies, an unpublish Master thesis.* http://essay.utwente.nl/66625/1/Eikenhout_MA_MB.pdf.

ELSamen, A. A., & Alshurideh, M. (2012). The impact of internal marketing on internal service quality: A case study in a Jordanian pharmaceutical company. *International Journal of Business and Management, 7*(19), 84–95.

Eniola, O. J., & Akinselure, O. P. (2016). Effect of Internal control on financial performance of firms in Nigeria. *IOSR Journal of Business and Management, 18*(10), 80–85.

Florio, C., & Leoni, G. (2017). Enterprise risk management and firm performance: The Italian case. *The British Accounting Review, 49*, 56–74.

Fraser, J. R., & Schoening-Thiessen, K. (2010). Who Reads What Most Often?: A Survey of Enterprise Risk Management Literature Read by Risk Executives. In: B. J. Simkins & J. Fraser (Eds.), *Enterprise risk management* (pp. 385–417). Wiley Online Library.

Fraser, J., & Simkins, B. (2010). *Enterprise risk management: Today's leading research and best practices for tomorrow's executives*. Wiley.

Gates, S., Nicolas, J. L., & Walker, P. L. (2012). *Enterprise risk management: A process for enhanced management and improved performance. Management Accounting Quarterly*, 28–38. https://hal.archives-ouvertes.fr/hal-00857435.

Golshan, N. M., & Rasid, S. Z. A. (2012). Determinants of enterprise risk management adoption: An empirical analysis of Malaysian public listed firms. *International Journal of Social and Human Sciences, 6*, 119–126.

Gordon, L. A., Loeb, M. P., & Tseng, C.-Y., (2009). Enterprise risk management and firm performance: A contingency perspective. *Journal of Accounting and Public Policy, 28*(4), 301–327.

Grace, M. F., et al. (2015). The value of investing in enterprise risk management. *Journal of Risk and Insurance, 82*(2), 289–316.

Group Corporate Governance Report, (2019). Corporate Governance Report Year 2019 submitted to Bursa Malaysia. https://www.bursamalaysia.com/market_information/announcements/company_announcement/announcement_details?ann_id=2974903.

Harrington, A., et al. (2009). Management of patients at risk for alcohol withdrawal in an acute adult psychiatry unit. *American Journal on Addictions, 18*(4), 327-327.

Hejase, A. J., & Hejase, H. J. (2013). *Research methods, a practical approach for business students* (2nd ed., p. 58). Masadir Inc.

Hoyt, R. E., & Liebenberg, A. P. (2011). The Value of Enterprise Risk Management. *Journal of Risk and Insurance, 78*(4), 795–822.

Khadash, H. A. A. (2017). COSO Enterprise Risk Management Implementation in Jordanian Commercial Banks and its Impact on Financial Performance, *1*, 5–23.

ISO 2009. 31000. (2018). Risk management principles and guidelines.

Kabrilyants, R., Obeidat, B., Alshurideh, M., & Masadeh, R. (2021). The role of organizational capabilities on e-business successful implementation. *International Journal of Data and Network Science, 5*(3), 417–432.

Kashif, S. M., & Fong, L. W. (2018). Enterprise risk management implementation and firm performance: Evidence from the Malaysian oil and gas industry. *14*(9), 47–53. https://doi.org/10.5539/ijbm.v14n9p47.

Kingsley, A. (2017). *The impact of enterprise risk management on firm performance : Evidence from Sri Lankan Banking and Finance Industry., 13*(1), 225–237. https://doi.org/10.5539/ijbm.v13n1p225

Kinyua, J. K., Gakure, R., Gekara, M., & Orwa, G. (2015). Effect of internal control environment on the financial performance of companies quoted in the Nairobi securities exchange. *International Journal of Innovative Finance and Economics Research, 3*(4), 29–48.

Kleffner, A. E., Lee, R. B., & McGannon, B. (2003). The effect of corporate governance on the use of enterprise risk management: Evidence from Canada. *Risk Management and Insurance Review, 6*(1), 53–73.

Liebenberg, A. P., & Hoyt, R. E. (2003). The determinants of enterprise risk management: Evidence from the appointment of chief risk officers. *Risk Management and Insurance Review, 6*(1), 37–52. https://doi.org/10.1111/1098-1616.00019

Likierman, A. (2007). Risk in performance measurement. In A. Neely (Ed.), *Business Performance Management* (pp. 261–277). Cambridge University Press.
McShane, M. K., Nair, A., & Rustambekov, E. (2011). Does enterprise risk management increase firm value. *Journal of Accounting, Auditing, and Finance, 26*(4), 641–658.
Merriam-webster. (2020). Conglomerate. Available from https://www.merriam-webster.com/dictionary/conglomerate#:~:text=Definition%20of%20conglomerate%20(Entry%203,conglomerate%20of%20some%20350%20businesses.
Mikes, A. (2009). Risk management and calculative cultures. *Management Accounting Research, 20*(1), 18–40.
Mikes, A., & Kaplan, R. S. (2013). *Managing risks: Towards a contingency theory of enterprise risk management.* [online]. Harvard Business School. http://ssrn.com/abstract=2311293.
Munene, M. J. (2013). *Effect of internal controls on the financial performance of technical training institutions in Kenya.* An un-published masters dissertation.
Neely, A. (2002). Business performance measurement. In Theory and practice (pp. 3–41). Cambridge University Press.
Olugbode, M., Elbeltagi, I., Simmons, M., & Biss, T. (2008). The Effect of information systems on firm performance and profitability using a case-study approach. *The Electronic Journal Information Systems Evaluation, 11*(1), 11–16.
Pagach, D., & Warr, R. (2010). The effects of enterprise risk management on firm performance.
Pagach, D., & Warr, R. (2011). The characteristics of firms that hire chief risk officers. *Journal of Risk and Insurance, 78*(1), 185–211.
Perera, A. A. S. (2019). Enterprise risk management-international standards and frameworks. *International Journal of Scientific and Research Publications, 9*(7), 211–216.
Pooser, D. M., (2012). An empirical examination of the interrelations of risks and the firm's relationships with enterprise risk management. 3539604 (Ph.D.). The Florida State University.
Power, M. (2013). The apparatus of fraud risk. *Accounting, Organizations and Society, 38*(6), 525–543.
Protiviti. (2006). *Protiviti guide to enterprise risk management—frequently asked questions, issued* [online]. Protiviti Consulting. www.protiviti.co.uk.
Quintero, D. (2012). Workload optimized systems tuning power7 for analytics. IBM. Ibm.com/redbook. P.14.
Rännar, S., et al. (1994). A PLS kernel algorithm for data sets with many variables and fewer objects. Part 1: Theory and algorithm. *Journal of Chemometrics, 8*(2), 111–125.
Raymond. H. (2016). The challenges and rewards of diversifying your business. https://excelbizsolutions.com.au/challenges-rewards-diversifying-business.
Razali, A. R., Yazid, A. S., & Tahir, I. M. (2011). The determinants of enterprise risk management (ERM) practices in Malaysian public listed companies. *Journal of Social and Development Sciences, 1*(5), 202–207.
Ryan, B. (2013). *An Empirical Investigation into the Association between Enterprise Risk Management and Firm Financial Performance., 1*(1), 17–184.
Solomon, D. C., & Muntean, M. (2012). Assessment of Financial Risk in Firm's Profitability Analysis. Economy *Transdisciplinarily Cognition, 15*(2).
Stroh, P. J. (2005). Enterprise risk management at UnitedHealth Group. Strategic Finance (July), 26–35.
Tariq, E., Alshurideh, M., Akour, I., & Al-Hawary, S. (2022). The effect of digital marketing capabilities on organizational ambidexterity of the information technology sector. *International Journal of Data and Network Science, 6*(2), 401–408.
Taouab, O., & Issor, Z. (2019). *Firm Performance : Definition and Measurement Models, 15*(1), 93–106. https://doi.org/10.19044/esj.2019.v15n1p93.
Teoh, P. A., & Rejendran, M. (2015). *The impact of enterprise risk management on firm performance: Evidence from Malaysia., 11*(22), 149–159. https://doi.org/10.5539/ass.v11n22p149
Trochim, W. (2007). The Research Methods Knowledge Base. Cornell University (1,828).
Vollmer, S. (2015). 6 steps to manage risks and drive performance. CGMA Magazine.

Wholey, J., Hatry, H., & Newcomer, K. (2010). *Handbook of practical program evaluation* (3rd ed.). Jossey-Bass.

Zou, X., & Hassan, C. H. (2017). Enterprise risk management in China: The impacts on organizational performance. *International Journal of Economic Policy in Emerging Economies, 10*, 226–239.

Examining Employee Performance During Covid-19 Pandemic: A Study of Aerospace Manufacturing Industry in Malaysia

Mohd Zulkhaizlan Sinor, Nurhanan Syafiah Abdul Razak, Mohamad Zamhari Tahir, and Muhammad Alshurideh

Abstract During Covid-19 pandemic situation, a lot of company facing problem in sustaining their businesses. Most organization has response to the current situation in maintaining the business sustainability. Unfortunately, the changes have been made by management has given significant impact on employee performance. This quantitative study is to examine the relationship between motivation, work environment and work from home towards employee performance. The target respondents of this study are among employee in an aerospace manufacturing company in Malaysia during Covid-19 pandemic situation. Statistical Product and Service Solutions (SPSS) and Analysis of a Moment Structures AMOS will be used in analyzing the data that has been collected to get the accuracy of result. The finding shows that motivation, work environment and WFH have a significant relationship with employee performance. The finding of this study is to provide insights to manufacturing company in managing strategic change management during Covid-19 pandemic as well as for future studies in revealing the factors that influence employee performance.

Keywords Covid-19 · Employee performance · Motivation · Work environment · Work from home

---

M. Z. Sinor · N. S. A. Razak · M. Z. Tahir
Faculty of Business and Management, DRB, HICOM University of Automotive Malaysia, 26607 Pahang, Malaysia
e-mail: zulkhaizlan.sinor@gmail.com

N. S. A. Razak
e-mail: hana.syafiah@gmail.com; nurhanan@meritus.edu.my

N. S. A. Razak
MERITUS University, Kuala Lumpur, Malaysia

M. Alshurideh (✉)
Department of Management, College of Business Administration, University of Sharjah, Sharjah, United Arab Emirates
e-mail: m.alshurideh@ju.edu.jo; malshurideh@sharjah.ac.ae

Marketing Department, School of Business, The University of Jordan, Amman, Jordan

M. Alshurideh et al. (eds.), *The Effect of Information Technology on Business and Marketing Intelligence Systems*, Studies in Computational Intelligence 1056,
https://doi.org/10.1007/978-3-031-12382-5_67

1 Introduction

The end of 2019 is the beginning of a kind of epidemic known as the novel corona virus (Covid-19). The virus has become a worldwide pandemic causing a high number of deaths. In addition, the pandemic has also hampered economic sectors around the world starting in 2020 (Alshurideh et al., 2021; Amarneh et al., 2021; Wang et al., 2020). This Covid-19 pandemic crisis has affected to the uncertain global economic, social and community that cause leadership to response and make significant changes in an organization to strengthen company financial stability and plan for future growth (Al-Dmour et al., 2021; Leo et al., 2021; Shah et al., 2021). When a significant change is required, leadership should quickly increase their sense of the situation and adjust their competency as well as responding to it by relying on their professional's instinct and Human Resource Development, HRD. Critical planning of strategic change management is required. Changes to a more transformational leadership style are needed in making effective decisions on the survival of the company. This is because, an organization will grow under leaders that gives a strong purpose and roles, excellent communications, share their leadership values, concern on employee's emotional stability, make sure employees can access to the technology, ensure the organizational health and resilience during the pandemic.

Readiness to change and careful planning can derive the effectiveness of transformational leadership. Periodic evaluations as well as appropriate incentives in making change plans are important for every management. To attract employee commitment and create a common vision, a comprehensive effort in changing the attitudes, values and beliefs of all employees is among the normative methods of using the hearts (empathy) and minds in making plans. With good communication and interaction between leadership and followers, providing knowledge and clear organizational goals is one of the adaptive planning methods. Change requires negotiation, consensus and agreement from the interactions that take place (Abbasi, 2017).

While change is taking place, the ability of transformational leaders to motivate their employees is important in shaping the culture of a fair organization. In other words, fair interactional and procedural become important factors in increasing intrinsic motivation and preventing demotivation from occurring among employees in the organization (Deschamps et al., 2016).

Strategic change management is very important in making strategic decisions in determining and planning the future of a company. It is more about planning for the long term to ensure the future growth of the company such as formulating strategies, implementing strategies and re-evaluating the strategies implemented as well as making improvements (Lynch, 2018). Effective strategic change management gives the company value added, seeks, creates and strengthens in overcoming its competitive position (Fuertes et al., 2020).

1.1 Overview of Selected Company

CTRM Aero Composites is a wholly-owned subsidiary of Composites Technology Research Malaysia Sdn Bhd (CTRM) of which CTRM Sdn Bhd is a member of DRB Hicom. CTRM is a composite component manufacturer in the aerospace manufacturing industry sector which is confined to the wings and engine cover of an aircraft, supply for main commercial aircraft manufacturer in the world such as Airbus SE and Boeing Co. As the pandemic strike down the world's economy, airliners delaying the aircraft ordering or even canceling order books due to high fusclage reserves. CTRM also did not escape from this crisis since customers have reduced their demand by approximately 40% and this has impacted in a reduction of company revenue. The company needs to survive to maintain in this sector of the industry, otherwise 2200 of employees will lose their jobs.

1.2 Background of Study

Covid-19 pandemic has swept away billions from the aviation industry where the global crisis has narrow down the profit margin as well as weakened the sector growth. The crisis has given a global impact on air traffic where a drastic drop in the number of world flights is due to the Movement Control Order (MCO) across the country to curb the virus from further spreading. International flights have a very bad impact compared to domestic flights. However, the domestic aviation market experienced a steady decline due to lack of demand for airlines (Suau-sanchez et al., 2020).

According to the International Civil Aviation Organization (ICAO) reported on 25th May 2021, there was a significant decline of 60% overall in passenger numbers and about USD 371 billion of airline revenue suffered losses on gross passenger operations, impact from the Covid-19 pandemic on world scheduled passenger traffic for 2020 compared to the previous year. Nevertheless, in 2021, the decline in passenger volume is estimated at 41–49% with approximately a fall in revenue on airlines from USD 274 billion to USD 322 billion of gross passenger operations (ICAO, 2021). The crisis has narrow down the profit margin as well as weakened the sector growth.

Some leaders will respond to the crisis by rising up and facing the challenge of strengthening the economy, society and community while there are also leaders who will disappear from the struggle. At that time, authentic leaders will help the entire system in an organization and strive to overcome limitations and fears as well as further increase their performance (Dirani et al., 2020). A crisis is something unexpected that can cause stress in an organization. It occurs when there is a strategic gap due to a lack of adaptation to the changes taking place in the organization. Strategic leadership plays important roles in crisis management process (Obeidat & Al Thani, 2020). In every change that leaders have made in setting goals, they will lead to the overall performance in the organization which is basically determined from the performance of the employees.

Employees will become depressed due to the financial pressures faced during the crisis which in turn affects their productivity. In the efforts of the management has been carried out in ensuring the stability of the company has resulted in a negative impact on employee morale. Employees will feel insecurity in their jobs, fear of a pandemic, unable to adapt in the new norm situation as well as feel anxious about the future of the company. This has caused employees to lose their focus in achieving company goals as a result in declining of employee performance and in turn have an impact on the organizational performance.

In existing studies, researchers study the effects of Covid-19 pandemic on employee performance in the global aviation industries were limited to aviation services providers as well as air transportation (Alkadash & Alamarin, 2021; Buhusayen et al., 2020; Liu et al., 2020; Melas & Melasová, 2020; Suau-sanchez et al., 2020). However, there is a limited study on the effect of employee performance in aerospace manufacturing sector. Therefore, this study will determine the factors that will influence employee performance through motivation, work performance and Work from Home.

2 Literature Review

2.1 Employee Performance

Employee Performance (EP) is dependent variable that need to focus on this study. In an organization, the organizational performance indicates a company's ability to achieve independent goals efficiently. The employee performance is among the elements that can assess the efficiency of an organization through its level of productivity (Almatrooshi et al., 2016).

Employee performance can be influenced by many factors, especially during Covid-19 pandemic where the new normal changes affect employee performance. Reactions to change during a pandemic have a negative impact on employee performance. As organizations try to adapt to new situations or crises, there are changes that need to be made especially for safety. The changes basically are the barrier for employees to perform on their daily task to meet organization goals (Hamid et al., 2020).

2.2 Change in Management

To adapt in current situation, leaders need to react quickly and do some changes in strategic management plan in order to survive in the industry. Due to changes in management decision have resulted in several factors that have influenced the performance of employees. There are three factors as dimension for independent variables that need to be study which give the significant impact to Employee Performance. They are motivation, working environment and work from home.

2.2.1 Motivation

Motivation in work is measured through three components, namely the intrinsic motivation component, the extrinsic motivation component, and the demotivation component. Extrinsic motivation refers to rewards for an achievement, annual increments, food subsidies, and other things that refer to short -term motivation. Intrinsic motivation refers to appreciation, job security, good work environment, job safety, willingness to work with high commitment and others that include motivation for the long term.

By providing extrinsic motivation in the form of rewards such as overtime pay during the Covid-19 pandemic, rewards to employees who innovate in the employment process as well as savings in operating costs, etc. which play an important role in helping meet the company's operational needs. Nevertheless, it should be noted that this short-term motivation will not last long and will put risk to the company such as retrenchment, absenteeism, turnover, as well as job dissatisfaction, especially during this pandemic situation. It must be followed by intrinsic motivation whereby employees will feel more valued for success and performance achieved as well as safety in the workplace and strong working environment (Christian et al., 2020).

2.2.2 Work Environment

Every employee needs a comfortable and safe work environment, either physically or mentally. During Covid-19 pandemic, safety and optimal work environment can help employees to increase their job performance (Susilo, 2020). Therefore, it is important for an organization to create better work environment to make sure employees maintain their performance. It is because a conducive work environment can increase the efficiency of employees to work which will affect employee performance (Aropah et al., 2020).

2.2.3 Work from Home

WFH is known as telework where it is a flexible work arrangement when the employee works from home and is not in the office or on the production premises, without contact with colleagues, but the ability to communicate with colleagues using information and communication technology (Aropah et al., 2020).

Throughout the Covid-19 pandemic, work from home (WFH) has become an alternative for every company in pursuing day-to-day work. WFH will have a positive effect on employee performance where performance will increase as employees are given the freedom to reorganize work from home. WFH mode also can increase stress on employees due to constraints that occur such as lack of resources and direction, disruption during WFH, fear of something unknown, technology issues, communication issues, time management, and so on (Sahni, 2020). However, there are advantages and disadvantages during WFH. The advantages of working from home are the working environment is more flexible, avoid transportation cost and traffic, no time constraint, very relaxing and not subject to the company rules. But the disadvantages are we need to pay electricity and internet cost. Besides, we will be more vulnerable to data security problems (Purwanto et al., 2020).

2.3 Underpinning Theory

Typically, employee performance will decline with organizational restructuring, innovation, mergers and downsizing. With effective leadership, employees' development, motivation, communication, organizational culture, and self-directed team can overcome employee performance problems (Daniel, 2019). In existing studies, researchers study the effects of Covid-19 on employee performance in global aviation industries were limited to aviation services providers as well as air transportation (Alkadash & Alamarin, 2021; Buhusayen et al., 2020; Liu et al., 2020; Melas & Melasová, 2020; Suau-sanchez et al., 2020). There are still limited of studies has been conducted on the effect of employee performance of the aero composite manufacturing sector.

There were a study shows the negative impact of Covid-19 and reaction to change on employee performance of Bangladesh. This is because they are unable to adapt to changes with new norms that affect the way they communicate and work concentration (Hamid et al., 2020). (Alkadash & Alamarin, 2021) recommend that SME's company need to consider motivation, work environment and job satisfaction factors where they show positive significant effect to the employee performance. In another study, employee motivation had a positive relationship to employee performance. Rewards and recognition are among the motivations that influence the positive relationship with employee performance. In addition, job security as well as a good work environment also have a positive relationship with employee performance. This hypothesis is as explained in Maslow's theory of Need Hierarchy because meeting their needs is more motivated and directed towards good output (Sandhu et al., 2017).

According to Christian et al. (2020), in the midst of the current Covid19 pandemics, the main needs of employees are focus on physiological needs in daily life as well as safety in employment, health and a strong work environment as other human needs are subject to current policies in curbing the current Covid-19 virus from spreading with the practice of physical distancing, self-quarantine as well as government movement control orders. The purpose is to maintain employee motivation throughout the pandemic.

Based on the results of previous studies, this research puts forward the following hypotheses:

H1: Motivation has significant impact to Employee Performance.

New normal has changed the work environment for any organization through Covid-19 pandemic. It has impact on daily activities where employees have limitation to perform daily tasks as they need to comply with government SOPs such as social distancing, wearing face masks, frequently wash their hands, sanitize workplace before and after working, no physical contact or attend any outside activities and implementation of MCO. From the study of the impact of Covid-19 on the employee performance of Bangladesh has shown the negative relationship where the fear of this pandemic has disrupted their daily routine, ways of communication and their behaviour (Hamid et al., 2020).

Based on the study, this study was made with hypothesis:

H2: Work Environment has significant impact to Employee Performance.

Based on a study wrote by Sahni (2020), among service industry employees, WFH can increase the level of stress and challenges. The findings of her study have supported the notion that the need for an effective organizational support system from management, society and psycho-social are very important. According to Susilo (2020) in his empirical study which are most of the respondents come from HR and Marketing sector shows, the people that work from home have a relationship with the job performance, job satisfaction, work environment and work motivation. However, they show that WFH has no significant relationship with job performance (Susilo, 2020). (Purwanto et al., 2020) showed the impact of WFH on Indonesian teachers where there are positive and negative impact on their performance.

The argument has come out with the next hypothesis:

H3: Work From Home has significant impact to Employee Performance.

2.4 Research Framework

(Fig. 1)

Fig. 1 Research framework

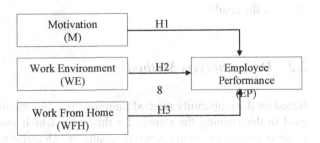

3 Methodology

3.1 Research Method

This study is a quantitative study where the respondent's coverage is among employees at CTRM, Batu Berendam, Malacca, Malaysia. This survey question has been adopted by previous studies and there have been modifications to the question to suit the scope of the study (Hamid et al., 2020; Susilo, 2020). This survey question was produced using the Likert scale in the section that focuses on dependent variable and independent variables, while in the demographic section using the category scale. There are five Likert scales used from 1 to 5 (1-strongly disagree, 2-disagree, 3-neutral, 4-agree, 5-strongly agree) are used to indicate the most appropriate answer from the respondents.

Questionnaires will be distributed to employees via Google form where the respondents are from executives and non-executives' level with experience more than 5 working years without consideration of gender, age, income & department. It is because, the majority of employees has 5 to 7 years working experience in the company. The contribution range is 10–15% of respondent for both executives and non-executives' level from the population (approximately 2200 of employees in the organization) and this study observation performed on post covid situation in FY2021. From the contribution range, 10% of senior management (Head of Department and above), 30% from middle management (Area Leader, Supervisor and Head of Section) and 60% from technicians had been involved in this study.

However, there are limitation of resource which sample population will not represent of total population of employee. Besides, high minimum range of experience employee might interrupt in data analysis to evaluate the effectiveness of change have been made. This study is conducted on only one company in the aerospace manufacturing sector in Malaysia. The small sociodemographic differences between our volunteer sample and the overall study population are not expected to greatly affect in the results.

3.2 Data Analysis Method

Based on the probability method sampling, random sampling techniques have been used in determining the sample for this study which involved full-time employee in aero composite manufacturing company. Questionnaires were distributed to employees in every department within the company via an online Google form. The study population is composing of employees working in the aero composite company in Malaysia. After receiving feedback from respondents, data will be extracted from the Google form for the evaluation and analysis process. Data obtained will be compiled using appropriate statistical analysis techniques. Statistical Product and

Service Solutions (SPSS) and Analysis of a Moment Structures (AMOS) will be used in analyzing the data that has been collected to get the accuracy of result.

4 Conclusion

The main objective of this study is to examine factors that have relationship with employee performance in an aerospace manufacturing company in Malaysia during Covid-19 pandemic. Review of the literature has shown the effects of motivation, work environment, and WFH on job satisfaction and consequently influence employee performance during the pandemic. The study has shown that motivation, work environment and WFH have a relationship with employee performance.

All other studies have agreed that motivation has a significant impact on employee performance during the Covid-19 pandemic where employees will be more diligent in achieving good performance if they are given appropriate rewards and appreciation. The work environment also has a significant impact on employee performance where if the workplace is safer, it will improve their performance. WFH shows a relationship to employee performance in different view. WFH will have a positive effect on employee performance where performance will increase as employees are given the freedom to reorganize work from home. But WFH also might increase stress among employees due to lack of resources, technology constrain and any other boundaries which caused mental health that leads to prolonged stress or even psychosomatic illness. Therefore, to sustain employee emotional, support by organization and community is require during crisis like Covid-19 pandemic and in turn will affect their mental health. However, motivation, work environment and WFH have a significant relationship to job satisfaction as a mediator and in turn have an impact on employee performance where if employees have satisfaction in doing work, then employee performance will increase.

Malaysia is in the process of providing vaccines to all citizens. In achieving herd immunity through the National COVID-19 Immunization Program (NCIP) implemented by the government will provide opportunities to all industries, especially in the aerospace sector in reviving the country's economic sector. Although the economic sector is opening up gradually, this does not mean that it is safe out there. The virus can spread through the airborne and can cause an increased case of Covid-19 infection. Thus, the results of this study will provide insights into future studies in revealing the factors that influence employee performance to help organizations give more focus to these factors during making company's strategic management changes when the organization is facing difficult situations such as this Covid-19 pandemic, in order to achieve the desired goal besides curbing the Covid-19 viruses from further spreading into the community. This can also give guidance to the company in developing a vision, mission or business strategy plan. Human Resources Department (HR) should also be involved in identifying factors that influence employee performance where HR involvement in increasing employee engagement is very important.

References

Abbasi, B. (2017). Transformational leadership and change readiness and a moderating role of perceived bureaucratic structure: An empirical investigation. *Problems and Perspectives in Management, 15*(1), 35–44.

Al-Dmour, R., AlShaar, F., Al-Dmour, H., Masa'deh, R., & Alshurideh, M. T. (2021). The effect of service recovery Justices strategies on online customer engagement via the role of 'customer satisfaction during the Covid-19 pandemic: An Empirical Study. *Effects of Coronavirus Disease Bus Intelligence, 334*, 346–325.

Alkadash, T. M., & Alamarin, F. (2021) An integrative conceptual framework on employee performance during COVID-19 pandemic for Bahrain SMEs. *Psychology and Education, 58*(2), 3812–3817.

Almatrooshi, B., Singh, S. K., & Farouk, S. (2016). Determinants of organizational performance: A proposed framework. *International Journal of Productivity and Performance Management, 65*(6), 844–859.

Alshurideh, M. T., et al. (2021). Factors affecting the use of smart mobile examination platforms by universities' postgraduate students during the COVID 19 pandemic: An empirical study. *Informatics, 8*(2), 32.

Amarneh, B. M., Alshurideh, M. T., Al Kurdi, B. H., Obeidat, Z. (2021). The Impact of COVID-19 on E-learning: Advantages and Challenges. In *The International Conference on Artificial Intelligence and Computer Vision* (pp. 75–89).

Aropah, V. D., Sarma, M., & Sumertajaya, I. M. (2020). Factors Affecting Employee Performance during Work from Home. *International Research Journal of Business Studies, 13*(2), 201–214.

Buhusayen, B., Seet, P. S., & Coetzer, A. (2020). Turnaround management of airport service providers operating during covid-19 restrictions. *Sustain, 12*(23), 1–24.

Christian, S., Susita, D., & Martono, S. (2020). How to maintain employee motivation amid the Covid-19 virus pandemic. *International Research Journal of Business Studies, VIII*(4), pp. 78–86.

Daniel, C. O. (2019). Effect of organizational change on employee job performance. *Asian Journal of Business and Management, 7*(1), 22–27.

Deschamps, C., Rinfret, N., Lagacé, M. C., & Privé, C. (2016). Transformational leadership and change: How leaders influence their followers' motivation through organizational justice. *Journal of Healthcare Management, 61*(3), 194–213.

Dirani, K. M., et al. (2020). Leadership competencies and the essential role of human resource development in times of crisis: A response to Covid-19 pandemic. *Human Resource Development International, 23*(4), 1–15.

Fuertes, G., Alfaro, M., Vargas, M., Gutierrez, S., Ternero, R., & Sabattin, J. (2020). Conceptual framework for the strategic management: A literature review—Descriptive. *Journal of Engineering (United Kingdom).*

Hamid, M., Wahab, S. A., Hosna, A. U., Hasanat, M. W., Kamruzzaman, M. (2020). Impact of coronavirus (COVID-19) and employees' reaction to changes on employee performance of Bangladesh. *International Journal of Business and Management, 8*(8).

ICAO. (2021). Effects of novel coronavirus (Covid-19) on civil aviation: Economic impact analysis. *International Civil Aviation Organization (ICAO), Montréal, Canada* (p. 125).

Leo, S., Alsharari, N. M., Abbas, J., & Alshurideh, M. T. (2021). From offline to online learning: A qualitative study of challenges and opportunities as a response to the COVID-19 pandemic in the UAE higher education context. *Effects of Coronavirus Disease Bus Intelligence, 334*, 203–217.

Liu, J., et al. (2020). Will the aviation industry have a bright future after the COVID-19 outbreak? Evidence from Chinese airport shipping sector. *Journal of Risk and Financial Management, 13*(11), 276.

Lynch, R. (2018). *Strategic management* (8th ed.). Pearson Education.

Melas, D., & Melasová, K. (2020). The early impact of Covid-19 pandemic on the aviation industry. *Acta Avion Journal, XXII*(1), 38–44.

Obeidat, A. M., & Al Thani, FBH. (2020). The impact of strategic leadership on crisis management. *International Journal of Asian Social Science, 10*(6), 307–326.

Purwanto, A., et al. (2020). Impact of work from home (WFH) on Indonesian teachers performance during the Covid-19 pandemic : An exploratory study. *International Journal of Advanced Science and Technology, 29*(5), 6235–6244.

Sahni, J. (2020). Impact of COVID-19 on employee behavior: Stress and coping mechanism during WFH (work from home) among service industry employees. *International Journal of Operations Management, 1*(1), 35–48.

Sandhu, M. A., Iqbal, J., Ali, W., & Tufail, M. S. (2017). Effect of employee motivation on employee performance. *Journal of Business and Social Review in Emerging Economies, 3*(1), 85–100.

Shah, S. F., Alshurideh, M. T., Al-Dmour, A., & Al-Dmour, R. (2021). Understanding the influences of cognitive biases on financial decision making during normal and COVID-19 pandemic situation in the United Arab Emirates. *Effects of Coronavirus Disease Business Intelligence, 334*, 274–257.

Suau-sanchez, P., Voltes-dorta, A., & Cugueró-escofet, N. (2020). An early assessment of the impact of COVID-19 on air transport: Just another crisis or the end of aviation as we know it ? *Journal of Transport Geography, 86*(May), 39–43.

Susilo, D. (2020). Revealing the effect of work-from-home on job performance during the Covid-19 Crisis: Empirical evidence from Indonesia. *The Journal of Contemporary Issues in Business and Government, 26*(01), 23–40.

Wang, C., Horby, P. W., Hayden, F. G., & Gao, G. F. (2020). A novel coronavirus outbreak of global health concern. *Lancet, 395*(10223), 470–473.

The Effect of Virtual Working Team on Job Performance: A Study of Aerospace Manufacturing Industry in Malaysia

Azmi Adam, Nurhanan Syafiah Abdul Razak, and Muhammad Turki Alshurideh

Abstract Job performance is one of most businesses' measures to ensure that their business is sustainable, lucrative, and expandable. It is also an important level indicator in Malaysia aerospace manufacturing industry to maintain competitiveness in the global market. By the end of 2019, when a pandemic had hit the world, the situation became more challenging where the government-imposed movement controls to prevent the disease from spreading. This situation led most aerospace manufacturing companies to explore alternative measures to maintain their business operations, including introducing work from home as a virtual working team for their indirect workers. As this strategy is just introduced in Malaysia aerospace manufacturing sector, the concern arises as to the extent of its impact on their job performance. This study purposely to examines the influence factors in the virtual working team environment and its relationship to job performance. Therefore, the findings hope to provide valuable insights to Malaysian manufacturing organizations in effectively managing their virtual working team and enhancing their employees' job performance.

Keywords Job performance · Virtual working team · Team coordination · Communicate frequently · Quality of work · Effective communication

A. Adam · N. S. A. Razak
Faculty of Business and Management, DRB, HICOM University of Automotive Malaysia, 26607 Pahang, Malaysia
e-mail: azmi.adam@ctrm.com.my

N. S. A. Razak
e-mail: hana.syafiah@gmail.com; nurhanan@meritus.edu.my

N. S. A. Razak
MERITUS University, Kuala Lumpur, Malaysia

M. T. Alshurideh (✉)
Department of Marketing, School of Business, The University of Jordan, Amman 11942, Jordan
e-mail: m.alshurideh@ju.edu.jo; malshurideh@sharjah.ac.ae

Department of Management, College of Business, University of Sharjah, 27272 Sharjah, United Arab Emirates

M. Alshurideh et al. (eds.), *The Effect of Information Technology on Business and Marketing Intelligence Systems*, Studies in Computational Intelligence 1056,
https://doi.org/10.1007/978-3-031-12382-5_68

1 Introduction

Malaysia's aerospace manufacturing industry has been critical in developing countries' efforts to expand and flourish, especially in the manufacturing and aviation sector. This industry has been predominantly export-oriented, which is directly contributes to Malaysia's economy. For the manufacturing industry, competent employees are the most important function for the sector to survive for a long business period. Understanding and exploiting the implications of the job performance, which is advantageous to both employers and employees. Employee performance refers to how efficiently and effectively a person does their work (Purbasari and Septian, 2017). An organization must understand how its employees perform at work to succeed in a worldwide competitive market. Employees' effort to achieve organizational objectives is determined by whether the individual succeeds or not in completing each strategy outlined and implemented effectively. It is attractive to know and discuss components of their working environment aspect related to their job performance. The issue of job performance becomes interesting to investigate in perspective of what factors can influence employees' performance.

Many factors can influence employees' performance in the manufacturing industry; however, this study will focus on a different situation where the global is dealing with a pandemic that impacts the entire socioeconomic system. To protect the community and prevent the spread of the disease, the Malaysian government has taken proactive strategies by imposing Movement Control Order (MCO). Only a few essential businesses categories are allowed to continue their operating after getting approval from the Ministry of International Trade and Industry, Malaysia (MITI). However, the approval is subject to stringent requirements with limited numbers of workers in the plant (not exceeding more than 60%). In addition, they must adhere to the government's Standard Operating Procedures (SOP). This situation gives the leadership a challenge in completing their task effectively to the given timeframe. When working from home as a virtual working team, the main concerns raised were how they managed their way of communicating effectively and the factors influencing their job performance.

1.1 Background of Study

Job performance refers to the standard to which an employee satisfies the variables specified in the job description. It also depends on an employee's abilities, experience, seriousness, and the amount of time available to perform the task (Raharjo & Sulistiasih, 2019). According to Bokhori and Halim, several factors are influencing on an employee's performance, such as personal issues, supervisory attitudes, work environment, general company culture, working conditions, job responsibilities, teamwork, employee relationships, non-financial and financial incentives rewards, and flexible working hours. Managers need to assess if their subordinates are doing their jobs effectively and efficiently or room for improvement (Robbins, 2009). The performance of each employee has a significant impact on the company's overall performance (Onyebuchi et al., 2019). As for the aerospace manufacturing sector, it is critical to guarantee that the global supply chain is adequate to deliver commercial aircraft components so as not to be disrupted. Due to the new Covid-19 pandemic situation, many authorized industries have minimized the number of workers in their manufacturing plants to curb the spread of disease. This led to most non-direct production employees working from home as a virtual working team to communicate among employees, suppliers, and customers.

The effective virtual working team on daily routine work becomes a critical aspect that needs to be focused on since it will influence job performance. However, there were limited studies on virtual working teams related to the manufacturing industry, especially in the aerospace manufacturing sector in Malaysia. Hence, this study aims to evaluate the factors influencing job performance for Malaysia's aerospace manufacturing companies working in virtual communication during a pandemic. There are two main objectives outlined in this study; to examine the factors influencing employees' job performance of effective virtual working communication and identifying strategies for employees to enhance their job performance by effectively working in communication during a pandemic.

By conducting this research systematically and ordered, it is important to show how the study is performed. This study is limited to the Malaysian aerospace manufacturing sector and its employees, which are specializing in the aero-structure manufacturing industry and dealing with global suppliers and customers. The study focuses on the employees who have worked as a virtual working team and communicate during the pandemic from the year 2020–2021.

2 Literature Review

At a company level, the organization's dedication to systematize the objectives and mission of every individual in the company and collaboration with high-level management is required to manage the company as effectively as possible (Raharjo & Sulistiasih, 2019). According to Kassim et al. (2009), solid organizational commitment will motivate people to work toward organizational goals and employee performance.

2.1 Job Performance

Job performance is defined as the result of an employee's best effort to complete a task depending on their skills, experience, and seriousness, and the amount of time available (Raharjo and Sulistiasih, 2019). It means the precise results that a person wants to achieve in their employment and the organization's desire to attain

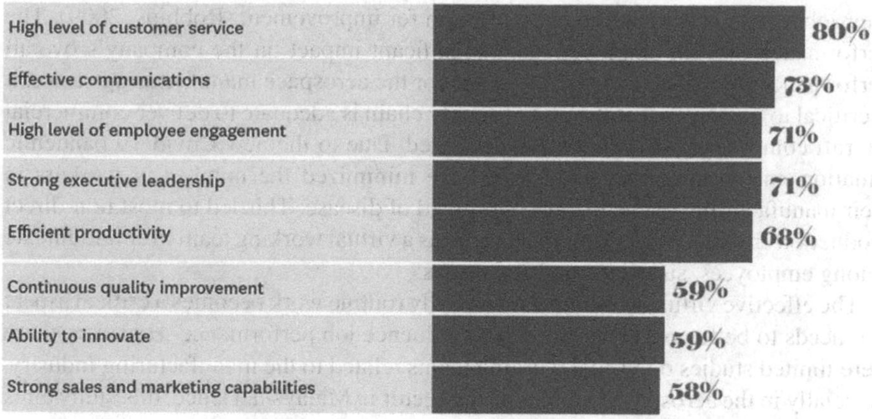

Fig. 1 Factors most likely to bring success. *Source* 2013 HBR Analytic Services Survey

those results. While Bokhori and Halim defined it as the capability and ability of an employee to accomplish their job is referred to as employee performance. The job requirement includes each employee's responsibilities, activities, and tasks in the organization. Leadership style also has a significant and beneficial impact on organizational success and employee performance (Simsek & Ozturk, 2018).

Meanwhile, three primary elements are measured for employee performance: job productivity, job quality, and job accomplishment (Iqbal et al., 2012). However, in the context of work from home or work remotely, on job performance of employees remains uncertain (Allen et al., 2015). According to Susilo (2020), the characteristics of the job itself significantly impact job performance when working in telecommuting. Telecommuting is defined as working on a computer connected within employees from home or wherever else. It is crucial for employees in advisory or supporting tasks, such as promptness in responding and good attitude attributes, such as being trustworthy (Judge et al., 2017). Management support, trust, and communication are essential success factors for a virtual working environment (Kowalski and Swanson, 2005). Based on findings from a global survey done by Harvard Business Review (HBR) in 2013 shows that the effective communication element provides the second-highest (Fig. 1) and most influential factor to enhancing employee's performance.

2.2 *Virtual Working Team*

Virtual communication becomes a common platform for every meeting, discussion, presentation, and reporting of all necessary information among employees as a daily routine works. Virtual communication is more complex than face-to-face, demanding, and challenging communication (Lockwood, 2015). Effective communication includes words, behaviours, quick and correct responses in delivering information (Morgan et al., 2014). With the pressures of global economic competition and supported by the rapid development in communication technology, most organizations are increasingly focusing on virtual team's communication capabilities to increase the effectiveness in communication on their operation (Nydegger & Nydegger, 2010). Madell (2019) claimed that working remotely has weaknesses, such as a lack of monitoring, which increases the risk of miscommunication. Based on Mlitz (2021), many employees who work remotely do not have a properly designated workspace; instead, their living and work environments are mixed. Thus, employees struggling to focus while working remotely, and it will be affecting their job performance (Fig. 2).

A virtual working team, also known as the remote team, refers to a group of employees who collaborate and work together to plan, execute, and assess the progress of the project (Salas et al., 2008). Each working assignment and communication are through technology such as email, voice, or video conferencing services in order to collaborate. Although technology has improved communication ability, the performance of virtual working teams faces the challenge of losing knowledge due to ineffective coordination and communications and lack of time in sharing the information (Karayaz, 2008).

Most previous studies show that virtual team communication is more effective when it is coordinated with the right style and traits of the leaders (Balthazard et al., 2009; Hoch & Kozlowski, 2014; Kelley & Kelloway, 2012). In addition, the practical key elements for a virtual working team include frequent communication and

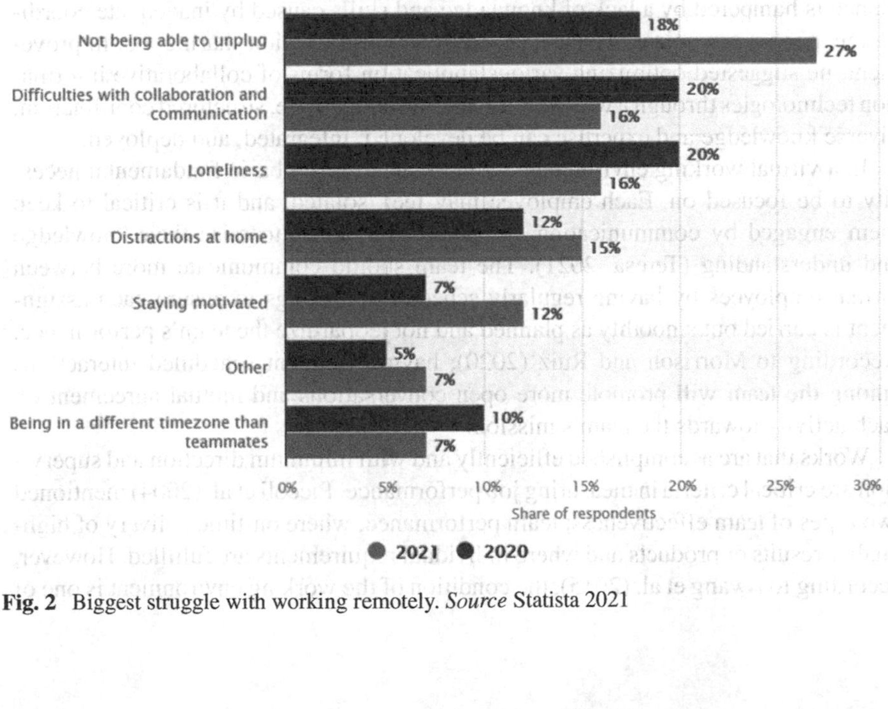

Fig. 2 Biggest struggle with working remotely. *Source* Statista 2021

have clear objectives among team members (Morgan et al., 2014). By practising regular communication positively influences the motivation of each member of the virtual team to concentrate and convey information effectively in order to complete the task perfectly according to the plan. Individual quality of work is also one of the essential parts of a virtual working team. Each employee is responsible for the output of quantity and quality required to achieve task objectives as agreed upon by the working team or their superior (Ivancevich & Matteson, 1996). According to Cullen & Sackett (2004), the concept of employee integrity itself directly impacts the quality of their job performance. Furthermore, to complete all of the tasks every day, employees must consistently have a certain level of commitment. Working as part of a team requires synchronization and subsequent interconnected activities across team members; small mistakes or delays will affect the team's goals and performance.

2.3 The Relationship Between Virtual Working Team and Job Performance

Virtual working teams depend heavily on technology communication channels to coordinate their daily work. It is crucial to examine the relation between the task's structure and coordination effectiveness as it affects individual and team performance (Victor et al., 2021). Working as a team must have strong coordination among employees in an organization. Iskandar et al. (2014) claimed its absence will show difficulty in operating business operations or, more importantly, cause serious harm between individuals. In addition, according to Karayaz (2008), virtual team performance is hampered by a lack of knowledge and skills caused by inadequate coordination, poor communication, and a lack of timely information sharing. For improvement, he suggested optimizing various application forms of collaborative information technologies through a well-coordinated system. Hence, structured coordination, diverse knowledge and expertise can be developed, integrated, and deployed.

In a virtual working environment, communication is a basic fundamental necessity to be focused on. Each employee may feel isolated, and it is critical to keep them engaged by communicating with them frequently to raise their knowledge and understanding (Teresa, 2021). The team should communicate more between virtual employees by having regularly scheduled meetings to ensure each assignment is carried out smoothly as planned and not jeopardize the team's performance. According to Morrison and Ruiz (2020), having frequent scheduled interactions among the team will promote more open conversations and mutual agreement on each activity towards the team's mission.

Works that are accomplished efficiently and with minimum direction and supervision are critical criteria in measuring job performance. Piccoli et al. (2004) mentioned two types of team effectiveness; team performance, where on-time delivery of high-quality results or products and where individual requirements are fulfilled. However, according to Awang et al. (2015), the condition of the working environment is one of the factors influencing employees' work speed and effectiveness. When working at home, it is necessary to create a good work environment that is both comfortable and conducive to acquire concentration and produce good outcomes. Understanding the concept of quality of work that adheres to the standards such as accuracy, neatness, attention to detail, consistency, and follows procedures will improve job and team performance.

Based on past studies in literature, several elements have been identified in a virtual working team and their relationships with job performance. Hence, there are three hypotheses have been constructed to verify the correlation between the dimension of the virtual working team and job performance as follows:

H1—Team coordination has a direct positive influencing on Job Performance.

H2—Communicate frequently has a direct positive influencing on Job Performance.

H3—Quality of work has a direct positive influencing on Job Performance.

2.4 *Conceptual Framework*

Virtual working teams are influenced by many factors of variables that have been studied in the literature and have resulted in a variety of models for studying and connecting them to job performance. However, the studied and theoretical concept established by Victor et al. (2021) is emphasized in this research. According to the study's findings, one of the essential variables in the model is task features. The main part of the task feature is effectively coordinate their work in communication technologies where the result gives positive influence with a high confidence level. Next, an important variable is the empowerment of the team members by the leader. Empowerment consists of sharing a vision and regularly communicating within the team to create a significant connection to the team performance. Another important factor identified in the study is working at a high-efficiency level for an extended period. It entails delivering high-quality results and solutions on time, ensuring all stakeholders' satisfaction, and increasing their job performance.

Based on this underpinning theory, the framework is designed to examine the relationship between three dimensions in a virtual working team as an independent variable and the dependent variable's job performance. The three dimensions are team coordination, communication frequently, and quality of work. The study's model was developed after a thorough evaluation of the scientific literature on the subject (Fig. 3).

3 Research Methodology

This study uses a quantitative causal method for analyzing the effect and relevance of independent variables on dependent variables. Quantitative analysis is based on

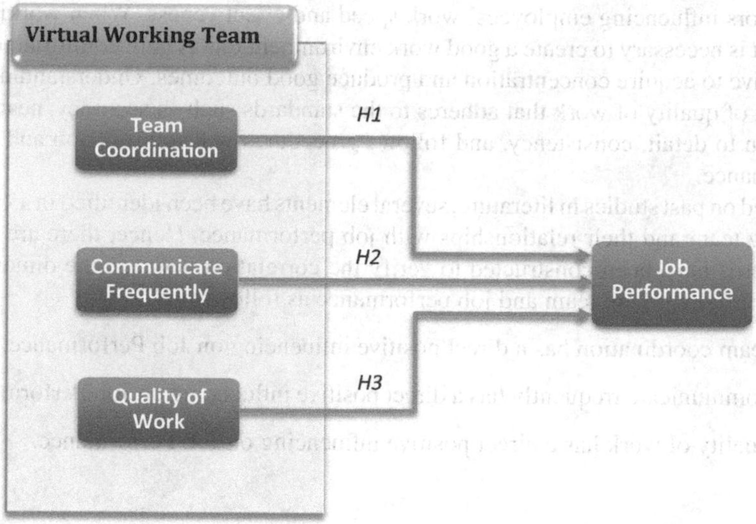

Fig. 3 Research conceptual framework

numerical results and delivers systematic conclusions covering a wide range of research topics while being more efficient, effective, and cost-effective (Ticehurst & Veal, 2000). The selected method is similar to the previous studies that have almost related to the topic (Iskandar et al., 2014; Susilo, 2020; Victor et al., 2021).

This study employs inferential statistics approaches, which means it aimed to answer the research questions by collecting and analyzing the samples and then generalizing the outcome to explain a population. Data collection will be conducted by distributing the questionnaire and completing online to the relevant aerospace manufacturing industry in Malaysia. For the population design, respondents' selection was based on an executive level in random departments, including managers, associate managers, and senior executives. Assuming that all respondents have work experience and can work from home, it is justifiable to ensure they have access to the internet. Therefore, the Purposive sampling method is being used because it is for employees who had access to the internet and had direct experience working in virtual communication. To help gather respondents, the questionnaire link will be distributed through any means of social media. The responses are measured using a 5-point Likert scale and evaluated accordingly: 1-strongly disagree; 2-disagree; 3-neutral; 4-agree; 5-strongly agree. The questionnaire targeted 500 respondents. All data will be check for response bias, validate the theory and evidence to support the interpretation. Reliable responses data will be kept and analyze further. For statistical analysis, SPSS and AMOS software will assist with the required analyses to effectively evaluate all reliable responses. The results will be presented in figures, tables, and charts, making them easier to interpret.

4 Conclusion

This study demonstrates how a realistic conceptual framework was developed to investigate and understand current factors existing in virtual working communication and their relationship to job performance. It is made a decisive contribution to the research of the influencing factors in virtual working teams on job performance in Malaysia's aerospace manufacturing industry. Furthermore, by analyzing this study's findings, the employees could improve their job performance by implementing appropriate and effective virtual working communication.

References

Allen, T. D., Golden, T. D., & Shockley, K. M. (2015). How effective is telecommuting? Assessing the status of our scientific findings. *Psychological Science in the Public Interest, 16*(2), 40–68.

Awang, N. A., Mahyuddin, N., & Kamaruzzaman, S. N. (2015). Indoor environmental quality assessment and users. *Journal of Building Perrformancce, 6*(1).

Balthazard, P., Waldman, D., & Warren, J. (2009). Predictors of the emergence of transformational leadership in virtual decision teams. *The Leadership Quarterly, 20*, 651–663.

Cullen, M. J., & Sackett, P. (2004). Integrity testing in the workplace. In J. C. Thomace (Ed). *The comprehensive handbook of psychological testing* (Vol. 4). Hoboken.

Duarte, D. L., & Snyder, N. T. (2006). *Mastering virtual teams: Strategies, tools, and techniques that succeed*. Wiley.

Hoch, J. E., & Kozlowski, S. W. (2014). Leading virtual teams: Hierarchical leadership, structural supports, and shared team leadership. *Journal of Applied Psychology, 99*(3), 390–403. https://doi.org/10.1037/a0030264.

Iqbal, J., Yusaf, A., Munawar, R., & Naheed, S. (2012). Employee motivation in modern organization: A review of 12 years. *Interdisciplinary Journal of Contemporary Research in Business, 4*(3), 692–708.

Iskandar, M. et al., (2014). Factors influencing employees's performance: A study on the islamic banks in Indonesia. *International Journal of Business and Social Science* 5(2).

Ivancevich, J. M., & Matteson, M. T. (1996). *Organizational behaviour and management* (4th ed). Chicago.

Judge, A. T., & Robbins, S. P. (2017). Investigating the perceptions of Iranian employees on teleworking. *Industrial and Commercial Training, 44*(4), 236–241. https://doi.org/10.1108/00197851211231513.

Karayaz, G. (2008). Utilizing knowledge management for effective virtual teams. *The Business Review, Cambridge, 10*(1), 294–299.

Kassim, S.H., et al., (2009). Retaining customers through relationship marketing in an Islamic financial institution in Malaysia. *International Journal of Marketing Studies, 1*(1), 66–71.

Kelley, E., & Kelloway, E. K. (2012). Testing a model of remote leadership. *Journal of Leadership and Organizational Studies, 19*(4), 437–449. https://doi.org/10.1177/1548051812454173.

Kowalski, K. B., & Swanson, J. A. (2005). *Critical success factors in developing teleworking programs*. https://doi.org/10.1108/14635770510600357.

Lee, C. (2019). Manufacturing performance and services inputs evidence from Malaysia. ISEAS–Yusof Ishak Institute No. 2019–02.

Lockwood, J. (2015). Virtual team management: what is causing communication breakdown?. *Language and Intercultural Communication, 15*(1). https://doi.org/10.1080/14708477.2014.985310.

Madell, R. (2019). *Pros and cons of working from home*. Retrieved from https://money.usnews.com/money/blogs/outside-voices-careers/articles/pros-and-cons-ofworking-from-home.

Mlitz, K. (2021). *Struggles with working remotely 2020–2021*. Retrieved from https://www.statista.com/statistics/1111316/biggest-struggles-to-remote-work.

Morgan, L., Wright, G., & Paucar-Caceres, A. (2014). Leading effective global virtual teams: The consequences of methods of communication. *Systemic Practice and Action Research, 27*, 607–624. https://doi.org/10.1007/s11213-014-9315-2.

Morrison, S., & Ruiz, J. (2020). Challenges and barriers in virtual teams. *A Literature Review. SN Applied Sciences, 2*(6), 1096. https://doi.org/10.1007/s42452-020-2801-5.

Nydegger, R., & Nydegger, L. (2010). Challenges in managing virtual teams. *Journal of Business and Economics Research, 8*(3), 69–82.

Onyebuchi, O., Obibhunun, L., & Omah, O. (2019). Impact of employee job satisfaction on organizational performance. *Article in International Journal of Current Research*.

Piccoli, G. et al., (2004). Virtual teams: Team control structure, work processes, and team effectiveness. *Information Technology and People, 17*(4), 359.

Purbasari, R. N., & Septian, T. A. (2017). Factors influencing on employee performance of production department on the manufacturing food industry in Indonesia. *Polish Journal of management studies, 16*(2).

Raharjo, D. S., & Sulistiasih, S. (2019). The Model of Manufacturing Industries Employee Performance. *International Review of Management and Marketing, 9*(5), 82–86.

Robbins, S. P., & Coulter, M. (2009). In C. Fernandes & E. Davis (Eds.), *Management* (Tenth-Inte). New Jersey Pearson Education Limited.

Salas, E., Cooke, N. J., & Rosen, M. A. (2008). On teams, teamwork, as well as team performance: Discoveries and developments. *Human Factors: The Journal of the Human Factors and Ergonomics Society., 50*(3), 540–547.

Simsek, A., & Ozturk, I. (2018). Kültürel Zekâ ve Liderlik Arasındaki İlişkinin Analizi: Hastane Yöneticileri Örneği. Ekonomi Bilimlerinde Güncel Akademik Çalışmalar-Gece Kitaplığı Turkey. Conference Paper (pp. 181–191).

Susilo, D. (2020). Revealing the effect of work-from-home on job performance during the Covid-19 crisis: empirical evidence from Indonesia. *The Journal of Contemporary Issues in Business and Government, 26*(1), 23–40.

Teresa, M. P. (2021). *Factors affecting the productivity and satisfaction of virtual workers*. Walden Dissertations and Doctoral Studies. Walden University (pp. 119).

Ticehurst, G. W., & Veal, A. J. (2000). *Business Research Methods. French Forest*. Longman.

Victor, G. A. et al., (2021). Virtual teams in times of pandemic: Factors that influence performance. *Frontiers in Psychology, 12*(624637). www.frontiersin.org.

Creating Organizational Culture that Compact Corruption in Local Government: The Role of Municipal Leadership

Ehap Alahmead, Susan Boser, Ra'ed Masa'deh, and Muhammad Turki Alshurideh

Abstract Corruption in U.S. local government is a phenomenon that boosted by the decentralization tendency. Most of the literature focus on reducing opportunity for corruption by monitoring and controlling the local governments' transactions. However, opportunity reduction faces limited success in combating local governments' corruption. This chapter highlights the need to focus on corruption rationalization. According to the Fraud Triangle Theory, committing a crime requires not only an opportunity but also a rationalization process that precedes or follows the crime. The rationalization is a defense mechanism that aims to normalize the crime and restore good self-image by claiming for instance that everybody practices the crime, the crime did not hurt anyone, or the crime was unavoidable etc. This chapter focuses on how corruption in local governments is normalized (rationalized). Particularly, the role of local governments' leadership in creating and/or maintaining an organizational culture that rationalize corruption.

Keywords Literature review · Corruption opportunities · Public sector

---

E. Alahmead (✉) · S. Boser
College of Health and Human Services, Indiana University of Pennsylvania, Indianapolis, USA
e-mail: YFPW@iup.edu

S. Boser
e-mail: sboser@iup.edu

R. Masa'deh
Department of Management Information Systems, School of Business, The University of Jordan, Amman, Jordan
e-mail: r.masadeh@ju.edu.jo

M. T. Alshurideh
Department of Marketing, The University of Jordan, Amman, Jordan
e-mail: malshurideh@sharjah.ac.ae

University of Sharjah, Sharjah, UAE

M. Alshurideh et al. (eds.), *The Effect of Information Technology on Business and Marketing Intelligence Systems*, Studies in Computational Intelligence 1056,
https://doi.org/10.1007/978-3-031-12382-5_69

1 Introduction

According to Agha et al. (2021), the interest in studying ethics in the U.S. government increased after the Watergate crisis. The Ethics in Government Act of 1978 established the federal Office of Government Ethics (OGE) to lead and direct ethics management in the executive branch (Agha et al., 2021). The OGE requires each federal agency to have an ethics officer to conduct training and keep paperwork of the ethics record (Agha et al., 2021); however, the OGE requirements are overwhelming to federal employees and reduce their performance (Yang, 2009). The OGE concerns with the integrity of the federal employees; when it comes to municipality officials, the federal government has no crucial role in managing their integrity (Mackenzie & Hafclen, 2002). At the federal level, power is distributed between three branches of government, and the legislative branch (Congress) is also divided between parties. If a decision is to be taken to govern municipalities, it requires the consensus of a large number of actors (Mackenzie & Hafclen, 2002). Moreover, due to the municipalities' independency, any federal prosecutions are perceived as a violation of the local government sovereignty; besides, Washington is not concerned if a municipality lacks on ethical standards or financially corrupts as long as no federal fund is involved (Berman & Carter, 2018; Zhao & Peters, 2009). Thus, addressing local governments' integrity is the role of the state government and the citizens.

Most of the literature reviewed for this study is focused on corruption opportunity. It seeks preventing (or reducing) the chance for corruption by monitoring and controlling the transactions of local government staff and hold them accountable. While reducing the opportunity for corruption is very important, it is not enough; as the Fraud Triangle theory (Brown, 2005) presumes, fraud requires three factors: pressure, opportunity, and rationalization. Some research such as Cressey (1973) added a fourth factor to the Fraud Triangle theory, the capability. Capability includes position within the organization to create or exploit an opportunity for fraud, being smart to understand and exploit internal control weaknesses, strong ego and great confidence that he will not be detected, ability to coerce others to conceal fraud. Capability as introduced by Cressey (1973) is part of the opportunity factor not a separate one. Without having predispositions like courage and cleverness to exploit a situation and without being in a position in the organization that enables corruption, we cannot say that there is an opportunity for corruption. The study Wolfe & Hermanson (2004) explores religiosity (commitment to anti-fraud religious values) as another factor to understand fraud. Religiosity also is not a new factor; it deals with religious values that influence rationalization. Religious values like honesty and integrity mark the absence of corruption rationalization for the individual who is committed to these values. Thus, the Fraud Triangle theory is stabilized at the three factors of pressure, opportunity, and rationalization (Said et al., 2018).

The incentive (pressure) for corruption does not require much focus because it usually exists whether to satisfy a need for higher income (Free, 2015); or to satisfy a want like living a lavish lifestyle, and abusing drugs (Liu & Lin, 2012). The corruption opportunity is heavily studied, and it may be less important than corruption rationalization (Muhtar et al., 2018). The research work (Muhtar et al., 2018) study corruption in Indonesia local governments focusing on the influence of opportunity factors like e-government (public easy access to information that improves transparency and combat corruption), and rationalization factors like the response on audit result (recommendations on audit result that received no follow-up from local government). The results show that the rationalization factors have effect on corruption; meanwhile, the opportunity factors like e-government and internal audit show no evidence of effect on corruption. In this research we will focus on corruption rationalization.

2 Literature Review

The literature review starts with discussing the three main approaches to eliminate corruption opportunity namely transparency, citizens involvement, and recentralization and explains why these approaches are problematic. Then, a discussion of corruption rationalization follows and illustrates why rationalization should be attributed to the organization culture (not only to individual characteristics). Finally, the organization culture section discusses how organizational culture emerged and maintained before ending with section on the role of municipal leadership in constructing the organization ethical culture (Ravisankar et al., 2011; Alshraideh et al., 2017; AlShurideh et al., 2019; Ashal et al., 2021; Hamadneh et al., 2021).

2.1 Opportunity

The literature in local government corruption focusses on combating corruption opportunity, which depends on the probability of being caught by auditing or monitoring systems (Nuseir et al., 2021; Al-bawaia et al., 2022). Eliminating corruption opportunity "is to deter and prevent corruption via a system of thorough and efficient observation and surveillance, including accounting, auditing, and layers of oversight" (Silva, 2010). Opportunity for corruption may depend on the economic activity in the local government area; when tourism is the major activity, there is more presence of pressure groups and lobbies who may directly participate in the political system and corruption may flourish more easily (Gong, 2015). Corruption opportunity may increase in crisis times when local officials are given relief aid and simultaneously the natural disaster brings disorder and infrastructure collapse, i.e. less chance for being caught (Anechiarico & Jacobs, 1996). Many approaches to eliminate corruption opportunity are discussed in the literature; for instance, having more women in the local government (Jimenez et al., 2017), restructuring anti-corruption agencies to produce more comprehensive integrity management (Al-bawaia et al., 2022), reducing the size of the city council (Nguyen et al., 2017), or reshaping incentive system and based it on citizens' perception of low corruption (Vijayalakshmi, 2008).

However, the three approaches to eliminate corruption opportunity that have the most attention in the literature are transparency, citizens' involvement, and recentralization (as will be discussed next).

2.2 *Transparency*

Members of the government are more prone to corruption because they have more control over the flow of information than citizens (Bergh et al., 2017). The more transparency (the more timely information delivered to citizens), the more the citizens power to monitor the government performance and combat corruption (Bergh et al., 2017; Schopf, 2018). Transparency improved by providing the public with easy access to information (Muhtar et al., 2018). This can be achieved through information technologies such as websites, social media, and mobile technologies; the citizens who interact with municipal government through information technology have better perception of the municipality's transparency and integrity (Kim et al., 2009; Benito et al., 2015; Al Dmour et al., 2014; Valle-Cruza et al., 2016; Akour et al., 2021; Al-Hamad et al., 2021; Alshurideh et al., 2021a). Moreover, according to Vijayalakshmi (2008), transparency can be improved by attaching the official's names to government documents to clarify responsibility and establishing a computerized information center to provide citizens with information about their application's progress and related administrative rules. In addition to using information technology, law should be amended to improve transparency. Law should be amended to require posting lease or sale of property with big font on the city hall, as well as establishing an independent database of lease or sale of property (Alshurideh et al., 2021b). Law should guarantee that records and meetings are open to public to hold the local government accountable and prevent elected officials from concealing their corrupt practices (Suleman et al., 2021). Budget law should include clear rules on the timing and specifics of disclosure requirements for key items such as payments, debt, departmental budgets, as well as legal liabilities for breaching the disclosure norms. Finally, to improve transparency, local media should be developed because underdeveloped media hardly puts any pressure on the local officials (Gonzalez et al., 2017); meanwhile, municipalities with strong media presence are more likely to abide with state-mandated ethics (Ferraz & Finan, 2011). The hope of promoting transparency is that it may invite public supervision (Al-bawaia et al., 2022; Fording et al., 2003), however, even when transparency makes corruption known to the public, the public may not confront it (Suleman et al., 2021).

2.3 *Citizens Involvement*

Central authorities encourage local agents to actively participate to improve anti-corruption effectiveness by holding the local government accountable to citizens

(Al-bawaia et al., 2022). To reduce opportunity for corruption, citizens need to volunteer at public hearings and committees, discuss public issues they know through the media, exchange their ideas/opinions, and stand up in county meetings or city council to air their complaints (Suleman et al., 2021). Citizens can also participate by establishing a committee of outside experts to receive appeals and investigate bureaucrats, and a citizen's court that is comprised of outside academics, lawyers, and journalists to decide corruption prone issue (Vijayalakshmi, 2008). There should be effective auditing programs that monitor the elected officials' corrupt actions and timely disseminate the auditing results, i.e., shortly before an election (Coxson, 2009). Otherwise, if the citizens lack meaningful information to cast a meaningful vote, they elect officials who lack accountability to them (Klasnja & Tucker, 2013). Citizens can discipline local officials rent seeking behavior by the power of their vote; mayors with eligibility to run again are less corrupted than the second term mayors specially when the probability of corruption being detected (by judiciary agents or media) is high (Bobonis et al., 2016); municipalities in cities with high levels of electoral competition are more likely to abide with state-mandated ethics (Ferraz & Finan, 2011). Citizen's report of corrupt behavior is also an anti-corruption tool; to improve the local officials' integrity, the reporting mechanisms should be enhanced (Al-bawaia et al., 2022). There is a need to increase the public awareness of corrupt behaviors and where to report them (Alshurideh et al., 2021b).

However, relying on local agents to report corruption may not be effective. Many people do not want to involve because strong evidence is required before reporting corruption, anonymity is not guaranteed, or they do not understand their rights; even if they take the decision to report corruption they may not know where to report it (Asthana, 2008). Reporting corruption requires two preconditions, first, positive perception of the quality of anti-corruption governance (i.e., believing that their report will be taken seriously and make difference); and second, a low level of tolerance towards corruption (Masters & Graycar, 2016), both conditions are not guaranteed. For the first condition, people may not believe in their report influence because as it moves from an agency to agency, the report diminishes, it changed from the hard language of allegation to a softer approach to administrative action which makes the act of corruption disappear (Asthana, 2008) Regarding the second condition, people may tolerate corruption, (Muhtar et al., 2018) find that while in 'low-corruption' countries like Sweden voters punish corrupted politicians regardless of the state of economy, in 'high-corruption' countries like Moldavia, voters punish corrupt officials only when the economy is bad. Thus, relying on citizens or local agents' engagement to reduce opportunity for corruption may be problematic because these citizens may not have the knowledge (Alshurideh et al., 2021b), the desire (Suleman et al., 2021), or the ability (Asthana, 2008) to stand against corruption in their counties or cities.

2.4 Recentralization

Government decentralization refers to the "redistribution of powers of decision making in favor of lower levels of administrative machinery" (Gonzalez et al., 2017), Governments decentralized authority to local level assuming that local governments can better tailor the public good to suit local populations (Assad & Alshurideh, 2020). The division of authority among local governments enhances the chance for corruption (Hess, 2017) because as the authority devolves from upper levels, local officials have fewer constraints and more opportunities to abuse their power (Masters & Graycar, 2016). For instance, fiscal decentralization (decentralize the government budget to local governments) provides local officials with opportunity for fraud and aggravate corruption in local government (Tiebout, 1956) Local officials expanded power and their deep involvement in economic affairs provide opportunity for corruption (Masters & Graycar, 2016). As Salloum et al. (2020) discussed, decentralization aggravates corruption and encourages local officials to collude against the central government in China. To solve these problems, the Chinese Communist Party linked the cadre promotion to quantifiable performance, but this policy failed so, corruption and budgetary crisis have grown in the last decades. Under the current Chinese President administration, the national leadership led a central anti-corruption campaign to constrain the local officials' authority (Bobonis et al., 2016). In the United States, decentralization of authority to local governments increases their control over policies (Berman & Carter, 2018). According to Klasnja & Tucker (2013), New Public Administration practices such as decentralization, deregulation, and privatization weaken the municipalities' accountability to the state authorities.

While decentralization contributes to local government corruption, tackling the problem by recentralizing the authority is not an effective approach (Bobonis et al., 2016). The recentralization tendency may reflect the center's desire to address the local government integrity through hierarchy (Al-bawaia et al., 2022) such as state ethics commissions that seek addressing the local government integrity problem by tracking election expenditures and establish rules for conflicts of financial interest (Agha et al., 2021). However, recentralization may be counterproductive; in New York City it led to inadequate managers' authority, increased delay, and low morals (Silva, 2010). By managing the integrity problem through the state hierarchy, the central government underemphasizes the role of public participation (Al-bawaia et al., 2022), emphasizes compliance with rules and regulations over ethics (Agha et al., 2021), treats public employees as potential corrupt (Silva, 2010), and slows down the investment attraction and development (Bobonis et al., 2016).

The previously mentioned articles represent the traditional approaches to public integrity, which focus on reducing the corruption opportunity by enforcing commissions, anti-corruption agencies, formalized guidelines, or whistle-blowing (Beeri & Navot, 2013). However, there is a need to understand corruption from a behavioral perspective that may explain the individual's response to the corruption opportunity and the organizational culture in which the corrupt behavior occurs (Ko & Zhi, 2013). The opportunity elimination approaches have met with limited success because they neglect the human factor and the need to embed virtue and values in the individual (Beeri & Navot, 2013). Individuals who engage in corruption usually work in environment that provides definitions favorable to corruption (Collins, 2012). Although the external controls (that seek eliminating corruption opportunity) are important, public administrators should have inner check (ethical values) and adhere to democratic principles while they execute their responsibilities. In conclusion, although technical solutions cannot be neglected, these solutions alone have not proved the success that their advocates had hoped; so, other approaches like developing the organizational values are essential to complement them (Beeri & Navot, 2013).

2.5 Rationalization

Rationalization is the justification made by the fraudsters to avoid feeling guilty (Cressey, 1973). Rationalization contributes to corruption because it helps the fraudster seeing himself as an honest person caught in a negative set of circumstances (Brown, 2005). It is the greatest factor causing tolerance of petty theft—which indicates a problematic organization culture. Rationalization is the factor leading to fraud in the sense that others also do it. So, it is positively related to employees' fraud (Wolfe & Hermanson, 2004).

Although understanding the nature of fraud rationalization is critical to properly formulating anti-fraud measures, it is still an under-researched issue (Said et al., 2018). Research is limited in rationalization area because it is a kind of mystery (O'Mally, 2002), it may not be observable (Kula et al., 2011), and it is difficult to measure (Hogan et al., 2008). The government employees who committed corruption usually rationalize it before they committed it with excuses that allowed them to "innocently" carry out the act like, we have no idea that the behavior is considered as bribe, bribery is our country's culture, or we are doing this as usual. Rationalization is important to justify corruption before and after committing the crime (Casabona & Grego, 2003).

Some research focus on corruption rationalization based on objective factors like low income. The public officials wage is an important factor in justifying corruption therefore "high salary for transparency" is an important policy in preventing corruption (Free, 2015). Meanwhile, other research finds that higher civil service salaries induce civil servants to demand higher bribes (Skousen et al., 2003). So, controlling corruption by increasing civil service pay is a myth because corruption is a complex social issue that cannot be tackled at the individual level (Mohd-sanusi et al., 2015).

Some researches such as (Foltz & Opoku-Agyemang, 2015) capture corruption rationalization based on age, education, and gender. The researcher in (Gong & Wu, 2012) emphasizes that younger managers want to achieve career progression quickly, so they take risks. Meanwhile, older managers seek more information, diagnosis the information precisely, and execute decisions carefully. Therefore, age may indicate the individual's moral development. Regarding the influence of education on corruption rationalization, Machado & Garter (2018) finds that individuals with knowledge in their work area are more aware of the unethical behavior consequences while without knowledge, individuals tend to rationalize fraud as an acceptable decision. Coming to gender, Zahara (2017) find that female executives can be more ethical in decision making than males. There are two problems with studying corruption rationalization based on age, education, or gender. First, these factors assess the likelihood of corruption rationalization not actual rationalization incidents. If the employee is an old female with good education in her work field, the researcher takes this case as a low possibility for rationalizing corrupt behavior; meanwhile, that woman may in fact rationalize her corrupt behavior. Second, studying corruption rationalization based on demographic factors (age, education, and gender) downplays the important role of individual values (Wolfe & Hermanson, 2004) and organizational values (Purcell, 2014) in explaining how individuals rationalize their corrupt behavior. This study focuses on how the municipality employees rationalize the corrupt behavior based on their organizational values (culture).

The study Troy et al. (2011) replaced corruption rationalization with "personal integrity" which shifted the focus to individualistic explanations of corruption. For instance, a person may not care about rationalizing corruption because he is an industrial psychopath who lacks empathy and behaves antisocial (Steffensmeier et al., 2013). Some individuals committed fraud because they are pathological gamblers who lost control on themselves and want money to continue gambling (Albrecht et al., 1984). Some individuals are fraud predators described as "born crooks" that continuously practice fraud (unlike accidental fraudsters) (Ramamoorti, 2008). For predators "pressure and rationalization play little or no role because the predator needs only opportunity" (Ramamoorti, 2008).

The previous discussion shows that attributing corruption rationalization to individual factors like the person's demographic characteristics (Foltz & Opoku-Agyemang, 2015) or his psychological disorders (Steffensmeier et al., 2013; Albrecht et al., 1984) may be insufficient. The violator does not invent rationalization to his violation but rather uses the verbalization from the culture he is in contact with (Brown, 2005). Rationalization "is the set of ethical values that allow certain parties to commit acts of fraud" (Kelly & Hartley, 2010). Thus, cultural mechanisms that influence fraud should be considered (Steffensmeier et al., 2013). Only few studies focused on the organizational culture role in corruption (Said, 2018). Some of these studies suggest establishing an ethics' commission that foster and support inner check like ethical standards and professional values on public officials and employees (Dorminey et al., 2012). Other study suggests adopting ethical code (a set of ethical norms and principles) to create awareness on standards of ethical behavior, guide public officials on ethically "grey areas", and help public officials internalize values (Manurung & Hadian, 2013). Some research suggests developing ethics office like the one in the city of Atlanta (established after mayor corruption scandal in 2001) that provides advice on ethical decision-making, educates city employees on ethics policy, and investigates reports on ethics violation and persecute these violations (Smith, 2003). These studies suggest solutions (ethical office, ethical commissions, or code of ethics) for a corrupted culture. My research though is concerned more with how a culture that tolerates corruption emerges.

3 Method

3.1 *Organizational Culture*

Based on Jerinic (2006), the following discussion traces the emergence and the development of organization culture (as a field of study). The British sociologist Elliot Jaques (1952) was the first theorist who described organizational culture. He argued that researchers ignore the human and emotional elements of organizational life because they focus on organizational structure. Jaques's work inspired organization theorists from the United Kingdom (Barry Turner), Italy (Pasquale Gagliardi), Canada (Peter Frost) and United States (Linda Smircich), who all studied organizational culture by focusing on the role of symbolism in organizational life. However, these authors viewed culture as an objective entity that can be managed or a tool to enhance performance.

According to Jerinic (2006), organizational culture researchers who adopted the symbolic perspective doubt the ease with which culture can be manipulated to managerial ends. These researchers used qualitative methods because it is difficult to define culture in operational terms that captured the nuances of meaning. Culture is constructed by interaction between individuals as they interpret their surrounding and collectively create meaning. If the meaning is to be understood, it should be studied in the location and situation in which it was created by observing how the organization members speak about this meaning. The study Green (1997) stated that "Believing with Max Weber, that man is an animal suspended in webs of significance he himself has spun; I take culture to be those webs, and the analysis of it to be therefore not an experimental science in search of law but an interpretive one in search of meaning." In other words, culture should not be studied as an objective matter that is governed by universal laws (regardless of social context); rather, culture is socially constructed and as such, the researcher should focus on how and why a certain meaning is constructed in a particular social context. This study depends on (Geertz, 1973) theory of *norm circle* to better understand how organizational culture is developed and maintained.

Geertz (1973) explains the emergence of culture because of interactions between individuals in a group. He called that group the *norm circle* and defined it as "the group of people who are committed to endorsing and enforcing a particular norm". The interaction between the group members has a causal power to produce a tendency in individuals to follow standardized practices (norms). An example is the norm of not blocking a person's route. If person A is blocking person B's path, B may encourage conformance to the norm, for example, by a glare or a push. On the other hand, if person A conforms to the norm and stands aside, B may indicate approval by a smile or thanking A verbally. By the repetition of endorsing and enforcing incidents, person A realizes that s/he is expected to stand aside from the others path and develops the tendency to do so. In other words, the norm circle creates a disposition in person A to conform to the 'stand aside' norm by the influence of the norm circle members who have endorsed and enforced the norm. Person A, as a member in a norm circle, feels committed to the other members and may act in their behalf to endorse and enforce the norm concerned and expect these members' support to do so.

Based on Geertz (1973) argument, culture has no ideational existence beyond its existence as beliefs or dispositions of individual human beings. It can only attain an objective aspect as the members of norm circles. If an organization has written documents of ethical codes or organizational values, it does not mean that the culture of this organization has an objective existence (as text on papers). Culture is not necessarily what is written, but rather what is actually practiced; and what is actually practiced is what the norm circle endorsed and enforced. The endorsement and enforcement of a norm by the norm circle is not only an *incentive* for an individual member to conform, it also serves as *feedback* that informs the individual as to whether his behavior was conforming with the norm or not. Consequently, the endorsement and enforcement mechanism (as feedback and behavior correction mechanism) explain how ambiguous shared values that may be understood differently by individuals still produce a relatively standardized behavior.

The study Geertz (1973) argues that culture is socially constructed because the individual's beliefs and behaviors arise from his/her communicative interactions with the other individuals in a particular social context. If these interactions changed, they produce different beliefs and then the normative environment may be constructed differently. Culture is composed of many cross-cutting norms, if these norms contradict; the individual conforms to the norm that is endorsed by the most powerful and resourceful group (i.e. the most powerful norm circle). Culture is maintained as long as the power of each norm circle is stable. However, cultural change occurs if a norm circle grows in size or influence on the expense of the other norm circles. In this case, the individuals conform to the norm that is endorsed by the newly more powerful group (instead of the norm they used to conform to).

3.2 Leadership

Culture is a group of shared beliefs and norms held by individuals; these norms are socially constructed, they are created and maintained through the individual's interactions with other members who endorse and enforce these norms. In organizational context, the leaders (by the virtue of their position and may be charisma) are powerful in endorsing and enforcing norms. The leading organizational psychologist, Edgar Schein, argues that leadership is critical to the creation and maintenance of culture; cultural norms emerge and change due to what leaders tend to focus on, the leaders role modeling, and their reactions to crises (Free et al., 2007).

According to Elder-Vass (2012), "leadership is a process whereby the leader influences others to reach a common goal". When an organization's leaders are perceived to be ethically positive, there are less reports of counterproductive employee behavior as in the studies (Northouse, 1999; Mayer et al., 2009; Al-dhuhouri et al., 2020; AlShehhi et al., 2020; Al Harahsheh, 2014). By contrast, when leaders become morally disengaged (underemphasizing the negative consequences of misconducts), employees' ethical behavior is negatively affected (Al-Lozi & Papazafeiropoulou, 2012; Bonner et al., 2016). The values exhibited by leaders may significantly influence the values exhibited by other individuals in the organization (Alyammahi et al., 2020; Schminke et al., 2002). The ethical leader exhibits a set of traits that will promote the shared understanding of what constitutes unethical behavior such as corruption and what constitutes ethical behavior in an organization (Alameeri et al., 2020).

According to Salloum et al. (2020), leaders may influence the employees' perceptions of the organization's ethical climate (i.e., the employees' perception of what is appropriate or not appropriate behavior in their organization). In an egoistic climate, people care for themselves, protect their own interest, and have no room for personal moral, which correlates positively with unethical behavior (Pertiwi, 2018). Meanwhile, a positive ethical climate (that focuses on others) has a positive influence on ethical behavior (Peterson, 2002). An organization's ethical climate affects which issues organization members consider ethically relevant, and whose interests they consider when deciding on moral issues (O'Fallon & Butterfield, 2012).

Leadership may create a culture that highlights financial performance over ethical considerations by (Said, 2018; Free et al. 2007) communicating inappropriate ethical standards while extremely focusing on financial earnings and growth rates (Kula et al., 2011). The example of Enron Company illustrates the role of leadership in creating a culture that rationalizes corruption. According to Free et al., (2007), Enron Company had adopted one of the best management control systems in the United States. The control system requires that any deal needs to be assessed by risk analysts and approved from various levels in several departments, including approval from the most senior levels, and the board of directors. Enron also had other corporate governance mechanisms including (among other mechanisms) credentialed board of directors, an audit and compliance committee, external auditor, financial disclosure office of director, and a finance committee. In the late 1990s, Enron had a new CEO, Jeffrey Skilling, who change the organizational culture into what (Free et al., 2007) describe as a culture that is permissible for corruption. Under Skilling, an extreme performance-oriented culture emerged which institutionalized and tolerated deviant behavior. The "creative risk-taking" led to circumventing and breaking legal and ethical boundaries. The internal competition and the huge incentives led to private information, and deceit short term performance. In 2001, Enron became one of the biggest company bankruptcies in the US history. According to Free et al. (2007), the case of Enron company illustrates that despite the sophisticated control system (which supposed to reduce opportunity for corruption), when Enron leadership created a culture that rationalizes corruption (by prioritizing performance over ethics), the company collapsed.

An important example of leadership successfully changing unethical culture in the local government is the case of Cookingham (Gorsira et al., 2018), the manager of Kansas City between 1940 and 1959 as discussed by (Cookingham, 2020). For two decades before Cookingham arrived, every aspect of Kansas City was controlled by the corrupted and unelected political boss Tom Pendergast (Perego, 2019). In Pendergast time, illegal gambling was $ 12 million enterprise (roughly $216 million in 2019), and the ghost voters were more than registered voters, which enabled Pendergast to control the city council elections. He developed a system of patronage hiring that bloated the payroll with illiterate police officers and employees who never show to work; a small note signed by Pendergast would guarantee a job in the city. Eventually, Pendergast was convicted of federal tax evasion and Cookingham was hired by the mayor and city council whose campaign symbol was a broom.

Realizing that cleaning the stairs starts from the top, the first challenge for Cookingham was to assess the ethical integrity of his team and determining who engaged in the unethical conduct. Cookingham himself was an example of ethical leader who does not compromise his principles. When he was rushed by the city council to fire the employees from Pendergast era quickly, he argued that not all employees were corrupted, and he defended them against the council with his letter of resignation ready in case he didn't prevail. Cookingham message to all employees was that the only allegiance which any employee owes is to all the people of Kansas City and that political action among the employees was discouraged. By these measures, Cookingham was a bright example of the leadership ability to influence the culture in local government.

The discussion above shows a significant influence for the organization leaders' values and ethical conduct on their subordinates. Geertz (1973) explained the individual's adherence to norm as a result of the norm circle (members) strength in endorsing/enforcing the norm. By the virtue of their position, leaders in municipalities may strongly enforce norms on subordinates. Thus, this study to recommend for asking the municipality employees about their perceptions of norms that rationalizes corruption that their leaders support/discourage, and how the leaders and subordinates construct meaning to these norms.

4 Conclusion

The literature review starts with discussing the three main approaches to eliminate corruption opportunity namely transparency, citizens involvement, and recentralization and explains why these approaches are problematic. Then, a discussion of corruption rationalization follows and illustrates why rationalization should be attributed to the organization culture (not only to individual characteristics). Finally, the organization culture section discusses how organizational culture emerged and maintained before ending with section on the role of municipal leadership in constructing the organization ethical culture.

Reference

Agha, K., Alzoubi, H. M., & Alshurideh, M. T. (2021). Measuring reliability and validity instruments of technologically driven cognitive intrusion towards work-life balance. In *The International Conference on Artificial Intelligence and Computer Vision* (pp. 601–614).

Akour, I., Alshurideh, M., Al Kurdi, B., Al Ali, A., & Salloum, S. (2021). Using machine learning algorithms to predict people's intention to use mobile learning platforms during COVID-19 pandemic: Machine learning approach. *JMIR Medical Education, 7*(1), 1–17.

Al-Hamad, M., Mbaidin, H., AlHamad, A., Alshurideh, M., Kurdi, B., & Al-Hamad, N. (2021). Investigating Students' behavioral intention to use mobile learning in higher education in UAE during Coronavirus-19 pandemic. *International Journal of Data and Network Science, 5*(3), 321–330.

Al-Lozi, E., & Papazafeiropoulou, A. (2012). Intention-based models: The theory of planned behavior within the context of IS. In *Information System Theory* (pp. 219–239). Springer, New York, NY.

Al-bawaia, E., Alshurideh, M., Obeidat, B., Masa'deh, R. (2022). The impact of corporate culture and employee motivation on organization effectiveness in Jordanian banking sector. *Academy of Strategic Management Journal, 21*(Special Issue 2), 1–18.

Al-dhuhouri, F. S., Alshurideh, M., Al Kurdi, B., & Salloum, S. A. (2020, October). Enhancing our understanding of the relationship between leadership, team characteristics, emotional intelligence and their effect on team performance: A critical review. In *International Conference on Advanced Intelligent Systems and Informatics* (pp. 644–655). Springer, Cham.

Al Dmour, H., Alshurideh, M., & Shishan, F. (2014). The influence of mobile application quality and attributes on the continuance intention of mobile shopping. *Life Science Journal, 11*(10), 172–181.

Al Harahsheh, M. A. (2014). The relationship between leadership skills and successful change management skills: A survey study on Police Sciences Academy in Sharjah, UAE. *Journal of Emerging Trends in Economics and Management Sciences, 5*(5), 490–497.

AlShehhi, H., Alshurideh, M., Kurdi, B. A. & Salloum, S. A. (2020, October). The impact of ethical leadership on employees performance: A systematic review. In *International Conference on Advanced Intelligence Systems and Informatics* (pp. 417–426). Springer, Cham.

AlShurideh, M., Alsharari, N. M., & Al Kurdi, B. (2019). Supply chain integration and customer relationship management in the airline logistics, *Theoretical Economics Letters, 9*(02), 392–414.

Alameeri, K., AlShurideh, M., Kurdi, B. A., & Salloum, S. A. (2020, October). The effect of work environment happiness on employee leadership. In *International Conference on Advanced Intelligent Systems and Informatics* (pp. 668–680). Springer, Cham.

Albrecht, W., Howe, K. and Romney, M. (1984). Deterring fraud: The internal auditor's perspective, *Institute of Internal Auditors Research Foundation*, Altamonte springs, FL.

Alshamsi, M., Salloum, S. A., Alshurideh, M., & Abdallah, S. (2021). Artificial intelligence and blockchain for transparency in governance. In *Artificial intelligence for sustainable development. Theory, practice and future applications* (pp. 219–230). Springer, Cham.

Alshraideh, A., Al_lozi, M., & Alshurideh M. (2017). The impact of training strategy on organizational loyalty via the mediating variables of organizational satisfaction and organizational performance: An empirical study on jordanian agricultural credit corporation staff, *Journal of Social and Science, 6*, 383–394.

Alshurideh, M. T., Kurdi, B., & Salloum, S. A. (2021a). The moderation effect of gender on accepting electronic payment technology: A study on United Arab Emirates consumers. *Review of International Business and Strategy, 31*(3), 375–396.

Alshurideh, M. T., Kurdi, B. A., AlHamad, A. Q., Salloum, S. A., Alkurdi, S., Dehghan, A., & Masa'deh, R. E. (2021b, June). Factors affecting the use of smart mobile examination platform by universities' postgraduate students during the COVID 19 pandemic: An empirical study. In Informatics, (Vol. 8, No. 2, pp. 1–21). Multidisciplinary Digital Publishing Institute.

Alyammahi, A., AlShurideh, M., Kurdi, B. A., & Salloum, S. A. (2020, October). The impact of communication ethics on workplace decision making and productivity. In *International conference on advanced intelligent systems and informatics* (pp. 488–500). Springer, Cham.
Anechiarico, F., & Jacobs, J. B. (1996). *The pursuit of absolute integrity: How corruption control makes government ineffective*. University of Chicago Press.
Ashal, N., Alshuraideh, M., Obeidat, B., Masa'deh, R. (2021) The impact of strategic orientation on organizational performance: Examining the mediating role of learning culture in Jordan telecommunication companies. *Academy of Strategic Management Journal, 21*(Special issue 6), 1–29.
Assad, N. F., & Alshurideh, M. T. (2020). Investment in context of financial reporting quality: A systematic review, *WAFFEN-UND Kostumkd. Journal, 11*(3), 255–286.
Asthana, A. N. (2008). Decentralization and corruption: Evidence from drinking water sector. *Public Administration Development, 28*(3), 181–189.
Beeri, I., & Navot, D. (2013). Local Political Corruption: Potential structural malfunctions at the central-local, local-local and intra-local levels. *Public Management Review, 15*(5), 712–739.
Benito, B., Guillamon, M-D., & Bastida, F. (2015). Determinants of urban political corruption in local governments. *Crime, Law & Social Change, 63*(3/4), 191–210.
Bergh, A., Fink, G., & Ohrvall, R. (2017). More politicians, more corruption: Evidence from Swedish municipalities. *Public Choice, 172*(3), 483–500.
Berman, E., & Carter, J., (2018). Policy analysis: Scientific integrity in federal policymaking under past and present administrations. *Journal of Science Policy Government 13*(1), p. 26.
Bobonis, G. J., Camara fuertes, L., R., & Schwabe, R. (2016). Monotoring corruptible politicians. *The American Economic Review, 106*(8), 2371–2405.
Bonner, J. M., Greenbaum, R. L., & Mayer, D. M. (2016). My boss is morally disengaged: The role of ethical leadership in explaining the interactive effect of supervisor and employee moral disengagement on employee behaviors. *Journal of Business Ethics, 137*(4), 731–742.
Brown, G.D. (2005). Carte Blanche: Federal prosecution of state and local officials after Sabri. *Catholic University Law review, 54*(2), 403–443.
Casabona, P. A., & Grego, M. J. (2003). SAS 99 – consideration of fraud in a financial statement audit: A revision of statement on auditing standards 82. *Reviews of Business, 24*(2), 16–20.
Collins, P. D. (2012). Introduction to the special issue: The global anti-corruption discourse towards integrity management? *Public Administration & Development, 32*(1), 1–10.
Cookingham, B. E. (2020). Advocacy and service delivery in the voluntary sector: Exploring the history of voluntary sector actvities for new minority and migrant groups in East London, 1970s–1990s. Voluntas. *International Journal of Voluntary and Nonprofit Organizations,* 1–11.
Coxson, S. L. (2009). Assessment of American local government corruption potential. *Public Administration and Development, 29*(3), 193–203.
Cressey, D. (1973). *Other people's money: A study in the social psychology of embezzlement*. Patterson Smith Series in Ciminology, Law Enforcement & Social Problems. Montclair, N.J.: Patterson Smith.
Dorminey, J., Fleming, A. S., Kranacher, M., & Richard, R. A. (2012). The evolution of fraud theory. *Issues in Accounting Education, 27*(2), 555–579.
Elder-Vass, D. (2012). The reality of social construction. Cambridge University Press. Experiment on West Africa's highway.
Ferraz, C., & Finan, F. (2011). Electoral accountability and corruption: Evidence from the audits of local governments. *American Economic Review, 101*(4), 1274–1311.
Foltz, J. D., & Opoku-Agyemang, K. A. (2015). Do higher salary low petty corruption? A policy.
Fording, R. C., Miller, P. M., & Patton, D. J. (2003). Reform or resistance? local government responses to state-mandated ethics reform in Kentucky. *Publius, 33*(2).
Free, C. (2015). Looking through the fraud triangle: A review and call for new directions.
Free, C., Macintosh, N., & Stein, M. (2007). Management controls: The organizational fraud triangle of leadership, culture and control in Enron. *Ivey Business Journal, 71*(6), 1–5.
Geertz, C. (1973). The interpretation of cultures: Selected essays. New York: Basic Books.

Gong, T. (2015). Managing government integrity under hierarchy: Anti-corruption efforts in local China. *Journal of Contemporary China, 24*(94), 684–700.
Gong, T., & Wu, A. M. (2012). Does increased civil service pay deter corruption? Evidence.
Gong, T. (2006). Corruption and local governance: The double identity of Chinese local governments in market reform. *Pacific Review, 19*(1), 85–102.
Gonzalez III, J. J., Kemp, R. L., & Rosenthal, J. (Eds.). (2017). Small town economic development: Reports on on growth strategies in practice. McFarland.
Gorsira, M., Steg, L., Denkers, A., & Huisman, W. (2018). Corruption in organization: Ethical climate and individual motives. *Administrative Sciences, 8*(1), 4.
Green, J. (1997). Organization theory: Modern symbolic and postmodern perspectives. Oxford: Oxford University Press.
Hamadneh, S., Hassan, J., Alshurrideh, M., Al Kurdi, B., & Aburayya, A. (2021). The effect of brand personality on consumer self-identity: The moderation effect of cultural orientation among British and Chinese consumers. *Journal of Legal, Ethical and Regulatory Issue, 24*, 1–14.
Hess, S. (2017). Decentralized meritocracy. *Problems of Post-Communism, 64*(1), 20–31.
Hogan, C., Rezaee, Z., Riley, R., & Velury, U. (2008). Financial statement fraud: Insights from the academic literature, Auditing. *A Journal of Practice and Theory, 27*(2), 231–252.
Jerinic, J. (2006). Development of codes of conduct for local government officials in Serbia: A beginner's case. Viesoji Politika Ir Administravimas.
Jimenez, J. L., Nombela, G., & Suarez-Aleman, A. (2017). Tourist municipalities ad local political corruption. *International Journal of Tourism Research, 19*(5), 515–523.
Kelly, P., & Hartley, C. A. (2010). Casino gambling and workplace fraud: A cautionary tale for managers. *Management Research Review.*
Kim, S., Kim, H. J., & Lee, H. (2009). An institutional analysis of an e-government system for anti-corruption: The case of OPEN. *Government Information Quarterly, 26*, 42–50.
Klasnja, M. & Tucker, J. A. (2013). The economy, corruption, and the vote: Evidence from experiments in Sweden and Moldova. *Electoral Studies, 32*,(3), 536–543.
Ko, K., & Zhi, H. (2013). Fiscal decentralization: Guilty of aggravating corruption in China? *Journal of Contemporary China, 22*(79), 35–55.
Kula, V., Yilmaz, C., Kaynar, B., & Kaymaz, A. R. (2011). Managerial assessment of employee fraud risk factors relating to misstatement arising from misappropriation of assets: A survey of ise companies. *International Journal of Business and Social Science, 2*(23), 171–180.
Liu, J., & Lin, B. (2012). Government auditing and corruption control: Evidence from China's provincial panel data. *China Journal of Accounting Research, 5*(2), 163–186.
Machado, M. R. R., & Gartner, I. R. (2018). The Cressey hypothesis (1953) and an investigation into the occurrence of corporate fraud: An empirical analysis conducted in Brazilian banking institutions. *Revista Contabilidade & Financas–USP, 29*(76), 60–81.
Mackenzie, G. C., & Haflcen, M. (2002). *Scandal proof do ethics laws make government ethical? Washington*, DC: Brookings Institution Press.
Manurung, D. T. H., & Hadian, N. (2013). *Detection fraud of financial statement with fraud triangle.* Paper presented at the 23rd International Business Research conference, Melbourne.
Masters, A. B., & Graycar, A. (2016). Making corruption disappear in local government. *Public Integrity, 18*(1), 42–58.
Mayer, D. M., et al. (2009). How low does ethical leadership flow? Test of a trickle-down model. *Organizational Behavior and Human Decision Processes 108*(1), 1–13.
Mohd-Sanusi, Z., Khalid, N. H., & Mahir, A. (2015). An evaluation of client's fraud reasoning, motives in assessing fraud risks: From the perspective of external and internal. New York: Free Press.
Muhtar., Sutaryo., & Sriyanto. (2018). Corruption in Indonesian local government: Study on triangle fraud theory. *International Journal of Business and Society, 19*(2), 536–552.
Nguyen, T. V., Bach, T. N., Le, T. Q., & Le, C. Q. (2017). Local governance, corruption, and public service quality: Evidence from a national survey in Vietnam. *International Journal of Public Sector Management.*

Northouse, P. G. (1999). Leadership: Theory and practice (p. xxiii). *E-Content Generic Vendor*.
Nuseir, M. T., Al Kurdi, B. H., Alshuridah, M. T., & Alzoubi, H. M. (2021, June). Gender discrimination at workplace: Do Artificial Intelligence (AI) and Machine Learning (ML) have opinions about It. In *The International Conference on Artificial Intelligence and Computer Vision* (pp. 301–316). Springer, Cham.
O'Mally, P. (2002). Criminology and fraud, fraud investigation course. La Trobe University, Bundoora, Victoria, Australia.
O'Fallon, M. J., & Butterfield, K. D. (2012). The influence of un ethical peer behavioron observers' unethical behavior: A social cognitive perspective. *Journal of Business Ethics, 109*(2), 117–131.
Perego, M. (2019). Honor cookingham's legacy. *Public Management, 101*(10), 2–3.
Pertiwi, K. (2018). Contextualizing corruption: A cross-disciplinary approach to studying corruption in organizations. *Administrative Sciences, 8*(2),12.
Peterson, D. k. (2002). The relationship between unethical behavior and the dimensions of the ethical climate questionnaire. *Journal of Business Ethics, 41*(4), 313–326.
Purcell, A. J. (2014). Corruption and misconduct: A behavioral reflection from investigative reports into local government. *Journal of Business Governance & Ethics, 9*(1), 1–22.
Ramamoorti, S. (2008). The psychology and sociology of fraud: Integrating the behavioral sciences component into fraud and forensic accounting curricula. *Issues in Accounting Education, 23*(4), 521–533.
Ravisankar, P., Ravi, V., Rao, G. R., & Bose, I. (2011). Detection of financial statement fraud and feature selection using data mining techniques. *Decision Support Systems, 50*(2), 491–500.
Said, J., Alam, M. M., Karim, Z. A., & Johari, R.J. (2018). Integrating religiosity into fraud triangle theory: Findings on Malaysian police officers. *Journal of Criminological Research, Policy and Practice.*
Salloum, S. A., Alshurideh, M., Elnsagar, A., & Shaalan, K. (2020). Machine learning and deep learning techniques for cybersecurity: A review. In *Joint European-US Workshops on Applications of Invariance in Computer Vision* (pp. 50–57).
Schminke, M., Cropanzano, R., & Rupp, D. E. (2002). Organization structure and fairness perceptions: The moderating effects of organizational level. *Organizational Behavior and Human Decision Processes, 89*(1), 881–905.
Schopf, J. C. (2018). Cutting corruption without institutionalized parties: The story of civic groups, elected local government, and administrative reform in Korea. *Korea Observer, 49*(4), 573–604.
Silva, P. (2010). Learning to fear the inspector-general: Measuring spillovers from anti-corruption politics. Conference papers. *American Political Science Association,* 1–52.
Skousen, C. J., Smith, K. R., & Wright, C. J. (2009). Detecting and predicting financial statement.
Smith, R. W. (2003). Enforcement or ethical capacity: Considering the role of state ethics commissions at the millennium. *Public Administration Review, 63*(3), 283–295.
Steffensmeier, D. J., Schwartz, J., & Rocche, M. (2013). Gender and twenty-1st-century corporate crime: Female involvment and the gender gap in E-era corporate frauds. *American Sociological Review, 78*(3), 448–476.
Suleman, M., Soomro, T. R., Ghazal, T. M., & Alshurideh, M. (2021, June). Combating against potentially harmful mobile apps. In *The International Conference on Artificial Intelligence and Computer Vision* (pp. 154–173). Springer Cham.
Tiebout, C. (1956). A pure theory of local expenditures. *Journal of Political Economy 64,* 416–424.
Tolchin, M., & Tolchin, S. J. (2015). *Pinstripe Patronage.* Routledge.
Troy, C., Smith, K. G., & Domino, M. A. (2011). CEO demographics and accounting fraud: Who is more likely to rationalize illegal acts? *Strategic Organization, 9*(4), 259–282.
Valle-Cruza, D., Sandoval-Almazan, R., & Gil-Garcia, J. R. (2016). Citizens perceptions of the impact of information technology use on transparency, efficiency and corruption in local governments. Information Policy. *The International Journal of Government & Democracy in the Information Age, 21*(3), 321–334.
Vijayalakshmi, V. (2008). Corruption and local governance. *Development in Karnataka: Challenges of governance, equity, and empowerment*, p. 167.

Wolfe, D. T., & Hermanson, D. R. (2004). The fraud diamond: Considering the four elements of fraud. *The CPA Journal, 74*(12), 38.

Yang, K. (2009). Institutional congruence, ideas, and anticorruption policy: The case of China and the United States. *Public Administration Review, 69*, S142–S150.

Zahara, Z. (2017). Role of resource-based entrepreneurship development to increase competitiveness of traditionally woven sargon creative industry. *Russian Journal of Agriculture and Socio-Economic Sciences, 67*(7).

Zhao, Y. and Peters, B.G., (2009). The state of state: Comparing governance in China and the United States. *Public Administration Review, 69*, S122–S128.

Reviewing the Literature of Internal Corporate Social Responsibility on Job Satisfaction

Sura Altheeb, Bader Obeidat, Muhammad Turki Alshurideh, and Ra'ed Masa'deh

Abstract In the last decade a social responsibility for the environment, local communities, working conditions, and the ethical practices have become necessary for any organization that is aiming to be sustainable and successful in the business world. Also, job satisfaction reflects the positive and favorable attitudes of the workers that have a great effect on their performance. The purpose of this paper is to reviewing the literature and the associations among internal corporate social responsibility, job satisfaction and their dimensions.

Keywords Internal corporate social responsibility · Skills development · Working environment · Employment stability · Work life balance · Empowerment · Intrinsic job satisfaction · Extrinsic job satisfaction

1 Introduction

Corporate social responsibility defines the relation between business and larger society components (Snider et al., 2003; Masa'deh et al., 2013; Obeidat et al., 2017a, b; Abu Zayyad et al., 2020). It also discusses the delivery of shareholder's value and

S. Altheeb · B. Obeidat
Department of Business Management, School of Business, The University of Jordan, Amman, Jordan
e-mail: Bader.obeidat@buid.ac.ae

B. Obeidat
Faculty of Business and Law, The British University in Dubai, Dubai, UAE

M. T. Alshurideh (✉)
Department of Marketing, School of Business, The University of Jordan, Amman 11942, Jordan
e-mail: m.alshurideh@ju.edu.jo

R. Masa'deh
Department of Management Information Systems, School of Business, The University of Jordan, Amman, Jordan
e-mail: r.masadeh@ju.edu.jo

M. Alshurideh et al. (eds.), *The Effect of Information Technology on Business and Marketing Intelligence Systems*, Studies in Computational Intelligence 1056,
https://doi.org/10.1007/978-3-031-12382-5_70

promoting for the societal value. Hopkins (2003) put the most appealing definition forward for corporate social responsibility (CSR):

"CSR is concerned with treating the stakeholders of the firm ethically or in a socially responsible manner. Stakeholders exist both within a firm and outside. The aim of social responsibility is to create higher and higher standards of living, while preserving the profitability of the corporation, for its stakeholders both within and outside the corporation". Also, Chuck Robbins, CEO Cisco stated:

> As I reflect on the progress we made in fiscal year 2017, I've never been more certain that what is good for people, society, and the planet is good for business.

Employees are considered one of the primary stakeholders in any company, who are directly responsible for the success and development of the company (Al-dalahmeh et al., 2018; Bauman & Skitka, 2012; Masa'deh et al., 2016; Shannak et al., 2012). Generally, being unhappy and unsatisfied at work will be reflected on the employee's personal and daily life (Schultz & Schultz, 2010; Abualoush et al., 2018a, b; Masa'deh et al., 2018; Abuhashesh et al., 2019). Hence, one of the most vital responsibilities of a company is to satisfy their employees by using both financial and non-financial incentives (Tsourela et al., 2008).

This paper is designed to review and explore the related literature in order to form a comprehensive understanding of internal corporate social responsibility and job satisfaction. Hence, the paper discusses the concepts of internal corporate social responsibility, job satisfaction and their dimensions.

2 Literature Review

2.1 Corporate Social Responsibilities

The concept of corporate social responsibility has been a very popular topic in business world and had brought many improvements to it, especially in the cases related to the environment, discrimination and abuse at workplace. The financial responsibility has been the sole bottom line driving force for the success of any organization. The corporate responsibility is considered as the second most important factor contributing to the company's reputation after the quality of its services and products (Kaufmann & Olaru, 2012). Bowen (1953) asked a vital question which is 'to what extent do the interests of business in the long run merge with the interests of society?'. Projects and external activities of any organization will have effect on the environment and society that the organization resides within. Some of these effects might be beneficial and some of them can be detrimental to it. Hence, in today's business market, companies are being put under a huge pressure to gain public trust and stay competitive in the global market (Jamali, 2007).

The social contract theory was used to explain the relationship between company and the society (Rousseau, 2008). Figure 1 shows the concept of interrelation between any organization and the components of the community.

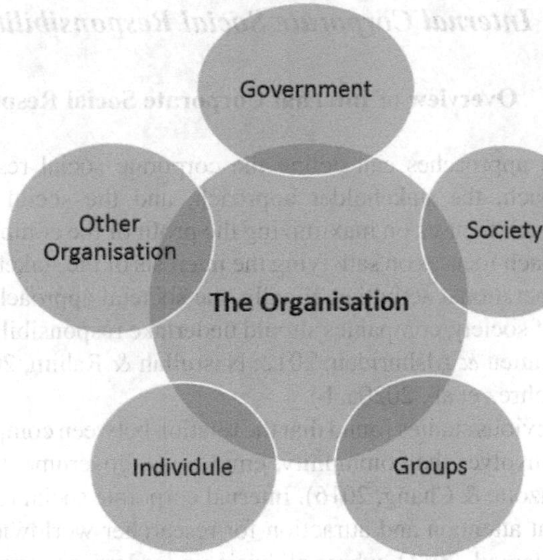

Fig. 1 Relationship between organization and society (Rousseau, 2008)

Corporate social responsibility is found on both the external and internal domain of any company (Tamm et al., 2010). The external social responsibility is the company's involvement in some types of external social project. On the other hand, internal responsibility involves human resources practices such as training and labor participation (Calveras, 2013), in addition to dealing with all issues that can affect the well-being of the employees (Tamm et al., 2010). The corporate social responsibility (CSR) affects the stakeholders of the company; the external CSR affects the outsider stakeholders who are the consumers of products or services, whereas the internal CSR have a social effect on the internal stakeholders represented by the employees (Alshurideh et al., 2012, 2021; Elfenbein et al., 2012). Both external and internal corporate social responsibilities are linked to the services and products quality (Calveras, 2013; Moh'd et al., 2013).

Dutch social and economic council (DSEC) categorized CSR into three dimensions: profit, people and planet. 'Profit' stands for profit maximization, 'Planet' for ecological quality and 'People' for well-being in and outside the organization. Wood (1991) concluded a framework of CSR that consists of three elements: (1) responsibility principles and the motivators of action and choice; (2) intra organizational responsiveness processes for determining action and choice; and (3) resulting outcomes of action and choice. The next section will discuss in more details the internal corporate social responsibility and its impact on corporations. Furthermore, the I-CSR's dimensions and measurements will be reviewed.

2.2 Internal Corporate Social Responsibility

2.2.1 Overview of Internal Corporate Social Responsibility

Three approaches can define the corporate social responsibility: the shareholder approach, the stakeholder approach, and the social approach. The shareholder approach focuses on maximizing the profit of the company, whereas the stakeholder approach focuses on satisfying the interests of the stakeholder that can be affected by the operational activities. Finally, the societal approach suggests that as an integral part of society, companies should undertake responsibilities to the society as a whole (ELSamen & Alshurideh, 2012; Nasrullah & Rahim, 2014; Alshurideh et al., 2015; Al Mehrez et al., 2020a, b).

Previous studies found that the relation between companies and society is complex as it involves the community, employees, governments and even other companies (Cavazotte & Chang, 2016). Internal corporate social responsibility (I-CSR) has got a great attention and attraction for researcher worldwide and from different industries (Ismail, 2011), where all business leaders, government officials and academics are focusing more and more on the concept of I-CSR due to its importance (Reinhardt et al., 2008). Many companies consider CSR as a strategy employed to achieve competitive advantage (Lee et al., 2013). Employees are considered as primary stakeholders, they have a high power to influence the firm, and they are another source of stakeholder demand for CSR (Mcwilliams & Siegel, 2001).

2.2.2 Impact of Internal Corporate Social Responsibility on Companies

Two studies by Aguilera et al. (2007), Heslin and Ochoa (2008) showed that I-CSR has a positive effect on turn over, recruitment, satisfaction, retention, loyalty and commitment of employees. Hence, that will be reflected on the performance and productivity of the employees to work harder and adapt the company's policies and strategies to increase the profit of the company (Hoskins, 2005). It also provides the employees with skills that allow them to be more sensitive toward the problems around them and deal with this problem in a way that serve the company and increase the financial reward. Furthermore, others suggested that CSR should be considered as an ethical behavior, and cares for the environment. Community involvement leads to organizational morality that leads to commitment and satisfaction between the employees. Recent studies showed that adapting the concepts of I-CSR is an effective method to motivate employees (Skudiene & Auruskevicine, 2012). Internal corporate social responsibility (I-CSR) refers to how firms respond to their responsibilities concerning their employees, and the work relation sphere (Cavazotte & Chang, 2016). It focusses on employees more directly and addresses their specific needs (Aguilera et al., 2007).

Pietersz (2011) referred to I-CSR as anything done inside the organization to improve the life of its employees that affect their productivity and it is directly reflected on the profitability of the company. It also affects the physical and psychological working environment of employees. I-CSR gives the company good reputation and promote the positive image for the employee which make them more satisfied and committed to the organization and improve the results and productivity of their work (Cavazotte & Chang, 2016). Moreover, Yousaf et al. (2016) mentioned that the most valuable benefit behind I-CSR is achieving a higher level of employee's job satisfaction and work engagement. The successful performance of any organization depends on a fully satisfied and engaged workforce (Santoso, 2014). Keraita et al. (2013) indicates that it is very important to have loyal employees for the success and stability in the business world. Therefore, investments in I-CSR encourage competent employees to stay with the company (Cooper & Wagman, 2009). This leads to retention of knowledge between the employees who are considered as an important asset for the development of the company and in reducing turnover cost (Kacmar et al., 2006).

2.2.3 Measurements of Internal Corporate Social Responsibility

Previous studies in the literature investigated various dimensions to measure the internal corporate social responsibility. Thang and Fassin (2017) stated that I-CSR comprises five components: labor relations, work life balance, social dialogue, health and safety, and training and development. Shibeika (2015) mentioned that I-CSR based on stakeholder theory is divided into eight dimensions: training and career development, health and safety, employees' rights, employees' welfare, vacation entitlement, social work environment, workplace diversity and disabled support. Calveras (2013) categorized I-CSR into labor stability, employee participation, and high commitment human resource practices such as training and high wages. Others categorize internal corporate social responsibility to work life balance, training and education, health and safety, human rights and workplace diversity. According to Longo et al. (2005), internal CSR can be classified to four categories: value classes' which include social equity, health and safety at workplace, well-being and satisfaction of the worker and quality of work.

Yousaf et al. (2016) focused on training and education, health and safety and human rights as the three components of I-CSR. Mory et al. (2015) stated that internal CSR comprises seven components derived based on social exchange theory: employment stability, working environment, skills development, work life balance, empowerment, workface diversity and tangible employee involvement. For the purpose of this study and incompatible with how the employees are being affected in pharmaceutical market in Jordan, the dimensions which were chosen to measure the I-CSR are: the skills development, working environment, employment stability, work life balance and empowerment.

Skills Development

Knowledge and skills development are considered the driving forces of economic growth and social development for any country. Countries with higher and better levels of skilled human capital are more likely to respond to the challenges facing it and having a better chance to benefits form the opportunities of globalization compared to the countries with low skilled workforce and employees (Planning Commission, 2007).

Skill development can be defined as all the efforts that allow somebody to learn to do something better than before or do something new that has not been done before. Mory et al. (2015) defines skills developments as to which extent the employees' skills are promoted within the organization. Due to globalization, skills development becomes an essential part of employee's life to meet the new demand of changing economic and new technology (Kawar, 2001; Tarhini et al., 2017). According to Bayley (2015), skills development is focusing on two different segments: education and vocational training. Good education and training can enhance people's capacities and creativity, empower people, raise productivity of workers, and contribute to boost future innovation. Some studies for European countries show that a one percent increase in training days leads to a three percent increase in productivity. Education and development have been associated with positive organizational outcomes such as performance and it promote greater level of efficiency through learning new skills used in the work place (Barney, 2000; Ferraz & Gallardo- Vázquez, 2016).

Trainings offer the employees the opportunity to learn new concepts and gain highly required skills. When these skills and concepts are put into practice, the outcomes from their work will be more qualified and their productivity will increase (Joyce & Showers, 2002).

Working Environment

Mory et al. (2015) refers the working conditions to the health and safety status at work place. Most countries have legislations to protect workers from hazards and accidents at work (Frick et al., 2000). The workplace environment is part of a collective perception that directly contribute to satisfaction and performance of employees (Davis & Newstrom, 1999). According to Terry Irwin the TCII strategic and Management consultant, a healthy work environment can be achieved when maintaining these three aspects: the ethics and value foundation upon which the organization rests, the policies that take those principles and convert them into day-to-day actions and the corporeal environment in which people work.

Some researchers indicate that workplace must be safe in order to reduce the accidents that have negative impact on employee's health. Health and safety practices should be effective and efficient to influence employee satisfaction positively and attract them to work for a longer time in the organization. Good working conditions can reduce any work-related illness and disease that can occur inside the organization. Most of the organizations care about health and safety because that will prevent the legal and financial issues associated with any accidents that might occur during the working hours (Hughes & Ferrett, 2011).

There is a strong relationship between improving health and safety and the satisfaction of the employees. Mostly, job satisfaction can be described as the extent to which the working environment meets the need and values of employees. Some researchers found a strong relationship between good working conditions at the organization and corporate social performance. Health and safety are considered a legal compliance leads to competitive advantage and world class business performance. Furthermore, health and safety have a strong relationship with intangible issue such as reputation, brand name, morale and customer satisfaction (Smallman & John, 2001).

Employment Stability

Employment stability is what the company can provide for the employees regarding job security. It must provide comfort, support, security and stability for the employees regarding their jobs (Mory et al., 2015). When providing employment stability, employees will be more flexible about their work roles, more involved in their jobs and more committed to the success of the organization (Al Kurdi et al., 2020; Gerhart, 1991). Lin and Wei (2006) focused on employment security and mentioned it as one of the most vital issues in the present time. Moreover, there is a strong positive relationship between job security and job satisfaction and with the commitment of the employees (Jandaghi et al., 2011).

The majority of companies are focusing on motivating their employees, to make them work hard and to be loyal to the company. Hence, they are offering the employees a job security as a part of their plans in motivating them to invest more time and effort (Leung, 2009).

Work Life Balance

'An employee must live two separate lives, one at work, and the second at home' (Lyness & Judiesch, 2014). Work life balance first appeared in 1986 (Byrne, 2005). Mory et al. (2015) explained that the balance in working life offered by the company could be achieved when there is no conflict between employee's family life and work. They measure this dimension as the time you spend in work compared to time spent in doing something you enjoy. Many companies have been helping their employees to balance between the demand of work and their personal life knowing that this will be reflected on their satisfaction and eventually on the company's plans and goals (Agha et al, 2021; Grady et al., 2008).

Welford (2008) considered the work life balance as an essential part of corporate social responsibility. Work in general causes stress and burnout for workers, which affects the performance and the ability to perform all tasks efficiently. Therefore, being able to balance between work and personal life is so important for individuals to relax and recharge. Companies that care about work life balance have a great ability in attracting and maintaining qualified and highly skilled employees.

Work life balance can be achieved by creating an internal culture that minimize negative norms and designing policies that support this issue such as flexible work hours, paid maternity leave, leave and time off, health assistance and vacations. Lazar et al. (2010) mentioned that work life balance could increase the participation of female personnel and make use of their capacities and capabilities. In order to keep the employees motivated and maintain an effective performance, the organization should include and work on the concept of work life balance in their corporate social responsibility strategy and plans.

Empowerment

Focusing on the empowerment concept as a part of the internal corporate social responsibility came after the huge escalation noticed in global competition and after facing a strong resistance for the change and the raise in the need for innovation (Drucker, 1988; Obeidat et al., 2019). Bowen and Lawler (1992) define the empowerment as sharing information with front line employees to enable them to understand the organization and act toward improving the performance of the company.

Empowerment can be classified into three categories: structural, motivational and leadership (Menon, 2001; Al Khayyal et al., 2020a). Employees' empowerment strengthens the self-efficacy and confidence to accomplish their tasks (Bettayeb et al., 2020; Ugboro & Obeng, 2000). Empowerment also increases the self-esteem of individuals, which is reflected positively on the performance of the employees and the organization (Spreitzer, 1995).

2.3 *Job Satisfaction*

2.3.1 Definition of Job Satisfaction

Some scholars defined the job satisfaction as the series of action and attitudes arise from emotional state of the employees towards their job experience and work environment. Moreover, job satisfaction reflects the feelings of the employees during doing their duties and responsibilities (Ardakani et al., 2013).

2.3.2 Impact of Job Satisfaction on Companies

As identified by Tewksbury and Higgins (2006), satisfied employees have a positive attitude toward their job with grater motivation and increased job performance. According to Suher et al., job satisfaction can affect the decision-making of employees regarding whether to leave the job and search for a new company or to stay at the same position in the same company. The idea of leaving the job will no longer exists if the HR department gives the employees the opportunity to achieve personal development, reward them, evaluate their performance and provide them with environment to express themselves freely (Alshurideh, 2022; Alshurideh et al., 2022). Organizations with satisfied employees have low turnover rate. In addition, satisfied employees are easy to retain in the organization, therefore the organization is able to cut hiring cost of new employees (Cho et al., 2009; Alshraideh et al., 2017; Aburayya et al., 2020a; Kurdi et al., 2020; Hayajneh et al., 2021; Sultan et al., 2021).

The job satisfaction degree can be measured by identifying the extent to which the needs and desires of employees are meeting their expectations. In recent years, achieving a high level of job satisfaction has been an important issue facing many companies (Kusku, 2001). Job satisfaction is an indicator of organizational effectiveness (Rothmann & Coetzer, 2002). Funmilola (2013) stated that organization with satisfied employees are more productive and effective than those with dissatisfied workers, it also reduces employee's turnover and enhance the job performance. Job satisfaction is the fundamental block that leads to recognition, income, promotion and achievement of other goals that leads to the feeling of fulfilment (Kaliski, 2007).

2.3.3 Measurements of Job Satisfaction

Regarding the measurement and types of job satisfaction, many classifications have been proposed over the years. Nash (1985) mentioned that promotions, salary, work environment, work group and work conditions are factors that highly affect the job satisfaction. Lee et al. (2012) figured four types of satisfaction: social such as occupational prestige, organizational reputation and corporate social responsibility, job itself, job environment and organizational characteristic. According to Funmilola (2013), pay, promotion, supervision, work itself and working condition are the dominating dimensions of job satisfaction. Moreover, Manafi et al. (2012) proposed that there are seven types of job satisfaction: workload, perceived control, reward system and recognition, sense of community and social support, perceived fairness on the job and conflicting values.

Some researchers argued that job satisfaction is considered as a bi-dimensional concept classified to intrinsic side that depends on individual characteristic of the person and extrinsic side that depends on environment such as salary and promotion. Herzberg (1966) in his Two-factor theory classified the job satisfaction into intrinsic and extrinsic dimensions, he explained the relationship between these factors and showed how they can increase the effective commitment of employees to their company and decreased their intention to quit. Figure 2 shows Herzberg model explaining the relationship between these factors.

Also, two types of job satisfaction are being studied by researchers: the intrinsic and extrinsic in accordance to the studies provided by Olorunsola (2012) and Herzberg (1966).

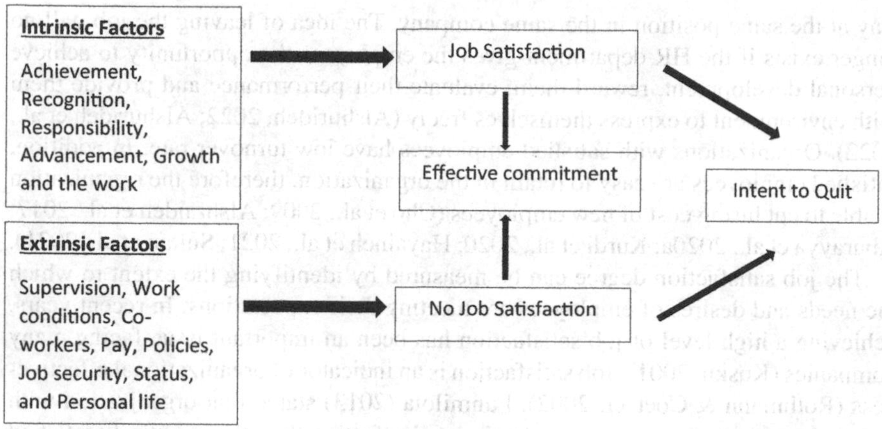

Fig. 2 Relationship between job satisfaction and the intrinsic and extrinsic factors (Herzberg, 1966)

Intrinsic Job Satisfaction

Intrinsic job satisfaction is a pivotal in job behavior (Decker et al., 2009). It refers to employees feeling about their job. Many factors can contribute in evaluating the intrinsic job satisfaction such as moral values, creativity, achievement, power and independence (Abdallah et al., 2017; Al-Dmour et al., 2021; Alzoubi et al., 2020). Herzberg (1966) defined the intrinsic factors as motivating factors that are dealing with achievement, recognition, responsibility, advancement, growth, and the work itself. Intrinsic job components are focusing on the idea of using personnel abilities, job participation, job involvement, and have the feeling of accomplishment (Yang, 2009). According to Herzberg (1966) intrinsic job satisfaction factors were described as motivating factors. In 20% of the cases, the presence of these factors leads to satisfied individuals. The motivation factors are comprised of the physiological need for growth, recognition, achievement, advancement, the job itself, and responsibility.

Extrinsic factors such as managerial behavior and the reward system in the work place affect intrinsic motivation (Deci et al., 1999). In other words, their absence was not necessarily dissatisfying, but when present, they could be a very effective motivational force (Herzberg, 1966).

Extrinsic Job Satisfaction

Extrinsic job satisfaction reflects the satisfaction of employee regarding the work tasks or the work itself (Abdallah et al., 2017; Ammari et al., 2017; Al Shebli et al., 2021; Harahsheh et al., 2021). It can be achieved by providing a positive supervision behavior, and paying the employee their benefits suitable with their work load and tasks (Decker et al., 2009). Extrinsic job satisfaction is considering the need of adequate remuneration, opportunity for advancement, and being praised for doing a good job (Yang, 2009). It can be related to "how employees feel about the portion of the work situation that is external to the work tasks or the work itself". Extrinsic satisfaction is usually influenced by environmental factors that are correlated with the work environment (Abdallah et al., 2017, p.32). Extrinsic factors show less impact on job satisfaction than the intrinsic ones. However, their absence may create dissatisfaction in the work environment (Baylor, 2010).

3 Conclusion

Recently, most of the researchers are trying to study and focus on internal corporate social responsibility as an essential factor for enhancing the employees' job satisfaction (Mory et al., 2015). The interest in internal corporate social responsibility first stemmed from questing to build strong company-employee relationship in order to achieve better level of employee engagement, which will be reflected on the organizational performance and strengthening its position in the competitive market and its reputation in the community. Internal CSR plays a vital role in increasing the productivity of the employees, it keeps them motivated to continuously achieve improvements in the quality of the product and it contributes to the research and development and the innovation in the organization (Hunaiti et al., 2009; Tarhini et al., 2015; Alshurideh et al., 2017; Abdallah et al., 2017; Alshurideh et al., 2019; Al-Khayyal et al., 2020b; Aburayya et al., 2020b; Alameeri et al., 2021; Nuseir et al., 2021).

Employees in any organization are considered the most valuable assets. Hence, it is important to keep all employees satisfied in order to increase their productivity and maintain them motivated. Many companies have been helping their employees to balance between the demand of work and their personal life knowing that this will be reflected on their satisfaction and eventually on the company's plans and goals (Grady et al., 2008).

References

Abdallah, A., Obeidat, B., Aqqad, N., Al Janini, M., & Dahiyat, S. (2017). An integrated model of job involvement, job satisfaction and organizational commitment: A structural analysis in Jordan's banking sector. *Communications and Network, 9*, 28–53.

Abualoush, S., Bataineh, K., & Alrowwad, A. (2018a). The role of knowledge management process and intellectual capital as intermediary variables between knowledge management infrastructure and organizational performance. *Interdisciplinary Journal of Information, Knowledge, and Management, 13*, 279–309.

Abualoush, S., Obeidat, A., & Tarhini, A. (2018b). The role of employees' empowerment as an intermediary variable between knowledge management and information systems on employees' performance. *VINE Journal of Information and Knowledge Management Systems, 48*(2), 217–237.

Aburayya, A., Alshurideh, M., Alawadhi, D., Alfarsi, A., Taryam, M., & Mubarak, S. (2020a). An investigation of the effect of lean six sigma practices on healthcare service quality and patient satisfaction: Testing the mediating role of service quality in Dubai primary healthcare sector. *Journal of Advanced Research in Dynamical and Control Systems, 12*(8), 56–72.

Aburayya, A., Alshurideh, M., Al Marzouqi, A., Al Diabat, O., Alfarsi, A., Suson, R., & Salloum, S. A. (2020b). An empirical examination of the effect of TQM practices on hospital service quality: An assessment study in UAE hospitals. *Systematic Reviews in Pharmacy, 11*(9), 347–362.

Abu Zayyad, H. M., Obeidat, Z. M., Alshurideh, M. T., Abuhashesh, M., & Maqableh, M. (2020). Corporate social responsibility and patronage intentions: The mediating effect of brand credibility. *Journal of Marketing Communications*. https://doi.org/10.1080/13527266.2020.1728565

Abuhashesh, M., Al-Khasawneh, M., & Al-Dmour, R. (2019). The impact of Facebook on Jordanian consumers' decision process in the hotel selection. *IBIMA Business Review* https://doi.org/10.5171/2019.928418.

Agha, K., Alzoubi, H. M., & Alshurideh, M. T. (2021). Measuring reliability and validity instruments of technologically driven cognitive intrusion towards work-life balance. In *The International Conference on Artificial Intelligence and Computer Vision* (pp. 601–614). Cham: Springer.

Aguilera, R., Rupp, D., Williams, C., & Ganapathi, J. (2007). Putting the S back in corporate social responsibility: A multilevel theory of social change in organizations. *Academy of Management Review, 32*(3), 836–863.

Alameeri, K. A., Alshurideh, M. T., & Al Kurdi, B. (2021). The effect of Covid-19 pandemic on business systems' innovation and entrepreneurship and how to cope with it: A theatrical view. *The effect of coronavirus disease (COVID-19) on business intelligence* (Vol. 334, pp. 288–275).

Al-dalahmeh, M., Masa'deh, R., Khalaf, R., & Obeidat, B. (2018). The effect of employee engagement on organizational performance via the mediating role of job satisfaction: The case of IT employees in Jordanian banking sector. *Modern Applied Science, 12*(6), 17–43.

Al-Dmour, R., AlShaar, F., Al-Dmour, H., Masa'deh, R., & Alshurideh, M. T. (2021). The effect of service recovery justices strategies on online customer engagement via the role of "customer satisfaction" during the Covid-19 pandemic: An empirical study. *The Effect of Coronavirus Disease (COVID-19) on Business Intelligence* (Vol. 334, p. 325).

Al-Khayyal, A., Alshurideh, M., Al Kurdi, B., & Aburayya, A. (2020). The impact of electronic service quality dimensions on customers' E-shopping and E-loyalty via the Impact of E-satisfaction and E-trust: A qualitative approach. *International Journal of Innovation, Creativity and Change, 14*(9), 257–281.

Al Khayyal, A. O., Alshurideh, M., Al Kurdi, B., & Salloum, S. A. (2020a). Women empowerment in UAE: A systematic review. In *International Conference on Advanced Intelligent Systems and Informatics* (pp. 742–755). Cham: Springer.

Al Kurdi, B., & Alshurideh, M., Al afaishat, T. (2020). Employee retention and organizational performance: Evidence from banking industry. *Management Science Letters, 10*(16), 3981–3990.

Al Mehrez, A. A., Alshurideh, M., Al Kurdi, B., & Salloum, S. A. (2020a). Internal factors affect knowledge management and firm performance: a systematic review. In *International Conference on Advanced Intelligent Systems and Informatics* (pp. 632–643). Cham: Springer.

AlMehrzi, A., Alshurideh, M., & Al Kurdi, B. (2020b). Investigation of the key internal factors influencing knowledge management, employment, and organisational performance: A qualitative study of the UAE hospitality sector. *International Journal of Innovation, Creativity and Change, 14*(1), 1369–1394.

Al Shebli, K., Said, R. A., Taleb, N., Ghazal, T. M., Alshurideh, M. T., & Alzoubi, H. M. (2021ne). RTA's employees' perceptions toward the efficiency of artificial intelligence and big data utilization in providing smart services to the residents of Dubai. In *The International Conference on Artificial Intelligence and Computer Vision* (pp. 573–585). Cham: Springer.

Alshraideh, A. T. R., Al-Lozi, M., & Alshurideh, M. T. (2017). The impact of training strategy on organizational loyalty via the mediating variables of organizational satisfaction and organizational performance: An empirical study on Jordanian agricultural credit corporation staff. *Journal of Social Sciences (COES&RJ-JSS), 6*(2), 383–394.

Alshurideh, M. (2022). Does electronic customer relationship management (E-CRM) affect service quality at private hospitals in Jordan? *Uncertain Supply Chain Management, 10*(2), 325–332.
Alshurideh, M., Alhadid, A. Y., & Barween, A. (2015). The effect of internal marketing on organizational citizenship behavior an applicable study on the University of Jordan employees. *International Journal of Marketing Studies, 7*(1), 138–145.
Alshurideh, M. T., Al-Hawary, S., Mohammad, A., Al-Hawary, A., & Al Kurdi, A. (2017). The impact of Islamic bank's service quality perception on Jordanian customer's loyalty. *Journal of Management Research, 9*(2), 139–159.
Alshurideh, M. T., Al Kurdi, B., AlHamad, A. Q., Salloum, S. A., Alkurdi, S., & Dehghan, A. (2021). Factors affecting the use of smart mobile examination platforms by universities' postgraduate students during the COVID-19 pandemic: An empirical study. *Informatics, 8*(2). https://doi.org/10.3390/informatics8020032.
Alshurideh, M. T., Al Kurdi, B., Alzoubi, H. M., Ghazal, T. M., Said, R. A., AlHamad, A. Q., & Al-kassem, A. H. (2022). Fuzzy assisted human resource management for supply chain management issues. *Annals of Operations Research*, 1–19.
Alshurideh, M., Masa'deh, R., & Alkurdi, B. (2012). The effect of customer satisfaction upon customer retention in the Jordanian mobile market: An empirical investigation. *European Journal of Economics, Finance and Administrative Sciences, 47*, 69–78.
Alshurideh, M., Salloum, S. A., Al Kurdi, B., Monem, A. A., & Shaalan, K. (2019). Understanding the quality determinants that influence the intention to use the mobile learning platforms: A practical study. *International Journal of Interactive Mobile Technologies, 13*(11), 183–157.
Alzoubi, H., Alshurideh, M., Kurdi, B., & Inairat, M. (2020). Do perceived service value, quality, price fairness and service recovery shape customer satisfaction and delight? A practical study in the service telecommunication context. *Uncertain Supply Chain Management, 8*(3), 579–588.
Ammari, G., Alkurdi, B., Alshurideh, A., & Alrowwad, A. (2017). Investigating the impact of communication satisfaction on organizational commitment: A practical approach to increase employees' loyalty. *International Journal of Marketing Studies, 9*(2), 113–133.
Ardakani, M., Zare, M., Mahdavi, S., Ghezavati, M., Fallah, H., Halvani, G., & Bagheraat, A. (2013). Relation between Job stress dimensions and job satisfaction in workers of a refinery control room. *Journal of Community Health Research, 1*(3), 198–208.
Barney, J. B. (2000). Firm resources and sustained competitive advantage. *Economics Meets Sociology in Strategic Management, 17*, 203–227.
Bauman, C. W., & Skitka, L. J. (2012). Corporate social responsibility as a source of employee satisfaction. *Research in Organizational Behavior, 32*, 63–86.
Bayley, D. H. (2015). *Police and political development in India*. Princeton University Press.
Baylor, K. M. (2010). The influence of intrinsic and extrinsic job satisfaction factors and affective commitment on the intention to quit for occupations characterized by high voluntary attrition. Nova Southeastern University.
Bettayeb, H., Alshurideh, M. T., & Al Kurdi, B. (2020). The effectiveness of mobile learning in UAE universities: A systematic review of motivation, self-efficacy, usability and usefulness. *The International Journal of Control, Automation, 13*(2), 1558–1579.
Bowen, D. E., & Lawler, E. E. (1992). The Empowerment of service workers: What, why, how and when. *Sloan Management Review, 33*(3), 31–39.
Bowen, H. R. (1953). *Social responsibilities of the businessman*. Harper & Row.
Byrne, U. (2005). Work-life balance. *Business Information Review, 22*(1), 53–59.
Calveras, A. (2013). External and internal corporate social responsibility: Complements through product quality. Evidence from the hotel industry. Retrieved December 2015, from http://www.webmeets.com/files/papers/earie/2013/339/CSR%20and%20product%20quality%20EARIE%202013.pdf.
Cavazotte, F., & Chang, N. C. (2016). Internal corporate social responsibility and performance: A study of publicly traded companies. *BAR-Brazilian Administration Review, 13*(4).

Cho, S., Johanson, M. M., & Guchait, P. (2009). Employees intent to leave: A comparison of determinants of intent to leave versus intent to stay. *International Journal of Hospitality Management, 28*, 374–381.

Cooper, S., & Wagman, G. (2009). Corporate social responsibility: A study of progression to the next level. *Journal of Business and Economics Research, 7*(5), 97–102.

Davis, J. W., & Newstrom, D. (1999). *Comportamiento Humano en el Trabajo* (10th ed.). McGraw-Hill.

Deci, E. L., Koestner, R., & Ryan, R. M. (1999). A meta-analytic review of experiments examining the effects of extrinsic rewards on intrinsic motivation. *Psychological Bulletin, 125*(6), 627.

Decker, F. H., Harris-Kojetin, L. D., & Bercovitz, A. (2009). Intrinsic job satisfaction, overall satisfaction, and intention to leave the job among nursing assistants in nursing homes. *The Gerontologist, 49*(5), 596–610.

Drucker, P. F. (1988). The coming of the new organization. *Harvard Business Review, 66*, 45–53.

Elfenbein, D., Fisman, R., & McManus, B. (2012). Charity as a substitute for reputation: Evidence from an online marketplace. *Review of Economic Studies, 79*(4), 1441–1468.

ELSamen, A. A., & Alshurideh, M. (2012). The impact of internal marketing on internal service quality: A case study in a Jordanian pharmaceutical company. *International Journal of Business and Management, 7*(19), 95–84.

Ferraz, F. A., & Gallardo-Vázquez, D. (2016). Measurement tool to assess the relationship between corporate social responsibility, training practices and business performance. *Journal of Cleaner Production, 129*, 659–672.

Ferreira, P., & Real de Oliveira, E. (2014). Does corporate social responsibility impact on employee engagement? *Journal of Workplace Learning, 26*(3/4), 232–247.

Frick, K., Jensen, P. L., Quinland, M., & Wilthagen, T. (2000). Systematic Occupational Health and Safety Management—An Introduction to a New Strategy for Occupational Safety, Health and Well-being. In K. Frick, P.L. Jensen, M. Quinland, T. Wilthagen (Eds.), *Systematic occupational health and safety management. Perspectives on an international development* (pp. 1–14). Pergamon.

Funmilola, O. F. (2013). Impact of job satisfaction dimensions on job performance in a small and medium enterprise in Ibadan, South Western, Nigeria. *Interdisciplinary Journal of Contemporary Research in Business, 4*(11).

Gerhart, B. A. (1991). Employment stability under different managerial compensation systems, (CAHRS Working Paper #91–02). Ithaca, NY: Cornell University, School of Industrial and Labor Relations, Center for Advanced Human Resource Studies.

Grady, G., McCarthy, A., Darcy, C., & Kirrane, M. (2008). *Work-life balance policies and initiatives in Irish organisations: A best practice management guide*. Oak Tress Press.

Harahsheh, A. A., Houssien, A. A., Alshurideh, M. T., & AlMontaserm, M. (2021). The effect of transformational leadership on achieving effective decisions in the presence of psychological capital as an intermediate variable in private Jordanian universities in light of the Corona pandemic. *The effect of coronavirus disease (COVID-19) on business intelligence*, (Vol. 334, pp. 243–221.

Hayajneh, N., Suifan, T., Obeidat, B., Abuhashesh, M., Alshurideh, M., & Masa'deh, R. (2021). The relationship between organizational changes and job satisfaction through the mediating role of job stress in the Jordanian telecommunication sector. *Management Science Letters, 11*(1), 315–326.

Herzberg, F. (1966). *Work and the nature of man*. World Publishing.

Heslin, P. A., & Ochoa, J. D. (2008). Understanding and developing strategic corporate social responsibility. *Organizational Dynamics, 37*(2), 125–144.

Hopkins, M. (2003). *The planetary bargain-corporate social responsibility matters*. Earthscan.

Hoskins, T. (2005). *The ICSA corporate social responsibility handbook: Making CSR work for business*. ICSA.

Hughes, P., & Ferrett, E. (2011). *Introduction to health and safety at work*. Routledge.

Hunaiti, Z., Mansour, M., & Al-Nawafleh, A. (2009). *Electronic commerce adoption barriers in small and medium-sized enterprises (SMEs) in developing countries: The case of Libya.* Paper presented at the innovation and knowledge management in twin track economies challenges and solutions—Proceedings of the 11th International Business Information Management Association Conference, IBIMA 2009, 1–3, (pp. 1375–1383).

Ismail, T. (2011). Corporate social responsibility: The influence of the silver book. *International Journal of Business and Management Studies, 3*(2), 371–383.

Jamali, D. (2007). The case for strategic corporate social responsibility in developing countries. *Business and Society Review, 112*(1), 1–27.

Jandaghi, G., Mokhles, A., & Bahrami, H. (2011). The impact of job security on employees' commitment and job satisfaction in Qom municipalities. *African Journal of Business Management, 5*(16), 6853–6858.

Joyce, B., & Showers, B. (2002). *Student achievement through staff development.* National College for School Leadership.

Kacmar, K., Andrews, M. C., Van Rooy, D. L., Chris Steilberg, R., & Cerrone, S. (2006). Sure everyone can be replaced… but at what cost? turnover as a predictor of unit-level performance. *Academy of Management Journal, 49*(1), 133–144.

Kaliski, B. S. (2007). *Encyclopaedia of business and finance* (2nd ed.). Thompson Gale.

Kaufmann, M., & Olaru, M. (2012). The impact of corporate social responsibility on business performance-can it be measured, and if so, how. *The Berlin International Economics Congress, 1*, 1–16.

Kawar, M. (2001). Skills development for job creation, economic growth and poverty reduction. Doha Forum on Decent Work and Poverty Reduction 25–26 October 2011, Doha, Qatar.

Keraita, J. M., Oloko, M. A., & Elijah, C. M. (2013). The influence of internal corporate social responsibility on employee commitment in the banking sector: A survey of commercial banks in Kisii Town, Kenya. *International Journal of Arts and Commerce, 2*(1), 59–76.

Kurdi, B., Alshurideh, M., & Alnaser, A. (2020). The impact of employee satisfaction on customer satisfaction: Theoretical and empirical underpinning. *Management Science Letters, 10*(15), 3561–3570.

Kusku, F. (2001). Dimensions of employee satisfaction: A state university example. *METU Studies in Development, 28*(3/4), 399–430.

Lazar, I., Osoian, C., & Ratiu, P. (2010). The role of work-life balance practices in order to improve organizational performance. *European Research Studies, 13*(1), 201.

Lee, C., An, M., & Noh, Y. (2012). The social dimension of service workers' job satisfaction: The perspective of flight attendants. *Journal of Service Science and Management, 5*(02), 160.

Lee, C., Song, H. J., Lee, H. M., Lee, S., & Bernhard, B. J. (2013). The impact of CSR on casino employees' organizational trust, job satisfaction, and customer orientation: An empirical examination of responsible gambling strategies. *International Journal of Hospitality Management, 33*, 406–415.

Leung, W. (2009). Job security and productivity: Evidence from academics. Berkeley, CA, 1–44.

Lin, C. Y., & Wei, Y. C. (2006). The role of business ethics in merger and acquisition success: An empirical study. *Journal of Business Ethics, 69*(1), 95–109.

Longo, M., Mura, M., & Bonoli, A. (2005). Corporate social responsibility and corporate performance: The case of Italian SMEs. *Corporate Governance: The International Journal of Business in Society, 5*, 28–42.

Lyness, K. S., & Judiesch, M. K. (2014). Gender egalitarianism and work-life balance for managers: multisource perspectives in 36 countries. *Applied Psychology, 63*(1), 96–129.

Manafi, M., Gheshmi, R., & Hojabri, R. (2012). The impact of different job dimensions toward job satisfaction and tendency to leave: A study of pharmaceutical industry in Iran. *International Journal of Business and Social Science, 3*(1).

Masa'deh, R., Alananzeh, O., Tarhini, A., & Algudah, O. (2018). The effect of promotional mix on hotel performance during the political crisis in the middle east. *Journal of Hospitality and Tourism Technology, 9*(1), 32–47. https://doi.org/10.1108/JHTT-02-2017-0010.

Masa'deh, R., Obeidat, B., & Tarhini, A. (2016). A Jordanian empirical study of the associations among transformational leadership, transactional leadership, knowledge sharing, job performance, and firm performance: A structural equation modelling approach. *Journal of Management Development, 35*(5), 681–705.

Masa'deh, R., Shannak, R., & Maqableh, M. (2013). A structural equation modeling approach for determining antecedents and outcomes of students' attitude toward mobile commerce adoption. *Life Science Journal, 10*(4), 2321–2333.

McWilliams, A., & Siegel, D. (2001). Corporate social responsibility: A theory of the firm perspective. *The Academy of Management Review, 26*(1), 117–127.

Menon, S. (2001). Employee empowerment: An integrative psychological approach. *Applied Psychology, 50*(1), 153–180.

Moh'd Taisir Masa'deh, R., Shannak, R. O., & Mohammad Maqableh, M. (2013). A structural equation modeling approach for determining antecedents and outcomes of students' attitude toward mobile commerce adoption. *Life Science Journal, 10*(4), 2321–2333.

Mory, L., Wirtz, B., & Göttel, V. (2015). Factors of internal corporate social responsibility and the effect on organizational commitment. *The International Journal of Human Resource Management, 27*(13), 1393–1425.

Nash, M. (1985). *Managing organizational performance*. Jossey-Bass.

Nasrullah, N., & Rahim, M. (2014). CSR in Private enterprises in developing countries. Evidences from the ready-made garments industry in Bangladesh. Springer Publications. https://link.springer.com/book/10.1007%2F978-3-319-02350-2.

Nuseir, M. T., Aljumah, A., & Alshurideh, M. T. (2021). How the business intelligence in the new startup performance in UAE during COVID-19: The mediating role of innovativeness. *The effect of coronavirus disease (COVID-19) on business intelligence* (Vol. 334, pp. 79–63).

Obeidat, B., Hadidi, A., & Tarhini, A. (2017a). Factors affecting strategy implementation: A case study of pharmaceutical companies in the middle East. *Review of International Business and Strategy, 27*(3), 386–408.

Obeidat, B., Tarhini, A., & Aqqad, N. (2017b). The impact of intellectual capital on innovation via the mediating role of knowledge management: A structural equation modeling approach. *International Journal of Knowledge Management Studies, 8*(3/4), 273–298.

Obeidat, Z. M., Alshurideh, M. T., & Al Dweeri, R. (2019). The Influence of Online Revenge Acts on Consumers Psychological and Emotional States: Does Revenge Taste Sweet? Paper presented at the Proceedings of the 33rd International Business Information Management Association Conference, IBIMA 2019: Education Excellence and Innovation Management through Vision 2020, 4797–4815.

Olorunsola, E. O. (2012). Job satisfaction and personal characteristics of administrative Staff in South West Nigerian University. *Journal of Emerging Trends in Educational Research and Policy Studies, 3*(1), 46–50.

Pietersz, G. (2011). *Corporate social responsibility is more than just donating money*. KPMG International.

Planning Commission. (2007). Task force on skill development, government of India, New Delhi.

Reinhardt, F., Stavins, R., & Vietor, R. (2008). Corporate social responsibility through an economic lens. *Review of Environmental Economics and Policy, 2*(2), 219–239.

Rothmann, S., & Coetzer, E. (2002). The relationship between personality dimensions and job satisfaction. *Business Dynamics, 14*(2), 213–226.

Rousseau, J. J. (2008). *Discourse on political economy and the social contract*. Oxford University Press.

Santoso, I. L. (2014). The impact of internal CSR towards employee engagement and affective commitment in XYZ hotel Surabaya. *Business Management, 2*(2), 79–88.

Schultz, D., & Schultz, S. (2010). *Psychology and work today* (9th ed.). Pearson Education Inc.

Shannak, R., Al-Zu'bi, Z., Obeidat, B., Alshurideh, M., & Altamony, H. (2012). A theoretical perspective on the relationship between knowledge management systems, customer knowledge management, and firm competitive advantage. *European Journal of Social Sciences, 32*(4), 520–532.

Shibeika, A. M. (2015). The impact of internal corporate social responsibility on job satisfaction within the banking sector in Sudan. *Khartoum University Journal of Management Studies, 9*(1), 65–87.

Skudiene, V., & Auruskeviciene, V. (2012). The contribution of corporate social responsibility to internal employee motivation. *Baltic Journal of Management, 7*(1), 49–67.

Smallman, C., & John, G. (2001). British directors perspectives on the impact of health and safety on corporate performance. *Safety Science, 38*(3), 227–239.

Snider, J., Ronald, P., & Martin, D. (2003). Corporate social responsibility in the 21st century: A view from the world's most successful firms. *Journal of Business Ethics, 48*(2), 175–187.

Spreitzer, G. M. (1995). Psychological empowerment in the workplace: dimensions, measurement, and validation. *Academy of Management Journal, 38*(5), 1442–1465.

Sultan, R. A., Alqallaf, A. K., Alzarooni, S. A., Alrahma, N. H., AlAli, M. A., & Alshurideh, M. T. (2021). How students influence faculty satisfaction with online courses and do the age of faculty matter. In *The International Conference on Artificial Intelligence and Computer Vision* (pp. 823–837). Cham: Springer.

Tamm, K., Eamets, R., & Mõtsmees, P. (2010). *Relationship between corporate social responsibility and job satisfaction: the case of Baltic countries.* The University of Tartu Faculty of Economics and Business Administration (Working Paper No. 76–2010). SSRN: https://ssrn.com/abstract=1717710 or http://dx.doi.org/https://doi.org/10.2139/ssrn.1717710.

Tarhini, A., Al-Busaidi, K., Maqableh, M., & Mohammed, A. B. (2017). Factors influencing students' adoption of E-learning: A structural equation modeling approach. *Journal of International Education in Business, 10*(2), 164–182.

Tarhini, A., Mgbemena, C., & Trab, M. S. A. (2015). User adoption of online banking in Nigeria: A qualitative study. *Journal of Internet Banking and Commerce, 20*(3). https://doi.org/10.4172/1204-5357.1000132.

Tewksbury, R., & Higgins, G. E. (2006). Prison staff and work stress: The role of organizational and emotional influences. *American Journal of Criminal Justice, 30*(2), 247–266.

Thang, N., & Fassin, Y. (2017). The impact of internal corporate social responsibility on organizational commitment: Evidence from vietnamese service firms. *Journal of Asia-Pacific Business*, 1–17.

Tsourela, M., Mouza, A. M., & Paschaloudis, D. (2008). Extrinsic job satisfaction of employees, regarding their intention to leave work position. A survey in small and medium enterprises.

Ugboro, I. O., & Obeng, K. (2000). Top management leadership, employee empowerment, job satisfaction, and customer satisfaction in TQM organizations: An empirical study. *Journal of Quality Management, 5*(2), 247–272.

Welford, R. (2008). Work life balance in Hong Kong: Survey results. Retrieved February, 16, 2015.

Wood, D. J. (1991). Corporate social performance revisited. *Academy of Management Review, 16*, 691–718.

Yang, Y. F. (2009). An investigation of group interaction functioning stimulated by transformational leadership on employee intrinsic and extrinsic job satisfaction: An extension of the resource-based theory perspective. *Social Behavior and Personality: An International Journal, 37*(9), 1259–1277.

Yousaf, H. Q., Ali, I., Sajjad, A., & Ilyas, M. (2016). Impact of internal corporate social responsibility on employee engagement a study of moderated mediation model. *International Journal of Sciences: Basic and Applied Research (IJSBAR)*, 1–17.

Can Better Capabilities Lead to Better Project and Program Governance? Cases from Dubai

Mounir El Khatib, Fatma Beshwari, Maryam Beshwari, Ayesha Beshwari, Haitham M. Alzoubi, and Muhammad Alshurideh

Abstract This paper addresses how to improve project and program governance through improving process and infrastructure, resources and stakeholders, organization and culture, and leadership and decision making. A brief overview of the theoretical background of these four governance capabilities is presented. The research approach is described along with three case studies from local government entities in the UAE have been analyzed. The data collected revealed that there is a gap in project/program governance implementation within the chosen organizations for this study. The paper suggested some approaches to improve the four capabilities that can contribute to creating and enhancing the current governance structure utilized across the organizations for achieving better project execution and program benefit outcomes.

Keywords Project management · Program management · Governance · Capabilities · UAE

M. El Khatib · F. Beshwari · M. Beshwari · A. Beshwari
Hamdan Bin Mohamad Smart University, Dubai, UAE

H. M. Alzoubi (✉)
School of Business, Skyline University College, Sharjah, UAE
e-mail: haitham.alzubi@skylineuniversity.ac.ae

M. Alshurideh
Department of Marketing, School of Business, University of Jordan, Amman, Jordan
e-mail: malshurideh@sharjah.ac.ae; m.alshurideh@ju.edu.jo

Department of Management, College of Business Administration, University of Sharjah, Sharjah, UAE

M. Alshurideh et al. (eds.), *The Effect of Information Technology on Business and Marketing Intelligence Systems*, Studies in Computational Intelligence 1056,
https://doi.org/10.1007/978-3-031-12382-5_71

1 Introduction

Program Management is a rapidly emerging profession fulfilling a key need to be used in establishing companies both big and small. Program management can be defined as the integration of skills, knowledge, techniques, and tools to meet program objectives (Project Management Institute, 2016). The programs are used to either better manage costs, maintain overall conditions, and successfully controls over conflicts. The program management generates value through linking different groups within an organization around common processes and goals. Therefore, there has been a strong interest in business market toward establishing project and program governance (Alaali et al., 2021; AlShamsi et al., 2021). The development and implementation of program governance are essential keys for project/program managers on how to manage a successful program (Al Alshurideh, 2019; Batayneh et al., 2021). This is because organizations are facing a huge problem when it comes to managing several projects, especially if they were across the geographies without having a formal program structure (Kabrilyants et al., 2021; Odeh et al., 2021). This made top management find difficulties in monitoring and controlling various projects execution in the absence of proper governance structure and formal channels. This leads to create a disconnection between project execution and organization strategies as well as no primary body like Steering Committee to provide oversights to programs and its components. Other challenges noticed with implementing an effective program governance are inadequately executive support, globalization, and rapid change in technology and business environment (Hasan et al., 2022; Project Management Institute, 2016; Tariq et al., 2022). The problem that many program managers has struggled with is how to determine making the program governance framework adaptable to different program requirements. Therefore, the aim of this paper is to address how to improve project/program governance through improving four capabilities which are process and infrastructure, resources and stakeholders, organization culture, and leadership and decision making.

2 Literature Review

2.1 Concept of Governance

According to Klakegg, the term "governance" is closely linked to words like governing, government, and control (Klakegg et al., 2008). In organizational settings, governance offers the necessary structure based on definite roles, accountability, and transparency for managerial actions and ethical decision-making within an organization (Müller, 2009). According to literatures, there are mainly two schools of thought about governance. The first school of thought postulated that different subunits of an organization require different forms of governance, such as project governance

(Abednego & Ogunlana, 2006; Miller & Hobbs, 2005; Winch, 2001), public governance (Du & Yin, 2010; Klakegg et al., 2008; Williams et al., 2010;), network governance, knowledge governance (Ghosh et al., 2012; Pemsel et al., 2016), and IT governance (Marnewick & Labuschagne, 2009; Martin & Gregor, 2006; Willson & Pollard, 2009). This school of thought about governance is likely developed by project managers, IT managers, government officials, and academics who work exclusively within these disciplines (Guergov & Radwan, 2021; Hamadneh et al., 2021; Obaid, 2021). Proponents of this school of thought viewed governance as a function of any entity or management responsible for decision making and/or overseeing the operations and projects of the organization. They also hold that there is no integrated governance practice as all governance practices operate independently from one another.

The second school of thought was proposed by organizations like agencies that govern stock exchanges, various Institutes of Directors (Institute of Directors Southern Africa and Australian Institute of Company Directors), and the OECD. According to this school of thought, governance is a single process but has different phases, which include viability and sustainability, financial governance, governing the people within the organization, governing changes, governing relationships, and the performance of individual directors and the Board of Directors. This school of thought holds that the core values of a well-governed organization include its vision, values and ethics, commitment to corporate social responsibility (CSR) and the way the board governs itself. While these core values are not absolute, they should exclusively be the responsibility of the organization's governing board or its equivalent.

2.2 Project Governance

Project governance has been explored by several practitioners and academicians (Abednego & Ogunlana, 2006; Sankaran et al., 2007; Yin et al., 2008). Project governance is a subset of corporate governance and focuses on the areas of corporate governance that relate to project activities, such as project and program disclosure and reporting, project and program management and efficiency, project sponsorship, and portfolio direction (Garland, 2009). Project governance has been around for several years and is the tool necessary to deliver strong and accurate project analysis. According to literatures, project governance is a general term used to encompass leadership, authority, direction and control, accountability, and stewardship to accomplish a project (Akhtar et al., 2021; Cruz, 2021; Khan, 2021). Project governance, especially at the pre-stage of a project procurement is very important to determine the success of any particular project. Project governance provides the right framework through which the goals and objectives of organizational projects are set. It also provides the structure though which achieving those goals and objectives are determined, and how monitoring project performance are determined (Australian Government, 2017). The essence of project governance is to ensure that project goals and objectives perfectly align with the organization objective and portfolio (Alzoubi et al., 2022; Aziz & Aftab, 2021; Mehmood, 2021). Hence, project governance has three main goals, which are selecting the right project, delivering the selected project effectively and efficiently, and ensuring that the selected projects can be sustained. Project governance helps in outlining the relationships between internal and external individuals involved in delivering a project, which include the project stakeholders.

2.3 *Organizational Responsibility for Project Governance*

As stated above, project governance is the process and framework that ensures that organizations properly set up their resources and investments to ensure that the resulting project delivery is carried out properly and as desired. Generally, project governance is implemented at different levels of the organization. For instance, project leaders receive directions from and report back to their project managers, project managers receive direction from and report back to the executive, and the executive to their director or board of directors. According to Selig (2008), effective project governance is built on three key elements, which are:

i. Leadership, organization and decision rights.
ii. Importance of flexible and scalable processes improvement; and
iii. The use of enabling technology.

Project governance starts with setting clear goals and objectives for the organization to provide the initial directions. After clear goals and objectives have been set, then strategic planning and execution of the goals and objectives ensue. The implementation of the project governance action plan, policy, and strategy will ensure that the project governance is effectively. Afterwards, a continuous loop is established to measure the project performance and compare it to the set goals and objectives of the project. This may lead to the redirection of project-related activities or changes to goals and objectives where appropriate.

2.4 *Project/Program Governance Overview*

Governance impacts every organization no matter how big or small, public, or private it is. So, understanding governance is an important factor to the organization because appropriate governance results in the success or failure of strategic initiatives (Project Management Institute, 2016). Program Governance can be defined as what a program is supposed to accomplish, describe how to keep the program on track, and detail how the organization can provide support. Project/program governance contains all key elements that lead to execute a successful project or program. It is important to adapt these key elements to the needs of the organization as project/program governance needs to be aligned with the business' governance (Ali et al., 2022; Alshurideh et al., 2022; Ben-Abdallah et al., 2022). When program governance framework well designed, it will provide good practices for effective decision-making and managing the program appropriately. Program governance support creating an environment for better communication and addressing program risks, uncertainties and opportunities that occur during the program performance. It assures stakeholder engagement by establishing clear expectations for program interaction and designing, authorizing the assurance when it is required by reviewing and health checks of program delivery (Project Management Institute, 2016).

Governance capabilities is the ability of providing an efficient form of governance for the organization to gain competitive advantages and to initiate, maintain its relationship with another organization (Doğru, 2019). The purpose of governance capabilities is to allow and assure the ability to control business, process, IT system capabilities and services offering of the enterprise with an acceptable limit that set by the organization.

2.5 *Process and Infrastructure*

The process in governance can be defined as how an organization consolidates their management, tasks, and services in order to ensure that project or program is executed properly based on the expectation of managers and stakeholders (Alsharari, 2021; Hanaysha et al., 2021a, b; Lee & Ahmed, 2021). Applying governance can make it possible by implementing standardized practices, process, and rules across the project. To get the successful performance in process, it is very important to establish process office, test process before mandating, defining team members, documentation, risk management and communication.

Infrastructure in program governance is a backbone for achieving a long-term inclusive development. Governance of infrastructure includes a range of tools, process, decision-making and monitoring by the organization. The Program Management Office (PMO) is part of the infrastructure governance and provides support along administrative, process, financial, and staff dimensions correlate with successful program implementation. In addition, Program Management Information System (PMIS) is another component of infrastructure governance which should play a critical role in integrating between all activities, phases and systems utilized with program execution (Alshurideh et al., 2020a; Ashal et al., 2021; Kurdi et al., 2021). Therefore, organizations tend to follow the latest technology such as, blockchain and big data to enhance their infrastructure governance's system. According to Risius and Spohrer (2017), blockchain technology is used as a tool to improve the project management success factor like monitoring, communication, traceability, and scalability.

2.6 Resources and Stakeholders

Stakeholders is individual or group that has interests in decision or activity in the organization and they can be an internal or external stakeholder (ASQ, 2020). Stakeholder can be people who are invested in the project or program and who will be affected at any phases along the way (Alhamad et al., 2021; Rui et al., 2022; Shakhour et al., 2021). In addition, their input can be directly impacted on the outcome of the project. It is very important to communicate with them in order to collaborate on the project (Rafiee & Sarabdeen, 2012). Good governance can influence taking the cooperation of stakeholder into account with the concentrating on interests of a wide range of constituencies as well as assuring rapid access to information and good communication with stakeholder which leads to better results. Resource on the other hand, it is all necessary asset that is utilized to carry out certain tasks or a project and it can be person, tool, team, finance. In addition, it is very important to allocate resources before starting the project or the program (Ghazal et al., 2021). Good governance framework is encouraging the efficient use of resources and equally distribute the accountability.

2.7 Organization and Culture

Governance and organization should work together on aligning strategies, objective, and risk tolerance. In addition, they can help in clear defining of roles, accountabilities, and empowering risk management (Ali et al., 2021; Alshurideh et al., 2021). Governance and the organization contribute to identify and promote all related matters across the organization such as centralized and decentralized model, the role of leadership and how to make decision making (Al-Dhuhouri et al., 2021; AlShehhi et al., 2021). Organizational structure is the outline how certain activities are directed to meet organization goals (Alshurideh et al., 2017; Murray & Christison, 2012). A poor governance structure will leave project/program in a continuous reactive state which lead to difficulty in catch up with changing conditions.

Culture represents organization personality and it compromised of assumptions, belief, value, norm, and organization team member behavior (Abuhashesh et al., 2021; Hamadneh & Al Kurdi, 2021). Understanding the culture of the organization is critical to run a successful project/program (Al-bawaia et al., 2022; Alzoubi et al., 2020). Culture can influence the organization policies through the value held by the decision maker. Moreover, culture contributes to the interpersonal relationship of the individual and the organization relationship which is leading to change a choice of governance structure (Alzoubi & Aziz, 2021). Several elements of the organization structure and culture has effects on how project/program managed. Therefore, it is important to identify them to help in creating organizational culture that incorporates project/program management (Rafiee & Sarabdeen, 2012).

2.8 Leadership and Decision Making

Leadership in project is the act of team toward successful completion of project/program. In fact, leadership requires skill in both leadership and management. Many studies and literature found that there is a significant impact between leadership and governance (Alzoubi et al., 2021). Governance plays an important role in providing planned direction for leaders and help them to foster commitment, accountability, and shared aims. In addition, governance supports leadership in project/program through arrangement and frameworks. Also, governance provides important boundaries for leadership in the project/program (Lord et al., 2009).

Decision making in project/program is on a daily basis and can lead to success or fail of the project/program. Engagement of governance in project/program can set the boundaries within people operate. In addition, it identifies who is responsible for decision making and define the process that should use to make legitimate decisions (Hanford, 2005).

2.9 Challenges of Program Management Capabilities Development

Despite the good potentials of program management capabilities, program management has been facing a failure also in the government sector (ALnuaimi et al., 2020). This issue has been studied for several years and many studies addressed that program management in the context of efforts to improve certain organization. A study has been conducted by Council for Excellence in US government in 2008, identified five major challenges of program management capability development across government sector (Alzoubi et al., 2020). First, policies have been developed overtimes to address a specific issue and do not holistically address the challenges of program management. Second, program management is not recognized as a management discipline that is essential to organization success, performance and results. Third, stakeholder and agency executive do not understand their responsibilities and roles (Joghee et al., 2020). Fourth, no consistency in government sector for training and development a program manager. Finally, the program manager is lacking professional community within the government organization as many program management roles in the organization are working in relative isolation (Marshall et al., 2015).

3 Research Methods

This is a conceptual paper that purposes to use case studies to address how to improve project/program governance through improving capabilities specifically Process and Infrastructure, Resources and Stakeholders, Organization and Culture, and Leadership and Decision Making. Three case studies are reviewed: two from Dubai Government entities and one of the Federal Government entities. A combination of primary and secondary data has been utilized to gather the information (Alshurideh et al., 2020b). Mainly an interview was used as primary data and.gov,.org, and official organization press used as secondary data. The interviews have been done through virtual meetings and they were based on open ended structure.

3.1 Data Collection

3.1.1 Case A: Dubai Electricity & Water Authority (DEWA)

DEWA is one of the big government entities in Dubai Government that provides utility services specifically water and electricity services. DEWA considers governance as a foundation of its operation since its establishment. Good implementation of governance in DEWA starts with the top management and leadership being an example. Organization structure, control system, and operation driven by policies and documented processes have led to greater efficiency and productivity, which is a good achievement for the world governance culture (Alzoubi & Yanamandra, 2020). DEWA program governance focus is to provide active direction, periodically review the results, identify and adjustments to ensure achievements of the planned outcome that contributes to the success of the overall business strategies.

The project framework covers the classic components of governance such as board oversights, a clear organization chart and span of control, and a well-documented strategy and proper delegation of financial and administrative authorities (Lee et al., 2022b). In keeping with international guidelines, DEWA adopts the three lines of defense approach with management and supervision being the first line, risk legal compliance governance being the second line, and internal audit as third line. A board base control review system is enabled through external auditors and the government audit. DEWA organizational model decomposes all management and oversight functions and describe the relation among them.

3.1.2 Case B: Dubai Municipality (DM)

DM is one of the biggest government's institutions in Dubai and one of those institutes who are leading the development of smart services. DM has a PMO that is reported directly to the director general office in the organizational hierarchy. PMO is responsible for defining and maintaining projects executions through adapting PMI methodology (Mehmood et al., 2019). Since 2009, DM has established governance by identifying and issuing general framework and manuals based on enhanced guidance and control, filling the right, relations, and obligations.

DM Project Governance framework provides a comprehensive application of regulations, procedures, and laws as well as it is a sharing process that involve stakeholders and through which vision and missions are achieved. It includes a network of organizational relationships that gives integrated system which contributes to finishing tasks and works professionally. In addition, Program Governance framework defines management principles and decision-making hierarchy (Alzoubi et al., 2019). A unified portal called PMWeb has been used as PMIS assigned to each department to allow project managers to monitor and control activities within project (Asem Alzoubi, 2021a, 2021a, b; Eli, 2021).

3.1.3 Case C: Ministry of Climate Change and Environment (MOCCAE)

MOCCAE is one of the federal authorities that was established in February 2006 as the Ministry of Environment and Water. In 2016, the UAE Cabinet reshuffle the Ministry and subsequent integration of the climate change function, as well to changing the name to the Ministry of Climate Change and Environment (Alzoubi & Ahmed, 2019). MOCCAE aims to strengthen the UAE's efforts in preserving the environment and promoting food diversity on the national level (Miller, 2021).

MOCCAE has recently adopted organizational governance based on the new decision of the UAE Cabinet on the Governance System for the Federal Government Boards (2020). The decision was made after the COVID-19 pandemic. However, MOCCAE had started in project governance in the late 2018. It is considered as a reform for the organization and a key element to maintain the sustainability of projects and initiatives (Alnazer et al., 2017). It is a technique and infrastructure that helps in adapting to different challenges that might occur in the future.

4 Data Analysis

Program governance capabilities	Case A	Case B	Case C
Process and Infrastructure	– Well-structured as project governance – Unclear about program governance – SAP PPM as PMIS	– Well-structured as project governance – Newly introduced program governance – PMWeb as PMIS	– Newly introduced project governance – Unclear about program governance – No PMIS and mainly rely on Basic MS office
Resources and Stakeholders	– Well defined resources allocations and utilization – Lack of external stakeholder involvement	– Well defined resources allocations and utilization – Lack of external stakeholder involvement	– Traditional methods in allocating resources and estimations – Traditional methods followed to identify and involve internal and external stakeholders
Organization and Culture	– Methods are communicated across organization – Regular awareness	– Familiar with projects knowledge – Regular awareness	– Unclear project culture
Leadership and Decision Making	– Automated process – Clear direction and involvement of leadership	– Programs are linked to strategy – Automated decision-making process	– Not structured decision-making process

4.1 *Case A: DEWA*

DEWA has a well-structured project governance, but the concept of program governance is not clear to them. Programs are considered as big projects and follows project governance framework that in some cases could not be tailored to meet the program requirements. This is because project focuses more on product and services executions while programs focus on the benefits which is not included in project governance framework (AlHamad et al., 2022; Lee et al., 2022a; Miller, 2021).

Decision making is based an automated process and follows three levels of approvements review, confirm, and approve in Smart Document Application as well as DEWA uses SAP PPM module as PMIS tool for project management, financial and resource allocations. However, there is lack of consideration of external stakeholders with existing governance framework and it is mainly focused on internal stakeholders like divisions, departments, and project teams. The main drawbacks found that many of big projects (programs) managers are not qualified enough to manage the execution of such projects where they usually are delayed and required many modifications and change requests due to inaccurate planning. Finally, DEWA organization culture has a shortfall of understanding program governance and its role to differentiate between projects and programs which impact the organization operational performance.

4.2 Case B: DM

Program governance in DM is still a new concept and challenging as well because it requires more efforts to establish well-structured processes with the applicable approval flow. The approval flow must be capable to develop, modify, or reconfirm the business case for the component (Mondol, 2021; Radwan & Farouk, 2021). Moreover, the management in DM has good resources and technologies. They have a PMIS called PMweb which is used for initiation, monitoring, and documenting projects within organization as well as it acts as a centralized source of all projects and programs that can help in successfully meeting the organization strategies. This system provides a cross functional system that combines many departments to work as one team in order to meet the goal and vision of the organization. In addition, top management has clear direction, roles, responsibilities toward stakeholder involvement to assure the success of the program, but there is a lack of external stakeholder involvement. Overall, DM built a professional ecosystem of confidence and credibility by actively promoting culture of governance, but still there are confusion between programs and initiatives.

4.3 Case C: MOCCAE

The ministry has recently adopted project governance. Therefore, there was some limitation in collecting the data and lack of knowledge about program governance. There is no clear approval authority for projects and PMO is still unstructured well. The programs are called initiatives and they have many initiatives that contains many project components, but there is no awareness about the correct terminologies, tools, and system. In addition, MOCCAE does not have a PMIS and mainly rely on basic MS Office documents to create project plans and share the information that create chances of creating human errors and lack of integrity. Moreover, they follow weak approach in stakeholders' identification and involvement that causes delays of the project and program (Al Ali, 2021). The resources estimations and allocation process have many drawbacks and can be biased in some cases because of inappropriate structure followed in this phase. Many of project managers are assigned based on their job roles not specialized in project management which leads to plan the project inaccurately and results in delaying or failing the project because of many changes are occurring in the scope before and within execution phase.

4.4 *Discussion of Results*

Establishment of program governance is not an easy task. Indeed, huge investments are needed when it comes to improvement of project/program governance framework. Program governance needs to be tailored to an organization specific need and there are four major components that have been identified can influence how effectively implement and improve project and program governance.

4.4.1 Process and Infrastructure

Organizations must establish strategies and support infrastructure with the vision and their capabilities. PgMO must be established since it is a core part of the program infrastructure and it supports the management and coordination of the program and component work. In addition, PMIS another key aspect that must be focused on because it is used for collecting, integrating, and communicating information (Alzoubi, 2021a, 2021a, b; Farouk, 2021). It should incorporate change management system, risk database, financial management system, earned value management system, configuration management tool, and knowledge repositories. The infrastructure should be optimized, transparent, and capable in managing complexity which will help in managing the project/program more efficiently and confidently. That is why, new technologies such as blockchain and AI can be utilized to enhance the tools and skills. In addition, it will make the organization more proactive through identifying, diagnosing, and resolve performance problem before they negatively impact the business (Al Alhashmi et al., 2020; Shebli et al., 2021; Yousuf et al., 2021).

The organizations need to introduce Business Process Management (BPM) that must focus on redesigning the current processes. With BPM, organization get the flexibility of making changes to processes with minimal costs and can easily be customized to suit the requirements of the organization (Kashif et al., 2021). In addition, BPM can simplify the automation of repetitive steps with regular workflow such as elimination of redundant steps, removal of bottlenecks and introduction of parallel processing. This will result in creating a sustainable profitability and growth.

4.4.2 Resources and Stakeholders

Governance is provided by the people and processes that guide them directly for the benefit of the organization and all stakeholders. This includes the shareholders, board of directors and executive team. The main aim of governance is to breach the gap between the managers and stakeholders. It is more concerned about the organization contribution of all the stakeholders. So, the main step for the program manager is to identify the key stakeholders and manage their concerns through providing clear program updates of the findings and ensure the continuous leadership alignment (Harahsheh et al., 2021; Kurikawa & Kaneko, 2016). The capability of the resources, external support, and experience are important aspects to the success of the program. Establishing trackers and reporting tool is one way to disciplined forecasting the resources.

4.4.3 Organization and Culture

The organization and program culture has unique aspects that need to be understood and developed appropriately to be able to match their complexity. Therefore, organizations need to search for the suitable ways to achieve the benefits from the program. Shifting the organization be a dynamic organization is one way that can be followed. Dynamic organization simulates the changes in people, processes, environment, and culture (Alshraideh et al., 2017; Zu'bi et al., 2012). It has the ability to grow the organization with less risk through dynamic strategies and operation reviews which help in removing the obstacle blocks and allows for actions that accelerate the growth (Allozi et al., 2022; Kurikawa & Kaneko, 2016).

4.4.4 Leadership and Decision Making

A critical aspect of program governance is delegated specific decision-making authority to each administrative and management role. Leadership should continuously linkage to enterprise business strategy and direction, provide administrative support and assign an executive control over program evolution and outcomes (Ali et al., 2021; Ghazal et al., 2021). Program managers can hold special group work meeting for this purpose and then establish and allocate a matrix for major decision areas and responsibility (Alameeri et al., 2021; Hanford, 2005). This will create a clear and well understood decision making authority as well as effective oversight of and insight program progress and direction (Ahmad et al., 2021; Ahmed et al., 2021; Alyammahi et al., 2020).

5 Conclusion

In conclusion, establishing an effective program governance framework is an important activity requiring concentration on all capabilities. A strong governance process provides a sustainable foundation for the organization business to support growth for the future (Alhamad et al., 2021; Kurdi et al., 2021). The three case studies analyzed in this paper revealed the need for a proper program governance framework because the program's success relies on best utilization of the governance framework, policies, and best practices. When governance is built properly, it provides sufficient reporting and control activities. This is because governance empowers program managers to execute their responsibilities in a structured and well-defined workflow. In addition, continuous feedback and communication are necessary components of a program governance and program team members must understand each organization's values (Alzoubi & Aziz, 2021; Alzoubi et al., 2021). For best implementation, organizational structure of each component within the program must be applicable to the enterprise overall management philosophy and methods.

6 Recommendations

- Organizations should establish a clear program governance framework with stakeholder involvements instead of relying on project governance.
- Create a dedicated team for Program Management Office.
- Evaluate several models of successful program governance implementations and adapt most applicable ones.
- Identify the program roles, decision making hierarchy, strategies, and best practices.
- Introduce new technologies such as big data and blockchain to enhance the lesson learned data and program knowledge management to be utilized as a source for future project/programs.

References

Abednego, M. P., & Ogunlana, S. O. (2006). Good project governance for proper risk allocation in public-private partnerships in Indonesia. *International Journal of Project Management, 24*, 622–634.

Abuhashesh, M. Y., Alshurideh, M. T., Ahmed, A., Sumadi, M., & Masa'deh, R. (2021). The effect of culture on customers' attitudes toward Facebook advertising: The moderating role of gender. *Review of International Business Strategy*, 31.

Ahmad, A., Alshurideh, M., Al Kurdi, B., Aburayya, A., & Hamadneh, S. (2021). Digital transformation metrics: Aconceptual view. *Journal of Management Information and Decision Science, 24*, 1–18.

Ahmed, A., Alshurideh, M., Al Kurdi, B., & Salloum, S. A., (2021). *Digital transformation and organizational operational decision making: A systematic review*. Advances in Intelligent Systems and Computing. Springer International Publishing.

Akhtar, A., Akhtar, S., Bakhtawar, B., Kashif, A. A., Aziz, N., & Javeid, M. S. (2021). COVID-19 Detection from CBC using Machine Learning Techniques. *International Journal of Technology Innovation Management, 1*, 65–78.

Al-bawaia, E., Alshurideh, M., Obeidat, B., & Masa'deh, R. (2022). The impact of corporate culture and employee motivation on organization effectiveness in Jordanian banking sector. *AcadamyStrategy Management Journal, 21*, 1–18.

Al-Dhuhouri, F. S., Alshurideh, M., Al Kurdi, B., & Salloum, S. A. (2021). Enhancing our understanding of the relationship between leadership, team characteristics, emotional intelligence and their effect on team performance: A critical review. *Advances in Intelligent Systems and Computing*.

Al Ali, A. (2021). The impact of information sharing and quality assurance on customer service at UAE banking sector. *International Journal of Technology Innovations Management., 1*, 01–17.

Al Batayneh, R. M., Taleb, N., Said, R. A., Alshurideh, M. T., Ghazal, T. M., & Alzoubi, H. M. (2021). IT governance framework and smart services integration for future development of Dubai infrastructure Utilizing AI and big data, its reflection on the citizens standard of living. In *The International Conference on Artificial Intelligence and Computer Vision*. Springer, pp. 235–247.

Al Shebli, K., Said, R. A., Taleb, N., Ghazal, T. M., Alshurideh, M. T., & Alzoubi, H. M., (2021). RTA's Employees' perceptions toward the efficiency of artificial intelligence and big data utilization in providing smart services to the residents of Dubai. In *The International Conference on Artificial Intelligence and Computer Vision*. Springer, pp. 573–585.

Alaali, N., Al Marzouqi, A., Albaqaeen, A., Dahabreh, F., Alshurideh, M., Mouzaek, E., Alrwashdh, S., Iyadeh, I., Salloum, S., & Aburayya, A. (2021). The impact of adopting corporate governance strategic performance in the tourism sector: A case study in the Kingdom of Bahrain. *Journal of Legal Ethical and Regulatory Issues, 24*.

Alameeri, K., Alshurideh, M., Al Kurdi, B., & Salloum, S. A. (2021). The effect of work environment happiness on employee leadership. *Advances in Intelligent Systems and Computing*.

AlHamad, A., Alshurideh, M., Alomari, K., Kurdi, B., Alzoubi, H., Hamouche, S., & Al-Hawary, S. (2022). The effect of electronic human resources management on organizational health of telecommuni-cations companies in Jordan. *International Journal of Data Networks Science, 6*, 429–438.

Alhamad, A. Q. M., Akour, I., Alshurideh, M., Al-Hamad, A. Q., Kurdi, B. A., & Alzoubi, H. (2021). Predicting the intention to use google glass: A comparative approach using machine learning models and PLS-SEM. *International Journal of Data Network Science, 5*, 311–320.

Alhashmi, S. F. S., Alshurideh, M., Al Kurdi, B., & Salloum, S. A. (2020). a systematic review of the factors affecting the artificial intelligence implementation in the health care sector. *Advances in Intelligent Systems and Computing*.

Ali, N., Ahmed, A., Anum, L., Ghazal, T. M., Abbas, S., Khan, M. A., Alzoubi, H. M., & Ahmad, M. (2021). Modelling supply chain information collaboration empowered with machine learning technique. *Intelligent Automation and Soft Computer, 30*, 243–257.

Ali, N., Ghazal, T. M., Ahmed, A., Abbas, S., Khan, M. A., Alzoubi, H., Farooq, U., Ahmad, M., & Adnan Khan, M. (2022). Fusion-based supply chain collaboration using machine learning techniques. *Intelligent Automation Soft Computer 31*, 1671–1687.

Allozi, A., Alshurideh, M., AlHamad, A., & Al Kurdi, B. (2022). Impact of transformational leadership on the job satisfaction with the moderating role of organizational commitment: Case of UAE and Jordan manufacturing companies. *Academy of Strategic Management Journal, 21*, 1–13.

Alnazer, N., Alnuaimi, M., & Alzoubi, H. (2017). Analyzing the appropriate cognitive styles and its effect on strategic innovation in Jordanian Universities. *International Journal of Business Excellence, 13*, 127–140.

ALnuaimi, M., Alzoubi, H., Dana Ajelat, & Alzoubi, A. (2020). toward intelligent organizations: An empirical investigation of learning orientation's role in technical innovation. *International Journal of Innovation Learning, 29*, 207–221.

AlShamsi, M., Salloum, S. A., Alshurideh, M., & Abdallah, S. (2021). Artificial intelligence and blockchain for transparency in governance. *Studies in Computational Intelligence*.

Alsharari, N. (2021). Integrating blockchain technology with internet of things to efficiency. *International Journal of Technology Innovation Management, 1*, 1–13.

AlShehhi, H., Alshurideh, M., Kurdi, B. A., & Salloum, S. A. (2021). The impact of ethical leadership on employees performance: A systematic review. *Advances in Intelligent Systems and Computing*.

Alshraideh, A. T. R., Al-Lozi, M., & Alshurideh, M. T. (2017). The impact of training strategy on organizational loyalty via the mediating variables of organizational satisfaction and organizational performance: An empirical study on Jordanian agricultural credit corporation staff. *Journal Society Science, 6*, 383–394.

Alshurideh, M. (2019). Do electronic loyalty programs still drive customer choice and repeat purchase behaviour? *International Journal of Electronics Customer Relationship Management*, 12.

Alshurideh, M., Al Kurdi, B., Abu Hussien, A., & Alshaar, H. (2017). Determining the main factors affecting consumers' acceptance of ethical advertising: A review of the Jordanian market. *Journal Marking Communication*, 23.

Alshurideh, M., Al Kurdi, B., & Salloum, S. A. (2020a). Examining the main mobile learning system drivers' effects: a mix empirical examination of both the expectation-confirmation model (ECM) and the technology acceptance model (TAM). *Advances in Intelligent Systems and Computing*.

Alshurideh, M., Gasaymeh, A., Ahmed, G., Alzoubi, H., & Kurd, B. A. (2020b). Loyalty program effectiveness: Theoretical reviews and practical proofs. *Uncertain Supply Chain Manag*, 8.

Alshurideh, M. T., Al Kurdi, B., AlHamad, A. Q., Salloum, S. A., Alkurdi, S., Dehghan, A., Abuhashesh, M., & Masa'deh, R. (2021). Factors affecting the use of smart mobile examination platforms by universities' postgraduate students during the COVID-19 pandemic: An empirical study. *Informatics*, 8.

Alshurideh, M. T., Al Kurdi, B., Alzoubi, H. M., Ghazal, T. M., Said, R. A., AlHamad, A. Q., Hamadneh, S., Sahawneh, N., & Al-kassem, A. H. (2022). Fuzzy assisted human resource management for supply chain management issues. *Annals of Opererations Research*, 1–19.

Alyammahi, A., Alshurideh, M., Kurdi, B. Al., & Salloum, S. A. (2020). The impacts of communication ethics on workplace decision making and productivity. In *International Conference on Advanced Intelligent Systems and Informatics*. Springer, pp. 488–500.

Alzoubi, A. (2021a). Renewable Green hydrogen energy impact on sustainability performance. *International Journal of Computer Integrated Manufacturing, 1*, 94–110.

Alzoubi, A. (2021b). The impact of process quality and quality control on organizational competitiveness at 5-star hotels in Dubai. *International Journal of Technology Innovation Management, 1*, 54–68.

Alzoubi, H., & Ahmed, G. (2019). Do TQM practices improve organisational success? A case study of electronics industry in the UAE. *International Journal. Economy Business Research, 17*, 459–472.

Alzoubi, H., Ahmed, G., Al-Gasaymeh, A., & Alkurdi, B. (2019). Empirical study on sustainable supply chain strategies and its impact on competitive priorities: the mediating role of supply chain collaboration. *Management Science Letter, 10*, 703–708.

Alzoubi, H., Alshurideh, M., Kurdi, B. A., & Inairat, M. (2020). Do perceived service value, quality, price fairness and service recovery shape customer satisfaction and delight? A practical study in the service telecommunication context. *Uncertain Supply Chain Manag., 8*, 579–588.

Alzoubi, H., Alshurideh, M., Kurdi, B., Akour, I., & Aziz, R. (2022). Does BLE technology contribute towards improving marketing strategies, customers' satisfaction and loyalty? The role of open innovation. *International Journal of Data Network Science, 6*, 449–460.

Alzoubi, H. M., & Aziz, R. (2021). Does Emotional intelligence contribute to quality of strategic decisions? the mediating role of open innovation. *Journal of Open Innovation Technology Market and Complexity, 7*, 130.

Alzoubi, H. M., Vij, M., Vij, A., & Hanaysha, J. R. (2021). What leads guests to satisfaction and loyalty in UAE five-star hotels? AHP analysis to service quality dimensions. *Enlightening Tourisum, 11*, 102–135.

Alzoubi, H. M., & Yanamandra, R. (2020). Investigating the mediating role of information sharing strategy on agile supply chain. *Uncertain Supply Chain Manag., 8*, 273–284.

Ashal, N., Alshurideh, M., Obeidat, B., & Masa'deh, R. (2021). The impact of strategic orientation on organizational performance: Examining the mediating role of learning culture in Jordanian telecommunication companies. *Academy of Strategic Management Journal Special Is*, 1–29.

ASQ, (2020). What are Stakeholders? Stakeholder Definition | ASQ [WWW Document].

Australian Government. (2017). Australian Government Assurance Reviews Resource Management Guide No. 106.

Aziz, N., & Aftab, S. (2021). Data mining framework for nutrition ranking: Methodology: SPSS modeller. *International Journal of Technology Innovation Management, 1*, 85–95.

Ben-Abdallah, R., Shamout, M., & Alshurideh, M. (2022). Business development strategy model using EFE, IFE and IE analysis in a high-tech company: An empirical study. *Academy of Strategic Management Journal, 21*, 1–9.

Cruz, A. (2021). convergence between blockchain and the internet of things. *International Journal of Technology Innovation Management, 1*, 35–56.

Doğru, C. (2019). *Handbook of research on contemporary approaches in management and organizational strategy*. IGI Global.

Du, Y., & Yin, Y. (2010). Governance-Management-Performance (GMP) Framework: A Fundamental Thinking for Improving the Management Performance of Public Projects. iBusiness 02, 282–294.

Eli, T. (2021). Studentsperspectives on the use of innovative and interactive teaching methods at the University of Nouakchott Al Aasriya, Mauritania: English department as a case study. *International Journal of Technology Innovation Management, 1*, 90–104.

Farouk, M. (2021). The universal artificial intelligence efforts to face coronavirus COVID-19. *International Journal of Computer Integrated Manufacturing, 1*, 77–93.

Garland, R. (2009). *Project governance : A practical guide to effective project decision making*, 210.

Ghazal, T. M., Hasan, M. K., Alshurideh, M. T., Alzoubi, H. M., Ahmad, M., Akbar, S. S., Al Kurdi, B., & Akour, I. A. (2021). IoT for smart cities: machine learning approaches in smart healthcare—a review. *Future Internet, 13*, 218.

Ghosh, S., Amaya, L., & Skibniewski, M. J. (2012). Identifying areas of knowledge governance for successful projects. *Journal of Civil Engineering and Management, 18*, 495–504.

Guergov, S., & Radwan, N. (2021). Blockchain convergence: Analysis of issues affecting IoT, AI and blockchain. *International Journal of Computer Integrated Manufacturing, 1*, 1–17.

Hamadneh, S., & Al Kurdi, B. (2021). The effect of brand personality on consumer self-identity: The moderation effect of cultural orientations among British and Chinese consumers. *Journal Legal Ethical Regulatory Issues, 24*, 1–14.

Hamadneh, S., Pedersen, O., & Al Kurdi, B. (2021). An investigation of the role of supply chain visibility into the Scottish Bood supply Chain. *Journal Legal Ethical Regulatory Issues, 24*, 1–12.

Hanaysha, J. R., Al-Shaikh, M. E., Joghee, S., & Alzoubi, H. (2021a). impact of innovation capabilities on business sustainability in small and medium enterprises. *FIIB Business Review*, 1–12.

Hanaysha, J. R., Al Shaikh, M. E., & Alzoubi, H. M. (2021b). Importance of marketing mix elements in determining consumer purchase decision in the retail market. *International Journal of Service Science Management Engineering Technology, 12*, 56–72.

Hanford, M. (2005). Defining program governance and structure. *Devsion Work*. IBM 1–12.

Harahsheh, A. A., Houssien, A. M. A., & Alshurideh, M. T. (2021). The effect of transformational leadership on achieving effective decisions in the presence of psychological capital as an intermediate variable in Private Jordanian. In: *The Effect of Coronavirus Disease (COVID-19) on Business Intelligence*. Springer Nature, pp. 243–221.

Hasan, O., McColl, J., Pfefferkorn, T., Hamadneh, S., Alshurideh, M., & Kurdi, B. (2022). Consumer attitudes towards the use of autonomous vehicles: Evidence from United Kingdom taxi services. *International Journal of Data Network Science, 6*, 537–550.

Joghee, S., Alzoubi, H. M., & Dubey, A. R. (2020). Decisions effectiveness of FDI investment biases at real estate industry: Empirical evidence from Dubai smart city projects. *International Journal of Scientific & Technology Research, 9*, 3499–3503.

Kabrilyants, R., Obeidat, B. Y., Alshurideh, M., & Masa'deh, R. (2021). The role of organizational capabilities on e-business successful implementation. *International Journal of Data Network Science*, 5.

Kashif, A. A., Bakhtawar, B., Akhtar, A., Akhtar, S., Aziz, N., & Javeid, M. S. (2021). Treatment response prediction in hepatitis c patients using machine learning techniques. *International Journal of Technology Innovation Management, 1*, 79–89.

Khan, M. A. (2021). Challenges facing the application of iot in medicine and healthcare. *International Journal of Computer Integrated Manufacturing, 1*, 39–55.

Klakegg, O. J., Williams, T., Magnussen, O. M., & Glasspool, H. (2008). Governance frameworks for public project development and estimation. *Project Management Journal, 39*, S27–S42.

Kurdi, B. A., Elrehail, H., Alzoubi, H. M., Alshurideh, M., & Al-adaileh, R. (2021). The interplay among HRM practices. *Job Satisfaction and Intention to Leave: An Empirical Investigation, 24*, 1–14.

Kurikawa, T., & Kaneko, K. (2016). *Dynamic Organization of Hierarchical Memories.*

Lee, C., & Ahmed, G. (2021). Improving IoT privacy, data protection and security concerns. *International Journal of Technology Innovation Management, 1*, 18–33.

Lee, K., Azmi, N., Hanaysha, J., Alzoubi, H., & Alshurideh, M. (2022a). The effect of digital supply chain on organizational performance: An empirical study in Malaysia manufacturing industry. *Uncertain Supply Chain Management, 10*, 495–510.

Lee, K., Romzi, P., Hanaysha, J., Alzoubi, H., & Alshurideh, M. (2022b). Investigating the impact of benefits and challenges of IOT adoption on supply chain performance and organizational performance: An empirical study in Malaysia. *Uncertain Supply Chain Management, 10*, 537–550.

Lord, P., Martin, K., Atkinson, M., & Mitchell, H. (2009). *Narrowing the gap in outcomes: What is the relationship between leadership and governance?* 8.

Marnewick, C., & Labuschagne, L. (2009). Deriving projects from the organisational vision using the Vision-to-Projects (V2P) Framework. *South. African Business Review, 13*, 119–146.

Marshall, P., Chenok, D., & Wholey, J. (2015). improving program management in the federal government. A White Pap. by a Panel Natl. Acad. PUBLIC Adm, 1–32.

Martin, N., Gregor, S., 2006. ICT Governance | Request PDF [WWW Document]. J. E-Government .

Mehmood, T. (2021). Does information technology competencies and fleet management practices lead to effective service delivery? empirical evidence from E-commerce industry. *International Journal of Technology Innovation Management, 1*, 14–41.

Mehmood, T., Alzoubi, H. M., Alshurideh, M., Al-Gasaymeh, A., & Ahmed, G. (2019). Schumpeterian entrepreneurship theory: Evolution and relevance. *Academy of Entrepreneurship Journal, 25*, 1–10.

Miller, D. (2021). The best practice of teach computer science students to use paper prototyping. *International Journal of Technology Innovation Management, 1*, 42–63.

Miller, R., & Hobbs, B. (2005). Governance regimes for large complex projects. *Project Management Journal, 36*, 42–50.

Mondol, E. P. (2021). The impact of block chain and smart inventory system on supply chain performance at retail industry. *International Journal of Computer Integrated Manufacturing, 1*, 56–76.

Müller, R. (2009). *Project governance*, 105.

Murray, D. E., & Christison, M. A. (2012). Organizational structure. In *Leadership in English Language Education: Theoretical Foundations and Practical Skills for Changing Times*, pp. 125–135.

Obaid, A. J. (2021). Assessment of smart home assistants as an IoT. *International Journal of Computer Integrated Manufacturing, 1*, 18–36.

Odeh, R., Obeidat, B. Y., Jaradat, M. O., Masa'deh, R., & Alshurideh, M. T. (2021). The transformational leadership role in achieving organizational resilience through adaptive cultures: the case of Dubai service sector. *International Journal of Production Performation Management*.

Pemsel, S., Müller, R., & Söderlund, J. (2016). Knowledge governance strategies in project-based organizations. *Long Range Planning, 49*, 648–660.

Project Management Institute, (PMI). (2016). Governance of Portfolios, Programs, and Projects. Project Management Institute.

Radwan, N., & Farouk, M. (2021). The growth of internet of things (IoT) in the management of healthcare issues and healthcare policy development. *International Journal of Technology Innovation Management, 1*, 69–84.

Rafiee, V., & Sarabdeen, J. (2012). Cultural Influence in the Practice of Corporate Governance in Emerging Markets. Commun. IBIMA 1–10.

Risius, M., & Spohrer, K. (2017). A blockchain research framework. *Business & Information System Engineering, 596*(59), 385–409.

Rui, L. S., Khai, L. L., & Alzoubi, H. M. (2022). Determinants of emerging technology adoption for safety among construction businesses. *Academy of Strategic Management Journal, 21*, 1–20.

Sankaran, S., Remington, K., & Turner, C. (2007). Relationship between project governance and Project performance: A multiple case study of shutdown maintenance projects in a maritime environment. Asia-Pacific PMI Glob. Congr. Proceedings, pp. 1–9.

Selig, G. J. (2008). *Implementing IT governance a practical guide to global best practices in IT management*. Van Haren Publ. 1–23.

Shakhour, N. H. T., Obeidat, B. Y., Jaradat, M. O., & Alshurideh, M. (2021). Agile-minded organizational excellence: Empirical investigation. *Academy of Strategic Management Journal, 20*, 1–25.

Tariq, E., Alshurideh, M., Akour, I., & Al-Hawary, S. (2022). The effect of digital marketing capabilities on organizational ambidexterity of the information technology sector. *International Journal of Data Network Science, 6*, 401–408.

Williams, T., Klakegg, O. J., Magnussen, O. M., & Glasspool, H. (2010). An investigation of governance frameworks for public projects in Norway and the UK. *International Journal of Project Management, 28*, 40–50.

Willson, P., & Pollard, C. (2009). *Exploring IT Governance in theory and practice in a large multi-national organisation in Australia*.https://doi.org/10.1080/1058053090279476026,98-109.

Winch, G. M. (2001). Governing the project process: A conceptual framework. *Construction Management and Economics, 19*, 799–808.

Yin, Y., Yan, L., Du, Y., & Wei, Z. (2008). Continuous improvement of public project management performance based on project governance. *2008 International Conference Wireless Communication Network Moblie Computer* WiCOM 2008.

Yousuf, H., Zainal, A. Y., Alshurideh, M., & Salloum, S. A. (2021). Artificial intelligence models in power system analysis. In *Artificial Intelligence for Sustainable Development: Theory, Practice and Future Applications*. Springer, pp. 231–242.

Zu'bi, Z., Al-Lozi, M., Dahiyat, S., Alshurideh, M., & Al Majali, A. (2012). Examining the effects of quality management practices on product variety. *European Journal of Economics Finance and Adminstrative Science, 51*, 123–139.

A Trial to Improve Program Management in Government Bodies Through Focusing on Program Resource Management: Cases from UAE

Mounir El khatib, Alia Mahmood, Amani Al Azizi, Ayesha Al Marzooqi, Khalil Al Abdooli, Saeed Al Marzooqi, Sumaya Al Jasmi, Haitham M. Alzoubi, and Muhammad Alshurideh

Abstract Improving a program management in government bodies is a critical aspect, and this research paper demonstrates a thesis on Program Resource Management and its process to the success of government bodies' roles. The qualitative methodology in this paper studies the program management several aspects in terms of resource management and its effect on the success of the government sector's programs in UAE. The study covers program management's impact on the sectors and issues in government sectors in relation to the lack of professionals and its effect on the availability and clarity of the senior management career paths. The research paper addressed the main problem in program resource management and its wide effects. Similarly, the data gathered to cover the same thesis to compare and analyze the government bodies' key results and finding. The researchers have been addressed the impacts of resource management on the program management in the government sector. Since resources are always allocated and utilized with a proper record being kept in the government sector, this research analyzes the issues in the government sector. It was estimated that the program managers lack a professional community within the federal government that can provide support and a voice on issues affecting the development of program management in the sector. This give rise to several concerns which are need to be addressed and the research will be on providing a solution to this problem. Alongside that, the research pointed out that

M. El khatib · A. Mahmood · A. Al Azizi · A. Al Marzooqi · K. Al Abdooli · S. Al Marzooqi · S. Al Jasmi
Hamdan Bin Mohamad Smart University, Dubai, UAE

H. M. Alzoubi (✉)
School of Business, Skyline University College, Sharjah, UAE
e-mail: haitham.alzubi@skylineuniversity.ac.ae

M. Alshurideh
Department of Marketing, School of Business, University of Jordan, Amman, Jordan
e-mail: m.alshurideh@ju.edu.jo; malshurideh@sharjah.ac.ae

Department of Management, College of Business Administration, University of Sharjah, Sharjah, UAE

M. Alshurideh et al. (eds.), *The Effect of Information Technology on Business and Marketing Intelligence Systems*, Studies in Computational Intelligence 1056,
https://doi.org/10.1007/978-3-031-12382-5_72

the government-wide job series for program managers that spans business functions with a career path that extends into senior career executive management ranks is full of flaws. There is a lack of credibility and quality at the level of the management which manages the resources. The research also allowed the researchers to carry out the SWOT analysis on program resource management. This will allow to identify the major threats and the opportunities which the field has for the young aspirants in the future. The researchers' main focus is on to identify the major problems in the program resource management. The aim has been to identify these issues and then propose proper and sustainable solutions for them to ensure employees and management make better use of the resources either in the program or project. The issues like managing resource allocations, choosing the right career path and lack of training and development are a few to name which the research has discussed in detail. The aim has been to evaluate these issues by ensuring that they are analyzed from the perspective of the employees as well as the management.

Keywords Project management · Program management · Resource management · Capabilities · Government bodies · UAE

1 Introduction

One of the most continues challenges for several programs across worldwide is maintaining resource management effectively (Alkalha et al., 2012; Alshurideh et al., 2019; Zu'bi et al., 2012). In the program environments, program resource management decides the success or failure of the program. It is very essential and critical that the program manager requires to work within the uncertainty boundary and assure the necessary elements and items for the projects (Al Ali, 2021; AlHamad et al., 2022; AlHamad et al., 2021; Ali et al., 2021). Program resource management plays a vital role in ensuring the availability of the needed resources to the project managers. This will lead to accomplish projects in an approved schedule date and get program benefits.

The researchers listed out the critical success factors to support program resource management. The listing out of CSF has allowed the organizations to significantly be able to identify the key factors responsible for the success of the organization and how the organization should focus on these key factors for sustaining success for long term period. The gaps and the issues in the organizations in terms of the program resource management was done by conducting interviews from the program managers of Emirates Group, Water and Power Producer Companies (IWPP) and The Executive Council of Dubai as a source of data gathering. Each manager made suggestions to improve the capacity and capability for the project and program management. It was suggested that the program manager job series must have a clear career path and policies to support the career development of program managers. The research aimed to solve the problems of the program managers for better resources use and allocation alongside effective management. This research explored the question under study by getting in touch with the professionals for effective results. At the end, there are a several recommendations for better use of the resources in program management.

On the other hand, program resource management faces several significant challenges, which are an insufficient correlation with the owners of the resource, poor project planning, inadequate HR planning, lack of career path, no consistency in the training of resource management, ...etc. (Al Kurdi et al., 2021; Ashal et al., 2021; Shamout et al., 2022).

In this research study, we will discuss the impact of resource management on the success of the program management alongside its issues. The gaps in resource management will be addressed, and solutions will be proposed. Qualitative data is gathered from three different organizations. Then, analyze data based on the literature review and data gathering interviews. Finally, drawing the conclusion and recommendations based on data analysis.

2 Literature Review

2.1 Program Resource Management Definition and Explanation

Program resource management is a very important organizational procedure that should be regulated to ensure that the critical inputs are used accordingly. This concept entails two distinct aspects, which are program and resource management. Programs refer to the projects organized by entities, whether in the private or public sector, to achieve certain goals and objectives (NAPA, 2015). The programs might range from training and development initiatives to the construction of public amenities to serve the local citizens and international corporations operating in the country. On the other hand, resource management refers to the specific strategies instituted by an organization to promote effective and efficient use of the available inputs (Petersen & Kumar, 2015). The practice is prevalent in both private and government agencies, which thrive to promote transparency and accountability in the utilization of core resources (Ali et al., 2022; Alnazer et al., 2017; Alnuaimi et al., 2021; Alsharari, 2021). Based on the breakdown of the constituent elements, program resource management can be defined as the specific initiatives incorporated within an organization, whether government institutions or corporations, to ensure that the resources are utilized effectively and efficiently in the core operations. However, the complexity of the process varies between organizations depending on the nature and the scope of operations. Therefore, it vital for the government organizations to adopt an effective resource management strategy to enhance its output.

2.2 Resource Management Impact on Program Management in the Government Sector

Resource management can be utilized by the government agencies to ensure that the implemented programs are managed effectively to deliver maximum utility to the members of the public. The government organizations and state-owned enterprises, such as agricultural, sports, tourism, and health agencies, are faced with the challenge of allocating the resources efficiently to yield optimal returns to the members of the public (Alshurideh et al., 2020; Alshurideh et al., 2022; Alzoubi, 2021a). The complex system of governance in the government entities renders the structure porous, hence encouraging inefficient use of resources, which includes duplication of core operations and embezzlement of the available funds. As such, the introduction of an effective resource management initiative would significantly combat the above-mentioned constraint and ensure that the government utilized the tax revenue effectively to benefit its citizens.

To start with, through resource allocation, the initiative would enable the government institutions to optimize the use of the available resources. It is vital for the government institutions to identify the resources that contribute to sustainability and allocate them efficiently (Braganza et al., 2017). Resource management enables organizations to organize their available assets, whether human, financial, or physical, to align with the overall goals and objectives of the projects (Patanakul et al., 2016). For example, the government agencies can divide work accordingly so as to match the competence of each individual in the organization. The process would promote effectiveness in operations, hence preventing the common operational inadequacies, which enhance wastages. In order to actualize the process, the managers and leaders in the government organizations should often use resource allocation reports. The documentation provides a detailed overview of resource availability (Scarlett & Boyd, 2015), hence preventing schedule delays or over budgeting. As such, the enhanced reporting capabilities would support transparency and efficiency over the government projects.

In addition, the government organizations can use resource leveling, a common type of resource management technique, to enhance the execution of their projects. The agencies can analyze the current and prior projects to identify the underutilized resources (De Carvalho et al., 2015). Timely identification of the inefficiently used resources would enable the agencies to apply the inputs to their advantage, which will ultimately increase the delivery of public goods to the citizens. For example, the government organizations can utilize the range of internal skills and expertise to acquire certain professional services rather than outsourcing the competency from private institutions. The interior employers within the institutions might have special talents and expertise, such as graphic design, auditing skills, and project evaluation, which could be utilized internally to foster productivity (Ayers, 2015). As such, the action would save some finances, which will be channeled towards other viable projects that would assist the members of the public.

Furthermore, the government institutions can utilize resource forecasting, another resource management practice, to improve the outcomes of their projects. The practice utilizes the historical data and level of current operations to predict the future resource requirements before the commencement of the project. It is a reasonable plan that can be used to optimize people, materials, and budget efficiency (Alzoubi, 2021b; Alzoubi & Yanamandra, 2020; Alzoubi et al., 2020a, b, 2022). At the initial stages of planning, the government institutions, led by the management, will have to identify the scope, potential constraints, and possible contingencies. However, it would be essential for the managers in the government institutions to be familiar with the public project lifecycle and the general overview of the requisite resources within the organization. The management can utilize project management software to provide the required project visibility (Serrador & Turner, 2015; Kabrilyants et al., 2021; Shakhour et al., 2021). The software will also enable the management to centrally access the projects and the corresponding resources. As such, having a comprehensive expectation of the future resource requirements enables the government institutions to plan accordingly, hence promoting efficient resource utilization.

2.2.1 Program Managers Lack a Professional Community Within the Federal Government that Can Provide Support and a Voice on Issues Affecting the Development of Program Management

However, irrespective of the advantages associated with the program resource management techniques mentioned in the prior section, the program managers in the government institutions might lack the requisite support to apply the measures effectively (Alzoubi et al., 2021; Alzoubi & Aziz, 2021; Alzoubi & Ahmed, 2019). Indeed, program managers lack a professional community and backing within the federal government that can provide the necessary support and reinforcement on fundamental issues that affect the development of program management. The government organizations accompanied by poor priority, especially in the government projects, might alienate the program managers and fail to offer the necessary financial, infrastructural, and human support to ensure that the resources are utilized efficiently and effectively. As such, most of the government projects are encompassed with issues of various magnitudes.

2.2.2 A Government-Wide Job Series for Program Managers that Spans Business Functions with a Career Path that Extends into Senior Career Executive Management Ranks

The application of resource management in government programs provides a culture that prioritizes efficiency, which ultimately strengthens career development. Particularly, a government-wide job series for program managers, accompanied by resource management and viable practices, bolsters business function efficiency, hence providing a pathway for career development into senior executive positions

(Alzoubi et al., 2020a, b; Aziz & Aftab, 2021; Cruz, 2021). The application of resource management provisions by the program managers in the government institutions bolsters their reputation greatly due to the high level of productivity. As such, the process creates a path upon which the project managers can elevate into higher managerial positions. As such, it is vital for the personnel within the government organizations with the aspiration of progressing in their careers to adhere to the resource management practices in their projects.

2.3 Program Resource Management SWOT Analysis

SWOT analysis can be used by the project and program managers to identify areas in resource management that need enhancement. The strategic tool enables the project and program managers to bolster program resource management to gain optimal efficiency (Hornstein, 2015). It also enables the leading personnel to mitigate the risks associated with the functions and optimizes the entire program. The constituent quadrants of the SWOT analysis matrix are discussed below.

- **Strengths**

Program resource management has certain strengths that make it appropriate in dealing with government-related projects. To start with, it aligns with the organization's goals and objectives formulated by the program and project managers, which is enhanced productivity. Program resource management ensures that the government organizations use the available resources optimally (Kogan et al., 2017; Rui et al., 2022), therefore bolstering the output. Second, the initiative is universal, hence compatible with different projects within an organization. Precisely, program resource management can be used in any organizational process to promote effective and efficient consumption of resources. Finally, program resource management encourages a culture of creativity and inventiveness, which enables the organizations to join optimal output from the project and program team members. The strengths indicate that the initiative would be appropriate in government organizations (Eli, 2021; Ghazal et al., 2021; Guergov & Radwan, 2021; Hamadneh, Pedersen, & Al Kurdi, 2021a, 2021b).

- **Weaknesses**

Irrespective of the strengths mentioned in the prior section, program resource management is associated with certain deficiencies, which if not addressed, might hamper the effectiveness of the operations set by the program and project managers alongside their team members. To start with, the process is very sensitive and disposed to potential errors, which might lead to further loses. Particularly, the forecasting of resource needs might be affected by environmental uncertainties, which might ultimately lead to either under-allocation or over-allocation of resources (Hanaysha et al. 2021a, b). Second, relying on internal expertise, which is obtained from the project and program team members, would lead to the acquisition of services that are not entirely professional. The reason is that the internal employees might not be as experienced as the seasoned professionals in various fields, including accounting and project evaluation. Finally, the initiative might be expensive, especially in cases where intensive forecast is required. As such, the weaknesses should be addressed to ensure that program is effective.

- **Opportunities**

Application of program resource management grants the program and project managers plenty of opportunities to flourish in their respective markets. The initiative provides an opportunity for the organizations to expand their investments to other viable projects. Particularly, through cost minimization (wastage reduction), the government organizations, which are steered by the project and program managers, would channel the saved finances to other projects that would yield optimal returns to the citizens. Second, through the government support, especially in relation to oversight, the program managers can plan effectively and acquire sufficient resources to execute the major initiatives (Patanakul et al., 2016). As such, it would be rational for the project managers to implement effective program resource management practices.

- **Threats**

There are some threats that emanate from the external surroundings, which might threaten the success of program resource management. Political instability might interrupt the government organizations operation (Patanakul et al., 2016), hence affecting their resource planning adversely. Furthermore, political realignment, especially in relation to the leadership positions of government organizations, might disorient the implementation of resource management practices. Therefore, it is essential for the government organizations to formulate viable corrective management practices to deal with the uncertainties.

2.4 Main Problem in Program Resource Management

2.4.1 Managing Resource Allocation

The project management resources require the organization to adopt the right set of procedures, techniques and philosophies in allocating these resources in order to determine success and growth in the long run. The allocation of these resources is most commonly perceived as the functional and cross-functional resources according to the requirement of the business organization. Project resource management is considered to be one of the most essential components of a business sector and requires the transparency to be maintained in terms of demand and supply of the resources to ensure the suitable business activities (Alshurideh, 2022; Dwivedula, 2019; Hamadneh et al., 2021a, b; Hasan et al., 2022).

2.4.2 Career Path

While stating a new project managing the availability of the employees and ensuring their capacity to meet the needs and requirements of the project is also extremely important. This requires the management to identify the capacity of the members involved and also in choosing the right career path for them which also supports their abilities to perform according to the project requirements. This would also support the job description of the team members based upon their previous experiences and current skills.

2.4.3 Training and Development

In most of the cases, the members of the team involved in a new project often lack certain set of skills for the completion of a project. This requires the management to design the training and development sessions for the employees in order to ensure success of the project. This would ensure that all tools and procedure are rightly utilized and monitored for ensuring success for the organization (Laslo, 2010).

2.5 CSF or Best Internal or External Factors Required to Support Program Resource Management

CSF or critical success factors are the activities which a business setup requires in order to ensure the success of the business; this supports the business in allocating its internal or external resources and managing them according to the requirement of the organization by conducting the data and business analysis and in reaching a practical approach for the business sector. This supports the organization to significantly be able to identify the key factors responsible for the success of the organization and how the organization should focus on these key factors for sustaining success for a longer period of time (Kastor & Sirakoulis, 2009).

There are a number of internal or external factors responsible for project resource management within a business sector. Some of these factors include in Table 1.

All these internal and external factors are responsible for acting as the driving force for an organization to manage and support its program resource management and allocating the right resources at the right time and conveying the performance of the organization to the stakeholders more effectively.

Table 1 The main internal and external factors responsible for project resource management

Internal factors	External factors
• Plans and policies of an organization • Human resources • Financial resources • Marketing resources • Corporate image • Labour management of the organization • PR relations • Quality of infrastructure • Project execution • IT management • Interdependency with the employees	• Customers • Suppliers • Competitors • Marketing and advertisement • Economic environment • Political and legal factors • Technological growth • Social factors • Public

2.6 *Assessment and Gaps in Program Resource Management to Support Program Management*

There are a number of gaps and issues observed while intending to support the project resource management for an organization. A few of the most common gaps include identifying whether the organization has the right set of resources available or not and how an organization should be able to identify which resources to request further and how. Moreover, the organization also needs to identify whether all the required resources are proactively involved within the teams and if these resources need to be explored further. Moreover, the organization also needs to effectively communicate the gaps among the team members involved in order to plan the right set of solutions for the team members (Smith, 2003).

In order to support the concept of project resource management, the organization needs to equip themselves with the resource the capabilities and processes being put to right use in terms of amount of time, money and manpower to ensure the most suitable project resource management.

3 Data Gathering

To accomplish the aims of this research, the researchers conducted three interviews among program managers from three different government organizations that are essential, huge and have a high reputation, which are;

(1) Emirates Group.
(2) Water and Power Producer Companies (IWPP).
(3) The Executive Council of Dubai.

The program managers were the purposive sampling in this study who are the most applicable professionals in the program management field. Also, they have experiences, knowledge, skills, and tools. The interviews were personal, structured, and contained qualitative data. The questions were to cover our research purpose in-depth and details. Thus, the selected program managers were from various industries, which will assist the researchers in gaining the required information from different perspectives and views.

To reach these organizations, the researchers contact the program managers' secretaries to take the permissions to meet the program managers through call conversations and emails. The importance of meeting and conducting the interviews with the program managers explained as well as the aims of the research discussed with the secretaries (Joghee et al., 2020; Kashif et al., 2021; Khan, 2021; Lee & Ahmed, 2021). After getting program managers' agreements, the researchers proposed an applicable date and time for conducting interviews for the project managers. One day before the agreed interview date, the researchers sent a gentle reminder to the program managers in order to assure and confirm their availability.

The researchers were available in the organizations before 15 min of the arranged time. At the beginning of the interviews, the researchers informed the program managers that their responses will be treated highly confidential and will be used for research purposes. The interview will take approximately 45 to 60 min. Also, the researchers took their agreement to record the interviews. Finally, the researchers proceeded to interviews and took the required information and notes in order to achieve the research goal.

4 Data Analysis

The data were gathered from three government bodies where each presented professionals' roles and how crucial are their responsibilities in the success of program management in the authority. This section will discuss key results from the research qualitative data and compare of these primary findings to the literature review.

Emirates Group suggested that organizations can ensure the four factors through effective management strategy where the top management provide the right leadership focused on these factors, facilitate supportive infrastructure and ensure change implementation through proven standards and practices, clearly defining the change and aligning it to the set goals and objectives, effective communication, and through practical training to provide knowledge, skills, and competencies needed for efficient operation as change is rolled out. IWPP suggested that all the employees must complete the Position Qualification Requirement (PQR), written instinct tests, and assessment interview. Doing so gives a good indicator that they are well-trained and competent (Lee, Azmi, et al., 2022a, b; Lee, Romzi, et al., 2022a, b; Mehmood, 2021; Mehmood et al., 2019). For experienced professionals, IWPP chooses them based on the size of the program and its requirements. To check how supportive the infrastructure is, organizations must review their past performance with what they have and how good their infrastructure compared to what available in the market. To ensure proven standards and practices, IWPP follows the federal law of power and Abu Dhabi financial law references, follows the Power and Water Purchase Agreement, refers to the international power and water practices in the Middle East and Europe, and attending the yearly user summits that takes place in different countries in the world to check the global standards. The Executive Council of Dubai believes that to ensure the four factors, employees should undergo job-related education programs with experienced professionals and being grouped in teams with similar skill sets. Moreover, it was suggested to create a well-structured framework to gives employees incentives to be more productive and maintain a higher quality of work ethics. The practices build bridges between members to acknowledge the importance of following standards allowing for constant skill growth. The three organizations agreed with the findings in the literature review of how important those four factors resource management.

To improve the capacity and capability for the project and program management, Emirates Group suggested a similar solution to the literature review of providing a capable workforce to support the organization. Emirates Group suggested ensuring consistent, senior-level support the discipline of program management, integrate program management as part of the organization's mission, and involve the staff in the planning for change initiatives (Miller, 2021; Mondol, 2021; Obaid, 2021; Radwan & Farouk, 2021). IWPP's suggestions were about the concept discussed in the literature review of using the right resources at the right time and the concept of priority. IWPP believes that fewer priority projects should be removed or delayed focusing on the projects that are directly supporting the strategy of the company. Also, using the right resource at the right time, the company will have more resources to work with. Similarly, to IWPP, the Executive Council of Dubai suggested that to increase the capacity and capability of the programs, and organizations must focus on their goals and standards for projects.

Emirates Group's implementation for improving the government-wide capability for project and program management is to make sure that executives and stakeholders clearly understand their roles and responsibilities as well as consistency in development and training. IWPP implementation is about using quality tools such as the Deming wheel, Kizan continues improvement, and the fishbone diagram. The Executive Council of Dubai suggested that a clear vision and plan are the key to successful implementation for enhancements.

To improve the government-wide capability for both project as well as program management, there are several recommendations raised from the interviewees. Emirates Group recommended using an integrated approach to develop policy. The integration of performance improvement, goal setting, and the strategic planning process play an essential role in program management. Also, leadership should support program management, as active support from top agency leadership has proved to be a critical factor in the success of change management initiatives. Besides that, establishing roles and responsibilities of the agency's stakeholders in the program management processes will assist in promoting accountability and achievement of set goals. IWPP suggestions were to listen and refer to the department of finance yearly governmental budget that gives clear limitations or priorities of the country's/government every year requirement. This gives the overall country's vision and mission toward the right project requirement, to avoid objecting and rejection of the projects after putting all the efforts and resources for years. Also, it is very significant that all the EMDs/CEOs of different sectors gather regularly to discuss the main agendas on the development of the country to use all the experience brains to improve all areas simultaneously. The Executive Council of Dubai recommended that government services should coop with the latest technology to serve the public faster, more comfortable, and for consistent improvements. Public awareness about their role in project sustainability. This can be achieved via caretaking and appreciating the projects to promote longevity and quality of projects.

To resolve the issue with the inconsistency in the training and development of program managers, Emirates Group suggested first to make sure that the program manager job series have a clear career path and policies to support the career development of program managers. Also, Emirates Group suggested that organizations should comply with the policies guiding the certification of personnel in program management and design a systematic approach to program managers' training and development. IWPP suggested enrolling program managers in courses that require a real competency/certification. Also, IWPP suggested enrolling the program managers in courses based on their yearly appraisal and direct supervisor recommendation. Emirates Group and IWPP responses were aligned with the findings in the literature review regarding how to choose the courses based on their skills and performances. On the other hand, the Executive Council of Dubai believes that it is impossible to create a consistent training and development program for such practice because of the different functions and services in each organization.

The observation shows a successful strategy followed by top management technique in Emirates Group to allocate the responsibility and skills of stakeholders. The results from Emirates Group interview show significant details in which VIP stakeholders' responsibility are determined and training programs to be conducted when required. The other result summarizes that the Authority priories the communication between stakeholders and consider the opinion of all parties where they believe how significant this sharing will change in the Resource Management requirement and delivering the strategy benefits. Emirates Group offer training program to ensure that all members have clear understanding of the roles of each. On the other hand, IWPP uses modern technique to evaluate the performance of stakeholders and assess the number of completed projects against planned projects. The Executive Council ensures the importance of relating the program topics and functions to the organization nature of works and top management ensures the curriculum of the topics. Similarly, the literature review emphasized that stakeholders should understand governance roles and the authority strategy by utilizing tools and procedures by management and introducing effective resource management initiatives to ensure the full understanding by stakeholders.

It's assessed that Emirates team support each other, and every professional maintain a good environment and IWPP team provide transparency in the meetings between public and private sectors professionals. And the Executive Council support developing team-based project structure that involves various members with different qualifications. This matches with the research literature review in clarifying that the availability of resource management application helps in strengthening career development and supports business function. Also, this provide a career series for the career development of senior level. This path puts the project managers into higher managerial positions.

The most important results from Emirates Group Interview emphasizes the important of Project Manager Career path, Project Manager Tasks and its effect to the Emirates Group Strategy. IWPP focus on business quality and clear career path to senior management to get the desired output and benefits. This doesn't reflect with literature review views where it highlighted that the government entities lack employees with experiences and background in program management where this may cause a failure in many aspects.

5 Recommendations

The research analyzed the primary data gathered from three government bodies where the analysis highlighted the main findings and finalized the interpretation of each. The main statement that summarized the analysis is addressing the point in which government bodies' lack of professionals in program management that would change in the prosperity of the entities. The three government bodies improve the issue in offering trainings and programs to qualify the organization Human Resources.

There are several essential recommendations to initiate strong program resource management, including:

1. Preparing a clear description of the program resource management roles and responsibilities for the executives and key stakeholders.
2. Creating an official job description, job series, as well as a career path for program managers based on theories that are applied for all organizations.
3. The consistency of the program resource management required to develop and evolve a standards-based model via the federal government.
4. Program resource managers required to initiate a professional community among the federal government. To create this new community, people should be incentives to join the community. They can be incentives with financial bonuses within their organizations, and with volunteer hours from the government.
5. Implementing an integrated approach for both program policy as well as a program road map.
6. Designating senior executives per each organization in order to be responsible for both program policy as well as strategy.
7. Creating integration between program management, strategic planning, goals, and performances.
8. In government bodies, it's recommended to promote successful entrepreneurs that see challenges as opportunities to achieve the success.

9. To consider recommendations in legal aspects, top management should designate more project and program managers for the sake of the success of program management.
10. Develop a high level training and work shop to support program and resource management
11. Set up a strategy to develop a creative lesson learned regards the resource in pervious projects and programs and share it will all program members to better plan future.
12. Create a portal for sharing the information and experience among involved employee in the project to have a better allocation of resource next time.

6 Research Limitation

In despite of the above provided details and information, there were several limitations faced during the research. For instance, we have selected 6 organizations to meet with their program managers and interview them to explore, have more accurate data about the subject and to achieve the research objectives. However, 3 out of 6 organizations have approved our request to have the same completed and which lead to the success of the research. Other limitation faced is the timeline. Due to the official holidays and year end, it was difficult to schedule a suitable time with the interviewees since the time requested is between 45–60 min. In addition, the interviewees were very discreet regarding the information provided during the interview and most of the questions were answered very generally. Which this led the researchers to ask more specific questions to gain the information needed related to the respective organization.

7 Conclusion

Before the start of any project, the resources requirements have to be well-defined including quantity and type since it's an important component in any project success. Resource management application has a huge impact on the success of any program or project if it's applied efficiently. The literature review has deal with the importance of program resource management and its important impact in the government sectors. One of its impacts it helps in reducing the number of outsourcing while utilizing the internal resources skills and experience in achieving the program goals and objectives. As per the interviews conducted with the program managers of Emirates Group, Water and Power Producer Companies (IWPP) and The Executive Council of Dubai, it was concluded that the program manager job series must have a clear career path and policies to support the career development of program managers. It was farther concluded that resource management could be made more effective if more emphasis is put on the importance of project manager career path, project manager tasks and their effect on the company overall. Further addition was made regarding the fact that resource management puts excess focus on business quality and clear career path to senior management to get the desired output and benefits. If these two are lacking, the organization will suffer in the long run. Applying resource management leads in reaching to a smooth project success. There should more adoption of the resource management in both government and private sectors. Which this will give the opportunity for skills development. Moreover, organizations should be more aware about the benefits of program resource management and arrange for more specific trainings for the respective teams who should apply resource management.

Appendix

Appendix 1: Interview 1 (Emirates Group)

1. **There is no guarantee of success in large-scale, complex change initiatives. However if program management is undertaken by:**

    1. **Well-trained,**
    2. **Experienced professionals**
    3. **Supportive infrastructure**
    4. **Proven standards and practices**

We believe that success will be more consistently achieved

How organizations can ensure these 4 factors?

Emirates Group, it can ensure these factors through effective management strategy where the top management provide the right leadership focused on these factors. The top leadership of the group should be equipped with the capacity to coach their direct reports and to facilitate supportive infrastructure and ensure change implementation through proven standards and practices. They should be in a position to engage in conversations with individual team members to find ways of making sure that these 4 factors are taken care of adequately thereby increasing the possibility of success of the change initiative.

Emirates Group can ensure these 4 factors by clearly defining the change and aligning it to the set goals and objectives. Doing so will provide an opportunity for the organization and its leaders and team members to critically review the aspects of the change initiative to ensure they integrate these 4 factors, thereby contributing to the success of the change process.

Emirates Group can ensure these 4 factors through effective communication. They should determine the most effective means of communication for the team members. The communication strategy should take into account aspects like the timeline regarding how change will be communicated incrementally, which channels to use and what would be the key messages.

Emirates Group can ensure these 4 factors is through effective training to provide knowledge, skills, and competencies needed for efficient operation as change is rolled out.

2. **Program management is not consistently recognized as a management discipline that is essential to government performance, success, and results.**

What *solutions* for improving government-wide *capability* for project and program management

The solutions include ensuring consistent, senior-level support the discipline of program management in Emirates Group; program management should be made an integral part of accomplishing the organization's mission rather than being considered as a technical or administrative specialty focused on the implementation; and, program management staff to be consistently involved in the planning for change initiatives.

What *Implementation* for improving government-wide capability for project and program management

The implementation for improving government-wide capability for project and program management is that which entails executives and stakeholders clearly understanding their roles and responsibilities. It also entails ensuring consistency across the government in both the development and training of the involved personnel, especially program managers.

What *recommendations* for improving government-wide capability for project and program management

(i) The development of government-wide program and project management policy should be done using an integrated approach.
(ii) The leadership should support program management, as active support from top agency leadership has proved to be a critical factor in the success of change management initiatives. Such support is vital in facilitating the alignment of diverse groups within the organization.
(iii) There should be clearly established roles and responsibilities of agency's stakeholders in the program management processes to promote accountability and achievement of set goals.
(iv) Program management should be integrated into performance improvement, goal-setting, and strategic planning process.

3. **There is no consistency across the government in the training and development of program managers. How this can be resolved?**

Considering that competent and experienced program managers are the most critical determinants of successful program, it is important to resolve the problem of lack of consistency across the government in their training and development. In the case of

Emirates Group, it can be resolved first by making sure that the program manager job series have a clear career path and policies to support the career development of program managers. The problem can also be solved by making sure that the Emirates Group comply with the policies guiding the certification of personnel in program management. Besides, Emirates Group should have systematic approach to program managers' training and development.

4. **Organizations executives and VIP stakeholder's governance role requires clear understanding of:**

The role of how programs are derived from strategies

And how programs are delivering strategy benefits

The roles and responsibilities of the program manager

How this can be ensured by training and development?

Training and development within the Emirates Group would ensure that organizations executives and VIP stakeholders have clear understanding of their governance roles by providing them with clear guidance on their roles and responsibilities within the program management processes. It would also enhance their skills, competencies, and knowledge regarding the various aspects of their governance roles, such as how programs are delivering strategy benefits.

5. **Program managers lack a professional community within the federal government that can provide support and a voice on issues affecting the development of program management.**

Give us your opinion in regards to your organization

While in most cases program managers lack a professional community, Emirates Group has made efforts to create a professional community. The efforts have been informed by the need to ensure that program managers work in collaboration with other professionals within the organization rather than working in relative isolation. It has created an environment where program managers are able to participate in discussions of policy, thus allowing their voice to be heard in building and improving program management capabilities within the organization.

6. **A government-wide job series for program managers that spans business functions with a career path that extends into senior career executive management ranks.**

Give us your opinion in regards to your organization

Emirates Group has made considerable efforts in establishing job series for program managers with a career path extending into senior management ranks. It has put a policy in place supporting the career development of these professionals. As they progress to senior ranks across a job series, their levels of responsibilities also increase.

Appendix 2: Interview 2 (Water and Power Producer Companies)

1. **There is no guarantee of success in large-scale, complex change initiatives.**

However, if program management is undertaken by:

1. **Well-trained,**
2. **Experienced professionals**
3. **Supportive infrastructure**
4. **Proven standards and practices**

We believe that success will be more consistently achieved

How organizations can ensure these 4 factors?

Well, all the employees must complete the Position Qualification Requirement (PQR) that includes all the essential competencies for every job. The same is assessed in written instinct tests and finally with assessment interview and any failure will lead to repeat the exercise till reaching to full understanding and passing all the previously mentioned assessment method. This PQR has different levels for every position in order to be confirmed for the targeted position/designation.

Regarding the experienced professionals, keep in mind that the members of the project are all having not less than 15 years of experience in a relevant field particularly in power and water sector. The members had worked in numerous power and desalination plants that are having different designs and complex changes.

Actually, there is a big supportive infrastructure; one of them was mentioned in the first interview which is using a stable standard simulator. This simulator can be feeded with all the necessary data in order to provide the feasibility of the project in terms of dividends, loses, capability to handle the capital and operating budgets. Additionally, EWEC and ADPC are very professional entities that compares the bid of any open or floating tender in terms of cost effectiveness, innovation, simplicity, flexibility and prompt support to the power grid and water network in case of shortages to achieve sustainability of resources.

I believe this is really very important point, as we keen on to follow the federal law of power and Abu Dhabi financial law references in order to move on a solid foundation. Besides that, we do follow the Power and Water Purchase Agreement which is a standard agreement and was signed and agreed between all the shareholders to manage technical, training, financial and management requirements in which all are compliant with. Regarding a standard practice, we do refer to the International power and water practices in the Middle East and Europe. This benchmark is regularly implemented between the 10 different Independent Water and Power Producer companies (IWPP) in UAE (From Fujairah till Al Dhafra). Not all that but also in Middle East and Europe as mentioned via attending the yearly user summits that takes place in different countries in the world.

2. **Program management is not consistently recognized as a management discipline that is essential to government performance, success, and results.**

- **What *solutions* for improving government-wide *capacity* for project and program management**

In fact, removing or delaying less priority projects and focusing on the projects that directly supporting the power and water demand that ensures security of the network grid is much better!

Allocating different teams for different projects in which everyone team is dedicated for one project from the day one kick off till handing over the projects. These dedicated teams must always respect the time via starting and completing the projects on time with minimum delays.

- **What *solutions* for improving government-wide *capability* for project and program management**

The capability part can be improved via increasing the capital budget of the project that needs to be handled by capable people. Those people must be result oriented, working as one team, understand the project well, know the quick wins via using simple, effective and innovative methods to during project execution.

- **What *Implementation* for improving government-wide capability for project and program management**

Use the management pioneers' methodologies to organize the projects and achieve the targets for instance, Deming wheel and KiZAN continues improvement. Furthermore, to use the right management tools, for instance fish bone diagram that illustrates the cause and effect actions to identify the bottle nicks and the remedial solutions. Finally, use the skill matrix, SWAT analysis and Pareto diagram that all can guide and lead the project to the desired targets as per the planed duration and achieve the necessary quality.

- **What *recommendations* for improving government-wide capability for project and program management**

Always listen and refer to the department of finance yearly governmental budget that gives a clear limitations or priorities of the country's/government every year requirement. Basically, it gives the overall country's vision and mission toward the right project requirement, to avoid objecting and rejection of the projects after putting all the efforts and resources for years. Additionally, I do recommend that all the EMDs/CEOs of different sectors gather every year twice or tries to discuss the main agendas pertaining to development of the country in order to use all the experience brains to improve all sectors simultansly.

3. **There is no consistency across the government in the training and development of program managers.**

How this can be resolved?

I addition to the yearly selected training courses by the employees based on the yearly appraisal, direct supervisor recommendation, the management shall select and enroll all the managers together to the one course that needs a real competency/certification in order to be specialist. In the same program to assess the managers in an either examination or interview or (both).

4. **Organizations executives and VIP stakeholder's governance role requires clear understanding of:**

The role of how programs are derived from strategies

And how programs are delivering strategy benefits

The roles and responsibilities of the program manager

How this can be ensured by training and development?

The training department shall use a tool named primavera that organize the project and send daily and then monthly update report to the sectors that shows the work progress in detail. Implement a KPI for individual stakeholder managers that assess total number of completed projects on time and as per requirement. In addition, to have weekly and monthly meetings to ensure alignment between the strategy and the role of the program.

5. **Program managers lack a professional community within the federal government that can provide support and a voice on issues affecting the development of program management.**

Give us your opinion in regard to your organization

In our case since we're a private sector subsidized by the government as we own 40% of the project while the government own 60%, we don't face problem. The reason for that is that it is a must for every stakeholder to provide transparency via having regular meetings. These meetings' agendas are related to all aspects of the project and how construct bridges of supports in technical, administrative and financial terms.

6. **A government-wide job series for program managers that spans business functions with a career path that extends into senior career executive management ranks.**

Give us your opinion in regard to your organization

In my point of view, this is a good approach that we are implementing not only to do the job in order to be completed on time, but also with the desired quality. Having a career path extended into senior career executive can lead to have the same understanding that consequently will produce the right project specification, as per all the procedures and agreements.

Having this approach can eliminate or minimize deviation of the project requirement from the tasks breakdown. This basically the right plan in order to retain the brains and having them to be loyal toward the work. This can generate proactive approaches and techniques to balance and align the program managers' ways of working into the senior staff way without misstatements that can effect negatively on the project progress or sometimes to the failure to complete the ongoing project.

Appendix 3: Interview 3 (The Executive Council of Dubai)

1. **There is no guarantee of success in large-scale, complex change initiatives.**

However, if program management is undertaken by:

1. **Well-trained,**
2. **Experienced professionals**
3. **Supportive infrastructure**
4. **Proven standards and practices**

We believe that success will be more consistently achieved

How organizations can ensure these 4 factors?

We can achieve that by scheduling consistent form of training for employees to undergo job related education programs in order to sustain an up-to-date level of experience of the relevant work With Experienced professionals, if grouped in teams with similar skills sets, can perform better especially when a suitable leader is involved in managing employees from different sectors and entities While being able to work under a well-structured framework gives employees incentives to be more productive and maintain a higher quality of work ethics mostly beneficial for group of teams with lower skill sets. The practices builds bridges between members to acknowledge the importance of following standards allowing for a constant skill growth.

2. **Program management is not consistently recognized as a management discipline that is essential to government performance, success, and results.**

- **What *solutions* for improving government-wide *capacity* for project and program management**

- **What *solutions* for improving government-wide *capability* for project and program management**
- **What *Implementation* for improving government-wide capability for project and program management**
- **What *recommendations* for improving government-wide capability for project and program management**

Setting goals and standards for projects. i.e. Dubai Government and Sheikh Mohammed Bin Rashid's vision has become the key benchmark unit for setting quality standards in executing projects.

The vision is followed by implementation. Again, His Highness's vision would not reach out without the energy put for implementing wide scale projects up to the completion.

Consistent improvements. With the ever growing age of technology, the overall government services should coop with the latest technology to service the public faster and easier.

Public awareness, the public should be more responsible to sustain the projects by caretaking and appreciating the projects to promote longevity and quality of projects. A simple example, never to throw chewing gums on pavements which can last stuck to a long time effecting the overall visuals of the street.

3. **There is no consistency across the government in the training and development of program managers.**

How this can be resolved?

It is impossible to create a nationwide consistency for such practices. In Dubai there are almost 60 entities. Each holding different functions and services.

4. **Organizations executives and VIP stakeholder's governance role requires clear understanding of:**

The role of how programs are derived from strategies

And how programs are delivering strategy benefits

The roles and responsibilities of the program manager

How this can be ensured by training and development?

To ensure proper exposure of the learning programs to the intended project managers. The topics of the programs should be relevant to the nature of works being exposed to the project managers. Executives should review the curriculum modules for each course.

5. **Program managers lack a professional community within the federal government that can provide support and a voice on issues affecting the development of program management.**

Give us your opinion in regard to your organization

The lack comes because of the absence of dedicated proper research and development program to become solely assigned to support the project managers. A simple solution is to use team based project structure (horizontal style) with members from different entities and backgrounds.

References

Al Ali, A. (2021). The impact of information sharing and quality assurance on customer service at UAE banking sector. *International Journal of Technology, Innovation and Management (IJTIM), 1*(1), 1–17. https://doi.org/10.54489/ijtim.v1i1.10.

Al Kurdi, B., Elrehail, H., Alzoubi, H., Alshurideh, M., & Al-adaileh, R. (2021). The interplay among HRM practices, job satisfaction and intention to leave: An empirical investigation. *Journal of Legal, Ethical and Regulatory, 24*(1), 1–14.

AlHamad, A., Alshurideh, M., Alomari, K., Kurdi, B., Alzoubi, H., Hamouche, S., & Al-Hawary, S. (2022). The effect of electronic human resources management on organizational health of telecommuni-cations companies in Jordan. *International Journal of Data and Network Science, 6*(2), 429–438.

Alhamad, A. Q. M., Akour, I., Alshurideh, M., Al-Hamad, A. Q., Kurdi, B. A., & Alzoubi, H. (2021). Predicting the intention to use google glass: A comparative approach using machine learning models and PLS-SEM. *International Journal of Data and Network Science, 5*(3). https://doi.org/10.5267/j.ijdns.2021.6.002.

Ali, N., Ahmed, A., Anum, L., Ghazal, T. M., Abbas, S., Khan, M. A., Alzoubi, H. M., & Ahmad, M. (2021). Modelling supply chain information collaboration empowered with machine learning technique. *Intelligent Automation and Soft Computing, 30*(1), 243–257. https://doi.org/10.32604/iasc.2021.018983.

Ali, N. M., Ghazal, T., Ahmed, A., Abbas, S., A. Khan, M., Alzoubi, H., Farooq, U., Ahmad, M., & Adnan Khan, M. (2022). Fusion-based supply chain collaboration using machine learning Techniques. *Intelligent Automation & Soft Computing, 31*(3), 1671–1687. https://doi.org/10.32604/iasc.2022.019892.

Alkalha, Z., Al-Zu'bi, Z., Al-Dmour, H., Alshurideh, M., & Masa'deh, R. (2012). Investigating the effects of human resource policies on organizational performance: An empirical study on commercial banks operating in Jordan. *European Journal of Economics, Finance and Administrative Sciences, 51*(1), 44–64.

Alnazer, N. N., Alnuaimi, M. A., & Alzoubi, H. M. (2017). Analysing the appropriate cognitive styles and its effect on strategic innovation in Jordanian universities. *International Journal of Business Excellence, 13*(1), 127–140. https://doi.org/10.1504/IJBEX.2017.085799

Alnuaimi, M., Alzoubi, H. M., Ajelat, D., & Alzoubi, A. A. (2021). Towards intelligent organisations: An empirical investigation of learning orientation's role in technical innovation. *International Journal of Innovation and Learning, 29*(2), 207–221. https://doi.org/10.1504/IJIL.2021.112996

Alsharari, N. (2021). Integrating blockchain technology with internet of things to efficiency. *International Journal of Technology, Innovation and Management (IJTIM), 1*(2), 1–13.

Alshurideh, M. (2022). Does electronic customer relationship management (E-CRM) affect service quality at private hospitals in Jordan? *Uncertain Supply Chain Management, 10*(2), 1–8.

Alshurideh, M., Alsharari, N., & Al Kurdi, B. (2019). Supply chain integration and customer relationship management in the airline logistics. *Theoretical Economics Letters, 9*(02), 392–414.

Alshurideh, M., Gasaymeh, A., Ahmed, G., Alzoubi, H., & Kurd, B. A. (2020). Loyalty program effectiveness: Theoretical reviews and practical proofs. *Uncertain Supply Chain Management, 8*(3). https://doi.org/10.5267/j.uscm.2020.2.003.

Alshurideh, M. T., Al Kurdi, B., Alzoubi, H. M., Ghazal, T. M., Said, R. A., AlHamad, A. Q., Hamadneh, S., Sahawneh, N., & Al-kassem, A. H. (2022). Fuzzy assisted human resource management for supply chain management issues. *Annals of Operations Research*, 1–19.

Alzoubi, Ali. (2021a). The impact of process quality and quality control on organizational competitiveness at 5-star hotels in Dubai. *International Journal of Technology, Innovation and Management (IJTIM), 1*(1), 54–68. https://doi.org/10.54489/ijtim.v1i1.14.

Alzoubi, Asem. (2021b). Renewable Green hydrogen energy impact on sustainability performance. *International Journal of Computations, Information and Manufacturing (IJCIM), 1*(1), 94–110. https://doi.org/10.54489/ijcim.v1i1.46.

Alzoubi, H. M., & Aziz, R. (2021). Does emotional intelligence contribute to quality of strategic decisions? the mediating role of open innovation. *Journal of Open Innovation: Technology, Market, and Complexity, 7*(2), 130. https://doi.org/10.3390/joitmc7020130

Alzoubi, H. M., Vij, M., Vij, A., & Hanaysha, J. R. (2021). What leads guests to satisfaction and loyalty in UAE five-star hotels? AHP analysis to service quality dimensions. *Enlightening Tourism, 11*(1), 102–135. https://doi.org/10.33776/et.v11i1.5056.

Alzoubi, H. M., & Yanamandra, R. (2020). Investigating the mediating role of information sharing strategy on agile supply chain. *Uncertain Supply Chain Management, 8*(2), 273–284. https://doi.org/10.5267/j.uscm.2019.12.004

Alzoubi, H., Ahmed, G., Al-Gasaymeh, A., & Kurdi, B. (2020a). Empirical study on sustainable supply chain strategies and its impact on competitive priorities: The mediating role of supply chain collaboration. *Management Science Letters, 10*(3), 703–708.

Alzoubi, H., Alshurideh, M., Kurdi, B., Akour, I., & Aziz, R. (2022). Does BLE technology contribute towards improving marketing strategies, customers' satisfaction and loyalty? The role of open innovation. *International Journal of Data and Network Science, 6*(2), 449–460.

Alzoubi, H., & Ahmed, G. (2019). Do TQM practices improve organisational success? A case study of electronics industry in the UAE. *International Journal of Economics and Business Research, 17*(4), 459–472. https://doi.org/10.1504/IJEBR.2019.099975

Alzoubi, H., Alshurideh, M., Kurdi, B. A., & Inairat, M. (2020b). Do perceived service value, quality, price fairness and service recovery shape customer satisfaction and delight? A practical study in the service telecommunication context. *Uncertain Supply Chain Management, 8*(3), 579–588. https://doi.org/10.5267/j.uscm.2020.2.005

Ashal, N., Alshurideh, M., Obeidat, B., & Masa'deh, R. (2021). The impact of strategic orientation on organizational performance: Examining the mediating role of learning culture in Jordanian telecommunication companies. *Academy of Strategic Management Journal, Special Is*(Special Issue 6), 1–29.

Ayers, R. S. (2015). Aligning individual and organizational performance: Goal alignment in federal government agency performance appraisal programs. *Public Personnel Management, 44*(2), 169–191.

Aziz, N., & Aftab, S. (2021). Data mining framework for nutrition ranking: methodology: SPSS modeller. *International Journal of Technology, Innovation and Management (IJTIM), 1*(1), 85–95.

Braganza, A., Brooks, L., Nepelski, D., Ali, M., & Moro, R. (2017). Resource management in big data initiatives: Processes and dynamic capabilities. *Journal of Business Research, 70*, 328–337.

Cruz, A. (2021). Convergence between blockchain and the internet of things. *International Journal of Technology, Innovation and Management (IJTIM), 1*(1), 35–56.

de Carvalho, M. M., Patah, L. A., & de Souza Bido, D. (2015). Project management and its effects on project success: Cross-country and cross-industry comparisons. *International Journal of Project Management, 33*(7), 1509–1522.

Dwivedula, R. (2019). Human resource management in project management: Ideas at the cusp. *European Project Management Journal, 9*(1), 34–41.

Eli, T. (2021). Studentsperspectives on the use of innovative and interactive teaching methods at the University of Nouakchott Al Aasriya, Mauritania: English Department as a Case Study. *International Journal of Technology, Innovation and Management (IJTIM), 1*(2), 90–104.

Ghazal, T. M., Hasan, M. K., Alshurideh, M. T., Alzoubi, H. M., Ahmad, M., Akbar, S. S., Al Kurdi, B., & Akour, I. A. (2021). IoT for smart cities: Machine learning approaches in smart healthcare—a review. *Future Internet, 13*(8), 218. https://doi.org/10.3390/fi13080218

Guergov, S., & Radwan, N. (2021). Blockchain convergence: Analysis of issues affecting IoT, AI and blockchain. *International Journal of Computations, Information and Manufacturing (IJCIM), 1*(1), 1–17. https://doi.org/10.54489/ijcim.v1i1.48.

Hamadneh, S., Pedersen, O., & Al Kurdi, B. (2021a). An investigation of the role of supply chain visibility into the scottish bood supply Chain. *Journal of Legal, Ethical and Regulatory Issues, 24*(Special Issue 1), 1–12.

Hamadneh, S., Pedersen, O., Alshurideh, M., Kurdi, B. Al, & Alzoubi, H. (2021b). An investigation of the role of supply Chain visibility into the scottish blood supply chain. *Journal of Legal, Ethical and Regulatory Issues, 24*(Special Issue 1), 1–12.

Hanaysha, J. R., Al-Shaikh, M. E., Joghee, S., & Alzoubi, H. (2021a). Impact of innovation capabilities on business sustainability in small and medium enterprises. *FIIB Business Review*, 1–12.https://doi.org/10.1177/23197145211042232.

Hanaysha, J. R., Al Shaikh, M. E., & Alzoubi, H. M. (2021b). Importance of marketing mix elements in determining consumer purchase decision in the retail market. *International Journal of Service Science, Management, Engineering, and Technology (IJSSMET), 12*(6), 56–72.

Hasan, O., McColl, J., Pfefferkorn, T., Hamadneh, S., Alshurideh, M., & Kurdi, B. (2022). Consumer attitudes towards the use of autonomous vehicles: Evidence from United Kingdom taxi services. *International Journal of Data and Network Science, 6*(2), 537–550.

Hornstein, H. A. (2015). The integration of project management and organizational change management is now a necessity. *International Journal of Project Management, 33*(2), 291–298.

Joghee, S., Alzoubi, H. M., & Dubey, A. R. (2020). Decisions effectiveness of FDI investment biases at real estate industry: Empirical evidence from Dubai smart city projects. *International Journal of Scientific and Technology Research, 9*(3), 3499–3503.

Kabrilyants, R., Obeidat, B. Y., Alshurideh, M., & Masa'deh, R. (2021). The role of organizational capabilities on e-business successful implementation. *International Journal of Data and Network Science, 5*(3). https://doi.org/10.5267/j.ijdns.2021.5.002.

Kashif, A. A., Bakhtawar, B., Akhtar, A., Akhtar, S., Aziz, N., & Javeid, M. S. (2021). Treatment response prediction in hepatitis C patients using machine learning techniques. *International Journal of Technology, Innovation and Management (IJTIM), 1*(2), 79–89. https://doi.org/10.54489/ijtim.v1i2.24.

Kastor, A., & Sirakoulis, K. (2009). The effectiveness of resource levelling tools for resource constraint project scheduling problem. *International Journal of Project Management, 27*, 493–500.

Khan, M. A. (2021). Challenges facing the application of IoT in medicine and healthcare. *International Journal of Computations, Information and Manufacturing (IJCIM), 1*(1), 39–55. https://doi.org/10.54489/ijcim.v1i1.32

Kogan, L., Papanikolaou, D., Seru, A., & Stoffman, N. (2017). Technological innovation, resource allocation, and growth. *The Quarterly Journal of Economics, 132*(2), 665–712.

Laslo, Z. (2010). Project portfolio management: An integrated method for resource planning and scheduling to minimize planning/scheduling-dependent expenses. *International Journal of Project Management, 28*(6), 609–618.

Lee, C., & Ahmed, G. (2021). Improving IoT privacy, data protection and security concerns. *International Journal of Technology, Innovation and Management (IJTIM), 1*(1), 18–33. https://doi.org/10.54489/ijtim.v1i1.12

Lee, K., Azmi, N., Hanaysha, J., Alzoubi, H., & Alshurideh, M. (2022a). The effect of digital supply chain on organizational performance: An empirical study in Malaysia manufacturing industry. *Uncertain Supply Chain Management, 10*(2), 495–510.

Lee, K., Romzi, P., Hanaysha, J., Alzoubi, H., & Alshurideh, M. (2022b). Investigating the impact of benefits and challenges of IOT adoption on supply chain performance and organizational performance: An empirical study in Malaysia. *Uncertain Supply Chain Management, 10*(2), 537–550.

Mehmood, T. (2021). Does information technology competencies and fleet management practices lead to effective service delivery? Empirical Evidence from E-Commerce Industry. *International Journal of Technology, Innovation and Management (IJTIM), 1*(2), 14–41.

Mehmood, T., Alzoubi, H. M., & Ahmed, G. (2019). Schumpeterian entrepreneurship theory: Evolution and relevance. *Academy of Entrepreneurship Journal, 25*(4).

Miller, D. (2021). The best practice of teach computer science students to use paper prototyping. *International Journal of Technology, Innovation and Management (IJTIM), 1*(2), 42–63. https://doi.org/10.54489/ijtim.v1i2.17

Mondol, E. P. (2021). The impact of block chain and smart inventory system on supply chain performance at retail industry. *International Journal of Computations, Information and Manufacturing (IJCIM), 1*(1), 56–76. https://doi.org/10.54489/ijcim.v1i1.30

NAPA. (2015). *Improving program management in the federal government.*

Obaid, A. J. (2021). Assessment of smart home assistants as an IoT. *International Journal of Computations, Information and Manufacturing (IJCIM), 1*(1), 18–36. https://doi.org/10.54489/ijcim.v1i1.34

Patanakul, P., Kwak, Y. H., Zwikael, O., & Liu, M. (2016). What impacts the performance of large-scale government projects? *International Journal of Project Management, 34*(3), 452–466.

Petersen, J. A., & Kumar, V. (2015). Perceived risk, product returns, and optimal resource allocation: Evidence from a field experiment. *Journal of Marketing Research, 52*(2), 268–285.

Radwan, N., & Farouk, M. (2021). The growth of internet of things (IoT) in the management of healthcare issues and healthcare policy development. *International Journal of Technology, Innovation and Management (IJTIM), 1*(1), 69–84. https://doi.org/10.54489/ijtim.v1i1.8

Rui, L. S., Khai, L. L., & Alzoubi, H. M. (2022). Determinants of emerging technology adoption for safety among construction businesses. *Academy of Strategic Management Journal, 21*(Special Issue 4), 1–20.

Scarlett, L., & Boyd, J. (2015). Ecosystem services and resource management: Institutional issues, challenges, and opportunities in the public sector. *Ecological Economics, 115*, 3–10.

Serrador, P., & Turner, R. (2015). The relationship between project success and project efficiency. *Project Management Journal, 46*(1), 30–39.

Shakhour, N. H. T., Obeidat, B. Y., Jaradat, M. O., & Alshurideh, M. (2021). Agile-minded organizational excellence: Empirical investigation. *Academy of Strategic Management Journal, 20*(Special Issue 6), 1–25.

Shamout, M., Elayan, M., Rawashdeh, A., Kurdi, B., & Alshurideh, M. (2022). E-HRM practices and sustainable competitive advantage from HR practitioner's perspective: A mediated moderation analysis. *International Journal of Data and Network Science, 6*(1), 165–178.

Smith, A. D. (2003). Surveying practicing project managers on curricular aspects of project management programs: A resource-based approach. *Project Management Journal, 34*(2), 26–33.

Zu'bi, Z., Al-Lozi, M., Dahiyat, S., Alshurideh, M., & Al Majali, A. (2012). Examining the effects of quality management practices on product variety. *European Journal of Economics, Finance and Administrative Sciences, 51*(1), 123–139.

Project Quality Management in the United Arab Emirates Mining and Construction Sector: A Literature Review

Mounir El Khatib, Haitham M. Alzoubi, Muhammad Alshurideh, and Ali A. Alzoubi

Abstract The United Arab Emirates (UAE) presents a special set of circumstances for sustainable long-term development and economic growth. In this paper, quality management in the mining and industries have been identified such circumstances. More specifically, this paper examines the level of integration of quality management practices in UAE's mining and construction sectors. Using a literature-based approach, the paper identifies the common factors that lead to failure during project management. Recommendations are also offered accordingly. Also, this paper concludes that although there is a general awareness of the need for quality management in the UAE, quality related issues are not uncommon in both industries. Moreover, delays and poorly structured processes are the main issues causes of low quality projects.

Keywords Project quality · Construction · Mining · UAE

M. El Khatib
Program Chair, Hamdan Bin Mohamad Smart University, Dubai, UAE

H. M. Alzoubi (✉)
School of Business, Skyline University College, Sharjah, UAE
e-mail: haitham.alzubi@skylineuniversity.ac.ae

M. Alshurideh
Department of Marketing, School of Business, University of Jordan, Amman, Jordan
e-mail: m.alshurideh@ju.edu.jo; malshurideh@sharjah.ac.ae

Department of Management, College of Business Administration, University of Sharjah, Sharjah, UAE

A. A. Alzoubi
Public Security Directorate, Amman, Jordan

M. Alshurideh et al. (eds.), *The Effect of Information Technology on Business and Marketing Intelligence Systems*, Studies in Computational Intelligence 1056,
https://doi.org/10.1007/978-3-031-12382-5_73

1 Introduction

For many years, crude oil has been the biggest contributor to the economy of countries in the Middle East, including United Arab Emirates (UAE). Oil mining is still an important activity in the UAE but the recent price instabilities and transition from overreliance on hydro fuels has forced the region to diversify its focus to other minerals and the construction industry as a way of bolstering the economy even more (Shahbaz et al., 2014). Modern highways, connecting high end cities have emerged within a short period of time. Today cities like Dubai that were once barren deserts have transformed into beautiful cities, characterized by modern hospitals, hotels, schools and airports (Aburayya et al., 2020; Delgado, 2016; Taryam et al., 2020). In fact, with the ongoing plans to transform Dubai into a major tourist destination, the construction industry might soon be ranked higher than oil as the leading contributor to the economy (Al Batayneh et al., 2021; Al Shebli et al., 2021; AlSuwaidi et al., 2021; Madi Odeh et al., 2021).

From an economic point of view, growth in the construction and mining industries is an entirely good thing. Yet, the complexity that is inherent in new projects presents significant huddles to researchers and project managers on the basis that determining the vital success factors is a continuous and complicated process. As Greeff and Barker (2014) declared, one of the common characteristics between the construction and mining industries is their fragmentation. Compared to the manufacturing sector, for example, construction projects involve multiple shareholders including contractors, the government and miners (Greeff & Barker, 2014). As a direct result of this complexity, each year the UAE reports considerable number of construction claims on the basis of low quality projects that fail to meet the stakeholders' requirements (Mishmish & El-Sayegh, 2018). In fact, the main reason for failure in many projects failure to understand stakeholder's needs and expectations as well as failing to consider the consumer environment or the end users of a service or product.

2 Problem Statement

Many construction industries in the United Arab Emirates underestimate the capacity of customers to identify service quality; they fail to apprehend that customers not only evaluate end service but are also affected by the way a service or product is delivered (Alshurideh, 2022; Awadhi et al., 2021; Mishmish & El-Sayegh, 2018). Moreover, the management of mining and construction projects requires a comprehensive understanding of modern management practices coupled with the understanding of the best practices in design and construction (Mehmood et al., 2019; Alshurideh et al., 2022; Shamout et al., 2022). In this paper the author will investigate the important elements of Project quality management in UAE's mining and construction sector.

3 Objectives of the Research

1. To Investigate the level of integration between quality management with projects in the United Arab Emirates mining and construction sector.
2. To study the success and deficiency factors in project quality management.
3. To come up with recommendations that will guide project managers and organizations on proper ways of delivering projects that satisfy customers throughout the project lifecycle.

4 Research Method

This research is based on the Organization Project Management Model (OPM3) as described by the Project Management Institute (PMI). This method is preferred because it not only provides a guideline for understanding the best practices in project management but also recommends measures to ensure that projects meet quality standards including the capability to perform predictably and consistently (Alzoubi et al., 2021; Hanaysha, Al-Shaikh, et al., 2021a, b). In this regard, OPM3 describes three key elements to successful quality Project Management: (PMI, 2017).

Knowledge acquired by investigating successful projects.
Assessment of existing needs (i.e. level of integration of quality management with projects in the UAE's mining and construction sector) and recommend improvements.
Improvement based on assessments to determine the necessary steps to attaining improvement goals.

The research employed a literature-based approach as the primary data collection method. In this regard, the authors reviewed key research papers and articles relating to quality management in UAE's construction and mining industries.

5 Background and Literature Review

5.1 The Concept of Quality Management

According to Rumane (2017), quality is a strategic imperative that helps organizations improve the bottom line, thus quality is defined on the basis of conformance to guidelines and requirements. In other words, quality is not gauged in terms of the common concepts of goodness, beautiful or excellent (Rumane, 2017). Instead, any service, process or product that is delivered in a manner that conforms to the client's requirements is seen as quality. This view is echoed by Pepall (2014) who stated that quality is "meeting the customer's (agreed) requirements, formal and informal, at the lowest cost, first time, every time." The Project Management Advisor (PMA) defines requirements in project management as specifications on the "capabilities, features or attributes of the project's deliverables" Requirements are drawn from an analysis of stakeholders' wishes and needs and thus is an expression of the mutual understanding and agreement between the parties involved (Ali et al., 2021; Ghazal et al., 2021). Hence, the definition of quality can be, more or less, said to be the ability to conform to the stakeholder's wishes (Ali et al., 2022; Alnazer et al., 2017).

Yang and Trewn (2004) asserted that although the concept of quality has been around for a longtime, it was not until the 1940s, during World War 2, that people started viewing quality as a statistical concept. During this time, the military introduced statistical sampling techniques to manage quality. Monitoring was also done using quality control charts were production process. Notably, Chung (2002) suggests that it was not until the 1960s that people viewed quality as an organization-wide concept i.e., quality was now not only used in production but in most business processes (Mukherjee, 2006).

5.2 Quality Management in the Construction and Mining Industries

As already established, the general aims of quality management is to achieve standardization (Pepall, 2014). The quality of a mineral or complete structure is an assimilation of all the relevant activities that ensure the results of a mining/construction project are satisfactory. According to the concept of quality management is more than reacting to events in the during site production. Rather, it is a strategic business function that governs all business processes in construction organizations so that organizations that fail to guarantee quality delivery are unable to compete for new projects (Alshamsi et al., 2021; Alshurideh et al., 2019; Alzoubi, 2021a; Farouk, 2021).

In his study Chung, (2002), asserted that the concept of quality management in the construction industry became popular in the 1980s and is still being practiced today. Generally, the ISO 9000 family of quality management standards is used in construction organizations to govern service delivery so that it meets the needs of all the stakeholders involved (Alzoubi et al., 2022; Obaid, 2021). This is possible on the basis that the ISO 9000 quality management standards establish the fundamentals of quality management systems and proposes "coordinated activities to direct and control an organization with regard to quality" (Peach, 2003). The activities of directing and controlling with relation to quality encompass establishing the necessary quality policies, objectives, quality planning, quality assurance, quality improvement: and quality control (ISO, 2009).

The quality planning function consists of specifying quality objectives and putting into place the necessary operational resources, and, processes to attain the quality management objectives (Al Hamad et al., 2021, 2022). One of the peculiar challenges faced by researchers and project managers alike is discerning the different in meanings between quality management and quality related activities including Quality Control (QC) and Quality Assurance (QA) (Alnuaimi et al., 2021; Alzoubi & Ahmed, 2019; Joghee et al., 2020). Mukherjee (2006) made the distinction between the terms, thus; QC and QA are essentially sub-elements of project quality management. In terms of functional relevance, QC and QA find meaning during the implementation phase while quality management represents the organization's strategically philosophy as regards project completion (Mukherjee, 2006). In this regard, quality management is relevant at all stages of project (Alketbi et al., 2020; AlMehrzi et al., 2020; Mashaqi et al., 2020). Finally, the aims of quality improvement, are to increase an organization's ability to fulfil all the quality requirements (Al Dmour et al., 2014; Pepall, 2014).

In both construction and mining projects the quality control function involves ensuring that the product specifications are met, inspection of tender documents, etc. (Alzoubi, 2021b; Hanaysha et al., 2021a). At the very basic, quality in construction and mining organizations translates to ensuring high level products that meet the customer's needs (Al Ali, 2021; Kashif et al., 2021). But quality also gives these organizations a competitive edge over competitors and thus safeguards business survival (Rumane, 2017). Similarly, it is equally important to manage the quality of services delivery during the pre-tender stage and this should be practiced through the entire lifetime of mining and construction projects (Guergov & Radwan, 2021; Khan, 2021; Mondol, 2021). As Taher (2009) noted that in construction, and generally all civil engineering projects, a satisfactory product is only essential but not all that is required. In fact, most defects in buildings arise from faulty design processes and inadequacies when incorporating the design in the construction (Akhtar et al., 2021; Eli, 2021).

Just like in most projects, QC consists of activities that govern the implementation of QA programs so that when implemented effectively, QC programs reduce the risk of mistakes and accidents. According to Greeff and Barker (2014), quality assurance in the construction and mining sectors involves ensuring that relevant activities are put in place to meet the quality management requirements of a project. Such activities include setting standards, training guidelines, procedures and policies (Alsharari, 2021; Mehmood, 2021). In essence, good QA practices serve as a shield against quality related issues in the sense that the defined procedures, for example, are a framework to gauge avoid accidents (Aziz & Aftab, 2021; Miller, 2021). Policies are equally important as they give early warnings when there is potential clash of internal and external processes.

6 Findings and Discussion

6.1 *Quality Management with Projects in the Construction Sector*

From a general perspective, quality management appears to be a well understood concept in the United Arab Emirates (Magd & Nabulsi, 2012). Yet, a more granular review of the literature indicates that construction claims are a leading hindrance in construction projects, mainly because claims cause delays (El-Sayegh & Mansour, 2015; Zaneldin, 2006). In a pilot study by Zaneldin (2006), the researcher studied the main causes of construction claims and analyze their frequency in Abu Dhabi and Dubai. While quality management systems have been adopted in the UAE, there is still a considerable number of quality related claims including changes claims, extra work claims and delay claims. As shown in the Table 1 and as claimed by Zaneldin, (2006), change related claims are the most frequent.

The assertions made by Zaneldin, (2006), conquer with the information gathered from general interviews and surveys conducted in this study. In fact, there was a general consensus among respondents that the most common quality management issues in UAE's construction industry are related to minor issues like not reproducing the design in the actual work and delays. A research by Ren et al. (2008) asserted that delays is generally the most common quality related issues that is experienced in UAE's construction industry (Fig. 1).

El-Sayegh and Mansour (2015) also identified the major causes of delays stating that "they {causes} vary from the unrealistic project duration, nominated sub-contractors, and the culture impacts". These factors have also been identified by other authors who emphasize that delays are a major failure factor and a leading reason for low quality in Saudi Arabia's construction sector (Cruz, 2021; Lee & Ahmed, 2021; Radwan & Farouk, 2021). The importance of time has also been emphasized in quality management literature due to the importance of time delays to both contractors and business owners (Jarkas & Younes, 2014; Mishmish & El-Sayegh, 2018).

Table 1 Causes of construction delays in the UAE

Types of claims	Weighted average	Importance index (%)	Rank
Changes claims	2.48	62.0	1
Extra-work claims	2.46	61.5	2
Delay claims	2.00	50.0	3
Non-performance claims	1.54	38.5	4
Different site conditions claims	1.46	36.5	5
Acceleration claims	1.39	34.8	6
Damage claims	1.14	28.5	7
Contract ambiguity claims	1.07	26.8	8

Factors, Reasons and Causes of delay

Client	Consultant	Contractor
Regular interference and poor communication	Incomplete contract documents	Inappropriate organization management
Variation order and late approval for payment	Incomplete drawings	Lack of technical professional in the organisation
Late supply of information and late decision making	Poor design management	Unsmooth external and internal communications
Project objectives are not very clear	Slow response	Lack coordination with sub-contractors
Nomination of Sub-contractors and suppliers	Delayed approval of drawings and BOQ for construction	Centralization with top management
Many provisional sums and prime cost	Inadequate duration for inspection	Delayed mobilization
Duration is not enough for constructing the project	Experience of staff in management and technical inspection	Incompetent contractor staff
Irregular payments and disturbed cash flow of main contractor	Delay in submittal and approval	Poor planning, scheduling or resource management
Routine of government authorities and approvals	Poor communication between consultant staff	Poor quality control
Irregular attending of weekly meetings	Poor quality control	Congested construction site

Fig. 1 Factors, reasons and Causes of Delays in UAE's construction sector (Ren et al., 2008)

6.2 *Quality Management with Projects in the Mining Sector*

Respected to the fact that there are limited studies that have studied the quality management topic in the United Arab Emirates. However, some existing literature show a high level of awareness among mining organizations as declared by Alhanshi and Albraiki (2015). Good to keep in mind that most project managers are aware of quality management and expect mining organizations to act in the best interests of their clients (Alzoubi & Aziz, 2021; Hamadneh et al., 2021). However, other quality managers are not aware of the granular details and activities of the individual functions in quality management (Alzoubi & Yanamandra, 2020; Alzoubi et al., 2020a, b; Lee et al., 2022a, b).

Existing literature asserts these findings. Elbadawi and Selim (2016), for example, suggested that managers of mining organizations in the UAE feel that quality is a critical element of any mining project. Moreover, organizations should play their part to ensure that by adhering to quality standards and supporting initiatives that are support quality management (Alshurideh et al., 2020; Alzoubi et al., 2020a, b). Existing literature also asserts that it is important to establish the requirements of a project at the inception stage so that project participants can attain the right balance the design requirements and the needs of the owner. In a study by Al Hashmi (2013), the author explored project critical failure factors in UAE's gas and oil industries (Fig. 2).

Using the likert methodology as demonstrated by Denni-fiberesima and Rani (2011) found out that scope of works, procurement, and communication are the critical factors that determine the success or failure of mining projects. Referring to

Most Five Important Scope of Work Factors		
Ranking No.	*Code No.*	*Procurement Factors*
1	S2	Absence of structured process for adding work
2	S9	Adding anytime work with overhaul work
3	S12	Change of scope due to absence of real cutoff date
4	S6	Unclear objectives for the project
5	S4	Undefined scope

Fig. 2 Project Scope related factors that result in Quality related problems in UAE's Mining Industry (Al Hashmi, 2013)

Most Five Important Scope of Work Factors		
Ranking No.	*Code No.*	*Procurement Factors*
1	P16	Unsuitable type of contract
2	P12	Inappropriate contractor selection criteria
3	P14	Incapable contractor
4	P17	Lack of experience of subcontractor
5	P13	Poor project material delivery system

Fig. 3 Procurement factors that result in Quality related problems in UAE's Mining Industry (Al Hashmi, 2013)

Most Five Important Scope of Work Factors		
Ranking No.	*Code No.*	*Procurement Factors*
1	C2	Un-established communication procedure within the project organization
2	C5	Lack of communication between stakeholders
3	C13	Lack of sharing information
4	C7	Incompetent project documentation management system
5	C12	Late and incomplete feedback

Fig. 4 Communication Factors that result in Quality related problems in UAE's Mining Industry (Al Hashmi, 2013)

scope, for example, at the very basic, this involves the owner's requirements so that quality consists of balancing the aesthetic needs with the owner's economic muscle. Scope may also include the professional's needs for time and adequate resources to complete a project. Figure 3, shows the common failures associated with the scope of construction projects. Al Hashmi (2013) also discussed the other two group of factors, identifying five areas of deficiency for each as seen in Fig. 4.

7 Conclusion and Recommendations

The mining and construction industries are two of the biggest contributors to UAE's development and economic growth. Because of the backward and forward linkages that these two industries have with each other and other industries, sustainability and quality are inescapable requirements. In essence, the construction and mining sectors have multiplier impact on larger economy. This research has shown that project participants in the UAE are generally aware of these facts and also the need for quality management. Yet, quality related issues are still an issue in UAE's construction and mining projects.

Based on the outcomes, the deficiency factors in terms of quality management evidently revolve around easily solvable issues such as delays during the inception stage and broken communication channels. In this regard, the authors believe that a shift in focus to structured process is the way forward for both industries. In essence, the absence of structured processes hinders the important functions of total quality management on the basis that there is inadequate control of work activities which eventually plays a key role in quality related issue, particularly time delays. On a practical level, the problem of unstructured problems can be overcome by developing a project steering committee which is involved in managing the project from start to finish and approves any additional work. Of course, proper communication channels are a vital requirement for such a committee to function appropriate. As discussed above, communication is also an important deficiency factor in UAE's mining and construction sector. To avoid breakdown in communication, it is recommended that organizations put in place new communication channels, encompassing all project participants. Preferably, such a system should be implemented at least three months prior to initiation of the project.

Finally, in the case when quality claims cannot be avoided claims should be settled and resolved swiftly to avoid delays. The paper proposes three methods of resolving claims in that regard; litigation, mediation, arbitration or negotiation. But the more important step is to analyze the causes of delays in projects. The results of this study indicate that this is an important success factor in UAE's mining and construction industry. Not only that, but quality assurance, quality control and timing can help organizations to improve quality management systems.

References

Aburayya, A., Alshurideh, M., Alawadhi, D., Alfarsi, A., Taryam, M., & Mubarak, S. (2020). An investigation of the effect of lean six sigma practices on healthcare service quality and patient satisfaction: Testing the mediating role of service quality in Dubai primary healthcare sector. *Journal of Advanced Research in Dynamical and Control Systems, 12*(8), 56–72.

Akhtar, A., Akhtar, S., Bakhtawar, B., Kashif, A. A., Aziz, N., & Javeid, M. S. (2021). COVID-19 Detection from CBC using machine learning techniques. *International Journal of Technology, Innovation and Management (IJTIM), 1*(2), 65–78. https://doi.org/10.54489/ijtim.v1i2.22.

Al Ali, A. (2021). The impact of information sharing and quality assurance on customer service at UAE banking sector. *International Journal of Technology, Innovation and Management (IJTIM), 1*(1), 1–17. https://doi.org/10.54489/ijtim.v1i1.10.

Al Batayneh, R. M., Taleb, N., Said, R. A., Alshurideh, M. T., Ghazal, T. M., & Alzoubi, H. M. (2021). IT governance framework and smart services integration for future development of Dubai infrastructure utilizing ai and big data, its reflection on the citizens standard of living. *The International Conference on Artificial Intelligence and Computer Vision*, 235–247.

Al Dmour, H., Alshurideh, M., & Shishan, F. (2014). The influence of mobile application quality and attributes on the continuance intention of mobile shopping. *Life Science Journal, 11*(10), 172–181.

Al Hashmi, S. A. S. (2013). *The exploration of projects' failure factors in oil & gas industry in UAE*. (Doctoral dissertation, The British University in Dubai (BUiD)).

Al Shebli, K., Said, R. A., Taleb, N., Ghazal, T. M., Alshurideh, M. T., & Alzoubi, H. M. (2021). RTA's Employees' perceptions toward the efficiency of artificial intelligence and big data utilization in providing smart services to the residents of Dubai. *The International Conference on Artificial Intelligence and Computer Vision*, 573–585.

AlHamad, A., Alshurideh, M., Alomari, K., Kurdi, B., Alzoubi, H., Hamouche, S., & Al-Hawary, S. (2022). The effect of electronic human resources management on organizational health of telecommunications companies in Jordan. *International Journal of Data and Network Science, 6*(2), 429–438.

AlHamad, M., Akour, I., Alshurideh, M., Al-Hamad, A., Kurdi, B., & Alzoubi, H. (2021). Predicting the intention to use google glass: A comparative approach using machine learning models and PLS-SEM. *International Journal of Data and Network Science, 5*(3), 311–320.

Alhanshi, M., & Albraiki, H. (2015). Knowledge sharing and employee development in oil and gas companies in the United Arab Emirates. *SPE Middle East Intelligent Oil and Gas Conference and Exhibition*.

Ali, N., Ahmed, A., Anum, L., Ghazal, T. M., Abbas, S., Khan, M. A., Alzoubi, H. M., & Ahmad, M. (2021). Modelling supply chain information collaboration empowered with machine learning technique. *Intelligent Automation and Soft Computing, 30*(1), 243–257. https://doi.org/10.32604/iasc.2021.018983.

Ali, N., M. Ghazal, T., Ahmed, A., Abbas, S., A. Khan, M., Alzoubi, H., Farooq, U., Ahmad, M., & Adnan Khan, M. (2022). Fusion-based supply chain collaboration using machine learning techniques. *Intelligent Automation & Soft Computing, 31*(3), 1671–1687. https://doi.org/10.32604/iasc.2022.019892.

Alketbi, S., Alshurideh, M., & Al Kurdi, B. (2020). The influence of service quality on customers'retention and loyalty in the UAE hotel sector with respect to the impact of customer' satisfaction, trust, and commitment: a qualitative study. *PalArch's Journal of Archaeology of Egypt/egyptology, 17*(4), 541–561.

AlMehrzi, A., Alshurideh, M., & Al Kurdi, B. (2020). Investigation of the key internal factors influencing knowledge management, employment, and organisational performance: A qualitative study of the UAE hospitality sector. *International Journal of Innovation Creativity Change, 14*(1), 1369–1394.

Alnazer, N. N., Alnuaimi, M. A., & Alzoubi, H. M. (2017). Analysing the appropriate cognitive styles and its effect on strategic innovation in Jordanian universities. *International Journal of Business Excellence, 13*(1), 127–140. https://doi.org/10.1504/IJBEX.2017.085799.

Alnuaimi, M., Alzoubi, H. M., Ajelat, D., & Alzoubi, A. A. (2021). Towards intelligent organisations: An empirical investigation of learning orientation's role in technical innovation. *International Journal of Innovation and Learning, 29*(2), 207–221. https://doi.org/10.1504/IJIL.2021.112996.

Alshamsi, A., Alshurideh, M., Kurdi, B. A., & Salloum, S. A. (2021). The influence of service quality on customer retention: A systematic review in the higher education. In *Advances in Intelligent Systems and Computing, Vol. 1261 AISC* (pp. 404–416). https://doi.org/10.1007/978-3-030-58669-0_37.

Alsharari, N. (2021). Integrating blockchain technology with internet of things to efficiency. *International Journal of Technology, Innovation and Management (IJTIM), 1*(2), 1–13.

Alshurideh, M. (2022). Does electronic customer relationship management (E-CRM) affect service quality at private hospitals in Jordan? *Uncertain Supply Chain Management, 10*(2), 1–8.

Alshurideh, M., Gasaymeh, A., Ahmed, G., Alzoubi, H., & Kurd, B. (2020). Loyalty program effectiveness: Theoretical reviews and practical proofs. *Uncertain Supply Chain Management, 8*(3), 599–612.

Alshurideh, M., Salloum, S. A., Al Kurdi, B., Monem, A. A., & Shaalan, K. (2019). Understanding the quality determinants that influence the intention to use the mobile learning platforms: A practical study. *International Journal of Interactive Mobile Technologies, 13*(11). https://doi.org/10.3991/ijim.v13i11.10300.

Alshurideh, M. T., Al Kurdi, B., Alzoubi, H. M., Ghazal, T. M., Said, R. A., AlHamad, A. Q., Hamadneh, S., Sahawneh, N., & Al-kassem, A. H. (2022). Fuzzy assisted human resource management for supply chain management issues. *Annals of Operations Research*, 1–19.

AlSuwaidi, S. R., Alshurideh, M., Al Kurdi, B., & Aburayya, A. (2021). The main catalysts for collaborative R&D projects in Dubai industrial sector. *The International Conference on Artificial Intelligence and Computer Vision*, 795–806.
Alzoubi, A. (2021a). The impact of process quality and quality control on organizational competitiveness at 5-star hotels in Dubai. *International Journal of Technology, Innovation and Management (IJTIM), 1*(1), 54–68. https://doi.org/10.54489/ijtim.v1i1.14.
Alzoubi, A. (2021b). Renewable Green hydrogen energy impact on sustainability performance. *International Journal of Computations, Information and Manufacturing (IJCIM), 1*(1), 94–110. https://doi.org/10.54489/ijcim.v1i1.46.
Alzoubi, H. M., & Aziz, R. (2021). Does emotional intelligence contribute to quality of strategic decisions? the mediating role of open innovation. *Journal of Open Innovation: Technology, Market, and Complexity, 7*(2), 130. https://doi.org/10.3390/joitmc7020130.
Alzoubi, H. M., Vij, M., Vij, A., & Hanaysha, J. R. (2021). What leads guests to satisfaction and loyalty in UAE five-star hotels? AHP analysis to service quality dimensions. *Enlightening Tourism, 11*(1), 102–135. https://doi.org/10.33776/et.v11i1.5056.
Alzoubi, H. M., & Yanamandra, R. (2020). Investigating the mediating role of information sharing strategy on agile supply chain. *Uncertain Supply Chain Management, 8*(2), 273–284. https://doi.org/10.5267/j.uscm.2019.12.004.
Alzoubi, H., Ahmed, G., Al-Gasaymeh, A., & Kurdi, B. (2020a). Empirical study on sustainable supply chain strategies and its impact on competitive priorities: The mediating role of supply chain collaboration. *Management Science Letters, 10*(3), 703–708.
Alzoubi, H., Alshurideh, M., Kurdi, B., Akour, I., & Aziz, R. (2022). Does BLE technology contribute towards improving marketing strategies, customers' satisfaction and loyalty? The role of open innovation. *International Journal of Data and Network Science, 6*(2), 449–460.
Alzoubi, H., & Ahmed, G. (2019). Do TQM practices improve organisational success? A case study of electronics industry in the UAE. *International Journal of Economics and Business Research, 17*(4), 459–472. https://doi.org/10.1504/IJEBR.2019.099975.
Alzoubi, H., Alshurideh, M., Kurdi, B. A., & Inairat, M. (2020b). Do perceived service value, quality, price fairness and service recovery shape customer satisfaction and delight? A practical study in the service telecommunication context. *Uncertain Supply Chain Management, 8*(3), 579–588. https://doi.org/10.5267/j.uscm.2020.2.005.
Awadhi, J., Obeidat, B., & Alshurideh, M. (2021). The impact of customer service digitalization on customer satisfaction: Evidence from telecommunication industry. *International Journal of Data and Network Science, 5*(4), 815–830.
Aziz, N., & Aftab, S. (2021). Data mining framework for nutrition ranking: Methodology: SPSS modeller. *International Journal of Technology, Innovation and Management (IJTIM), 1*(1), 85–95.
Chung, H. W. (2002). *Understanding quality assurance in construction: A practical guide to ISO 9000 for contractors*. Routledge.
Cruz, A. (2021). Convergence between blockchain and the internet of things. *International Journal of Technology, Innovation and Management (IJTIM), 1*(1), 35–56.
Delgado, P. A. A. D. L. (2016). *The United Arab Emirates case of economic success: The federal government economic policies*. (Doctoral dissertation).
Denni-fiberesima, D., & Rani, N. S. A. (2011). *An Evaluation of Critical Success Factors in Deepwater Oil & Gas Project Portfolios in Nigeria*. https://doi.org/10.4043/21336-ms.
El-Sayegh, S. M., & Mansour, M. H. (2015). Risk assessment and allocation in highway construction projects in the UAE. *Journal of Management in Engineering, 31*(6), 4015004.
Elbadawi, I., & Selim, H. (2016). *Understanding and avoiding the oil curse in resource-rich Arab economies*. Cambridge University Press.
Eli, T. (2021). Students perspectives on the use of innovative and interactive teaching methods at the University of Nouakchott Al Aasriya, Mauritania: English Department as a Case Study. *International Journal of Technology, Innovation and Management (IJTIM), 1*(2), 90–104.

Farouk, M. (2021). The universal artificial intelligence efforts to face coronavirus COVID-19. *International Journal of Computations, Information and Manufacturing (IJCIM), 1*(1), 77–93. https://doi.org/10.54489/ijcim.v1i1.47.

Ghazal, T. M., Hasan, M. K., Alshurideh, M. T., Alzoubi, H. M., Ahmad, M., Akbar, S. S., Al Kurdi, B., & Akour, I. A. (2021). IoT for smart cities: Machine learning approaches in smart healthcare—a review. *Future Internet, 13*(8), 218. https://doi.org/10.3390/fi13080218.

Greeff, W. J., & Barker, R. (2014). Communicating for survival in the mining and construction industries: Northern conversations and Southern contextualisations. *Communicatio, 40*(2), 191–205.

Guergov, S., & Radwan, N. (2021). Blockchain convergence: analysis of issues affecting IoT, AI and blockchain. *International Journal of Computations, Information and Manufacturing (IJCIM), 1*(1), 1–17. https://doi.org/10.54489/ijcim.v1i1.48.

Hamadneh, S., Pedersen, O., & Al Kurdi, B. (2021). An investigation of the role of supply chain visibility into the scottish bood supply chain. *Journal of Legal, Ethical and Regulatory Issues, 24*(Special Issue 1), 1–12.

Hanaysha, J. R., Al-Shaikh, M. E., Joghee, S., & Alzoubi, H. (2021a). Impact of innovation capabilities on business sustainability in small and medium enterprises. *FIIB Business Review*, 1–12.https://doi.org/10.1177/23197145211042232.

Hanaysha, J. R., Al Shaikh, M. E., & Alzoubi, H. M. (2021b). Importance of marketing mix elements in determining consumer purchase decision in the retail market. *International Journal of Service Science, Management, Engineering, and Technology (IJSSMET), 12*(6), 56–72.

International Organization for Standardization. (2009). *ISO 9000 quality management*. Geneva, Switzerland: International Organization for Standardization (ISO).

Jarkas, A. M., & Younes, J. H. (2014). Principle factors contributing to construction delays in the State of Qatar. *International Journal of Construction Project Management, 6*(1), 39.

Joghee, S., Alzoubi, H. M., & Dubey, A. R. (2020). Decisions effectiveness of FDI investment biases at real estate industry: Empirical evidence from Dubai smart city projects. *International Journal of Scientific and Technology Research, 9*(3), 3499–3503.

Kashif, A. A., Bakhtawar, B., Akhtar, A., Akhtar, S., Aziz, N., & Javeid, M. S. (2021). Treatment response prediction in hepatitis C patients using machine learning techniques. *International Journal of Technology, Innovation and Management (IJTIM), 1*(2), 79–89. https://doi.org/10.54489/ijtim.v1i2.24.

Khan, M. A. (2021). Challenges facing the application of IoT in medicine and healthcare. *International Journal of Computations, Information and Manufacturing (IJCIM), 1*(1), 39–55. https://doi.org/10.54489/ijcim.v1i1.32.

Lee, C., & Ahmed, G. (2021). Improving IoT privacy, data protection and security concerns. *International Journal of Technology, Innovation and Management (IJTIM), 1*(1), 18–33. https://doi.org/10.54489/ijtim.v1i1.12.

Lee, K., Azmi, N., Hanaysha, J., Alzoubi, H., & Alshurideh, M. (2022a). The effect of digital supply chain on organizational performance: An empirical study in Malaysia manufacturing industry. *Uncertain Supply Chain Management, 10*(2), 495–510.

Lee, K., Romzi, P., Hanaysha, J., Alzoubi, H., & Alshurideh, M. (2022b). Investigating the impact of benefits and challenges of IOT adoption on supply chain performance and organizational performance: An empirical study in Malaysia. *Uncertain Supply Chain Management, 10*(2), 537–550.

Madi Odeh, R. B. S., Obeidat, B. Y., Jaradat, M. O., Masa'deh, R., & Alshurideh, M. T. (2021). The transformational leadership role in achieving organizational resilience through adaptive cultures: the case of Dubai service sector. *International Journal of Productivity and Performance Management*. https://doi.org/10.1108/IJPPM-02-2021-0093.

Magd, H., & Nabulsi, F. (2012). The effectiveness of ISO 9000 in an emerging market as a business process management tool: The case of the UAE. *Procedia Economics and Finance, 3*, 158–165.

Mashaqi, E., Al-Hajri, S., Alshurideh, M., & Al Kurdi, B. (2020). The impact of e-service quality, e-recovery services on e-loyalty in online shopping: theoretical foundation and qualitative proof. *PalArch's Journal of Archaeology of Egypt/egyptology, 17*(10), 2291–2316.

Mehmood, T. (2021). Does information technology competencies and fleet management practices lead to effective service delivery? Empirical evidence from E-commerce industry. *International Journal of Technology, Innovation and Management (IJTIM), 1*(2), 14–41.

Mehmood, T., Alzoubi, H. M., & Ahmed, G. (2019). Schumpeterian entrepreneurship theory: evolution and relevance. *Academy of Entrepreneurship Journal, 25*(4).

Miller, D. (2021). The best practice of teach computer science students to use paper prototyping. *International Journal of Technology, Innovation and Management (IJTIM), 1*(2), 42–63. https://doi.org/10.54489/ijtim.v1i2.17.

Mishmish, M., & El-Sayegh, S. M. (2018). Causes of claims in road construction projects in the UAE. *International Journal of Construction Management, 18*(1), 26–33.

Mondol, E. P. (2021). The impact of block chain and smart inventory system on supply chain performance at retail industry. *International Journal of Computations, Information and Manufacturing (IJCIM), 1*(1), 56–76. https://doi.org/10.54489/ijcim.v1i1.30.

Mukherjee, P. N. (2006). *Total quality management*. PHI Learning Pvt. Ltd.

Obaid, A. J. (2021). Assessment of smart home assistants as an IoT. *International Journal of Computations, Information and Manufacturing (IJCIM), 1*(1), 18–36. https://doi.org/10.54489/ijcim.v1i1.34.

Peach, R. W. (2003). *The ISO 9000 Handbook*.

Pepall, D. (2014). *Total Quality Management: Case Studies*.

PMI. (2017). *Organizational Project Management*.

Radwan, N., & Farouk, M. (2021). The growth of internet of things (IoT) in the management of healthcare issues and healthcare policy development. *International Journal of Technology, Innovation and Management (IJTIM), 1*(1), 69–84. https://doi.org/10.54489/ijtim.v1i1.8.

Ren, Z., Atout, M., & Jones, J. (2008). Root causes of construction project delays in Dubai. *Procs 24th Annual ARCOM Conference*, 1–3.

Rumane, A. R. (2017). *Quality management in construction projects*. CRC Press.

Shahbaz, M., Sbia, R., Hamdi, H., & Ozturk, I. (2014). Economic growth, electricity consumption, urbanization and environmental degradation relationship in United Arab Emirates. *Ecological Indicators, 45*, 622–631.

Shamout, M., Elayan, M., Rawashdeh, A., Kurdi, B., & Alshurideh, M. (2022). E-HRM practices and sustainable competitive advantage from HR practitioner's perspective: A mediated moderation analysis. *International Journal of Data and Network Science, 6*(1), 165–178.

Taher, N. A. B. A. (2009). *Understanding and preventing construction conflict, claims and disputes: a critical in-depth study into their causes and recommendations to control in the United Arab Emirates*. University of South Wales (United Kingdom).

Taryam, M., Alawadhi, D., Aburayya, A., Albaqa'een, A., Alfarsi, A., Makki, I., Rahmani, N., Alshurideh, M., & Salloum, S. A. (2020). Effectiveness of not quarantining passengers after having a negative COVID-19 PCR test at arrival to dubai airports. *Systematic Reviews in Pharmacy, 11*(11). https://doi.org/10.31838/srp.2020.11.197.

Yang, K., & Trewn, J. (2004). *Multivariate statistical methods in quality management*. McGraw-Hill Education.

Zaneldin, E. K. (2006). Construction claims in United Arab Emirates: Types, causes, and frequency. *International Journal of Project Management, 24*(5), 453–459.

E-Government Implementation: A Case Study of Jordanian e-Government Program

Mua'ad Abu-Faraj, Ra'ed Masa'deh, and Muhammad Turki Alshurideh

Abstract The diffusion of Information and Communication Technologies (ICTs) including the high-speed Internet within the private sector and the exemplar of the private sector's implementation and use of e-commerce and e-business have caused the public sector around the world to start thinking about this phenomenon become aware of its potentials and consequently utilize them. Therefore, governments around the world started investing heavily in e-Government projects, along with e-services in order to exploit the expected benefits of this phenomenon. E-Government is a new approach followed by nations in providing better services to people and businesses. It increases the efficiency of departments, reduces the cost, streamlines the processes and thus provides several benefits to all stakeholders involved. In this paper, we describe a case study on the successful implementation of e-Government in Jordan using Chan, Lau, and Pan's e-Government Implementation Framework. The various e-Government initiatives undertaken by the Jordanian government have been described using the four components, namely information content, ICT infrastructure, e-Government Infostructure, and e-Government promotion.

Keywords Jordan · MoICT · e-Government · m-Government · ICT infrastructure · Information content · e-Government infostructure · e-Government promotion

M. Abu-Faraj (✉)
Department of Computer Information Systems, School of Information Technology and Systems, The University of Jordan, Aqaba, Jordan
e-mail: m.abufaraj@ju.edu.jo

R. Masa'deh
Department of Management Information Systems, School of Business, The University of Jordan, Amman, Jordan
e-mail: r.masadeh@ju.edu.jo

M. T. Alshurideh
Department of Marketing, School of Business, The University of Jordan, Amman, Jordan
e-mail: m.alshurideh@ju.edu.jo
University of Sharjah, Sharjah, UAE

M. Alshurideh et al. (eds.), *The Effect of Information Technology on Business and Marketing Intelligence Systems*, Studies in Computational Intelligence 1056,
https://doi.org/10.1007/978-3-031-12382-5_74

1 Introduction

Many definitions for Electronic Government (e-Government) have been adopted. For example, (Heeks, 2001) defined e-Government as the initiative of the public sector to use ICTs to improve existing processes and good governance, build good relationships within society, provide citizens and businesses with more convenient access to online government services, and increase effectiveness and efficiency in the public sector entities. E-Government services can enhance a country's economic and social development by releasing citizens' time and cost, providing citizens with more convenient access to government information and services. Government e-services also encourage foreign investment that will contribute to a country's economy and social development.

E-Government has been recognized as a tool for radically improving the way governments interact with their citizens. While the obvious benefits of e-Government include providing convenience and easy access to customers thereby resulting in improved services to citizens, reduction of costs (by re-deploying resources from back-end processing to front-end customer service); providing easier access to information, increasing transparency and communication between government departments and with the public, the long term benefit of e-Government is also to make it easier for people to conduct their daily lives and business in the nation, thereby making it an attractive location to attract global talent. On the other hand, introducing e-Government requires a transformation in the way the government structure functions—thereby requiring a change in the organizational culture, managerial styles, systems and procedures apart from large financial investments towards technology upgrades. In order to ensure that e-Government is actually bringing about the changes that are desired (via the benefits) it is vital that strict performance measures be instituted on a continuous basis to check the effectiveness of e-Government.

In this paper, we describe a case study on the successful implementation of e-Government in Jordan. This study is done based on Chan, Lau, and Pan's e-Government Implementation Framework which has been published in by Chan et al. (2008). The various e-government initiatives undertaken by Jordan government have been described using the four components, namely information content, ICT infrastructure, e-Government infostructure, and e-Government promotion.

2 Jordan e-Government

E-Government in Jordan is a national program initiated by his Majesty King Abdullah II. The purpose of the e-Government program is to improve the performance of government backend processes that generate services with the end users in terms of efficiency, accuracy, time and cost effectiveness, transparency, and accountability. Services are at the heart of e-Government in Jordan. E-Government in Jordan is dedicated to delivering services to people across society, irrespective of location, economic status, education or ICT ability. Government e-services should be delivered to citizens in a seamless manner, citizens should have the ability to use the government e-services and also feel confident using them as declared by MoICT (2013) and Blakemore and Dutton (2003).

In 2017, the Jordanian government unveiled its Economic Growth Plan (EGP) that aims to automate procedures to reach an e-government by 2020. In 2018, citizens and businesses have access to more than 100 e-Services offered by various institutions, with the number expected to reach 131 e-services in 2019 and 293 e-services in 2020 as given by Ghazal (2018). According to the United Nations (UN) e-Government survey 2018 as mentioned by United Nations (2018); Jordan ranked 98th globally and 8th regionally. The purpose of this survey is to measure countries' use of information and communications technologies to deliver public services. The Index captures the scope and quality of online services, status of telecommunication infrastructure and existing human capacity and is rated between zero (lowest) and one (highest). Despite Jordan's e-Government Development Indcx (EGDI) value was 0.5575 compared with 0.5123 in 2016, Jordan drops seven places globally.

The key emphasis of the Jordanian e-Government program has started with the re-engineering of processes to become more efficient, as well as human performance development and eventually, deployment of best practices using the latest technologies as a tool to enable government stakeholders to implement the new processes. The Ministry of Communications and Information Technology (MoICT) has been assigned to take the lead in coordinating the efforts of implementing the e-Government program and providing support where needed in the implementation MoICT (MoICT, 2013).

The deployment of e-Government in Jordan can be achieved by consolidating government resources, engaging greater citizen participation in the local economic development and facilitating citizen access to government services demonstrating more citizen empowerment and less government control over public data. e-Inclusion, the participation of all in e-Government is imperative to promote economic and social empowerment through ICT for all citizens including vulnerable groups pre-identified by the United Nations as the poor, illiterate, blind, old, young, immigrants and women as mentioned by the United Nations in United Nations (2012). The Strategy for development and implementation of e-Government in Jordan crystallizes this vision into objectives; it presents priority e-Government initiatives, tools and projects as shown in Fig. 1; and it identifies targets and milestones to facilitate performance control and raises accountability by clearly defining the responsibilities of key stakeholders as given by Majdalawi et al. (2015).

This strategy emphasizes the e-Government role of encouraging and motivating government entities to deliver high-quality customer-centric and performance-driven services to e-Government beneficiaries while transforming from traditional service delivery to more effective and efficient service providers to their beneficiaries (Citizens, Residents, Visitors, Businesses, Government Entities and Government Employees).

The four stages of e-transformation are evolving around the maturity of service delivery (emerging, enhanced, transactional, Connected). Government of Jordan

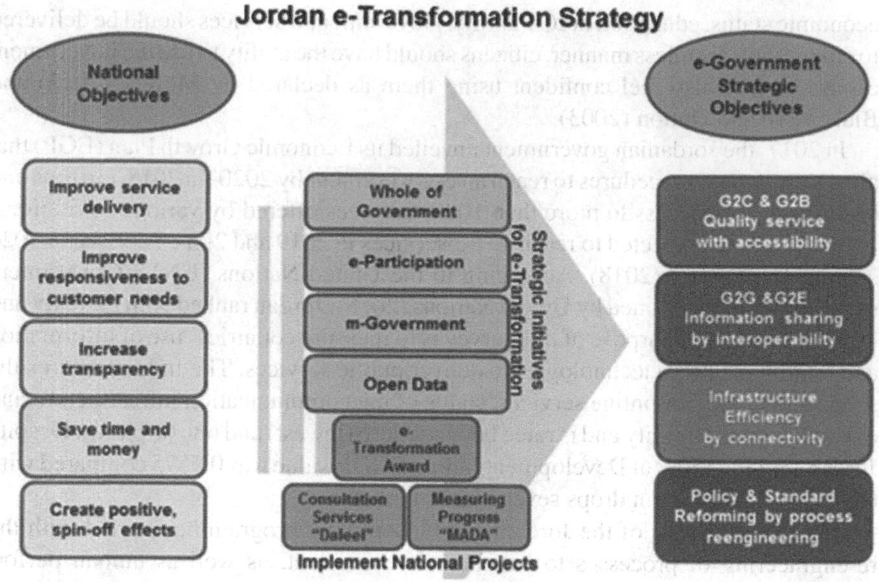

Fig. 1 Jordan e-Transformation Strategy (MoICT, 2018)

is aiming to achieve the transactional stage by end of this Strategy term. Jordan is currently in the late enhanced stage given that Government of Jordan offer more sources of information through the National Government Portal (www.jordan.gov.jo), the National Contact Center (NCC), the National Mobile Portal and National SMS Gateway as given by Jordan e-Government (2017).

3 E-Government Implementation Framework

Abu-Shanab and Abu Baker (2011) have proposed a model for evaluating the Jordanian e-Government website from users' perspective, this study was conducted in 2011. This model was proposed based on study a sample of 300 Jordanian students, they were asked to perform specific tasks and answered questions related to these tasks. The study focused on usability, accessibility and privacy/security as factors in predicting the adoption of e-government website.

Chan et al., (2008) offer a macro perspective of the various activities involved in the implementation of e-Government through an interpretive analysis of the various e-Government-related initiatives undertaken by the Singapore Government. The analysis led to the identification of four main components in the implementation of e-Government, namely (i) information content, (ii) ICT infrastructure, (iii) e-Government Infostructure, and (iv) e-Government promotion. These four components were then conceptually integrated into the e-Government Implementation

Framework. Sethi et al., 2008) have applied the same framework proposed by Chan et al. (2008) to study the case of the e-Government in Dubai. In this paper, the same framework implemented by Chan et al. (2008) and (Sethi et al., 2008), will be used as a descriptive tool to organize and coordinate various e-Government initiatives in Jordan. This framework will also be used as a prescriptive structure to plan and strategize e-Government implementation in Jordan.

4 ICT Infrastructure

Jordan is already a regional leader when it comes to the ICT sector and one of the main launch pads of entrepreneurship in the Middle East and North Africa. The sector is one of Jordan's fastest growing industries, contributing 12% of the GDP and creating over 80,000 value-added jobs (6% of the labor force). Jordan's ICT exports grew eight-fold since 2001 and currently reach over 40 countries around the globe. Amman is recognized as one of the 10 best destinations to start a technology firm worldwide and a leading center for developing Arabic internet content as mentioned by Oxford Business Group (2018).

The ICT sector in Jordan consists of the IT industry (software and content providers), the telecommunication industry with three main telecom companies (Orange, Zain and Umniah), and the Business Process Outsourcing (BPO) industry (call centers). The ICT sector accounts for approximately 12 percent of GDP direct and indirect, and 84,000 jobs (direct employment figures for the IT industry reached 16,000, with indirect and induced employment for ICT estimated at around 68,000). The sector attracts an average of USD 150 million in investments annually as given by MoICT (2017). Competition is strongest in the broadband and mobile markets. Jordan has emerged as a regional tech startup hub due to an ICT-focused educational system, low startup costs, and business-friendly environment. Its growing reputation is increasingly attracting international capital eager to tap into the region's underserved, but growing online market. Due to the country's small market, most start-ups in Jordan either focus exclusively on, or grow with, the expectation of expanding into the wealthier markets of the Middle East Gulf Region as mentioned by MoICT (2009).

The three network operators compete in a mobile market, with a majority of their subscribers being prepaid users. All three have launched LTE networks, which is underpinning mobile broadband growth. The potential for large-scale mobile internet penetration due to competition has captured the attention of Jordan's technology start-up companies (Al-Omari, 2006). Many are developing applications to target the higher spending demographic traditionally associated with mobile internet early adopters. Jordan's initiative to develop an e-payment infrastructure coincides with a rollout of Near Field Communications (NFC) mobile payment terminals across Jordan by MasterCard and a number of partners.

In March 2015, the Central Bank of Jordan (CBJ) announced the launch of the eFAWATEERcom portal (www.eFAWATEERcom.jo) that will enable clients inquire about, review and settle their bills on-line. The portal was established in partnership with Madfoo3atCom for Electronic Payments Company as the main operator, MasterCard Internet Gateway Service (MiGS), and Emerging Markets Payments Group (EMP Group) as the hosting company (Times, 2017). Payments through eFAWATEERcom, an online bill payment service, have increased to JD3.643 billion by the end of July, 2018, compared with JD1.208 billion in the same period of 2017, marking a 200 per cent increase, the number of bills paid through the service in this year's January-July period stood at 4.869 million, compared with 2.303 million invoices that were paid in the same period of 2017, as given by Times (2018).

The Jordanian government is committed to completing its National Broadband Network (NBN) project, which is owned by MoICT. The objective of the National Broadband Network is to connect all public schools, governmental entities, health entities, community colleges, and knowledge stations. The network has been broken up into 3 phases, Middle phase, North phase and South phase. These phases are connected to each other by an Overlay Network which will provide physical connectivity between the Point of Presences (POPs) for each module and the National Data Centers. Public facilities will be connected to the National Broadband Network by the end of 2019, this will greatly boost e-services, e-commerce, e-health and e-education in Jordan. The network will provide high speed connectivity between public facilities, hospitals, schools and government agencies (Ghazal, 2016).

5 Information Content

The United Nations has identified "Five Stage e-Government Model", this model is useful in developing the e-services that are being offered by the various government departments in Jordan as shown in Table 1 as published by United Nations (2002).

In 2018, citizens and businesses have access more than 100 e-Services on the National Government Portal four of which are transactional and three of those services offer online payment through the Jordan Payment Gateway (JoPAY). Many more online information and interactional services are offered on government official websites. The National Mobile Gateway and SMS Gateway offer 40 Government

Table 1 The Stages of e-Government

The Stages of e-Government	
Emerging	An official government online presence is established
Enhanced	Government sites increase; information becomes more dynamic
Interactive	Users can download forms, e-mail officials and interact through the web
Transactional	Users can actually pay for services and other transactions online
Seamless	Full integration of e-services across administrative boundaries

e-Services that are offered by 22 government agencies while the National Contact Center (NCC) is currently serving citizens and businesses by answering calls for 22 government agencies regarding over 200 services in addition to receiving citizens' complaints and suggestions against government services. In addition, as of today, 77 government Agencies have subscribed to National SMS Gateway push services (MoICT, 2017).

At the time of the portal launch in 2003, the Jordanian e-Government was in "Emerging Stage" with an Internet presence through its official portal with 15 e-services, in addition to launching the strategic plan of the e-Government in Jordan. Also, during the emerging stage, the government started working on the National Broadband Network and the Secured Government Network (SGN). E-government efforts accelerated in Jordan since 2003, entering the "Enhanced" stage three years later. By 2006 the number of e-services rose to more than 30, also the government launched the National SMS gateway.

The Jordanian e-Government project has entered "Interactive Stage" in 2009, with 55 online e-service also the government focused on providing newer official websites and enhanced national portal. Furthermore, the e-Government integrated many systems such as Government Finance Management Information System (GFMIS) by Ministry of Finance, Human Resource Management Information System (HRMIS) being developed by the Civil Status Bureau and Inventory Management System that is developed by the General Supply Department. In 2013, the government has activated the e-Government Steering Committee (EGSC) which is responsible of setting the strategic directions for e-Government in Jordan as announced by the Jordan e-Government (Jordan e-Government, 2017). The Jordanian e Government promoted to "Transactional Stage" in 2013, were 96 e-services have been released including mobile services. The government also activated e-Procurement in addition to the e-participation policy. The government utilized the existing shared and composite services, integrate related vertical services and integrate government systems through Enterprise Service Bus (ESB) for the purpose of providing cross governmental services. The Jordanian e-Government is supposed to enter the "Seamless Stage" by 2020, were the number of e-services will be more than 290 e-services, and all public facilities are connected to the National Broadband Network.

To improve government efficiency and enable each entity to focus on its core functionality rather than managing administrative and financial systems, government must work as a holistic government by organizing and standardizing processes across government entities around user needs. This will solve the current silos of public sector which create redundancy and fragmentation in government. This approach can also enable different entities to reap the benefit of information sharing to facilitate service provision and reduce the burden on citizens in term of running around different agencies to authenticate documents and clearances. Table 2 shows the development stages of the e-Government in Jordan.

The governmental departments in Jordan have huge differences in its technical competencies, core business process and each of them is equipped with uniform applications, this forced the e-Government management to use a single centralized entity with well-defined common tools known as National e-Government tools. These

Table 2 Jordan e-Government Transformation into Maturity

e-Service	2003–2005	2006–2008	2009–2012	2013–2015	2016–2020
G2C and G2B	15 online e-services	30 online e-services on National Portal and Official websites	55 online e-services on enhanced National Portal and newer official website	96 online e-services on enhanced National Portal and newer official website launching Mobile services	More than 100 online e-services on 2018 Launch Bekhedmetkom e-services to reach 131 e-services in 2019 and 293 in 2020
G2G and G2E			GFMIS HRMIS IMIS	e-Procurement	
Institutional	Launch the initial strategy of e-Government program	National Strategy (2006–2009) Roadmap EGSC 70 CIO units	National Strategy (2014–2016) Roadmap 37 additional CIO units Launching eFAWATEERcom portal	Activate EGSC CIO Council MADA Daleel e-Government award	
Business		Training 10,000 government employees	Training 3000 government employees		Training 5000 government employees
Legal			e-Transaction Amendments Draft	e-Participation Policy	
Infrastructure	Launch SGN to include 14 entities Start working on the National Broadband Network	Launch National Portal SMS gateway	Launch Mobile Gateway, Payment Gateway ESB SGN to include 73 entities with enhanced portal	Transactional National portal PKI Cloud Computing Data Model and Open Data Standards	Public facilities will be connected to the National Broadband Network

tools which are shown in Table 3 are considered as expedited service delivery by other government departments and helped to achieve cost savings by providing commonly utilized e-services to other departments. These tools will help other departments to pay more efforts toward e-Transformation to better serve their beneficiaries and improve the overall progress of Jordan achievements compared with regional and global countries (Navarra, 2007).

Table 3 National e-Government tools

Synergistic Tool	Description
eFAWATEERcom	eFAWATEERcom is an electronic bill presentment and payment system that is owned by the Central Bank of Jordan and aims at providing this service to its customers through different banking channels and payment service providers
MADA	Jordan e-Government Program mandate to follow up on the progress of e-Transformation in Jordan
Daleel	initiative guiding tool to promote e-Government consulting services toward business development and employ accumulative knowledge and experience in enabling government entities to achieve e-Transformation
Bekhedmetkom	Interactive platform to communicate with the government (Ask government, Suggestion, Complement, Complaint, Reporting)
Profiling	Is a portal provided to deliver highly personalized services to users allowing a differentiated personalized experience for each user, by unifying registration procedures and applying high security and privacy calibrations?

6 E-Government Infostructure

At the early stages of the e-Government program in Jordan, there was no national portal for the e-Government services. During the years from 2003 till 2005, the e-services provided by the e-Government was considered as separate services with no major hub to accommodate these services to other participants. The e-Government website (www.jordan.gov.jo), launched in 2008. The number of e-services at that time was around 30 e-services. The aim of the e-Government portal was to be a central hub for all the services provided by the government. The e-Government program started creating new e-services and then integrating these e-services to the e-Government portal.

In 2010, the portal was revamped with the addition of more e-services and new implementation approach. This development phase included an important approach to give the governmental departments the freedom to create their own services and content and publishing these services at their own portals while remaining an integral part of the portal e-Government. This approach can be described in other words by following a decentralized approach with the governmental department and centralized approach in the main e-Government portal. At that time, the portal offered 55 e-services targeting various segments of the community including the individuals, business sector, and the private sector. In 2014, the e-Government program launched mobile services and it added its portal to the e-Government website. These mobile services include launching SMS gateway and launching mobile applications, which are available on iOS and Google Play stores.

In 2016 the e-Government portal was again revamped with more user-friendly features. The e-Government portal provides 96 online e-services on enhanced national Portal and newer official website. The bilingual portal was divided into four groups: citizen, resident, investor, and visitor/tourist. The relevant e-services were listed under each group. In 2018, The portal is positioned as central hub and it is considered as the gateway to all government departments and their services, in addition to the portal of each governmental department. The portal provides more than 100 e-services with the aim to provide 131 e-services in 2019 and 293 in 2020. The e-Government official website provides the Government Entities which includes prime ministry, ministries, and government agencies. Also, it provides "Bekhedmetkom", which is interactive platform to communicate with the government (Ask government, Suggestion, Complement, Complaint, Reporting). In 2018, the mobile applications portal includes more than 25 different mobile applications.

The fast evolvement of ICT has offered government multiple service delivery channels. Since beneficiaries vary in their literacy, location and accessibility to those channels, they require diverse service provision. Internet has proven to be the least effective channel to reach Jordanians. Mobile penetration has exceeded by far all expectations and should be utilized as one of the main delivery channels in Jordan. Ultimately, e-Government services should transform the way all beneficiaries interact with government.

7 E-Government Promotion

The e-Government Promotion concept focuses on enhancing governance and achieving efficient, convenient, more responsive citizen-oriented public services delivery to provide better access to online public services. The e-Government Promotion can be achieved through:

- Support in improving e-Government interoperability, and applying an effective business process reengineering mechanism (Awareness);
- Assistance in enhancing government online services delivery (Assistance);
- Enhancing e-Government institutional (Assurance).

The e-government promotion involves a three-prong approach of awareness, assistance and assurance (Chan et al., 2008). The first prong of awareness referred to the various publicity activities and strategies employed by the Jordanian e-Government to raise the public's awareness of e-services. Jordanian e-government has conducted different approaches to promote the participants to use e-services, such as online marketing, marketing with government departments, announcement on public TV stations and newspapers. For example, The Jordanian e-Government has launched two tools: "Daleel", and "Bekhedmetkom". Daleel is initiative guiding tool to promote e-Government consulting services toward business development and employ accumulative knowledge and experience in enabling government entities to achieve e-Transformation. Bekhedmetkom is an interactive platform to communicate with the government. The government also has launched a national award for content development which evaluates the creativity and innovation in the content as well as the creativity and innovation in the usage of the content as mentioned by Odat and Khazaaleh (2012).

Creating awareness about government e-services was not enough; the next task was to improve the computer literacy rates. This was addressed by the second prong of assistance, where Jordanian e-Government provided assistance though the different initiatives that targeted the use and diffusion of ICT in schools in Jordan. The government also has launched a national project to connect 3,000 public schools, ministry directorates and military education schools to an integrated network, which enabled more than 1.5 million students and 80,000 teachers to benefit from this project. The communications network will provide teachers and students with an interactive learning environment, and will also allow administrators to communicate more easily, expediting decision making. The Jordanian government also worked on integrating ICT use into university curriculum in areas such as journalism (facebook use, twitter use, etc.), drama and the arts which are becoming ICT and digital content heavy as given by Al-Yaseen et al. (2013).

The final prong of assurance in e-Government promotion is very crucial as it provides assurance on privacy and security issues of e-services (Chan et al., 2008). The quality evaluation of government departments' services by Jordanian e-Government was based on a number of well-defined criteria, one of which was security and privacy of e-services. It was extremely important to Jordanian e-Government that the security and privacy policy of e-services was created in both English and Arabic by the government departments and made known to the users. In order to increase users' confidence in the website, all government departments were advised to add a statement around security of the website and how information was shared with other government organizations. Furthermore, to enhance the reliability of the statement, government departments were instructed to add a link or reference to any kind of legislation in the security and privacy statement. At the same time, online payment security of centralized service, eFAWATEERcom was beefed up to match the world class security standard.

Other than the three prongs of e-Government promotion, Jordanian e-Government also launched a "Bekhedmetkom" through Jordan.gov.jo portal in 2016 to measure customer satisfaction level and enhance the quality of e-services. It invited the public to submit their suggestions, complains, compliments, and views on Jordanian government's official portal, Jordan.gov.jo.

8 Discussion

The Jordanian e-Government provides a variety of online services for both individuals and businesses. The Jordanian e-Government has managed to add more than 100 e-services to its portal, there is still much work to be done in this filed. The percent of e-services that the Jordanian government managed to get online compared to the services it provides is very low, also this percent compared to the other countries in the region is also considered very low.

The ICT infrastructure in Jordan is developing very fast, also Jordan has enjoyed a strong leadership, this enabled Jordan to move from the informational government websites to transactional websites and currently the aim is to fully integrate of e-services across the administrative boundaries (reaching seamless phase) by 2020.

Strong Leadership, Commitment and Vision: e-Government is a complex initiative involving multiple stakeholders and entities. The strong leadership with unified and coherent vision is extremely important for e-Government success. The Jordanian e-Government project aims to ease the lives of people and businesses interacting with the government.

Flexible and Robust Infrastructure: The successful implementation of e-Government initiative depends on the underlying ICT infrastructure comprising of secure servers, routers, firewalls, internet connectivity etc. In order to incorporate changes and additions in the technology infrastructure, Jordanian e-Government had formulated a technology master plan (National Strategy Roadmap) right from start for the required infrastructure and its related architecture.

Central Flexible Model: To better manage and monitor its e-Government initiative, Jordanian government chose a hybrid approach which accelerated the release of new e-services. Jordanian e-Government is centrally controlled and monitored the e-services development of various government departments, the government departments were given the freedom to creatively build their own e-services in the earlier phase of e-Government initiative.

Development Strategies: Although the overall vision of Jordanian e-Government remained the same, the development strategies kept changing to meet the vision. The biggest challenge faced by Jordanian e-Government in the initial phase of the e-Government initiative was difference in the technical competence of various government departments. Some departments were technically qualified and demonstrated high competency, while others lacked even the minimum technological infrastructure. To bridge the digital divide among the government departments in the initial phase, Jordanian e-Government established a number of strategies such as providing support to government departments through work teams sharing e-services, and wide training programs.

Development of Human Resources: The e-Government initiative can't be successful without the development of human resources. An individual (citizen or employee) is the basis of the success or failure of this initiative. Without citizens or employees able to accept electronic systems, the e-Government model can collapse. Realizing the importance of raising the technical qualification of both citizens and employees, The Jordanian e-Government launched several training programs for government.

Public–Private Partnerships: e-Government initiatives require a variety of business and IT skills such as business process reengineering, systems analysis and design, networks, software design and implementation etc. In order to address the IT skills shortage, the Jordanian e-Government had formed strategic alliances with private sector to complement its competencies in various areas such as portal hosting and management, customer care, electronic payment, and mobile services among others. These strategic partnerships had been a critical step in achieving both short-term and long-term strategic objectives of Jordanian e-Government.

To conclude, this is believed to be the most up-to-date and comprehensive analysis of Jordan's plans and assessment of its level of readiness for delivery of e-Government services. The findings present a number of key factors that hinder Jordan's e-Government development. These findings can be useful for researchers and practitioners, as they provide rich insights on e-Government development. The findings can be also useful to other developing countries, as they can help them in understanding citizen related challenges when designing, planning and implementing their e-Government initiatives.

References

Abu-Shanab, E., & Abu Baker, A. (2011). Evaluating Jordan's e-government website: A case study. *Electronic Government an International Journal, 8*(4), 271–289. https://doi.org/10.1504/EG.2011.042807

Al-Omari, H., (2006). E-government architecture in Jordan: A comparative analysis. *Journal of Computer Science, 2*(11), 846–852.

Al-Yaseen, H., Al-Soud, A., & Al-Jaghoub, S. (2013). Assessing Jordan's e-government maturity level: Citizen's perspective on awareness, acceptability and usage of e-government services. *International Journal Electronic Government Research, 9*(4), 1–18.

Blakemore, M., & Dutton, R., (2003). E-Government, e-Society and Jordan: Strategy, theory, practice, and assessment. *First Monday, 8*(10). https://doi.org/10.5210/fm.v8i11.1095

Chan, C. M. L., Lau, Y. M., & Pan, S. L. (2008). E-government implementation: A macro analysis of Singapore's e-government initiatives. *Government Information Quarterly, 25*(2), 239–255.

Ghazal, M. (2016, June 28). Public facilities to be connected to broadband network by 2019, Jordan Times.

Ghazal, M. (2018, July 23). Jordan drops seven places in global e-gov't ranking, Jordan Times.

Heeks, R. (2001). Understanding e-Government for Development, Institute for Development Policy and Management (IDPM). Retrieved from: http://unpan1.un.org/intradoc/groups/public/documents/NISPAcee/UNPAN015484.pdf

Jordan e-Government. (2017). Jordan National Information and Communications Technology Strategy (2013–2017). Retrieved from: http://inform.gov.jo/Portals/0/Report%20PDFs/6.%20Infrastructure%20&%20Utilities/ii.%20ICT/2013-2017%20National%20ICT%20Strategy.pdf

Majdalawi, Y., Almarabeh, T., Mohammad, H., & Quteshate, W. (2015). E-government strategy and plans in Jordan. *Journal of Software EngiNeering and Applications, 8*, 211–223. https://doi.org/10.4236/jsea.2015.84022

MoICT. (2009). Assessment of the economic impacts of ICT in the Hashemite Kingdom of Jordan Project. Retrieved from: http://moict.gov.jo/uploads/Policies-and-Strategies-Directorate/Studies/Assessment-of-the-Economic-Impact-of-ICT/Hashemite-Kingdom-of-Jordan-2009.pdf

MoICT. (2013). e-Government strategy 2014–2016. Retrieved from: https://jordan.gov.jo/wps/wcm/connect/56d75661-abb5-4ecb-8826-67a1c3ee30df/e-Government_StrategyJO_Draft.pdf?MOD=AJPERES

MoICT. (2017). Jordan-information and communication technology. Retrieved from: https://www.export.gov/article?id=Jordan-Information-and-Communication-Technology.

MoICT. (2018). Jordan's e-Government program. Retrieved from: https://jordan.gov.jo/wps/wcm/connect/gov/eGov/Government+Ministries+_+Entities/Ministry+of+Information+and+Communications+Technology/FAQ/

Navarra, D. D. (May 2007). The Architecture of Global ICT Programs: A case Study of E-Governance in Jordan. In *Paper presented at the 9th International conference on social implications of computers in developing countries*, São Paulo, Brazil.

Odat, A., & Khazaaleh, M. (2012). E-government challenges and opportunities: A case study of Jordan. *International Journal of Computer Science Issues, 9*(5), no 2, 361–367.

Oxford Business Group. (2018). Expansion of Jordan's ICT sector remains a government priority. Retrieved from: https://oxfordbusinessgroup.com/overview/dynamic-market-private-sector-ambitions-rising-internet-penetration-and-shift-4g-are-tran-0

Sethi, N., & Sethi, V., (2008, December). E-government implementation: A case study of Dubai e-Government. In *Paper presented at the 6th International Conference on E-Governance (ICEG 2008)*. New Delhi, India.

Jordan Times. (2017). eFAWATEERcom online payments 'exceed JD2 billion since 2015, Jordan Times.

Jordan Times. (2018). 200 per cent increase in eFAWATEERcom e-payments, Jordan Times.

United Nations. (2002). Benchmarking E-government: A global perspective. Retrieved from: https://publicadministration.un.org/egovkb/portals/egovkb/documents/un/english.pdf

United Nations. (2012). Survey: Towards a more citizen-centric approach. Retrieved from: http://unpan1.un.org/intradoc/groups/public/documents/un/unpan047965.pdf

United Nations. (2018). UN E-Government Survey 2018. Retrieved from: https://publicadministration.un.org/egovkb/en-us/Reports/UN-E-Government-Survey-2018

The Impact of the Motivational Culture on the Job Satisfaction of Aqaba Special Economic Zone Authority Employee's in Jordan

Shaker Habis Nawafleh, Ra'ed Masa'deh, and Muhammad Alshurideh

Abstract The aim of this study was to examine the impact of Motivational Culture on the Job Satisfaction among the employees in the Aqaba Special Economic Zone Authority. Descriptive, correlational, cross-sectional design were used by distributing of (115) self-reporting questionnaire, which was consist of three parts; the first part was for the demographical characteristics of the employees, the second part was for the Motivation at Work Scale (MAWS) and the third part was the Short Form Minnesota Satisfaction Questionnaire (SFMSQ). The results revealed that motivational culture and the job satisfaction levels were in a moderate level {38.4 (SD = 4.392)} and {62.6 (SD = 14.87)}. Moreover, the regression results reveals that there was a strong significance positive impact of the motivational culture on the job satisfaction (r = 0.78); and (61%) of motivational culture for the employee could be explained from motivational culture. The positive effect of motivational culture in the promoting a high level of job satisfaction, need more focus by the human resources departments to maximum the benefit of the motivational culture.

Keywords Motivational culture · Job satisfaction · Aqaba special economic zone authority employee's · Jordan

S. H. Nawafleh (✉)
Aqaba Special Economic Zone Authority, Aqaba, Jordan
e-mail: Shalnawafleh@aseza.jo

R. Masa'deh
Department of Management Information Systems, School of Business, The University of Jordan, Amman, Jordan

M. Alshurideh
Department of Marketing, School of Business, The University of Jordan, Amman, Jordan
e-mail: m.alshurideh@ju.edu.jo

Department of Marketing, School of Business, University of Sharjah, Sharjah, UAE

M. Alshurideh et al. (eds.), *The Effect of Information Technology on Business and Marketing Intelligence Systems*, Studies in Computational Intelligence 1056,
https://doi.org/10.1007/978-3-031-12382-5_75

1 Introduction

Human resources are considered the key to the success of any organization, they are the most valuable and essential part of any successful operating system (Alrowwad & Abualoush, 2020; Lorincová et al., 2019; Rapsanjani & Johannes, 2019). Without Qualified human resources, other resources will be Ineffective (Pancasila et al., 2020), so Organizational goal achievement couldn't depend on the technological excellence alone because the quality of the employees is a crucial foundation stone (Aldmour & Obeidat, 2017; Bernanthos, 2018; Rozi & Sunarsi, 2020; Vanesa et al., 2019). Each organization Seeks to have special characteristics that recognize it from other organizations, these characteristics help the organization to keep the best employees within their team, this characteristic is called motivation (Abuhashesh & Al-Dmour, 2019; Bernanthos, 2018; Hayajneh et al., 2021; Paais & Pattiruhu, 2020; Rozi & Sunarsi, 2020). The presence of expert and high-quality employees will enhance the productivity of the organization (Al-dalahmeh et al., 2018; Obeidat & Altheeb, 2018).

1.1 Background

Human resources are considered the key to the success of any organization, they are the most valuable and essential part of any successful operating system (Alrowwad & Abualoush, 2020; Lorincová et al., 2019; Rapsanjani & Johannes, 2019). Without Qualified human resources, other resources will be Ineffective (Pancasila et al., 2020), so Organizational goal achievement couldn't depend on the technological excellence alone because the quality of the employees is a crucial foundation stone (Aldmour & Obeidat, 2017; Bernanthos, 2018; Rozi & Sunarsi, 2020; Vanesa et al., 2019). Each organization Seeks to have special characteristics that recognize it from other organizations, these characteristics help the organization to keep the best employees within their team, this characteristic is called motivation (Abuhashesh & Al-Dmour, 2019; Bernanthos, 2018; Hayajneh et al., 2021; Paais & Pattiruhu, 2020; Rozi & Sunarsi, 2020; Vanesa et al., 2019). The presence of expert and high-quality employees will enhance the productivity of the organization (Al-dalahmeh et al., 2018; Obeidat & Altheeb, 2018).

1.2 Problem Statement

Organizational employee satisfaction affects their performance which has a negative effect on the organization's performance and its ability to perform the desired goals (Paais & Pattiruhu, 2020), and it is difficult to suppose that we can have an effective business without professional teamwork which is hard to have without a reward and motivation system (Ali & Anwar, 2021), each organization needs to give more attention to the employees' needs and talents in order to help them in their future plans which help in the improvement of the company's productive (Pancasila et al., 2020).

1.3 Significance of the Study

Human resources management department in the Aqaba Special Economic Zone Authority intended to enhance the job satisfaction levels for its employees. The researcher highlights in this study the positive impact of the motivational culture on the job satisfaction levels for the Aqaba Special Economic Zone Authority employees and helps the Human resources management department work hard to achieve their aim to increase job satisfaction levels of their employees.

1.4 Purpose and Research Questions

The objective of this study was to examine the impact of motivational culture in the Aqaba Special Economic Zone Authority on job satisfaction levels for its employees. To accomplish this goal, the researcher focuses on a number of objectives:

- To recognize the levels of motivational culture and job satisfaction among the employees in the Aqaba Special Economic Zone Authority.
- To investigate the impact of motivational culture on job satisfaction among the employees in the Aqaba Special Economic Zone Authority.

These objectives guided the researcher to identify the following research questions:

- What are the levels of motivational culture and job satisfaction among the employees in the Aqaba Special Economic Zone Authority?
- Is there an impact of motivational culture on job satisfaction among the employees in the Aqaba Special Economic Zone Authority?

Accordingly, the following hypothesis is presented. In fact, in this study the decision to accept or reject the hypothesis is based on statistical significance. Statistical significance considered as a description for a result or experiment when the probability is less than the significance level. The current study objective to examine the following hypothesis:

H1: There is an impact of motivational culture on job satisfaction among the employees in the Aqaba Special Economic Zone Authority.

1.5 Study Variables and Framework

This study was involved two major variables:

(1) Motivational Culture

The word motivation is originating from the Latin word "movere", which means to "transfer" or "push" (Pang & Lu, 2018).

Motivation is a group of situations and worth's to influence the individual to realize the specific goal in correspondence with the individual goals (Sudiardhita et al., 2018). It could be a form of organizational philosophy and values adopted by human resources within the organization, but the formation of motivation is enhanced by the organizational managers (Bernanthos, 2018).

Developing a well-structured motivation system is an essential element for the organization to authorize employees to produce the maximum benefits for the organization (Pang & Lu, 2018).

Work motivation is the force that remains within an individual, which, fosters his attitude to take action (Barasa et al., 2018). It is the gift of mobility that inspired a person's affection for them to work simultaneously, work productively, and incorporated with every attempt to achieve satisfaction (Vanesa et al., 2019).

Motivational culture is a set of phenomena of physical and spiritual survival of the organization's personnel: the controlling moral rules and values, customs that emerge from the moment of induction of the project and are shared by the majority of its employees (Lebedynets & Zhurakivska, 2021).

(2) Job Satisfaction

Job satisfaction reflects the emotional state (either pleasant or unpleasant) in which employees saw their work. Job satisfaction reflects an employee's point of view across his work, job satisfaction is a serious agent to be considered in the perception of the organizational behavior (Barasa et al., 2018; Masa'deh, 2016; Vratskikh et al., 2016).

Job Satisfaction is a fundamental standard that cannot be overestimated, representing an integration of psychological, physiological, and environmental situations that mark an employee's satisfaction with their job, emotional stability, and judgment (Davidescu et al., 2020; Masa'deh et al., 2019; Sudiardhita et al., 2018).

Job satisfaction is one of the most significant factors to get the best work result. When a person feels satisfaction in the work of course he will attempt as much as possible with all the skills that have to accomplish the job task (Barasa et al., 2018).

In order to achieve the study goal, the researcher develops the following Study frame work:

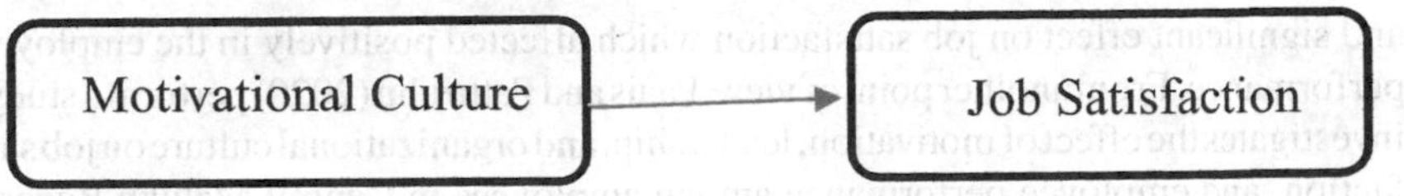

Source: Authors

2 Literature Review

2.1 *Motivational Culture*

The influence of the motivation and job satisfaction against employees' performance among Indonesian employees were studied in a descriptive study by Bernanthos (2018) depending in the results of an (114) employees' respond which revealed that there was moderate levels of a motivation and job satisfaction, and positive, significant effect of the motivation process on the employees' performance. Furthermore, a study was conducted by Rozi and Sunarsi (2020) in South of Indonesia to determine the effect of motivation and work experience on employee performance; the researchers involved (43 respondent) in their study and the result reflect that (47.6%) of the employee performance are influenced by motivational attitude.

2.2 *Job Satisfaction*

Job satisfaction considered as the most influential variable on the employee's performance, as a conclusion of the results of a descriptive study of Rapsanjani and Johannes (2019) which was aimed to examine the effect of remuneration, work motivation and job satisfaction on employee's performance among Indonesian employees in housekeeping department in Jakarta. Moreover, Ali and Anwar (2021) conduct a study in Iraq among 128 bank employees to analyze the effect of the culture on employee satisfaction in different banks in 2021, and the result show that compensation as motivation has significant positive influence on job satisfaction in the analysis of the randomly collected questionnaire which was developed by researchers previously.

2.3 *The Impact of Motivational Culture on Job Satisfaction*

The effect of compensation, motivation of employee and work satisfaction to employee performance pt. Bank in Indonesia were studied by Sudiardhita et al. (2018) among 346 Bank's employees, and the results shows that the work motivation has a positive and significant effect on job satisfaction which affected positively in the employees' performance. From another point of view, Paais and Pattiruhu (2020) perform a study to investigates the effect of motivation, leadership, and organizational culture on job satisfaction, and employee performance among employees in Central Maluku Regency, Indonesia, the result showed that work motivation and organizational culture had a positive and significant effect on performance, but did not significantly influence employee job satisfaction.

3 Methodology

3.1 Research Design

This study used a descriptive, correlational, cross-sectional design to assess impact of the motivational culture on the job satisfaction among Aqaba Special Economic Zone Authority Employee's, Jordan, and evaluate the levels of motivational culture among employees.

3.2 Population and Sample

The accessible population of the study was all employees in the Aqaba Special Economic Zone Authority. The sample size was calculated using G-power program with α probability of (5%) and power of (95%), with a medium effect size of (0.15), so the calculating program indicated the minimum number of sample is (107) employees.

3.3 Instrument of Data Collection

In order to achieve the goal of the study and based on previous literature, the researcher designed the study questionnaire, which was consisted of three parts; the first part was for the demographical characteristics of the employees, the second part was for the Motivation at Work Scale (MAWS) and the third part was Short Form Minnesota Satisfaction Questionnaire (SFMSQ).

The Motivation at Work Scale (MAWS) was used on its English version for data collection. It was developed by Gagne et al. (2010). It contains 12 items. First three items were related to intrinsic (indicating inner motivation), next three to identification (self-actualization), next three for introjections (getting inspiration) and the last three items were related to extrinsic (motivated from external sources, i.e., promotion, money etc.) level of motivation. A 5-point Likert system was used to respond

Table 1 The instrument reliability

	Variables	No of items	Alpha value
1	Motivation at Work Scale (MAWS)	12	0.88
2	Short Form Minnesota Satisfaction Questionnaire (SFMSQ)	20	0.78

Prepared by researcher depend on pilot study results

to the items, (1) for strongly disagree, (2) for disagree, (3) for Neutral, (4) for agree, and (5) for strongly agree. The increased of the scale sub-total indicated increase Motivational level (Gagne et al., 2010).

The Short Form Minnesota Satisfaction Questionnaire (SFMSQ) in an English form, was used to assess job satisfaction using a 5-point Likert scale ranging from one (extremely dissatisfied) to five (extremely satisfied). It consists of 20 items, and each item represents a feature in the work environment. The possible scores for SFMSQ range from 20 to 100. The level of job satisfaction classified to high satisfaction for score of (>75), low level for score (<25), while the score in the middle range of percentile (26–74) would indicate average satisfaction level (Al-Dmour et al., 2021).

For collecting the data, convenience sampling technique was used to distribute the questionnaire for the employees. But before conducting the study, pilot testing is performed by taking (15) employees for the purpose of checking the reliability of the questionnaire. The responses of the employees were scored and the reliability of the instrument was determined using Cronbach's Alpha. The statistically acceptable value of the coefficient of alpha responses to the questionnaire for all Cronbach is 70%. The result shows that the Cronbach's alpha ranges from (0.78) to (0.88) which show the scale is reliable, as in Table 1.

3.4 Data Analysis

After collecting the valid questionnaires and calculating the response rate, the data has been coded manually and entered to Statistical Package for Social Sciences (SSPS version 21) data base. Descriptive statistics (mean and standard deviation) were used to the employee's demographical characteristics. Another descriptive statistic (mean and standard deviation) used to calculate the motivational levels and job satisfaction level for research question number one. Inferential statistics (Pearson correlation and simple liner regression) were used for testing the impact of motivational culture on job satisfaction level for the employees in order to answer the question number two.

3.5 Ethical Considerations

The approval of the Aqaba Special Economic Zone Authority to conduct this study among the Authority employees was confirmed. Informed consent has been obtained for each participant and the confidentiality and anonymity where guaranteed, and emphasizing that the participating is voluntary; the participant could withdraw at any time during the data collection time. There is no risk on the involved employees and no financial issues in participating in this study. The researcher's phone numbers were mentioned, for any questions, or any help for the participants in filling out the questionnaire.

4 Results

4.1 Demographical Information

The researcher distributed (115) questionnaires among the employees in the Aqaba Special Economic Zone Authority. Only (110) questionnaires were valid for analysis, making the response rate of 96%. The analysis of the demographical characteristic of the sample shows that the mean age of the sample was 27.3 years (SD = 4.96) ranged from (22–45) years. And the average monthly income was ranged between (700 – 2000) JD. With the sample mean 1250 JD. (SD = 350). While the mean of the experience level for the participant was 10 years (SD = 4.5) as represented in Table 2. Furthermore, the Categorical Characteristic of the sample showed in Table 3; which revealed that out of (110) employees participated in this study; almost more than the half of the participants (57%) were male. Almost the half of the employees (47%) has BCS degree in the view of educational levels, and (29%) of the participants were has higher education.

Table 2 Demographical characteristic of the sample (N = 110)

Category	Mean	SD	Rang
Age	27.3 years	4.96	22–45 years
Monthly income	1250 JD	350	700–2000 JD
Experience level	10 years	4.5	2–19 years

Prepared by researcher depend on SPSS results

Table 3 Categorical Socio-demographic information of about the respondents (N = 110)

Category	Component	Frequency	Percentage (%)
Gander	Male	63	57
	Female	47	43
Educational level	School	26	24
	BSC	53	47
	M.A	28	26
	Ph.D.	3	3

Prepared by researcher depend on SPSS results

Table 4 Levels of MAWS and SFMSQ (N = 110)

	Scale	Mean	SD.	Range
1	Motivation at Work Scale (MAWS)	38.4	4.392	33–48
2	Short Form Minnesota Satisfaction Questionnaire (SFMSQ)	62.6	14.87	48–82

Prepared by researcher depend on pilot study results

4.2 The Answers of Research Questions

4.2.1 The First Research Question

To answer the first research question (What are the levels of motivational culture and job satisfaction among the employees in the Aqaba Special Economic Zone Authority?); the researcher analyzes the responses of the participant for the questionnaire of "Motivation of the Work Scale (MAWS)" and "Short Form Minnesota Satisfaction Questionnaire (SFMSQ)", by using the descriptive statistics (mean, standard deviation and frequency). Analyzing the responses of the participating employees to ward "MAWS" shows that the scores were ranged from (33) to (48) and the mean was {38.4 (SD = 4.392)}, which revealed a moderate level of work motivation among the employees in the Aqaba Special Economic Zone Authority. In the anther point of view, the result of participant respondent to ward "SFMSQ" revealed that the scores ranged from (48) to (82) and the mean was {62.6 (SD = 14.87)}, which revealed a moderate level of job satisfaction among the employees in the Aqaba Special Economic Zone Authority, as in Table 4.

4.2.2 The Second Research Question

To answer the second research question (Is there an impact of motivational culture on job satisfaction among the employees in the Aqaba Special Economic Zone

Table 5 Result of regression model for the impact of motivational culture on job satisfaction

Model	F	Sig. F	*R*	R^2
1	20.74	0.001**	0.78	0.61

Prepared by researcher depend on SPSS results

Authority?); the researcher applied simple liner Regression analysis. The regression result Table 5; reveals that the regression equation reaches the significance (F = 20.74, p < 0.001**). And there was a positive strong significance impact for motivational culture on job satisfaction (r = 0.78); the results conclude that (61%) of the job satisfaction could be explained from motivational culture.

5 Discussion and Conclusions

Studying the levels of work motivation among employees in the Aqaba Special Economic Zone Authority reported a moderate level of work motivation {M = 38.4 (SD = 4.392)}. Such a supported result was reported in Indonesia by Bernanthos (2018) during studying the levels of work motivation among employee questionnaire data in Private Universities and report moderate levels of work motivation with mean {39.3(SD = 3.796)} (Bernanthos, 2018). The results of job satisfaction levels indicate a moderate level of job satisfaction among the employees; the mean was {62.6 (SD = 14.87)}. These results were supported by previous studying of job satisfaction levels among employees working at Housekeeping Department in Indonesia and revealed the job satisfaction mean (58.3 (SD = 35.3%) (Rapsanjani & Johannes, 2019).

Furthermore, the results of this study indicate that there a there was a positive strong significance impact for motivational culture on job satisfaction among the employees in the Aqaba Special Economic Zone Authority, this impact can be explaining (61%) of the employee's job satisfaction a result from motivational culture. Such a strongly positive impact motivational culture was reported by Ali and Anwar (2021), by studying the impact of motivational culture on job satisfaction.

This study highlights on the impact of motivational culture on job satisfaction among the employees in the Aqaba Special Economic Zone Authority, and reported that the motivational culture can explain (61%) of the job satisfaction level and indicate that there a significance relationship between motivational culture and the job satisfaction level. This relationship helps the organizations to work more to promote employee job satisfaction high. In addition, the Aqaba Special Economic Zone Authority should offer better services in alignment with the latest Information Systems as suggested by researchers (e.g., Abuhashesh & Al-Dmour, 2018; Al-Dmour et al., 2021; Alshurideh et al., 2021; Davidescu et al., 2020; Lorincová et al., 2019; Madi Odeh et al., 2021; Obeidat et al., 2019; Paais & Pattiruhu, 2020; Pancasila et al., 2020; Pang & Lu, 2018; Rozi & Sunarsi, 2020; Shannak et al., 2012; Sudiardhita et al., 2018; Vanesa et al., 2019; Zureik & Al-Sondos, 2021).

According to the result of the study, employee's motivational level for the whole sample was "moderate", and in the aim to increase this level, the researcher recommended to the human resources managers to increase the behaviors that promote motivational culture by taking into consideration employees' desires and expectations, reinforcing communication within the authority (Al Kurdi et al., 2020; Alameeri et al., 2020; Al-bawaia et al., 2022; AlShehhi et al., 2020; Alsuwaidi et al., 2020; Bettayeb et al., 2020; Kurdi et al., 2020), and establishing open communication channels directly with the employees (Aljumah et al., 2021; Alshurideh, 2022; Alshurideh et al., 2014; Alyammahi et al., 2020; Sweiss et al., 2021). Similar studies can be carried out by comparing governmental and private organizations in the aim of co-operations between the different organizations to increase the levels of motivational level and job satisfaction among the employee (Aburayya et al., 2020; Al-Khayyal et al., 2020; Alshurideh et al., 2012; Alzoubi et al., 2020, 2022).

References

Abuhashesh, M., & Al-Dmour, R. (2019). Factors that affect employees job satisfaction and performance to increase customers' satisfactions. *Journal of Human Resources Management Research, 2019*, 1–23.

Abuhashesh, M., & Al-Dmour, R. (2018). Factors that impact job satisfaction and performance among employees in the Jordanian industrial sector. In *Proceedings of the 32nd international business information management association conference, IBIMA 2018-vision 2020: Sustainable economic development and application of innovation management from regional expansion to global growth, 15–16 November, 2018, Seville, Spain* (pp. 4285–4305).

Aburayya, A., Alshurideh, M., Alawadhi, D., Alfarsi, A., Taryam, M., & Mubarak, S. (2020). An investigation of the effect of lean six sigma practices on healthcare service quality and patient satisfaction: Testing the mediating role of service quality in Dubai primary healthcare sector. *Journal of Advanced Research in Dynamical and Control Systems, 12*(8), 56–72.

Al Kurdi, B., Alshurideh, M., & Al afaishat, T. (2020). Employee retention and organizational performance: Evidence from banking industry. *Management Science Letters, 10*(16), 3981–3990.

Alameeri, K., Alshurideh, M., Al Kurdi, B., & Salloum, S. A. (2020, October). The effect of work environment happiness on employee leadership. In *International conference on advanced intelligent systems and informatics* (pp. 668–680). Springer, Cham.

Al-bawaia, E., Alshurideh, M., Obeidat, B., Masa'deh, R. (2022) The impact of corporate culture and employee motivation on organization effectiveness in Jordanian banking sector. *Academy of Strategic Management Journal, 21*(Special Issue 2), 1–18.

Al-dalahmeh, M., Abu Khalaf, R., & Obeidat, B. (2018). The effect of employee engagement on organizational performance via the mediating role of job satisfaction: The case of IT employees in Jordanian banking sector. *Modern Applied Science, 12*(6), 17–43.

Al-Dmour, R., AlShaar, F., Al-Dmour, H., & Alshurideh, M. T. (2021). The effect of service recovery justices strategies on online customer engagement via the role of "Customer satisfaction" during the covid-19 pandemic: An empirical study. https://doi.org/10.1007/978-3-030-67151-8_19

Aldmour, R., & Obeidat, B. (2017). Factors influencing the adoption and implementation of HRIS applications: Are they similar. *International Journal of Business Innovation and Research, 14*(2), 139–167.

Ali, B. J., & Anwar, G. (2021). An empirical study of employees' motivation and its influence job satisfaction. *International Journal of Engineering, Business and Management, 5*(2), 21–30.

Aljumah, A., Nuseir, M. T., & Alshurideh, M. T. (2021). The impact of social media marketing communications on consumer response during the COVID-19: Does the brand equity of a University Matter. In *The effect of coronavirus disease (COVID-19) on business intelligence* (pp. 367–384).

Al-Khayyal, A., Alshurideh, M., Al Kurdi, B., & Aburayya, A. (2020). The impact of electronic service quality dimensions on customers' E-shopping and E-loyalty via the Impact of E-satisfaction and E-trust: A qualitative approach. *International Journal of Innovation, Creativity and Change, 14*(9), 257–281.

Alrowwad, A., & Abualoush, S. (2020). Innovation and intellectual capital as intermediary variables among transformational leadership, transactional leadership, and organizational performance. *Journal of Management Development, 39*(2), 196–222.

AlShehhi, H., Alshurideh, M., Al Kurdi, B., & Salloum, S. A. (2020, October). The impact of ethical leadership on employees performance: A systematic review. In *International conference on advanced intelligent systems and informatics* (pp. 417–426). Cham: Springer.

Alshurideh, M. (2022). Does electronic customer relationship management (E-CRM) affect service quality at private hospitals in Jordan? *Uncertain Supply Chain Management, 10*(2), 325–332.

Alshurideh, M., Masa'deh, R. M. D. T., & Alkurdi, B. (2012). The effect of customer satisfaction upon customer retention in the Jordanian mobile market: An empirical investigation. *European Journal of Economics, Finance and Administrative Sciences, 47*(12), 69–78.

Alshurideh, M. T., Shaltoni, A., & Hijawi, D. (2014). Marketing communications role in shaping consumer awareness of cause-related marketing campaigns. *International Journal of Marketing Studies, 6*(2), 163–168.

Alshurideh, M. T., Hassanien, A. E., & Masa'deh, R. (2021). *Preface: The effect of coronavirus disease (COVID-19) on business intelligent systems*, 1st Eds., Springer International Publishing, eBook ISBN: 978–3–030–67151–8, Hardcover ISBN: 978–3–030–67150–1, Series ISSN: 2198–4182. https://doi.org/10.1007/978-3-030-67151-8

Alsuwaidi, M., Alshurideh, M., Al Kurdi, B., & Salloum, S. A. (2020, October). Performance appraisal on employees' motivation: a comprehensive analysis. In *International conference on advanced intelligent systems and informatics* (pp. 681–693). Cham: Springer.

Alyammahi, A., Alshurideh, M., Al Kurdi, B., & Salloum, S. A. (2020, October). The impacts of communication ethics on workplace decision making and productivity. In *International conference on advanced intelligent systems and informatics* (pp. 488–500). Cham: Springer.

Alzoubi, H. M., Alshurideh, M., Al Kurdi, B., & Inairat, M. (2020). Do perceived service value, quality, price fairness and service recovery shape customer satisfaction and delight? A practical study in the service telecommunication context. *Uncertain Supply Chain Management, 8*(3), 579–588.

Alzoubi, H., Alshurideh, M., Kurdi, B., Akour, I., & Aziz, R. (2022). Does BLE technology contribute towards improving marketing strategies, customers' satisfaction and loyalty? The role of open innovation. *International Journal of Data and Network Science, 6*(2), 449–460.

Barasa, L., Gunawan, A., & Sumali, B. (2018). Determinants of job satisfaction and it's implication on employee performance of port enterprises in DKI Jakarta. *International Review of Management and Marketing, 8*(5), 43.

Bernanthos, B. (2018). The direct and indirect influence of leadership, motivation and job satisfaction against employees' performance. *European Research Studies Journal, 2*, 236–243.

Bettayeb, H., Alshurideh, M. T., & Al Kurdi, B. (2020). The effectiveness of mobile learning in UAE universities: A systematic review of motivation, self-efficacy, usability and usefulness. *International Journal of Control and Automation, 13*(2), 1558–1579.

Davidescu, A. A., Apostu, S. A., Paul, A., & Casuneanu, I. (2020). Work flexibility, job satisfaction, and job performance among Romanian employees—Implications for sustainable human resource management. *Sustainability, 12*(15), 6086.

Gagne, M., Forest, J., Gilbert, M. H., Aubé, C., Morin, E., & Malorni, A. (2010). The motivation at work scale: Validation evidence in two languages. *Educational and Psychological Measurement, 70*(4), 628–646.

Hayajneh, N., Suifan, T., Obeidat, B., Abuhashesh, M., & Alshurideh, M. (2021). The relationship between organizational changes and job satisfaction through the mediating role of job stress in the Jordanian telecommunication sector. *Management Science Letters, 11*(1), 315–326.

Kurdi, B., Alshurideh, M., & Alnaser, A. (2020). The impact of employee satisfaction on customer satisfaction: Theoretical and empirical underpinning. *Management Science Letters, 10*(15), 3561–3570.

Lebedynets, I. S., & Zhurakivska, Y. M. (2021). Features of establishment and development of motivational culture at enterprises in modern economic conditions. *НАУКОВИЙ ВІСНИК, 8*(3), 92.

Lorincová, S., Štarchoň, P., Weberova, D., Hitka, M., & Lipoldová, M. (2019). Employee motivation as a tool to achieve sustainability of business processes. *Sustainability, 11*(13), 3509.

Madi Odeh, R. B. S., Obeidat, B. Y., Jaradat, M. O., & Alshurideh, M. T. (2021). The transformational leadership role in achieving organizational resilience through adaptive cultures: The case of dubai service sector. *International Journal of Productivity and Performance Management*. https://doi.org/10.1108/IJPPM-02-2021-0093

Masa'deh, R. (2016). The role of knowledge management infrastructure in enhancing job satisfaction at Aqaba five star hotels in Jordan. *Communications and Network, 8*(4), 219–240.

Masa'deh, R., Almajali, D., Alrowwad, A., & Obeidat, B. (2019). The role of knowledge management infrastructure in enhancing job satisfaction: A developing country perspective. *Interdisciplinary Journal of Information, Knowledge, and Management, 14*, 1–25.

Obeidat, B., & Altheeb, S. (2018). The impact of internal corporate social responsibility on job satisfaction in Jordanian pharmaceutical companies. *Modern Applied Science, 12*(11), 105–120.

Obeidat, Z. M., Alshurideh, M. T., Al Dweeri, R., & Masa'deh, R. (2019). The influence of online revenge acts on consumers psychological and emotional states: Does revenge taste sweet? In *Paper presented at the proceedings of the 33rd international business information management association conference, IBIMA 2019: Education excellence and innovation management through vision 2020* (pp. 4797–4815).

Paais, M., & Pattiruhu, J. R. (2020). Effect of motivation, leadership, and organizational culture on satisfaction and employee performance. *The Journal of Asian Finance, Economics, and Business, 7*(8), 577–588.

Pancasila, I., Haryono, S., & Sulistyo, B. A. (2020). Effects of work motivation and leadership toward work satisfaction and employee performance: Evidence from Indonesia. *The Journal of Asian Finance, Economics and Business, 7*(6), 387–397.

Pang, K., & Lu, C. S. (2018). Organizational motivation, employee job satisfaction and organizational performance: An empirical study of container shipping companies in Taiwan. *Maritime Business Review, 3*(1), 36–52.

Rapsanjani, A., & Johannes, S. (2019). The effect of remuneration, work motivation and job satisfaction on employee's performance. *International Humanities and Applied Sciences Journal (IHSJ), 2*(2), 23–29.

Rozi, A., & Sunarsi, D. (2020). The influence of motivation and work experience on employee performance at PT. Yamaha Saka Motor in South Tangerang. *Journal Office, 5*(2), 65–74.

Shannak, R. O., Al-Zu'bi, Z. M. F., Obeidat, B. Y., Alshurideh, M., & Altamony, H. (2012). A theoretical perspective on the relationship between knowledge management systems, customer knowledge management, and firm competitive advantage. *European Journal of Social Sciences, 32*(4), 520–532.

Sudiardhita, K. I., Mukhtar, S., Hartono, B., Sariwulan, T., & Nikensari, S. I. (2018). The effect of compensation, motivation of employee and work satisfaction to employee performance Pt. Bank Xyz (Persero) Tbk. *Academy of Strategic Management Journal, 17*(4), 1–14.

Sweiss, N., Obeidat, Z. M., Al-Dweeri, R. M., Mohammad Khalaf Ahmad, A., M. Obeidat, A., & Alshurideh, M. (2021). The moderating role of perceived company effort in mitigating customer misconduct within Online Brand Communities (OBC). *Journal of Marketing Communications*, 1–24.

Vanesa, Y. Y., Matondang, R., Sadalia, I., & Daulay, M. T. (2019). *The influence of organizational culture, work environment and work motivation on employee discipline in PT Jasa Marga (Persero) TBK, Medan Branch, North Sumatra, Indonesia*. American International Journal of Business Management (AIJBM) (pp. 37–45).
Vratskikh, I., Al-Lozi, M., & Maqableh, M. (2016). The impact of emotional intelligence on job performance via the mediating role of job satisfaction. *International Journal of Business and Management, 11*(2), 69–91.
Zureik, A. Y., & Al-Sondos, I. A. S. A. (2021). The effect of negative incentives on job stability–An applied study on Saudi Post in the Asir region. https://doi.org/10.26389/AJSRP.R040520.

Innovation, Entrepreneurship and Leadership

Strategic Leadership and Its Role on Implementing Public Policies in the Government Departments in Karak Governorate

Kamel Mohammad Al-hawajreh, Alaa Radwan Al-Nawaiseh, Reham Zuhier Qasim Almomani, Menahi Mosallam Alqahtani, Basem Yousef Ahmad Barqawi, Muhammad Turki Alshurideh, Sulieman Ibraheem Shelash Al-Hawary, Ayat Mohammad, and Anber Abraheem Shlash Mohammad

Abstract The major aim of the study was to examine the impact of strategic leadership on implementing public policies in government departments in Karak Governorate. Therefore, it was focused on government departments in Karak Governorate in Jordan. Data were primarily gathered through questionnaires which were distributed to a random sample of (175) respondent. In total, (175) responses were

K. M. Al-hawajreh
Business faculty, Mu'tah University, Karak, Jordan

A. R. Al-Nawaiseh
Business faculty, Mu'tah University, Karak, Jordan

R. Z. Q. Almomani
Business Administration, Amman, Jordan

M. M. Alqahtani
Administration Department, Community College of Qatar, Doha, Qatar

B. Y. A. Barqawi
Business Administration department, Faculty of Administration and Financial Sciences, Petra University, Amman, Jordan

M. T. Alshurideh
Department of Marketing, School of Business, The University of Jordan, Amman 11942, Jordan
e-mail: m.alshurideh@ju.edu.jo; malshurideh@sharjah.ac.ae

Department of Management, College of Business, University of Sharjah, 27272 Sharjah, United Arab Emirates

S. I. S. Al-Hawary (✉)
Department of Business Administration, School of Business, Al al-Bayt University, P.O. BOX 130040, Mafraq 25113, Jordan
e-mail: dr_sliman73@aabu.edu.jo; dr_sliman@yahoo.com

A. Mohammad
Business and Finance Faculty, the World Islamic Science and Education University (WISE), P.O Box 1101, Amman 11947, Jordan

M. Alshurideh et al. (eds.), *The Effect of Information Technology on Business and Marketing Intelligence Systems*, Studies in Computational Intelligence 1056,
https://doi.org/10.1007/978-3-031-12382-5_76

received including (5) invalid to statistical analysis due to uncompleted or inaccurate. Hence, the final sample contained (170) responses. Structural equation modeling (SEM) was conducted to test hypotheses. The study reached the presence of a statistically significant effect of strategic leadership on the implementation of public policies. The study recommended the necessity of paying attention to human capital in government departments in Karak governorate and enhancing interest in ethical practices from major implications in planning activities and processes and participation with stakeholders in providing services and preserving government departments'resources in Karak governorate.

Keywords Strategic leadership · Public policies · Government departments · Karak Governorate · Jordan

1 Introduction

The relationship between the state and the individual has been characterized since its existence by complexity and effective change, which was reflected in the diversity of ideas, and theories that sought to clarify and explain them, and this is represented through the basic role played by the leadership in its strategic concept in paying attention to the organizations development in order to achieve its vision, mission and overall performance (AlTaweel & Al-Hawary, 2021; Mohammad et al., 2020; Al-Hawary & Al-Syasneh, 2020; Alhalalmeh et al., 2020; Al-Hawary & Obiadat, 2021; Al-Quran et al., 2020; Allahow et al., 2018; Alolayyan et al., 2018; Alshurideh et al., 2017; Al-Lozi et al., 2017; Al-Nady et al., 2016; Al-Hawary & Hadad, 2016; Al-Nady et al., 2013; Al-Hawary et al., 2013).

Organizations are greatly affected by the strategic leadership practices (Alameeri et al., 2020; AlShehhi et al., 2020). Indeed, those who chronicle the organizations appear to be chronicling the leadership figures who led these organizations (Al-Dhuhouri et al., 2020; Odeh et al., 2021). Strategic leadership is also concerned with forming strategic decisions that represent the core administrative process in response to the adaptation reality to bring about the required change, taking into account the expectations of individuals and society (Al-Hawary & Mohammed, 2017; Al-Hawary et al., 2012; Al-Hawary, 2009; Harahsheh et al., 2021). The implementation of public policies differs between countries according to production factors reality and the variables related to economic and social ...etc. Based on that, public policies implementation must be based on the executive government apparatus and the administrative component in the different regions of the country. This requires local councils to generate energies and resources to activate this implementation process. Also, achieving the set goals and objectives requires a solid formulation of these policies, in line with the possibility of implementing these policies.

A. A. S. Mohammad
Marketing, Marketing Department, Faculty of Administrative and Financial Sciences, Petra University, B.O. Box: 961343, 11196 Amman, Jordan

The importance of the role of strategic leaders in implementing policies by linking the two stages of formulation and implementation of public policies from the reality of local councils in the governorates is not hidden (Allozi et al., 2022; Tariq et al., 2022). The importance of strategic leader's role in implementing policies by linking the two stages of public policies formulation and implementation from the reality of local councils in the governorates is become clear (Alkalha et al., 2012; Zu'bi et al., 2012). The implementing public policies process is considered one of the main pillars of all government departments (Alaali et al., 2021; Alzoubi et al., 2022). Because it is directly related to improving the efficiency and effectiveness of these departments, As a result of the current rapid developments, these developments require the adoption of a management approach based on providing unfamiliar services of high quality; to reach outstanding performance, Jordanian government departments in Karak governorate work like other departments in other governorates, in the face of changing environmental conditions with a high degree of complexity, which calls for the need to adopt effective strategic leadership practices, according to an administrative methodology, focused on achieving excellence in sustainable performance, which is based on creating lower cost services,

Relying on the formulation and implementation of the general policy related to each department and its dissemination, public resources planning and mobilization, especially the scarce ones, and stakeholders' participation and achievement that enable them to adapt to the emerging changes that affect government departments in light of the threats they face (Alshurideh et al., 2019a, 2022). In general, these departments success requires keeping pace with strategic leadership practices in their developments. Because it is one of the important tools that enable these departments to deal with the environmental uncertainty, and it is considered one of the important factors in supporting, coordinating and unifying all efforts in order to achieve the strategic goals efficiently and effectively (Aburayya et al., 2020; Alshurideh et al., 2019b). Accordingly, the purpose consistent with this study is to show the role of strategic leadership practices on implementing public policies in government departments in Karak Governorate.

2 Literature Review and Hypotheses Development

2.1 Strategic Leadership

There are many definitions and concepts according to the researchers view to leadership and its concepts, but there is no specific agreement about its concept, or even general agreement about leadership component, as some researchers focused on leadership as the outcome of interaction between the leader, subordinates and the situation as a continuous process (Meuser et al., 2016; Batayneh et al., 2021), There are many definitions of strategic leadership with the multiplicity of functions it performs, where strategic leadership is defined as the leader's ability to visualize future visions, deep strategic thinking, identify environmental changes that will affect the level of organization continuous improvement, and work with others to take the necessary actions to achieve a better future., while maintaining flexibility in implementation, reviewing and evaluating tasks and procedures (Alyammahi et al., 2020). It is also defined by Ratanasongtham and Ussahawanitchakit (2015) as leadership that possesses the ability to predict, visualize, maintain flexibility, empower others, develop competencies and capabilities, improve organizational structures, and choose to develop the next leader's generation of leader, sustain an effective organizational culture, and emphasize ethical practices. While Al-Nafar (2015) referred to strategic leadership as the leadership capable of developing a clear and appropriate strategic vision based on strategic objectives in accordance with an appropriate organizational culture in order to develop a human cadre that is characterized by values, ethics, innovation, and flexibility that works to bring about change and continuous development that leads to excellence and uniqueness. Looking at the previous definitions of strategic leadership concept, it becomes clear that strategic leadership focuses on a set of activities and operations, including determining the organization current situation, formulating the organization strategic direction, strategic alliances, implementing change and making strategic decisions to ensure the organization's growth and continuity, improving its outputs, and achieving its desired goals (Ahmad et al., 2021; Al-Dmour et al., 2021).

According to what was indicated by Muwafaq, the strategic leadership importance lies in the fact that it contributes to: building the organization's strategy, achieving long-term success and prosperity, and enhancing the continuous competitive advantage. Poor strategic leadership practices cause the organization to lose focus on defining long-term direction. Strategic leadership enhances the ability to deal with different types of individuals personalities inside and outside the organization, and coordinate with parties that did not interact in the first place, in addition to enhancing commitment and participation, and confirming the ethical model in organizational relations, with consideration of strategic leadership as the organization representative and its negotiator in dealing with relevant external agencies, entities, and organizations (Ashal et al., 2021; Ben-Abdallah et al., 2022).

What distinguishes strategic leadership is that it has an impact that extends outside the functional scope of the leader and business units, and sometimes outside the organization, which means that the organization is a strongly interconnected system, as decisions taken in one part of the business units affect other parts of the organization, and the impact of strategic leadership extends to long period of time, The strategic leader keeps long-term goals in his mind, and works to achieve short-term goals derived from long-term goals, which requires balancing and harmonizing these two types of organizational goals (Njeri, 2017). The strategic leader keeps long-term goals in his mind, and works to achieve short-term goals derived from long-term goals, which requires balancing and harmonizing these two types of organizational goals (Njeri, 2017).

There are a number of strategic leadership practices that were dealt with by researchers and constituted a basic requirement for this type of leadership. The researcher decided to address a number of them in agreement with the study population, which are as follows: Strategic direction: It refers to the general path chosen by the organization to achieve its strategic goals, taking into account the external environment conditions in which it is active, and its available resources while adhering to the available capabilities. Human capital: It refers to the processes related to training, education, and other professional initiatives in order to increase the knowledge level, skills, abilities, values, and employee's social assets that leads to employee satisfaction and performance (Ukenna et al., 2010). Ethical practices: They refer to a set of foundations, standards and principles that are the basis for the proper behavior that individuals must undertake to abide by at work (Al-Aqili, 2014). It is also referred to as a practical framework for decision-making and it serves as a means to refine and evaluate practices, because it reflects a set of standards, values, and ethical principles that guide the behavior and actions of employees. Strategic control: it is defined as an administrative process that seeks to detect deviations, whether positive or negative, with the aim of correcting them, taking into account if everything is done in accordance with established plans, instructions issued, and applicable principles. It includes all the organization moral and material parts (Reza, 2011).

2.2 *Public Policies Implementation*

Public policies have kept pace with developments in the social sciences fields such as economics, management and politics, as a scientific field that intersects with all these sciences. Therefore, researchers, experts, academics and leaders have been interested in the issue of public policies because of its great importance at the institutions, public organizations and government's levels. There are many definitions of public policies to include all aspects of decisions issued by the government, or which it authorizes in order to fulfill societal demands. The public policies approved and implemented by the political system are characterized by diversity and comprehensiveness that affects all community life. Therefore, public policy has been defined as a set of plans, programs, decisions and activities resulting from the official authorities to solve current and future problems in all sectors. It is issued in the form of decisions, laws and regulations with the allocation of financial and technical resources to achieve the public interest (Salihi, 2019). Public policy is what the government does or intends to do to solve a general problem facing the community in order to provide the community required needs, It is an organized effort aimed at analyzing, understanding and evaluating how the government dose its role in serving the community, and improving the government performance efficiency (Altamony et al., 2012; Farghaly, 2019). implementing public policies importance came from the fact that it helps in obtaining more knowledge about its sources, stages, active contributors, its importance, and its effects on society, and a deeper understanding of society and its needs through knowledge of the foundations and results of public policy programs, and this deepening course increases the understanding of the relationship between the political system and its society. There is a set of dimensions that constitute the implementation of public policies structure, and in many cases they are closely interconnected, and the separation of one of them constitutes a flaw in the implementation of the rest of the dimensions in a way that achieves their implementation effectiveness. The researcher decided to address a number of dimensions that, the researcher formed dimensions that are closely related to the study population. They were addressed as follows: public policy Formulation and dissemination: This process represents the activities that take place before they are announced by the government, and according to what (Ikelegbe mentioned, 2005), said it includes defining the problem related to the policy, developing, analyzing alternatives and selecting the appropriate alternatives for implementation. Planning for implementation and moving resources): Planning and resource mobilization are referred to as contemplation, reflection and insight into the past, present and future, and study the size of required policies and required resources availability through which this can be implemented (Hussain and Jassim, 2017). Stakeholders Participation: (Freeman, 1984) define stakeholders as an individual or group that influences by the organization's achievement of its objectives. It is noteworthy that the policy that is effectively developed without the intervention of the capital owner may lead to difficulty in implementation because it does not reflect the actual needs and does not take into account the feelings of the shareholders and owners who may contribute to the policy implementation (Klein & Knight, 2005).

2.3 *Strategic Leadership and Its Role in Implementing Public Policies*

Boal and Schultzy (2007) emphasizes that strategic leadership is a critical element in developing the organizational capabilities through opportunities and avoiding threats in a timely manner, Strategic leadership helps in making strategic decisions, creating and communicating the future vision, developing structures, activating oversight and organizational processes, maintaining an effective organizational culture, and spreading the ethical values system in the organization culture. According to Muwafaq, the strategic leadership importance lies in the fact that it contributes to: building the organization's strategy, achieving long-term success, and enhancing the continuous competitive advantage. And that the strategic leadership practices weakness makes the organization lose focus on determining the long-term direction. Strategic leadership enhances the ability to deal with different types of individuals personalities inside and outside the organization, and coordinate with parties that did not interact in the first place, in addition to enhancing commitment and participation, and confirming the ethical model in organizational relations, with consideration of strategic leadership as the representative of the organization and its negotiator in dealing with relevant external agencies, bodies, and organizations. Al-Arif pointed to strategic leadership importance in the process of evaluating organizational performance, through balanced control of decisions, directions, and what it does in providing advice and guidance. As for Al-Qaryouti (2006), says that public policy helps in identifying the activities and programs undertaken by the government and their results, and public policy helps in analyzing and formulating policies, providing alternatives and solutions to the problems presented, and implementing them. In addition to assisting the government in adopting successful policies to achieve the planned goals, it also works to improve the knowledge of the problems facing society through scientific diagnosis of these problems and educating politicians with the latest methods used in the process of developing, implementing and evaluating public policies. In addition to identifying the determinants that benefit the government in the process of drawing up and implementing public policy, in order to seek a greater understanding of its role, assistance in predicting foreign public policies that affect internal policies, taking precautions for them, and assisting the government in setting programs and public policies that can receive affirmation and support. Based on what was discussed, the study hypothesis can be formulated as follows:

There is a statistically significant impact of the strategic leadership practices on the implementation of public policies in government departments in Karak Governorate.

3 Study Model

See Fig. 1.

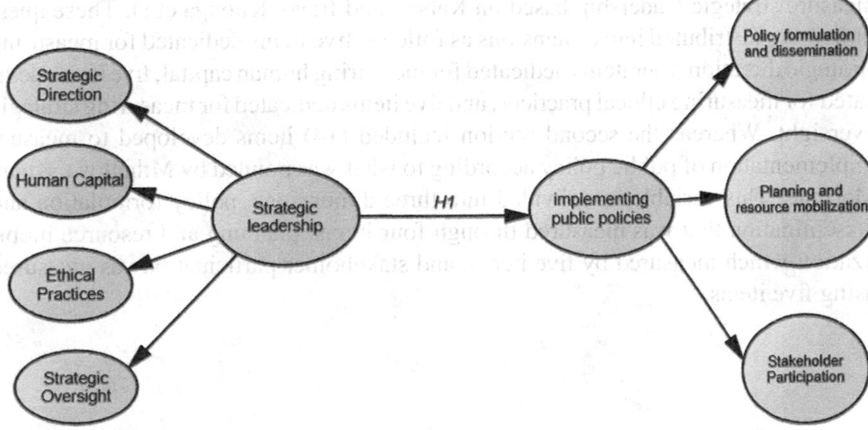

Fig. 1 Research model

4 Methodology

4.1 Population and Sample Selection

A qualitative method based on a questionnaire was used in this study for data collection and sample selection. The major aim of the study was to examine the impact of strategic leadership on implementing public policies in government departments in Karak Governorate. Therefore, it was focused on government departments in Karak Governorate in Jordan. Data were primarily gathered through questionnaires which were distributed to a random sample of (175) respondent. In total, (175) responses were received including (5) invalid to statistical analysis due to uncompleted or inaccurate. Hence, the final sample contained (170) responses suitable to analysis requirements that were formed a response rate of 97.86) %) where it proved to be sufficient to the extent that was predictable and allowed for a presumption of data saturation (Sekaran & Bougie, 2016).

4.2 Measurement Instrument

A questionnaire consists of two main sections along with a section regarding control variables was used as the measurement instrument. Control variables considered as categorical measures were composed of gender, age, educational level, and experience. The two main sections were dealt with a five-point Likert scale (from 1 = strongly disagree to 5 = strongly agree). The first section contained (19) items to measure strategic leadership based on Kabetu and Iravo, Kitonga et al. These questions were distributed into dimensions as follows: five items dedicated for measuring strategic direction, four items dedicated for measuring human capital, five items dedicated for measuring ethical practices, and five items dedicated for measuring strategic oversight. Whereas the second section included (14) items developed to measure implementation of public policy according to what was pointed by Mthethwa, Khan, Marume. This variable was divided into three dimensions: policy formulation and dissemination that was measured through four items, planning and resource mobilization which measured by five items, and stakeholder participationwas measured using five items.

5 Findings

5.1 Measurement Model Evaluation

This study was conducted structural equation modeling (SEM) to test hypotheses, which represents a contemporary statistical technique for testing and estimating the relationship between factors and variables (Wang & Rhemtulla, 2021). Accordingly, the reliability and validity of the constructs were tested using confirmatory factor analysis (CFA) through the statistical program AMOSv25. Table 1 summarizes the results of convergent and discriminant validity, as well the indicators of reliability.

Table 1 shows that the standard loading values for the individual items were within the domain (0.620–0.895), these values greater than the minimum retention of the elements based on their standard loads (Al-Lozi et al., 2018; Sung et al., 2019). Average variance extracted (AVE) is a summary indicator of the convergent validity of constructs that must be above 0.50 (Howard, 2018). The results indicate that the AVE values were greater than 0.50 for all constructs, thus the used measurement model has an appropriate convergent validity. Rimkeviciene et al. (2017) suggested the comparison approach as a way to deal with discriminant validity assessment in covariance-based SEM. This approach is based on comparing the values of maximum shared variance (MSV) with the values of AVE, as well as comparing the values of square root of AVE ($\sqrt{}$AVE) with the correlation between the rest of the structures. The results show that the values of MSV were smaller than the values of AVE, and that the values of $\sqrt{}$AVE were higher than the correlation values among the rest of the constructs. Therefore, the measurement model used is characterized by discriminative validity. The internal consistency measured through Cronbach's Alpha coefficient (α) and compound reliability by McDonald's Omega coefficient (ω) was conducted as indicators to evaluate measurement model. The results listed in Table 1 demonstrated that both values of Cronbach's Alpha coefficient and McDonald's Omega coefficient were greater than 0.70, which is the lowest limit for judging on measurement reliability (De Leeuw et al., 2019).

5.2 Structural Model

The structural model illustrated no multicollinearity issue among predictor constructs because variance inflation factor (VIF) values are below the threshold of 5, as shown in Table 1 (Hair et al., 2017). This result is supported by the values of model fit indices shown in (Figs. 1 and 2).

The results in Fig. 1 indicated that the chi-square to degrees of freedom (CMIN/DF) was 2.199, which is less than 3 the upper limit of this indicator. The values of the goodness of fit index (GFI), the comparative fit index (CFI), and the Tucker-Lewis index (TLI) were upper than the minimum accepted threshold of 0.90.

Table 1 Results of validity and reliability tests

Constructs	1	2	3	4	5	6	7
1. SD	**0.766**						
2. HC	0.556	**0.782**					
3. EP	0.587	0.633	**0.768**				
4. SO	0.598	0.678	0.521	**0.785**			
5. PFAD	0.578	0.574	0.673	0.699	**0.757**		
6. PARM	0.784	0.682	0.695	0.752	0.683	**0.759**	
7. SP	0.794	0.622	0.605	0.759	0.761	0.709	**0.787**
VIF	3.195	3.017	2.960	2.779	–	–	–
Loadingsrange	0.805–0.849	0.782–0.822	0.758–0.847	0.705–0.884	0.826–0.891	0.620–0.895	0.750–0.890
AVE	0.868	0.731	0.850	0.860	0.839	0.826	0.813
MSV	0.614	0.549	0.543	0.497	0.556	0.485	0.469
Internal consistency	0.884	0.823	0.858	0.883	0.894	0.693	0.753
Composite reliability	0.748	0.865	0.874	0.905	0.911	0.945	0.877

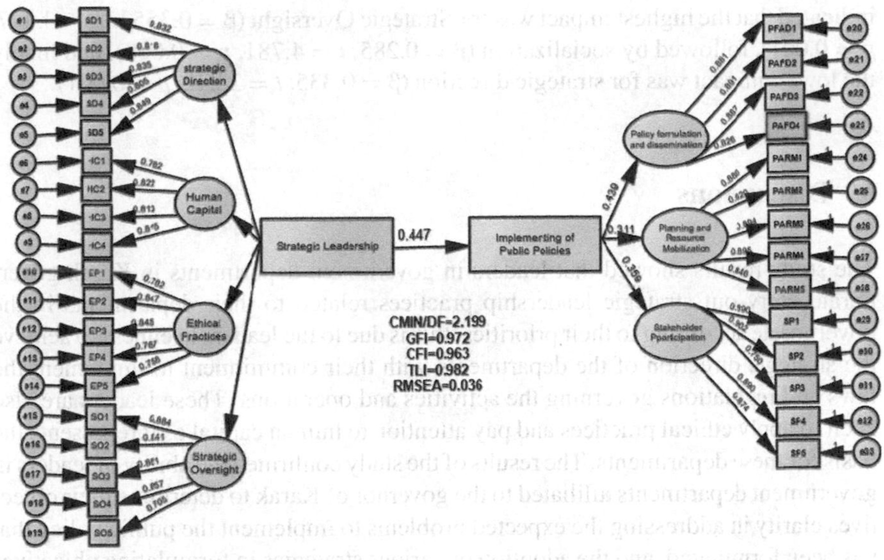

Fig. 2 SEM results of the digital marketing effect on mental image

Moreover, the result of root mean square error of approximation (RMSEA) indicated to value 0.036, this value is a reasonable error of approximation because it is less than the higher limit of 0.08. Consequently, the structural model used in this study was recognized as a fit model for predicting the DEP and generalization of its result (Ahmad et al., 2016; Shi et al., 2019). To verify the results of testing the study hypotheses, structural equation modeling (SEM) was used, the results of which are listed in Table 2.

The results demonstrated in Table 2 show that all strategic leadership practices have a positive impact relationship on implementing public policies except human capital and ethical practices which have no impact on implementation public policies (β = 121, $t = 1.207$, $p = 0.229$ and β = 0.068, $t = 0.682$, $p = 0.496$). However, the results

Table 2 Hypothesis testing

Hypothesis	Relation	Standard beta	*t* value	*p* value
H1	strategic direction→implementing public policies	0.355	3.242	0.001
H2	Human capital→implementing public policies	0.121	1.207	0.229
H3	Ethical practices→implementing public policies	0.068	0.682	0.496
H4	Strategic Oversight→implementing public policies	0.333	3.447	0.001

Note * $p < 0.05$, ** $p < 0.01$, *** $p < 0.001$

indicated that the highest impact was for Strategic Oversight ($\beta = 0.35533, t = 3.447, p = 0.001$), followed by socialization ($\beta = 0.285, t = 4.781, p = 0.000$), and finally the lowest impact was for strategic direction ($\beta = 0.335, t = 3.242, p = 0.001$).

6 Discussions

The study results showed that leaders in government departments in Karak governorate carry out strategic leadership practices related to their departments in the governorate according to their priorities, this is due to the leaders keenness to achieve the strategic direction of the departments with their commitment to implement the laws and regulations governing the activities and operations. These leaders are also keen to apply ethical practices and pay attention to human capital as it represents the basis for these departments, The results of the study confirmed the ability of leaders in government departments affiliated to the governor of Karak to determine their objectives clarity in addressing the expected problems to implement the public policy that has been formulated, and the adoption of various strategies in formulating objectives when formulating public policies in the governorate so that these public policies adopted by leaders lead In the government departments in Karak governorate, new programs should be set up so that the beneficiaries are aware of their benefits. It also shows the leaders keenness in government departments in Karak governorate to plan for potential changes in public policies, work systems and procedures, with their ability to draw plans in their departments according to the capabilities to implement public policy.

The results indicated the leaders' interest in the aspects related to the stakeholder's participation and related to their keenness on effective communication with the audience and employees, with the leaders' awareness of the public needs analyzing importance according to time, and their keenness to receive complaints and work to resolve them with the general policy unconfident. The results revealed a statistically significant impact of the general strategic leadership practices on public policies in government departments in Karak Governorate. This result can be explained by the adoption of government departments in Karak governorate to the strategic leadership practices represented in strategic direction and strategic control, This reinforces two dimensions of strategic leadership, one related to the strategic vision of government departments, and the other dimension is related to the strategic control process, which means the process of control and oversight in government departments in the governorate so that they are able to face challenges, anticipate the future and take advantage of opportunities.

The results also indicated that the government departments in the Karak governorate were able to draw up, formulate and publish the general policy, in line with the complementary and coordinating viewpoint between the various government departments in the Karak governorate. The results indicated that the government adoption of strategic direction dimensions and strategic control ensures the achievement of these departments aspirations to exploit resources in a high degree, based on their scarcity and decline, which requires careful vision and organizational control over all available resources and the possibility of using them to meet citizens needs and desires in Karak Governorate. The government departments in the Karak governorate does their internal functions and operations and link them to citizen's needs, desires and tendencies, in accordance with the stakeholders in the governorate, and based on strategic plans prepared in partnership with stakeholders and civil society organizations in Karak governorate. This result is consistent with Al-Amri and Al-Janabi (2020), which indicated that there is an impact of strategic leadership practices in avoiding strategic pitfalls, and Juma's study (2019), which showed that there is an impact of strategic leadership practices in enhancing strategic performance, and also agreed with the study.

7 Recommendations

Based on the previous results, the study recommends the need to pay attention to human capital in government departments in Karak governorate, as it represents one of the strategic leadership practices because of its reflection on the quality of services provided by employees and that its efficiency enhances interest in public policy formulation, planning, resource mobilization and the participation of stakeholders in Karak governorate. The study also recommends the need to enhance attention to ethical practices in government departments in Karak governorate because of their great repercussions in activities and planning processes and participation with stakeholders in providing services and preserving government departments' materials in Karak governorate, also the need to enhance stakeholders participation when formulating public policies, as these parties are part of the work environment of these departments, and these departments success lies in the extent of their stakeholders participation.

References

Aburayya, A., Alshurideh, M., Alawadhi, D., Alfarsi, A., Taryam, M., & Mubarak, S. (2020). An investigation of the effect of lean six sigma practices on healthcare service quality and patient satisfaction: Testing the mediating role of service quality in Dubai primary healthcare sector. *Journal of Advanced Research in Dynamical and Control Systems, 12*(8), 56–72.

Ahmad, S., Zulkurnain, N., & Khairushalimi, F. (2016). Assessing the validity and reliability of a measurement model in structural equation modeling (SEM). *British Journal of Mathematics and Computer Science, 15*(3), 1–8. https://doi.org/10.9734/BJMCS/2016/25183

Ahmad, A., Alshurideh, M. T., Al Kurdi, B. H., & Alzoubi, H. M. (2021, June). Digital strategies: A systematic literature review. In *The international conference on artificial intelligence and computer vision* (pp. 807–822). Cham: Springer.

Al- Quran, A. Z., Alhalalmeh, M. I., Eldahamsheh, M. M., Mohammad, A. A., Hijjawi, G. S., Almomani, H. M., & Al-Hawary, S. I. (2020). Determinants of the green purchase intention in

Jordan: The moderating effect of environmental concern. *International Journal of Supply Chain Management, 9*(5), 366–371.

Alaali, N., Al Marzouqi, A., Albaqaeen, A., Dahabreh, F., Alshurideh, M., Alrwashdh, S., Iyadeh, I., Salloum, S., & Aburayya, A. (2021) The impact of adopting corporate governance strategic performance in the tourism sector: A case study in the Kingdom of Bahrain. *Journal of Legal, Ethical and Regulatory, 24*(Special Issue 1), 1–18.

Al-Afifi, B., & Madi, K. (2016). *Ethical practices and their role in developing organizational commitment among workers in Palestinian universities in the Gaza Strip, Master's thesis (unpublished)*. Al-Azhar University, Gaza, Palestine.

Alameeri, K., Alshurideh, M., Al Kurdi, B., & Salloum, S. A. (2020, October). The effect of work environment happiness on employee leadership. In *International conference on advanced intelligent systems and informatics* (pp. 668–680). Cham: Springer.

Al-Amiri, B. R. A., & Al-Janabi, M. H. A. H. (2020). The effect of strategic control in avoiding strategic pitfalls. *Journal of Economics and Administrative Sciences, 26*(117), 128–151.

Al-Aqili, Come On. (2014). *Work ethics and its relationship to job satisfaction for female administrative staff at King Saud University in Riyadh, an unpublished master's thesis*. Naif Arab University for Security Sciences.

Al-Dhuhouri, F. S., Alshurideh, M., Al Kurdi, B., & Salloum, S. A. (2020, October). Enhancing our understanding of the relationship between leadership, team characteristics, emotional intelligence and their effect on team performance: A Critical Review. In *International conference on advanced intelligent systems and informatics* (pp. 644–655). Cham: Springer.

Al-Dmour, R., AlShaar, F., Al-Dmour, H., Masa'deh, R., & Alshurideh, M. T. (2021). The effect of service recovery justices strategies on online customer engagement via the role of "customer satisfaction" during the Covid-19 pandemic: An empirical study. *The Effect of Coronavirus Disease (COVID-19) on Business Intelligence, 334*, 325–346.

Alhalalmeh, M. I., Almomani, H. M., Altarifi, S., Al- Quran, A. Z., Mohammad, A. A., & Al-Hawary, S. I. (2020). The nexus between corporate social responsibility and organizational performance in Jordan: The mediating role of organizational commitment and organizational citizenship behaviour. *Test Engineering and Management, 83*(July), 6391–6410.

Al-Hawary, S. I. (2009). The effect of the leadership style on the effectiveness of the organization: A field study at Zarqa Private University. *The Egyptian Journal for Commercial Studies, 33*(1), 361–393.

Al-Hawary, S. I., & AL-Awawdeh, W., & Abden, M. A. (2012). The impact of the leadership style on organizational commitment: A field study on Kuwaiti telecommunications companies. *ALEDARI, 130*, 53–102.

Al-Hawary, S. I., & Hadad, T. F. (2016). The effect of strategic thinking styles on the enhancement competitive capabilities of commercial banks in Jordan. *International Journal of Business and Social Science, 7*(10), 133–144.

Al-Hawary, S. I. S., & Obiadat, A. A. (2021). Does mobile marketing affect customer loyalty in Jordan? *International Journal of Business Excellence, 23*(2), 226–250.

Al-Hawary, S. I., Al-Hawajreh, K., & AL-Zeaud, H., & Mohammad, A. (2013). The impact of market orientation strategy on performance of commercial banks in Jordan. *International Journal of Business Information Systems, 14*(3), 261–279.

Al-Hawary, S. I., & Al-Syasneh, M. S. (2020). Impact of dynamic strategic capabilities on strategic entrepreneurship in presence of outsourcing of five stars hotels in Jordan. *Business: Theory and Practice, 21*(2), 578–587.

Al-Hawary, S. I. S., & Mohammed, A. K. (2017). Impact of team work traits on organizational citizenship behavior from the viewpoint of the employees in the education directorates in North Region of Jordan. *Global Journal of Management and Business, 17*(2-A), 23–40.

Alkalha, Z., & Al-Zu'bi, Z., Al-Dmour, H., Alshurideh, M., & Masa'deh, R. (2012). Investigating the effects of human resource policies on organizational performance: An empirical study on commercial banks operating in Jordan. *European Journal of Economics, Finance and Administrative Sciences, 51*(1), 44–64.

Allahow, T. J. A. A., Al-Hawary, S. I. S., & Aldaihani, F. M. F. (2018). Information technology and administrative innovation of the central agency for information technology in Kuwait. *Global Journal of Management and Business, 18*(11-A), 1–16.

Al-Lozi, M., Almomani, R. Z., & Al-Hawary, S. I. (2018a). Talent Management strategies as a critical success factor for effectiveness of Human Resources Information Systems in commercial banks working in Jordan. *Global Journal of Management and Business Research: A Administration and Management, 18*(1), 30–43.

Allozi, A., Alshurideh, M., AlHamad, A., & Al Kurdi, B. (2022). Impact of transformational leadership on the job satisfaction with the moderating role of organizational commitment: Case of UAE and Jordan manufacturing companies. *Academy of Strategic Management Journal, 21*, 1–13.

Al-Lozi, M., Almomani, R. Z., & Al-Hawary, S. I. (2017). Impact of talent management on achieving organizational excellence in Arab Potash Company in Jordan. *Global Journal of Management and Business Research: A Administration and Management, 17*(7), 15–25.

Al-Nady, B. A., Al-Hawary, S. I., & Alolayyan, M. (2013). Strategic management as a key for superior competitive advantage of sanitary ware suppliers in Kingdom of Saudi Arabia. *International Journal of Management and Information Technology, 7*(2), 1042–1058.

Al-Nady, B. A., Al-Hawary, S. I., & Alolayyan, M. (2016). The role of time, communication, and cost management on project management success: An empirical study on sample of construction projects customers in Makkah City, Kingdom of Saudi Arabia. *International Journal of Services and Operations Management, 23*(1), 76–112.

Al-Nafar, H. N. (2015). Strategic leadership practices and their role in the application of total quality: a field study by application to Palestinian universities in Gaza Governorate, unpublished Ph.D. thesis, Suez Canal University, Faculty of Commerce.

Alolayyan, M., Al-Hawary, S. I., Mohammad, A. A., & Al-Nady, B. A. (2018). Banking service quality provided by commercial banks and customer satisfaction. A structural equation modelling approaches. *International Journal of Productivity and Quality Management, 24*(4), 543–565.

Al-Qaryouti, M. Q. (2006). *Drawing, implementing, evaluating and analyzing public policy, 1st floor*, Kuwait, Al Falah Library.

AlShehhi, H., Alshurideh, M., Al Kurdi, B., & Salloum, S. A. (2020, October). The impact of ethical leadership on employees performance: A systematic review. In International Conference on Advanced Intelligent Systems and Informatics (pp. 417–426). Cham: Springer.

Alshurideh, M., Alsharari, N. M., & Al Kurdi, B. (2019a). Supply chain integration and customer relationship management in the airline logistics. *Theoretical Economics Letters, 9*(02), 392–414.

Alshurideh, M., Kurdi, B. A., Shaltoni, A. M., & Ghuff, S. S. (2019b). Determinants of pro-environmental behaviour in the context of emerging economies. *International Journal of Sustainable Society, 11*(4), 257–277.

Alshurideh, M., Al-Hawary, S. I., Batayneh, A. M., Mohammad, A., & Al-Kurdi, B. (2017). The impact of Islamic Banks' service quality perception on Jordanian customers loyalty. *Journal of Management Research, 9*(2), 139–159.

Alshurideh, M. T., Al Kurdi, B., Alzoubi, H. M., Ghazal, T. M., Said, R. A., AlHamad, A. Q., & Al-kassem, A. H. (2022). Fuzzy assisted human resource management for supply chain management issues. *Annals of Operations Research*, 1–19.tr

Altamony, H., Masa'deh, R., Alshurideh, M., Obeidat, B. (2012) Information systems for competitive advantage: Implementation of an organisational strategic management process. *Innovation and Sustainable Competitive Advantage: From Regional Development to World Economies*, 583–592.

AlTaweel, I. R., & Al-Hawary, S. I. (2021). The mediating role of innovation capability on the relationship between strategic agility and organizational performance. *Sustainability, 13*(14), 7564.

Alyammahi, A., Alshurideh, M., Al Kurdi, B., & Salloum, S. A. (2020, October). The impacts of communication ethics on workplace decision making and productivity. In *International conference on advanced intelligent systems and informatics* (pp. 488–500). Cham: Springer.

Alzoubi, H., Alshurideh, M., Kurdi, B., Akour, I., & Aziz, R. (2022). Does BLE technology contribute towards improving marketing strategies, customers' satisfaction and loyalty? The role of open innovation. *International Journal of Data and Network Science, 6*(2), 449–460.

Ashal, N., Alshurideh, M., Obeidat, B., Masa'deh, R. (2021). The impact of strategic orientation on organizational performance: Examining the mediating role of learning culture in Jordanian telecommunication companies. *Academy of Strategic Management Journal, 21*(Special Issue 6), 1–29.

Batayneh, R. M. A., Taleb, N., Said, R. A., Alshurideh, M. T., Ghazal, T. M., & Alzoubi, H. M. (2021, June). IT governance framework and smart services integration for future development of Dubai infrastructure utilizing AI and big data, its reflection on the citizens standard of living. In *The international conference on artificial intelligence and computer vision* (pp. 235–247). Cham: Springer.

Ben-Abdallah, R., Shamout, M., & Alshurideh, M. (2022) Business development strategy model using EFE, IFE and IE analysis in a high-tech company: An empirical study. *Academy of Strategic Management Journal, 21*(Special Issue 2), 1–9.

Boal, B., & Schultz, L. (2007). Storytelling, time, and evolution: The role of strategic leadership in complex adaptive systems. *The Leadership Quarterly, 18*(4), 411–428.

de Leeuw, E., Hox, J., Silber, H., Struminskaya, B., & Vis, C. (2019). Development of an international survey attitude scale: Measurement equivalence, reliability, and predictive validity. *Measurement Instruments for the Social Sciences, 1*(1), 9. https://doi.org/10.1186/s42409-019-0012-x

Farghali, A. Z. (2019). The effectiveness of the role of local administration in public policies: A comparative study. *Journal of the College of Economics and Political Science, 20*(2), 265–297.

Freeman, E. (1984). *Strategic management: A stakeholder approach.* Bitman.

Gomaa, M. H., & Gomaa,. (2019). The role of strategic leadership in enhancing strategic performance: A exploratory study of the views of administrative leaders at the University of Diyala. *Journal of the College of Administration and Economics for Economic, Administrative and Financial Studies, 11*(4), 502–522.

Hair, J. F., Babin, B. J., & Krey, N. (2017). Covariance-based structural equation modeling in the journal of advertising: Review and recommendations. *Journal of Advertising, 46*(1), 163–177. https://doi.org/10.1080/00913367.2017.1281777

Harahsheh, A., Houssien, A., Alshurideh, M. & AlMontaser, M. (2021). The effect of transformational leadership on achieving effective decisions in the presence of psychological capital as an intermediate variable in private Jordanian Universities in light of the corona pandemic. *The Effect of Coronavirus Disease (COVID-19) on Business Intelligence, 334*, 221–243.

Howard, M. C. (2018). The convergent validity and nomological net of two methods to measure retroactive influences. *Psychology of Consciousness: Theory, Research, and Practice, 5*(3), 324–337. https://doi.org/10.1037/cns0000149

Hussein, F. Z., & Jassem, H. F. (2017). The effectiveness of administrative leaders in implementing public policies: Analytical research in the Baghdad Provincial Council. *Journal of Economic and Administrative Sciences, 23*(98), 128–148.

Ikelegbe, A. (2005). *Public policy analysis: Concepts, issues and case*. Lagos: Imprint Services.

Klein, K. J., & Knight, A. P. (2005). Innovation implementation: overcoming the challenge. *Current Directions in Psychological Science, 14*(5):243–246.

Kusumawati, A. (2019). Impact of digital marketing on student decision-making process of higher education institution: A case of Indonesia. *Journal of E-Learning and Higher Education, 2019*(2019), 1–11. https://doi.org/10.5171/2019.267057

Makrash, F. (2015). The impact of management with intelligence on the strategic direction "A case study of the Air Algerie company". Unpublished Ph.D. Thesis, Faculty of Economics, Commercial and Management Sciences, University of Mohamed Khider, Biskra, Algeria.

Meuser, J. D., Gardner, W. L., Dinh, J. E., Hu, J., Liden, R. C., & Lord, R. G. (2016). A network analysis of leadership theory: The infancy of integration. *Journal of Management, 42*(5), 1374–1403.

Mohammad, A. A., Alshura, M. S., Al-Hawary, S. I. S., Al-Syasneh, M. S., & Alhajri, T. M. (2020). The influence of Internal Marketing Practices on the employees' intention to leave: A study of the private hospitals in Jordan. *International Journal of Advanced Science and Technology, 29*(5), 1174–1189.

Mowaffaq, S. (2013): The contribution of strategic leadership to achieving the competitive advantage of the institution, a master's thesis (published), Mohamed Khedir University, Biskra, the People's Democratic Republic of Algeria.

Njeri, B. N. (2017). Influence of strategic leadership on strategy implementation in the Kenyan Motor vehicle industry. *European Journal of Business and Strategic Management, 2*(9), 29–44.

Odeh, R. B. M., Obeidat, B. Y., Jaradat, M. O., & Alshurideh, M. T. (2021). The transformational leadership role in achieving organizational resilience through adaptive cultures: the case of Dubai service sector. *International Journal of Productivity and Performance Management*. https://doi.org/10.1108/IJPPM-02-2021-0093

Ratanasongtham, W., & Ussahawanitchakit, P. (2015). Strategic audit planning and audit quality: An empirical research of CPAs in Thailand. *The Business and Management Review, 7*(1), 384–398.

Reda, H. H. (2011). *Administrative reform* (1st ed.). Amman - Jordan.

Rimkeviciene, J., Hawgood, J., O'Gorman, J., & De Leo, D. (2017). Construct validity of the acquired capability for suicide scale: Factor structure, convergent and discriminant validity. *Journal of Psychopathology and Behavioral Assessment, 39*(2), 291–302. https://doi.org/10.1007/s10862-016-9576-4

Salehi, A. (2019). Evaluating public policies between qualitative and quantitative use: A study of concepts and models. *Al-Naqid Journal for Political Studies, 3*(1), 192–208.

Samra, A., & Khamis, H. (2019). The role of strategic leadership in developing managerial skills for employees of the Palestinian Ministry of Labor, unpublished doctoral thesis, Al-Aqsa University.

Sekaran, U., & Bougie, R. (2016). *Research methods for business: A skill-building approach* (7th edn.). Wiley.

Shi, D., Lee, T., & Maydeu-Olivares, A. (2019). Understanding the model size effect on SEM fit indices. *Educational and Psychological Measurement, 79*(2), 310–334. https://doi.org/10.1177/0013164418783530

Sung, K.-S., Yi, Y. G., & Shin, H.-I. (2019). Reliability and validity of knee extensor strength measurements using a portable dynamometer anchoring system in a supine position. *BMC Musculoskeletal Disorders, 20*(1), 1–8. https://doi.org/10.1186/s12891-019-2703-0

Tariq, E., Alshurideh, M., Akour, E., Al-Hawaryd, S., & Al Kurdi, B. (2022). The role of digital marketing, CSR policy and green marketing in brand development at UK. *International Journal of Data and Network Science, 6*(3), 1–10.

Ukenna, S., Ijeoma, N., Anionwu, C., & Olise, C. (2010). Effect of investment in human capital development on organisational performance: empirical examination of the perception of small business owners in Nigeria. *European Journal of Economics, Finance and Administrative Sciences*, 93–107.

Wang, Y. A., & Rhemtulla, M. (2021). Power analysis for parameter estimation in structural equation modeling: A discussion and tutorial. *Advances in Methods and Practices in Psychological Science, 4*(1), 1–17. https://doi.org/10.1177/2515245920918253

Warokka, A. (2020). Digital marketing support and business development using online marketing tools: aAn experimental analysis. *International Journal of Psychosocial Rehabilitation, 24*(1), 1181–1188. https://doi.org/10.37200/IJPR/V24I1/PR200219

Zu'bi, Z., Al-Lozi, M., Dahiyat, S., Alshurideh, M., & Al Majali, A. (2012). Examining the effects of quality management practices on product variety. *European Journal of Economics, Finance and Administrative Sciences, 51*(1), 123–139.

The Impact of Innovative Leadership on the Strategic Intelligence in the Insurance Companies in Jordan

Basem Yousef Ahmad Barqawi, Mohammad Mousa Eldahamsheh, Menahi Mosallam Alqahtani, Kamel Mohammad Al-hawajreh, Nancy Abdullah Shamaileh, Anber Abraheem Shlash Mohammad, Muhammad Turki Alshurideh, Ayat Mohammad, and Mohammed Saleem Khlif Alshura

Abstract The aim of the study was to examine the impact of innovative leadership practices on strategic intelligence in insurance companies in Jordan. Data were primarily gathered through the questionnaires which were distributed to the managers of the different levels. In total, (256) responses were received including (18) invalid

B. Y. A. Barqawi
Faculty of Administration and Financial Sciences, Business Administration Department, Petra University, Amman, Jordan

M. M. Eldahamsheh
Researcher Strategic management, Amman, Jordan

M. M. Alqahtani
Administration Department, Community College of Qatar, Doha, Qatar
e-mail: mena7i@icloud.com

K. M. Al-hawajreh
Business Faculty, Mu'tah University, Kerak, Jordan

N. A. Shamaileh
Department of Business Administration, School of Business, Al Al-Bayt University, P.O. BOX 130040, Mafraq 25113, Jordan

A. A. S. Mohammad (✉)
Faculty of Administrative and Financial Sciences, Marketing, Marketing Department, Petra University, B.O. Box: 961343, Amman 11196, Jordan
e-mail: mohammad197119@yahoo.com

M. T. Alshurideh
Department of Marketing, School of Business, The University of Jordan, Amman 11942, Jordan
e-mail: malshurideh@sharjah.ac.ae; m.alshurideh@ju.edu.jo

Department of Management, College of Business, University of Sharjah, 27272 Sharjah, United Arab Emirates

A. Mohammad
Business and Finance Faculty, The World Islamic Science and Education University (WISE), Postal Code 11947, P.O Box 1101, Amman, Jordan
e-mail: dr_ayatt@yahoo.com

M. Alshurideh et al. (eds.), *The Effect of Information Technology on Business and Marketing Intelligence Systems*, Studies in Computational Intelligence 1056,
https://doi.org/10.1007/978-3-031-12382-5_77

to statistical analysis due to uncompleted or inaccurate. Hence, the final questionnaires contained (238) responses suitable to analysis requirements. Structural equation modeling (SEM) was conducted to test hypotheses. The results showed that all innovative leadership dimensions have a positive impact relationship on strategic intelligence except changeability which has no impact on Strategic Intelligence. Based on the study results, managers and decision makers of insurance companies have to instill enthusiasm and courage in employees, allowing them to express themselves more freely, with follow-up to identify complaints that may arise from a lack of knowledge or a lack of coordination or integration between departments and units at the institutional level.

Keywords Innovative leadership · The strategic intelligence · Insurance companies · Jordan

1 Introduction

The link between leadership and innovation is important in the success of the continuation of innovative processes within society and in the development and creation of the creative climate (Alameeri et al., 2020; AlShehhi et al., 2020; Odeh et al., 2021). Innovative leaders are not only satisfied with rearranging existing structures but are motivated by the continuous quest to find a better way (Al-Dhuhouri et al., 2020; Harahsheh et al., 2021). The innovative leadership style is discovery and self-renewal and their impact contributes to achieving the broader interest and raises the awareness of leadership, followers and the institution as a whole (Al-Hawary, 2009; Al-Hawary et al., 2012; Alzoubi et al., 2022). Innovation gives leadership characteristics that distinguish it from other leaders, such as openness towards change, the desire to identify internal problems and issues, in addition to giving the ability to control the environment in which he lives, confidence in the ability of others to achieve and bear responsibility, respect for others' choices, and the ability to rush towards learning and using culture, appreciating the achievements of others, and moving towards the future with full force (Mohammad et al., 2020; Al-Hawary & Mohammed, 2017; Al-Zubaidi et al., 2015). The communications and information technology revolution has contributed to unprecedented scientific, cultural, and technological developments in human history. To keep up with these developments, institutions of all kinds and orientations are reviewing their policies, objectives, technologies, and activities related to creativity, change, and continuous improvement of their services and products (Al-Hawary & Ismael, 2010; Al-Hawary et al., 2013; Al-Nady et al., 2013; Al-Hawary & Hadad, 2016; Al-Nady et al., 2016; Al-Lozi et al., 2017; Allahow et al., 2018; Al-Hawary & Al-Syasneh, 2020; Al-Hawary et al., 2020a,

M. S. K. Alshura
Faculty of Money and Management, Management Department, The World Islamic Science University, P.O. Box1101, Amman 11947, Jordan

2020b; Al-Hawary & Alhajri, 2020; Al- Quran et al., 2020; Alhalalmeh et al., 2020; Al-Hawary & Obiadat, 2021; AlTaweel & Al-Hawary, 2021; Alshurideh et al., 2022).

Organizations must prepare leaders and provide them with the skills they need to complete the responsibilities entrusted to them to the best of their abilities and achieve the organization's goals and objectives (Mohammad et al., 2020; Al-Hawary et al., 2012; Al-Hawary, 2009; Al-Hawary & Mohammed, 2017; Al-Hawary & Al-Hamwan, 2017). Monitoring the developments in developed countries shows that institutions were able to overcome all obstacles by encouraging creative practices and putting them into practice on the ground, as well as updating the various methods of work that were a reason for countries' superiority and progress, provided that these methods and tools were adapted to suit Arab environments.

Strategic intelligence was named one of the types of intelligence that anchors leadership traits, then finding creative leaders, especially in light of the growing importance of leadership in institutions and reliance on traits (intelligence, mind, and personality) in judging the efficiency of leaders (Al-Hawary & Hadad, 2016; Allozi et al., 2022). The true responsibility of the innovative leader is to uncover and invest his followers' innovative energies, as well as to work to ensure that his organization learns innovation and creativity on a constant basis. In general, the performance of these institutions depends on keeping up with new strategic management advancements. Strategic intelligence is also considered an advanced field; it is one of the most important tools for these institutions to deal with the environmental uncertainty they face, and it is a key factor in supporting and coordinating all efforts to unite them in order to achieve strategic goals efficiently and effectively (Al Kurdi et al., 2020; Alshurideh, 2022; Alsuwaidi et al., 2020; Kurdi et al., 2020). The study mainly aimed to determine the role that innovative Leadership plays on enhancing strategic intelligence.

2 Theoretical Framework and Hypotheses Development

2.1 Innovative Leadership

It is clear that the innovative leadership in the organization is the innovator who translates knowledge into choices and who has ideas that are new and useful and related to solving specific problems or re-combining known patterns in unique forms. Innovative leadership is the ability to change, renew or create a new approach or style, and use it with techniques. It is noticeable that innovative leadership contains all the characteristics of these styles, but each of them quotes a characteristic or advantage of the innovative leadership. Innovative leadership is also defined as working continuously to influence employees and convince them to accept work in order to achieve the goals of the organization according to the method determined by the innovative leader who can Re-dissolving the differences and contradictions between work goals and the goals of the institution (Bilton, 2010; Al Mehrez et al., 2020; Al Naqbi et al., 2020; Al Suwaidi et al., 2020; Nuseir et al., 2021; Lee et al., 2022a, 2022b).

Carr defined innovative leadership as the ability to change, renew or introduce a new approach or method, and use it with modern technologies that fit with the requirements of the environment and the aspirations of the modern era and meet the needs of institutions and society together. Al-Baz (2012) defines it as the innovator who translates knowledge into new choices and who has ideas that are new, useful and related to solving specific problems or re-combining known patterns of knowledge in unique forms, while Eid refers to it as "leadership that pursues an unconventional approach." in the development of the institution, in order to achieve continuous adaptation to the changes accompanying the field of work at the local and international levels, and those who practice it possess special characteristics and skills that are compatible with the requirements of applying modern trends in management." Innovative leadership has a set of basic features that appear to influence the behavior of leaders, and for the purposes of this study the following dimensions were adopted:

Problem solving and decision-making: This refers to the ability to comprehend and comprehend what is truly going on in a scenario, including other people's behaviors and feelings. An understanding of what is going on in daily events in a way that aids productivity, followed by the ability to act successfully in response to those conditions. **Changeability**: by generating novel ideas and recommendations for scenarios in which the institution must follow certain administrative guidelines in order to accomplish self-renewal and gain a competitive edge (Al-Jafra, 2013). **Initiative** is a trait of a creative leader that fosters initiative, hires creative individuals, encourages healthy and ethical competition, and does not suffocate excitement and originality with oppression and professional envy. **The ability to communicate** entails devising the most effective methods for identifying ideas, information, and directions by establishing a social structure in which opinions, trends, and knowledge are formed (Alyammahi et al., 2020; Sweiss et al., 2021). **Stimulating creativity**: a paragraph about a mental trait that allows an individual to think in unique ways, and this necessitates thorough preparation by the institution's upper management (Al-Zubaidi et al., 2015). **Creative knowledge**: This is the knowledge that enables a company to lead its industry.

2.2 *Strategic Intelligence*

Strategic intelligence is defined as "a systematic and ongoing process of producing smart information necessary for an organization's activity that is characterized by its strategic value and in an actionable form to aid long-term decision-making" (Ahmadi et al., 2020). It's also known as the essential competences required to make decisions that affect an institution's future and capacity to compete with others. This process qualifies its owner to have a set of synergistic analytical skills that leaders in organizations use in general to improve their ability to identify and understand systems, predict their behavior, and devise modifications to them in order to produce the desired effects, as these skills work together as a system, to help them with future aspirations that determine the nature of business future that the organization wants to achieve based on the environment i.e. It encourages firms to form cooperative partnerships in order to find investment possibilities, improve the quality of their products, or increase the efficiency of their operations.

Strategic intelligence is the delivery of the correct information to the appropriate people at the right time so that they may make the best and most practical decisions for the organization's future (Muslim & Qader, 2015). Basadur (2014) defined strategic intelligence as an aggregate formula for multiple sorts of skills in the context of determining the most efficient and effective ways for the institution's position to deal with future challenges and opportunities. Several studies looked into the various aspects of the idea of strategic intelligence, with the following being the most common: **Systems thinking**: It refers to the individuals' ability to deal with a group of variables, address them or research their relationship, and analyze them in a way that helps them explain their causes and outcomes and profit from them in attaining the overall goals that the institution tries to achieve is referred to as systems thinking (Al-Sheikhly, and Ibrahim, 2015). **Strategic vision** is a set of concepts for driving company development operations that is usually contained in a document or statement so that all management of the organization can share the same vision and make decisions based on common principles and the firm's mission (Yanez, Uruburu, Moreno & Lumbreras, 2019). An agreement between two or more institutions to collaborate and exchange material and/or intellectual resources is referred to as a **strategic partnership**, and their relationship is formalized through an employment contract. This form of agreement is designed to assist both parties in achieving their objectives (Kaur et al., 2019).

2.3 Innovative Leadership and Strategic Intelligence

Strategic intelligence is seen as the organization's protective shield, as it is involved in every stage of the organization's strategies and activities., and relying on the traits (intelligence, mind, and personality) in judging the efficiency of its leaders, as well as the leaders' need to enjoy different types of intelligence and employ them in adapting to new technology, the emergence of employment and information patterns (such as the internet). Innovative leaders no longer rely on a single sort of intelligence; instead, they must diversify the types of intelligence rather than relying on a single model to assess intelligence efficiency.

The responsibility of leading the institution in an innovative manner is placed on the shoulders of higher leadership, which necessitates all of the characteristics of sharp and strategic intelligence. No single leader can take on all issues and topics and manage them from the top, so the true role of the creative leader is to reveal creative energies to subordinates, invest them, and work to teach his organization to learn innovation. Organizations must adopt the concept of a good and modern leadership style, which is accomplished through attracting and employing highly skilled leaders, as well as establishing fair processes for evaluating their performance and providing suitable incentives.

In general, organizational success necessitates keeping up with changes in the world of leadership and implementing them efficiently and successfully as a work strategy. Obtaining advanced positions in leadership, global competitiveness indicators, liberating employees' creative and intellectual powers, allowing for positive engagement, and developing the spirit of collaboration and, within organizations, innovative leadership is regarded as a critical component that must not be overlooked. In one way or another, innovative leadership helps the institution improve its performance. The relevance of innovative leadership in working on the organization's continual development and progress by promoting and supporting the leader's presentation of new ideas, facilitating the institution's work with external parties, given that the leader has special skills that enable him to understand with the various external parties that the institution deals with, read expectations distinctly from competitors, and possess a vision that supports creativity, which provides opportunities and the ability to discover products or new services (Al-Nashar, 2018).

Strategic intelligence is a tool for providing comprehensive information on the external environment to senior decision-makers in a timely manner to support the strategy development process, and the reasons for strategic intelligence's increasing importance in supporting the decision-making process are discussed (Al Abdali et al., 2012). Zhang, Siribanpitak, and Charoenkul (2018) discovered a link between creative leadership strategies and encouraging innovation among Chinese teachers, while Abd, Abbas, and Khudair discovered an effect of strategic management practices on organizational leadership through strategic intelligence, which was confirmed by Tham and Kim (2002) that strategic intelligence is important because it is part of the organization's culture, and it improves the culture of teamwork and sharing information by preparing information for all departments and allowing management and employees to contribute their perspectives on the future by involving them in decision-making.

Organizations that apply the most strategic intelligence will undoubtedly achieve excellence, sustainability, and performance excellence, as strategic intelligence plays an important role in solving problems, removing barriers and restrictions, and structural reorganization, which aids in development and change to keep pace with the demands of the environment, and plays a fundamental role in improving the situation, the growth and development of intangible assets, which are among the most essential resources of any firm, the competitiveness of companies. Those interested in strategic intelligence have also done a lot of research and study in the subject of intelligence. Because of its significance, previous research has found a statistically significant relationship between strategic intelligence practices and organizational ingenuity, as well as a relationship between senior leaders' strategic intelligence and promoting organizational excellence among academic leaders at Hamdania University. Accordingly, the study hypothesis can be formulated as follows:

There is a statistically significant effect of creative leadership on strategic intelligence.

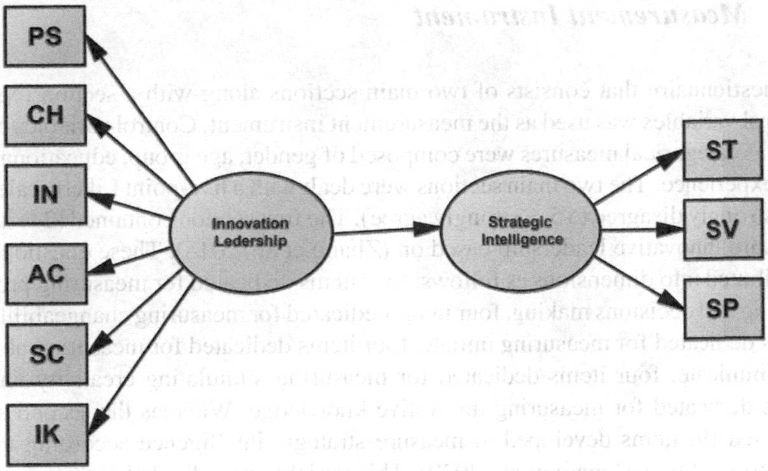

Fig. 1 Research Model

3 Study Model

See (Fig. 1).

4 Methodology

4.1 Population and Sample Selection

A quantitative method based on a questionnaire was used in this study for data collection and sample selection. The aim of the study was to examine the impact of innovative leadership practices on strategic intelligence in insurance companies in Jordan. Data were primarily gathered through the questionnaires which were distributed to the managers of the different levels. In total, (256) responses were received including (18) invalid to statistical analysis due to uncompleted or inaccurate. Hence, the final questionnaires contained (238) responses suitable to analysis requirements., where it proved to be sufficient to the extent that was predictable and allowed for a presumption of data saturation (Sekaran & Bougie, 2016).

4.2 Measurement Instrument

A questionnaire that consists of two main sections along with a section regarding control variables was used as the measurement instrument. Control variables considered as categorical measures were composed of gender, age group, educational level, and experience. The two main sections were dealt with a five-point Likert scale (from 1 = strongly disagree to 5 = strongly agree). The first section contained (24) items to measure innovative leadership based on (Zhang et al., 2018). These questions were distributed into dimensions as follows: four items dedicated for measuring problems solving and decisions making, four items dedicated for measuring changeability, four items dedicated for measuring initiate, four items dedicated for measuring ability to communicate, four items dedicated for measuring stimulating creativity and four items dedicated for measuring innovative knowledge. Whereas the second section included (9) items developed to measure strategic intelligence according to what was pointed by (Ahmadi et al., 2020). This variable was divided into three dimensions: systems thinking that was measured through three items, strategic vision which measured by three items and strategic partnership was measured using three items.

5 Findings

5.1 Measurement Model Evaluation

This study was conducted structural equation modeling (SEM) to test hypotheses, which represents a contemporary statistical technique for testing and estimating the relationship between factors and variables (Wang & Rhemtulla, 2021). Accordingly, the reliability and validity of the constructs were tested using confirmatory factor analysis (CFA) through the statistical program AMOSv24. Table 1 summarizes the results of convergent and discriminant validity, as well the indicators of reliability.

Table 1 shows that the standard loading values for the individual items were within the domain (0.518–0.661), these values greater than the minimum retention of the elements based on their standard loads (Al-Lozi et al., 2018; Sung et al., 2019). Average variance extracted (AVE) is a summary indicator of the convergent validity of constructs that must be above 0.50 (Howard, 2018). The results indicate that the AVE values were greater than 0.50 for all constructs, thus the used measurement model has an appropriate convergent validity. Rimkeviciene et al. (2017) suggested the comparison approach as a way to deal with discriminant validity assessment in covariance-based SEM. This approach is based on comparing the values of maximum shared variance (MSV) with the values of AVE, as well as comparing the values of square root of AVE ($\sqrt{AVE}$) with the correlation between the rest of the structures. The results show that the values of MSV were smaller than the values of AVE, and that the values of $\sqrt{AVE}$ were higher than the correlation values among the rest of the constructs. Therefore, the measurement model used is characterized by

Table 1 Results of validity and reliability tests

Constructs	1	2	3	4	5	6	7	8	9
1. PS	–								
2. CH	0.54	–							
3. IN	0.62	0.57	–						
4. AC	0.82	0.80	0.84	–					
5. SC	0.59	0.62	0.55	0.69	–				
6. IK	0.61	0.48	0.61	0.67	0.57	–			
7. ST	0.62	0.51	0.63	0.63	0.69	0.58	–		
8. SV	0.69	0.61	0.68	0.78	0.82	0.76	0.765	–	
9. SP	0.45	0.55	0.42	0.62	0.74	0.44	0.749	0.782	–
VIF	2.580	2.384	1.777	1.460	1.670	1.884	–	–	
Loadings range	0.578–0.827	0.644–0.858	0.665–0.750	0.518–0.661	0.674–0.752	0.568–0.777	0.599–0.887	0.758–0.859	0.634–0.735
AVE	0.733	0.759	0.719	0695	0.789	0.755	0.719	0.733	0.795
MSV	0.511	0.588	0.596	0.522	0.644	0.616	0.559	0.537	0.619
Internal consistency	0.866	0.915	0.845	0.844	0.839	0.822	0.849	0.833	0.852
Composite reliability	0.829	0.844	0.856	0.878	0.899	0.883	0.878	0.859	0.849

Note PS: problems solving and decisions making, CH: changeability, IN: initiate, AC: ability to communicate, SC: stimulating creativity, IK: innovative knowledge, ST: systems thinking, SV: strategic vision, SP:strategic partnership, bold fonts in the table indicate to root square of AVE discriminative validity. The internal consistency measured through Cronbach's Alpha coefficient (α) and compound reliability by McDonald's Omega coefficient (ω) was conducted as indicators to evaluate measurement model. The results listed in Table 1 demonstrated that both values of Cronbach's Alpha coefficient and McDonald's Omega coefficient were greater than 0.70, which is the lowest limit for judging on measurement reliability (de Leeuw et al., 2019).

5.2 *Structural Model*

The structural model illustrated no multicollinearity issue among predictor constructs because variance inflation factor (VIF) values are below the threshold of 5, as shown in Table 1 (Hair et al., 2017).This result is supported by the values of model fit indices shown in Fig. 1.

The results in Fig. 1 indicated that the chi-square to degrees of freedom (CMIN/DF) was 2.198 which is less than 3 the upper limit of this indicator. The values of the goodness of fit index (GFI), the comparative fit index (CFI), and the Tucker-Lewis index (TLI) were upper than the minimum accepted threshold of 0.90. Moreover, the result of root mean square error of approximation (RMSEA) indicated to value 0.046, this value is a reasonable error of approximation because it is less than the higher limit of 0.08. Consequently, the structural model used in this study was recognized as a fit model for predicting the DEP and generalization of its result (Ahmad et al., 2016; Shi et al., 2019).To verify the results of testing the study hypotheses, structural equation modeling (SEM) was used, the results of which are listed in Table 2 (Fig. 2).

The results demonstrated in Table 2 show that all innovative leadership dimensions have a positive impact relationship on strategic intelligenceexcept changeability which has no impact on Strategic Intelligence ($\beta = 0.048$, $t = 1.002$, $p = 0.317$)

Table 2 Hypothesis testing

Hypothesis	Relation	Standard beta	*t* value	*p* value
H1	Problems solving and decisions making → Strategic Intelligence	0.214	*5.335	0.000
H2	Changeability → Strategic Intelligence	0.048	1.002	0.317
H3	Initiate → Strategic Intelligence	0.134	*2.913	0.004
H4	Ability to communicate → Strategic Intelligence	0.093	1.557	0.061
H5	Stimulating creativity → Strategic Intelligence	0.124	*2.446	0.016
H6	Innovative knowledge → Strategic Intelligence	0.259	*5.540	0.000

Note * $p < 0.05$, ** $p < 0.01$, *** $p < 0.001$

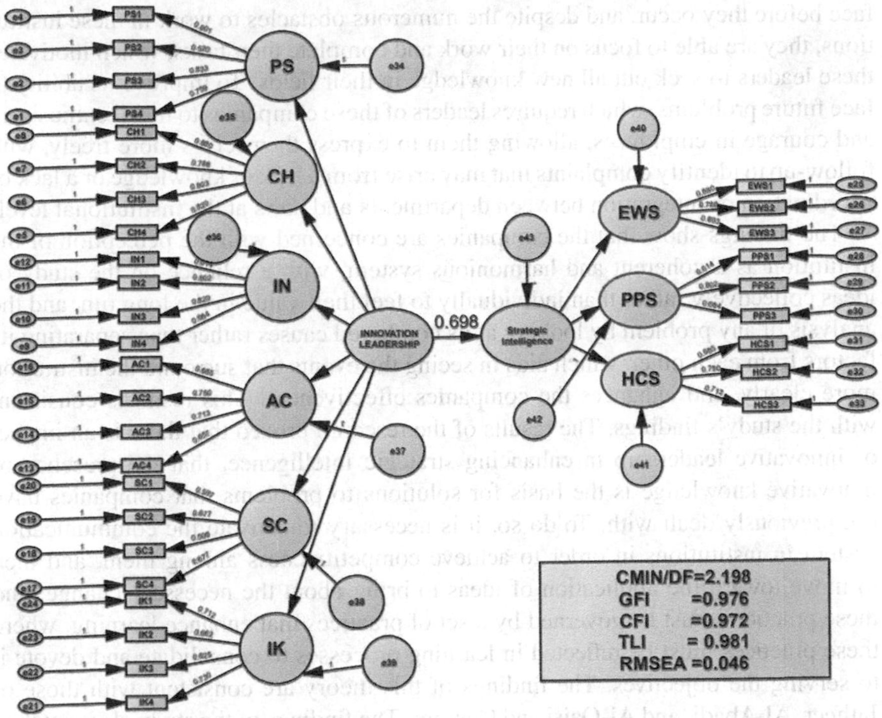

Fig. 2 SEM results of the digital marketing effect on mental image

and Ability to communicate (β = 0.093, $t = 1.557, p = 0.061$). However, the results indicated that the highest impact was for innovative knowledge (β = 0.259, $t = 5.540, p = 0.000$), followed by problems solving and decisions making (β = 0.214, $t = 5.335, p = 0.00$), and finally the lowest impact was for stimulating creativity (β = 0.124, $t = 2.446, p = 0.016$).

6 Discussion and Recommendations

Leadership in insurance companies in Jordan has features related to creative leadership with the ability to solve problems and make decisions, and what distinguishes it is also its active role in taking the initiative, and to achieve all the features, leadership needs knowledge because knowledge represents strength in the institution, and knowledge needs it all the operations that the company wants to accomplish, in addition, the communication system in the companies plays a major role in facilitating the delivery of knowledge that enhances the possibility of solving problems and making decisions, Leaders have the ability to anticipate problems that their institutions may face before they occur, and despite the numerous obstacles to work in these institutions, they are able to focus on their work and complete their tasks, which motivates these leaders to seek out all new knowledge in their fields. To improve its ability to face future problems, which requires leaders of these companies to instill enthusiasm and courage in employees, allowing them to express themselves more freely, with follow-up to identify complaints that may arise from a lack of knowledge or a lack of coordination or integration between departments and units at the institutional level.

The findings show that the companies are concerned with the perception of the institution as a coherent and harmonious system, with a reliance on the study of ideas collectively rather than individually to feel their value in the long run, and the analysis of any problem by looking at its combined causes rather than separating its factors from each other, which aids in seeing the events that surround the institution more clearly, and enhances the companies effectiveness. This result is consistent with the study's findings. The results of the research proved that there is an impact of innovative leadership in enhancing strategic intelligence, that the presence of innovative knowledge is the basis for solutions to problems that companies have not previously dealt with. To do so, it is necessary to activate the communication system in institutions in order to achieve competitiveness among them, and then to move toward the application of ideas to bring about the necessary change, and these practices must be governed by a set of practices that enhance learning, where these practices must be reflected in learning processes to consolidate and devote it to serving the objectives. The findings of this theory are consistent with those of Jatheer, Al-Abadi, and Al-Qaisi and Qassem. The findings of the study also matched those of Al-Quwaz, who used the Asia Company in Iraq as a case study to validate the impact of strategic knowledge in boosting strategic intelligence.

References

Abdali, S., Abdel-Al, M. S., & Al-Tayeb, A.-A. (2012). The role of information and communication technology in enhancing strategic intelligence, an exploratory study of a sample of directors of departments and administrative units in Al-Salam Hospital in Mosul, the eleventh annual practical conference, in business intelligence and knowledge economy, College Economics and Administrative Sciences, Public Institutions in Zaytuna.

Ahmad, S., Zulkurnain, N., & Khairushalimi, F. (2016). Assessing the validity and reliability of a measurement model in structural equation modeling (SEM). *British Journal of Mathematics & Computer Science, 15*(3), 1–8. https://doi.org/10.9734/BJMCS/2016/25183

Ahmadi, M., Baei, F., Hosseini-Amiri, S.-M., Moarefi, A., Suifan, T.S., & Sweis, R. (2020). Proposing a model of manager's strategic intelligence, organization development, and entrepreneurial behavior in organizations. *Journal of Management Development*, ahead-of-print No. ahead-of-print.

Al Kurdi, B., Alshurideh, M., & Al Afaishat, T. (2020). Employee retention and organizational performance: Evidence from banking industry. *Management Science Letters, 10*(16), 3981–3990.

Al Mehrez, A. A., Alshurideh, M., Al Kurdi, B., & Salloum, S. A. (2020). Internal factors affect knowledge management and firm performance: A systematic review. In *International Conference on Advanced Intelligent Systems and Informatics* (pp. 632–643). Springer, Cham. (2020, October).

Al Naqbi, E., Alshurideh, M., AlHamad, A., Al Kurdi, B. (2020). The impact of innovation on firm performance: A systematic review. *International Journal of Innovation, Creativity and Change, 14*(5), 31–58.

Al-Quran, A. Z., Alhalalmeh, M. I., Eldahamsheh, M. M., Mohammad, A. A., Hijjawi, G. S., Almomani, H. M., & Al-Hawary, S. I. (2020). Determinants of the green purchase intention in Jordan: The moderating effect of environmental concern. *International Journal of Supply Chain Management, 9*(5), 366–371.

Al Suwaidi, F., Alshurideh, M., Al Kurdi, B., & Salloum, S. A. (2020). The impact of innovation management in SMEs performance: a systematic review. In *International Conference on Advanced Intelligent Systems and Informatics* (pp. 720–730). Springer, Cham. (2020, October).

Alameeri, K., Alshurideh, M., Al Kurdi, B., & Salloum, S. A. (2020). The effect of work environment happiness on employee leadership. In *International Conference on Advanced Intelligent Systems and Informatics* (pp. 668–680). Springer, Cham. (2020, October).

Al-Baz, A. (2012). The role of creative leadership in crisis management. *Al-Nahda Journal, 11*(3), 54–82.

Al-Dhuhouri, F. S., Alshurideh, M., Al Kurdi, B., & Salloum, S. A. (2020). Enhancing our understanding of the relationship between leadership, team characteristics, emotional intelligence and their effect on team performance: A Critical Review. In *International Conference on Advanced Intelligent Systems and Informatics* (pp. 644–655). Springer, Cham. (2020, October).

Alhalalmeh, M. I., Almomani, H. M., Altarifi, S., Al-Quran, A. Z., Mohammad, A. A., & Al-Hawary, S. I. (2020) The nexus between corporate social responsibilty and organizational performance in Jordan: The mediating role of organizational commitment and organizational citizenship behavior. *Test Engineering and Management, 83*(July), 6391–6410.

Al-Hawary, S. I. (2009). The effect of the leadership style on the effectiveness of the organization: A field study at Zarqa Private University. *The Egyptian Journal for Commercial Studies, 33*(1), 361–393.

Al-Hawary, S. I., & Al-Hamwan, A. (2017). Environmental analysis and its impact on the competitive capabilities of the commercial banks operating in Jordan. *International Journal of Academic Research in Accounting, Finance and Management Sciences, 7*(1), 277–290.

Al-Hawary, S. I., & Hadad, T. F. (2016). The effect of strategic thinking styles on the enhancement competitive capabilities of commercial banks in Jordan. *International Journal of Business and Social Science, 7*(10), 133–144.

Al-Hawary, S. I., & Ismael, M. (2010). The effect of using information technology in achieving competitive advantage strategies: A field study on the Jordanian pharmaceutical companies. *Al Manara for Research and Studies, 16*(4), 196–203.

Al-Hawary, S. I., AL-Awawdeh, W., & Abden, M. A. (2012). The impact of the leadership style on organizational commitment: A field study on Kuwaiti telecommunications companies. *ALEDARI*, (130), 53–102.

Al-Hawary, S. I., Al-Qudah, K., Abutayeh, P., Abutayeh, S., Al-Zyadat, D. (2013). The impact of internal marketing on employee's job satisfaction of commercial banks in Jordan. *Interdisciplinary Journal of Contemporary Research in Business*, 4(9), 811–826.

Al-Hawary, S. I. S., & Alhajri, T. M. S. (2020).Effect of electronic customer relationship management on customers' electronic satisfaction of communication companies in Kuwait. *Calitatea, 21*(175), 97–102.

Al-Hawary, S. I. S., & Mohammed, A. K. (2017). Impact of team work traits on organizational citizenship behavior from the viewpoint of the employees in the education directorates in North Region of Jordan. *Global Journal of Management and Business, 17*(2-A), 23–40.

Al-Hawary, S. I. S., & Obiadat, A. A. (2021). Does mobile marketing affect customer loyalty in Jordan? *International Journal of Business Excellence, 23*(2), 226–250.

Al-Hawary, S. I. S., Mohammad, A. S., Al-Syasneh, M. S., Qandah, M. S. F., &Alhajri, T. M. S. (2020a). Organisational learning capabilities of the commercial banks in Jordan: do electronic human resources management practices matter? *International Journal of Learning and Intellectual Capital, 17*(3), 242–266.

Al-Hawary, S. I. S., Mohammad, A. S., Al-Syasneh, M. S., Qandah, M. S. F., & Alhajri, T. M. S. (2020b). Organisational learning capabilities of the commercial banks in Jordan: Do electronic human resources management practices matter? *International Journal of Learning and Intellectual Capital, 17*(3), 242–266.
Al-Hawary, S. I., & Al-Syasneh, M. S. (2020). Impact of dynamic strategic capabilities on strategic entrepreneurship in presence of outsourcing of five stars hotels in Jordan. *Business: Theory and Practice, 21*(2), 578–587.
Allahow, T. J. A. A., Al-Hawary, S. I. S., & Aldaihani, F. M. F. (2018). Information technology and administrative innovation of the central agency for information technology in Kuwait. *Global Journal of Management and Business, 18*(11-A), 1–16.
Allozi, A., Alshurideh, M., AlHamad, A., & Al Kurdi, B. (2022). Impact of transformational leadership on the job satisfaction with the moderating role of organizational commitment: Case of UAE and Jordan manufacturing companies. *Academy of Strategic Management Journal, 21*, 1–13.
Al-Lozi, M., Almomani, R. Z., & Al-Hawary, S. I. (2018a). Talent management strategies as a critical success factor for effectiveness of human resources information systems in commercial banks working in Jordan. *Global Journal of Management and Business Research: A Administration and Management, 18*(1), 30–43.
Al-Lozi, M., Almomani, R. Z., & Al-Hawary, S. I. (2017). Impact of talent management on achieving organizational excellence in Arab Potash company in Jordan. *Global Journal of Management and Business Research: A Administration and Management, 17*(7), 15–25.
Al-Nady, B. A., Al-Hawary, S. I., & Alolayyan, M. (2013). Strategic management as a key for superior competitive advantage of sanitary ware suppliers in Kingdom of Saudi Arabia. *International Journal of Management and Information Technology, 7*(2), 1042–1058.
Al-Nady, B. A., Al-Hawary, S. I., & Alolayyan, M. (2016). The role of time, communication, and cost management on project management success: An empirical study on sample of construction projects customers in Makkah City, Kingdom of Saudi Arabia. *International Journal of Services and Operations Management, 23*(1), 76–112.
Al-Nashar, A. I. (2018). The role of creative leadership in promoting organizational health in the Ministry of Interior and National Security in the southern governorates of Palestine. Unpublished Master's Thesis, Public Institutions in Al-Aqsa.
AlShehhi, H., Alshurideh, M., Al Kurdi, B., & Salloum, S. A. (2020). The impact of ethical leadership on employees performance: A systematic review. In *International Conference on Advanced Intelligent Systems and Informatics* (pp. 417–426). Springer, Cham. (2020, October).
Sheikhly, A. R. I. (2015). The impact of strategic intelligence on decision-making methods. *Journal of Economic and Administrative Sciences, 11*(85), 1–18.
Alshurideh, M. (2022). Does electronic customer relationship management (E-CRM) affect service quality at private hospitals in Jordan? *Uncertain Supply Chain Management, 10*(2), 1–8.
Alsuwaidi, M., Alshurideh, M., Al Kurdi, B., & Salloum, S. A. (2020). Performance appraisal on employees' motivation: a comprehensive analysis. In *International Conference on Advanced Intelligent Systems and Informatics* (pp. 681–693). Springer, Cham. (2020).
AlTaweel, I. R., & Al-Hawary, S. I. (2021). The mediating role of innovation capability on the relationship between strategic agility and organizational performance. *Sustainability, 13*(14), 7564.
Alyammahi, A., Alshurideh, M., Al Kurdi, B., & Salloum, S. A. (2020). The impacts of communication ethics on workplace decision making and productivity. In *International Conference on Advanced Intelligent Systems and Informatics* (pp. 488–500). Springer, Cham. (2020, October).
Alzoubi, H., Alshurideh, M., Kurdi, B., Akour, I., & Aziz, R. (2022). Does BLE technology contribute towards improving marketing strategies, customers' satisfaction and loyalty? The role of open innovation. *International Journal of Data and Network Science, 6*(2), 449–460.
Al-Zubaidi, Lamia Salman Abd Ali and Al-Amouri, Kawkab Aziz Hamoudi. (2015). The impact of focusing on creativity engines in crisis management, published research paper. *Journal of Management and Economics, 38*(103).

Basadur, M. (2014). Leading others to think innovatively together: Creative leadership. *The Leadership Quarterly, 15*, 103–121.
Bilton, D. (2010). Manageable creativity. *International Journal of Cultural Policy, 16*(3), 255–269.
de Leeuw, E., Hox, J., Silber, H., Struminskaya, B., & Vis, C. (2019). Development of an international survey attitude scale: Measurement equivalence, reliability, and predictive validity. *Measurement Instruments for the Social Sciences, 1*(1), 9. https://doi.org/10.1186/s42409-019-0012-x
Hair, J. F., Babin, B. J., & Krey, N. (2017). Covariance-based structural equation modeling in the journal of advertising: Review and recommendations. *Journal of Advertising, 46*(1), 163–177. https://doi.org/10.1080/00913367.2017.1281777
Harahsheh, A., Houssien, A., Alshurideh, M., & AlMontaser, M. (2021). The effect of transformational leadership on achieving effective decisions in the presence of psychological capital as an intermediate variable in Private Jordanian universities in light of the Corona Pandemic. In *The Effect of Coronavirus Disease (COVID-19) on Business Intelligence* (Vol. 334, pp. 221–243).
Howard, M. C. (2018). The convergent validity and nomological net of two methods to measure retroactive influences. *Psychology of Consciousness: Theory, Research, and Practice, 5*(3), 324–337. https://doi.org/10.1037/cns0000149
Kaur, S., Gupta, S., Singh, S. K., & Perano, M. (2019). Organizational ambidexterity through global strategic partnerships: A cognitive computing perspective. *Technological Forecasting and Social Change, 145*, 43–54.
Kurdi, B., Alshurideh, M., & Alnaser, A. (2020). The impact of employee satisfaction on customer satisfaction: Theoretical and empirical underpinning. *Management Science Letters, 10*(15), 3561–3570.
Lee, K., Azmi, N., Hanaysha, J., Alshurideh, M., & Alzoubi, H. (2022a). The effect of digital supply chain on organizational performance: An empirical study in Malaysia manufacturing industry. *Uncertain Supply Chain Management, 10*(2), 1–16.
Lee, K., Ramiz, P., Hanaysha, J., Alzoubi, H., & Alshurideh, M. (2022b). Investigating the impact of benefits and challenges of IOT adoption on supply chain performance and organizational performance: An empirical study in Malaysia. *Uncertain Supply Chain Management, 10*(2), 1–14.
Mohammad, A. A., Alshura, M. S., Al-Hawary, S. I. S., Al-Syasneh, M. S., & Alhajri, T. M. (2020). The influence of Internal Marketing Practices on the employees' intention to leave: A study of the private hospitals in Jordan. *International Journal of Advanced Science and Technology, 29*(5), 1174–1189.
Muslim, T. H., & Qader, A. (2015). *The Impact of Strategic Intelligence on Leadership, unpublished MA thesis, presented for the purpose of obtaining a degree*. Public Institutions in Az-Zahr, Gaza.
Nuseir, M. T., Aljumah, A., & Alshurideh, M. T. (2021). How the Business Intelligence in the New Startup Performance in UAE During COVID-19: The Mediating Role of Innovativeness. In *The Effect of Coronavirus Disease (COVID-19) on Business Intelligence, 334*, 63–79.
Odeh, R. B. M., Obeidat, B. Y., Jaradat, M. O., & Alshurideh, M. T. (2021). The transformational leadership role in achieving organizational resilience through adaptive cultures: the case of Dubai service sector. *International Journal of Productivity and Performance Management*, Vol. ahead-of-print No. ahead-of-print. https://doi.org/10.1108/IJPPM-02-2021-0093
Rimkeviciene, J., Hawgood, J., O'Gorman, J., & De Leo, D. (2017). Construct validity of the acquired capability for suicide scale: factor structure, convergent and discriminant validity. *Journal of Psychopathology and Behavioral Assessment, 39*(2), 291–302. https://doi.org/10.1007/s10862-016-9576-4
Sekaran, U., & Bougie, R. (2016). *Research methods for business: A skill-building approach* (7th ed.). Wiley.
Shi, D., Lee, T., & Maydeu-Olivares, A. (2019). Understanding the model size effect on SEM fit indices. *Educational and Psychological Measurement, 79*(2), 310–334. https://doi.org/10.1177/0013164418783530

Sung, K.-S., Yi, Y. G., & Shin, H.-I. (2019). Reliability and validity of knee extensor strength measurements using a portable dynamometer anchoring system in a supine position. *BMC Musculoskeletal Disorders, 20*(1), 1–8. https://doi.org/10.1186/s12891-019-2703-0

Sweiss, N., Obeidat, Z. M., Al-Dweeri, R. M., Mohammad Khalaf Ahmad, A., M. Obeidat, A., & Alshurideh, M. (2021). The moderating role of perceived company effort in mitigating customer misconduct within Online Brand Communities (OBC). *Journal of Marketing Communications*, 1–24.

Tham, K., & Kim, M. (2002). Towards Strategies intelligence with Anthology based Enterprise Modeling & ABS. In *Proceeding of the IBER Conference,2002 .The Society of Competitive Intelligence Professionals, 1999, An Introduction to Competitive Intelligence.*

Wang, Y. A., & Rhemtulla, M. (2021). Power analysis for parameter estimation in structural equation modeling: A discussion and tutorial. *Advances in Methods and Practices in Psychological Science, 4*(1), 1–17. https://doi.org/10.1177/2515245920918253

Yanez, S., Uruburu, A., Moreno, A., & Lumbreras, J. (2019). The sustainability report as an essential tool for the holistic and strategic vision of higher education institutions. *Journal of Cleaner Production, 207*, 57–66.

Zhang, Q., Siribanpitak, P., & Charoenkul, N. (2018). Creative leadership strategies for primary school principals to promote teachers' creativity in Guangxi, China. *Kasetsart Journal of Social Sciences.*

The Impact of Organizational Innovation Capabilities on Sustainable Performance: The Mediating Role of Organizational Commitment

Mohammed Saleem Khlif Alshura, Faisal Khaleefah Jasem Alsabah, Raed Ismael Ababneh, Muhammad Turki Alshurideh, Mohammad Issa Ghafel Alkhawaldeh, Faraj Mazyed Faraj Aldaihani, Ayat Mohammad, Sulieman Ibraheem Shelash Al-Hawary, and Anber Abraheem Shlash Mohammad

Abstract The major aim of the study was to examine the impact of organizational ambidexterity on sustainable performance in the presence of organizational commitment as a mediator. Therefore, it focused on Kuwaiti Industrial sector. Data were primarily gathered through self-reported questionnaires created by Google Forms which were distributed to a sample of (450) administrators via email. In total, (402)

M. S. K. Alshura · F. K. J. Alsabah
Management Department, Faculty of Money and Management, the World Islamic Science University, P.O. Box1101, Amman 11947, Jordan

R. I. Ababneh
Policy, Planning, and Development Program, Department of International Affairs, College of Arts and Sciences, Qatar University, 2713 Doha, Qatar

M. T. Alshurideh
Department of Marketing, School of Business, The University of Jordan, Amman 11942, Jordan
e-mail: m.alshurideh@ju.edu.jo; malshurideh@sharjah.ac.ae

Department of Management, College of Business, University of Sharjah, 27272 Sharjah, United Arab Emirates

M. I. G. Alkhawaldeh
Directorates of Building and Land Tax, Ministry of Local Administration, Zarqa Municipality, Amman, Jordan

F. M. F. Aldaihani
Kuwait Civil Aviation, Ishbiliyah Bloch 1, Street 122, Home 1, Safat, Kuwait

A. Mohammad
Business and Finance Faculty, The World Islamic Science and Education University, Postal Code 11947, P.O Box 1101, Amman, Jordan

S. I. S. Al-Hawary (✉)
Department of Business Administration, School of Business, Al Al-Bayt University, P.O BOX 130040, Al-Mafraq 25113, Jordan
e-mail: dr_sliman73@aabu.edu.jo

M. Alshurideh et al. (eds.), *The Effect of Information Technology on Business and Marketing Intelligence Systems*, Studies in Computational Intelligence 1056,
https://doi.org/10.1007/978-3-031-12382-5_78

responses were received including (10) invalid to statistical analysis due to uncompleted or inaccurate. Hence, the final sample contained (392) responses suitable to analysis requirements that were formed a response rate of (87.11%). structural equation modeling (SEM) was conducted to test hypotheses. The results showed that organizational commitment mediated the relationship between organizational capabilities and sustainable performance. Based on the research results the researchers recommend the oil sector to follow up the information related to the latest technological methods used in its sector, providing attention to setting the plans that aim at improving the human resources and train them to support exploration and exploitation skills.

Keywords Authentic leadership · Sustainable performance · Knowledge ability · Jordan customs department

1 Introduction

Sustainable performance concept is based on development concept (social and economic development in line with environmental constraints), needs concept (resources redistribution to ensure the life quality for all) and future generations' concept (the possibility of using resources in the long term to ensure the necessary quality of life for future generations). Where the essence of sustainable performance concept stems from the triple bottom line concept, which includes a balance between three pillars of sustainability, namely environmental sustainability, which focuses on maintaining the quality of the environment necessary to carry out economic activities and the quality of people's lives, and social sustainability, which strives to ensure human rights and equality, and preserve cultural identity., respect for cultural diversity, ethnicity and religion, and economic sustainability necessary to maintain the natural, social and human capital necessary for income and standards of living (AlTaweel & Al-Hawary, 2021; Al-Hawary & Al-Syasneh, 2020; Al-Hawary & Mohammed, 2017; Al-Hawary et al., 2012; Al-Hawary, 2009). Ambidexterity has become a popular concept for understanding certain aspects of organizational planning, and it plays an important role in competition because it supports organizational growth and enhances organizations' future vision for survival in addition to the challenges faced by organizations that have forced them to adopt organizational ambidexterity as an integral part of their strategy, as it is considered essential that be aware of these challenges in the local and global business environment (Al-Nady et al., 2013, 2016; Al-Hawary & Aldaihani, 2016; Allahow et al., 2018; Mohammad et al., 2020; Al-Hawary & Hadad, 2016; Al-Hawary & Obiadat, 2021; Alhalalmeh et al., 2020; Al-Hawary & Alhajri, 2020; Alshurideh et al., 2017; Alolayyan et al., 2018; Metabis & Al-Hawary, 2013). This highlights the human element role in achieving this and what it requires of the employee's stay in the organization in order to achieve the goals set

A. A. S. Mohammad
Marketing Department, Faculty of Administrative and Financial Sciences, Petra University, B.O Box: 961343, Amman 11196, Jordan

and maintains the presence of the workforce committed to the organization in order to achieve a high level of organizational commitment (Al-Hawary & Alajmi, 2017; Al-Hawary & Nusair, 2017; Al-Hawary et al., 2020; Al-Lozi et al., 2017, 2018). This study is an objective attempt to shed light on the rapid development of modern administrative concepts, especially the organizational ambidexterity concept, which has become an effective management method and its role in promoting sustainable performance, as well as achieving the oil companies' goals through theoretical knowledge to achieve a clear understanding and define the concepts and objectives of the variables under discussion. Study importance also lies in applying the study to one of the most important economic sectors in Kuwait capital, which is the petroleum sector. The petroleum sector in the Kuwait capital is one of the most important economic sectors that contribute to the gross domestic production. As oil companies operate in a very dynamic and complex environment, and this is what prompted them to adapt to work in the administrative, economic, technological and societal fields, they are obliged to respond quickly to environmental changes, because the lack of a rapid response may lead to damage their work and their retreat from their competitors. Thus, these companies must effectively manage the changing business conditions, trying to retain their customers as well as attract new ones (Alshurideh, 2016, 2022; Alkitbi et al., 2020; Alsharari & Alshurideh, 2020), and this is what made the capabilities and skills that help them to succeed through achieving organizational ambidexterity of great importance to them. It is also necessary to ensure, through this study, a high level of organizational commitment that may contribute to enhancing sustainable performance in the Kuwaiti petroleum sector (Oyewobi et al., 2019).

2 Literature Review and Hypotheses Development

2.1 *Organizational Ambidexterity*

Organizational ambidexterity broadly refers to the organization's ability to pursue two different things at the same time, that is, the organization can exploit existing competencies in addition to exploring new opportunities with equal skill based on adapting to changing environmental conditions and obtaining a sustainable competitive advantage (Altamony et al., 2012; Obeidat et al., 2021; Shamout et al., 2022). It also depends on both exploiting existing resources and exploring new opportunities in order to make an innovation (Al Naqbi et al., 2020; Alameeri et al., 2021). The concept of ambidexterity is the ability to use both hands equally effectively. In the field of organizations ambidexterity is related to balancing the capabilities that exist in the organization and the way to use them, and at the same time searching for other possibilities, which are exploitation and exploration. Booth-Kewley et al. (2017) emphasize the same idea, noting that from a terminological perspective, the word "ambidextrous" means the ability to use both hands equally at the same time. Kafetz opoulos (19,2020) defines organizational ambidexterity as the organization ability to implement and adopt incremental and revolutionary changes, while Du et al., (2020, 50) defined them as organizations that are able to exploit existing competencies as well as explore new opportunities. Gürlek (2021) defined organizational ambidexterity as the organization's ability to implement new changes in its environment in an orderly and correct manner. Organizational ambidexterity is also defined by Bresciani et al. (2018) as the organization ability to exploit existing competencies and explore new opportunities at the same time. For long-term survival, the organization needs to exploit existing capabilities and explore alternative opportunities (Martin et al., 2019). The ability to do both activities simultaneously is called organizational ambidexterity. It is the ability to combine contradictory, inconsistent and sometimes incompatible organizational elements such as exploratory and exploitative activities, strategies, structures, processes, mindsets, behaviors, etc. (Posch & Garaus, 2020). At the core of ambidexterity, there are two different activities–exploration and exploitation. The exploitation purpose is to improve and increase performance. While exploration goal, in turn, is to look for new opportunities and experiment. Both exploration and exploitation are necessary if the company aim to remain sustainable and to continue for the long term (Alshurideh et al., 2020; Alshurideh et al., 2022; Lee et al., 2022a, 2022b). But at the same time, many organizations struggle to strike a balance between exploration and exploitation, discovery and exploitation are the most important capabilities of organizational ambidexterity (López-Sánchez & Santos-Vijande, 2020).

2.2 *Sustainable Performance*

Sustainability is a broad term, as it draws on politics, economics, philosophy and other social sciences in addition to the natural sciences. Sustainability skills and environmental awareness are a priority in many organizations, as organizations seek to comply with new legislation related to urban planning, environmental consulting (the built and natural environment), and agriculture, not for profit. In addition to corporate strategies, health assessment and planning, and even in law and decision-making (Alzoubi et al., 2022; Ben-Abdallah et al., 2022; Ögmundarson et al., 2020). Rosenberg et al. (2021) indicates that sustainability practices support health, environmental, human and economic practices, where the sustainability principle assumes that resources are limited and must be used wisely, taking into account long-term priorities and the consequences of the ways in which resources are used.

Pislaru et al. (2019) defined sustainable performance as the competitiveness of an organization that ensures a sustainable presence in the marketplace and is able to reconcile all partners expectations, create value for shareholders and customers, and recognize the sustainable development importance. On the other hand, Huo et al. (2019) viewed sustainable performance as the alignment of financial, environmental and social objectives in delivering an organization's core business activities in order to maximize value, including operational excellence, risk management, product innovation, growth, and governance to transform business performance and radically improve it and create stakeholders value. Among the types of sustainable performance on which previous studies focused.

Among the most prominent types of sustainable performance mentioned in previous studies, **Economic performance**: It is a practical and actionable toolkit that designs strategies to act on unexpected and unpredictable economic events (Al-Jarrah et al., 2012; Poltronieri et al., 2019), **Social performance** is the effective translation of organization's mission into practical reality in line with accepted social values related to improving the individuals lives in society and giving them the opportunity to work and develop. **Environmental Performance**: Ojo and Fauzi (2020) find that organizations' environmental performance helps individual governments assess progress in achieving their environmental goals; In addition to promoting continuous policy dialogue and peer learning; and stimulate greater governments accountability to each other and to public opinion.

2.3 *Organizational Commitment*

Organizational commitment takes an important role in employees' retention and turnover in organizations, and organizational commitment appears when employees are committed to their work in the organization and are motivated to achieve high performance levels (Al Kurdi et al., 2020; Alameeri et al., 2020; Al-Hawary & Alajmi, 2017; Kurdi et al., 2020). Karami et al. (2017) noted that workplace commitment can be divided into different aspects, where an employee may be committed to the job, profession, goals, teams, leaders, or organization, However, it is well recognized that employees develop more than one commitment related to work, and while all of these can be seen in the workplace and affect the overall employee commitment in the workplace, they all have their own characteristics. While Kaplan and Kaplan (2018) assure that just as commitment is difficult to comprehensively define, the same difficulty lies in organizational commitment, however, it has been possible to find common factors for different definitions, Alhalalmeh et al. (2020) define organizational commitment as the organization member psychological view towards his association with the organization in which he works, and organizational commitment plays a vital role in determining whether the employee will stay with the organization for a longer period of time and work enthusiastically towards achieving organizational goals. As for Imamoglu et al. (2019) viewed organizational commitment as the employee's relationship size with the organization and the extent of the strength of this relationship, and organizational commitment is related to many things, including the employee's belief in the organization's goals and values, and the employee's position in making effort for the organization and the desire to stay.

Novitasari et al. (2020) define organizational commitment as the employee's attachment to the organization based on the intention to remain in it; adherence to the organization values and goals; And the willingness to make an extra effort in its favor, that is, for individuals to look at the extent to which their values and goals are related to the organization values as part of the organizational commitment, so it is considered the link between the individual employee and the organization. Labrague et al. (2018) argue that the important theories of organizational commitment are a Three Component Model (TCM), and according to this theory, there are three distinct components of organizational commitment: Emotional commitment is the emotional attachment that an employee has to the organization, ongoing commitment which is the employee's belief that leaving the organization will be costly. The last dimension of the organizational commitment model is normative commitment, which is seen as the obligation sense to keep going because it is the right thing to do (AlShehhi et al., 2020; Alsuwaidi et al., 2020; Gopinath et al., 2020; Allozi et al., 2022).

2.4 *Organizational Ambidexterity, Sustainable Performance and Organizational Commitment*

Kraner (2018) indicates that there are many previous studies that proved that organizational ambidexterity has a significant impact on organizational excellence, where the researcher confirmed that organizational ambidexterity with its complete concept and capabilities can contribute significantly to increasing the organization growth despite the difficulty of achieving it on a practical level. As Backović and Djurić (2019) pointed out that the achieving organizational ambidexterity at the practical level lies in "balancing", finding and creating a balance between discovery and exploitation in an equal manner at the same time maintaining the highest possible performance level, while Campanella et al. (2020) agreed with the same opinion, pointing out that organizational ambidexterity appears to be a management term of some kind, but in reality there is a great conflict between the organizational ambidexterity abilities in terms of exploration and exploitation. Limaj and Bernroider (2019) argue that organizational ambidexterity has a positive effect on organizational excellence, where researchers found that ambidexterity had a positive impact on the organizations growth by creating an appropriate balance in the adaptation attempts in a turbulent environment, The previous literature–Felício et al. (2019) and Rialti et al. (2019) highlights some of achieving ambidexterity in organizations outcomes, where organizational ambidexterity can lead to better performance, better founders, long-term success and sustainable competitive advantage. Sustainability is now viewed as a key business imperative that leads to cost savings and value creation. While many organizations already have a sustainability strategy in place, there is a growing need to make this business more focused, in terms of operational excellence, risk management, and product innovation, growth, and governance (Almaazmi et al., 2020; Al

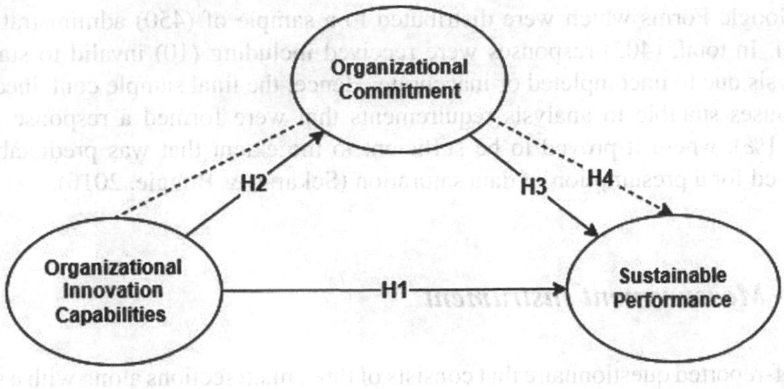

Fig. 1 Research model

Suwaidi et al., 2020; Belhadi et al., 2021). Rialti et al. (2019) study was found that organizational commitment has a positive and significant impact on economic performance, and organizational commitment has a positive and important impact on environmental performance, and a positive and significant impact on social performance, and organizational commitment has a positive and important impact on the company's performance in sustainability. Accordingly, the study hypothesis can be formulated as follows:

There is a significant impact of organizational ambidexterity on achieving sustainable performance in the presence of organizational commitment in oil companies in the capital of Kuwait.

3 Study Model

See (Fig. 1).

4 Methodology

4.1 Population and Sample Selection

A qualitative method based on a questionnaire was used in this study for data collection and sample selection. The major aim of the study was to examine the impact of organizational ambidexterity on sustainable performance in the presence of organizational commitment as a mediator. Therefore, it focused on Kuwaiti Industrial sector. Data were primarily gathered through self-reported questionnaires created by Google Forms which were distributed to a sample of (450) administrators via email. In total, (402) responses were received including (10) invalid to statistical analysis due to uncompleted or inaccurate. Hence, the final sample contained (392) responses suitable to analysis requirements that were formed a response rate of (87.11%), where it proved to be sufficient to the extent that was predictable and allowed for a presumption of data saturation (Sekaran & Bougie, 2016).

4.2 *Measurement Instrument*

A self-reported questionnaire that consists of three main sections along with a section regarding control variables was used as the measurement instrument. Control variables considered as categorical measures were composed of gender, age group, educational level, and experience. The three main sections were dealt with a five-point Likert scale (from 1 = strongly disagree to 5 = strongly agree). The first section contained (15) items to measure organizational ambidexterity capabilities based on (López-Sánchez & Santos-Vijande, 2020). These items were distributed into dimensions as follows: five items dedicated for measuring acknowledgment, five items dedicated for measuring exploitation, and five items dedicated for measuring re-engineering. The second section related to organizational commitment variable which contained eight items developed according to (Al-Hawary & Alajmi, 2017). Whereas the third section included (15) items developed to measure sustainable performance according to what was pointed by (Ojo & Fauzi, 2020). This variable was a second-order construct divided into three first-order constructs. Economic performance measured by five items, environmental performance measured through five items, and social performance measured using five items.

5 Findings

5.1 *Measurement Model Evaluation*

This study was conducted structural equation modeling (SEM) to test hypotheses, which represents a contemporary statistical technique for testing and estimating the relationship between factors and variables (Wang & Rhemtulla, 2021). Accordingly, the reliability and validity of the constructs were tested using confirmatory factor analysis (CFA) through the statistical program AMOSv24. Table 1 summarizes the results of convergent and discriminant validity, as well the indicators of reliability.

Table 1 shows that the standard loading values for the individual items were within the domain (0.582–0.902), these values greater than the minimum retention of the elements based on their standard loads (Al-Lozi et al., 2018; Sung et al., 2019). Average variance extracted (AVE) is a summary indicator of the convergent validity

Table 1 Results of validity and reliability tests

Constructs	1	2	3	4	5	6	7
1. Acknowledgment	**0.747**						
2. Exploitation	0.312	**0.768**					
3. Re-Engineering	0.297	0.344	**0.748**				
4. Organizational Commitment	0.415	0.281	0.461	**0.760**			
5. Economic Performance	0.364	0.384	0.320	0.566	**0.761**		
6. Environmental Performance	0.481	0.465	0.405	0.502	0.468	**0.777**	
7. Social Performance	0.375	0.597	0.554	0.495	0.566	0.591	**0.742**
VIF	1.805	1.225	1.648	–	–	–	–
Loadings range	0.643–0.823	0.671–0.902	0.705–0.813	0.582–0.871	0.705–0.824	0.726–0.825	0.681–0.792
AVE	0.558	0.589	0.559	0.578	0.579	0.603	0.551
MSV	0.415	0.387	0.501	0.495	0.461	0.513	0.466
Internal consistency	0.860	0.875	0.861	0.913	0.870	0.883	0.857
Composite reliability	0.862	0.878	0.863	0.915	0.873	0.884	0.859

Note All correlations have a significance level less than 0.05, bold fonts refer to the square root of AVE of constructs that must be above 0.50 (Howard, 2018). The results indicate that the AVE values were greater than 0.50 for all constructs, thus the used measurement model has an appropriate convergent validity. Rimkeviciene et al. (2017) suggested the comparison approach as a way to deal with discriminant validity assessment in covariance-based SEM. This approach is based on comparing the values of maximum shared variance (MSV) with the values of AVE, as well as comparing the values of square root of AVE ($\sqrt{}$AVE) with the correlation between the rest of the structures. The results show that the values of MSV were smaller than the values of AVE, and that the values of $\sqrt{}$AVE were higher than the correlation values among the rest of the constructs. Therefore, the measurement model used is characterized by discriminative validity. The internal consistency measured through Cronbach's Alpha coefficient (α) and compound reliability by McDonald's Omega coefficient (ω) was conducted as indicators to evaluate measurement model. The results listed in Table 1 demonstrated that both values of Cronbach's Alpha coefficient and McDonald's Omega coefficient were greater than 0.70, which is the lowest limit for judging on measurement reliability (de Leeuw et al., 2019).

5.2 *Structural Model*

The structural model illustrated no multicollinearity issue among predictor constructs because variance inflation factor (VIF) values are below the threshold of 5, as shown in Table 1 (Hair et al., 2017). This result is supported by the values of model fit indices shown in Fig. 1. (Fig. 2)

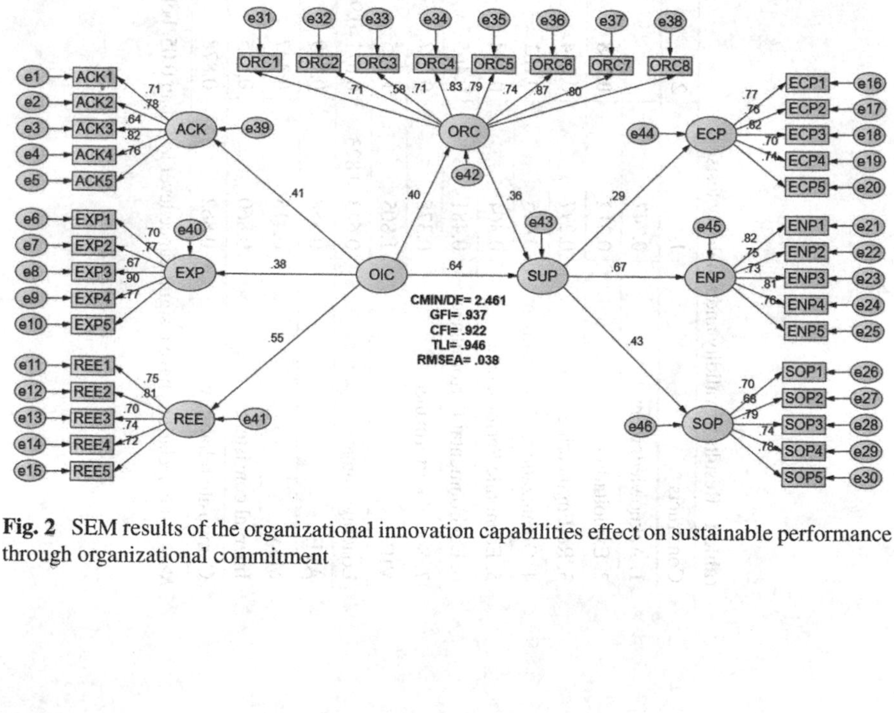

Fig. 2 SEM results of the organizational innovation capabilities effect on sustainable performance through organizational commitment

Table 2 Hypothesis testing

Relation	Direct effect		Indirect effect		Total effect		
	β	*p* value	β	*p* value	β	*t* value	*p* value
OIC → SUP	0.642***	0.000			0.642***	18.609	0.000
OIC → ORC	0.403**	0.002			0.403**	16.224	0.002
ORC → SUP	0.364***	0.000			0.364***	12.166	0.000
OIC → ORC → SUP	0.642***	0.000	0.147*	0.02	0.789*	21.547	0.02

Note OIC: Organizational Innovation Capabilities, SUP: Sustainable Performance, ORC: Organizational Commitment, * $p < 0.05$, ** $p < 0.01$, *** $p < 0.001$

The results in Fig. 1 indicated that the chi-square to degrees of freedom (CMIN/DF) was 2.461, which is less than 3 the upper limit of this indicator. The values of the goodness of fit index (GFI), the comparative fit index (CFI), and the Tucker-Lewis index (TLI) were upper than the minimum accepted threshold of 0.90. Moreover, the result of root mean square error of approximation (RMSEA) indicated to value 0.038, this value is a reasonable error of approximation because it is less than the higher limit of 0.08. Consequently, the structural model used in this study was recognized as a fit model for predicting the sustainable performance and generalization of its result (Ahmad et al., 2016; Shi et al., 2019). To verify the results of testing the study hypotheses, structural equation modeling (SEM) was used, the results of which are listed in Table 2.

The results demonstrated in Table 2 show that organizational ambidexterity have a positive direct effect on sustainable performance ($\beta = 0.642$, $t = 18.609$, $p = 0.000$) and organizational commitment($\beta = 0.403$, $t = 16.224$, $p = 0.002$). Organizational commitment has also positive direct effect on sustainable performance ($\beta = 0.364$, $t = 12.166$, $p = 0.000$). Furthermore, organizational commitment mediated the relationship between organizational capabilities and sustainable performance with total effect ($\beta = 0.789$, $t = 21.547$, $p = 0.02$) and indirect effect ($\beta = 0.147$, $p = 0.02$).

6 Discussion

The study results showed an impact of organizational ambidexterity on sustainable performance through organizational commitment, study results showed that the organizational ambidexterity strategy application achieves cost-effectiveness and helps organizations meet different customer's needs. In addition, organizational ambidexterity is critical to organizational survival and success. Also, reaching a state of organizational excellence requires providing constructive feedback, not negative criticism, and employees must be provided with constructive comments whenever needed, study results showed that organizational commitment can lead to a stable and productive workforce and help employees to increase their creativity and contribute to organizational development initiatives. Also, highly committed employees do not leave the organization because they are dissatisfied and tend to undertake difficult work activities, as committed members are usually achievers and innovative mentors with the ultimate goal of participating and improving performance. When an employee commits to an organization, many positive outcomes can be seen for the individual as well as for the organization itself. One of the advantages that it can affect the customer's well-being, additionally, increases employee's job satisfaction increases commitment. An organization can benefit from committed employees in a variety of ways, at scale, in terms of increasing their ability to influence the organization's effectiveness, employees are also less likely to leave the organization, which reduces employee turnover. Equally important, committed employees can often make things work even without good systems, and they are key to increased productivity in organizations. This result agrees with Muhammad, which confirmed the existence of a significant correlation between organizational ambidexterity and sustainable performance, it also agrees with the study of Hijjawi and Mohammad, which confirmed that organizational ambidexterity had a significant effect on organizational conflict.

7 Recommendations

Based on the research results the researchers recommend the oil sector to follow up the information related to the latest technological methods used in its sector, providing attention to setting the plans that aim at improving the human resources and train them to support exploration and exploitation skills Support the organizational culture the promote organizational commitment. Reducing the amount of wastage that cause harm to the environment to the least possible level, reducing the risks associated with work to the least possible level.

References

Ahmad, S., Zulkurnain, N., & Khairushalimi, F. (2016). Assessing the validity and reliability of a measurement model in structural equation modeling (SEM). *British Journal of Mathematics & Computer Science, 15*(3), 1–8. https://doi.org/10.9734/BJMCS/2016/25183

Al Kurdi, B., Alshurideh, M., & Al afaishat, T. (2020). Employee retention and organizational performance: Evidence from banking industry. *Management Science Letters, 10*(16), 3981–3990.

Al Naqbi, E., Alshurideh, M., AlHamad, A., & Al Kurdi, B. (2020). The impact of innovation on firm performance: A systematic review. *International Journal of Innovation, Creativity and Change, 14*(5), 31–58.

Al Suwaidi, F., Alshurideh, M., Al Kurdi, B., & Salloum, S. A. (2020). The impact of innovation management in SMEs performance: a systematic review. In *International Conference on Advanced Intelligent Systems and Informatics* (pp. 720–730). Springer, Cham. (2020, October).

Alameeri, K. A., Alshurideh, M. T., & Al Kurdi, B. (2021). The effect of covid-19 pandemic on business systems' innovation and entrepreneurship and how to cope with it: A theatrical view. In *The effect of coronavirus disease (COVID-19) on Business Intelligence* (Vol. 334, pp. 275–288).

Alameeri, K., Alshurideh, M., Al Kurdi, B., & Salloum, S. A. (2020). The effect of work environment happiness on employee leadership. In *International Conference on Advanced Intelligent Systems and Informatics* (pp. 668–680). Springer, Cham. (2020, October).

Alhalalmeh, M. I., Almomani, H. M., Altarifi, S., Al- Quran, A. Z., Mohammad, A. A., & Al-Hawary, S. I. (2020). The nexus between corporate social responsibilty and organizational performance in Jordan: The mediating role of organizational commitment and organizational citizenship behavior. *Test Engineering and Management, 83*(July), 6391–6410.

Al-Hawary, S. I. (2009). The effect of the leadership style on the effectiveness of the organization: A field study at Zarqa Private University. *The Egyptian Journal for Commercial Studies, 33*(1), 361–393.

Al-Hawary, S. I. S., & Alhajri, T. M. S. (2020). Effect of electronic customer relationship management on customers' electronic satisfaction of communication companies in Kuwait. *Calitatea, 21*(175), 97–102.

Al-Hawary, S. I. S., & Mohammed, A. K. (2017). Impact of team work traits on organizational citizenship behavior from the viewpoint of the employees in the education directorates in North Region of Jordan. *Global Journal of Management and Business, 17*(2-A), 23–40.

Al-Hawary, S. I. S., & Obiadat, A. A. (2021). Does mobile marketing affect customer loyalty in Jordan? *International Journal of Business Excellence, 23*(2), 226–250.

Al-Hawary, S. I. S., Mohammad, A. S., Al-Syasneh, M. S., Qandah, M. S. F., & Alhajri, T. M. S. (2020). Organisational learning capabilities of the commercial banks in Jordan: Do electronic human resources management practices matter? *International Journal of Learning and Intellectual Capital, 17*(3), 242–266.

Al-Hawary, S. I., & Alajmi, H. M. (2017). Organizational commitment of the employees of the ports security affairs of the state of Kuwait: The impact of human recourses management practices. *International Journal of Academic Research in Economics and Management Sciences, 6*(1), 52–78.

Al-Hawary, S. I., & Aldaihani, F. M. (2016). Customer relationship management and innovation capabilities of Kuwait airways. *International Journal of Academic Research in Economics and Management Sciences, 5*(4), 201–226.

Al-Hawary, S. I., & Al-Syasneh, M. S. (2020). Impact of dynamic strategic capabilities on strategic entrepreneurship in presence of outsourcing of five stars hotels in Jordan. *Business: Theory and Practice, 21*(2), 578–587.

Al-Hawary, S. I., & Hadad, T. F. (2016). The effect of strategic thinking styles on the enhancement competitive capabilities of commercial banks in Jordan. *International Journal of Business and Social Science, 7*(10), 133–144.

Al-Hawary, S. I., & Nusair, W. (2017). Impact of human resource strategies on perceived organizational support at Jordanian Public Universities. *Global Journal of Management and Business Research: A Administration and Management, 17*(1), 68–82.

Al-Hawary, S. I., AL-Awawdeh, W., & Abden, M. A. (2012). The impact of the leadership style on organizational commitment: a field study on Kuwaiti Telecommunications Companies. *ALEDARI*, (130), 53–102.

Al-Jarrah, I., Al-Zu'bi, M. F., Jaara, O., & Alshurideh, M. (2012). Evaluating the impact of financial development on economic growth in Jordan. *International Research Journal of Finance and Economics, 94*, 123–139.

Alkitbi, S. S., Alshurideh, M., Al Kurdi, B., & Salloum, S. A. (2020). Factors affect customer retention: A systematic review. In *International Conference on Advanced Intelligent Systems and Informatics* (pp. 656–667). Springer, Cham. (2020, October).

Allahow, T. J. A. A., Al-Hawary, S. I. S., & Aldaihani, F. M. F. (2018). Information technology and administrative innovation of the central agency for information technology in Kuwait. *Global Journal of Management and Business, 18*(11-A), 1–16.

Allozi, A., Alshurideh, M., AlHamad, A., & Al Kurdi, B. (2022). Impact of transformational leadership on the job satisfaction with the moderating role of organizational commitment: Case of UAE and Jordan manufacturing companies. *Academy of Strategic Management Journal, 21*, 1–13.

Al-Lozi, M., Almomani, R. Z., & Al-Hawary, S. I. (2017). Impact of talent management on achieving organizational excellence in Arab Potash company in Jordan. *Global Journal of Management and Business Research: A Administration and Management, 17*(7), 15–25.

Al-Lozi, M., Almomani, R. Z., & Al-Hawary, S. I. (2018). Talent Management strategies as a critical success factor for effectiveness of human resources information systems in commercial banks working in Jordan. *Global Journal of Management and Business Research: A Administration and Management, 18*(1), 30–43.

Almaazmi, J., Alshurideh, M., Al Kurdi, B., & Salloum, S. A. (2020). The effect of digital transformation on product innovation: a critical review. In *International Conference on Advanced Intelligent Systems and Informatics* (pp. 731–741). Springer, Cham. (2020, October).

Al-Nady, B. A., Al-Hawary, S. I., & Alolayyan, M. (2013). Strategic management as a key for superior competitive advantage of sanitary ware suppliers in Kingdom of Saudi Arabia. *International Journal of Management and Information Technology, 7*(2), 1042–1058.

Al-Nady, B. A., Al-Hawary, S. I., & Alolayyan, M. (2016). The role of time, communication, and cost management on project management success: An empirical study on sample of construction projects customers in Makkah City, Kingdom of Saudi Arabia. *International Journal of Services and Operations Management, 23*(1), 76–112.

Alolayyan, M., Al-Hawary, S. I., Mohammad, A. A., & Al-Nady, B. A. (2018). Banking service quality provided by commercial banks and customer satisfaction. A structural equation modelling approaches. *International Journal of Productivity and Quality Management, 24*(4), 543–565.

Alsharari, N. M., & Alshurideh, M. T. (2020). Student retention in higher education: the role of creativity, emotional intelligence and learner autonomy. *International Journal of Educational Management, 35*(1), 233–247.

AlShehhi, H., Alshurideh, M., Al Kurdi, B., & Salloum, S. A. (2020). The impact of ethical leadership on employees performance: A systematic review. In *International Conference on Advanced Intelligent Systems and Informatics* (pp. 417–426). Springer, Cham. (2020, October).

Alshurideh, M. (2016). Scope of customer retention problem in the mobile phone sector: A theoretical perspective. *Journal of Marketing and Consumer Research, 20*(2), 64–69.

Alshurideh, M. (2022). Does electronic customer relationship management (E-CRM) affect service quality at private hospitals in Jordan? *Uncertain Supply Chain Management, 10*(2), 325–332.

Alshurideh, M. T., Al Kurdi, B., Alzoubi, H. M., Ghazal, T. M., Said, R. A., AlHamad, A. Q., Hamadneh, S., Sahawneh, N., & Al-kassem, A. H. (2022). Fuzzy assisted human resource management for supply chain management issues. *Annals of Operations Research*, 1–19.

Alshurideh, M., Al-Hawary, S. I., Batayneh, A. M., Mohammad, A., & Al-Kurdi, B. (2017). The impact of Islamic banks' service quality perception on Jordanian customers loyalty. *Journal of Management Research, 9*(2), 139–159.

Alshurideh, M., Gasaymeh, A., Ahmed, G., Alzoubi, H., & Kurd, B. (2020). Loyalty program effectiveness: Theoretical reviews and practical proofs. *Uncertain Supply Chain Management, 8*(3), 599–612.

Alsuwaidi, M., Alshurideh, M., Al Kurdi, B., & Salloum, S. A. (2020). Performance appraisal on employees' motivation: a comprehensive analysis. In *International Conference on Advanced Intelligent Systems and Informatics* (pp. 681–693). Springer, Cham. (2020, October).

Altamony, H., Masa'deh, R., Alshurideh, M., Obeidat, B. (2012) Information systems for competitive advantage: Implementation of an organisational strategic management process. In *Innovation and sustainable competitive advantage: From regional development to world economies* (pp. 583–592).

AlTaweel, I. R., & Al-Hawary, S. I. (2021). The mediating role of innovation capability on the relationship between strategic agility and organizational performance. *Sustainability, 13*(14), 7564.

Alzoubi, H., Alshurideh, M., Kurdi, B., Akour, I., & Aziz, R. (2022). Does BLE technology contribute towards improving marketing strategies, customers' satisfaction and loyalty? The role of open innovation. *International Journal of Data and Network Science, 6*(2), 449–460.

Backović, N., & Djurić, M. (2019). Managing critical tasks within ambidextrous organizations. In *RSEP CONFERENCES* (p. 83).

Belhadi, A., Kamble, S., Gunasekaran, A., & Mani, V. (2021). Analyzing the mediating role of organizational ambidexterity and digital business transformation on industry 4.0 capabilities and sustainable supply chain performance. *Supply Chain Management: An International Journal.*

Ben-Abdallah, R., Shamout, M., & Alshurideh, M. (2022). Business development strategy model using EFE, IFE and IE analysis in a high-tech company: An empirical study. *Academy of Strategic Management Journal, 21*(Special Issue 2), 1–9.

Booth-Kewley, S., Dell'Acqua, R. G., & Thomsen, C. J. (2017). Factors affecting organizational commitment in Navy Corpsmen. *Military Medicine, 182*(7), e1794–e1800.

Bresciani, S., Ferraris, A., & Del Giudice, M. (2018). The management of organizational ambidexterity through alliances in a new context of analysis: Internet of Things (IoT) smart city projects. *Technological Forecasting and Social Change, 136*, 331–338.

Campanella, F., Del Giudice, M., Thrassou, A., & Vrontis, D. (2020). Ambidextrous organizations in the banking sector: An empirical verification of banks' performance and conceptual development. *The International Journal of Human Resource Management, 31*(2), 272–302.

De Leeuw, E., Hox, J., Silber, H., Struminskaya, B., & Vis, C. (2019). Development of an international survey attitude scale: Measurement equivalence, reliability, and predictive validity. *Measurement Instruments for the Social Sciences, 1*(1), 9. https://doi.org/10.1186/s42409-019-0012-x

Du, W., Pan, S. L., & Wu, J. (2020). How do IT outsourcing vendors develop capabilities? An organizational ambidexterity perspective on a multi-case study. *Journal of Information Technology, 35*(1), 49–65.

Felício, J. A., Caldeirinha, V., & Dutra, A. (2019). Ambidextrous capacity in small and medium-sized enterprises. *Journal of Business Research, 101*, 607–614.

Gopinath, R. (2020). Impact of job satisfaction on organizational commitment among the academic leaders of Tamil Nadu Universities. *GEDRAG & Organisatie Review, 33*(2), 2337–2349.

Gürlek, M. (2021). Effects of high-performance work systems (HPWSs) on intellectual capital, organizational ambidexterity and knowledge absorptive capacity: Evidence from the hotel industry. *Journal of Hospitality Marketing & Management, 30*(1), 38–70.

Hair, J. F., Babin, B. J., & Krey, N. (2017). Covariance-based structural equation modeling in the journal of advertising: review and recommendations. *Journal of Advertising, 46*(1), 163–177. https://doi.org/10.1080/00913367.2017.1281777

Howard, M. C. (2018). The convergent validity and nomological net of two methods to measure retroactive influences. *Psychology of Consciousness: Theory, Research, and Practice, 5*(3), 324–337. https://doi.org/10.1037/cns0000149

Huo, B., Gu, M., & Wang, Z. (2019). Green or lean? A supply chain approach to sustainable performance. *Journal of Cleaner Production, 216*, 152–166.

Imamoglu, S. Z., Ince, H., Turkcan, H., & Atakay, B. (2019). The effect of organizational justice and organizational commitment on knowledge sharing and firm performance. *Procedia Computer Science, 158*, 899–906.

Kaplan, M., & Kaplan, A. (2018). The relationship between organizational commitment and work performance: a case of industrial enterprises.

Karami, A., Farokhzadian, J., & Foroughameri, G. (2017). Nurses' professional competency and organizational commitment: Is it important for human resource management? *PLoS ONE, 12*(11), e0187863.

Kraner, J. (2018). Literature Review and Theoretical Propositions. *Innovation in High Reliability Ambidextrous Organizations*, 9–53.

Kurdi, B., Alshurideh, M., & Alnaser, A. (2020). The impact of employee satisfaction on customer satisfaction: Theoretical and empirical underpinning. *Management Science Letters, 10*(15), 3561–3570.

Labrague, L. J., McEnroe–Petitte, D. M., Tsaras, K., Cruz, J. P., Colet, P. C., & Gloe, D. S. (2018). Organizational commitment and turnover intention among rural nurses in the Philippines: Implications for nursing management. *International Journal of Nursing Sciences, 5*(4), 403–408.

Lee, K., Azmi, N., Hanaysha, J., Alshurideh, M., & Alzoubi, H. (2022a). The effect of digital supply chain on organizational performance: An empirical study in Malaysia manufacturing industry. *Uncertain Supply Chain Management, 10*(2), 1–16.

Lee, K., Ramiz, P., Hanaysha, J., Alzoubi, H., & Alshurideh, M. (2022b). Investigating the impact of benefits and challenges of IOT adoption on supply chain performance and organizational performance: An empirical study in Malaysia. *Uncertain Supply Chain Management, 10*(2), 1–14.

Limaj, E., & Bernroider, E. W. (2019). The roles of absorptive capacity and cultural balance for exploratory and exploitative innovation in SMEs. *Journal of Business Research, 94*, 137–153.

López-Sánchez, J. Á., & Santos-Vijande, M. L. (2022). Key capabilities for frugal innovation in developed economies: insights into the current transition towards sustainability. *Sustainability Science*, 1-

Martin, A., Keller, A., & Fortwengel, J. (2019). Introducing conflict as the microfoundation of organizational ambidexterity. *Strategic Organization, 17*(1), 38–61.

Metabis, A., & Al-Hawary, S. I. (2013). The impact of internal marketing practices on services quality of commercial banks in Jordan. *International Journal of Services and Operations Management, 15*(3), 313–337.

Mohammad, A. A., Alshura, M. S., Al-Hawary, S. I. S., Al-Syasneh, M. S., & Alhajri, T. M. (2020). The influence of internal marketing practices on the employees' intention to leave: A study of the private hospitals in Jordan. *International Journal of Advanced Science and Technology, 29*(5), 1174–1189.

Novitasari, D., Asbari, M., Wijaya, M. R., & Yuwono, T. (2020). Effect of organizational justice on organizational commitment: mediating role of intrinsic and extrinsic satisfaction. *International Journal of Science and Management Studies (IJSMS), 3*(3), 96–112.

Obeidat, U., Obeidat, B., Alrowwad, A., Alshurideh, M., Masadeh, R., & Abuhashesh, M. (2021). The effect of intellectual capital on competitive advantage: The mediating role of innovation. *Management Science Letters, 11*(4), 1331–1344.

Ögmundarson, Ó., Sukumara, S., Herrgård, M. J., & Fantke, P. (2020). Combining environmental and economic performance for bioprocess optimization. *Trends in Biotechnology, 38*(11), 1203–1214.

Ojo, A. O., & Fauzi, M. A. (2020). Environmental awareness and leadership commitment as determinants of IT professionals engagement in Green IT practices for environmental performance. *Sustainable Production and Consumption, 24*, 298–307.

Oyewobi, L. O., Oke, A. E., Adeneye, T. D., & Jimoh, R. A. (2019). Influence of organizational commitment on work–life balance and organizational performance of female construction professionals. *Engineering, Construction and Architectural Management.*

Pislaru, M., Herghiligiu, I. V., & Robu, I. B. (2019). Corporate sustainable performance assessment based on fuzzy logic. *Journal of Cleaner Production, 223*, 998–1013.

Poltronieri, C. F., Ganga, G. M. D., & Gerolamo, M. C. (2019). Maturity in management system integration and its relationship with sustainable performance. *Journal of Cleaner Production, 207*, 236–247.

Posch, A., & Garaus, C. (2020). Boon or curse? A contingent view on the relationship between strategic planning and organizational ambidexterity. *Long Range Planning, 53*(6), 101878.

Rialti, R., Zollo, L., Ferraris, A., & Alon, I. (2019). Big data analytics capabilities and performance: Evidence from a moderated multi-mediation model. *Technological Forecasting and Social Change, 149*, 119781.

Rimkeviciene, J., Hawgood, J., O'Gorman, J., & De Leo, D. (2017). Construct validity of the acquired capability for suicide scale: Factor structure, convergent and discriminant validity.

Journal of Psychopathology and Behavioral Assessment, 39(2), 291–302. https://doi.org/10.1007/s10862-016-9576-4

Rosenberg, A., Lynch, P. M., & Radmann, A. (2021). Sustainability comes to life. Nature-based Adventure Tourism in Norway. *Frontiers in Sports and Active Living, 3*, 154.

Sekaran, U., & Bougie, R. (2016). *Research methods for business: A skill-building approach* (7th edn.). Wiley.

Shamout, M., Elayan, M., Rawashdeh, A., Kurdi, B., & Alshurideh, M. (2022). E-HRM practices and sustainable competitive advantage from HR practitioner's perspective: A mediated moderation analysis. *International Journal of Data and Network Science, 6*(1), 165–178.

Shi, D., Lee, T., & Maydeu-Olivares, A. (2019). Understanding the model size effect on SEM fit indices. *Educational and Psychological Measurement, 79*(2), 310–334. https://doi.org/10.1177/0013164418783530

Sung, K.-S., Yi, Y. G., & Shin, H.-I. (2019). Reliability and validity of knee extensor strength measurements using a portable dynamometer anchoring system in a supine position. *BMC Musculoskeletal Disorders, 20*(1), 1–8. https://doi.org/10.1186/s12891-019-2703-0

Wang, Y. A., & Rhemtulla, M. (2021). Power analysis for parameter estimation in structural equation modeling: A discussion and tutorial. *Advances in Methods and Practices in Psychological Science, 4*(1), 1–17. https://doi.org/10.1177/2515245920918253

Authentic Leadership and Its Impact on Sustainable Performance: The Mediating Role of Knowledge Ability in Jordan Customs Department

Mohammed Saleem Khlif Alshura, Saud Saleh Alloush Abu Tayeh, Yahia Salim Melhem, Fuad N. Al-Shaikh, Hanan Mohammad Almomani, Fatima Lahcen Yachou Aityassine, Reham Zuhier Qasim Almomani, Sulieman Ibraheem Shelash Al-Hawary, and Anber Abraheem Shlash Mohammad

Abstract The study aimed to identify the impact of authentic leadership on sustainable performance through knowledge Ability from the point of view of employees in Jordan customs department. The study population consisted of all 678 employees in the border centers of the Jordan Customs Service, where a random sample was taken proportional represented by the study population by 271 individuals, and the study

M. S. K. Alshura
Faculty of Money and Management, Management Department, The World Islamic Science University, P.O. Box 1101, Amman 11947, Jordan

S. S. A. A. Tayeh
Researcher, Management Department, Faculty of Money and Management, The World Islamic Science University, P.O. Box 1101, Amman 11947, Jordan

Y. S. Melhem
Professor of Business Management, Business Administration Department, School of Business, Yarmouk University, Irbid, Jordan
e-mail: ymelhem@yu.edu.jo

F. N. Al-Shaikh
Department of Business Administration-Faculty of Economics and Administrative Sciences, Yarmouk University, P.O Box 566-Zip Code 21163, Irbid, Jordan
e-mail: eco_fshaikh@yu.edu.jo

H. M. Almomani · S. I. S. Al-Hawary (✉)
Department of Business Administration, School of Business, Al Al-Bayt University, P.O. BOX 130040, Mafraq 25113, Jordan
e-mail: dr_sliman73@aabu.edu.jo; dr_sliman@yahoo.com

F. L. Y. Aityassine
Department of Financial and Administrative Sciences, Irbid University College, Al-Balqa'Applied University, As-Salt, Jordan
e-mail: Fatima.yassin@bau.edu.jo

R. Z. Q. Almomani
Assistant Professor, Business Administration, Amman, Jordan

M. Alshurideh et al. (eds.), *The Effect of Information Technology on Business and Marketing Intelligence Systems*, Studies in Computational Intelligence 1056,
https://doi.org/10.1007/978-3-031-12382-5_79

used the statistical package for social sciences (SPSS), in addition to using the Amos program. The study concluded that there is a statistically significant effect of authentic leadership in sustainable performance through knowledge Ability from the point of view of employees in border centers in the Jordan Customs Department. The study recommended a set of recommendations and suggestions that contribute to maintaining the sustainability of performance and enhancing growth and development in the Jordanian Customs Department.

Keywords Authentic leadership · Sustainable performance · Knowledge ability · Customs department · Jordan

1 Introduction

Leadership plays an important role in the individuals, societies and organizations level, as history has witnessed over the past ages nations and societies that have risen and developed thanks to their great leaders, and many organizations have achieved success and development at the global level through their leaders work, and on the contrary, leaders have been a cause of organizations failure and extinction (Alameeri et al., 2020; Al-Hawary, 2009; Al-Hawary et al., 2012, 2013; Metabis & Al-Hawary, 2013). On the other hand, recent years have witnessed many conferences dealing with ethics especially in work ethics, most of which centered on the necessity of creating a global ethical charter and limiting scandals and violations in the business field (Al-Hawary, 2015; Al-Hawary & Abu-Laimon, 2013; Al-Lozi et al., 2017; AlShehhi et al., 2020; Alyammahi et al., 2020). Organizations leadership style is the basis of organizations work, as it is mainly responsible for their progress or decline in front of competing organizations, as the issue of leadership is considered one of the most important topics in management science (Al-Dhuhouri et al., 2020; Odeh et al., 2021). Authentic leadership theory clarifies the behaviors that leaders must exhibit as well as addressing the absence of the most important of these behaviors that leaders may lack, and as a reaction to the business environment that is characterized by technological developments and changes and intense competition; Organizations have found new ways to counter these developments, Organizations must pay attention to the employees positive state and their sense of vitality and optimism towards achieving personal and organizational goals, which is better and positively reflected on the organizations overall performance continuity and sustainability (Alhalalmeh et al., 2020; Al-Hawary & Alajmi, 2017; Al-Hawary & Al-Namlan, 2018; Al-Hawary & Hadad, 2016; Al-Hawary & Obiadat, 2021; Al-Hawary et al., 2020; Al-Nady et al., 2013; Harahsheh et al., 2021; Mohammad et al., 2020). The authentic leadership style has emerged recently, in which leaders apply the principles and values of honesty, transparency and ethical standards. The term sustainable performance has emerged

A. A. S. Mohammad
Faculty of Administrative and Financial Sciences, Marketing Department, Petra University, B.O. Box: 961343, Amman 11196, Jordan

as a modern management concept concerned with society and the environment as basic pillars for achieving sustainable growth and development. It is not limited to achieving financial and economic performance only, but also integrates all social and environmental conditions within the organization's strategic plan (Al- Quran et al., 2020). As well as the term knowledge competence, which requires more time and work to understand and clarify, as it represents all the capabilities that contribute to building a pioneering knowledge organization capable of keeping pace with all developments in the surrounding environment, and providing the best services with the least time and effort, through the optimal use of modern technology and workers attention and motivating towards creativity and excellence (Allahow et al., 2018; Al-Hawary & Al-Syasneh, 2020; AlTaweel & Al-Hawary, 2021; Al-Hawary & Alwan, 2016; Al-Hawary & Aldaihani, 2016).

Study importance stems from the importance of the effect achieved by authentic leadership on sustainable performance through the knowledge ability in the Jordan Customs Department. Where this study linked between modern variables and dealt with modern sub-dimensions of those variables which are of great importance in organizations growth and development study problem is the Jordanian Customs Department sustainable performance and how to maintain this performance in light of the surrounding environmental changes, through authentic leadership, and with the emergence of contemporary global challenges and the spread of the global epidemic of covid-19, All this led to a decline in revenues and customs collections, due to border closures and global closures of trade outlets, here, authentic leadership appeared to be the best solution in overcoming these difficulties and transforming challenges into opportunities (Almazrouei et al., 2020; Allozi et al., 2022). However, the problem lies in the existence of different standards in the Customs Department regarding the extent of plans implementation, either due to negligence or shortcomings in legislation or changes in regulations and laws that affect the implementation process (Al Kurdi et al., 2021; Alshurideh, 2022). While this gap remains, the performance outputs and the quality of provided services are affected, and this gap may lead to a decrease in revenues levels, and this was evident during the year (2020), where revenues decreased from the year (2019) by (39) million dinars, and the department also witnessed a decrease in the customs data registered numbers as in 2019 (819) thousand customs declarations, In 2020, it amounted to (699) thousand statements.

This study came to clarify the relationship between authentic leadership as a modern leadership style and sustainable performance, which ensures the organization permanence and survival. This study attempts to provide a qualitative addition to knowledge that helps decision makers to identify this modern leadership style, which helps in making change decisions and facing challenges. The study focused on authentic leadership in terms of concept, application requirements, and advantages achieved by organizations in which there is authentic leadership, and also dealt with the relationship and role of authentic leadership in maintaining the performance sustainability and continuity. Also, many studies focused on the authentic leadership concept, its components, its importance and its relationship to the organizational, market and economic performance of business organizations. This study comes to shed light on the components of authentic leadership in the Jordanian

Customs Department, and this study contributes to measuring the extent of authentic leadership influence the Jordanian Customs Department overall performance.

2 Literature Review and Hypotheses Development

2.1 Authentic Leadership

Organizations dominant leadership style is the main focus of their work, as it is responsible for their progress or decline extent with competing organizations (Al-Hawary & Nusair, 2017). The issue of leadership is considered one of the most important topics in management science, and perhaps clarifying the most important behaviors that leaders must have was the share of the original leadership theory, as well as addressing the absence of the most important of these behaviors that leaders may lack, As organizations are operating in a business environment characterized by technological developments and changes, and intense competition; Therefore, it is necessary to adopt new methods to confront these developments, especially the crises that the world has faced as a result of the Corona pandemic and the subsequent economic crises, and the breakdown of organizations (Alshurideh et al., 2021; Pendleton and Jonathan 2021, 18).

Authentic leadership includes, at its roots, all positive qualities, such as self-confidence, integrity, and credibility, and seeks to achieve the goal in order to achieve psychological and social well-being, through the application of positive features of leadership, based on authentic leader values, believes, and the consistent behaviors (Muhammed & Zaim, 2020, 14–15). Authentic leadership focuses on building followers' strengths, developing their capabilities, expanding their thinking, and creating a positive organizational context. The authenticity approach indicates that it is linked to followers' attitudes and behaviors and that the most important effects of authentic leadership are those whose results are due to the working and subordinate individual's behavior and actions. The authentic leadership effect arises through the individual sense of social affiliation between the leader and the follower (Chalon, 2018, 36). In light of this, studies and research have clarified multiple definitions of authentic leadership. Studies have shown that the authentic leadership essence is the availability of frankness, integrity and transparency, and positive psychological capabilities, such as (trust, hope and optimism), high moral behaviors and trust between the leaders and subordinates.

Studies also emphasized that authentic leadership focuses on positive behaviors that include enhancing self-confidence, sincerity, credibility, merit, and high moral values that are reflected in improving the relationship between the leader and his subordinates (Linda & Chester, 2013). Nair and Shrekumar (2021) believes that authentic leadership is formed as a result of the stimulating events that the individual goes through, during the various stages of his life, which generates experiences and capabilities in facing circumstances, and these experiences formed lead to the positive psychological capabilities development (e.g., hope, optimism and confidence). Which leads to the development of self-awareness within the individual, and positive self-development leads to authentic leadership formation with a high degree of self-awareness and the ability to express values and behaviors with full force and determination. Authentic leadership is also defined as a leader's behavior pattern based on promoting positive self-development, a transparent and ethical leadership behavior that encourages openness in sharing information necessary for decision-making while accepting subordinates' views (Sparks, 2021, 55).

Self-awareness is one of the most essential aspects of authentic leadership, because it includes special knowledge of authentic leader values, beliefs, motives and experiences, and it refers to the successful leader behavior according to his motives and desires, as it helps the leader to control his emotions and deal based on value, Self-awareness represents a guarantee that the leader will take full responsibility for his behavior and his work, as it is not a point he reach, but rather it is an ongoing process. As for the Balanced Proccssing, it refers to the leader collecting the data and information necessary to make decisions, analyzing them accurately and objectively, extracting the results and discussing them before making the decision. With regard to the Internalized Moral Perspective, it means that managers possess and demonstrate internal moral standards and values rather than recognizing external pressures, however, originality reflects managers' ability to achieve harmony with themselves and their followers broadly, to achieve cooperation within and outside the organization.

Therefore, managers' values should be consistent with those generally accepted in the environment (e.g. society, organization) (Alshurideh et al., 2017, 2019). Individuals who have high levels of self-regulation manage their behavior to conform to ethical standards, which, on the one hand, represent the consistency of managers' behavior with their own values, and on the other hand, the managers' behavior consistency with the ethical rules that prevail within society and the organization (Alshurideh et al., 2017). And finally, relations transparency, that through transparency, ideas and beliefs are exchanged between the leader and the followers, as it works to present the authentic personality to others, and thus more confidence and reduce inappropriate emotions, which contributes to reducing organizational conflict intensity. Transparency in relationships also includes revealing the leader's authentic inner self and not the illusion or artificial personality. By showing the true self, the leader can gain the subordinates confidence and enhance their organizational commitment. And this also includes exchanging information and ideas with credibility and clarity (Al-Mansi, 2019).

2.2 *Sustainable Performance*

Organizations and companies seek to develop their competencies and resources, by noting the discrepancy in talents, aspirations and performance levels, and seek to build programs and policies to improve and raise organizations performance levels and encourage the transfer of knowledge and expertise within the work environment, as this requires the leaders with experience, knowledge and ability to face difficulties (Chalon, 2018, 15). Organizations looking for sustainability have goals that go beyond traditional concepts. The emergence of sustainable development concept has forced organizations to integrate the sustainable development dimensions and making sustainability a part of the organization's strategy, and this requires radical changes in the organization's performance. Sustainable performance is real performance and gives indications that the organization is moving in the right direction because it is a continuous and it can be expected in the future (Majid, 2020). Hence, sustainable performance is a combination of economic, social and environmental performance as it extends beyond the organization boundaries, and does not only address financial results, but extends to include taking into account working individuals' interests, the surrounding community and the natural environment (AlTaweel & Al-Hawary, 2021).

Ziemak and Katarzyna (2020) defines sustainable performance as the process of interaction between the organization work, the environment and society requirements, and profitability levels, and includes the extent to which the organization has succeeded in translating its goals into practical reality to maintain sustainability and continuity, to achieve more positives and reduce negatives, taking into account satisfying customers and employees and maintaining workplace safety. Rasool et al. refers to sustainable performance as the process of coordinating financial, environmental and social goals in order to maximize value, and it is based on four pillars that include achieving goals, use of resources, internal processes, and stakeholders.

Sustainable performance importance is based on the great responsibility that falls on organizations shoulders in the global economy, so performance sustainability has become a must, as it achieves a state of integration between the decisions taken daily by the organizations in order to facilitate business and enters the organizations in a new phase represented by how to transform into entities that serve the environmental, social and economic interests while preserving the stakeholders at the same time (Al-Mansi, 2019; Shamout et al., 2022).

2.3 *Knowledge Ability*

Al-Mansi (2019) indicated that knowledge ability represents the individual's ability to have a great knowledge of their working conditions and understand the procedures and rules they follow, that enable them to act in the appropriate way away from traditional organizational structures, and they have the ability to make better changes and use cognitive thinking to analyze problems and develop appropriate solutions (Al-Hawary & Al-Namlan, 2018; Al-Maroof et al., 2021; Al Mehrez et al., 2020). This means that they have an integrated cognitive knowledge that generates for them the ability to achieve high performance. Knowledge ability includes all cognitive capabilities and modern institutional methods that keep pace with development in order to be able to process the business content and tasks assigned to them; Cognitive competence is one of the modern concepts that has attracted the researchers' attention and has become the most prominent guarantee for the future building requirements by focusing on the sustainable work efficiency and the employing the mental energies of working individuals efficiency (Agha et al., 2021; Al-Mawla & Harbi, 2019).

Knowledge ability importance is represented by what was stated by some writers and researchers. It has been pointed out that knowledge ability is linked to knowledge and experience, it absorbs its components and benefit from its potential and leads to policies and practice development and the means and strategies adoption, and it is capable of overcoming all obstacles and difficulties that prevent the goals achievement. To catch up with progress requires scientific and knowledge capabilities that give the organization and society the reasons for solid strength, and it also promotes working individuals to higher positions in their specialization field, and they are those who have the ability to assume responsibility and build ideas that serve the production or service process and have a spirit of responsibility (Al Suwaidi et al., 2020; Al-Mawla & Harbi, 2019).

The researchers tried directly or indirectly to describe the knowledge ability through a set of dimensions that came after the processes of analysis and interpretation of all its related aspects, the purpose of accurately defining knowledge ability dimensions is that it represents a great importance to the organization and reflects a distinguished level of human competencies, personal excellence, and commitment to moral guidance. As defined by the researchers, knowledge ability dimensions are as follows: Information Technology which is defined as a set of software, hardware, communications, database management and various data processing technologies that are used in computer-based information systems (Ibarra, 2015). Creative thinking which represents the ability to organize ideas in a different way, in order to invent new things, and includes restructuring knowledge to form a new vision (Proctor, 2021, 8).

2.4 *Authentic Leadership, Sustainable Performance and Knowledge Ability*

By reviewing books and studies on authentic leadership, the researcher concluded that it is based on the leader individual capabilities, where decisions are made based on clear and accurate information and with the others participation. It enhances trust between the leader and subordinates, which is reflected in the organization overall performance. Authentic leadership is closely related to the organization internal quality (organizational culture and organizational climate), and it is linked to job satisfaction, and this ultimately leads to raising the productivity levels, The researchers found that whenever there was an improvement in the level of job satisfaction by 5%, there was an increase of 1.2% in customer satisfaction, which led to an increase in the performance level by 0.5%, leadership influence was clear in employees performance, which increases the productivity levels (Pendleton and Jonathan 2021).

Locared refers to authentic leadership as working to build the general welfare of humanity and achieve society interest, and defines an authentic leader as that individual who is fully aware of how to think and how to act, and is seen by others as aware of his own values and the values of others, and by integrating positive qualities such as hope and optimism, he motivates the followers to perform well and identify the individuals needs and desires in order to meet them and ensure the achievement of organization general objectives. Knowledge ability is an important resource for organization creativity and innovation processes, through the acquisition of external knowledge and the maximization of internal knowledge in order to raise the organization's ability to achieve competitive advantage (Duan et al., 2020). The term knowledge ability is one of the modern terms that requires more time and work to understand it, as it represents the possibility that works to re-characterize the organization from a traditional organization to a knowledge organization, so that it turns into a pioneering organization that has a high ability to face challenges through modern technological knowledge. In this way, knowledge ability can transform the organization through the type, activity or theories used, and then transform it into a knowledge organization. It represents the characteristic that responsible individuals carry outside the traditional role, as they have knowledge ability, technological skills and institutional methods that make them capable of knowing and treating their business content (Majid, 2020). Based on the above, study hypothesis can be formulated as follows:

There is no statistically significant impact of authentic leadership on sustainable performance with the presence of knowledge ability as a mediating variable in the Jordan Customs Department.

3 Study Model

See (Fig. 1).

Fig. 1 Research model

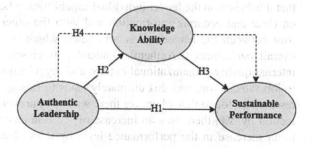

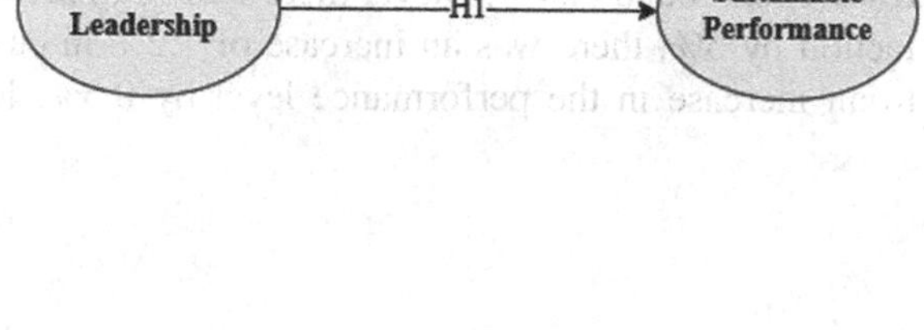

4 Methodology

4.1 Population and Sample Selection

A qualitative method based on a questionnaire was used in this study for data collection and sample selection. The major aim of the study was to examine the impact of authentic leadership on sustainable performance through the mediating role of knowledge ability. Therefore, it focused on Jordanian customs centers. Data were primarily gathered through self-reported questionnaires creating by Google Forms which were distributed to a random sample of (320) employees via email. In total, (283) responses were received including (12) invalid to statistical analysis due to uncompleted or inaccurate. Hence, the final sample contained (270) responses suitable to analysis requirements that were formed a response rate of (84.68%), where it proved to be sufficient to the extent that was predictable and allowed for a presumption of data saturation (Sekaran & Bougie, 2016).

4.2 Measurement Instrument

A self-reported questionnaire that consists of three main sections along with a section regarding control variables was used as the measurement instrument. Control variables considered as categorical measures were composed of gender, age group, educational level, and experience. The three main sections were dealt with a five-point Likert scale (from 1 = strongly disagree to 5 = strongly agree). The first section contained (16) items to measure authentic leadership based on. These items were distributed into dimensions as follows: four items dedicated for measuring self-awareness, four items dedicated for measuring balanced processing, four items dedicated for measuring internalized moral perspective, and four items dedicated for measuring relational transparency. The second section related to knowledge ability variable which contained nine items developed according to (Al-Hawary & Alwan, 2016). These items distributed into two first-order constructs: five items for information technology and four items for creative thinking. Whereas the third section included (15) items developed to measure sustainable performance according to what was pointed by Afum et al. and Kamble et al. This variable was a second-order construct divided into three first-order constructs. Economic performance measured by five items, environmental performance measured through five items, and social performance measured using five items.

5 Findings

5.1 Measurement Model Evaluation

This study was conducted structural equation modeling (SEM) to test hypotheses, which represents a contemporary statistical technique for testing and estimating the relationship between factors and variables (Wang & Rhemtulla, 2021; Al-Adamat et al., 2020; Al-Gasawneh & Al-Adamat, 2020). Accordingly, the reliability and validity of the constructs were tested using confirmatory factor analysis (CFA) through the statistical program AMOSv24. Table 1 summarizes the results of convergent and discriminant validity, as well the indicators of reliability.

Table 1 shows that the standard loading values for the individual items were within the domain (0.627–0.902), these values greater than the minimum retention of the elements based on their standard loads (Al-Lozi et al., 2018; Sung et al., 2019). Average variance extracted (AVE) is a summary indicator of the convergent validity of constructs that must be above 0.50 (Howard, 2018). The results indicate that the AVE values were greater than 0.50 for all constructs, thus the used measurement model has an appropriate convergent validity. Rimkeviciene et al. (2017) suggested the comparison approach as a way to deal with discriminant validity assessment in covariance-based SEM. This approach is based on comparing the values of maximum shared variance (MSV) with the values of AVE, as well as comparing the values of square root of AVE ($\sqrt{AVE}$) with the correlation between the rest of the structures. The results show that the values of MSV were smaller than the values of AVE, and that the values of $\sqrt{AVE}$ were higher than the correlation values among the rest of the constructs. Therefore, the measurement model used is characterized by discriminative validity. The internal consistency measured through Cronbach's Alpha coefficient (α) and compound reliability by McDonald's Omega coefficient (ω) was conducted as indicators to evaluate measurement model. The results listed in Table 1 demonstrated that both values of Cronbach's Alpha coefficient and McDonald's Omega coefficient were greater than 0.70, which is the lowest limit for judging on measurement reliability (De Leeuw et al., 2019).

5.2 Structural Model

The structural model illustrated no multicollinearity issue among predictor constructs because variance inflation factor (VIF) values are below the threshold of 5, as shown in Table 1 (Hair et al., 2017). This result is supported by the values of model fit indices shown in Fig. 1 (Fig. 2).

The results in Fig. 1 indicated that the chi-square to degrees of freedom (CMIN/DF) was 2.348, which is less than 3 the upper limit of this indicator. The values of the goodness of fit index (GFI), the comparative fit index (CFI), and the Tucker-Lewis index (TLI) were upper than the minimum accepted threshold of 0.90.

Table 1 Results of validity and reliability tests

Constructs	1	2	3	4	5	6	7	8	9
1. SAW	**0.786**								
2. BPR	0.347	**0.755**							
3. IMP	0.471	0.297	**0.774**						
4. RTR	0.445	0.364	0.415	**0.756**					
5. ITE	0.534	0.428	0.547	0.581	**0.744**				
6. CTH	0.394	0.497	0.552	0.496	0.418	**0.766**			
7. ECP	0.405	0.537	0.374	0.335	0.532	0.614	**0.759**		
8. ENP	0.534	0.637	0.622	0.579	0.674	0.597	0.574	**0.765**	
9. SOP	0.614	0.569	0.637	0.510	0.458	0.602	0.614	0.574	**0.781**
VIF	3.458	2.664	1.858	1.374	–	–	–	–	–
Loadings range	0.644–0.854	0.694–0.883	0.716–0.839	0.627–0.894	0.681–0.854	0.653–0.902	0.661–0.881	0.645–0.837	0.691–0.854
AVE	0.618	0.570	0.599	0.572	0.554	0.586	0.577	0.585	0.610
MSV	0.501	0.415	0.508	0.467	0.513	0.406	0.515	0.497	0.522
Internal consistency	0.862	0.837	0.855	0.839	0.857	0.846	0.869	0.873	0.882
Composite reliability	0.865	0.840	0.856	0.841	0.861	0.848	0.871	0.875	0.886

Note SAW: self-awareness, BPR: balanced processing, IMP: internalized moral perspective, RTR: relational transparency, ITE: information technology, CTH: creative thinking, ECP: economic performance, ENP: environmental performance, SOP: social performance

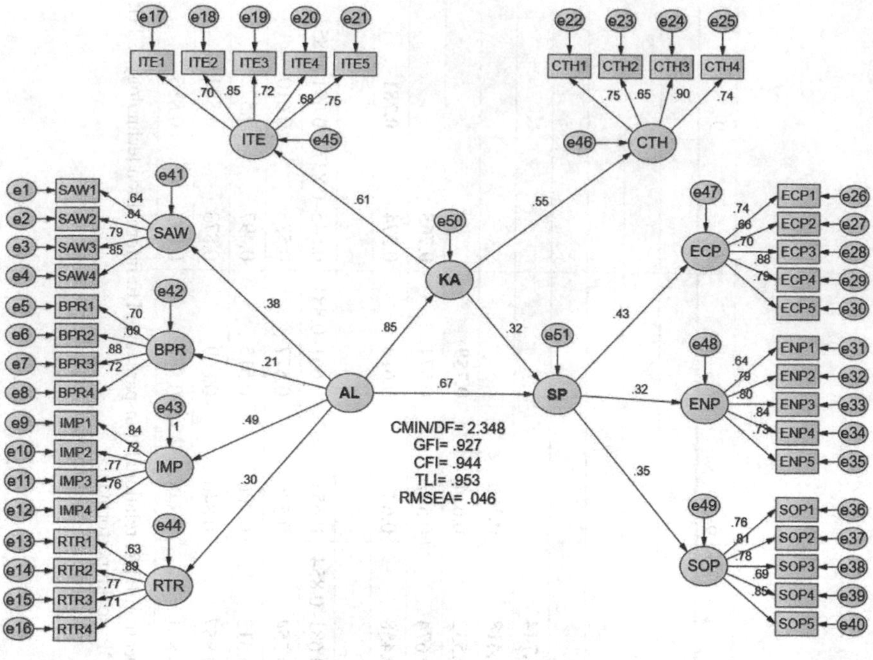

Fig. 2 SEM results of the ALimpact on SP through KA

Moreover, the result of root mean square error of approximation (RMSEA) indicated to value 0.046, this value is a reasonable error of approximation because it is less than the higher limit of 0.08. Consequently, the structural model used in this study was recognized as a fit model for predicting the sustainable performance and generalization of its result (Ahmad et al., 2016; Shi et al., 2019). To verify the results of testing the study hypotheses, structural equation modeling (SEM) was used, the results of which are listed in Table 2.

Table 2 Hypothesis testing

Relation	Direct		Indirect		Total		
	β	p-value	β	p-value	β	t-value	p-value
Authentic Leadership → Sustainable Performance	0.672	0.000			0.672	19.367	0.000
Authentic Leadership → Knowledge Ability	0.848	0.001			0.848	21.228	0.001
Knowledge Ability → Sustainable Performance	0.325	0.03			0.325	16.015	0.03
Authentic Leadership → Knowledge Ability → Sustainable Performance	0.672	0.000	0.275	0.000	0.947	28.641	0.000

The results demonstrated in Table 2 show that authentic leadership has a positive direct impact on sustainable performance (β = 0.672, t = 19.367, p = 0.000) and knowledge ability (β = 0.848, t = 21.228, p = 0.001). Moreover, it indicated that knowledge ability has a positive impact on sustainable performance (β = 0.325, t = 16.015, p = 0.03). Hence, knowledge ability has a partial mediation effect in the relationship between authentic leadership and sustainable performance, where the total effect was (β = 0.947, t = 28.641, p = 0.000) with indirect effect of (β = 0.275, p = 0.000).

6 Discussion

Study results indicated that there is a statistically significant impact of the authentic leadership on sustainable performance in the presence of knowledge ability, and this explains the Customs Department interests in developing leaders that contribute to growth and development and have a future vision that keeps pace with developments in addition to modernization in individuals and equipment that contribute to raising the performance levels and achieving high growth indicators. Study results showed that the mediating role played by knowledge ability between authentic leadership and sustainable performance achieves positive results, and this is an indicator that confirms that the three main variables achieve a strong correlation, in that the Customs Department works to implement the authentic leadership style and maintains performance sustainability through it. Companies and individuals dealing with the Jordanian Customs Department need facilitated services and facilities to protect investment and the national economy, facilitate trade movement, achieve the highest quality levels in providing services and facilities, and ensure goods smooth arrival.

The results showed that the study variables dimensions represented in authentic leadership, sustainable performance and knowledge ability, reached high levels of application, which indicates the awareness extent of the Customs Department of these modern managerial concepts importance, and their role in achieving performance sustainability and growth in light of the surrounding changes and difficulties that the economy suffers from world and the international trade system. Study results showed the strategic role played by the Customs Department in supplying the state treasury with revenues through the high level of economic performance. This result is consistent with the study (Balogun et al., 2020), which showed that there is a significant and clear impact of authentic leadership on the performance level.

7 Recommendations

According to the study results and conclusions, the researcher presents a set of suggestions and recommendations, which in turn can benefit the researched organization, through the study results, it is clear that it has become necessary for the

Customs Department to promote the environmental sustainability concept and to practice activities and policies that are reflected on the surrounding environment by preventing pollution, preventing the entry of products harmful to the environment and society, and promoting a culture of green environment among employees. The Customs Department is considered one of the institutions based on knowledge and technology. Therefore, the Customs Department must develop strategies based on knowledge transfer process and ease of access and conduct more knowledge transfer courses and enhance the growth and self-development levels. The study recommends also for increasing the space of participation in decision-making by listening to different viewpoints, and this is done through holding meetings and seminars that contribute to the views transfer on the performance level and strengths and weaknesses in order to overcome problems and find solutions to all customs work obstacles. Through the study results, which reflected the development and growth level reached by the Customs Department, the researcher recommends increasing the participants numbers in foreign courses to learn about global customs experiences in order to gain more experience and knowledge, as the participants number in foreign courses is few in light of the surrounding environment developments and the global competition increase.

References

Agha, K., Alzoubi, H. M., & Alshurideh, M. T. (2021). Measuring reliability and validity instruments of technologically driven cognitive intrusion towards work-life balance. In *The International Conference on Artificial Intelligence and Computer Vision* (pp. 601–614). Springer, Cham. (2021, June).

Ahmad, S., Zulkurnain, N., & Khairushalimi, F. (2016). Assessing the validity and reliability of a measurement model in structural equation modeling (SEM). *British Journal of Mathematics & Computer Science, 15*(3), 1–8. https://doi.org/10.9734/BJMCS/2016/25183

Al- Quran, A. Z., Alhalalmeh, M. I., Eldahamsheh, M. M., Mohammad, A. A., Hijjawi, G. S., Almomani, H. M., & Al-Hawary, S. I. (2020). Determinants of the green purchase intention in Jordan: The moderating effect of environmental concern. *International Journal of Supply Chain Management, 9*(5), 366–371.

Al-Adamat, A., Al-Gasawneh, J., & Al-Adamat, O. (2020). The impact of moral intelligence on green purchase intention. *Management Science Letters, 10*(9), 2063–2070.

Alameeri, K., Alshurideh, M., Al Kurdi, B., & Salloum, S. A. (2020). The effect of work environment happiness on employee leadership. In *International Conference on Advanced Intelligent Systems and Informatics* (pp. 668–680). Springer, Cham. (2020, October).

Al-Dhuhouri, F. S., Alshurideh, M., Al Kurdi, B., & Salloum, S. A. (2020). Enhancing our understanding of the relationship between leadership, team characteristics, emotional intelligence and their effect on team performance: A critical review. In *International Conference on Advanced Intelligent Systems and Informatics* (pp. 644–655). Springer, Cham. (2020, October).

Al-Gasawneh, J. A., & Al-Adamat, A. M. (2020). The relationship between perceived destination image, social media interaction and travel intentions relating to Neom city. *Academy of Strategic Management Journal, 19*(2), 1–12.

Alhalalmeh, M. I., Almomani, H. M., Altarifi, S., Al- Quran, A. Z., Mohammad, A. A., & Al-Hawary, S. I. (2020). The nexus between corporate social responsibility and organizational performance in

Jordan: The mediating role of organizational commitment and organizational citizenship behavior. *Test Engineering and Management, 83*(July), 6391–6410.

Al-Hawary, S. I. (2015). Human resource management practices as a success factor of knowledge management implementation at health care sector in Jordan. *International Journal of Business and Social Science, 6*(11/1), 83–98.

Al-Hawary, S. I. (2009). The effect of the leadership style on the effectiveness of the organization: A field study at Zarqa Private University. *The Egyptian Journal for Commercial Studies, 33*(1), 361–393.

Al-Hawary, S. I., & Alajmi, H. M. (2017). Organizational commitment of the employees of the ports security affairs of the State of Kuwait: The impact of human recourses management practices. *International Journal of Academic Research in Economics and Management Sciences, 6*(1), 52–78.

Al-Hawary, S. I., & Aldaihani, F. M. (2016). Customer relationship management and innovation capabilities of Kuwait Airways. *International Journal of Academic Research in Economics and Management Sciences, 5*(4), 201–226.

Al-Hawary, S. I., & Al-Namlan, A. (2018). Impact of electronic human resources management on the organizational learning at the private hospitals in the State of Qatar. *Global Journal of Management and Business Research: A Administration and Management, 18*(7), 1–11.

Al-Hawary, S. I., & Hadad, T. F. (2016). The effect of strategic thinking styles on the enhancement competitive capabilities of commercial banks in Jordan. *International Journal of Business and Social Science, 7*(10), 133–144.

Al-Hawary, S. I., & Nusair, W. (2017). Impact of human resource strategies on perceived organizational support at Jordanian Public Universities. *Global Journal of Management and Business Research: A Administration and Management, 17*(1), 68–82.

Al-Hawary, S. I., AL-Awawdeh, W., & Abden, M. A. (2012). The impact of the leadership style on organizational commitment: A field study on Kuwaiti Telecommunications Companies. *ALEDARI* (130), 53–102.

Al-Hawary, S. I., Al-Qudah, K., Abutayeh, P., Abutayeh, S., & Al-Zyadat, D. (2013). The impact of internal marketing on employee's job satisfaction of commercial banks in Jordan. *Interdisciplinary Journal of Contemporary Research in Business, 4*(9), 811–826.

Al-Hawary, S. I. S., & Alwan, A. M. (2016). Knowledge management and its effect on strategic decisions of Jordanian Public Universities. *Journal of Accounting-Business & Management, 23*(2), 24–44.

Al-Hawary, S. I. S., & Obiadat, A. A. (2021). Does mobile marketing affect customer loyalty in Jordan? *International Journal of Business Excellence, 23*(2), 226–250.

Al-Hawary, S. I. S., Mohammad, A. S., Al-Syasneh, M. S., Qandah, M. S. F., & Alhajri, T. M. S. (2020). Organisational learning capabilities of the commercial banks in Jordan: Do electronic human resources management practices matter? *International Journal of Learning and Intellectual Capital, 17*(3), 242–266.

Al-Hawary, S. I., & Abu-Laimon, A. A. (2013). The impact of TQM practices on service quality in cellular communication companies in Jordan. *International Journal of Productivity and Quality Management, 11*(4), 446–474.

Al-Hawary, S. I., & Al-Syasneh, M. S. (2020). Impact of dynamic strategic capabilities on strategic entrepreneurship in presence of outsourcing of five stars hotels in Jordan. *Business: Theory and Practice, 21*(2), 578–587.

Al Kurdi, B., Elrehail, H., Alzoubi, H., Alshurideh, M., & Al-Adaila, R. (2021). The interplay among HRM practices, job satisfaction and intention to leave: An empirical investigation. *Journal of Legal, Ethical and Regulatory, 24*(1), 1–14.

Allahow, T. J. A. A., Al-Hawary, S. I. S., & Aldaihani, F. M. F. (2018). Information technology and administrative innovation of the central agency for information technology in Kuwait. *Global Journal of Management and Business, 18*(11-A), 1–16.

Allozi, A., Alshurideh, M., AlHamad, A., & Al Kurdi, B. (2022). Impact of transformational leadership on the job satisfaction with the moderating role of organizational commitment: Case of UAE and Jordan manufacturing companies. *Academy of Strategic Management Journal, 21*, 1–13.

Al-Lozi, M., Almomani, R. Z., & Al-Hawary, S. I. (2018). Talent Management strategies as a critical success factor for effectiveness of Human Resources Information Systems in commercial banks working in Jordan. *Global Journal of Management and Business Research: A Administration and Management, 18*(1), 30–43.

Al-Lozi, M., Almomani, R. Z., & Al-Hawary, S. I. (2017). Impact of talent management on achieving organizational excellence in Arab Potash Company in Jordan. *Global Journal of Management and Business Research: A Administration and Management, 17*(7), 15–25.

Al-Maroof, R., Ayoubi, K., Alhumaid, K., Aburayya, A., Alshurideh, M., Alfaisal, R., & Salloum, S. (2021). The acceptance of social media video for knowledge acquisition, sharing and application: A comparative study among YouYube users and TikTok users' for medical purposes. *International Journal of Data and Network Science, 5*(3), 197–214.

Al-Mansi, M. (2019). Authentic leadership and its impact on bullying behaviors in the workplace: An applied study on the Egyptian Tax Authority. *Journal of Business Research, 41*(1), Egypt.

Al-Mawla, S., & Harbi, A. (2019). The effect of cognitive competence in promoting future Orientalism. *Journal of Administration and Economics, 42*(120), Faddad.

Almazrouei, F. A., Alshurideh, M., Al Kurdi, B., & Salloum, S. A. (2020). Social media impact on business: a systematic review. In *International Conference on Advanced Intelligent Systems and Informatics* (pp. 697–707). Springer, Cham. (2020, October).

Al Mehrez, A. A., Alshurideh, M., Al Kurdi, B., & Salloum, S. A. (2020, October). Internal factors affect knowledge management and firm performance: a systematic review. In International Conference on Advanced Intelligent Systems and Informatics (pp. 632–643). Springer, Cham.

Al-Nady, B. A., Al-Hawary, S. I., & Alolayyan, M. (2013). Strategic management as a key for superior competitive advantage of sanitary ware suppliers in Kingdom of Saudi Arabia. *International Journal of Management and Information Technology, 7*(2), 1042–1058.

AlShehhi, H., Alshurideh, M., Al Kurdi, B., & Salloum, S. A. (2020). The impact of ethical leadership on employees performance: A systematic review. In *International Conference on Advanced Intelligent Systems and Informatics* (pp. 417–426). Springer, Cham. (2020, October).

Alshurideh, M., Al Kurdi, B., Abu Hussien, A., & Alshaar, H. (2017). Determining the main factors affecting consumers' acceptance of ethical advertising: A review of the Jordanian market. *Journal of Marketing Communications, 23*(5), 513–532.

Alshurideh, M., Kurdi, B. A., Shaltoni, A. M., & Ghuff, S. S. (2019). Determinants of proenvironmental behaviour in the context of emerging economies. *International Journal of Sustainable Society, 11*(4), 257–277.

Alshurideh, M. T., Hassanien, A. E., & Ra'ed Masa'deh. (2021). *The effect of coronavirus disease (COVID-19) on business intelligence*. Springer.

Alshurideh, M. (2022). Does electronic customer relationship management (E-CRM) affect service quality at private hospitals in Jordan? *Uncertain Supply Chain Management, 10*(2), 1–8.

Al Suwaidi, F., Alshurideh, M., Al Kurdi, B., & Salloum, S. A. (2020). The impact of innovation management in SMEs performance: a systematic review. In *International Conference on Advanced Intelligent Systems and Informatics* (pp. 720–730). Springer, Cham. (2020, October).

AlTaweel, I. R., & Al-Hawary, S. I. (2021). The mediating role of innovation capability on the relationship between strategic agility and organizational performance. *Sustainability, 13*(14), 1–14.

Alyammahi, A., Alshurideh, M., Al Kurdi, B., & Salloum, S. A. (2020). The impacts of communication ethics on workplace decision making and productivity. In *International Conference on Advanced Intelligent Systems and Informatics* (pp. 488–500). Springer, Cham. (2020, October).

Balogun, V., Bright, M., Charles, A. (2020). A confirmatory factor analytic study of an authentic leadership measure in Nigeria. *Journal of Human Resource Management*. ISSN: (Online) 2071-078X, (Print) 1683–7584

Chalon, C. (2018). *Achieving world class performance* (1st edn.). This book was professionally typeset on Reeds Find out more at reedsy.com.

De Leeuw, E., Hox, J., Silber, H., Struminskaya, B., & Vis, C. (2019). Development of an international survey attitude scale: Measurement equivalence, reliability, and predictive validity. *Measurement Instruments for the Social Sciences, 1*(1), 9. https://doi.org/10.1186/s42409-019-0012-x

Duan, K., Zhang, C., Zhang, R., & Zhang, Y. (2020). Boundary-spanning search for knowledge, knowledge reconstruction and the sustainable innovation ability of agricultural enterprises: A Chinese perspective. *Agriculture, 10*, 39. https://doi.org/10.3390/agriculture10020039

Hair, J. F., Babin, B. J., & Krey, N. (2017). Covariance-based structural equation modeling in the journal of advertising: Review and recommendations. *Journal of Advertising, 46*(1), 163–177. https://doi.org/10.1080/00913367.2017.1281777

Harahsheh, A., Houssien, A., Alshurideh, M., & AlMontaser, M. (2021). The effect of transformational leadership on achieving effective decisions in the presence of psychological capital as an intermediate variable in private Jordanian universities in light of the corona pandemic. In *The Effect of Coronavirus Disease (COVID-19) on Business Intelligence* (Vol. 334, pp. 221–243).

Howard, M. C. (2018). The convergent validity and nomological net of two methods to measure retroactive influences. *Psychology of Consciousness: Theory, Research, and Practice, 5*(3), 324–337. https://doi.org/10.1037/cns0000149, https://doi.org/10.1007/978-3-319-653075

Ibarra, H. (2015). *The Authenticity Paradox*. Harvard Business Review. https://hbr.org/2015/01/the-authenticity-paradox

Linda, L. N., & Chester, A. S. (2013). Advances in authentic and ethical leadership. *A volume in Research in Management, 10*.

Majid, Z. (2020). Authentic leadership and its role in promoting organizational creativity -a field study in the State Company for Petrochemical Industries. *Anbar University Journal of Economic and Administrative Sciences, 12*(29).

Metabis, A., & Al-Hawary, S. I. (2013). The impact of internal marketing practices on services quality of commercial banks in Jordan. *International Journal of Services and Operations Management, 15*(3), 313–337.

Mohammad, A. A., Alshura, M. S., Al-Hawary, S. I. S., Al-Syasneh, M. S., & Alhajri, T. M. (2020). The influence of internal marketing practices on the employees' intention to leave: A study of the private hospitals in Jordan. *International Journal of Advanced Science and Technology, 29*(5), 1174–1189.

Muhammed, S., & Zaim, H. (2020). Peer knowledge sharing and organizational performance: The role of leadership support and knowledge management success. *Journal of Knowledge Management*. Emerald Publishing Limited, ISSN 1367-3270. https://doi.org/10.1108/JKM-03-2020-0227.

Nair, B., Shrekumar, N. (2021). Exploring authentic leadership through leadership. *Journey of Gandhi, The Qualitative Report*, 26(3), 714–733. https://doi.org/10.46743/2160-3715/2021.4004

Odeh, R. B. M., Obeidat, B. Y., Jaradat, M. O., & Alshurideh, M. T. (2021). The transformational leadership role in achieving organizational resilience through adaptive cultures: the case of Dubai service sector. *International Journal of Productivity and Performance Management. ahead-of-print*(ahead-of-print). https://doi.org/10.1108/IJPPM-02-2021-0093

Pendleton, F., & Jonathan, C. (2021). Leadership. *Springer Nature*. https://doi.org/10.1007/978-3-030

Proctor, T. (2021). *Absolute essentials of creative thinking and problem solving*. Routledge, 605 Third Avenue, New York, NY 10158.

Rimkeviciene, J., Hawgood, J., O'Gorman, J., & De Leo, D. (2017). Construct validity of the acquired capability for suicide scale: factor structure, convergent and discriminant validity. *Journal of Psychopathology and Behavioral Assessment, 39*(2), 291–302. https://doi.org/10.1007/s10862-016-9576-4

Sekaran, U., & Bougie, R. (2016). *Research methods for business: A skill-building approach* (7th edn.). Wiley.

Shamout, M., Elayan, M., Rawashdeh, A., Kurdi, B., & Alshurideh, M. (2022). E-HRM practices and sustainable competitive advantage from HR practitioner's perspective: A mediated moderation analysis. *International Journal of Data and Network Science, 6*(1), 165–178.

Shi, D., Lee, T., & Maydeu-Olivares, A. (2019). Understanding the model size effect on SEM fit indices. *Educational and Psychological Measurement, 79*(2), 310–334. https://doi.org/10.1177/0013164418783530

Sparks, R. (2021). *The authentic leader using the Meisner technique for embracing the values of truthful leadership*. Routledge/Productivity Press 52 Vanderbilt Avenue.

Sung, K.-S., Yi, Y. G., & Shin, H.-I. (2019). Reliability and validity of knee extensor strength measurements using a portable dynamometer anchoring system in a supine position. *BMC Musculoskeletal Disorders, 20*(1), 1–8. https://doi.org/10.1186/s12891-019-2703-0

Wang, Y. A., & Rhemtulla, M. (2021). Power analysis for parameter estimation in structural equation modeling: A discussion and tutorial. *Advances in Methods and Practices in Psychological Science, 4*(1), 1–17. https://doi.org/10.1177/2515245920918253

Ziemak, A., & Katarzyna, W. (2020). The relationship between organizational learning and sustainable performance: an empirical examination. *Journal of Workplace Learning Emerald Publishing* 1366–5626.

The Impact of Innovative Leadership on Crisis Management Strategies in Public Institutions in the State of Qatar

Kamel Mohammad Al-Hawajreh, Abdullah Matar Al-Adamat, Snaid Saleh Al-Daiya Al-Marri, Zaki Abdellateef Khalaf Khalaylah, Menahi Mosallam Alqahtani, Yahia Salim Melhem, Ziad Mohd Ali Smadi, Ayat Mohammad, and Sulieman Ibraheem Shelash Al-Hawary

Abstract The study aimed to examine the impact of innovative leadership practices on crisis management strategies in public institutions in the State of Qatar. The study population consisted of public institutions in the State of Qatar. In order to achieve the objectives of the study, a questionnaire was developed to collect data from the analytical unit, the Social Statistical Sciences Package (SPSS V.25) was adopted. The

K. M. Al-Hawajreh · S. S. Al-Daiya Al-Marri
Business Faculty, Mu'tah University, Karak, Jordan

A. M. Al-Adamat
Department of Business Administration and Public Administration, School of Business, Al al-Bayt University Jordan, P.O. Box 130040, Mafraq 25113, Jordan
e-mail: aaladamat@aabu.edu.jo

Z. A. K. Khalaylah
Faculty of Money and Management, Management Department, The World Islamic Science University, P.O. Box 1101, Amman 11947, Jordan

M. M. Alqahtani
Administration Department, Community College of Qatar, Doha, Qatar
e-mail: mena7i@icloud.com

Y. S. Melhem
Business Administration Department, School of Business, Yarmouk University, Irbid, Jordan
e-mail: ymelhem@yu.edu.jo

Z. M. A. Smadi
Department of Business Administration, School of Business, Al al-Bayt University, P.O. Box, Mafraq 13004025113, Jordan
e-mail: ziad38in@aabu.edu.jo

A. Mohammad
Business and Finance Faculty, The World Islamic Science and Education University (WISE), 11947, P.O Box 1101, Amman, Jordan

S. I. S. Al-Hawary (✉)
Department of Business Administration, School of Business, Al al-Bayt University, P.O. Box 130040, Mafraq 25113, Jordan
e-mail: dr_sliman73@aabu.edu.jo; dr_sliman@yahoo.com

M. Alshurideh et al. (eds.), *The Effect of Information Technology on Business and Marketing Intelligence Systems*, Studies in Computational Intelligence 1056,
https://doi.org/10.1007/978-3-031-12382-5_80

AMOS V.25 program were adopted for path analysis. The study showed an impact of Innovative leadership on crisis management strategies. The study recommended the need for Qatari public institutions to focus on creating strong relationships with strategic partners in the internal Qatari environment and the external environment.

Keywords Innovative leadership · Crisis management strategies · Public institutions · Qatar state

1 Introduction

The communications and information technology revolution has contributed to the scientific, cultural and technological developments that are unprecedented in human history, which made institutions of all kinds and orientations seek to keep pace with these developments, and review their policies, objectives, technologies and activities related to creativity, change and continuous improvement of their services and products (Al-Lozi et al., 2017; Al-Hawary & Al-Syasneh, 2020; Al-Quran et al., 2020; AlTaweel & Al-Hawary, 2021; Al-Hawary & Obiadat, 2021).

Organizations must prepare leaders and provide them with the skills that enable them to carry out the tasks entrusted to them to the fullest and to achieve the goals and objectives of the organization (Al-Hawary et al., 2012; Al-Hawary & Al-Hamwan, 2017; Al-Hawary & Mohammed, 2017; Al-Hawary, 2009; Mohammad et al., 2020). It is evident from monitoring the developments that occurred in the developed countries that the institutions were able to overcome all the obstacles they were facing by encouraging creative practices and applying them on the ground and updating the different methods of work that was a reason for the superiority and progress of countries, provided that these methods and tools are adapted to suit Arab environments (Al Kurdi et al., 2020; AlShehhi et al., 2020; Odeh et al., 2021; Alshurideh, 2022).

Since the crisis has a renewed and developed character, it needs the presence of creative leaders to deal with it by formulating a vision, a message and strategic goals to deal with crises as they occur, as most of the crises are considered emergency and sudden events that often have no introduction, and therefore the matter requires the presence of creative leaders able to take decisive decisions at the appropriate times and seize the initiative in leading, influencing and directing events as appropriate. The Ambidexterity of leadership lies in envisioning the possibility of transforming the crisis and the risks it carries into opportunities to unleash the innovative capabilities that exploit the crisis as an opportunity to reformulate conditions and find appropriate solutions (Hijjawi & Mohammad, 2019; Alameeri et al., 2020; Al-Dhuhouri et al., 2020). The real role of the innovative leader lies in revealing and investing the innovative energies of his followers, and working to make his organization learn innovation and creativity continuously (Al-Hawary et al., 2013; Al-Hawary & Hadad, 2016; Al-Hawary & Alhajri, 2020; Harahsheh et al., 2021).

Public institutions in the State of Qatar work like other institutions, facing environmental conditions, with a high degree of uncertainty and severe environmental complexity, which leads It calls for the necessity of a creative leadership approach, according to a management methodology, that focuses on developing crisis management strategies, so that this leads to relying on finding products/services at a lower cost, and with flexibility that enables them to adapt to emerging changes that affect them in light of the threats they face, also, adopting an approach of creativity to develop the products/services provided by public institutions in the State of Qatar, and to improve the quality of the products offered to customers. Through the researcher's extrapolation of the general climate in the State of Qatar, and the sector of public sector institutions in particular, he realized that public institutions are among the most prominent sectors that have witnessed changes and new challenges, in the crisis of the blockade, which posed a major challenge to these institutions; Therefore, public institutions in the State of Qatar are required to improve their products and enhance their competitiveness indicators, especially in the midst of the challenges the world is witnessing.

In general, the success of these institutions requires keeping pace with the new developments of strategic management. Strategic intelligence is also considered an advanced field (Al-Hawary & Ismael, 2010; Al-Hawary et al., 2020; Al-Nady et al., 2013; Alzoubi et al., 2022). It is considered one of the important tools that enable these institutions to deal with the environmental uncertainty they face, and it is an important factor in supporting and coordinating all efforts to unify them in order to achieve the strategic goals efficiently and effectively (Al-Dmour et al., 2021; Ashal et al., 2021; Shakhour et al., 2021). Due to the scarcity of studies that dealt with the impact of innovative leadership on crisis management strategies, especially in the work environment in the State of Qatar, the problem of the current study appears in analyzing the impact of innovative leadership in its dimensions (problem solving and decision-making, susceptibility to change, initiative, ability to communicate, stimulating creativity, and innovative knowledge), in crisis management strategies in its dimensions (early warning strategy, preparedness and prevention strategy, damage containment strategy, recovery strategy, learning and growth strategy) in public institutions in the State of Qatar.

2 Theoretical Framework and Hypothesis Building

2.1 *Innovative Leadership*

Leadership is one of the most important approaches to developing strategic performance. Because of its important role in influencing employees and directing their ideas and behavior in order to achieve the goals sought by the institution, because it has a significant impact on their behavior, performance and productivity at work. Innovative leaders are satisfied not only by rearranging existing organizational structures, but are motivated by the constant quest for a better way (Al-Hawary, 2015; Al-Hawary & Abu-Laimon, 2013; Al-Hawary & Alajmi, 2017).

The innovative leadership style is discovery and self-renewal and their impact contributes to achieving the general interest and raises the awareness of the leadership, followers and the institution as a whole, and the innovative leader is able to collect new ideas with each other in an individual way that organizes their unconnected relationship, and make them a focus of innovation (Al-Hawary & Al-Namlan, 2018; Allozi et al., 2022; Metabis & Al-Hawary, 2013). The term leadership is one of the concepts that have received great interest by researchers and those interested, as this concept has appeared in several different definitions due to the different views of researchers and the different approaches to studying this concept. Leadership is one of the most important topics that researchers have studied in depth, and despite the diversity and multiplicity of research and studies, they did not reach agreement on the concept of leadership.

The meaning of innovation in the business dictionary is to provide a new method, or a new service, which is the ability to find solutions to problems and crises through a new tool or a new technical effect, and it has been defined as the process that leads to the creation of a new idea and its output through a useful product or service, or methods of Operations (Al-Hawary & Aldaihani, 2016; Al-Hawary & Alwan, 2016; Al-Hawary & Nusair, 2017; Dogan, 2019). Other researchers also defined innovation it as a combination of abilities, preparations, and characteristics of a person which, if found in an appropriate environment, can elevate mental processes to lead to original and useful products, whether in relation to the individual's previous experiences or to the experiences of the institution, society or the world if the results are on a level of creative breakthroughs in one of the fields of human life (Hamdi, 2018).

The link between leadership and innovation is important in the success of the continuation of creative processes within society and in the development and creation of the creative climate. Innovative leaders are not only satisfied with rearranging the existing structures but are motivated by the continuous quest to find a better way. The creative leadership style is discovery and self-renewal, and their impact contributes to achieving the broader interest and raises the awareness of leadership, followers and the institution as a whole (Al Dukan, 2010). Innovation gives leadership characteristics that distinguish it from other leaders, such as openness towards change, the desire to identify internal problems and issues, in addition to giving the ability to control the environment in which he lives, confidence in the ability of others to achieve and bear responsibility, respect for others' choices, and the ability to rush towards learning and using culture, appreciating the achievements of others, and moving towards the future with full force (Al-Zubaidi & Al-Mamouri, 2015; Abuhashesh et al., 2021; Hamadneh et al., 2021).

It becomes clear that the innovative leadership in the organization is the innovator who translates knowledge into choices and who has ideas that are new and useful and related to solving specific problems or re-combining known patterns in unique forms (Al-Nady et al., 2016; Allahow et al., 2018; Alhalalmeh et al., 2020). Al-Saghir and Muhammad (2017) referred to creative leadership as the ability to change, renew or The development of a new approach or method, and its use of modern technologies that are compatible with the requirements of the environment and the aspirations of the modern era and meet the needs of the institution, on the other hand is the ability of the leader to use familiar things in an unfamiliar way, and to propose innovative ideas and solutions to the problems and issues he deals with, characterized by novelty, quality and feasibility. The traits of creators may be available, all or some of them, in the leader of the organization who has the ability to be innovative, and the absence of some of these traits does not mean her inability to lead the organization in a innovative way, but rather they are helpful and influential traits that develop creativity and creativity in the leader, and those in charge of the organization must take appropriate methods to develop these features. Innovative leadership has a set of basic features that appear to influence the behavior of leaders.

For the purposes of this study, the following dimensions were adopted: **Problem-solving and decision-making**: The sensitivity to problems and decision-making is one of the most important characteristics of leadership, as a innovative leader has the ability to face a situation that involves a problem or several problems that need to be solved or change, **Changeability**: The emergence of many changes and the many challenges facing the institution, which need a creative leader who can face those challenges, and come up with innovative solutions and proposals about them, especially after the institution's application of some administrative trends, which gave the institution more empowerment and independence to achieve self-renewal to enable the institution to Achieving a competitive advantage (Al-Jaafira, 2013). **Initiative**: It refers to the ability to be innovating and to act consciously and quickly in various situations. It is a characteristic of innovation and an administrative principle. Initiatives appear in a healthy climate that motivates workers and encourages them to take initiatives. It is the characteristic of an innovative leader who encourages the spirit of initiative. He takes the initiative and shows his initiative in making decisions that bring about important changes (Abdul Qader, 2012). **Ability to communicate**: Communication aims to identify ideas, information and directions through a social system and the ways in which opinions, trends and knowledge are formed. **Stimulating innovation**: It is a paragraph about a mental characteristic that enables the individual to think in unconventional ways, or as it is known and popularized "thinking outside the box, which contributes in one way or another to creativity of all kinds or the use of different, unusual methods while dealing with a specific task or issue. This requires good preparation by the higher management of the institution (Al-Zubaidi & Al-Mamouri, 2015). **Innovative knowledge**: It is the knowledge that gives the institution the ability to lead the sector, and in the case the institution's differentiation is clear compared to competitors, which enables it to change the rules of the game in the style and timing that it determines, and through the knowledge pyramid, it can be said that it contributes well to clarifying how to form and build the knowledge system.

2.2 *Crisis Management Strategies*

The concept of crisis has several meanings, including a sudden change to the worst and the risks and opportunities it entails to restore conditions to their normal state and find constructive solutions, or in other words, it is a turning point in unstable conditions, or a series of events that have a negative impact on the institution (Al-Zubaidi & Al-Mamouri, 2015). As for crisis management, it is the method of dealing with the crisis by the administrative and methodological processes that leaders in institutions adopt to confront the crisis before, during and after its occurrence, control and prevent it, and try to predict it by developing and implementing plans, and taking appropriate decisions through the necessary information and providing an effective communication system, achieving the desired goals, in minimal losses, cost and effort (Bobyleva & Sidorova, 2015; Alzoubi et al., 2021; Tariq et al., 2022).

The concept of crisis management refers to how to overcome the crisis by using scientific administrative methods in order to avoid its negative aspects as much as possible and maximize its answers, and crisis management is a purposeful activity based on research and obtaining the necessary information that enables management to predict the locations and trends of the expected crisis and create the appropriate climate To deal with it by taking measures to control the expected crisis.

The administrative literature has included a number of strategies for dealing with crises according to their nature, where the crisis differs in terms of its type, severity and causes, the goal of facing crises is to seek the available human and material capabilities to manage the situation. Therefore, the method of dealing with crises is the determinant of the management's efficiency in facing them. Accordingly, dealing with crises had to be subject to the scientific administrative approach to confirm the success factors and protect the administrative entity from any uncalculated change and secretions, present or future. Contemporary administrative thought has resulted in a set of strategies, and these strategies are consistent with the contents of technological and administrative developments, and are in agreement to some extent with the modern trends adopted by institutions towards workers and all stakeholders, taking into account the positive image of the institution, taking into account the achievement of community interest and deepening the institutional role of the organization in society (Al-Lami & Al-Issawi, 2016).

2.3 *The Relationship Between Innovative Leadership and Crisis Management*

The personality of the leader is clearly reflected on the institution and its management, especially when emergency matters or crises occur in the institution. The leader's adoption of a certain behavior inevitably includes the performance and satisfaction of employees to directly affect the effectiveness of the institution and depending on the nature of the crisis as an opportunity or a threat, managers and leaders of institutions may resort to changing their behavior between centralization and decentralization of crisis management, he resorts to the first type when the cause of the crisis is internal and when the institution's ability to control external factors weakens, the leader resorts to (decentralization) and because the crisis started outside the institution, the leader needs information about the crisis and it is natural for him to resort to middle and executive management, and thus (decentralization) is a preferred method for such a situation.

Researchers believe that many institutions are exposed to different crises in which the leader exhibits different patterns and behaviors. It is obvious that the leader deals with every crisis that occurs in different patterns, so what works for managing a crisis may not work for another crisis. The crisis is a catalyst for the emergence of different types of behavior and therefore different responses to crises. Also, cooperation and respect between the leader and the workers and between them facilitates the process of containing the crisis, in order to reach the optimal leadership style, more than one style can be used at the same time. As there is no fixed leadership style that the leader follows in all situations, and the strategies used to solve the crisis are determined according to the type and severity of crises and the characteristics of leadership qualified to overcome crises, and an attempt to diagnose the nature of the relationship between the strategy used by the administrative leader in the face of the crisis and the type, severity and reasons for the occurrence of the crisis.

Al-Kaabi (2014) said that there is a significant effect relationship between successful administrative leadership and crisis management. The study of Svejenova reached several results, stating the importance of the concept of innovative leadership in the context of social change, as it increases the understanding of creative industries by proposing a social purpose as a distinctive and open engine for innovation and a different source of budget, as well as significant limits to research and practice in the creative industries. The study of Ghosh (2016) confirmed that the innovation of leaders is subject to several influences, the most important of which are technical support at work and their level of creativity, and the leadership style has positive effects on innovation. **Based on the above literature, the study hypotheses may be formulated as:**

There is a significant effect of innovative leadership on the crisis management strategies.

3 Study Model

See Fig. 1.

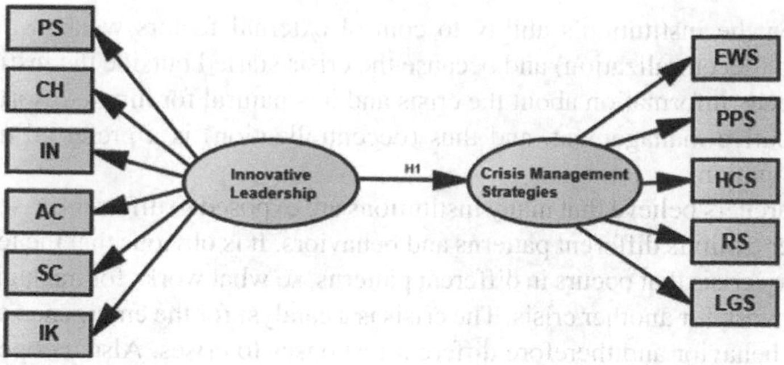

Fig. 1 Research model

4 Methodology

4.1 Population and Sample Selection

A quantitative method based on a questionnaire was used in this study for data collection and sample selection. The major aim of the study was to examine the impact of innovative leadership practices on crisis management strategies in public institutions in the State of Qatar. Data were primarily gathered through the questionnaires which were distributed to a random sample of (380) postgraduate students. In total, (327) responses were received including (53) invalid to statistical analysis due to uncompleted or inaccurate. Hence, the final sample contained (346) responses suitable to analysis requirements that were formed a response rate of (86.05%), where it proved to be sufficient to the extent that was predictable and allowed for a presumption of data saturation (Sekaran & Bougie, 2016).

4.2 Measurement Instrument

A questionnaire that consists of two main sections along with a section regarding control variables was used as the measurement instrument. Control variables considered as categorical measures were composed of gender, age group, educational level, and experience. The two main sections were dealt with a five-point Likert scale (from 1 = strongly disagree to 5 = strongly agree). The first section contained (24) items to measure innovative leadership based on Zhang et al. These questions were distributed into dimensions as follows: four items dedicated for measuring problems solving and decisions making, four items dedicated for measuring changeability, four items dedicated for measuring initiate, four items dedicated for measuring ability to communicate, four items dedicated for measuring stimulating creativity and four items dedicated for measuring innovative knowledge. Whereas the second section included (15) items developed to measure crisis management strategies according to what was pointed by Eliadis. This variable was divided into five dimensions: early warning strategy that was measured through five items, preparedness and prevention strategy which measured by five items, Harm containment strategy was measured using five items, recovery strategy that was measured by five items and learning and growth strategy was measured using five items.

5 Findings

5.1 Measurement Model Evaluation

This study was conducted structural equation modeling (SEM) to test hypotheses, which represents a contemporary statistical technique for testing and estimating the relationship between factors and variables (Wang & Rhemtulla, 2021). Accordingly, the reliability and validity of the constructs were tested using confirmatory factor analysis (CFA) through the statistical program AMOSv24. Table 1 summarizes the results of convergent and discriminant validity, as well the indicators of reliability.

Table 1 shows that the standard loading values for the individual items were within the domain (0.520–0.820), these values greater than the minimum retention of the elements based on their standard loads (Al-Lozi et al., 2018; Sung et al., 2019). Average variance extracted (AVE) is a summary indicator of the convergent validity of constructs that must be above 0.50 (Howard, 2018). The results indicate that the AVE values were greater than 0.50 for all constructs, thus the used measurement model has an appropriate convergent validity. Rimkeviciene et al. (2017) suggested the comparison approach as a way to deal with discriminant validity assessment in covariance-based SEM. This approach is based on comparing the values of maximum shared variance (MSV) with the values of AVE, as well as comparing the values of square root of AVE ($\sqrt{AVE}$) with the correlation between the rest of the structures. The results show that the values of MSV were smaller than the values of AVE, and that the values of $\sqrt{AVE}$ were higher than the correlation values among the rest of the constructs. Therefore, the measurement model used is characterized by discriminative validity. The internal consistency measured through Cronbach's Alpha coefficient (α) and compound reliability by McDonald's Omega coefficient (ω) was conducted as indicators to evaluate measurement model. The results listed in Table 1 demonstrated that both values of Cronbach's Alpha coefficient and McDonald's Omega coefficient were greater than 0.70, which is the lowest limit for judging on measurement reliability (De Leeuw et al., 2019).

Table 1 Results of validity and reliability tests

Constructs	1	2	3	4	5	6	7	8	9	10
1. PS	–									
2. CH	0.54	–								
3. IN	0.62	0.57	–							
4. AC	0.82	0.80	0.84	–						
5. SC	0.59	0.62	0.55	0.69	–					
6. IK	0.61	0.48	0.61	0.67	0.57	–				
7. EWS	0.62	0.51	0.63	0.63	0.69	0.58	–			
8. PPS	0.69	0.61	0.68	0.78	0.82	0.76	0.65	–		
9. HCS	0.45	0.55	0.42	0.62	0.74	0.44	0.59	0.74	–	
10.RS	0.57	0.74	0.65	0.55	0.74	0.55	0.66	0.53	0.66	–
11. LGS	0.77	0.80	0.74	0.65	0.69	0.53	0.45	0.48	0.59	0.70
VIF	2.580	2.384	1.777	1.460	1.670	1.884	–	–		
Loadings range	0.520–0.820	0.614–0.880	0.685–0.750	0.508–0.721	0.694–0.772	0.518–0.787	0.568–0.819	0.788–0.892	0.644–0.795	0.577–0.835
AVE	0.741	0.692	0.684	0.669	0.609	0.685	0.593	0.709	0.633	0.684
MSV	0.522	0.587	0.594	0.588	0.627	0.628	0.522	0.574	0.609	0.622
Internal consistency	0.844	0.957	0.883	0.795	0.869	0.922	0.912	0.888	0.844	0.839
Composite reliability	0.827	0.798	0.843	0.877	0.819	0.887	0.778	0.759	0.7490.738	0.833

5.2 Structural Model

The structural model illustrated no multicollinearity issue among predictor constructs because variance inflation factor (VIF) values are below the threshold of 5, as shown in Table 1 (Hair et al., 2017). This result is supported by the values of model fit indices shown in Fig. 1.

The results in Fig. 2 indicated that the chi-square to degrees of freedom (CMIN/DF) was 2.0002, which is less than 3 the upper limit of this indicator. The values of the goodness of fit index (GFI), the comparative fit index (CFI), and the Tucker-Lewis index (TLI) were upper than the minimum accepted threshold of 0.90. Moreover, the result of root mean square error of approximation (RMSEA) indicated to value 0.038, this value is a reasonable error of approximation because it is less than the higher limit of 0.08. Consequently, the structural model used in this study was recognized as a fit model for predicting the DEP and generalization of its result (Ahmad et al., 2016; Shi et al., 2019). To verify the results of testing the study hypotheses, structural equation modeling (SEM) was used, the results of which are listed in Table 2.

The results demonstrated in Table 2 show that all innovative leadership dimensions have a positive impact relationship on crisis management strategies. However, the

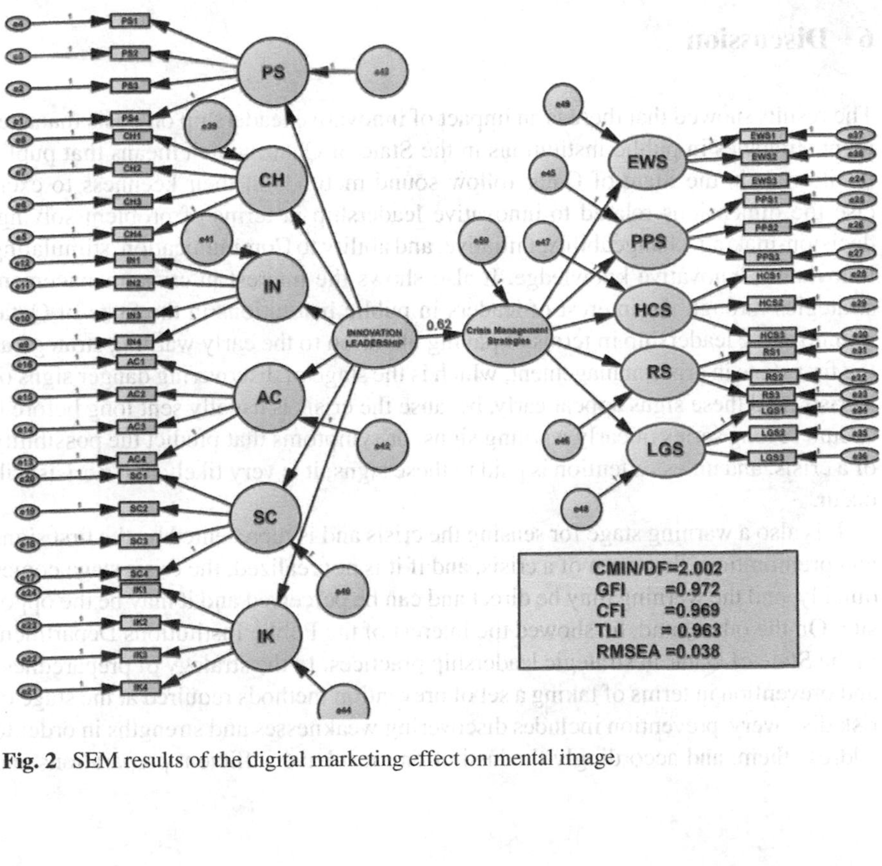

Fig. 2 SEM results of the digital marketing effect on mental image

Table 2 Hypothesis testing

Hypothesis	Relation	Standard beta	*t* value	*p* value
H1	Problems solving and decisions making → crisis management strategies	0.106	*8.912	0.000
H2	Changeability → crisis management strategies	0.109	*9.909	0.000
H3	Initiate → crisis management strategies	0.148	*13.910	0.000
H4	Ability to communicate → crisis management strategies	0.250	*14.176	0.000
H5	Stimulating creativity → crisis management strategies	0.247	*13.368	0.000
H6	Innovative knowledge → crisis management strategies	0.246	*11.900	0.000

results indicated that the highest impact was for combination (β = 0.250, $t = 14.176$, $p = 0.000$), followed by stimulating creativity (β = 0.247, $t = 14.368$, $p = 0.000$), and finally the lowest impact was for problems solving and decisions making (β = 0.106, $t = 8.912$, $p = 0.000$).

6 Discussion

The results showed that there is an impact of innovative leadership on crisis management strategies in public institutions in the State of Qatar, which means that public institutions in the State of Qatar follow sound methods in their keenness to exercise the dimensions related to innovative leadership in terms of problem solving, decision-making, changeability, initiative, and ability to Communication, stimulating innovation, innovative knowledge. It also shows the interest in crisis management strategies through the interest of leaders in public institutions in the State of Qatar in innovative leadership in terms of paying attention to the early warning strategy as the first stage in crisis management, which is the stage of discovering danger signs of a crisis, and these signs appear early, because the crisis is usually sent long before it occurs A long series of early warning signs, or symptoms that predict the possibility of a crisis, and unless attention is paid to these signs, it is very likely that a crisis will occur.

It is also a warning stage for sensing the crisis and is represented by the first signs and premonitions that warn of a crisis, and if it is not realized, the crisis stage comes quickly, and the warning may be direct and can be perceived and it may be the opposite. On the other hand, he showed the interest of the Public Institutions Department in the State of Qatar in strategic leadership practices. In the strategy of preparedness and prevention in terms of taking a set of prevention methods required at the stage of risk discovery, prevention includes discovering weaknesses and strengths in order to address them, and accordingly the institution must have sufficient preparations and methods to prevent crises. The results of the current study is consistent with Hamdi (2018), which dealt with the impact of innovative leadership on crisis management in the Jordanian telecommunications sector, and Al-Qahtani (2012), which dealt with the reality of the practice of innovative leadership among secondary school principals in Riyadh from the point of view of principals and supervisors, and the results of Saed (2016), which dealt with the role of innovative leadership in improving the level of organizational culture in the Palestinian Ministry of Health.

7 Recommendations

According to the findings of the analysis, the researchers present a number of recommendations to decision-makers to enhance the desired results from the application of the dimensions of creative leadership in adopting crisis management strategies in public institutions in the State of Qatar through interest in creating strong and solid strategic partnership relationships with strategic partners in the internal environment of a Qatari state, as well as the external environment at the level of countries and regional and global institutions, and work on developing the capabilities of the leaders of public institutions in the State of Qatar through training in skills and methods of solving problems in scientific ways through scientific thinking and focusing on predicting problems before they occur. And the need for public institutions in the State of Qatar to pay attention to indicators of growth and leadership, such as training on brainstorming sessions, new ways and methods of work, generating creative ideas, enhancing innovative knowledge, and trying to inform them of what is being developed of creative methods and thinking outside the box.

References

Abdul Qadir, S. A. Q. (2012). Creative leadership and its relationship to the organizational climate, Saudi Arabia, Umm Al-Qura, Master's Thesis, Kingdom of Saudi Arabia.

Abuhashesh, M. Y., Alshurideh, M. T., & Sumadi, M. (2021). The effect of culture on customers' attitudes toward Facebook advertising: The moderating role of gender. *Review of International Business and Strategy, 31*(3), 416–437.

Ahmad, S., Zulkurnain, N., & Khairushalimi, F. (2016). Assessing the Validity and reliability of a measurement model in structural equation modeling (SEM). *British Journal of Mathematics & Computer Science, 15*(3), 1–8. https://doi.org/10.9734/BJMCS/2016/25183

Al Kurdi, B., Alshurideh, M., & Al Afaishat, T. (2020). Employee retention and organizational performance: Evidence from banking industry. *Management Science Letters, 10*(16), 3981–3990.

Al-Quran, A. Z., Alhalalmeh, M. I., Eldahamsheh, M. M., Mohammad, A. A., Hijjawi, G. S., Almomani, H. M., & Al-Hawary, S. I. (2020). Determinants of the green purchase intention in Jordan: The moderating effect of environmental concern. *International Journal of Supply Chain Management, 9*(5), 366–371.

Alameeri, K., Alshurideh, M., Al Kurdi, B., & Salloum, S. A. (2020, October). The effect of work environment happiness on employee leadership. In *International conference on advanced intelligent systems and informatics* (pp. 668–680). Cham: Springer.

Al-Dhuhouri, F. S., Alshurideh, M., Al Kurdi, B., & Salloum, S. A. (2020, October). Enhancing our understanding of the relationship between leadership, team characteristics, emotional intelligence and their effect on team performance: A Critical Review. In *International conference on advanced intelligent systems and informatics* (pp. 644–655). Cham: Springer.

Al-Dmour, R., AlShaar, F., Al-Dmour, H., Masa'deh, R., & Alshurideh, M. T. (2021). The effect of service recovery Justices strategies on online customer engagement via the role of "Customer Satisfaction" during the Covid-19 pandemic: An empirical study. In *The effect of coronavirus disease (COVID-19) on business intelligence* (Vol. 334, pp. 325–346).

Alhalalmeh, M. I., Almomani, H. M., Altarifi, S., Al- Quran, A. Z., Mohammad, A. A., & Al-Hawary, S. I. (2020). The nexus between corporate social responsibilty and organizational performance in Jordan: The mediating role of organizational commitment and organizational citizenship behavior. *Test Engineering and Management, 83*(July), 6391–6410.

Al-Hawary, S. I. (2009). The effect of the leadership style on the effectiveness of the organization: A field study at Zarqa Private University. *The Egyptian Journal for Commercial Studies, 33*(1), 361–393.

Al-Hawary, S. I. (2015). Human resource management practices as a success factor of knowledge management implementation at health care sector in Jordan. *International Journal of Business and Social Science, 6*(11/1), 83–98.

Al-Hawary, S. I. S., & Alhajri, T. M. S. (2020). Effect of electronic customer relationship management on customers' electronic satisfaction of communication companies in Kuwait. *Calitatea, 21*(175), 97–102.

Al-Hawary, S. I. S., & Alwan, A. M. (2016). Knowledge management and its effect on strategic decisions of Jordanian Public Universities. *Journal of Accounting-Business & Management, 23*(2), 24–44.

Al-Hawary, S. I. S., & Mohammed, A. K. (2017). Impact of team work traits on organizational citizenship behavior from the viewpoint of the employees in the education directorates in north region of Jordan. *Global Journal of Management and Business, 17*(2-A), 23–40.

Al-Hawary, S. I. S., & Obiadat, A. A. (2021). Does mobile marketing affect customer loyalty in Jordan? *International Journal of Business Excellence, 23*(2), 226–250.

Al-Hawary, S. I. S., Mohammad, A. S., Al-Syasneh, M. S., Qandah, M. S. F., & Alhajri, T. M. S. (2020). Organisational learning capabilities of the commercial banks in Jordan: Do electronic human resources management practices matter? *International Journal of Learning and Intellectual Capital, 17*(3), 242–266.

Al-Hawary, S. I., & Abu-Laimon, A. A. (2013). The impact of TQM practices on service quality in cellular communication companies in Jordan. *International Journal of Productivity and Quality Management, 11*(4), 446–474.

Al-Hawary, S. I., & Alajmi, H. M. (2017). Organizational commitment of the employees of the ports security affairs of the state of Kuwait: The impact of human recourses management practices. *International Journal of Academic Research in Economics and Management Sciences, 6*(1), 52–78.

Al-Hawary, S. I., & Aldaihani, F. M. (2016). Customer relationship management and innovation capabilities of Kuwait airways. *International Journal of Academic Research in Economics and Management Sciences, 5*(4), 201–226.

Al-Hawary, S. I., & Al-Hamwan, A. (2017). Environmental analysis and its impact on the competitive capabilities of the commercial banks operating in Jordan. *International Journal of Academic Research in Accounting, Finance and Management Sciences, 7*(1), 277–290.

Al-Hawary, S. I., & Al-Namlan, A. (2018). Impact of electronic human resources management on the organizational learning at the private hospitals in the state of Qatar. *Global Journal of Management and Business Research: A Administration and Management, 18*(7), 1–11.

Al-Hawary, S. I., & Al-Syasneh, M. S. (2020). Impact of dynamic strategic capabilities on strategic entrepreneurship in presence of outsourcing of five stars hotels in Jordan. *Business: Theory and Practice, 21*(2), 578–587.
Al-Hawary, S. I., & Hadad, T. F. (2016). The effect of strategic thinking styles on the enhancement competitive capabilities of commercial banks in Jordan. *International Journal of Business and Social Science, 7*(10), 133–144.
Al-Hawary, S. I., & Ismael, M. (2010). The effect of using information technology in achieving competitive advantage strategies: A field study on the Jordanian pharmaceutical companies. *Al Manara for Research and Studies, 16*(4), 196–203.
Al-Hawary, S. I., & Nusair, W. (2017). Impact of human resource strategies on perceived organizational support at Jordanian Public Universities. *Global Journal of Management and Business Research: A Administration and Management, 17*(1), 68–82.
Al-Hawary, S. I., AL-Awawdeh, W., & Abden, M. A. (2012). The impact of the leadership style on organizational commitment: A field study on kuwaiti telecommunications companies. *ALEDARI*, (130), 53–102.
Al-Hawary, S. I., Al-Qudah, K., Abutayeh, P., Abutayeh, S., & Al-Zyadat, D. (2013). The impact of internal marketing on employee's job satisfaction of commercial banks in Jordan. *Interdisciplinary Journal of Contemporary Research in Business, 4*(9), 811–826.
Al-Jaafira, S. (2013). Methods of managing organizational conflict and its relationship to administrative creativity among the principals of public schools in the Karak Governorate from their point of view. *Dirasat Journal, 40*(2), 1663–1687.
Al-Kaabi, H. S. (2014). The role of successful administrative leaders in crisis management, an analytical study of some companies in the Ministry of Transport and Communications, Al-Rafidain College, public institutions in the Department of Business Administration, Iraq.
Allahow, T. J. A. A., Al-Hawary, S. I. S., & Aldaihani, F. M. F. (2018). Information technology and administrative innovation of the central agency for information technology in Kuwait. *Global Journal of Management and Business, 18*(11-A), 1–16.
Al-Lami, G. Q., & Al-Issawi, K. A. (2016). *Crisis Management, foundations and applications* (1st ed.). Amman, Jordan: House of Methodology for Publishing and Distribution.
Allozi, A., Alshurideh, M., AlHamad, A., & Al Kurdi, B. (2022). Impact of transformational leadership on the job satisfaction with the moderating role of organizational commitment: Case of UAE and Jordan manufacturing companies. *Academy of Strategic Management Journal, 21*, 1–13.
Al-Lozi, M., Almomani, R. Z., & Al-Hawary, S. I. (2017). Impact of talent management on achieving organizational excellence in Arab potash company in Jordan. *Global Journal of Management and Business Research: A Administration and Management, 17*(7), 15–25.
Al-Lozi, M., Almomani, R. Z., & Al-Hawary, S. I. (2018). Talent Management strategies as a critical success factor for effectiveness of Human Resources Information Systems in commercial banks working in Jordan. *Global Journal of Management and Business Research: A Administration and Management, 18*(1), 30–43.
Al-Nady, B. A., Al-Hawary, S. I., & Alolayyan, M. (2013). Strategic management as a key for superior competitive advantage of sanitary ware suppliers in Kingdom of Saudi Arabia. *International Journal of Management and Information Technology, 7*(2), 1042–1058.
Al-Nady, B. A., Al-Hawary, S. I., & Alolayyan, M. (2016). The role of time, communication, and cost management on project management success: An empirical study on sample of construction projects customers in Makkah City, Kingdom of Saudi Arabia. *International Journal of Services and Operations Management, 23*(1), 76–112.
Al-Qahtani, Jaafar is enough. (2012). The reality of the practice of creative leadership among secondary school principals in Riyadh from the point of view of principals and supervisors. Unpublished Ph.D. Thesis, Riyadh: Public Institutions in King Saud.
Al-Saghir, G., & Muhammad, H. (2017). Creative leadership and its role in the development of security institutions, an unpublished Ph.D. Thesis, public institutions in Khartoum.

AlShehhi, H., Alshurideh, M., Al Kurdi, B., & Salloum, S. A. (2020, October). The impact of ethical leadership on employees performance: A systematic review. In *International conference on advanced intelligent systems and informatics* (pp. 417–426). Cham: Springer.

Alshurideh, M. (2022). Does electronic customer relationship management (E-CRM) affect service quality at private hospitals in Jordan? *Uncertain Supply Chain Management, 10*(2), 325–332.

AlTaweel, I. R., & Al-Hawary, S. I. (2021). The Mediating Role of Innovation Capability on the Relationship between Strategic Agility and Organizational Performance. *Sustainability, 13*(14), 7564.

Alzoubi, H. M., Alshurideh, M., & Ghazal, T. M. (2021, June). Integrating BLE beacon technology with intelligent information systems IIS for operations' performance: A managerial perspective. In *The international conference on artificial intelligence and computer vision* (pp. 527–538). Cham: Springer.

Alzoubi, H., Alshurideh, M., Kurdi, B., Akour, I., & Aziz, R. (2022). Does BLE technology contribute towards improving marketing strategies, customers' satisfaction and loyalty? The role of open innovation. *International Journal of Data and Network Science, 6*(2), 449–460.

Al-Zubaidi, L. S. A. A., & Al-Amouri, K. A. H. (2015). The impact of focusing on creativity engines in crisis management, published research paper. *Journal of Management and Economics, 38*(103).

Ashal, N., Alshurideh, M., Obeidat, B., & Masa'deh, R. (2021) The impact of strategic orientation on organizational performance: Examining the mediating role of learning culture in Jordanian telecommunication companies. *Academy of Strategic Management Journal, 21*(Special Issue 6), 1–29.

Bobyleva, A., & Sidorova, A. (2015). Crisis management in higher education in Russia. *Internationalization in Higher Education. Management of Higher Education and Research, 3*(1), 23–35.

de Leeuw, E., Hox, J., Silber, H., Struminskaya, B., & Vis, C. (2019). Development of an international survey attitude scale: Measurement equivalence, reliability, and predictive validity. *Measurement Instruments for the Social Sciences, 1*(1), 9. https://doi.org/10.1186/s42409-019-0012-x

Dogan, M. (2019). *Creative marginality: Innovation at the intersections of social sciences*. Routledge.

Dukan, A. (2010). Creative leadership. http://www.manhal.net/articles.php?action=show&id=2794.

Ghosh, K. (2016). Creative leadership for workplace innovation: An applied SAP-LAP framework. *Development and Learning in Organizations: An International Journal, 30*(1), 10–14.

Hair, J. F., Babin, B. J., & Krey, N. (2017). Covariance-based structural equation modeling in the journal of advertising: Review and recommendations. *Journal of Advertising, 46*(1), 163–177. https://doi.org/10.1080/00913367.2017.1281777

Hamadneh, S., Hassan, J., Alshurideh, M., Al Kurdi, B., & Aburayya, A. (2021). The effect of brand personality on consumer self-identity: The moderation effect of cultural orientations among British and Chinese consumers. *Journal of Legal, Ethical and Regulatory Issues, 24*, 1–14.

Hamdi, K. W. (2018). *The impact of creative leadership on crisis management: A field study on the Jordanian telecommunications sector*. Unpublished Master's Thesis, Public Institutions in Al-Bayt, Jordan.

Harahsheh, A., Houssien, A., Alshurideh, M. & AlMontaser, M. (2021). The effect of transformational leadership on achieving effective decisions in the presence of psychological capital as an intermediate variable in Private Jordanian Universities in light of the corona pandemic. In: *The effect of coronavirus disease (COVID-19) on business intelligence* (Vol. 334, pp. 221–243).

Hijjawi, G. S., & Mohammad, A. (2019). Impact of organizational ambidexterity on organizational conflict of zain telecommunication company in Jordan. *Indian Journal of Science and Technology, 12*, 26.

Howard, M. C. (2018). The convergent validity and nomological net of two methods to measure retroactive influences. *Psychology of Consciousness: Theory, Research, and Practice, 5*(3), 324–337. https://doi.org/10.1037/cns0000149

Metabis, A., & Al-Hawary, S. I. (2013). The impact of internal marketing practices on services quality of commercial banks in Jordan. *International Journal of Services and Operations Management, 15*(3), 313–337.

Mohammad, A. A., Alshura, M. S., Al-Hawary, S. I. S., Al-Syasneh, M. S., & Alhajri, T. M. (2020). The influence of internal marketing practices on the employees' intention to leave: A study of the private hospitals in Jordan. *International Journal of Advanced Science and Technology, 29*(5), 1174–1189.

Odeh, R. B. M., Obeidat, B. Y., Jaradat, M. O., & Alshurideh, M. T. (2021). The transformational leadership role in achieving organizational resilience through adaptive cultures: The case of Dubai service sector. *International Journal of Productivity and Performance Management*. Vol. ahead-of-print No. ahead-of-print. https://doi.org/10.1108/IJPPM-02-2021-0093.

Rimkeviciene, J., Hawgood, J., O'Gorman, J., & De Leo, D. (2017). Construct validity of the acquired capability for suicide scale: Factor structure, convergent and discriminant validity. *Journal of Psychopathology and Behavioral Assessment, 39*(2), 291–302. https://doi.org/10.1007/s10862-016-9576-4

Saed, N. A. (2016). The role of creative leadership in improving the level of organizational culture in the Palestinian Ministry of Health, Palestine Administration and Politics Academy for Graduate Studies, unpublished Master's Thesis, Palestine.

Sekaran, U., & Bougie, R. (2016). *Research methods for business: A skill-building approach* (7th ed.). Wiley.

Shakhour, R., Obeidat, B., Jaradat, M., Alshurideh, M., & Masa'deh, R. (2021) Agile-minded Organizational Excellence: Empirical investigation. *Academy of Strategic Management Journal*, 20(Special Issue 6), 1–25.

Shi, D., Lee, T., & Maydeu-Olivares, A. (2019). Understanding the model size effect on SEM fit indices. *Educational and Psychological Measurement, 79*(2), 310–334. https://doi.org/10.1177/0013164418783530

Sung, K.-S., Yi, Y. G., & Shin, H.-I. (2019). Reliability and validity of knee extensor strength measurements using a portable dynamometer anchoring system in a supine position. *BMC Musculoskeletal Disorders, 20*(1), 1–8. https://doi.org/10.1186/s12891-019-2703-0

Tariq, E., Alshurideh, M., Akour, I., & Al-Hawary, S. (2022). The effect of digital marketing capabilities on organizational ambidexterity of the information technology sector. *International Journal of Data and Network Science, 6*(2), 401–408.

Wang, Y. A., & Rhemtulla, M. (2021). Power analysis for parameter estimation in structural equation modeling: A discussion and tutorial. *Advances in Methods and Practices in Psychological Science, 4*(1), 1–17. https://doi.org/10.1177/2515245920918253

A Systematic Review on the Influence of Entrepreneurial Leadership on Social Capital and Change Propensity

Khadija Alameeri, Muhammad Turki Alshurideh, and Barween Al Kurdi

Abstract The last three decades have been characterized by an increased number of research studies examining how the management of organizations has become a critical area of concern in modern literature. This paper presents a systematic literature review of the influence of entrepreneurial leadership on social capital and change propensity. The work clearly states the research question on how these variables affect a firm's success. The methodology involved drawing ideas from recent studies that have explored the association between leadership styles and an organization's ability to achieve consistent social capital and effective change management. Relevant literature was accessed on Google scholar search and used to examine the relationships between the variables. This study seeks to fill a gap in the literature by presenting a theoretically-grounded methodology that helps leaders design and adopt approaches that focus on achieving and integrating social capital in managing change in organizations.

Keywords Systematic review · Entrepreneurial · Leadership · Entrepreneurial leadership · Change propensity · Social capital

---

K. Alameeri
University of Sharjah, Sharjah, UAE

M. T. Alshurideh (✉)
Department of Management, College of Business Administration, University of Sharjah, Sharjah, UAE
e-mail: malshurideh@sharjah.ac.ae; m.alshurideh@ju.edu.jo

Department of Marketing, School of Business, The University of Jordan, Amman, Jordan

B. Al Kurdi
Faculty of Economics and Administrative Sciences, Department of Marketing, The Hashemite University, Zarqa, Jordan
e-mail: barween@hu.edu.jo

M. Alshurideh et al. (eds.), *The Effect of Information Technology on Business and Marketing Intelligence Systems*, Studies in Computational Intelligence 1056,
https://doi.org/10.1007/978-3-031-12382-5_81

1 Introduction

Entrepreneurial leadership is the cornerstone of development for any organization that aims at achieving outstanding milestones in business. While various researchers have focused on the importance of proper management of companies, there is a need for methodical approaches to assess the significance of actualizing social capital to initiate and implement change processes (Chitsaz et al., 2019; Dean & Ford, 2017; Ximenes et al., 2019). The paper provides the literature review based on the role of entrepreneurial leadership in articulating common organizational values, vision, risk-taking, effective communication, and creativity.

A number of scholars have drawn parallels between social capital and change propensity historically and conceptually (Clark et al., 2019). In this emerging concept, some scholars define entrepreneurial leadership within a narrow view, specifically within the context of small businesses (Alameeri et al., 2021; Dost et al., 2018; Felix et al., 2019; Leitch & Volery, 2017; Mehmood et al., 2019a; Yani et al., 2020). Although it may appear as if there is a single leadership approach to the actualization of social capital, a closer examination would reveal that numerous strategies may be necessary to align these to favor the propensity of change within an organization. Alameeri et al. (2021), Al-Dhuhouri et al. (2021), AlShehhi et al. (2021), Harahsheh et al. (2021), Madi Odeh et al. (2021), Mehmood et al. (2019b) argued that this new paradigm of management stretches beyond the convergence of fields and is linked to the characteristics of leadership styles across diverse conditions and contexts.

The purpose of this paper is to provide an insight into the development of entrepreneurial leadership. The resulting systematic literature review and discussion delve into providing a conceptual map of the prevailing state of research in the new paradigm with a view of identifying directions into future studies. It provides a benchmark for further investigation into the influence of entrepreneurial leadership on social capital and inclination to change.

2 Research Question

Does entrepreneurial leadership influence the development of social capital and enable organizations to have a high propensity to change?

3 The Motivation to Investigate the Research Question

Current studies have explored the link between leadership style and the organizations' ability to attain social capital and effective change management. Nevertheless, there is growing confusion because these studies claim that such a relationship depends on the leadership approach. Despite its growing influence on the two aspects of business, most of these studies undermine the role of entrepreneurial leadership in developing social capital and the ability of the organizations to cope with emerging changes in the business environment. Besides, studies attempting to explore the link between entrepreneurial leadership and social capital have appeared to ignore this leadership strategy's role by contending that social capital only associates with transformational and democratic leadership approaches (Abu Zayyad et al., 2021; Al Kurdi et al., 2021; Allozi et al., 2022; Al-Maroof, et al., 2021; Alshurideh et al., 2019). Other researchers have argued that social capital depends on multiple factors, including individual attitudes and behaviors in the workplace (Al Khasawneh et al., 2021a, 2021b; Aljumah et al., 2021; Almazrouei et al., 2021; Nuseir et al., 2021; Refae et al., 2021). Therefore, there is a need to determine the role of entrepreneurial leadership in developing social capital and effective change management.

4 Research Methodology

The primary method that would be applied in the literature review is the systematic review methodology. It involves a critical description and appraisal of current and previous studies associated with a particular research inquiry to determine relationships between themes and ideas (Ahmad et al., 2021; Al Naqbia et al., 2020; Al Suwaidi et al., 2021; Alhashmi et al., 2020; Alshamsi et al., 2021; Bazan et al., 2020; Linares-Espinós et al., 2018; Mehmood et al., 2019c; Mehrez et al., 2021; Mishra & Misra, 2017). The systematic review strategy will involve two significant steps: identifying the research process and determining which secondary research sources would help provide the study with the resources (Ahmad et al., 2021; Alhashmi et al., 2020; Assad & Alshurideh, 2020; Bettayeb et al., 2020). The study will adopt two search strategies. The first one will involve using Google Search Engine to locate studies relevant to the systematic review. The second one will include using specific searching terms and phrases such as "influence of entrepreneurial leadership on social capital" or influence of entrepreneurial leadership on change management."

5 Tabulated Systematic Literature Review

Author/s	Year	Title	Sample	Country of origin	Main aspect	Method	Main findings	Future directions
Mamun et al. (2018)	2018	Entrepreneurial leadership, performance, and sustainability of micro-enterprises in Malaysia	100 people	Malaysia	Establishing the link between entrepreneurial leadership, performance d sustainability of micro-enterprises	Cross-sectional research design	Entrepreneurial leadership contains emotional intelligence that sustains micro-enterprises	Future studies should incorporate relevant constructs in the present model to explain detailed aspects of entrepreneurial leadership in relation to achieving social capital and change mobility
Leitch and Volery (2017)	2017	Entrepreneurial leadership: Insights and directions	Sampled previous studies/no Primary study	Not specified	Providing insights and directions about the link entrepreneurial leadership and its impact on social capital		Entrepreneurial leadership is still evolving. It lacks appropriate tools to determine its traits and nature	Should focus on determining the key attributes of entrepreneurial leadership

(continued)

(continued)

Author/s	Year	Title	Sample	Country of origin	Main aspect	Method	Main findings	Future directions
Mishra and Misra (2017)	2015	Entrepreneurial leadership and organizational effectiveness: A comparative study of executives and non-executives	410 workers for India's firms	India	Determining the relationship between entrepreneurial leadership and organizational effectiveness		Executives and non-executives differed on the role of this leadership style on a firm's effectiveness	Future studies should include other organizational variables to draw possible links
Mehmood et al. (2019b)	2019	Impact of Entrepreneurial Leadership on Employee's Innovative Behavior: Mediating role of Psychological Empowerment	301 managers and employees of SMEs	Pakistan	The mediating role of psychological empowerment that entrepreneurial leaders have on workers		Entrepreneurial leadership has a direct impact on intelligent behaviors in workers and an indirect impact on intelligent behavior through psychological empowerment	Future researchers must adopt longitudinal to validate the existing link between entrepreneurial leadership and intelligent behavior

(continued)

(continued)

Author/s	Year	Title	Sample	Country of origin	Main aspect	Method	Main findings	Future directions
Ranjan (2018)	2018	Entrepreneurial leadership: a review of measures, antecedents, outcomes, and moderators	Review of 50 studies	India	Reviewing measures, antecedents, outcomes, and moderators. Examining the dimensions of Entrepreneurial leadership (strategic factors, communicative factors, motivational factors, personal factors, and leadership behaviors		The research discovered EL traits, including ambitiousness, performance-oriented, diplomatic behaviors, inspiration, positive attitude, intellectual stimulation	Extend future research on the possible link between the mentioned constructions and employee performance
Yang et al. (2019)	2019	Entrepreneurial Leadership and Turnover Intention of Employees: The Role of Affective Commitment and Person-job Fit	427 workers	China	The mediating role of affective commitment created by EL on-employee turnover intentions		The study discovered that EL promotes effective commitment, which eventually reduces employee turnover intentions	Future research can collect data from more companies to gather more insights into the relationship

(continued)

(continued)

Author/s	Year	Title	Sample	Country of origin	Main aspect	Method	Main findings	Future directions
Ximenes et al. (2019)	2019	Entrepreneurial leadership moderating high-performance work system and employee creativity on employee performance	200 workers	Indonesia	Entrepreneurial leadership as a moderator of performance work systems, employee creativity, and performance		The research discovered that EL influences high-performance work systems, which eventually influence employee creativity and performance	The results could not be generalized. Hence, future studies should expand the scope of the study beyond cooperatives
Miao et al. (2018)	2018	How Leadership and Public Service Motivation Enhance Innovative Behavior	281 Chinese civil servants	China	How various leadership approaches influenced innovativeness in employees		Entrepreneurial leadership influences innovation in employees	Future researchers should use objective data to determine innovative behavior

(continued)

(continued)

Author/s	Year	Title	Sample	Country of origin	Main aspect	Method	Main findings	Future directions
Cai et al. (2019)	2019	Does entrepreneurial leadership foster creativity among employees and teams? The mediating role of creative efficacy beliefs?	43 leaders and 237 employees	China	Using social cognitive theory, the researchers wanted to establish the link between EL, individual creativity, and team creativity		There is entrepreneurial leadership creativity generated by creative efficacy beliefs	Future research on this subject should focus on other countries that nurture collectivist cultural norms
Li et al. (2020)	2020	Impact of Entrepreneurial Leadership on Innovative Work Behaviour: Examining Mediation and Moderation Mechanisms	350	China	The influence of entrepreneurial leadership on innovative work behavior and culture among employees		There was a significant relationship between entrepreneurial leadership and innovative work behavior	The study relied on cross-sectional data of a single country. Future researchers should focus on more countries

6 Results

6.1 *Entrepreneurial Leadership, Social Capital, and Change Propensity*

Despite some studies ignoring the influence of entrepreneurial leadership on the organization's social capital, others have captured the issue and provided evidence for such relationships. First, it is vital to ascertain the nature and characteristics of entrepreneurial leadership before drawing the connection between the two subjects. To evaluate the critical attributes of entrepreneurial leadership, Leitch and Volery (2017) defined this leadership approach as one that involves influencing other individuals in a workplace to engage in collective efforts towards achieving shared goals. In a different study that explored entrepreneurial leadership in the context of organizational performance and sustainability, entrepreneurial leadership generally involves a combination of leadership and entrepreneurial capabilities (Mamun et al., 2018). Such combinations portray a hybrid leader who can effectively influence others to achieve a common goal, optimize risks, encourage innovation, and take advantage of presenting opportunities in a dynamic environment.

6.2 *Entrepreneurial Leadership and Change Propensity*

A propensity to change is the likelihood of a company to change its structure and functioning in response to emerging issues (Harrison et al., 2018; Sarabi et al., 2020; Sawaean & Ali, 2020). The research associates this unique potential of companies to high levels of innovativeness and creativity often nurtured through a careful application of the entrepreneurial leadership approach. As discussed by Dean and Ford (2017), entrepreneurial leaders discover and tap opportunities in a dynamic business environment to alleviate risks, create new ways of thriving, and increasing organizational innovativeness.

7 Limitations of the Literature Review

The research method used in this systematic literature review has numerous merits and demerits. A distinguished usefulness of the methodical search is the use of up-to-date and peer-reviewed sources to provide applicable responses to the research question. Nonetheless, the literature review has been restrained by the phrases "influence of entrepreneurial leadership on social capital" and the "influence of entrepreneurial leadership on change management" used to identify and retrieve the information. As a result, further studies should look for better sources that have not been included in the synthesis. The approach to selecting the vital terms and keywords was intentionally meant for the identification of reliable literature on the relationship between entrepreneurial leadership and achievement of social capital in organizations. The second limitation of the study was that the search method was constrained to publications written in the English language, which led to a bias inclining to research conducted in English-speaking countries.

8 Directions for Future Research

As highlighted by the systematic literature review, the concept of entrepreneurial leadership opens vast opportunities for further research. Although this area of study has drawn increased attention recently, the influence of entrepreneurial leadership on social capital and change propensity has received little consideration (Ha et al., 2020). Intellectual developments in this concept have presented researchers with opportunities to explore the role and influence of leaders in shaping the abilities of workers to accept change processes (Dabić et al., 2021; Obeidat et al., 2021; Pasricha & Rao, 2018). Since entrepreneurs from different countries face unique challenges due to discrepancies in GDP per capita, policies and conditions suitable for business may not favor them equally (Al-Gasaymeh et al., 2015, 2020; Al-Jarrah et al., 2012; Hsiao et al., 2016; Hsu & Chen, 2019; Setini et al., 2020; Theodoraki et al., 2018). The review has raised questions that would be useful in conducting future research. For instance, how can entrepreneurial leadership attributes be developed to realize social capital and change propensity? Consequently, there is a need to carry out further surveys to come up with a definitive model of entrepreneurial leadership that clearly expresses the link between the mentioned variables.

9 Conclusion

Many studies undermine the role of leadership styles in actualizing social capital and effective change management in organizations. While many of the authors acknowledge that little research has been done to draw relationships between social capital and change propensity, the systematic literature review proves that entrepreneurial leadership is a relatively new area of study. Nonetheless, facts from the review indicate that it is an essential factor in the development of an enterprise through capitalization on human resources and active management of change. The findings of the study underpinned the conceptual development of management frameworks that promote shared values, vision, risk-taking, effective communication, and creativity. A concern that arose from the literature review is that some studies overlooked the influence of entrepreneurial leadership on the organization's social capital, while others adequately covered the issue by providing evidence for such relationships.

References

Abu Zayyad, H. M., Obeidat, Z. M., Alshurideh, M. T., Abuhashesh, M., Maqableh, M., & Masa'deh, R. (2021). Corporate social responsibility and patronage intentions: The mediating effect of brand credibility. *Journal of Marketing Communications, 27*(5). https://doi.org/10.1080/13527266.2020.1728565.

Ahmad, A., Alshurideh, M. T., Al Kurdi, B. H., & Alzoubi, H. M. (2021). Digital strategies: A systematic literature review. In *The international conference on artificial intelligence and computer vision* (pp. 807–822).

Al Khasawneh, M., Abuhashesh, M., Ahmad, A., Masa'deh, R., & Alshurideh, M. T. (2021a). *Customers online engagement with social media influencers' content related to COVID 19* (Vol. 334).

Al Khasawneh, M., Abuhashesh, M., Ahmad, A., Alshurideh, M. T., & Masa'deh, R. (2021b). *Determinants of E-word of mouth on social media during COVID-19 outbreaks: An empirical study* (Vol. 334).

Al Kurdi, B., Alshurideh, M., Nuseir, M., Aburayya, A., & Salloum, S. A. (2021).*The effects of subjective norm on the intention to use social media networks: An exploratory study using PLS-SEM and machine learning approach* (Vol. 1339).

Al Naqbia, E., Alshuridehb, M., AlHamadc, A., & Al, B. (2020). The impact of innovation on firm performance: A systematic review. *International Journal of Innovation, Creativity and Change, 14*(5), 31–58.

Al Suwaidi, F., Alshurideh, M., Al Kurdi, B., & Salloum, S. A. (2021). *The impact of innovation management in SMEs performance: A systematic review* (Vol. 1261). AISC.

Alameeri, K., Alshurideh, M., Al Kurdi, B., & Salloum, S. A. (2021). *The effect of work environment happiness on employee leadership* (Vol. 1261). AISC.

Alameeri, K. A., Alshurideh, M. T., & Al Kurdi, B. (2021). *The effect of Covid-19 pandemic on business systems' innovation and entrepreneurship and how to cope with it: A theatrical view* (Vol. 334).

Al-Dhuhouri, F. S., Alshurideh, M., Al Kurdi, B., & Salloum, S. A. (2021). *Enhancing our understanding of the relationship between leadership, team characteristics, emotional intelligence and their effect on team performance: A critical review* (Vol. 1261). AISC.

Al-Gasaymeh, A., Kasem, J., & Alshurideh, M. (2015). Real exchange rate and purchasing power parity hypothesis: Evidence from ADF unit root test. *International Research Journal of Finance and Economics, 14*, 450–2887.

Al-Gasaymeh, A., Almahadin, A., Alshurideh, M., Al-Zoubid, N., & Alzoubi, H. (2020). The role of economic freedom in economic growth: Evidence from the MENA region. *International Journal of Innovation, Creativity and Change, 13*(10), 759–774.

Alhashmi, S. F. S., Alshurideh, M., Al Kurdi, B., & Salloum, S. A. (2020). *A systematic review of the factors affecting the artificial intelligence implementation in the health care sector* (Vol. 1153). AISC.

Al-Jarrah, I. M., Al-Zu'bi, Z. M. F., Jaara, O. O., & Alshurideh, M. (2012). Evaluating the impact of financial development on economic growth in Jordan. *International Research Journal of Finance and Economics, 94*, 123–139.

Aljumah, A., Nuseir, M. T., & Alshurideh, M. T. (2021). *The impact of social media marketing communications on consumer response during the COVID-19: Does the brand equity of a university matter?* (Vol. 334).

Allozi, A., Alshurideh, M., AlHamad, A., & Al Kurdi, B. (2022). Impact of transformational leadership on the job satisfaction with the moderating role of organizational commitment: Case of UAE and Jordan manufacturing companies. *Academy of Strategic Management Journal, 21*, 1–13.

Al-Maroof, R., et al.: The acceptance of social media video for knowledge acquisition, sharing and application: A comparative study among YouYube users and TikTok users' for medical purposes. *nternational Journal of Data and Network Science, 5*(3). https://doi.org/10.5267/j.ijdns.2021.6.013.

Almazrouei, F. A., Alshurideh, M., Al Kurdi, B., & Salloum, S. A. (2021). *Social media impact on business: A systematic review* (Vol. 1261). AISC.

Alshamsi, A., Alshurideh, M., Kurdi, B. A., & Salloum, S. A. (2021). *The influence of service quality on customer retention: A systematic review in the higher education* (Vol. 1261). AISC.

AlShehhi, H., Alshuridehl, M., Kurdi, B. A., & Salloum, S. A. (2021). *The impact of ethical leadership on employees performance: A systematic review* (Vol. 1261). AISC.

Alshurideh, M., Salloum, S. A., Al Kurdi, B., & Al-Emran, M. (2019).Factors affecting the social networks acceptance: An empirical study using PLS-SEM approach. In *Proceedings of the 2019 8th International Conference on Software and Computer Applications* (pp. 414–418).

Assad, N. F., & Alshurideh, M. T. (2020). Investment in context of financial reporting quality: A systematic review. *Waffen-und Kostumkunde Journal, 11*(3), 255–286.

Bazan, C., et al. (2020). A systematic literature review of the influence of the university's environment and support system on the precursors of social entrepreneurial intention of students. *Journal of Innovation and Entrepreneurship, 9*(1), 4.

Bettayeb, H., Alshurideh, M. T., & Al Kurdi, B. (2020). The effectiveness of mobile learning in UAE universities: A systematic review of motivation, self-efficacy, usability and usefulness. *International Journal of Control and Automation, 13*(2), 1558–1579

Cai, W., Lysova, E. I., Khapova, S. N., & Bossink, B. A. G. (2019). Does entrepreneurial leadership foster creativity among employees and teams? The mediating role of creative efficacy beliefs. *Journal of Business and Psychology, 34*(2), 203–217.

Chitsaz, E., Tajpour, M., Hosseini, E., Khorram, H., & Zorrieh, S. (2019). The effect of human and social capital on entrepreneurial activities: A case study of Iran and implications. *Entrepreneurship and Sustainability Issues, 6*(3), 1393.

Clark, C. M., Harrison, C., & Gibb, S. (2019). Developing a conceptual framework of entrepreneurial leadership: A systematic literature review and thematic analysis. *International Review of Entrepreneurship, 17*(3).

Dabić, M., Stojčić, N., Simić, M., Potocan, V., Slavković, M., & Nedelko, Z. (2021). Intellectual agility and innovation in micro and small businesses: The mediating role of entrepreneurial leadership. *Journal of Business Research, 123*, 683–695.

Dean, H., & Ford, J. (2017). Discourses of entrepreneurial leadership: Exposing myths and exploring new approaches. *International Small Business Journal, 35*(2), 178–196.

Dost, M., Arshad, M., & Afsar, B. (2018). The influence of entrepreneurial orientation on types of process innovation capabilities and moderating role of social capital. *Entrepreneurship Research Journal, 8*(4).

Felix, C., Aparicio, S., & Urbano, D. (2019). Leadership as a driver of entrepreneurship: An international exploratory study. *Journal of Small Business and Enterprise Development*.

Ha, N. T., Doan, X. H., Vu, T. N., Nguyen, T. P. L., Phan, T. H., & Duong, C. D. (2020). The effect of social capital on social entrepreneurial intention among Vietnamese Students. *The Journal of Asian Finance, Economics and Business, 7*(8), 671–680.

Harahsheh, A. A., Houssien, A. M. A., Alshurideh, M. T., & Mohammad, A. M. (2021). *The effect of transformational leadership on achieving effective decisions in the presence of psychological capital as an intermediate variable in private Jordanian universities in light of the corona pandemic* (Vol. 334).

Harrison, R. T., Leitch, C. M., & McAdam, M. (2018). Breaking glass: Towards a gendered analysis of entrepreneurial leadership. In *Research Handbook on Entrepreneurship and Leadership*. Edward Elgar Publishing.

Hsiao, C., Lee, Y.-H., & Chen, H.-H. (2016). The effects of internal locus of control on entrepreneurship: The mediating mechanisms of social capital and human capital. *International Journal of Human Resource Management, 27*(11), 1158–1172.

Hsu, B.-X., & Chen, Y.-M. (2019). Industrial policy, social capital, human capital, and firm-level competitive advantage. *The International Entrepreneurship and Management Journal, 15*(3), 883–903.

Leitch, C. M., & Volery, T. (2017). Entrepreneurial leadership: Insights and directions. *International Small Business Journal, 35*(2), 147–156.

Li, C., Makhdoom, H. U. R., & Asim, S. (2020). Impact of entrepreneurial leadership on innovative work behavior: Examining mediation and moderation mechanisms. *Psychology Research and Behavior Management, 13*, 105.

Linares-Espinós, E., et al. (2018). Metodología de una revisión sistemática. *Actas Urológicas Españolas, 42*(8), 499–506.

Madi Odeh, R. B. S., Obeidat, B. Y., Jaradat, M. O., Masa'deh, R., & Alshurideh, M. T. (2021). The transformational leadership role in achieving organizational resilience through adaptive cultures: The case of Dubai service sector. *International Journal of Productivity and Performance Management*. https://doi.org/10.1108/IJPPM-02-2021-0093.

Mamun, A. A., Ibrahim, D. M., Yosoff, M. H., & Fazal, S. A. (2018). Entrepreneurial leadership, performance, and sustainability of micro-enterprises in Malaysia. *Sustainability, 10*(5), 1591.

Mehmood, T., Alzoubi, H. M., Alshurideh, M., Al-Gasaymeh, A., & Ahmed, G. (2019a). Schumpeterian entrepreneurship theory: Evolution and relevance. *Academy of Entrepreneurship Journal, 25*(4), 1–10.

Mehmood, M. S., Jian, Z., Waheed, A., Younas, A., & Khan, S. Z. (2019b). Impact of entrepreneurial leadership on employee's innovative behavior: Mediating role of psychological empowerment. In *Proceedings of the 2019 3rd International Conference on Management Engineering, Software Engineering and Service Sciences* (pp. 223–229).

Mehmood, M. S., Jian, Z., & Waheed, A. (2019c). The influence of entrepreneurial leadership on organisational innovation: Mediating role of innovation climate. *International Journal of Information Systems and Change Management, 11*(1), 70–89.

Mehrez, A. A. A., Alshurideh, M., Kurdi, B. A., & Salloum, S. A. (2021). *Internal factors affect knowledge management and firm performance: A systematic review* (Vol. 1261). AISC.

Miao, Q., Newman, A., Schwarz, G., & Cooper, B. (2018). How leadership and public service motivation enhance innovative behavior. *Public Administration Review, 78*(1), 71–81.

Mishra, P., & Misra, R. K. (2017). Entrepreneurial leadership and organizational effectiveness: A comparative study of executives and non-executives. *Procedia Computer Science, 122*, 71–78.

Nuseir, M. T., El Refae, G. A., & Alshurideh, M. (2021). The impact of social media power on the social commerce intentions: Double mediating role of economic and social satisfaction. *Journal of Legal, Ethical and Regulatory Issues, 24*(Special Issue 6), 1–15.

Obeidat, U., Obeidat, B., Alrowwad, A., Alshurideh, M., Masadeh, R., & Abuhashesh, M. (2021). The effect of intellectual capital on competitive advantage: The mediating role of innovation. *Management Science Letters, 11*(4), 1331–1344.

Pasricha, P., & Rao, M. K. (2018). The effect of ethical leadership on employee social innovation tendency in social enterprises: Mediating role of perceived social capital. *Creativity and Innovation Management, 27*(3), 270–280.

Ranjan, S. (2018). Entrepreneurial leadership: A review of measures, antecedents, outcomes and moderators. *Asian Social Science, 14*(12), 104–114.

El Refae, G. A., Nuseir, M. T., & Alshurideh, M. (2021) The influence of social media regulations boundary on marketing and commerce of industries in Uae. *Journal of Legal, Ethical and Regulatory Issues, 24*(Special Issue 6), 1–14.

Sarabi, A., Froese, F. J., Chng, D. H. M., & Meyer, K. E. (2020). Entrepreneurial leadership and MNE subsidiary performance: The moderating role of subsidiary context. *International Business Review, 29*(3), 101672.

Sawaean, F., & Ali, K. (2020). The impact of entrepreneurial leadership and learning orientation on organizational performance of SMEs: The mediating role of innovation capacity. *Management Science Letters, 10*(2), 369–380.

Setini, M., Yasa, N. N. K., Gede Supartha, I. W., Ketut Giantari, I., & Rajiani, I. (2020). The passway of women entrepreneurship: Starting from social capital with open innovation, through to knowledge sharing and innovative performance. *Journal of Open Innovation: Technology, Market, and Complexity, 6*(2), 25.

Theodoraki, C., Messeghem, K., & Rice, M. P. (2018). A social capital approach to the development of sustainable entrepreneurial ecosystems: An explorative study. *Small Business Economics, 51*(1), 153–170.

Ximenes, M., Supartha, W. G., Dewi, I. G. A. M., & Sintaasih, D. K. (2019). Entrepreneurial leadership moderating high performance work system and employee creativity on employee performance. *Cogent Business & Management*, *6*(1), 1697512.

Yang, J., Pu, B., & Guan, Z. (2019). Entrepreneurial leadership and turnover intention of employees: The role of affective commitment and person-job fit. *International Journal of Environmental Research and Public Health, 16*(13), 2380.

Yani, A., Eliyana, A., Hamidah, I. K. R., & Buchdadi, A. D. (2020). The impact of social capital, entrepreneurial competence on business performance: An empirical study of SMEs. *Systematic Reviews in Pharmacy, 11*(9), 779–787.

Factors and Challenges Influencing Women Leadership in Management: A Systematic Review

Khadija Alameeri, Muhammad Alshurideh, and Barween Al Kurdi

Abstract Leadership and management have always been common among men, yet women are stepping and succeeding into these roles too. However, with any new change that is projected to a system, challenges arise to reach adaptation or even acceptance of the change. This paper aims in pointing out the challenges and difficulties faced by women in their leadership and management roles. Six factors were studied which are competence, perception, stereotyping, harassment, organizational culture, and gender inequality on the performance and participation of a person, and thus how they impact gender discrimination. The results have presented a set of impressive challenges faced by women in their leadership roles such as gender and sex-based discrimination, stereotyping, harassment, gender inequality and perceived incompetence. This paper articulates the importance of paying attention to the factors affecting women in leadership and management in order to empower them and decrease gender discrimination in the workspace.

Keywords Women leadership · Gender inequality · Stereotyping · Lack of competence · Discrimination · Harassment · Gender inequality

K. Alameeri
University of Sharjah, Sharjah, UAE

M. Alshurideh (✉)
Department of Management, College of Business Administration, University of Sharjah, Sharjah, UAE
e-mail: malshurideh@sharjah.ac.ae; m.alshurideh@ju.edu.jo

Marketing Department, School of Business, The University of Jordan, Amman, Jordan

B. Al Kurdi
Faculty of Economics and Administrative Sciences, Department of Marketing, The Hashemite University, Zarqa, Jordan
e-mail: barween@hu.edu.jo

M. Alshurideh et al. (eds.), *The Effect of Information Technology on Business and Marketing Intelligence Systems*, Studies in Computational Intelligence 1056,
https://doi.org/10.1007/978-3-031-12382-5_82

1 Introduction

Nowadays, it is increasingly evident that women hold leadership positions, and this is always a topic of discussion. In the modern period, many stereotypes formed over the centuries are beginning to be rethought in society, in particular, the idea that women are destined. The paper aims to consider certain difficulties that female leaders face. Regardless of gender, a successful leader should have a certain set of personal and professional qualities. Strong character, ability to lead people, responsibility, business sense—all that is necessary for the leader. Currently, both men and women can possess these traits. Despite this, there are differences between male and female forms of management. A woman leader rarely resorts to an authoritarian style of management, she most often uses persuasion methods, tries to motivate staff, and does not force them to fulfill their goals and objectives. In conflict situations and negotiations, she might often seek a compromise, rather than coercion, carefully considers decisions, taking into account the opinions of others.

Modern working people and women in management need to make maximum personal efforts to create real competition for men in the managerial labor market. The success of professional self-realization depends not only on competencies, creative and creative thinking; but also on the gender characteristics of the individual. They also have competitive qualities such as high educational potential, sociability, responsibility, integrity, and hard work. Some features of the career strategy of women are presented, the leadership style of which is individual. Each female leader uses the experience of other leaders, including men, often shows traditionally masculine qualities, of course, not forgetting about feminine qualities, but she always introduces a personal, individual component in her professional activity, which is always tuned for success.

2 Literature Review

There are a number of key factors and challenges that directly or indirectly affect women's leadership in management. In the manufacturing industry, it is important to note that women are influenced by their level of competence, family support, and emotional stability (Al Khayyal et al., 2021; Harahsheh et al., 2021; Klenke, 2017). It is mainly due to the fact that the given field puts a significant amount of stress on women. The study conducted in New York shows that women in the automotive industry are primarily affected by appointment criteria, work environment, and skills and knowledge, and all these factors are manifested in their overall performance (Glass & Cook, 2016). In addition, stereotyping plays a major role in determining a woman's success as a leader, because it hinders career progression (Brookes, 2010). Gender inequality and stereotyping are also critical factors that influence women of color substantially due to two main components, such as racism and sexism (Abuhashesh et al., 2021; Alshurideh et al., 2021; Sanchez-Hucles & Davis,

2010). In the context of trait, contingency, and behavioral leadership, women are more likely to be subjects of gender inequality and stereotyping, and harassment (Al Kurdi & Alshurideh, 2021; Ayman & Korabik, 2010; Nuseir et al., 2021). Therefore, various leadership approaches become more challenging for female managers (Alameeri et al., 2021; Allozi et al., 2022; AlShehhi et al., 2021).

Organizational culture is another important challenge to consider due to the fact that traditional leaders cannot fully adapt to the modern concept of diversity. Therefore, they tend to exclude women from key positions (Al-Dhuhouri et al., 2021; Eagly & Chin, 2010; Madi Odeh et al., 2021). The given factor also creates a thick glass ceiling which makes it difficult for women to climb to upper management positions (Glass & Cook, 2016). On certain occasions, women may lack competence in specific fields due to the absence of proper supervision and guidance (Klenke, 2017). The given factor might be manifested in the fact that there is gender inequality in educating and training female workers (Flick, 2015). Women are prone to experience significant challenges in the context of business and entrepreneurship (Alameeri et al., 2021; Gray, 2019; Mehmood et al., 2019). The factor of stereotyping women can hinder their ability to express outstanding leadership behavior (Chin, 2011). In counseling and negotiations, there are also elements of stereotyping that constrain them in leading a successful negotiation process (Levitt, 2010). Thus, the latter notion decreases women's overall performance in a number of industries.

The factor of discrimination against women is present across many nations. In India, gender inequality in the organizational environment alongside the local culture acts as a major factor diminishing the role of a woman as a leader (Swarup, 2014). In addition, a New Zealand study shows that specific programs allow women to regain their self-confidence in order to properly perform in managerial positions in an organization (Harris & Leberman, 2012). There is a problem of objectification of women's bodies which can be a key major factor in setting the perception of women's inferiority (O'Neill, 2019). The given faulty perception is a form of sexism, which prevents women to acquire upper-management occupations (Ibarra et al., 2010). Gender inequality is also affected by women's work and family problems, where they are in a vulnerable position of duality, where they need to care for their family and perform their male counterparts (Hideg & Shen, 2019). This issue is strengthened by the fact that stereotyping based on cultures and traditions is prevalent across all fields, such as politics and business (McLean & Beigi, 2016). These factors play a detrimental role in hindering women's performance in management and their ability to reach higher positions.

3 Methods

The methodology used was primarily manifested in a systematic review of recent literature on women's performance and participation in management and the workplace. A number of sources were searched for the most relevant and important keywords regarding the subject (Ahmad et al., 2021; Ahmed et al., 2021; Al Naqbia

Table 1 Search cretira

Keyword search
"Women" AND "Leadership"
"Women" AND "Management"
"Women" AND "Discrimination"
"Women" AND "Career"

Table 2 Web search engines

Databases	Frequency
Google Scholar	51
SAGE Journals	32
Elsevier	17
Wiley Online Library	14
APA PsycNet	9

et al., 2020; Al Suwaidi et al., 2021; Alhashmi et al., 2020; Alshamsi et al., 2021; Assad & Alshurideh, 2020a; Bettayeb et al., 2020; Mehrez et al., 2021; Shah et al., 2021). These keyword searches are summarized and presented in Table 1. A total of 123 articles and books on the subject were screened and overviewed, from which 16 sources were selected. Table 2 shows the frequency and type of databases used in the literature search as done by other studies (Alketbi et al., 2020; Al-khayyal et al., 2020; Almaazmi et al., 2021; Almazrouei et al., 2021; AlMehrzi et al., 2020; Alsuwaidi et al., 2021, 2021; Alyammahi et al., 2021; Assad & Alshurideh, 2020b; Yousuf et al., 2021).

4 Results

In order to obtain and analyze major challenges affecting women in management, the primary external factors were summarized and categorized into six classes, which are competence, perception, stereotyping, harassment, organizational culture, and gender inequality. The detailed information on each source and the external factor discussed are demonstrated in Table 3. In addition, Table 4 shows how the given external factors and databases are correlated with each other. Inclusion/exclusion criteria are represented and demonstrated in Table 5 and the critical inclusion questions are listed in Table 6. Figure 1 demonstrates the flowchart of the study selection process. The result of the quality assessment process is presented in Table 7 and the research model is presented in Fig. 2.

The following hypotheses were generated:

H1a: Stereotyping has a negative effect on women's participation in management.
H1b: Stereotyping has a negative effect on women's performance in management.

Table 3 The papers' sources chosen and the external factors found

Authors	Dependent factors	Sample size	Place	Subjects	External factors
Ayman and Korabik (2010)	Performance	180 women	India	Employees	Harassment
Brookes (2010)	Participation	135 women	Nigeria	Employees	Stereotyping
Chin (2011)	Participation	85 women	Florida	Employees	Stereotyping
Eagly and Chin (2010)	Performance	80 women	New York	Employees	Organizational culture
Flick (2015)	Performance	85 women	Arizona	Employees	Gender inequality
Glass and Cook (2016)	Performance	120 women	New York	Employees	Organizational culture
Gray (2019)	Performance	90 women	Italy	Employees	Organizational culture
Harris and Leberman (2012)	Participation	70 women	New Zealand	Employees	Organizational culture
Hideg and Shen (2019)	Performance	95 women	US	Employees	Gender inequality
Ibarra et al. (2010)	Participation	150 women	New York	Employees	Perception
Klenke (2017)	Performance	110 women	Virginia	Employees	Competence
Levitt (2010)	Performance	115 women	US	Employees	Stereotyping
McLean and Beigi (2016)	Performance	100 women	US	Employees	Stereotyping
O'Neil (2019)	Performance	110 women	Chicago	Employees	Perception
Sanchez-Hucles and Davis (2010)	Performance	110 women	US	Employees	Gender inequality
Swarup (2014)	Performance	85 women	India	Employees	Gender inequality

H2a: Discriminatory organizational culture hinders women's participation in management.
H2b: Discriminatory organizational culture hinders women's performance in management.
H3a: Gender inequality acts as barrier for women's participation in management.
H3b: Gender inequality acts as barrier for women's performance in management.
H4a: Discriminatory perception negatively affects women's participation in management.
H4b: Discriminatory perception negatively affects women's performance in management.
H5a: The lack of competence hinders women's participation in management.
H5b: The lack of competence hinders women's performance in management.

Table 4 The external factors found connected to each database used

External factors	Databases					
	SAGE journals	Wiley online library	Google scholar	APA PsycNet	Elsevier	Total
Competence	0	0	1	0	0	1
Perception	1	0	1	0	0	2
Stereotyping	1	1	1	1	0	4
Harassment	0	0	0	1	0	1
Organizational culture	1	0	1	1	1	4
Gender inequality	2	0	1	1	0	4

Table 5 The used inclusion/exclusion criteria

Inclusion criteria	Exclusion criteria
Should involve the factors and challenges that influence women in management	Factors and challenges affecting women in management are not present
Should include possible causes of women discrimination in the workplace or organization	Possible causes of discrimination are not present
The study must be in English	Non-English sources

Table 6 The critical inclusion questions used

#	Questions
1	Are the subject and goals being specifically targeted women in management?
2	Does the study specify factors and challenges affecting women in management?
3	Are data collection and analysis methods adequate?
4	Does the study contribute to the current knowledge?

H6a: Harassment negatively affects women's participation in management.
H6b: Harassment negatively affects women's performance in management.
H7: Women's participation in management faces gender discrimination.
H8: Women's performance in management faces gender discrimination.

5 Discussion

Gender discrimination in the world of work is a multilateral phenomenon with a large number of different manifestations. Nevertheless, some researchers are convinced that discrimination is becoming the norm, while others, on the contrary, deny the very fact of its existence on the basis of sex. It is necessary to analyze the degree of

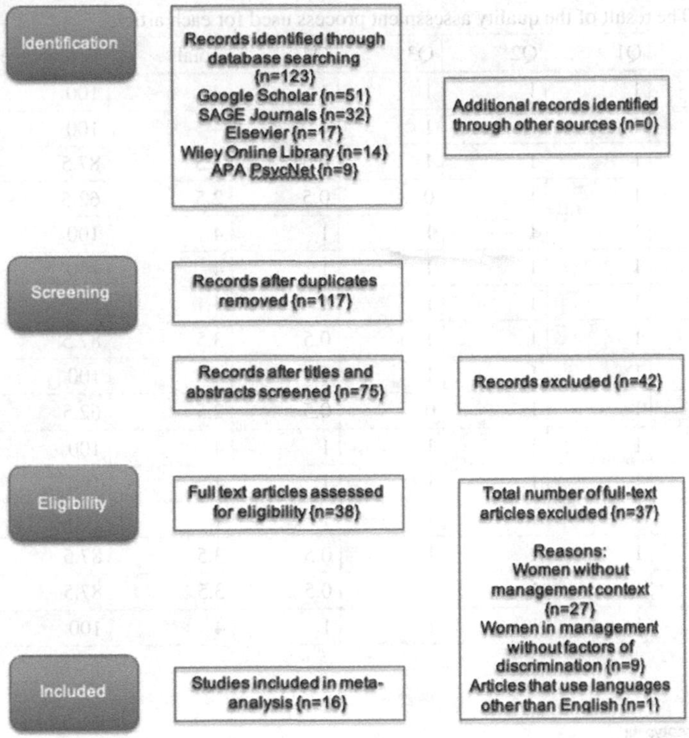

Fig. 1 Flow chart of selected studies

discrimination against women, that is, the extent and form of restriction of their rights in any sphere of public life, primarily industrial, but also family, social, political, and spiritual. This approach aims to explore the mechanism of integration of women in the entire system of government, society, and the economy. In fact, any social institution whose activity, the principle of gender equality, is deliberately violated can act as a subject of violence. Social discrimination of women is most expediently investigated in a double plan. On one hand, this is a social phenomenon that encompasses a large demographic community, consisting of various social, professional, age, and status groups. On the other hand, it is a social process with a successive change in the state of an object. Under the influence of external and internal conditions, objective and subjective factors, the process inherent in the sustainable interaction of female society with various social institutions and communities in order to achieve equal rights and opportunities, eliminate discrimination based on gender.

The forms of social discrimination of women, depending on the sphere of their activity, can be different. By its type, discrimination can be both violent and non-violent. In any case, the basis of discriminatory actions is violence against a woman's personality. A sociological study of social discrimination against women is based on data from a number of particular sociological factors. Among them, the concept

Table 7 The result of the quality assessment process used for each article

Study	Q1	Q2	Q3	Q4	Total	Percentage (%)
S1	1	1	1	1	4	100
S2	1	1	1	1	4	100
S3	1	1	1	0.5	3.5	87.5
S4	1	1	0	0.5	2.5	62.5
S5	1	1	1	1	4	100
S6	1	1	1	1	4	100
S7	1	1	1	1	4	100
S8	1	1	1	0.5	3.5	87.5
S9	1	1	1	1	4	100
S10	1	1	0	0.5	2.5	62.5
S11	1	1	1	1	4	100
S12	1	1	1	1	4	100
S13	1	1	1	1	4	100
S14	1	1	1	0.5	3.5	87.5
S15	1	1	1	0.5	3.5	87.5
S16	1	1	1	1	4	100

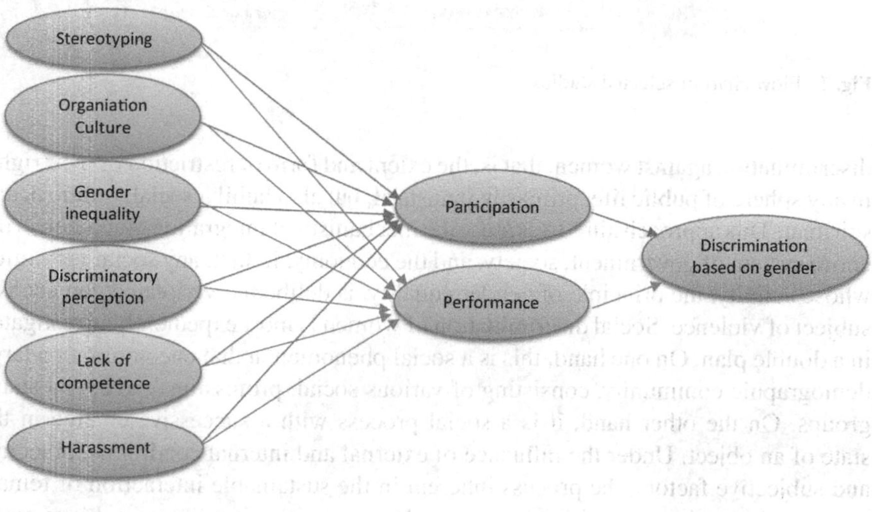

Fig. 2 The study proposed model

of gender identity reflects ideas about the field, and through their consciousness identified with it. Awareness of one's gender does not always correspond to the biological characteristics of an individual. The concept of gender ideals reflects public ideas about male and female behavior, in recent decades they have changed significantly. The category of biological sex means taking into account the primary and secondary characteristics typical of men and women. The concept of sexual role is associated with the division of labor, with the rights and obligations of the sexes.

The female leader feels the surrounding world more subtly, does not go to an open conflict, which can harm. In relation to the staff, she is more open than men, she does not seek to keep her distance. She may be interested not only in the work process, but also in her personal life. This allows you to make the relationship more trusting, more friendly, which has a beneficial effect on the results of the work. A woman leader treats subordinates with warmth and attention, since maternal instinct is inherent in any woman. Women govern with instruments of influence based on horizontal connections, and men with instruments of power. Literary analysis has shown that there are few publications on gender management issues, all the more so when it comes to comparative analysis of gender theories. At the same time, the situation of women, who make up more than half of the population, reflects the state of society as a whole. Only recently, women began to actively occupy leadership positions, becoming the first persons of corporations, and even states.

The success of a woman as a leader and manager depends primarily on two factors—objective and subjective. The first is the efforts that the woman herself will make in order to achieve success in her career. The second is from stereotypes and views on it from the side and often from the side of men. That is, subjectively, a woman's career will depend on the social and cultural environment in which the woman rotates, and in which her personality and business environment are formed. As noted above, in general, assessments of the perception of women were the result of many factors—religious, political, social. The feminine principle was ontogenetically secondary and often subordinate to the masculine principle. In addition, it was the male sex who was the landowner and breadwinner, and this was also reflected in the legislation of that time. The birth of girls in the family was a painful and joyful event for the family and the financial conditions. The woman was completely subordinate to the man, both psychologically and economically.

However, over time, everything began to change. Nevertheless, the introduction of legal equality and a certain change in social roles could not completely distort the traditional patriarchal views. From the results, it seems that men continue to occupy top management positions, and women could be content only with grass-roots management. No one will deny the fact that a working woman is an everyday occurrence. However, the female leader sometimes causes some inhabitants some doubt and not many women manage to stand on the top of the career ladder. The reason for this difference is that female and male labor is evaluated differently, due to stereotyping. Perhaps with leadership, one must combine intuition, rigidity, and diplomacy, adhere to ethical standards and be philanthropic.

Gender features of the management style are very relevant issues in the psychology of management due to the fact that positive trends are currently emerging. When hiring a manager, they began to give preference to women, because, unlike men, they are characterized by a low level of empathy, communication skills, conservatism, and aggressiveness. Women not only unite employees, show sympathy, but also know how to motivate them to perform difficult tasks in critical situations (Gray, 2019).

Gender features of managerial activity are the characteristics of men and women leaders, manifested in the implementation of the main managerial functions, in the types and types of managerial decision-making, and the specificity of response to the impact of psychological factors of activity. Management style is a relatively stable system of methods, methods and forms of influence of a leader on subordinates in accordance with the goals of joint activities.

At the same time, in the analysis of managerial literature, the overwhelming majority of works clearly show latent discrimination against women. It is believed that the manager and specialist is necessarily a man, and the assessment of personality behavior, analysis of managerial situations, recommendations for the development of managerial decisions are given precisely from male positions. Moreover, the peculiarities of a woman's reaction to the external environment, the motivation of her actions, and personal characteristics are considered extremely limited. The reasons lie in the patriarchal views on the role of women, rooted in the mass consciousness. For centuries, a man raised the primary perception of a woman as the guardian of the hearth, the teacher of children, and the hostess in the kitchen. Each new appearance of a woman in management is presented as a sensation. At the same time, economic instability, periods of recessions, and booms predetermine the inevitability of a change in ideas about effective management.

It can be speculated that women quickly gain experience managing firms in situations of uncertainty. To counter stiff competition in modern conditions has become possible thanks to the involvement of women leaders practicing managerial style. This was mainly expressed in the development of more flexible forms of management orientation to an ever-changing situation. The introduction of a new mentality, style, methods inherent in the female management style into the renewed management has contributed to increasing the competitiveness of products through the search for the optimal combination of male and female management. Differences in the hierarchy of values, behavior, motivation between men and women are manifested in the desire to organize the management process in different ways. A woman is considered a more effective leader in the areas of staff motivation, communication, and decision-making processes. Speaking about the potential of women managers in management, it is important to note a number of features that confirm that the female management style is much richer. Firstly, a woman reacts more subtly to the moral and psychological climate in the team, relies on attention technology. Secondly, a woman is inclined to resolve a conflict situation, to experience the possible outcome of events.

The ever-changing change in social relations in the economy has led to serious problems in the employment of women, especially with regard to their career growth. Thus, the ongoing exclusion of women from the field of intellectual work, the employment of women in various sectors of the economy without formal employment, and discrimination against women in the workplace have determined the low level of women in the management sphere, and this despite the fact that working women make up a significant part of the working population. It is well known that professional activity is one of the main areas of personal self-realization. In professional activities, a person reveals and shows his skills, communicative and professional qualities. It is this activity that allows a person to be a special individuality, to achieve recognition of their originality and originality, significance for other people, and the society as a whole.

A professional career should be considered as a process of self-realization of an individual in his main and socially significant sphere - in the form of a person moving from one professional status to another. This shows the special role of the individual and the mandatory coordination of the needs and capabilities of the individual with the competencies and qualification requirements of the profession. On the one hand, this means that a person was able to realize himself, and on the other, the fact that his activity was highly appreciated. In this regard, a caree—r is of great importance in the professional development of an individual. Career is an activity of conscious implementation of professional advancement, as a result of which a status appears that guarantees an acceptable level of social and professional recognition. The specifics of women's professional careers are the presence of sex-role stereotypes that affect the degree of manifestation of their career activity, as well as the personal characteristics of women, which result in the efforts made by them in the process of professional and career growth.

Moreover, these efforts are the main components of women's readiness to pursue a career. However, it is necessary to take into account the fact that not all women strive for leadership in the professional sphere. For some women, professional career and professional growth are in first place among life strategies and values, while for another part of women, work is supportive. This category of women does not seek to occupy a higher position, leadership positions, they self-actualize themselves in the family, they are quite satisfied with their position and salary. Thus, it is not possible to consider all women as a social group that has certain characteristics.

Modern employers prefer career-oriented women to work for them, and of course, it is not beneficial for them to have workers like mothers. The image of a housewife in society is supported mainly by men and is firmly fixed by the media. On the contrary, in professional activities, a woman overcomes a huge number of difficulties. These can be difficulties of both a socio-economic and political nature, such as a personnel management policy that limits the professional growth of women. This includes non-recognition of professional qualities by colleagues, unemployment, and fierce competition from men. Considering professional activity, it should be said about the self-realization of the personality, which is perceived as the realization of individual and personal capabilities of a person in their creative activity. A person realizes their own potential, a combination of knowledge, skills, creative abilities and abilities, as they realize their significance, goals of the activity, and methods of its implementation. The problem of professional self-realization is determined by the growth of requirements for professional activity and the growth of human attention to the expansion of their capabilities.

A professional career for men and women is built in different ways, therefore, it is necessary to determine the competitive qualities of women, the style of their leadership, and the urgent problems of women leaders. Currently, a woman has the right to choose the most attractive orientation to her family, professional or social activities. However, women's leadership in real production continues to be quite low, due to some problems. Firstly, gender asymmetry has developed in the economy, with guaranteed equal rights for men and women in management, women, unfortunately, do not have equal opportunities with men to engage in leadership activities. On one hand, asymmetry is associated with objective difficulties in managing enterprises, and on the other, it is determined by patriarchal stereotypes of society.

Secondly, it is an entrenched patriarchal-conservative view of the leadership profession as a purely masculine one. It is worth paying attention to the fact that all the names of leadership positions are masculine and it becomes clear why a woman has to overcome a great many barriers to career growth in order to occupy the position of head of the organization. The modern world of business involves a masculine manner of behavior and lifestyle, and it is difficult for men to perceive women with power. Professionally independent women constantly come to grips with the male desire for excellence and the stereotypes that exist in society about female abilities. It turns out that women are forced to show their business qualities a hundred times more to be taken seriously. In this regard, the man himself is changing, with whom the fate of a prosperous wife is connected and there is a change in the principles for the redistribution of responsibilities in the household. Consequently, it can be stated that women have great difficulties in promoting their ranks. The success of a career in many respects depends on herself, on her personal qualities, on the ability to deal with stereotypical thinking in society. The reality is that women achieve success in work not as a result of copying the male management style, but through the use of their abilities, which provide a competitive advantage for women leaders over men.

One of the most important qualities is the industriousness of women is the ability to complete the work that has been started and to take responsibility, which is explained by the maternal instinct. Female executives might be better oriented towards users of the products of their enterprise, towards new services that the market needs (Aburayya et al., 2020; Al-Dmour et al., 2021; Al-Khayyal et al., 2021; Alshurideh et al., 2019; Alzoubi et al., 2020). The flexibility and diplomacy of women, combined with perseverance, allows them to more effectively build relationships with business partners and find mutual understanding and support. Femininity, charm, logic, developed intuition are often used by women leaders to solve the most complex and promising areas. Women managers can also be distinguished from men by greater democracy, willingness to cooperate, and collective decision-making with a skillful distribution of responsibilities between subordinates. Women leaders are probably more law-abiding, which leads to the establishment of constructive relations with authorities (Alshurideh, 2019; Alshurideh et al., 2016; Levitt, 2010; Svoboda et al., 2021). A detailed and thoughtful approach to solving problems, attention to detail, excellent friendship with numbers—all these qualities help female leaders to perceive and analyze details perfectly.

Thus, modern working women need to make maximum personal efforts to create real competition for men in the managerial labor market. An important role in the promotion of a specialist in the career ladder is played by the style of his communication with colleagues, and in the future, the style of leadership over them. Researchers study the similarities and differences between male and female leadership styles, and they often disagree (Gray, 2019). The traditional view is that women choose relationship-oriented leadership because they are more emotional in nature and male leadership is characterized by perseverance and determination. There are only some features of the male and female leadership styles, which are determined by the difference in the psychological structure of personality in men and women. A woman effectively implements both survival strategies and development strategies, determining more cautious relationships with her colleagues and partners and, most importantly, avoiding risk strategies. The success of a woman in her career almost always changes the psychological microclimate of the family, as her authority as a mother and wife increases.

The first and most significant advantage is the high educational potential of women. In terms of education, women can lead in both higher and secondary vocational education. Female intuition, combined with a tendency to double-check information through various sources, allows you to choose the most favorable from a variety of decisions, understand the partnership, and evaluate the prospects of starting work. Compared to men, women can be distinguished by greater organization skills, determination, and consistency, which provides female leaders with the opportunity to perfectly notice and take into account any trifles in their work. Sociability is expressed in the fact that female managers use feedback opportunities better than men, they tend to trust behavior, they quickly adapt to changing conditions.

6 Conclusion

In conclusion, it is highly important to understand that factors and challenges affecting women's performance and participation in management are comprised of a multitude of forces. These can include gender and sex-based discrimination, such as stereotyping, harassment, gender inequality, perceived incompetence, and organizational culture. Most industries either do not provide sufficient knowledge and training to women or do not allow them to reach the upper-management and supervisory positions.

References

Aburayya, A., et al. (2020). An empirical examination of the effect of TQM practices on hospital service quality: An assessment study in uae hospitals. *Systematic Reviews in Pharmacy, 11*(9). https://doi.org/10.31838/srp.2020.9.51.

Ahmad, A., Alshurideh, M. T., Al Kurdi, B. H., & Alzoubi, H. M. (2021). Digital strategies: A systematic literature review. In *The international conference on artificial intelligence and computer vision* (pp. 807–822).

Ahmed, A., Alshurideh, M., Al Kurdi, B., & Salloum, S. A. (2021). *Digital transformation and organizational operational decision making: A systematic review* (Vol. 1261). AISC.

Al-Dhuhouri, F. S., Alshurideh, M., Al Kurdi, B., & Salloum, S. A. (2021). *Enhancing our understanding of the relationship between leadership, team characteristics, emotional intelligence and their effect on team performance: A critical review* (Vol. 1261). AISC.

Al-Dmour, R., AlShaar, F., Al-Dmour, H., Masa'deh, R., & Alshurideh, M. T. (2021). *The effect of service recovery justices strategies on online customer engagement via the role of "Customer Satisfaction" during the Covid-19 pandemic: An empirical study* (vol. 334).

Al-Khayyal, A., Alshurideh, M., Al Kurdi, B., & Salloum, S. A. (2021). Factors influencing electronic service quality on electronic loyalty in online shopping context: Data analysis approach. In *Enabling AI applications in data science* (pp. 367–378). Springer.

Al-khayyal, A., Alshurideh, M., & Al, B. (2020). The impact of electronic service quality dimensions on customers' E-shopping and E-loyalty via the impact of E-satisfaction and E-trust: A qualitative approach. *International Journal of Innovation, Creativity and Change, 14*(9).

Al Kurdi, B. H., & Alshurideh, M. T. (2021). Facebook advertising as a marketing tool: Examining the influence on female cosmetic purchasing behaviour. *International Journal of Online Marketing, 11*(2), 52–74.

Al Khayyal, A. O., Alshurideh, M., Al Kurdi, B., & Salloum, S. A. (2021). *Women empowerment in UAE: A systematic review* (Vol. 1261). AISC.

Al Suwaidi, F., Alshurideh, M., Al Kurdi, B., Salloum, S. A. (2021). *The impact of innovation management in SMEs performance: A systematic review* (Vol. 1261). AISC.

Al Naqbia, E., Alshuridehb, M., AlHamadc, A., & Al, B. (2020). The impact of innovation on firm performance: a systematic review. *International Journal of Innovation, Creativity and Change, 14*(5), 31–58.

AlMehrzi, A., Alshurideh, M., & Al Kurdi, B. (2020). Investigation of the key internal factors influencing knowledge management, employment, and organisational performance: A qualitative study of the UAE hospitality sector. *Journal of Innovation, Creativity and Change, 14*(1), 1369–1394.

AlShehhi, H., Alshurideh, M., Kurdi, B. A., & Salloum, S. A. (2021). *The impact of ethical leadership on employees performance: A systematic review* (Vol. 1261). AISC.

AlSuwaidi, S. R., Alshurideh, M., Al Kurdi, B., & Aburayya, A. (2021). The main catalysts for collaborative R&d projects in Dubai industrial sector. In *The international conference on artificial intelligence and computer vision* (pp. 795–806).

Alameeri, K., Alshurideh, M., Al Kurdi, B., & Salloum, S. A. (2021). *The effect of work environment happiness on employee leadership* (Vol. 1261). AISC.

Alameeri, K. A., Alshurideh, M. T., & Al Kurdi, B. (2021). *The effect of Covid-19 pandemic on business systems' innovation and entrepreneurship and how to cope with it: A theatrical view* (Vol. 334).

Alhashmi, S. F. S., Alshurideh, M., Al Kurdi, B., & Salloum, S. A. (2020). *A systematic review of the factors affecting the artificial intelligence implementation in the health care sector* (Vol. 1153). AISC.

Alketbi, S., Alshurideh, M., & Al Kurdi, B. (2020). The influence of service quality on customers'retention and loyalty in the uae hotel sector with respect to the impact of customer'satisfaction, trust, and commitment: A qualitative study. *PalArch's Journal of Archaeology of Egypt/Egyptology, 17*(4), 541–561.

Allozi, A., Alshurideh, M., AlHamad, A., & Al Kurdi, B. (2022). Impact of transformational leadership on the job satisfaction with the moderating role of organizational commitment: Case of UAE and Jordan manufacturing companies. *Academy of Strategic Management Journal, 21*, 1–13.

Almaazmi, J., Alshurideh, M., Al Kurdi, B., & Salloum, S. A. (2021). *The effect of digital transformation on product innovation: A critical review* (Vol. 1261). AISC.

Almazrouei, F. A., Alshurideh, M., Al Kurdi, B., & Salloum, S. A. (2021). *Social media impact on business: A systematic review* (Vol. 1261). AISC.

Alshamsi, A., Alshurideh, M., Kurdi, B. A., & Salloum, S. A. (2021). *The influence of service quality on customer retention: A systematic review in the higher education* (vol. 1261). AISC.

Alshurideh, M., Alsharari, N. M., & Al Kurdi, B. (2019). Supply chain integration and customer relationship management in the airline logistics. *Theoretical Economics Letters*, *9*(02), 392–414.

Alshurideh, M., Al Kurdi, B. H., Vij, A., Obiedat, Z., & Naser, A. (2016). Marketing ethics and relationship marketing-An empirical study that measure the effect of ethics practices application on maintaining relationships with customers. *International Business Research*, *9*(9), 78–90.

Alshurideh, D. M. (2019). Do electronic loyalty programs still drive customer choice and repeat purchase behaviour? *International Journal of Electronic Customer Relationship Management*, *12*(1). https://doi.org/10.1504/IJECRM.2019.098980.

Alshurideh, M. T., Al Kurdi, B., Masa'deh, R., Salloum, S. A. (2021). The moderation effect of gender on accepting electronic payment technology: A study on United Arab Emirates consumers. *Review of International Business and Strategy*, *31*(3). https://doi.org/10.1108/RIBS-08-2020-0102.

Alsuwaidi, M., Alshurideh, M., Al Kurdi, B., & Salloum, S. A. (2021). *Performance appraisal on employees' motivation: A comprehensive analysis* (Vol. 1261). AISC.

Alyammahi, A., Alshurideh, M., Kurdi, B. A., & Salloum, S. A. (2021). *The impacts of communication ethics on workplace decision making and productivity* (Vol. 1261). AISC.

Alzoubi, H., Alshurideh, M., Kurdi, B. A., Inairat, M. (2020). Do perceived service value, quality, price fairness and service recovery shape customer satisfaction and delight? A practical study in the service telecommunication context. *Uncertain Supply Chain Management*, *8*(3). https://doi.org/10.5267/j.uscm.2020.2.005.

Assad, N. F., & Alshurideh, M. T. (2020a). Investment in context of financial reporting quality: A systematic review. *Waffen-und Kostumkunde Journal*, *11*(3), 255–286.

Assad, N. F., & Alshurideh, M. T. (2020b). Financial reporting quality, audit quality, and investment efficiency: Evidence from GCC economies. *Waffen-und Kostumkunde Journal*, *11*(3), 194–208.

Ayman, R., & Korabik, K. (2010). Leadership: Why gender and culture matter. *American Psychologist*, *65*(3), 157.

Bettayeb, H., Alshurideh, M. T., & Al Kurdi, B. (2020). The effectiveness of mobile learning in UAE universities: A systematic review of motivation, self-efficacy, usability and usefulness. *International Journal of Control and Automation*, *13*(2), 1558–1579.

Brookes, S. (2010). Telling the story of place: The role of community leadership. In *The new public leadership challenge* (pp. 150–168). Springer.

Chin, J. L. (2011). Women and leadership: Transforming visions and current contexts. In *Forum on public policy online* (Vol. 2011, No. 2).

Eagly, A. H., & Chin, J. L. (2010). Diversity and leadership in a changing world. *American Psychologist*, *65*(3), 216.

Flick, U. (2015). *Introducing research methodology: A beginner's guide to doing a research project*.

Glass, C., & Cook, A. (2016). Leading at the top: Understanding women's challenges above the glass ceiling. *The Leadership Quarterly*, *27*(1), 51–63.

Gray, D. E. (2019). *Doing research in the business world*. Sage.

Harahsheh, A. A., Houssien, A. M. A., Alshurideh, M. T., & Mohammad, A. M. (2021). *The effect of transformational leadership on achieving effective decisions in the presence of psychological capital as an intermediate variable in private Jordanian universities in light of the corona pandemic* (Vol. 334).

Harris, C. A., & Leberman, S. I. (2012). Leadership development for women in New Zealand universities: Learning from the New Zealand women in leadership program. *Advances in Developing Human Resources*, *14*(1), 28–44.

Hideg, I., & Shen, W. (2019). Why still so few? A theoretical model of the role of benevolent sexism and career support in the continued underrepresentation of women in leadership positions. *Journal of Leadership & Organizational Studie*, *26*(3), 287–303.

Ibarra, H., Carter, N. M., Silva, C., et al. (2010). Why men still get more promotions than women. *Harvard Business Review*, *88*(9), 80–85.

Klenke, K. (2017). *Women in leadership: Contextual dynamics and boundaries*. Emerald Group Publishing.

Levitt, D. H. (2010). Women and leadership: A developmental paradox? *Adultspan Journal, 9*(2), 66–75.
M. Y. Abuhashesh, M. T. Alshurideh, A. Ahmed, M. Sumadi, and R. Masa'deh, "The effect of culture on customers' attitudes toward Facebook advertising: the moderating role of gender," *Review of International Business and Strategy, 31*(3). https://doi.org/10.1108/RIBS-04-2020-0045.
Madi Odeh, R. B. S., Obeidat, B. Y., Jaradat, M. O., Masa'deh, R., & Alshurideh, M. T. (2021). The transformational leadership role in achieving organizational resilience through adaptive cultures: The case of Dubai service sector. *International Journal of Productivity and Performance Management*. https://doi.org/10.1108/IJPPM-02-2021-0093.
McLean, G. N., & Beigi, M. (2016). The importance of worldviews on women's leadership to HRD. *Advances in Developing Human Resources, 18*(2), 260–270.
Mehmood, T., Alzoubi, H. M., Alshurideh, M., Al-Gasaymeh, A., & Ahmed, G. (2019). Schumpeterian entrepreneurship theory: Evolution and relevance. *Academy of Entrepreneurship Journal, 25*(4), 1–10.
Mehrez, A. A. A., Alshurideh, M., Kurdi, B. A., & Salloum, S. A. (2021). *Internal factors affect knowledge management and firm performance: A systematic review* (Vol. 1261). AISC.
Nuseir, M. T., Al Kurdi, B. H., Alshurideh, M. T., & Alzoubi, H. M. (2021). Gender discrimination at workplace: Do artificial intelligence (AI) and machine learning (ML) Have opinions about it. In *The international conference on artificial intelligence and computer vision* (pp. 301–316).
O'Neill, C. (2019). Unwanted appearances and self-objectification: The phenomenology of alterity for women in leadership. *Leadership, 15*(3), 296–318.
Sanchez-Hucles, J. V., & Davis, D. D. (2010). Women and women of color in leadership: Complexity, identity, and intersectionality. *American Psychologist, 65*(3), 171.
Shah, S. F., Alshurideh, M., Kurdi, B. A., & Salloum, S. A. (2021). *The impact of the behavioral factors on investment decision-making: A systemic review on financial institutions* (Vol. 1261). AISC.
Svoboda, P., Ghazal, T. M., Afifi, M. A. M., Kalra, D., Alshurideh, M. T., & Alzoubi, H. M. (2021). Information systems integration to enhance operational customer relationship management in the pharmaceutical industry. In *The international conference on artificial intelligence and computer vision* (pp. 553–572).
Swarup, V. (2014). Women in leadership roles in public sector. *NHRD Network Journal, 7*(3), 82–85.
Yousuf, H., Zainal, A. Y., Alshurideh, M., & Salloum, S. A. (2021). *Artificial intelligence models in power system analysis* (Vol. 912).

The Relation Between Creative Leadership and Crisis Management Among Faculty Members at Imam Abdulrahman Bin Faisal University in Light of the Corona Pandemic from the Perspective of Department Heads

Saddam Rateb Darawsheh, Anwar Saud Al-Shaar, Muhammad Alshurideh, Nabila Ali Alomari, Amira Mansour Elsayed, Asma Khaleel Abdallah, and Tareq Alkhasawneh

Abstract The current study aimed to reveal the relation between the practice of creative leadership style and crisis management among faculty members at Imam Abdulrahman bin Faisal University (IAU), from the perspective of the heads of academic departments. A descriptive correlative approach was used by applying a questionnaire to a random sample of (100) Head of Department, during the second semester of the academic year 1442/1443. The findings revealed that faculty members practice both creative leadership and crisis management to a high degree. The study recommended holding training courses for faculty members, creating incentives systems and rewards, stimulating creative leadership among faculty members in universities, and spreading a culture of creativity.

S. R. Darawsheh (✉)
Department of Administrative Sciences, The Applied College, Imam Abdulrahman Bin Faisal University, P.O. Box: 1982, Dammam 43212, Saudi Arabia
e-mail: srdarawsehe@iau.edu.sa

A. S. Al-Shaar
Deanship of Preparatory Year and Supporting Studies, Department of Self Development, (Imam Abdulrahman Bin Faisal University), P.O. Box: 1982, Dammam 43212, Saudi Arabia

M. Alshurideh
Marketing Department, School of Business, The University of Jordan, Amman, Jordan
e-mail: m.alshurideh@ju.edu.jo; malshurideh@sharjah.ac.ae

Department of Management, College of Business Administration, University of Sharjah, Sharjah, UAE

N. A. Alomari · A. M. Elsayed
Assistant Professor, Department of Administrative Sciences, The Applied College, Imam Abdulrahman Bin Faisal University, P.O. Box: 1982, Dammam 43212, Saudi Arabia
e-mail: naalomari@iau.edu.sa

M. Alshurideh et al. (eds.), *The Effect of Information Technology on Business and Marketing Intelligence Systems*, Studies in Computational Intelligence 1056,
https://doi.org/10.1007/978-3-031-12382-5_83

Keywords Originality · Fluency · Containment · Escape

1 Introduction

The educational and academic sectors as any other sectors in all countries are influenced by emergent crises. However, these crises differ in terms of their quality, severity, and consequences, necessitating the development of effective ways for containing and managing them (Khairullah, 2010). Indeed, since we live in a constantly changing era, the emergence of crises has become commonplace in all business sectors; as a result, being attentive and aware of the effects of these crises has become a prerequisite to successfully managing their positive and negative impacts (Al-Khamees & Al-Salihi, 2019; Joghee et al., 2021; Svoboda et al., 2021). In this context, it is critical for decision-makers in Saudi Arabian universities to prioritize crisis management during the planning and organizing processes, as well as to invest in creative leadership thinking to solve problems and make decisions that reduce the likelihood of crises (Ahmad et al., 2021a, b, c; Al Naqbia et al., 2020; Alshurideh et al., 2020a; Harahsheh et al., 2021; Kahwagi, 2015; Shah et al., 2021).

Several studies have dealt with the topic of creative leadership and crisis management, such as the study of Hamdi (2018), Al-Dhuhouri et al. (2020), Alameeri et al. (1261), AlShehhi et al. (2020), Madi Odeh et al. (2021), Allozi et al. (2021), which aimed to identify the impact of creative leadership on crisis management in Jordanian telecommunications companies. The study variables included (problem-solving and decision-making, changeability, initiative, and risk-taking, ability to communicate and motivate creativity) (Naqvi et al., 2021; Shah et al., 2020). The findings of the study revealed an impact of creative leadership on the employees' crisis management. The study recommended stimulating employees who can innovate and be creative in Jordanian telecom companies.

Also, (Abu Zaytoun, 2015) cited that the employees in the Jordanian Civil Defense Directorates in the North Region possess a highly creative leadership, in his study that aimed to identify the role of creative leadership in crisis and disaster management in the Jordanian Civil Defense Directorates in the North Region. It also revealed a statistically significant relationship between the Innovative leadership dimensions and indicators of crisis management.

According to Sohmen (2015), Al-Salami, 2012, Shakhour et al., 2021), creative leadership comprises the leader's ability to bring about managerial style growth;

A. M. Elsayed
e-mail: amabdelkhalik@iau.edu.sa

A. K. Abdallah
Foundation of Education Department, College of Education, United Arab University, Al Ain, UAE
e-mail: asma.abdallah@uaeu.ac.ae

T. Alkhasawneh
Humanities and Social Sciences, College of Education, Al Ain University, Al Ain, UAE
e-mail: Tareq.alkhasawneh@aau.ac.ae

Working by activating and utilizing current technologies that are appropriate for the educational environment, conforming to modern era language, and satisfying the demands of the beneficiaries (Al Kurdi et al., 2020a, b; Al-Hamad et al., 2021; Alshamsi et al., 2020; Alsharari & Alshurideh, 2020; Alshurideh et al., 2020, 2019a, b; Bettayeb et al., 2020; Leo et al., 2021). They also emphasized the leader's ability to use innovative work methods for change and improvement, as well as qualitative and quantitative development in the workplace; to solve problems, improve the workplace environment, serve the requirements of subordinates, and influence others (Al Kurdi et al., 2020c; Alshurideh et al., 2019b; Kurdi et al., 2020; Zu'bi et al., 2012). In addition to good manners in dealing with employees, motivating them, and investing their talents and mental abilities in achieving the goals of the organization.

Other scholars add that creative leadership is the idea or suggestion that contributes to increasing the productivity of the organization (Abbas, 2015; Al-Abdullah et al., 2019; Peterson et al., 2015; Shawish & F, 2018).

Al-Hayyaf (2015) proposed a program to develop the administrative performance of department heads in education departments in northern Saudi Arabia in light of their training needs. The results of the study indicated that the training needs of administrative performance of department heads in education departments in northern Saudi Arabia came to a high degree.

While in a study aimed at building a training program in creative leadership for deans of faculties and heads of departments in Jordanian universities in the light of reality and contemporary administrative trend, (Tawiqat, 2007) results revealed that the degree of availability of creative leadership characteristics in the respondents was generally medium. Based on this, a training program in creative leadership was designed for deans of faculties and heads of departments.

Some believe that educational institutions, especially universities, must work to reduce the severity and risks of crises and support creative leadership (Al-Hussein, 2018; Al-Mashaqbeh, 2018; Battah, 2016; Hamdouneh, 2012; Maqableh, 2019) by developing targeted plans that help to address these crises based on the capabilities, as well as the material and human resources that make up effective forces that work and participate in the plans' implementation so that they operate as an immune shield against crises.

It should be noted that, in terms of crisis management, the majority of studies conducted on university education in the Kingdom of Saudi Arabia recommended the establishment of a crisis management unit in university administration to predict the foreseeable future in terms of crises and plan to avoid the risks and effects that they entail.

1.1 Problem Statement

All sectors, including the academic sector, faced an urgent global crisis represented by the outbreak of the Corona pandemic. In fact, the world is still trying to control the consequences of this crisis and plan to reduce its risks, which affected all fields.

Regarding education fields, this crisis led to the closure of schools and to shift to distance learning, and its spread also resulted in governments taking preventive measures that had social and economic consequences. Among the most prominent negative consequences of this crisis are the cessation of learning, the unwillingness of parents for distance learning, the unequal access to educational platforms, the high economic cost, the high dropout rate, and social isolation (Obeidat et al., 2004) which was confirmed by the study of Nairokh (2020) as the results demonstrated the need to prepare for crisis management beforehand through future planning and support the capabilities of the unique faculty members to prevent its occurrence or mitigate its effects. Within the limits of the researchers' knowledge, there are no previous studies on the subject of the current study. Furthermore, the current study is one of several recent studies that have addressed an ongoing crisis.

1.2 Study Questions

The study seeks to answer the following questions:

1. What is the degree of creative leadership practice in crisis management among faculty members at IAU, concerning (originality, flexibility, problem-solving, fluency) from the point of view of department heads, in light of the Corona pandemic?
2. What is the degree of creative leadership practice in crisis management among faculty members at IAU, about (escape, confrontation, cooperation, containment) from the point of view of department heads, in light of the Corona pandemic?

1.3 Study Objectives

The study aimed to achieve the following main goals: To reveal the relation between the practice of creative leadership style and crisis management among faculty members at IAU, from the heads of academic departments' viewpoint in the light of the Corona pandemic.

1.4 The Importance of the Study

The importance of this study is stemmed from the significance of the topic it deals with. The results of the current study may add scientific facts regarding the relation between creative leadership and crisis management in educational institutions. Moreover, the results of this study may benefit faculty members in identifying the creative traits that could be practiced effectively during crises that could afflict or befall academic institutions.

1.5 The Study Limits

This study was limited to an attempt to conclude the relationship between creative leadership and crisis management among faculty members at Imam Abdul Rahman bin Faisal University from the point of view of department heads, during the second semester of the year 1442/1443.

1.6 Study Procedural Terminology

Creative leadership indicates the faculty members' ability to accomplish desired changes to overcome crises by engaging in decision-making and driving solution creation to attain excellence in goal implementation.

*Crisis Management i*medicates what the university's academic leadership and faculty members do in response to present and future crises, following particular preventive measures.

2 Methods and Procedures

2.1 Study Approach

The correlative descriptive approach was used in this study as it helps in studying the phenomenon or research problem and describing it scientifically, to reach logical explanations, which gives the researcher the ability to define the frameworks of the research problem and helps in deducing the research results.

2.2 Population and Sample

The study population consisted of all department heads at Imam Abdul Rahman bin Faisal University in 2020/2021. The study population was chosen from department heads because they are the backbone of the educational process, especially in the field of creative leadership and crisis management. Where the study sample consisted of 100 department heads, (80 males and 20 females), who were chosen through a simple random method from the study population.

2.3 Study Instrument

The researchers benefited from the theoretical literature and previous studies to design a questionnaire including two main axes to meet the study's goals. The questionnaire was composed of two sub-axes: the first axis was for measuring creative leadership, it included (24) items distributed over four domains: originality (7) statements, flexibility (5) statements, problem-solving (5) statements, fluency (7) statements. Where the second axis was concerned with crisis management, including (30) items distributed into four areas: the escape method (8) statements, the confrontation method (7) statements, the cooperation method (7) statements, The fourth area is the manner of containment (8) statements.

2.4 Instrument Validity and Reliability

To check the instrument validity, the correlation coefficients of the scale items with the domain to which it belongs were extracted. The scale items were also analyzed, and the correlation coefficient of each item was calculated. Cronbach's alpha reliability coefficient was applied to check the instrument reliability through the Likert scale. The reliability value of the creative leadership axis was (0.950) which indicates a very high degree of reliability. And the reliability value of the crisis management axis was (0.940) which also indicates a very high degree of reliability.

2.5 Procedures

1. Reviewing the literature on the subject, and previous studies.
2. Preparing the study tool as explained previously.
3. The questionnaire was examined by specialists and modified until it reached the final form.
4. The researchers selected the faculty members using the simple random method from the different faculties of the university.
5. Applying the questionnaire to faculty members in the second semester of the academic year 2020/2021,
6. Entering the questionnaire data into special tables for statistical analysis.
7. Extracting the results of the study and linking them with previous studies to answer the study questions.

2.6 *Statistical Analysis*

The Statistical Package for Social Sciences (SPSS) program was employed to analyze the data according to the study problem and its questions. The researchers used statistical methods, including arithmetic mean, standard deviation, and one-way analysis of variance (ANOVA).

3 Results and Discussion

3.1 *Results of the First Question*

'What *is the degree of creative leadership practice in crisis management among faculty members at IAU, concerning (originality, flexibility, problem-solving, fluency) from the point of view of department heads, in light of the Corona pandemic?*' Descriptive analysis was used to answer this question, including computing the mean, standard deviation, degree of consistency, and order, for each of the sub-domains of the creative leadership, and the overall axis as shown in Table 1.

Table 1 shows that the degree of faculty members' practice of creative leadership was high, where the mean of their responses on this axis was (3.65), with a standard deviation (0.85) and a (high) degree. The means of the responses of the faculty members on thc sub-domains of the axis ranged between (3.58–3.67). The researchers attribute these findings to the faculty members' possession of the scientific and practical skills that qualify them to be creative in the field of academic leadership, and they also attribute this to their benefiting from the academic development provided to them in this area and to the academic freedom that allows them to be creative in the academically. These results are consistent with what was revealed by Al-Hussein (2018) study, which showed that the primary school leaders in the Hawtah Bani Tamim governorate practice all creative leadership processes, while it differed from the results of the study of Maqableh (2019).

Table 1 The degree of creative leadership practice among faculty members at the university ranked in descending order

Field	Mean	Standard deviation	Degree	Rank
Fluency	3.67	0.93	High	1
Flexibility	3.60	0.87	High	2
Originality	3.61	0.88	High	3
Problem solving	3.58	0.90	High	4
Creative leadership	3.65	0.85	High	

Table 2 The degree to which faculty members practice crisis management

Field	Mean	Standard deviation	Degree	Rank
Containment	3.77	0.86	High	1
Cooperation	3.75	0.90	High	2
Confrontation	3.69	0.85	High	3
Escape	3.45	0.71	High	4
Crisis management	3.69	0.77	High	

3.2 *Results of the Second Question*

'What is the degree of creative leadership practice in crisis management among faculty members at IAU, about (escape, confrontation, cooperation, containment) from the point of view of department heads, in light of the Corona pandemic?' Descriptive analysis was used to answer this question, including calculating the mean, standard deviation, degree of consistency, and items rank, for each of the sub-domains of the crisis management axis, and the overall axis as shown in Table 2.

The practice of the faculty members in crisis management came to a high degree, as the mean of their total responses on this axis was (3.69) with a (high) degree, and the standard deviation was (0.77). Where the means of the responses of the faculty members ranged between (3.77–3.45). These results can be explained by the fact that the faculty members are aware of the academic crises that the university may face during the Corona pandemic, emergencies, and various problems that necessitate preparing for it and preparing the available resources. In addition to the respondents' and society's perceptions of the crisis's threats and the necessity to provide the requisite capacities to deal with them appropriately when they occur. The findings coincide with the results of Hamdouneh (2012), Al-Mashaqbeh (2018), Almazrouei et al. (2020), which revealed that school principals practice crisis management skills to a high degree, and they are also consistent with the study of Al-Khamees and Al-Salihi (2019), which showed that the application of secondary school leaders to decisions concerning managing school crises was high.

4 Recommendations

Considering the findings, the researchers recommend the following:

1. Holding training and development workshops for university faculty members that emphasize the importance of creative leadership and how it can improve members' abilities to manage and face crises.
2. Creating a system of incentives and rewards to encourage creative leadership among university faculty members and to establish a culture of innovation.

3. Developing educational technology, communication, and informatics infrastructure at universities to improve creative leadership and raise the likelihood of success in dealing with academic crises.

References

Abbas, K. F. (2015). *The relationship of crisis management with administrative creativity among the heads of scientific departments in Iraqi universities*. Mustansiriya University.

Abu Zaytoun, H. H. M. (2015). *Creative leadership and crisis and disaster management in the directorates of civil defense in the northern region*. Faculty of Economics and Administrative Sciences.

Ahmad, A., Alshurideh, M. T., & Al Kurdi, B. H. (2021a). The four streams of decision making approaches: Brief summary and discussion. In *Advanced Machine Learning Technologies and Applications: Proceedings of AMLTA* (pp. 570–580).

Ahmad, A., Alshurideh, M. T., Al Kurdi, B. H., & Salloum, S. A. (2021b). Factors impacts organization digital transformation and organization decision making during Covid19 pandemic. *Effect of Coronavirus Disease on Business Intelligence* (p. 95).

Ahmad, A., Alshurideh, M., Al Kurdi, B., Aburayya, A., & Hamadneh, S. (2021c). Digital transformation metrics: a conceptual view. *Journal of Management Information and Decision Science, 24*(7), 1–18.

Al Kurdi, B., Alshurideh, M., Salloum, S. A., Obeidat, Z. M., & Al-dweeri, R. M. (2020a). An empirical investigation into examination of factors influencing university students' behavior towards eLearning acceptance using SEM approach. *International Journal of Interactive Mobile Technologies, 14*(2), 19–41.

Al Kurdi, B., Alshurideh, M., & Salloum, S. (2020b). Investigating a theoretical framework for e-learning technology acceptance, *International Journal of Electrical and Computer Engineering, 10*(6), 6484–6496.

Al Kurdi, B., Alshurideh, M., & Al afaishata, T. (2020c). Employee retention and organizational performance: Evidence from banking industry. *Management Science Letters, 10*(16), 3981–3990.

Al Naqbia, E., Alshuridehb, M., AlHamadc, A., & Al, B. (2020). The impact of innovation on firm performance: A systematic review. *International Journal of Innovation, Creativity and Change, 14*(5), 31–58.

Al-Abdullah, M., Abolfadl, A., Ibrahim, S., & Al-Ansari, A. (2019). Developing the role of middle school principals in light of the educational crisis management approach: A field study in the State of Kuwait. Culture and development. *Cultural Development Association, 2*, 388–339.

Alameeri, K., Alshurideh, M., Al Kurdi, B., & Salloum, S. A. (2021). The effect of work environment happiness on employee leadership. *Advances in Intelligent Systems and Computing, 261* AISC, 668–680.

Al-Dhuhouri, F. S., Alshurideh, M., Al Kurdi, B., & Salloum, S. A. (2020). Enhancing our understanding of the relationship between leadership, team characteristics, emotional intelligence and their effect on team performance: A Critical Review. In *International Conference on Advanced Intelligent Systems and Informatics* (pp. 644–655).

Al-Hamad, M., Mbaidin, H., AlHamad, A., Alshurideh, M., Kurdi, B., & Al-Hamad, N. (2021). Investigating students' behavioral intention to use mobile learning in higher education in UAE during Coronavirus-19 pandemic. *International Journal of Data and Network Science, 5*(3), 321–330.

Al-Hayyaf, H. (2015). *A proposed program to develop the administrative per-formance of department heads in education departments in northern Saudi Arabia in light of their training needs*. Yarmouk University, Ir-bid.

Al-Hussein, A. S. (2018). Creative leadership among primary school leaders. *International. Journal of Educational Psychology and Science Arab Found Science Research and Human Development, 15*, 79–179.

Al-Khamees, K., Al-Salihi, I. (2019). *The reality of the application of secondary school leaders for decision-making in managing school crises in the Qassim region.* Saudi Arabia: Qassim University.

Allozi, B., Alshurideh, A., AlHamad, M., & Al Kurdi. (2021). Impact of transformational leadership on the job satisfaction with the moderating role of organizational commitment: case of UAE and Jordan manufacturing companies. *Academy of Strategic Management Journal, 20*(1), 1–13.

Al-Mashaqbeh, M. (2018). The degree to which government school principals in Zarqa Governorate possess the skill of managing school crises from their point of view. *Journal of Educational and Psychological Science and National Research, 2*(29), 68–83.

Almazrouei, F. A., Alshurideh, M., Al Kurdi, B., & Salloum, S. A. (2020). Social media impact on business: A systematic review. In *International Conference on Advanced Intelligent Systems and Informatics* (pp. 697–707).

Al-Salami, B. (2012). *Creative leadership and its relationship to the organizational climate in intermediate public schools in Jeddah.*

Alshamsi, A., Alshurideh, M., Al Kurdi, B., & Salloum, S. A. (2020). The influence of service quality on customer retention: A systematic review in the higher education. In *International Conference on Advanced Intelligent Systems and Informatics* (pp. 404–416).

Alsharari, N. M., & Alshurideh, M. T. (2020). Student retention in higher education: The role of creativity, emotional intelligence and learner autonomy. *International Journal of Educational Management, 35*(1), 233–247.

AlShehhi, H., Alshurideh, M., Al Kurdi, B., & Salloum, S. A. (2020). The impact of ethical leadership on employees performance: A systematic review. In *International Conference on Advanced Intelligent Systems and Informatics* (pp. 417–426).

Alshurideh, M., Al Kurdi, B., & Salloum, S. A. (2020). *Examining the Main Mobile Learning System Drivers' Effects: A Mix Empirical Examination of Both the Expectation-Confirmation Model (ECM) and the Technology Acceptance Model (TAM)*, 1058.

Alshurideh, M., Salloum, S. A., Al Kurdi, B., Monem, A. A., & Shaalan, K. (2019a). Understanding the quality determinants that influence the intention to use the mobile learning platforms: A practical study. *International Journal of Interactive Mobile Technologies, 13*(11).

Alshurideh, M., Al Kurdi, B., Shaltoni, A. M., & Ghuff, S. S. (2019b). Determinants of pro-environmental behaviour in the context of emerging economies. *International Journal of Sustainable Society*, 11(4), 257–277.

Alshurideh, M., Al Kurdi, B., & Salloum, S. A. (2020a). Digital transformation and organizational operational decision making: A systematic review. In *International Conference on Advanced Intelligent Systems and Informatics* (pp. 708–719).

Alshurideh, M., Al Kurdi, B., Salloum, S. A., Arpaci, I., & Al-Emran, M. (2020b). Predicting the actual use of m-learning systems: A comparative approach using PLS-SEM and machine learning algorithms. *Interactive Learning Environment*, 1–15.

Battah, M. (2016). *A proposed scenario for crisis management among government secondary school principals in the State of Kuwait.* Al al-Bayt University, College of Educational Sciences.

Bettayeb, H., Alshurideh, M. T., & Al Kurdi, B. (2020). The effectiveness of mobile learning in UAE Universities: A systematic review of Motivation, Self-efficacy, Usability and Usefulness. *International journal of control Automation, 13*(2), 558–1579.

Hamdi, K. (2018). *The impact of creative leadership on crisis management.* Al al-Bayt University.

Hamdouneh, H. (2012). *The exercise of the secondary school principal's crisis management skill in Gaza Governorate, Faculty of Education.* Islamic University.

Harahsheh, A. A., Houssien, A. M. A., Alshurideh, M. T., & Mohammad, A. M. (2021). The Effect of Transformational leadership on achieving effective decisions in the presence of psychological capital as an intermediate variable in private Jordanian Universities in Light of the Corona Pandemic. *Studies in Systems, Decision and Control, 334*, 221–243.

Joghee, S., Alzoubi, H. M., Alshurideh, M., & Al Kurdi, B. (2021). The role of business intelligence systems on green supply chain management: Empirical analysis of FMCG in the UAE. In *The International Conference on Artificial Intelligence and Computer Vision* (pp. 539–552).

Kahwagi, A. (2015). *The role of administrative leadership in crisis management*. Economic dimensions magazine: University of M'hamed Bouguerra Boumerdes.

Khairullah, J., *Administration creativity* (3ed). Jordan: Dar Osama for Publishing and Distribution.

Kurdi, B., Alshurideh, M., & Alnaser, A. (2020). The impact of employee satisfaction on customer satisfaction: Theoretical and empirical underpinning. *Management Science Letters, 10*(15), 3561–3570.

Leo, S., Alsharari, N. M., Abbas, J., & Alshurideh, M. T. (2021). From offline to online learning: A qualitative study of challenges and opportunities as a response to the COVID-19 Pandemic in the UAE higher education context. *Effect of Coronavirus Disease on Business Intelligence, 334*, 203–217.

Madi Odeh, R. B. S., Obeidat, B. Y., Jaradat, M. O., Masa'deh, R., & Alshurideh, M. T. (2021). The transformational leadership role in achieving organizational resilience through adaptive cultures: the case of Dubai service sector. *International Journal of Productivity and Performance Management*.

Maqableh, W. (2019). *The degree of innovative leadership practice and its relationship to creativity in performance among secondary school administrators in Jerash Governorate from the point of view of female teachers*. Deanship of Scientific Research and Graduate Studies, University of Jerash.

Nairokh, R. (2020). *The degree of school crisis management practice among government school principals in the Hebron governorate*. Hebron University.

Naqvi, R., Soomro, T. R., Alzoubi, H. M., Ghazal, T. M., & Alshurideh, M. T. (2021). The nexus between big data and decision-making: A study of big data techniques and technologies. In *The International Conference on Artificial Intelligence and Computer Vision* (pp. 838–853).

Obeidat, K., Adass, T., & Abdelhak, A. (2004). *Scientific research concept, tools, and methods*. Jordan: Dar alfiker.

Peterson et al. 2015Peterson, J. F., Frankham, N., McWhinnie, L., & Forsyth, G. (2015). Leading creative practice pedagogy futures. *Art, Design and. Communication Higher Education, 14*(1), 71–86.

Shah, S. F., Alshurideh, M. T., Al-Dmour, A., & Al-Dmour, R. (2021). Understanding the influences of cognitive biases on financial decision making during normal and COVID-19 pandemic situation in the United Arab Emirates. *Effect of Coronavirus Disease on Business Intelligence, 334*, 274–257.

Shah, S. F., Alshurideh, M., Al Kurdi, B., & Salloum, S. A. (2020). The impact of the behavioral factors on investment decision-making: A systemic review on financial institutions. In *International Conference on Advanced Intelligent Systems and Informatics* (pp. 100–112).

Shakhour, M. R., Obeidat, B., Jaradat, N., Alshurideh, M. (2021). Agile-minded organizational excellence: Empirical investigation. *Academy of Strategic Management Journal, 20*(6), 1–25.

Shawish, A. F. (2018). *A proposed conception for developing creative leadership among Palestinian secondary school principals*. The Islamic University of Gaza.

Sohmen, V. S. (2015). Reflections on creative leadership. *International Journal of Globalisation Business, 8*(1), 1–14.

Svoboda, P., Ghazal, T. M., Afifi, M. A. M., Kalra, D., Alshurideh, M. T., & Alzoubi, H. M. (2021). Information systems integration to enhance operational customer relationship management in the pharmaceutical industry. In *The International Conference on Artificial Intelligence and Computer Vision* (pp. 553–572).

Tawiqat, M. (2007). *Developing a training program in creative leadership for deans of faculties and heads of academic departments in Jordanian universities in the light of contemporary administrative trends Unpublished*. Amman Arab University, Amman.

Zu'bi, Z., Al-Lozi, M., Dahiyat, S., Alshurideh, M., & Al Majali, A. (2012). Examining the effects of quality management practices on product variety, *European Journal of Economics, Finance and Administrative Science*, *51*(1),123–139.

The Effect of Transformational Leadership Toward Organizational Innovation: A Study of Aerospace Manufacturing Industry in Malaysia

Mohd Fahmi Ahmad, Nurhanan Syafiah Abdul Razak, Musmuliadi Kamaruding, and Muhammad Alshurideh

Abstract This study objectives are to enhance the understanding of relationship between the transformational leadership on organizational innovation. This framework is proposing to identify the transformational leadership influencing effect in the organizational innovation. A sample of 120 respondent from one of the established Aero Composites Company at Malaysia participated in this research. The result of the study is expected positive support relationship between transformational leadership and organizational innovation. The result of the transformational leadership not only give a significant impact to the process innovation within the organization but in the same time the sustainability of the innovation in driven the better organization management in the competitive market. Only three transformational leadership is considered in the research project which is idealized influence, inspirational motivation, and individualized consideration in the context of aero composites industries generally. This quantitative study used SPSS method for the statistical analysis measurement.

Keywords Transformational leadership · Organizational innovation · Idealized influence · Inspirational motivation · Intellectual simulation

M. F. Ahmad · N. S. Abdul Razak (✉)
Faculty of Business and Administration, DRB-HICOM University of Automotive Malaysia (DHUAM), Pahang, Malaysia
e-mail: hana.syafiah@gmail.com; nurhanan@meritus.edu.my

N. S. Abdul Razak
MERITUS University, Kuala Lumpur, Malaysia

M. Kamaruding
Engineering Department, Faculty of Civil Engineering, University Teknologi MARA, Shah Alam, Selangor, Malaysia

M. Alshurideh
Marketing Department, School of Business, The University of Jordan, Amman, Jordan
e-mail: m.alshurideh@ju.edu.jo; malshurideh@sharjah.ac.ae

Department of Management, College of Business Administration, University of Sharjah, Sharjah, UAE

M. Alshurideh et al. (eds.), *The Effect of Information Technology on Business and Marketing Intelligence Systems*, Studies in Computational Intelligence 1056,
https://doi.org/10.1007/978-3-031-12382-5_84

1 Introduction

In the era of dynamic environment with the high demand for new products and amenities almost company in the world are facing the tremendously pressure on sustaining the product, services, competitiveness in the market faced by rapid technology change, shortening product cycle life and globalization itself. The new borderless environment are created enforcement to leader become creative in managing the organization for sustainability, competing and growth the company in a good manner (Mumford & Gustafson, 1988). Innovation through creativity is essential for the success and competitive advantage of organizations as well as for strong markets leader in these new century.

Thus, this situation need creativity and innovation in leadership style (Shin & Zhou, 2003). This passion also facing at aero composite industry since the main player are limited to technology dependency and extensive research and development. Internal and external factor like COVID-19 pandemic also make contribution in today leadership style in managing crisis with digital transformation platform, spirit of changes and comprehensive financial analysis in managing the operation. Current situation shows the demand from aviation industry declining and give a tremendously pressure in the aero composites industries. The decreasing number of flights up to 20% from the world airlines give a big impact to the entire supply chain. In this case, the aero composite company are not excluding from this impact. The reduction of demand almost 40% from MYR945 to MYR540 million by the end of financial year 2020 need to be restrained to ensure the business sustainability. Even the future still uncertain the trend of world Passenger Load Factor (PLF) for commercial aircraft are increasing over the time as per (Fig. 1).

The increasing demand with the limited resource needs to mitigate, in managing these challenges, leader must be strategizing and clear direction to drive the organization sustainability. The unstable and complex future lead the aero composited industries into consideration of fracturing backlog of delivery to customer especially Airbus as a main player in commercial aircraft passenger industries. With the fragile

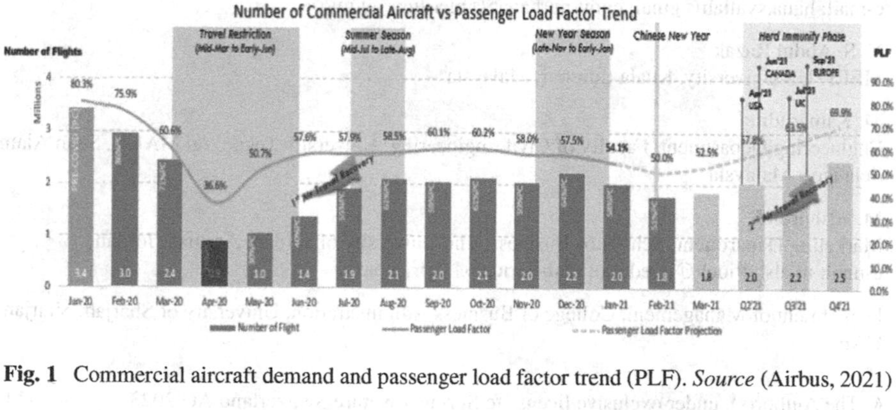

Fig. 1 Commercial aircraft demand and passenger load factor trend (PLF). *Source* (Airbus, 2021)

manufacturing environment, the recovery phases must be handled in a proper manner to make sure the customer need are materialized. The transformational leadership one of leadership style are proven can bring new ideas and possibilities into reality (Bass, 1985). Then, the organizational innovation aims to improve business outcomes by enhancing the current services, process or product to be success in the market (Scott & Bruce, 1994). It is also important for companies to constantly evolve in order to remain sustainable in the face of growing competition from businesses. However, further analysis about the connection between organizational innovation and transformational leadership are essential (Avolio et al., 1999). The strong literature influence and optimistic result of transformational leadership combined with the organizational innovation capability to respond on the demands in the changing of corporate world to makes sure this transformational leadership an attractive strategized in the future. Thus, this research hopefully can contribute additional understanding to the current literature on transformational leadership influencing to organizational innovation especially in the context of aero composite company in Malaysia.

2 Background

The new economic environment and the increasing demand from customer for varieties of product landscapes, services and quality are challenge the companies facing. This situation changes the way of respond and functionary of leadership. An innovation effort is required for organization to stay competitive in the industry. A part of that, organization need effective leaders to change the people. In order to manage the complex pressure, the leader need to come out the best strategies in ensuring business sustainability. The function of leadership as to determinant of organizational inspiration and innovation. This responsible become more important with the growing complexity of job and processes competitiveness in the market. The attribution of innovation as central to competitiveness has been driven by the technological advancement, emergence knowledge economy, competitive element and reliable service in the market. In fact, organizational innovation are more important than before in organization for survival, growth, competing, and remain relevant in the industry (Avolio et al., 1999). Thus, this required an organization to become innovative to gain and survive.

Transformational leadership is knowing as a major determinant of innovation. It can be defined as some or leader share vision and goal inspiring followers in an organization or unit. The leader functional as to challenge the sub accessor to become innovative, problem solver and develop the capability of follower to become a leader by talent development by coaching, mentoring, and providing the challenges and support in materializing the strategic overview (Bass, 1995). In others perceptive, leadership also involves communication among more than one member of a group that regularly engage in managing the situations, perceptions, and expectations of team members.

Innovation can be expressed as an execution of a concept, new item, service, process, marketing strategy, and organizational method. Organizational innovation is essential to current complex organization, this method result from the strategic decision that management has taken. In others words organizational innovation was define as the tendency of an organization or people to changes in developing a new product, services, process and improve it to the new level with its impact on economic development (Gumusluoglu & Ilsev, 2009). Moving forward, the organization must have unique approach to attract their customer, an innovation can play a significant role in this development model. Innovation must be adopted individually and in organizational level to succeed (Jung et al., 2003). Over the years, researchers explored the significant of leadership in influence organizational innovation and identified the related characteristics that significantly affect the organizational innovation (Shin, 1996). Previous studies recommend that transformational leadership has a resilient correlation with innovation both at the organizational level than the individual level (Rosing et al., 2011). Leader are key important roles towards the efficiency of organizational innovation by influencing secure strategic decisions, procedure and policies that are functional as change agent in supporting innovation (Prasad & Junni, 2016). Thus, the right type of leadership are important in driving the effectiveness of organizational innovation (Oke et al., 2009).

Among the empirical study the transformational leadership are most suitable leadership types to enhance the ability of group or organization to change in innovative and creative way in developing the new designs of solutions, organizational structures changes, processes improvement, system practicality and efficiency (Prasad & Junni, 2016).

This study attempts to magnify the understanding of transformational leadership influence of transformational leadership on organizational innovation through resource supply. There are requirement on additional empirical study to support the theoretical proposal analysis on the relationship between transformational leadership and organizational innovation especially on the mediating and mediator variable or determinant need to be examine systematically (Jaskyte, 2004). According to Choi et al. (2016), previous study findings indicate that further research is needed to explicitly focus on the relationship between transformational leadership and organizational innovation, include the limited factor to successes this both relationship. As a finding of previous studies on leadership and innovation have a many opinions and constraint and not drive a simple decision and come out with variation of result and findings. While different leadership styles are positively related to innovation, but most of study show a wide range of correlations based on the setup of moderating conditions. This are lead to arguments that the relationship between transformational leadership and organizational innovation are not always same understanding, which drive for discovering mediating and moderating variable between the theory in detail (Choi et al., 2016). The other studies also suggest, a leader with a transformational leadership style should focus on many areas of the innovation process, such as an overall supportive environment to enhance leader influences. This supportive environment is regarded as an organization's supporting contribution to its employees in order to increase engagement and improve performance (Choi et al., 2016). As a result, through supporting processes inside the organization, it can encourage people to give their time and efforts to innovation initiatives (Amabile, 1998). Therefore, transformational leadership has a substantial positive relationship with organizational innovation, both directly and indirectly through the level of the organization's climate for innovation. Empirical study highlighted the importance of expanding and refining the measurement of organizational innovation because the variables in the study used need to be quantity in materializing organizational innovation were inadequate enough in generating the success outcomes (Gumusluoglu & Ilsev, 2009a). Hence enhancing this research has an significant contribution to understanding and more practicality systematic approach of the relationship between transformational leadership and organizational innovation (Jong & Hartog, 2007), in aero composites industries generally.

3 Literature Review

3.1 Organizational Innovation

Organizational innovation are refer to tendency or the organization to developed a new product or product improvement either in term of competitive price, good services and the successful in bringing value added products or services to the market (Gumusluoglu & Ilsev, 2009a).

It can be defined as the organizational capability to revamp ideas and understanding into new products, services or processes constantly for the benefit to the organization. Accordingly to the empirical research mention creativity as the production of inspired, valuable ideas, and innovation as the successful reality of innovative ideas neither in the group or organization, from the variance of study several researchers have rather defined organization innovation in concurrence with the individual creativity and recognizing individuals talent are the ultimate underpinning of new improvement (Shalley et al., 2004).

3.2 Transformational Leadership

There has innumerable definition of leader by different researcher in various aspect. According to empirical study leadership can be define as the process of providing a significant track or determination for collective exertion, encouraging and willingness to commitment in materializing the target. Based on previous research transformational leaders have been identified give an influence on innovation, which are the ability leader enhance the tendency to innovate within the corporate organization. Normally the transformational leader use inspirational motivation and intellectual simulation which are the most critical factor in influencing organizational innovation (Elkins & Keller, 2003). Transformational leadership style are proven effective through the managerial level (Howell & Avolio, 1993), working environment (Bass, 1985), national culture (Jung & Avolio, 2000), computer related setting (Sosik et al., 1998), innovative developmental environment (Howell & Avolio, 1993), and TQM programs (Sosik & Doinne, 1997). The transformational leadership characteristic are answer to the competitive business environment adaptive leadership is considered to be an important tool (Bass et al., 2003). These adaptive leadership behaviours are characterized as transformational leadership. The four component consist of idealized influence, inspirational motivation, individualized consideration, and intellectual stimulation (Bass et al., 2003). In this research only three main components are considered which are, idealize influence (II), inspirational motivation (IM), and intellectual simulation (IS).

3.3 Idealize Influence

Idealized influence is referred to the type of leader with trusted and well respect behaviour by the follower in the organization or group of people. Typically, this type of leader is stimulus to take a new risk and frequently show the act as role model. Another named of this type of leadership are known as known a charismatic. This type of leader is highly regarded on the moral and ethical value. Previous study showed the effect of transformational leadership characteristic among supporter's shows most of the enthusiast are able to take possibility risk after being organizing with transformational leaders. The confident personality and self-efficacy of such leaders in turn affect how the groups feel about their own capabilities (Bass & Riggio, 2006).

Even the idealized influence are was considered as the most important dominant component of transformational leadership style but this component does not be enough to measure the understanding of transformational leadership (Bass, 1995) individually.

3.4 Inspirational Motivation

Inspirational motivation main characteristic are including enunciating a convincing vision, articulating confidence in accomplishing goals, talking positively and passion about the future along with the desires to be accomplished (Jung & Avolio, 2000).

The functional of inspirational motivation are to enhancing and improve team understanding with the organization vision and mission. This technique will lead to increase group member or organization initiatives and awareness to perform beyond the expectations. The supported levels of motivation is linked towards the higher levels of performance which likely to enhance organizational innovation (Shamir et al., 1993). In these technique leaders encounter followers in reminiscent ways and demonstrate commitment in attaining goals and shared visions. Apart from being consistently included in inspirational motivation in transformational leadership, the establishment of organizational vision and mission is also stated to have relationship in promoting organizational innovation.

As the previous article cited, the successful organization improvement are dependent on the preparation of the followers to achieve the vision through implementation change element, expressing optimism characteristic, confidence and passionate (Jung & Avolio, 2000).

3.5 Intellectual Simulation

An intellectual stimulation is define as a cause's emotion encouragement behaviour among resilient followers. These technique inspire the followers to be innovative and creative in handling problem solution, handling new challenges, and promoting the new way of thinking as a change agent (Jung & Avolio, 2000).

Intellectual stimulation refers to leader who are promotes the intelligence technique, knowledge and learning in developing innovation and solution in problem solving behaviour in materializing the organization improvement planning. These leaders will display environmental awareness and recognize innovation opportunities through cautious evaluation of environment impact (Conger & Kanungo, 1988). According to Woodman et al. (1993), the leader as a founder of different planning and own the obligation on the project by comprehensive strategic planning. At the same time react as inventors and visionaries who inspire team or group of people. The creation of competitive new products within an organization framework is the results of organizational innovation. Commonly, transformational leadership are related to a higher level of performance (Woodman et al., 1993). Transformational leadership have positive impact on employee performance in group and organization (Sosik et al., 1998).

Meanwhile at previous research study have confidence in transformational leadership influences the organizational innovation creatively, these also support by research on actuality and nature of this relationship (Mumford et al., 2002). Finally, the team factor such as intellectual simulation behaviour also influenced by transformational leadership (Waldman & Atwater, 1994).

3.6 *Relationship Between Transformational Leadership and Organizational Innovation*

From the selected previous research, transformational leadership can be define as the process of providing a significant track or determination for collective exertion, encouraging and willingness to commitment in materializing the target (Avolio et al., 1999).

Transformational leadership are recommended have significant impact in organizational innovation process by using the four element behaviour (Bass, 1995), in this study consideration which are:

1. Idealize Influence (II)—focus on leader charismatic action on value, mission, vision and belief.
2. Inspirational Motivation (IM)—the technique of the leader to gain up the follower in belief on future growth and determinant goal.
3. Intellectual Stimulation (IS)—refer to how the leader challenge their follower in materializing the improvement and creative thinking and solution will be implemented.

While the organizational innovation can be define as the organizational capability to improve the new idea and information in to a new product services or process enhancement continuously give beneficial to the firm (Gumusluoglu & Ilsev, 2009a) and tendency of organization to improve, develop, and sustain the product, process or services to the market (Damanpour, 1991). To relate the transformational leadership with organizational innovation the previous research shows the leadership style will affect the organizational tendency to innovate (Gumusluoglu & Ilsev, 2009a) and in the same time will drive the result beyond expectation by enhancing level of self-confident and motivation (Mumford, 2002). Thus the direct and indirect impact of transformational leadership have correlation to organizational innovation and drive the positive and significant relationship between the two constructions (Gumusluoglu & Ilsev, 2009a). Thus, the various definition of transformational leadership and organization innovation are shows at (Table 1) as summary of the various section overview.

The selection of variable for this research are:

1. Variable 1—Transformational Leadership (IV)

   Transformational leadership style is being used as independent variable (IV) in this research.
2. Variable 2—Organizational Innovation (DV)

   Organizational innovation process is being used as dependent variable (DV) in this research.

Previous empirical case study is had been used to understand the conceptual of the research as per below justification.

Study Title: Transformational leadership and organizational innovation: Moderated by organizational size.

Table 1 Various definitions of transformational leadership and organizational innovation

Definition	Source
The technique decision making units are structured within the organization, the way decision making, power, and skills are distributed within the organization and between decision making elements and the type of information and communication structures that are well organized	Greenan (2003)
Organizational innovation is the introduction of new organizational systems of work and collaboration	Mulej (2002)
Managerial innovations occur in the social system of an organization. It includes rules, characters, procedures and structures that are related to the communication and discussion among people and environment	Damanpour (1991)
Transformational leadership emerged to be effective across managerial levels	Howell and Avolio (1993)
Process of providing a significant track or determination for collective exertion, encouraging and willingness to commitment in materializing the target	Bass (1995)
The special characteristic of organizational innovation can be obtain as it's bring the change to the firm social system, while the change might be not technical and intangible but in the same time it might affect the technical aspects as well	Edquist et al. (2001)
The words of organizational innovation itself in general are applied by the organization by adapting the changing conditions of competition, technical innovation and business growth through introducing better products, techniques and systems	Utterback (1994)
The leader functional as to challenge the sub accessor to become innovative, problem solver and develop the capability of follower to become a leader by talent development via coaching, mentoring, and provision of both challenge and support in materializing the strategic overview	Bass (1985)
Transformation leader are able to bring change in their follower personal normal and encourage the organization performance beyond the expectation at the same time increased the level of motivation similarly with self-motivation	Jung and Avolio (2000)
Organizational innovation can be defined as the organizational capability to improve the new idea and information into a new product services or process enhancement continuously give beneficial to the firm	Bass (1995)

4 Summary

This case study emphasizes on the impact of transformational leadership on organizational innovation. A sample of 296 managers are used from the telecommunication sector at Pakistan. The average ages of participant used within 25 to 60 years.

Hierarchical regression models are used to demonstrate organizational size significantly moderating the relationship between transformational leadership. This study proof that organizational population size is moderated significantly the relationship between all aspects of transformational leadership and shows the positive impact with organizational innovation.

Keywords: Organizational innovation, transformational leadership, attribute charisma, organizational size, inspirational motivation, intellectual stimulation and individualized consideration, idealized influence (Khan et al., 2019).

Based on study model the construction of hypothesis development for this research are:

1. Hypothesis One—Idealize Influence
   H01—Idealized influences is not related to organizational innovation at Aero Composites Company.
   H1—Idealized influences is positively related to organizational innovation at Aero Composites Company.
2. Hypothesis Two—Inspirational Motivation
   H02—Inspirational motivation is not related to organizational innovation at Aero Composites Company.
   H2—Inspirational motivation is positively related to organizational innovation at Aero Composites Company.
3. Hypothesis Three—Intellectual Simulation
   H03—Intellectual simulation is not related to organizational innovation at Aero Composites Company.
   H3—Intellectual simulation is positively related to organizational innovation at Aero Composites Company.

5 Theoretical Framework

The dependent variable (DV) for this study is refer to the Contingency Theory (CT) initiated by "The work of Weber" (1947). Under the scope of contingency theory, the most suitable structure for an organization is the one that best matches a particular operational contingency, such as size of operation (Pugh et al., 1969), technology (Woodward, 1965) or environment. This theory underpins the knowledge of the interactions between nature, technology, environment, structure, and performance (Burns & Stalker, 1961). The demonstrated of differentiate technology and market environment in the perspective of rate of change and complexity will affect the organizational structure and innovation are shown in the (Box 1) (Burns & Stalker, 1961).

Box 1 Mechanistic and Organic Structures. ***Source*** **Burns and Stalker (1961)**

Bums and Stalker (1961) set out to explore whether differences in the technological and market environments affect the structure and management processes in firms. They investigated twenty manufacturing films in depth, and classified environments into 'stable and predictable' and 'unstable and unpredictable'. They found that firms could be grouped into one of the Wo main types, mechanistic and organic forms, with management practices and structures that Rums and Stalker considered to be logical responses lo environmental conditions.

The *Mechanistic Organization* has a more rigid structure and is typically found where the environment is stable and predictable. Its characteristics are:

(a) tasks required by the organization arc broken down into specialised, functionally differentiated duties and individual tasks arc pursued in an abstract way, that is more or less distinct from the organization as a whole;
(b) the precise definition of rights, obligations and technical methods is attached to roles, and these arc translated into the responsibilities of a functional position, and there is a hierarchical structure of control, authority and communication;
(c) Knowledge of the whole organization is located exclusively at the top of the hierarchy, with greater importance and prestige being attached to internal and local knowledge, experience and skill rather than that which is general lo the whole organization;
(d) a tendency for interactions between members of the organization to be vertical, i.e. between superior and subordinate.

The Organic Organization has a much more fluid set of arrangements and is art appropriate form to changing environmental conditions which require emergent and innovative responses Its characteristics are:

(a) individuals contribute to the common task of the organization and there is continual adjustment and re-definition of individual tasks through interaction with others;
(b) the spread of commitment to the organization beyond any technical definition, a network structure of control authority and communication, and the direction of communication is lateral rather vertical;
(c) knowledge may be located anywhere in the network, with this ad hoc location becoming the centre of authority and communication;
(d) importance and prestige attach lo affiliations and expertise valid in industrial and technical and commercial milieu external to the firm.

Mechanistic and organic forms are polar types at the opposite ends of a continuum and. in some organizations, a mixture of both types could be found.

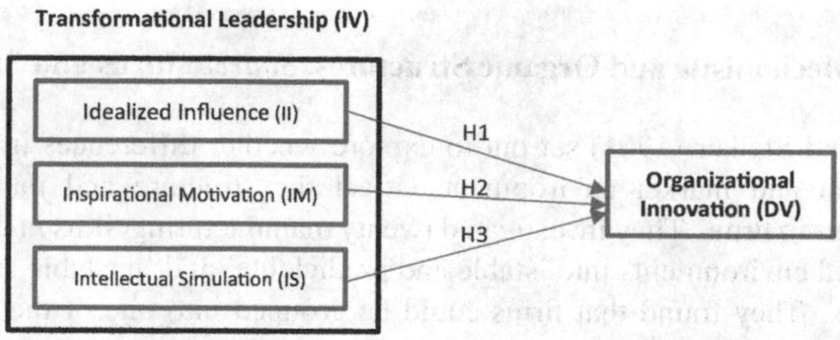

Fig. 2 Research conceptual framework

Discussing on theory of the study, the research framework is develop to demonstrate the relationship between the external determinant and internal determinant in transformational leadership style as the independent variable (IV) with the organizational innovation as dependent variable (DV) as per (Fig. 2).

6 Research Methodology

This research design used quantitative approach to support the hypothesis. The questionnaire will be used to pertaining the leadership style and to excess the full range of leadership scale, the influence factor and the relation with organizational innovation. The study will use data measurement Likert Scale. The scale weightage used logic numbers 1 through 5 with the format indicate (Armstrong, & Robert, 1987). The collection data by conducting an internet web survey trough SPSS application will be presenting the descriptive analysis which provide simple summaries about the sample and the measures. The analysis result will be present using simple graphics analysis in the basis of virtual. Every quantitative analysis of data as part of data analysis and measurement.

7 Conclusion

The ability to continuously innovate by implementing creative efforts as well as the necessary skills and attitudes is critical to an organization's competitiveness. Leadership become major determinant of an organization ability to innovate, with its significant impact on the creation of the organizational vision and the methods to achieve the goal, the benefits of transformational leadership in stimulating organizational innovation can be secure by organizational in the context that support innovation and increased the overall performance of the organization.

References

Amabile, T. M. (1998). How to kill creativity. *Harvard Business Review, 76*(9), 77–87.Google Scholar Pub Med.

Armstrong, & Robert. (1987). The midpoint on a five-point likert-type scale. *Perceptual and Motor Skills, 64*(2), 359–362.

Avolio, B., Bass, B., & Jung, D. (1999). Re-examining the components of transformational and transactional leadership using the multifactor leadership questionnaire. *Journal of Occupational and Organizational Psychology, 72*(4), 441–463.

Bass, B. (1985). Transformational leadership. *Journal of Management Inquiry, 4*(3), 293–298.

Bass, B. (1995). Transformational leadership. *Journal of Management Inquiry, 4*(3), 293–298.

Bass, B., Avolio, B., Jung, D., & Berson, Y. (2003). Predicting unit performance by assessing transformational and transactional leadership. *Journal Applied Psychology, 88*, 207–218.

Bass, B., & Riggio, R. E. (2006). *Transformational leadership. Mahwah.* NJ: L. Erlbaum Associates.Google Scholar.

Burns, T., & Stalker, G. M. (1961). *The management of innovation.* Tavistock.

Choi, S. B., Kim, K., Ullah, S. E., & Kang, S. W. (2016). How transformational leadership facilitates innovative behavior of Korean workers: Examining mediating and moderating processes. *Personnel Review, 45*(3), 459–479.Cross Ref Google Scholar.

Conger, J., & Kanungo, R. (1988). The Empowerment process: Integrating Theory and Practice. *The Academy of Management Review, 13*(3), 471–482.

Damanpour, F. (1991). Organizational innovation: A meta-analysis of effects of determinants and moderators. *Acadamy of Management Journal., 34*(3), 555–590.

Edquist, C., Hommen, L., & McKelvey, M. (2001). *Innovation and employment.* Process innovation versus Product Innovation. Edward elgar.

Elkins, T., & Keller, R. T. (2003). Leadership in research and development organizations: A literature review and conceptual framework. *Leadership Quarterly, 14*, 587–606.

Greenan, N. (2003). Organisational change, technology, employment and skills: An empirical study of French manufacturing. *Cambridge Journal of Economics, 27*(27), 287–316.

Gumusluoglu, L., & Ilsev, A. (2009a). Transformational leadership and organizational innovation: The roles of internal and external support for innovation. *Journal of Product Innovation Management, 26*, 264–277.Cross Ref Google Scholar.

Gumusluoglu, T., & Ilsev, A. (2009). Transformational leadership, creativity and organizational innovation. *Journal of Business Research, 62*, 461–473.

Howell, J., & Avolio, B. (1993). Transformational leadership, transactional leadership, locus of control and support for innovation: Key predictors of consolidated-business-unit performance. *Journal of Applied Psychology, 78*, 891–902.

Jaskyte, K. (2004). Transformational leadership, organizational culture, and innovativeness in nonprofit organizations. *Nonprofit Management and Leadership, 15*(2), 153–168. https://doi.org/10.1002/nml.59.CrossRefGoogleScholar.

Jong, J. P. J. D., & Hartog, D. N. D. (2007). How leaders influence employees' innovative behaviour. *European Journal of Innovation Management, 10*(1), 41–64.https://doi.org/10.1108/14601060710720546.CrossRefGoogleScholar.

Jung, D., & Avolio, B. (2000). Opening the black box: An experimental investigation of the mediating effects of trust and value congruence on transformational and transactional leadership. *Journal of Organizational Behavior, 21*, 949–964.

Jung, D., Chow, C., & Wu, A. (2003). The role of transformational leadership in enhancing organizational innovation: Hypotheses and some preliminary findings. *The Leadership Quarterly, 14*(4–5), 525–544. https://doi.org/10.1016/s1048-9843(03)00050-x.Cross Ref Google Scholar https://doi.org/10.1016/j.orgdyn.2008.10.005.Cross Ref Google Scholar.

Khan, R., Rehman, A. U., & Fatima, A. (2019).Transformational leadership and organizational innovation: Moderated by organizational size. *African Journal of Business Management, 3* (11), 678–684.

Mulej. (2002). How innovative are the Business and Management of the Slovenian manufacturing enterprises? Our economy No. 3-4, 217–237.

Mumford, M., Scott, G., Gaddis, B., & Strange, J. (2002). Leading creative people: Orchestrating expertise and relationships. *The Leadership Quarterly, 13*, 705–750.

Mumford, M. D. (2002). Leading creative people: Orchestrating expertise and relationships. *The Leadership Quarterly, 13*(6), 705–750.Cross Ref Google Scholar.

Mumford, M., & Gustafson, S. (1988). Creativity syndrome: Integration, application, and innovation. *Psychological Bulletin*, 27–43.

Oke, A., Munshi, N., & Walumbwa, F. (2009). The influence of leadership on innovation processes and activities. *Organizational Dynamics, 38*(1), 64–72. https://doi.org/10.1016/j.orgdyn.2008.10.005.CrossRefGoogleScholar.

Prasad, B., & Junni, P. (2016). CEO transformational and transactional leadership and organizational innovation: The moderating role of environmental dynamism. *Management Decision, 54*(7), 1542–1568.Cross Ref Google Scholar.

Pugh, D. S., Hickson, D. J., & Hinings, C. R. (1969). The context of organization structures. *Administrative Science Quarterly, 14*, 47–61.

Rosing, K., Frese, M., & Bausch, A. (2011). Explaining the heterogeneity of the leadership-innovation relationship: Ambidextrous leadership. *The Leadership Quarterly, 22*(5), 956–974.Cross Ref Google Scholar.

Scott, S., & Bruce, R. (1994). Determinants of innovative behavior: A path model of individual innovation in the workplace. *Academy of Management Journal, 37*(3), 580–607.

Shalley, C., Zhou, J., & Oldham, G. (2004). The effects of personal and contextual characteristics on creativity: Where should we go from here? *Journal of Management, 30*(6), 933–958.

Shamir, B., House, R., & Arthur, M. (1993). The motivational effects of charismatic leadership: A selfconcept based theory. *Organization Science, 4*, 577–594.

Shin, J. (1996). *The effects of executive leadership on organizational innovation in nonprofit*. University of Pittsburgh. Google Scholar.

Shin, S., & Zhou, J. (2003). Transformational leadership, conservation and creativity: Evidence from Korea. *Academy of Management Journal, 46*(6), 703–714.

Sosik, J., & Doinne, S. (1997). Self concept based aspects of charismatic leader: More than meets the eye. *The Leadership Quarterly, 9*, 503–526.

Sosik, J., Kahai, S., & Avolio, B. (1998). Transformational leadership and dimensions of creativity: Motivating idea generation in computer-mediated groups. *Creativity Research Journal, 11*(2), 11–121.

Utterback, J. (1994). *Mastering the dynamics of innovation: how companies can seize opportunities in the face of technological change*. Harvard Business School Press.

Waldman, D., & Atwater, L. (1994). The Nature of Effective leadership and championing processes at different levels in an R&D hierarchy. *Journal of High Technology Management Research, 5*(2), 233–245.

Woodman, R., Sawyer, J., & Griffin, R. (1993). Toward a theory of organizational creativity. *Academy of Management Review, 18*(2), 293–321.

Woodward, J. (1965). *Industrial Organization, theory and practice*. Oxford University Press.

Entrepreneurial Marketing: An Approach-Based Paradigm Shift to Marketing

Hanin Damer, Shafig Al-Haddad, Ra'ed Masa'deh, and Muhammad Turki Alshurideh

Abstract This paper aims to augment the research field of Entrepreneurial Marketing (EM) as a concept and approach, and the substantial differences it holds in comparison to the prevailing giving of the marketing. In doing so, the research aims to explore and analyze the causes and drivers that resulted in the emergence of EM as a new school of thought and conceptualize the paradigm shift EM is providing to the prevailing giving of the marketing field. Grounded theory and logical argumentation based on the literature review including both qualitative and quantitative research, to develop the argument leading to conceptualizing the proposed approach-based paradigm. The research conceptualizes a proposed paradigm to formalized marketing with four main components comprising it: resilient management; scalable/gradual matter of sub-frames including multi-dimensional orientation effectual-causal, proactiveness-reactiveness; living strategic process, and Entrepreneurial Innovation. This paper represents the first bold proposal on the need for a paradigm shift to the prevailing marketing giving and introduces a frame for conceptualized paradigm. Specifically, does the body of literature in EM provide a base toward a paradigm shift of the prevailing of marketing giving or only a new school of thought?

H. Damer · S. Al-Haddad (✉)
Princess Sumaya University for Technology (PSUT), Amman, Jordan
e-mail: s.haddad@psut.edu.jo

H. Damer
e-mail: hanindamer@gmail.com

R. Masa'deh
Department of Management Information Systems, School of Business, The University of Jordan, Amman, Jordan
e-mail: r.masadeh@ju.edu.jo

M. T. Alshurideh
Department of Marketing, School of Business, The University of Jordan, Amman, Jordan
e-mail: m.alshurideh@ju.edu.jo

M. Alshurideh et al. (eds.), *The Effect of Information Technology on Business and Marketing Intelligence Systems*, Studies in Computational Intelligence 1056,
https://doi.org/10.1007/978-3-031-12382-5_85

Keywords Entrepreneurial marketing · Termed-labeled marketing · The paradigm shift

1 Introduction

Research in entrepreneurial marketing has developed since its inception involving the number of views, lenses, perspectives, theories and models. That aim at defining, investigating, understanding and even measuring and assessing validity and impact. And this holds the main themes of viewing EM. Along with this growth in research comes the notice of labeled or termed marketing, with the adjectives attached to marketing including "conventional", "conventional wisdom", "traditional", "standard", "administrative", "textbook marketing", "corporate traditional marketing" and others. Mostly in the context of explaining the difference between EM and labeled marketing, in addition to other notices on the critique side concerning being no longer suitable to address the changing of the business, economic and market landscape. This labeling of marketing is the result of the emergence of new definitions or models that differs noticeably from the giving of the formalized marketing discipline and curricula. And that proved to be more suitable in the current market landscape. The discipline is still there, but the approach to these elements developed noticeably with applied models differing from the current giving's. While "Marketing" remains as a discipline, this discipline had witnessed development until shaping the formalized principles and theoretical frames deriving the giving of the main textbooks. And it shall also be subject to continual development with the advancement and changes requiring a response. Moreover, research indicates a lag between this discipline as a research body and academia and the business and practice world, which raises the need for such evolvement rather than development.

This paper augments both the research in EM and critique to the formalized marketing to propose the concept of an approach-based paradigm shift in marketing. Arguing that this provides a basis and a call for a paradigm shift, which shifts the approach and concept interpretation providing a frame that, responds to the changes shaping the market, business and economy. In proposing so, this paper also aims to investigate the main drivers behind in link with those addressed in EM; environmental landscape, mindset, activities, resources landscape, etc. Noticing that all stress the need for a novel approach.

Although vast researches on EM indicates the insufficiency of the formalized market giving's to the Entrepreneurial Ventures (EV), SMEs and also on the general level with the changing in business and market landscape, none augmented on the body available to promote it concerning paradigm shift. Some tended to indicate the need for a new school of marketing thought (Abu Zayyad et al., 2021; Al-Dmour et al., 2020; Mort et al., 2012). Under this paper, we refer to the existing prevailing course of marketing as formalized market giving's indicating the standards, we refer to also to avoid muddling with terms used in labeling marketing in the reviewed papers. This covers the "marketing activities of planning, formal marketing research, design and implementation of the marketing mix or program, that are the core of the traditional textbooks in marketing" including Brassington & Pettitt, Jobber and Kotler et al. as indicated in Amjad et al. (2020a), Alwan and Alshurideh (2022).

Still, marketing is evidenced to be a key to the success of new ventures (Mort et al., 2012; Al Kurdi et al., 1339). Furthermore (Miles & Darroch, 2006) argues that EMP could be leveraged in large firms as equal to SMEs to establish a sustainable competitive advantage. Especially in the context of dramatic changes on social and technological bases, where EM is seen as an approach toward establishing sustained customer relationships (Miles & Darroch, 2006).

2 Entrepreneurship and Marketing

This paper does not aim to provide a literature review for the two disciplines of Entrepreneurship and Marketing, however, below aims to review key definitions and concepts on these two disciplines. To establish the base for the argument and augmentation on research.

2.1 Entrepreneurship

The core focus of this paper is entrepreneurship in the business field. Moreover, this provides the base required, with the review of literature linked to it covering behavioral and activity perspectives of entrepreneurship. Thus, starting by the entrepreneur definition as proposed by Drucker as cited in, defining an entrepreneur from a behavioral and activity-based perspective as the person who looks for the change, but differs in being responding to the change in an innovative way, thus utilizing it as an opportunity (Alameeri et al., 2021; Mehmood et al., 2019; Stokes, 2000). Building on Drucker's ideas entrepreneur is defined from the process viewpoint as "a process, an action-oriented management style which takes innovation and change as the focus of thinking and behaviour" (Stokes, 2000). Morrish et al. (2010) indicates a definition for an entrepreneur as "engaging in opportunity creation or discovery, assessment and exploitation of attractive opportunities" from the perspective of what the meaning of being such.

Moving to firm context, the entrepreneurial firm is defined as "one that engages in product-market innovation, undertakes somewhat risky ventures, and is first to come up with proactive innovations" (Kraus et al., 2010). Attached to this definition is entrepreneurial management defined as "the process by which individuals—either on their own or inside organizations—pursue opportunities without regard to the resources they currently control" (Kraus et al., 2010). Phua and Jones (2010) considers that entrepreneurship is better to be understood from theoretical and practical views, to help to understand what actually in practice it is. Under this view, the authors adopt the process perspective definition of EM considered by Morris et al.

(2002), where the key elements composing the process are "the entrepreneur, the opportunity, and the acquisition and management of resources".

Also from a process view, entrepreneurship is defined as "the process of creating value by bringing together a unique package of resources to exploit an opportunity." Where it entails the activities required for opportunity identification, business concept definition, assessment of required resources and acquiring them, in addition to managing and harvesting the venture (Morris et al., 2002). Moving to the concept, the entrepreneurship concept is found to have three fundamental dimensions, "innovativeness, risk-taking/management, and proactiveness" (Morris et al., 2001). The author defines concept and process perspective as "Entrepreneurship is how firms discover, create or assemble resource assortments that allow them to produce valued market offerings".

From a firm contextual frame, (Collinson & Shaw, 2001) sees that the entrepreneurship area holds relevancy for large and small size firms as well. Explaining the cause behind being attached more to SMEs to two points: commonly visible more in smaller sized firms and with firms growing in size, the complex structure of management makes it more challenging to maintain an entrepreneurial focus. The authors continue to explain, most of the research in this field is around the SME, is due to the difficulty of spotting the true entrepreneurial large firms. However, the number of large-size firms developing this approach is increasing. The author states that the number of large firms through their approach to marketing activities can be entrepreneurial.

Even though literature relates EM to SME context, (Hills & Hultman, 2011) notes the discrepancy between EM and small business marketing that is reflected in a conceptual separation between both. Attaching the difference to the behaviors of the person running the firm; owner mindset running the business as opposed to the entrepreneur mindset. The note is valid, however, in this paper, the author relates SME to the EM in the commonalities of the organization context of size, constrained resources, newness in the market and other attributes. The same extends to the references reviewed that tackle SMEs from the same perspective. In this view, we attribute to the notion "A small business is not a little big one" observation of *Welsh/White*'s (Gruber, 2004) provides an insightful piece to shift the understanding of new ventures as in comparison to larger ones. Gruber (2004) identifies three main characteristics differentiating new ventures namely: newness, small size, Uncertainty, and turbulence.

Collinson and Shaw (2001) brings another perspective of defining entrepreneurship within the context of this paper with "entrepreneurial effort" linked with "entrepreneurial scenario" encountered with a focus on firm and management context. Where "entrepreneurial effort" is described to hold several characteristics demonstrated by the entrepreneur or those of the management team, these include "energy, zeal, commitment, determination, persistence, opportunity and focus". On the other hand, the "entrepreneurial scenario" is referred to as a level of risk encountered by the firm, along with resources available in addition to the need among individuals for "skills, knowledge, experience and personal independence". The consequence of these two factors is found to be affecting the degree of firms' development and growth over (Collinson & Shaw, 2001). This explains why with the growth marinating the same level of "entrepreneurial effort" becomes more difficult, along with this challenge comes the tendency toward activities lowering risk levels.

2.2 Marketing

Marketing is considered by several researchers as one of the crucial elements of firm success (Alshurideh, 2022; Lee et al., 2022a, 2022b; Tariq et al., 2022). Where it is also considered as one of the vital pillars for the realization of both objectives and success (Franco et al., 2014). Definitions cover both scholarly work and textbooks of academia, with different views mainly of concept and process perspectives. As overall briefed definition marketing is viewed as, "meeting needs shortly" where needs refer to human and social needs and marketing to include identification of those needs (Kotler & Keller, 2016).

The recent revision for the definition of the American Marketing Association (AMA) dated 2007 defines it as" Marketing is the activity, set of institutions, and processes for creating, communicating, delivering, and exchanging offerings that have value for customers, clients, partners, and society at large" (Hills & Hultman, 2011). Continuing on concept level definition, five components are indicated to be included in any definition of marketing, these are "two or more parties, something that is given up by each party, something that is received by each party, some level of communication between the parties, and some mechanism to perform the exchange" (Morris et al., 2001). Stokes (2000) from an organizational point of view defines four components underlying the varied definitions of marketing including an imperative one represented by organizational culture and the three of the strategic process, tactical methods, and marketing intelligence. The integration of these components suggests the comprehensive definition of marketing as an "organizational philosophy of market orientation, guided by segmentation, targeting and positioning strategies, operationalised through the marketing mix and underpinned throughout by market intelligence" (Stokes, 2000).

More specifically at the process level (Morris et al., 2001) views it as "the process of planning and executing the conception, pricing, promotion and distribution of ideas, goods and services to create exchanges that satisfy individual and organizational goals". The authors indicate also four categories of interrelated linkages that are introduced in most definitions of marketing: "product", "price", "promotion", and "distribution", these four combined results in numerous scenarios of decisions each denoted as the "Marketing Mix". The mix decisions shall be designed to reflect the needs identified of targeted customers, allowing at the same time provision of the firm with sustainably based differentiation edge as compared to competitors.

2.3 *New Approaches/Recent Approaches*

In searching for new methods, that are more suitable for different business environments, new marketing approaches emerged in practice and academic literature. With adjectives terming these as new approaches to the marketing, among these main are "Environmental marketing management", "Subversive marketing", "proactive marketing", "expeditionary marketing", "guerrilla marketing", "Viral marketing", "radical marketing" and "disruptive marketing", and "viral marketing" (Morris et al., 2001, 2002). Some authors consider the emergence of these as an informal rise of entrepreneurial marketing, including the new models /methods /perspectives that emerged both in the trade and in academic literature. The common factor seen among these is being concerned about an entrepreneurial role for marketing in leading-edge companies (Morris et al., 2001).

Below provides a highlight on the key aspects of the main new approaches indicated earlier to provide the understanding of the difference between this and the formalized market giving and to provide the base for the next section on Entrepreneurial marketing (Table 1).

Assessment of these indicates the commonalities briefed by Morris et al. (2001) as including efficient marketing budgets, where with constrained resources extensive amount is conducted; different modes of resource leveraging, where people in charge capable of employing resources, not under ownership or control, refusal of "conventional approaches" to marketing variable management; continuing product/service innovation; leading customers along with the capability of affecting change in the

Table 1 Explanation of new approaches/models/perspectives to marketing

Guerrilla marketing: relies on bootstrapping, creative use of available resources, and a highly targeted mix of innovative communications techniques
Radical Marketing: describe a set of approaches that they argue challenge the status quo, or immutable laws, of marketing
Expeditionary Marketing" describes the "role of marketers in creating markets ahead of competitors
Subversive marketing: refers to the need for marketers to undermine company structure and process to implement innovative marketing practices
Environmental marketing management: where marketing theory explicitly adopts a proactive, entrepreneurial orientation to the management of environmental conditions
Proactive marketing: includes ongoing responsibility for redefining the product and market context within which the firm operates, identifying novel sources of customer value, and emphasizing unproven wants, new market segments, new technologies, and continuous innovation in all areas of the marketing mix
Viral Marketing encourages individuals to pass on a marketing message to others, creating the potential for exponential growth in the message's exposure and influence
Buzz marketing is the aggregate of all communication about a certain product that is propelled through grassroots activities among people at any given time

Source Adapted from Morris et al. (2001)

environment of context. Morris et al. (2001) argues that these different approaches and perspectives, advocate the base for an "alternative concept" of marketing. Supported by the point that these emerged with the changes occurring in the market and business environment and the need for new approaches to respond, in the light of insufficiency of existing ones.

3 Entrepreneurial Marketing

3.1 Marketing Within an Entrepreneurial Context

"In each small entrepreneurial firm, marketing is done" (Gross et al., 2014). Even if not through the involvement of the entrepreneur or the team in the formal or theoretical process of it, it remains to be practiced on a day to day with the intuition, formulation and implementation (Gross et al., 2014). Stokes (2000) notes the increased recognition of the marketing as practiced by entrepreneurs to be different than those concepts offered in what is referred to as conventional marketing textbooks and give the example of Kotler. Stokes and Stokes (2000) make the note that "a small firm is not simply a scaled-down version of a large firm" clarifying that not the same rule of large firms can presumably assume to be valid for new ventures. On the other hand (Collinson & Shaw, 2001) sees that EM is also valid for a larger firm, and provides the example of Virgin.

Entrepreneurial firms are deemed to be driving markets, through influencing the marketplace directly or indirectly and supporting shaping it (Morrish, 2011). Though celebrated as far different from conventional, small business holds several difficulties when it came to marketing, such as a slim customer base, marketing activities of both limited scope and impact, efforts of unplanned and diversified nature (Stokes, 2000). Marketing is considered among the utmost challenges faced in SMEs, yet it is also of crucial importance for the survival and success of SMEs (Franco et al., 2014). Although not all entrepreneurial firms are SMEs, (Becherer et al., 2012) considers marketing to have a higher significance in the context of the size of SMEs, where one customer lost or gained has a high impact on the survival of these firms. Becherer et al. (2012) confirms the same point and considers marketing to be a primary success determinant among all new ventures. Also within the entrepreneurial context, venture capitalist rates marketing as of extreme importance (Becherer et al., 2012).

Morrish (2011) proposed that EM firms still perform the processes of traditional marketing, but in a way that is influenced profoundly by the main attributes of entrepreneurs guiding the process from the core. These attributes are innovation, reactiveness, and risk-taking. Furthermore, (Morrish, 2011) views EM firms as having a twofold core of both customer-centric, being similar to AM and entrepreneur-centric. Providing the EM firms with the flexibility of navigating Market-Oriented strategy and Entrepreneurially Oriented strategy as deemed suitable for the different environments. Morrish et al. (2010) also suggests that EM could represent one of the evolution stages of marketing sophistication within the firm. With the possibility of occurrence at the initial or last stages, implying the EM is still considered as "fully embedded by the traditional processes of marketing management and strategy". Morrish et al. (2010) is a support to this stage model of Kotler, proposing that EM could be very effective along all the stages of the model, at any time the organization is in a need of creating or discovering new and attractive opportunities and exploiting them. In proposing so, the authors see that marketing's core functions and processes remain the same. However, the difference occurs in the approach within which where these functions and processes are conducted while adapting an approach of an "entrepreneur/customer opportunity-driven" to the three elements of marketing namely: culture/philosophy, strategy and tactics. Affecting consequently the implementation of the process of segmenting, targeting and positioning.

This view sees the difference in EM being in the influencing attributes, while the process remains the same as the referred to "traditional marketing" or formalized marketing giving's. And appears to conflict with the mainstream literature on EM viewing it as substantially different in approach, strategy and process from the formalized marketing giving's such as (Gruber, 2004; Morris et al., 2002; Stokes, 2000). It also appears to have a sort of conflict with the stage model of Kotler dated 2003 adopted by the author in their definition of EM, (Morrish et al., 2010) concludes that EM organizations "have a core of AM [Administrative Marketing] functions, augmented by the entrepreneur's desires, needs and motivations [….], nested and overlayed by the multiple layers of EM processes". In indicating EM processes, the author defines a new set of processes other than those of AM, where at the same position sees that marketing core functions and processes remain the same, with the difference in the adopted approach to these processes.

Another point is attached to the notion of EM organizations having a core of AM functions, which comes in conflict with several works indicating that this function is not available at EV in most cases (Stokes & Stokes, 2000). This conflict is better understood within the perspective adopted by Morrish et al. (2010) focusing on the interrelationship among AM and EM. In addition, not the integrative perspective. Entrepreneurial marketing is defined as "the marketing of small businesses which grows through entrepreneurship" (Yadav & Bansal, 2002). This view limits the definition to the contextual practice rather than the concept-based definition.

3.2 *Theoretical Foundation and Development*

Collinson and Shaw (2001) draws three main focus areas in the development of EM seen as emerging from the two management disciplines of marketing and entrepreneurship, considered traditionally as two distinct areas of study (Stokes & Stokes, 2000). After which research evolved around the overlap between both, identifying the areas of similarity and differentiation while investigating the role of each in the other and how these can benefit from each other. The interface comes as research advances, with the focus on the attempts to combine the two disciplines and provide an integrative area (Collinson & Shaw, 2001; Stokes & Stokes, 2000).

Uslay and Teach (2009) highlights the issue raised by other scholars on the fast emergence of the MEI, with the potential of becoming a "distinct marketing school of thought". Notwithstanding this, this remains insufficiently developed with segmented works, while a single paradigm has not been developed yet. EM (2012) concludes four main different approaches that were developed along the time of researching the field of EM, these are commonalities between the two main disciplines, "entrepreneurship in marketing" focusing on marketing framework within which entrepreneurship elements are introduced; third is "marketing in entrepreneurship"; and the fourth comes in opposite to the first approach in identifying what is unique about EM rather than similarities, arguing that combination of these two disciplines yields in a new distinctive thing. Collinson and Shaw (2001) sees three key areas of interface between marketing and entrepreneurship: both being change focused; both in nature are opportunistic, and both are innovative in approaching management. Another views build on the interface role and define EM research as focusing on exploring the interfaces of three disciplines and not only the two of marketing and entrepreneurship, with the addition of Innovation. Jones and Rowley (2009). This evolvement in the approach to understanding EM provides a rational sequence of approaching this field. It also on the other hand reflects an extent an implicit evolvement in the marketing discipline too, that is affected by this evolution. The fact that EM is viewed as the first approach of studying it under the commonalities between Entrepreneurship and marketing and later as the role of it in Entrepreneurship or the other way of role relationship provided a base toward the evolvement of marketing school of thought. While if EM is considered as integrative concept from the onset this would have seen limitations.

EM (2012) indicates the first emergence of EM to be dated 1982 at a conference at the University of Illinois. Later in 1987, the first formal research convention on Entrepreneurship and marketing was held at the (AMA), turning later on to the annual event of "Research at the marketing-entrepreneurship interface conference" (Amjad et al., 2020b; Stokes & Stokes, 2000). Amjad et al. (2020b) in researching the current development is "Entrepreneurial Marketing Theory" indicates that research in the EM witnessed rapid progress over the last decade, attributed to the effectiveness of EM in the uncertain conditions and markets of high competition. Figure 1 from the same study shows the first literature on EM to be in 1976, with slow growth following; and rapid one starting from 2000, specifically from 2010. Totaling of 160 articles were examined in their study between the duration of 1976–2018, among these 140 are in the ten years between 2008–2018.

Despite the progress in the research and the rapid one as shown in Fig. 1, the EM area is considered to be still substantially lacking theoretical and academic models (Amjad et al., 2020b). The author also indicates that EM is identified as a "much under-researched area" in a vast range of the literature, with interventions and practices in this area being sprinted away ahead of the work on in the research and development of theories needed to vindicate and explain (Amjad et al., 2020b). Despite this situation, new models are developed during the progress in the research in EM. These are identified by Amjad et al. (2020b) based on a systematic review and

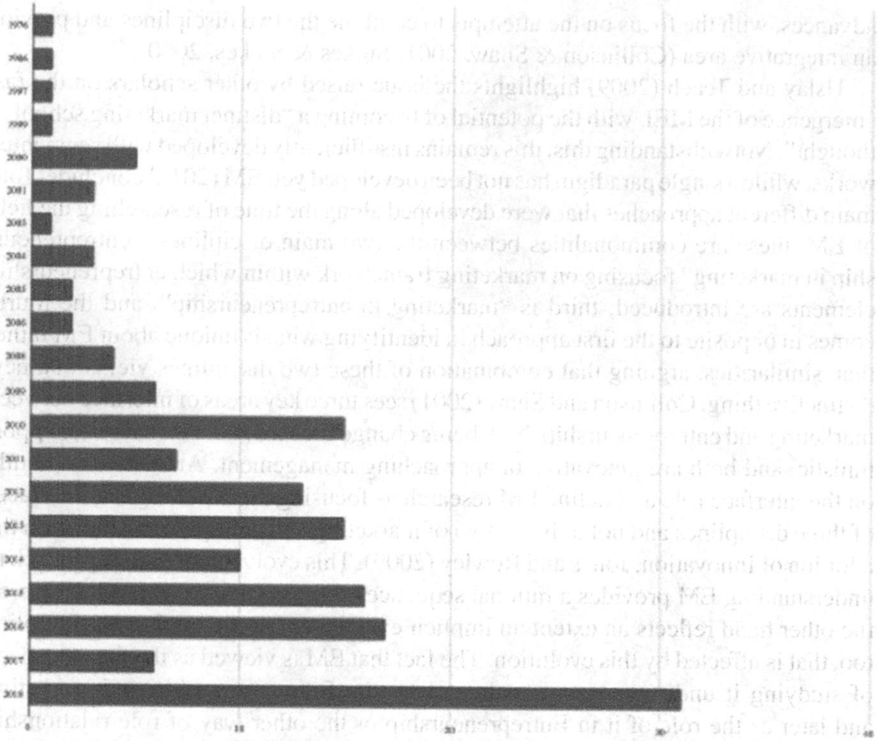

Fig. 1 Number of EM articles per year (Amjad et al., 2020b)

synthetization of all peer-reviewed literature in the period of 2008–2018, considered to be the era of much progress in the literature. These newly developed models, although provides further knowledge on the EM, are restricted by the controllers of the research behind them. Özdemir (2013) investigated the process of 'social value creation' in EM and presented a model of the whole process of creation. Furthermore, these represent only ten-out of the one hundred forty articles identified and examined-who have contributed to developing substantive models. This validates the viewpoints of this field to be substantially lacking theoretical models.

Another note made by the author on the models developed to be mostly based on deductive methods, where there is more need for inductive methods to support the development of fundamental theories and models that can stand as principles for the EM (Amjad et al., 2020b). Under the same state of art paper (Amjad et al., 2020b). Listed the varied theoretical lenses within which the EM has been examined in 20 articles from the 140 identified from 2008 to 2018. Among these lenses, two are shown to be of most frequently used, and thus tested. Moreover, are also referred to in some other papers to support the author's point of view. These are the Seven Dimension Model of Morris et al. (2002) and the EMICO framework by Jones and

Rowley, (2009). In building the seven dimensions, (Morris et al., 2001) suggests the EM to be particularly consistent with Resource-Advantage Theory.

These two lenses are tackling the dimensions of the EM, providing a brief on the nature of the EM phenomena and an explanation of the dimension comprising the theoretical lens/model. Thus, both are considered basic and descriptive. Amjad et al. (2020b) Seven Dimensions of Morris et al. (2002) are indicated to provide an overall primary model, while EMICO of Jones and Rowley (2009) are considered to be more comprehensive and inclusive of those of Morris et al. (2002). However, looking at the EMICO framework's fifteen dimensions, it is noticeable that some are more of functions and processes, such as understanding and delivering customer value, sales and promotion, research and development. The author indicates also that both do not cover the relationship between the dimensions. However, (Morris et al., 2002) indicates an overall frame for this relationship indicating that these dimensions are not independent and need not be available together to form the EM.

3.3 *Multifaceted Understanding of EM*

In this research, the author presents a multifaceted understanding of EM-based on the research development in supporting the explanation and exploration of EM. This multifaceted understanding is presented in the layering of EM as an Interface-influence between composing disciplines, as concepts and definitions, a Mindset, an Orientation, a Concept, a dimension, a strategy, and a process followed by a distinct view of marketing from the entrepreneur's point.

3.3.1 As an Interface—Influence

Franco et al. (2014) in their study on the exploration of the influence of the founder-entrepreneur and the role, this one plays in EM in SMEs and found that marketing is recognized as an important element for business growth and sustainability. However, the two cases studied showed that its role is not activated through the formal planning process, but rather through the informal and reactive process: mainly due to the restricted resources. A noteworthy conclusion claims that founder-entrepreneur have a high influence on the marketing in SMEs and is closely dependent on his/her decision. Being in charge of the whole business. Stokes and Stokes (2000) supports this view, highlighting that characteristics of the owner or manager have a dominant influence in SMEs. On the same point, (Martin, 2009) considers "the active role of an entrepreneur" as a factor distinguishing "formal' marketing from 'entrepreneurial' one in addition to the factor of "networks" that are considered to provide the link behind the phenomena.

3.3.2 Mindset

Some studies suggest an alternative conceptualization, with EM understood as "marketing with an entrepreneurial mindset, irrespective of firm size or age" (Kraus et al., 2010). Adopting this mindset approach, they define EM as an "organizational function of marketing by taking into account innovativeness, risk-taking, reactiveness and the pursuit of opportunities without regard for the resources currently controlled", which compared to that provided on Fario, shows more toward entrepreneurship. The authors in presenting their definition indicate that it applies to large firms and shall not be limited to small ventures.

3.3.3 Orientation

EM as an orientation lies under the interface role of marketing in entrepreneurship and the other way. A stream of research is found to be focusing on orientation perspective with a focus on the marketing orientation and entrepreneurial orientation. These orientations also are considered based on the most frequently referred models of seven dimensions of Morris et al. (2002) and the EMICO of Jones and Rowley (2009). Morris et al. (2001) propose a relationship between the two orientations as "Coupling a market orientation with entrepreneurial values provides the necessary focus for the firm's information processing efforts, while it also encourages frame-breaking action, thus greatly increasing the prospects for generative learning". Morris et al. (2001) suggest another interpretation for this relation, where the authors consider that these two orientations may be "be part of a single, overriding organizational philosophy" and not only holding relationship among them.

3.3.4 As Concept and Definition

Yadav and Bansal (2002) indicates that the concept of 'Entrepreneurial marketing' emerged in 1982. In addition, since then this concept is defined from different perspectives and contexts. To define it, some tend to go in a straightforward position by describing it as "marketing carried out by entrepreneurs or owner-managers of entrepreneurial ventures" (Stokes, 2000). Kraus et al. (2010) views it from a mindset view and defines it as "the marketing activities with the entrepreneurial mindset". Franco et al. (2014) presents a description of its nature as being" informal and ad hoc in nature" and defines it as "a process with an Entrepreneurial Spirit".

Morris et al. (2001) could be considered as the first one introducing the term of entrepreneurial marketing as an integrative concept. That emerged as a response in the time of information intensity and the continuing change faced by the firms in the environmental landscape within which they operate. Authors introduced this integrative concept, to conceptualize marketing within this context, and define it as "the proactive identification and exploitation of opportunities for acquiring and retaining profitable customers through innovative approaches to risk management, resource leveraging and value creation" (Amjad et al., 2020a).

This definition attributes to entrepreneurship from an activity perspective, with the similarities in the three main definition components if compared to the scale of Covin and Slevin on entrepreneurial activity, considered to be the most used one and is established about "risk-taking, innovation and proactive response" as the behavioral dimensions of the scale (Stokes, 2000). Some tend to view SME marketing as a synonym to EM. Franco et al. (2014) considers that Marketing in SMEs should be labeled as EM with attachment to similar contextual environments, making SME modeling an aggressive and unconventional approach, that enables the provision of the means toward success notwithstanding the limitation of resources. Although this contextual attachment holds relevancy, there are other main dimensions of entrepreneurial ventures that put limitations on this generalization, among these is the discard of resources when pursuing the opportunity, mindset, entrepreneurial activities, and others.

EM is defined by Becherer et al. (2012) as "effective action or adaptation of marketing theory to the specific needs of SMEs" and uses the term to describe the marketing process adopted by firms pursuing opportunities in an inexact circumstance of the market. As a contextual definition EM is used mostly in describing marketing conducted at firms of small and medium sizes, more often at the stages of the start-up or the following early growth one (Amjad et al., 2020a). These SMEs entrepreneurial ventures are mostly characterized by limitations on their human and financial resources, which as a response necessitates creative and innovative methods in marketing (Alshurideh et al., 2021, 2022; Amjad et al., 2020a).

Mort et al. (2012) adapts the seven dimensions of Morris et al. and defines EM accordingly as "the proactive identification and exploitation of opportunities for acquiring and retaining profitable customers through innovative approaches to risk management and resource leveraging for value creation". Morrish (2011) differs from the viewpoint of EM as a nexus between the two composing disciplines and tends to view EM as "an augmented process, where both the entrepreneur and customer are the core actors, co-creating value within the marketing environment". In this supporting, (Morrish et al., 2010) view of an entrepreneur being a central actor in the marketing process and equal to that of customer-defined as a focal point in traditional marketing.

EM is also viewed as "the Result of Entrepreneurial Interpretation of Information, Decision-making and Marketing Actions" (Hills & Hultman, 2011), where the entrepreneurial interpretation of the business environment influences the decision making and with actions, these are resulting in noticeable EM outcomes. Morrish et al. (2010) provides a definition for EM, that extends on (Morris et al., 2002) definition and incorporates AMA definition of marketing with "a spirit, an orientation as well as a process of passionately pursuing opportunities and launching and growing ventures that create perceived customer value through relationships by employing innovativeness, creativity, selling, market immersion, networking and flexibility". This is seen as an attempt to define how an organization can adopt EM rather than conceptualizing EM.

Fiore et al. (2013) propose a comprehensive definition as "Entrepreneurial marketing is an organizational function and a set of processes for creating, communicating, and delivering value to customers and for managing customer relationships in ways that benefit the organization and its stakeholders, and that is characterized by innovativeness, risk-taking, proactiveness, and may be performed without resources currently controlled". Although this definition is provided as an extension to Morris et al. (2002) and the AMA definition. This stands as putting the two definitions together rather than defining it as a standalone one. Martin (2009) suggests the 4P's of: "person, process, purpose, and practices" as a frame better suited to understand marketing in entrepreneurial firms, where each of these new P's "is grounded in relationships and networks".

3.3.5 Dimensions

The most frequently referred work on the identification of the dimensions underlying the EM is Morris et al. (2002), proposing seven dimensions. These are namely, "Opportunity driven", "Proactiveness", "Innovation focused", "Risk Management", "Customer Intensity", "Value Creation" and "Resource Leveraging". The first four of these are based on the work of the number of scholars on the 'entrepreneurial orientation' of the firm and the following two on the market orientation, resource leveraging is relevant to entrepreneurship and marketing literature and the new marketing approaches. Although the concept of EM as introduced by Morris evolves around the role of marketing within the organization. The design and the seven dimensions of the new approach affects also the business represented by the organization/firm and their relationship with the marketplace (Morris et al., 2001).

The proposal of these seven dimensions provided a theoretical base and has been since then as the main underpinning theory used in several studies (Amjad et al., 2020b). Jones and Rowley (2009) built on these seven dimensions and number of "23 characteristics of EM" that are identified by Hills and Hultman (2011) and developed EMICO framework that evolves around the four of "entrepreneurial orientation (EO)", "market orientation (MO)", Innovation Orientation (IO)" and "Customer/Sales Orientation CO". The fifteen dimensions of EMICO are:" research and development, speed to market, risk-taking, proactiveness, market intelligence generation, responsiveness towards competitors, integration of business processes, networks and relationships, knowledge infrastructure, propensity to innovate, responsiveness towards customers, communication with customers, proactively exploiting markets, understanding and delivering customer value, and sales and promotion". Some views these to be more comprehensive than the seven dimensions, however these dimensions relate to function orientation more than dimensional one. Below further elaboration of each of the seven dimensions, clarifies further both the design of the concept and the new approach proposed for EM as an integrative concept (Table 2).

Table 2 Seven dimension of EM-(Morris et al., 2002)

Customer intensity" is proposed to "capture a sense of conviction, passion, zeal, enthusiasm and belief in where marketing is attempting to take the firm and how it plans to get there
Sustainable Innovation involves the ability at an organizational level to maintain a flow of internally and externally motivated new ideas that are translatable into new products, services, processes, technology applications, and/or markets
Strategic Flexibility: involves a willingness to continuously rethink and make adjustments to the firm's strategies, action plans, and resource allocations, as well as to company structure, culture, and managerial systems
Calculated Risk-taking involves a willingness to pursue opportunities that have a reasonable chance of producing losses or significant performance discrepancies
Environmental Proactiveness: redefine elements of the external environment in ways that reduce environmental uncertainty, lessen the firm's dependency and vulnerability, and/or modify the task environment in which the firm operates
Resource Leveraging: being not constrained by the resources currently controlled or owned

Morris et al. (2002) clarifies that these seven dimensions are not independent, where interventions in one of them affect the others. Furthermore, the operation of the whole of the seven dimensions together is not a condition for the EM occurrence, where EM could be in place with a number of these dimensions only. In stressing so, the authors clarify the gradation nature of EM, where the different amalgamations resulting from these seven dimensions show a degree of less or more EM. And when the combination considers these seven dimensions collectively, the resulting marketing does differ considerably from the formalized marketing giving.

Amjad et al. (2020a) proposed another dimension to be added which is 'legitimation'. Although this work is the first to introduce it as the eighth dimension, 'legitimation' is identified to be integral to entrepreneurial success in the study of Amjad et al. (2020a). Mort et al. (2012) identifies legitimacy as one among four key strategies identified for EM. Amjad et al. (2020a) defines legitimation from the perspective of EM as "legitimation exhibits the reliability of a firm/entrepreneur, which positively influences the trading decisions of the immediate audience". In advancement of Morris et al. (2002) work, (Fiore et al., 2013) provide scale validation for the seven dimensions in small firms, with proposed scale validated showing the influence of the entrepreneurial intentions on the EM and the role the latter plays on 5Ps and 4Es affecting the Brand distinctiveness.

3.3.6 EM as a Strategy

Three entrepreneurial marketing strategies are highlighted in Morrish (2011) and are concerning market driving approach under market-oriented overall one. These include deconstruction, construction and functional modification, and are applied by EV aiming to "reshape customer preference and behaviour" covering anticipation of future ones as well.

Mort et al. (2012) identify four core strategies for EM, where the authors find that it is through "purposeful strategy based on effectuation approach" that EM contributes to the outstanding performance in small firms. In their work, Mort et al. examined EM in "born global firms" or "International new venture INV" covering high- and low-tech industries, where this internationalization and globalization context within different industry profiles provides a wide perspective to examine the EM process and its impact on performance, leading to the understanding of EM strategies.

Core strategies of EM identified that are found to be contributing to the firm performance when applied strategically are "opportunity creation; customer intimacy based innovative products; resource enhancement; and legitimacy" (Mort et al., 2012). These core strategies when viewed from the concept and dimensional frameworks presented above are found to be augmenting on these. With the dominance of the key elements of opportunity, customer intensity, innovation, and resources (Table 3).

Morrish et al. (2010) under their viewpoint of EM as being both Customer and entrepreneur-centric, provides a frame for EM strategy suggesting "EM firms focus on the needs, desires and motives of both the customer and the entrepreneur to shape the firm's concept of strategy and its tactical processes". Swenson et al. (2012) suggests an entrepreneurial marketing framework that helps in creating opportunities from a competitive angle base. and that also supports marketing strategy development. Their framework provides a systematic approach for the evaluation and execution of the EM and is composed of 5 elements: Opportunity Creation; effect multiplication; relationships leveraging; process acceleration; and profit-making. Where opportunity creation is attached to the choices on competitive angles and is considered to be of importance in identifying losers and winners in the startup context. Competitive angles are composed of five facets (1) need to believe, (2) reason to believe, (3) blows away expectations, (4) quantifiable support, and (5) unique product claim (Swenson et al., 2012).

In a contrast to the research proposing EM as having a different strategy frame, (Phua & Jones, 2010) finds that entrepreneurs indeed use existing marketing tools in the daily management of their venture. They do also use marketing strategies in a notion to the exaggerated myth of the informality nature of planning activities in small firms. However, the authors stress the limitations of the study are based on four NES entrepreneurs and that cannot be generalized as a rule. And that this doesn't exclude informal marketing practices from being adopted. A notion on this is the pattern matching method used in matching the studies practices in the marketing strategy model adopted by the authors, which results in a tendency to a framed analysis of the activities within the model.

3.3.7 Process

Fiore et al. (2013) suggests that "EM is a complex process as well as an orientation for how entrepreneurs behave in the marketplace". The term Entrepreneurial Marketing Processes (EMP) is conceptualized by Miles and Darroch (2006) as the

Table 3 Core Strategies of EM from Concept and Process Relevance

Core strategies	Concept/dimensional relevance	Process relevance
"Opportunity Creation": where opportunities that do not previously exist are created actively. Rather than the identification of pre-existing ones and exploitation of them This is valid with the emergence of new challenges needed to be tackled An ongoing multifaceted process	Advancement on opportunity orientation, where opportunities are to be discovered	
"Customer intimacy based innovative products": the ability to develop marketable innovative products based on ongoing relationships with customers and intimacy	Nexus amid sustainable innovation and "customers intimacy", linking both in one strategic dimension	Contrasting the process of a new product (Mort et al., 2012) Advancing EM in empirical clarification of the Iterative process of customer intimacy and innovative product development
Resource enhancement: highly developed ability to identify and mobilize resources from external sources, enrich and extend existing internal resources and recombine these in a novel, elaborated ways with a strategic purpose	Extending on resource leveraging Mort et al. (2012) sees that (Morris et al., 2002) considers resource leveraging as a dimension as "the ability to use other people's resources" making them able to overcome the limitations, where they consider it as developing new combinations	
Legitimacy Strategies and tactics used to get marketing acceptance and trust for the firm and its products and services		

Source Author based on (Mort et al., 2012)

marketing processes emphasizing" "opportunity creation and/or discovery, evaluation and exploitation" (Kraus et al., 2010). Entrepreneurs are likely to have a different view of marketing. Stokes (2000) investigated the definition of marketing from the perspective of entrepreneurs and with a focus on the strategic level of process, his work is based on data from face-to-face interviews with forty entrepreneurial characteristics, along with following focus groups in two series. Results from this show that entrepreneurs tend to define marketing from the tactics component-third level in marketing after culture and strategy—with the goal of new business attracting.

His work also reported less awareness of the strategic and culture-related levels. In some parts, this is related to not being aware of the terminology, rather than ignoring it. Furthermore, focus groups show that considerable resources, effort and time are spared on marketing but are flagged by another name. His work defines the Entrepreneurial marketing process, through the differences between the "giving's of marketing theory", and insights provided by what successful small businesses and entrepreneurs do. This definition is better understood as a comparison along the four components composing marketing (concept, strategy, tactic and marketing intelligence) each compared between the two sides of conventional marketing and entrepreneurial marketing and resulting in four main aspects defining the EM.

At concept or cultural components, the key difference is in the orientation directing the new product or service development. "The idea comes first and the check for market acceptance second" (Stokes, 2000). Thus, under EM the process is more toward Innovation oriented, implying that that idea, intuition or competition pressures stimulate the process followed by finding the market. This stands as the other way, compared to the customer orientation in conventional marketing, where study, assessment, or research of market needs establish and commence the process of new product development. Stokes (2000) also indicates that under EM, customer needs remain as a key issue, but the core difference is that in most cases ideas or intuitions are transformed through innovation activity to new products or in some cases into defining total new markets in the case of disruptive innovation.

Under the strategic process component, a key difference is attached to the direction or route of strategizing. Entrepreneurial founders-manager, while developing strategies, lean toward a "bottom-up" direction, where a majority of marketing textbooks advise "top-down". Formalized Strategy involves processes toward attaining customers with applying segmentation, targeting and positioning sequentially ordered; segmentation is developed using different criteria to develop segmented profiles, followed by the assessment of these segments and their attractiveness to inform the selection of targeted one; position is then developed and communicated resulting in the differentiated offering.

Under EM "bottom-up", the direction is more of reversed one in process order, where targeting starts at a smaller scale base of customers and then widens up to attract more. This process is articulated in three main stages; market opportunities detection through non-formal, ad-hoc actions and intuitive based testing; initial customer base attraction, and through the regular contact with this initial base during this stage, entrepreneurs become more knowledgeable by their needs; followed by the extension of this base, with attracting more customers of similar profile that is confirmed in the prior stage. Also, this expansion is more of non-formal nature and involves methods such as word of the mouth (Stokes, 2000). The small-scale base at the initial stage and the regular contact enables a wider definition of customers, beyond the end-user, consumer, or the one who buys, to cover the whole network of supportive.

Under Tactical methods, a key variance lies in "4P's" versus "Interactive and Word-of-Mouth Marketing"; where the marketing mix of product, price, place, and promotion design the marketing program for the product or service, developed based on the understanding of the marketplace. On the Entrepreneurs' side, their activities seem to have different methods of themes, focusing on 'interactive marketing' that comes in harmony with the "bottom-up" approach, and that involves direct interaction and development of personal relationships with their customer. This conversational relationship enables entrepreneurs to have a deep understanding of their customers, built through the process of listening and responding.

However, this is enabled with the small initial base of customers and could be of higher challenge to maintain with the expansion of this base while moving toward top level in their approach. Notwithstanding this challenge, this "meaningful dialogue" remains one of the key differentiators. To overcome this, these interactions are tied with "Word-of-Mouth" marketing that enables spreading the message and expanding the customer base. In addition, is defined as "Oral, person-to-person communication between a perceived non-commercial communicator and a receiver concerning a brand, a product or a service offered for sale" (Stokes, 2000). Finally, concerning Marketing intelligence "Market Research" versus "Networking" delineates the key difference, whereas in traditional marketing formal research and market intelligence inform the different stages of marketing such as market segmentation and targeting. Entrepreneurs tend for lateral and non-formal methods, where information is gathered through their contacts' network of those involved in the industry or trade (Stokes, 2000).

Although seems confusing to compare traditional marketing to EM on a concept versus processes base, it shows to be a valid comparison to understand and define. Where processes are referred to about the primary source of forty face-to-face and focus groups, aiming in a different method to identify the EM and define it from how Entrepreneurs see it and practice it through a process of a set of activities. Stokes (2000) propose a framework or entrepreneurial marketing process through the combination of four strategic processes differentiating it from the formalized giving of marketing. As a process, another dimension is the management of the EM. Collinson and Shaw (2001) identifies three themes differentiating management of EM. These are the process, the position of marketing within an organization and the approach. The process of managing EM is described to involve a shorter process of decision–making headed more often by minute formal planning. This differs from managing regular approaches to marketing, main reason lies in the contextual environment of EM with fluctuating and changing one, where time consumed for the regular planning process is not valid and may end in obsolete outputs with the change occurring.

Under the organizational position of marketing, this position is more often adopted as the guiding philosophy in entrepreneurial ventures. With the organizational activities orientating around marketplace and customer. This is valid, even though in most EVs no identified market section is in place, and no individuals are assigned responsible solely for this. Gearing the above two themes is the informal approach most often adopted for managing marketing in EV. Portraying this approach is the organizational-level commitment of grasping, foreseeing, and responding to the needs of a continually changing marketplace. This makes marketing activity under the ownership of the whole organization and managed by the whole, decreasing the

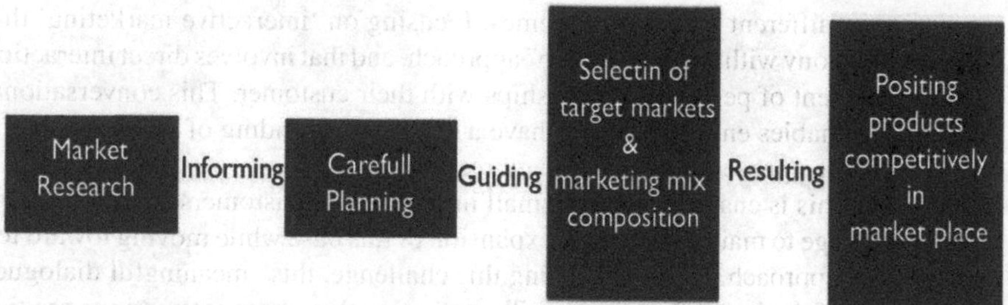

Fig. 2 Formalized Process of Marketing Giving's. *Source* Author contribution based on Collinson and Shaw (2001)

importance of managing it in a formal approach. The above allows for more flexibility, where formal approaches cannot hold this nature within the sequential order of process and the gates at each required to activate the others. The below graph illustrates the sequential order of formalized process of marketing gives (Fig. 2).

The key issue distinguishing EM is related to the effectual logic or frame. Within this logic of decision-making, entrepreneurs are found not to believe in the possibility of predicting the future, which explains why they do not attach great importance to set objectives as the case in the causation frame. Instead, they are found to be starting with the means and what they have "tangible and intangible assets", the things they can do "capabilities", and those whom they know "networks" and with these, they establish variant options of diversified ends (EM, 2012). In their work of investigating the EM in practice and from a social practice theory perspective, (Gross et al., 2014) notes that particularly in small firms, the activities of marketing extends beyond the mandate and scope of the marketing department designated, furthermore it is more likely not to have a dedicated role of marketing. In their work, they emphasize that firms adopting EMPs are "better suited to discover and create, assess, and exploit attractive entrepreneurial opportunities and that this enhanced level of corporate entrepreneurship enables the EMP firm to more effectively and efficiently create and renew competitive advantage" (Miles & Darroch, 2006).

4 Why the Paradigm Shift

The practice in marketing witnessed raising critique of various issues, with the main notion of on the criticism of traditional marketing related to the "one size fits all" assumptions of marketing theory (Martin, 2009). In addition to this comes "an over-reliance on established rules of thumb, encouragement of formula-based thinking [....] an emphasis on the promotion elements of the marketing mix [....] the tendencies to imitate instead of innovating and serve existing markets instead of creating new ones, a concentration on short-term, low-risk payoffs, and marketing as a silo with static and reactive approaches (Morris et al., 2002). Along with this comes the notion of increasing evidence highlighting the insufficiency of "conventional marketing approaches" in addressing and tackling the needs of the firms working in contemporary environments with the changes these environments are witnessing. France et al. (2014) sees that this competitive paradigm witnesses continuous change. In reflection of current changes and mega trends, this is further forced with the issues of technological advancement and the fourth industrial revolution, competition on local, regional and global affecting new and emerging needs of the customers. The market is furthermore described as context-dependent by Morris et al. (2002), and nowadays more than before this context is continually changing, with multidimensional changes and a larger number of forces driving this change. Since 1999, suggestions on the need for a reconceptualization of marketing are proposed. Arguing that development of the discipline building on the extension of the existing theoretical frameworks possibly will not be enough to reflect the changes and shifts in the marketplace, and consequently guiding the practice in the new context (Morris et al., 2002). The same extends to academia and the gap in contribution to marketing practices.

In another strategic critique, (Miles & Darroch, 2006) noted that "Much concern has been expressed about the relationship of marketing to innovation. The marketing concept, it is argued, has contributed to the death of true product innovation in North America." Morrish et al. (2010) also, mark that entrepreneurship is needed for firms to grow. Along with this, EM emerged as providing a more effective and efficient model substituting the formalized marketing giving. The literature body in the EM is showing increasing consensus that EM differs substantially from the conventional one (Mort et al., 2012). This is supported by the works of Stokes (2000), Morris et al., 2002). Some scholars developed their work responding to the challenge made to EM scholars "to [more] fully develop EM as a school of marketing thought," (Mort et al., 2012).

Mort et al. (2012) indicates that noted contributions are provided to the 'emerging paradigm' of EM by several authors. With implicit propositions on the superior role of EM on value creation. The author also disagrees with the point of labeling EM as unplanned, rather they send the view of EM being non-linear activity in harmony with characteristics of effectuation proposed by Sarasvathy as indicated by Morrish (2011). This represents one of the key contributions of Mort et al. (2012), in the raise of EM being an effectuation rather than causal. Under effectuation (Mort et al., 2012) explains that entrepreneurial ventures have broad known goals however the ways to get to them cannot be approached in a logical process following causal setting. This causal setting seems to be more appropriate for the case of existing firms aiming at increasing market share in their existing markets (Morrish (2011) defines the effectuation processes as that which "take a set of means as given and focus on selecting any of the multiple possible outcomes or effects".

Collinson and Shaw (2001) identifies structured management and its impacts on the approach for marketing as one of the key issues making entrepreneurs apart from adopting it. With the preference for a more flexible, non-sequential approach. On the same point, the focus of most textbooks in marketing and management is focused around this structured and ordered approach. Morrish (2011) make a good referral on the omission of the marketing mix from the then-recent AMA definition, raising the point of the development of functional focus toward process focus with increasing the importance of this process and the network. Another force driving the change is the issue related to the changes at the consumer side, where studies show more focus on the characteristics of a contemporary consumer being more knowledgeable and discriminating, and the consumer tendency to focus on experiential and intangible factors (Fiore et al., 2013).

The paradigm shift is raised by Hill and Wright in a note that a "paradigmatic shift would allow for the full expression of the entrepreneurial personality in the management and marketing activities of the SME" (Martin, 2009). In their note Hill and wright bring up the argument that a new entrepreneurial paradigm of marketing could be created from the further complete adjoining of both marketing and entrepreneurship. Behind this, several scholars question the major elements of textbook marketing. Gronroos debates the high questionability of the usefulness of the general theory of the 4P's, while others raise the issue of missing the "fundamental point of marketing—adaptability, flexibility and responsiveness" with strict adherence to the 4P's" (Martin, 2009).

Hills and Hultman (2011) investigated the relation of EM to mainstream marketing, to the 'paradigm discussion' on the relation of EM to marketing discipline described as one of the EM's parents. In their analysis, they note that the U.S marketing stream dominating of the discipline for decades, with the AMA revisions for definition made in 1935,1982,2004 and 2007. These revisions reflect the evolution on the focus from the performance of business activities to the focus on the process of planning and executing, followed by the focus on marketing as an organization function, in addition, to set of processes, and recently on the comprehensiveness of activity, institutions, and process with the exchanging brought up as part of value creation. However, (Hills & Hultman, 2011) sees EM as a research field to be supplementing the market mainstream theories, however through focusing on marketing as done by entrepreneurs. Renton et al. (2015) notes the recognition in the latter literature in EM of entrepreneurs "doing" traditional marketing, but the form of doing this is not understood fully. Entrepreneurs are found to use practices that are recognized as traditional marketing but in a different approach and themes. They "do so in adaptive, intuitive, opportunistic and pragmatic ways, fitted to the needs of the business" (Renton et al., 2015).

With this in mind, the social practice theory emerges as a strong base, and with the emergence of the research stream of marketing-as-practice, that is focusing on the "actors, practical activities, performances, representations and tasks involved in the complex performance of marketing" (Gross et al., 2014). This provides a response to nature required. The social practice also provides the potential of investigating the "practical marketing practice (doing or praxis) of the firm, which is permeated and characterized by the behavioral and cognitive attributes associated with entrepreneurship" (Gross et al., 2014).

4.1 The Recent Business Environment and Drivers for a Shift

Morris et al. (2002) describes the business environment, which is still valid in parts of it in describing today's environment. Being consisting of "increased risk, decreased ability to forecast, fluid firm and industry boundaries […] and new structural forms that not only allow for change but also help create it" this fluid, rapidly changing nature is further augmented with the advancement in data and information, the fourth industrial revolution, global and regional market exposure and other issues. Amjad et al. (2020b) attaches specific importance to the economic and social challenges in developing countries, resulting in further uncertainty and higher risks compared to the conditions in developed countries. These conditions are among the main drivers for entrepreneurs toward EM, being more suitable at such conditions (Amjad et al., 2020b).

At the global level, entrepreneurship as a means for economic development represents a key concern (Amjad et al., 2020a). This concern has raised the importance of entrepreneurship role. However, this is challenged at the entrepreneur's levels with some issues they face in the field including market dynamics, rapid changes, competition, uncertainty and at the top of it is the pandemic caused (Amjad et al., 2020a). The market can no longer be considered as a "given" issue or place for transactions as considered by the positivist school of thinking. It is being shaped more toward "potential" things according to the constructivism school of thinking. Thus, it is more a process of socio-relational nature, where providers or producers and customers or consumers engage in a co-consumption and co-production of lifestyle, identity in addition to the product (EM, 2012; Al-Dmour et al., 2020).

4.2 Labeled-Termed Marketing

Labeling or terming of the existing principles of marketing is noticed along the course of the emergence of EM research and the new methods of marketing. This is evident with the various adjectives attached to marketing such as "conventional", "conventional wisdom", "traditional", "standard", "administrative", "textbook marketing", "corporate traditional marketing" in referral to the existing principles and the course of marketing textbook givings. This labeling is used to refer to the "formalized market giving's", that the author refers to -under this paper- to indicate the existing prevailing course of marketing. And this terming holds two folds first; the terming for comparison purposes with the EM or new approaches to identify the differences; the second one relates to the description and exploration of the mainstream and consequently understanding the reasons behind being no longer sufficient to respond and tackle the new shape of the environment. All papers studied under the work for this paper include this labeling/terming except for two of them that are (Kraus et al., 2010) and (Uslay & Teach, 2009). Where the most used labeling terms are found to be traditionally followed by conventional. And these are used interchangeably in most of the papers.

Morris et al. (2001) use the terms of traditional marketing and conventional marketing interchangeably and introduce a detailed comparison contrasting conventional marketing from EM. Traditional marketing is the one more used by Stokes (2000) and refers to as "marketing as presented in standard textbooks such as Kotler" the author also presents a comparison between the traditional marketing process and the EMP. The author also uses the term conventional marketing but less frequently than traditional. Amjad et al. (2020a) use the term traditional marketing, with the examples of "traditional marketing textbooks" to include "Brassington & Pettitt dated 2007, Jobber dated 2009 and Kotler et al. dated 2008". The same author (Amjad et al., 2020b) uses the term standard marketing about marketing in textbooks noting it"was originally developed for larger enterprises, and it widely ignores the marketing particularities of SMEs". Collinson and Shaw (2001) use traditional marketing indicating being operated in a consistent environment, characterized by continuous market conditions and clear satisfaction of perceived needs of customers. Morrish et al. (2010) use traditional, conventional, and administrative in addition to presenting a comparison between administrative marketing and EM on the strategy and tactic levels. EM (2012) use traditional marketing with a dedicated section comparing EM to it. The traditional market is described by the author as "exclusively customer-centric focused" also refers to it as administrative marketing.

Farsi et al. (2018), Fiore et al. (2013), Franco et al. (2014), Gross et al. (2014), Miles and Darroch (2006), Phua and Jones (2010), Stokes and Stokes (2000), Swenson et al. (2012) use the term traditional marketing. The latter also uses the term administrative marketing. Becherer et al. (2012), Mort et al. (2012), Renton et al. (2015), Yadav and Bansal (2002) use both traditional and conventional marketing interchangeably. While (Jones & Rowley, 2009) uses conventional only. Some authors use other less frequently used terms where (Martin, 2009) use the term corporate/traditional marketing (CTM) composed of the culture, strategy, and tactic. Morrish (2011) use the term traditional marketing wisdom in addition to administrative marketing. Hills and Hultman (2011) use the term traditional mainstream marketing. Gruber (2004) uses the more generic term of conventional wisdom referring to textbooks. This labeling/terming shows the growth of the views toward formalized market giving and the main textbook giving's as no longer sufficient, and lacking the tools and approaches required especially in the new ventures and SMEs. The literature also indicates that formality and over-reliance on given rules and methods resulted in lower levels of innovation and proactiveness.

5 An-Approach Based Paradigm Shift-Discussion

A paradigm shift is defined in the dictionary as "an important change that happens when the usual way of thinking about or doing something is replaced by a new and different way". The term is used for the first time by Thomas S. Kuhn in his writings between (1922–96), with the view of a paradigm shift as" a fundamental change in approach or underlying assumptions". Three main forces are defined to be driving for a paradigm shift of the prevailing marketing giving, these are the critique of the giving and principles deemed to be no longer sufficient in the contemporary market; strong emergence of the EM as a field, and new marketing school of thought with the views toward EM as a concept valid for the large size firm as well as the new ventures and SMEs; nature of the landscape composing the market and business environment at the different levels with the rapidly changing and fluid boundaries being nature of this market rather than occurrences; change at customer side with the globalization and open markets making the customer more knowledgeable. All under the umbrella of information technology advancements and the fourth industrial revolution, raising disruption almost across all industries.

While marketing remains as the main discipline and one of the key issues of the success for any firm, of any size, within any context, across the full spectrum of industries and markets. These forces raised since the -90's in some of them- elevate the need for a bold evolution in the prevailing school of marketing. An evolution that builds on it but views it in other frames and theories to yield a novel one, and no extension of the existing ones to develop theories for new challenges. The extension of the existing theoretical frames is seen to be no longer sufficient in reflecting shifts in the market landscape (Morris et al., 2002). EM with the development in the research field provides a strong groundwork for this evolution seen as a paradigm shift. A shift on the approach-based layer or frame and not on the detailed levels of strategy, tactic, or process. As detailed levels do not allow for a comprehensive frame of revision, that addresses the paradigm at the school of thought. The frame for this approach-based paradigm is composed of the main four elements/dimensions, derived from the main forces calling for the change. These are resilient management; scalable/gradual matter of sub-frames including multi-dimensional orientation effectual-causal, proactiveness-reactiveness; living strategic process, and Entrepreneurial Innovation. This shape the approach-based paradigm to be resilience in taking and managing risks without severe impacts, dynamic with flexible scalability allowing for the activation and deactivation of the sub-dimension according to the landscape of operation, and with living strategic process that allows for smooth capturing of changes and seamless reflection on the processes all with entrepreneurial innovation that is also gradual but activated as deemed necessary for the growth of the firms. Ironically, marketing concepts are also considered to be about change, but different than entrepreneurial activity/process in a further methodical and structured way (Stokes, 2000).

Effectuation theory provides a supportive frame for resilience. Effectuation allows the changing goals and shaping and re-constructing them over time and making use of the contingencies arising over time (Hills & Hultman, 2011). The proposed paradigm, support deliberating the marketing as a discipline from being organizational or context defined and allow for a flexible frame instead that augments on the theories and practices of new approaches and models and provide a conceptual frame that overrides the different views supporting the formal or informal planning, sequential or fluid and the other various points on the sort of orientation better suited.

Another note is made on the conceptual nature of the paradigm proposed this links with the academia and the practice field. Where the definition and structure of the framework provide the understanding of the term or issue of study, while concept proposals allow for the overall perception and further case-level interpretation. This allows for increasing innovation that is not restricted in the structured forms or given order of thinking. Concepts shall give the overall understanding without putting rigid frames to enable flexibility required more than before in the current landscape, and also allow for more creativity and innovation in projecting these concepts under different contexts, contents, cultures, etc. this is the core of the proposed paradigm.

6 Conclusions and Future Work

The objective of this research is to augment the research field of Entrepreneurial Marketing as a concept and approach, and the substantial differences it holds in comparison to the prevailing giving of the marketing. In doing so the research aims to explore and analyze the causes and drivers that resulted in the emergence of EM as a new school of thought in addition to other new approaches, and how these are understood as better suited for the ventures in the contemporary market and the shifts the marketplace is witnessing. When compared to the formalized giving's of marketing referred to frequently with "traditional marketing" and "conventional marketing". It is noted that EM is viewed to be fundamentally different than formalized giving of marketing, with this difference examined at the different levels of concept definitions, mindset, influence, dimension, strategy, and process. With the recent tendency for addressing EM as suitable for large firms as well as EV and SMEs.

Along with the emergence of the EM, it is also found to have an emerging criticism for the formalized giving of marketing and the rules set. With the infer that these giving's requires development if to be sufficient in addressing the market shift needs in the long run. It is concluded that this is also implicitly rooted in the changing in the landscape of operation and the customer side. Covering market, business, and economic landscape. Augmenting on the above, this research proposes an approach-based paradigm shift frame for the formalized givings of marketing. The proposed paradigm responds to the drivers of evolution concluded in the study and expands on the evolution in the EM in defining the main dimensions conceptualizing this proposed paradigm.

The main contribution lies in the proposed paradigm linked with the driving forces for the need of the shift. However, this research also provides a contribution in the EM literature in exploring EM in a multi-faceted perspective integrating the advancement and development in the field. It also contributes under the labeled/termed marketing introduction, aiming to have a bold move toward responding to the change needed to the marketing discipline.

Limitations on the work are represented by the literature focusing on EM, while other new approaches require study to aid the development of the paradigm. In addition to the literature specifically on the critique of formalized giving's of marketing to view other possible forces of change. This paper is a concept paper, proposing a conceptual paradigm. Where the proposed paradigm requires further work to develop and examine within inductive methods.

References

Abu Zayyad, H. M., Obeidat, Z. M., Alshurideha, M. T., Abuhasheshc, M., Maqableh, M., & Masa'deh, R. (2021). Corporate social responsibility and patronage intentions: The mediating effect of brand credibility. *Journal of Marketing Communication, 27* (5).

Al Kurdi, B., Alshurideh, M., Nuseir, M., Aburayya, A., & Salloum, S. A. (2021). *The Effects of Subjective Norm on the Intention to Use Social Media Networks: An Exploratory Study Using PLS-SEM and Machine Learning Approach* (Vol.1339) .

Alameeri, K. A., Alshurideh, M. T., & Al Kurdi, B. (2021). *The Effect of Covid-19 Pandemic on Business Systems' Innovation and Entrepreneurship and How to Cope with It: A Theatrical View* (Vol. 334).

Al-Dmour, R., Dawood, E. A. H., Al-Dmour, H., & Masa'deh, R. (2020). The effect of customer lifestyle patterns on the use of mobile banking applications in Jordan. *International Journal of Electronic Marketing and Retailing, 11*(3).

Alshurideh, M. (2022). Does electronic customer relationship management (E-CRM) affect service quality at private hospitals in Jordan? *Uncertain Supply Chain Management, 10*(2), 1–8.

Alshurideh, M. T., Al Kurdi, B., Masa'deh, R., & Salloum, S. A. (2021). The moderation effect of gender on accepting electronic payment technology: A study on United Arab Emirates consumers. *Review of International Business and Strategy, 31*(3). https://doi.org/10.1108/RIBS-08-2020-0102.

Alshurideh, M. et al. (2022). Fuzzy assisted human resource management for supply chain management issues. *Annals of Operations Research*, 1–19.

Alwan, M., & Alshurideh, M. T. (2022). *The effect of digital marketing on purchase intention : Moderating effect of brand equity.* https://doi.org/10.5267/j.ijdns.2022.2.012.

Amjad, T., Abdul Rani, S. H., & Sa'atar, S. (2020a). A new dimension of entrepreneurial marketing and key challenges: A case study from Pakistan. *SEISENSE Journal of. Management, 3*(1), 1–14. https://doi.org/10.33215/sjom.v3i1.272.

Amjad, T., Abdul Rani, S. H., & Sa'atar, S. (2020b). Entrepreneurial marketing theory: Current developments and future research agenda. *SEISENSE Journal Management, 3*(1), 27–46. https://doi.org/10.33215/sjom.v3i1.274.

Becherer, R. C., Helms, M. M., & McDonald, J. P. (2012). The effect of entrepreneurial marketing on outcome goals in SMEs. *New England Journal of Entrepreneurship, 15*(1), 7–18. https://doi.org/10.1108/neje-15-01-2012-b001.

Collinson, E., & Shaw, E. (2001). Entrepreneurial marketing—a historical perspective on development and practice. *Management Decision, 39*(9), 761–766. https://doi.org/10.1108/EUM0000000006221.

Entrepreneurial marketing: A new approach for challenging times. *Entrepreneurial marketing A New Approach Challenging Times, 7*(1), 131.

Farsi, J. Y., Mobaraki, M. H., Toghraee, M. T., & Rezvani, M. (2018). Entrepreneurial marketing in creative art based businesses. *International Journal of Management Practice, 11*(4), 448. https://doi.org/10.1504/ijmp.2018.10014325.

Fiore, A. M., Niehm, L., Hurst, J., Son, J., & Sadachar, A. (2013). Entrepreneurial marketing: Scale validation with small, independently-owned businesses. *Journal of Marketing Development and Competitiveness, 7*(4), 63.

Franco, M., Santos, M. de F., Ramalho, I., & Nunes, C. (2014). An exploratory study of entrepreneurial marketing in SMEs: The role of the founder-entrepreneur. *Journal of Small*

Business and Enterprise Development, 21(2), 265–283. https://doi.org/10.1108/JSBED-10-2012-0112.

Gross, N., Carson, D., & Jones, R. (2014). Beyond rhetoric: Re-thinking entrepreneurial marketing from a practice perspective. *Journal of Research in Marketing and Entrepreneurship, 16*(2), 105–127. https://doi.org/10.1108/JRME-01-2014-0003.

Gruber, M. (2004). Marketing in new ventures: Theory and empirical evidence. *Schmalenbach Business Review, 56*(2), 164–199. https://doi.org/10.1007/bf03396691.

Hills, G. E., & Hultman, C. M. (2011). Academic Roots: The past and present of entrepreneurial marketing. *Journal of Small Business and Entrepreneurship, 24*(1), 1–10. https://doi.org/10.1080/08276331.2011.10593521.

Jones, R., & Rowley, J. (2009). Presentation of a generic 'EMICO' framework for research exploration of entrepreneurial marketing in SMEs. *Journal of Research in Marketing and Entrepreneurship, 11*(1), 5–21. https://doi.org/10.1108/14715200911014112.

Kotler, P., & Keller, K. L. (2016). *Marketing management*. Global Edi. Pearson Education Limited.

Kraus, S., Harms, R., & Fink, M. (2010). Entrepreneurial marketing: Moving beyond marketing in new ventures. *International Journal of Entrepreneurship and Innovation Management, 11*(1), 19–34. https://doi.org/10.1504/IJEIM.2010.029766.

Lee, K., Azmi, N., Hanaysha, J., Alzoubi, H., & Alshurideh, M. (2022). The effect of digital supply chain on organizational performance: An empirical study in Malaysia manufacturing industry. *Uncertain Supply Chain Management, 10*(2), 495–510.

Lee, M., Ramiz, K., Hanaysha, P., Alzoubi, J., & Alshurideh, H. (2022). Investigating the impact of benefits and challenges of IOT adoption on supply chain performance and organizational performance: An empirical study in Malaysia. *Uncertain Supply Chain Management* (10), 1–14.

Martin, D. M. (2009). The entrepreneurial marketing mix. *Qualitative Market Research, 12*(4), 391–403. https://doi.org/10.1108/13522750910993310.

Mehmood, T., Alzoubi, H. M., Alshurideh, M., Al-Gasaymeh, A., & Ahmed, G. (2019). Schumpeterian entrepreneurship theory: Evolution and relevance. *Academy of Entrepreneurship Journal, 25*(4).

Miles, M. P., & Darroch, J. (2006). Large firms, entrepreneurial marketing processes, and the cycle of competitive advantage. *European Journal of Marketing, 40*(5–6), 485–501. https://doi.org/10.1108/03090560610657804.

Morris, M. H., Schindehutte, M., & LaForge, R. W. (2002). Entrepreneurial marketing: A construct for integrating emerging entrepreneurship and marketing perspectives. *Journal of Marketing Theory and Practice, 10*(4), 1–19. https://doi.org/10.1080/10696679.2002.11501922.

Morris, M. H., Schindehutte, M., & LaForge, R. W. (2001). The emergence of entrepreneurial marketing: Nature and meaning. *Entrepreneurship the Way Ahead*, 91–104, 2003. https://doi.org/10.4324/9780203356821.

Morrish, S. C. (2011). Entrepreneurial marketing: A strategy for the twenty-first century? *Journal of Research in Marketing and Entrepreneurship, 13*(2), 110–119. https://doi.org/10.1108/14715201111176390.

Morrish, S. C., Miles, M. P., & Deacon, J. H. (2010). Entrepreneurial marketing: Acknowledging the entrepreneur and customer-centric interrelationship. *Journal of Strategic Marketing, 18*(4), 303–316. https://doi.org/10.1080/09652541003768087.

Mort, G. S., Weerawardena, J., & Liesch, P. (2012). Advancing entrepreneurial marketing: Evidence from born global firms. *European Journal of Marketing, 46*(3–4), 542–561. https://doi.org/10.1108/03090561211202602.

Özdemir, Ö. G. (2013). Entrepreneurial marketing and social value creation in Turkish art industry: An ambidextrous perspective. *Journal of Research in Marketing and Entrepreneurship, 15*(1), 39–60. https://doi.org/10.1108/JRME-03-2013-0012.

Phua, S., & Jones, O. (2010). Marketing in new business ventures: Examining the myth of informality. *International Journal of Entrepreneurship and Innovation Management, 11*(1), 35–55. https://doi.org/10.1504/IJEIM.2010.029767.

Renton, M., Daellenbach, U., Davenport, S., & Richard, J. (2015). Small but sophisticated: Entrepreneurial marketing and SME approaches to brand management. *Journal of Research in Marketing and Entrepreneurship, 17*(2), 149–164. https://doi.org/10.1108/JRME-05-2014-0008.

Stokes, D. (2000). Putting entrepreneurship into marketing. *Journal of Research in Marketing and Entrepreneurship, 2*(1), 1–16.

Stokes, D., & Stokes, D. (2000). Selected papers from the academy of marketing conference entrepreneurial marketing : A conceptualisation from qualitative research. *Qualitative Market Research an International Journal, 3*(1), 47–54. https://doi.org/10.1108/14715200080001536.

Swenson, M., Rhoads, G., & Whitlark, D. (2012). Entrepreneurial marketing: A framework for creating opportunity with competitive angles. *Journal of Applied Business and Economics, 13*(1), 47–52.

Tariq, E., Alshurideh, M., Akour, I., Al-Hawary, S., & Al, B. (2022). The role of digital marketing, CSR policy and green marketing in brand development. *International Journal Data Network Science, 6*(3), 1–10.

Uslay, C., & Teach, R. D. (2009). Marketing/entrepreneurship interface research priorities (2010/2012). *Journal of Research in Marketing and Entrepreneurship, 10*(1), 70–75. https://doi.org/10.1108/01443571010996244.

Yadav, A., & Bansal, S. (2002). Viewing marketing through entrepreneurial mindset: A systematic review. *International Journal of Emerging Markets*. https://doi.org/10.1108/IJOEM-03-2019-0163.

A Development of a Newly Constructed Model Related to the Impact of Entrepreneurial Motivation on Entrepreneurial Intention

C. Al Deir, M. Al Khasawneh, M. Abuhashesh, R. Masa'deh, and A. M. Ahmad

Abstract The current research developed a model which included Entrepreneurship Motivation (EM) as a higher order construct that consists of twelve first order indicators (need for achievement, need for autonomy, need for affiliation, need for power, subjective norms, attitude toward becoming entrepreneur, perceived behavioral control, governance motive, financial motive, economic motive, creativity and risk taking) and considered EM as an independent variable for Entrepreneurial Intention (EI). The current research methodology employed a quantitative approach, in which data were administrated and gathered via an online and face-to-face survey with entrepreneurs in Jordan; 221 surveys were collected, the data were later tested using a (SEM) via AMOS program. The findings of this research showed that need for achievement, need for autonomy, need for affiliation, need for power, subjective norms, attitude toward becoming entrepreneur, PBC, financial motive, creativity, and risk taking are considered as indicators for EM, whereas governance motive and economic motive were not indicators for EM.

Keywords Entrepreneurship · Entrepreneurship motivation · McClelland's theory · Theory of planned behavior · Entrepreneurial intentions · Jordanian context

C. Al Deir · M. Al Khasawneh · M. Abuhashesh (✉) · A. M. Ahmad
E-Marketing and Social Media Department, Princess Sumaya University for Technology (PSUT), Amman, Jordan
e-mail: m.abuhashesh@psut.edu.jo

M. Al Khasawneh
e-mail: m.alkhasaawneh@psut.edu.jo

A. M. Ahmad
e-mail: a.ahmed@psut.edu.jo

R. Masa'deh
Department of Management Information Systems, School of Business, The University of Jordan, Amman, Jordan
e-mail: r.masadeh@ju.edu.jo

M. Alshurideh et al. (eds.), *The Effect of Information Technology on Business and Marketing Intelligence Systems*, Studies in Computational Intelligence 1056,
https://doi.org/10.1007/978-3-031-12382-5_86

1 Introduction

The significance of entrepreneurship as a contributor and supporter to venture creation, economic development, and innovation is widely recognized (Alameeri et al., 2021; Hopp & Stephan, 2012; Nabi & Liñán, 2011; Qandah et al., 2020). Therefore, it is essential to define Entrepreneurship as a concept. In particular, entrepreneurship is the creation and application of new opportunities in an uncertain and complex environment (Neck & Greene, 2011) that drives sustainability and growth in an economy (Rasmussen & Sorheim, 2006). In addition, it is considered as a phenomenon encompassing the action of humans; it means that the entrepreneurial process exists because opportunities are being exploited by individuals, and as similar, human roles cannot be overlooked (Shane et al., 2003). Taking this point further, starting a new business relies largely on intentions as suggested by a growing body of relevant literature (Bird, 1988; Liñán & Chen, 2009a; Schlaegel & Koenig, 2014). Several previous studies confirmed the aforementioned assertions by stating that the development of entrepreneurial intention is followed by entrepreneurial activities (Douglas, 2013; Shook et al., 2003).

Furthermore, it is important to be motivated to insist and continue until the work is done and goals are achieved (Farhangmehr et al., 2016). Hence, Entrepreneurial Motivation (EM) is viewed as observable and behavioral and is connected to the organizational growth, survival and origin (Bird, 1995). More specifically, EM represents relevant particular goals to which potential entrepreneurs believe and aspire they can accomplish with new ventures, and accordingly their beliefs will motivate them to implement an entrepreneurial action (Kuratko et al., 1997).

After examining the previous studies about EM, it is noted that these studies did not developed a comprehensive model to measure the indicators affecting EM. For that reason, this research aims to examine the indicators that affect EM, and the effect of EM on EI. In addition and based on reviewing previous studies, it is found that different theories were used to measure EM, where needs theory (McClelland, 1961) was used in various studies (Jayachandra, 2011; Lee, 1996; Sethi & Saxena, 2006), while theory of planned behavior was used in several other EM studies (Malebana, 2014; Solesvik, 2013). In addition to that, there was no agreement in the natures or the number of the indicators to measure EM, such as, some studies used the three needs theory (McClelland, 1961) to measure EM whereby four indicators were included (Sethi & Saxena, 2006), while seven indicators were examined in another stream of research (Jayachandra, 2011). Whereas the theory of planned behavior related studies examined EM with eight indicators (Malebana, 2014) and others with only three indicators (Solesvik, 2013).

For this reason, this study developed a model with an extended theories and variables from different studies to measure EM by adopting the needs theory (McClelland, 1961), and theory of planned behavior (Ajzen, 1991), in addition to other influencing variables as recommended in the existing relevant literature such as governance motive (Aziza et al., 2013), financial motive (Aziza et al., 2013; Hessels et al., 2008; Sarri & Trihopoulou, 2004) economic core motive (Aziza et al., 2013; Vijaya,

1998) creativity (Estay et al., 2013), and risk taking (Estay et al., 2013; Jayachandra, 2011; Nishantha, 2009; Segal et al., 2005). Further, the current study will add to the current knowledge regarding the relationship between EM and EI by examining the influence of EM as a multidimensional construct on EI in the Jordanian context.

EM described briefly in this background discussion, and it focuses on the need of having deeper knowledge and understanding the indicators of EM and its effect on EI in a wide range approach with the following objectives:

(1) To understand the key dimensions of EM from a theoretical perspective.
(2) To understand the key dimensions of EM from a statistical perspective in the Jordanian context.
(3) To investigate the level to which EM influences EI in the Jordanian context.

2 Literature Review

To provide a comprehensive understanding of this research, an intensive revision of the relevant literatures related to EM and EI has been conducted. EM topic is very broad and explored by many researchers (Shane et al., 2003; Jayachandra 2011; Lee 1996; Malebana 2014; McClelland, 1961; Sethi & Saxena, 2006; Solesvik, 2013; Aziza et al., 2013; Antonioli et al., 2016; Batchelor et al., 2014; Fereidouni et al., 2010; Hackman & Oldham, 1975). EM is defined as the willingness and propensity to manipulate and control companies, thoughts and human beings as independently and rapidly as possible (Johnson, 1990). Previous research found that EM impacts the growth of small business (Delmar & Wiklund, 2008). On the other hand, a significant relationship was found between EI and EM (Achchuthan & Nimalathasan, 2013; Solesvik, 2013). Some researchers found that there is a link between EI, motivation and behavior in which they showed that since intentions cannot transform immediately into action, then motivation is considered as link between EI and actions; motivation will translate an intention into real action (Carsrud & Brännback, 2011).

2.1 Three Needs Theory

Various theories and indicators have been used to measure EM (Collins et al., 2004; Shane et al., 2003). For example, many studies have examined the Three Needs Theory developed by McClelland to measure the EM. Taking this point further, (McClelland, 1961) suggested that the high need for achievement was the personal characteristics common to entrepreneurs, in support to this, a numerous research has focused on entrepreneurs' characteristics (Churchill & Lewis, 1986; Shaver & Scott, 1991). In particular, by duplicating and extending the Three Needs Theory, several studies included this theory in their investigation such a study which was conducted by Lee (1996) to examine the indicators that influence the motivational needs of entrepreneurs. These indicators include need for achievement, need for affiliation, need for autonomy and need for dominance and the results demonstrated that entrepreneurs will be motivated by a high need for achievement, a moderate need for autonomy and affiliation and a little high need for dominance. Moreover, it was found that entrepreneurs need a higher need for dominance and achievement than employees, and a significant difference in the needs for autonomy and affiliation (Lee, 1996). In contrast, another study used the same theory with the same variables but they found that need for achievement have a crucial role in EM (Sethi & Saxena, 2006).

Other studies have focused on investigating EM using psychological traits (Bygrave, 1989; Robinson et al., 1991). For instance, (Bygrave, 1989), developed a model and confirmed that need for achievement, tendency toward risk-taking, ambiguity toleration and inner locus of control act determinants of the entrepreneurial motivation. However, (Robinson et al., 1991) found in their study that the accomplishment, self-confidence and the locus of control are the most significant indicators that are predicting entrepreneurial behavior.

2.2 Theory of Planned Behavior

Other researchers have emphasized on examining and measuring EM using the Theory Planned Behavior (Ajzen, 2012; Malebana, 2014). This theory suggested that an individual's intention to perform or not to perform an action is the most important determinant of action (Ajzen, 2012). In particular, (Malebana, 2014) found that EM is a significant link between EI and entrepreneurial action; hence, it is suggested that the theory of planned behavior should be utilized to explore the determinants of EM and EI. Depending on the theory of planned behavior, there are several studies used this theory to determine the effect of the following three variables (attitude toward entrepreneurs, subjective norms, and PBC) on EM and their role in forming EI (Solesvik, 2013). These three particular variables were found to affect EM by mediating the relationship between EM and EI. While, (Locke et al., 2001) extended the Theory Planned Behavior theory in his research by adding five variables (EI, entrepreneurial role models, social valuation of entrepreneurship, knowledge of entrepreneurial support and perceived barriers to start a business) to determine their effect on EM. The findings showed that the attitude towards becoming an entrepreneur, entrepreneurial support, subjective norms, social valuation of entrepreneurship, EI and knowledge of entrepreneurial role models have a great influence on EM (Baum et al., 2001).

Various approaches and methods have been used to measure EM (Collins et al., 2004; Shane et al., 2003). For example, one of the studies suggested that need for achievement, locus of control, risk taking and goal settings are major indicators contributing to the formation of EM. While other studies measured EM with passion, independence and drive factors (Collins et al., 2004; Shane et al., 2003). In addition to that, other related studies to EM developed their models based on key previous studies within the EM context. For example, some researchers examined need for achievement, locus of control, vision, desire for independence, passion, drive, goal settings and self-efficacy as an indicator that will impact EM (Baum et al., 2001; Hisrich, 1985; Locke and Latham, 1990; McClelland, 1961; Tracy et al., 1998). Whereas another study conducted by Nishantha (2009) measured EM including the following indicators: personality traits (internal locus of control, need for achievement, and risk taking), and socio demographic background (parents' occupation, gender, self-employment, and (Nishantha, 2009). One more research made their model as a comparison to determine where the focus should be to be motivated in order to start a business between psychological behavior variables, which includes achievement striving, social networking and optimism, with external environment variables which is the perceived importance of a favorable business environment (Alameeri et al., 2021; Taormina & Lao, 2007). They concluded that both were significant predictors, but the psychological factors influence was stronger on potential entrepreneurs, whereas the external environment was stronger for successful entrepreneurs (Taormina & Lao, 2007). Whereas the variable perceived importance of business environment was used in another study with other several variables. These indicators perceived importance of business environment, social status, and country external conflicts. They revealed that the perceived importance of business environment is the most important variable in predicting EM, while the perceived social status doesn't affect the motivation of an entrepreneur to start a new business (Abu Zayyad et al., 2020; Fereidouni et al., 2010).

A revision of other streams of existing studies concluded that the motivation of an individual is complicated and multi-faceted (Marlow & Strange, 1994; Shane et al., 1991) and relied mosly on push and pull theory. In particular, individuals who think they are able to create or to grow in a business are stronger and more motivated to create new venture and to develop in certain business successfully (Malebana, 2014). The push and pull indicators play a key role in EM, it helps in understanding why people decide to create a new venture. Pull indicators contain independence, achievement, personal growth and recognition, in contrast push indicators are lack of innovation and alternatives, insecurity of work and unemployment (Wickham, 2006). Individuals who depend on pull indicators perceive the entrepreneurial job opportunity more attractive, so they are considered as entrepreneurs, while others who depend on push indicators choose to follow entrepreneurship and decide to create new business just because the traditional work is less attractive (Malebana, 2014). Taking this point further, (Reynolds et al., 2002) suggested that individuals who are opportunity driven are more motivated to exploit opportunities for economic gain in order to achieve success, and those who are necessity driven are motivated just for the need of survival.

In conclusion, several indicators have been identified as determinants of EM within the EM literature and there was no agreement upon which theories most applicable to EM determinants and which are the most important indicators that affect EM. To address this gap in the entrepreneurship literature, the current study seeks to; firstly, present a theoretical model of the most important determinants of EM by incorporating variables synthesized from the EM and EI literature. Secondly, this study will empirically validate the model from the entrepreneurial perspective. The final result will be the development and empirical validation of a comprehensive model of determinants of EM and EI.

2.3 Research Model

Previous studies about EM have been examined, it is noted that most of the studies did not develop a comprehensive model to measure EM as the results showed that there are several indicators forming EM construct. In reviewing the existing relevant literature, several studies attempted to examine EM using various indicators separately such as; need for achievement (Antonioli et al., 2016; Begley & Boyd, 1987; Johnson, 1990; Lee, 1996; Nishantha, 2009; Rauch & Frese, 2000; Sethi & Saxena, 2006; Shane et al., 2003), need for affiliation (Lee, 1996; Sethi & Saxena, 2006), need for autonomy (Estay et al., 2013; Lee, 1996; McClelland, 1985; Oosterbeek et al., 2010; Sethi & Saxena, 2006) need for power (McClelland, 1985; Sethi & Saxena, 2006) attitude toward becoming an entrepreneur (Malebana, 2014; Solesvik, 2013; Segal et al., 2005; Ajzen, 2005; Fishbein, 1975; Linan 2004), subjective norms (Malebana, 2014; Segal et al., 2005; Solesvik, 2013) perceived behavioral Control (Malebana, 2014; Segal et al., 2005; Solesvik, 2013). Based on theory of planned behavior (Fishbein & Ajzen, 1975), creativity (Caird, 1991; Cromie & O'Donoghue 1992; Zhu et al., 2016), risk taking (Estay et al., 2013; Jayachandra, 2011; Koh, 1996; Nishantha, 2009; Segal et al., 2005) economic Core (Aziza et al., 2013; Vijaya, 1998) financial motive (Aziza et al., 2013; Hessels et al., 2008; Sarri & Trihopoulou, 2004) and governance (Audretsch, 2002; Stevenson & Lundström, 2001; OECD, 2010). Therefore, the current research proposed that all of these indicators may act as significant indictors to measure EM, as well as the impact of EM on EI (Liñán & Chen, 2009b).

The elements of this research are established based on preceding literature review either theoretically or empirically. Indeed, this study uses variables that are common Entrepreneurship Motivation and Entrepreneurial Intention literature. Figure 1 represents a model of the study that shows the independent variables (in which twelve indicators have been identified as the key indicators of EM as shown below in Fig. 1), the dependent variable and the proposed relationship between them.

2.3.1 Need for Achievement

Need for achievement is defined as an individual drive to exceed with respect to some settled arrangement of norms and standards (Lee, 1996; McClelland, 1985). Much of the literature have examined how entrepreneurs are motivated, by a higher need for achievement than employees (Lee, 1996; Nishantha, 2009; Oosterbeek et al., 2010; Sethi & Saxena, 2006). Taking this point further, individuals with higher need for achievement are more satisfied in jobs that have hard challenges and high skill levels (Eisenberger et al., 2005) and are more frequently demanded for a feedback on their advance towards goal fulfillment (Emmons, 1997; McAdams, 1994). Furthermore, individuals with high need for achievement are more likely to be motivated in order to become an entrepreneur (Antonioli et al., 2016; Lee, 1996; Nishantha, 2009). Therefore, based on the previous discussion this study supposed that, need for achievement is highly correlated for the higher-order construct EM.

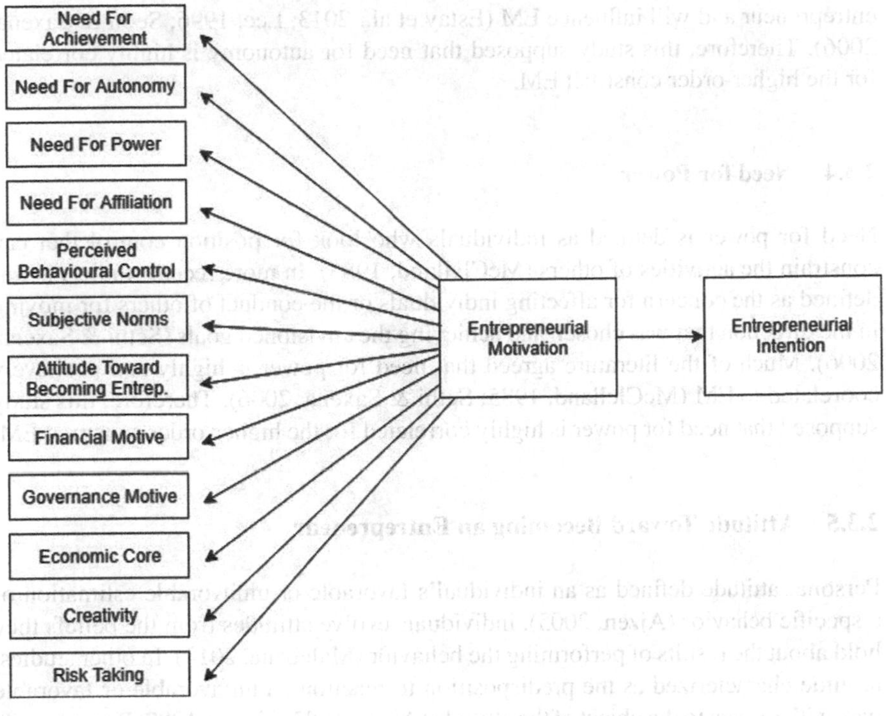

Fig. 1 Research model

2.3.2 Need for Affiliation

Need for affiliation is viewed as the desire to have warm, friendly relationships with others (Robbins, 2003; McClelland, 1985; Veroff, 1980). Several studies have applied need for affiliation in various fields such as entrepreneurship (Lee, 1996; Sethi & Saxena, 2006), and managemnet (Lee, 1996). Studies have shown that people with higher need for affiliation tend to be more successful managers and leaders (Lee, 1996). In addition to that, people with high need for affiliation were more likely to become an entrepreneur and it is highly correlated with EM (Lee, 1996; Sethi & Saxena, 2006). Therefore, this study supposed that need for affiliation is highly correlated for the higher-order construct EM.

2.3.3 Need for Autonomy

Need for autonomy is defined as a want for independence, (Sethi & Saxena, 2006) and to be able get things done without taking into consideration to what others may think and to avoid commitment and responsibility (Lee, 1996). Several studies indicated that individuals with high need for autonomy were more motivated to become an entrepreneur and will influence EM (Estay et al., 2013; Lee, 1996; Sethi & Saxena, 2006). Therefore, this study supposed that need for autonomy is highly correlated for the higher-order construct EM.

2.3.4 Need for Power

Need for power is defined as individuals who look for position control that can constrain the activities of others (McClelland, 1985). In more recent studies, it was defined as the concern for affecting individuals or the conduct of others for moving in the direction that was chosen and achieving the envisioned goals (Sethi & Saxena, 2006). Much of the literature agreed that need for power is highly and positevely coorelated to EM (McClelland, 1985; Sethi & Saxena, 2006). Therefore, this study supposed that need for power is highly correlated for the higher-order construct EM.

2.3.5 Attitude Toward Becoming an Entrepreneur

Personal attitude defined as an individual's favorable or unfavorable estimation of a specific behavior (Ajzen, 2005), individuals evolve attitudes from the beliefs they hold about the results of performing the behavior (Malebana, 2014). In other studies, attitude characterized as the predisposition to reaction an unfavorable or favorable way with respect to the object of the attitude (Ajzen, 1982; Shaver, 1987; Rosenberg & Hovland, 1960). Taking this point further, attitude toward a particular object has been found to be positively correlated to motivation and behavioral intentions (Bauer et al., 2005). Attitudes have a great influence on EM (Fishbein & Ajzen, 1975, Linan, 2004; Malebana, 2014; Solesvik, 2013), for that reason we hypothesize that attitude is highly correlated for the higher-order construct EM.

2.3.6 Subjective Norms

Subjective norms refer to the perceived social compression to perform or not the behavior (Malebana, 2014). In other studies, subjective norms defined as the social pressure to implement the behavior being referred to, moreover subjective norms are a measure of social help of the behavior by significant others, for example, family, connections and other considered as role models and associates (Bauer et al., 2005; Segal et al., 2005; Yau & Ho, 2015). In reviewing the related literature, subjective norms were found to be positively related to student behavioral intention to use e-learning in their education (Park, 2009; Yau & Ho, 2015). Moving to the entrepreneurial context, it is found that subjective norms have a great influence on the intention to become an entrepreneur (Malebana, 2014; Segal et al., 2005; Solesvik, 2013), for that reason, we hypothesize that subjective norms are correlated for the higher-order construct EM.

2.3.7 Perceived Behavioral Control (PBC)

PBC refers to people's appraisals of how much they are able of performing a specific behavior (Chiou, 1998; Malebana, 2014; Segal et al., 2005). Several researches have examined attitude in many fields, for example, entrepreneurship context (Malebana, 2014; Solesvik, 2013) and behavioral intention (Ajzen, 2005). PBC will influence the behavior of intention when individuals possess the accurate resources and skills to implement it (Ajzen, 2005). In addition to that, PBC have a great influence on individual's intention to become entrepreneurs and start their own ventures thus it will influence EM (Malebana, 2014; Segal et al., 2005; Solesvik, 2013), For that reason we hypothesize that PBC is highly correlated for the higher-order construct EM.

2.3.8 Creativity

Creativity defined as evolving new methods instead of utilizing standard procedures (Estay et al., 2013). In other studies, creativity characterized as a method for seeing or doing things that described by familiarity; to create many ideas, adaptability; to shift point of view effectively, elaboration; to expand on others thoughts, and, originality; to imagine something new (Alvino, 1990; Cromie & O'Donoghue, 1992). Several researches have examined creativity in many fields such as management and marketing context (Caird, 1991; Cromie & O'Donoghue, 1992; Zhu et al., 2016) and entrepreneurship and technology (Estay et al., 2013). Creativity influences owner—managers higher than managers (Caird, 1991), moreover, creative entrepreneurs are more likely to be to create new ideas or developing existing ideas in order to create more value and to be more innovative (Cromie & O'Donoghue, 1992). Creativity was found to have a great influence on individuals to start their new ventures and become entrepreneurs (Estay et al., 2013; Jaillot, 2017). For that reason, we suppose that creativity is correlated for the higher-order construct EM.

2.3.9 Financial Motive

Financial motive is viewed as the desire to earn wealth (Brett et al., 1995; Doran et al., 1991; George & Brief, 1990). Several researches have examined financial motive in many fields, for example, in organizations (Burton, 2012) and in entrepreneurship (Aziza et al., 2013; Hessels et al., 2008; Sarri & Trihopoulou, 2004). Financial motive has a great influence on employees, in which employees who are compensated financially will be more motivated and their performance will be higher (Burton, 2012). Entrepreneurs begin their businesses in order to gain more money. In particular, financial motive is considered as one of the indicators of EM in order to become an entrepreneur and start a business (Aziza et al., 2013; Hessels et al., 2008; Sarri & Trihopoulou, 2004), and for that reason we suppose that financial motive is correlated for the higher-order construct EM.

2.3.10 Risk Taking

Risk taking is characterized as an individual tendency towards taking opportunities in uncertain decision-making contexts (Estay et al., 2013; March & Shapira, 1987; Nishantha, 2009). Several researches have studied risk taking in many fields, for example, management context (Crimmon & Wehrung, 1986; March & Shapira, 1987) and entrepreneurship (Estay et al., 2013; Jayachandra, 2011; Koh, 1996; Nishantha, 2009; Segal et al., 2005). It has been previously found that risk taking influences managers' willingness to take risks (Crimmon & Wehrung, 1986) in which they connect risk taking with the expectation of their work and it influences managers' decision making (Abuhashesh et al., 2019; March & Shapira, 1987). In particular, risk taking is considered as a trait to distinguish founders from non-founders (Abuhashesh et al., 2019). Risk taking is considered high for individuals and has an influence for individuals who are entrepreneurially inclined than non-entrepreneurially inclined (Koh, 1996). Individuals with high-risk taking have an influence toward entrepreneurship (Nishantha, 2009). Risk taking has an influence on EI, in addition to that, risk taking have a great influence on EM (Estay et al., 2013; Nishantha, 2009), and for that reason we suppose that risk taking is highly correlated for the higher-order construct EM.

2.3.11 Economic Motive

Economic motive is defined as the investigation of how people use knowledge to recognize resources and use these scarce resources to make, utilizing knowledge, and commodities and disseminate them among individuals (Ayers & Collinge, 2005; Khumalo, 2012; McConnell, 1969). Several researches have studied economic core, such as, entrepreneurship (Aziza et al., 2013; Vijaya, 1998) and economic development (Deaton, 2010). Based on these studies, it is noted that economic motive makes entrepreneurs more motivated when they feel independence and self-achievement (Vijaya, 1998). Furthermore, economic motive effects saving rates, in which saving rates are higher in rapidly developing and growing economies (Deaton, 2010). Economic motive considered as one of the indicators of EM (Aziza et al., 2013; Vijaya, 1998). Therefore, we hypothesize that economic motive is correlated for the higher-order construct EM.

2.3.12 Governance Motive

Governance is defined as the activity of political, economic and executive authority to deal with a country's affairs. It is the complex systems, procedures and establishments, through which citizens and groups express their interests, practice their legitimate rights and commitments, and arrange their differences (Lusaka, 2013). For the current research, governance is defined by Ha et al. (2016), as "the traditions and institutions by which authority in a country is exercised. This includes (1) the process by which governments are selected, monitored and replaced; (2) the capacity of the government to effectively formulate and implement sound policies; and (3) the respect of citizens and the state for the institutions that govern economic and social interactions among them" (P. 517).

Many researches have studied governance in several contexts, such as, corporate governance (Albourini et al., 2020) and entrepreneurship (Aziza et al., 2013; Yalcin et al., 2008). Good corporate governance is a way to make market confidence and business integrity, which then will be basic for organizations that need access to value capital for long term venture (Albourini et al., 2020). Governance is considered as one of the indicators of EM (Aziza et al., 2013; Yalcin et al., 2008) suggested that specific government policies influence people's desire to be entrepreneurs. Bureaucracy, exaggerated taxation, and limited access to capital are considered as barriers to entrepreneurship. Alternatively, their study proposes that entrepreneurs' perception that business laws and regulations are ideal to begin and manage new businesses are positively related to entrepreneurial motives. Limited studies discussed the influence of governance on EM, so more knowledge is needed to examine such influence in the Jordanian context. Therefore, this research hypothesize that governance is highly correlated for the higher-order construct EM. Based on the previous literature, we have developed the following hypothesis.

H1: EM is a higher order construct that consist from twelve first order indicators (need for achievement, need for autonomy, need for affiliation, need for power, subjective norms, attitude toward becoming entrepreneur, PBC, governance motive, creativity, financial motive, economic core motive, and risk taking).

2.4 Entrepreneurial Motivation

EM is defined as the willingness and propensity to manipulate and control companies, thoughts and human beings as independently and rapidly as possible (Johnson, 1990). Several studies have discussed EM in the business context (Delmar & Wiklund, 2008), education context (Ferreira et al., 2007; Raposo et al., 2008) and entrepreneurship context (Achchuthan & Nimalathasan, 2013; Carsrud & Brännback, 2011; Ajzen, 2012; Malebana, 2014; Solesvik, 2013). Some researchers focused on the significance of entrepreneurship education in the advancement of the EM and EI. For instance, (Ferreira et al., 2007; Raposo & A. Pac¸o & J. Ferreira, 2008) asserted that education has the most significant impact on the tendency to be motivated to create a new business among students. (Delmar & Wiklund, 2008) found that EM impacts the growth of small business. On the other hand, a significant relationship was found between EI and EM (Achchuthan & Nimalathasan, 2013; Solesvik, 2013). Some researchers found that there is a link between entrepreneurial intentions, motivation and behavior, they showed that since intentions cannot transform immediately into action, then motivation is considered as link between EI and actions; motivation will translate an intention into real action (Carsrud & Brännback, 2011). While other researchers have emphasized on examining and measuring EM using the Theory of Planned Behavior (Ajzen, 2012; Malebana, 2014). This theory suggested that an individual's intention to perform or not to perform an action is the most important determinant of action (Ajzen, 2012). In particular, (Malebana, 2014) found that EM is a significant link between EI and entrepreneurial action; hence, it is suggested that the theory of planned behavior should be utilized to explore the determinants of EM and EI. According to Solesvik (2013), his empirical result depending on the theory of planned behavior showed that the three variables (attitude toward entrepreneurs, subjective norms, and PBC) affect EM and they mediate the relationship between EM and EI. Therefore, EM will have a significance influence on EI: Based on the previous studies, we have developed the following hypothesis.

H2: EM has a significant positive influence on EI.

3 Methodology

3.1 Population and Sample of the Study

The aim of this study to develop and validate a newly constructed model related to the EM construct and its impact on EI in Jordan. This research utilized the survey questionnaire as the fundamental method for gathering information. Online self-administered questionnaires were created based on prior literature. As this research is considering the indicators of EM that influences EI. The population of the research comprises entrepreneurs running their businesses in Jordan. Therefore, it is important for individuals who aim to start a business in Jordan to understand the indicators of EM that will influence their decision to run a business. The current study adopted a non-probability sampling procedure in which a convenience sampling technique is used. Thus, convenience sampling is incorporated into the research, 221 surveys were completed. The data was collected with support from the business incubators which operates in Jordan (Zinc, Oasis 500, The tank, Adamtec VC, QRCE, Business development center, Business and professional women, Shaml start, Migretmena, Big, I park, Injaz, Intaj).

The survey composed of two parts; the first part is information about demographics characteristics, which include gender, age, level of education, size of the company and legal form. The second part consists of EM indicators: need for achievement, (Culp, 2015) need for Affiliation (Culp, 2015) need for power (Lee, 1996), need for autonomy (Broeck et al., 2010), PBC (Malebana, 2014) subjective norms (Malebana, 2014) attitude toward becoming an entrepreneur (Malebana, 2014), governance motive (Ahmed et al., 2021), economic motive (Ahmed et al., 2021), financial motive (Ahmed et al., 2021), risk taking (Perry, 2014), creativity (Ala'eddin Ahmed et al., 2021), and EI (Liñán & Chen, 2009b). The measures in the questionnaire were based on prior studies and used the five-point Likert scale, the extent from "1" for strongly disagree to "5" for strongly agree (Farhangmehr et al., 2016), The survey questions were pre-tested by 30 entrepreneurs and ten university professors to take their feedback. The order of some questions after the feedback were reorganized, the content and wording of some questions were changed, furthermore some questions were complex, so rewording has been done in order to make them easier to be understood. Based on their review, some questions were deleted.

4 Data Analysis

The research model was tested a structural equation modeling via AMOS. The reason for using AMOS is that the study model contains a high order construct that consists of first order indicators which makes AMOS as the most appropriate statistical technique to test the model.

4.1 Reliability Analysis

Cronbach alpha (α) was used to assess the reliability of each scale (Hair, 1995). Table 1 shows that the Cronbach alpha of the items of each scale was above the threshold of 0.35 and the Cronbach alpha of each construct was above the threshold of 0.70.

4.2 Model Validation

It should be noted that AMOS 23 was used to assess convergent validity and discriminant validity. To do so, a model of 12 first-order correlated variables was specified. However, two of the 12 variables namely (economic motive and governance motive) did not correlate with other ten variables, which indicate that they are not part of the construct of entrepreneurial motivation. Therefore, these two variables were omitted from the model. Their inclusion actually produced a bad model-fit: CMIN/DF (3.248), IFI (0.701), TLI (0.627), CFI (0.692) and NFI (0.619). However, after deleting the two variables, the model fit has become as follows: CMIN/DF (2.357), IFI (0.878), TLI (0.824), CFI (0.864) and NFI (0.848), which is within the acceptable range (see Table 2).

4.3 Convergent and Discriminant Validities

Average variance extracted (AVE) was calculated in order to assess the convergent validity of each scale. Table 3 shows that the values of the AVE of each scale was

Table 1 Construct reliability

Construct	Cronbach alpha
Need for achievement	0.848
Need for affiliation	0.844
Need for power	0.85
Need for autonomy	0.847
PBC	0.844
Subjective norms	0.845
Creativity	0.847
Risk taking	0.847
Financial motive	0.843
Economic core	0.851
Attitude toward becoming entrepreneur	0.844
Governance motive	0.849
Entrepreneurial intention	0.843

Table 2 Rule of thumbs of goodness-fit-indices

Type of fit indices	Cut-off point
CMIN (X^2)	p-value $\geq$ 0.05
Normed CMIN (CMIN/DF)	1.0 < CMIN/DF < 3.0
Normed Fit Index (NFI)	0.8 < NFI < 0.9 (acceptable fit)
	0.9 $\leq$ NFI > 1 (satisfactory fit)
	NFI = 1 (perfect fit)
Competitive Fit Index (CFI)	0.8 < CFI < 0.9 (acceptable fit)
	0.9 $\leq$ CFI > 1 (satisfactory fit)
	CFI = 1 (perfect fit)
Tucker-Lewis Index (TLI)	0.8 < TLI < 0.9 (acceptable fit)
	0.9 $\leq$ TLI > 1 (satisfactory fit)
	TLI = 1 (perfect fit)
Incremental Fit Index (IFI)	0.8 < IFI < 0.9 (acceptable fit)
	0.9 $\leq$ IFI > 1 (satisfactory fit)
	IFI = 1 (perfect fit)

above than the cut-off-point of 0.50. Discriminant validity is characterized as two variables, which are not, correlated with each other the items that measures each variable is poorly correlated with each other in the same model.

Table 3 Convergent validity and discriminant validity

	AVE	Financial motive	Need for affiliation	Need for achievement	Need for autonomy	Creativity	Power	Subject norms	PBC	Risk taking	Attitude	Entrepreneurial intention
Financial Motive	0.63	**0.794**										
Affiliation	0.585	0.284	**0.765**									
Achievement	0.534	0.465	0.573	**0.731**								
Autonomy	0.671	0.458	0.522	0.67	**0.819**							
Creativity	0.528	0.407	0.594	0.703	0.579	**0.726**						
Power	0.618	0.495	0.495	0.484	0.408	0.46	**0.786**					
Subject norm	0.567	0.423	0.472	0.482	0.448	0.595	0.529	**0.753**				
Perceived behavioral control	0.532	0.598	0.524	0.481	0.541	0.681	0.559	0.491	**0.729**			
Risk taking	0.628	0.434	0.515	0.548	0.62	0.534	0.565	0.412	0.588	**0.793**		
Attitude	0.605	0.504	0.564	0.514	0.443	0.581	0.468	0.52	0.615	0.589	**0.778**	
Entrepreneurial intention	0.564	0.492	0.552	0.502	0.431	0.569	0.456	0.508	0.603	0.577	0.517	**0.751**

4.4 Model Testing

AMOS 23 was also used to test the thesis's hypotheses. To do so, a high order construct of entrepreneurial motivation that consists of ten first-order indicators was specified. The results of the hypotheses are reported below.

H1: EM is a high order construct that consist from twelve first order indicators (Need for achievement, Need for Autonomy, Need for Affiliation, need for power, Subjective norms, Attitude, perceived behavioral control, Governance, creativity, financial motive, Economic core, and Risk Taking).

This hypothesis was rejected as economic core motive and governance motive were deleted, which indicates the EM is a high order construct that consists of ten rather than twelve first order variables. These variables are Need for achievement, need for autonomy, need for affiliation, need for power, Subjective norms, Attitude, PBC, Creativity, Financial motive, and Risk taking.

The fit indices of this model are as follows: CMIN/DF (2.357), IFI (0.878), TLI (0.824) and CFI (0.864). Table 4 also shows the relative importance of each of variables in forming the construct of EM. All the variables were significant at (0.001), (***).

H2: EM is positively associated with EI.

This hypothesis is accepted ($\beta = 0.743$, $p < 0.001$). Additionally, the variance in EI that was explained by entrepreneurial motivation was substantial ($R^2 = 0.745$), All the results are reported in Table 5.

Table 4 Regression weights of each indicator

			Regression weights
Financial motive	←	EM	0.912
Need for affiliation	←	EM	0.436
Need for achievement	←	EM	0.49
Need for autonomy	←	EM	0.449
Creativity	←	EM	0.337
Need For power	←	EM	0.596
Subjective norms	←	EM	0.48
Perceived behavioral control	←	EM	0.838
Risk taking	←	EM	0.758
Attitude toward becoming entrepreneur	←	EM	0.716

Table 5 Regression weights of each indicator with EI

			Regression weights
Financial motive	←	EM	0.912
Need for affiliation	←	EM	0.436
Need for achievement	←	EM	0.49
Need for autonomy	←	EM	0.449
Creativity	←	EM	0.337
Need For power	←	EM	0.596
Subjective norms	←	EM	0.48
Perceived behavioral control	←	EM	0.838
Risk taking	←	EM	0.758
Attitude toward becoming entrepreneur	←	EM	0.716
Entrepreneurial intention	←	EM	0.743

5 Discussion and Conclusion

5.1 Discussion

The analysis results showed a strong evidence for the current research model. In particular, the research proposed that EM is a higher order construct, that consists from twelve first order indicators (need for achievement, need for autonomy, need for affiliation, need for power, subjective norms, attitude, PBC, governance, creativity, financial motive, economic core, and risk taking). The results of the data analysis showed that there is a correlation with the high order construct by these indicators in a descending order as follows; financial motive (β = 0.912); PBC (β = 0.838); risk taking (β = 0.758); attitude toward becoming entrepreneur (β = 0.716); need for power (β = 0. 596); need for achievement (β = 0.490); subjective norms (β = 0.480); need for autonomy (β = 0.449); need for affiliation (β = 0.436); creativity (β = 0.337), With a significance level of 0.000.

Concerning McClelland needs theory which consists of need for achievement, need for affiliation, need for autonomy and need for power, the current research added a significant contribution for the existing literature by enhancing our knowledge about the EM indicators. Some researchers found that entrepreneurs with high need for achievement are motivated more than employees (Lee, 1996; Nishantha, 2009; Oosterbeek et al., 2010; Sethi & Saxena, 2006). While according to Eisenberger et al. (2005) who indicated that individuals with higher need for achievement are more satisfied in jobs that have hard challenges and high skill levels. As well as need for affiliation, (Lee, 1996) found that people with need for affiliation tend to be more successful managers and helpers. As well as, need for affiliation has a significant influence onorganizational identification. Researchesrs found that individuals with need for autonomy are highly motivated to start a new ventures (Estay et al., 2013).

While people with high need for power influenced EM. However, people with high need for achievement, need for affiliation, need for autonomy and need for power influenced the decision to be an entrepreneur as well as EM (Antonioli et al., 2016; Lee, 1996; Nishantha, 2009; Sethi & Saxena, 2006; Shane et al., 2003). The results of the current research data analysis are consistent with prior researches in which McClelland Needs Theory variables considered as indicators of EM.

With regard to theory of planned behavior which consists of attitude toward becoming entrepreneur, subjective norms, and PBC, the current research found that variables related to this theory are considered as significant indicator of EM, which is consistent with the finding of Malebana (2014), Solesvik (2013), Segal et al. (2005), Fishbein and Ajzen (1975), Linan (2004) in which they found that the theory of planned behavior affects the EM. For Malebana (2014), found attitude toward becoming entrepreneur is significantly influenced EM. While according to Bauer et al. (2005), found attitude toward becoming an entrepreneur influenced the motivation and behavioral intentions. As well as, some studies found subjective norms as an influencer toward student behavioral intention towards e learning to be used in their education (Park, 2009; Yau & Ho, 2015). Others found that subjective norms affected the intention to become an entrepreneur (Malebana, 2014; Segal et al., 2005; Solesvik, 2013). In addition to that, (Ajzen, 2005) found that PBC influenced the behavioral intention when individuals possess the accurate resources and skills to implement it (Ajzen, 2005). While others found that PBC influenced the intention of individuals to be entrepreneurs in order to start their own ventures (Malebana, 2014; Segal et al., 2005; Solesvik, 2013).

The findings of this research showed that creativity is a significant indicator in EM, which is consistent with prior studies. For example, creativity has higher influence on managers (Caird, 1991). While according to Cromie and O'Donoghue (1992), found that creativity influenced entrepreneurs in order to create new ideas or develop existing ideas for creating more value and to be more innovative. Others found that creativity influenced individuals to start new ventures and become entrepreneurs (Estay et al., 2013; Jaillot, 2017). For Zhu et al. (2016) found that creativity is a significant influencer on EM. Based on the previous discussion, the current research is consistent with prior studies and it enhanced our knowledge by determining creativity as a significant indicator in EM in the Jordanian context.

Regarding financial motive, previous studies found an influence of financial motive on EM, for Burton (2012), found that financial motive have a great influence on employees, in particularly, employees who are compensated financially are more motivated and their performance are higher. Whereas others found that entrepreneurs begin their businesses to gain more money, in which they found that financial motive is considered as one of EM indicators that influenced individuals to become entrepreneurs and start new business (Aziza et al., 2013; Hessels et al., 2008; Sarri & Trihopoulou, 2004). For the current research it was found that financial motive is the most highly correlated indicator for the higher-order construct EM, this result was consistent with the previous discussed studies.

Concerning risk taking, for Crimmon and Wehrung (1986) found that risk taking influenced managers' willingness to take risks. Moreover, March and Shapira (1987) found that risk taking is connected to the work expectations as well as, risk taking influenced the decision making of managers. Whereas, Nishantha (2009) found individuals that have high risk-taking influenced entrepreneurship. In addition, others found that risk taking have influenced the intention to become an entrepreneur and have influenced EM too (Estay et al., 2013; Nishantha, 2009). For the current research, it was found that risk taking as an indicator has a significant correlation towards EM in the Jordanian context, which was consistent with the previous literatures.

Regarding governance, previous studies found that governance in corporation influenced the confidence of the market and business integrity, in which governance is a basic need for organizations that need access to value capital for long-term venture (Albourini et al., 2020). Whereas, (Yalcin et al., 2008) found that specific government policies influenced people's desire to be entrepreneurs, they found that bureaucracy and limited access to capital are barriers to entrepreneurship, as well as they also found that business laws and regulations are ideal to begin and manage new businesses for entrepreneurs thus entrepreneurs are highly motivated. While according to Aziza et al. (2013), they found that governance influence EM. However, it is noteworthy to mention that limited studies have been done to the best of our knowledge, about governance and its correlation with EM. For the current research, it was found that there's an insignificant correlation with other 10 variables, which indicate that it is not part of the construct of EM in the Jordanian context. This result was inconsistent with the previous discussed studies. However, it should be noted that the current research was conducted in Jordan, while previous studies performed their studies in different countries, such as in China (Cullen et al., 2014) and in the Kingdom of Saudi Arabia (Salem, 2014).

Concerning economic core motive, (Deaton, 2010) found that economic core effects saving rates that are higher in rapidly developing and growing economies. According to Vijaya (1998), found that economic motive makes entrepreneurs more motivated when they feel independence and self-achievement. Economic core is one of the indicators of EM (Aziza et al., 2013; Vijaya, 1998). The current research found that economic motive did not correlate with other 10 variables, which indicate that it is not part of the construct of EM in the Jordanian context. However, Limited studies have been done to the best of our knowledge, about economic influence on EM. The result of this research is inconsistent with the previous studies. However, it should be noted that this research was conducted in Jordan, while previous studies performed their studies in Malaysia (Obaji, 2014) and in Nigeria (Oni, 2012).

Concerning the key findings of the higher order construct EM and its impact on EI, the standardized coefficients ($\beta = 0.743$), as well as, ($R^2 = 0.729$) and ($p < 0.001$). This research added a significant contribution for the existing literature by enhancing our knowledge about the influence of EM on EI. Some researchers found that entrepreneurship education have a great influence in the advancement of EM and EI, in which they discovered that education influenced students towards becoming motivated and creating new business (Ferreira et al., 2007; Raposo et al., 2008). According to Delmar and Wiklund (2008), found that EM impacts the growth of small business. On the other hand, some researchers found a significant positively relationship between EI and EM (Achchuthan & Nimalathasan, 2013; Solesvik, 2013). Whereas, others found that there is a link between EIs, motivation and behavior, they showed that since intentions cannot transform immediately into action, then motivation is linked between EI and actions; motivation translated an intention into real action (Carsrud & Brännback, 2011). While other researchers have found that, an individual's intention to perform or not to perform an action is the most important determinant of action (Ajzen, 2012). In particular, (Malebana, 2014) found that EM connects EI and entrepreneurial action. According to Solesvik (2013), his empirical result showed that theory of planned behavior influence and it mediates the relationship between EM and EI. The results of the current research are consistent with previous literature, which showed that there is a great significance influence of EM on EI.

5.2 Theoretical Implications

The current research attempted to fill the gap in the literature, as well as, to develop a comprehensive model to measure EM and its impact on EI, also to detect the indicators of EM and to recognize its effect on EI. Previous studies have concentrated on a specific theories or variables in developing their model, while the current research has concentrated on developing a comprehensive model. The model consists of twelve indicators to measure EM, from a different previous studies, using theories and key variables, the current research used McClelland Needs theory (need for achievement, need for affiliation, need for autonomy, need for power), theory of planned behavior (PBC, subjective norms, attitude toward becoming entrepreneur) and key variables such as: financial motive, governance motive, economic motive, creativity and risk taking, in addition to that, the current study enhanced the understanding of EM construct through using rigorous statistical techniques. The developed model can be applied in other countries. This research aimed to provide better understanding related to the indicators of EM and its influence on EI.

5.3 Implications for Asian Business Contexts

The value of this research regarding entrepreneurship context can be applied in practice. It can be applied by schools and universities by developing for students an educational material for each indicator and provide them with training courses, or to make a mandatory course for them, thus, will lead to increase the number of entrepreneurs in the future. Furthermore, governance and economic motives are considered as entrepreneurial motivators in other countries, but when we implement the scale in the Jordanian context we found that they are not correlated to EM for Jordanian entrepreneurs, thus, these results suggest the need for Jordanian policy makers, planners and economist to consider additional methods to improve governance motive (i.e., policies and regulations needed to run a business), and improve the economic motive (i.e., taxation, material cost and interest rate). In addition to that, Jordanian government can create favorable business laws, offer tax reductions to entrepreneurs, support entrepreneurs and be willing to give government loans.

5.4 Limitation

A number of limitations were found within this research, the research was conducted only in the Jordanian context, as well as this research was influenced by money and time constraints. Furthermore, it was not easy to access the business incubators, and it was difficult to differentiate between entrepreneurs and SMEs owner. As well as, this research found lack of interest for entrepreneurs in the scientific researches.

5.5 Future Studies

The current study examined the EM indicators and its impact on EI in the Jordanian context, for future researchers can examine the same model in other countries to see how entrepreneurs from different regions perceive the dimensions of EM and it impact on EI. In addition to that, mixed methodology can be used in future studies including quantitative and qualitative together such as focus groups or interviews, to further elaborate on key issues of such topic.

5.6 Conclusion

This research aims to examine the factors that considered as indicators of EM and the effect of EM on EI in a comprehensive model, based on previous studies, there was no comprehensive model used to measure EM. This current research contribution is to fill this gap in the literature, finding comprehensive model to measure EM and its impact on EI, Understanding EM indicators and their effect on EI, can be helpful for entrepreneurs and their SME's because it might influence the increase of percentage of succesfull businesess (Stefanovic et al., 2010).

The developed model for the current research included EM which is a high order construct that consists from twelve first order indicators (need for achievement, need for autonomy, need for affiliation, need for power, subjective norms, attitude toward becoming entrepreneur, PBC, governance, financial motive, economic core, creativity, and risk taking) and considered EM as an independent variable for the dependent variable EI.

A quantitative approach was used in this research; data were administrated via an online and face-to-face survey. Survey was filled by entrepreneurs in Jordan; the data were tested using a structural equation modeling via AMOS. The findings of this research showed that need for achievement, need for autonomy, need for affiliation, need for power, subjective norms, attitude toward becoming entrepreneur, PBC, financial motive, creativity, and risk taking are considered as indicators for EM, whereas, governance and economic core is not considered as indicators for EM in the Jordanian context. Furthermore, the research found that EM is a highly significant influencer of EI.

References

Abuhashesh, M., Al-Khasawneh, M., & Al-Dmour, R. (2019). The impact of facebook on Jordanian consumers' decision process in the hotel selection. *IBIMA Business Review*, 1–16.

Abu Zayyad, H. M., Obeidat, Z. M., Alshurideh, M. T., Abuhashesh, M., Maqableh, M., & Masa'deh, R. (2020). Corporate social responsibility and patronage intentions: The mediating effect of Brand Credibility. *Journal of Marketing Communications, 27*(5), 510–533.

Achchuthan, S., & Nimalathasan, B. (2013). Relationship between entrepreneurial motivation and entrepreneurial intention: A case study of management undergraduates of the university of jaffana, Sri Lanka. http://www.academia.edu/2951985.

Ahmed, A., Aldahabreh, N., Nusairat, N. M., Abuhashesh, M., Rawashdeh, L., Al-Gasawneh, J. A. (2021). The Impact of Entrepreneurs' Emotional Intelligence on Creativity: The Moderating Role of Personal Traits. *Annals of the Romanian Society for Cell Biology*, 5626–5640. Retrieved from.

Ajzen, I. (1982). On behaving in accordance with one's attitudes. *Hillsdale, NJ: Erlbaum., 2*, 3–15.

Ajzen, I. (1991). Theory of planned behaviour. *Organizational Behaviour and Human Decision Processes, 50*, 179–211.

Ajzen. (2005). Attitudes, personality and behaviour. Berkshire, England: Open University Press.

Ajzen, I. (2012). The theory of planned behaviour. In Lange, P. A. M., Kruglanski, A. W., & Higgins, E. T. (Eds)., *Handbook of theories of social psychology, 1*, 438–459, London, UK: Sage.

Alameeri, K. A., Alshurideh, M. T., & Al Kurdi, B. (2021). The effect of covid-19 pandemic on business systems' innovation and entrepreneurship and how to cope with it: A theatrical view. *Studies in Systems, Decision and Control*, pp. 275–288.

Albourini, F., Ahmad, A., Abuhashesh, M., & Nusairat, N. (2020). The effect of networking behaviors on the success of entrepreneurial startups. *Management Science Letters, 10*(11), 2521–2532.

Alvino, J. (1990). A glossary of thinking-skills terms. *Learning, 16*(50).

Antonioli, D., Ramaciotti, L., & Rizzo, U. (2016). *The effect of intrinsic and extrinsic motivations on academics' entrepreneurial intention.*

Audretsch, D. (2002). The dynamic role of small firms.

Ayers, R. M., & Collinge, R. A. (2005). *Economics; Explore and Study*. Upper Saddle River: Pearson Education.

Aziza, N., Friedmanb, B. A., Bopievac, A., & Kelesd, I. (2013). Entrepreneurial motives and perceived problems: An empirical study of entrepreneurs in Kyrgyzstan. *International Journey of Business, 18*(2).

Bandura, A. (1997). *Self-efficacy: The exercise of self control*. New York: Freeman.

Batchelor, J. H., Abston, K. A., Lawlor, K. B., & Burch, G. F. (2014). The job characteristics model: An extension to entrepreneurial motivation. *Small Business Institute Journal, 10*(1), 1–10.

Bauer, H. H., Barnes, S. J., Reichardt, T., & Neumann, M. M. (2005). Driving consumer acceptance of mobile driving marketing: A theoretical framework and empirical study. *Journal of Electronic Commerce Research, 6*(3).
Baum, J. R., Locke, E. A., & Smith, K. G. (2001). A multi-dimensional model of venture growth. *Academy of Management Journal, 44*(2), 292–303.
Baum, J. R., Locke, E. A., & Smith, K. G. (2001). A multi-dimensional model of venture growth. *Academy of Management Journal, 44*(2), 292–303.
Begley, T., & Boyd, D. (1987). A comparison of entrepreneurs and managers of small business firms. *Journal of Management, 13*, 99–108.
Bird, B. (1988). Implementing entrepreneurial ideas: The case for intention. *Academy of Management Review, 13*(3), 442–453.
Bird, B. (1995). Towards a theory of entrepreneurial competency. *Advances in Entrepreneurship, Firm Emergence and Growth, 2*, 51–72.
Brett, J., Cron, W., & Slocum, J. (1995). Economic dependency on work: Amoderator of the relationship between organizational commitment and performance. *Academy of Management Journal, 38*, 261–271.
Broeck, A. V., Vansteenkiste, M., Witte, H. D., BartSoenens & Lens, W. (2010). Capturing autonomy, competence, and relatedness at work: Construction and initial validation of the Work-related Basic Need Satisfaction scale. *Journal of Occupational and Organizational Psychology* (Vol. 83, pp. 981–1002).
Burton. (2012). *A study of motivation: How to get your employees moving*. Indiana University.
Bygrave, W. (1989). The entrepreneurship paradigm (I): A philosophical look at its research methodologies. *Entrepreneurship Theory and Practice, 14*, 7–26.
Caird, S. (1991). Testing enterprising tendency in occupational groups. *British Journal of Management, 2*, 177–186.
Carsrud, A., & Brännback, M. (2011). Entrepreneurial motivations: What do we still need to know? *Journal of Small Business Management, 49*(1), 9–26.
Chiou, J.-S. (1998). The Effects of Attitude, Subjective Norm, and Perceived Behavioral Control on Consumers' Purchase Intentions: The Moderating Effects of Product Knowledge and Attention to Social Comparison Information. *Proc. Natl. Sci. Counc. ROC, 9*(2), 298–308.
Churchill, N., & Lewis, V. (1986). *Entrepreneurship Research*. Cambridge: Ballinger Publishing.
Collins, C., Hanges, P., & Locke, E. (2004). The relationship of achievement motivation to entrepreneurial behavior: A meta-analysis. *Human Performance, 17*(1), 95–117.
Crimmon, M., R. K. R., & Wehrung, D. A. (1986). The management of uncertainty, taking risks. *Free Press*, New York.
Cromie, S., & O'Donoghue, J. (1992). Assessing entrepreneurial inclinations. *International Small Business Journal, 10*, 66–73.
Culp, K. (2015). Principal specialist for volunteerism, Dept. of 4-H Youth Development, Adjunct Associate Professor, Dept. of Family Sciences, University of Kentucky, Brown, J, Culp, III.
Cullen, M., Calitz, A., & Chandler. (2014). Business incubation in the eastern Cape: A case study. *International Journal for Innovation Education and Research, 2*(05), 76–89.
Deaton, A. (2010). Understanding the mechanisms of economic development. *Journal of Economic Perspectives, 24*(3), 3–16.
Delmar, F., & Wiklund, J. (2008). The effect of small business managers' growth motivation on firm growth: A longitudinal study. *Entrepreneurship Theory & Practice, 5*, 437–457.
Doran, L., Stone, V., Brief, A., & George, J. (1991). Behavioral intentions as predictors of job attitudes: The role of economic choice. *Journal of Applied Psychology, 76*, 40–46.
Douglas, E. J. (2013). Reconstructing entrepreneurial intentions to identify predisposition for growth. *Journal of Business Venturing, 28*, 633–651.
Emmons, R. A. (1997). *Motives and life goals*.
Eisenberger, R., Jones, J. R., Stinglhamber, F., Shanock, L., & Randall, A. T. (2005). Flow experiences at work: For high need achievers alone. *Journal of Organizational Behavior, 26*, 755–775.

Estay, C., Durrieu, F., & Akhter, M. (2013). Entrepreneurship: From motivation to start-up. *Journal of International Entrepreneurship, 11*, 243–267.
Farhangmehr, M., Gonçalves, P., & Sarmento, M. (2016). Predicting entrepreneurial motivation among university students: The role of entrepreneurship education. *Education + Training, 58*(7/8), 861–881.
Fereidouni, H. G., Masron, T. A., Nikbin, D., & Amiri, R. E. (2010). Consequences of External Environment on Entrepreneurial Motivation in Iran. *Asian Academy of Management Journal, 15*(2), 175–196.
Ferreira, J., Pac¸o, A., Raposo, M., & Rodrigues, R. (2007). Entrepreneurship education and business creation propensity: Testing a structural model. In Proceedings of IntEnt 2007—17th Global Conference, Internationalizing Entrepreneurship Education and Training,Gdansk.
Fishbein, M., & Ajzen, I. (1975). Belief, attitude, intention, and behavior. *An Introduction to Theory and Research.*
George, J., & Brief, A. (1990). The economic in strumentality of work: An examination of the eVects of nancial requirements and sex on the pay—life satisfaction relationship. *Journal of Vocational Behavior, 37*, 357–368.
Hopp, C., & Stephan, U. (2012). The influence of socio-cultural environments on the performance of nascent entrepreneurs: Community culture, motivation, self-efficacy and start-up success. *Entrepreneurship & Regional Development: An International Journal.*
Ha, T. T., Chau, N. N., & Hieu, N. T. (2016). The Impact of Governance on Entrepreneurship Development in ASEAN+1 Countries: Evidence from World Bank Datasets. *Modern Economy, 7*, 515–525.
Hackman, J. R., & Oldham, G. R. (1975). Development of the job diagnostic survey., 60, 159–170. *Journal of Applied Psychology, 60*, 159–170.
Hair, J., Anderson, R. E. Jr., Ta tham, R. L., & Black, W. C. (1995). Multivariate data analysis with readings (4th ed.). Englewood Cliffs, NJ: Prentice-Hall.
Hessels, J., Van Gelderen, M., & Thurik, R. (2008). Drivers of entrepreneurial aspirations at the country level: The start-up motivations and social security. *International Entrepreneurship & Management Journal, 4*, 401–417.
Hisrich, R. D. (1985). The woman entrepreneur in the United States and Puerto Rico: A comparative study. *Leadership and Organizational Development Journal, 5*, 3–8.
Jayachandra, B. (2011). Entrepreneurial motivation: A conceptual analysis. *Unleashing Entrepreneurship in India: Opportunities and challenges* (pp. 265–268).
Johnson, B. (1990). Toward a multidimensional model of entrepreneurship: The case of achievement motivation and the entrepreneur. *Entrepreneurship Theory and Practice, 14*(3), 39–54.
Khumalo, B. (2012). Defining Economics in the Twenty First Century. *Modern Economy, 3*, 597–607.
Koh, H. (1996). "Testing hypotheses of entrepreneurial characteristics," a study of Hong Kong MBA students. *J Man Psychol, 11*, 12–25.
Kuratko, D., Hornsby, J., & Naffziger, D. (1997). An examination of owner's goals in sustaining entrepreneurship. *Journal of Small Business Management, 35*(1), 24–33.
Jaillot, M. (2017). The importance of entrepreneurial creativity, innovation engines, 73–90.
Lee, J. (1996). The motivation of women entrepreneurs in Singapore. *Women in Management Review, 11*(2), 18–29.
Liñán, F., & Chen, Y. W. (2009b). Development and cross–cultural application of a specific instrument to measure entrepreneurial intentions. *Entrepreneurship Theory and Practice, 33*(3), 593–617.
Linan, F. (2004). Intention-based models of entrepreneurship education. Research Gate.
Liñán, F., & Chen, Y. (2009a). Development and cross-cultural application of a specific instrument to measure entrepreneurial intentions. *Entrepreneurship Theory & Practice*, 593–617.
Locke, E. A., & Latham, G. P. (1990). A theory of goal setting and performance. Englewood Cliffs: NJ: Prentice-Hall.
Lusaka. (2013). *Political governance study.*

Malebana, M. J. (2014). Entrepreneurial Intentions and Entrepreneurial Motivation of South African Rural University Students. *Journal of Economics and Behavioral Studies, 6*(9), 709–726, September.

March, J. G., & Shapira, Z. (1987). *Managerial perspectives on risk and risk taking*, 33(11), 1404–1418.

Marlow, S., & Strange, A. (1994). Female entrepreneurs—success by whose standards?.

McAdams, D. P. (1994). The person: An introduction to personality psychology (2nd ed.). Fort Worth, TX: Harcourt Brace.

McClelland, D. C. (1961). The achieving society. Princeton, NJ: Van Nostrand.

McClelland, D. C. (1985). *Human motivation*, Glenview, IL.

McConnell, C. R. (1969). *Economics: Principles, problems and policies*, p. 23. New York: McGraw-Hillp.

Neck, H., & Greene, P. (2011). Entrepreneurship education: Known worlds and new frontiers. *Journal of Small Business Management, 49*(1), 55–70.

Nabi, G., & Liñán, F. (2011). Graduate entrepreneurship in the developing world: Intentions, education and development. *Education & Training, 53*(3), 325–334.

Nishantha, B. (2009). Influence of personality traits and socio-demographic background of undergraduate students on motivation for entrepreneurial career: The case. of Sri Lanka. *ResearchGate, 49,*(2), october.

Obaji, N. O., & Olugu, M. U. (2014). The role of government policy in entrepreneurship development. *Science Journal of Business and Management, 2*(4), 109–115.

OECD. (2010). *Regulatory policy and the road to sustainable growth.*

Oni, E., & Daniya, A. (2012). Development of small and medium scale enterprises: The role of government and other financial institutions. *Arabian Journal of Business and Management Review, 1*(7), 16–29.

Oosterbeek, H., Praag, M. v., & Ijsselstein, A. (2010). The impact of entrepreneurship education on entrepreneurship skills and motivation. *European Economic Review.*

Park, Y. (2009). An analysis of the technology acceptance model in understanding university students' behavioral intention. *Educational Technology & Society, 12*(3), 150–162.

Perry, A. L. (2014). *Creativity and its antecedents; An investigation of different assessments and training effects*, ProQuest LLC.

Qandah, R., Suifan, T. S., Masa'deh, R., & Obeidat, B. Y. (2020). The impact of knowledge management capabilities on innovation in entrepreneurial companies in Jordan. *International Journal of Organizational Analysis, 29*(4), 989–1014.

Raposo, M., Pac¸o, A., & Ferreira, J. (2008). Entrepreneur's profile: A taxonomy of attributes and motivations of university students. *Journal of Small Business and Enterprise Development, 15* (2), 405–18.

Rasmussen, E., & Sorheim, R. (2006). Action-based entrepreneurship education. *Technovation, 26*(2), 185–194.

Rauch, A., & Frese, M. (2000). Psychological approaches to entrepreneurial success. A general model and an overview of findings. In: Cooper CL, Robertson IT (eds) *International review of industrial and organizational psychology*. Chichester: Wiley.

Reynolds, P., Bygrave, W., Autio, E., Cox, L., & Hay, M. (2002). Global entrepreneurship monitor: 2002 executive report. Kauffman Center for Entrepreneurial Leadership, Kansas City, MO.

Robbins, S. P. (2003). Organizational behavior (10th ed.). NJ: Prentice-Hall, Englewood Cliffs.

Robinson, P., Stimpson, D., Huefner, J., & Hunt, H. (1991). An attitude approach to the prediction of entrepreneurship. *Entrepren Theor Pract, 15*, 13–32.

Rosenberg, M. J., & Hovland, C. I. (1960). *Cognitive, affective, and behavioral components of attitudes.* Yale University.

Rotter, J. B. (1996). Generalized expectancies for internal versus external control of reinforcemen. *Psychological Monographs: General and Applied.*

Salem, M. (2014). The role of business incubators in the economic development of Saudi Arabia. *International Business and Economics Research Journal, 13*(4), 853–860.

Sarri, K., & Trihopoulou, A. (2004). Female entrepreneurs' personal characteristics and motivation: A review of the Greek situation. *Women in Management Review, 20*(1), 24–36.
Schlaegel, C., & Koenig, M. (2014). Determinants of entrepreneurial intent: A meta-analytic test and integration of competing models. Entrepreneurship Theory & Practice, 291–332.
Segal, G., Bogia, D., & Schoenfeld, J. (2005). The motivation to become an entrepreneur. *International Journal of Entrepreneurial Behaviour and Research, 11*(1), 42–57.
Sethi, J., & Saxena, A. (2006). Enterpreneurial competencies, motivation, performance and rewards.
Shane, S., Kolvereid, L., & Westhead, P. (1991). An exploratory examination of the reasons leading to new firm formation across country and gender. *Journal of Business Venturing, 6*(6).
Shane, S., Locke, E., & Collins, C. (2003). Entrepreneurial motivation. *Human Resource Management Review, 13*(2), 257–279.
Shaver, K. G. (1987). Principles of social psychology (3rd ed.). Cambridge, MA: Winthrop.
Shaver, K., & Scott, L. (1991). Person, process, choice: The psychology of new venture creation. *Entrepreneurship Theory and Practice, 16*(2), 23–45.
Shook, C., Priem, R., & McGee, J. (2003). Venture creation and the enterprising individual: A review and synthesis. *Journal of Management, 29*(3), 370–399.
Solesvik, M. Z. (2013). Entrepreneurial motivations and intentions: Investigating the role of education major. *Education & Training, 55*(3), 253–271.
Stevenson, L., & Lundström, A. (2001). Patterns and trends in entrepreneurship/SME policy and practice in ten economies, entrepreneurship policy for the future series. *Swedish Foundation for Small Business Research, 3.*
Stefanovic, I., Prokic, S., & Rankovic, L. (2010). Motivational and success factors of entrepreneurs: The evidence from a developing country. *Zbornik Radova Ekonomskog Fakulteta u Rijeci, 28*(2), 251–269.
Taormina, J. R., & Lao, S. K. (2007). Measuring Chinese entrepreneurial motivation-Personality and environmental variables. *International Journal of Entrepreneurial Behaviour & Research, 13*(4), 200–221.
Tracy, K., Locke, E., & Renard, M. (1998). Conscious goal setting versus subconscious motives: Longitudinal and concurrent effects on the performance of entrepreneurial firms.
Veroff, J. B. (1980). Social incentives: A life span developmental approach. New York: Academic Press.
Vijaya, V., & Kamalanabhan, T. (1998). A scale to assess entrepreneurial motivation. *Journal of Entrepreneurship*, September.
Wickham, P. A. (2006). Strategic entrepreneurship, 4th ed., England: Pearson Education.
Yalcin, S., & Kapu, H. (2008). Entrepreneurial dimensions in transitional economies: A review of relevant literature and the case of Kyrgyzstan. *Journal of Developmental Entrepreneurship, 13*(2), 185–204.
Yau, H. K., & Ho, T. C. (2015). The influence of subjective norm on behavioral intention in using e-learning: An empirical study in Hong Kong higher education. *The International MultiConference of Engineers and Computer Scientists, 2.*
Zhu, Y.-Q., Gardner, D. G., & Chen, H.-G. (2016). Relationships between work team climate, individual motivation, and creativity. *Journal of Management.*

Exploring the Relationship Between Open Innovation, Procurement Sustainability and Organisational Performance: The Mediating Role of Procurement Agility

Nawaf Al Awadhi and Muhammad Turki Alshurideh

1 Introduction

Open innovation concept within the highly dynamic global socio-economic environment is based on the idea that corporate entities should take advantage of the various external and internal ideas for doing things differently in an organization (Almaazmi et al., 2020; Nuseir et al., 2021). Open innovation in this context should be understood to mean the utilization of combination of knowledge outflows and attendant inflows, to accelerate internal innovation while at the same time expanding opportunities for external utilization of innovation (Al Suwaidi et al., 2020; Alameeri et al., 2021). A robust open innovation model for any enterprise within the global socio-economic and political landscape should be rather strategic in managing the exchange of valuable information with is irrelevant to the boundaries of classic organizations (Alzoubi et al., 2022; Ghannajeh et al., 2015). Such exchange of information is critical to the growth of an enterprise, particularly because it is aimed at influencing and integrating individual corporate resources and related knowledge into the internal channels of an organization's innovative processes (Al Naqbi et al., 2020; Obeidat et al., 2021). In order to critically analyze the concept of open innovation in the modern business environment, it would be important to analyze how such an idea helps improve growth of typical enterprises (AlMehrzi et al., 2020; Odeh et al., 2021). Such an analysis would regularly span a discussion on the different mediators in the link between open

N. Al Awadhi (✉)
DBA Candidate, College of Business Administration, University of Sharjah, Sharjah, United Arab Emirates
e-mail: U19106220@sharjah.ac.ae

M. T. Alshurideh
Department of Management, College of Business Administration, University of Sharjah, Sharjah, United Arab Emirates
e-mail: malshurideh@sharjah.ac.ae

M. Alshurideh et al. (eds.), *The Effect of Information Technology on Business and Marketing Intelligence Systems*, Studies in Computational Intelligence 1056,
https://doi.org/10.1007/978-3-031-12382-5_87

innovation and organizational performance. Of specific and extreme importance in the following discussion is an analysis on whether the concept of procurement agility is explored in any possible research as a critical mediating factor in exploring the relationship between open innovation, procurement sustainability and organizational performance. The two questions the research will attempt to answer are:

Q1: Does open innovation impact the achievement of procurement sustainability and effect organizational performance?

Q2: What is the influence of procurement sustainability on organizational performance?

2 Research Variables and Methodology

The following study will adopt a qualitative research methodology, primarily through exploration of the relevant literature on the link between open innovation and organization growth, while also taking into account the various possible mediating factor like procurement agility. A qualitative research study in this context involves the ability of a researcher to pursue an in-depth and deeper exploring of various factors affecting a phenomenon trough adopting a specific and focused line of questioning. By focusing through the available literature on the connection between open innovation and organization growth, the research study is prepared to gather evidence that will prove the existence of such a relationship or otherwise.

The key research variables in the ensuing research study include, organizational performance, procurement sustainability and the concept of open innovation. Other variables that will suffice in the research study will largely span the different mediating factors in the relationship between open innovation and firm growth with the modern global business environment. Open innovation as a key variable in the research study has been dissected by researchers such as Parveen et al. (2015) who averred that the use of invention and technology creates value and plays a major role in enhancing social well-being. Other authors such as Islam et al. (2017) argued that the concept of open innovation presumes that corporates apply an open system that is based on internal and external factors. The variable is termed as a distributed process founded on purposeful managed information that flows across institutions using non-pecuniary and pecuniary mechanisms (Parveen et al., 2015). **Open innovation** is defined as an approach for organizations to achieve sustainable growth. The second variable examined in the current research is **procurement sustainability**. The rise trend and incorporation of tenders within institutions has enhanced the need to achieve sustainable approaches in procurement methods. Markedly, the capacity to meet the requirements is founded on the level of innovation used to promote the process. Openness, accountability, and timing are critical factors while assessing the sustainability of the procurement process. As organizations strive to achieve their set future objectives, procurement plays a significant role in determining how they attain consistent growth (Starzyńska, 2019). The procurement process serves as a factor facilitating innovation diffusion to reduce the risks of losses and exit from markets. The research defines procurement sustainability is the level where institutions incorporate environmental and social factors alongside the financial aspects of acquiring goods or services (Blind et al., 2020). It is therefore, an essential part of a business that ensures enhanced performance.

Organizational performance is the third variable examined in the report. According to Sethibe and Steyn (2016), it is an aspect measured in the operational and financial sustainability of a firm. The measures of institutional performance depend on the related question and correlate to the independent variables. In this report, organizational performance relies on the adoption of open innovation in procurement processes and the achievement of sustainability in the overall procedures. Therefore, it is a critical element within the institution that defines the success in business functions but relies on different factors. Also, Hameed (2018) investigated the main drivers of an archetypal firm's open innovation performance om small and medium entreprises in Malaysia. The study found that a decline in open innovation practices in Malaysian firms the potential to hamper overall performance. The study suggested a raft of measures designed to correct the anomaly through a cross-sectional study fo Malaysina firms. The research adopted a quantitative research design, and the subsequent research results collated using a 5-point likert scale from mail surveys. The study listed internal innovation, Research and Development (R&D), and external knowledge as critical determinant's of a typical enterprsise's open innovation performance. Moreover, Asha (2008) dissected the issue the impact of various styles of open innovation and sectorl organizational performance. The study found tow critical variables linked to the construct of open innovation that includes, an understanding of open styles innovation by firms, and the question as to whether the ability of a firm to discern open innovation design capabilities has a bearing on individual corporation's performance. The research study was largely distinctive particuallry because of its design and attendant conceptualization. The research tested the hypothesis from a vast cros-sectional data of 16,445 businesses across the United Kingdom, derived from the UK innovation suvey database. The study found that design capability is an important construct of open innovation practices, and thus managers should give it greater attention.

Stanisławski (2020) conducted a study on open innovation as a value chain for small and medium enterprises. The study mainly focused on the various determinants of the use of open innovation for firms in Poland. The study identified both internal and external determinats of open innovation. In this regard, the study found that larger entities are more poised to use open innovation than comparatively smaller entities. The study further asserted that both external and internal determinants of open innovation have a significant impact on the general application of open innovation concepts in companies across Poland. Also, Canh et al. (2019) conducted a study on the general impact of open innovation on firm performance and corporate social responbilities for firms across Vietnam. The study focused on individual and process innovation, and the manner in which such costructs interact with an entitys internal collaborations, and the attendant impact on corporate social responsibility and firm performance. The findinsg of the study were that consistent investments in open innovation by entities have the potential to enhance firm performance. To add more, Durst and Ståhle (2014) conducted a study on the various critical success factors of open innovation. Through a systematic review of 29 studies on open innovation, the study identified a total of 9 themes deemed to be critical to the success of open innovation in manufacturing entities. Such themes include; provision of resources, leadership, culture, governance, relational aspects, strategy, process management, and the pepple involved in open innovation processes.

3 Review Approach

The current report uses a systematic review to assess and examine the relationship between open innovation, sustainable procurement, and organizational performance. The approach identifies, chooses, and critically appraises various researches as an approach to answer the formulated questions (Perényi & Losoncz, 2018; Al Mehrez et al., 2020; AlShehhi et al., 2020; Ahmad et al., 2021). According to Cooper et al. (2018), a systematic review incorporates searching studies to undertake a transparent report. It leaves a straightforward approach about how the studies were identified and how the researcher arrived at the findings in relevance to the research questions (Ahmed et al., 2020; Alkitbi et al., 2020; Alshamsi et al., 2020). The major advantage of using a systematic review in this report is to enhance reliability and transparency on the research findings in relation to how the variables correlate to each other (Al Khayyal et al., 2020; Almazrouei et al., 2020). The study focused on 35 articles that consisted the literature review. Out of the 35 articles that comprised the study, a total of 19 articles qualified for an indepth analysis. The initial article searches in the entire research study was more than 40,000 articles. Subsequently, the universe of the research article searches related to the study was cut down to 35 articles. A total of 16 articles were related to innovation, while 19 articles were related to procurement.

4 Main Finding Table

Articles	Related to innovation	Related to procurement	Total
Number	16	19	35
Variables	Open innovaition	Procurement sustainability	Procurement agility
	16	16	3
Dabase	Number of articles related to open innovation	Number of articles related to procurement sustainability	Number of articles related to procurement agility
Pro quest	7	4	2
Initial search	8,279	79	3
Google scholar	9	12	1
Initial search	37,000	497	30

S.No	Title	Authors	Year	Place	Context	Independent variable	Mediating variable	Depend variable	Data collection methods	Sample size	Journal title	Findings/main aspects	Limitations	Future study
1	Revitalizing the concept of Public Procurement for Innovation (PPI) from a systemic perspective: objectives, policy types, and impact mechanisms	Shin, Kiyoon, Yeongjun Yeo, and Jeong-Dong Lee	2019	Korea	Public sector	Innovativeness	Policies	Procurement efficiency	Qualita-tive	N/A	Systemic practice and action research	Innovativeness in procurement is critical for efficiency and transparency	The research fails to provide a suitable policy implication	Future studies need to focus on economic variables
2	Innovation and standardization as drivers of companies' success in public procurement: an empirical analysis	Blind, Knut, Jakob Pohlisch, and Anne Rainville	2020	N/A	Public sector	Innovation	Standardiza-tion	Procurement success	Qualita-tive	N/A	The journal of technology transfer	Innovation influencing procurement outcomes by enhancing efficiency and transparency	Possibility of endogeneity	There are no future recommen-dations

(continued)

(continued)

S.No	Title	Authors	Year	Place	Context	Independent variable	Mediating variable	Depend variable	Data collection methods	Sample size	Journal title	Findings/main aspects	Limitations	Future study
3	The benefits and challenges of E-procurement implementation: a case study of Malaysian company	Mohd Nasrun Mohd Nawi1*, Saniah Roslan2, Nurul Azita Salleh3, Faisal Zulhumadi4, and Aizul Nahar Harun5	2016	Malaysia	Telecommunication industry	E-procurement adoption	N/A	Institution's performance	Qualitative case study	N/A	International journal of economics and financial issues	Adoption of e-procurement has eminent benefits, which include saving cost and enhancing efficiency in the procurement process. innovation is a recent development that lacks benchmarking	The study considered one institution as a case, which limits the generalizability of the findings	There are no recommendations for future studies
4	Measuring the link between public procurement and innovation	Appelt, Silvia, and Fernando Galindo-Rueda	2016	Multi-country	OECD project	Innovation	N/A	Procurement	Qualitative	N/A	OECD science technology and industry working papers	Innovation plays a major role in enhancing the procurement process	There are no limitations to the study	There are no future recommendations given

(continued)

(continued)

S.No	Title	Authors	Year	Place	Context	Independent variable	Mediating variable	Depend variable	Data collection methods	Sample size	Journal title	Findings/main aspects	Limitations	Future study
5	Public procurement for innovation and civil preparedness: a policy-practice gap	Isabell Therese Storsjö and Hlekiwe Kachali	2017	Finland	Healthcare sector	Policies relating to Innovation and civil preparedness	N/A	Procurement effectiveness	Quantitative	92	International journal of public sector management	Policies guide the process of innovation in the procurement process, which enhance the ability to achieve sustainable development	The data collected were for the Finland health sector	The future direction is to map supply networks and establish points that require innovation as an approach to enhance the effectiveness
6	Sustainable innovation in public procurement: The decisive role of the individual	Martin E. Eikelboom	2018	N/A	Large public institutions	Individual initiative	N/A	Sustainable innovativeness	Quantitative	283	Journal of public procurement	From the study, innovation is critical in enhancing sustainable innovations	The study was conducted within one institution	Future research should expand on the model and assess the various connections between the constructs
7	The effect of innovation capability on business performance: a focus on IT and business service companies	Seung Hoo Jin and Sang Ok Choi	2019	Korea	IT sector	Innovation capability	N/A	Business performance	Quantitative	160	Sustainability basel	Innovation plays a major role in enhancing sustainability within firms, which include procurement processes	The results are analyzed based on productivity and revenue	

(continued)

(continued)

S.No	Title	Authors	Year	Place	Context	Independent variable	Mediating variable	Depend variable	Data collection methods	Sample size	Journal title	Findings/main aspects	Limitations	Future study
8	Public procurement as a policy tool: using procurement to reach desired outcomes in society	Grandia, Jolien, and Joanne Meehan	2017	Europe	Sectors of the economy	Innovation	N/A	Procurement performance	Qualita-tive	N/A	International journal of public sector management	Procurement procedures are supported through advancements in technology	N/A	N/A
9	A field-level examination of the adoption of sustainable procurement in the social housing sector	Meehan, Joanne, and David J. Bryde	2015	N/A	Social housing sector	Sustainable procurement	N/A	Institution performance	Quantita-tive	116	International journal of operations and production management	Provisions for sustainable procurement processes enable firms to achieve high performance	The findings give an insight into the research topic	Longitu-dinal studies need to be undertaken across different countries
10	Modeling the adoption of sustainable procurement in construction organizations	Agbesi, Kwaku, Frank D. Fugar, and Theophilus Adjei-Kumi	2018	N/A	Construction sector	Sustainable procurement	N/A	Institutional performance	Quantita-tive	193	Built environment project and asset management	To attain consistent growth, institutions need to adopt sustainable procurement as a vessel to incorporate innovation	The study is limited in scope	Future research should enhance the scope of the topic

(continued)

(continued)

S.No	Title	Authors	Year	Place	Context	Independent variable	Mediating variable	Depend variable	Data collection methods	Sample size	Journal title	Findings/main aspects	Limitations	Future study
11	Key factors hindering sustainable procurement in the Brazilian public sector: A Delphi study	Da Costa, Bruno BF, and Ana Lúcia TS Da Motta	2019	Brazil	Public sector	Sustainable procurement	External and internal factors	Organizational performance	Qualitative	N/A	International journal of sustainable development and planning	Sustainable procurement within the public sector enhances accountability and improved performance	The findings focus on the Brazilian public sector	Future studies could test the applicability of results in under-developed countries
12	**Cite**Eriksson, Per Erik. "Procurement strategies for enhancing exploration and exploitation in construction projects	Eriksson, Per Erik	2017	N/A	Construction industry	Procurement strategies	N/A	Exploitation of projects	Qualitative	N/A	Journal of financial management of property and construction	Procurement strategies enhance the management and implementation of projects	The article focuses on the construction sector	Future studies need to focus on diverse industries to enhance generalizability
13	E-Procurement adoption: A case study about the role of two Italian advisory services	Belisari, Sara, Daniele Binci, and Andrea Appolloni	2020	Italy	Legal sector	Innovation adoption	N/A	Achievement if sustainable procurement processes	Quantitative	51 contracting authorities	*Sustainability basel*	The study found that adoption of innovaton is critical to frm's sustainablity	The study was inconclusive due to inadequate literature	Future studies should explore recent literatures

(continued)

(continued)

S.No	Title	Authors	Year	Place	Context	Independent variable	Mediating variable	Depend variable	Data collection methods	Sample size	Journal title	Findings/main aspects	Limitations	Future study
14	factors influencing the digitization of procurement and supply chains	Bien-haus,Florian., Haddud, Abu	2018	UK	Supply Chain	Procurement and supply chain	N/A	Digitaiztion factors in procurement	Quantita-tive	30	*Business Process Management Journal*	The study found that digitization is critical to development of procurement in an typical firms	The study was not broad in terms of coverage of firms in the UK	Future studies should focus on technology as a mediating factor
15	Innovation and standardization as drivers of companies' success in public procurement: an empirical analysis	Knu Blid,Jacob Pohlisch,Anne Rainville	2019	Poland	Procurement	innovation	N/A	Innovative factors	Qualita-tive	N/A	*The Journal of Technology Transfer volume*	Innovation plays a major role in enhancing sustainability within firms, which include procurement processes	The study covered a narrow view of the topic	Future studies should focus on bootlenecks that hinder innovation in the public sector
16	Defining the process to literature searching in systematic reviews: A literature review of guidance and supporting studies	Chris Cooper, Andrew Booth,Jo Varley-campbell, Nicky Britten,Ruth Garside	2018	UK	Systematic reviews	Innovation	N/A	N/A	Qualita-tive	N/A	*BMC medical research methodology*	The study found that systematic review of literarure is critical to understanding open innovation	The study was limited in terms of scope	Future studies should focus on wide range of recent literature on the topic

(continued)

(continued)

S.No	Title	Authors	Year	Place	Context	Independent variable	Mediating variable	Depend variable	Data collection methods	Sample size	Journal title	Findings/main aspects	Limitations	Future study
17	Key factors hindering sustainable procurement in the Brazilian public sector: A Delphi study		2019	Bazil	Procurement	Sustainable development	N/A	N/A	Qualita-tive	N/A	*International journal of sustainable development and planning*	Thes study foundnout that governemnt bureucarcy and related bottleneck are a hindrance to achievemnet of sustainabel procuremnt in Brazilian public sector	The study narrowly analyzed the Brazilina public sector which is Markedly unique within the global procurement landscape	Future research study should avoid a narrow view of public sector procure-ment
18	Implementation of sustainable public procurement in local governments: A measurement approach	Lily Hsueh, Stuart Bretschneider, JustinStritch	2020	UK	Procurement	Policy implementation	N/A	Procuremnt policies	Quntatitve	10	*International journal of public sector management*	Public policy implementtaion is crtical in procuremnet	The studu could not effectively measure sustainable development in the UK public sector	Future studies should focus on harmoniza-tion policies in the public sector
19	E-procurement use in the Nigerian building industry	Ibem, E. O., Aduwo, E. B., Ayo-Vaughan, E. A., Uwakonye, U. O., Owolabi, J. D	2017	Nigeria	E-procurement	Technological enhancements	N/A	Procurment	Quntatitve	9	*International journal of electronic commerce studies*	Innovativeness in procurement is critical for efficiency and transparency	The nascent nature of E-procuremnt in Nigeria was a major limitation	Future studies should focus on more developed economies with regards to E-procurment procedures

(continued)

(continued)

S.No	Title	Authors	Year	Place	Context	Independent variable	Mediating variable	Depend variable	Data collection methods	Sample size	Journal title	Findings/main aspects	Limitations	Future study
20	Do sustainable procurement practices improve organizational performance	Mazharul IslamAbalala Turki Md. Wahid MuradAzharul Karim	2017	UAE	Procurment	Sustainable procurment	N/A	Sustain-ability	Quntatitve	25	*Sustainability journal*	The study found that sustainable development practices have a direct impact on frim performance	The study was not conclusive in defining and contextual-izing sustainable development	Future studies should focus on conclusive definition and contex-tualizing of various terms
21	A method to encourage and assess innovations in public tenders for infrastructure and construction projects	Bart Lenderink, Johannes I.M. Halman, Hans Boes and Hans Voordijk	2020	UK	Infrastructure	Procurment	N/A	Infrastruc-ture	Quntatitve	24	*Construction innovation*	The study found that innovation in public sector has a direct impact in successful delivery of governmnet projects	The study was inconclusve on the nature of impact of innovation on positive delivery of government projects	Future studies should focus on nature of impact of innovation on government projects
22	Preparing today for prosperity tomorrow: Using sustainable public procurement as a tool for development in Ghana	A.B. Adjei	2019	Ghana	Procuremnt laws	Sustainable procurement	N/A	Sustain-ability in preocure-ment	Quntatitve	5	*Public contract law journal*	The study found that sustainable public sector procurment in Ghana is widely embraced	The study was qualitative rather than quantitatiev with apparent gaps in findings	Future studies should adopt a quantative methods of study

(continued)

(continued)

S.No	Title	Authors	Year	Place	Context	Independent variable	Mediating variable	Depend variable	Data collection methods	Sample size	Journal title	Findings/main aspects	Limitations	Future study
23	Organization culture and open innovation: A quadruple helix open innovation model approach	*Shazia Parveen, Aslan Amat Senin, Arslan Umar*	2015	India	Organizational culture	Open innovation	N/A	Organization culture in oen innovation	Qualita-tive	N/A	*International journal of economics and financial issues,*	The study found that organizational culture is critical in the ability of an entreprise to embrace open innovation	The study made a feeble connection between orgnaiza-tional culture and the desire to embrace open innovation	Future studies should be conclusive on what is organiza-tional culture with regards to open innovation
24	A systematic review of international entrepreneur-ship special issue articles	Aaron Perenyi, Miklos Losonzc	2018	Australia	Entrepreneur-ship	Entrepreneurshi	N/A	Systematic review	Qualita-tive	N/A	*Sustainability masel*	The studay found motley of articles that advocated for adoption of interantional entreprenurship	The study was inconclsuive with regards to the findings	Future research should focus on one aspect of interna-tional entreprenur-ship
26	Implementation of sustainable public procurement practices and policies: A sorting framework	Eric Prier, Edward Schwerin, Clifford P. McCue	2016	N/A	Pocurement	Sustainable procurement	N/A	Policies	Qualita-tive	N/A	*Journal of public procurement*	The article found a prevalent use of sustainable public procuremnt	The article was inconclsuive on the drivers of sustainable public procuremnt in public sector	Future studies should focus on the drivers of sustainable procure-ment in the public sector

(continued)

(continued)

S.No	Title	Authors	Year	Place	Context	Independent variable	Mediating variable	Depend variable	Data collection methods	Sample size	Journal title	Findings/main aspects	Limitations	Future study
27	Developing strategies to improve agility in the project procurement management (PPM) process	Santosh B. Rane, Shivangi Viral Thakker	2019	N/A	Sustainable Busines	Sustainability in busines processes	N/A	Policies	Qualita-tive	N/A	*Business process management journal*	The article found a major drive among corporates in embarcing the concept of agilty in project procuremnt	The artocel didi not properely define the concept of agility in procuremnet	Future research should delev more deeply on the concept of agility in procure-ment
28	Perspectives of introduction sustainable procurement in public procurement in Russia	Irina Romodina Maxim Silin	2016	Russia	Public procurement	Sustainable procurement	N/A	Sustainable policies	Qualita-tive	N/A	*Oeconomia Copernicana*	The study found that the introduction of sustainable procuremnt in Russia positively influenced organizational perfoemnace	The article ocused on Russia which has a less robust public sector	Future studies should focus on the main drivesr of sustainable procure-ment
29	Innovation and organizational performance: A critical review of the instruments used to measure organizational performance	Tebogo Sethibe, Renier Steyn	2016	N/A	Organizational performance	Measurement of organizational performence	N/A	Internal processes	Qualita-tive	N/A	*The southern arican journal of entrepreneur-ship and small business management*	From the study, innovation is critical in enhancing sustainable innovations	The study was conducted within one institution	Future studies should cover a wide range of sectors for conclusive research

(continued)

(continued)

S.No	Title	Authors	Year	Place	Context	Independent variable	Mediating variable	Depend variable	Data collection methods	Sample size	Journal title	Findings/main aspects	Limitations	Future study
30	Innovative procurement in Poland in the light of the report on the evaluation of public procurement functioning after the amendment of the public procurement law of 2016	Wacława Starzyńska	2016	Poland	Innovation	Procurement	Innovation in procurement	Technology	Qualitative	N/A	*Folia Oeconomica Stetinensia, Journal of Global Entrepreneurship Research*	The study found that innovative procurment is widely embraced in Polish public sector	The study focused on legal constructs of innovative procurment in Polish public sector which is a rather narrow view	Future studies should focus on various hindrances to the implementation of procurement policies in the pulic sector
31	Determinants of firm's open innovation performance and the role of R & D department: an empirical evidence from Malaysian SME's	Waseem Ul Hameed, Mumammd Basher Jawad Iqbal, Ayesha Anwar, Hafiz Ahmad	2018	Malaysia	Open innovation	Innovation performance	Firm performance	Technology	Qualitative	N/A	*Folia Oeconomica Stetinensia,*	The study found out that technology, organizational cultures, and organizational structure are critical determinants of firm'spredilections to open innovation	The study was no emprirically backedwith regards to the findings	Future roles should focus on internal structure of open design

(continued)

(continued)

S.No	Title	Authors	Year	Place	Context	Independent variable	Mediating variable	Depend variable	Data collection methods	Sample size	Journal title	Findings/main aspects	Limitations	Future study
32	Open by design: The role Of design In open innovation	Virginia Acha	2008	UK	Open innovation	Innovation design	Mediating factors	N/A	Qualita-tive	N/A	*Folia Oeconomica Stetinensia,*	The study found that open design as a construct of organizational structure has a positive impact on the abiity of a firm to embrace open innovation	The study was mainly focused on the organiza-tional design of typical enterprise which is rather narrow	Future studies should focus on the role of market as mediating factors in open innovation
33	Open innovation as a value chain for small and medium-sized enterprises: determinants of the useof open innovation	Robert Stanisławski	2020	Poland	Open innovation	Dterminants of open innovation	External and internal factors	N/A	Qualita-tive	N/A	*Folia MDPI jOURNAL*	The study found out that open innovation is a critical process funnel in a typical SME's value chain	The study was inconclusive on the impact of enhanced value chain on organiza-tional performnce	Future studies should provide a link between open innovation and improved value cahin in organiza-tions
34	The impact of innovation on the firm perfor-manceand corporate social responsibility of Vietnamese manufacturing firms	Nguyen Thi Canh1, Nguyen Thanh Liem1,*, Phung Anh Thu2,*andNguyen Vinh Khuong	2019	Vietnam	Open innovation	Dterminants of open innovation	Impact of open innovation on firm performance	N/A	Qualita-tive	N/A	*Folia MDPI jOURNAL*	The study found that open innovation has a direct impact on a firm's ability to embrace and sustain corporate social responsibility	The study provided no link between organiza-tional performance and corporate social respeonbility	Future studies shoud focus on other drivers of firm's perfor-mance, which mediate the concept of open innovation

(continued)

(continued)

S.No	Title	Authors	Year	Place	Context	Independent variable	Mediating variable	Depend variable	Data collection methods	Sample size	Journal title	Findings/main aspects	Limitations	Future study
35	Success factors of open innovation—A literature review	Sussanne, Durst, Pirjo Ståhle	2014	Sweden	Open innovation	Citical success factors	Fcators influencing open innovation	N/a	Qualitative	N/A	*Finland Futures Research Centre and Center for Knowledge and Innovation Research*	The study found that technology, and organizational culture are critical in the ability of a firm to embrace open innovation	The study was hampered by voluminous nature of the research, that sometimes threw the findings of the study off-tangent	Future studies should focus more on the different factors that drive open innovation

5 Discussion

Research Q1: Does open innovation impact the achievement of procurement sustainability and effect organizational performance?

The review articles answer the question of how open innovation influences the achievement of sustainable procurement in organizations. The focus of the review centers on the application of innovation as a source of enhancing procurement processes within organizations. The study is centered on the research question and purpose to establish the relationship between open innovation and sustainable procurement. It is essential to acknowledge that few reviewed articles do not directly show the relationship but demonstrate the link between the two variables. Consistent with the question, the independent variable in this case is open innovation, while the dependent variable is sustainable procurement. There are positive outcomes in all the reviewed articles, with all the scholars agreeing with the concept that innovation has a major role in ensuring sustainable procurement practices. From the literature, it can be stated that open innovation is a tool introducing new technology in the procurement process, thus enabling firms to achieve high value.

Various studies examined in the research shows a significant correlation between open innovation, sustainable procurement, and organizational performance. According to Parveen, Senin and Umar (2015), open innovation is a strategic tool for enhancing sustainability within businesses. There are justifications to link open innovation to the attainment of an efficient and transparent procurement process within institutions from the seven reviewed articles. Rane et al. (2019) envision that the capacity of institutions to enhance efficiency is founded on the ability to attain procurement agility as a tool for managing the acquisition of goods and services. This stand is supported by Nyantakyi's (2019) study that exemplifies the role of innovation in enhancing sustainable practices. Acquisition of goods and services as a critical component of business growth has in the past decade been observed to lean on firm's innovation capabilities. A firm's innovation capability has the ultimate effect on the attendant supply chain agility. Iddris et al. (2014) define agility as the ability of an arceytpal firm to adapt, sense, perform and respond in light of an increasingly dynamic and competitive environment. Other scholars such as Cristopher (2012) defines agility as the technical capability of firms to match the demand of goods and services with the attendant supply, mediated by downstream and upstream supply chains. Iddris et al. (2014) avers that a typical firm's agility is a factor open innovation, technology, trust, with the ultimate goal of enhanced competitiveness. The principles of technology applied in acquisition processes play a major role in improving value for money. Prier et al. (2016) posit that procurement processes are considered to attain sustainability when they integrate criteria, specifications, and requirements that favor and align to protecting the environment and supporting constant growth (Yevu & Yu, 2019). This exemplifies the notion that open innovation is a critical factor while striving to achieve sustainable procurement in companies.

The review of past studies shows a significant similarity in establishing the relationship between open innovation and sustainable procurement. From the thirteen reviewed articles, seven scholars agree that innovation significantly enhances efficiency in acquisition processes. With all articles having a positive relationship, it exemplifies the idea that adoption of technology has enhanced the achievement of procurement agility. With no conflicting findings from the review, it is evident that institutions consistently use open innovation to achieve sustainability and value to stakeholders.

Research Q2: What is the influence of procurement sustainability on organizational performance?

The systematic review involved six articles selected to examine the relationship between sustainable procurement and organizational performance. With respect to the research question, sustainable procurement is an independent variable, while organizational performance is the dependent variable. Few of the articles from the table do not give a direct link between two variables. Nevertheless, there are strategic assessments that correlate sustainable procurement to institutional performance. The review shows positive findings from all the sources used in the current report. This exemplifies the significant relationship between procurement processes and their ability to enhance performance. Also, Yevu and Yu's (2019) findings show that procurement has a significant role in enhancing institutional performance. A similar argument is illustrated by Vluggen et al. (2019) study that focused on examining the role of sustainable procurement in enhancing accountability. Findings from the studies show that for institutions to create an enabling environment characterized by reduced cost and responsibility, procurement agility must be achieved. Stritch, Bretschneider, Darmall, Hsueh and Chen (2020) affirm the role of sustainable procurement processes in enhancing efficiency and addressing the dynamic external and internal factors.

The six articles reviewed to assess the relationship between sustainable procurement and organizational performance show a positive correlation. None of the article's findings show a contradicting argument with respect to the two variables. This illustrates the idea that procurement agility has a mediating role of blending innovation as a tool for promoting organizational performance.

6 Summary

Open innovation is described as a critical aspect that ensures firms apply technology, processes, and inventions that satisfy the need for an economical cost. From the research, it is evident that innovation is an approach for organizations to create value for investment by ensuring consistent growth. Studies show that open innovation is a strategy to ensure firms adapt to the changing business environment, thus enhancing their ability to respond to shifting challenges (Hsueh et al., 2020). Adoption of technology encompasses the integration of critical development and business models, which enhance social responsibilities to stakeholders (Ibem et al., 2017). The application of open innovation in procurement mandates paradigm shifts in focus, strategy and mindsets while assimilating the process's internal and external factors (Vluggen et al., 2019). While open innovation is a way of introducing new terminology, it plays a key role in enhancing institution's capacity to achieve sustainability (Romodina & Silin, 2016). Its adoption increases the possibility of attaining sustainable procurement processes within organizations.

The current business environment compels institutions to remain adaptive to the changing factors (Lenderinket al., 2020). Markedly, attaining sustainability in the procurement process is a critical aspect while creating value for stakeholders. Open innovation in the procurement process transforms the practice and makes provision for reduced costs, accountability, and efficiency while managing suppliers electronically (Lee et al., 2022a; Rane et al., 2019). Markedly, it exemplifies the mediating role of procurement agility in aligning technological advancement in enhancing institutional performance (Shakhour et al., 2021). It is imperative to state that there is a significant relationship between open innovation, sustainable procurement, and organizational performance (Stritch et al., 2020). Achieving institutional growth relies on the ability to attain effective procurement procedures while, in turn, depends on the invention adopted. This shows a linear relationship between the three variables, thus exemplifying the need for investing in technology as a way of initiating sustainable development in firms (Prier et al., 2016). Innovation streamlines the process of procurement which has a significant impact on performance and the ability to safeguard resources within public and private companies.

7 Research Limitations

The current findings are limited by the drawbacks of the research. The current study relies on secondary data from past researches to affirm the significant relationship between open innovation, sustainable procurement, and organizational performance. One of the drawbacks is the large nature of the search results, triggering a possibility of the researcher taking a different line form the main focus of the study. Similarly, there was no enough literature available on the mediator factor "Procurement Agility". The current report recommends the use of primary data as an approach to verify the findings.

8 Research Future Directions

Future studies are required to exemplify the relationship between open innovation, sustainable procurement, and organizational performance. This can be achieved through a preliminary study or an intense review incorporating data from various libraries and regions. Future studies should also deeply explore the impact of the concept of procurement agility mediated by procurement sustainability or openinnovation. Or the impact procurement Agility could have on organization performance. There was very little literature availbe on this subject matter.

9 Conclusion

The mediating role of procurement agility is to align the benefits of open innovation as tools for enhancing organizational performance. The process of attaining sustainable procurement is driven by the need to improve institutional performance. Through the invention and adoption of technology, companies have promoted social responsibilities about procuring goods and services. From the current report, it is affirmed that innovation plays a major role in enhancing transparency and accuracy in procurement departments. Besides the limitations of the study, it is conclusive that the role of achieving procurement agility is to ensure new technology is strategically affiliated to address organizational challenges. Additionally, it is also conclusive that there is a positive relation between the variables open innovation, procurement sutainability and organizational performance.

References

Ahmad, A., Alshurideh, M. T., Al Kurdi, B. H., & Alzoubi, H. M. (2021). Digital strategies: A systematic literature review. In The International Conference on Artificial Intelligence and Computer Vision (pp. 807–822), June. Cham: Springer.

Ahmed, A., Alshurideh, M., Al Kurdi, B., & Salloum, S. A. (2020). Digital transformation and organizational operational decision making: A systematic review. In International Conference on Advanced Intelligent Systems and Informatics (pp. 708–719), October. Cham: Springer.

Al Naqbi, E., Alshurideh, M., AlHamad, A., & Al Kurdi, B. (2020). The impact of innovation on firm performance: A systematic review. *Int. J. Innov. Creat. Change, 14*(5), 31–58.

Al Khayyal, A. O., Alshurideh, M., Al Kurdi, B., & Salloum, S. A. (2020). Women empowerment in UAE: A systematic review. In International Conference on Advanced Intelligent Systems and Informatics (pp. 742–755), October. Cham: Springer.

Al Mehrez, A. A., Alshurideh, M., Al Kurdi, B., & Salloum, S. A. (2020). Internal factors affect knowledge management and firm performance: A systematic review. In International Conference on Advanced Intelligent Systems and Informatics (pp. 632–643). Cham: Springer.

Al Suwaidi, F., Alshurideh, M., Al Kurdi, B., & Salloum, S. A. (2020). The impact of innovation management in SMEs performance: a systematic review. In International Conference on Advanced Intelligent Systems and Informatics (pp. 720–730), October. Cham: Springer.

AlMehrzi, A., Alshurideh, M., & Al Kurdi, B. (2020). Investigation of the key internal factors influencing knowledge management, employment, and organisational performance: A qualitative study of the UAE hospitality sector. *Int. J. Innov. Creat. Chang, 14*(1), 1369–1394.

AlShehhi, H., Alshurideh, M., Al Kurdi, B., & Salloum, S. A. (2020). The impact of ethical leadership on employees performance: A systematic review. In International Conference on Advanced Intelligent Systems and Informatics (pp. 417–426), October. Cham: Springer.

Alameeri, K. A., Alshurideh, M. T., & Al Kurdi, B. (2021). The effect of covid-19 pandemic on business systems' Innovation and entrepreneurship and how to cope with it: A theatrical view. *The Effect of Coronavirus Disease (COVID-19) on Business Intelligence, 334*, 275–288.

Alkitbi, S. S., Alshurideh, M., Al Kurdi, B., & Salloum, S. A. (2020). Factors affect customer retention: A systematic review. In International Conference on Advanced Intelligent Systems and Informatics (pp. 656–667), October. Cham: Springer.

Almaazmi, J., Alshurideh, M., Al Kurdi, B., & Salloum, S. A. (2020). The effect of digital transformation on product innovation: A critical review. In International Conference on Advanced Intelligent Systems and Informatics (pp. 731–741), October. Cham: Springer.

Almazrouei, F. A., Alshurideh, M., Kurdi, B. A., & Salloum, S. A. (2020). Social media impact on business: a systematic review. In International conference on advanced intelligent systems and informatics (pp. 697–707), October. Cham: Springer.

Alshamsi, A., Alshurideh, M., Al Kurdi, B., & Salloum, S. A. (2020). The influence of service quality on customer retention: A systematic review in the higher education. In International Conference on Advanced Intelligent Systems and Informatics (pp. 404–416), October. Cham: Springer.

Alzoubi, H., Alshurideh, M., Kurdi, B., Akour, I., & Aziz, R. (2022). Does BLE technology contribute towards improving marketing strategies, customers' satisfaction and loyalty? The role of open innovation. *International Journal of Data and Network Science, 6*(2), 449–460.

Asha, V. (2008). Open by design: The role of design in open innovation. *Academy of Management, 1*(6), 1–45.

Aziz, K. (2014). Determinants for supply chain agility practices: An empirical studies on Malaysian Electrical and Electronics (E&E) firms. *Asian Academy of Management Journal, 1*(1), 1–25.

Bienhaus, F., & Haddud, A. (2018). Procurement 4.0: Factors influencing the digitisation of procurement and supply chains. *Business Process Management Journal, 24*(4), 965–984.

Blind, K., Pohlisch, J., & Rainville, A. (2020). Innovation and standardization as drivers of companies' success in public procurement: An empirical analysis. *The Journal of Technology Transfer, 45*(3), 664–693.

Canh, N., Liem, N., Thu, P., & Khuong, N. (2019). The Impact of innovation on the firm performance and corporate social responsibility of Vietnamese manufacturing firms. *MDPI Journal, 1*(6), 1–14.

Cooper, C., Booth, A., Varley-Campbell, J., Britten, N., & Garside, R. (2018). Defining the process to literature searching in systematic reviews: A literature review of guidance and supporting studies. *BMC Medical Research Methodology, 18*(1), 1–14.

Da Costa, B. B., & Da Motta, A. L. T. (2019). Key factors hindering sustainable procurement in the Brazilian public sector: A Delphi study. *International Journal of Sustainable Development and Planning, 14*(2), 152–171.

Durst, S., & Ståhle, P. (2014). Success factors of open innovation—A literature review. *International Journal of Business Research and Management, 4*(4), 1–21.

Eikelboom. M. E. (2018). Sustainable innovation in public procurement: The decisive role of the individual. *Journal of Public Procurement, 18*(3), 190–201.

Eriksson, P. E. (2017). Procurement strategies for enhancing exploration and exploitation in construction projects. *Journal of Financial Management of Property and Construction, 22*(2), 211–230.

Ghannajeh, A. M., AlShurideh, M., Zu'bi, M. F., Abuhamad, A., Rumman, G. A., Suifan, T., & Akhorshaideh, A. H. O. (2015). A qualitative analysis of product innovation in Jordan's pharmaceutical sector. *European Scientific Journal, 11*(4), 474–503.

Grandia, J., & Meehan, J. (2017). Public procurement as a policy tool: Using procurement to reach desired outcomes in society. *International Journal of Public Sector Management, 30*(4), 302–309.

Hameed, W., Basheer, M., Iqbal, J., Anwar, A., & Ahmad, H. (2018). Determinants of firm's open innovation performance and the role of R & D department: An empirical evidence from Malaysian SME's. *Journal of Global Entrepreneurship Research, 1*(1), 1–6.

Hsueh, L., Bretschneider, S., Stritch, J. M., & Darnall, N. (2020). Implementation of sustainable public procurement in local governments: A measurement approach. *International Journal of Public Sector Management, 33*(6/7), 697–712.

Ibem, E. O., Aduwo, E. B., Ayo-Vaughan, E. A., Uwakonye, U. O., & Owolabi, J. D. (2017). E-procurement use in the Nigerian building industry. *International Journal of Electronic Commerce Studies, 8*(2), 219–254.

Iddris, F., Awuah, G., & Gebrekidah, D. (2014). The role of innovation capability in achieving supply chain agility. *International Journal of Management and Computing Sciences, 4*(2), 104–112.

Islam, M., Turki, A., Murad, W., & Karim, A. (2017). Do sustainable procurement practices improve organizational performance? *Sustainability, 9*, 1–17.

Jin, S. H., & Choi, S. O. (2019). The effect of innovation capability on business performance: A focus on IT and business service companies. *Sustainability, 11*(19), 1–15.

Jing, S., Hu, K., Yan, J., Ping-Ho, Z., & Han, L. (2020). Investigating the effect of value stream mapping on procurement effectiveness: A case study. *Journal of Intelligent Manufacturing, 12*(1), 1–19.

Lee, K., Azmi, N., Hanaysha, J., Alshurideh, M., & Alzoubi, H. (2022a). The effect of digital supply chain on organizational performance: An empirical study in Malaysia manufacturing industry. *Uncertain Supply Chain Management, 10*(2), 1–16.

Lee, K., Ramiz, P., Hanaysha, J., Alzoubi, H., & Alshurideh, M. (2022b). Investigating the impact of benefits and challenges of IOT adoption on supply chain performance and organizational performance: An empirical study in Malaysia. *Uncertain Supply Chain Management, 10*(2), 1–14.

Lenderink, B., Halman, J. I., Boes, H., & Voordijk, H. (2020). A method to encourage and assess innovations in public tenders for infrastructure and construction projects. *Construction Innovation, 20*(2), 171–189.

Meehan, J., & Bryde, D. J. (2015). A field-level examination of the adoption of sustainable procurement in the social housing sector. *International Journal of Operations & Production Management, 35*(7), 982–1004.

Nawi, M. N. M., Roslan, S., Salleh, N. A., Zulhumadi, F., & Harun, A. N. (2016). The benefits and challenges of E-procurement implementation: A case study of Malaysian company. *International Journal of Economics and Financial Issues, 6*(7S), 329–332.

Nuseir, M. T., Aljumah, A., & Alshurideh, M. T. (2021). How the business intelligence in the new startup performance in UAE during COVID-19: The mediating role of innovativeness. *The Effect of Coronavirus Disease (COVID-19) on Business Intelligence, 334*, 63–79.

Nyantakyi, A. (2019). Preparing today for prosperity tomorrow: Using sustainable public procurement as a tool for development in Ghana. *Public Contract Law Journal, 48*(2), 377–396.

Obeidat, U., Obeidat, B., Alrowwad, A., Alshurideh, M., Masadeh, R., & Abuhashesh, M. (2021). The effect of intellectual capital on competitive advantage: The mediating role of innovation. *Management Science Letters, 11*(4), 1331–1344.

Odeh, R. B. M., Obeidat, B. Y., Jaradat, M. O., & Alshurideh, M. T. (2021). The transformational leadership role in achieving organizational resilience through adaptive cultures: The case of Dubai service sector. *International Journal of Productivity and Performance Management*. Vol. ahead-of-print No. ahead-of-print. https://doi.org/10.1108/IJPPM-02-2021-0093.

Parveen, S., Senin, A. A., & Arslan, U. (2015). Organization culture and open innovation: A quadruple helix open innovation model approach. *International Journal of Economics and Financial Issues, 5*(1), 335–342.

Perényi, A., & Losoncz, M. (2018). A systematic review of international entrepreneurship special issue articles. *Sustainability, 10*(10), 3476, 1–26.

Prier, E., Schwerin, E., & McCue, C. P. (2016). Implementation of sustainable public procurement practices and policies: A sorting framework. *Journal of Public Procurement, 16*(3), 312–346.

Rane, S. B., Narvel, Y. A. M., & Bhandarkar, B. M. (2019). Developing strategies to improve agility in the project procurement management (PPM) process. *Business Process Management Journal, 16*(3), 312–346.

Romodina, I., & Silin, M. (2016). Perspectives of introduction sustainable procurement in public procurement in Russia. *Oeconomia Copernicana, 7*(1), 35–48.

Sethibe, T., & Steyn, R. (2016). Innovation and organizational performance: A critical review of the instruments used to measure organizational performance. *The Southern African Journal of Entrepreneurship and Small Business Management, 8*(1), 1–12.

Shakhour, R., Obeidat, B., Jaradat, M., Alshurideh, M., Masa'deh, R. (2021). Agile-minded organizational excellence: Empirical investigation. *Academy of Strategic Management Journal, 20*(Special Issue 6), 1–25.

Shin, K., Yeo, Y., & Lee, J. D. (2019). Revitalizing the concept of public procurement for innovation (PPI) from a systemic perspective: Objectives, policy types, and impact mechanisms. *Systemic Practice and Action Research, 33*(2), 187–211.

Stanisławski, R. (2020). Open innovation as a value chain for small and medium-sized enterprises: Determinants of the use of open innovation. *MDPI Journal, 1*(6), 1–24.

Starzyńska, W. (2019). Innovative procurement in Poland in the light of the report on the evaluation of public procurement functioning after the amendment of the public procurement law of 2016. *Folia Oeconomica Stetinensia, 19*(2), 149–159.

Storsjö, I. T., & Kachali, H. (2017). Public procurement for innovation and civil preparedness: A policy-practice gap. *International Journal of Public Sector Management., 30*(4), 342–356.

Stritch, J. M., Bretschneider, S., Darnall, N., Hsueh, L., & Chen, Y. (2020). Sustainability policy objectives, Centralized decision making, and efficiency in public procurement processes in US local governments. *Sustainability, 12*(17), 1–17.

Vluggen, R., Gelderman, C. J., Semeijn, J., & Van Pelt, M. (2019). Sustainable public procurement: External forces and accountability. *Sustainability, 11*(20), 1–16.

Yevu, S. K., & Yu, A. T. W. (2019). The ecosystem of drivers for electronic procurement adoption for construction project procurement. *Engineering, Construction and Architectural Management., 27*(2), 411–440.

The Effect of High Commitment Management Requirements in Achieving Strategic Entrepreneurship Through the Perceived Organizational Support in Iraqi Ministry of Health

Waleed Radeef Al-Janabi and Barween Al Kurdi

Abstract This study aimed at identifying the effect of the requirements of the administration of high commitment and its dimensions (normative commitment, career enrichment, job security) on achieving strategic entrepreneurship and its dimensions (entrepreneurial leadership, entrepreneurial culture, risk taking) through perceived organizational support in the Iraqi Ministry of Health. To achieve the objectives of the study, a questionnaire was designed and distributed to a sample of (181) of employees working in the Anbar Health Department / the Iraqi Ministry of Health. The descriptive approach was used to describe the analytical procedures, data collection, classification, processing, and analysis to find out the results, while the heuristic approach was used to analyse, interpret and draw conclusions based on the sample of the study. The results of the study showed that there is a statistically significant impact of high commitment and its dimensions (normative commitment, career enrichment, job security) on strategic entrepreneurship in its dimensions (entrepreneurial leadership, entrepreneurial culture, risk taking) through perceived organizational support. The study recommended to increase the level of high management commitment to develop organisational committeemen amongst the employees as this will lead to positive working experience.

Keywords Commitment management requirements · High commitment · Strategic entrepreneurship · Perceived organizational support

W. R. Al-Janabi
Iraqi Ministry of Health/Anbar Health Department, Baghdad, Iraq

B. Al Kurdi (✉)
Faculty of Economics and Administrative Sciences, Department of Marketing, The Hashemite University, Zarqa, Jordan
e-mail: barween@hu.edu.jo

M. Alshurideh et al. (eds.), *The Effect of Information Technology on Business and Marketing Intelligence Systems*, Studies in Computational Intelligence 1056,
https://doi.org/10.1007/978-3-031-12382-5_88

1 Introduction

Organizations are undergoing cultural, economic, environmental and political changes and challenges, as well as rapid technological changes in the light of the globalization of modern times. Modern technological developments are a motivation for increasing competition among organizations, especially hospitals, in achieving competitive advantage that contributes to increasing health awareness, which maintains growth, continuity and achievement of the desired goals.

The present rapid changes, technological and environmental developments have intensified competition between organizations, leaving many organizations exposed to real risks that have almost ended their life cycle or that of other organizations due to poor risk foreseeing and poor human resource management practices represented by poor polarization, selection, recruitment, incentives and other such practices.

The exposure of many organizations to a wave of rapid changes in various technological, economic and social fields has led them to the need to awaken the spirit of the strategic entrepreneurship within it. The public administration has had few contributions to entrepreneurial activity as an input to the development of organizations, at least if the individual identifies entrepreneurial activity as more than just an investment, as strategic entrepreneurship is linked to the overall strategy of the organization and you may see, for example, that it opens up new branches working on programs and systems with new modes such as e-government (Ghazi & Parvaneh, 2014).

High commitment management practices and perceived organizational support have a clear and important impact on the life of every individual working in the organization through their participation in the organization's policy-making in order to achieve goals and their participation in decision-making, Providing them with opportunities, training and developing them and distributing rewards fairly among them, all this improves performance and excellence in this performance (Said & Al-Nasrawi, 2016).

Hence this study aims at identifying the effect of high commitment management requirements in achieving strategic entrepreneurship through perceived organizational support as an intermediary in health department of Anbar/Iraqi Ministry of Health.

This research also comes as a response to several previous studies (Al-Nuaimi & Hamid, 2015; Said & Al-Nasrawi, 2016; Howard, 2013; Park et al., 2019) which recommended studying the requirements of high commitment management in achieving strategic entrepreneurship.

2 Problem of the Study and the Study Questions

The development of management thought has seen a lot of discussions generated by the requirements and conditions of the work environment and its challenges and dynamic change in the internal and external environment and the opportunities and threats it faces. Therefore, a constant change became a necessary characteristic attach to them. The problem of the study lies in the poor employment of high commitment management requirements. In order to achieve strategic entrepreneurship within the organizations, including Anbar Health Service, there are many competent human resources, but they lack the systematic support of the level of awareness on the part of the administrations. This compels these departments to adopt targeted practices in order to invest and develop these human resources to enable them to properly think about opportunities and commit themselves to confronting threats in a way that will lead to the strategic entrepreneurship of their organizations.

Hence, the problem of the study is reflected in the answer to the following questions:

Do the requirements of managing high-level commitment to its dimensions (normative commitment, job enrichment, job security) have any effect on strategic entrepreneurship in its dimensions (entrepreneurial leadership, entrepreneurial culture, taking risks) in Anbar Health Directorate/Iraqi Ministry of Health?

It gives rise to the following sub-questions:

The first Sub-question: Do the requirements of the high commitment management in its dimensions (normative commitment, job enrichment, job security) have an effect on the entrepreneurial entrepreneurship of the Anbar Health Directorate/Iraqi Ministry of Health?

The second sub-question: Do the requirements of the management of high commitment to its dimensions (standard commitment, job enrichment, job security) have an effect on the entrepreneurial culture at Anbar Health Directorate/Iraqi Ministry of Health?

The third sub-question: Is there an effect of risk taking on the requirements of high commitment management (standard commitment, job enrichment, job security) in Anbar Health Directorate/Iraqi Ministry of Health?

The second main question: Is there an effect of the requirements of high commitment management in their dimensions (normative commitment, job enrichment, job security) on the perceived organizational support in Anbar Health Directorate/Iraqi Ministry of Health?

The third main question: Is there an effect of perceived organizational support on strategic entrepreneurship in its dimensions (entrepreneurial leadership, entrepreneurial culture, risk taking) in Anbar Health Directorate/Iraqi Ministry of Health?

The fourth main question: Do the requirements of high commitment management (normative commitment, job enrichment, job security) have an effect on the strategic entrepreneurship (entrepreneurial leadership, entrepreneurial culture, risk taking)?

3 Significance of the Study

This study contributes in its attempt to fill the deficit of the Arab Research Group on the concept of high commitment management requirements in achieving strategic entrepreneurship, where most of the previous studies focused on studying the relationship between independent and dependent variable only. Whereas this study takes a third variable, which is the perceived organizational support medium variable which will add a cognitive accumulation to the previous studies and scientific libraries. The importance of this study is also reflected in the recommendations that will be proposed to the Anbar Health Directorate at the Iraqi Ministry of Health, which illustrates the effect of the requirements of high commitment management via focusing on their dimensions (normative commitment, job enrichment, job security), develop the inspection unit's understanding of the strategic entrepreneurship of the workers at department of health, and how to grow and develop it in the way that achieves the continuity, efficiency and effectiveness of the organization via the organizational support to all parties.

4 Literature Review

This aspect includes a presentation of the theoretical framework and literature related to the subject and variables of the study, through a review of Arabic and English previous studies related to the current study.

4.1 High Commitment Management

The concept of a high level of commitment in its various forms is one of the management practices that has received increasing attention from researchers. This interest has led to its widespread perception among managers and academics that a high level commitment represents a crucial element to link the organization with the individuals working in it. Harem (2004) indicates that the commitment of individuals to their organizations is an important factor in ensuring the success of those organizations.

Although more and more researchers are interested in the topic of high commitment, there is no agreement among them on a specific definition of this concept due to the multiplicity of perspectives and angles that researchers have looked at and the difficulty in defining this behavior input. Al-Sameeh (2010) defines it as an emotional connection between the organization and its values and goals. Al-Shayab and Abu Hammour (2014) focus on the psychological perspective of commitment and define it as the psychological bond that binds the individual to the organization, which leads him to integrate into work and to embrace the values of the organization. Qassimi (2011) defines it as "a strong belief and acceptance by individuals of the organization's goals and values, and their desire to make a greater contribution to it, with the strong desire to continue its membership." Flayiah and Abdul Majid (2014) add that "The commitment is the positive feeling generated by the employee towards his organization, his commitment to it, his loyalty to it, his conformity with its values and objectives, his commitment to its survival through exerting efforts and favoring it from other organizations, and be proud of its effects, which enhances its success." According to the perspective of the authors of this study, the commitment is considered a mutual investment between the individual and the organization via the continuation of the contractual relationship between them, where the individual has the desire to give more in order to contribute to the success and continuation of the organization.

The previous studies have discussed three dimensions of high commitment: Normative commitment, job enrichment and job security (Howard, 2013; Al-Gharbawi, 2014; Al-Nuaimi & Hamid, 2015; Said & Al-Nasrawi, 2016; Park et al., 2019).

Concerning the concept of normative commitment, it refers to the individual's feeling of being committed to remain in the organization because of the pressures of others. Individuals with a strong normative commitment take into account what others will say if they leave the organizations in which they work, so these individuals don't want to make a bad impression on their colleagues by leaving their organizations, due to their literary obligation even if it is at the expense of themselves (Park et al., 2019).

On the other hand, career enrichment is what makes working individuals more responsible and independent in decision-making, and this, in turn, strengthens organizational commitment to work. Career enrichment is achieved through the workers' sense that the benefits of the organization also benefit them (Howard, 2013) because this feeling would strengthen their commitment to the organization. Some organizations are trying to do this directly through incentive schemes, particularly profit-sharing programs. Such plans and programs, if managed in a fair manner, will play an effective role in supporting individuals' organizational loyalty.

As far as the third dimension of high commitment is job security. This concept refers to the power of one's desire to remain in an organization because he believes that leaving would cost him dearly. The longer an individual lasts in the organization, the more likely an individual will lose much of what he or she has invested in overtime such as: (Pension plans, intimate friendships for some individuals). Many individuals are unwilling to sacrifice such things, and they are said to have a high degree of continuing loyalty (Howard, 2013).

4.2 The Strategic Entrepreneurship

There have been a number of definitions related to the concept and nature of entrepreneurship in recent times. Some definitions refer to leadership as the ability and desire to organize and manage related businesses, in addition to including some definitions on new concepts, innovation, innovation, and risk (Agostini & Nosella,

2017). Entrepreneurship was defined as the process of creating something a new and valuable thing by devoting the time and necessary effort by assuming accompanying physical, psychological, and social risks, and reaping the resulting financial returns (Oviat & McDougall, 2005). Mandel and Barnes (2014) defined it as "activities involving new product or processes, or entering new markets, or creating new enterprises". Seitovirta (2011) defined it as "the process of creating a new valuable thing via allocating the time, effort and money needed for the project, bearing the accompanying risks, and receiving the rewards for what happens when wealth is accumulated."

All in all, there is a consensus among researchers on the importance of strategic entrepreneurship and its role in developing and supporting the economies of countries, especially after the major shift in the capital economy to the knowledge economy, in the hope of achieving the economic stability that all peoples seek to ensure for themselves economic and political security (Sultan, 2016). The technological and scientific development in modern business organizations has helped to develop many entrepreneurial organizations in various sectors of business, especially with the increase of global competition and the emergence of other factors to make organizations more pioneering, and the exploitation of investment opportunities in the market through innovation and innovation, as well as the entrepreneurial processes that have become part of the strategic management of the work of these organizations (Plummer, 2016). Workers entrepreneurship is used to demonstrate innovators and inventors in all fields via showing their achievements and successes despite the lack of resources, capabilities and difficulties they have faced. In light of the ongoing and increasing economic globalization, the idea of entrepreneurship on the subject has become the focus of attention of leaders and managers, not only in their understanding of the subject, but also in the ability to develop strategies and appropriate planning for the success of organizations and the persistence of institutional excellence. Accordingly, entrepreneurship is one of the competitive advantages of today's organizations. In order for the organization to succeed, it must develop a vision that encourages growth through entrepreneurship (Ismail, 2010). Based on the above, it can be said that leadership plays a pivotal role in achieving success. It pursues the task of defining and articulating the goals and objectives of each phase of the implementation of the organization's strategy, which identifies the necessary resources, manages them and distributes them to the different business units and job areas of the organization.

4.3 Perceived Organizational Support

Business organizations are currently facing a complex and rapid set of global and local challenges. It is imperative that organizations address these challenges in all ways. To achieve this, the organization, represented by its management, has to undertake many planned and targeted activities that provide financial, psychological and social support to workers, reflecting its concern for human resources and its interest in meeting their needs and motivations (Anomneze et al., 2016). Hence the importance of the concept of conscious organizational support, which refers to the level of moral and material support provided by the management to employees, through which the employees feel that the management appreciates and appreciates their efforts. In return, the Organization obtains high performance and commitment (Mansour & Ashour, 2016).The theory of organizational support (perceived organizational support) derives from the theory of social exchange and exchange values. According to the theory of social exchange, individuals who benefit from the actions of one party or another feel a commitment towards reciprocity in the form of certain patterns and responses (Drover & Ariel, 2015). A high perception of organizational support can, therefore, create a sense of commitment on the part of the individual that makes him/her feel that he/she needs not only to belong to the organization but also to feel that he/she has a commitment to reinstate the organization's affiliation by engaging in behaviors that support organizational goals.

4.4 Previous Related Studies

Researchers have drawn from many Arab and foreign references to formulate the study's assumptions. For example, Al-Nuaimi and Hameed study (2015) aims at clarifying the relationship between high commitment management activities and strategic entrepreneurship in Iraqi constructions activity. The study has found out a number of results, the most important ones are the effective relationship and a link between high commitment management and strategic entrepreneurship indicating that providing high commitment management requirements contributes to building strategic entrepreneurship in studied companies. In another study, Said and Al-Nasrawi (2016) aims at clarifying the nature of the relationship and the impact between the management of high commitment and organizational excellence et al.-Furat General Company for Chemical Industries. One of the most important results of this study is that the management of high commitment has a positive effect on the organizational excellence. On the other hand, Mansour and Ashour (2016) tested the mediating role of perceived organizational support between the High Containment Department and the company's organizational commitment. The results of this study indicated a positive and moral impact between global containment management and organizational commitment through perceived organizational support. In contrast, Khudayr and Ahmed (2017) conducted a study to examine the relationship between the perceived organizational support and professional compatibility and their effect on reducing organizational satire found that cognitive organizational support and professional consensus combined had an impact on reducing organizational satire. As for Al-Hawajira study (2018), it aimed at measuring the impact of the organizational entrepreneurship on the strategic success through business intelligence capabilities in universities and found that organizational leadership has an impact on strategic success, and that organizational leadership has an impact on business intelligence capabilities. Al-Tahan (2018) also conducted a research to understand the impact of future vision on achieving the strategic entrepreneurship by organizations in the Egyptian telecommunications sector and found that it has a positive impact on looking ahead in achieving the strategic entrepreneurship by the organizations.

On the other hand, Farndale et al. (2011) study aimed at knowing the perceptions of workers on managing commitment, high performance, roles of justice and trust in the organization. The study community and its sample of workers consisted of four organizations in the United Kingdom. The most important finding of the study was that there was a correlation between the expertise of highly committed, high-performing, fair and trusting employees that influenced human resources management practices. Kraus and his colleagues' study (2011) aimed at organizing and synthesizing the current scientific work on the strategic entrepreneurship using the formation approach, finding that strategic entrepreneurship contributed to understanding how companies were creating a new enterprise that could be moved to the greater level of corporate strategies, whether small or medium. Howard (2013) in his study discussed the relationship between the high commitment to human resources management and the employee's well-being in three large organizations in South Africa and found a strong correlation between the human resources management content and the employee's well-being. The study also reached the most important recommendations on the need to have positive effects on organizations, human resources management activities and employees in order to achieve high commitment in organizations. The study of Ghazi and Parvaneh (2014) examined the different dimensions of strategic entrepreneurship as well as understanding the dimensions of strategic entrepreneurship from the theory to practice through the employment of many startup companies in Iran. The most crucial findings of this study are the emphasis on business leadership and innovation, capital mobilization, entrepreneurship and leadership management, the simultaneous attention to the strategic entrepreneurship, management and growth, and finally the profitability.

4.5 *What Distinguishes the Current Study From Previous Studies?*

Several issues exist that are important to highlight to demonstrate the merit of this study over previous studies. First, the objective of the study, it aims at measuring the impact of high commitment management requirements in achieving strategic entrepreneurship through the perceived organizational support in Anbar Health Directorate/Iraqi Ministry of Health. Most of the previous studies addressed only two variables, such as the study of (Al-Gharbawi, 2014), the study of (Al-Nuaimi & Hamid, 2015), the study of (Said & Al-Nasrawi, 2016), and the study of (Park et al., 2019). This study addressed a third variable, the organizational support that is aware of its impact on the relationship between the dependent and independent variable. Second, study variables: The study looked at the dimensions of their independent variable, the requirements of managing high commitment (normative commitment, job enrichment, job security). Their dependent variable was strategic entrepreneurship, represented by dimensions (entrepreneurial leadership, entrepreneurial culture, risk taking) through their intermediate variants and perceived organizational support. Third: Study environment: Most of the previous studies that were examined varied in their environment, and the studies that were conducted were applied to address them. However, most of these studies did not take into account the Iraqi environment. The importance of this study results its attempts to add an in-depth understanding of the concepts of managing high-level commitment and strategic entrepreneurship in an environment that was not previously addressed in the researches.

5 Hypotheses of the Study

The following hypotheses have been formulated in the preceding paragraphs in terms of the object and importance of the study, and as an indication from previous scientific studies.

The first main hypothesis H01: There is no statistically significant impact at the level ($\alpha \leq 0.05$) of the requirements for managing the global commitment in its combined dimensions (normative commitment, job enrichment, job security) on strategic entrepreneurship in its combined dimensions (leadership, leadership culture, risk taking) in the Anbar Health Directorate/Iraqi Ministry of Health.

A number of sub-hypotheses emerge from this main hypothesis:

The first sub-hypothesis: H01-1: There is no statistically significant effect at significance level ($\alpha \leq 0.05$) of the requirements of the high management Commitment in its dimensions (normative commitment, job enrichment, job security) are placed on the entrepreneurial leadership of Anbar Health Directorate/Iraqi Ministry of Health.

The second sub-hypothesis: H01-2: There is no statistically significant effect at significance level ($\alpha \leq 0.05$) of the requirements of the high management commitment in its dimensions (normative commitment, job enrichment, job security), over the entrepreneurial culture in Anbar Health Directorate/Iraqi Ministry of Health.

The third sub-hypothesis: H01-3 There is no statistically significant impact at the level of the ($\alpha \leq 0.05$) of the requirements of the high management commitment in its dimensions (normative commitment, job enrichment, job security) on the risk taking in the Anbar Health Directorate/Iraqi Ministry of Health.

The second main hypothesis: H02: There is no statistically significant effect at significance level ($\alpha \leq 0.05$) of the requirements of the management of high commitment in its combined dimensions (normative commitment, job enrichment, job security) depend on the perceived organizational support at Anbar Health Directorate/Iraqi Ministry of Health.

The third main hypothesis: H03: There is no statistically significant impact at the significance level ($\alpha \leq 0.05$) of the perceived organizational support on the strategic entrepreneurship in all its dimensions (entrepreneurial leadership, entrepreneurial culture, risk taking) in Anbar Health Directorate/Iraqi Ministry of Health.

The fourth main hypothesis: H04: There is no statistically significant effect at significance level ($\alpha \leq 0.05$) for the requirements of high managing commitment in its combined dimensions (normative commitment, job enrichment, job security), focusing on strategic entrepreneurship in its combined dimensions (entrepreneurial leadership, entrepreneurial culture, risk taking) through the perceived organizational support at the Anbar Health Directorate /Iraqi Ministry of Health.

6 Model of the Study

To achieve the objective of the study, a special model was developed that was shown in Fig. 1 as the following (Table 1).

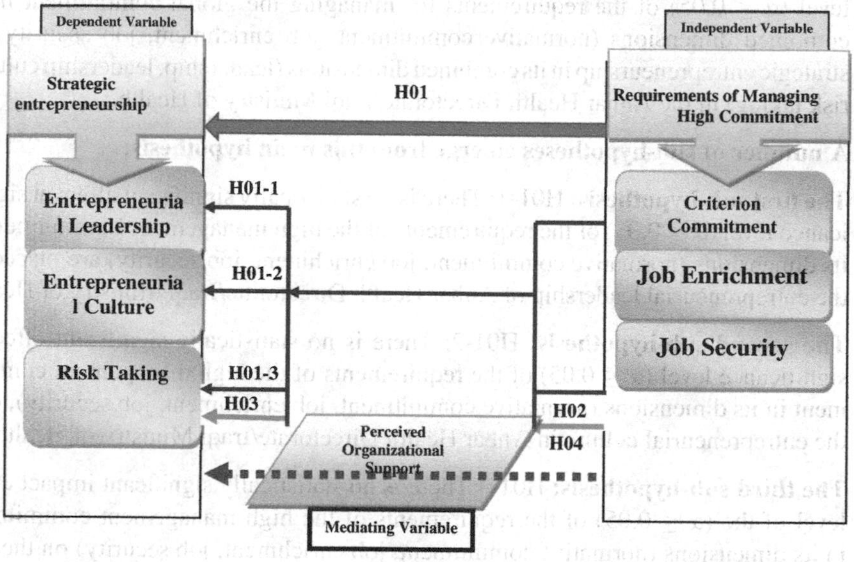

Fig. 1 Model of the study

Table 1 References and studies that support the model of the study

Variable	Studies depended on
Requirements of managing high commitment	Al-Gharbawi (2014), Al-Nuaimi and Hamid (2015), Said and Al-Nasrawi (2016), Howard (2013), Park et al. (2019)
Perceived organizational support	Al-Shanty (2015), Mansour and Ashour (2016), Khudayr and Ahmad (2017), Worku (2015), Gigliotti et al. (2019)
Entrepreneurial leadership	Al-Nuaimi and Hamid (2015), Al-Hawajira (2018), Al-Tahan (2018), Kraus et al. (2011), Ghazi and Parvaneh (2014)

7 Methodology

7.1 *Method of the Study*

To achieve the objective of the study which used the descriptive analytical approach, two approaches were used: The first approach is theoretical, and I use desk surveys to look at existing books, references, resources, and studies relevant to the subject of the study, to construct a theoretical background for the study. Concerning the second field approach, I used the study tool (the questionnaire), which was developed to collect data from the research category and to work on analyzing them statistically to achieve the study's objective.

7.2 *Population and Sample of the Study*

The population of the study consisted of all the approximately 347 employees working at various administrative levels in the Anbar Health Directorate/Iraqi Ministry of Health. Based on Sekaran and Bougie (2012), a random sample of employees was collected through the distribution of (181) questionnaires, of which (176) were retrieved, and (5) were excluded because the respondents did not complete the answers that the number of questionnaires subject to analysis was to (171), rated 94% of the total distributed questionnaires.

7.3 *Instrument of the Study*

In order to achieve the study's objectives, its questionnaire was developed as a study tool, and it was developed in a way that corresponds to the study variables by looking at previous studies related to the study topic. The questionnaire consisted of three

Table 2 Likert pentagonal graded scale

Mark	1	2	3	4	5
Agreement level	Strongly agree	Agree	Moderately agree	Disagree	Strongly disagree

parts: The first part aims at collecting demographic data for study sample individuals (age, education qualification, number of years of experience, job title), and the second part is related to the independent variable (requirements for managing high commitment) in its dimensions (normative commitment, job enrichment, job security). In order to develop this part, previous studies were made use of via using Al-Gharbawi (2014), Al-Nuaimi and Hamid (2015), Said and Al-Nasrawi (2016), Howard (2013), Park et al. (2019). The third part is the dependent variable (strategic entrepreneurship) in its dimensions (entrepreneurial leadership, entrepreneurial culture, risk taking) through the studies of Kraus et al. (2011), Ghazi and Parvaneh (2014), Al-Nuaimi and Hamid (2015), Al-Hawajira (2018) and Al-Tahan (2018). The intermediate variant (perceived organizational support), via using the Al-Shanty (2015), Worku (2015), Mansour and Ashour (2016), Khudayr and Ahmed (2017) and Gigliotti et al. (2019) studies. The study tool was developed via using the exploratory factor analysis and the confirmatory factor analysis.

A progressive Likert scale was adopted in the resolution to give more flexibility to the sample members in the selection as shown in Table 2.

Table 3 shows the related limits adopted by this study, when commenting on the arithmetic mean of the variables in the study model and determining the degree of approval, the researcher has determined three levels (high, medium, low) based on the following equation.

Length of period = (upper limit of alternative−lower limit of alternative)/number of levels.

(5–1)/3 = 4/3 = 1.33 so the levels are as follows:

A low approval score of 1-less than 2.34.

An average approval score of 2.34-less than 3.68.

A high approval score of 3.68–5.

Table 3 shows the measure in determining the level of fit for the arithmetic mean to be used when commenting on arithmetic means.

Table 3 Likert scale treatment

Mean	Degree evaluation
1. Lower than 2.34	Low
2. 34- Lower than 3.68	Middle
3. 68–5	High

Table 4 Reliability coefficient of the study instrument

	Cronbach's alpha
Requirements of managing high commitment	0.773
Perceived organizational support	1000
Entrepreneurial leadership	0.841

7.4 Validity and Reliability of the Study Istrument

The study instrument was validated by presenting it in its initial form to a group of academic arbitrators with expertise and competence to verify the veracity of the contents of its paragraphs and to ascertain their validity to measure what should be measured and to ascertain the clarity of its paragraphs and the verbal and moral validity of their words. Their observations on the study instrument were taken and paragraphs were modified, deleted and added until the tool was finalized. Cronbach's alpha was also used to check internal consistency. It measures the consistency of the researchers' answers to the questions in the full scale, and its high value indicates a persistence between 0 and 1, acceptable at (0.60) and above. According to other Vogt study, the value of Cronbach alpha is acceptable if it is greater than (0.70) (Table 4).

8 Study Findings and Discussion

8.1 Description of Study Sample Properties

This part of the study aims at showing the frequency and percentages of demographic variables for respondents regarding the first section of resolution, in terms of age, gender, educational level, and number of years of experience. Table 5 shows the responses of the study sample.

Table 5 showed that the majority of study sample members were in the age group (35-under-45) at 63.2% (n = 108), followed by the sample members in the age group (25-under-35) at 18.1% (n = 31), and then the sample members were in the age group (45 years and older) at 16.4% (n = 28). Finally, the sample population was 2.3% (n = 4) in the age group (under 25 years). It also showed that (77.8) of the study sample were male and (22.2) of the study sample were female. This finding reflects the fact that the proportion of male health workers is higher than that of female workers. The data also show that the majority of the study sample had a bachelor's degree of 69.5% (n = 119), followed by those with a diploma score of 12.3% (n = 21). Then those with a master's degree of 11.7% (n = 20) and finally those with a doctorate of 2.9% (n = 5). This is an indication of the interest of the research sector in recruiting those with scientific qualifications. As for the number of years of experience, the data in Table 5 indicates that the majority of the sample members were 66.1% (n

Table 5 Frequencies and percentages of sample responses of the study based on their demographic variables

Variable	Group	Frequency	Percentage
Age	Lower than 25 Years	4	2.3
	25 Years–Lower than 35 Years	31	18.1
	35 Years–Lower than 45 Years	108	63.2
	45 years and above	28	16.4
	Total	171	100.0
Sex	Male	133	77.8
	Female	38	22.2
	Total	171	100.0
Academic qualification	Diploma or below	6	3.5
	Diploma	21	12.3
	Bachelor	119	69.5
	Master	20	11.7
	Doctorate	5	2.9
	Total	171	100.0
Experience period	Lower than 5 Years	3	1.8
	5 Years–Lower than 10 Years	14	8.2
	10 Years–Lower than 15 Years	41	24.0
	15 Years of above	113	66.1
	Total	171	100.0

= 113), whose years of experience ranged from (15 years and older), followed by those whose years of experience were 10 and younger than 15 years old by 24% (n = 41), those whose years of experience (5-less than 10 years) by 8.2% (n = 14), and finally those whose years of experience were less than 5 years by 1.8% (n = 3). This is an indication that researchers have reached the appropriate categories to answer the study tool's paragraphs.

8.2 *Description of Study Variables Results*

The mission of this part of the study is to describe the basic study dimensions according to the estimates of the study sample members, through their answers to the elements of questionnaire with their sub-variables.

First: Describing the high commitment management variable in its different dimensions: Normative commitment, job enrichment and job security

Table 6 shows that the arithmetic averages of responses of sample members to high compliance management requirements between (4.35 and 4.00) with a high rating for all dimensions ranked first after "standard commitment" with an arithmetic mean (4.35), second after "job enrichment" with an arithmetic mean (4.13), third after "job security" with an average of 4.00), and third after "computational average" with high compliance management requirements (4.16) with a high rating.

Second: Describing of the mediating variable: the Perceived organizational support

Arithmetic means and standard deviations were extracted for the responses of the sample individuals to the dimensions of the perceived organizational support, Table 7 shows this.

Table 7 shows that the arithmetic means of the workers' sample responses around the perceived organizational support variant ranged from (4.21–2.35), hence we note that the sample of workers gave consent to all paragraphs of the perceived organizational support variant. Paragraph 22, which reads: "management cares about

Table 6 Comparison of means, standard deviations, and level of significance of the dimensions of the requirements of managing high commitment

No	Item	Mean	Rank	Level of significance
1	Criterion commitment	4.372	1	High
2	Job enrichment	4.13	2	High
3	Job security	4.00	3	High
Requirements of managing high commitment		4.16	–	High

Table 7 Means, standard deviations, and of the sample responses of the perceived organizational support

Item no	Mean	Standard deviation	Rank	Level of significance
22	4.21	0.836	1	High
23	4.04	0.777	2	High
24	3.97	0.928	3	High
25	3.92	0.926	4	High
26	3.87	0.889	7	High
27	3.87	0.864	8	High
28	3.75	0.987	9	High
29	2.35	1.26	10	High
30	3.91	0.941	5	High
31	3.89	0.936	6	High

Table 8 Means, standard deviations, and significance level of strategic entrepreneurship dimensions

No	Item	Mean	Rank	Level of significance
1	Entrepreneurial leadership	3.78	3	High
2	Entrepreneurial culture	4.03	1	High
3	Risk taking	3.97	2	High
Requirements of strategic entrepreneurship		3.92	–	High

meeting my job needs and desires", gave the highest of the paragraphs on average (4.21) and with a standard deviation (0.836), and at a high level of importance, while paragraph (29), which states that: "I'm looking for another organization to work on seriously," at a minimum, with an average score of 2.35, a standard deviation of 1.26, and of average relative importance.

In general, the total sum of the means of this variable was 3.78, indicating that the relative importance of the perceived organizational support was high, with a standard deviation (539), which reflects a lack of dispersion and homogeneity of study sample responses to the study paragraphs.

Third: Description of the dependent variable: Strategic entrepreneurship in its various dimensions: entrepreneurship, entrepreneurial culture, and taking risks

Arithmetic means and standard deviations were extracted for the respondents' answers to the dimensions of strategic entrepreneurship, Table 8 shows this. The table shows that the arithmetic averages of the respondents' responses to strategic entrepreneurship between (4.03 and 3.78) with high score for all dimensions came first after "entrepreneurial culture" with an arithmetic average (4.03), second after "risk taking" with an arithmetic mean (3.97), third after "entrepreneurial culture" with an arithmetic mean (4.03), and the arithmetic average of strategic entrepreneurship (3.92) with high score.

8.3 *Testing the Hypotheses of the Study*

This part of the study pertains to the testing of the main study hypotheses and their sub-hypotheses, in order to arrive at the presentation of results and clarify the behavior of the relationships between the study variables.

Results of the main hypothesis (H01) test

There is no statistically significant impact at the level (a $\leq$ 0.05) of the requirements for managing the global commitment in its combined dimensions (normative commitment, job enrichment, job security) on strategic entrepreneurship in its combined dimensions (leadership, leadership culture, risk taking) in the Anbar Health Directorate/Iraqi Ministry of Health.

Table 9 shows the statistical results of testing the first main hypothesis (direct effect) as follows.

Via extrapolating the results in Table 9, we note that the value of the marker (R2) is 658, which means that the high-commitment management requirements variant accounts for 65% of the variation in the dependent variable (strategic entrepreneurship). The results indicate a statistically significant effect of the high-commitment management requirements variant in strategic entrepreneurship, with beta (811.) and t (25.411) being less than 0.005.

Result of testing the first sub- Hypothesis (H01-1)

Table 10 shows the result of the test for the first sub-hypothesis from the first major hypothesis that there is a statistically significant effect of the requirements of the Department of High Commitment on leadership, with a beta value (0.757) and a t value (15.186) with a significance level (0.00) less than (0.05).

Result for testing the second sub-hypothesis (H01-2)

Table 11 Result of the test of the second sub-hypothesis from the first main hypothesis, showing statistically significant effect of the requirements of the management of high Commitment on entrepreneurial confidence, with beta (B) with value of (640.) and (t) with value of (12.220) with a significance level of (0.00) less than (0.05).

Table 9 Results of (SmartPLS) of measuring the effect of the requirements of managing high commitment on strategic entrepreneurship

R-squared value of the model	Beta value	Calculated T- value	Level of significance
0.658	0.811	25.411	0.000

* Statistically significant at ($\alpha \leq 0.05$)

Table 10 Results of (SmartPLS) of measuring the effect of the requirements of managing high commitment on entrepreneurial leadership

Beta value	Calculated T value	Level of significance
0.757	15.186	0.00

* Statistically significant at ($\alpha \leq 0.05$)

Table 11 Results of (SmartPLS) of measuring the effect of the requirements of managing high commitment on entrepreneurial confidence

Beta value	Calculated T value	Level of significance
640	12.220	0.00

* Statistically significant at ($\alpha \leq 0.05$)

Table 12 Results of (SmartPLS) of measuring the effect of the requirements of managing high commitment on risk taking

Beta value	Calculated T value	Level of significance
759	15.186	0.00

* Statistically significant at ($\alpha \leq 0.05$)

Result of testing the third sub-hypothesis (H01-3)

The result of the third sub-test derived from the first major hypothesis shows a significant effect of the requirements of the High commitment Management on the risk taking, with a value of beta (B) of (759) and a value of (t) of (15.186) with a significance level of (0.000) less than (0.05) (Table 12).

Results of testing the second main hypothesis (H02)

The second main hypothesis was tested which states that there is no statistically significant effect at a significant level of (a ≤ 0.05) for the management requirements of high commitment in their combined dimensions (normative commitment, functional enrichment, job security) on the perceived organizational support in Anbar Health Directorate/Iraqi Ministry of Health.

Via extrapolating the results in Table 13, we note that the R2 coefficient is (563). (This means that the high commitment management requirements variant accounts for 56.3% of the variation in the median variance and perceived organizational support.

A statistically significant effect of the requirement management variant for high commitment management is evident in perceived organizational support, with a value of beta (B) of (730.) and a value of (t) of (18.805) with a significance level of (0.000) less than (0.05).

Table 13 Results of (SmartPLS) of measuring the effect of the requirements of managing high commitment on the perceived organizational support

R-squared value of the model	Beta value	Calculated T-value	Level of significance
0563	0.0730	18.805	0.000

* Statistically significant at ($\alpha \leq 0.05$)

Table 14 Results of (SmartPLS) of measuring the effect of the perceived organizational support on the strategic entrepreneurship

R-squared value of the model	Beta value	Calculated T- value	Level of significance
0.660	0.179	2.263	0.024

* Statistically significant at ($\alpha \leq 0.05$)

Results of testing the third main hypothesis (H03)

The third main hypothesis was tested, showing that "there is no statistically significant impact at the significance level $\alpha \leq 0.05$) of organizational support recognized for strategic entrepreneurship in all its dimensions (leadership, leadership, leadership culture, risk taking) in the Anbar Health Department/Iraqi Ministry of Health".

Via extrapolating the results in Table 14, we note that the value of the marker (R2) is (0.660), which means that the intermediate variant (perceived organizational support) accounts for 66% of the variance in the dependent variable (strategic entrepreneurship). A statistically significant effect show a statistically significant effect of the perceived variant, with a value of beta (B) of (179.) and a value of (t) of (2.263) with a significance level of (0.024) less than (0.05).

Result of testing the fourth main hypothesis (H04)

The fourth key hypothesis that there is no statistically significant effect was tested at a significance level ($\alpha \leq 0.05$) for the requirements of managing the high commitment in its combined dimensions (normative commitment, job enrichment, job security), on the strategic entrepreneurship in its combined dimensions (entrepreneurial leadership, entrepreneurial culture, risk taking) via the perceived organizational support of Anbar Health Directorate/Iraqi Ministry of Health.

There is a statistically significant effect of the perceived organizational Support variable, which is recognized in the relationship between the requirements of high commitment management and the strategic entrepreneurship in both courses:

First direct course: The direct course value of beta(B) is 672 and the direct course value of (t) is 9.748 with a significance level of (0.000) less than (0.05).

The second indirect course: The indication of beta values (B) in the indirect course are respectively (0.730) and (179), and (t) values, which were (16.805) and (2.263), with a significance level (0.000) for both courses, less than (0.05). These values indicate that the medium resulting from the perceived organizational variable is considered a partial medium, as the results demonstrated a direct effect of the requirements of the high commitment management without the intermediate variant. However, the effect is considered as the median variable, since the value of the marker (R2) before the intervention of the mediator (perceived organizational support) was (0.658) With regard to the first hypothesis, after the entry of the mediator, this percentage rose slightly by (0.02), which means that the variant of requirements for high commitment management in its combined dimensions refers to a value (66.0%) of the variation in the dependent variable (strategic entrepreneurship) (Table 15).

Table 15 Results of track analysis (Smart PLS) to measure the effect of the requirements of managing high commitment in their combined dimensions on the strategic entrepreneurship via perceived organizational support as a mediating variable

Direction	Beta value for indirect direction	Beta value for direct direction	The calculated T-value	Significance level
Requirements of managing high commitment -> The perceived organizational support		0.730	16.805	0.000
The perceived organizational support-> The strategic entrepreneurship		0.179	2.263	0.024
Requirements of managing high commitment -> The perceived organizational support-> The Strategic entrepreneurship	0131		2.234	0.000
Requirements of managing high commitment - > The strategic entrepreneurship		0.672	9.748	0.000

9 Results' Discussion

9.1 Discussing the Results of the First Main Hypothesis

That there is a statistically significant impact at the significance level ($\alpha \leq 0.05$) of the requirements for managing the high-level commitment to its dimensions on strategic entrepreneurship. The results indicate a statistically significant effect of the variant in the demands of managing high commitment in strategic entrepreneurship, reaching a beta value (B) (811) and a (t) value (25.411) with a significance level (0.000) less than (0.05). This finding was consistent with Howard study (2013) and Al-Naimi and Hamid (2015) studies in which it was stated that there is an effect of high commitment management requirements on the strategic entrepreneurship.

9.2 *Discussing the Results of the Sub-hypotheses Arising from the First Main Hypothesis*

1. Discussing the results of the first sub-hypothesis, which states that there is a statistically significant impact at the significant level of ($\alpha \leq 0.05$) of the requirements of the high Commitment management in its dimensions (normative commitment, job enrichment, job security) on the entrepreneurship in "Anbar Health Directorate/Iraqi Ministry of Health."

   As it is proved by the value of beta (ß), which amounted to (0.757) and the value of (t) which amounted to (15.186) with a significance level of (0.00) less than (0.05). This result is consistent with a study (Farndale et al., 2011), which concluded with some results that indicate that there is a correlation between the experiences of employees with commitment, high performance, fairness and confidence, which affect the activities of human resources management, and accordingly the researcher believes that the high level of commitment of working individuals helps entrepreneurial leadership in the Health office.
2. Discussing the results of the second sub-hypothesis, which states that there is a statistically significant impact at the level of significance ($\alpha \leq 0.05$) of the requirements of the management of high compliance in its dimensions (normative commitment, job enrichment, job security) on the entrepreneurial culture in the "Anbar health department/ Iraqi Ministry of Health."

   It is evidenced by the value of beta (B) of (640.) and (t) of (12.220) with a significance level of (0.00) less than (0.05). This finding is consistent with a study of Al-Hawajira (2018) which concluded that some findings suggest that enhancing a the entrepreneurial culture, linking it to strategy, goals and short and long-term performance outcomes, and attempting to transform the organizational structure from a bureaucratic one into a flat structure based on intelligence capabilities.
3. Discussing the results of the third sub-hypothesis, which states that there is a statistically significant effect at the level of significance ($\alpha \leq 0.05$) of the high commitment management requirements in their dimensions (normative commitment, job enrichment, job security) on risk taking in Anbar Health Directorate/ Iraqi Ministry of Health.

   As it is proved by the value of beta (ß), which amounted to (0.759) and the value of (t) which amounted to (15.186) at the level of significance (0.000) less than (0.05). The result is therefore considered compatible with a study of Said and Al-Nasrawi (2016).

9.3 *Discussing the Results of the Second Main Hypothesis*

There is a statistically significant effect at the significant level of ($\alpha \leq 0.05$) for the requirements of the high commitment management in all its combined dimensions (normative commitment, job enrichment, job security) on the perceived organizational support in Anbar Health Directorate/Iraqi Ministry of Health. "the result of the second main hypothesis indicates that there is a statistically significant effect of the high commitment management requirements variable in perceived organizational support, with a beta value (ß) of (0.730). and the value of (t) which amounted to (18.805) with a level of significance (0.000) which is less than (0.05).

The results also showed that the R2 value is (0.563). This means that the high commitment management requirements variable accounts for (56.3%) of the variance in the intermediate variable perceived organizational support. This result is similar to the results of the studuies of Worku (2015) and Mansour and Ashour (2016), and in terms of the presence of a variable effect in the high commitment in the perceived organizational support.

9.4 Discussing the Results of the Third Key Hypothesis

There is a statistically significant effect at the significance level ($\alpha \leq 0.05$) of the organizational support that recognizes the strategic entrepreneurship in its dimensions (entrepreneurial leadership, entrepreneurial culture, risk taking) in Anbar Health Directorate/Iraqi Ministry of Health. "The outcome of the third major sub-hypothesis indicates a statistically significant effect of the organizational support variable in strategic entrepreneurship, with a beta (B) value of (0.179) and a (t) value of (2.263) with a significance level of (0.024) which is less than (0.05).

Besides the value of the marker (R2) is (0.860), which means that the median variant (perceived organizational support) accounts for 66% of the variation in the dependent variable (strategic entrepreneurship), and these results are considered supportive of the studies of Khudayr and Ahmed (2017) and Giglioti et al. (2019).

9.5 Discussing the Results of the Fourth Main Hypothesis

There's a statistically significant effect at the significance level ($\alpha \leq 0.05$) for the requirements of high commitment management in its combined dimensions (normative commitment, job enrichment, job security) on strategic entrepreneurship in its combined dimensions (leadership, leadership culture, risk taking) through perceived organizational support in the Anbar Health Directorate/Iraqi Ministry of Health. The results showed a statistically significant effect of the organizational support variable perceiving the relationship between the requirements of high commitment management and the strategic entrepreneurship in both courses.

The first direct course: The direct course value of beta (B) is (0.672) and the direct course value of (t) is (9.748) with a significant level of (0.000) which is less than (0.05).

The second indirect course: The significant values of beta (B) in the indirect course are respectively (0.730) and (0.179) and (t) values are (16.805) and (2.263) with a significance level equal to (0.000) for both courses less than (0.05).

These values indicate that the mediation resulting from the perceived organizational variable is considered to be a partial medium, as the results demonstrated a direct effect of the requirements of the high commitment management without the intermediate variable. However, the effect of the intermediate variable is positive, since the value of the coefficient of determination (R^2) prior to the intervention of the mediator (perceived organizational support) was (0.658); after the mediator's entry this percentage rose slightly by (0.02) This means that the variant of the requirements for managing the high commitment in its combined dimensions has come to explain (0.66) the variation in the dependent variable (strategic entrepreneurship).

Hence, the result is considered consistent with a study (Mansour & Ashour, 2016) that found a positive and significant effect between high commitment management and organizational commitment through perceived organizational support. Therefore, the research company must work continuously to improve employees' outcomes, especially the level of organizational commitment due to its positive effect on the organizational performance.

10 Recommendations

Based on the findings, this study recommends the following:

1. Formulating a future vision with a leadership orientation of the department depends on the requirements of high commitment management and how to recruit in operations to attract, select and recruit people who are able to explore opportunities and expected threats with the results of performance analysis to recognize and strengthen strengths and weaknesses to overcome them.
2. The directorate uses the quantitative and qualitative methods and resorts to numbers and statistics to put a clear picture of the future of the directorate within a scientific and practical methodology aimed at adopting appropriate methodologies for its working environment to achieve strategic entrepreneurship.
3. The need to invest the research directorate for high commitment management practices to enhance the perceived organizational support of the employees in order to keep strengthening the strategic entrepreneurship and distinguishing it from other organizations. it should realize the experiences of the developed countries in applying these practices to benefit from them in the best possible form.
4. The high management in the health directorate should develop the capabilities and skills of the service providers, take the interest of the service providers, increase organizational support for them and meet the expectations of the employees about attention to their personal needs, self-esteem and work efforts.

References

Agostini, L., & Nosella, A. (2017). Enhancing radical innovation performance through intellectual capital components. *Journal of Intellectual Capital.*

Al-Gharbawi, M. (2014). *Role of organizational commitment in improving quality of service: An applied study on the civilian sector at the ministry of interior and national security–Gaza bank.* Master's Thesis, Islamic University-Gaza, Palestine.

Al-Hawajira, K. (2018). Business intelligence mediates between organizational entrepreneurship and strategic success at Jordanian government universities. *Jordan Journal in Business Administration, 14*(3), 413–444.

Al-Nuaimi, S., & Hamid, R. (2015). Requirements for high commitment management to achieve strategic entrepreneurship in business organizations field research in Iraqi contracting companies. *Journal of Economic and Administrative Sciences, 21*(84), 36–59.

Al-Sameeh, 2010 Al-Sameeh, A. M. (2010). *Studies in school administration.* Amman, Jordan: Dar Al-Hamid for Publishing and Distribution.

Al-Shanty, M. (2015). The role of perceived organizational support as a mediating variable in the relationship between organizational justice and organizational citizenship behavior: An applied study on the employees of the Ministry of the Interior–Civil Sector–Gaza bank. *Journal of the Islamic University for Economic and Administrative Studies, 23*(2), 31–59.

Al-Shayab, A., & Abu Hammour, A. (2011). *Modern Management Concepts,* Dar Al-Manhal Publishing House.

Al-Tahan, I. (2018). The Effect of future in achieving strategic organizational entrepreneurship: A field study on the Egyptian Company for Communications (MECSI). *Comprehensive Multidisciplinary Electronic Journal for Scientific and Educational Research Dissemination, 1*(6), 24–31.

Anomneze, A., Ugwu, I., Enwereuzor, K., Leonard, I., & Ugwu, I. (2016). Teachers' emotional labour and burnout: Does perceived organizational support matter? *Asian Social Science, 12*(2), 9–16.

Drover, P., & Ariel, B. (2015). Leading an experiment in police body-worn video cameras. *International Criminal Justice Review, 25*(1), 80–97.

Farndale, E., Hope-Hailey, V., & Kelliher, C. (2011). High commitment performance management: The roles of justice and trust. *Personnel Review.*

Flayiah, F., Abdul Majid, M. (2014). *Organizational behavior in management of educational institutions.* Amman, Jordan: T1, Dar Al-Massira Publishing, Distribution and Printing.

Ghazi, E., & Parvaneh, G. (2014). Strategic entrepreneurship element from theory to practice. *International Journal of Business and Technopreneurship, 4*(2), 205–219.

Gigliotti, R., Vardaman, J., Marshall, R., & Gonzalez, K. (2019). The role of perceived organizational support in individual change readiness. *Journal of Change Management, 19*(2), 1–16.

Harem, H. (2004). *Organizational behavior: Individual behaviour in Organizations* (4th ed.). Amman, Jordan: 4, Zahran Publishing and Distribution House.

Howard, T. (2013). *High commitment human resource management and employee wellbeing.* Master's thesis, University of Cape Town.

Ismail, O. (2010). Entrepreneurial characteristics of industrial organizations and their impact on technical innovation. *Al-Qadissiyah Journal of Administrative and Economic Sciences, 4*(12).

Khudayr, A., & Ahmad, S. (2017). perceived organizational support and professional compatibility and their effect in reducing the phenomenon of organizational satire, field research of the views of a sample of employees in the Ministry of Youth. *Journal of Al-Dananir, 1*(10), 286–325.

Kraus, S., Kauranen, I., & Henning Reschke, C. (2011). Identification of domains for a new conceptual model of strategic entrepreneurship using the configuration approach. *Management Research Review, 34*(1), 58–74.

Mandel, D. R., & Barnes, A. (2014). Accuracy of forecasts in strategic intelligence. *Proceedings of the National Academy of Sciences, 111*(30).

Mansour, T., & Ashour, M. (2016). The perceived organizational support mediated between high containment management and organizational commitment: A survey of the opinions of a sample of employees of the South Oil Company. *Gulf Economic Journal, 32*(30), 31–77.

Oviatt, B. M., & McDougall, P. P. (2005). The internationalization of entrepreneurship. *Journal of International Business Studies, 36*(1), 2–8.

Park, J., Jung, D., & Lee, P. (2019). How to make a sustainable manufacturing process: A high-commitment HRM system. *Sustainability, 11*(8), 2309.

Plummer, P. (2016). Examining the relationship between entrepreneurial orientation and the use of business advisory services by small and medium-sized businesses. Doctoral Dissertation, Saint Leo University, USA.

Qassimi, N. (2011). *A guide of sociology terminologies: Organization and work.* University Press Department.

Said, H., & Al-Nasrawi, H. (2016). Management of high commitment and its role in achieving organizational excellence. Prospective research of the views of a sample of directors in the Al-Furat State company for chemical industries/babil. *Journal of Economic and Administrative Sciences, 22*(89), 195–230.

Seitovirta, L. (2011). The role of strategic intelligence services in corporate decision making. Master Thesis, The Aalto University.

Sekaran, U., & Bougie, R. (2012). *Research methods for business: A skill-building approach* (6th ed.). John Wiley & Sons Inc.

Sultan, M. S. S. (2016). Level of availability of entrepreneurial characteristics and its association with certain personality variables: An applied study for undergraduate students specializing in "business administration" at universities in the southern West Bank. *Journal of the Islamic University of Economic and Administrative Studies*, 24(2).

Worku, A. (2015). An investigation of the relationship among perceived organizational support, perceived supervisor support, job satisfaction and turnover intention. *Journal of Marketing and Consumer Research, 13*(1), 1–8.

Knowledge Management

Interdependencies and Integration of Smart Buildings and Smart Cities: A Case of Dubai

Mounir El Khatib, Gouher Ahmed, Muhammad Alshurideh, and Ahmad Al-Nakeeb

Abstract The literary meaning of the word 'smart' is bright, intelligent, high performing, which, by and large, is also true in the case of 'smart' buildings and cities. The present digital age (seems) to be a smart age and smart times, where everything is expected to be smart, including cities and their dwellings or buildings as living and working spaces. This study addresses Dubai, well known for its smart service initiatives, through Dubai Municipality (DM), and Dubai Electricity & Water Authority (DEWA). The Results reveal smart power and water management efficiencies and smart and effective delivery of civic services through the use of ICT. Dubai has plan to turn the sustainable city into power, water, infrastructure, information and communication efficient one by the 2021 Golden Jubilee year of the founding of UAE. The message of it is that the future is of smart dwellings and smart cities, and the concept of 'smart' in the case of ultramodern Dubai too should extend to the citizens.

Keywords Smart buildings · Smart Cities · Sustainability · Eco-system · Dubai

---

M. El Khatib · A. Al-Nakeeb
School of Business and Quality Management, Hamdan Bin Mohammed Smart University, Dubai, UAE

G. Ahmed (✉)
School of Business, Skyline University College, Sharjah, UAE
e-mail: gouher@usa.net

M. Alshurideh
Department of Marketing, School of Business, The University of Jordan, Amman, Jordan
e-mail: m.alshurideh@ju.edu.jo; malshurideh@sharjah.ac.ae

Department of Management, College of Business Administration, University of Sharjah, Sharjah, United Arab Emirates

M. Alshurideh et al. (eds.), *The Effect of Information Technology on Business and Marketing Intelligence Systems*, Studies in Computational Intelligence 1056,
https://doi.org/10.1007/978-3-031-12382-5_89

1 Introduction

To start with, a smart city is a community in which citizens, business firms, knowledge institutions, and municipal agencies collaborate with one another to achieve system integration and efficiency, citizen engagement, and a continually improving quality of life (Al Batayneh et al., 2021; Al Shebli et al., 2021; Ghazal et al., 2021a). The term Smart Cities or Digital Cities are urban development, with the help/aid of implementing information and communication technologies. They are cities of the future which are forward-looking and resource-efficient providing a high quality of life, and are innovative in respect of infrastructure and homes, offices, business premises, besides being energy efficient and nature friendly (AlSuwaidi et al., 2021; Madi Odeh et al., 2021; Taryam, et al., 2020). They are made-up of new energy and transport system. Government and public Partnerships are essential for smart cities, and to make the cities to be smart and intelligent, decisions need to be taken at the strategic level to see the decision makers' commitment to smart city development implementations (Aburayya et al., 2020a; Alhashmi et al., 1153; Kabrilyants et al., 2021).

The concept of smart building is elastic and evolving in a constant way. This flexibility supports the facility to add and insert various functions and components to smart buildings and smart cities programmes. Smart buildings are generally kept to be flexible, energy efficient, interconnected and automated to enhance comforts of life (Aburayya et al., 2020c; Alshurideh et al., 2016; Al-Zu'bi et al., 2012). "Smart home refers to an intelligent home concept in which the user can work on sensors and devices to make the house respond in a way to stimuli". Smart buildings have been of scholarly attention for over thirty years. The phenomena are present in almost every smart materials, smart detectors, and smart constructs, with the aim of creating efficient buildings, making use of advanced and efficient building construction technologies, as the 'green' one (Alshurideh et al., 2019a; Joghee et al., 2021; Lazaroiu et al., 2018; Shishan et al., 2022).

This paper review Smart Cities as the future of urban settlements, with large populations that require, first, smart dwellings make-up good city to live-in buildings block of smart cities. Smart Cities has an average technology size, sustainable, comfortable and also secure (Lazaroiu & Roscia, 2012). Dubai, which has reputation as an ultra-modern city (Aburayya et al., 2020b; Ahmed, 2021; Alzoubi et al., 2021; Krane & Dubai, 2009), is taken as a case study of the problem. A comparison made of 'dazzling' Dubai with other smart cities, to see how they differ Dubai, a city state of the Emirate is in part of the 7- emirate of the United Arab Emirates (1971), which has come to be an up-to-date state of the world. The basic question addressed in the study is whether smart buildings lead to smart cities?

The research objectives are to (i) define the constituents of smart buildings from both international and local point of view, and also (ii) to see whether smart buildings are enough to build a smart city or not, and (iii) to study Dubai's initiatives in this regard. And the study is organized around the following questions: (a) What constitutes a smart building? (b) Are smart buildings milestones that lead to smart cities? (c) Are smart buildings the only key component that lead to smart cities? (d) What is Dubai government's strategy and future plan towards smart buildings?

2 Literature Review

2.1 Smart Building

The concept of Smart Building has been in existence for about two decades by now. Its definition, as well as implementation, has been evolving and varied in line with development of knowledge and technology over the years. (Kroner, 1997) asserts that the contemporary smart buildings are ones that have been enhanced electronically. Accordingly, Smart Buildings can be defined as buildings that are capable of enhancing the productivity of their occupants, optimize environmental performance, enhance satisfaction and safety by promoting comfort and adopting morbidity and health features.

Arkin and Paciuk (1997) have suggested that the smartness of a building is not only based on the sophistication of the technology used but the integration that occurs between different systems. According to Atif and Galasiu (2003), the factors that influence the increased use of Smart buildings in the world and more importantly in the United Arab Emirates, (UAE) are ensuring cost effectiveness through cost reduction measures and enhancing the technological capabilities that characterize them (Aburayya et al., 2020d; Allozi et al., 2022; Nuseir et al., 2021a). Other major drives include industry and government sustainability policies, the need for intense security, the changing market trends in real estate, the ongoing business pressures, the triple bottom line requirements for reporting, the needs to outsource the building services and the shifting nature of work environments (Al Naqbia et al., 2020; Al Suwaidi et al., 1261; Alameeri et al., 1261). (Amponsah & Ahmed, 2017a) and other researchers further note that it is vital to analyze the drivers to determine the applications of technologies that would yield more beneficial results.

2.1.1 Features of Smart Buildings

- *The buildings are flexible:* The buildings make use of information collected from different sources to prepare for future events. They can adjust to their operations and physical nature so that they can suit these surroundings. A smart building can account for people's different needs and changes in occupancy rates (Akram et al., 2016).
- *Smart buildings have control:* They are either automated or possess human control. The buildings regularly warn occupants of the current temperature and direct them to their comfort areas.
- *Environmental preservation:* They use the real-time environmental information to guide the house owners on ways to preserve the environment. The buildings also have good sewage systems that ensure efficient disposal of waste.
- *Possess an enterprising feature:* The buildings make use of information gathered to advise visitors on the nature of a house. This enterprising nature makes a tremendous potential for the buildings to provide a comfortable atmosphere (Buckman et al., 2014).

- *The buildings are comfortable:* According to Das et al. (2015), the buildings have advanced technologies, which make the life of the owners easy. They have a remote control which helps the owners to perform their duties comfortably.

Smart buildings in sum are airy, bright, energy and water conserving. It enabled, light but should rowed attention, and social and environment friendly, well maintained and clean and green.

2.2 Smart Cities

The literature on Smart Cities includes a broad spectrum of multidimensional components and a diversity of conceptual relatives, as "knowledge-based development," "intelligent city," and "skilled cities," to name a few (Nam & Pardo, 2011). The term smart city covers a considerable ground and includes a range of concepts related to urban development conflated under the "smart" label (Hollands, 2020).

Chourabi et al. (2012) focus on smart governance, ICT, and/or smart people, while (Caragliu et al., 2011) a smart city has a progressive economy and outlook, ICT infrastructure, an attitude toward positive development, a focus on creative solutions to management complexity, and the use of technology to streamline city services. Further, a smart city must contain "a smart economy; smart mobility; a smart environment; smart people; smart living; and smart governance." Moss Kanter and Litow (2009) claim that all of aforesaid systems must come together to create an "organic whole," so that all elements of the city must function as a single network.

By the way, literally 'smart' means *Stylish,* up to date, (Webster Dictionary) by which smart cities are Bright and Stylish cities. Of course, every city, with the passage of time, has to get updated. For example, compare Dubai of 1971 with Dubai of 2019, between which there is a sea of difference. The opposite of smart is Dull, Faded. Dubai is said to dazzle its ever-flowing tourist visitors, who come in millions, to see the wonder city of the twenty-first century and shop to their hearts full or over flowing satisfactory. There is no doubt about Dubai being a very Smart City. What is more, it is credited with Smart and forward looking leadership (Moss Kanter & Litow, 2009), without which cities cannot take a smart shape.

All smart city working definitions include most or all of the attributes of sustainability, advanced information communication technologies (ICT), smart/good governance, and human capital development (Khatib et al., 2019). According to the researchers, the four principal elements that operate in an integrated way in smart buildings are actuators, sensors, performance models and integrated information management system.

A smart city is not only technologically savvy, but also culturally and intellectually oriented. It is an integrated and holistic city in which people love to live and work and indulge in some creative activity (Amponsah & Ahmed, 2017b). It, in a way, is of advanced global city, as one can easily visualize. In this respect, Dubai is an established and acclaimed Smart City with a smart and dynamic and sophisticated feel-at-home tourism industry (Khatib & Ahmed, 2018).

2.3 Smart Buildings and Smart Cities Relationship

The influence of smart buildings in contributing to development of smart cities is visible in the ability of these buildings to accommodate a high number of residents and creating sustainable real estate solutions. Urban populations are constantly growing, requiring the application of information and communication technology to develop buildings that will offer improved quality as well as interactivity of services in urban regions (Intel, 2016).

Buildings interact with other economic, social, and environmental aspects in the community. For instance, buildings consume approximately two-thirds of electrical energy, therefore through installation of appropriate sensors, control systems and actuators electrical energy could be saved and environmental preservation would be achieved (Morvaj et al., 2011). Smart buildings are hence a significant influence in development of smart cities. The key elements that create a framework for a smart city include energy, water, waste, infrastructure, public safety, education, green buildings, transportations as well as citizen services and buildings are associated with a majority of these elements, illustrating the potential that smart buildings have in creation of smart cities (Doherty xxxx).

2.4 Information Technology and Urban Growth

The translation of a smart building acknowledges the need for an integrated and universal model. They are structures which combine and consider business management, intelligence, and adaptability, so as to meet the directors of development. Smart cities are urban growth insights to combine the Internet of things (IoT) and IT uniformly to handle a town's properties (Deakin, 2013). These resources may include schools, power plants, hospitals, and community services amongst others. The cities utilize IT to improve their effectiveness, connectivity, quality of work, costs decrease, resources use, and enhance the contact between the state and its nationals. IoT is developing a new range of smart buildings which are more aligned with the organizations' strategies (Ghazal et al., 2021b; Krane & Dubai, 2009; Lee et al., 2022; Nuseir et al., 2021b). It allows operational processes to produce precise and sufficient information for enhancing functions and offers the best to its clients.

2.5 The IT Alignment with Smart City Strategy

Information Technology is gradually transforming the cities across the globe, making the cities digital cities with swing on-line transitions. The internet is modifying the conventional urban strategy design and compelling developers to consider the utilization of IT to enhance the mobility, administration, finances, and the overall environment of a city. Even from the definition of the smart cities, IT plays a vital function in creating a city that is adaptable to the modern needs of its inhabitants. Further, IT offers a technology and trade review of the latest trends in logistics, water supply, and energy in urban developments (Khatib & Ahmed, 2018). Thus, IT appears to be the vehicle propelling the development of Smart Cities.

In the vital and worrying energy sector, 'smart power' relates to the any type of energy, whose purpose is to decrease energy losses, enhancing provider services, raise effectiveness, and reduce wastage. The plans rely on the advanced IT facilities and services. Smart logistics target information-grounded traffic regulation and people. This sector depends on smart transportation processes for the checking and management of the fleet and traffic services on real-time data basis. Smart water relates to the programs that aim at encouraging water conservation and efficient use of energy in the water arena by the implementation of the machine-to-machine tech, and data review.

In a firm where IT and business strategies are aligned, the organization understands the goals and objectives of the IT department and ensures every change takes into account the influence it has on this department (Alshurideh et al., 2021a, 2021b; Leo et al., 2021). The IT manager comprehends the overall aim and objective of the firm and thereby acts according to its terms. The misalignment of IT and business plan takes place when the firm never incorporated IT into the planning or has slowly become disconnected as the company's requirements and priorities evolved. As the modifications take place, most managers update their plans but fail to incorporate the effect of these changes to IT.

The supply chain, sale, and marketing departments may have evolved over a period. However, these changes usually occur with no observation of what IT must change so as to support such modifications. Similarly, technology changes at a high rate. Firms notice the significant capital input they made in particular technologies and are cautious to leave out legacy processes so as to favour new and outsourced services due to the increased costs and difficulties. Consequently, in such firms the conflicts arise as the business plans change and IT strategies remain stagnant or even worse. Now it is time to take up Dubai and its plans of Development, to make the city a more renowned global metro or mega-polis.

2.6 Dubai Plan 2021

Dubai 2021 plan aims at ensuring safety standards within the built environment of the city. The plan focuses on guaranteeing building safety for the residents and visitors living in the city by offering disaster management responses. It also plans to provide high-quality housing, to all socio-economic groups. The plan involves integration of new technologies that ensure optimum facilities, and also focus on optimum energy management and sustainability, in all matters of the city development to put the stellar city on a for-ever smart footing. The buildings are also to be made ecosystem friendly, to ensure a clean and healthy environment. To be able to achieve the 2021 plan, Dubai needs to make sure that the buildings have no negative impact on their surroundings and that they maximize economic benefits of the city. Dubai Plan 2021 is an example of keeping the reputed city dynamically smart pleasing in appearance and most efficient in functioning and administration.

3 Research Methodology

3.1 Research Design and Data Collection

This research is descriptive research that uses the qualitative method for data collection to identify, categorize publicly available information about international Smart Cities including, Dubai and some others to compare their smartness in terms of building and other standards. Data collection was done through interviews, and Surveys, concerning select Dubai civic bodies of DM & DEWA.

3.2 Interviews

Person-to-person interviews of the head of each department (IT & Strategy) in both DEWA and Dubai Municipality Sr. manager, manager, sr. specialist level in the know-how of the City matter's as their administrators, in their organizations. They had provided with required amount of information for the study being experts in the current state of affairs and the plans Dubai plan 2021, the Golden Jubilee Year of founding the United Arab Emirates (UAE), by which date the Emirates, intend to 'achieve, say a very 'smart' high and steady state of development. Three questions were put to the heads in both the organisations, they are (a) What is DM's/DEWA Information Technology strategy? Especially in smart building field. (b) Do DEWA forecast or predict for future technology? How? (c) How they utilize future technology in their plans/projects? giving a good deal of information/Data. It is significant, Dubai's urban/municipal organizations and authorities operate on a minute state of Information. So, 'Smart' in the matter of departmental information.

3.3 Online Resources

The study had drawn from the DEWA and Dubai Municipality official websites for the organizations' initiatives and real projects. Also, other related websites to collect more data about the specified topics, which had got buttressed by articles and researchers have been used and utilized to support the research.

Even as it is Dubai seems to be a smart global city by appearances, and functioning, which doesn't mean it doesn't have un-smart features like over per capita water consumption. In a desert city like Dubai, water has to be used with utmost care and ZERO or near zero wastage Water is the nectar of life, it is needless to say.

3.4 Sample Interviewed Organisations (DEWA & DM)

3.4.1 Dubai Electricity and Water Authority (DEWA)

In 1992, Dubai Electricity Company and the Dubai Water Department were merged by the Sheikh Maktoum bin Rashid Al Maktoum, Vice President & Ruler of Dubai, to be Dubai Electricity and Water Authority (DEWA), the main provider of electricity and water service in Dubai. DEWA is one of the authorities that has e-Government service which provides the online users with up to date information. In addition, DEWA has launched three initiatives programs to effort sustainable development in Dubai as a part of smart Dubai initiatives such as connecting solar energy to houses and buildings, Smart Applications & your Home and Infrastructure and Electrical vehicles charging stations. It is important that these initiatives were quite ahead of them in the Arab World, by and large, and elsewhere too.

3.4.2 Dubai Municipality (DM)

Dubai Municipality is a government organization which provides municipal services. It is a well acclaimed municipal body. It is made up of 34 units under six sectors (Sector's of Engineering and Planning/Environment, Public Safety and Control Sector/Environmental Services and Public Safety Sector/Institutional Support Sector/Public Support Sector/Communication Sector and Society) with multiple authorities & responsibilities. The municipality's vision is "Developing a happy and sustainable city", and the mission is to plan, develop and manage an excellent city that provides the essence of success and sustainable living–Dubai Municipality is considered to be one of the largest governmental institutions in terms of the services it provides and the projects it carries out, in the Emirate of Dubai and UAE.

3.5 *Limitations*

In spite of researcher's best efforts and the respondent cooperative, the study may have some limitations, which however do impact upon the soundness and Results. (a) Difficulties to get the required answers from the organizations in Dubai as the majority of projects are not announced yet. (b) Time limitation. (c) The Variety of the criteria that define and measure the smart buildings all over the "Smart cities" made the comparison a challenge. (d) Lack of enough research in this field, as most of the research are based on smart building from an environmental aspect.

4 Analysis and Results

By interviewing people from Strategy department in both DEWA and DM, it was found that the important part of any project is the strategy used to achieve the organizations' goals. The strategy is to primarily facilitate positive energy and by doing this DEWA and DM have provided a positive and optimistic work environment. In return they expect to enhance the people's happiness, and the individual happiness leads to corporate efficiency.

Just to give a glimpse, they have incorporated a library, nursery, cafeteria, outdoor jogging track, and lots and lots of green spaces. In addition, DEWA and DM pride them self on the implementation of technology to reach their goals. For example, as of 2017, DEWA had adopted Microsoft's HoloLens technology, which is absolutely ground-breaking. Through this piece of technology, DEWA will be able to efficiently boost its operations, and greatly improve overall productivity and increase the process of innovation. By doing all this, they are able to provide services, with maximum potential to their customers. Through this, DEWA is able to assert its position at international level, as an efficient electricity and water dispenser of world repute. It is noteworthy, both the Dubai Emirates and the Emirates aim at international working standards for their organizations, of which both DM and DEWA have proved themselves to service organizational of global distinction.

The DEWA and DM belief is that future planning is a must to the success of any project or any of its related task. With this in mind, 2020 is likely to be a challenging as well as a great opportunity moment for the United Arab Emirates. DEWA and DM have programmes that lay the pathway for the achievement of a smartest possible city of Dubai. For instance, in 2016, DM had implemented a'10-week' programme, during which a number of technologies were undertaken and implemented to open further strategies. This programme was also called the ExO Sprint Programme, meaning a quick results program.

IT manager at DEWA had discussed about Al Sheraa or 'The Sail' project to be completed by 2019. DEWA had planned to design the building solely based on solar energy as the future un-bounded energy was planned to be shaped as a sail to capture solar energy. It had plan to build a 'net zero' type building where the energy used will be much lesser than that of a normal building. It was calculated and found that the building will be utilizing about 70 kWh/sq. This helps DEWA to achieve the vision of an environmentally friendly approach and innovative planning.

DEWA and DM's primary concept is innovation, which form a greater part of their vision. In relation to it, they had recently entered into cooperation with a number of companies from India. All these companies operate in the energy sector. It was all a part of the 'Dubai Clean Energy Strategy 2050'. Through this cooperation, a number of projects are kept in order, which would transform the United Arab Emirates to be the leading and the most environmentally sustainable country. For example, Al Sheraa is a huge and ambitious project, to be implemented by DEWA. But there are other projects in place, which aims at sustainability, and integration with innovation. DEWA had initially planned a project called the 'Mohammed bin Rashid Al Maktoum Solar Park', which is planned to achieve 100% completion by 2020.

These are all 'smart' energy initiatives and plans which are to give Dubai a smart energy city image, as the solar energy is the cleanest and near to nature energy. The project is a part of Dubai Clean Energy Strategy. With this, the Al Sheraa project is also a sustainable related project, as, control centre would help keep track of all the building systems and switch off lights and AC, when they are not in use. Sustainability and innovations take part of each and every aspect of a project of DEWA projects.

The finding of it is that the entities in Dubai are involved and aware of Dubai vision, which is building up of a smart Dubai City that stands out in the country of global smart cities, through the medium of ICT. However, surprisingly, ICT was found not involved the premier organizations business strategy. It is however, a different that the researchers had not gone into the Question of Smart staff.

From the interviews it was found that, the organisations were still not involving the IT in their business strategies. And, without IT, smart organizations remain a distant dream, with IT standing for 'smartness' as the smart phones.

Al-Sheraa Building—The Smartest and Largest Government building around the world. Al- Sheraa, in Arabic means 'The Sail'. The name has a close correlation with the structural design of the building which shaped like a sail, and is covered with solar panels to produce over 3,500 kilowatt hours (KW/h). In addition, the building using technologies such as, the Internet on Things, Big Data and Open Data, and most importantly Artificial intelligence (AI). These robots are helpful in a lot of operations within the building. An application is given to the employees, which advises them on the traffic condition, and also assists in the process of booking meeting rooms and parking spots. The Al-Sheraa building is essentially a Zero Energy Building, which utilizes a way less energy as compared to other city structures.

4.1 Smart Water Management

As said, water is a very precious gift of mother nature, which however is in a declining supply Dubai being a desert land, the water problem is very acute, with the problem compounded by a meagre annual average rainfall of 80–140 mm (Sherif et al., 2014). With this, the need is for the Smartest water management of the expanding City's water resources.

The DEWA is at the Job of smart management of the city's water resources, wherein the role of the citizenry is very vital, as they need to cut down their excess per capita water use, which is noted to be the highest of any country in the world. Smart cities, whatever they are supposed to be first of all, call for smart citizenry. This give reference point, however, is seldom come across in the smart cities' literature. As such, more than anything else, smart people in the sense of socially very responsible ones, is the first requirement of Smart Cities.

Al-Sheera is minutely managed with the latest technologies, the model of which should extend to homes, offices, businesses premises and the whole city. Dubai Municipality Smart Buildings exercise is quite an elaborative one and is mostly technologically driven. Dubai Municipality, which is reputed municipal body across the world, 'smartness' pervades all the big building initiatives (7) of the Dubai Municipality. Infrastructure, communications, energy, health, waste material, elevators, printers, building management, security and safety. From all buildings appear to be independent, yet inter-connected, building system. They are like little islands is the smart sea of municipality.

Smart Buildings mean energy, technology, and environmental benefits smart buildings are, say, the way forward to smart cities. Smart buildings confer, on the whole environmental health, economical, and technological advantages on Smart Cities and thus help building up of Smart Cities. In other words, it is a wider public perception in Dubai that Smart Buildings are akin to the bricks that go into the making of the edifice of smart city. In a word, public perception in Dubai is well in favour of Smart Buildings through which Smart Cities is presumed to be next step. Dubai is felt to be on the road to a smarter future.

4.2 Dubai's March Towards a Smart Future

Dubai is on the way to a large-scale digital transformation. The goal is possible through the adoption of the approaches of cutting-edge digital transformation, information, and communication utilizations. Adjustments, in the policies as well as regulations facilitate the adoption of prompt technological advancements which together with societal revolution, enable infrastructures for the internet-based businesses as well as the creation of a knowledge economy are being restored. Dubai city intends to expand the usage of innovative digital means to achieve sustainability in development, and enhanced governance, and growth as well as public well-being.

The ambitious endeavour of Dubai entails the creation of a "smartest" and "happiest" city by 2017 (Alshurideh et al., 2019b; Alzoubi et al., 2020; Sherif et al., 2014). The achievement of their vision relies on the maintenance of the digital-age entrepreneurial approach to leadership, networked and collaborative style of governance. Other indictors can be a positive sign to put Dubai in the right way towards smart transformation ICT infrastructure and economic development and business. Such an achievement can lead to the minimization of the future barriers together with the expedite efforts that facilitate the attainment of their ambitious goals of a 'Smart'/Smartest and happier/happiest city of international distinction and standing.

The strategic plan of digital or smartest city transformation comprises of six dimensions that include Economy, Governance, Environment, Living, Mobility, and People, and the areas of concern include the quality of life, lifestyle, ICT, communications, information, infrastructure, services, and people, together with the social, environment, sustainability, Governance and administration, economy and finance and Mobility and transportation (Sherif et al., 2014).

4.3 Dubai Plan Toward Transforming into a Smart City

The findings from reading and analyzing is explained in this part. Dubai plan 2021 to transform the acclaimed city of Dubai into a smart city by achieving several accomplishments. The plan is an output of the forward-looking government of Dubai, headed by very concerned leadership, which intends to put their Emirate of Dubai in the first ranks of the developed countries. The plan is a model of smart-city development (Sherif et al., 2014).

Firstly, Dubai plans on building an infrastructure, which will support the economy and social life of the city. The infrastructure will be wholly integrated and connected, to ensure easy mobility for the people in the city have access to the economic and social sectors.

Incidentally, 2021 is the Golden Jubilee year (50th) of the founding the new Arab state of the United Arab Emirates (UAE) (1971), which has achieved many records of development, in which Dubai as the second largest Emirate after Abu Dhabi has a stellar tale.

Secondly, it plans on ensuring that the city uses the available resources sustainably. Sustainability means that it must pay attention to the future by providing sound management and use of renewable energy sources.

Thirdly, it plans on ensuring a healthy and clean environment, to ensure sustainability of the city in the long-term.

Lastly, it plans on managing safe and resilient built surroundings. The program aims at making Dubai stand it on par with the world-class cities, where there are high levels of efficiency and facilities for future growth.

4.4 Comparison of Dubai and Singapore

Singapore recognizes the influence of smart buildings in contributing to the development of smart cities and smart nations, and in this regard, the country highly prioritizes integration of technological as well as automated systems in buildings to enhance control and monitoring, in addition to energy saving and improvement of human comfort. Comparison of efforts and plans of Dubai Municipality and the City Council of Singapore illustrates that both bodies are focused on transforming the public sector. One indicator is the degree(s) provided for various Smart buildings initiatives. For instance, governmental buildings are provided 15 points in Dubai regarding installation and the same holds for Singapore.

However, Singapore's total building degree for governmental buildings is higher at 100, compared to 95 in Dubai. The inference is that Singapore's efforts and resources allocated to this initiative would be higher. In 2016, the Building Construction Authority (BCA) in the country of Singapore was signed a partnership agreement with Microsoft aiming to integrate technological and automated systems in 30 commercial buildings (Sherif et al., 2014).

Singapore', allocation of a higher degree to Industrial Buildings have implications for Dubai's efforts, because Dubai may fail to achieve optimal outcomes. By allocating 25 to 'total building degree', Dubai fails to transform industrial buildings entirely and since these buildings often consume high energy amounts, they need to be monitored in the same capacity (perhaps even more) compared to government buildings, public buildings or other building types, most variables significantly receive the same degree in both Dubai and Singapore, including Energy Control Systems (15), Security and Monitoring Systems (10) and Water Control Systems (10). While both Dubai and Singapore invest significantly on tourism and hospitality, the degree provided to Hotels and Hotel Apartments is higher (at 100) in Dubai compared to Singapore which has provided a degree of 90 to this building type. The inference is that Dubai's popularity as a tourism destination encourages the municipality to improve human comfort through transformation of Hotels and Hotel Apartments into smart buildings.

As per the rankings stats, infrastructure from Table 1, Dubai is doing better in terms of ranking, the ranking has improved from 43rd out of 118 in 2020 to 29th out of 118 in 2021 when it comes to smart city ratings, Singapore is rated higher when compared to Dubai in terms of Population, HDI, GNI per capita. Dubai Ranks 'BB' in Smart City Global Rating and Singapore holds the top position with the 'AAA' in the year 2020 (IMD Smart City 2021 Index). Dubai has to rise over its under ratings in respect of health, utilities, education, and public safety in its journey towards an Emirates future.

As per the rankings stats, infrastructure from Table 2, Dubai is doing better than Singapore when it comes to the development of smart buildings, Singapore is rated higher in smart health care delivery. Adoption of smart transport is higher in Dubai than in Singapore while adoption of smart technology seems to be higher in Singapore as compared to Dubai. Dubai, nevertheless, is rated lower when it comes to electricity, education and public safety as compared to Singapore (KPMG Revenue (Annual Report), 2017). Dubai Ranks 45 in in Smart City Global Ranking and Singapore holds the top position with the 1st Rank in the year 2019 (Sherif et al., 2014). Dubai has to rise over its under ratings in respect of health, utilities, education, and public safety in its journey towards an Emirates future.

Table 1 Dubai and Singapore comparison

Component	Dubai	Singapore
Population	2,880,000	5,940,000
HDI (2019)	0.890	0.938
GNI per capita (PPP $)	67,462	88,155
Smart city ranking (in 2021)	29 out of 118 (43rd in 2020)	1 out of 118 1st in (2020)
Smart city rating (in 2020)	BB	AAA
Factor ratings (Structures) (in 2020)	BB	AAA
Factor ratings (Technologies) (in 2020)	BB	AAA

Smart city index (2021)–All ratings range from AAA to D.

Table 2 Ratings

Component	Dubai (%)	Singapore (%)
Smart buildings	70	60
Smart urban health care	50	75
Smart transport	75	65
Smart technology	70	70
Tourism	75	60
Utility	45	50
Education	60	70
Public safety	56	70

Source KPMG Revenue (Annual Report 2017)

5 Discussion and Findings

From the theoretical data and the collected information, a smart building is a construction that takes advantage of improved materials and technology in structural terms, electrical systems, HVACR, plumbing, and appliances. The compilation of all these factors enables such a building to achieve a considerable amount of energy saving. Certain elements of these type of buildings are true and tried, a familiar example being weatherization. Other elements include new technologies in lighting among other aspects that are new to the market. Some largely depend on the availability of renewable energy. Landscaping is also incorporated with the inclusion of green walls and green roofs. Advanced systems of control connecting these buildings are yet another major element, which is a way to optimize savings. A smart building uses information technology in the integration of every aspect.

Smart buildings are the key ingredients to the creation of smart cities. As such, constructing a smart building in any city is a sure way of improving the citizen's quality of life. Improved quality, in turn, is the base of creation of smart cities. Smart buildings enable a city to efficiently and frequently collect utility data. An example of this is replacing the tallying of the bills with smart meters to do the collection. Doing this would make the identification of patterns of usage and effectively plan the usage of energy based on the amount of demand. An efficient city regarding service delivery translates into saving of energy, simplified processes and thus the city can identify the capital expenditure that is more helping without the risk of breaking the budget.

As for resident happiness, a smart city program enabled by smart buildings can improve the lifestyle of residents more than any other initiative tailored towards a smart city. In the long run, residents are empowered with better working conditions, safer environment, which entitles them to do more meaningful goals in life through efficiency, comfort, and safety of the systems around them. This is aside from the happiness induced by belonging to a sustainable city that practices fair distribution of capital income over other important aspects of life such as education systems and economic growth.

Is Dubai in the right way to be a smart city? The finding from the research is that a smart city must create a high quality of life and possess sustainable economic development. Smart buildings, good infrastructure, digital technologies, good governance, sustainable environment and well-educated residents, characterize a smart city. In the 2021 Dubai Plan, the city aims at building an infrastructure that supports economic development. It plans on creating a sustainable setting with a clean and safe environment, good governance, educated individuals and a vibrant and rich culture. It is evident that Dubai is in the right way to be a smart city since its plan is in line with features of a smart city. The program focuses on creating a high-quality lifestyle for the people of Dubai and building a sustainable economy, which are the key determinants of a smart city.

The government of Dubai has a plan known as "Smart Dubai Gov or SDG" the main aim of which is to create the happiest and smartest city in the world. The mission of the government is to create and deliver a city that is smart, and of the world class. This includes smart services and infrastructure, which comprise smart buildings. According to the government, their end goal is not technology, but rather to create an extraordinary life quality. The goal is to create sustainability among the citizen residents and ensure visitors are satisfied and happy. The city is benchmarking itself against other similar cities geographically, socially, economically, and culturally to compete to be a global leader.

Smart buildings are, no doubt, the key component to the creation of smart cities. For instance, the power use in such buildings enables the balancing of electricity demand and supply. The advancement in these buildings can be said to be like having a brain; that can reconcile and balance competing interests like occupant comfort, minimize energy and get grid stability. All these factors make smart buildings the only and best factor to the realization of smart cities anywhere in the world. They are efficient in every aspect, majorly solving most of the issues that are currently facing many cities. The best thing about such cities is that they are automatic and so will need minimal training to operate on the systems to get the cities up and running. From the data analysis, Dubai has the qualifications and the ability to be a smart city which however may need more efforts in some phases but with government's commitment and knowledgeable people the city can step forward to start more smart initiatives with supporting innovations and provide the needed technology.

But, there are certain hurdles in the way of realisation smart city goals by Dubai, such as the over-use of water and electricity and excessively taking to outsource ownership and use. For after all, it is citizens who make homes, cities and states. Hence, there is a need, above all, for 'smart' or public and nature minded citizen behaviour.

6 Conclusion, Recommendations and Policy Implications

6.1 Conclusion

The field of the smart buildings has grown worldwide, including the United Arab Emirates and its constituent emirate of Dubai. There are several activities, initiatives and real projects that reflect the growing interest and awareness of smart buildings in Dubai. Dubai's experience in the field is still in its early stages. Evidently, smart buildings are considered to be the cornerstones for successful smart city program. But the smart city programs are driven home to re-integrated to be effective and wholesome, as no price meal approach to the problem couldn't be well paying.

The benefits of smart buildings tap into the fundamental necessities of human life, such as safety, health, and energy, that culminate in the happiness of a city. Smart cities do not come without, substantial or capital expenditure on infrastructure, environmental, etc., which is justified in terms of its many benefits. Above all, for reducing global warming, the starting points of which are cities and their dwellings in whom a behavioural change is to be brought over.

In sum, Smart Buildings and Smart Cities are tidy, neat and green, well- functioning and well governed and orderly, advanced technologically, economically well off in which the residents are happy to live in and work. Dubai is inching towards to be the world's smarter city. However, Smart Cities, have also been noted to suffer from some wicked problems as acute class and housing differences (Colding et al., 2019).

6.2 Recommendation and Policy Implications

In the light of the results of this modest but significant exercise, the following suggestions are put-forth with policy implications.

- To start with, smart buildings as suggested by very concerned scholars should be built with advanced materials which are nature friendly and nanotechnology for energy and water use efficiency (Sherif et al., 2014).

- While many cities can easily be labelled as 'smart' for adopting new technologies, what is more necessary is that to truly establish the smart factor in every building, we must take the right approach towards the construction of new buildings and the transformation of the existing ones.
- Find and Set international features or standardized criteria to define and recognize the smart buildings and smart cities.
- The planners must learn how to overcome the barriers that hinder the progress of building Smart Cities. The common hurdles are the limitations of the infrastructure, gaps between the technological potential and the developmental requirements, public concerns, policymakers', technological determinism, obsolete regulatory frameworks, the government silos, resistance to change, and capacity deficit.
- Align the IT plan with the business strategy to start smart buildings. Make managers understand the features and the future of the IT and the technology supporting the smart building infrastructures, for which the IT involvement in planning smart cities instruction is important.
- Government and higher political involve leadership in Building Smart cities is essential.
- Citizens and people who get the beneficial from smart buildings should be involve as well through public awareness and more involvement via social media.
- Support and educational facility are important to encourage innovators that new ideas and creative touches to improve the smart initiatives in the cities are aiming to be smarter.
- Smart Buildings and Smart Cities movement should be a people's movement, involving all sections of the people.
- There are 'wicked' problems of Smart Cities like the demolition of poor dwellings for the purpose of the high-rise buildings for the rich and super rich.
- Not the least, Smart Cites call for socially responsible citizens who are active in the city affairs.

References

Aburayya, A., Alshurideh, M., Alawadhi, D., Alfarsi, A., Taryam, M., & Mubarak, S. (2020a). An investigation of the effect of lean six sigma practices on healthcare service quality and patient satisfaction: Testing the mediating role of service quality in Dubai primary healthcare sector. *Journal of Advanced Research in Dynamical and Control Systems, 12*(8), 56–72.

Aburayya, A., Alshurideh, M., Albqaeen, A., Alawadhi, D., Al A'yadeh, I. (2020b). An investigation of factors affecting patients waiting time in primary health care centers: An assessment study in Dubai. *Management Science Letters, 10*(6), 1265–1276. https://doi.org/10.5267/j.msl.2019.11.031.

Aburayya, A., et al. (2020c). Critical success factors affecting the implementation of tqm in public hospitals: A case study in UAE Hospitals. *Systematic Reviews in Pharmacy, 11*(10). https://doi.org/10.31838/srp.2020.10.39.

Aburayya, A., et al. (2020d). An empirical examination of the effect of TQM practices on hospital service quality: An assessment study in uae hospitals. *Systematic Reviews in Pharmacy, 11*(9). https://doi.org/10.31838/srp.2020.9.51.

Ahmed, G. (2015). "DESTINATION 2021" Forbes Middle East Guide, pp. 46–47.

Akram, O. K., Franco, D. J., & Ismail, S. (2016). Smart buildings–A new environment (Theoretical approach). *International Journal of Engineering Technology, Management and Applied Sciences, 4*(4), 1–5.

Al Batayneh, R. M., Taleb, N., Said, R. A., Alshurideh, M. T., Ghazal, T. M., & Alzoubi, H. M. (2021). IT governance framework and smart services integration for future development of Dubai infrastructure utilizing AI and Big Data, its reflection on the citizens standard of living. In *The International Conference on Artificial Intelligence and Computer Vision* (pp. 235–247).

Al Naqbia, E., Alshuridehb, M., AlHamadc, A., & Al, B. (2020). The impact of innovation on firm performance: a systematic review. *International Journal of Innovation, Creativity and Change, 14*(5), 31–58.

Al Shebli, K., Said, R. A., Taleb, N., Ghazal, T. M., Alshurideh, M. T., & Alzoubi, H. M. (2021). RTA's employees' perceptions toward the efficiency of artificial intelligence and big data utilization in providing smart services to the residents of Dubai. In *The International Conference on Artificial Intelligence and Computer Vision* (pp. 573–585).

Al Suwaidi, F., Alshurideh, M., Al Kurdi, B., & Salloum, S. A. (2021). *The impact of innovation management in SMEs performance: a systematic review*, vol. 1261. AISC.

Alameeri, K., Alshurideh, M., Al Kurdi, B., & Salloum, S. A. (2021). *The effect of work environment happiness on employee leadership*, vol. 1261. AISC.

Alhashmi, S. F. S., Alshurideh, M., Al Kurdi, B., Salloum, S. A. (2020). A systematic review of the factors affecting the artificial intelligence implementation in the health care sector, vol. 1153. In *AISC*.

Allozi, A., Alshurideh, M., AlHamad, A., & Al Kurdi, B. (2022). Impact of transformational leadership on the job satisfaction with the moderating role of organizational commitment: case of UAE and Jordan manufacturing companies. *Academy of Strategic Management Journal*, 21, 1–13.

Alshurideh, M., Al Kurdi, B. H., Vij, A., Obiedat, Z., & Naser, A. (2016). Marketing ethics and relationship marketing-An empirical study that measure the effect of ethics practices application on maintaining relationships with customers. *International Business Research, 9*(9), 78–90.

Alshurideh, M., Al Kurdi, B., Shaltoni, A. M., & Ghuff, S. S. (2019a). Determinants of pro-environmental behaviour in the context of emerging economies. *International Journal of Sustainable Society, 11*(4). https://doi.org/10.1504/IJSSOC.2019.104563.

Alshurideh, M., Salloum, S. A., Al Kurdi, B., & Al-Emran, M. (2019b). Factors affecting the social networks acceptance: an empirical study using PLS-SEM approach. In *Proceedings of the 2019b 8th International Conference on Software and Computer Applications* (pp. 414–418).

Alshurideh, M. T., Al Kurdi, B., Masa'deh, R., & Salloum, S. A. (2021b). The moderation effect of gender on accepting electronic payment technology: a study on United Arab Emirates consumers. *Review of International Business and Strategy, 31*(3). https://doi.org/10.1108/RIBS-08-2020-0102.

Alshurideh, M. T., et al. (2021a). Factors affecting the use of smart mobile examination platforms by universities' postgraduate students during the COVID-19 pandemic: An empirical study. *Informatics, 8*(2). https://doi.org/10.3390/informatics8020032.

AlSuwaidi, S. R., Alshurideh, M., Al Kurdi, B., & Aburayya, A. (2021). The main catalysts for collaborative R&D projects in Dubai industrial sector. In *The International Conference on Artificial Intelligence and Computer Vision* (pp. 795–806).

Alzoubi, H., Alshurideh, M., Kurdi, B. A., & Inairat, M. (2020). Do perceived service value, quality, price fairness and service recovery shape customer satisfaction and delight? A practical study in the service telecommunication context. *Uncertain Supply Chain Management, 8*(3). https://doi.org/10.5267/j.uscm.2020.2.005.

Alzoubi, B., Alshurideh, H., Akour, M., Shishan, I., Aziz, F., & Al Kurdi, R. (2021). Adaptive intelligence and emotional intelligence as the new determinant of success in organizations. An empirical study in Dubai's real estate. *Journal of Legal, Ethical and Regulatory Issues, 24*(6), 1–15.

Al-Zu'bi, Z. M. F., Al-Lozi, M., Dahiyat, S. E., Alshurideh, M., Al Majali, A. (2012). Examining the effects of quality management practices on product variety. *European Journal of Economics, Finance and Administrative Sciences, 51*(1).

Amponsah, C. T., & Ahmed, G. (2017a). New global dimensions of business excellence. *International Journal of Business Excellence, 13*(1), 60–78.

Amponsah, C. T., & Ahmed, G. (2017b). Factors affecting entrepreneurship in emerging economies: A case of Dubai. *Journal for International Business and Entrepreneurship Development, 10*(2), 120–137.

Arkin, H., & Paciuk, M. (1997). Evaluating intelligent buildings according to level of service systems integration. *Automation in Construction, 6*(5–6), 471–479.

Atif, M. R., & Galasiu, A. D. (2003). Energy performance of daylight-linked automatic lighting control systems in large atrium spaces: Report on two field-monitored case studies. *Energy and Buildings, 35*(5), 441–461.

Buckman, A. H., Mayfield, M., & Beck, S. B. M. (2014). What is a smart building? *Smart and Sustainable Built Environment.*

Caragliu, A., Del, B., C-Nijkamp, P. (2011). Smart cities in Europe Journal of Urban Technology, *18*(2), 65–82.

Chourabi, H., et al. (2012). Understanding smart cities: An integrative framework. In *2012 45th Hawaii International Conference on System Sciences* (pp. 2289–2297).

Colding, J., Barthel, S., & Sörqvist, P. (2019). Wicked problems of smart cities. *Smart Cities, 2*(4), 512–521.

Das, R., Tuna, G., & Tuna, A. (2015). Design and Implementation of a Smart Home for the Elderly and Disabled. *Environment, 1*, 3.

Deakin, M. (2013). *Smart cities: governing, modelling and analysing the transition.* Routledge.

Doherty, P. From Smart Buildings to Smart Cities. *Intelligent Building, 12*(49), 1–4.

El Khatib, M. M., & Ahmed, G. (2018). Improving efficiency in IBM asset management software system 'Maximo': A case study of Dubai Airports and Abu Dhabi national energy company. *Theoretical Economics Letters, 8*(10), 1816–1829.

El Khatib, M. M., Al-Nakeeb, A., & Ahmed, G. (2019). Integration of cloud computing with artificial intelligence and Its impact on telecom sector—A case study. *iBusiness, 11*(01), 1.

Ghazal, T. M., et al. (2021a). IoT for smart cities: Machine learning approaches in smart healthcare—a review. *Future Internet, 13*(8), 218. https://doi.org/10.3390/fi13080218

Ghazal, T. M., Alshurideh, M. T., & Alzoubi, H. M. (2021b). Blockchain-enabled internet of things (IoT) platforms for pharmaceutical and biomedical research. In *The International Conference on Artificial Intelligence and Computer Vision* (pp. 589–600).

Hollands, R. G. (2020). Will the real smart city please stand up?: Intelligent, progressive or entrepreneurial? In *The Routledge companion to smart cities* (pp. 179–199). Routledge.

Intel, D. (2016). *Smart cities start with smart buildings.* Harvard business review.

Joghee, S., Alzoubi, H. M., Alshurideh, M., Al Kurdi, B. (2021). The role of business intelligence systems on green supply chain management: Empirical analysis of FMCG in the UAE. In *The International Conference on Artificial Intelligence and Computer Vision* (pp. 539–552).

Kabrilyants, R., Obeidat, B. Y., Alshurideh, M., Masa'deh, R. (2021). The role of organizational capabilities on e-business successful implementation. *International Journal of Data and Network Science, 5*(3). https://doi.org/10.5267/j.ijdns.2021.5.002.

KPMG 2017 Revenue (Annual Report). (2017). Retrieved, from https://big4accountingfirms.com/the-blog/kpmg-2017-revenue-annual-report/.

Krane, J. (2009). *Dubai: The story of the world's fastest city.* Atlantic Books Ltd.

Kroner, W. M. (1997). An intelligent and responsive architecture. *Automation in Construction, 6*(5–6), 381–393.

Lazaroiu, C., Roscia, M., Zaninelli, D. (2018). Fuzzy logic to improve prosumer experience into a smart city. In *2018 International Conference on Smart Grid (icSmartGrid)* (pp. 52–57).
Lazaroiu, G. C., & Roscia, M. (2012). Definition methodology for the smart cities model. *Energy, 47*(1), 326–332.
Lee, M., Ramiz, K., Hanaysha, P., Alzoubi, J., Alshurideh, H. (2022). Investigating the impact of benefits and challenges of IOT adoption on supply chain performance and organizational performance: An empirical study in Malaysia. *Uncertain Supply Chain Management, 10*, 1–14.
Leo, S., Alsharari, N. M., Abbas, J., & Alshurideh, M. T. (2021). *From offline to online learning: A qualitative study of challenges and opportunities as a response to the COVID-19 Pandemic in the UAE higher education context*, vol. 334.
Madi Odeh, R. B. S., Obeidat, B. Y., Jaradat, M. O., Masa'deh, R., & Alshurideh, M. T. (2021). The transformational leadership role in achieving organizational resilience through adaptive cultures: the case of Dubai service sector. *International Journal of Productivity and Performance Management*. https://doi.org/10.1108/IJPPM-02-2021-0093.
Morvaj, B., Lugaric, L., & Krajcar, S. (2011). Demonstrating smart buildings and smart grid features in a smart energy city. In *Proceedings of the 2011 3rd International Youth Conference On Energetics (IYCE)* (pp. 1–8).
Moss Kanter, R., & Litow, S. S. (2009). *Informed and interconnected: A manifesto for smarter cities*. Harvard Business School General Management Unit Working Paper.
Nam, T., & Pardo, T. A. (2011). Conceptualizing smart city with dimensions of technology, people, and institutions. In *Proceedings of the 12th Annual International Digital Government Research Conference: Digital Government Innovation in Challenging Times* (pp. 282–291).
Nuseir, M. T., Aljumah, A., & Alshurideh, M. T. (2021a). *How the business intelligence in the new startup performance in UAE during COVID-19: The mediating role of innovativeness*, vol. 334.
Nuseir, M. T., Al Kurdi, B. H., Alshurideh, M. T., & Alzoubi, H. M. (2021b). Gender discrimination at workplace: do artificial intelligence (AI) and machine learning (ML) have opinions about it. In *The International Conference on Artificial Intelligence and Computer Vision* (pp. 301–316).
Sherif, M., Almulla, M., Shetty, A., & Chowdhury, R. K. (2014). Analysis of rainfall, PMP and drought in the United Arab Emirates. *International Journal of Climatology, 34*(4), 1318–1328.
Shishan, F., Mahshi, R., Al Kurdi, B., Alotoum, F. J., & Alshurideh, M. T. (2022). Does the past affect the future? An analysis of consumers' dining intentions towards green restaurants in the UK. *Sustainability, 14*(1), 1–14.
Smart City Index. (2021). A tool for action, an instrument for better lives for all citizens. Retrieved, from https://imd.cld.bz/SmartCity20211
Taryam, M., et al. (2020). Effectiveness of not quarantining passengers after having a negative COVID-19 PCR test at arrival to dubai airports. *Systematic Reviews in Pharmacy, 11*(11). https://doi.org/10.31838/srp.2020.11.197.

Dubai Smart City as a Knowledge Based Economy

Mounir M. El Khatib, Naseem Abidi, Ahmad Al-Nakeeb, Muhammad Alshurideh, and Gouher Ahmed

Abstract Innovation, emerging technologies and digitalization have significant impacts on personal, social, environmental, and business aspects. This study explores the initiatives taken by the United Arab Emirates (UAE) to accelerate its knowledge-based economy to ensure the best living experience for its people. Knowledge is the key factor in growth and labor market expansion where skills, knowledge, and experience possessed by individuals, are the driver of creativity and innovation, with dependence on information and communication technology (ICT) as an enabler. The UAE is moving towards a knowledge-based economy in order to maintain a diversified economy in line with "The Abu Dhabi Economic Vision 2030" which was launched in 2009. Through a systematic literature review, researchers have developed a perspective on transformation of economy and society through innovation, use of Information and Communication Technology (ICT), as well as government performance management frameworks. Theory of sense data, nomothetic explanations, predictive and prescriptive analytics, along with Smart Dubai as a case study, are used to assess the achievements, emerging trends and expectations of knowledge-based economy in the UAE. Smart Dubai is a unique example of a high-level quality system that strives towards making Dubai the happiest city on earth. Smart Dubai has been adopting many projects that enhance the life of people and support the

M. M. El Khatib · A. Al-Nakeeb
School of Business and Quality Management, Hamdan Bin Mohammed Smart University, Dubai, United Arab Emirates
e-mail: a.alnakeeb@hbmsu.ac.ae

N. Abidi · G. Ahmed
Skyline University College, Sharjah, United Arab Emirates
e-mail: gouher@usa.net

M. Alshurideh (✉)
Department of Marketing, School of Business, The University of Jordan, Amman, Jordan
e-mail: m.alshurideh@ju.edu.jo; malshurideh@sharjah.ac.ae

Department of Management, College of Business Administration, University of Sharjah, Sharjah, United Arab Emirates

M. Alshurideh et al. (eds.), *The Effect of Information Technology on Business and Marketing Intelligence Systems*, Studies in Computational Intelligence 1056,
https://doi.org/10.1007/978-3-031-12382-5_90

economy of the UAE. This study highlights the challenges related to the implementation of the Smart Dubai project. The study also provides a clear view of smart cities transformation into a knowledge-based economy and utilizes predictive and prescriptive analytics tools that can assist other researches in the future to improve and develop research regarding smart cities and gain a better understanding. Ten recommendations are made in this study, for the Smart Dubai to consider, in order to maintain high-quality performance and accelerate the transformation process of Knowledge-based economy.

Keywords Knowledge-based economy · Innovation · Analytics · Digitalization · United Arab Emirates · Sustainable development

1 Introduction

The OECD defines Knowledge-based economies as "economies which are directly based on the production, distribution, and use of knowledge and information" (OECD, 1996, p.7). In the Knowledge Economy (KE), possession, use, and transfer of knowledge is essential (Al Mehrez et al., 2020; AlMehrzi et al., 2020; Altamony et al., 2012; Shannak et al., 2012). So, it is imperative to combine and synergize people, knowledge and technology to facilitate enhancement of value addition at micro and macro-economic levels. Powell & Snellman (2004) defined knowledge economy as "production and services based on knowledge-intensive activities that contribute to an accelerated pace of technological and scientific advance as well as equally rapid obsolescence". According to (Powell and Snellman 2004; Obeidat et al., 2021; Alsuwaidi et al., 2020; Al Naqbia et al., 2020; Al-Dhuhouri et al., 2020), the main components of a knowledge economy include a greater dependence on intellectual capabilities than on physical inputs or natural resources, combined with efforts to integrate improvements in various stages of the production process, from the R&D lab to the factory floor to the interface with customers (Ahmad et al., 2021; AlSuwaidi et al., 2021; Harahsheh et al., 2021; Svoboda et al., 2021).

Technological advancements and Information and communications technology (ICT) largely effect the globalization and knowledge economy because of the imbalances, and instability (Aljumah et al., 2021; Alshurideh et al., 2014; Alyammahi et al., 2020; Alzoubi et al., 2020; Ammari et al., 2017; Suciu, 2004). Dahlman and Utz (2005) highlighted that a knowledge-based economy is successful when there is a link between industry, science and technology. The empowerment comes from increased lifelong and education learning with higher investments in software and research and development. The dissemination of knowledge and technology is important and requires the "knowledge networks and national innovation systems" (Chartrand, 2006). According to Brinkley (2006), knowledge economy is, when firms bring technology and talent to create wealth. Also, According to Ahmed and Alfaki (2013), a knowledge-based economy is wider than the new economy or high technology. The foundation of a knowledge-based economy is to create, disseminate and use knowledge, and knowledge assets are given more importance than labor and capital assets. Social and economic activities reach to a very high level through the use of quality of knowledge.

The economic growth and competitiveness are added to the earlier definitions of knowledge-based economy (Huggins et al., 2014), which is the value addition to the science and technology, and also to the social, economic, and educational progress. In this context, (Skrodzka, 2016), suggested that the knowledge-based economy promotes innovation and entrepreneurship. Similarly, (Khanmirzaee et al., 2018), noted that science and technology parks play a major role in the development of technology and are able to boost the economic growth of countries. (Al-Gasaymeh et al., 2015, 2020; Assad & Alshurideh, 2020a; López-Cabarcos et al., 2020), study based on knowledge-based theory, organizational learning theory and dynamics theory, investigates how knowledge transforms into profitability, and their results show the importance of both tacit and explicit knowledge for achieving sustainable competitive advantages. According to LaFayette et al., 2019a, 2019b; Shah et al., 2021, 2020; Assad & Alshurideh, 2020b; Al-Jarrah et al., 2012), the world economy is passing through a major transformation from an industrial economy to a knowledge economy, rendering knowledge as a primary factor in the production.

This study presents the perspective of knowledge-based economy, predictive and prescriptive analytics, theory of sense data, and nomothetic explanations for Smart Dubai as a case study. The Smart Dubai case study was examined on three perspectives such as knowledge-based economy, the use of the analytical tools, and Dubai pulse initiative, and an insight was developed to examine the transformation of the UAE's economy to knowledge-based economy. Secondary sources data were collected from World Bank Report, Global Innovation Index Report, ICT-Global Competitiveness Index Report and websites of government of the UAE.

According to Snowdon (2015), 'Sense-data' is a modern term for a long-standing idea, earlier articulated in the language of 'ideas' and 'impressions'. "On the dominant use of 'sense-data' it stands for an entity present in perceptual (and other) experiences which fundamentally has a mental nature and possesses certain qualitative properties which the subject apprehends" (Snowdon 2015). Those authors who involved in the project use sense data to foresee challenges associated with the Smart Dubai project to make nomothetic explanations and to provide necessary recommendations.

Predictive analytics tools are used to analyze the current data available at the organization's databases to come up with certain conclusions that will assist the management to make decisions based on knowledge rather than speculations and assumptions. According to Gartner official website, predictive analysis "is a form of advanced analytics which examines data or content to answer the question "What is going to happen?" or more precisely, "What is likely to happen?", and is characterized by techniques such as regression analysis, forecasting, multivariate statistics, pattern matching, predictive modeling, and forecasting." (Gartner Glossary, 2019a).

Prescriptive analytics tools help to anticipate what will happen with the reason and the timing of the events in order to reduce future risks and make better decisions. According to Gartner official website "Prescriptive Analytics is a form of advanced analytics which examines data or content to answer the question "What should be done?" or "What can we do to make it happen?", and is characterized by techniques such as graph analysis, simulation, complex event processing, neural networks, recommendation engines, heuristics, and machine learning (Gartner Glossary, 2019b).

Prescriptive tools applied for two approaches, optimization model and rules model. The optimization model chooses the best alternative or decision to maximize/minimize objective function under constraints. Optimization model is done through an analytical code using a software with constraints to optimize the planning, scheduling, and assigning limited resources to minimize the cost. For example, deciding on the best route for oil-transporting vehicles from the depo to the petrol station, the model will be based on many constraints like vehicle size, petrol type, destination, distance, driver and the expected traffic. The second approach is the rules model. This model uses the inputs, judgment and business rules to assign the best action. For example, providing customers with certain treatments or offers (Gartner Research, 2015).

2 Rationale and Objectives

UAE desires to maintain a diversified economy in line with "The Abu Dhabi Economic Vision 2030" which was launched in 2009 (Ahmed & Alfaki, 2013). It is obvious that the UAE is moving towards knowledge-based economy (Aswad et al., 2011). Accordingly, the Government of the UAE aims to build a knowledge-based economy with sustainability that will include process, product, and distribution technologies to meet the needs for the UAE economy diversification while overcoming any challenges or conflicts and acquiring all opportunities available during a knowledge-based economy transformation (Mina, 2014). There is a need for the use of the analytics tools to ensure that a knowledge-based economy adds value to the economy of the UAE.

Considering the initiatives taken by the government of UAE to transform its traditional oil-based (resource-based) economy to a knowledge-based economy, the main objective of this research paper is to assess the achievements, emerging trends and expectations of knowledge-based economy. It is also an objective to review the Smart Dubai project as a case study to foresee challenges and make recommendations to accelerate the process of its knowledge-based economy.

3 Knowledge Economy in Brief

Literature has outlined key characteristics of knowledge economy with a common thread that knowledge plays an important role in it. According to Karlsson et al. (2009), the Knowledge Economy's main characteristics are continuous increase in knowledge investments such as education and knowledge production, and widening application of Knowledge for development, production, distribution, and use of goods and services. White et al. (2013) proposed open innovation, education, knowledge management, and creativity as characteristics for the knowledge-based economy. Similarly, Tapscott (2014), puts forth that knowledge is the basic production factor, knowledge economy is a digital economy and visualization plays an important role in the knowledge economy as major characteristics of the knowledge economy.

The particular nature and high quality level of the knowledge-based economy provides new prospects for companies, countries and local communities to create a proper environment for a sustainable development (Sirbu et al., 2009).

Knowledge-based economy and sustainable development are targeted by many economies worldwide. "Knowledge-based economy can promote sustainable development through innovation sustained by economic growth" (Jednak & Kragulj, 2015).

Researchers accept that knowledge economy includes the advancement of knowledge creation, diffusion and the positive effects on economy; hence, there is a need to measure the knowledge economy at the global level. To measure the knowledge economy, the World Bank has Knowledge Management assessment methodology, which is based on 109 items, and is used in 146 countries for performance measurements based on 4 pillars (Debnath, 2002; The World Bank, 2012).

The four pillars are grouped into two indexes: Knowledge Index (KI), which includes Education Index, Innovation Index and ICT Index and Knowledge Economy Index (KEI), which comprises of Economic and Institutional Regime Index and Education Index. These two KI and KEI are based on a scale from 0 to 10, whereas the 10 is the highest level (Chen & Dahlman, 2005; Sundać & Fatur Krmpotić, 2011). According to Parcero and Ryan (Parcero & Ryan, 2017), the four pillars of a knowledge-based economy are (1) information and communication technology, (2) education, (3) innovation, and (4) economy and regime.

4 UAE Economy

After its independence in the year 1971, the economy of UAE was mainly driven by oil and global finance for over three decades. Early in the twenty-first century, UAE started initiatives to diversify its economy to reduce its dependence on oil. One of the national priorities of UAE is to become a competitive knowledge economy, driven by innovation (Khatib & Ahmed, 2019). In 2014, H.H. Sheikh Mohammed bin Rashid Al Maktoum, Vice President and Prime Minister of the UAE and the Ruler of Dubai, launched the 'National Agenda' to guide efforts towards the country's Vision 2021, which is aimed to position the UAE among the best countries in the world by the Golden Jubilee of the Union in 2021. The agenda specified a wide-range of work programs centered around 6 national priorities and 52 National Key Performance Indicators (NKPIs). The third national priority of UAE is "Competitive

Knowledge Economy driven by Innovation", and the government is putting all its efforts to achieve it under vision 2021.

The UAE is known as a 'trailblazer' in the Arab World, setting pace to economic development in the Arab region through innovations and excellence, which is the watch-word of the country in any economic domain (Aburayya et al., 2020a; Ahmed, 2021; Amponsah & Ahmed, 2017a, 2017b). In fact, the UAE seems to be a Renaissance Arab State, standing for the Arab World's resurgence (Ahmed, 2018). The incredible economic development of the UAE is greatly due to its eminent leaders at the helm of the affairs of the state and their exemplary leadership (Aburayya et al., 2020b; Alzoubi & Ahmed, 2019; Khatib et al., 2019; Taryam et al., 2020).

The Nations vision 2021, in the golden jubilee year, intends to take the economy far to the ranks of the developed countries and world-class in every respect. The UAE had a population of 9.63 million and commanded a GDP worth of US$ 414.18bn in year 2018 from its $2 billion in year 1971, and mobile phone subscription of 208 per 100 people with around 99% internet users (The World Bank, 2019). It seems a fairy tale-like transformation, attracting tourists and shoppers not only from the Arab region but also across the world.

The UAE's economy has a dynamic change in all the indicators from 1990 to 2018, except the agricultural sector which is less than 1% of GDP and the fertility rate has decreased drastically from 4.5 to 1.5 births per women, which is due to the women participating in the UAE work force and the UAE's government encouragement towards the Emiratization. GDP has increased to about 8 times from 1990 to 2018. At present, the UAE is the third largest economy in the MENA region, after Saudi Arabia and Iran (The World Bank, 2019).

5 UAE Knowledge Economy

If KEI rankings of Middle East and North Africa (MENA) countries are compared in 2012 with respect to their ranking in the year 2000, it can be observed that 10 countries of the region have risen upward, whereas KEI rankings of 13 countries fell back. On the knowledge economy index, UAE stood the second-best ranked country in the MENA region after Israel, which slipped down 8 ranks in the past 12 years. Its leading position reflected strong performance on the pillars of innovation, education and ICT. Apart from the UAE, Oman, Saudi Arabia, Azerbaijan, Algeria and Tunisia had good performance on the KEI, as Oman climbed 18 positions, Saudi Arabia 26 positions and Azerbaijan 15 position, Algeria 14 positions and Tunisia 9 positions as compared to 2000. Since World Bank stopped publishing KEI ranking from 2012, the authors considered Global Innovation Index (GII) report to gauge the progress of UAE on knowledge-based economy.

The GII ranks the innovation performance of nearly 130 economies around the world. Since 2011, Switzerland has ranked 1st in the GII ranking every year. UAE was ranked 47 in the year 2015 and has improved to 36 in the year 2019. In North America, U.S and Canada ranked No.1 & No.2, in Europe Switzerland, Sweden and

Netherlands ranked 1, 2 and 3, whereas in Northern Africa and Western Asia, Israel, Cyprus and the UAE ranked 1, 2, and 3 respectively. The UAE achieved good success in knowledge utilization due to availability of technological infrastructure and moved up 6 points towards knowledge-based economy (Dutta et al., 2019), as it had one of the strategic goals to expand in the non-oil sector with diversification strategies for competing with emerging markets (Mehmood et al., 2019). Furthermore, it has also been observed in the report that the UAE produce lower level of output relative to their innovation inputs and comes under below expectations for level of developments under high income category.

It can be noted that, UAE needs to focus on improving its ranks on business environment, knowledge workers, knowledge creation and knowledge impact, which are key pillars of knowledge-based economy.

6 Case of Smart Dubai

Smart Dubai was first established in 2000 and was known as Dubai e-Government until 2013 when H.H Sheikh Mohammed Bin Rashid Al Maktoum changed its name to Smart Dubai government department. The vision of Smart Dubai is aligned with the vision of H.H Sheikh Mohammed bin Rashid Al Maktoum to make Dubai the happiest city on Earth and since then Smart Dubai has been pioneering smart city solutions and has been focusing on transforming and empowering Dubai through disruptive innovative initiatives, large scale projects by collaborating with the public and private sector, leading technological projects and many other initiatives (Smart Dubai, 2019d).

Smart Dubai has several initiatives that made a big impression regionally and even globally.

- Dubai Paperless initiative targeted towards the environmental sustainability aspect to preserve the environment by making Dubai a fully paperless government where governmental departments will no longer issue or ask for any papers or printed documents to be handed over across all operations (Smart Dubai, 2019c).
- Artificial Intelligence Ethical toolkit that includes the guidelines, self-assessment tool for developers, ethical considerations and common agreements. This initiative is targeted toward government entities, private sector, and individuals (Smart Dubai, 2019a).
- Dubai Blockchain Strategy project, which is a joint initiative between Smart Dubai and the Dubai Future Foundation that will lead the introduction of the blockchain technology in Dubai which is expected to "unlock 5.5 billion dirhams in savings annually in document processing alone" according to smart Dubai official website (Smart Dubai, 2019b).

Smart Dubai had a particular initiative called Dubai Pulse, which is a platform designed to provide all stakeholders, including government entities, private sector, and entrepreneurs, access to one location of shared open data. This initiative includes core analytical services as well as big data services that can be used by various entities to enhance the living conditions and the economy in the UAE by promoting the vision of making Dubai the smartest and happiest city on earth.

The following insights are developed from the Smart Dubai Case Study;

1. Smart Dubai is supporting the knowledge-based economy through empowering the government entities with the technological tools and the reliable infrastructure of data sharing, storage, processing, and visualization. This enables the government entities to make better knowledge-based decisions in order to serve the society more efficiently and effectively.
2. Smart Dubai is focused on building collaboration between the public and the private sectors to achieve several targets like smart transportation, smart society, smart economy, smart governance, and smart environment.
3. Smart Dubai has released the data law in 2015 "Law No. (26) of 2015 on the Organization of Dubai Data Publication and Sharing" (Data, 2019), which makes Dubai the first city globally to release a Law to state guidelines for sharing data between government entities and the open data available for the public.
4. The Dubai Data initiative will benefit the city by "increasing efficiencies in spending, job creation and innovation investment enhancement through creating new products and services and raise the level of citizen engagement. These benefits promote government transparency and the ultimate happiness of Dubai residents and citizens" (Pulse, 2016).
5. The main challenge that Smart Dubai is facing with building a knowledge-based economy, is the culture as people relay on printed papers and the usual routine to perform tasks. Providing tools that can replace people jobs can make some people feel uncomfortable.
6. Smart Dubai vision is to make Dubai the happiest city worldwide, although it may seem simple but it is rather a large scope and challenging vision as stated by Hamarneh (2019).
7. The Artificial Intelligence (AI) initiative by Smart Dubai that is considered as a predictive analysis project, which is used by Smart Dubai. According to Hamarneh (2019), this tool can be used in so many ways like:
    a. An AI service that will match children to schools, so it will make it easier for parents to choose the right school for their children, and reduces the burden of this task on parents.
    b. Another AI application is at Dubai Customs where thousands of containers are arriving to the port every day. AI will help them to decide what to inspect and why and when by analyzing the volume, history and country the container came from.
8. The AI can be of a great use in reducing or automating the repetitive tasks, like validating customer papers or the inserting of forms information or detecting certain trends that will enable the government entities to perform better.

9. Smart Dubai does not apply the prescriptive analytics tools, although (Hamarneh, 2019) mentioned that applying these tools can be of a great benefit to the government and help them in making better decisions that will lower the cost, and enable them expect the demand through supply and demand mathematical modules.
10. Dubai pulse is an initiative provided to all stakeholders to make different analytical tools by the different data sets available for the use of the public that are regulated through Dubai data law. Many government entities are participating in this platform like Dubai Customs, Dubai Municipality, Dubai Silicon Oasis Authority, Department of Dubai Economic Development, Roads and Transport Authority (RTA), Dubai Health Authority and many other departments.
11. Dubai Pulse provides different data views like city flow, property land distribution, people flow, and property buildings distribution. These data views are available to provide stakeholders with insights through interactive displays to enhance the stakeholder's decision-making process and create business value. Therefore, building a better knowledge-based economy (Pulse, 2019).
12. Smart Dubai project is basically an Information Technology enabled Services (ITeS) to stakeholders and need to be examined on its sustainability. The project needs to be viewed on six indicators to measure e-environmental sustainability performance of ITeS organizations, based on consumption rather than cost as suggested by Yadav et al. (2017).

Therefore, it is obvious that Smart Dubai is a unique example of a high-level quality system that strives towards making Dubai the happiest city on earth. Smart Dubai has been adopting many projects that enhance the life of people and support the economy of the UAE (Aburayya et al., 2020c; Al Batayneh et al., 2021; Al Shebli et al., 2021; Taryam et al., 2020). One of the main projects that Smart Dubai has initiated is the Dubai pulse project which is a platform to gather different data and visualize it in data views. This project is considered as a descriptive tool of data analytics, which can later be used for predictive and prescriptive analytics to assist in the decision-making process of the various stakeholders. Another project is the Artificial Intelligence and machine-learning project, which is a form of the predictive analytics tools, this project aims to eliminate any repetitive tasks and optimize the resources in order to provide better services for the citizens. As the use of the analytical tools can aid in the decision-making process on different managerial levels, hence establishing better services for the citizens and building a knowledge-based economy. Smart Dubai has successfully established a collaborative data platform but still needs to establish projects that will have the different forms of the analytical tools like regression analysis, predictive modeling, pattern matching, data mining and machine learning.

6.1 Challenges

The implementation of the tools in the government can encounter many challenges concerning infrastructure as well as the issue of technological compatibility. Some of the key challenges related to the implementation of Smart Dubai project are;

- Successful implementation of the tools depends on purchasing and deploying software for analytics. It also requires the engagement of the right people and processes to leverage, implement, maintain, and govern the models and tools.
- Government data is scattered across many departments making the gathering process a challenging task for the smart government. Furthermore, it can also be challenging to decide on what is actually important from the massive amount of data.
- Government data can consist of unneeded and redundant data and, therefore, cleaning and reconstruction of data will be required.
- The data available can consist of different data formats or still manual paper information or even scanned documents with no stored fields making it hard to rely on for analytical tools.
- The models that are generated from the analytical tools must be embedded in the operations and in the decision-making process to utilize their benefits.
- Testing and validating the models, based on different scenarios, can be time consuming and strenuous.
- Changing the culture and mentality of the people also can be considered as a challenge. Shifting from the regular decision-making process to advanced analytics tools can be challenging and requires a lot of change management.
- A systems approach to management of innovations in supply chains has two categories (value chain and supply chain) and has six levels. These levels of innovations are interlinked with different actors in the supply chain and different functions of value chain. The role of each actor in the supply chain for respective levels of innovations needs to be identified, incubated and managed professionally to utilize the benefits for growth and success (Gupta, 2013).

6.2 Global Cases

The advanced analytics tools are still considered as a new technological advancement; however, many governments and companies have decided to add the use of the tools to their strategies. The following cases can be compared with the case of Smart Dubai.

- A successful implementation of the advanced data analytical tools at US Immigration and Customs Enforcement (ICE), to manage the overseas offices effectively. The ICE manages more than 60 field offices in 45 nations. Until 2014 the decisions of choosing the ICE international offices were made based on subjective evidence or periodical surveys. However, in 2015 they decided to use the data of the workloads and the activities of the offices to decide on where to expand or where to shut offices without affecting the mission performance (Kelkar et al., 2016).

- Another case to benchmark with is the Virginia state case in building an integrated system for benefits eligibility of medical assistance. The offices that processed those applications were often understaffed and overloaded with work. However, in 2015, Virginia Department of Social Services (DSS) decided to take a different approach in addressing this problem. The information was gathered and analyzed by the use of the analytical tools to identify issues in processing the applications and, therefore, enabling the DSS to provide better social services to the citizens. (Kelkar et al., 2016).

7 Recommendations

Insights developed in the case study in the field of smart cities, may help and support other cities to become smart, discuss the conflict, challenges and opportunities of knowledge-based economy and add value to their economies. It also provides a clear view of smart cities transformation into a knowledge-based economy in addition to using predictive and prescriptive analytics tools that help other researchers in the future to improve and develop studies regarding smart cities and get a better understanding. The following recommendations are made to the Smart Dubai to consider in order to maintain the high-quality performance and accelerate the transformation process of a Knowledge-based economy:

- It is important to establish an information system and a collaborative platform of data for digital transformation of a smart city.
- Increase the utilization of technology in order to improve the economy.
- All stakeholders should be engaged in the transformation to smart city and the building of the platform for analytical tools.
- Create a roadmap to have a successful use of the analytical tools by monitoring, experimenting and implementation of these tools.
- Increase investments in education, technology and communication in order to achieve competitive advantages against other economies regionally and globally.
- Focus on training programs for employees to develop, improve and enhance their skills and experience.
- Train the government sector leaders on the new data analytics and how it can boost and sustain the economy.
- Identify the obstacles that are preventing the government from using the full potentials of the disruptive innovations.
- Establish a policy and a foundation of security, and accountability that protects privacy and promotes digital ethics.
- Identification of the most important data relevant to the decision-making process and the elimination of redundant and unneeded data.

8 Conclusion

Due to strong competition between cities and economies worldwide, Dubai's economy is perceived to be an exemplary example. His highness Sheikh Mohammed Bin Rashid Al Maktoum Vice President and Prime Minister of UAE, and Ruler of Dubai, envisioned economy of the UAE to be transformed from oil-dependent economy to a knowledge-based sustainable economy with the use of innovation and technology, and the creation of knowledge. Several initiatives in the UAE in general, and particularly in Dubai, made the economy of UAE stable and more flexible. The strategies of His Highness attracted some of the best and largest organizations, from different industries worldwide, to invest in Dubai in order to make it the best city in the world as well as the happiest city on earth by leveraging emerging technologies such as Blockchain, Artificial Intelligence, along with harnessing Data Science capabilities (Alshurideh et al., 2020, 2021; Alzoubi et al., 2021; Dubai, 2021; Ghazal et al., 2021; Joghee et al., 2021; Naqvi et al., 2021; Nuseir et al., 2021a, 2021b). Since the world bank stopped publishing reports on knowledge economy index, countries performance on KE parameters are now gauged from other international reports and indices. The UAE is ranked 36 globally, in the Global Innovation Index (GII) report in the year 2019, 2 ranks above from its position in the year 2018. UAE enjoys the first rank globally in the rate of mobile-broadband subscriptions, rate of the population covered by a mobile-cellular network, as well as the population covered by at least a 3G mobile network as per the Global Competitiveness Index, published by the International Telecommunication Union (ITU) in the year 2018. The UAE also ranked second in Mobile-cellular subscriptions, fourth in the rate of households with a computer, seventh in rate of households with Internet access at home, and eighth in internet users globally (TDRA, 2018). The UAE with such an excellent ICT infrastructure and, over 90 percent of expatriate population from all over the world as users and contributors to the economy, is an ideal setting that should to be utilized for accelerating knowledge-based economy with the use of predictive and prescriptive analytics tools and strong support from the government.

To accelerate the transformation process of knowledge-based economy, the UAE needs to improve its ranking on business environment, knowledge workers, knowledge creation and knowledge impact.

References

Aburayya, A., Alshurideh, M., Alawadhi, D., Alfarsi, A., Taryam, M., & Mubarak, S. (2020a). An investigation of the effect of lean six sigma practices on healthcare service quality and patient satisfaction: testing the mediating role of service quality in Dubai primary healthcare sector. *Journal of Advanced Research in Dynamical and Control Systems, 12*(8), 56–72.

Aburayya, A., Alshurideh, M., Albqaeen, A., Alawadhi, D., & Al A'yadeh, I. (2020b). An investigation of factors affecting patients waiting time in primary health care centers: An assessment study in Dubai. *Management Science Letters, 10*(6), 1265–1276.

Ahmad, A., Alshurideh, M. T., Al Kurdi, B. H., & Salloum, S. A. (2021). Factors Impacts Organization Digital Transformation and Organization Decision Making During Covid19 Pandemic. In *The Effect of Coronavirus Disease (COVID-19) on Business Intelligence* (p. 95).

Ahmed, G. (2015). *Destination 2021*. Forbes Middle East, pp. 46–47.

Ahmed, A., & Alfaki, I. M. A. (2013). Transforming the United Arab Emirates into a knowledge-based economy: The role of science, technology and innovation. *World Journal of Science, Technology and Sustainable Development*.

Al Batayneh, R. M., Taleb, N., Said, R. A., Alshurideh, M. T., Ghazal, T. M., & Alzoubi, H. M. (2021). IT Governance Framework and Smart Services Integration for Future Development of Dubai Infrastructure Utilizing AI and Big Data, Its Reflection on the Citizens Standard of Living. In *The International Conference on Artificial Intelligence and Computer Vision* (pp. 235–247).

Al Mehrez, A. A., Alshurideh, M., Al Kurdi, B., & Salloum, S. A. (2020). Internal factors affect knowledge management and firm performance: A systematic review. In *International Conference on Advanced Intelligent Systems and Informatics* (pp. 632–643).

Al Naqbia, E., Alshuridehb, M., AlHamadc, A., & Al, B. (2020). The impact of innovation on firm performance: A systematic review. *International Journal of Innovation, Creativity and Change*, 14(5), 31–58.

Al Shebli, K., Said, R. A., Taleb, N., Ghazal, T. M., Alshurideh, M. T., & Alzoubi, H. M. (2021). RTA's employees' perceptions toward the efficiency of artificial intelligence and big data utilization in providing smart services to the residents of Dubai. In *The International Conference on Artificial Intelligence and Computer Vision* (pp. 573–585).

Al-Dhuhouri, F. S., Alshuridch, M., Al Kurdi, B., & Salloum, S. A. (2020). Enhancing our understanding of the relationship between leadership, team characteristics, emotional intelligence and their effect on team performance: A Critical Review. In *International Conference on Advanced Intelligent Systems and Informatics* (pp. 644–655).

Al-Gasaymeh, A., Kasem, J., & Alshurideh, M. (2015). Real exchange rate and purchasing power parity hypothesis: Evidence from ADF unit root test. *International Research Journal of Finance and Economics, 14*, 450–2887.

Al-Gasaymeh, A., Almahadin, A., Alshurideh, M., Al-Zoubid, N., & Alzoubi, H. (2020). The role of economic freedom in economic growth: Evidence from the MENA region. *International Journal of Innovation Creative Change, 13*(10), 759–774.

Al-Jarrah, I., Al-Zu'bi, M. F., Jaara, O., & Alshurideh, M. (2012). Evaluating the impact of financial development on economic growth in Jordan. *International Research Journal of Finance and Economics, 94*, 123–139.

Aljumah, A., Nuseir, M. T., Alshurideh, M. T. (2021). The impact of social media marketing communications on consumer response during the COVID-19: Does the brand equity of a University Matter. In *The effect of Coronavirus Disease (COVID-19) on business intelligence* (vol 334, pp 367–384).

AlMehrzi, A., Alshurideh, M., & Al Kurdi, B. (2020). Investigation of the key internal factors influencing knowledge management, employment, and organisational performance: A qualitative study of the UAE hospitality sector. *International Journal of Innovation, Creativity and Change*, 14(1), 1369–1394.

Alshurideh, M., Shaltoni, A., & Hijawi, D. (2014). Marketing communications role in shaping consumer awareness of cause-related marketing campaigns. *International Journal of Marketing Studies, 6*(2), 163.

Alshurideh, M., Al Kurdi, B., & Salloum, S. A. (2020). Digital transformation and organizational operational decision making: A systematic review. In *International Conference on Advanced Intelligent Systems and Informatics* (pp. 708–719).

Alshurideh, M. T., Al Kurdi, B., & Salloum, S. A. (2021). The moderation effect of gender on accepting electronic payment technology: a study on United Arab Emirates consumers. *Review of International Business and Strategy*.

Alsuwaidi, M., Alshurideh, M., Al Kurdi, B., & Salloum, S. A. (2020). Performance appraisal on employees' motivation: A comprehensive analysis. In *International Conference on Advanced Intelligent Systems and Informatics* (pp. 681–693).

AlSuwaidi, S. R., Alshurideh, M., Al Kurdi, B., & Aburayya, A. (2021). The main catalysts for collaborative R&D projects in Dubai industrial sector. In *The International Conference on Artificial Intelligence and Computer Vision* (pp. 795–806).

Altamony, H., Masa'deh, R. M. T., Alshurideh, M., & Obeidat, B. Y. (2012). Information systems for competitive advantage: Implementation of an organisational strategic management process. In *Innovation and Sustainable Competitive Advantage: From Regional Development to World Economies—Proceedings of the 18th International Business Information Management Association Conference* (vol. 1, pp. 583–592).

Alyammahi, A., Alshurideh, M., Al Kurdi, B., & Salloum, S. A. (2020). The impacts of communication ethics on workplace decision making and productivity. In *International Conference on Advanced Intelligent Systems and Informatics* (pp. 488–500).

Alzoubi, H., & Ahmed, G. (2019). Do TQM practices improve organisational success? A case study of electronics industry in the UAE. *International Journal of Economics and Business Research, 17*(4), 459–472.

Alzoubi, H., Alshurideh, M., Kurdi, B. A., & Inairat, M. (2020). Do perceived service value, quality, price fairness and service recovery shape customer satisfaction and delight? A practical study in the service telecommunication context. *Uncertain Supply Chain Management, 8*(3), 579–588.

Alzoubi, H. M., Alshurideh, M., & Ghazal, T. M. (2021). Integrating BLE Beacon technology with intelligent information systems IIS for operations' performance: A managerial perspective. In *The International Conference on Artificial Intelligence and Computer Vision* (pp. 527–538).

Ammari, G., Al kurdi, B., Alshurideh, M., & Alrowwad, A. (2017). Investigating the impact of communication satisfaction on organizational commitment: A practical approach to increase employees' loyalty. *International Journal of Marketing Studies, 9*(2), 113–133.

Amponsah, C. T., & Ahmed, G. (2017a). Factors affecting entrepreneurship in emerging economies: A case of Dubai. *Journal for International Business and Entrepreneurship Development, 10*(2), 120–137.

Amponsah, C. T., & Ahmed, G. (2017b). New global dimensions of business excellence. *International Journal of Business Excellence, 13*(1), 60–78.

Assad, N. F., & Alshurideh, M. T. (2020a). Financial reporting quality, audit quality, and investment efficiency: Evidence from GCC economies. *WAFFEN-UND Kostumkd Journal, 11*(3), 194–208.

Assad, N. F., & Alshurideh, M. T. (2020b). Investment in context of financial reporting quality: A systematic review. *WAFFEN-UND Kostumkd Journal, 11*(3), 255–286.

Aswad, N. G., Vidican, G., & Samulewicz, D. (2011). Creating a knowledge-based economy in the United Arab Emirates: Realising the unfulfilled potential of women in the science, technology and engineering fields. *European Journal of Engineering Education, 36*(6), 559–570.

Brinkley, I. (2006). *Defining the knowledge economy*. The Work Foundation.

Chartrand, H. H. (2006). *Ideological evolution: The competitiveness of nations in a global knowledge-based economy*.

Chen, D. H. C., & Dahlman, C. J. (2005). The knowledge economy, the KAM methodology and World Bank operations. *World Bank Institute Working Paper* no. 37256.

Dahlman, C. J., & Utz, A. (2005). *India and the knowledge economy: leveraging strengths and opportunities*. World Bank Publications.

Data, D. (2019). Data Law. dubaidata.ae

Debnath, S. C. (2002). *Creating the knowledge-based economy in Kingdom of Saudi Arabia to solve the current unemployment crisis*.

Dubai Pulse. (2016). Dubai data initiative- governance, policy and engagement. dubaipulse.gov.ae

Dubai Pulse. (2019). *About Dubai Pulse*.

Dutta, S., Reynoso, R. E., Wunsch-Vincent, S., León, L. R., & Hardman, C. (2019). Creating the future healthy of medical lives–innovation. In *Global Innovation Index 2019: Creating Healthy Lives—The Future of Medical Innovation* (p. 201941).

El Khatib, M. M., & Ahmed, G. (2019). Management of artificial intelligence enabled smart wearable devices for early diagnosis and continuous monitoring of CVDS. *International Journal of Innovative Technology and Exploring Engineering, 9*(1), 1211–1215.
Gartner Glossary. (2019a). Predictive analytics (2). Gartner.com
Gartner Glossary. (2019b). Predictive analytics. Gartner.com
Gartner Research. (2015). *How to get started with prescriptive analytics.* Gartner.com.
Ghazal, T. M., Alshurideh, M. T., & Alzoubi, H. M. (2021). Blockchain-enabled internet of things (IoT) platforms for pharmaceutical and biomedical research. In *The International Conference on Artificial Intelligence and Computer Vision* (pp. 589–600).
Gupta, V. (2013). *Framework for managing innovations in supply chains of ICT products authors.*
Hamarneh, H. (2019). *How Smart Dubai is building a Knowledge-based Economy (interview).*
Harahsheh, A. A., Houssien, A. M. A., & Alshurideh, M. T. (2021). The effect of transformational leadership on achieving effective decisions in the presence of psychological capital as an intermediate variable in private Jordanian. In *The Effect of Coronavirus Disease (COVID-19) on Business Intelligence* (pp. 243–221). Springer Nature.
Huggins, R., Izushi, H., Prokop, D., & Thompson, P. (2014). *The global competitiveness of regions.* Routledge.
Jednak, S., & Kragulj, D. (2015). Achieving sustainable development and knowledge-based economy in Serbia. *Management: Journal of Sustainable Business and Management Solutions in Emerging Economies, 20*(75), 1–12.
Joghee, S., Alzoubi, H. M., Alshurideh, M., & Al Kurdi, B. (2021). The Role of business intelligence systems on green supply chain management: empirical analysis of FMCG in the UAE. In *The International Conference on Artificial Intelligence and Computer Vision* (pp. 539–552).
Karlsson, C., Johansson, B., & Stough, R. R. (2009). Human capital, talent, agglomeration and regional growth. KTH Royal Institute of Technology.
Kelkar, M., Viechnicki, P., Conlin, S., Frey, R., & Strickland, F. (2016). *Mission analytics* (pp. 1–28).
Khanmirzaee, S., Jafari, M., & Akhavan, P. (2018). A study on the role of science and technology parks in development of knowledge-based economy. *World journal of entrepreneurship, management and sustainable development.*
El Khatib, M. M., Al-Nakeeb, A., Ahmed, G. (2019). Integration of cloud computing with artificial intelligence and Its impact on telecom sector—A case study. *iBusiness 11*(1), 1.
LaFayette, B., Curtis, W., Bedford, D., & Iyer, S. (2019a). *Knowledge economies and knowledge work.* Emerald Group Publishing.
LaFayette, B., Curtis, W., Bedford, D., & Iyer, S. (2019b). How the economic landscape is changing. In *Knowledge Economies and Knowledge Work.* Emerald Publishing Limited.
López-Cabarcos, M. Á., Srinivasan, S., & Vázquez-Rodr\'\iguez, P. "The role of product innovation and customer centricity in transforming tacit and explicit knowledge into profitability. *Journal of Knowledge Management.*
Mehmood, T., Alzoubi, H. M., Alshurideh, M., Al-Gasaymeh, A., & Ahmed, G. (2019). Schumpeterian entrepreneurship theory: Evolution and relevance. *Acad. Entrep. J., 25*(4), 1–10.
Mina, W. (2014). United Arab emirates FDI outlook. *The World Economy, 37*(12), 1716–1730.
Naqvi, R., Soomro, T. R., Alzoubi, H. M., Ghazal, T. M., & Alshurideh, M. T. (2021). The Nexus between big data and decision-making: A study of big data techniques and technologies. In *The International Conference on Artificial Intelligence and Computer Vision* (pp. 838–853).
Nuseir, M. T., Al Kurdi, B. H., Alshurideh, M. T., & Alzoubi, H. M. (2021a). Gender discrimination at workplace: Do Artificial Intelligence (AI) and Machine Learning (ML) have opinions about it. In *The International Conference on Artificial Intelligence and Computer Vision* (pp. 301–316).
Nuseir, M. T., Aljumah, A., & Alshurideh, M. T. (2021b). How the business intelligence in the new startup performance in UAE during COVID-19: The mediating role of innovativeness. In *The effect of coronavirus disease (covid-19) on business intelligence* (pp. 63–79).
Obeidat, U., Obeidat, B., Alrowwad, A., Alshurideh, M., Masadeh, R., & Abuhashesh, M. (2021). The effect of intellectual capital on competitive advantage: The mediating role of innovation. *Management Science Letters, 11*(4), 1331–1344.

Parcero, O. J., & Ryan, J. C. (2017). Becoming a knowledge economy: The case of Qatar, UAE, and 17 benchmark countries. *Journal of the Knowledge Economy, 8*(4), 1146–1173.
Powell, W. W., & Snellman, K. (2004). The knowledge economy. *Annual Review of Sociology, 30*, 199–220.
Shah, S. F., Alshurideh, M., Al Kurdi, B., & Salloum, S. A. (2020). The impact of the behavioral factors on investment decision-making: A systemic review on financial institutions. In *International Conference on Advanced Intelligent Systems and Informatics* (pp. 100–112).
SF Shah MT Alshurideh A Al-Dmour R Al-Dmour 2021 Understanding the influences of cognitive biases on financial decision making during normal and COVID-19 pandemic situation in the United Arab Emirates. In *The Effect of Coronavirus Disease (COVID-19) on Business Intelligence* (pp. 257-274).
Shannak, R. O., Zu'bi, M. F., & Alshurideh, M. T. (2012). A theoretical perspective on the relationship between knowledge management systems, customer knowledge management, and firm competitive advantage. *European Journal of Social Sciences.*
Sirbu, M., Doinea, O., Mangra, M. G., and others, "Knowledge based economy-the basis for insuring a sustainable development. *Annals of the University of Petrosani, Economics, 9*(4), 227–232.
Skrodzka, I. (2016). Knowledge-based economy in the European Union--Cross-country analysis. *Statistics in Transition new series, 17*(2), 281–294.
Smart Dubai. (2019a). Artificial intelligence principles & ethics. smartdubai.ae
Smart Dubai. (2019b). Blockchain. smartdubai.ae.
Smart Dubai. (2019c). Paperless. smartdubai.ae.
Smart Dubai. (2019d). The vision of H.H. Sheikh Mohammed Bin Rashid Al Maktoum is to make dubai the happiest city on earth. smartdubai.ae
Snowdon, P. F. (2015). Sense-data. In *Oxford Handbook of Philosophy of Perception* (pp. 118–135).
Suciu, M. C. (2004). Noua economie şi societatea cunoaşterii [New Economics and Knowledge based society]. *Bucharest ASE*.
Sundać, D., & Fatur Krmpotić, I. (2011). Knowledge economy factors and the development of knowledge-based economy. *Croatian Economic Survey*, (13), 105–141.
Svoboda, P., Ghazal, T. M., Afifi, M. A. M., Kalra, D., Alshurideh, M. T., & Alzoubi, H. M. (2021). Information Systems Integration to Enhance Operational Customer Relationship Management in the Pharmaceutical Industry. In *The International Conference on Artificial Intelligence and Computer Vision* (pp. 553–572).
Tapscott, D. (2014). The Digital Economy Anniversary Edition: Rethinking Promise and Peril in the Age of Networked Intelligence New York, NY: McGraw-Hill, 2014 Reviewed by Howard A. Doughty. Citeseer.
Taryam, M., Alawadhi, D., & Aburayya, A. (2020). Effectiveness of not quarantining passengers after having a negative COVID-19 PCR test at arrival to Dubai airports. *Systematic Reviews in Pharmacy, 11*(11), 1384–1395.
TDRA. (2018). *UAE leads the ICT global competitiveness index*. tdra.gov.ae
The World Bank. (2012). *Knowledge assessment methodology*. Knoema.com
The World Bank (2019). *Country United Arab Emirates*. data.theworldbank.org
White, D. S., Gunasekaran, A., & Ariguzo, G. C. (2013). The structural components of a knowledge-based economy. *International Journal of Business Innovation and Research, 7*(4), 504–518.
Yadav, S. S. K., Abidi, N., & Bandyopadhayay, A. (2017). Development of the Environmental Sustainability Indicator Profile for ITeS Industry. *Procedia Computer Science, 122*, 423–430.

Impact of Organizational Learning Capabilities on Service Quality of Islamic Banks Operating in Jordan

Suleiman Ibraheem Shelash Al-Hawary, Ibrahim Rashed Soliaman AlTaweel, Nida'a Al-Husban, Mohammad Fathi Almaaitah, Faraj Mazyed Faraj Aldaihani, Muhammad Turki Alshurideh, Doa'a Ahmad Odeh Al-Husban, and Rania Ibrahim Mohammad

Abstract The major aim of the study was to examine the impact of Impact of organizational learning capabilities on service quality. Therefore, it focused on Islamic banks operating in Jordan. Data were primarily gathered through self-reported questionnaires creating by Google Forms which were distributed to a purposive sample

S. I. S. Al-Hawary (✉)
Faculty of Economics and Administrative Sciences, Department of Business Administration, Al Al-Bayt University, P.O. Box 130040, Mafraq 25113, Jordan
e-mail: dr_sliman73@aabu.edu.jo; dr_sliman@yahoo.com

I. R. S. AlTaweel
Faculty of Business School, Department of Business Administration, Qussim University, P.O. Box 6502, Al Russ City 51452, Saudi Arabia
e-mail: toiel@qu.edu.sa

N. Al-Husban
Faculty of Economics and Administrative Sciences, Department of Business Administration, Al Al-Bayt University, P.O. Box 130040, Mafraq 25113, Jordan

M. F. Almaaitah
Faculty of Economic and Administration Sciences, Department of Business Administration and Public Administration, Al Al-Bayt University Jordan, P.O. Box 130040, Mafraq 25113, Jordan
e-mail: m.maaitah@aabu.edu.jo

F. M. F. Aldaihani
Kuwait Civil Aviation, ishbiliyah bloch 1, street 122, home 1, Istanbul, Kuwait

M. T. Alshurideh
School of Business, Department of Marketing, The University of Jordan, Amman 11942, Jordan
e-mail: malshurideh@sharjah.ac.ae

Department of Management, College of Business, University of Sharjah, 27272 Sharjah, United Arab Emirates

D. A. O. Al-Husban
Faculty of Alia College, Department of Human Fundamental Sciences, Al-Balqa Applied University, Amman, Jordan
e-mail: D_husban@bau.edu.jo

M. Alshurideh et al. (eds.), *The Effect of Information Technology on Business and Marketing Intelligence Systems*, Studies in Computational Intelligence 1056,
https://doi.org/10.1007/978-3-031-12382-5_91

of managers via email. This study was conducted using structural equation modeling (SEM) to test hypotheses. The results show that interaction with external environment has the highest positive impact relationship on service quality. Upon study findings, the researcher recommends managers and decision makers to invest the human element continuously by involving them in the decision-making process to acquire administrative skills along with technical skills they already have.

Keywords Organizational learning capabilities · Service quality · Islamic banks · Jordan

1 Introduction

Organizations operating in the local and global fields of all sizes and activities face great challenges in a competitive world, and the rapid change of the client's wants (Al-Hawary & Al-Syasneh, 2020). Organizations tended to race to invest in information technology in order to enable and enhance the organization's ability to learn, in order to reach and satisfy customer expectations; the thing which is reflected in achieving the organization's goal and then survival and stability in a rapidly changing environment (Al-Hawary & Ismael, 2010; Al-Hawary & AlDafiri, 2017; Al-Hawary et al., 2018; Al-Hawary & Al-Hamwan, 2017). Learning is an effective tool to achieve strength and survival for organizations in light of the information revolution that the world is witnessing, and it is a basic requirement for the success of any organization nowadays (Al-Hawary, 2015; Al-Hawary & Aldaihani, 2016). Lukas (1996) shows the importance of organizational learning by saying that "Organizational learning is the key to future organizational success". Therefore, organizations need to increase the pace of organizational learning to be able to keep pace with the massive changes in the contemporary business environment, and to achieve this, it needs more modern ideas and concepts (Al-Hawary & Al-Namlan, 2018; Al-Hawary & Alwan, 2016; Al-Hawary & Nusair, 2017). It also needs to innovate products and modern techniques in training the human resource and encourage innovation and creativity during work, and organizations are now dealing with human resources as mental and intellectual energy, a source of information and knowledge, and innovation and creativity (Alshurideh et al., 2019a, 2019b; Al-Hawary et al., 2020; Al-Hawary & Al-Rasheedy, 2021; AlHamad et al., 2022; Alshurideh et al., 2022; Shamout et al., 2022).

With the continued fierce competition in the market, the quality of service has become an essential factor in maintaining competitive advantage (Alzoubi et al., 2020; Alshurideh, 2022; Hasan et al., 2022) in addition, it is an urgent and necessary need to establish and maintain customer satisfaction. And according to Mansour the quality of service is an essential approach to total quality management. Service quality has become the focus of many organizations due to the increased intensity of competition, especially between service organizations, and because the purchase

R. I. Mohammad
Ministry of Education, Amman, Jordan

decision has become more complex (Aburayya et al., 2020a; Alshamsi et al., 2020; Alshurideh et al., 2019a, 2019b; Zu'bi et al., 2012). Despite the importance of service quality, there is a difficulty in evaluating service quality compared to goods (Parasurainan et al., 1985) because it varies with different people and times, in addition to the fact that the service is intangible. Therefore, this study came to test the impact of job satisfaction between organizational learning capabilities and service quality in telecommunications companies in Jordan.

The issue of service quality is one of the topics that preoccupied researchers in accessing practical and conceptual models through which organizations can reach an actual reality of quality commensurate with market requirements, as there are many models for measuring quality that differ according to the environment and activity (Metabis & Al-Hawary, 2013; Alshurideh et al., 2017; Alolayyan et al., 2018; Al-Nady et al., 2016; Al-Hawary & Metabis, 2012; Al-Hawary & Hussien, 2017; Al-Hawary & Harahsheh, 2014; Al-Hawary & Abu-Laimon, 2013; Al-Hawary, 2013a, 2013b). Moreover, the increasing in intensity of competition is a result of globalization and the rapid development of information technology, and the huge diversity in the number and methods of providing the service, the thing that reflected in the organizations' abilities to strive towards continuous improvement in the quality of their services, to reach a high level of the quality of their services, which is reflected in their abilities to improve the level of their customer satisfaction, in order to maintain their customers and increase their market share (Abbad & Al-Hawary, 2011; Abu Qaaud et al., 2011; Al-Hawajreh et al., 2011; Al-Hawary et al., 2012; Al-Hawary, 2012; Al-Hawary, 2013a, 2013b; Al-Dmour et al., 2021; Alzoubi et al., 2022).

2 Theoretical Framework and Hypotheses Development

2.1 Organizational Learning Capabilities

Learning carries multiple factors of human activities, these factors prompted researchers to discover what is called organizational learning in contemporary organizations. Like any system, learning is a vital process that helps the growth and development of the organization and increases its organizational abilities; accordingly, learning refers to the organized transformation in the behavior of organizational groups from old procedures to others with competitive priority.

Researchers and practitioners have been paying close attention to the topic of organizational learning since the end of the 1970s, this revealed to the research the organizations' interest in new competitive methods that help them to survive in the markets and adapt to environmental changes in order to perform their work related to information technology and modern knowledge. Communication in the learning process is the main source of sustainable competitive advantage in light of the changing environment conditions (Al Kurdi et al., 2020; Alshurideh et al., 2020; Ashal et al., 2021). If the organization wants to raise its level of performance, and accordingly, organizations will not be able to achieve that goal except through human staff capable of learning and willing to learn, only then will they be able to succeed and adapt to the renewable environment variables.

How can the organization today achieve competitive advantage? In light of the great competitive challenges faced by organizations, the answer is the organization's ability to learn enables it to achieve competitive advantage, therefore learning is an effective tool that gives strength and survival to organizations, it also is one of the basic requirements for the success of any organization these days, Lukas (1996) says that organizational learning is, according to many researchers, the key of organizational success for the future, emphasizing the importance of organizational learning.

Organizational learning is one of the dynamic capabilities of the organization, through which human resources can absorb strategic knowledge in a way that enables them to improve their conditions and competitive position. Organizational learning capabilities represent the organization's ability to extract knowledge to implement strategic actions and then benefit from that knowledge when needed in order to adjust the organization's strategy. (Thite, 2004) confirms that organizational learning is a tactic aimed at learning dynamic capabilities, and (Garcia et al., 2006) confirms that implementing organizational learning processes provides the organization with a series of mechanisms that lead to many benefits such as outperforming organizational performance and increasing its competitiveness. (Berghman, 2006) said that organizational learning leads to an Increase in the dynamism of organizations and increase strategic innovation initiatives.

Learning capabilities have developed to become one of the most important managerial topics, The term organizational learning capabilities is influenced by a number of fields such as: organization theory, production management, strategic management, and management science (Beheshtifar et al., 2012). The learning capabilities encourage the organization and organizational process to learn effectively (Ahmed et al., 2011), it also increases organizational knowledge.

Researchers addressed the concept of organizational learning in different ways, and according to researchers and the models used, the concept of organizational learning differed among them, and by reviewing the management literature, the researcher did not find a unified definition of organizational learning capabilities, although many studies have dealt with this term, researchers didn't agree on a specific definition, Yueh defined organizational learning as: a process an organization undertakes to acquire and develop knowledge, this process has four basic components that are: getting knowledge from its sources, distributing information to those who deserve it, translating information into useful sentences, and finally storing it in the organizational memory. Tohidi and Mandegari (2012) referd to organizational learning capabilities as organizational and managerial characteristics that facilitate the process of organizational learning and allow the organization to learn, while Bahadori et al. defined organizational learning abilities as a set of tangible and intangible resources and tasks necessary to achieve competitive advantage, (Rashidi et al., 2010) considers organizational learning capabilities as an evidence that confirms the organization's ability to create and integrate ideas in a distinctive way and the ability to communicate with organizational boundaries using creative management methods.

From the previous argument, a definition of organizational learning capabilities can be deduced which is a group of managerial and organizational components (tangible and non-tangible) that plays an important role in the organizational learning process and encourages the organization to learn greatly. Organizational learning capabilities play an important role in helping an organization learn, and therefore improve the organization's performance, and helps it to achieve a sustainable competitive advantage, plus a lot of reasons, Bess et al. (2011) pointed that organizational learning capabilities can be measured in seven dimensions represented by knowledge sharing, dialogue, participation in decision-making, managerial commitment, expertise and openness, knowledge transfer, and risk, as every one of indicated (Alegre & Chiva, 2008; Goh & Richards, 1997; Jerez-Gomez et al., 2005; Onağ et al., 2014).

The organizational learning capabilities are represented in eleven dimensions, which are: Openness and interaction with the external environment, experimentation, managerial commitment, participative decision making, empowerment and leadership commitment, clarity of purpose and vision, knowledge transfer and integration, work teams, dialogue, risk taking, and systems perspective. While Chiva and Alegre (2009) pointed out that organizational learning capabilities are represented by five dimensions, which are interaction with the external environment, experience, participation in decision-making, dialogue, and risk-taking. External environment, experience, participation in decision-making, dialogue, and risk-taking.

Experimentation: Many researchers have relied on experimentation as a dimension of organizational learning capabilities (Nevis et al., 1995), Alegre and Chiva defines it as "the extent to which suggestions and new ideas are presented and how to deal with them." Researchers assured that experimentation guarantees that the organization comes up with new ideas, clarifies how things work, helps to make changes in work procedures, and suggests innovative solutions to problems, based on the use of clear methods and procedures (Garvin, 1993). Mat and Razak (2011) sees that experimentation results in ideas flow and proposals that guarantee a change in the existing system, so that experimentation is a symptoms of the creative environment. Accordingly, management should encourage and support freedom to conduct experiments using new and innovative methods of work (Garvin, 1993; McGill & Lei, 1992).

Risk Taking: Taking risks is one of the organizational learning capabilities dimensions, as many researchers addressed (Onağ et al., 2014). Risk taking is the degree to which ambiguity, uncertainty, and error are tolerated (Chiva & Alegre, 2009). And Liles defined risk taking as futile probability outcome due to different procedures. The business organization is supposed to encourage risk taking and accept mistakes (Rafiq et al., 2011).

Interaction with the external environment: interaction with the external environment dimension represents one of the organizational learning capabilities dimension that many researchers have included in their studies (Bess et al. 2011; Onağ et al., 2014; Alameeri et al., 2020) and Chiva and Alegre (2009) defined interaction with the external environment by the extent of which organization's interact with the environment and its adaptation to changes and the ability to influence the environment requirements." Environmental characteristics affect the learning process relationships therefore communication with the organizations external environment are important, especially with changing environments.

Dialogue: Researchers such as Bess et al. (2011) and Alegre and Chiva (2008) saw that dialogue as organizational learning capabilities dimension and a tool to provide communication and collective learning within the organization and between groups, dialogue has become an important process in business organizations as an aspect of understanding difficulties and possibilities of learning and change (Gear et al., 2003). Organization good communication network enhances learning as it provides an entrance to the appared knowledge, which in turn leads to new knowledge. Communication between management and workers allows for knowledge development and increase experience within the organization.

Participative decision making: Participation in decision-making is one of the organizational learning capacities as discussed by researchers (Alegre & Chiva, 2008). The process of giving employees the opportunity to participate in decision-making contribute to improve quality of the decision (Cotton et al., 1988). As it is considered as an incentive for employees, and involves achieving job satisfaction and employees organizational commitment (Daniels & Bailey, 2012).

2.2 Service Quality

Due to the competition intensity between service organizations and the fact that the purchase decision has become complicated, in addition to the increase of technological development that has helped organizations to provide electronic services with traditional service, organizations has increased its interests in service quality concept. Despite the service quality importance, there is a difficulty in evaluating it compared to goods (Parasurainan et al., 1985) because it varies according to people, time and place, in addition to the fact that the service is intangible (Al-Hawary & Al-Syasneh, 2020; Al-Hawary et al., 2017; Al-Hawary & Al-Menhaly, 2016; Al-Hawary et al., 2011)). There are many definitions of the concept of service quality; Where Gefan (2002) defined service quality as customer's subjective comparison to find the differences between the wanted service characteristics and the actual provided service. Parasuraman et al. define service quality as the comparison made by the customer between the current experience of receiving the service and the previous experiences of obtaining similar services. Also, Lehtinen and Lehtinen (1982) emphasized that service quality got three dimensions: physical quality, which focuses on the organization equipment and buildings, interactive quality, which result from organization employees and customers interaction that provide services, and organization quality, which is similar to organization image dimension that we referred to earlier. On the other hand, Parasurainan et al. (1985) detailed ten dimensions of service quality:

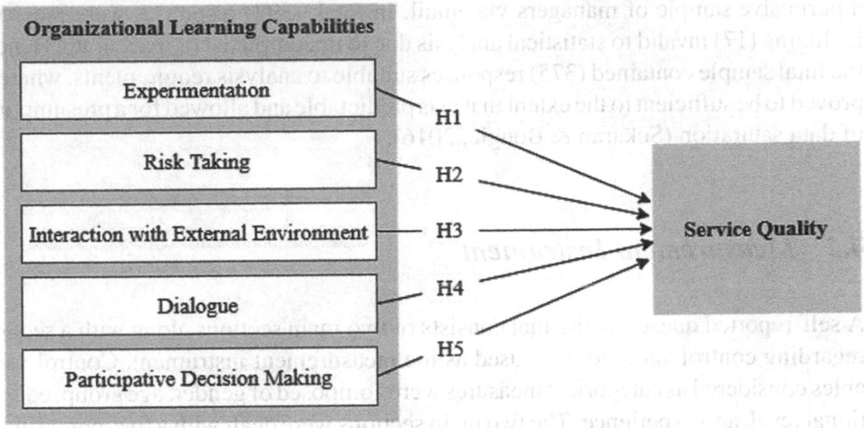

Fig. 1 Conceptual model

reliability, empathy, responsiveness, credibility, access, communication, courtesy, safety, tangibility, and the ability to understand the consumer.

Parasuraman et al. developed a 22-item tool to measure service quality, and when filtering service quality determinants, it now includes only five dimensions: tangible things, credibility, response, safety, and empathy, which is the individual care and attention that the organization shows towards its customers.

3 Research Model

See Fig. 1.

H1: There is a statistically significant impact of Organizational Learning Capabilities on Service Quality of Islamic banks operating in Jordan.

4 Methodology

4.1 Population and Sample Selection

A qualitative method based on a questionnaire was used in this study for data collection and sample selection. The major aim of the study was to examine the impact of Impact of organizational learning capabilities on service quality. Therefore, it focused on Islamic banks operating in Jordan. Data were primarily gathered through self-reported questionnaires creating by Google Forms which were distributed to a purposive sample of managers via email. In total, (392) responses were received including (17) invalid to statistical analysis due to uncompleted or inaccurate. Hence, the final sample contained (375) responses suitable to analysis requirements, where it proved to be sufficient to the extent that was predictable and allowed for a presumption of data saturation (Sekaran & Bougie, 2016).

4.2 Measurement Instrument

A self-reported questionnaire that consists of two main sections along with a section regarding control variables was used as the measurement instrument. Control variables considered as categorical measures were composed of gender, age group, educational level, and experience. The two main sections were dealt with a five-point Likert scale (from 1 = strongly disagree to 5 = strongly agree). The first section contained (21) items to measure organizational learning capabilities based on (Alegre & Chiva, 2008; Bess et al., 2011). These questions were distributed into dimensions as follows: five items dedicated for measuring experimentation, four items used for measuring risk taking, four items for measuring interaction with external environment, four items for measuring dialogue, and five items dedicated for measuring participative decision making. Whereas the second section included eight items used to measure service quality according to what was pointed by Aburayya et al. (2020b), Al-Hawary and Al-Menhaly (2016), Al-Hawary and Al-Smeran (2017), Al-Hawary et al. (2011), Al-Khayyal et al. (2020).

5 Findings

5.1 Measurement Model Evaluation

This study was conducted structural equation modeling (SEM) to test hypotheses, which represents a contemporary statistical technique for testing and estimating the relationship between factors and variables (Wang & Rhemtulla, 2021). Accordingly, the reliability and validity of the constructs were tested using confirmatory factor analysis (CFA) through the statistical program AMOSv24. Table 1 summarizes the results of convergent and discriminant validity, as well the indicators of reliability.

Table 1 shows that the standard loading values for the individual items were within the domain (0.637–0.831), these values greater than the minimum retention of the elements based on their standard loads (Al-Lozi et al., 2018; Sung et al., 2019). Average variance extracted (AVE) is a summary indicator of the convergent validity of constructs that must be above 0.50 (Howard, 2018). The results indicate that the AVE values were greater than 0.50 for all constructs, thus the used measurement

Table 1 Results of validity and reliability tests

Constructs	1	2	3	4	5	6
1.EXP	**0.717**					
2.RTA	0.428	**0.739**				
3.IEE	0.406	0.331	**0.737**			
4.DIA	0.527	0.468	0.442	**0.759**		
5.PDM	0.397	0.528	0.439	0.578	**0.731**	
6.SEQ	0.681	0.609	0.655	0.661	0.624	**0.728**
VIF	1.894	2.667	1.530	2.472	3.015	–
Loadings range	0.682–0.761	0.671–0.791	0.642–0.803	0.724–0.792	0.637–0.822	0.644–0.831
AVE	0.514	0.547	0.542	0.576	0.535	0.530
MSV	0.483	0.328	0.439	0.451	0.397	0.499
Internal consistency	0.835	0.826	0.822	0.841	0.847	0.881
Composite reliability	0.840	0.828	0.825	0.845	0.851	0.900

Note EXP: experimentation, RTA: risk taking, IEE: interaction with external environment, DIA: dialogue, PDM: participative decision making, SEQ: service quality, Bold fonts express to square root of AVE

model has an appropriate convergent validity. Rimkeviciene et al. (2017) suggested the comparison approach as a way to deal with discriminant validity assessment in covariance-based SEM. This approachis based on comparing the values of maximum shared variance (MSV) with the values of AVE, as well as comparing the values of square root of AVE ($\sqrt{AVE}$) with the correlation between the rest of the structures. The results show that the values of MSV were smaller than the values of AVE, and that the values of $\sqrt{AVE}$ were higher than the correlation values among the rest of the constructs. Therefore, the measurement model used is characterized by discriminative validity. The internal consistency measured through Cronbach's Alpha coefficient (α) and compound reliability by McDonald's Omega coefficient (ω) was conducted as indicators to evaluate measurement model. The results listed in Table 1 demonstrated that both values of Cronbach's Alpha coefficient and McDonald's Omega coefficient were greater than 0.70, which is the lowest limit for judging on measurement reliability (De Leeuw et al., 2019).

5.2 *Structural Model*

The structural model illustrated no multicollinearity issue among predictor constructs because variance inflation factor (VIF) values are below the threshold of 5, as shown

in Table 1 (Hair et al., 2017).This result is supported by the values of model fit indices shown in Figs. 1 and 2.

The results in Fig. 1 indicated that the chi-square to degrees of freedom (CMIN/DF) was 2.173, which is less than 3 the upper limit of this indicator. The values of the goodness of fit index (GFI), the comparative fit index (CFI), and the Tucker-Lewis index (TLI) were upper than the minimum accepted threshold of 0.90. Moreover, the result of root mean square error of approximation (RMSEA) indicated to value 0.068, this value is a reasonable error of approximation because it is less than the higher limit of 0.08. Consequently, the structural model used in this study was recognized as a fit model for predicting the DEP and generalization of its result (Ahmad et al., 2016; Shi et al., 2019). To verify the results of testing the study hypotheses, structural equation modeling (SEM) was used, the results of which are listed in Table 2.

The results demonstrated in Table 2 show that interaction with external environment has the highest positive impact relationship on service quality($\beta = 0.708$, $t = 29.10$, $p = 0.000$), followed by experimentation ($\beta = 0.672$, $t = 27.20$, $p = 0.000$), then dialogue($\beta = 0.614$, $t = 23.48$, $p = 0.003$),then participative decision making($\beta = 0.562$, $t = 20.81$, $p = 0.005$), and finally the lowest impact was for risk taking(β

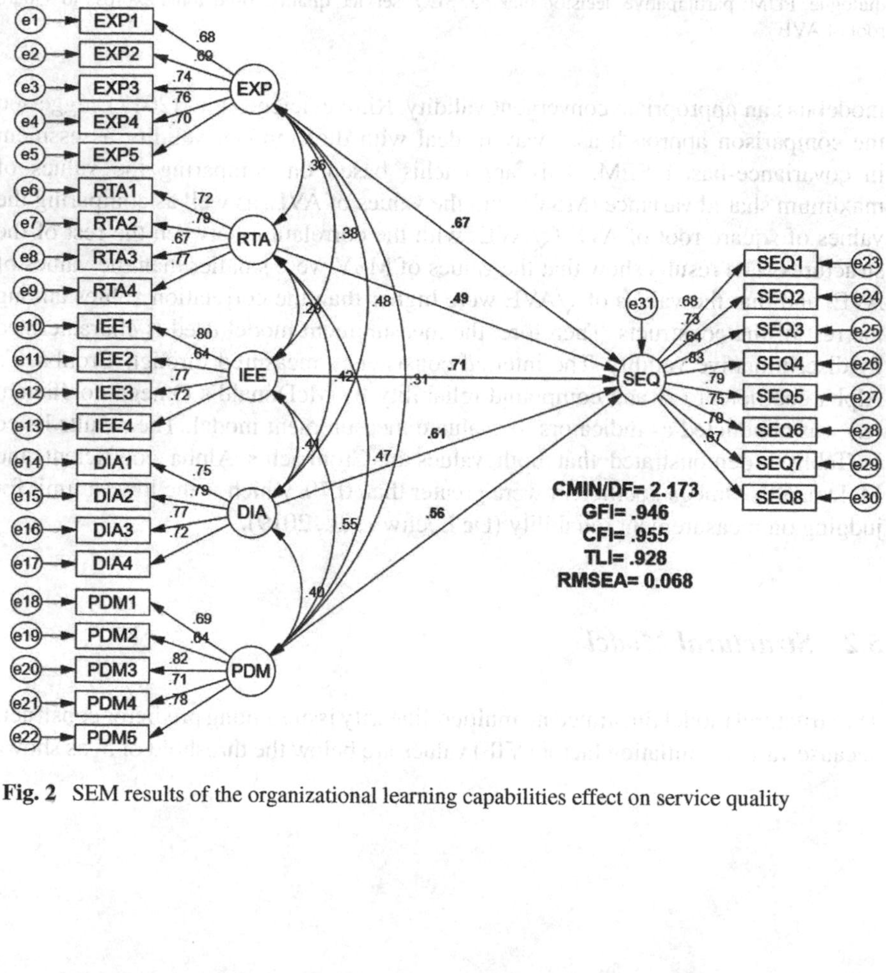

Fig. 2 SEM results of the organizational learning capabilities effect on service quality

Table 2 Hypothesis testing

Hypothesis	Relation	Standard beta	*t* value	*p* value
H1	EXP → SEQ	0.672	27.20***	0.000
H2	RTA → SEQ	0.493	18.45*	0.02
H3	IEE → SEQ	0.708	29.10***	0.000
H4	DIA → SEQ	0.614	23.48**	0.003
H5	PDM → SEQ	0.562	20.81**	0.005

Note EXP: experimentation, RTA: risk taking, IEE: interaction with external environment, DIA: dialogue, PDM: participative decision making, SEQ: service Quality, * $p < 0.05$, ** $p < 0.01$, *** $p < 0.001$

$= 0.493$, $t = 18.45$, $p = 0.02$). Thus, all the study hypotheses were supported based on these results.

6 Discussions and Recommendations

Study results showed that there is a significant impact of organizational learning capabilities (experimentation, risk taking, interaction with the external environment, dialogue, and participation in decision-making) on the service quality of Islamic banks operating in Jordan. this result is rational, as skills, knowledge and human resources optimal availability and the existence of learning work environment enables company to create and integrate ideas effectively and communicate with various organizational boundaries to improve knowledge sharing through communication and interaction and create an appropriate environment through which human efforts gather in order to create creative ways contribute in improving service quality level. This result agree with Darwaza et al. who concluded that there is an impact of organizational learning on the elements of quality culture, this also agree with Ajilat, who assured the impact of learning orientation on technical creativity, as well as the study of Mat and Razak (2011), which found a relationship between organizational learning capabilities dimensions (participation in decision-making, Experimentation, interaction with the external environment and risk) on the technological product innovation successful implementation.

Upon study findings, the researcher recommends managers and decision makers to invest the human element continuously by involving them in the decision-making process to acquire administrative skills along with technical skills they already have. It's better for organizations to work on consolidating the relationship with the external environment to facilitate linking it with the internal environment and to provide a greater opportunity for workers to learn from competitors' experiences, and learn more about the latest developments in their sector technology. In addition meeting employees annually to open dialogue and discussion with them, listening to them, and arguing them to reach executive solutions that suit both parties

References

Abbad, J., & Al-Hawary, S. I. (2011). Measuring banking service quality in Jordan: A case study of Arab bank. *Abhath Al-Yarmouk, 27*(3), 2179–2196.

Abu Qaaud, F., Al-Shoura, M., & Al-Hawary, S. I. (2011). The impact of the service marketing mix in the service quality of health services from the viewpoint of patients in government hospitals in Amman "A Field study." *Abhath Al-Yarmouk, 27*(1B), 417–441.

Aburayya, A., Alshurideh, M., Alawadhi, D., Alfarsi, A., Taryam, M., & Mubarak, S. (2020b). An investigation of the effect of lean six sigma practices on healthcare service quality and patient satisfaction: Testing the mediating role of service quality in Dubai primary healthcare sector. *Journal of Advanced Research in Dynamical and Control Systems, 12*(8), 56–72.

Aburayya, A., Alshurideh, M., Al Marzouqi, A., Al Diabat, O., Alfarsi, A., Suson, R., et al. (2020a). An empirical examination of the effect of TQM practices on hospital service quality: An assessment study in UAE hospitals. *Systematic Reviews in Pharmacy, 11*(9), 347–362.

Ahmad, S., Zulkurnain, N., & Khairushalimi, F. (2016). Assessing the validity and reliability of a measurement model in structural equation modeling (SEM). *British Journal of Mathematics & Computer Science, 15*(3), 1–8. https://doi.org/10.9734/BJMCS/2016/25183

Ahmed, B., Sabir, H. M., Sohail, N., & Mumtaz, R. (2011). Does corporate entrepreneurship matter for organizational learning capability? A study on textile sector in Pakistan. *European Journal of Business and Management, 3*(7), 53–58.

Al Kurdi, B., Alshurideh, M., Salloum, S., Obeidat, Z., & Al-dweeri, R. (2020). An empirical investigation into examination of factors influencing university students' behavior towards elearning acceptance using SEM approach. *International Journal of Interactive Mobile Technologies, 14*(2), 19–24.

Alameeri, K., Alshurideh, M., Al Kurdi, B., & Salloum, S. A. (2020, October). The effect of work environment happiness on employee leadership. In *International conference on advanced intelligent systems and informatics* (pp. 668–680). Cham: Springer.

Al-Dmour, R., AlShaar, F., Al-Dmour, H., Masa'deh, R., & Alshurideh, M. T. (2021). The effect of service recovery justices strategies on online customer engagement via the role of "Customer Satisfaction" during the Covid-19 pandemic: An empirical study. In *The effect of coronavirus disease (COVID-19) on business intelligence* (Vol. 334, pp. 325–346).

Alegre, J., & Chiva, R. (2008). Assessing the impact of organizational learning capability on product innovation performance: An empirical test. *Technovation, 28*, 315–326.

AlHamad, A., Alshurideh, M., Alomari, K., Kurdi, B., Alzoubi, H., Hamouche, S., & Al-Hawary, S. (2022). The effect of electronic human resources management on organizational health of telecommunications companies in Jordan. *International Journal of Data and Network Science, 6*(2), 429–438.

Al-Hawajreh, K., AL-Zeaud, H., Al-Hawary, S. I., & Mohammad, A. A. (2011). The influence of top management support and commitment on total quality management indicators from managers and heads of departments viewpoint: A case study of Sahab industrial city. *Jordan Journal of Business Administration, 7*(4), 557–576.

Al-Hawary, S. I. (2012). Health care services quality at private hospitals, from patient's perspective: A comparative study between Jordan and Saudi Arabia. *African Journal of Business Management, 6*(22), 6516–6529.

Al-Hawary, S. I. (2013a). The role of perceived quality and satisfaction in explaining customer brand loyalty: Mobile phone service in Jordan. *International Journal of Business Innovation and Research, 7*(4), 393–413.

Al-Hawary, S. I. (2013b). The roles of perceived quality, trust, and satisfaction in predicting brand loyalty: The empirical research on automobile brands in Jordan market. *International Journal of Business Excellence, 6*(6), 656–686.

Al-Hawary, S. I. (2015). Human resource management practices as a success factor of knowledge management implementation at health care sector in Jordan. *International Journal of Business and Social Science, 6*(11/1), 83–98.

Al-Hawary, S. I., & Abu-Laimon, A. A. (2013). The impact of TQM practices on service quality in cellular communication companies in Jordan. *International Journal of Productivity and Quality Management, 11*(4), 446–474.

Al-Hawary, S. I. S., & AlDafiri, M. F. S. (2017). Effect of the components of information technology adoption on employees performance of interior ministry of Kuwait state. *International Journal of Academic Research in Economics and Management Sciences, 6*(2), 149–169.

Al-Hawary, S. I., & Aldaihani, F. M. (2016). Customer relationship management and innovation capabilities of Kuwait airways. *International Journal of Academic Research in Economics and Management Sciences, 5*(4), 201–226.

Al-Hawary, S. I., & Al-Hamwan, A. (2017). Environmental analysis and its impact on the competitive capabilities of the commercial banks operating in Jordan. *International Journal of Academic Research in Accounting, Finance and Management Sciences, 7*(1), 277–290.

Al-Hawary, S. I., & Al-Menhaly, S. (2016). The quality of E-government services and its role on achieving beneficiaries satisfaction. *Global Journal of Management and Business Research: A Administration and Management, 16*(11), 1–11.

Al-Hawary, S. I., & Al-Namlan, A. (2018). Impact of electronic human resources management on the organizational learning at the private hospitals in the state of Qatar. *Global Journal of Management and Business Research: A Administration and Management, 18*(7), 1–11.

Al-Hawary, S. I. S., & Al-Rasheedy, H. H. (2021). The effect of strategic learning for human resources on dynamic capabilities of airlines companies in Kuwait. *International Journal of Business Information Systems, 37*(4), 421–441.

Al-Hawary, S. I., & Al-Smeran, W. (2017). Impact of electronic service quality on customers satisfaction of Islamic banks in Jordan. *International Journal of Academic Research in Accounting, Finance and Management Sciences, 7*(1), 170–188.

Al-Hawary, S. I. S., & Alwan, A. M. (2016). Knowledge management and its effect on strategic decisions of Jordanian public universities. *Journal of Accounting-Business & Management, 23*(2), 24–44.

Al-Hawary, S. I., & Harahsheh, S. (2014). Factors affecting jordanian consumer loyalty toward cellular phone brand. *International Journal of Economics and Business Research, 7*(3), 349–375.

Al-Hawary, S. I., & Hussien, A. J. (2017). The impact of electronic banking services on the customers loyalty of commercial banks in Jordan. *International Journal of Academic Research in Accounting, Finance and Management Sciences, 7*(1), 50–63.

Al-Hawary, S. I., & Ismael, M. (2010). The effect of using information technology in achieving competitive advantage strategies: A field study on the Jordanian pharmaceutical companies. *Al Manara for Research and Studies, 16*(4), 196–203.

Al-Hawary, S. I., & Metabis, A. (2012). Service quality at Jordanian commercial banks: What do their customers say? *International Journal of Productivity and Quality Management, 10*(3), 307–334.

Al-Hawary, S. I., & Nusair, W. (2017). Impact of human resource strategies on perceived organizational support at Jordanian public universities. *Global Journal of Management and Business Research: A Administration and Management, 17*(1), 68–82.

Al-Hawary, S. I., Alghanim, S., & Mohammad, A. (2011). Quality level of health care service provided by King Abdullah educational hospital from patient's viewpoint. *Interdisciplinary Journal of Contemporary Research in Business, 2*(11), 552–572.

Al-Hawary, S. I. S., Mohammad, A. S., Al-Syasneh, M. S., Qandah, M. S. F., & Alhajri, T. M. S. (2020). Organisational learning capabilities of the commercial banks in Jordan: Do electronic human resources management practices matter? *International Journal of Learning and Intellectual Capital, 17*(3), 242–266.

Al-Hawary, S. I., & Al-Syasneh, M. S. (2020). Impact of dynamic strategic capabilities on strategic entrepreneurship in presence of outsourcing of five stars hotels in Jordan. *Business: Theory and Practice, 21*(2), 578–587.

Al-Hawary, S. I., AL-Zeaud, H., & Matabes, A. (2012). Measuring the quality of educational services offered to postgraduate students at the faculty of business and finance: A field study on the universities of the north region. *Al Manara for Research and Studies, 18*(1), 241–278.

Al-Hawary, S. I. S., Abdul Aziz Allahow, T. J., & Aldaihani, F. M. F. (2018). Information technology and administrative innovation of the central agency for information technology in Kuwait. *Global Journal of Management and Business, 18*(11-A), 1–16.

Al-Khayyal, A., Alshurideh, M., Al Kurdi, B., & Aburayya, A. (2020). The impact of electronic service quality dimensions on customers' E-shopping and E-loyalty via the impact of E-satisfaction and E-trust: A qualitative approach. *International Journal of Innovation, Creativity and Change, 14*(9), 257–281.

Al-Lozi, M. S., Almomani, R. Z. Q., & Al-Hawary, S. I. S. (2018). Talent management strategies as a critical success factor for effectiveness of human resources information systems in commercial banks working in Jordan. *Global Journal of Management and Business Research: A Administration and Management, 18*(1), 30–43.

Al-Nady, B. A., Al-Hawary, S. I., & Alolayyan, M. (2016). The role of time, communication, and cost management on project management success: An empirical study on sample of construction projects customers in Makkah City, Kingdom of Saudi Arabia. *International Journal of Services and Operations Management, 23*(1), 76–112.

Alolayyan, M., Al-Hawary, S. I., Mohammad, A. A., & Al-Nady, B. A. (2018). Banking service quality provided by commercial banks and customer satisfaction. A structural equation modelling approaches. *International Journal of Productivity and Quality Management, 24*(4), 543–565.

Alshamsi, A., Alshurideh, M., Al Kurdi, B., & Salloum, S. A. (2020, October). The influence of service quality on customer retention: A systematic review in the higher education. In *International conference on advanced intelligent systems and informatics* (pp. 404–416). Cham: Springer.

Alshurideh, M. (2022). Does electronic customer relationship management (E-CRM) affect service quality at private hospitals in Jordan? *Uncertain Supply Chain Management, 10*(2), 1–8.

Alshurideh, M., Al-Hawary, S. I., Batayneh, A. M., Mohammad, A., & Al-Kurdi, B. (2017). The impact of Islamic banks' service quality perception on Jordanian customers loyalty. *Journal of Management Research, 9*(2), 139–159.

Alshurideh, M., Salloum, S. A., Al Kurdi, B., Monem, A. A., & Shaalan, K. (2019b). Understanding the quality determinants that influence the intention to use the mobile learning platforms: A practical study. *International Journal of Interactive Mobile Technologies, 13*(11), 183–157.

Alshurideh, M., Al Kurdi, B., & Salloum, S. A. (2019a, October). Examining the main mobile learning system drivers' effects: A mix empirical examination of both the expectation-confirmation model (ECM) and the technology acceptance model (TAM). In *International conference on advanced intelligent systems and informatics* (pp. 406–417). Cham: Springer.

Alshurideh, M., Al Kurdi, B., Salloum, S. A., Arpaci, I., & Al-Emran, M. (2020). Predicting the actual use of m-learning systems: A comparative approach using PLS-SEM and machine learning algorithms. *Interactive Learning Environments*, 1–15.

Alshurideh, M. T., Al Kurdi, B., Alzoubi, H. M., Ghazal, T. M., Said, R. A., AlHamad, A. Q., et al. (2022). Fuzzy assisted human resource management for supply chain management issues. *Annals of Operations Research*, 1–19.

Alzoubi, H. M., Alshurideh, M., Al Kurdi, B., & Inairat, M. (2020). Do perceived service value, quality, price fairness and service recovery shape customer satisfaction and delight? A practical study in the service telecommunication context. *Uncertain Supply Chain Management, 8*(3), 579–588.

Alzoubi, H., Alshurideh, M., Kurdi, B., Akour, I., & Aziz, R. (2022). Does BLE technology contribute towards improving marketing strategies, customers' satisfaction and loyalty? The role of open innovation. *International Journal of Data and Network Science, 6*(2), 449–460.

Ashal, N., Alshurideh, M., Obeidat, B., & Masa'deh, R. (2021). The impact of strategic orientation on organizational performance: Examining the mediating role of learning culture in Jordanian telecommunication companies. *Academy of Strategic Management Journal, 21*(Special Issue 6), 1–29.

Beheshtifar, M., Mohammad-Rafiei, R., & Nekoie-Moghadam, M. (2012). Role of career competencies in organizational learning capability. *International Journal of Contemporary Research in Business, 4*(8), 563–569.

Berghman, L. (2006). *Strategic innovation capacity: A mixed method study on deliberate strategic learning mechanisms.* Erasmus Research Institute of Management.

Bess, K. D., Perkins, D. D., & McCown, D. L. (2011). Testing a measure of organizational learning capacity and readiness for transformational change in human services. *Journal of Prevention and Intervention in the Community, 39*, 35–49.

Chiva, R., & Alegre, J. (2009). Organizational learning capability and job satisfaction: An empirical assessment in the ceramictile industry. *British Journal of Management, 20*, 323–340.

Cotton, J. L., Vollrath, D. A., Foggat, K. L., Lengnick-Hall, M. L., & Jennings, K. R. (1988). Employee participation: Diverse forms and different outcomes. *Academy of Management Review, 13*(1), 8–22.

Daniels, K., & Bailey, A. (2012). Strategy development processes and participation in decision making: Predictors of role stressors and job satisfaction. *Journal of Applied Management Studies, 8*, 27–42.

De Leeuw, E., Hox, J., Silber, H., Struminskaya, B., & Vis, C. (2019). Development of an international survey attitude scale: Measurement equivalence, reliability, and predictive validity. *Measurement Instruments for the Social Sciences, 1*(1), 9. https://doi.org/10.1186/s42409-019-0012-x

Garcia, M., Llorens, M., & Verd, J. (2006). Antecedents and consequences of organizational innovation and organizational learning in entrepreneurship. *Industrial Management & Data Systems, 106*(1), 21–42.

Garvin, D. A. (1993). Building a learning organization. *Harvard Business Review, 71*(4), 78–91.

Gear, T. R., Vince, M. R., & Minkes, A. L. (2003). Group enquiry for collective learning in organizations. *Journal of Management Development, 22*(2), 88–102.

Gefan, D. (2002). Customer loyalty in e-commerce. *Journal of the Association of Information System, 3*(1), 27–51.

Goh, S., & Richards, G. (1997). Benchmarking the learning capability of organizations. *European Management Journal, 15*(5), 757–583.

Hair, J. F., Babin, B. J., & Krey, N. (2017). Covariance-based structural equation modeling in the journal of advertising: Review and recommendations. *Journal of Advertising, 46*(1), 163–177. https://doi.org/10.1080/00913367.2017.1281777

Hasan, O., McColl, J., Pfefferkorn, T., Hamadneh, S., Alshurideh, M., & Kurdi, B. (2022). Consumer attitudes towards the use of autonomous vehicles: Evidence from United Kingdom taxi services. *International Journal of Data and Network Science, 6*(2), 537–550.

Howard, M. C. (2018). The convergent validity and nomological net of two methods to measure retroactive influences. *Psychology of Consciousness: Theory, Research, and Practice, 5*(3), 324–337. https://doi.org/10.1037/cns0000149

Jerez-Gomez, P., Spedes-Lorente, J., & Valle-Cabrera, R. (2005). Organizational learning capability: A proposal of measurement. *Journal of Business Research, 58*(6), 715–725.

Lehtinen, J. R., & Lehtinen, U. (1982). *Service quality: A study of quality dimensions.* Unpublished Working Paper, Helsinki: Service Management Institute.

Lukas, B. A. (1996). Striving for quality: The key role of internal and external customers. *Journal of Market Focused Management, 1*(2), 87–175.

Mat, A., & Razak, R. C. (2011). The influence of organizational learning capability on success of technological innovation (Product) implementation with moderating effect of knowledge complexity. *International Journal of Business and Social Science, 2*(17), 217–225.

McGill, W. S., & Lei, D. (1992). Management practice in learning organization. *Organizational Dynamics, 21*(1), 5–17.

Metabis, A., & Al-Hawary, S. I. (2013). The impact of internal marketing practices on services quality of commercial banks in Jordan. *International Journal of Services and Operations Management, 15*(3), 313–337.

Nevis, E., DiBella, A. J., & Gould, J. M. (1995). Understanding organizational as learning systems. *Sloan Management Review, 36*(2), 73–85.

Onağ, A. O., et al. (2014). Organizational learning capability and its impact on firm innovativeness. *Procedia: Social and Behavioral.*

Parasurainan, A., Berry, L. L., & Zeithaini, V. A. (1985). A conceptual model of service quality and its implications for future research. *Journal of Marketing, 49*(Fall).

Rafiq, M., Naseer, Z., & Ali, B. (2011). Impact of emotional intelligence on organizational learning capability. *International Journal of Academic Research, 3*(4), 321–325.

Rashidi, M. M., Habibi, M., & JafariFarsani, J. (2010). The relationship between intellectual assets organizational learning capability at the institute for international energy studies. *Management and Human Resource in the Oil Industry, 11*(4), 59–76.

Rimkeviciene, J., Hawgood, J., O'Gorman, J., & De Leo, D. (2017). Construct validity of the acquired capability for suicide scale: Factor structure, convergent and discriminant validity. *Journal of Psychopathology and Behavioral Assessment, 39*(2), 291–302. https://doi.org/10.1007/s10862-016-9576-4

Sekaran, U., & Bougie, R. (2016). *Research methods for business: A skill-building approach* (7th ed.). Wiley.

Shamout, M., Elayan, M., Rawashdeh, A., Kurdi, B., & Alshurideh, M. (2022). E-HRM practices and sustainable competitive advantage from HR practitioner's perspective: A mediated moderation analysis. *International Journal of Data and Network Science, 6*(1), 165–178.

Shi, D., Lee, T., & Maydeu-Olivares, A. (2019). Understanding the model size effect on SEM fit indices. *Educational and Psychological Measurement, 79*(2), 310–334. https://doi.org/10.1177/0013164418783530

Sung, K.-S., Yi, Y. G., & Shin, H.-I. (2019). Reliability and validity of knee extensor strength measurements using a portable dynamometer anchoring system in a supine position. *BMC Musculoskeletal Disorders, 20*(1), 1–8. https://doi.org/10.1186/s12891-019-2703-0

Thite, M. (2004). Strategic positioning of HRM in knowledge-based organizations. *The Learning Organization, 11*(1), 28–44.

Tohidi, H., & Mandegari, M. (2012). Assessing the impact of organizational learning capability on firm innovation. *African Journal of Business Management, 6*(12), 4522–4535.

Wang, Y. A., & Rhemtulla, M. (2021). Power analysis for parameter estimation in structural equation modeling: A discussion and tutorial. *Advances in Methods and Practices in Psychological Science, 4*(1), 1–17. https://doi.org/10.1177/2515245920918253

Zu'bi, Z., Al-Lozi, M., Dahiyat, S., Alshurideh, M., & Al Majali, A. (2012). Examining the effects of quality management practices on product variety. *European Journal of Economics, Finance and Administrative Sciences, 51*(1), 123–139.

Impact of Knowledge Management on Administrative Innovation of Software Companies in Jordan

Riad Ahmad Mohammed Abazeed, Mohammad Fathi Almaaitah, Ayat Mohammad, Doa'a Ahmad Odeh Al-Husban, Ibrahim Rashed Soliaman AlTaweel, Sulieman Ibraheem Shelash Al-Hawary, Nida'a Al-Husban, Abdullah Ibrahim Mohammad, and Rana Ibrahim Mohammad

Abstract The major aim of the study was to examine the impact of knowledge management on administrative innovation. A qualitative method based on a questionnaire was used in this study for data collection and sample selection. Therefore, it focused on of Software Companies in Jordan. Data were primarily gathered through self-reported questionnaires creating by Google Forms which were distributed to a

R. A. M. Abazeed
Faculty of Finance and
Business Administration, Business Management and Public Administration, Department of Business Administration, Al Al-Bayt University, P.O. Box 130040, Mafraq 25113, Jordan

M. F. Almaaitah
Faculty of Economic and Administration Sciences, Department of Business Administration & Public Administration, Al Al-Bayt University Jordan, P.O. Box 130040, Mafraq 25113, Jordan
e-mail: m.maaitah@aabu.edu.jo

A. Mohammad
Business and Finance Faculty, the World Islamic Science and Education University (WISE), P.O Box 1101, Amman 11947, Jordan

D. A. O. Al-Husban
Faculty of Alia College, Department of Human Fundamental Sciences, Al-Balqa Applied University, Amman, Jordan
e-mail: D_husban@bau.edu.jo

I. R. S. AlTaweel
Faculty of Business School, Department of Business Administration, Qussim University, P.O. Box 6502, Al Russ City 51452, Saudi Arabia
e-mail: toiel@qu.edu.sa

S. I. S. Al-Hawary (✉) · N. Al-Husban
Faculty of Economics and Administrative Sciences, Department of Business Administration, Al Al-Bayt University, P.O. Box 130040, Mafraq 25113, Jordan
e-mail: dr_sliman73@aabu.edu.jo; dr_sliman@yahoo.com

A. I. Mohammad
Department of Basic Scientific Sciences, Al-Huson University College, Al-Husun, Jordan

Department of Basic Scientific Sciences, Al-Balqa Applied University, Salt, Jordan

M. Alshurideh et al. (eds.), *The Effect of Information Technology on Business and Marketing Intelligence Systems*, Studies in Computational Intelligence 1056,
https://doi.org/10.1007/978-3-031-12382-5_92

purposive sample of managers via email. Statistical program AMOSv24 was used to test the study hypotheses. The study results show that the highest impact on administrative innovation was for knowledge generation, followed by knowledge applying, then knowledge storing, and finally the lowest impact was for knowledge dissemination. Researchers recommend decision-makers to pay attention to the tacit knowledge and experiences of employees, and turn them into tangible ideas and methods that contribute to the development of work performance, and to refine the personalities of their employees by supporting their ideas and suggestions and activating their role in solving problems.

Keywords Knowledge management · Administrative innovation · Software companies · Jordan

1 Introduction

There have become challenges facing organizations in general worldwide; Because of the developments and changes that occur rapidly, organizations must keep pace with these developments and work to find the best ways to ensure their survival, and their continuity to achieve goals such as increasing profits, which led to the organizations' interest in managing the knowledge represented by the tacit and explicit knowledge that individuals have to take advantage of that knowledge, and employ it to achieve administrative creativity (Al-Hawary, 2015; Al-Hawary & Al-Hamwan, 2017; Al Mehrez et al., 2020).

The current era is described as the era of knowledge, as knowledge is the real capital that outweighs in its importance and values other natural resources (Al-Maroof et al., 2021; Al Mehrez et al., 2020). This era is also characterized by intense interest in the human resource and his development and investment of his mental abilities, and considering it the basis for achieving any progress in the effectiveness of the organization, as well as interest in the integration of the knowledge sources, organizing and development (Al-Hawary & Al-Namlan, 2018; Al-Lozi et al., 2017; Al-Nady et al., 2013; Metabis & Al-Hawary, 2013). Knowledge has become the highest percentage in the composition of the gross national product of the economically, technically and cognitively advanced countries, and for the importance of knowledge in building and developing individuals, groups, organizations and countries (Al-Hawary et al., 2016). The managements have interested in building programs for the development and employment of knowledge to achieve continuous improvement in the main and supportive operations and activities of the organization. The management also pointed out the importance of distinguishing between the latent or self-knowledge of the individual and the external knowledge that comes to him from environmental sources, and the need to mix them, as one is indispensable to the other (Al-Hawary & Alwan, 2016).

R. I. Mohammad
Ministry of Education, Amman, Jordan

The need to pay attention to the human element and work to invest in it positively began, as human societies do not advance or develop, without the presence of creators and innovators in various fields who are considered basic pillars in the structure of the human society as they work to produce, and develop, they represent the hope of solving problems that hinder civilized and human progress (Alhalalmeh et al., 2020; Al-Hawajreh et al., 2011; Al-Hawary et al., 2012, 2020; Mohammad et al., 2020). Creativity in all fields has become an urgent and essential necessity, so that society, including individuals and institutions, is able to keep pace with the times, which is characterized by the nature of technical progress, knowledge explosion, the abundance of inventions and the multiplicity of cultures, each of them is trying to impose itself on the other in the era of globalization. Today, we live in a world characterized by complexity and problems that explode day after day, so that creativity and innovation are the solution that makes the individual and society able to keep pace with the requirements of this era (Ghannajeh et al., 2015; Al-Hawary & Aldaihani, 2016; Almaazmi et al., 2020).

Many researchers have revealed that humans invent, create and innovate, since the inception of creation until now. Most human civilizations, throughout the ages and time, have cared for their children and worked to develop them and increase mental abilities in all fields (Al-Hawary & Al-Rasheedy, 2021; Al-Lozi et al, 2017). Creativity is important in the modern era because it is the real bridge on which the theoretical ideas of peoples and individuals cross to the righteousness of practical creative works, in addition to that creativity is an effective and practical test for measuring excellence because it is easy to see and evaluate the works of creators accurately. (Al-Hawary & Alwan, 2016).

2 Theoretical Framework and Hypotheses Development

2.1 *Knowledge Management*

Knowledge is one of the most important resources of the most interest in business organizations, and the most important fact is that development and progress in advanced economic societies has become dependent on knowledge and intellectual capital (Al-Hawary & AL-Zeaud, 2010; Al-Hawary, 2009; Al-Hawary & Shdefat, 2016; Al-Hawary & Nusair, 2017), and the exploitation of intellectual assets as is the case in other assets and with the same degree of importance, but rather in some sectors that depend heavily Directly on the intellectual assets in the nature of its business. The abilities, experiences and skills of humans are not equal among them, there is a group of them that possess the mentioned components more than others (Allahow et al., 2018; AlHamad et al., 2022; Alshurideh et al., 2022).

Knowledge has become one of the critical driving forces for business success regardless of the nature of its work, whether commercial, industrial or service, organizations have become more knowledge-intensive, and the belief of most managers to achieve their strategic goals has become the extent of their possession of knowledge (Aldaihani & Ali, 2018, 2019; Alshura et al., 2016), so they relied on minds more than hands and the need to benefit increases of the value of knowledge, and as a result knowledge has been treated systematically and considered as an important resource like capital and other tangible resources (Al-Hawary & Metabis, 2012).

Many organizations are exploring the field of knowledge management in order to improve their competitiveness, and sustainability, which increases company returns, employee satisfaction and loyalty, and improves the competitive position by focusing on intangible assets (Al-Hawary & Hadad, 2016; Al-Hawary & Ismael, 2010), which necessitates the need for a more systematic and in-depth study concerning the critical success factors for the implementation of knowledge management is critical, organizations must be aware of the factors that will affect the success of the knowledge management practice.

Knowledge is one of the old renewable terms, and interest in it has been since ancient times, and in the present time, organizations have come to view knowledge as an effective basis for achieving competitive advantage and creativity, as it has become the competitive tool for many organizations in various sectors, so knowledge is one of the most important factors that organizations need. To achieve success, and keep pace with the changes that the world is witnessing from a technological revolution as a result of the development in the field of technology and communications and its use in the field of obtaining knowledge and sending information (Al-Hawary & Aldaihani, 2016).

Knowledge is the key to outstanding competitiveness, better customer service, and faster access to markets, performance development, innovation and creativity (Allahow et al., 2018; Alshurideh, 2019; Alshurideh et al., 2020). This necessitated the need to organize and manage that process to take advantage of it to reach the goals and help these organizations in making the appropriate decision. Knowledge management has become one of the solutions used in organizations to face the risks and challenges that these organizations may be exposed to, and therefore the knowledge balance of those organizations must be invested (Al-Hawary & AL-Zeaud, 2010; Shannak et al., 2012; Al-Hawary & Nusair, 2017). The use of ICT is one of the important factors that helped and stimulated knowledge management. The challenge is no longer limited to finding information for the organization, but rather how to search for it through the information available in the organization to employ it and enhance its competitive position (Allahow et al., 2018). In the following, we will mention some of the definitions of knowledge management (see Table 1).

Through the previous definitions, knowledge management can be defined as a set of operations and activities by the management of the organization aimed at identifying and finding the required knowledge and following different administrative methods in order to help in investing it well to achieve a competitive advantage that is unique from other organizations and gives it an opportunity for survival, continuity and development. Knowledge is the set of processes and activities that begin with the dissemination of knowledge among individuals and groups in the organization to achieve the goals of raising the level of performance, making decisions, solving problems and survival. Researchers noted that knowledge management measures varied,

Table 1 Definitions of knowledge management

Author	Definitions
Girard and Girard (2015)	The processes of creating, sharing, using and managing information and knowledge in an organization
Wiig (2013)	A set of clear and well-defined approaches and processes aimed at discovering Knowledge, both positive and negative, of various types of operations, managing them, defining new products or strategies, enhancing human resource management, and achieving a number of other objectives
Laudon and Laudon (2007)	That process that provides knowledge to workers in various sectors by providing information and disseminating knowledge and information

Table 2 Knowledge management measures in previous studies

Author	Dimensions
Parisa (2015)	Knowledge production, knowledge sharing, knowledge development and continuous improvement
Bognar and Bansal	Knowledge generation, knowledge building, knowledge efficiency
William et al. (2008)	Knowledge generation, knowledge acquisition and sharing Knowledge, knowledge transfer
Nasiruzzaman and Dahlan (2013)	Explicit knowledge and tacit knowledge

due to the different methods and visions of researchers about knowledge management and its objectives. Table 2 is illustrated knowledge management measures as mentioned by researchers:

Knowledge acquisition: The process of obtaining knowledge is the first stage of learning at the individual and collective level in the organization and it is the cornerstone of this process (Marquardt, 1996). Knowledge is gained and obtained through internal and external sources. Examples of external sources are: marketing intelligence, industry and academic research, as well as from mergers and acquisitions and through the recruitment of new employees. As for the internal sources of knowledge, they are represented by dialogue sessions and communication between group members, in addition to the availability of basic data such as the financial and economic data of the company.

Knowledge Sharing: Sinkula emphasized that communication between the departments of the organization is necessary and is one of the most important dimensions of organizational learning (Sinkula, 1994; Argyris & Schön, 1996; Teo et al., 2006). Whereas, the communication between the departments will lead to the dissemination of knowledge among them, and makes the knowledge available throughout the organization. Workers share it, which leads to an increase in their experiences, and the transfer of tacit knowledge inside their minds to their colleagues. This was confirmed by Herschel, who said that the process of transferring productive knowledge requires transforming it from tacit knowledge to explicit knowledge (Herschel, 2000). Heisig and Rorbeek (2000) indicated that there are several methods for distributing and disseminating knowledge such as internal information networks (intranets), training by old colleagues, experience teams, learning circles, and knowledge circles. These methods provide knowledge throughout the organization, which helps to increase effectiveness, access to new technology, and identify the factors in the organization's surrounding environment, and over time it becomes a culture in the organization (Teo et al., 2006).

Knowledge Implementation: At this stage, the greatest benefit is evident from all previous stages, as knowledge moves into application and benefit from what has been learned to reflect on the performance of the organization, and allow its employees to solve the problems they face through the knowledge that has been acquired, and thus becomes the knowledge acquired and spread in all parts of the organization have meaning and value. The use of knowledge spread in the organization in all its forms of documents, procedures and databases leads to an increase in the experience and capabilities of employees. This helps the organization to innovate, which will eventually lead to the continuation of the organization and increase its competitiveness.

Knowledge storage: The acquired knowledge is stored in repositories and knowledge stores within the organization, and the aim is to preserve the acquired knowledge so that all members of the organization can access and benefit from it, especially the knowledge that was used as a way to solve previous problems, to try to solve the current problems in the same way or avoid the failed attempts that were previously used to solve it. Knowledge can be stored in several ways, including organizational memory, which contains existing knowledge in its various forms, including written documents, and information stored in electronic databases. Human knowledge is stored in expert systems as well as tacit knowledge gained by individuals and business networks. Knowledge repositories constitute an essential technology in knowledge-based learning organizations, as they are keen to manage their knowledge content to meet the opportunities of future changes using the necessary supporting technology.

2.2 Administration Innovation

Today, the world is witnessing a lot of changes and developments in various fields of economic, social, cultural, political, technical, and information life, which prompted organizations, whether public or private, to seek to develop internal working methods, on a permanent and continuous basis, so that these organizations can keep pace with the tremendous knowledge developments that occur in today's world, and achieve this by investing the energies of innovators within them to the fullest extent, and benefiting from them so that these innovators can perform the roles entrusted to them to the fullest (Al Suwaidi et al., 2020; Al-Hawary & Aldaihani, 2016).

Administrative innovation is an important basic means for the growth and development of organizations, for their sustainability on the one hand, and in order to achieve the desired goals on the other hand, as an organization that does not innovate or develop is destined to decline and decay, but may be demise. Administrative innovation helps organizations adapt to various changes, and helps them to face challenges, whether political, economic, technical or informational, and thus achieve the goals of the organization that it seeks to achieve, and creativity leads to renewal and innovation, which makes the organization ahead of other competing organizations, it guarantees them access to advanced positions in the competition (Allahow et al., 2018; Alameeri et al., 2021; Alzoubi et al., 2022). Because of the importance that innovation achieves, it is necessary to provide the appropriate atmosphere for work in addition to allocating the necessary means for developing innovation by following stages and strategies that enable the organization to reach the highest levels of administrative creativity, which leads to an increase in the efficiency of the human element as the main factor for increasing productivity within organizations (Altamony et al., 2012; Alameeri et al., 2020; Al-Hawary & Al-Rasheedy, 2021).

Administrative innovation is divided into incremental innovation and radical innovation based on the size of the change it causes. Radical innovation the amount of change is large, in other words, "innovation can be embodied in a new technology that leads to changing the current structure of the market" (Claver et al., 1998; Alyammahi et al., 2020) and contains new products and services that did not exist previously. Incremental innovation includes gradual change, is not significant, such as developing the current incentive system and not completely abandoning it. There are clear differences between the two types. Incremental innovation is related to exploitative learning, which is the acquisition of new behavioral capabilities, but within the same current visions of the organization. On the other hand, radical innovation is related to exploratory learning, which occurs when the organization acquires new behavioral capabilities that lead to changing the current visions of the organization and forming new ones (March, 1991; AlShehhi et al., 2020; Agha et al., 2021; Batayneh et al., 2021; Kurdi et al., 2020; Nuseir et al., 2021).

It is not permissible to talk about administrative innovation without the availability of its basic elements and components, as the importance of the elements and components of innovation is to define and measure it at the level of the group, the individual and the organization. Accordingly, management scholars and researchers unanimously agreed on the existence of the following elements of innovation:

Originality: This is intended to be renewal or singling out with ideas. The innovator person has an original thinking, that is, he moves away from the ordinary or the common. He does not repeat the ideas of others. It also refers to the ability of the innovator to produce original ideas with few repetitions within the group to which he belongs and the criterion for judging the idea as original. It is not subject to circulated ideas and out of the ordinary (Al-Hawary & Al-Rasheedy, 2021).

Fluency: It is intended to produce the largest number of synonyms, ideas or uses when responding to a specific stimulus, and the speed in generating them, and it also means the ability to recall the largest possible number of ideas for a particular situation within a relatively short period of time, and also means the multiplicity of ideas that can come with it. The innovator individual, characterized by its suitability to the requirements of the real environment (Al-Hawary & Alwan, 2016).

Flexibility: It means the diversity of ideas that the innovator individual brings, and his ability to change or transform his thinking or his point of view according to the requirements of the situation. The degree of ease with which the innovator changes a certain position or mental point of view, and it also means looking at things from several angles (Al-Hawary & Al-Rasheedy, 2021).

Dealing with risks: It means how brave the innovator individual is in exposing himself to failure or criticism, making a lot of guesses, working under ambiguous circumstances, and defending his own ideas. It also means taking the initiative to adopt new ideas and methods and studying for solutions and at the same time being the individual is capable of bearing the risks resulting from the work he undertakes, and he is ready to face the responsibilities.

Sensitivity to problems: the ability to sense problems and realize their nature and identify them and the innovator person notices that there is something wrong that others do not notice (Allahow et al., 2018).

2.3 Knowledge Management and Administrative Innovation

Knowledge is the key to outstanding competitiveness, better customer service, faster access to markets, performance development, innovation and creativity (Al-Hawary & Alwan, 2016; Alshurideh, 2022). The use of knowledge spread in the organization in all its forms of documents, procedures and databases leads to an increase in the experience and capabilities of employees, which helps the organization to innovate, which will eventually lead to the continuation of the organization and increase its competitiveness. Hajir notes that there is a significant positive impact of knowledge management infrastructure on innovation. The results strongly increase from previous studies, which reveal that the dimension of knowledge management infrastructure has the greatest impact on innovation. The results of the current study have several managerial implications for organizations. If mobile telecommunications companies in Jordan wish to enhance their innovation capabilities, they must work to maintain an effective knowledge management infrastructure that invests mainly in information technology. Ling also showed that the effectiveness of knowledge acquisition has a significant impact on managerial innovation. Accordingly, the study hypothesis can be formulated as follows:

There is a statistically significant effect of knowledge management on administrative innovation.

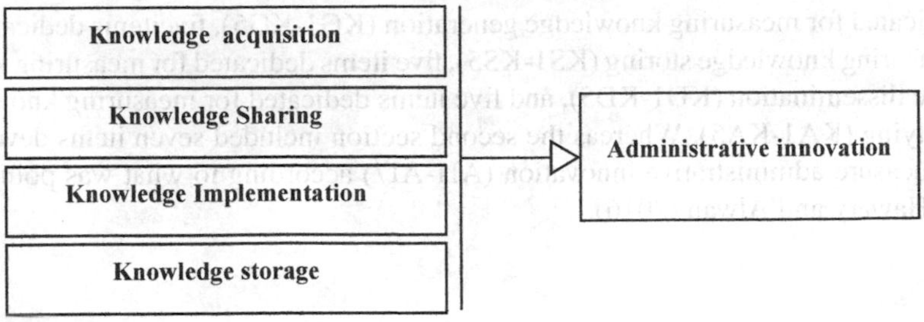

Fig. 1 Theoretical model

3 Research Model

See Fig. 1.

4 Methodology

4.1 Population and Sample Selection

A qualitative method based on a questionnaire was used in this study for data collection and sample selection. The major aim of the study was to examine the impact of knowledge management on administrative innovation. Therefore, it focused on of Software Companies in Jordan. Data were primarily gathered through self-reported questionnaires creating by Google Forms which were distributed to a purposive sample of managers via email. In total, (148) responses were received including (13) invalid to statistical analysis due to uncompleted or inaccurate. Hence, the final sample contained (135) responses suitable to analysis requirements, where it proved to be sufficient to the extent that was predictable and allowed for a presumption of data saturation (Sekaran & Bougie, 2016).

4.2 Measurement Instrument

A self-reported questionnaire that consists of two main sections along with a section regarding control variables was used as the measurement instrument. Control variables considered as categorical measures were composed of gender, age group, educational level, and experience. The two main sections were dealt with a five-point Likert scale (from 1 = strongly disagree to 5 = strongly agree). The first section contained (20) items to measure knowledge management based on (Al-Hawary & Aldaihani, 2016). These items were distributed into dimensions as follows: fiveitems dedicated for measuring knowledge generation (KG1-KG5), fiveitems dedicated for measuring knowledge storing (KS1-KS5), five items dedicated for measuring knowledge dissemination (KD1-KD5), and five items dedicated for measuring knowledge applying (KA1-KA5). Whereas the second section included seven items developed to measure administrative innovation (AI1-AI7) according to what was pointed by Al-Hawary and Alwan (2016).

5 Findings

5.1 *Measurement Model Evaluation*

This study was conducted structural equation modeling (SEM) to test hypotheses, which represents a contemporary statistical technique for testing and estimating the relationship between factors and variables (Wang & Rhemtulla, 2021). Accordingly, the reliability and validity of the constructs were tested using confirmatory factor analysis (CFA) through the statistical program AMOSv24. Table 3 summarizes the results of convergent and discriminant validity, as well the indicators of reliability.

Table 3 shows that the standard loading values for the individual items were within the domain (0.657–0.857), these values greater than the minimum retention of the elements based on their standard loads (Al-Lozi et al., 2018; Sung et al., 2019). Average variance extracted (AVE) is a summary indicator of the convergent validity of constructs that must be above 0.50 (Howard, 2018). The results indicate that the AVE values were greater than 0.50 for all constructs, thus the used measurement model has an appropriate convergent validity. Rimkeviciene et al. (2017) suggested

Table 3 Results of validity and reliability tests

Constructs	KG	KS	KD	KA	AI
KG	**0.791**				
KS	0.517	**0.760**			
KD	0.467	0.614	**0.773**		
KA	0.539	0.468	0.544	**0.769**	
AI	0.715	0.705	0.721	0.734	**0.760**
VIF	1.674	2.197	2.371	1.937	–
Loadings range	0.668–0.857	0.681–0.824	0.657–0.843	0.693–0.827	0.673–0.817
AVE	0.625	0.577	0.597	0.591	0.578
MSV	0.514	0.497	0.502	0.535	0.473
Internal consistency	0.890	0.869	0.876	0.873	0.901
Composite reliability	0.892	0.872	0.880	0.878	0.905

Note KG: knowledge generation, KS: knowledge storing, KD: knowledge dissemination, KA: knowledge applying, AI: administrative innovation, Bold fonts indicate to square root of AVE

the comparison approach as a way to deal with discriminant validity assessment in covariance-based SEM. This approachis based on comparing the values of maximum shared variance (MSV) with the values of AVE, as well as comparing the values of square root of AVE ($\sqrt{AVE}$) with the correlation between the rest of the structures. The results show that the values of MSV were smaller than the values of AVE, and that the values of $\sqrt{AVE}$ were higher than the correlation values among the rest of the constructs. Therefore, the measurement model used is characterized by discriminative validity. The internal consistency measured through Cronbach's Alpha coefficient (α) and compound reliability by McDonald's Omega coefficient (ω) was conducted as indicators to evaluate measurement model. The results listed in Table 3 demonstrated that both values of Cronbach's Alpha coefficient and McDonald's Omega coefficient were greater than 0.70, which is the lowest limit for judging on measurement reliability (De Leeuw et al., 2019).

5.2 *Structural Model*

The structural model illustrated no multicollinearity issue among predictor constructs because variance inflation factor (VIF) values are below the threshold of 5, as shown in Table 3 (Hair et al., 2017). This result is supported by the values of model fit indices shown in Fig. 2.

The results in Fig. 2 indicated that the chi-square to degrees of freedom (CMIN/DF) was 2.468, which is less than 3 the upper limit of this indicator. The values of the goodness of fit index (GFI), the comparative fit index (CFI), and the Tucker-Lewis index (TLI) were upper than the minimum accepted threshold of 0.90. Moreover, the result of root mean square error of approximation (RMSEA) indicated to value 0.038, this value is a reasonable error of approximation because it is less than the higher limit of 0.08. Consequently, the structural model used in this study was recognized as a fit model for predicting the DEP and generalization of its result (Ahmad et al., 2016; Shi et al., 2019). To verify the results of testing the study hypotheses, structural equation modeling (SEM) was used, the results of which are listed in Table 4.

The results demonstrated in Table 4 show that the highest impact on administrative innovation was for knowledge generation ($\beta = 0.579, t = 24.21, p = 0.000$), followed by knowledge applying ($\beta = 0.521, t = 22.96, p = 0.000$), then knowledge storing ($\beta = 0.462, t = 19.88, p = 0.003$), and finally the lowest impact was for knowledge dissemination ($\beta = 0.388, t = 17.38, p = 0.008$). Thus, all the research hypotheses were supported based on these results.

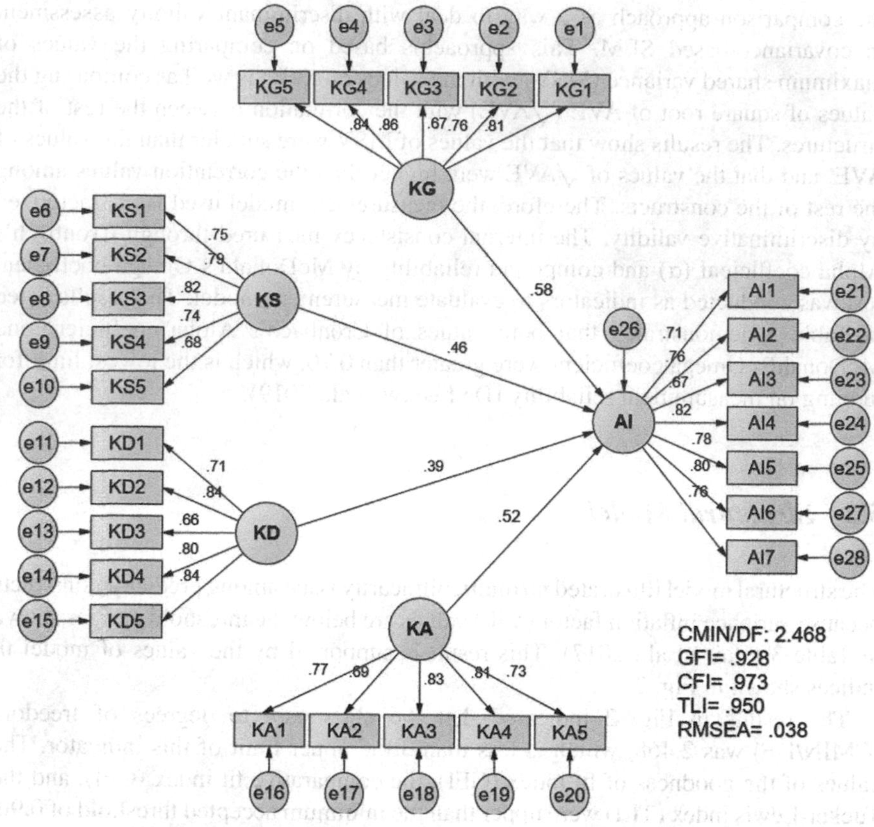

Fig. 2 SEM results of the knowledge management effect on administrative innovation

Table 4 Hypothesis testing

Hypothesis	Relation	Standard beta	*t* value	*p* value
H1	KG → AI	0.579	24.21***	0.000
H2	KS → AI	0.462	19.88**	0.003
H3	KD → AI	0.388	17.38**	0.008
H4	KA → AI	0.521	22.96***	0.000

Note KG: knowledge generation, KS: knowledge storing, KD: knowledge dissemination, KA: knowledge applying, AI: administrative innovation, * $p < 0.05$, ** $p < 0.01$, *** $p < 0.001$

6 Conclusion

The study results showed that the study population adopts strategies that contribute to achieving the maximum benefit from all intellectual and information resources through the collection and sharing of ideas and information among all work-related parties, including employees, administrators and various business partners, on the other hand, providing all the needs of technology and equipment that enable workers to increasing their stock of various information and knowledge that helps them to develop their skills and abilities to accomplish various work tasks and duties, and provides them with a flexible environment that enables them to unleash their creative ideas which contribute to inventing advanced ways and methods of work, and motivates them to participate in developing solutions to the problems that hinder the performance of work as required, and provides their employees with tools and equipment that help them to retain all information and data related to work, and enables them to refer to it when needed, and motivates them to share ideas and information, and provides them with modern communication channels to receive and send all work-related activities between them. It also gives workers complete freedom in how to perform their work, which enables them to take the approach and method as required by the current stage of work in accordance with its policies in performing work activities, which leads to their possession of capabilities and skills that enable them to analyze problems which affect work, and what are the appropriate solutions to them.

7 Discussion

The results of the study showed that there is an impact of knowledge management on administrative innovation, the researchers attribute this result to the importance of the role played by knowledge management by putting forward ideas and methods that enable employees to accomplish their work in innovative and modern ways which affect the work better, on the other hand, increasing the ability of management to discover all the talents and skills of employees. In addition, the workers possess the skills, expertise and experience related to work by creating information that enables them to devise advanced methods and methods to perform work activities and tasks, and provide a system of incentives for each individual who presents ideas and suggestions that will contribute to the development and prosperity of work, and their keenness to provide tools which enables them to keep all suggestions and constructive ideas that serve the objectives and policies of the work in an appropriate manner, and which enables all employees to easily and quickly refer to them when needed through the use of an advanced system of information systems and databases. Also, providing all the means and capabilities that help them implement their ideas and knowledge, and supervise them by specialists to ensure that they are applied in the correct and appropriate manner. On the other hand, workers are provided with the ability to study and analyze the implementation mechanisms, requirements and expected obstacles, enabling them to apply their knowledge without falling into problems or errors.

8 Recommendations

Researchers recommend decision-makers to pay attention to the tacit knowledge and experiences of employees and turn them into tangible ideas and methods that contribute to the development of work performance, and to refine the personalities of their employees by supporting their ideas and suggestions and activating their role in solving problems, which makes them able to be creative and innovative in accomplishing work tasks. In addition to keeping pace with technological developments by providing modern equipment that enables them to create ideas and information, preserve and distribute them in ways that enable employees to make optimal use of knowledge, and pay attention to supporting the positive qualities of workers and administrators represented in their feeling that they are rewarded for their work, and play a major role in solving problems and maintaining the survival of and business boom. And finally, attention to individuals with outstanding creative abilities, providing them with the appropriate atmosphere and conditions, and motivating them through an appropriate reward system.

References

Agha, K., Alzoubi, H. M., & Alshurideh, M. T. (2021, June). Measuring reliability and validity instruments of technologically driven cognitive intrusion towards work-life balance. In *The international conference on artificial intelligence and computer vision* (pp. 601–614). Cham: Springer.

Ahmad, S., Zulkurnain, N., & Khairushalimi, F. (2016). Assessing the validity and reliability of a measurement model in structural equation modeling (SEM). *British Journal of Mathematics & Computer Science, 15*(3), 1–8. https://doi.org/10.9734/BJMCS/2016/25183

Al Mehrez, A. A., Alshurideh, M., Al Kurdi, B., & Salloum, S. A. (2020). Internal factors affect knowledge management and firm performance: A systematic review. In *International conference on advanced intelligent systems and informatics* (pp. 632–643). Cham: Springer.

Al Suwaidi, F., Alshurideh, M., Al Kurdi, B., & Salloum, S. A. (2020, October). The impact of innovation management in SMEs performance: A systematic review. In *International conference on advanced intelligent systems and informatics* (pp. 720–730). Cham: Springer.

Alameeri, K. A., Alshurideh, M. T., & Al Kurdi, B. (2021). The effect of Covid-19 pandemic on business systems' innovation and entrepreneurship and how to cope with it: A theatrical view. In *The effect of coronavirus disease (COVID-19) on business intelligence* (Vol. 334, pp. 275–288).

Alameeri, K., Alshurideh, M., Al Kurdi, B., & Salloum, S. A. (2020, October). The effect of work environment happiness on employee leadership. In *International conference on advanced intelligent systems and informatics* (pp. 668–680). Cham: Springer.

Aldaihani, F. M. F., & Ali, N. A. B. (2018). Impact of social customer relationship management on customer satisfaction through customer empowerment: A study of Islamic Banks in Kuwait. *International Research Journal of Finance and Economics, 170*(170), 41–53.

Aldaihani, F. M. F., & Ali, N. A. B. (2019). Impact of relationship marketing on customers loyalty of Islamic banks in the State of Kuwait. *International Journal of Scientific & Technology Research, 8*(11), 788–802.

Alhalalmeh, M. I., Almomani, H. M., Altarifi, S., Al-Quran, A. Z., Mohammad, A. A., & Al-Hawary, S. I. (2020). The nexus between corporate social responsibilty and organizational performance in

Jordan: The mediating role of organizational commitment and organizational citizenship behavior. *Test Engineering and Management, 83*, 6391–6410.

AlHamad, A., Alshurideh, M., Alomari, K., Kurdi, B., Alzoubi, H., Hamouche, S., & Al-Hawary, S. (2022). The effect of electronic human resources management on organizational health of telecommunications companies in Jordan. *International Journal of Data and Network Science, 6*(2), 429–438.

Al-Hawajreh, K., AL-Zeaud, H., Al-Hawary, S. I., & Mohammad, A. A. (2011). The influence of top management support and commitment on total quality management indicators from managers and heads of departments viewpoint: A case study of Sahab industrial city. *Jordan Journal of Business Administration, 7*(4), 557–576.

Al-Hawary, S. I. AL-Awawdeh, W., & Abden, M. A. (2012). The impact of the leadership style on organizational commitment: A field study on Kuwaiti telecommunications companies. *ALEDARI*, (130), 53–102.

Al-Hawary, S. I. (2009). The effect of the leadership style on the effectiveness of the organization: A field study at Zarqa private university. *The Egyptian Journal for Commercial Studies, 33*(1), 361–393.

Al-Hawary, S. I. (2015). Human resource management practices as a success factor of knowledge management implementation at health care sector in Jordan. *International Journal of Business and Social Science, 6*(11/1), 83–98.

Al-Hawary, S. I. S., & Al-Rasheedy, H. H. (2021). The effect of strategic learning for human resources on dynamic capabilities of airlines companies in Kuwait. *International Journal of Business Information Systems, 37*(4), 421–441.

Al-Hawary, S. I. S., & Alwan, A. M. (2016). Knowledge management and its effect on strategic decisions of Jordanian public universities. *Journal of Accounting-Business & Management, 23*(2), 24–44.

Al-Hawary, S. I. S., Mohammad, A. S., Al-Syasneh, M. S., Qandah, M. S. F., & Alhajri, T. M. S. (2020). Organisational learning capabilities of the commercial banks in Jordan: Do electronic human resources management practices matter? *International Journal of Learning and Intellectual Capital, 17*(3), 242–266.

Al-Hawary, S. I., & Aldaihani, F. M. (2016). Customer relationship management and innovation capabilities of Kuwait airways. *International Journal of Academic Research in Economics and Management Sciences, 5*(4), 201–226.

Al-Hawary, S. I., & Al-Hamwan, A. (2017). Environmental analysis and its impact on the competitive capabilities of the commercial banks operating in Jordan. *International Journal of Academic Research in Accounting, Finance and Management Sciences, 7*(1), 277–290.

Al-Hawary, S. I., & Al-Namlan, A. (2018). Impact of electronic human resources management on the organizational learning at the private hospitals in the state of Qatar. *Global Journal of Management and Business Research: A Administration and Management, 18*(7), 1–11.

Al-Hawary, S. I., & AL-Zeaud, H. (2010). The relationship between job satisfaction and organizational commitment in the cellular communications companies in Jordan: A field study. *Al Manara for Research and Studies, 17*(5), 109–147.

Al-Hawary, S. I., & Hadad, T. F. (2016). The effect of strategic thinking styles on the enhancement competitive capabilities of commercial banks in Jordan. *International Journal of Business and Social Science, 7*(10), 133–144.

Al-Hawary, S. I., & Ismael, M. (2010). The effect of using information technology in achieving competitive advantage strategies: A field study on the Jordanian pharmaceutical companies. *Al Manara for Research and Studies, 16*(4), 196–203.

Al-Hawary, S. I., & Metabis, A. (2012). Implementation of internal marketing in Jordan banks. *International Journal of Data Analysis and Information, 4*(1), 37–53.

Al-Hawary, S. I., & Nusair, W. (2017). Impact of human resource strategies on perceived organizational support at Jordanian public universities. *Global Journal of Management and Business Research: A Administration and Management, 17*(1), 68–82.

Al-Hawary, S. I., & Shdefat, F. (2016). Impact of human resources management practices on employees' satisfaction a field study on the Rajhi cement factory. *International Journal of Academic Research in Accounting, Finance and Management Sciences, 6*(4), 274–286.

Allahow, T. J. A. A., Al-Hawary, S. I. S., & Aldaihani, F. M. F. (2018). Information technology and administrative innovation of the central agency for information technology in Kuwait. *Global Journal of Management and Business, 18*(11-A), 1–16.

Al-Lozi, M., Almomani, R. Z., & Al-Hawary, S. I. (2017). Impact of talent management on achieving organizational excellence in Arab potash company in Jordan. *Global Journal of Management and Business Research: A Administration and Management, 17*(7), 15–25.

Al-Lozi, M., Almomani, R. Z., & Al-Hawary, S. I. (2018). Talent Management strategies as a critical success factor for effectiveness of Human Resources Information Systems in commercial banks working in Jordan. *Global Journal of Management and Business Research: A Administration and Management, 18*(1), 30–43.

Almaazmi, J., Alshurideh, M., Al Kurdi, B., & Salloum, S. A. (2020, October). The effect of digital transformation on product innovation: A critical review. In *International conference on advanced intelligent systems and informatics* (pp. 731–741). Cham: Springer.

Al-Maroof, R., Ayoubi, K., Alhumaid, K., Aburayya, A., Alshurideh, M., Alfaisal, R., & Salloum, S. (2021). The acceptance of social media video for knowledge acquisition, sharing and application: A comparative study among YouYube users and TikTok users' for medical purposes. *International Journal of Data and Network Science, 5*(3), 197–214.

Al-Nady, B. A., Al-Hawary, S. I., & Alolayyan, M. (2013). Strategic management as a key for superior competitive advantage of sanitary ware suppliers in kingdom of Saudi Arabia. *International Journal of Management and Information Technology, 7*(2), 1042–1058.

AlShehhi, H., Alshurideh, M., Al Kurdi, B., & Salloum, S. A. (2020, October). The impact of ethical leadership on employees performance: A systematic review. In *International conference on advanced intelligent systems and informatics* (pp. 417–426). Cham: Springer.

Alshura, M. S. K., Nusair, W. K. I., & Aldaihani, F. M. F. (2016). Impact of internal marketing practices on the organizational commitment of the employees of the insurance companies in Jordan. *International Journal of Academic Research in Economics and Management Sciences, 5*(4), 168–187.

Alshurideh, D. M. (2019). Do electronic loyalty programs still drive customer choice and repeat purchase behaviour? *International Journal of Electronic Customer Relationship Management, 12*(1), 40–57.

Alshurideh, M. (2022). Does electronic customer relationship management (E-CRM) affect service quality at private hospitals in Jordan? *Uncertain Supply Chain Management, 10*(2), 325–332.

Alshurideh, M. T., Al Kurdi, B., Alzoubi, H. M., Ghazal, T. M., Said, R. A., AlHamad, A. Q., et al. (2022). Fuzzy assisted human resource management for supply chain management issues. *Annals of Operations Research*, 1–19.

Alshurideh, M., Gasaymeh, A., Ahmed, G., Alzoubi, H., & Kurd, B. (2020). Loyalty program effectiveness: Theoretical reviews and practical proofs. *Uncertain Supply Chain Management, 8*(3), 599–612.

Altamony, H., Masa'deh, R., Alshurideh, M., & Obeidat, B. (2012) Information systems for competitive advantage: Implementation of an organisational strategic management process. *Innovation and Sustainable Competitive Advantage: From Regional Development to World Economies*, 583–592.

Alyammahi, A., Alshurideh, M., Al Kurdi, B., & Salloum, S. A. (2020, October). The impacts of communication ethics on workplace decision making and productivity. In *International conference on advanced intelligent systems and informatics* (pp. 488–500). Cham: Springer.

Alzoubi, H., Alshurideh, M., Kurdi, B., Akour, I., & Aziz, R. (2022). Does BLE technology contribute towards improving marketing strategies, customers' satisfaction and loyalty? The role of open innovation. *International Journal of Data and Network Science, 6*(2), 449–460.

Argyris C., & Schön D. (1996). Organizational learning II: Theory, method and practice. MA: Addison-Wesley.

Batayneh, R. M. A., Taleb, N., Said, R. A., Alshurideh, M. T., Ghazal, T. M., & Alzoubi, H. M. (2021, June). IT governance framework and smart services integration for future development of Dubai infrastructure utilizing AI and big data, its reflection on the citizens standard of living. In *The international conference on artificial intelligence and computer vision* (pp. 235–247). Cham: Springer.

Claver, E., Llopis, J., Garcia, D., & Molina, H. (1998). Organizational culture for innovation and new technological behavior. *The Journal of High Technology Management Research, 9*(1), 55–68.

de Leeuw, E., Hox, J., Silber, H., Struminskaya, B., & Vis, C. (2019). Development of an international survey attitude scale: Measurement equivalence, reliability, and predictive validity. *Measurement Instruments for the Social Sciences, 1*(1), 9. https://doi.org/10.1186/s42409-019-0012-x

Ghannajeh, A. M., AlShurideh, M., Zu'bi, M. F., Abuhamad, A., Rumman, G. A., Suifan, T., & Akhorshaideh, A. H. O. (2015). A qualitative analysis of product innovation in Jordan's pharmaceutical sector. *European Scientific Journal, 11*(4).

Girard, J., & Girard, J. (2015). Defining knowledge management: Toward an applied compendium. *Online Journal of Applied Knowledge Management, 3*(1), 1–20.

Hair, J. F., Babin, B. J., & Krey, N. (2017). Covariance-based structural equation modeling in the journal of advertising: Review and recommendations. *Journal of Advertising, 46*(1), 163–177. https://doi.org/10.1080/00913367.2017.1281777

Heisig, A., & Rorbeck (2000). *Knowledge management an integrated approach*, England, Prentice Hall.

Herschel, S. (2000). *Managing knowledge work*. New York: Palgrave MacCmillan.

Howard, M. C. (2018). The convergent validity and nomological net of two methods to measure retroactive influences. *Psychology of Consciousness: Theory, Research, and Practice, 5*(3), 324–337. https://doi.org/10.1037/cns0000149

Kurdi, B., Alshurideh, M., & Alnaser, A. (2020). The impact of employee satisfaction on customer satisfaction: Theoretical and empirical underpinning. *Management Science Letters, 10*(15), 3561–3570.

Laudon, K. C., & Laudon, J. P. (2007). *Essentials and management Information system*, Upper Saddle River New Jersey: Prentice Hall, Inc. Ltd.

March, J. G. (1991). Exploration and exploitation in organizational learning. *Organization Science, 2*(1), 71–87.

Marquardt, M. J. (1996). *Building the learning organization: A systems approach to quantum improvement and global success*. New York: McGraw-Hill.

Metabis, A., & Al-Hawary, S. I. (2013). The impact of internal marketing practices on services quality of commercial banks in Jordan. *International Journal of Services and Operations Management, 15*(3), 313–337.

Mohammad, A. A., Alshura, M. S., Al-Hawary, S. I. S., Al-Syasneh, M. S., & Alhajri, T. M. (2020). The influence of Internal Marketing Practices on the employees' intention to leave: A study of the private hospitals in Jordan. *International Journal of Advanced Science and Technology, 29*(5), 1174–1189.

Nasiruzzaman, M., & Dahlan, A. R. A. (2013). Project success and knowledge management (KM) practices in Malaysian institution of higher learning (IHL). *Journal of Education and Vocational Research*.

Nuseir, M. T., Al Kurdi, B. H., Alshurideh, M. T., & Alzoubi, H. M. (2021, June). Gender discrimination at workplace: Do artificial intelligence (AI) and machine learning (ML) have opinions about it. In *The international conference on artificial intelligence and computer vision* (pp. 301–316). Cham: Springer.

Parisa, G. (2015). Relationship between knowledge management and quality management in insurance companies. *International Journal of Academic Research*.

Rimkeviciene, J., Hawgood, J., O'Gorman, J., & De Leo, D. (2017). Construct validity of the acquired capability for suicide scale: Factor structure, convergent and discriminant validity.

Journal of Psychopathology and Behavioral Assessment, 39(2), 291–302. https://doi.org/10.1007/s10862-016-9576-4

Sekaran, U., & Bougie, R. (2016). *Research methods for business: A skill-building approach* (7th ed.). Wiley.

Shannak, R. O., Zu'bi, M. F., & Alshurideh, M. T. (2012). A theoretical perspective on the relationship between knowledge management systems, customer knowledge... *European Journal of Social Sciences, 32*(4), 520–532.

Shi, D., Lee, T., & Maydeu-Olivares, A. (2019). Understanding the model size effect on SEM fit indices. *Educational and Psychological Measurement, 79*(2), 310–334. https://doi.org/10.1177/0013164418783530

Sinkula, J. M. (1994). Market information processing and organizational learning. *Journal of Marketing, 58*(1), 35–45.

Sung, K.-S., Yi, Y. G., & Shin, H.-I. (2019). Reliability and validity of knee extensor strength measurements using a portable dynamometer anchoring system in a supine position. *BMC Musculoskeletal Disorders, 20*(1), 1–8. https://doi.org/10.1186/s12891-019-2703-0

Teo, H. H., Wang, X., Wei, K. K., Sia, C. L., & Lee, M. K. (2006). Organizational learning capacity and attitude toward complex technological innovations: An empirical study. *Journal of the American Society for Information Science and Technology, 57*(2), 264–279.

Wang, Y. A., & Rhemtulla, M. (2021). Power analysis for parameter estimation in structural equation modeling: A discussion and tutorial. *Advances in Methods and Practices in Psychological Science, 4*(1), 1–17. https://doi.org/10.1177/2515245920918253

Wiig, K. (2013). *Knowledge management foundations thinking about thinking*. USA: Schema Press.

William, R. K., Chung, T. R., & Honey, M. N. (2008). Knowledge management and organizational learning. *Omega, 36*(2), 167–172.

The Impact of Intellectual Capital on Competitive Capabilities: Evidence from Firms Listed in ASE

Doa'a Ahmad Odeh Al-Husban, Sulieman Ibraheem Shelash Al-Hawary, Ibrahim Rashed Soliaman AlTaweel, Nida'a Al-Husban, Mohammad Fathi Almaaitah, Faraj Mazyed Faraj Aldaihani, Anber Abraheem Shlash Mohammad, Ayat Mohammad, and Dheifallah Ibrahim Mohammad

Abstract This study aims at examining the impact of intellectual capital on achieving competitive capabilities. A qualitative method based on a questionnaire was used in this study for data collection and sample selection. Data were primarily gathered through self-reported questionnaires creating by Google Forms which were

---

D. A. O. Al-Husban
Faculty of Alia College, Department of Human Fundamental Sciences, Al-Balqa Applied University, Amman, Jordan
e-mail: D_husban@bau.edu.jo

S. I. S. Al-Hawary (✉) · N. Al-Husban (✉)
Faculty of Economics and Administrative Sciences, Department of Business Administration, Al Al-Bayt University, P.O. Box 130040, Mafraq 25113, Jordan
e-mail: dr_sliman73@aabu.edu.jo; dr_sliman@yahoo.com

I. R. S. AlTaweel
Faculty of Business School, Department of Business Administration, Qussim University, P.O. Box 6502, Al Russ City 51452, Saudi Arabia
e-mail: toiel@qu.edu.sa

M. F. Almaaitah
Faculty of Economic and Administration Sciences, Department of Business Administration & Public Administration, Al Al-Bayt University Jordan, P.O. Box 130040, Mafraq 25113, Jordan
e-mail: m.maaitah@aabu.edu.jo

F. M. F. Aldaihani
Kuwait Civil Aviation, ishbiliyah bloch 1, street 122, home 1, Kuwait, Kuwait

A. A. S. Mohammad
Faculty of Administrative and Financial Sciences, Marketing Department, Petra University, P.O. Box 961343, Amman 11196, Jordan

A. Mohammad
Business and Finance Faculty, The World Islamic Science and Education University (WISE), P.O Box 1101, Amman 11947, Jordan

D. I. Mohammad
Ministry of Education, Amman, Jordan

M. Alshurideh et al. (eds.), *The Effect of Information Technology on Business and Marketing Intelligence Systems*, Studies in Computational Intelligence 1056,
https://doi.org/10.1007/978-3-031-12382-5_93

distributed to a purposive sample of senior managers via email. Structural equation modeling (SEM) was used to test hypotheses. The study results showed that the highest impact on competitive capabilities was for structural capital, followed by intellectual capital, and finally the lowest impact was for human capital. In light of the findings of the study, researchers recommend paying attention to intellectual capital by paying attention to scientific competencies and preserving them by all means, providing the appropriate atmosphere to support talents and providing opportunities for ideas that are put forward by workers and put into practice.

Keywords Intellectual · Capital · Competitive capabilities · Firm · ASE

1 Introduction

The importance of information escalated with the emergence of the technology and computer revolution, so that information and knowledge became a source and basis for information technology, and the importance of knowledge based in the minds of individuals, focusing on individuals' skills and abilities to innovate and creativity (Allahow et al., 2018; Mohammad, 2020; Al-Hawary & Obiadat, 2021; Al-Hawary & AlDafiri, 2017; Bebba et al., 2017), generated the concept of intellectual capital (Al-Hawary et al., 2020; Alshura et al., 2016; Metabis & Al-Hawary, 2013; Mohammad et al., 2020). And it enjoys great and growing interest from the managers and leaders in organizations that seek to develop and succeed in an era which has become characterized by speed, technological development and intense competition between business organizations (Al-Hawary & Al-Syasneh, 2020; Al-Lozi et al., 2018; Al-Nady et al., 2016). Thus, the organizations looked at the concept of intellectual capital after the physical capital was dominant and whoever owns it owns the organization. The one who owns the knowledge becomes the one who owns the organization, and the organizations are now more accommodating of knowledge and more using it in an attempt to achieve their goals, satisfy customers and distinguish between their competitors (Al-Hawary & Nusair, 2017; Al-Hawary & Shdefat, 2016).

It has become necessary for organizations to work on evaluating their intellectual capital, and to work on developing it and transforming it into gains and a competitive advantage that distinguishes it from others, he higher the value of human resources, the greater the value of the organization (Al-Lozi et al., 2017; Al-Hawary et al., 2011; Mohammad et al., 2020; Al-Hawary, 2015; Alhalalmeh et al., 2020; Al-Hawary, 2013). Human capital has become the real capital and is responsible for the process of renewal and innovation in the organization, and since the current era is characterized by openness, rapid development and globalization, business organizations, if they want to live in this environment, must possess the strength that enables them to grow, continue and survive, this strength is represented by intellectual capital (Al-Hawary & Al-Syasneh, 2020; Al-Hawary & Alajmi, 2017; Al-Hawajreh et al., 2011; Al-Hawary & Al-Namlan, 2018; Al-Hawary et al., 2013a, 2013b). Therefore, the organization must realize the importance of intellectual capital and try to realize it and work with it in order to achieve its goals and mission, and therefore it needs workers who have knowledge and are good and always strive to provide what is best, and seek to find an internal organizing to perform work flexibly and effectively, which must be built a distinguished intellectual capital that serves it in achieving its strategy and guarantees it continuity in the work performance of its work (Al- Quran et al., 2020; Al-Hawary & Al-Kumait, 2017; Al-Hawary et al., 2011). Contemporary organizations live in a highly competitive environment, an environment characterized by speed and technological change, and these organizations are trying to catch up with technology and competition between companies and try to gain an advantage over other organizations (Al-Hawary & Batayneh, 2015). Leaders, management and workers in these organizations are accused of neglecting the issue of intellectual capital and not delving into it and realizing its importance despite its role in making them strong competitive organizations in the era of speed and technological change, which is the force that enables them to achieve what they aspire to. This study seeks to identify the impact of intellectual capital on competitive capabilities.

2 Theoretical Framework and Hypotheses Development

2.1 Intellectual Capital

With the emergence of globalization and the increasing importance of information and knowledge, and in light of the challenges and economic changes, organizations' interest in intellectual capital increased, and these organizations realized that the human element is the most important type of investment and that all its physical assets are useless if there is no human element that has the capabilities, abilities, and the necessary advanced and renewable knowledge to manage these assets (Al-Hawary & Ismael, 2010; Al-Hawary & Metabis, 2012). Thus, the focus of organizations shifted to intangible assets and considered them a source of wealth. Intellectual capital became the indicator of the organization's possession of a competitive advantage, but its value depends on the extent of interest in it, its development and its effective management (Al-Hawary & Haddad, 2016; Al-Hawary & Al-Hamwan, 2017).

Ungerer (2004) stated that researchers, analysts, and managers became increasingly interested in intellectual capital in the 1980s when intangible assets often emerged as a major determinant of company profits. Since the concept of intellectual capital appeared and its importance increased, many definitions of it have been put forward (Al-Hawary & Hadad, 2016).

Fen et al. (2012) argues that intellectual capital is only evidence that an organization is operating effectively or not. Auer described it as: "The difference between the market and the book value of the company, and there is no doubt that the intellectual capital represents the most important stock of knowledge on an organized basis, this value is usually not announced in the annual reports and does not appear in the analysis of traditional models." Bounfour and Edvinsson (2005) see it as a process of integrating human capital with strategic structural capital to obtain additional results for the future performance of employees. Rehman et al. (2011) see that human capital is the skill and experience of employees that increase with training and the higher the efficiency of the employees, the higher the efficiency of the organization. Leaniz and Bpsque (2013: 266) defined it as "knowledge that can be turned into profits in the future comprising resources such as ideas, inventions, technologies, designs, processes, and information programmes".

The subject of measurement is one of the most important topics that take into account when mentioning intellectual capital. Its importance stems from the importance of the subject it measures. Villanueva states that valuation of intangible assets is an important process for evaluating the quality of performance. Abdel Moneim believes that modern methods of measuring intellectual capital depend on the present and the future in measuring the value of the company and focus on quality, while traditional accounting methods focus on the past in the measurement process and on the results of previous transactions. There is a great difference between researchers about the ways in which intellectual capital is measured, and the reason for this difference is due to the intangible nature of intellectual capital and therefore difficult to measure and identify.

Stewart Model is the most common model in measuring intellectual capital. Stewart (1997) classifies intellectual capital into human capital which represents the knowledge, experience and skills possessed by workers, the ability to be creative and everything that distinguishes workers in the organization from others. Structural capital represents systems, procedures, culture, patents, copyrights, and anything that remains in the organization when employees leave. And relational capital represents the rate of loyalty and relationships with allies and suppliers.

Human capital: Human capital is the most important component of intellectual capital. Whatever the organization possesses of material assets, it still needs human minds to manage these assets, benefit from and maximize them (Alkalha et al., 2012; AlHamad et al., 2022; Alshurideh et al., 2022). Hence the importance of human capital stems from the knowledge that workers possess of skills, experiences, and the ability to innovate and deal with problems, looking to the future and predicting it, the value of this knowledge increases with development and training (Al Kurdi et al., 2021; Shamout et al., 2022a, 2022b). Banany believes that it is one of the most important production tools that contribute to raising the level of quality of services provided to customers and thus lead to a competitive advantage in the market, which increases the value of the company. As for Joia (2007), he saw it as abilities, skills, knowledge, and ability to solve problems, ability to be creative, work with the group, ability to learn, influence others, and ability to adapt to changing circumstances.

Structural capital: Everything that remains in the organization after the employees leave it and forms the infrastructure of the organization such as policies, procedures, databases, equipment, buildings, trademarks and copyrights (Alshurideh et al., 2015; Alameeri et al., 2020; Al Kurdi et al., 2020a, 2020b). Despite the importance of the human element, physical assets are also important for organizations (Alsuwaidi et al., 2020; Al Shebli et al., 2021). When minds create, they work to develop and benefit from these assets and add values to them (Aburayya et al., 2020; Alshurideh, 2022; Taryam et al., 2020). Luthy (1998) also mentions that infrastructure that supports and enables human capital to function, is owned by the organization and remains in the organization when employees leave and includes hardware, software, buildings, processes, patents and trademarks. Roos and Mcdonald (2003) defines it as "consisting of administrative processes, information systems, organizational structure, intellectual property, and any intangible assets owned by a company that do not appear on balance sheet."

Relational capital: The organization is always interested in its external relations, whether it is with customers, partners or suppliers, and from here the importance of relational capital has grown as an important part of intellectual capital (Lee et al., 2022a, 2022b; Shamout et al., 2022a, 2022b). It is the value that results from the level of customer satisfaction, allies, suppliers and all external parties that the organization deals with Alshurideh et al. (2012), Al-Dmour et al. (2021), Alzoubi et al. (2022). Luthy (1998) defined relational capital as the extent of loyalty to customer relationships. As for Mazlan (2005), he worked on identifying the components of relationship capital and defined it as "all the relationships that link the organization with its stakeholders such as suppliers, government agencies, investors and customers, distribution channels, and strategic alliances that the organization establishes."

2.2 Competitive Capabilities

In recent years, the topic of competitive capabilities has received great attention from researchers as a result of rapid developments, economic changes and the emergence of technology, which intensified competition between companies and increased organizations' interest in competitive capabilities (Al-Tarawneh et al., 2012; Al-Hawary et al., 2013a, 2013b; Al-Hawary, 2011a, 2011b; Al-Nady et al., 2013; Al-Hawary & Aldaihani, 2016). For great opportunities that make it compete and rise in the labor market over its competitors, and then it must maintain these capabilities and develop them continuously through creativity, innovation and continuous renewal (Alameeri et al., 2021; Almaazmi et al., 2020; Nuseir et al., 2021). Covin et al. (2010) see that competitive capabilities may be considered as a specific feature or set of advantages that an organization possesses and distinguishes it from other organizations and that the basic work of any organization does not depend only on the process of producing and providing products, but in the ability to continue production and satisfy the continuous, changing and different needs of the market as well. Stevenson (2007) considered competitiveness as the firm's ability to discover and use new methods of production and to develop new products as well.

2.3 Intellectual Capital and Competitive Capabilities

Intellectual capital is one of the most important recent topics that have begun to focus and pay attention to it because of its great importance in increasing the competitive capabilities of organizations (Altamony et al., 2012; Obeidat et al., 2021). Al-Hawajrah (2010) defined intellectual capital as "a group of intangible assets that enable the organization to perform its functions according to a competitive advantage." Both Al-Rousan and Al-Ajlouni (2010: 44) believe that intellectual capital is the most important source of wealth for the organization, as it constitutes a competitive advantage that distinguishes it from other organizations in an environment in which competition has become based on knowledge and skills. Al-Hamdani and Ali (2010) see that intellectual capital as "the value extracted from transforming the tacit knowledge, information, experiences and skills that employees have into tangible forms that can be traded (copyright) and that enable the organization to build a mental stature and distinct relationships with beneficiaries with the aim of competitive advantage." From the point of view of Sharabati et al. (2013) they are organized intangible assets and knowledge of value or those that are used to create value. Guthrie (2001) sees that all the individual capabilities that characterize a specific number of employees in the organization enable them to contribute to increasing the productivity of the organization and raise the level of performance compared to its competitors. The importance of intellectual capital lies in the fact that it enables the organization to possess competitive capabilities that enable it to face competitors. It helps the organization to innovate, achieve what is always better and keep pace with continuous changes in an era characterized by change and rapid development. Based on the above, the hypothesis of the study can be formulated as follows:

There is a statistically significant effect of intellectual capital on competitive capabilities.

3 Study Model

See Fig. 1.

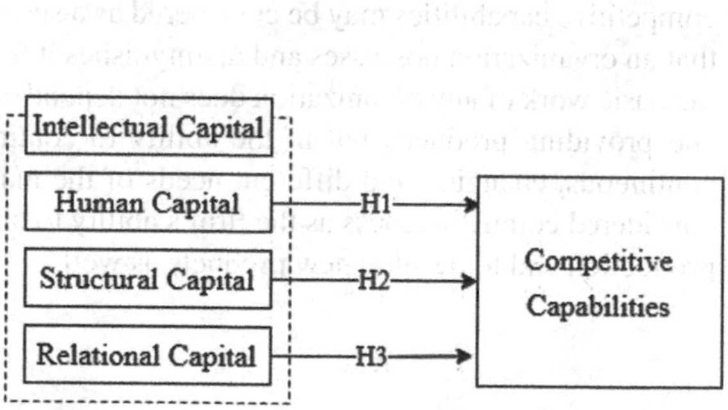

Fig. 1 Research model

4 Methodology

4.1 Population and Sample Selection

A qualitative method based on a questionnaire was used in this study for data collection and sample selection. The major aim of the study was to examine the impact of intellectual capital on competitive capabilities. Therefore, it focused on industrial firms listed in ASE. Data were primarily gathered through self-reported questionnaires creating by Google Forms which were distributed to a purposive sample of senior managers via email. In total, (167) responses were received including (21) invalid to statistical analysis due to uncompleted or inaccurate. Hence, the final sample contained (146) responses suitable to analysis requirements, where it proved to be sufficient to the extent that was predictable and allowed for a presumption of data saturation (Sekaran & Bougie, 2016).

4.2 Measurement Instrument

A self-reported questionnaire that consists of two main sections along with a section regarding control variables was used as the measurement instrument. Control variables considered as categorical measures were composed of gender, age group, educational level, and experience. The two main sections were dealt with a five-point Likert scale (from 1 = strongly disagree to 5 = strongly agree). The first section contained (14) items to measure intellectual capital based on Stewart (1997). These items were distributed into dimensions as follows: (HC1-HC5) items dedicated for measuring human capital, (SC1-SC4) items dedicated for measuring structural capital, and (IC1-IC5) items dedicated for measuring intellectual capital. Whereas the second section included seven items (CC1-CC7) developed to measure competitive capabilities according to what was pointed by Al-Hawary and Hadad (2016).

5 Findings

5.1 Measurement Model Evaluation

This study was conducted structural equation modeling (SEM) to test hypotheses, which represents a contemporary statistical technique for testing and estimating the relationship between factors and variables (Wang & Rhemtulla, 2021). Accordingly, the reliability and validity of the constructs were tested using confirmatory factor

Table 1 Results of validity and reliability tests

Constructs	HC	SC	IC	CC
HC	**0.769**			
SC	0.625	**0.752**		
IC	0.538	0.498	**0.749**	
CC	0.712	0.703	0.724	**0.787**
VIF	2.418	1.946	2.044	–
Loadings range	0.674–0.835	0.653–0.844	0.698–0.801	0.642–0.883
AVE	0.591	0.566	0.561	0.620
MSV	0.502	0.518	0.497	0.537
Internal consistency	0.875	0.834	0.862	0.916
Composite reliability	0.878	0.838	0.864	0.919

Note HC: human capital, SC: structural capital, IC: intellectual capital, CC: competitive capabilities, Bold fonts indicates to square root of AVE

analysis (CFA) through the statistical program AMOSv24. Table 1 summarizes the results of convergent and discriminant validity, as well the indicators of reliability.

Table 1 shows that the standard loading values for the individual items were within the domain (0.642–0.883), these values greater than the minimum retention of the elements based on their standard loads (Al-Lozi et al., 2018; Sung et al., 2019). Average variance extracted (AVE) is a summary indicator of the convergent validity of constructs that must be above 0.50 (Howard, 2018). The results indicate that the AVE values were greater than 0.50 for all constructs, thus the used measurement model has an appropriate convergent validity. Rimkeviciene et al. (2017) suggested the comparison approach as a way to deal with discriminant validity assessment in covariance-based SEM. This approach is based on comparing the values of maximum shared variance (MSV) with the values of AVE, as well as comparing the values of square root of AVE ($\sqrt{AVE}$) with the correlation between the rest of the structures. The results show that the values of MSV were smaller than the values of AVE, and that the values of $\sqrt{AVE}$ were higher than the correlation values among the rest of the constructs. Therefore, the measurement model used is characterized by discriminative validity. The internal consistency measured through Cronbach's Alpha coefficient (α) and compound reliability by McDonald's Omega coefficient (ω) was conducted as indicators to evaluate measurement model. The results listed in Table 1 demonstrated that both values of Cronbach's Alpha coefficient and McDonald's Omega coefficient were greater than 0.70, which is the lowest limit for judging on measurement reliability (De Leeuw et al., 2019).

5.2 Structural Model

The structural model illustrated no multicollinearity issue among predictor constructs because variance inflation factor (VIF) values are below the threshold of 5, as shown in Table 1 (Hair et al., 2017). This result is supported by the values of model fit indices shown in Fig. 1.

The results in Fig. 2 indicated that the chi-square to degrees of freedom (CMIN/DF) was 2.443, which is less than 3 the upper limit of this indicator. The values of the goodness of fit index (GFI), the comparative fit index (CFI), and the Tucker-Lewis index (TLI) were upper than the minimum accepted threshold of 0.90. Moreover, the result of root mean square error of approximation (RMSEA) indicated to value 0.033, this value is a reasonable error of approximation because it is less than the higher limit of 0.08. Consequently, the structural model used in this study was recognized as a fit model for predicting the DEP and generalization of its result (Ahmad et al., 2016; Shi et al., 2019). To verify the results of testing the study hypotheses, structural equation modeling (SEM) was used, the results of which are listed in Table 2.

The results demonstrated in Table 2 show that the highest impact on competitive capabilitieswas for structural capital ($\beta = 0.543$, $t = 22.96$, $p = 0.000$), followed by intellectual capital ($\beta = 0.477$, $t = 19.65$, $p = 0.002$), and finally the lowest impact was for human capital ($\beta = 0.396$, $t = 18.12$, $p = 0.006$). Thus, all the minor hypotheses of the study were supported based on these results.

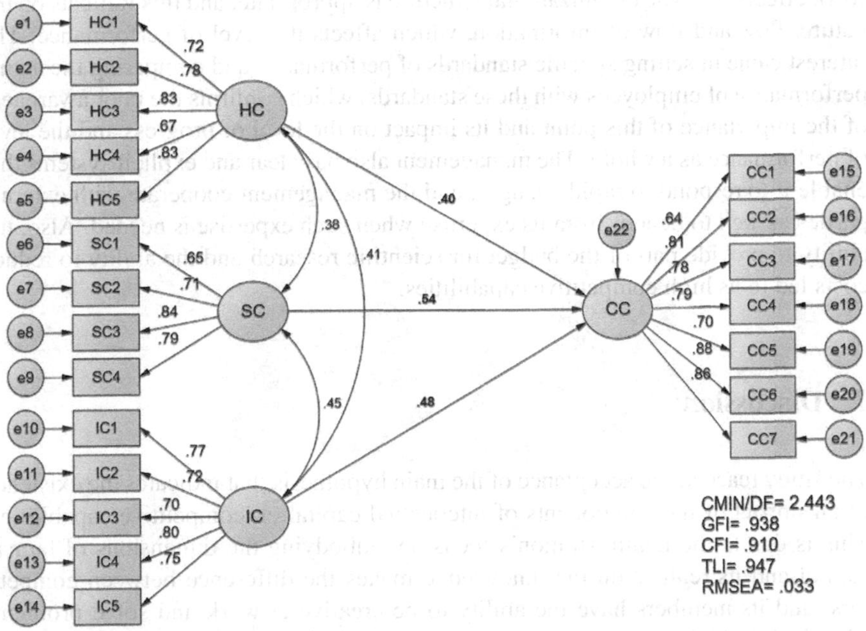

Fig. 2 SEM results of the intellectual capital effect on competitive capabilities

Table 2 Hypothesis testing

Hypothesis	Relation	Standard Beta	*t* value	*p* value
H1	HC → CC	0.396	18.12**	0.006
H2	SC → CC	0.543	22.96***	0.000
H3	IC → CC	0.477	19.65**	0.002

Note HC: human capital, SC: structural capital, IC: intellectual capital, CC: competitive capabilities
* $p < 0.05$, ** $p < 0.01$, *** $p < 0.001$

6 Conclusion

The results of the study indicate the management interest in structural capital through paying attention to developing its managerial processes, paying attention to its organizational structure and its continuous maintenance of infrastructure, which provides a suitable environment for employees to perform work in the best way, which is reflected in the overall performance of the organization. It also indicates that the management pays great attention to its managerial operations and works to develop them, and information systems and databases which are characterized by their integration and development. The management interest in maintaining its infrastructure was appropriate, which means that employees are not sufficiently satisfied with the infrastructure provided by the organization, despite the importance of providing a good infrastructure and maintaining it on an ongoing basis, and thus providing the appropriate environment for work, which encourages workers to perform their work effectively. The organizational structure is appropriate, and this Reflects on the nature, flow and flow of information, which affects the level of performance. The interest came in setting specific standards of performance and comparing the actual performance of employees with these standards, which confirms the poor awareness of the importance of this point and its impact on the level of progress and the level of performance as a whole. The management also has clear and explicit systems that enable it to respond to rapid changes, and the management cooperate with external parties as well to benefit from its expertise when such expertise is needed. Also, the ability to provide part of the budget for scientific research and the ability to reduce costs led to its high competitive capabilities.

7 Discussion

The study reached the acceptance of the main hypothesis that indicates the existence of an impact of the components of intellectual capital on competitive capabilities. This is due to the administration's focus on embodying the dimensions of human capital and its realization that knowledge makes the difference between competitors, and its members have the ability to be creative at work and solve problems effectively, which supports meet human capital requirements, the company's policy has also contributed to encouraging employees to acquire skills through rotation and development programmes, and thus employees have a great deal of skills and experience, and this is reflected in their competitive capabilities. It was found that there is an impact of structural capital on competitive capabilities, due to the interest in the field of information systems and databases that provide the necessary information at the right time, which supports the decision-making process. It also seeks to adopt policies, procedures, and methods that facilitate the implementation of operations efficiently and effectively and works to continuously review and evaluate its managerial operations to increase the effectiveness of the company. It adopts a flexible organizational structure that improves the performance of operations and individuals and enhances its competitive capabilities. It was also found that there is an impact of relational capital on competitive capabilities, due to the company's interest in its external relations, establishing relations with other organizations, and exchanging information and knowledge with them, which is beneficial and raises its competitive capabilities.

8 Recommendations

In light of the findings of the study, researchers recommend paying attention to intellectual capital by paying attention to scientific competencies and preserving them by all means, providing the appropriate atmosphere to support talents and providing opportunities for ideas that are put forward by workers and put into practice. In addition to the interest in structural capital, it is considered the supporting infrastructure for the organization through the provision of all modern technologies, the continuous development of information systems and databases, and the continuous maintenance of the infrastructure. If the organization has a flexible and strong organizational structure, this enhances its performance to obtain good outputs. Constant keenness to support human capital through the establishment of training and development programs, encouragement of teamwork and the exchange of knowledge, and the appointment must be according to the efficiency standard to reach the desired goals. And strengthening the relational capital, by paying attention to cultural activities, and providing the opportunity to establish human relations between the organization's family, establishing development and communication centers between the organization and the local community, consolidating the relationship between the organization and other organizations working in the same field, exchanging experiences among them, and training workers on technology and means different contact.

References

Aburayya, A., Alshurideh, M., Al Marzouqi, A., Al Diabat, O., Alfarsi, A., Suson, R., et al. (2020). An empirical examination of the effect of TQM practices on hospital service quality: An assessment study in UAE hospitals. *Systematic Reviews in Pharmacy, 11*(9), 347–362.

Ahmad, S., Zulkurnain, N., & Khairushalimi, F. (2016). Assessing the validity and reliability of a measurement model in structural equation modeling (SEM). *British Journal of Mathematics & Computer Science, 15*(3), 1–8. https://doi.org/10.9734/BJMCS/2016/25183

Al Kurdi, B., Alshurideh, M., & Al Afaishat, T. (2020a). Employee retention and organizational performance: Evidence from banking industry. *Management Science Letters, 10*(16), 3981–3990.

Al Kurdi, B., Alshurideh, M., & Alnaser, A. (2020b). The impact of employee satisfaction on customer satisfaction: Theoretical and empirical underpinning. *Management Science Letters, 10*(15), 3561–3570.

Al Kurdi, B., Elrehail, H., Alzoubi, H., Alshurideh, M., & Al-Adaila, R. (2021). The interplay among HRM practices, job satisfaction and intention to leave: An empirical investigation. *Journal of Legal, Ethical and Regulatory, 24*(1), 1–14.

Al- Quran, A. Z., Alhalalmeh, M. I., Eldahamsheh, M. M., Mohammad, A. A., Hijjawi, G. S., Almomani, H. M., & Al-Hawary, S. I. (2020). Determinants of the green purchase intention in Jordan: The moderating effect of environmental concern. *International Journal of Supply Chain Management, 9*(5), 366–371.

Al Shebli, K., Said, R. A., Taleb, N., Ghazal, T. M., Alshurideh, M. T., & Alzoubi, H. M. (2021, June). RTA's employees' perceptions toward the efficiency of artificial intelligence and big data utilization in providing smart services to the residents of Dubai. In *The international conference on artificial intelligence and computer vision* (pp. 573–585). Cham: Springer.

Alameeri, K. A., Alshurideh, M. T., & Al Kurdi, B. (2021). The effect of Covid-19 pandemic on business systems' innovation and entrepreneurship and how to cope with it: A theatrical view. In: *The effect of coronavirus disease (COVID-19) on business intelligence* (Vol. 334, pp. 275–288).

Alameeri, K., Alshurideh, M., Al Kurdi, B., & Salloum, S. A. (2020, October). The effect of work environment happiness on employee leadership. In *International conference on advanced intelligent systems and informatics* (pp. 668–680). Cham: Springer.

Al-Dmour, R., AlShaar, F., Al-Dmour, H., Masa'deh, R., & Alshurideh, M. T. (2021). The effect of service recovery Justices strategies on online customer engagement via the role of "customer satisfaction" during the Covid-19 pandemic: An empirical study. In *The effect of coronavirus disease (COVID-19) on business intelligence* (Vol. 334, pp. 325–346).

Alhalalmeh, M. I., Almomani, H. M., Altarifi, S., Al- Quran, A. Z., Mohammad, A. A., & Al-Hawary, S. I. (2020). The nexus between corporate social responsibilty and organizational performance in Jordan: The mediating role of organizational commitment and organizational citizenship behavior. *Test Engineering and Management, 83*(July), 6391–6410.

AlHamad, A., Alshurideh, M., Alomari, K., Kurdi, B., Alzoubi, H., Hamouche, S., & Al-Hawary, S. (2022). The effect of electronic human resources management on organizational health of telecommunications companies in Jordan. *International Journal of Data and Network Science, 6*(2), 429–438.

Al-Hamdani, N., & Ali, A. (2010). Intellectual capital and its impact on employee performance management: An analytical study of the opinions of a sample of heads of scientific departments at the university of Mosul. *Rafidain Development, 32*(98), 119–145.

Al-Hawajrah, K. (2010). a study of the relationship of knowledge capital investment strategies to the competitive performance of institutions. *University of Sharjah Journal of Humanities and Social Sciences, 7*(2), 279–309.

Al-Hawajreh, K., AL-Zeaud, H., Al-Hawary, S. I., & Mohammad, A. A. (2011). The influence of top management support and commitment on total quality management indicators from managers and heads of departments viewpoint: A case study of Sahab industrial city. *Jordan Journal of Business Administration, 7*(4), 557–576.

Al-Hawary, S. I. (2011a). Human resource management practices in ZAIN cellular communications company operating in Jordan. *Perspectives of Innovations, Economics and Business, 8*(2), 26–34.

Al-Hawary, S. I. (2011b). The effect of banks governance on banking performance of the Jordanian commercial banks: Tobin's Q Model" an applied study. *International Research Journal of Finance and Economics, 71*, 34–47.

Al-Hawary, S. I. (2013). The roles of perceived quality, trust, and satisfaction in predicting brand loyalty: The empirical research on automobile brands in Jordan market. *International Journal of Business Excellence, 6*(6), 656–686.

Al-Hawary, S. I. (2015). Human resource management practices as a success factor of knowledge management implementation at health care sector in Jordan. *International Journal of Business and Social Science, 6*(11/1), 83–98.

Al-Hawary, S. I., & Alajmi, H. M. (2017). Organizational commitment of the employees of the ports security affairs of the state of Kuwait: The impact of human recourses management practices. *International Journal of Academic Research in Economics and Management Sciences, 6*(1), 52–78.

Al-Hawary, S. I., & Aldaihani, F. M. (2016). Customer relationship management and innovation capabilities of Kuwait airways. *International Journal of Academic Research in Economics and Management Sciences, 5*(4), 201–226.

Al-Hawary, S. I., & Al-Hamwan, A. (2017). Environmental analysis and its impact on the competitive capabilities of the commercial banks operating in Jordan. *International Journal of Academic Research in Accounting, Finance and Management Sciences, 7*(1), 277–290.

Al-Hawary, S. I., & Al-Kumait, Z. (2017). Training programs and their effect on the employees performance at King Hussain Bin Talal development area at Al-Mafraq Governate in Jordan. *International Journal of Academic Research in Economics and Management Sciences, 6*(1), 258–274.

Al-Hawary, S. I., & Al-Namlan, A. (2018). Impact of electronic human resources management on the organizational learning at the private hospitals in the state of Qatar. *Global Journal of Management and Business Research: A Administration and Management, 18*(7), 1–11.

Al-Hawary, S. I., & Hadad, T. F. (2016). The effect of strategic thinking styles on the enhancement competitive capabilities of commercial banks in Jordan. *International Journal of Business and Social Science, 7*(10), 133–144.

Al-Hawary, S. I., & Haddad, I. (2016). Level of employers' satisfaction on the employees' performance at the Irbid industrial zone in Jordan. *International Journal of Academic Research in Economics and Management Sciences, 5*(4), 228–248.

Al-Hawary, S. I., & Ismael, M. (2010). The effect of using information technology in achieving competitive advantage strategies: A field study on the jordanian pharmaceutical companies. *Al Manara for Research and Studies, 16*(4), 196–203.

Al-Hawary, S. I., & Metabis, A. (2012). implementation of internal marketing in Jordan banks. *International Journal of Data Analysis and Information, 4*(1), 37–53.

Al-Hawary, S. I., & Nusair, W. (2017). Impact of human resource strategies on perceived organizational support at Jordanian public universities. *Global Journal of Management and Business Research: A Administration and Management, 17*(1), 68–82.

Al-Hawary, S. I., & Shdefat, F. (2016). Impact of human resources management practices on employees' satisfaction a field study on the Rajhi cement factory. *International Journal of Academic Research in Accounting, Finance and Management Sciences, 6*(4), 274–286.

Al-Hawary, S. I., Al-Hawajreh, K., AL-Zeaud, H., & Mohammad, A. (2013a). The impact of market orientation strategy on performance of commercial banks in Jordan. *International Journal of Business Information Systems, 14*(3), 261–279.

Al-Hawary, S. I., Al-Qudah, K., Abutayeh, P., Abutayeh, S., & Al-Zyadat, D. (2013b). The impact of internal marketing on employee's job satisfaction of commercial banks in Jordan. *Interdisciplinary Journal of Contemporary Research in Business, 4*(9), 811–826.

Al-Hawary, S. I., AL-Zeaud, H., & Batayneh, A. M. (2011). The relationship between transformational leadership and employee's satisfaction at Jordanian private hospitals. *Business and Economic Horizons, 5*(2), 35–46.
Al-Hawary, S. I. S., & AlDafiri, M. F. S. (2017). Effect of the components of information technology adoption on employees performance of interior ministry of Kuwait state. *International Journal of Academic Research in Economics and Management Sciences, 6*(2), 149–169.
Al-Hawary, S. I. S., & Batayneh, A. M. I. (2015). A Study of the strategic performance of shareholding industrial organizations in Jordan: Using Z- score model. *International Journal of Business and Social Science, 6*(11–1), 177–182.
Al-Hawary, S. I. S., & Obiadat, A. A. (2021). Does mobile marketing affect customer loyalty in Jordan? *International Journal of Business Excellence, 23*(2), 226–250.
Al-Hawary, S. I. S., Mohammad, A. S., Al-Syasneh, M. S., Qandah, M. S. F., & Alhajri, T. M. S. (2020). Organisational learning capabilities of the commercial banks in Jordan: Do electronic human resources management practices matter? *International Journal of Learning and Intellectual Capital, 17*(3), 242–266.
Al-Hawary, S. I., & Al-Syasneh, M. S. (2020). Impact of dynamic strategic capabilities on strategic entrepreneurship in presence of outsourcing of five stars hotels in Jordan. *Business: Theory and Practice, 21*(2), 578–587.
Alkalha, Z., Al-Zu'bi, Z., Al-Dmour, H., Alshurideh, M., & Masa'deh, R. (2012). Investigating the effects of human resource policies on organizational performance: An empirical study on commercial banks operating in Jordan. *European Journal of Economics, Finance and Administrative Sciences, 51*(1), 44–64.
Allahow, T. J. A. A., Al-Hawary, S. I. S., & Aldaihani, F. M. F. (2018). Information technology and administrative innovation of the central agency for information technology in Kuwait. *Global Journal of Management and Business, 18*(11-A), 1–16.
Al-Lozi, M., Almomani, R. Z., & Al-Hawary, S. I. (2017). Impact of talent management on achieving organizational excellence in Arab potash company in Jordan. *Global Journal of Management and Business Research: A Administration and Management, 17*(7), 15–25.
Al-Lozi, M., Almomani, R. Z., & Al-Hawary, S. I. (2018). Talent Management strategies as a critical success factor for effectiveness of Human Resources Information Systems in commercial banks working in Jordan. *Global Journal of Management and Business Research: A Administration and Management, 18*(1), 30–43.
Almaazmi, J., Alshurideh, M., Al Kurdi, B., & Salloum, S. A. (2020, October). The effect of digital transformation on product innovation: A critical review. In *International conference on advanced intelligent systems and informatics* (pp. 731–741). Cham: Springer.
Al-Nady, B. A., Al-Hawary, S. I., & Alolayyan, M. (2013). Strategic management as a key for superior competitive advantage of sanitary ware suppliers in kingdom of Saudi Arabia. *International Journal of Management and Information Technology, 7*(2), 1042–1058.
Al-Nady, B. A., Al-Hawary, S. I., & Alolayyan, M. (2016). The role of time, communication, and cost management on project management success: An empirical study on sample of construction projects customers in Makkah city, kingdom of Saudi Arabia. *International Journal of Services and Operations Management, 23*(1), 76–112.
Al-Rousan, M. A., & Al-Ajlouni, M. M. (2010). the impact of intellectual capital on creativity in Jordanian banks, a field study. *Damascus University Journal of Economic and Legal Sciences, 26*(2), 37–57.
Alshura, M. S. K., Nusair, W. K. I., & Aldaihani, F. M. F. (2016). Impact of internal marketing practices on the organizational commitment of the employees of the insurance companies in Jordan. *International Journal of Academic Research in Economics and Management Sciences, 5*(4), 168–187.
Alshurideh, M. (2022). Does electronic customer relationship management (E-CRM) affect service quality at private hospitals in Jordan? *Uncertain Supply Chain Management, 10*(2), 325–332.

Alshurideh, M. T., Al Kurdi, B., Alzoubi, H. M., Ghazal, T. M., Said, R. A., AlHamad, A. Q., et al. (2022). Fuzzy assisted human resource management for supply chain management issues. *Annals of Operations Research*, 1–19.

Alshurideh, M., Alhadid, A. Y., & Barween, A. (2015). The effect of internal marketing on organizational citizenship behavior an applicable study on the university of Jordan employees. *International Journal of Marketing Studies, 7*(1), 138–145.

Alshurideh, M., Masa'deh, R. M. D. T., & Alkurdi, B. (2012). The effect of customer satisfaction upon customer retention in the Jordanian mobile market: An empirical investigation. *European Journal of Economics, Finance and Administrative Sciences, 47*(12), 69–78.

Alsuwaidi, M., Alshurideh, M., Al Kurdi, B., & Salloum, S. A. (2020, October). Performance appraisal on employees' motivation: A comprehensive analysis. In *International conference on advanced intelligent systems and informatics* (pp. 681–693). Cham: Springer.

Altamony, H., Masa'deh, R., Alshurideh, M., Obeidat, B. (2012) Information systems for competitive advantage: Implementation of an organisational strategic management process. *Innovation and Sustainable Competitive Advantage: From Regional Development to World Economies*, 583–592.

Al-Tarawneh, K. A., Mohammad Alhamadani, S. Y., & Mohammad, A. A. (2012). Transformational leadership and marketing effectiveness in commercial banks in Jordan. *European Journal of Economics, Finance and Administrative Sciences, 46*, 71–87.

Alzoubi, H., Alshurideh, M., Kurdi, B., Akour, I., & Aziz, R. (2022). Does BLE technology contribute towards improving marketing strategies, customers' satisfaction and loyalty? The role of open innovation. *International Journal of Data and Network Science, 6*(2), 449–460.

Bebba, I., Bentafat, A., & Al-Hawary, S. I. S. (2017). An evaluation of the performance of higher educational institutions using data envelopment analysis: An empirical study on Algerian higher educational institutions. *Global Journal of Human Social Science Research, 17*(8), 21–30.

Bounfour, A., & Edvinsson, L. (2005). *Intellectual capital for communities: Nations- regions- and cities*. USA: Elsevier Butterworth-Heinemann.

Covin, J. G., Slevin, D. P., & Heeley, M. B. (2010). Pioneers and followers: Competitive tactics, environment, and firm growth. *Journal of Business Venturing, 15*(2), 175–210.

De Leeuw, E., Hox, J., Silber, H., Struminskaya, B., & Vis, C. (2019). Development of an international survey attitude scale: Measurement equivalence, reliability, and predictive validity. *Measurement Instruments for the Social Sciences, 1*(1), 9. https://doi.org/10.1186/s42409-019-0012-x

Guthrie, J. (2001). Study: Measurement and reporting of intellectual capital. *Journal of Intellectual Capital, 1*(2), 27–41.

Hair, J. F., Babin, B. J., & Krey, N. (2017). Covariance-based structural equation modeling in the journal of advertising: Review and recommendations. *Journal of Advertising, 46*(1), 163–177. https://doi.org/10.1080/00913367.2017.1281777

Howard, M. C. (2018). The convergent validity and nomological net of two methods to measure retroactive influences. *Psychology of Consciousness: Theory, Research, and Practice, 5*(3), 324–337. https://doi.org/10.1037/cns0000149

Joia, L. A. (2007). *Strategies for information technology and intellectual capital challenges and opportunities* (1st ed.). Brazil: Yurchak.

Leaniz, M., & Bpsque, R. (2013). *Intellectual capital and relational capital: The role of sustainability in developing corporate reputation Patricia*. Spain: Omnia Science.

Lee, K., Azmi, N., Hanaysha, J., Alshurideh, M., & Alzoubi, H. (2022a). The effect of digital supply chain on organizational performance: An empirical study in Malaysia manufacturing industry. *Uncertain Supply Chain Management, 10*(2), 1–16.

Lee, K., Ramiz, P., Hanaysha, J., Alzoubi, H., & Alshurideh, M. (2022b). Investigating the impact of benefits and challenges of IOT adoption on supply chain performance and organizational performance: An empirical study in Malaysia. *Uncertain Supply Chain Management, 10*(2), 1–14.

Luthy, D. H. (1998). Intellectual capital and its measurement. In *Proceedings of the Asian Pacific. Interdisciplinary Research in Accounting Conference (APIRA), Osaka, Japan* (pp. 16–17).

Mazlan, I. (2005). The role of employee development in the growth of intellectual capital. *Personnel Review, 29*(4), 521–533.

Metabis, A., & Al-Hawary, S. I. (2013). The impact of internal marketing practices on services quality of commercial banks in Jordan. *International Journal of Services and Operations Management, 15*(3), 313–337.

Mohammad, A. A. (2020). The effect of customer empowerment and customer engagement on marketing performance: The mediating effect of brand community membership. *Business: Theory and Practice, 21*(1), 30–38.

Mohammad, A. A. S., Saleem Khlif Alshura, M., Al-Hawary, S. I. S., Al-Syasneh, M. S., & Alhajri, T. M. S. (2020). The influence of internal marketing practices on the employees' intention to leave: A study of the private hospitals in Jordan. *International Journal of Advanced Science and Technology, 29*(5), 1174–1189.

Nuseir, M. T., Aljumah, A., & Alshurideh, M. T. (2021). How the business intelligence in the new startup performance in UAE during COVID-19: The mediating role of innovativeness. In *The effect of coronavirus disease (COVID-19) on business intelligence* (Vol. 334, pp. 63–79).

Obeidat, U., Obeidat, B., Alrowwad, A., Alshurideh, M., Masadeh, R., & Abuhashesh, M. (2021). The effect of intellectual capital on competitive advantage: The mediating role of innovation. *Management Science Letters, 11*(4), 1331–1344.

Rehman, W., Ghaudhary, A., Rehman, H. & Zahid, A. (2011). Intellectual capital performance and its impact on corporater performance: An empirical evidence from Modaraba sector of Pakistan. *Australian Journal of Business Management Research, 1*(5), 8–16.

Rimkeviciene, J., Hawgood, J., O'Gorman, J., & De Leo, D. (2017). Construct validity of the acquired capability for suicide scale: Factor structure, convergent and discriminant validity. *Journal of Psychopathology and Behavioral Assessment, 39*(2), 291–302. https://doi.org/10.1007/s10862-016-9576-4

Roos, G. & Mcdonald, S. (2003). *An intellectual capital primer*. This Paper is an Extension and Modification of an Unpublished Paper, Centre for Business Performance Cranfield University.

Sekaran, U., & Bougie, R. (2016). *Research methods for business: A skill-building approach* (7th ed.). Wiley.

Shamout, M., Ben-Abdallah, R., Alshurideh, M., Alzoubi, H., Kurdi, B., & Hamadneh, S. (2022a). A conceptual model for the adoption of autonomous robots in supply chain and logistics industry. *Uncertain Supply Chain Management, 10*(2), 577–592.

Shamout, M., Elayan, M., Rawashdeh, A., Kurdi, B., & Alshurideh, M. (2022b). E-HRM practices and sustainable competitive advantage from HR practitioner's perspective: A mediated moderation analysis. *International Journal of Data and Network Science, 6*(1), 165–178.

Sharabati, A. A., Nour, A. A. N., & NaserEddin, Y. N. (2013). Intellectual capital development: A case study of Middle East university. *Jordan Journal of Business Administration, 9*(3), 567–602.

Shi, D., Lee, T., & Maydeu-Olivares, A. (2019). Understanding the model size effect on SEM fit indices. *Educational and Psychological Measurement, 79*(2), 310–334. https://doi.org/10.1177/0013164418783530

Stevenson, W. J. (2007). *Production: Operations management* (8th ed.). Von Hoffmann Press.

Stewart, T. (1997). *Intellectual capital: The new wealth of organization*. London: Nicholas Brealey.

Sung, K.-S., Yi, Y. G., & Shin, H.-I. (2019). Reliability and validity of knee extensor strength measurements using a portable dynamometer anchoring system in a supine position. *BMC Musculoskeletal Disorders, 20*(1), 1–8. https://doi.org/10.1186/s12891-019-2703-0

Taryam, M., Alawadhi, D., Aburayya, A., Albaqa'een, A., Alfarsi, A., Makki, I., et al. (2020). Effectiveness of not quarantining passengers after having a negative COVID-19 PCR test at arrival to Dubai airports. *Systematic Reviews in Pharmacy, 11*(11), 1384–1395.

Ungerer, M. (2004). *Developing core capabilities in a financial services firm: An intellectual capital perspective*. Doctor desertion in industrial psychology, Faculty of Economic and Management Sciences, Rand Afrikans University.

Wang, Y. A., & Rhemtulla, M. (2021). Power analysis for parameter estimation in structural equation modeling: A discussion and tutorial. *Advances in Methods and Practices in Psychological Science, 4*(1), 1–17. https://doi.org/10.1177/2515245920918253

Wu, M. F., Lee, Y. J., & Wang, G. L. (2012). To verivy how intellectual capital affects organizational performance in listed Taiwan IC design companies with considering the moderator of corporate governance. *Journal of Global Business Management*, *88*(1), 20–32.

Impact of Knowledge Management on Total Quality Management at Private Universities in Jordan

Ali Zakariya Al-Quran, Rehab Osama Abu Dalbouh, Mohammed Saleem Khlif Alshura, Majed Kamel Ali Al-Azzam, Faraj Mazyed Faraj Aldaihani, Ziad Mohd Ali Smadi, Kamel Mohammad Al-hawajreh, Sulieman Ibraheem Shelash Al-Hawary, and Muhammad Turki Alshurideh

Abstract This study aims to know the impact of knowledge management on total quality management in Jordanian private universities. To achieve the objectives of the study, descriptive and analytical approach was adopted, where the researcher

A. Z. Al-Quran · R. O. A. Dalbouh · Z. M. A. Smadi · S. I. S. Al-Hawary (✉)
Department of Business Administration, School of Business, Al al-Bayt University, P.O. Box 130040, Mafraq 25113, Jordan
e-mail: dr_sliman73@aabu.edu.jo; dr_sliman@yahoo.com

A. Z. Al-Quran
e-mail: ali.z.al-quran@aabu.edu.jo

R. O. A. Dalbouh
e-mail: ali.z.al-quran@aabu.edu.jo

Z. M. A. Smadi
e-mail: ziad38in@aabu.edu.jo

M. S. K. Alshura
Faculty of Money and Management, Management Department, The World Islamic Science University, P.O. Box 1101, Amman 11947, Jordan

M. K. A. Al-Azzam
Department of Business Administration-Faculty of Economics and Administrative Sciences, Yarmouk University, P.O. Box 566-Zip Code 21163, Irbid, Jordan
e-mail: Majedaz@yu.edu.jo

F. M. F. Aldaihani
Kuwait Civil Aviation, Ishbiliyah bloch 1, street 122, home 1, Kuwait City, Kuwait

K. M. Al-hawajreh
Business Faculty, Mu'tah University, Karak, Jordan

M. T. Alshurideh
Department of Management, College of Business, University of Sharjah, 27272 Sharjah, United Arab Emirates
e-mail: m.alshurideh@ju.edu.jo

Department of Marketing, School of Business, The University of Jordan, Amman 11942, Jordan

M. Alshurideh et al. (eds.), *The Effect of Information Technology on Business and Marketing Intelligence Systems*, Studies in Computational Intelligence 1056,
https://doi.org/10.1007/978-3-031-12382-5_94

developed a questionnaire consisting of (35) items. The study population consisted of all (directors of departments, deans of colleges, department heads, administrative staff, and faculty members), where the study followed the simple random method. The results showed a statistically significant effect of knowledge management on total quality management. The study recommended the necessity for universities to organize meetings to exchange knowledge, opinions, experiences and ideas, address problems facing them at work, and promote a culture of openness, sharing, teamwork and dialogue, in a manner that ensures benefit and learning from each other to improve and implement quality programs.

Keywords Knowledge management · Total quality management · Private universities · Jordan

1 Introduction

At the end of the twentieth century, the world witnessed a great momentum in knowledge that led to rapid progress in science and technology, and it was called the fortune of information and communication technology (Alhalalmeh et al., 2020; Allahow et al., 2018; AlTaweel & Al-Hawary, 2021; Al-Nady et al., 2013). After that, knowledge became a necessity rather a strategy because of its impact on human life (Al Mehrez et al., 2020; Shannak et al., 2012). Because of the growing interest in knowledge management and the speed of its application in universities, as it is the most productive for knowledge and the challenges it faces and motivation to reach the competitive advantage (Al-Hawary & Alwan, 2016; Al-Hawary & Al-Hamwan, 2017). Universities tended to follow Total quality management, which is considered as a mechanism to improve organization and individual performance and enhance competitive advantage (Al-Hawary, 2015). It also enhances customer and employee satisfaction, and most importantly, its constant focus on continuous improvement (Abbas, 2019; Al-Hawary & Abu-Laimon, 2013; Al-Hawary & Batayneh, 2010).

The need for knowledge management has recently emerged as a way to address the problem of knowledge and information explosion to use it in solving problems. The ability of organizations, including Jordanian private universities in the northern region, faces a real challenge in gaining and maintaining a competitive advantage, which requires continuous improvement in all its operations and services, aiming to achieve the maximum satisfaction of the requirements and needs of students and employees which forced universities to follow the Total Quality Management along with knowledge management, which will help it achieve a high level of services. Many studies, including Berrish (2016) and Kahreh et al. (2014), have indicated in their recommendations that there is a strong correlation between knowledge management and total quality management and the impact of this integration on improving the performance of services provided in universities.

Looking at the previous theoretical literature, the researcher found many studies that recommend more studies that examine the issue of knowledge management and total quality, especially in the university sector. Thus, this study contributes to enrich the managerial literature with more studies; the results of the study may help universities to identify what is applied in terms of knowledge management, total quality, and what needs to be applied, as well as to benefit from the results in order to provide them with the best services (Al Kurdi et al., 2020a, 2020b; Al-Maroof et al., 2021; Alshurideh et al., 2021). Hence, this study came to focus on knowledge management and total quality management as one of the modern intellectual advances and modern perspectives as it helps the institutions, they follow to create the required development and improvement, as well as raise the effectiveness and efficiency of their various activities (Aljumah et al., 2021; AlMehrzi et al., 2020; Altamony et al., 2012). In order to answer the study's question, what is the impact of knowledge management on total quality management in Jordanian private universities?

2 Theoretical Framework and Making Hypothesis

2.1 *Knowledge Management*

Contemporary institutions, such as higher education institutions, witness a flow of rapid developments in the world today most importantly the information and technological revolution that relies on scientific knowledge and the best use of information which resulted from the great development in the Internet. Because of these developments, knowledge has become the strategic source in creating advantage and the most powerful and effective factor in the success or failure of the institution (Schwandt & Marqurdt, 2003).

Organizations have proactively engaged in knowledge management in the hope of improving performance through better management of what they know and possess (Miković et al., 2020). Knowledge management discusses a systematic and integrative procedure that helps organizations find, organize, distribute and transfer important evidence, knowledge, and experiences that are essential to actions such as problem solving, self-motivated education, strategic planning, and decision-making to achieve organizational goals (Danish et al., 2014; Lawson, 2013).

Al-Hawary (2015) defined knowledge management as the process of capturing, coordinating and preserving information and experiences of individuals and groups within an organization and providing it to others. Knowledge management has also been defined as a policy of giving the right knowledge to the right people at the right time to help people share that knowledge and put it into practice, in ways that attempt to expand the performance of the organization (Al-Hawary & Al-Namlan, 2018; Al-Hawary et al., 2020; Al-Lozi et al., 2017, 2018; Tayal et al., 2015).

Kor and Maden (2013) defined knowledge management as a process that allows the organization to create new knowledge and facilitate the use of existing knowledge when needed. Moreover, Knowledge management was defined by Obeid and Rabay'a (2016) as the processes that help the organization to acquire, organize, generate and share knowledge, as well as transfer important relevant information to various managerial activities by the organization such as decision-making, problem solving, learning and strategic planning. It was also defined as a process that allows individuals within the organization to obtain the correct data and information at the right time and coordination (Bolisani & Bratianu, 2018).

Knowledge management includes a set of sub-processes that help the organization in acquiring, developing, organizing, applying, sharing and transforming the knowledge, information and experience into management activities that improve the decision-making process, strategic planning and competitive advantage. In addition, it aims to improve the efficiency and effectiveness of the productivity and performance of managers, achieve organizational goals and enhance the competitive position of the institution (Al-Qatawneh, 2019).

Knowledge management practices or processes help organizations to use internal knowledge and take advantage of external resources, which leads to create value and improve innovative performance (Ferraris et al., 2017). Briefly, we assume that knowledge management has its own life cycle and processes, and therefore we need to manage it according to the stages of this cycle (Mikovic, 2019).

Knowledge management processes or practices are originally defined as the application of a systematic approach to the activities of obtaining, structuring, and managing and sharing knowledge throughout the organization in order to achieve faster and better work (Alameeri et al., 2020; Nonaka & Takeuchi, 1995). Many previous studies have attempted to define and classify knowledge management practices, as (Alavi & Leidner, 2001) classified knowledge management processes into four dimensions, which are knowledge acquisition, knowledge storage, knowledge sharing and knowledge application. (Becerra-Fernandez and Sabherwal, 2014) indicated that knowledge management processes include four dimensions, which are knowledge discovery, knowledge acquisition, knowledge exchange and knowledge application. (Malkawi & Abu Rumman, 2016) used six dimensions to describe knowledge management practices, which are knowledge acquisition, knowledge innovation, knowledge storage, knowledge exchange, knowledge application, and finally knowledge protection. According to Ramadan et al. (2017) it includes five processes, which are knowledge acquisition, knowledge innovation, knowledge sharing, knowledge application, and knowledge documentation. In this study, four dimensions were used (knowledge acquisition, knowledge sharing, knowledge storage, and knowledge application).

Knowledge Sharing: The process of knowledge sharing refers to the stage of transformation or productivity that ensures the dissemination and record of knowledge (Akour et al., 2021; Wong & Aspinwall, 2005). Knowledge sharing also refers to the process through which individuals, experiences, and information are shared together; thus, this increases the organization's resources and reduces the waste of time in tries and errors. In addition, (Birasnav, 2014) said it is a process by which explicit or tacit knowledge is transferred to other people through the communication that occurs between them.

Knowledge Application: Knowledge application is the vital element in knowledge management processes, as it is considered an output aspect in knowledge management. The process of knowledge application can be defined as the process of implementing knowledge for the beneficiary of the facility. The perspective which is based on knowledge considers the process of applying knowledge the last step that attempts to solve problems and achieve a comparative advantage over competitors (Alshurideh, 2016; Choi et al., 2010). Moreover, it is considered as an activity that can enhance the application of knowledge of a new scenario and learning from it.

Knowledge Storage: It is not enough to generate and acquire new knowledge for decision-making purposes, as mechanisms are needed to store and retrieve it when necessary. So, we can define knowledge storage as storing existing and acquired in properly indexed knowledge. According to (Obeidat et al., 2016), it is the process of identifying new information as it is relevant and necessary for current and future use, as well as store it in reasonable forms to be easily accessible in the organization.

Knowledge Acquisition: Al-Sa'di et al. (2017) defines knowledge acquisition as the process through which the organization obtains knowledge from internal and external sources, which is the process of taking knowledge from different people such as suppliers, customers, employees, etc. for continuous improvement in operations, products and services (Johnson et al, 2019).

2.2 *Total Quality Management*

Because of technological, social, political and environmental changes that have occurred over the past few years, the ability of the organization to gain and maintain a competitive advantage has become a real challenge (Al-Hawary & Hadad, 2016; Al-Hawary & Ismael, 2010; Al-Hawary & Obiadat, 2021; Mohammad et al., 2020). These changes have not only led to more choices for customers, but also led to a change in their preferences and demands (Al-Hawary, 2013; Al-Hawary & Aldaihani, 2016; Al-Hawary & Al-Smeran, 2017; Al-Hawary et al., 2013, 2017; Alolayyan et al., 2018; Alshurideh et al., 2017; Metabis & Al-Hawary, 2013). Hence, total quality management has been identified as a mechanism that has the ability to improve organization and individual performance, and strengthen competitive advantage (Alzoubi et al., 2021; Li et al., 2018).

Total quality management is one of the many quality-based methods, which began in Japan after World War II and spread to the United States of America in the eighties, mostly in American manufacturing companies that face competition from Japan. Moreover, total quality management is a management view that seeks to integrate all organizational purposes from marketing to finance, design, engineering, production and customer service to emphasize all customer needs and organizational goals (Alzoubi et al., 2020; Obeidat et al., 2016). In addition, Wang et al. (2012) said that it is a continuous procedure for the development of individuals and the entire organization, and the continuous development of processes inside organizations to provide greater value to customers and meet their needs. It also includes company processes that focus on acceptance, participation and response to customers through the concept of marketing.

Bouranta et al. (2017) defined total quality management as a science that depends on a set of rules, methods, equipment and theoretical methodologies. Modgil and Sharma (2016) defined it as a philosophy that aims to change for the better and protect the quality of the commodity and the process on an ongoing basis. This continuous improvement is the outcome of the he committed work of senior management and the participation of suppliers, customers and employees. As well as, it is considered the management philosophy that focuses on continuous improvement for all operations of the institution. Marefal and Fraidfathi said that it is a customer-based philosophy in which the sourcing business, business processes and institution culture are managed and blended in a good way to ensure customer satisfaction.

Wang et al. (2012) defined it as a consistent measure of development for individuals and the entire organization. Total quality management can be defined as a philosophy or a tool aims to increase the quality of products, services and process, as well as achieve competitive advantage and align quality with the company's mission, vision and goal through the application and commitment of senior management, employee empowerment, continuous improvement, resource management and customers.

2.3 *Knowledge Management and Total Quality Management*

A few researchers have recently expressed their interest in the relationship between total quality management and knowledge management, but they did not agree on the perception of this relationship. Some researchers consider knowledge management as a facilitator for Total Quality Management (Stewart & Waddell, 2008), at which from one perspective, knowledge management is identified as an activator of total quality management. Stewart and Waddell (2008) stated that expanding the concept of quality to product and service specifications as a quick response to customer needs clarifies the relationship between knowledge management and Total Quality Management.

The acquisition and sharing of knowledge also lead to finding a quality culture which create an understandable framework according to quality and knowledge management programs (Al-Salti & Hackney, 2011; Alshamsi et al., 2020; Alshurideh, 2022). On the other hand, there are many approaches assume that total quality management agree with knowledge management. Lin and Lee (2005) introduced ISO9000 processes based on knowledge management system, which support the flow of knowledge in the organization. In a case study revealed that Total Quality Management practices are factors that facilitate knowledge acquisition and sharing. Choo et al. (2007) presented a conceptual framework based on quality and knowledge management programs. Based on this study, quality programs are effective factors for knowledge management. Jayawarana and Holt (2009) analyzed the relationship between knowledge creation and transformation in the context of research and development; and based on their case study, they concluded that total quality management practices improve knowledge creation and transformation. In addition, Molina et al. examined the relationship between total quality management and knowledge transformation.

According to Hsu and Shen (2005), the relationship between total quality management and knowledge management as management practices appears close, as they share some processes such as results direction, people-based management, leadership and customer satisfaction; whereas similarities form the interactive relationship between the two practices (Leonard & Mccadam, 2001). Ooi et al. (2010) conducted an empirical study to measure the impact of total quality management on knowledge management. The results showed that total quality management is highly associated with knowledge sharing. Moreover, the results showed that there is a positive relationship between training, teamwork and financial analysis with knowledge exchange among middle management employees (Alshurideh et al., 2019a, b; Al Kurdi et al., 2020a, b). In addition, (Zwain et al., 2011) evaluated the relationship between the basic elements of total quality management and knowledge sharing. They also stressed that there is a positive relationship between the elements of total quality management and knowledge sharing, and that the basic elements must be implemented entirely and not individually to obtain the best results by sharing knowledge (Aburayya et al., 2020; Al-Khayyal et al., 2021; Zu'bi et al., 2012). Based on the above, the hypothesis of the study can be as follows:

There is a statistically significant effect of knowledge management on total quality management in private universities in Jordan.

3 Study Model

See Fig. 1.

4 Methodology

4.1 Population and Sample Selection

A qualitative method based on a questionnaire was used in this study for data collection and sample selection. The major aim of the study was to examine the impact of knowledge management on total quality management. Therefore, it focused on Jordan-based universities operating in the north region. Data were primarily gathered through self-reported questionnaires created by Google Forms which were distributed

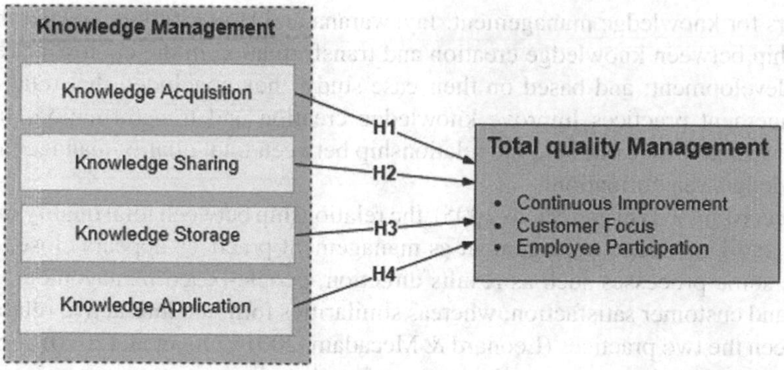

Fig. 1 Research model

to a random sample of (520) employees via email. In total, (331) responses were received including (28) invalid to statistical analysis due to uncompleted or inaccurate. Hence, the final sample contained (303) responses suitable to analysis requirements that were formed a response rate of (58.3%), where it proved to be sufficient to the extent that was predictable and allowed for a presumption of data saturation (Sekaran & Bougie, 2016).

4.2 Measurement Instrument

A self-reported questionnaire that consists of two main sections along with a section regarding control variables was used as the measurement instrument. Control variables considered as categorical measures were composed of gender, age group, educational level, and experience. The two main sections were dealt with a five-point Likert scale (from 1 = strongly disagree to 5 = strongly agree). The first section contained (20) items to measure knowledge management based on (Al-Hawary, 2015). These questions were distributed into dimensions as follows: five items dedicated for measuring knowledge acquisition, five items dedicated for measuring knowledge sharing, five items dedicated for measuring knowledge storage, and five items dedicated for measuring knowledge application. Whereas the second section included (15) items developed to measure total quality management according to what was pointed by (Al-Hawary & Abu-Laimon, 2013). This construct divided into three first-order constructs. Continuous improvement was measured through five items, customer focus was measured by five items, and employee participation was measured using five items.

5 Findings

5.1 Measurement Model Evaluation

This study was conducted structural equation modeling (SEM) to test hypotheses, which represents a contemporary statistical technique for testing and estimating the relationship between factors and variables (Wang & Rhemtulla, 2021). Accordingly, the reliability and validity of the constructs were tested using confirmatory factor analysis (CFA) through the statistical program AMOSv24. Table 1 summarizes the results of convergent and discriminant validity, as well the indicators of reliability.

Table 1 shows that the standard loading values for the individual items were within the domain (0.654–0.891), these values greater than the minimum retention of the elements based on their standard loads (Al-Lozi et al., 2018; Sung et al., 2019). Average variance extracted (AVE) is a summary indicator of the convergent validity of constructs that must be above 0.50 (Howard, 2018). The results indicate that the AVE values were greater than 0.50 for all constructs, thus the used measurement model has an appropriate convergent validity. Rimkeviciene et al. (2017) suggested the comparison approach as a way to deal with discriminant validity assessment in covariance-based SEM. This approachis based on comparing the values of maximum shared variance (MSV) with the values of AVE, as well as comparing the values of square root of AVE ($\sqrt{AVE}$) with the correlation between the rest of the structures. The results show that the values of MSV were smaller than the values of AVE, and that the values of $\sqrt{AVE}$ were higher than the correlation values among the rest of the constructs. Therefore, the measurement model used is characterized by discriminative validity. The internal consistency measured through Cronbach's Alpha coefficient (α) and compound reliability by McDonald's Omega coefficient (ω) was conducted as indicators to evaluate measurement model. The results listed in Table 1 demonstrated that both values of Cronbach's Alpha coefficient and McDonald's Omega coefficient were greater than 0.70, which is the lowest limit for judging on measurement reliability (de Leeuw et al., 2019).

5.2 Structural Model

The structural model illustrated no multicollinearity issue among predictor constructs because variance inflation factor (VIF) values are below the threshold of 5, as shown in Table 1 (Hair et al., 2017).This result is supported by the values of model fit indices shown in (Figs. 1 and 2).

The results in Fig. 1 indicated that the chi-square to degrees of freedom (CMIN/DF) was 1.973, which is less than 3 the upper limit of this indicator. The values of the goodness of fit index (GFI), the comparative fit index (CFI), and the Tucker-Lewis index (TLI) were upper than the minimum accepted threshold of 0.90.

Table 1 Results of validity and reliability tests

Constructs	1	2	3	4	5	6	7
1. KAC	*0.773*						
2. KSH	0.517	*0.765*					
3. KST	0.492	0.439	*0.762*				
4. KAP	0.395	0.547	0.517	*0.768*			
5. CIM	0.553	0.622	0.598	0.625	*0.773*		
6. CFO	0.617	0.602	0.638	0.589	0.628	*0.786*	
7. EPA	0.587	0.638	0.608	0.661	0.637	0.667	*0.774*
VIF	1.978	2.648	2.335	1.879	–	–	–
Loadings range	0.731–0.817	0.654–0.864	0.702–0.872	0.691–0.836	0.685–0.891	0.725–0.834	0.664–0.846
AVE	0.598	0.586	0.581	0.589	0.597	0.618	0.600
MSV	0.501	0.428	0.487	0.521	0.497	0.517	0.521
Internal consistency	0.879	0.872	0.870	0.875	0.877	0.888	0.878
Composite reliability	0.881	0.875	0.873	0.877	0.880	0.890	0.881

Note KAC: knowledge acquisition, KSH: knowledge sharing, KST: knowledge storage, KAP: knowledge application, CIM: continuous improvement, CFO: customer focus, EPA: employee participation, italic fonts indicate to root square of AVE

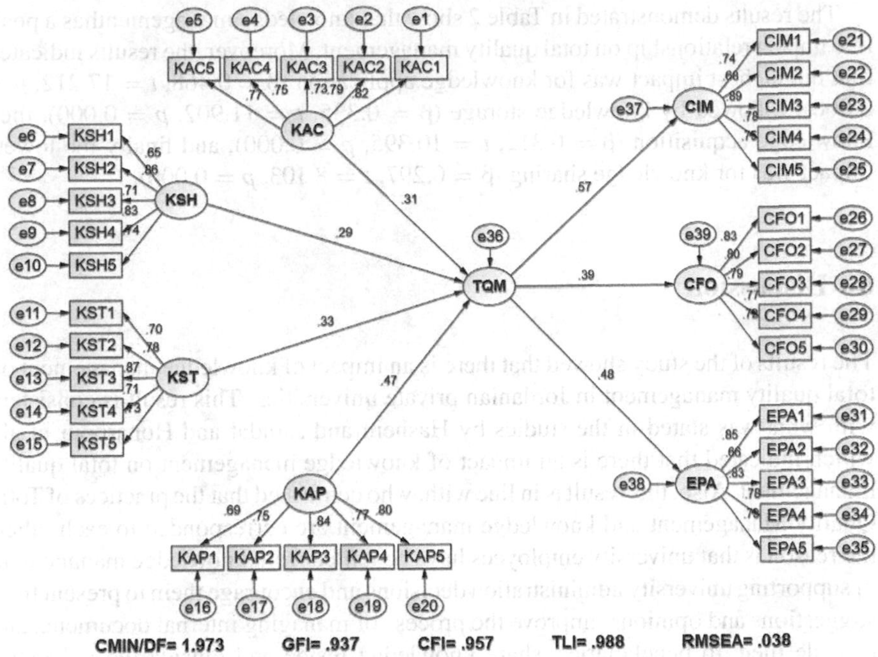

Fig. 2 SEM results of the knowledge management effect on total quality management

Moreover, the result of root mean square error of approximation (RMSEA) indicated to value 0.038, this value is a reasonable error of approximation because it is less than the higher limit of 0.08. Consequently, the structural model used in this study was recognized as a fit model for predicting the DEP and generalization of its result (Ahmad et al., 2016; Shi et al., 2019).To verify the results of testing the study hypotheses, structural equation modeling (SEM) was used, the results of which are listed in Table 2.

Table 2 Impact of knowledge management dimensions on total quality management

Hypothesis	Relation	Standard Beta	*t* value	*p* value	Result
H1	KAC→TQM	0.312	10.395	0.000	Supported
H2	KSH→TQM	0.297	8.103	0.000	Supported
H3	KST→TQM	0.335	11.902	0.000	Supported
H4	KAP→TQM	0.468	17.212	0.000	Supported

Note KAC: knowledge acquisition, KSH: knowledge sharing, KST: knowledge storage, KAP: knowledge application, TQM: total quality management
$* p < 0.05$, $** p < 0.01$, $*** p < 0.001$

The results demonstrated in Table 2 show that knowledge managementhas a positive impact relationship on total quality management. Moreover, the results indicated that the highest impact was for knowledge application (β = 0.468, $t = 17.212$, $p = 0.000$), followed by knowledge storage (β = 0.335, $t = 11.902$, $p = 0.000$), then knowledge acquisition (β = 0.312, $t = 10.395$, $p = 0.000$), and finally the lowest impact was for knowledge sharing (β = 0.297, $t = 8.103$, $p = 0.000$).

6 Discussion

The results of the study showed that there is an impact of knowledge management on total quality management in Jordanian private universities. This result is consistent with what was stated in the studies by Hashem and Jaradat and Honarpour et al., which indicated that there is an impact of knowledge management on total quality management. Also, this result is in line with, who confirmed that the practices of Total Quality Management and knowledge management are corresponded to each other; the reason is that university employees have benefited from knowledge management in supporting university administration decisions and encourage them to present their suggestions and opinions, improve the process of managing internal documents and provide them to beneficiaries, share knowledge inside and outside the university and make qualitative improvement in the educational process because of its great importance for improving work; this is because total quality is a flexible participatory process.

The results indicated that there is an impact of the acquisition of knowledge on total quality management, so that universities integrate the components of knowledge they have with knowledge they acquired from abroad, which led to an improvement in the quality of knowledge provided to students and satisfy their educational requirements. The university's interest in motivating creative and innovative employees led to accept and interact with the wise ideas proposed by the employees after evaluating and implementing these proposals, which contribute to the process of continuous improvement in the university. The university also provides programs that simplify and facilitate the provision, development and improvement of services that create an advantage over other competitive universities. The results indicated that there is an impact of knowledge storage on total quality management in Jordanian private universities in the north region, whereas university workers can easily access the information and documents stored in the university databases, and document the acquired knowledge of each project they accomplish. In addition, reliance on electronic documents led to the ease of finding documents, completing transactions, and taking appropriate procedures at the right time, making it easier for workers to improve the quality of services in a high and correct manner.

The results indicated that there is an impact of sharing knowledge on total quality management in Jordanian private universities in the North Region, as the university employees share information related to their work because there are valid and easy communication ways between the various sites and departments, which leads to finding appropriate solutions to the problems they face and taking preventive immediate procedures constantly. The results indicated that there is an impact of knowledge application on total quality management in Jordanian private universities in the north region. University workers use their knowledge in their work field which leads to improve the efficiency and effectiveness of their work quality. Moreover, the university administration discussing ideas and the outcomes of its work in formal meetings leads to involve employees at all levels, express their opinions, focus on students' satisfaction and respond quickly to their requirements, listen to their comments and complains which will help the university avoid shortcomings and increase the efficiency of their work. This will enhance student satisfaction as they are the core of the university's work.

7 Recommendations

Based on the results of this study, the study recommends Jordanian private universities to pay attention to the application of knowledge management and total quality by reviewing the best international practices in their application, and benefiting from that in improving and developing the quality of services they provide. It is also preferable for universities to organize meetings between them and the employees to exchange knowledge, opinions, experiences and ideas, address problems they face at work, promote the culture of openness partnership, teamwork and dialogue in order to ensure benefit and learning from each other to improve and implement quality programs. Moreover, it should establish a rapid mechanism for responding to the requirements of total quality and continue working on meeting the needs and requirements of students in the best way.

References

Abbas, J. (2019). Impact of total quality management on corporate sustainability through the mediating effect of knowledge management. *Journal of Cleaner Production, 118806*.

Aburayya, A., Alshurideh, M., Al Marzouqi, A., Al Diabat, O., Alfarsi, A., Suson, R., & Salloum, S. A. (2020). An empirical examination of the effect of TQM practices on hospital service quality: An assessment study in UAE hospitals. *Systematic Reviews in Pharmacy, 11*(9), 347–362.

Ahmad, S., Zulkurnain, N., & Khairushalimi, F. (2016). Assessing the validity and reliability of a measurement model in structural equation modeling (SEM). *British Journal of Mathematics & Computer Science, 15*(3), 1–8. https://doi.org/10.9734/BJMCS/2016/25183

Akour, I., Alshurideh, M., Al Kurdi, B., Al Ali, A., & Salloum, S. (2021). Using machine learning algorithms to predict people's intention to use mobile learning platforms during the COVID-19 pandemic: Machine learning approach. *JMIR Medical Education, 7*(1), 1–17.

Al Kurdi, B., Alshurideh, M., & Salloum, S. A. (2020a). Investigating a theoretical framework for e-learning technology acceptance. *International Journal of Electrical and Computer Engineering (IJECE), 10*(6), 6484–6496.

Al Kurdi, B., Alshurideh, M., Salloum, S., Obeidat, Z., & Al-dweeri, R. (2020b). An empirical investigation into examination of factors influencing university students' behavior towards elearning acceptance using SEM approach. *International Journal of Interactive Mobile Technologies, 14*(2), 19–24.

Al Mehrez, A. A., Alshurideh, M., Al Kurdi, B., & Salloum, S. A. (2020). Internal factors affect knowledge management and firm performance: A systematic review. In *International Conference on Advanced Intelligent Systems and Informatics* (pp. 632–643). Cham: Springer.

Alameeri, K., Alshurideh, M., Al Kurdi, B., & Salloum, S. A. (2020). The effect of work environment happiness on employee leadership. In *International Conference on Advanced Intelligent Systems and Informatics* (pp. 668–680). Cham: Springer.

Alavi, M., & Leidner, D. E. (2001). Review: Knowledge management and knowledge management system: Conceptual foundations and research issues. *MIS Quarterly Journal, 25*(1), 107–136.

Alhalalmeh, M. I., Almomani, H. M., Altarifi, S., Al- Quran, A. Z., Mohammad, A. A., & Al-Hawary, S. I. (2020). The nexus between corporate social responsibilty and organizational performance in Jordan: The mediating role of organizational commitment and organizational citizenship behavior. *Test Engineering and Management, 83*(July), 6391–6410.

Al-Hawary, S. I. (2013). The role of perceived quality and satisfaction in explaining customer brand loyalty: Mobile phone service in Jordan. *International Journal of Business Innovation and Research, 7*(4), 393–413.

Al-Hawary, S. I. (2015). Human resource management practices as a success factor of knowledge management implementation at health care sector in Jordan. *International Journal of Business and Social Science, 6*(11/1), 83–98.

Al-Hawary, S. I., & Aldaihani, F. M. (2016). Customer relationship management and innovation capabilities of Kuwait Airways. *International Journal of Academic Research in Economics and Management Sciences, 5*(4), 201–226.

Al-Hawary, S. I., & Al-Hamwan, A. (2017). Environmental analysis and its impact on the competitive capabilities of the commercial banks operating in Jordan. *International Journal of Academic Research in Accounting, Finance and Management Sciences, 7*(1), 277–290.

Al-Hawary, S. I., & Al-Namlan, A. (2018). Impact of electronic human resources management on the organizational learning at the private hospitals in the state of Qatar. *Global Journal of Management and Business Research: A Administration and Management, 18*(7), 1–11.

Al-Hawary, S. I., & Batayneh, A. M. (2010). The effect of marketing communication tools on non-Jordanian students' choice of Jordanian public universities: A field study. *International Management Review, 6*(2), 90–99.

Al-Hawary, S. I., & Hadad, T. F. (2016). The effect of strategic thinking styles on the enhancement competitive capabilities of commercial banks in Jordan. *International Journal of Business and Social Science, 7*(10), 133–144.

Al-Hawary, S. I., & Ismael, M. (2010). The effect of using information technology in achieving competitive advantage strategies: A field study on the Jordanian pharmaceutical companies. *Al Manara for Research and Studies, 16*(4), 196–203.

Al-Hawary, S. I., Al-Qudah, K., Abutayeh, P., Abutayeh, S., & Al-Zyadat, D. (2013). The impact of internal marketing on employee's job satisfaction of commercial banks in Jordan. *Interdisciplinary Journal of Contemporary Research in Business, 4*(9), 811–826.

Al-Hawary, S. I. S., & Alwan, A. M. (2016). Knowledge management and its effect on strategic decisions of Jordanian Public Universities. *Journal of Accounting-Business & Management, 23*(2), 24–44.

Al-Hawary, S. I. S., & Obiadat, A. A. (2021). Does mobile marketing affect customer loyalty in Jordan? *International Journal of Business Excellence, 23*(2), 226–250.

Al-Hawary, S. I. S., Mohammad, A. S., Al-Syasneh, M. S., Qandah, M. S. F., & Alhajri, T. M. S. (2020). Organisational learning capabilities of the commercial banks in Jordan: Do electronic human resources management practices matter? *International Journal of Learning and Intellectual Capital, 17*(3), 242–266.

Al-Hawary, S. I., & Abu-Laimon, A. A. (2013). The impact of TQM practices on service quality in cellular communication companies in Jordan. *International Journal of Productivity and Quality Management, 11*(4), 446–474.

Al-Hawary, S. I., & Al-Smeran, W. (2017). Impact of electronic service quality on customers satisfaction of Islamic banks in Jordan. *International Journal of Academic Research in Accounting, Finance and Management Sciences, 7*(1), 170–188.

Al-Hawary, S. I., Batayneh, A. M., Mohammad, A. A., & Alsarahni, A. H. (2017). Supply chain flexibility aspects and their impact on customers satisfaction of pharmaceutical industry in Jordan. *International Journal of Business Performance and Supply Chain Modelling, 9*(4), 326–343.

Aljumah, A., Nuseir, M. T., & Alshurideh, M. T. (2021). The Impact of Social Media Marketing Communications on Consumer Response During the COVID-19: Does the Brand Equity of a University Matter. The Effect of Coronavirus Disease (COVID-19) on Business Intelligence, pp. 367–384.

Al-Khayyal, A., Alshurideh, M., Al Kurdi, B., & Salloum, S. A. (2021). Factors influencing electronic service quality on electronic loyalty in online shopping context: Data analysis approach. In Enabling AI Applications in Data Science (pp. 367–378). Springer, Cham.

Allahow, T. J. A. A., Al-Hawary, S. I. S., & Aldaihani, F. M. F. (2018). Information technology and administrative innovation of the central agency for information technology in Kuwait. *Global Journal of Management and Business, 18*(11-A), 1–16.

Al-Lozi, M. S., Almomani, R. Z. Q., & Al-Hawary, S. I. S. (2018). Talent Management strategies as a critical success factor for effectiveness of human resources information systems in commercial banks working in Jordan. *Global Journal of Management and Business Research: A Administration and Management, 18*(1), 30–43.

Al-Lozi, M., Almomani, R. Z., & Al-Hawary, S. I. (2017). Impact of talent management on achieving organizational excellence in Arab Potash Company in Jordan. *Global Journal of Management and Business Research: A Administration and Management, 17*(7), 15–25.

Al-Maroof, R., Ayoubi, K., Alhumaid, K., Aburayya, A., Alshurideh, M., Alfaisal, R., & Salloum, S. (2021). The acceptance of social media video for knowledge acquisition, sharing and application: A comparative study among YouYube users and TikTok users' for medical purposes. *International Journal of Data and Network Science, 5*(3), 197–214.

AlMehrzi, A., Alshurideh, M., & Al Kurdi, B. (2020). Investigation of the key internal factors influencing knowledge management, employment, and organisational performance: A qualitative study of the UAE hospitality sector. *International Journal of Innovation, Creativity and Change, 14*(1), 1369–1394.

Al-Nady, B. A., Al-Hawary, S. I., & Alolayyan, M. (2013). Strategic management as a key for superior competitive advantage of sanitary ware suppliers in kingdom of Saudi Arabia. *International Journal of Management and Information Technology, 7*(2), 1042–1058.

Alolayyan, M., Al-Hawary, S. I., Mohammad, A. A., & Al-Nady, B. A. (2018). Banking service quality provided by commercial banks and customer satisfaction. A structural equation modelling approaches. *International Journal of Productivity and Quality Management, 24*(4), 543–565.

Al-Qatawneh, N. A.-W. (2019). The impact of knowledge management process on entriprenurial practices: The context of Jordanian manufacturing sector, published MA thesis. Mu'tah University, Karak.

Al-Sa'di, A. F., Abdallah, A. B., & Dahiyat, S. E. (2017). The mediating role of product and process innovations on the relationship between knowledge management and operational performance in manufacturing companies in Jordan. *Business Process Management Journal, 23*(2), 349–376

Al-Salti, Z., & Hackney, R. (2011). Factors impacting knowledge transfer success in information systems outsourcing. *Journal of Enterprise Information Management, 24*(5), 455–468.

Alshamsi, A., Alshurideh, M., Al Kurdi, B., & Salloum, S. A. (2020). The influence of service quality on customer retention: A systematic review in the higher education. In *International Conference on Advanced Intelligent Systems and Informatics* (pp. 404–416). Springer, Cham.

Alshurideh, M. (2016). Scope of customer retention problem in the mobile phone sector: A theoretical perspective. *Journal of Marketing and Consumer Research, 20*(2), 64–69.

Alshurideh, M. (2022). Does electronic customer relationship management (E-CRM) affect service quality at private hospitals in Jordan? *Uncertain Supply Chain Management, 10*(2), 325–332.
Alshurideh, M. T., Kurdi, B. A., AlHamad, A. Q., Salloum, S. A., Alkurdi, S., Dehghan, A., ... & Masa'deh, R. E. (2021). Factors affecting the use of smart mobile examination platforms by universities' postgraduate students during the COVID 19 pandemic: An empirical study. In Informatics, 8 2), 1–21. Multidisciplinary Digital Publishing Institute.
Alshurideh, M., Al-Hawary, S. I., Batayneh, A. M., Mohammad, A., & Al-Kurdi, B. (2017). the impact of islamic banks' service quality perception on Jordanian customers loyalty. *Journal of Management Research, 9*(2), 139–159.
Alshurideh, M., Salloum, S. A., Al Kurdi, B., & Al-Emran, M. (2019a). Factors affecting the social networks acceptance: An empirical study using PLS-SEM approach. In *Proceedings of the 2019a 8th International Conference on Software and Computer Applications* (pp. 414–418).
Alshurideh, M., Salloum, S. A., Al Kurdi, B., Monem, A. A., & Shaalan, K. (2019b). Understanding the quality determinants that influence the intention to use the mobile learning platforms: A practical study. *International Journal of Interactive Mobile Technologies, 13*(11), 183–157.
Altamony, H., Masa'deh, R., Alshurideh, M., Obeidat, B. (2012) Information systems for competitive advantage: Implementation of an organisational strategic management process. Innovation and sustainable competitive advantage: From regional development to world economies. 583–592.
AlTaweel, I. R., & Al-Hawary, S. I. (2021). The mediating role of innovation capability on the relationship between strategic agility and organizational performance. *Sustainability, 13*(14), 7564.
Alzoubi, H. M., Alshurideh, M., Al Kurdi, B., & Inairat, M. (2020). Do perceived service value, quality, price fairness and service recovery shape customer satisfaction and delight? A practical study in the service telecommunication context. *Uncertain Supply Chain Management, 8*(3), 579–588.
Alzoubi, H., Alshurideh, M., Akour, I., Al Shraah, A., & Ahmed, G. (2021) Impact of information systems capabilities and total quality management on the cost of quality. *Journal of Legal, Ethical and Regulatory Issues, 24*(Special Issue 6), 1–11.
Becerra-Fernandez, I., & Sabherwal, R. (2014). Knowledge management: Systems and processes, 2nd edition. New york and London Routledge.
Berrish, M. (2016). *The integration between knowledge management and total quality management and its impact on educational performance*. Published doctoral dissertation, University of Salford, Salford, UK.
Birasnav, M. (2014). Knowledge management and organizational performance in the service industry: The role of transformational leadership beyond the effects of transactional leadership. *Journal of Business Research, 67*, 1622–1629.
Bolisani, E., & Bratianu, C. (2018). *Emergent knowledge strategies: Strategic thinking in knowledge management* (pp. 1–22). Springer International Publishing.
Bouranta, N., Psomas E. L., & Pantouvakis, A. (2017). Identifying the critical determinants of TQM and their impact on company performance Evidence from the hotel industry of Greece. *The TOM Journal, 29*(1), 147–166.
Choi, S. Y., Lee, H., & Yoo, Y. (2010). The impact of information technology and transactive memory systems on knowledge sharing, application, and team performance: A field study. *MIS Quarterly, 34*(4), 855–870.
Choo, A. S., Linderman, K. W., & Schroeder, R. G. (2007). Method and context perspectives on learning and knowledge creation in quality management. *Journal of Operations Management, 25*(4), 918–931.
Danish, R., Asghar, A., & Asghar, S. (2014). Factors of knowledge management in banking sector of Pakistan. *Journal of Management Information System and Ecommerce, 1*(1), 41–49.
de Leeuw, E., Hox, J., Silber, H., Struminskaya, B., & Vis, C. (2019). Development of an international survey attitude scale: Measurement equivalence, reliability, and predictive validity. *Measurement Instruments for the Social Sciences, 1*(1), 9. https://doi.org/10.1186/s42409-019-0012-x

Ferraris, A., Santoro, G., & Dezi, L. (2017). How MNC's subsidiaries may improve their innovative performance? The role of external sources and knowledge management capabilities. *Journal of Knowledge Management, 21*(3), 540–552.

Hair, J. F., Babin, B. J., & Krey, N. (2017). Covariance-based structural equation modeling in the journal of advertising: Review and recommendations. *Journal of Advertising, 46*(1), 163–177. https://doi.org/10.1080/00913367.2017.1281777

Howard, M. C. (2018). The convergent validity and nomological net of two methods to measure retroactive influences. *Psychology of Consciousness: Theory, Research, and Practice, 5*(3), 324–337. https://doi.org/10.1037/cns0000149

Hsu, S.-H., & Shen, H.-P. (2005). Knowledge management and its relationship with TQM. *Total Quality Management, 16*(3), 351–361.

Jayawarana, D., & Holt, R. (2009). Knowledge and quality management: An R&D perspective. *Technovation, 29*(11), 775–785.

Johnson, T. L., Fletcher, S. R., Baker, W., & Charles, R. L. (2019). How and why we need to capture tacit knowledge in manufacturing: Case studies of visual inspection. *Applied Ergonomics, 74*, 1–9.

Kahreh, Z. S., Shirmohammadi, A., & Kahreh, M. S. (2014). Explanatory study towards analysis the relationship between total quality management and knowledge management. *Procedia-Social and Behavioral Sciences, 109*, 600–604.

Kor, B., & Maden, C. (2013). The relationship between knowledge management and innovation in Turkish service and high-tech firms. *International Journal of Business and Social Science, 4*(4), 293–304.

Lawson, S. (2013). Examining the relationship between organizational culture and knowledge management. Unpublished doctoral dissertation, Nova Southern University.

Leonard, D., & Mcadam, R. (2001). Grounded theory methodology and practitioner reflexivity in TQM research. *International Journal of Quality & Reliability Management, 18*(2), 180–194.

Li, D., Zhao, Y., Zhang, L., Chen, X., Cao, C., (2018). Impact of quality management on green innovation. *Journal of Cleaner Production, 170*, 462–470.

Lin, H. F., & Lee, G. G. (2005). Impact of organizational learning and knowledge management factors on e-business adoption. *Management Decision, 43*(2), 171–188.

Malkawi, M. S., & Abu Rumman, H. B. (2016). Knowledge management capabilities and its impact on product innovation in SME's. *International Business Research, 9*(5), 76–85.

Metabis, A., & Al-Hawary, S. I. (2013). The impact of internal marketing practices on services quality of commercial banks in Jordan. *International Journal of Services and Operations Management, 15*(3), 313–337.

Mikovic, R. (2019). *Integrated model of knowledge management based on social capital of the organization Doctoral Thesis.* Faculty of Organizational Sciences, University of Belgrade.

Miković, R., Petrović, D., Mihić, M., Obradović, V., & Todorović, M. (2020). The integration of social capital and knowledge management–The key challenge for international development and cooperation projects of nonprofit organizations, International Journal of Project Management.

Modgil, S., & Sharma, S. (2016). Impact of hard and soft TQM on supply chain performance: Empirical investigation of pharmaceutical industry. *International Journal of Productivity and Quality Management, 20*(4), 513–533.

Mohammad, A. A., Alshura, M. S., Al-Hawary, S. I. S., Al-Syasneh, M. S., & Alhajri, T. M. (2020). The influence of Internal Marketing Practices on the employees' intention to leave: A study of the private hospitals in Jordan. *International Journal of Advanced Science and Technology, 29*(5), 1174–1189.

Nonaka, I., & Takeuchi, H. (1995). *The knowledge-creating company.* Oxford University Press, Oxford.

Obeid, S. M., & Rabay'a, S. (2016). The impact of knowledge management dimensions in the learning organiza-tion from the perspective of the Arab American University's (AAU) faculty–Palestine. *Jordan Journal of Business Administration, 12*(4), 813–840.

Obeidat, B. Y., Al-Sarayrah, S., Tarhini, A., Al-Dmour, R. H., & Al-Salti, Z. (2016). Cultural influence on strategic human resource management practices: A Jordanian case study. *International Business Research, 9*(10), 33–50.

Ooi, K.-B., Cheah, W.-C., Lin, B., & Teh, P.-L. (2010). TQM practices and knowledge sharing: An empirical study of Malaysia's manufacturing organizations. *Asia Pacific Journal of Management, 29*(1), 59–78.

Ramadan, B. M., Dahiyat, S. E., Bontis, N., & Al-Dalahmeh, M. A. (2017). Intellectual capital, knowledge management and social capital within the ICT sector in Jordan. *Journal of Intellectual Capital, 18*(2), 437–462.

Rimkeviciene, J., Hawgood, J., O'Gorman, J., & De Leo, D. (2017). Construct validity of the acquired capability for suicide scale: Factor structure, convergent and discriminant validity. *Journal of Psychopathology and Behavioral Assessment, 39*(2), 291–302. https://doi.org/10.1007/s10862-016-9576-4

Schwandt, D., & Marqurdt, M. J. (2003). *Organizational learning: From world–class theories to global best practices*. St.Lucie Press.

Sekaran, U., & Bougie, R. (2016). *Research methods for business: A skill-building approach* (Seventh edition). Wiley.

Shannak, R. O., Zu'bi, M. F., & Alshurideh, M. T. (2012). A theoretical perspective on the relationship between knowledge management systems, customer knowledge. *European Journal of Social Sciences, 32*(4), 520–532.

Shi, D., Lee, T., & Maydeu-Olivares, A. (2019). Understanding the model size effect on SEM fit indices. *Educational and Psychological Measurement, 79*(2), 310–334. https://doi.org/10.1177/0013164418783530

Stewart, D., & Waddell, D. (2008). Knowledge management: The fundamental component for delivery of quality. *Total Quality Management & Business Excellence, 19*(9), 987–996.

Sung, K.-S., Yi, Y. G., & Shin, H.-I. (2019). Reliability and validity of knee extensor strength measurements using a portable dynamometer anchoring system in a supine position. *BMC Musculoskeletal Disorders, 20*(1), 1–8. https://doi.org/10.1186/s12891-019-2703-0

Tayal, A., Coleman, T. F., & Li, Y. (2015). Rankrc: Large-scale nonlinear rare class ranking. *IEEE Transactions on Knowledge and Data Engineering, 27*(12), 3347–3359.

Wang, C. H., Chen, K. Y., & Chen, S. C. (2012). Total quality management, market orientation and hotel performance: The moderating effects of external environmental factors. *International Journal of Hospitality Management, 31*(1), 119–129.

Wang, Y. A., & Rhemtulla, M. (2021). Power analysis for parameter estimation in structural equation modeling: A discussion and tutorial. *Advances in Methods and Practices in Psychological Science, 4*(1), 1–17. https://doi.org/10.1177/2515245920918253

Wong, K., & Aspinwall, E. (2005). An empirical study of the important factors for knowledge management adoption in the SME sector. *Journal of Knowledge Management, 9*(3), 64–82.

Zu'bi, Z., Al-Lozi, M., Dahiyat, S., Alshurideh, M., & Al Majali, A. (2012). Examining the effects of quality management practices on product variety. *European Journal of Economics, Finance and Administrative Sciences, 51*(1), 123–139.

Zwain, A. A. A., Lim, K. T., & Othman, S. N. (2011). The impact of total quality management (TQM) on knowledge sharing in Iraqi higher education institutions: an empirical study. In: *The 9th Asian Academy of Management International Conference, AAMC, Organized by Asian Academy of Management*, Penang, October 14–16.

The Effect Knowledge Creation Process on Organizational Innovation in Social Security Corporation in Jordan

Ali Zakariya Al-Quran, Raed Ismael Ababneh, Mohammad Hamzeh Hassan Al-Safadi, Mohammed saleem khlif Alshura, Mohammad Mousa Eldahamsheh, Majed Kamel Ali Al-Azzam, Main Naser Alolayyan, Muhammad Turki Alshurideh, and Sulieman Ibraheem Shelash Al-Hawary

Abstract The aim of this study is to identify the effect knowledge creation process on organizational innovation in Social Security Corporation in Jordan. To achieve the objectives of the study, a questionnaire was prepared consisting of (47) items. The

A. Z. Al-Quran · M. H. H. Al-Safadi · S. I. S. Al-Hawary (✉)
Department of Business Administration, School of Business, Al al-Bayt University, P.O. Box 130040, Mafraq 25113, Jordan
e-mail: dr_sliman73@aabu.edu.jo; dr_sliman@yahoo.com

A. Z. Al-Quran
e-mail: ali.z.al-quran@aabu.edu.jo

R. I. Ababneh
Policy, Planning, and Development Program, Department of International Affairs. College of Arts and Sciences, Qatar University, 2713 Doha, Qatar

M. Alshura
Faculty of Money and Management, Management Department, The World Islamic Science University, P.O. Box 1101, Amman 11947, Jordan

M. M. Eldahamsheh
Strategic Management, Irbid, Jordan

M. K. A. Al-Azzam
Department of Business Administration-Faculty of Economics and Administrative Sciences, Yarmouk University, P.O. Box 566-Zip Code 21163, Irbid, Jordan
e-mail: Majedaz@yu.edu.jo

M. N. Alolayyan
Faculty of Medicine, Health Management and Policy Department, Jordan University of Science and Technology, Ar-Ramtha, Jordan
e-mail: mnalolayyan@just.edu.jo

M. T. Alshurideh
Department of Marketing, School of Business, The University of Jordan, Amman 11942, Jordan
e-mail: malshurideh@sharjah.ac.ae

Department of Management, College of Business, University of Sharjah, 27272 Sharjah, United Arab Emirates

M. Alshurideh et al. (eds.), *The Effect of Information Technology on Business and Marketing Intelligence Systems*, Studies in Computational Intelligence 1056,
https://doi.org/10.1007/978-3-031-12382-5_95

study population consists of all (Managers of departments, Directors of directorates, Heads of departments and staff), who perform their duties in Social Security Corporation in Jordan. The number of employees was about (435) employees. Where the study followed the method of comprehensive survey to withdraw the sample study. The study reached there is an effect of the process of creating knowledge on organizational innovation in the Social Security Corporation in Jordan. The study recommended that knowledge management should be adopted as an input to improve the levels of individual and organizational innovation, through the establishment of an independent department in Social Security Corporation in.

Keywords Knowledge creation process · Organizational innovation · Social security corporation · Jordan

1 Introducing

The current era is characterized by rapid changes in globalization, information and communication technology, which presents many opportunities and challenges for all organizations in order to obtain the competitive strength to survive, continue to work efficiently and effectively to face these changes (Allahow et al., 2018; Al-Hawary & Al-Syasneh, 2020; AlTaweel & Al-Hawary, 2021; Al-Maroof et al., 2021). Perhaps it is distinguished by the emergence of the power of knowledge; so whoever can use it well possesses the competitive power that enables him to survive in light of these changes and challenges, and in the same context, the concept of the rule of the era of knowledge led to increase its role and importance by providing information and thinking about it (Al-Hawary et al., 2020; Al Mehrez et al., 2020). Therefore, the administrative thought focused on knowledge processes and their management to ensure that the organization achieves its desired goals, responds to continuous changes and adapts easily. In addition to training, developing and motivating human resources as it is considered an effective and important factor to enhance levels of organizational innovation, production capabilities and excellence of organizations, as the priority occupied by human capital in the knowledge economy has led to an intense struggle for the talented, the creative, and the brilliant minds (Al-Hawary & Alwan, 2016; Kabrilyants et al., 2021; Tariq et al., 2022).

Organizational innovation is considered a goal for organizations that are interested in knowledge management and its operations, due to the intensity of local and global competition in most fields (Al-Hawary, 2015; Al-Hawary & Aldaihani, 2016; Al-Hawary & Al-Namlan, 2018; Al-Hawary & Nusair, 2017; Al-Lozi et al., 2017, 2018; Alzoubi et al., 2022). And because the value of organizations lies in the available human resources, this prompts them to pay attention to their knowledge to reach the creativity they seek. As developed countries have made progress and superiority through talented and creative minds where their abilities, skills and energies have been utilized; so the specialists in measuring the progress of countries are based on the number of scholars and thinkers, as well as the resulting degree of knowledge and creative accumulation (Jawad et al., 2006; AlHamad et al., 2022; Alshurideh et al., 2022).

The most prominent problems that public organizations in Jordan suffer from are the lack of an appropriate climate, an appropriate work environment, and a culture that supports knowledge management as well as its operations that contribute to support and improve levels of organizational innovation in them. Moreover, these organizations face many challenges such as the difficulty of strategic prediction for their future, mission and goals. In addition to the financial pressures, bureaucratic complications, regulations and laws they face, and all of the foregoing severely impede the effective implementation of knowledge management initiatives and operations, looking for an increase in the levels of innovation, which is a vital and effective element in Raising the performance levels of employees, thus achieving success and excellence.

In light of the foregoing, the General Organization for Social Security in Jordan needs a comprehensive knowledge management system because of its importance, as it plays a vital and effective role in making public administration jobs more efficient by raising the levels of innovation among employees, it also contributes in supporting social goals for the public sector through searching for opportunities and large projects that allow society to flourish, progress and prosperity, in addition to enabling organizations and their members to work at high levels of performance, in order to improve the quality and quality of community life. Therefore, this study is a starting point for future training and development programs, the results and recommendations also help in interpreting the results of previous researches that dealt with some aspects of the subject, and use the findings of the research to support its results, as well as providing researchers with documented information derived from the field study, in addition to trying to attract the attention of researchers to direct their future research towards the variables of this study. So, this study came to test the impact of the knowledge creation process in achieving organizational innovation.

2 Theoretical Framework and Creating Hypotheses

2.1 Knowledge Creation

Nonaka (1994) pointed out that the process of creating knowledge lies mainly in the difference between tacit knowledge and explicit knowledge, as it made available and expanded entirely new knowledge by individuals, as well as that it had formed and documented knowledge with the knowledge of the organization; despite the formation of ideas in the minds of individuals, the interaction among these individuals plays an important role in its development, and this indicates that close cooperation and coordination between individuals within groups or work teams contributes to the expansion and development of completely new knowledge that extends across the borders of the organization. Toyama et al. (2000) added, it is a process by which the individual excels himself by transcending his old ideas, thinking more broadly and with a holistic thought to move to new ideas in order to acquire new content that keep pace with the changes and developments at the local and global levels. Gottschal (2008) stressed that this process leads to the creation of valuable ideas that contribute to solving problems and offering opportunities for development and superiority, as it includes developing new content or replacing the available content with new content through the processes of cooperation, consultation and effective coordination between individuals or groups; In addition, creativity is the basis in the process of creating knowledge.

Many researchers point out that creating and sharing knowledge is not a natural and automatic act, because knowledge often exists in the minds of individuals, but they face difficulties in transferring and sharing it with others, especially when facilities and support are not available from the management of the organization (AlMehrzi et al. 2020; Shannak et al., 2012). Magnier-Watanabe et al. (2011) and Elinfoo (2005) emphasized that the essence of knowledge management processes focuses on the process of creating and forming knowledge by employing the existing knowledge accumulation in the organization to create new knowledge or taking advantage of the organization's employees' interactions with its markets and competitive environment to create new knowledge.

Researchers in the field of knowledge management were interested in studying and analyzing the mechanisms of creating and forming organizational knowledge through the interaction of individuals, work teams, and groups in the internal environment of the organization, or in the external work environment or both, where they presented several models for the transformations of knowledge and the stages of its creation, formation and development. (Nonaka and Takeuchi) model for the process of creating and creating knowledge remains one of the most important and influential models on the development of the literature of knowledge transfer and management.

Therefore, this model is considered one of the most prominent models used in the process of creating knowledge, and it is called the Nonaka spiral, and it includes four stages of transformation of knowledge: (Socialization, Externalization, Combination, and Internalization), to represent a model designed to understand the dynamics of creating and forming knowledge in organizations through the availability of two types of Knowledge, which is tacit knowledge and explicit knowledge, as well as the dynamic interaction between them within three main levels of interaction, namely individuals, groups and the organization (Al-Hawary & Alwan, 2016; Alshurideh, 2022; Alzoubi et al., 2021). It is worth noting that Nonaka et al. (2000), presented a description of each stage of the four knowledge creation process as follows:

- **Socialization**: It is the stage at which new tacit knowledge is created through the exchange of experiences, ideas and technical skills between individuals. Socialization appears in human relationships at work, outside work or both, where mental and theoretical models are produced to the world.
- **Externalization**: It is the stage, at which tacit knowledge is highlighted and transformed into explicit knowledge, and it happens through oral communication using language in discussions and collective thinking, as the transformation of tacit

knowledge into explicit knowledge is logically and systematically crystallized to be easily shared.

- **Combination**: It is the stage at which explicit knowledge is transformed into more complex and systematic explicit knowledge, such as: (conducting research and preparing reports), where explicit knowledge is collected from a variety of different sources, and then combined, organized and coordinated to form new explicit knowledge, to be shared between many groups through the means and methods of distributing and sharing written documents, and electronic publishing.
- **Internalization**: It is the stage at which explicit knowledge is self-characterized, and transformed into tacit knowledge, where knowledge is understood by all members of the organization, and this is done in several ways such as: (training, experiments, and repeating the performance of tasks).

2.2 *Organizational Innovation*

Innovation is a vital and important element for organizations, because the skills and abilities required for their survival and to perform their role to the fullest, will be different from what they were in the past, as organizations at the present time and in the future will face a wave of many challenges and opportunities, as well as rapid and sharp changes, and this requires them to respond quickly and find appropriate ways and methods to adapt to them in a flexible manner (Almaazmi et al., 2020; Al Suwaidi et al., 2020; Alameeri et al., 2021; Nuseir et al., 2021). Therefore, perceptive organizations will seek to employ talented individuals and groups, in order to find ideas and creative solutions that will elevate the organization and raise its levels of performance and ability to exploit opportunities and face various challenges and threats (Al-Nady et al., 2013, 2016; Alshurideh et al., 2017; Alolayyan et al., 2018; Al-Hawary & Alhajri, 2020; Mohammad et al., 2020; Alhalalmeh et al., 2020; Al-Hawary & Obiadat, 2021).

Organizational innovation, it is a process through which the organization uses all its knowledge, skills, and technical systems in order to develop completely new goods or services or to improve and develop its operating systems and programs to raise the level of response to the needs and desires of the beneficiaries and to gain their satisfaction and loyalty to them (Jones, 2001). Guilford (1959) defined organizational innovation as a set of preparatory features that include (fluency in thinking, flexibility, sensitivity to problems and authenticity), as well as redefining the problem and clarifying it in detail.

The researchers differed on defining the elements of administrative innovation, but they agreed on a group of them (Altamony et al., 2012; Allahow et al., 2018). Fluency, means the individual's ability to present or produce a large number of synonyms, alternatives, uses or ideas, when confronting and responding to a specific event, situation or stimulus, as well as speed in generating it. Flexibility, means the ability of the individual to present various and different ideas in addition to his ability to change or transform his point of view, vision, or path of thinking according to the course and requirements of the event or problem, and sensitivity to problems; it also means the individual's ability to discover various problems, weaknesses and gaps in the problem, the issue or situation. Internalization means the individual's ability to present or produce something unconventional, unusual, amazing or rare and unique in the long run, where scientists point out that the idea is not completely original and new unless no one preceded it and was unfamiliar and far-reaching.

2.3 Knowledge Management Processes and Organizational Innovation

The basics of the knowledge creation process (Socialization, Externalization, Combination, and Internalization) affect innovation in the organization, so organizations must focus in their main activities on learning and knowledge management processes, and then use knowledge theory to find a specific direction to coordinate activities related to transformation and knowledge creation processes, in order to prove the value of brilliant minds for the active and continuous existence in society (Rowley, 2000). Chapman and Magnusson (2006) pointed out that knowledge is an essential component of all forms of creativity, as well as it is a widely accepted principle of innovation management. Stacey (2000) emphasizes that knowledge is one of the most vital assets as it is the main source of innovation and excellence, but there is no simple and generally accepted definition of knowledge, as knowledge differs from data and information, but it is related to both, unlike data and information, where knowledge emerges from human interpretations and their complex interactions. Rahimi et al. (2011) explained that the knowledge creation process can be seen as a spiral or ascending process from the individual level to the group level, then to the organizational level, and sometimes to the inter-organizational level; but innovation is the basis of problem solving, and to occur, it must not be limited to the exchange of knowledge only, but must be used and recombined, and sometimes the distinction between the creation and use of knowledge appears to be particularly reflected when confronted with complex systems such as innovation because it is the result of a combination of existing and new knowledge, and therefore the process of compiling new and current knowledge depends largely on the creation and use of knowledge; and in the next generation of knowledge management the importance of knowledge creation and use has been greatly emphasized to create value and became critical factors in the creative process. Alberto (2000), points out that the relationship between knowledge management processes and organizational innovation is of great importance to the success of competitive organizations, as well as its impact on the formulation of competition strategies, because the development of knowledge is linked to the characteristics and capabilities of individuals, and this supports the innovation of the organization and its competitive capabilities. Awad and Ghaziri (2004) noted the relationship between knowledge management processes and organizational innovation, which lies in the fact that the knowledge

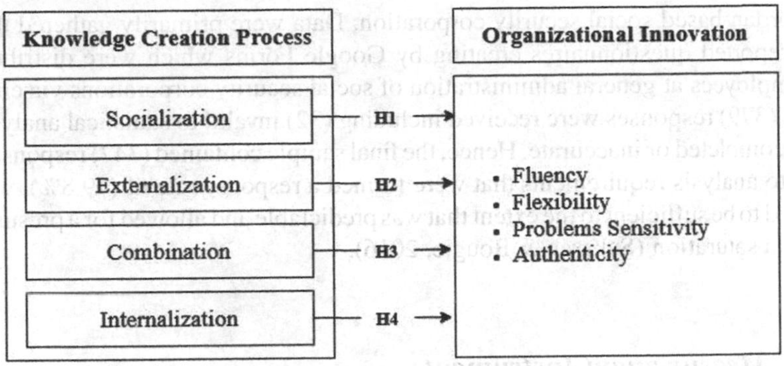

Fig. 1 Research model

capabilities of organizations have turned into processes that allowed them to create new products and services very quickly, or to modify and develop them quickly as well, in addition to their speed in providing them to beneficiaries in a distinctive way; this is because knowledge management processes, especially the process of creating knowledge, play an important role in influencing and advancing the levels of organizational innovation, as innovation cannot take place without it, especially in the current era, as it supports and provides individuals with new knowledge that will enhance their creative ideas in various fields, as well as It lights their way to original and unique ideas. Accordingly, the study hypothesis can be formulated as follows:

There is a statistically significant effect at the significance level (0.05 ≥ α) of the knowledge creation process on organizational innovation in the Public Institution for Social Security in Jordan.

3 Study Model

See Fig. 1.

4 Methodology

4.1 Population and Sample Selection

A qualitative method based on a questionnaire was used in this study for data collection and sample selection. The major aim of the study was to examine the impact of knowledge creation process on organizational innovation. Therefore, it focused on Jordan-based social security corporation. Data were primarily gathered through self-reported questionnaires creating by Google Forms which were distributed to all employees at general administration of social security corporations via email. In total, (379) responses were received including (32) invalid to statistical analysis due to uncompleted or inaccurate. Hence, the final sample contained (347) responses suitable to analysis requirements that were formed a response rate of (79.8%), where it proved to be sufficient to the extent that was predictable and allowed for a presumption of data saturation (Sekaran & Bougie, 2016).

4.2 *Measurement Instrument*

A self-reported questionnaire that consists of two main sections along with a section regarding control variables was used as the measurement instrument. Control variables considered as categorical measures were composed of gender, age group, educational level, and experience. The two main sections were dealt with a five-point Likert scale (from 1 = strongly disagree to 5 = strongly agree). The first section contained (24) items to measure knowledge creation processbased on (Chatterjee et al., 2018; Li et al., 2018; Al-Adamat et al., 2020; Prompreing & Hu, 2021). These questions were distributed into dimensions as follows: seven items dedicated for measuring socialization, six items dedicated for measuring externalization, five items dedicated for measuring combination, and six items dedicated for measuring internalization. Whereas the second section included (23) items developed to measure organizational innovation according to what was pointed by Silva and Cirani (2020). This variable was a second-order construct divvied into four first-order constructs. Fluency was measured through five items, flexibility was measured through six items, problems sensitivity was measured by six items, and authenticity was measured using six items.

5 Findings

5.1 *Measurement Model Evaluation*

This study was conducted structural equation modeling (SEM) to test hypotheses, which represents a contemporary statistical technique for testing and estimating the relationship between factors and variables (Wang & Rhemtulla, 2021). Accordingly, the reliability and validity of the constructs were tested using confirmatory factor analysis (CFA) through the statistical program AMOSv24. Table 1 summarizes the results of convergent and discriminant validity, as well the indicators of reliability.

Table 1 shows that the standard loading values for the individual items were within the domain (0.537–0.945), these values greater than the minimum retention of the elements based on their standard loads (Al-Lozi et al., 2018; Sung et al., 2019). Average variance extracted (AVE) is a summary indicator of the convergent validity

Table 1 Results of validity and reliability tests

Constructs	1	2	3	4	5	6	7	8
1. SOC	**0.784**							
2. EXT	0.387	**0.764**						
3. COM	0.485	0.527	**0.732**					
4. INT	0.594	0.439	0.521	**0.723**				
5. FLU	0.625	0.511	0.618	0.438	**0.763**			
6: FLE	0.581	0.637	0.594	0.494	0.374	**0.748**		
7. PSE	0.499	0.552	0.541	0.567	0.439	0.492	**0.769**	
8. AUT	0.614	0.482	0.558	0.634	0.552	0.502	0.513	**0.725**
VIF	1.624	1.513	2.334	1.504	–	–	–	–
Loadings range	0.684–0.846	0.642–0.945	0.681–0.775	0.657–0.774	0.622–0.841	0.713–0.791	0.537–0.882	0.674–0.781
AVE	0.614	0.584	0.536	0.523	0.583	0.560	0.591	0.522
MSV	0.554	0.429	0.374	0.415	0.502	0.487	0.515	0.497
Internal consistency	0.905	0.889	0.848	0.864	0.871	0.880	0.892	0.865
Composite reliability	0.917	0.892	0.852	0.867	0.874	0.884	0.894	0.867

Note SOC: socialization, EXT: externalization, COM: combination, INT: internalization, FLU: Fluency, FLE: flexibility, PSE: problems sensitivity, AUT: authenticity, Bold fonts refers to $\sqrt{AVE}$

of constructs that must be above 0.50 (Howard, 2018). The results indicate that the AVE values were greater than 0.50 for all constructs, thus the used measurement model has an appropriate convergent validity. Rimkeviciene et al. (2017) suggested the comparison approach as a way to deal with discriminant validity assessment in covariance-based SEM. This approach is based on comparing the values of maximum shared variance (MSV) with the values of AVE, as well as comparing the values of square root of AVE ($\sqrt{AVE}$) with the correlation between the rest of the structures. The results show that the values of MSV were smaller than the values of AVE, and that the values of $\sqrt{AVE}$ were higher than the correlation values among the rest of the constructs. Therefore, the measurement model used is characterized by discriminative validity. The internal consistency measured through Cronbach's Alpha coefficient (α) and compound reliability by McDonald's Omega coefficient (ω) was conducted as indicators to evaluate measurement model. The results listed in Table 1 demonstrated that both values of Cronbach's Alpha coefficient and McDonald's Omega coefficient were greater than 0.70, which is the lowest limit for judging on measurement reliability (De Leeuw et al., 2019).

5.2 *Structural Model*

The structural model illustrated no multicollinearity issue among predictor constructs because variance inflation factor (VIF) values are below the threshold of 5, as shown in Table 1 (Hair et al., 2017). This result is supported by the values of model fit indices shown in (Figs. 1 and 2).

The results in Fig. 1 indicated that the chi-square to degrees of freedom (CMIN/DF) was 1.287, which is less than 3 the upper limit of this indicator. The values of the goodness of fit index (GFI), the comparative fit index (CFI), and the Tucker-Lewis index (TLI) were upper than the minimum accepted threshold of 0.90. Moreover, the result of root mean square error of approximation (RMSEA) indicated to value 0.045, this value is a reasonable error of approximation because it is less than the higher limit of 0.08. Consequently, the structural model used in this study was recognized as a fit model for predicting the DEP and generalization of its result (Ahmad et al., 2016; Shi et al., 2019). To verify the results of testing the study hypotheses, structural equation modeling (SEM) was used, the results of which are listed in Table 2.

The results demonstrated in Table 2 show that all knowledge creation process dimensions have a positive impact relationship on organizational innovation except externalization which has no impact on organizational innovation ($\beta = 0.031$, $t = 0.498$, $p = 0.624$). However, the results indicated that the highest impact was for combination ($\beta = 0.338$, $t = 5.657$, $p = 0.000$), followed by socialization ($\beta = 0.285$, $t = 4.781$, $p = 0.000$), and finally the lowest impact was for internalization ($\beta = 0.196$, $t = 3.735$, $p = 0.000$).

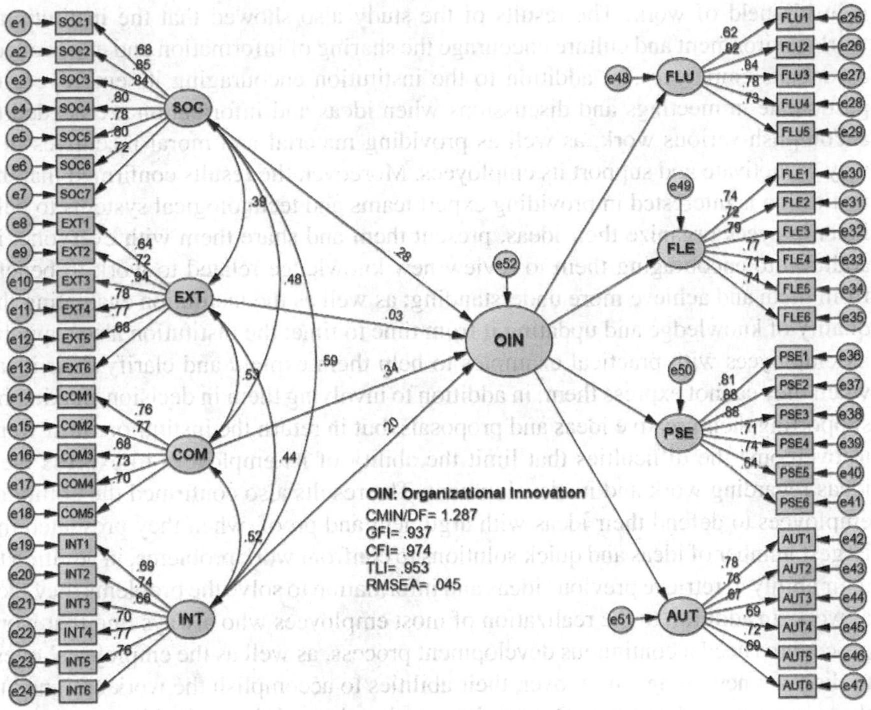

Fig. 2 SEM results of the knowledge creation process effect on organizational innovation

Table 2 Hypothesis testing

Hypothesis	Relation	Standard beta	*t* value	*p* value
H1	Socialization→Organizational Innovation	0.285	4.781	0.000
H2	Externalization→Organizational Innovation	0.031	0.498	0.624
H3	Combination→Organizational Innovation	0.338	5.657	0.000
H4	Internalization→Organizational Innovation	0.196	3.735	0.000

Note * $p < 0.05$, ** $p < 0.01$, *** $p < 0.001$

6 Discussion

The results showed the interest of the Social Security Corporation in the availability of experts in knowledge to train its employees on new knowledge to use in their work, in addition to giving them opportunities to think about what was discussed and support it with ideas and creative proposals, and to encourage everyone who supports knowledge with practical additions that can be benefited from and improved in work, in addition to the institution's interpretation of knowledge results obtained and saved in a comprehensive knowledge base so that everyone can access it to benefit from it in his field of work. The results of the study also showed that the institution's work environment and culture encourage the sharing of information and experiences among its employees, in addition to the institution encouraging its employees to participate in meetings and discussions when ideas and information are needed to accomplish various work, as well as providing material and moral incentives in a way to motivate and support its employees. Moreover, the results confirmed that the institution is interested in providing expert teams and technological systems to help its employees organize their ideas, present them and share them with everyone, in addition to encouraging them to review new knowledge related to work to benefit from them and achieve more understanding; as well as the institution evaluating the quality of knowledge and updating it from time to time; the institution also supports its employees with practical examples to help them explain and clarify their ideas when they cannot express them, in addition to involving them in decision-making by supporting their creative ideas and proposals, but in return the institution must work to overcome the difficulties that limit the ability of its employees to express their ideas regarding work and its development. The results also confirmed the ability of employees to defend their ideas with argument and proof, when they presented the largest number of ideas and quick solutions to confront work problems, in addition to their ability to retrieve previous ideas and information to solve the problems they face at work, in addition to the realization of most employees who always find that work procedures need a continuous development process, as well as the employees' quest to discover new things; moreover, their abilities to accomplish the work assigned to them in a renewed manner. The employees also showed the high ability to propose solutions to face work problems after knowing the shortcomings or weaknesses in the work they do. In addition to their medium ability to discover problems experienced by others and plan to solve them, as well as sometimes predict them before they occur, and their ability to make some changes in work methods from time to time, in addition to their medium ability to adapt to new work variables, see things from different sides and think quickly about working conditions and requirements.

The results of the study showed and found an impact of the knowledge creation process on organizational innovation in the Social Security Corporation in Jordan, because the Social Security Corporation is interested in (socialization) through the process of sharing experiences and information in order to create tacit knowledge such as mental models and joint technical skills; (Combination) through a process of systematically compiling and organizing the explicit knowledge that is available with explicit knowledge related to the work of the institution; (Internalization) by transforming explicit knowledge into tacit knowledge; (Personality) through training and repeating the performance of tasks, taking suggestions and ideas and supporting them, so that explicit knowledge becomes assimilated as tacit knowledge. But on the other hand, there is a clear weakness and less interest from the institution in (Externalization) through the process of accurately identifying and improving the tacit knowledge related to the work and then transforming it into explicit knowledge through interviews, discussions and meetings, and this result agreed to some extent with the studies by Young, Berraies and Chaher, and Huang and Wang.

7 Recommendations

In light of the results that have been reached, the study recommends the need to adopt knowledge management as an entrance to improve the levels of individual and organizational innovation, through the establishment of an independent department in the Public Institution for Social Security in Jordan, under the name (Knowledge Management) appearing in the organizational structure; and the necessity for the Social Security Corporation in Jordan to provide systems, methods and ways that support the process of sharing and exchanging tacit (personal) knowledge related to work among its employees, Which would enhance their innovation levels, thus raising the level of their performance at work; as well as the necessity of adopting the Public Institution for Social Security in Jordan, strategies that contribute to supporting employees and opening the way for them to express their ideas and creative proposals related to work and its development with ease. As well as providing specialized technological systems and tools to help its employees organize their information, ideas and suggestions, display them in exemplary ways, and share them with their colleagues, in addition, the institution train them on these systems through expert teams.

References

Ahmad, S., Zulkurnain, N., & Khairushalimi, F. (2016). Assessing the validity and reliability of a measurement model in structural equation modeling (SEM). *British Journal of Mathematics & Computer Science, 15*(3), 1–8. https://doi.org/10.9734/BJMCS/2016/25183

Al Mehrez, A. A., Alshurideh, M., Al Kurdi, B., & Salloum, S. A. (2020). Internal factors affect knowledge management and firm performance: a systematic review. In *International Conference on Advanced Intelligent Systems and Informatics* (pp. 632–643). Springer, Cham.

Al Suwaidi, F., Alshurideh, M., Al Kurdi, B., & Salloum, S. A. (2020). The impact of innovation management in SMEs performance: a systematic review. In *International Conference on Advanced Intelligent Systems and Informatics* (pp. 720–730). Springer, Cham.

Al-Adamat, A., Al-Gasawneh, J., & Al-Adamat, O. (2020). The impact of moral intelligence on green purchase intention. *Management Science Letters, 10*(9), 2063–2070.

Alameeri, K. A., Alshurideh, M. T., & Al Kurdi, B. (2021). The effect of Covid-19 pandemic on business systems' innovation and entrepreneurship and how to cope with it: A theatrical view. The Effect of Coronavirus Disease (COVID-19) on Business Intelligence, 334, 275–288.

Alberto, C. (2000). How does knowledge management influence innovation and competitiveness? *Journal of Knowledge Management, 4*(2).

Alhalalmeh, M. I., Almomani, H. M., Altarifi, S., Al- Quran, A. Z., Mohammad, A. A., & Al-Hawary, S. I. (2020). The nexus between corporate social responsibilty and organizational performance in Jordan: The mediating role of organizational commitment and organizational citizenship behavior. *Test Engineering and Management, 83*(July), 6391–6410.

AlHamad, A., Alshurideh, M., Alomari, K., Kurdi, B., Alzoubi, H., Hamouche, S., & Al-Hawary, S. (2022). The effect of electronic human resources management on organizational health of telecommunications companies in Jordan. *International Journal of Data and Network Science, 6*(2), 429–438.

Al-Hawary, S. I. (2015). Human resource management practices as a success factor of knowledge management implementation at health care sector in Jordan. *International Journal of Business and Social Science, 6*(11/1), 83–98.

Al-Hawary, S. I., & Aldaihani, F. M. (2016). Customer relationship management and innovation capabilities of Kuwait Airways. *International Journal of Academic Research in Economics and Management Sciences, 5*(4), 201–226.

Al-Hawary, S. I., & Al-Namlan, A. (2018). Impact of electronic human resources management on the organizational learning at the private hospitals in the state of Qatar. *Global Journal of Management and Business Research: A Administration and Management, 18*(7), 1–11.

Al-Hawary, S. I., & Nusair, W. (2017). Impact of human resource strategies on perceived organizational support at Jordanian Public Universities. *Global Journal of Management and Business Research: A Administration and Management, 17*(1), 68–82.

Al-Hawary, S. I. S., & Alhajri, T. M. S. (2020). Effect of electronic customer relationship management on customers' electronic satisfaction of communication companies in Kuwait. *Calitatea, 21*(175), 97–102.

Al-Hawary, S. I. S., & Alwan, A. M. (2016). Knowledge management and its effect on strategic decisions of Jordanian Public Universities. *Journal of Accounting-Business & Management, 23*(2), 24–44.

Al-Hawary, S. I. S., & Obiadat, A. A. (2021). Does mobile marketing affect customer loyalty in Jordan? *International Journal of Business Excellence, 23*(2), 226–250.

Al-Hawary, S. I. S., Mohammad, A. S., Al-Syasneh, M. S., Qandah, M. S. F., & Alhajri, T. M. S. (2020). Organisational learning capabilities of the commercial banks in Jordan: Do electronic human resources management practices matter? *International Journal of Learning and Intellectual Capital, 17*(3), 242–266.

Al-Hawary, S. I., & Al-Syasneh, M. S. (2020). Impact of dynamic strategic capabilities on strategic entrepreneurship in presence of outsourcing of five stars hotels in Jordan. *Business: Theory and Practice, 21*(2), 578–587.

Allahow, T. J. A. A., Al-Hawary, S. I. S., & Aldaihani, F. M. F. (2018). Information technology and administrative innovation of the central agency for information technology in Kuwait. *Global Journal of Management and Business, 18*(11-A), 1–16.

Al-Lozi, M. S., Almomani, R. Z. Q., & Al-Hawary, S. I. S. (2018). Talent Management strategies as a critical success factor for effectiveness of Human Resources Information Systems in commercial banks working in Jordan. *Global Journal of Management and Business Research: A Administration and Management, 18*(1), 30–43.

Al-Lozi, M., Almomani, R. Z., & Al-Hawary, S. I. (2017). Impact of talent management on achieving organizational excellence in Arab Potash Company in Jordan. *Global Journal of Management and Business Research: A Administration and Management, 17*(7), 15–25.

Almaazmi, J., Alshurideh, M., Al Kurdi, B., & Salloum, S. A. (2020). The effect of digital transformation on product innovation: a critical review. In *International Conference on Advanced Intelligent Systems and Informatics* (pp. 731–741). Springer, Cham.

Al-Maroof, R., Ayoubi, K., Alhumaid, K., Aburayya, A., Alshurideh, M., Alfaisal, R., & Salloum, S. (2021). The acceptance of social media video for knowledge acquisition, sharing and application: A comparative study among YouYube users and TikTok users' for medical purposes. *International Journal of Data and Network Science, 5*(3), 197–214.

AlMehrzi, A., Alshurideh, M., & Al Kurdi, B. (2020). Investigation of the key internal factors influencing knowledge management, employment, and organisational performance: A qualitative study of the UAE hospitality sector. *International Journal of Innovation Creativity and Change, 14*(1), 1369–1394.

Al-Nady, B. A., Al-Hawary, S. I., & Alolayyan, M. (2013). Strategic management as a key for superior competitive advantage of sanitary ware suppliers in Kingdom of Saudi Arabia. *International Journal of Management and Information Technology, 7*(2), 1042–1058. and Technology, 29(5), 1174–1189.

Al-Nady, B. A., Al-Hawary, S. I., & Alolayyan, M. (2016). The role of time, communication, and cost management on project management success: An empirical study on sample of construction projects customers in Makkah City, Kingdom of Saudi Arabia. *International Journal of Services and Operations Management, 23*(1), 76–112.

Alolayyan, M., Al-Hawary, S. I., Mohammad, A. A., & Al-Nady, B. A. (2018). Banking service quality provided by commercial banks and customer satisfaction. A structural equation modelling approaches. *International Journal of Productivity and Quality Management, 24*(4), 543–565.

Alshurideh, M. (2022). Does electronic customer relationship management (E-CRM) affect service quality at private hospitals in Jordan? *Uncertain Supply Chain Management, 10*(2), 325–332.

Alshurideh, M. T., Al Kurdi, B., Alzoubi, H. M., Ghazal, T. M., Said, R. A., AlHamad, A. Q., ... & Al-kassem, A. H. (2022). Fuzzy assisted human resource management for supply chain management issues. *Annals of Operations Research*, 1–19.

Alshurideh, M., Al-Hawary, S. I., Batayneh, A. M., Mohammad, A., & Al-Kurdi, B.(2017). The impact of Islamic Banks' service quality perception on Jordanian customers loyalty. *Journal of Management Research, 9*(2),139–159.

Altamony, H., Masa'deh, R., Alshurideh, M., Obeidat, B. (2012) Information systems for competitive advantage: Implementation of an organisational strategic management process. Innovation and sustainable competitive advantage: From regional development to world economies. 583–592.

AlTaweel, I. R., & Al-Hawary, S. I. (2021). The mediating role of innovation capability on the relationship between strategic agility and organizational performance. *Sustainability, 13*(14), 7564.

Alzoubi, H., Alshurideh, M., Akour, I., Al Shraah, A., & Ahmed, G. (2021). Impact of information systems capabilities and total quality management on the cost of quality. *Journal of Legal, Ethical and Regulatory Issues, 24*(Special Issue 6), 1–11.

Alzoubi, H., Alshurideh, M., Kurdi, B., Akour, I., & Aziz, R. (2022). Does BLE technology contribute towards improving marketing strategies, customers' satisfaction and loyalty? The role of open innovation. *International Journal of Data and Network Science, 6*(2), 449–460.

Awad, E., & Ghaziri, H. (2004). *Knowledge Management* (1st ed.). Prentice-Hall.

Chapman, R., & Magnusson, M. (2006). Continuous innovation, performance and knowledge Management. *Knowledge and Process Management, 13*(3), 129–131.

Chatterjee, A., Pereira, A., & Sarkar, B. (2018). Learning transfer system inventory (LTSI) and knowledge creation in organizations. *The Learning Organization, 25*(5), 305–319. https://doi.org/10.1108/TLO-06-2016-0039

De Leeuw, E., Hox, J., Silber, H., Struminskaya, B., & Vis, C. (2019). Development of an international survey attitude scale: Measurement equivalence, reliability, and predictive validity. *Measurement Instruments for the Social Sciences, 1*(1), 9. https://doi.org/10.1186/s42409-019-0012-x

Elinfoo. K. (2005). Knowledge creation and knowledge transfer in construction organization in Tanzania, Doctoral Thesis, Royal Stitue of Technology, Stokholm, Sweden.

Gottschal, P. (2008). *It in Knowledge Management, Knowledge Management: Concepts, Methodologies, Tools and Applications*. Jennex, Published in United Nations, By IGI Global.

Guilford, J. P. (1959). Traits of creativity. In H. H. Anderson (Ed.), *Creativity and its cultivation* (pp. 142–151). Harper and Row.

Hair, J. F., Babin, B. J., & Krey, N. (2017). Covariance-based structural equation modeling in the journal of advertising: Review and recommendations. *Journal of Advertising, 46*(1), 163–177. https://doi.org/10.1080/00913367.2017.1281777

Howard, M. C. (2018). The convergent validity and nomological net of two methods to measure retroactive influences. *Psychology of Consciousness: Theory, Research, and Practice, 5*(3), 324–337. https://doi.org/10.1037/cns0000149

Jawad, M., Hussein, A., & Abdul Hamid, K. (2006). The impact of intellectual capital on organizational creativity. Ahl al-Bayt Magazine, (4), pp. 20–49.

Jones, G. (2001). *Organizational theory: Text and cases* (3rd ed.). Prentice Hall.

Kabrilyants, R., Obeidat, B., Alshurideh, M., & Masadeh, R. (2021). The role of organizational capabilities on e-business successful implementation. *International Journal of Data and Network Science, 5*(3), 417–432.

Li, M., Liu, H., & Zhou, J. (2018). G-SECI model-based knowledge creation for CoPS innovation: The role of grey knowledge. *Journal of Knowledge Management, 22*(4), 887–911. https://doi.org/10.1108/JKM-10-2016-0458

Magnier-Watanabe, R., Benton, C., Senoo, D. (2011). A study of knowledge management enablers across countries. *Knowledge Management Research and Practice, 1*(9), 17–28.

Mohammad, A. A., Alshura, M.S., Al-Hawary, S. I. S., Al-Syasneh, M. S., & Alhajri, T. M. (2020). The influence of Internal Marketing Practices on the employees' intention to leave: A study of the private hospitals in Jordan. *International Journal of Advanced Science*

Nonaka, I. (1994). A dynamic theory of organizational knowledge creation. *Organization Science, 5*(1).

Nonaka, I., Toyama, R., & Konno, N. (2000). SECI, Ba and leadership: A unified model of dynamic knowledge creation. *Long Range Planning, 33*, 5–34.

Nuseir, M. T., Aljumah, A., & Alshurideh, M. T. (2021). How the business intelligence in the new startup performance in UAE During COVID-19: The mediating role of innovativeness. The Effect of Coronavirus Disease (COVID-19) on Business Intelligence, 334, 63–79.

Prompreing, K., & Hu, C. (2021). The role of knowledge-sharing behaviour in the relationship between the knowledge creation process and employee goal orientation. *International Journal of Business Science & Applied Management, 16*(2), 46–63.

Rahimi, H., Arbabisarjou, A., Allameh, A., & Razieh, S. (2011). Relationship between knowledge management process and creativity among faculty members in the university. *Interdisciplinary Journal of Information, Knowledge, and Management, 4*, 204–205.

Rimkeviciene, J., Hawgood, J., O'Gorman, J., & De Leo, D. (2017). Construct validity of the acquired capability for suicide scale: Factor structure, convergent and discriminant validity. *Journal of Psychopathology and Behavioral Assessment, 39*(2), 291–302. https://doi.org/10.1007/s10862-016-9576-4

Rowley, J. (2000). Is higher education ready for knowledge management? *International Journal of Educational Management, 14*(7), 325–333.

Sekaran, U., & Bougie, R. (2016). *Research methods for business: A skill-building approach* (Seventh edition). Wiley.

Shannak, R. O., Zu'bi, M. F., & Alshurideh, M. T. (2012). A theoretical perspective on the relationship between knowledge management systems, customer knowledge. *European Journal of Social Sciences, 32*(4), 520–532.

Shi, D., Lee, T., & Maydeu-Olivares, A. (2019). Understanding the model size effect on SEM Fit indices. *Educational and Psychological Measurement, 79*(2), 310–334. https://doi.org/10.1177/0013164418783530

da Silva, J. J., & Cirani, C. B. S. (2020). The capability of organizational innovation: Systematic review of literature and research proposals. *Gestão&produção, 27*(4), e4819. https://doi.org/10.1590/0104-530x4819-20

Stacey, R. D. (2000). The emergence of knowledge in organizations. *Emergence, 2*(4), pp: 23–39.

Sung, K.-S., Yi, Y. G., & Shin, H.-I. (2019). Reliability and validity of knee extensor strength measurements using a portable dynamometer anchoring system in a supine position. *BMC Musculoskeletal Disorders, 20*(1), 1–8. https://doi.org/10.1186/s12891-019-2703-0

Tariq, E., Alshurideh, M., Akour, I., & Al-Hawary, S. (2022). The effect of digital marketing capabilities on organizational ambidexterity of the information technology sector. *International Journal of Data and Network Science, 6*(2), 401–408.

Toyama, R., Nonaka, I., Konno, N. (2000). SECI, Ba and leadership: A unified model of dynamic knowledge creation, long range planning, Vol (33). www.elsevier.com.

Wang, Y. A., & Rhemtulla, M. (2021). Power analysis for parameter estimation in structural equation modeling: A discussion and tutorial. *Advances in Methods and Practices in Psychological Science, 4*(1), 1–17. https://doi.org/10.1177/2515245920918253

The Effect of Marketing Knowledge on Competitive Advantage in the Food Industries Companies in Jordan

Faraj Mazyed Faraj Aldaihani, Ali Zakariya Al- Quran, Laith Al-hourani, Mohammad Issa Ghafel Alkhawaldeh, Abdullah Matar Al-Adamat, Anber Abraheem Shlash Mohammad, Sulieman Ibraheem Shelash Al-Hawary, Muhammad Turki Alshurideh, and Barween Al Kurdi

Abstract This study aimed to identify the impact of marketing knowledge practices on the competitive advantage of the food industries companies in Jordan. In order to achieve the objectives of the study, the questionnaire was used to collect primary data, whereby the study population consisted of (160) administrators of all levels (senior management, middle management) and who numbered (16) food industry

F. M. F. Aldaihani
Kuwait Civil Aviation, Ishbiliyah bloch 1, street 122, home 1, Kuwait City, Kuwait

A. Z. Al- Quran · L. Al-hourani · S. I. S. Al-Hawary (✉)
Department of Business Administration, School of Business, Al al-Bayt University, P.O.BOX 130040, Mafraq 25113, Jordan
e-mail: dr_sliman73@aabu.edu.jo; dr_sliman@yahoo.com

A. Z. Al- Quran
e-mail: ali.z.al-quran@aabu.edu.jo

L. Al-hourani
e-mail: ali.z.al-quran@aabu.edu.jo

M. I. G. Alkhawaldeh
Directorates of Building and Land Tax, Ministry of Local Administration, Zarqa municipality, Jordan

A. M. Al-Adamat
School of Business, Department of Business Administration and Public Administration, Al Al-Bayt University Jordan, P.O. Box, (130040), Mafraq 25113, Jordan
e-mail: aaladamat@aabu.edu.jo

A. A. S. Mohammad
Marketing Department, Faculty of Administrative and Financial Sciences, Petra University, P.O. Box: 961343, Amman 11196, Jordan

M. T. Alshurideh
Department of Management, College of Business, University of Sharjah, 27272 Sharjah, United Arab Emirates
e-mail: malshurideh@sharjah.ac.ae; m.alshurideh@ju.edu.jo

Department of Marketing, School of Business, The University of Jordan, Amman 11942, Jordan

M. Alshurideh et al. (eds.), *The Effect of Information Technology on Business and Marketing Intelligence Systems*, Studies in Computational Intelligence 1056,
https://doi.org/10.1007/978-3-031-12382-5_96

companies. The questionnaire was distributed to the study sample, which included (160) administrators, and the researcher approved the sample the comprehensive survey and the descriptive analytical approach as it is appropriate to achieve the objectives of the study. One of the most important findings of the study is that there is a positive effect of marketing knowledge on the competitive. The study recommended the management of foods industries companies to provide the ability to process and analyze the information they possess in order to correctly forecast market demands, as well as employ delegates to study different markets, and worked to increase cooperation with suppliers and customers to follow up new developments in the market.

Keywords Marketing knowledge · Competitive advantage food industries companies · Jordan

1 Introduction

Business organizations use many factors of power in the context of facing the external environment variables, which are changing rapidly, and among these factors is knowledge, where organizations of all kinds can perform their work in a way that achieves competitiveness and entrepreneurship and is ahead of competitors in the field of activity that the organization is engaged in (Altamony et al., 2012; Al-Hawary & Al-Syasneh, 2020; AlTaweel & Al-Hawary, 2021; Obeidat et al., 2021). Marketing activity is the basis on which the organization relies to serve its markets appropriately, and this can only be achieved through the possession of those who carry out this activity with sufficient knowledge of market factors and variables from competitors and customers (Al Mehrez et al., 2020; Al-Hawary et al., 2011; Alshurideh, 2022; Alshurideh et al., 2017). Knowledge enables business organizations to understand the market, and how to deal with competitors by meeting the needs and desires of customers, which made there a need for marketing knowledge that enables the appropriate identification of the target market sector in light of the increasing competitiveness among business organizations, and thus directing the marketing mix that suits these sectors, which leads to enhancing the target market share (Al-Hawary & Alwan, 2016; Al-Lozi et al., 2017; Alshurideh, 2019; Alshurideh et al., 2020; Sweiss et al., 2021).

Competitive advantage expresses the organization's central characteristics, capabilities and efficiency, and also identifies the aspects in which the organization excels from other organizations. Dimensions of competitive advantage include improving internal efficiency, customer orientation, alliances and response to customers, cost reduction, awareness, flexibility and creativity (Al-Nady et al., 2016; Allahow et al.,

B. Al Kurdi
Department of Marketing, Faculty of Economics and Administrative Sciences, The Hashemite University, Zarqa, Jordan
e-mail: barween@hu.edu.jo

2018; Alhalalmeh et al., 2020; Al-Hawary et al., 2020; Mohammad et al., 2020). Business organizations can achieve and maintain competitive advantage in the presence of high competitiveness through innovation, improvement and development through continuous dynamic processes and not through individual events.

Marketing knowledge is an effective pillar in formulating various marketing plans that will achieve a competitive advantage, and given the many great challenges facing business organizations that stand in front of their aspirations to achieve development and growth, this imposes the need to fortify the competitive position of these organizations through the knowledge they possess with all competitors, customers, as well as suppliers who deal with these organizations, and knowing the current and future needs of the market so that these organizations can keep pace with what is happening in the market according to a new vision derived from its accurate diagnosis of the current reality (Al-Hawary & Alhajri, 2020; Al-Hawary & Obiadat, 2021; Al-Maroof et al., 2021). From the reality of the researcher's work in one of the food industries companies, he found that these companies lack appropriate marketing knowledge to enable them to know the market and the needs of customers, which creates several challenges for these companies to achieve competitive advantage (Al Kurdi et al., 2020a, 2020b; Khasawneh et al., 2021; Alzoubi et al., 2021). Therefore, this study comes to reveal the impact of marketing knowledge on the competitive advantage in food industries companies in Jordan.

2 Literature Review and Hypothesis Development

2.1 *Marketing Knowledge*

In light of the rapid changes in the movement of markets worldwide, marketing takes great importance at the level of work of business units in organizations, as without effective marketing strategic planning, companies cannot achieve the desired goals of marketing strategies, except by relying on knowledge that is characterized as a the only resource that builds up cumulatively and is used to generate and develop new ideas. Recent studies have shown the multiple role of knowledge in all fields, and the conviction of individuals and organizations of the importance of knowledge has increased. Hence, industrial companies seek to obtain knowledge from its sources, arrange, classify, store, and make it available in an easy and easy way to benefit from it (Alshurideh et al., 2012; Al-Hawary, 2015; Lee et al., 2022a, 2022b; Shamout et al., 2022).

There are several definitions of knowledge, Al-Hawary and Alwan (2016) defined knowledge as "the individual's ability to recognize and distinguish things" or "the ability that the individual possesses and stores in his mind in the form of cognitive maps". Al-Benaa and Qassem (2012) indicates that there are several approaches to clarify the concept of Knowledge, some of them focus on the economic approach, considering the intellectual capital owned by the organization, and others focus on the information approach, considering that knowledge is the extent of the ability to deal with information to transform it into permanent knowledge in the organization in order to employ it to achieve goals, while the technical approach considers them as technological capabilities that enable the organization to achieve excellence, and finally portrayed by the social approach as a social structure based on communication between individuals.

Al-Hawary et al. (2020) focused on creating, collecting, organizing and storing this knowledge for dissemination and exploitation by transforming knowledge into collaborative knowledge that can be shared within the organization. In order to implement it correctly, knowledge management and its processes within the organization must be properly adopted through an integrative, cumulative process that consists over long periods of time to become available for application and use in order to address specific problems and conditions. Organizations rely on knowledge to interpret the available information about a particular case, and to make a decision about how to manage and treat this case (Allahow et al., 2018; Alshurideh et al., 2021; Alwan & Alshurideh, 2022).

Macintosh (1998), Laudon and Laudon (2015), Becerra-Fernandez and Sabherwal (2014) and Campeanu-Sonea and Sonea (2016) dealt with several definitions of knowledge management, most of which were based on the fact that it is a discipline that focuses on analyzing the knowledge that can be obtained so as to link the knowledge assets (explicit and implicit) in all its internal and external operations, as knowledge management is based on the discovery, acquisition, sharing and application of knowledge in all units, in order to ensure the achievement of organizational goals. From here, we find that knowledge management is an essential aspect in permanent learning companies, as this supports the mechanism of sharing and reusing the knowledge of individuals or companies' knowledge through information technology such as the document management system, work groups, e-mail, and various databases that make up the knowledge sharing system integrates with additions of individuals and teams who use this system (Al-Hawary et al., 2013; Al-Dhuhouri et al., 2020).

Marketing knowledge management is based on being one of the most important skills for companies to acquire, create, retain and share knowledge based on the marketing environment, in order to build an information base on competitive market variables and identify target markets for use in making strategic marketing decisions (Alaali et al., 2021; Al-Hawary et al., 2017). Al-Hawary and Al-Namlan (2018) believes that the management of marketing knowledge in the organization is largely based on knowing all the information about the market, what are the movements of competitors, what are the desires and needs of customers, and the changes that have occurred, all of which prompted an attempt to integrate marketing activities into an organizational framework that leads to the human dimension which is represented in the tacit and explicit knowledge of employees, and linking this to information technology to form basic pillars in creating a marketing knowledge management system and promoting participatory, which enhances the accumulation of knowledge, retains and develops it, and enhances knowledge sharing among workers at all managerial levels, and thus develops the skills of workers, this, in turn, affects the support of the relationship with the customer and the identification of his desires and needs and how these needs and desires can be satisfied, which leads to achieving the competitive advantage of the organization in the target markets that it aspires to reach.

To enable the company to focus on the markets; You must generate information about customer needs, and the factors that influence customers, by collecting and responding to information by designing and implementing products and services that meet customer needs (Al-Hawary & Abu-Laimon, 2013; Al-Hawary & Harahsheh, 2014; Al-Hawary et al., 2017; Al-Lozi et al., 2018; Zebal & Others, 2019) indicated that marketing knowledge represents a graphic knowledge "know-what" that exists independently and must be distinguished and known from marketing skills or procedural knowledge "know-how". In the same direction, indicated that marketing knowledge represents both the codified information in the marketing databases of companies and the development of research skills that allow the application of this information for the benefit of the company. It is not only the codified information that managers possess, but also the refinement of personal skills for finding, evaluating, and using this information.

Marketing knowledge enables organizations to link better with existing customers, and to define the target customer more accurately. The dimensions of marketing knowledge in this research are indicated according to the following classification. First: Knowledge of the market: Knowledge of the market enables business organizations to identify the needs and desires of customers, their characteristics and interests about the nature of the products and services provided to benefit from them when designing the marketing mix for various products and services. Second: Knowledge of Suppliers: Knowledge of suppliers is important, as it includes knowledge of organizations that equip and provide the organization with the necessary resources that are important in producing goods and providing various services. Third: Knowledge of the customer: Knowledge of customers is based on understanding the needs and desires and discovering them in a renewed way by interacting with him and tracking his behavior, providing the value that satisfies him and acquiring the value that satisfies the organization, which includes knowledge from the customer and knowledge about the customer and knowledge of the customer, and the success and continuity of the company depends on accurate understanding For the customer as he is the cornerstone of the marketing process at the present time, so the organization must find many channels of mutual communication with him to identify his needs and desires and work to produce and provide goods and services of high quality, the lowest price, the greatest benefit and the safest and to align it with his needs and desires in a way that achieves his satisfaction to improve the image of the organization in his mind. Fourth: Product knowledge: Product knowledge includes clarity of the characteristics of the products that customer's desire and those characteristics that the organization desires to present, so that the organization can satisfy the needs of customers and the market." Fifth: Knowledge of competitors: Knowledge of competitors represents the source of the ability to confront competitors, and although competitors seek to make their knowledge difficult to imitate, however, competitors' businesses, products, and services provide opportunities for organizations to learn by analyzing their strengths and weaknesses (Alolayyan et al., 2018; Al-Dmour et al., 2021).

2.2 *Competitive Advantage*

The concept of competitive advantage emerged as one of the important topics in the field of business management, and the competitive advantage refers to the organization's ability to formulate and apply strategies that achieve distinction in front of competing companies, which operate in the same activity of this organization, and this can only be done through an optimal exploitation of the capabilities and resources of the organization; Financial, technical and organizational, in addition to the capabilities of competencies and knowledge of workers in this organization, through which they can apply and design competitive strategies, and this concept is related to the concept of perceived value to customers, and the extent of the organization's ability to achieve distinction among other organizations (Munizu, 2013). Competitive advantage has been defined as: "a set of factors that have a direct and indirect relationship to the organization's stability in the market, which includes the organization's effective participation in economic fields that affect stability and the development of profits, by actively employing material and human resources (Baroto et al., 2012). Al-Nady et al. (2013) defined competitive advantage as the most appropriate management that aims to discover new, innovative and creative ways to produce and provide goods and services more effectively than those offered by competitors in the market, where it is able to embody these methods on the ground, creating a innovative process that outperforms its competitors.

The competitive advantage of companies expresses their ability to attract customers, build a good image for them, improve their satisfaction and add serious value to their products (Baah & Jin, 2019). There are two types of models related to competitive advantage: the resource-based model, and the market-based model. These two models complement each other, and each of these two models includes variables, as the competitive advantage model that is based on the market includes variables such as differentiation, cost, effectiveness and evaluation of competitors, and the risks and threats facing the organization, while the model of competitive advantage, which is based on resources, it includes All the resources owned by the organization are material, financial or human, so that the movement and development are from within the organization (Korankey, 2013; AlHamad et al., 2022; Alshurideh et al., 2022; Hasan et al., 2022).

Due to developments and environmental changes and the changing needs and desires of the customer over time, the competitive dimensions have changed and developed. Thus, the success of the organization depends on determining the appropriate dimension on the basis of which it competes, which directly affects the organization's strategy and its entire performance. These are the main factors that achieve the organization excellence and superiority over competitors in the long term through its ability to identify the needs and desires of its target market and the possibility of satisfying these needs and desires in a way better than its competitors. The organization works on translating those needs and desires into certain indications that constitute the dimensions on which the organization will compete, and these dimensions constitute the competitive advantages (Al-Nady et al., 2013). The researchers differed in defining these dimensions, some of them identified them with five dimensions, and some identified them with six or more. However, they mostly agreed in four dimensions; delivery, quality, cost, flexibility (Al-Hawary et al., 2011; Alzoubi et al., 2021).

2.3 Marketing Knowledge and Competitive Advantage

Marketing knowledge has been studied by many researchers, Munyoro and Nyereyemhuka emphasized that the culture of customer relationship management leads to better performance, with strategies that use the needs and expectations of customers as a way to create a continuous culture to manage customer relationship. According to Abubakar and Mohammad (2019), companies looking for ways to improve the quality of decisions to improve their production lines in order to enhance competitive advantage. While Zebal and others: made five main proposals for business organizations to take advantage to improve the performance of their business in the markets, by adopting the internal and external approach to the management of tacit and explicit knowledge. In an attempt to understand market orientation to accelerate market innovation, define competitive advantage, and build performance. Na et al. (2019) researched marketing innovation, sustainable competitive advantage, and performance and confirmed that functional coordination of market cultural orientation increases product innovation, according to Hasniaty, Asram and Basmar (2018), improving marketing performance improves competitive advantage; as efficiency, innovation, information technology, facilities and infrastructure improve the company's performance and in terms of new products from which a competitive advantage can be built.

Husseini et al. (2018) refered to the existence of a relationship between efficiency, product quality, innovation and new product development. Competitive advantage may be achieved as a result of developing a new product strategy, and marketing knowledge management may create a sustainable competitive advantage because there is a positive role for marketing knowledge management to improve performance and reach a sustainable competitive advantage (Rezaee & Jafari, 2015). Marketing knowledge can also be used to improve the marketing communications of retailers in small and medium-sized businesses to obtain appropriate knowledge from customers and feedback to meet their needs and help win new customers (Marcek & Mucha, 2015). As knowledge management can affect the competitiveness of projects, knowledge management can help to make the company in a better position in terms of the effectiveness and efficiency of its activity, and thus enable the company to achieve a competitive advantage with the need to harmonize all characteristics of marketing knowledge depending on the environment and activity of the company (Semaskiene & Stancikiene, 2014). Based on Based on the above theoretical literature, the study hypothesis can be formulated as follows:

There is a statistically significant effect at the significance level (α < 0.05) of marketing knowledge on the competitive advantage of food industries companies in Jordan.

3 Study Model

See Fig. 1.

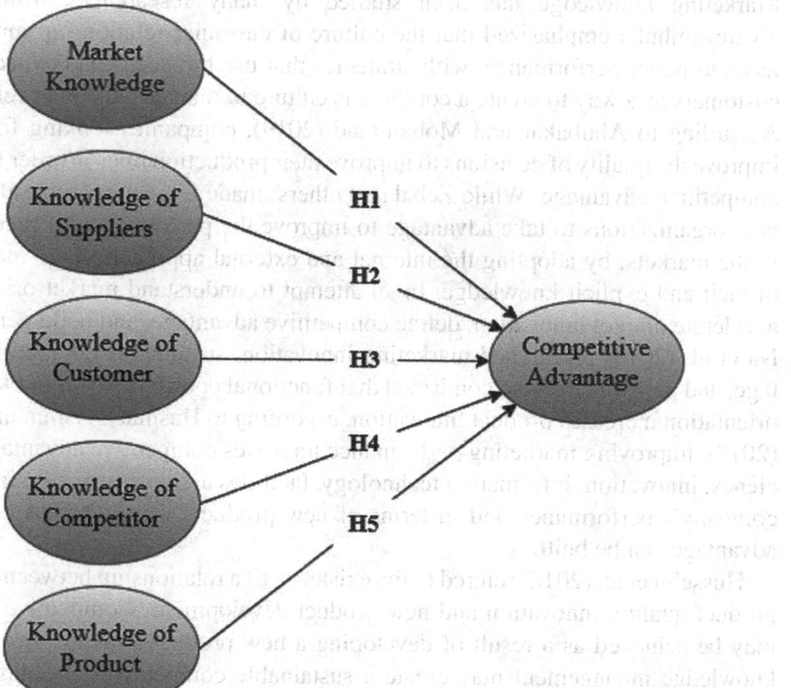

Fig. 1 Research model

4 Methodology

4.1 Population and Sample Selection

A qualitative method based on a questionnaire was used in this study for data collection and sample selection. The major aim of the study was to examine the impact of marketing knowledge on competitive advantage. Therefore, it focused on Jordan-based companies operating in the food industry sector in Abdullah II Ibn Al-Hussein Industrial City. Data were primarily gathered through self-reported questionnaires created by Google Forms which were distributed to a purposive sample of managers at high and middle levels via email. In total, (160) responses were received including (10) invalid to statistical analysis due to uncompleted or inaccurate. Hence, the final sample contained (150) responses suitable to analysis requirements that were formed a response rate of (93.7%), where it proved to be sufficient to the extent that was predictable and allowed for a presumption of data saturation (Sekaran & Bougie, 2016).

4.2 Measurement Instrument

A self-reported questionnaire that consists of two main sections along with a section regarding control variables was used as the measurement instrument. Control variables considered as categorical measures were composed of gender, age group, educational level, and experience. The two main sections were dealt with a five-point Likert scale (from 1 = strongly disagree to 5 = strongly agree). The first section contained (26) items to measure marketing knowledge based on (Al-Hawary & Alwan, 2016). These questions were distributed into dimensions as follows: five items dedicated for measuring market knowledge, five items dedicated for measuring knowledge of suppliers, five items dedicated for measuring knowledge of customers, six items dedicated for measuring knowledge of competitors, and five items dedicated for measuring knowledge of products. Whereas the second section included eight items developed to measure competitive advantage according to what was pointed by ((Distanont & Khongmalai, 2018; Yasa et al., 2020).

5 Findings

5.1 Measurement Model Evaluation

This study was conducted structural equation modeling (SEM) to test hypotheses, which represents a contemporary statistical technique for testing and estimating the relationship between factors and variables (Wang & Rhemtulla, 2021; Al-Adamat et al., 2020; Al-Gasawneh et al., 2020). Accordingly, the reliability and validity of the constructs were tested using confirmatory factor analysis (CFA) through the statistical program AMOSv24. Table 1 summarizes the results of convergent and discriminant validity, as well the indicators of reliability.

Table 1 shows that **the** standard loading values for the individual items were within the domain (0.673–0.854), these values greater than the minimum retention of the elements based on their standard loads (Al-Lozi et al., 2018; Sung et al., 2019). Average variance extracted (AVE) is a summary indicator of the convergent validity of constructs that must be above 0.50 (Howard, 2018). The results indicate that the AVE values were greater than 0.50 for all constructs, thus the used measurement model has an appropriate convergent validity. Rimkeviciene et al. (2017) suggested the comparison approach as a way to deal with discriminant validity assessment in covariance-based SEM. This approach is based on comparing the values of maximum shared variance (MSV) with the values of AVE, as well as comparing the

Table 1 Results of validity and reliability tests

Constructs	1	2	3	4	5	6
1. KMA	**0.755**					
2. KSU	0.527	**0.752**				
3. KCU	0.374	0.438	**0.743**			
4. KCO	0.422	0.527	0.502	**0.734**		
5. KPR	0.502	0.498	0.496	0.449	**0.774**	
6. COA	0.624	0.597	0.602	0.625	0.597	**0.761**
VIF	3.727	1397	2.311	2.646	2.648	–
Loadings range	0.681–0.825	0.718–0.831	0.682–0.802	0.685–0.792	0.687–0.837	0.673–0.854
AVE	0.570	0.566	0.553	0.538	0.599	0.579
MSV	0.418	0.502	0.468	0.513	0.484	0.511
Internal consistency	0.866	0.861	0.857	0.872	0.880	0.911
Composite reliability	0.869	0.867	0.860	0.875	0.882	0.916

Note KMA: market knowledge, KSU: knowledge of suppliers, KCU: knowledge of customers, KCO: knowledge of competitors, KPR: knowledge of products, COA: competitive advantage, Bold font indicates to root square of AVE values of square root of AVE ($\sqrt{AVE}$) with the correlation between the rest of the structures. The results show that the values of MSV were smaller than the values of AVE, and that the values of $\sqrt{AVE}$ were higher than the correlation values among the rest of the constructs. Therefore, the measurement model used is characterized by discriminative validity. The internal consistency measured through Cronbach's Alpha coefficient (α) and compound reliability by McDonald's Omega coefficient (ω) was conducted as indicators to evaluate measurement model. The results listed in Table 1 demonstrated that both values of Cronbach's Alpha coefficient and McDonald's Omega coefficient were greater than 0.70, which is the lowest limit for judging on measurement reliability (de Leeuw et al., 2019).

5.2 *Structural Model*

The structural model illustrated no multicollinearity issue among predictor constructs because variance inflation factor (VIF) values are below the threshold of 5, as shown in Table 1 (Hair et al., 2017).This result is supported by the values of model fit indices shown in (Figs. 1 and 2).

The results in Fig. 1 indicated that the chi-square to degrees of freedom (CMIN/DF) was 2.514, which is less than 3 the upper limit of this indicator. The values of the goodness of fit index (GFI), the comparative fit index (CFI), and the Tucker-Lewis index (TLI) were upper than the minimum accepted threshold of 0.90. Moreover, the result of root mean square error of approximation (RMSEA) indicated to value 0.036, this value is a reasonable error of approximation because it is less than the higher limit of 0.08. Consequently, the structural model used in this study was recognized as a fit model for predicting the DEP and generalization of its result (Ahmad et al., 2016; Shi et al., 2019).To verify the results of testing the study hypotheses, structural equation modeling (SEM) was used, the results of which are listed in Table 2.

The results demonstrated in Table 2 show that marketing knowledge dimensions had a positive impact relationship on competitive advantage except knowledge of customers ($\beta = 0.054, t = 0.267, p = 0.092$). However, the results indicated that the highest impact was for market knowledge ($\beta = 0.363, t = 2.418, p = 0.001$), followed by knowledge of products ($\beta = 0.251$, $t = 2.161$, $p = 0.003$), then knowledge of suppliers ($\beta = 0.190$, $t = 2.209$, $p = 0.002$), and finally the lowest impact was for knowledge of competitors ($\beta = 0.102$, $t = 1.474$, $p = 0.041$).

6 Discussion

The management of food industries companies is keen to deliver their products at a specific time to achieve financial return and gain customer satisfaction and loyalty, and to ensure quality and the ability to control activities and optimal use

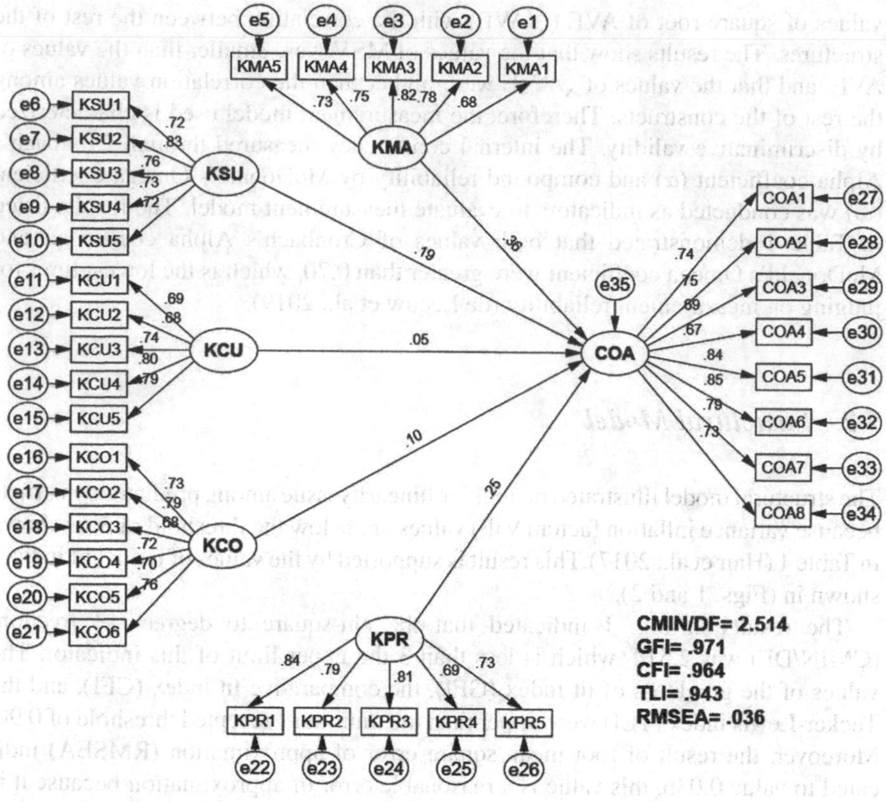

Fig. 2 SEM results of the marketing knowledge effect on competitive advantage

Table 2 Hypothesis testing

Hypothesis	Relation	Standard Beta	*t* value	*p* value	Result
H1	KMA→COA	0.363	2.418	0.001	Supported
H2	KSU→COA	0.190	2.209	0.002	Supported
H3	KCU→COA	0.054	0.267	0.092	Not supported
H4	KCO→COA	0.102	1.474	0.041	Supported
H5	KPR→COA	0.251	2.161	0.003	Supported

Note KMA: market knowledge, KSU: knowledge of suppliers, KCU: knowledge of customers, KCO: knowledge of competitors, KPR: knowledge of products, COA: competitive advantage
* $p < 0.05$, ** $p < 0.01$, *** $p < 0.001$

of its available resources to reach the largest segment, and to ensure that tasks are completed at the specified cost, and its ability to reduce prices and speed to respond to customers' requests, flexibility, simplification of transactions, and speed of response to customer requirements, which may create a competitive advantage over competitors. The results of the study showed that there is an impact of marketing knowledge on the competitive advantage in food industries companies in Jordan. It is clear to the researcher the impact of marketing knowledge on the competitive advantage, and that this effect stems from the keenness of the food industry companies to review, develop and constantly improve knowledge of the market, suppliers, customer, competitor, and product because of their conviction that these improvements with knowledge will benefit them in improving competitive advantage through accuracy of delivery, quality control, cost reduction and flexibility. This view is supported by Al-Mardini, Karthikeyan, Razaee and Jafari (2015). According to the results of the study, food industries companies in Jordan, through their officials, recognized the importance of providing effective channels of communication with the customer in the success of delivery and quality control of the various procedures. The keenness and commitment of the food industry companies to work with high efficiency reduce costs and achieve flexible procedures without delay.

7 Recommendations

Based on the findings of the study, researchers recommend managers and decision makers in food industries companies to provide the ability to process and analyze the information they possess in order to correctly predict market demands and study different markets. It also recommends to increase cooperation with suppliers and customers to follow up on new developments in the market. And maintain a database on suppliers and competitors for the purpose of constantly analyzing their environment and information. The study also recommends the need to train workers with multiple skills and abilities that make them able to perform multiple tasks, and work on continuous change in the design of production processes.

References

Abubakar, T., & Mohammad, H. (2019). Learning product line strategies to competitive advantage. *SEISENSE Journal of Management.*, *2*(4), 65–78.

Ahmad, S., Zulkurnain, N., & Khairushalimi, F. (2016). Assessing the validity and reliability of a measurement model in structural equation modeling (SEM). *British Journal of Mathematics & Computer Science, 15*(3), 1–8. https://doi.org/10.9734/BJMCS/2016/25183

Al Kurdi, B., Alshurideh, M., & Al afaishat, T. (2020a). Employee retention and organizational performance: Evidence from banking industry. *Management Science Letters, 10*(16), 3981–3990.

Al Kurdi, B., Alshurideh, M., & Alnaser, A. (2020b). The impact of employee satisfaction on customer satisfaction: Theoretical and empirical underpinning. *Management Science Letters, 10*(15), 3561–3570.

Al Mehrez, A. A., Alshurideh, M., Al Kurdi, B., & Salloum, S. A. (2020). Internal factors affect knowledge management and firm performance: a systematic review. In *International Conference on Advanced Intelligent Systems and Informatics* (pp. 632–643). Springer, Cham.

Alaali, N., Al Marzouqi, A., Albaqaeen, A., Dahabreh, F., Alshurideh, M., Alrwashdh, S., Iyadeh, I., Salloum, S., Aburayya, A. (2021). The impact of adopting corporate governance strategic performance in the tourism sector: A case study in the kingdom of Bahrain. *Journal of Legal, Ethical and Regulatory, 24* (Special Issue 1), 1–18.

Al-Adamat, A., Al-Gasawneh, J., & Al-Adamat, O. (2020). The impact of moral intelligence on green purchase intention. *Management Science Letters, 10*(9), 2063–2070.

Al-Benaa, H. M. Q. (2012). *The impact of organizational empowerment strategy on knowledge management processes in Jordanian commercial banks, unpublished doctoral thesis*. Jordan, Amman Arab University.

Al-Dhuhouri, F. S., Alshurideh, M., Al Kurdi, B., & Salloum, S. A. (2020). Enhancing our understanding of the relationship between leadership, team characteristics, emotional intelligence and their effect on team performance: A critical review. In *International Conference on Advanced Intelligent Systems and Informatics* (pp. 644–655). Springer, Cham.

Al-Dmour, R., AlShaar, F., Al-Dmour, H., Masa'deh, R., & Alshurideh, M. T. (2021). The effect of service recovery justices strategies on online customer engagement via the role of "Customer Satisfaction" During the Covid-19 Pandemic: An empirical study. *The Effect of Coronavirus Disease (COVID-19) on Business Intelligence, 334*, 325–346.

Al-Gasawneh, J. A., & Al-Adamat, A. M. (2020). The relationship between perceived destination image, social media interaction and travel intentions relating to Neom city. *Academy of Strategic Management Journal, 19*(2), 1–12.

Alhalalmeh, M. I., Almomani, H. M., Altarifi, S., Al- Quran, A. Z., Mohammad, A. A., & Al-Hawary, S. I. (2020). The nexus between corporate social responsibilty and organizational performance in Jordan: The mediating role of organizational commitment and organizational citizenship behavior. *Test Engineering and Management, 83*(July), 6391–6410.

AlHamad, A., Alshurideh, M., Alomari, K., Kurdi, B., Alzoubi, H., Hamouche, S., & Al-Hawary, S. (2022). The effect of electronic human resources management on organizational health of telecommunications companies in Jordan. *International Journal of Data and Network Science, 6*(2), 429–438.

Al-Hawary, S. I. (2015). Human resource management practices as a success factor of knowledge management implementation at health care sector in Jordan. *International Journal of Business and Social Science, 6*(11/1), 83–98.

Al-Hawary, S. I., & Al-Namlan, A. (2018). Impact of electronic human resources management on the organizational learning at the private hospitals in the State of Qatar. *Global Journal of Management and Business Research: A Administration and Management, 18*(7), 1–11.

Al-Hawary, S. I. S., & Alhajri, T. M. S. (2020). Effect of electronic customer relationship management on customers' electronic satisfaction of communication companies in Kuwait. *Calitatea, 21*(175), 97–102.

Al-Hawary, S. I. S., & Alwan, A. M. (2016). Knowledge management and its effect on strategic decisions of Jordanian Public Universities. *Journal of Accounting-Business & Management, 23*(2), 24–44.

Al-Hawary, S. I. S., & Obiadat, A. A. (2021). Does mobile marketing affect customer loyalty in Jordan? *International Journal of Business Excellence, 23*(2), 226–250.

Al-Hawary, S. I. S., Mohammad, A. S., Al-Syasneh, M. S., Qandah, M. S. F., & Alhajri, T. M. S. (2020). Organisational learning capabilities of the commercial banks in Jordan: Do electronic human resources management practices matter? *International Journal of Learning and Intellectual Capital, 17*(3), 242–266.

Al-Hawary, S. I., & Abu-Laimon, A. A. (2013). The impact of TQM practices on service quality in cellular communication companies in Jordan. *International Journal of Productivity and Quality Management, 11*(4), 446–474.

Al-Hawary, S. I., & Al-Syasneh, M. S. (2020). Impact of dynamic strategic capabilities on strategic entrepreneurship in presence of outsourcing of five stars hotels in Jordan. *Business: Theory and Practice, 21*(2), 578–587.

Al-Hawary, S. I., & Harahsheh, S. (2014). Factors affecting jordanian consumer loyalty toward cellular phone brand. *International Journal of Economics and Business Research, 7*(3), 349–375.

Al-Hawary, S. I., Al-Hawajreh, K., AL-Zeaud, H., & Mohammad, A. (2013). The impact of market orientation strategy on performance of commercial banks in Jordan. *International Journal of Business Information Systems, 14*(3), 261–279.

Al-Hawary, S. I., Batayneh, A. M., Mohammad, A. A., & Alsarahni, A. H. (2017). Supply chain flexibility aspects and their impact on customers satisfaction of pharmaceutical industry in Jordan. *International Journal of Business Performance and Supply Chain Modelling, 9*(4), 326–343.

Al-Hawary, S. I., Mohammad, A. A., & Al-Shoura, M. (2011). The impact of E-marketing on achieving competitive advantage by the Jordanian Pharmaceutical Firms. *DIRASAT*, *38*(1), 143–160.

Allahow, T. J. A. A., Al-Hawary, S. I. S., & Aldaihani, F. M. F. (2018). Information technology and administrative innovation of the central agency for information technology in Kuwait. *Global Journal of Management and Business, 18*(11-A), 1–16.

Al-Lozi, M. S., Almomani, R. Z. Q., & Al-Hawary, S. I. S. (2018). Talent Management strategies as a critical success factor for effectiveness of Human Resources Information Systems in commercial banks working in Jordan. *Global Journal of Management and Business Research: A Administration and Management, 18*(1), 30–43.

Al-Lozi, M., Almomani, R. Z., & Al-Hawary, S. I. (2017). Impact of talent management on achieving organizational excellence in Arab Potash Company in Jordan. *Global Journal of Management and Business Research: A Administration and Management, 17*(7), 15–25.

Al-Maroof, R., Ayoubi, K., Alhumaid, K., Aburayya, A., Alshurideh, M., Alfaisal, R., & Salloum, S. (2021). The acceptance of social media video for knowledge acquisition, sharing and application: A comparative study among YouYube users and TikTok users' for medical purposes. *International Journal of Data and Network Science, 5*(3), 197–214.

Al-Nady, B. A., Al-Hawary, S. I., & Alolayyan, M. (2013). Strategic management as a key for superior competitive advantage of sanitary ware suppliers in kingdom of Saudi Arabia. *International Journal of Management and Information Technology, 7*(2), 1042–1058.

Al-Nady, B. A., Al-Hawary, S. I., & Alolayyan, M. (2016). The role of time, communication, and cost management on project management success: An empirical study on sample of construction projects customers in Makkah City, Kingdom of Saudi Arabia. *International Journal of Services and Operations Management, 23*(1), 76–112.

Alolayyan, M., Al-Hawary, S. I., Mohammad, A. A., & Al-Nady, B. A. (2018). Banking service quality provided by commercial banks and customer satisfaction. A structural equation modelling approaches. *International Journal of Productivity and Quality Management*, *24*(4), 543–565.

Alshurideh, D. M. (2019). Do electronic loyalty programs still drive customer choice and repeat purchase behaviour? *International Journal of Electronic Customer Relationship Management, 12*(1), 40–57.

Alshurideh, M. (2022). Does electronic customer relationship management (E-CRM) affect service quality at private hospitals in Jordan? *Uncertain Supply Chain Management, 10*(2), 325–332.

Alshurideh, M. T., Al Kurdi, B., & Salloum, S. A. (2021). The moderation effect of gender on accepting electronic payment technology: A study on United Arab Emirates consumers. *Review of International Business and Strategy, 31*(3), 375–396.

Alshurideh, M. T., Al Kurdi, B., Alzoubi, H. M., Ghazal, T. M., Said, R. A., AlHamad, A. Q., ... & Al-kassem, A. H. (2022). Fuzzy assisted human resource management for supply chain management issues. *Annals of Operations Research*, 1–19.

Alshurideh, M., Al-Hawary, S. I., Batayneh, A. M., Mohammad, A. A., & Al-Kurdi, B.(2017). The impact of Islamic banks' service quality perception on Jordanian customers loyalty. *Journal of Management Research, 9*(2), 139–159.

Alshurideh, M., Gasaymeh, A., Ahmed, G., Alzoubi, H., & Kurd, B. (2020). Loyalty program effectiveness: Theoretical reviews and practical proofs. *Uncertain Supply Chain Management, 8*(3), 599–612.

Alshurideh, M., Nicholson, M., & Xiao, S. (2012). The effect of previous experience on mobile subscribers' repeat purchase behaviour. *European Journal of Social Sciences, 30*(3), 366–376.

Altamony, H., Masa'deh, R., Alshurideh, M., Obeidat, B. (2012). Information systems for competitive advantage: Implementation of an organisational strategic management process. Innovation and Sustainable Competitive Advantage: From Regional Development to World Economies, 583–592.

AlTaweel, I. R., & Al-Hawary, S. I. (2021). The mediating role of innovation capability on the relationship between strategic agility and organizational performance. *Sustainability, 13*(14), 7564.

Alwan, M., & Alshurideh, M. (2022). The effect of digital marketing on purchase intention: Moderating effect of brand equity. *International Journal of Data and Network Science, 10*(3), 1–12.

Alzoubi, H., Alshurideh, M., Akour, I., Al Shraah, A., & Ahmed, G. (2021) Impact of information systems capabilities and total quality management on the cost of quality. *Journal of Legal, Ethical and Regulatory Issues, 24*(Special Issue 6), 1–11.

Alzoubi, H., Alshurideh, M., Kurdi, B., Akour, I., & Aziz, R. (2022). Does BLE technology contribute towards improving marketing strategies, customers' satisfaction and loyalty? The role of open innovation. *International Journal of Data and Network Science, 6*(2), 449–460.

Baah, C., & Jin, Z. (2019). Sustainable supply chain management and organizational performance: The intermediary role of competitive advantage. *Journal of Management & Sustainability, 9*, 119.

Baroto, M., Abdullah, M., & Wan, H. (2012). Hybrid strategy: A new strategy for competitive advantage. *International Journal Business and Management, 7*(20), 120–133.

Becerra-Fernandez, I., & Sabherwal, R. (2014). Knowledge management: Systems and processes. Routledge.

Campeanu-Sonea, E., & Sonea, A. (2016). Knowledge management competence and introduction to the organization. Managerial challenges of the contemporary society. *Proceedings, 9*(1), 84-91.

de Leeuw, E., Hox, J., Silber, H., Struminskaya, B., & Vis, C. (2019). Development of an international survey attitude scale: Measurement equivalence, reliability, and predictive validity. *Measurement Instruments for the Social Sciences, 1*(1), 9. https://doi.org/10.1186/s42409-019-0012-x

Distanont, A., & Khongmalai, O. (2018). The role of innovation in creating a competitive advantage. *Kasetsart Journal of Social Sciences*, S2452315118300080https://doi.org/10.1016/j.kjss.2018.07.009

Hair, J. F., Babin, B. J., & Krey, N. (2017). Covariance-based structural equation modeling in the journal of advertising: Review and recommendations. *Journal of Advertising, 46*(1), 163–177. https://doi.org/10.1080/00913367.2017.1281777

Hasan, O., McColl, J., Pfefferkorn, T., Hamadneh, S., Alshurideh, M., & Kurdi, B. (2022). Consumer attitudes towards the use of autonomous vehicles: Evidence from United Kingdom taxi services. *International Journal of Data and Network Science, 6*(2), 537–550.

Hasniaty, O, Asran, E, Basmar, E (2018). Marketing performance in improving competitive advantages of macro business. 1st International Conference on Materials Engineering and Management Section (ICMEMm 2018) Advances in Economics, Business and Management Research, 1(75): 1–30.

Howard, M. C. (2018). The convergent validity and nomological net of two methods to measure retroactive influences. *Psychology of Consciousness: Theory, Research, and Practice, 5*(3), 324–337. https://doi.org/10.1037/cns0000149

Husseini, A., Soltani, S., & Mehdizadeh, M. (2018). Competitive advantage and its impact on new product development strategy (case study: Toos niro technical firms. *Journal of Open Innovation: Technology, Market and Complexity, 4*(17), 1–12.

Khasawneh, M. A., Abuhashesh, M., Ahmad, A., Masa'deh, R., & Alshurideh, M. T. (2021). Customers online engagement with social media influencers' content related to COVID 19. In The Effect of Coronavirus Disease (COVID-19) on Business Intelligence (pp. 385–404). Springer, Cham.

Korankye, A. (2013). Total quality management a source of competitive advantage, a comparative study of manufacturing and service firms in Ghana. *International Journal of Asian Social Science, 3*(6), 1293–1305.

Laudon, K. C., & Laudon, J. P. (2015). Management information systems .Upper Saddle River: Pearson.

Lee, K., Azmi, N., Hanaysha, J., Alshurideh, M., & Alzoubi, H. (2022a). The effect of digital supply chain on organizational performance: An empirical study in Malaysia manufacturing industry. *Uncertain Supply Chain Management, 10*(2), 1–16.

Lee, K., Ramiz, P., Hanaysha, J., Alzoubi, H., & Alshurideh, M. (2022b). Investigating the impact of benefits and challenges of IOT adoption on supply chain performance and organizational performance: An empirical study in Malaysia. *Uncertain Supply Chain Management, 10*(2), 1–14.

Macintosh, A. (1998). Position paper on knowledge asset management. www.aiai.edu.ac.uk/,alm/kam.html. Cited on 25/11/2020.

Marcek, P., & Mucha, M. (2015). The use of knowledge management in marketing communication of small and medium–sized companies, procedia. *Social and Behavioral Sciences, 175*, 182–192.

Mohammad and saleem khlif Alshura, M., Al-Hawary, S. I. S., Al-Syasneh, M. S., & Alhajri, T. M. S. , 2020Mohammad, A. A. S., saleem khlif Alshura, M., Al-Hawary, S. I. S., Al-Syasneh, M. S., & Alhajri, T. M. S. (2020). The influence of Internal Marketing Practices on the employees' intention to leave: A study of the private hospitals in Jordan. *International Journal of Advanced Science and Technology, 29*(5), 1174–1189

Munizu, M. (2013). The impact of total quality management practices towards competitive advantage and organizational performance: Case of fishery industry in South. Sulawesi Province of Indonesia, Pakistan. *Journal of Commerce and Social Sciences, 7*(1), 184–197.

Na, Y. K., Kang, S., & Jeong, H. Y. (2019). The effect of market orientation on performance of sharing economy business: Focusing on marketing innovation and sustainable competitive advantage. *Sustainability, 11*(3), 729.

Obeidat, U., Obeidat, B., Alrowwad, A., Alshurideh, M., Masadeh, R., & Abuhashesh, M. (2021). The effect of intellectual capital on competitive advantage: The mediating role of innovation. *Management Science Letters, 11*(4), 1331–1344.

Rezaee, F., & Jafari, M. (2015). The effect of marketing knowledge management on sustainable competitive advantage: Evidence from banking industry. *Accounting, 1*(2), 69–88.

Rimkeviciene, J., Hawgood, J., O'Gorman, J., & De Leo, D. (2017). Construct validity of the acquired capability for suicide scale: Factor structure, convergent and discriminant validity. *Journal of Psychopathology and Behavioral Assessment, 39*(2), 291–302. https://doi.org/10.1007/s10862-016-9576-4

Sekaran, U., & Bougie, R. (2016). *Research methods for business: A skill-building approach* (Seventh edition). Wiley.

Semaskiene, T., & Stancikiene, A. (2014). Influence of knowledge management on the competitiveness of enterprises. *Journal of Social Studies, 6*(3), 557–578.

Shamout, M., Ben-Abdallah, R., Alshurideh, M., Alzoubi, H., Kurdi, B., & Hamadneh, S. (2022). A conceptual model for the adoption of autonomous robots in supply chain and logistics industry. *Uncertain Supply Chain Management, 10*(2), 577–592.

Shi, D., Lee, T., & Maydeu-Olivares, A. (2019). Understanding the model size effect on SEM Fit indices. *Educational and Psychological Measurement, 79*(2), 310–334. https://doi.org/10.1177/0013164418783530

Sung, K.-S., Yi, Y. G., & Shin, H.-I. (2019). Reliability and validity of knee extensor strength measurements using a portable dynamometer anchoring system in a supine position. *BMC Musculoskeletal Disorders, 20*(1), 1–8. https://doi.org/10.1186/s12891-019-2703-0

Sweiss, N., Obeidat, Z. M., Al-Dweeri, R. M., Mohammad Khalaf Ahmad, A., M. Obeidat, A., & Alshurideh, M. (2021). The moderating role of perceived company effort in mitigating customer misconduct within Online Brand Communities (OBC). *Journal of Marketing Communications*, 1–24.

Wang, Y. A., & Rhemtulla, M. (2021). Power analysis for parameter estimation in structural equation modeling: A discussion and tutorial. *Advances in Methods and Practices in Psychological Science, 4*(1), 1–17. https://doi.org/10.1177/2515245920918253

Yasa, N. N. K., Giantari, I. G. A. K., Setini, M., & Rahmayanti, P. L. D. (2020). The role of competitive advantage in mediating the effect of promotional strategy on marketing performance. *Management Science Letters*, 2845–2848https://doi.org/10.5267/j.msl.2020.4.024

Zebal, M., Ferdous, A., & Chambers, C. (2019). An integrated model of marketing knowledge–a tacit knowledge perspective. *Journal of Research in Marketing and Entrepreneurship, 21*(1), 2–18.

The Effect of Talent Management on Organizational Innovation of the Telecommunications Companies in Jordan

Reham Zuhier Qasim Almomani, Saleem Sameeh Saleem AL-khaldi, Ali Zakariya Al-Quran, Hanan Mohammad Almomani, Fatima Lahcen Yachou Aityassine, Mohammad Mousa Eldahamsheh, Fuad N. Al-Shaikh, Muhammad Turki Alshurideh, Anber Abraheem Shlash Mohammad, and Sulieman Ibraheem Shelash Al-Hawary

Abstract This study aimed to identify the impact of talent management on Organizational Innovation. The study population consisted of 370 employees distributed over three companies representing the telecommunications sector in Jordan. An electronic

R. Z. Q. Almomani
Business Administration, Amman, Jordan

S. S. S. AL-khaldi · A. Z. Al-Quran · H. M. Almomani · S. I. S. Al-Hawary (✉)
Department of Business Administration, School of Business, Al al-Bayt University, P.O. Box 130040, Mafraq 25113, Jordan
e-mail: dr_sliman73@aabu.edu.jo; dr_sliman@yahoo.com

S. S. S. AL-khaldi
e-mail: ali.z.al-quran@aabu.edu.jo

A. Z. Al-Quran
e-mail: ali.z.al-quran@aabu.edu.jo

H. M. Almomani
e-mail: mohammad197119@yahoo.com

F. L. Y. Aityassine
Department of Financial and Administrative Sciences, Irbid University College, Al-Balqa' Applied University, As-Salt, Jordan
e-mail: Fatima.yassin@bau.edu.jo

M. M. Eldahamsheh
Strategic Management, Aqaba, Jordan

F. N. Al-Shaikh
Faculty of Economics and Administrative Sciences, Department of Business Administration, Yarmouk University, P.O. Box 566, Irbid 21163, Jordan
e-mail: eco_fshaikh@yu.edu.jo

M. T. Alshurideh
Department of Marketing, School of Business, The University of Jordan, Amman 11942, Jordan
e-mail: m.alshurideh@ju.edu.jo; malshurideh@sharjah.ac.ae

M. Alshurideh et al. (eds.), *The Effect of Information Technology on Business and Marketing Intelligence Systems*, Studies in Computational Intelligence 1056,
https://doi.org/10.1007/978-3-031-12382-5_97

questionnaire was distributed via (Google Forms) and shared by talent managers at Jordanian telecommunications companies, and 257 responses were valid for statistical analysis. After conducting a statistical analysis, the study reached the presence of a statistically significant effect of talent management has a statistically significant effect on Organizational Innovation in Jordanian Telecommunications companies. The study recommended giving attention to talent management, as it represents the first driver of all work activities.

Keywords Talent management · Organizational innovation · Mobile communications companies · Jordan

1 Introduction

Talent management as one of human resources field seeks to build and develop the idea and concept of focus on talent culture as a major competitive advantage source, as companies that possess talented and distinguished work force can develop their organizational culture to be an attractive place to work, a source of employees' pride, build loyalty and trust between existing employees (Abuhashesh et al., 2021; Odeh et al., 2021). Talent is an organizational important issue, as organizations are competing to include distinguished talent in light of the supply of job seekers increase and talent in qualitative terms scarcity (Al-Lozi et al, 2017; Al-bawaia et al., 2022). The human element, talent, was one of the main organizational characteristics that influence innovative behaviors and capabilities in service-oriented organizations (Almaazmi et al., 2020; AlTaweel & Al-Hawary, 2021). These talents include all the acquired skills that employees display from all of their training experiences (Al Suwaidi et al., 2020; Nuseir et al., 2021). Talent management represents the energy that moves and pushes talents towards creative and distinguished work and increases organizational attractiveness. Organizations' adoption of talent management leads to building of employee's capabilities, and at the same time, effective performance that builds competitive advantage depends on the competence and capabilities of these individuals (Al-Hawary & Nusair, 2017; Al-Hawary & Al-Namlan, 2018; Mohammad et al., 2020). Organizations strive to excel constantly, and this requires them to continuously improve their capabilities to suit their client's needs and expectations (Al-Hawary & Abu-Laimon, 2013; Al-Hawary & Al-Smeran, 2017; Al-Hawary & Obiadat, 2021; Al-Hawary et al., 2013; Alolayyan et al., 2018; Alshurideh et al., 2017). This improvement requires skilled, distinguished, trained and talented human resources, in addition to the presence of specific strategies that help them complete their work, which indicates talent management strategies importance in high levels

Department of Management, College of Business, University of Sharjah, 27272 Sharjah, United Arab Emirates

A. A. S. Mohammad
Faculty of Administrative and Financial Sciences, Marketing Department, Petra University, P.O. Box 961343, Amman 11196, Jordan

of achievement, which is positively reflected on the organizational innovation and its members in order to achieve high levels of strategic creativity in the short and long term (Alhalalmeh et al., 2020; Cappelli, 2008; Al-Hawary & Al-Syasneh, 2020; Al Kurdi et al., 2020a, 2020b; Alshurideh, 2022).

The researcher noted the discrepancy in telecommunications companies provided services quality and consequently the diversity processes involved in providing the service, and because the telecommunications sector targets all segments of Jordanian society, the circumstance must find human cadres with high flexibility capable of dealing with all the conditions created by the surrounding environment and finding the best ways to reach excellence at the level of service delivery and operations. In view of the fierce competition between Jordanian telecom networks, they strives to own what distinguishes them and makes it unique, whether it is a customer service that represents a superior response and meets the needs or a distinct process that achieves the customers value. Or strategies that are difficult for competitors to imitate. Therefore, we will start here to see the impact of talent management on the organizational innovation.

2 Literature Review and Hypotheses Development

2.1 Talent Management

The term "talent" has different definitions according to different fields of study. Nurhadi and zahro (2016) define a set of different definitions of "talent" such as mastery of abilities and skills, individual ideas to generate creative ideas, set of values, intrinsic endowment, and cognitive skills or enhanced competencies that allow employees to perform in an excellent manner. The term talent management is defined as the process of attracting, selecting, developing and retaining talent for the best employees in a strategic organizational position (Sculllion & Collings, 2011). As Cunningham (2007) says that talent management typically means high performance and a resource that must be managed mainly according to performance levels, therefore talent individuals must be sought, appointed and rewarded differently, regardless of the special roles they play or even organization special needs.

As Bhatnagar (2007) points out that talent management is the priority for rapidly attracting organization attention across the world, talent management is primarily designed to improve individuals recruitment and development process in order to meet the current organizational needs as the various characteristics of talent management are recruitment and selection Supervision, performance management, career development, leadership development, planning alternatives, excellence and reward. talent management importance appears in what it achieves for the organization, as it is considered a source of business organizations excellence, and the organizations' practice of talent management processes from attracting, developing, motivating and sustaining talented employees helps in achieving the organization's current and future goals, in addition to talent management importance appears in its ability to provide departments and sections needs of human resources who working responsibly, flexibly and quickly in decisions making. (Alameeri et al., 2020; AlShehhi et al., 2020; Grapragesem et al., 2014).

As indicated by Jones (2010) that talent management aims to build a culture that helps generate commitment and build the capabilities and competencies necessary to possess the integrated talent, using the distribution and development process that aligns with organizations objectives. Researchers have differed about talent management dimensions due to the multiplicity of views on this concept, as Stahl et al. (2007) sees that it includes talents recruitment, succession planning, training and talents development, and talent retention, while Collings and Mellahi (2009) indicates that it includes identifying vital jobs, develop and grow the talent pool, create a structure of diverse human resources, to measuring talent management Kumari and Bahuguna (2012) relied on several dimensions: identifying vital job positions, aggregating talent, attracting talent, managing talent performance, developing talent, disseminating talent, and retaining talent. Buthelezi (2010) identified several dimensions: Manpower planning, talent attraction and selection, career path management, performance management, talent development, reward and recognition.

As for this study, the study relied on a set of dimensions to measure talent management: talent attraction, talent education and development, talent training, talent performance management, and talent retention. Attracting talents: the process of identifying and selecting talent requires the use of multiple techniques to collect data on talented individuals, including the central assessment methodology, psychometric tests, talent data integration, and finally talent maps that represent the result of the talent identification and selection process (Al-Hawary, 2011; Egerova et al., 2013). Talent training: training refers to achieving the competency, information, and ability needed to carry out activities through training (Al-Hawary, 2015; Kum et al., 2014; Shamout et al., 2022; Alshurideh et al., 2022). It is evident in the literature that organizations that invests in talent training, talent development and continuous learning programs is more likely to attract and retain employees through its intent to eliminate employee turnover, it also has an economic and performance advantage from these employees. Talent education and development is the high potential individuals and future senior managers' development for what it wants from its senior leaders in the future, talent performance management (Al-Hawary et al., 2020; Alshamsi et al., 2020; Metabis & Al-Hawary, 2013). This step refers to performance measurement, that is, the process of planning, managing and evaluating employee's performance over time. This step is important in a talent management system for simple reason that management does not want to promote people whose current jobs are not performing effectively, in order to maintain credibility (Al-Hawary & Alajmi, 2017; Alkitbi et al., 2020). Finally, retaining talents by organizing an encouraging environment for talented people to present their suggestions through holding seminars, exchanging creative ideas and innovative skills, and forming development plans to support employee performance. It helps create a stronger working relationship between the employee and his manager that enhances loyalty and retention.

2.2 *Organizational Innovation*

Innovation refers to the ability to find new things that may be ideas, solutions, products or services, or useful ways and methods of work (Alameeri et al., 2021; AlTaweel & Al-Hawary, 2021). "Innovation is the ability to provide unique answers to posed problems and exploit opportunities." Organizational innovation becomes the main factor in the existence of the company in a competitive environment where the successful introduction of new products is the lifeblood of most organizations (Al-Hawary & Aldaihani, 2016; Tohidi & Jabbari, 2012). The importance of organizational innovation appears in that it helps the organization work well by improving coordination, internal control and the organizational structure (Alzoubi et al., 2022; Ghannajeh et al., 2015). It also leads to facilitating administrative processes that enable the organization to continue its work and find creative solutions to the problems it faces efficiently and effectively, and to bring positive changes in the institution's and its administrative processes, and helps it adapt and interact with all the surrounding environmental variables, improve its productivity, raise its performance levels and employees performance, and find discoveries, proposals and ideas for the development of new, innovative and creative systems, regulations, procedures and work methods, which results in the emergence of programs and services outside the main enterprise activities, improving the products and services quality provided to customers benefiting from the service, and increasing its financial input (Al-Nady et al., 2013; Allahow et al., 2018; Alzoubi et al., 2020). Organizational innovation is an integral part of the culture of any organization that seeks success, as it occupies the heart of its activities and operations (Al Naqbi et al., 2020; Al-Hawary & Alwan, 2016). "Innovation in organizations creates the appropriate climate that enables the organization to be able to develop new products to satisfy customers' needs in the market on the one hand, and the ability to achieve the growth goals sought by the organization on the other hand."

According to Walker (2008) organization innovations are innovations in structure, strategy and managerial processes. While Afuah (1998) classified organizational innovation into three categories: The first is technological, including products, services, and processes. The second is the market including product, price, place and promotion and the third is organizational and managerial including strategy, structure, systems and people. This research is based on the use of the following dimensions of organizational innovation, service innovation, which is new idea introduction that focus on services that provide new ways to provide benefit, new service concepts, or new service business models through continuous operational improvement (Enz, 2012).

Product innovation: introducing a new product, introducing new functions, improving performance, or adding new features to existing products; Consumers do not know it (Susman & Warren, 2006). Process innovation is new production methods introduction, management methods and techniques that can be used to improve production and management processes. And managerial innovation, which is often specifically related to strategies, structure, policies, or people within the organization (Popadiuk & Choo, 2006). It is a set of clear and well-defined approaches and processes aimed at discovering and managing knowledge functions, in various types of operations, identifying new products or strategies, strengthening human resource management, and achieving a number of other goals to be achieved.

2.3 The Relationship Between Talent Management and Organizational Innovation

Many studies have emphasized the relationship of talent management and organizational innovation. Attracting talent is one of the important things that determine organizational performance, and thus attracting talent has a significant positive impact on organizational performance. Abou-moghli (2019) emphasized that innovation strategies contribute significantly to enhancing organizational competitive advantage. Talent management is a catalyst for organizational performance improvement because the process of innovation, retention and attraction is a significant influence on organizational effectiveness. Retention, participation, and attraction influence the innovation process and organizational effectiveness. The investment in talent management and development enriches the intellectual capital, which is a great balance of competitive advantage and organizational creativity (Abd El Rahman & Farghaly, 2019). Mohammed et al. (2018) confirmed with his study on talent management that there is great competition between organizations in modern technology area, which has led to an increase in knowledgeable employees in addition to huge changes in the market. Consequently, academic institutions have begun to rethink their procedures and policies to best attract, develop and retain these employees. Olaka determines the extent of the relationship between talent management and creativity, and the results of his study showed that there are positive relationships between the variables of talent management and the variables corresponding to creativity. The researcher adds that the proper identification, development, use and retention of talent is likely to translate into higher innovations in terms of services, processes, marketing and products and this will also lead to improved organizations performance and profitability. Based on the foregoing, study hypothesis can be formulated as follows.

There is a statistically significant effect of talent management on organizational innovation of Jordanian telecom companies.

3 Study Model

See Fig. 1.

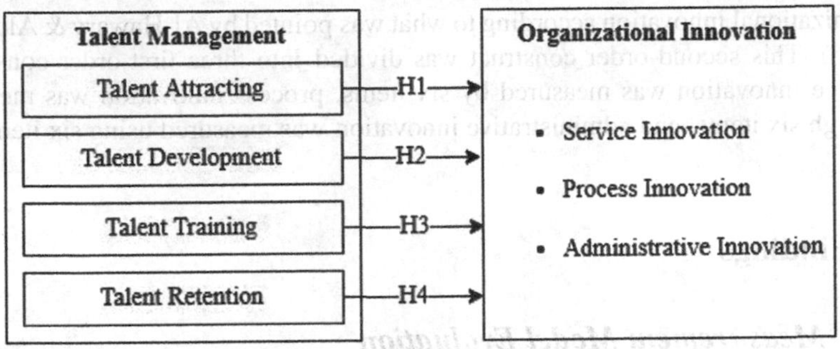

Fig. 1 Research model

4 Methodology

4.1 Population and Sample Selection

A qualitative method based on a questionnaire was used in this study for data collection and sample selection. The major aim of the study was to examine the impact of talent management on organizational innovation. Therefore, it focused on Jordan-based companies operating in the telecommunications sector. Data were primarily gathered through self-reported questionnaires creating by Google Forms which were distributed to a purposive sample of (370) HR managers via email. In total, (282) responses were received including (25) invalid to statistical analysis due to uncompleted or inaccurate. Hence, the final sample contained (257) responses suitable to analysis requirements that were formed a response rate of (69.4%), where it proved to be sufficient to the extent that was predictable and allowed for a presumption of data saturation (Sekaran & Bougie, 2016).

4.2 Measurement Instrument

A self-reported questionnaire that consists of two main sections along with a section regarding control variables was used as the measurement instrument. Control variables considered as categorical measures were composed of gender, age group, educational level, and experience. The two main sections were dealt with a five-point Likert scale (from 1 = strongly disagree to 5 = strongly agree). The first section contained (20) items to measure talent management based on Al-Lozi et al. (2018). These questions were distributed into dimensions as follows: five items dedicated for measuring talent attracting, five items dedicated for measuring talent development, five items dedicated for measuring talent training, and five items dedicated for measuring talent retention. Whereas the second section included (18) items developed to measure organizational innovation according to what was pointed by Al-Hawary & Aldaihani (2016). This second-order construct was divided into three first-order constructs. Service innovation was measured by six items, process innovation was measured through six items, and administrative innovation was measured using six items.

5 Findings

5.1 *Measurement Model Evaluation*

This study was conducted structural equation modeling (SEM) to test hypotheses, which represents a contemporary statistical technique for testing and estimating the relationship between factors and variables (Al-Adamat et al., 2020; Al-Gasawneh & Al-Adamat, 2020; Wang & Rhemtulla, 2021). Accordingly, the reliability and validity of the constructs were tested using confirmatory factor analysis (CFA) through the statistical program AMOSv24. Table 1 summarizes the results of convergent and discriminant validity, as well the indicators of reliability.

Table 1 shows that the standard loading values for the individual items were within the domain (0.651–0.901), these values greater than the minimum retention of the elements based on their standard loads (Al-Lozi et al., 2018; Sung et al., 2019). Average variance extracted (AVE) is a summary indicator of the convergent validity of constructs that must be above 0.50 (Howard, 2018). The results indicate that the AVE values were greater than 0.50 for all constructs, thus the used measurement model has an appropriate convergent validity. Rimkeviciene et al. (2017) suggested the comparison approach as a way to deal with discriminant validity assessment in covariance-based SEM. This approach is based on comparing the values of maximum shared variance (MSV) with the values of AVE, as well as comparing the values of square root of AVE ($\sqrt{AVE}$) with the correlation between the rest of the structures. The results show that the values of MSV were smaller than the values of AVE, and that the values of $\sqrt{AVE}$ were higher than the correlation values among the rest of the constructs. Therefore, the measurement model used is characterized by discriminative validity. The internal consistency measured through Cronbach's Alpha coefficient (α) and compound reliability by McDonald's Omega coefficient (ω) was conducted as indicators to evaluate measurement model. The results listed in Table 1 demonstrated that both values of Cronbach's Alpha coefficient and McDonald's Omega coefficient were greater than 0.70, which is the lowest limit for judging on measurement reliability (De Leeuw et al., 2019).

Table 1 Results of validity and reliability tests

Constructs	1	2	3	4	5	6	7
1. Talent attracting	**0.777**						
2. Talent development	0.517	**0.769**					
3. Talent training	0.497	0.471	**0.783**				
4. Talent retention	0.438	0.442	0.502	**0.754**			
5. Service innovation	0.628	0.588	0.556	0.567	**0.766**		
6. Process innovation	0.678	0.625	0.637	0.598	0.671	**0.779**	
7. Administrative innovation	0.597	0.567	0.697	0.684	0.662	0.604	**0.744**
VIF	1.763	2.357	1.746	2.156	–	–	–
Loadings range	0.681–0.881	0.705–0.835	0.651–0.901	0.693–0.827	0.683–0.864	0.662–0.897	0.655–0.825
AVE	0.604	0.591	0.614	0.569	0.586	0.607	0.554
MSV	0.502	0.472	0.513	0.520	0.483	0.516	0.409
Internal consistency	0.881	0.875	0.884	0.866	0.892	0.899	0.878
Composite reliability	0.883	0.878	0.887	0.868	0.894	0.902	0.881

Note Bold fonts in the table refer to the square root of AVE

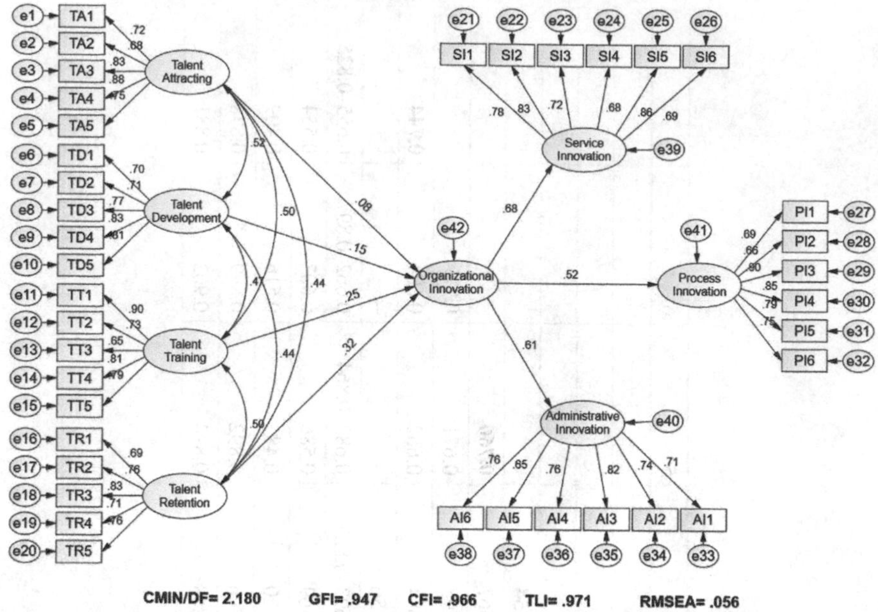

Fig. 2 SEM results of the talent management effect on organizational innovation

5.2 *Structural Model*

The structural model illustrated no multicollinearity issue among predictor constructs because variance inflation factor (VIF) values are below the threshold of 5, as shown in Table 1 (Hair et al., 2017). This result is supported by the values of model fit indices shown in Fig. 2.

The results in Fig. 2 indicated that the chi-square to degrees of freedom (CMIN/DF) was 2.180, which is <3 the upper limit of this indicator. The values of the goodness of fit index (GFI), the comparative fit index (CFI), and the Tucker-Lewis index (TLI) were upper than the minimum accepted threshold of 0.90. Moreover, the result of root mean square error of approximation (RMSEA) indicated to value 0.056, this value is a reasonable error of approximation because it is less than the higher limit of 0.08. Consequently, the structural model used in this study was recognized as a fit model for predicting the DEP and generalization of its result (Ahmad et al., 2016; Shi et al., 2019).To verify the results of testing the study hypotheses, structural equation modeling (SEM) was used, the results of which are listed in Table 2.

The results demonstrated in Table 2 show that all talent management dimensions had a positive impact relationship on organizational innovation except talent attracting ($\beta = 0.081$, $t = 1.302$, $p = 0.195$). However, the results indicated that the highest impact was for talent retention ($\beta = 0.316$, $t = 6.045$, $p = 0.000$), followed by talent training ($\beta = 0.248$, $t = 5.368$, $p = 0.000$), and finally the lowest impact was for talent development ($\beta = 0.148$, $t = 3.235$, $p = 0.040$).

Table 2 Hypothesis testing

Hypothesis	Relation	Standard Beta	*t* value	*p* value
H1	Talent Attracting → Organizational Innovation	0.081	1.302	0.195
H2	Talent Development → Organizational Innovation	0.148	3.235	0.040
H3	Talent Training → Organizational Innovation	0.248	5.368	0.000
H4	Talent Retention → Organizational Innovation	0.316	6.045	0.000

Note $*p < 0.05$, $**p < 0.01$, $***p < 0.001$

6 Discussion

It has been evident from the study results that the company under study is keen on developing, improving and training talents through its continuous endeavor to encourage talents to present ideas and proposals, involve them in educational workshops and conferences, and provide a distinctive training environment. Telecom companies seek to retain what they have of talent by providing material and moral incentives and opening doors for career development. study results also confirmed that company management concerned with talent management is not aware of the of attracting talent importance that they have not clear strategies for attracting employees and evaluate them carefully to discover and exploit their strengths. The results showed the interest of telecommunications companies in supporting talents to provide a high quality service that satisfies customers. This results agreed with the study of Emmy & Hillary, which showed that there is a significant positive impact of the variables (talent attraction and its dimensions are selection and polarization, employee promotion, job security) on job performance. And it appeared that Jordanian telecom companies are interested in supporting talents to present new ideas related to improving service delivery by automating procedures, modifying and improving their processes, and finding innovative ideas to reduce the operational cost of operations while raising their outputs. This result agreed with Al-Hiyasat study, where the study showed an impact of the talent management strategy (talent attraction, talent investment, talent development, talent development, talent retention) in achieving the competitive advantage. Study results showed that supporting, developing and retaining talent contributes greatly to innovation and creativity in presenting work methods, in addition to creating distinctive and renewable administrative policies. So that Jordanian telecommunications companies constantly seek to involve talented people in the decision-making process related to work goals. This result agreed with the study of Abou-Moghli (2019). The results showed an important and positive impact of the innovation process, retention and attraction on organizational effectiveness. In addition, the study found a positive and significant role in moderation of retention, engagement, and attraction in the innovation process and organizational effectiveness. Talent management in Jordanian telecommunications companies pays attention to the human resource for innovation and creation of creative ideas to provide high quality services. Talent management role is enhanced by attracting, developing and retaining talent in finding innovative and unique ways to enhance its competitive strength.

7 Recommendations

Based on the results, the study recommends that Jordanian telecommunications companies adopt clear employment strategies to attract talented people internally or externally and place them in the right place to increase creativity and enrich companies with unique and distinctive ideas. Encouraging talented people who are able to think innovatively by providing practical suggestions for creating new services, improving new services, and a new process. Emphasis on the education and development of talents and their development to enhance their creativity, allocate a budget to develop the capabilities of the gifted, and involve the gifted in educational workshops or conferences dealing with enhancing the quality of services and operations provided. As well as following up on employees strengths and weaknesses to fill the knowledge gap at all levels and qualify them to participate in presenting distinctive and unique ideas.

References

Abd El Rahman. R., & Farghaly, S. (2019). Application of optimis' talent management model for head nurses on nurses' job crafting and innovation. *Journal of Nursing and Health Science (IOSR-JNHS)*, 5(8), 81–95.

Abou-Moghli, A. A. (2019). Competitive innovation strategies and their effect on enhancing organizational effectiveness: Talent management as a moderator. *International Journal of Business and Management, 14*(4), 24–34.

Abuhashesh, M. Y., Alshurideh, M. T., & Sumadi, M. (2021). The effect of culture on customers' attitudes toward Facebook advertising: The moderating role of gender. *Review of International Business and Strategy, 31*(3), 416–437.

Afuah, A. (1998). *Innovation management: Strategies, implementation, and profits*. New York: Oxford University Press.

Ahmad, S., Zulkurnain, N., & Khairushalimi, F. (2016). Assessing the validity and reliability of a measurement model in structural equation modeling (SEM). *British Journal of Mathematics & Computer Science, 15*(3), 1–8. https://doi.org/10.9734/BJMCS/2016/25183

Al Kurdi, B., Alshurideh, M., & Al afaishat, T. (2020a). Employee retention and organizational performance: Evidence from banking industry. *Management Science Letters, 10*(16), 3981–3990.

Al Kurdi, B., Alshurideh, M., & Alnaser, A. (2020b). The impact of employee satisfaction on customer satisfaction: Theoretical and empirical underpinning. *Management Science Letters, 10*(15), 3561–3570.

Al Naqbi, E., Alshurideh, M., AlHamad, A., & Al Kurdi, B. (2020). The impact of innovation on firm performance: A systematic review. *International Journal of Innovation, Creativity and Change, 14*(5), 31–58.

Al Suwaidi, F., Alshurideh, M., Al Kurdi, B., & Salloum, S. A. (2020, October). The impact of innovation management in SMEs performance: A systematic review. In *International Conference on Advanced Intelligent Systems and Informatics* (pp. 720–730). Cham: Springer.

Al-Adamat, A., Al-Gasawneh, J., & Al-Adamat, O. (2020). The impact of moral intelligence on green purchase intention. *Management Science Letters, 10*(9), 2063–2070.

Alameeri, K. A., Alshurideh, M. T., & Al Kurdi, B. (2021). The effect of Covid-19 pandemic on business systems' innovation and entrepreneurship and how to cope with it: A theatrical view. *The Effect of Coronavirus Disease (COVID-19) on Business Intelligence, 334*, 275–288.

Alameeri, K., Alshurideh, M., Al Kurdi, B., & Salloum, S. A. (2020, October). The effect of work environment happiness on employee leadership. In *International Conference on Advanced Intelligent Systems and Informatics* (pp. 668–680). Cham: Springer.

Al-bawaia, E., Alshurideh, M., Obeidat, B., & Masa'deh, R. (2022). The impact of corporate culture and employee motivation on organization effectiveness in Jordanian banking sector. *Academy of Strategic Management Journal, 21*(Special Issue 2), 1–18.

Al-Gasawneh, J. A., & Al-Adamat, A. M. (2020). The relationship between perceived destination image, social media interaction and travel intentions relating to Neom city. *Academy of Strategic Management Journal, 19*(2), 1–12.

Alhalalmeh, M. I., Almomani, H. M., Altarifi, S., Al- Quran, A. Z., Mohammad, A. A., & Al-Hawary, S. I. (2020). The nexus between corporate social responsibilty and organizational performance in Jordan: The mediating role of organizational commitment and organizational citizenship behavior. *Test Engineering and Management, 83*, 6391–6410.

Al-Hawary, S. I., & Nusair, W. (2017). Impact of human resource strategies on perceived organizational support at Jordanian Public Universities. *Global Journal of Management and Business Research: A Administration and Management, 17*(1), 68–82.

Al-Hawary, S. I. (2015). Human resource management practices as a success factor of knowledge management implementation at health care sector in Jordan. *International Journal of Business and Social Science, 6*(11/1), 83–98.

Al-Hawary, S. I. (2011). Human resource management practices in ZAIN cellular communications company operating in Jordan. *Perspectives of Innovations, Economics and Business, 8*(2), 26–34.

Al-Hawary, S. I., & Alajmi, H. M. (2017). Organizational commitment of the employees of the ports security affairs of the state of Kuwait: The impact of human recourses management practices. *International Journal of Academic Research in Economics and Management Sciences, 6*(1), 52–78.

Al-Hawary, S. I., & Aldaihani, F. M. (2016). Customer relationship management and innovation capabilities of Kuwait Airways. *International Journal of Academic Research in Economics and Management Sciences, 5*(4), 201–226.

Al-Hawary, S. I., & Al-Namlan, A. (2018). Impact of electronic human resources management on the organizational learning at the private hospitals in the State of Qatar. *Global Journal of Management and Business Research: A Administration and Management, 18*(7), 1–11.

Al-Hawary, S. I. S., & Alwan, A. M. (2016). Knowledge management and its effect on strategic decisions of Jordanian Public Universities. *Journal of Accounting-Business & Management, 23*(2), 24–44.

Al-Hawary, S. I. S., & Obiadat, A. A. (2021). Does mobile marketing affect customer loyalty in Jordan? *International Journal of Business Excellence, 23*(2), 226–250.

Al-Hawary, S. I. S., Mohammad, A. S., Al-Syasneh, M. S., Qandah, M. S. F., & Alhajri, T. M. S. (2020). Organisational learning capabilities of the commercial banks in Jordan: Do electronic human resources management practices matter? *International Journal of Learning and Intellectual Capital, 17*(3), 242–266.

Al-Hawary, S. I., & Abu-Laimon, A. A. (2013). The impact of TQM practices on service quality in cellular communication companies in Jordan. *International Journal of Productivity and Quality Management, 11*(4), 446–474.

Al-Hawary, S. I., & Al-Smeran, W. (2017). Impact of electronic service quality on customers satisfaction of Islamic Banks in Jordan. *International Journal of Academic Research in Accounting, Finance and Management Sciences, 7*(1), 170–188.

Al-Hawary, S. I., & Al-Syasneh, M. S. (2020). Impact of dynamic strategic capabilities on strategic entrepreneurship in presence of outsourcing of five stars hotels in Jordan. *Business: Theory and Practice, 21*(2), 578–587.

Al-Hawary, S. I., Al-Hawajreh, K., AL-Zeaud, H., & Mohammad, A. (2013). The impact of market orientation strategy on performance of commercial banks in Jordan. *International Journal of Business Information Systems, 14*(3), 261–279.

Alkitbi, S. S., Alshurideh, M., Al Kurdi, B., & Salloum, S. A. (2020, October). Factors affect customer retention: A systematic review. In *International Conference on Advanced Intelligent Systems and Informatics* (pp. 656–667). Cham: Springer.

Allahow, T. J. A. A., Al-Hawary, S. I. S., & Aldaihani, F. M. F. (2018). Information technology and administrative innovation of the central agency for information technology in Kuwait. *Global Journal of Management and Business, 18*(11-A), 1–16.

Al-Lozi, M. S., Almomani, R. Z. Q., & Al-Hawary, S. I. S. (2018). Talent Management strategies as a critical success factor for effectiveness of Human Resources Information Systems in commercial banks working in Jordan. *Global Journal of Management and Business Research: A Administration and Management, 18*(1), 30–43.

Al-Lozi, M., Almomani, R. Z., & Al-Hawary, S. I. (2017). Impact of talent management on achieving organizational excellence in Arab Potash Company in Jordan. *Global Journal of Management and Business Research: A Administration and Management, 17*(7), 15–25.

Almaazmi, J., Alshurideh, M., Al Kurdi, B., & Salloum, S. A. (2020, October). The effect of digital transformation on product innovation: a critical review. In *International Conference on Advanced Intelligent Systems and Informatics* (pp. 731–741). Cham: Springer.

Al-Nady, B. A., Al-Hawary, S. I., & Alolayyan, M. (2013). Strategic management as a key for superior competitive advantage of sanitary ware suppliers in Kingdom of Saudi Arabia. *International Journal of Management and Information Technology, 7*(2), 1042–1058, and *Technology, 29*(5), 1174–1189.

Alolayyan, M., Al-Hawary, S. I., Mohammad, A. A., & Al-Nady, B. A. (2018). Banking service quality provided by commercial banks and customer satisfaction. A structural equation modelling approaches. *International Journal of Productivity and Quality Management, 24*(4), 543–565.

Alshamsi, A., Alshurideh, M., Al Kurdi, B., & Salloum, S. A. (2020, October). The influence of service quality on customer retention: A systematic review in the higher education. In *International Conference on Advanced Intelligent Systems and Informatics* (pp. 404–416). Cham: Springer.

AlShehhi, H., Alshurideh, M., Al Kurdi, B., & Salloum, S. A. (2020, October). The impact of ethical leadership on employees performance: A systematic review. In *International Conference on Advanced Intelligent Systems and Informatics* (pp. 417–426). Cham: Springer.

Alshurideh, M. (2022). Does electronic customer relationship management (E-CRM) affect service quality at private hospitals in Jordan? *Uncertain Supply Chain Management, 10*(2), 325–332.

Alshurideh, M. T., Al Kurdi, B., Alzoubi, H. M., Ghazal, T. M., Said, R. A., AlHamad, A. Q., … & Al-kassem, A. H. (2022). Fuzzy assisted human resource management for supply chain management issues. *Annals of Operations Research*, 1–19.

Alshurideh, M., Al-Hawary, S. I., Batayneh, A. M., Mohammad, A., & Al-Kurdi, B. (2017). The impact of Islamic Banks' service quality perception on Jordanian customers loyalty. *Journal of Management Research, 9*(2), 139–159.

AlTaweel, I. R., & Al-Hawary, S. I. (2021). The mediating role of innovation capability on the relationship between strategic agility and organizational performance. *Sustainability, 13*(14), 7564.

Alzoubi, H. M., Alshurideh, M., Al Kurdi, B., & Inairat, M. (2020). Do perceived service value, quality, price fairness and service recovery shape customer satisfaction and delight? A practical study in the service telecommunication context. *Uncertain Supply Chain Management, 8*(3), 579–588.

Alzoubi, H., Alshurideh, M., Kurdi, B., Akour, I., & Aziz, R. (2022). Does BLE technology contribute towards improving marketing strategies, customers' satisfaction and loyalty? The role of open innovation. *International Journal of Data and Network Science, 6*(2), 449–460.

Bhatnagar, J. (2007). Talent management strategy of employee engagement in Indian ITES employees: Key to retention. *Employee Relations, 29*(6), 640–663.

Buthelezi, N. C. (2010). *Developing a talent management framework for a South African sectorial education and training authority*. Doctoral dissertation, University of Stellenbosch, Stellenbosch.

Cappelli, P. (2008). Talent management for the twenty-first century. *Harvard Business Review, 86*(3), 74.

Collings, D. G., & Mellahi, K. (2009). Strategic talent management: A review and research agenda. *Human Resource Management Review, 19*(4), 304–313.

Cunningham, I. (2007). Talent management: Making it real. *Development and Learning in Organizations: An International Journal, 21*(2), 4–6.

De Leeuw, E., Hox, J., Silber, H., Struminskaya, B., & Vis, C. (2019). Development of an international survey attitude scale: Measurement equivalence, reliability, and predictive validity. *Measurement Instruments for the Social Sciences, 1*(1), 9. https://doi.org/10.1186/s42409-019-0012-x

Egerova, D., Eger, L., Jirincova, M., & Ali Taha, V. (2013). *Integrated talent management. Challenge and future for organizations in Visegrad countries*. Czech Republic: NAVA, Plzen.

Enz, C. A. (2012). Strategies for the implementation of service innovations. *Cornell Hospitality Quarterly, 53*(3), 187–195.

Ghannajeh, A. M., AlShurideh, M., Zu'bi, M. F., Abuhamad, A., Rumman, G. A., Suifan, T., & Akhorshaideh, A. H. O. (2015). A qualitative analysis of product innovation in Jordan's pharmaceutical sector. *European Scientific Journal, 11*(4), 474–503.

Grapragesem, S., Krishan, A., & Mansor, A. N. (2014). Current trends in Malaysian higher education and the effect on education policy and practice: An overview. *International Journal of Higher Education, 3*(1), 85–93.

Hair, J. F., Babin, B. J., & Krey, N. (2017). Covariance-based structural equation modeling in the journal of advertising: Review and recommendations. *Journal of Advertising, 46*(1), 163–177. https://doi.org/10.1080/00913367.2017.1281777

Howard, M. C. (2018). The convergent validity and nomological net of two methods to measure retroactive influences. *Psychology of Consciousness: Theory, Research, and Practice, 5*(3), 324–337. https://doi.org/10.1037/cns0000149

Jones, R. (2010). Talent Management in the new economy: Applying lessons learned from knowledge workers. In *Talent Management of Knowledge Workers* (pp. 23–32). London: Palgrave Macmillan.

Kum, F. D., Cowden, R., & Karodia, A. M. (2014). The impact of training and development on employee performance: A case study of ESCON consulting. *Singaporean Journal of Business Economics, and Management Studies, 3*(3), 72–105.

Kumari, P., & Bahuguna, P. C. (2012). Measuring the impact of talent management on employee behavior: An empirical study of oil and gas industry in India. *Journal of Human Resource Management and Development, 2*(2), 65–85.

Metabis, A., & Al-Hawary, S. I. (2013). The impact of internal marketing practices on services quality of commercial banks in Jordan. *International Journal of Services and Operations Management, 15*(3), 313–337.

Mohammad, A. A. S., Saleem Khlif Alshura, M., Al-Hawary, S. I. S., Al-Syasneh, M. S., & Alhajri, T. M. S. (2020). The influence of internal marketing practices on the employees' intention to leave: A study of the private hospitals in Jordan. *International Journal of Advanced Science and Technology, 29*(5), 1174–1189.

Mohammed, A. A., Hafeez-Baig, A., & Gururajan, R. (2018). Talent management as a core source of innovation and social development in higher education. *Innovations in Higher Education-Cases on Transforming and Advancing Practice* (pp. 1–31). IntechOpen.

Nurhadi, D., & Zahro, S. (2016). Integrating the talent management program as a new concept to develop a sustainable human resource at higher educational institutions. *International Journal of Organizational Innovation, 8*(4), 146–160.

Nuseir, M. T., Aljumah, A., & Alshurideh, M. T. (2021). How the business intelligence in the new startup performance in UAE during COVID-19: The mediating role of innovativeness. *The Effect of Coronavirus Disease (COVID-19) on Business Intelligence, 334*, 63–79.

Odeh, R. B. M., Obeidat, B. Y., Jaradat, M. O., & Alshurideh, M. T. (2021). The transformational leadership role in achieving organizational resilience through adaptive cultures: The case of Dubai service sector. *International Journal of Productivity and Performance Management*. Vol. ahead-of-print No. ahead-of-print. https://doi.org/10.1108/IJPPM-02-2021-0093.

Popadiuka, S., & Choob, C. (2006). Innovation and knowledge creation: How are these concepts related? *International Journal of Information Management, 26*, 302–312.

Rimkeviciene, J., Hawgood, J., O'Gorman, J., & De Leo, D. (2017). Construct validity of the acquired capability for suicide scale: Factor structure, convergent and discriminant validity. *Journal of Psychopathology and Behavioral Assessment, 39*(2), 291–302. https://doi.org/10.1007/s10862-016-9576-4

Scullion, H., & Collings, D. (2011). *Global talent management*. New York, NY: Routledge.

Sekaran, U., & Bougie, R. (2016). *Research methods for business: A skill-building approach* (7th edn.). Wiley.

Shamout, M., Elayan, M., Rawashdeh, A., Kurdi, B., & Alshurideh, M. (2022). E-HRM practices and sustainable competitive advantage from HR practitioner's perspective: A mediated moderation analysis. *International Journal of Data and Network Science, 6*(1), 165–178.

Shi, D., Lee, T., & Maydeu-Olivares, A. (2019). Understanding the model size effect on SEM fit indices. *Educational and Psychological Measurement, 79*(2), 310–334. https://doi.org/10.1177/0013164418783530

Stahl, G. K., Björkman, I., Farndale, E., Morris, S. S., Paauwe, J., Stiles, P., … & Wright, P. M. (2007). *Global talent management: How leading multinationals build and sustain their talent pipeline*. INSEAD Faculty and Research Working Papers. Industrial Management Review Association at the Alfred P. Sloan School of Management, Massachusetts Institute of Technology (2007/34).

Sung, K.-S., Yi, Y. G., & Shin, H.-I. (2019). Reliability and validity of knee extensor strength measurements using a portable dynamometer anchoring system in a supine position. *BMC Musculoskeletal Disorders, 20*(1), 1–8. https://doi.org/10.1186/s12891-019-2703-0

Susman, G., Warren, A., Ding, M., & Stites, J. P. (2006). *Product and service innovation in small and medium-sized enterprises*. Research Sponsored by United States Department of Commerce-The National Institute of Standards and Technology Manufacturing Extension Partnership.

Tohidi, H., & Jabbari, M. M. (2012). Innovation as a success key for organizations. *Procedia Technology, 1*, 560–564.

Walker, R. M. (2008). An empirical evaluation of innovation types and organizational and environmental characteristics: Towards a configuration framework. *Journal of Public Administration Research and Theory, 18*(4), 591–615.

Wang, Y. A., & Rhemtulla, M. (2021). Power analysis for parameter estimation in structural equation modeling: A discussion and tutorial. *Advances in Methods and Practices in Psychological Science, 4*(1), 1–17. https://doi.org/10.1177/2515245920918253

Running Head: Impact of Smart Buildings on Accounting and Management Control

Ala'a Yahya Ahmad, Nedal Fawzi Assad, and Muhammad Turki Alshurideh

Abstract This study investigates the impact of smart buildings on accounting and management controls. By analyzing the innovative building structures, this study will find out the accounting and management controls patterns and their solutions or help improve management control or reduce costs. Studying different articles for this study hardly find any relevant article or research paper on this study, but some documents are available on management and controls and intelligent buildings. So, the research gap of this study is that no investigation was done on this topic. As earlier, no research was done on this topic, so my analysis will be based on primary qualitative or exploratory research. Based on this, some previous papers will be analyzed and based on that information; thematic solutions will be provided.

1 Introduction

An intelligent building should provide its occupants and owners with a wide range of high-value features and benefits. However, after some time, expectations for a building shift while the structure itself is at risk of deterioration and degeneration. To ensure that intelligent buildings meet all their requirements throughout their lifecycle, careful attention to detail is required. In addition to thorough planning throughout the construction and execution phases, the building's operating term needs high-quality administration and upkeep. All the way through, the facilities manager is a

A. Y. Ahmad · N. F. Assad · M. T. Alshurideh
Department of Management, College of Business, University of Sharjah, 27272 Sharjah, United Arab Emirates
e-mail: u19106242@sharjah.ac.ae

N. F. Assad
e-mail: nassad@sharjah.ac.ae

M. T. Alshurideh (✉)
Department of Marketing, School of Business, The University of Jordan, 11942 Amman, Jordan
e-mail: m.alshurideh@ju.edu.jo; malshurideh@sharjah.ac.ae

M. Alshurideh et al. (eds.), *The Effect of Information Technology on Business and Marketing Intelligence Systems*, Studies in Computational Intelligence 1056,
https://doi.org/10.1007/978-3-031-12382-5_98

key actor. The contemporary building of the twenty-first century must show a high degree of adaptability and usefulness in addition to good architectural, technical, and environmental characteristics. Keeping an eye on all these features is essential. As a facilities manager, you may be responsible for installing and analyzing building monitoring systems. As the level of relevance of building optimization grows, so does the number of solutions available for improvement. As a result, the best chances of successful project implementation are found in the early stages of project planning. It is estimated that up to 80% of all operating expenses are established at this point, just before construction begins. At most, only around 20% of a building's operating expenditures may be modified throughout its life (while 80% of the overall costs are regularly attached to the operational phase). Although it is not common practice in many countries, the participation of facility managers in the early stages of building design should be the norm. As a result, existing buildings may be reconstructed or altered without compromising the integrity of the original design. When designing smart buildings, having a facility manager on board is essential. This means that the facility manager is tasked with representing the interests of both the investor (the future owner) and the building's future occupants. For example, suppose a facility manager assumes control of an already-completed construction project. In that case, they will have only a limited number of alternatives for improving the long-term cost-effectiveness of the building. For a facility manager devoted to their profession, such practices are difficult to bear. As a result of the inability of both the building owner and potential tenants to complete critical optimization efforts that affect operating expenses, this has happened in this instance. A significant deficiency in the building's use phase is that the concept of planned (preventive) maintenance has not yet been adequately extended. There is still a great deal of building and equipment maintenance done out of necessity rather than a genuine desire to address the root causes of any given manifestation or issue. As a result, both the frequency of repairs and maintenance costs have increased. Preventative maintenance may save you up to four times the cost of repairs if you don't undertake it. In addition, uncoordinated interventions in building management and necessary disturbances to the facility's functioning harm building users' well-being. In the context of smart buildings, this is an absolute no–no. Instead, facilities managers must work with other components (such as suppliers of innovative technologies) to mitigate such risks and prevent such eventualities.

2 Literature Review

2.1 *Smart Building Policy*

Building owners and users' needs are the core emphasis of "smart building," which may be defined in various ways. The primary focus is on how technology, services, and system solutions can be leveraged to meet these needs. Operational costs and energy usage, the interior atmosphere, and the preservation of the building's technical condition and market value are all factors that must be taken into account to accomplish this. To suit the growing demands of both owners and users, subgroups of criteria must be established for each of these factors (Stamatescu et al., 2019). This implies that the building's inherent intelligence must be preserved by not changing its qualities through time but rather by improving its quality. All kinds of buildings, from residential to administrative to public, may benefit from this approach, which considers the unique purpose for which they were designed. Some of the demands made by owners and renters of intelligent buildings are compatible, while others are at odds. Owners prefer the following alternatives:

- minimizing acquisition costs.
- minimizing operating costs.
- minimizing energy costs.
- minimizing repair and reconstruction costs.
- maximizing the return on investment of the building.
- continuous conservation or increase in construction value (investment protection).
- optimizing the level of security of the building and its users.

It was evident from this review that the financial demands of buildings are of crucial importance. However, although the user owner's) interests are vital, the owner does not need to interfere with those interests.

- The flexibility of leased spaces.
- The quality of the indoor environment that contributes to achieving the desired comfort of use or to increasing work productivity.
- Maximizing security level.

Renting out the interior of a building to satisfy tenants' demands is a frequent practice for building owners. The owner's expenditures throughout the building's construction, existence, and operation should be monitored less for "minimization" and more for "optimization" in light of the three requirements of the building. While the facility is being utilized, the facility manager's role is to keep an eye on and safeguard the interests of all parties involved throughout the planning, projection, and final realization phases. Many vendors and suppliers often give particular technologies or even partial facets of the building's intelligence as part of its implementation or upkeep. However, whatever technology is utilized to administrate sub-groups or to manage the building's general operational management must run smoothly at all times. Therefore, certain technologies are allocated to sub-groups to satisfy the building's specific functional demands. For example:

- optimization of energy management.
- power system management, heating, ventilation, cooling, air conditioning, lighting, shading technology, external sources (weather forecasting, etc.).
- fire safety of buildings and users—fire safety system, ventilation, lighting, lifts, security system, power system control.

- user safety—security system, person entry system, indoor CCTV circuit, charges, lighting.

The BMS (Building Management System) platform is used to regulate all of the building's systems and offer real-time access to information to establish an integrated multi-operating structure. BMS keeps track of individual technologies and raises warnings when something out of the ordinary occurs while archiving any intermediate data. The operator has to be subservient to the facility management to respond to the data supplied (Gonçalves et al., 2020). Human variables like security and cleaning, in addition to the BMS platform's technologies, must be managed by facility managers to guarantee that the intelligent building works optimally for its owners and users. Starting with the design phase and continuing throughout the life of the building, BIM (Building Information Modeling/Management) should be an integral part of any innovative building project. Stanford University's Center for Integrated Facilities Engineering studied 32 BIM-based projects. The design and implementation duration were significantly reduced because of the contributing components' more effective data management and coordination. The reduction of collisions throughout the pre-implementation and implementation phases saves up to 40% off-budget and 10% of total expenses. For intelligent buildings, BIM's financial advantages are anticipated to be far more significant.

2.2 *Facility Management in Smart Buildings*

In today's building management paradigm, facility managers perform a wide variety of tasks (in the case of smart buildings). The administration of technological assets, coordination, and integrated management are often included in these packages (Dong et al., 2019). Therefore, FM may be divided into two primary divisions that do not have a clear border, based on the demands of the client and the needs of the company:

- **Space and Infrastructure**—so-called "hard services", i.e., space management, space utilization, management and optimization of the workplace, technical management of buildings, energy management, waste management, indoor and outdoor cleaning.
- **People and Organizations**—so-called "soft services", i.e., health, hygiene, safety and security, internal services (boarding, receptions, meeting rooms, secretarial services, etc.), ICT, internal logistics (archive services, internal mail, mail service, transport services, car service, etc.).

Management of facilities is defined by the International Facility Management Association (IFMA) as including operations management, design and architecture, and the arts and sciences. All systems in the building must be functional and efficient, as well as the structure's interior space must be efficiently used to guarantee the building's long-term survival (Dong et al., 2019).

To summaries, the ultimate purpose of facility management is to provide a high-quality interior environment while also maximizing the building's energy and resource efficiency. Whether it's a single structure, a collection of buildings, or a complex, the basics remain the same. Effective use of client assets is crucial to provide the most excellent possible support for the client's primary business activity. As part of the design team, a facilities manager should accept these requirements throughout the planning phases for a new building or restoration of an existing structure (Ostadijafari et al., 2019). However, smart buildings have higher operational system accountability than traditional buildings. This is why they provide more advantages than conventional buildings, such as more flexibility and usefulness, better interior environment quality, and higher operational efficiency. More than a hundred different technologies exist for these features, and their network is far more extensive and better connected than a conventional building's. But if technology fails, the structure's functionality and security are damaged, which is not necessarily fatal for a typical installation. Non-financial ramifications, such as a sullied reputation, might accompany the financial ones. Competent building facilities managers are not one person but rather a team of experts who are directly responsible to building owners or CEOs. There is a good chance that an intelligent building's management staff and its owners will have conflicting interests (Ostadijafari et al., 2019). This means that in addition to having an excellent professional name, the facility manager must also have the ability to speak up for the owner's interests and convey their own opinions.

2.3 Management Controls

The conceptual framework from Balaji et al. (2018) addresses the limitations of researching MCS in organizations. Additionally, the model provides an overview of the many controls, enabling complete investigation since each regime has its research path. This approach and attitude are increasingly used in research (Le et al., 2019). MCS is called "systems" because they are "systems for regulating team member behaviour, not systems simply utilized to obtain information for decision-making" (Balaji et al. 2018). This description does not include accounting systems used for decision support, which instead emphasizes behaviour. To be considered an MCS, cost accounting must also align team member behaviour. Managerial control may be used for various purposes, including but not limited to team member behaviour.

- **Cultural Controls**: Clans, crests, and customs are all part of the package. In an informal form of command and control, a company's culture influences the behaviour of its employees. As an informal control, it is challenging to study culture.
- **Planning**: Malmi and Brown do not consider long-term and action planning forms of control (2008). To control employees, you must have them actively participate in

the process. Individuals and organizations may use it to set goals and measure their progress. Budgeting and planning are inseparable and should be done together.

- **Cybernetic Controls**: Financial and non-financial metrics and a mix of both are all subject to the same level of scrutiny under this system. Measurements may be used to hold employees or organizations responsible for their performance. They are the foundation of MCS, and every firm relies on them. Cybernetic systems may be classed as either information systems or control systems, depending on their function.
- **Reward and Compensation**: They might be used as a control to inspire individuals or groups. Cybernetic controls, team member retention, and cultural control promotion are all part of this strategy. Business plans must be aligned with incentive schemes (Epstein et al., 2010).
- **Administrative Controls**: The organization's structure and design, governance, and rules and procedures are examples of administrative controls that may affect team member behaviour.

To ensure that changes are gradual, cultural controls have been placed at the top. Thus, they provide a framework for other authorities to operate in. Commands begin with planning, followed by measurement both during and after the process, and then possible reward and pay based on performance. There are three sequential controls. To exert authority, one must first establish a base. We can see a significant difference in the design and use of rules (Liew, 2019). MCS may now be studied in more depth, thanks to (Pašek & Sojková, 2018). Two primary topics may be found in this chapter. The setup of an MCS package will be the first step. Only a limited amount of research has been undertaken on the combined effects of these two factors. More evidence from MCS packages that have been deployed in practice is needed to understand better the most crucial systems and the contents of containers. This section focuses on the way systems interact with one another and how it influences their results. As part of the intelligent environment projects, environmental sustainability is a primary focus for each of the five control kinds. Therefore, an interview model is utilized to organize and analyze and summarise the results when conducting interviews (Wei et al., 2018). MCS package design is the most critical factor in determining what controls the city will apply in its MCS package. As a collective, respondents are more inclined to discuss controls than individual controls while answering the survey's questions.

> *One of the world's fastest-growing developing countries, the United Arab Emirates is no exception. One of the critical reasons for Dubai's rapid development is that the government started to plan for a smart city and gradually reflected on other emirates to promote Dubai's future city of Dubai.*

The need of constructing a long-term, ecologically friendly environment could not be overstated. Another motivation to create a green metropolis is the UAE's reputation as one of the world's most polluting countries. According to WWF's Living Planet Report 2014, the United Arab Emirates has the third-largest Ecological Footprint per capita in the world. That's why it's essential to the government that the physical environment, rules and public awareness all undergo significant changes (promoting sustainability assessment methods, like LEED and ESTIDAMA, with stakeholders who are the decision-makers). The already-existing New York University Abu Dhabi campus was chosen for this study. The institution has 15.4 ha of property. Saadiyat Island in Abu Dhabi, the capital city of the United Arab Emirates, is home to the campus (Zou et al., 2018). On the property, you'll find classrooms, labs, and other resources for academic pursuits. In addition, students and lecturers can both find housing in the area. The building's structure is supposed to adapt to its surroundings, creating a setting that stimulates user-to-user contact. Because of its creative arrangement, the campus's many components, including the buildings themselves, interiors, and exteriors, form a unified whole. Students, faculty, staff, and administrators worked together to build a cooperative institution. This was reflected in the building's physical design and the school's academic offerings another noteworthy characteristic of the site in its landscape design that connects the many institutions on campus. For the ESTIDAMA sustainability accreditation, the NYUAD campus has been created to meet all of the requirements for its creation. Consequently, the case study will incorporate the interior and outside of campus characteristics to emphasize and analyze the components of the outside campus. The sustainable attributes of the contemporary building are divided into two categories:

- Outdoors features: Based on observed site visit.
- Interior Features: Based on a questionnaire survey with staff and students.
- Internal side streets are relatively narrow and shaded by the surrounding buildings.
- Use of local plants/palm trees and water channels that do not require excessive water while offering a soothing atmosphere during summertime and pleasant scenery for the people.
- The High Line is a must-see since it is the campus's most recognizable icon. The High Line is a system of bridges and stairways on the second level of every university campus that enables visitors to move between buildings freely. As seen in, a green canopy above the High Line works as a heat insulator. The advent of the High Line has made the university's campus more dynamic and exciting. In addition, students, instructors, and staff will be more likely to stroll between buildings if the walkway is well-designed. Because of the gardens and vistas from the tops of the buildings, students were more likely to walk between campuses. All across the school, solar panels have been installed on the roofs.

3 Thematic Results

3.1 *Impact of Smart Buildings*

To keep up with the growth of "smart buildings," facilities management must give way to real estate management. However, the smart (intelligent) building is a drawback because of the increasing reliance on embedded technology. In addition, the building's general operational performance and usability are harmed, putting the productivity and health of its inhabitants in danger and raising operating costs. Owners are responsible for ensuring that their facilities are adequately maintained by appointing a top-notch facility manager (Gholamzadehmir et al., 2020). Technical and non-technical abilities are necessary for a facilities manager's work, but so too is the ability to analyze, interpret, and apply massive volumes of data to real-world circumstances. We should reasonably expect significant technical progress over the next 80 years, given the expected lifespan of buildings. The development of these systems requires continuous updates and integration with the building's central management system to meet the changing needs of the building's owners and users. There is a trade-off between the flexibility, investment protection, convenience of presenting traffic information, and freedom of decision-making that comes with both BMS platforms and building owners through facility managers. This is the fundamental difference between conventional approaches and individual attention to the demands of each structure and its residents. As a result, the procedure has been judged to be useless. They deliver superior customer service by integrating a variety of systems into a cohesive whole. An example of this is SEC and LS integration, a security and access control system. Providing smart access to multiple university departments via a single card is comparable to delivering smart cards and smart access to all students everywhere (library, classrooms, parking. etc.). The smart card may also be used by teaching staff to access classrooms in their department. Energy management and automation provider Schneider Electric cooperated with the University of West England in the United Kingdom to demonstrate a successful building integration. There will also be Schneider Electric's integrated security system on campus, in addition to new departments. Schneider Electric's Andover Continuum building management system and Juniper switch systems were used for both heating and lighting (Gholamzadehmir et al., 2020).

3.2 *Management Controls*

Planning for a city involves several steps, each of which takes a significant amount of time to accomplish. An essential consideration in the planning phase is the project's budget. The same holds when it comes to finding a business partner.

3.3 *Cybernetic Controls*

As required by the, the city keeps track of CO_2 emissions to keep tabs on the progress of the city's climate action plan. Nevertheless, this analysis falls short since it fails to account for all relevant factors. Moreover, the city can only have a limited impact on these traits since they are more than simply its boundaries. It is hoped that a dashboard would allow progress toward Liège 2025's strategic objectives to be tracked and shown. Unfortunately, there was just a brief written account of the progress made at the time of the research, with no mention of the results or an overview of how well the objectives were met (Ramon-Jeronimo et al., 2019). Transversal strategic plan aims include the establishment of a dashboard, increased focus on a formal evaluation, and environmental effect measurement implementation in administration.—There must be a way to track progress toward goals and convey this information internally, as well as externally. According to individuals questioned, the city is working toward these goals and has a strong belief in the benefit of having more information available to residents. The alderman's goal is to have the scorecard in place by 2021. This dashboard may be used to keep tabs on and assess how far you are in achieving your objectives (Ramon-Jeronimo et al., 2019). If necessary, the project might be redirected during or after completion. Nevertheless, evaluating the long-term impact of many programmers is difficult since actions like spreading knowledge and teaching people are challenging to track. Objectivists may also benefit from the information. For example, they educate the public to make more informed decisions and get political support for their initiatives. In addition, real-time data management might help the city improve itself daily. Trash management already had a cleanliness dashboard in place, but it was marked as a work in progress. This dashboard displays data on things like resource use and rubbish collection by neighborhood. This information came from sensors and other sources (Neves et al., 2020). After never having attempted something like this, everyone in the city of Liège had to modify their mindsets. According to one responder, performance evaluations should improve performance, not detract from it. Keeping track of the city's finances is another critical cybernetic control. As a limited but vital resource, money is a primary cause of stress for the city's budget. To request a budget, there is a set procedure. The creation of an annual budget necessitates preparing one year in advance. To go forward, approval is required from the city council and the city financial supervisor (Neves et al., 2020). As a consequence, creating a budget is going to take some time. Nevertheless, the project must be finished on schedule to minimize financial losses at the end of the year. The transversal strategic plan states that enhancing budget management efficiency is a priority.

3.4 Reward and Compensation

For environmental activities in the Belgian city of Liège, there are no extra financial incentives or payments. However, other non-financial considerations were also listed to motivate people in the workplace. For example, workers' self-motivation was considered one of the essential variables in inspiring employees. However, the selection procedure did not include this as a consideration. Internally, the city sought to make its employees' jobs meaningful and express that it was improving people's lives due to this initiative. For this reason, employees are also trained and made aware of the company's impact on sustainability. Motivating and winning support for projects may be achieved in part via team member engagement. It has been established that reducing the workload negatively impacts team members well-being, which is critical for a motivated workforce.

3.5 Administrative Controls

This section explains the major essential components of the city's organizational structure without going into depth. The political body comprises the mayor, older persons, and cabinet members. In contrast, the administrative body includes the remainder of the city council members and their staff. As a link between the people and the government, the elected body acts as a conduit for the people's will. For the most part, the tasks of these politicians fall outside the purview of other governments. It is the responsibility of the administration to implement the government's policies and provide critical services to the people. Organizations like these have boards of directors, managers for each area of responsibility and regional heads for each sector or neighborhood. Between the two entities, the alderman and his cabinet serve as a vital point of contact, monitoring the progress of many projects and objectives (Parker & Chung, 2018). Officials from both parties meet regularly to examine the progress of environmental programmes and see whether they are accomplishing their aims. The alderman claims that the two organizations work together in harmony. There is an ecological department. However, it must coordinate with every other municipal division. Managing data is a vital IT department responsibility, for example, in intelligent environment efforts like the Internet of Things. Different departments may call upon the environmental expert. When working with others or making purchases, the government must use public procurement. This method entails drawing up the project's specs, issuing an RFP, and comparing at least three offers. It's not only the project's budget that expands inexorably; the legal procedure that governs it does, too. For example, environmental issues may be considered when the city decides, and the transversal strategic plan have the reform of public procurement as a goal. Include ethical, social, and environmental clauses in contracts is the goal of this project as previously said, budgets play a crucial role in this strategy.

3.6 Cultural Controls

Our conversation focused on culture and environmental sustainability since culture spans various themes. They believe that a plethora of little gestures may have a significant impact. Various sectors and generations need different strategies to promote interdisciplinarity but creating a feeling of togetherness is necessary. It is possible to make individuals feel more responsible for their behaviour by informing them of the environmental repercussions. Also, it attempts to emphasize the need of including all essential stakeholders. The city uses eco-teams as role models for its personnel.

There is an intranet, a letter called "info perse," and a quarterly journal sent internally in addition to the intranet. You may use these methods to get the word out about new projects or other departments.

3.7 Accounting

Some of the demands made by owners and renters of intelligent buildings are compatible, while others are at odds. As a general rule, it's pretty rare for owners to wish to lower their out-of-pocket expenditures in these four areas: acquiring the building at the lowest price, operating it at its most efficient level, and minimizing the amount of money they have to spend on repairs and rebuilds. Financial demands on structures are essential in this case. However, even though the user owner's interests are vital, the owner does not need to interfere with those interests. improving production, maximizing safety, and providing more adaptability in leased facilities are all examples of a company's "internal environment quality (Haseeb et al., 2019)."

4 Conclusion

To keep up with the growth of "smart buildings," facilities management must give way to real estate management. However, the smart (intelligent) building is a drawback because of the increasing reliance on embedded technology. In addition, the building's general operational performance and usability are harmed, putting the productivity and health of its inhabitants in danger and raising operating costs. Owners are responsible for ensuring that their facilities are adequately maintained by appointing a top-notch facility manager. Technical and non-technical abilities are necessary for a facilities manager's work, but so too is the ability to analyze, interpret, and apply massive volumes of data to real-world circumstances. We should reasonably expect significant technical progress over the next 80 years, given the expected lifespan of buildings. The development of these systems requires continuous updates and integration with the building's central management system to meet the changing needs of the building's owners and users. There is a trade-off between the flexibility, investment protection, and convenience of presenting traffic information, as well as the freedom of decision-making that comes with both BMS platforms and building owners through facility managers. This is the fundamental difference between conventional approaches and individual attention to the demands of each structure and its residents. As a result, the procedure has been judged to be useless.

References

Balaji, B., Bhattacharya, A., Fierro, G., Gao, J., Gluck, J., Hong, D., ... & Whitehouse, K. (2018). Brick: Metadata schema for portable smart building applications. *Applied Energy, 226*, 1273–1292.

Dong, B., Prakash, V., Feng, F., & O'Neill, Z. (2019). A review of smart building sensing system for better indoor environment control. *Energy and Buildings, 199*, 29–46.

Gholamzadehmir, M., Del Pero, C., Buffa, S., & Fedrizzi, R. (2020). Adaptive-predictive control strategy for HVAC systems in smart buildings—A review. *Sustainable Cities and Society*, 102480.

Gonçalves, D., Sheikhnejad, Y., Oliveira, M., & Martins, N. (2020). One step forward toward smart city Utopia: Smart building energy management based on adaptive surrogate modelling. *Energy and Buildings, 223*, 110146.

Haseeb, M., Lis, M., Haouas, I., & WW Mihardjo, L. (2019). The mediating role of business strategies between management control systems package and firms stability: Evidence from SMEs in Malaysia. *Sustainability, 11*(17), 4705.

Le, D. N., Le Tuan, L., & Tuan, M. N. D. (2019). Smart-building management system: An Internet-of-Things (IoT) application business model in Vietnam. *Technological Forecasting and Social Change, 141*, 22–35.

Liew, A. (2019). Enhancing and enabling management control systems through information technology: The essential roles of internal transparency and global transparency. *International Journal of Accounting Information Systems, 33*, 16–31.

Malmi, T., & Brown, D. A. (2008). Management control systems as a package—Opportunities, challenges and research directions. *Management Accounting Research, 19*(4), 287–300.

Neves, F. T., de Castro Neto, M., & Aparicio, M. (2020). The impacts of open data initiatives on smart cities: A framework for evaluation and monitoring. *Cities, 106*, 102860.

Ostadijafari, M., Dubey, A., Liu, Y., Shi, J., & Yu, N. (2019, August). Smart building energy management using nonlinear economic model predictive control. In *2019 IEEE Power & Energy Society General Meeting (PESGM)* (pp. 1–5). IEEE.

Parker, L. D., & Chung, L. H. (2018). Structuring social and environmental management control and accountability: Behind the hotel doors. *Accounting, Auditing & Accountability Journal*.

Pašek, J., & Sojková, V. (2018). Facility management of smart buildings. *International Review of Applied Sciences and Engineering, 9*(2), 181–187.

Ramon-Jeronimo, J. M., Florez-Lopez, R., & Araujo-Pinzon, P. (2019). Resource-based view and SMEs performance exporting through foreign intermediaries: The mediating effect of management controls. *Sustainability, 11*(12), 3241.

Stamatescu, G., Stamatescu, I., Arghira, N., & Fagarasan, I. (2019). Data-driven modelling of smart building ventilation subsystem. *Journal of Sensors*.

Wei, F., Li, Y., Sui, Q., Lin, X., Chen, L., Chen, Z., & Li, Z. (2018). A novel thermal energy storage system in smart building based on phase change material. *IEEE Transactions on Smart Grid, 10*(3), 2846–2857.

Zou, H., Zhou, Y., Jiang, H., Chien, S. C., Xie, L., & Spanos, C. J. (2018). WinLight: A WiFi-based occupancy-driven lighting control system for smart building. *Energy and Buildings, 158*, 924–938.

Predictive and Prescriptive Analytics Tools, How to Add Value to Knowledge-Based Economy: Dubai Case Study

Mounir El Khatib, Moza Abdalla Al Shamsi, Khalid Al Buraimi, Fatima Al Mansouri, Haitham M. Alzoubi, and Muhammad Alshurideh

Abstract Innovation and technology continue to be the driving force of growing economies. The need for constant change and improvement is driven by competing economies fueled by a more educated workforce and backed by proper governance structures. Dubai which has earned immense wealth in the petroleum industry is now venturing into these industries. Smart Dubai was open in 2000 as a smart city solution as an innovative initiative, large scale project for collaboration of private and public sector. These initiatives of Dubai Pulse which involves all stakeholders and entrepreneurs in accessing and sharing of data. It is having been touted for its analytical services. It will be instrumental to the future of uncovering Dubai from the veil of oil money. Dubai banks on technology and innovation particularly predictive and prescriptive analysis. The research study was conducted with the goals of analyzing the usage of both predictive and perspective analysis in the knowledge-based economy. The study also hopes to understand the value created through the use of predictive and perspective analysis in Smart Dubai and DEWA. The study raises conclusions on why the smart city project should be considered a success. Furthermore, there are areas that have been attributed to this success. The information was primarily collected through an interview with employees who heads the

M. E. Khatib
Program Chair, Hamdan Bin Mohamad Smart University, Dubai, UAE

M. A. Al Shamsi · K. Al Buraimi · F. Al Mansouri
Hamdan Bin Mohamad Smart University, Dubai, UAE

H. M. Alzoubi (✉)
School of Business, Skyline University College, Sharjah, UAE
e-mail: haitham.alzubi@skylineuniversity.ac.ae

M. Alshurideh
Department of Marketing, School of Business, University of Jordan, Amman, Jordan
e-mail: malshurideh@sharjah.ac.ae; m.alshurideh@ju.edu.jo

Department of Management, College of Business Administration, University of Sharjah, Sharjah, UAE

M. Alshurideh et al. (eds.), *The Effect of Information Technology on Business and Marketing Intelligence Systems*, Studies in Computational Intelligence 1056,
https://doi.org/10.1007/978-3-031-12382-5_99

data analytics department at Smart Dubai and employees in the middle management level at DEWA. The feedback from the interviews were elaborate and thorough to the research being gathered. The interview was divided into three concerning knowledge-based economy, use of analytical tools and Dubai Pulse Initiative. The use of the interview arises from the fact that the subject of study needed to be informed by firsthand knowledge. However, this primary source was complemented by other secondary sources. The data from Smart Dubai and DEWA were analyzed to determine the opportunities that the use of the analytical tools will have along with the challenges and benchmarking with global successful cases. Then the recommendations part provides similar governments with the roadmap of how to implement the analytical tools and overcome the challenges.

Keywords Predictive · Prescriptive · Analytics · Knowledge based economy · Smart Dubai · UAE

1 Introduction

UAE desires to maintain a diversified economy with "The Abu Dhabi Economic Vision 2030" and it was launched in 2009 (Ahmed & Alfaki, 2013). It shows that the UAE moves towards knowledge-based economy (Alshurideh et al., 2020a, 2020b; Aswad et al., 2011; Nuseir et al., 2021a, 2021b). Accordingly, the Government of the UAE aims to build a knowledge-based economy with sustainability that shall include process, product and distribution technology with the need for UAE economy diversification while overcoming any challenges or conflicts and acquiring all opportunities available during a knowledge-based economy transformation (Aburayya et al., 2020a, 2020b; AlMehrzi et al., 2020; Mina, 2014). There shall be need for the use of the analytics tools to ensure that a knowledge-based economy adds value to the economy of the UAE and that will be done in this study (Al-Maroof et al., 2021; Leo et al., 2021; Mehrez et al., 2021).

In this research study, literature review will be presented across two perspectives, the first one is the knowledge-based economy, then the analytical tools definition and how it can add value to the economy. Secondly, the research methodology and the data analysis will be presented based on the data gathering and analysis. The analyzed data then will be compared to the literature to identify gaps and opportunities of improvement. By the end of this research, we will conclude and make recommendations. The main objective of this assignment to analyze the usage of both perspective and predictive analysis tools and determine how it adds value to the UAE's knowledge-based economy. In order to meet this research objective, the research questions are the following:

1. How can the usage of perspective analysis tools and value to knowledge-based economy in Dubai?
2. How can the usage of predictive analysis tools add value to knowledge-based economy in Dubai?

2 Brief About the Organizations

2.1 Smart Dubai

Smart Dubai was first established in the year of 2000 but back then it was known as Dubai E-Government till 2013 when H.H Sheikh Mohammed Bin Rashid Al Maktoum changed its name to Dubai Smart government department. The vision of smart Dubai is aligned with the vision of H.H Sheikh Mohammed bin Rashid Al Maktoum to make Dubai the happiest city on earth and since then Smart Dubai has been pioneering smart city solutions and has been focused on transforming and empowering Dubai through disruptive innovative initiatives, large scale projects by collaborating with the public and private sector, leading technological projects and many more (Ahmad Aburayya et al., 2020a, 2020b; Taryam et al., 2020; Al Shebli et al., 2021; AlSuwaidi et al., 2021; Digital Dubai, 2022b).

To name a few, Smart Dubai had several initiatives that made quite an impression across the region and globally.

- Dubai Paperless initiative targeted towards the environmental aspect to preserve the environment by making Dubai a fully paperless government where the governmental department will no longer issue or ask for any papers or printed documents to be handed over across all operations (Digital Dubai, 2022d).
- Artificial Intelligence Ethical toolkit that includes the guidelines, self-assessment tool for developers, ethical considerations and common agreements. This initiative is targeted toward government entities, private sector, and individuals (Digital Dubai, 2022a).
- Dubai Blockchain Strategy project, which is a joint initiative between Smart Dubai and the Dubai Future Foundation that will lead the introduction of the blockchain technology to Dubai that is expected to "unlock 5.5 billion dirhams in savings annually in document processing alone" according to smart Dubai official website (Digital Dubai, 2022c).

This report will highlight a particular initiative called Dubai Pulse, this platform is designed to provide all the stakeholders including government entities, private sector, and entrepreneurs to access one location of shared open data. This initiative includes a core analytical services as well as big data services that can be used by various entities to enhance the living conditions and the economy in the UAE by promoting the vision of making Dubai the smartest and happiest city on earth (DubaiPulse, 2022).

2.2 Dubai Water and Electricity Department

Dewa was first established in 1992 as a result of the merge of Dubai Electricity Company and Dubai Water Department. DEWA provides its services to about

882,000 customers across Emirate of Dubai as the solo provider of electricity and water. DEWA has maintained a high customer happiness rate at a remarkably 95% in 2018 (Government of Dubai, 2022). DEWA strategy consists of five main pillars infrastructure, legislation, funding, building capacities and skills, and having an environmentally friendly energy mix. These pillars were supported of many innovative initiatives to transform Dubai to be the smartest city in the world (Government of Dubai, 2022).

3 Literature Review

3.1 Predictive and Prescriptive Analytics Tools

Predictive analytics are tools that are used to predict the future events through multiple techniques like data mining, statistics, modeling, machine learning and artificial intelligence. Those tools can help analyze the current data available at the organization's databases to come up with certain conclusions that will assist the management into making decisions based on knowledge rather than speculations (Abid et al., 2020; Alnazer et al., 2017; Alshurideh et al., 2020a, 2020b; Alzoubi & Ahmed, 2019; Mehmood et al., 2019). According to Gartner official website, predictive analysis "is a form of advanced analytics which examines data or content to answer the question "What is going to happen?" or more precisely, "What is likely to happen?", and is characterized by techniques such as regression analysis, forecasting, multivariate statistics, pattern matching, predictive modeling, and forecasting" (Gartner, 2022).

Moreover, prescriptive analytics are tools that helps the organizations to anticipate what will happen with the reason and the timing of the events in order to reduce future risks and make better decisions. According to Gartner official website "Prescriptive Analytics is a form of advanced analytics which examines data or content to answer the question "What should be done?" or "What can we do to make it happen?", and is characterized by techniques such as graph analysis, simulation, complex event processing, neural networks, recommendation engines, heuristics, and machine learning" (Gartner, 2022).

Prescriptive tools can be done by following two approaches, the first approach is by using the optimization model. This model chooses the best alternative or decision by running through a maximum objective function (Alzoubi & Yanamandra, 2020; Alzoubi et al., 2020a, 2020b; Joghee et al., 2020). This model is done through an analytical coded model using a software with many constraints to optimize the planning, scheduling, assigning limited resources to minimize the cost. For example, deciding on the best route for oil transporting vehicles from the depo to the petrol station, the model will be based on many constraints like vehicle size, petrol type, destination, distance, driver and traffic expected. The second approach is by using the rules model. This model uses the inputs, judgment and business rules to assign the best action. For example, providing customers with certain treatments or offers (Al Batayneh et al., 2021; Kart, 2015; Odeh et al., 2021).

3.2 The Value Added from the Use of the Analytics Tools

CIOs consider Data analytics as a "game changing" tools where it can unleash so many opportunities in fraud detection, analysis on transactional data, decision making process, reduce uncertainty and eliminate risk. The analytics tools vary in the process of decision making and the input they provide. As shown in the graph below the different analytics tools leads to better decision making with minimizing the human input (Fig. 3).

The use if the analytics tools can aid in the different layers of decision making, starting from the strategic, tactical, and operational level. Strategic decision happen less frequently but often has a wide range of impact, for example, building of a new school, a construction of a bridge, a drilling of oil location. These decisions often have a great amount of uncertainty, hence the application of the analytical tools specially the perspective tools can reduce the uncertainty and lead the government to make better decisions that will affect the citizens and the overall economy (Alshurideh, 2022; Hamadneh et al., 2021a, 2021b; Kart, 2015).

Another level of decisions is the tactical layer, these decisions often happen more frequently like on a daily, monthly, quarterly, or annual basis. The use of the analytical tools can lead the government to allocate the resources and the time properly therefore reducing cost and risk. An example of an approach of application is capital planning to allocate resources across tasks (Alaali et al., 2021; Alkalha et al., 2012; Kart, 2015).

The last level of decision is related to the day to day operations and translate the tactical decisions to actionable tasks. These decisions often carry less risk and the analytical tools can be of less impact here.

3.3 Combining of Predictive and Prescriptive Tools

To enable organizations to make optimum decisions, they must be quick, accurate in assisting complicated and time sensitive matters. Although making accurate predictions through the use of the predictive tools is essential but to truly add value the organizations must go even further by making decisions that can affect the predicted outcome through analyzing the likelihoods of a certain solution by the use of the prescriptive tools (Hasan et al., 2022; Herschel & Idoine, 2016).

Bringing together forecasts (predictive) with optimizations (prescriptive) lets an organization explore how changes to different variables are likely to change the outcomes or alter the relative trade-offs. It gets to the heart of the task of adding business value, of proactively making decisions that drive action and influence the future course of an organization.

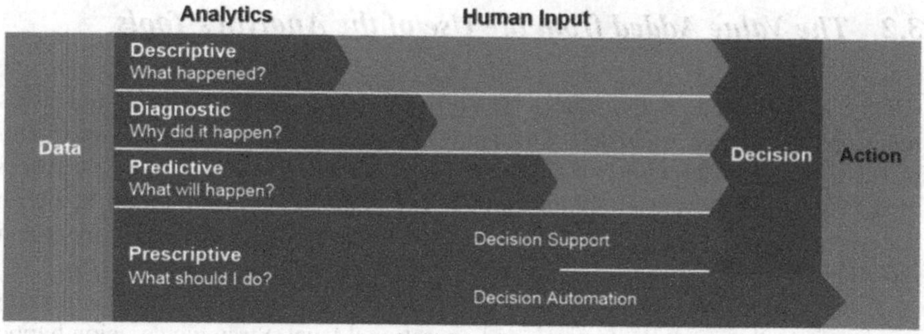

Fig. 1 Analytics capabilities

3.4 Knowledge Based Economy

According to Ahmed and Alfaki (2013), a knowledge-based economy is wider than the new economy or high technology and they are much related to the internet that is used often in the information society. The foundation of a knowledge-based economy is to create, disseminate and use knowledge and knowledge assets are given more importance than labor and capital assets (Al-Maroof et al., 2021; AlMehrzi et al., 2020; Shannak et al., 2012). Therefore, the societal and economic activities reach a very high level through sophistication and quantity of knowledge (Alzoubi et al., 2021a; Svoboda et al., 2021). Parcero and Ryan (2017) illustrates and explains the four pillars of a knowledge-based economy (see Fig. 1). It consists of four pillars and they are communication and information technology, innovation, education, and economy and regime (Alshurideh et al., 2019a, 2019b; Ashal et al., 2021; Awadhi et al., 2021).

The concept of a knowledge-based economy is used so an economy in the country can be described (Ahmed & Alfaki, 2013; Al-Jarrah et al., 2012; Alshurideh, Al Kurdi, et al., 2019a, 2019b). Dahlman and Utz (2005) states a knowledge-based economy is successful when there is link between industrial and science technology. The empowerment comes from increased lifelong and education learning with higher investments in software and research and development.

3.5 Knowledge Based Economy in UAE

According to Alfaki (2014), the UAE achieves good success in knowledge utilization due to technological infrastructure availability so they are moving towards knowledge-based economy. One of the strategic goals of the UAE is to expand in the non-oil sector with diversification strategy so competition with emerging markets can be made possible. In addition, the UAE federal strategy of 2011–2013 and UAE vision of 2021 and advocates in the increase of investments in research and development, science and technology so the economy of the UAE can be based on knowledge

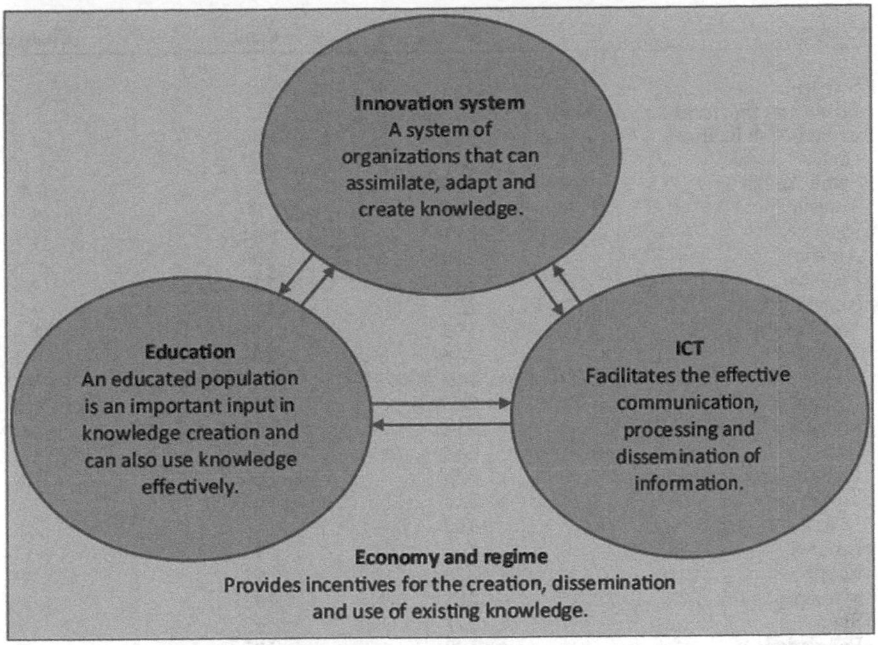

Fig. 2 The four pillars of a knowledge economy (Parcero & Ryan, 2017)

(Akhtar et al., 2021; Al Ali, 2021; Alzoubi, 2021a; Eli, 2021; Kashif et al., 2021). Figure 2 illustrates the performance of the UAE when comparing with Muslim countries as given by World Bank's KEI 2012 cited in Alfaki (2014). The performance of the UAE varies **with** other Muslim countries and the progress is quite impressive ranking 42nd, score 6.94 and the change is +6. In the UAE, there is good progress of a knowledge-based economy's implementation along with innovation, quality of infrastructure and macro-economic environment (Ahmed & Alfaki, 2013).

Alfaki (2014) finds that the UAE experienced major improvement in the ICT sector and ranks top when it comes to good performance in organizational innovation systems and ICT components of knowledge as given in Figs. 3 and 4. It shows a careful investigation of the four pillars such as ICT, innovation, education, and economic incentive (Fig. 5).

4 Research Methodology

The research methodology of this report was done through two methods, the first one used to gather the primary information was a qualitative method done thorough conducting a semi structured interview, this interview was done with an employee working at Smart Dubai and is expert in the data analysis field. Hamzeh Hamaraneh

Country	KEI	Rank	Change[a]
Sweden	9.43	1	–
Economies with large improvements in KEI rankings since 2000			
United Arab Emirates	6.94	42	+6
Oman	6.14	47	+18
Saudi Arabia	5.96	50	+26
Tunisia	4.56	80	+9
Iran	3.91	94	+1
Algeria	3.79	96	+14
Pakistan	2.45	117	+5
Nigeria	2.2	119	+5
Yemen	1.92	122	+6
Sudan	1.48	138	+1
Economies with decreases in KEI rankings since 2000			
Bahrain	6.9	43	−2
Malaysia	6.1	48	−3
Qatar	5.84	54	−5
Kuwait	5.33	64	−18
Turkey	5.16	69	−7
Jordan	4.95	75	−18
Lebanon	4.56	81	−13
Egypt	3.78	97	−9
Morocco	3.61	102	−10
Syria	2.77	112	−1
Bangladesh	1.49	137	−3
Myanmar	0.96	145	−8

Fig. 3 Muslim countries KEIs (Alfaki, 2014)

Index	UAE	Qatar	Bahrain	Kuwait	Oman	KSA	Korea	Singapore
KEI	6.73	6.73	6.04	5.85	5.36	5.31	7.82	8.44
KI	6.72	6.63	5.80	5.63	4.77	5.10	8.43	8.03
Economic incentive and institutional regime	6.75	7.05	6.75	6.50	7.15	5.94	6.00	9.68
Education	4.90	5.37	5.82	4.93	4.47	4.89	8.09	5.29
Innovation	6.69	6.45	4.29	4.98	4.94	3.97	8.60	9.58
ICT	8.59	8.06	7.30	6.96	4.90	6.43	8.60	9.22

Fig. 4 KEI, KI and the four pillar (UAE vs. Other countries) (Alfaki, 2014)

who the Head of Data Analytics has provided a thorough insight into the data analytics field at Smart Dubai, the interview questions were divided across three perspectives including knowledge-based economy, the use of the analytical tools, and Dubai pulse initiative.

The second method was to gather the secondary information of this report, though finding relevant publications, journals, and books from reliable sources like the University library, electronic databases like SSCI or EBSCO, recognized periodical journals and the organization main website.

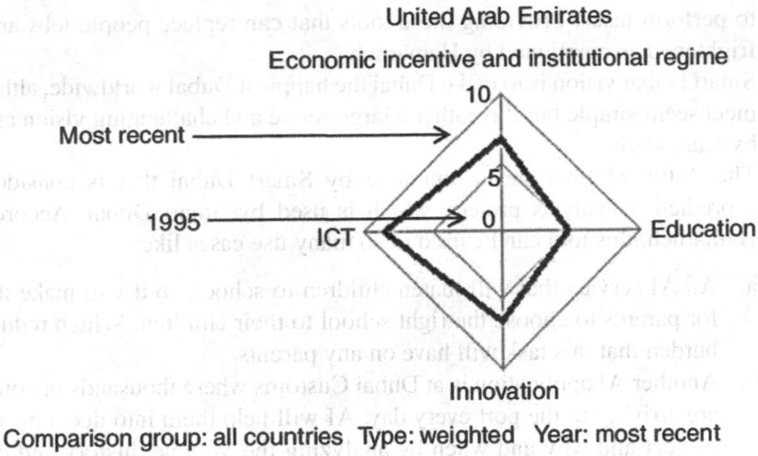

Fig. 5 KEI of the four pillars (UAE) (Alfaki, 2014)

5 Data Analysis

5.1 Smart Dubai Data Analysis

The following analysis points were concluded after a thorough analysis on the interview results and the related resources found:

1. Smart Dubai is supporting the knowledge-based economy through empowering the government entities with the technological tools and the reliable infrastructure of data sharing, storage, processing, and visualization. Which enables the government entities to make better and knowledge-based decisions in order to server the society better.
2. Smart Dubai is focused on building a collaboration between the public and the private sector to achieve several targets like smart transportation, smart society, smart economy, smart governance, and smart environment.
3. Smart Dubai has released the data law in 2015 "Law No. (26) Of 2015 on the Organization of Dubai Data Publication and Sharing" (Dubai Data, 2022), which makes Dubai the first city globally to release a Law to state the guidelines of the sharing of data between government entities and the open data available for the public.
4. The Dubai Data initiative will benefit the city by "increasing efficiencies in spending, job creation and innovation investment enhancement through creating new products and services and raise the level of citizen engagement. These benefits promotes government transparency and the ultimate happiness of Dubai residents and citizens" (DubaiPulse, 2018).
5. The main challenge that Smart Dubai is facing with building a knowledge-based economy, is the culture, people relay on printed papers and the usual routine

to perform tasks. Providing these tools that can replace people jobs are often frightened as mentioned by Hamarneh.

6. Smart Dubai vision is to make Dubai the happiest Dubai worldwide, although it meet seem simple but it is rather a large scope and challenging vision as stated by Hamarneh.
7. The Artificial Intelligence initiative by Smart Dubai that is considered as a predictive analysis project, which is used by Smart Dubai. According to Hamarneh, this tool can be used in so many use cases like:

   a. An AI service that will match children to school, so it will make it easier for parents to choose the right school to their children. Which reduces the burden that this task will have on any parents
   b. Another AI application is at Dubai Customs where thousands of containers are arriving to the port every day. AI will help them into deciding what to inspect and why and when by analyzing the Volume, history and country the container came from.

8. The AI can be of a great use in reducing or automating the repetitive tasks, like validating customer papers or inserting of forms information or detecting certain trends that will enable the government entity to perform better (Alzoubi et al., 2022; Hamadneh et al., 2021a, 2021b; Hanaysha et al., 2021a, 2021b; Lee et al., 2022a, 2022b).
9. Smart Dubai does not apply the prescriptive analysis tools, although that Hamarneh has mentioned that applying these tools can be of a great benefit to the government into making better decisions that will lower the cost, expect the demand through supply and demand mathematical modules (AlHamad et al., 2022; Ali et al., 2022; Alshurideh et al., 2022; Lee et al., 2022a, 2022b).
10. Dubai pulse is an initiative provided to all the stakeholders to make different analytical tools by the different data sets available for the use of the public that are regulated through Dubai data law. Many government entities are participating in this platform like Dubai Customs, Dubai Municipality, Dubai Silicon oasis Authority, Department of Dubai Economic Development, RTA, Dubai Health Authority and many other departments.
11. Dubai Pulse provides different data views like city flow, property land distribution, people flow, and property buildings distribution. These data views are available to provide the stakeholders with insights through interactive displays to enhance the stakeholder's decision-making process and create business value. Therefore, building a better and a knowledge-based economy (DubaiPulse, 2022).

5.2 DEWA Data Analysis

The following analysis points were concluded after a thorough analysis on the interview results and the related resources found:

1. DEWA is focused in making Dubai the smartest city on earth, and that can be accomplished through using the latest technologies and innovations to increase the happiness metric of Dubai citizens.
2. DEWA has adopted many initiatives to increase the reliance on Data in the decision-making process inclusive of the descriptive, predictive and prescriptive tools.
3. DEWA has adopted General Electric's industry wide Internet of Things platform, this platform enables the connection between machines, data, and people.
4. This platform will enable data to be collected and analyzed in real time to enhance the speed and efficiency of its operations.
5. This platform is designed to embed certain applications like built in Artificial intelligence to predict the assets lifecycle and prevent future malfunctions by using a continuous asset monitoring service to pre-identify and analyze root causes (AlShamsi et al., 2021; Holland, 2018; Nuseir et al., 2021a, 2021b; Yousuf et al., 2021).
6. DEWA currently uses dashboards to support the decision making process by the use of descriptive analytics.
7. The initiation of the GE platform will provide DEWA with many abilities to enhance their utilization of equipment's, better services to customers, attracting more beneficial opportunities and lowering cost.

6 Discussions

6.1 Comparison of Dubai Experience with the Analytical Tools' Literature

Smart Dubai is a unique example of high-level quality system that strive towards making Dubai the happiest city on earth. Smart Dubai has been adopting many projects that enhances the life of people and support the economy of the UAE. One of the main projects that Smart Dubai has initiated is the Dubai pulse project which is a platform to gather different data and visualize it in data views (Alsharari, 2021; Aziz & Aftab, 2021; Mehmood, 2021; Miller, 2021). Although that this project is roughly considered as a descriptive tool of data analytics rather than the prescriptive and predictive analytics. But it can assist in the decision-making process of the various stakeholders. Another project is the Artificial intelligence and machine learning project, which is a form of the predictive analytics tools, this project aims to eliminate any repetitive tasks and optimize the resources in order to provide better services to the citizens.

It is shown in the literature, that the use of the analytical tools can aid in the decision-making process on different managerial levels, hence establishing better services to the citizens and building a knowledge-based economy (Cruz, 2021; Khan, 2021; Lee & Ahmed, 2021; Mondol, 2021; Radwan & Farouk, 2021). Smart Dubai, in this matter lacks the proper use of the analytical tools where it can be considered still seen as at the earliest stages of building a data driven added value projects, Smart Dubai has successfully established a collaborative data platform but still needs to establish projects that will have the different forms of the analytical tools like regression analysis, predictive modeling, pattern matching, data mining and machine learning.

6.2 Opportunities for Adding Value

Predictive analytics can provide the following opportunities to add value:

- Predictions: this analytical feature can predict the probability of a specific outcome (Idoine & Brethenoux, 2018). For example, possible repetitive congestion in any of the country main roads by predicting the outcome the government can make a diversion in the roads or build a new bridge that will aid in eliminating this congestion and keep the smoothness of the transportation. Another example of the prediction feature of the analytical tool is the maintenance scheduling of the government buildings or any public areas, this tool can predict the maintenance schedule of the area instead of having a problem reported and then responding based on it.
- Forecasting: the predicting of a series of outcomes over time (Idoine & Brethenoux, 2018).This feature can forecast a certain event in the future. Like the expected tourists in a certain season.
- Simulation: predicting multiple outcomes and highlighting uncertainties (Idoine & Brethenoux, 2018).

Prescriptive analytics can provide the following opportunities to add value:

- Rules: predefined framework for choosing between alternatives (Idoine & Brethenoux, 2018), for example, calculating the best way in performing an action or building an optimization model to ensure the optimization of the resources across the government. Also, it can aid in reducing any unnecessary costs and risks.
- Optimization: outcome-driven, constraint-based evaluation of an interdependent set of options (Fig. 6).

Moreover, the tools can increase the speed of the planning analysis, decisions that took weeks to be decided can be done within hours. Improved executive insights, more projects and opportunities can be exploited by the use of the analytical tools therefore creating and adding more value to the economy and the citizens (Ali et al., 2021; Alzoubi, 2021b; Alzoubi et al., 2021b; Farouk, 2021; Guergov & Radwan, 2021; Obaid, 2021).

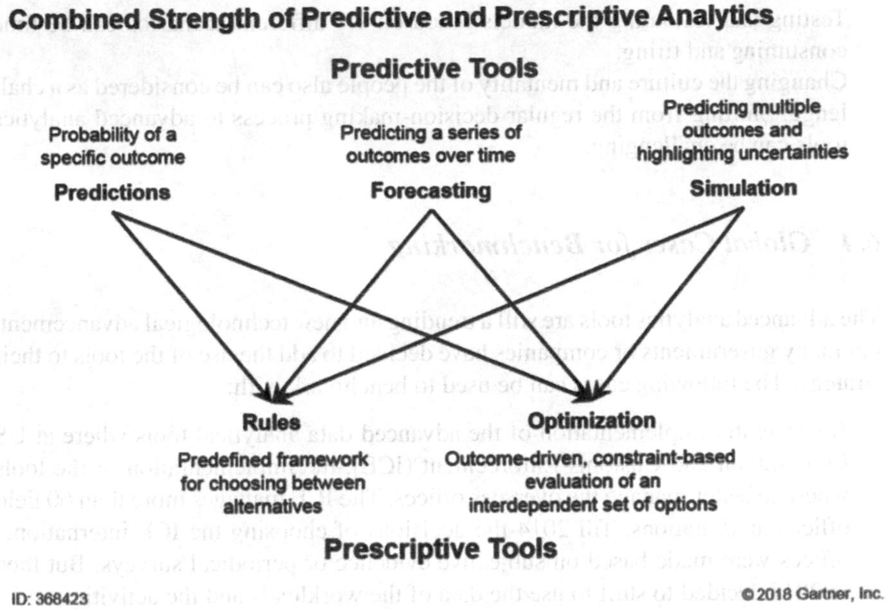

Fig. 6 Combine predictive and prescriptive analytics to exploit their respective strengths (Idoine & Brethenoux, 2018)

6.3 Challenges

The implementation of the tools in the government can encounter many challenges, the challenges can vary between infrastructure challenges to technological compatibility challenges. A few are mentioned below:

- Successful implementation of the tools comes from more than purchasing and deploying the software of the analytics, it also require the engagement of the right people, processes to leverage, implement, maintain, and govern the models and tools (Idoine & Brethenoux, 2018).
- Government data is scattered across many departments making the collection and the gathering process challenging to the smart government. Also, it can be challenging to decide on what is actually important from the massive amount of data.
- Government data can consist of unneeded and redundant data, therefore cleaning and reconstruction of data might be required.
- The data available can consist of different data formats or still manual paper information or even scanned documents with no stored fields making it hard to rely on it for analytical tools.
- The models that are generated from the analytical tools must be embedded in the operations and in the decision-making process to obtain the most use out of it.

- Testing and validating the models based on the different scenarios can be time consuming and tiring.
- Changing the culture and mentality of the people also can be considered as a challenge. Shifting from the regular decision-making process to advanced analytics tools can be challenging.

6.4 Global Cases for Benchmarking

The advanced analytics tools are still a trending and new technological advancement, but many governments or companies have decided to add the use of the tools to their strategy. The following cases can be used to benchmark with:

- A successful implementation of the advanced data analytical tools where at US Immigration and Customs Enforcement (ICE), the implementation of the tools where to better manage the overseas offices. The ICE manages more than 60 field offices in 45 nations. Till 2014 the decisions of choosing the ICE international offices were made based on subjective evidence or periodical surveys. But then in 2015 decided to shift to use the data of the workloads and the activities of the offices to decide on where to expand or where to shut offices without affecting the mission performance (Mahesh et al., 2016).
- Another case to benchmark with Virginia state case in building an integrated system of benefits eligibility of medical assistance. The offices that process those applications are often understaffed and overloaded with work (Alhamad et al., 2021; Alnuaimi et al., 2021; Alzoubi & Aziz, 2021; Ghazal et al., 2021; Hanaysha et al., 2021b). So, in 2015, Virginia Department of Social Services decided to take a different approach in addressing this problem. The information where gathered and analyzed by the use of the analytical tools to identify issues in processing the applications therefore enabling the DSS to provide better social services to the citizens (Mahesh et al., 2016).

7 Recommendations

This research paper aims to come out with new results in the field of smart cities that would help and support other cities to become smart and discuss the conflict, challenges and opportunities of knowledge-based economy and add value to their economies. Furthermore, this study provides a clear view of smart cities transformation and knowledge-based economy and the use of perspective and predictive analysis tools that help other researchers in the future to improve and develop studies regarding smart cities and get new results. The following are some points for smart Dubai to maintain the high-quality performance:

1. The result shows it is important to establish an information system and a collaborative platform of data for digital transformation of smart city.

.

2. Increase the benefits of technology in order to improve the economy.
3. All stakeholders should be engaged in the transformation to smart city and the build of the analytical tools' platform.
4. Create a roadmap to have a successful use of the analytical tools by monitoring, experimenting and implementation of these tools.
5. Investment in education, technology and communication in order to achieve competitive advantage against different economies (local and global).
6. Training program for the employees to develop, improve and enhance their skills and experience.
7. Engage and train the government sector leaders about the new data analytics and how it can benefit the economy.
8. Identify the gaps that are preventing the government from using the full capabilities of the use of this disruptive innovation tools.
9. Building a policy and a foundation of security, accountability that keeps privacy and promotes digital ethics.
10. Identification of the most important data relevant to the decision-making process and eliminating of redundant and unneeded data.

8 Conclusion

Due to the strong competition between cities and economies worldwide, Dubai became one of the most successful economies in the current century. By the vision of his highness Sheikh Mohammed Bin Rashid Al Maktoum vice president and prime minister of UAE, and ruler of Dubai, he builds up Dubai economy from oil economy to a non-oil economy and changed the economy to a diversified economy with various resources. There is numerous businesses in Dubai that made the economy stronger and more flexible, the strategies of his highness attracted most of the big companies and organizations worldwide to invest in Dubai and see Dubai as a very important business city in MENA region.

Dubai used the predictive analysis tools to know on what to depend in the future the use of that tools changed Dubai economy from oil to non-oil economy because they knew that in the future the world will not depend on the oil anymore, new resources are available now regardless oil and gas. The prescriptive analysis tools to know where Dubai economy want to go and what to do in order to increase and maintain its high performances and quality that attract business owners across the world.

Appendix

Smart Dubai—Interview Questions

As part of the information gathering of the Information Technology Management Course project, the following interview questions were prepared to be answered by an employee at Smart Dubai.

The questions were divided into sections as shown below.

Interview was conducted on 08/05/2019 at 10:45 AM with **Hamzeh Hamaraneh** who is the head of DATA Analytics in Smart Dubai.

Section 1: UAE Based Economy and Smart Dubai Contribution to it

1. **How can Smart Dubai empower or support UAE economy?**

   - We are empowering the government by using the tools and technology to represent the data and information.
   - By providing human resources to use the data in different uses of decision-making and across the city, while having the legislative and the authority to access such an information.
   - The Dubai data law that was released 2015. To be the first city to release such law, to regulate how the data used, stored, processed, and presented.
   - AI is filled by DATA. Not only we issued the law in 2015 but also we have the ethics of AI.

2. **What is the influence of smart Dubai on people of Dubai (residents, visitors, business owners, parents and families)?**

Smart Dubai has direct and indirect influence, by supporting the government and the non-government entities through our services so we can enable them to serve the society better like Dubai police, RTA, etc.

Another sector, government to citizens directly. Through channels like Dubai now through all the segregated government entities and bodies.

Traditionally, people will go through the normal norm of business. Now through the eservices we are providing better services by using DUBAI now by having a journey rather than the service itself. Customer focused.

3. **What challenges does Smart Dubai faces in the local and global market?**

   1. The most challenging area, or critical is the culture, people will feel that the paper is more reliable than the digital products. So changing the culture into knowledge society and paperless society is the challenge.
   2. Making people relay on the digital projects, technology advancements is moving into the society much higher than the culture will embrace it.

4. **What is Smart Dubai trying to achieve in the future?**

The goal is clear and very simple, to making Dubai the happiest city in the world; it is very simple but very big and challenging.

5. What new services, innovations and applications smart Dubai may provide to customer and Dubai society?

We have a lot of products and services, I think the main services are Dubai now, AI, and block chain.

Example of AI use case

An AI service that will match children to school, so we will make it easier for parents to choose the right school to their children. To reduce the burden that this task will have on any parents.

Use cases of AI

Dubai customs, thousands of containers are arriving to the port every day. AI will help them into deciding What to inspect and why and when. By analyzing the Volume, history and country the container came from.

In a nutshell, we have a lot of repetitive tasks and we tend to do it repetitively which can be resolved by the use of AI so we can focus on creative tasks.

To define any repetitive tasks to be automated.

Section 2: Analytics Tools Currently used in Smart Dubai

Does smart Dubai uses Predictive analysis, like machine learning, AI tools, and simulations.

Yes, they do.

Can you name few initiatives?

Many tools are used, our biggest initiative is the paperless initiative, by 2021 Dubai government will be the first government of paperless. AI technology can be used by eliminating any repetitive tasks. To call for the same documents repeatedly.

Do you think that those tools add value to the smart government and the knowledge-based economy?

Yes it does.

Does smart Dubai uses Prescriptive Analytics, like math programming models, optimization models, heuristics.

Hint: Those kind of tools often minimize the risks in making certain decisions related to costs and often important to use when there is time sensitive decisions to be made.

Exploring those tools, currently they are not using them.

Does smart Dubai uses those tools, and do you think they are adding any benefits to the government?

Of course it will add the benefits, like lowering costs, having those tools will benefit the government to make better decisions, like the supply and demand modules.

Can you give me a brief about Dubai Pulse?

Dubai pulse is a market place, we have tools and technologies and we are acquiring data sets and providing to the customers to make any analytical tools. So this service is provided for the public.

In addition, do you think it is related to the tools mentioned previously?

Yes it is, it is a representation of DATA to allow the different stakeholders to make better decisions.

References

Abid, K., Bari, Y. A., Younas, M., Tahir Javaid, S., & Imran, A. (2020). Progress of COVID-19 epidemic in Pakistan. *Asia-Pacific Journal of Public Health, 32*(4), 154–156. https://doi.org/10.1177/1010539520927259

Aburayya, A., Alshurideh, M., Al Marzouqi, A., Al Diabat, O., Alfarsi, A., Suson, R., Salloum, S. A., Alawadhi, D., & Alzarouni, A. (2020a). Critical success factors affecting the implementation of tqm in public hospitals: A case study in UAE Hospitals. *Systematic Reviews in Pharmacy, 11*(10). 10. 31838/srp.2020a.10.39.

Aburayya, A., Alshurideh, M., Alawadhi, D., Alfarsi, A., Taryam, M., & Mubarak, S. (2020b). An Investigation of the effect of lean six sigma practices on healthcare service quality and patient satisfaction: Testing the mediating role of service quality in Dubai primary healthcare sector. *Journal of Advanced Research in Dynamical and Control Systems, 12*(8), 56–72.

Ahmed, A., & Alfaki, I. M. A. (2013). Transforming the United Arab Emirates into a knowledge-based economy: The role of science, technology and innovation. *World Journal of Science, Technology and Sustainable Development*.

Akhtar, A., Akhtar, S., Bakhtawar, B., Kashif, A. A., Aziz, N., & Javeid, M. S. (2021). COVID-19 detection from CBC using machine learning techniques. *International Journal of Technology, Innovation and Management (IJTIM), 1*(2), 65–78. https://doi.org/10.54489/ijtim.v1i2.22.

Al-Jarrah, I. M., Al-Zu'bi, Z. M. F., Jaara, O. O., & Alshurideh, M. (2012). Evaluating the impact of financial development on economic growth in Jordan. *International Research Journal of Finance and Economics, 94*, 123–139.

Al-Maroof, R., Ayoubi, K., Alhumaid, K., Aburayya, A., Alshurideh, M., Alfaisal, R., & Salloum, S. (2021). The acceptance of social media video for knowledge acquisition, sharing and application: A comparative study among YouYube users and TikTok users' for medical purposes. *International Journal of Data and Network Science, 5*(3). https://doi.org/10.5267/j.ijdns.2021.6.013.

Al Ali, A. (2021). The impact of information sharing and quality assurance on customer service at UAE banking sector. *International Journal of Technology, Innovation and Management (IJTIM), 1*(1), 1–17. https://doi.org/10.54489/ijtim.v1i1.10.

Al Batayneh, R. M., Taleb, N., Said, R. A., Alshurideh, M. T., Ghazal, T. M., & Alzoubi, H. M. (2021). IT governance framework and smart services integration for future development of Dubai infrastructure utilizing AI and big data, Its reflection on the citizens standard of living. In *The International Conference on Artificial Intelligence and Computer Vision* (pp. 235–247).

Al Shebli, K., Said, R. A., Taleb, N., Ghazal, T. M., Alshurideh, M. T., & Alzoubi, H. M. (2021). RTA's employees' perceptions toward the efficiency of artificial intelligence and big data utilization in providing smart services to the residents of Dubai. In *The International Conference on Artificial Intelligence and Computer Vision* (pp. 573–585).

Alaali, N., Al Marzouqi, A., Albaqaeen, A., Dahabreh, F., Alshurideh, M., Mouzaek, E., Alrwashdh, S., Iyadeh, I., Salloum, S., & Aburayya, A. (2021). The impact of adopting corporate governance strategic performance in the tourism sector: A case study in the Kingdom of Bahrain. *Journal of Legal, Ethical and Regulatory Issues, 24*(Special Issue 1).

Alfaki, I. M. A. (2014). Evaluating UAE success in utilizing knowledge and technological infrastructure. *International Journal of Innovation and Knowledge Management in the Middle East and North Africa, 3*(1), 33.

AlHamad, A., Alshurideh, M., Alomari, K., Kurdi, B., Alzoubi, H., Hamouche, S., & Al-Hawary, S. (2022). The effect of electronic human resources management on organizational health of telecommunications companies in Jordan. *International Journal of Data and Network Science, 6*(2), 429–438.

Alhamad, A. Q. M., Akour, I., Alshurideh, M., Al-Hamad, A. Q., Kurdi, B. A., & Alzoubi, H. (2021). Predicting the intention to use google glass: A comparative approach using machine learning models and PLS-SEM. *International Journal of Data and Network Science, 5*(3). https://doi.org/10.5267/j.ijdns.2021.6.002.

Ali, N., Ahmed, A., Anum, L., Ghazal, T. M., Abbas, S., Khan, M. A., Alzoubi, H. M., & Ahmad, M. (2021). Modelling supply chain information collaboration empowered with machine learning technique. *Intelligent Automation and Soft Computing, 30*(1), 243–257. https://doi.org/10.32604/iasc.2021.018983.

Ali, N., M. Ghazal, T., Ahmed, A., Abbas, S., A. Khan, M., Alzoubi, H., Farooq, U., Ahmad, M., & Adnan Khan, M. (2022). Fusion-based supply chain collaboration using machine learning techniques. *Intelligent Automation & Soft Computing, 31*(3), 1671–1687. https://doi.org/10.32604/iasc.2022.019892.

Alkalha, Z., Al-Zu'bi, Z., Al-Dmour, H., Alshurideh, M., & Masa'deh, R. (2012). Investigating the effects of human resource policies on organizational performance: An empirical study on commercial banks operating in Jordan. *European Journal of Economics, Finance and Administrative Sciences, 51*(1), 44–64.

AlMehrzi, A., Alshurideh, M., & Al Kurdi, B. (2020). Investigation of the key internal factors influencing knowledge management, employment, and organisational performance: A qualitative study of the UAE hospitality sector. *The International Journal of Innovation, Creativity and Change, 14*(1), 1369–1394.

Alnazer, N. N., Alnuaimi, M. A., & Alzoubi, H. M. (2017). Analysing the appropriate cognitive styles and its effect on strategic innovation in Jordanian universities. *International Journal of Business Excellence, 13*(1), 127–140. https://doi.org/10.1504/IJBEX.2017.085799

Alnuaimi, M., Alzoubi, H. M., Ajelat, D., & Alzoubi, A. A. (2021). Towards intelligent organisations: An empirical investigation of learning orientation's role in technical innovation. *International Journal of Innovation and Learning, 29*(2), 207–221. https://doi.org/10.1504/IJIL.2021.112996

AlShamsi, M., Salloum, S. A., Alshurideh, M., & Abdallah, S. (2021). Artificial intelligence and blockchain for transparency in governance. In *Studies in computational intelligence* (Vol. 912). https://doi.org/10.1007/978-3-030-51920-9_11.

Alsharari, N. (2021). Integrating blockchain technology with internet of things to efficiency. *International Journal of Technology, Innovation and Management (IJTIM), 1*(2), 1–13.

Alshurideh, M. (2022). Does electronic customer relationship management (E-CRM) affect service quality at private hospitals in Jordan? *Uncertain Supply Chain Management, 10*(2), 1–8.

Alshurideh, M., Al Kurdi, B., Shaltoni, A. M., & Ghuff, S. S. (2019a). Determinants of pro-environmental behaviour in the context of emerging economies. *International Journal of Sustainable Society, 11*(4). https://doi.org/10.1504/IJSSOC.2019.104563.

Alshurideh, M., Salloum, S. A., Al Kurdi, B., Monem, A. A., & Shaalan, K. (2019b). Understanding the quality determinants that influence the intention to use the mobile learning platforms: A practical study. *International Journal of Interactive Mobile Technologies, 13*(11). https://doi.org/10.3991/ijim.v13i11.10300.

Alshurideh, M., Al Kurdi, B., Salloum, S. A., Arpaci, I., & Al-Emran, M. (2020a). Predicting the actual use of m-learning systems: A comparative approach using PLS-SEM and machine learning algorithms. *Interactive Learning Environments*. https://doi.org/10.1080/10494820.2020.1826982

Alshurideh, M., Gasaymeh, A., Ahmed, G., Alzoubi, H., & Kurd, B. A. (2020b). Loyalty program effectiveness: Theoretical reviews and practical proofs. *Uncertain Supply Chain Management*, *8*(3). https://doi.org/10.5267/j.uscm.2020.2.003.

Alshurideh, M. T., Al Kurdi, B., Alzoubi, H. M., Ghazal, T. M., Said, R. A., AlHamad, A. Q., Hamadneh, S., Sahawneh, N., & Al-kassem, A. H. (2022). Fuzzy assisted human resource management for supply chain management issues. *Annals of Operations Research*, 1–19.

AlSuwaidi, S. R., Alshurideh, M., Al Kurdi, B., & Aburayya, A. (2021). The main catalysts for collaborative R&D projects in Dubai industrial sector. In *The International Conference on Artificial Intelligence and Computer Vision* (pp. 795–806).

Alzoubi, A. (2021a). The impact of process quality and quality control on organizational competitiveness at 5-star hotels in Dubai. *International Journal of Technology, Innovation and Management (IJTIM)*, *1*(1), 54–68. https://doi.org/10.54489/ijtim.v1i1.14.

Alzoubi, A. (2021b). Renewable Green hydrogen energy impact on sustainability performance. *International Journal of Computations, Information and Manufacturing (IJCIM)*, *1*(1), 94–110. https://doi.org/10.54489/ijcim.v1i1.46.

Alzoubi, H., & Ahmed, G. (2019). Do TQM practices improve organisational success? A case study of electronics industry in the UAE. *International Journal of Economics and Business Research*, *17*(4), 459–472. https://doi.org/10.1504/IJEBR.2019.099975

Alzoubi, H. M., & Aziz, R. (2021). Does emotional intelligence contribute to quality of strategic decisions? The mediating role of open innovation. *Journal of Open Innovation: Technology, Market, and Complexity*, *7*(2), 130. https://doi.org/10.3390/joitmc7020130

Alzoubi, H. M., & Yanamandra, R. (2020). Investigating the mediating role of information sharing strategy on agile supply chain. *Uncertain Supply Chain Management*, *8*(2), 273–284. https://doi.org/10.5267/j.uscm.2019.12.004

Alzoubi, H. M., Alshurideh, M., & Ghazal, T. M. (2021a). Integrating BLE beacon technology with intelligent information systems IIS for operations' performance: A managerial perspective. In *The International Conference on Artificial Intelligence and Computer Vision* (pp. 527–538).

Alzoubi, H. M., Vij, M., Vij, A., & Hanaysha, J. R. (2021b). What leads guests to satisfaction and loyalty in UAE five-star hotels? AHP analysis to service quality dimensions. *Enlightening Tourism*, *11*(1), 102–135. https://doi.org/10.33776/et.v11i1.5056.

Alzoubi, H., Ahmed, G., Al-Gasaymeh, A., & Kurdi, B. (2020a). Empirical study on sustainable supply chain strategies and its impact on competitive priorities: The mediating role of supply chain collaboration. *Management Science Letters*, *10*(3), 703–708.

Alzoubi, H., Alshurideh, M., Kurdi, B., Akour, I., & Aziz, R. (2022). Does BLE technology contribute towards improving marketing strategies, customers' satisfaction and loyalty? The role of open innovation. *International Journal of Data and Network Science*, *6*(2), 449–460.

Alzoubi, H., Alshurideh, M., Kurdi, B. A., & Inairat, M. (2020b). Do perceived service value, quality, price fairness and service recovery shape customer satisfaction and delight? A practical study in the service telecommunication context. *Uncertain Supply Chain Management*, *8*(3), 579–588. https://doi.org/10.5267/j.uscm.2020.2.005

Ashal, N., Alshurideh, M., Obeidat, B., & Masa'deh, R. (2021). The impact of strategic orientation on organizational performance: Examining the mediating role of learning culture in Jordanian telecommunication companies. *Academy of Strategic Management Journal*, (Special Issue 6), 1–29.

Aswad, N. G., Vidican, G., & Samulewicz, D. (2011). Creating a knowledge-based economy in the United Arab Emirates: Realising the unfulfilled potential of women in the science, technology and engineering fields. *European Journal of Engineering Education*, *36*(6), 559–570.

Awadhi, J., Obeidat, B., & Alshurideh, M. (2021). The impact of customer service digitalization on customer satisfaction: Evidence from telecommunication industry. *International Journal of Data and Network Science*, *5*(4), 815–830.

Aziz, N., & Aftab, S. (2021). Data mining framework for nutrition ranking: Methodology: SPSS modeller. *International Journal of Technology, Innovation and Management (IJTIM), 1*(1), 85–95.
Cruz, A. (2021). Convergence between blockchain and the Internet of Things. *International Journal of Technology, Innovation and Management (IJTIM), 1*(1), 35–56.
Dahlman, C. J., & Utz, A. (2005). *India and the knowledge economy: leveraging strengths and opportunities*. World Bank Publications.
Digital Dubai. (2022a). *Artificial intelligence principles & ethics*.
Digital Dubai. (2022b). *Ditial Dubai—About us*.
Digital Dubai. (2022c). *Dubai blockchain strategy*.
Digital Dubai. (2022d). *Dubai paperless strategy*.
Dubai Data. (2022). *Dubai data regulations*.
DubaiPulse. (2018). *Dubai data initiative-governance, policy and engagement*.
DubaiPulse. (2022). *About DubaiPulse*.
Eli, T. (2021). Students perspectives on the use of innovative and interactive teaching methods at the University of Nouakchott Al Aasriya, Mauritania: English Department as a Case Study. *International Journal of Technology, Innovation and Management (IJTIM), 1*(2), 90–104.
Farouk, M. (2021). The universal artificial intelligence efforts to face coronavirus COVID-19. *International Journal of Computations, Information and Manufacturing (IJCIM), 1*(1), 77–93. https://doi.org/10.54489/ijcim.v1i1.47.
Gartner. (2022). *Predictive analytics*.
Ghazal, T. M., Hasan, M. K., Alshurideh, M. T., Alzoubi, H. M., Ahmad, M., Akbar, S. S., Al Kurdi, B., & Akour, I. A. (2021). IoT for smart cities: Machine learning approaches in smart healthcare—A review. *Future Internet, 13*(8), 218. https://doi.org/10.3390/fi13080218
Government of Dubai. (2022). *DEWA the brand*.
Guergov, S., & Radwan, N. (2021). Blockchain convergence: Analysis of issues affecting IoT, AI and blockchain. *International Journal of Computations, Information and Manufacturing (IJCIM), 1*(1), 1–17. https://doi.org/10.54489/ijcim.v1i1.48.
Hamadneh, S., Keskin, E., Alshurideh, M., Al-Masri, Y., & Al Kurdi, B. (2021a). The benefits and challenges of RFID technology implementation in supply chain: A case study from the Turkish construction sector. *Uncertain Supply Chain Management, 9*(4), 1071–1080.
Hamadneh, Samer, Pedersen, O., & Al Kurdi, B. (2021b). An investigation of the role of supply chain visibility into the Scottish bood supply chain. *Journal of Legal, Ethical and Regulatory Issues, 24*(Special Issue 1), 1–12.
Hanaysha, J. R., Al-Shaikh, M. E., Joghee, S., & Alzoubi, H. (2021a). Impact of innovation capabilities on business sustainability in small and medium enterprises. *FIIB Business Review*, 1–12. https://doi.org/10.1177/23197145211042232.
Hanaysha, J. R., Al Shaikh, M. E., & Alzoubi, H. M. (2021b). Importance of marketing mix elements in determining consumer purchase decision in the retail market. *International Journal of Service Science, Management, Engineering, and Technology (IJSSMET), 12*(6), 56–72.
Hasan, O., McColl, J., Pfefferkorn, T., Hamadneh, S., Alshurideh, M., & Kurdi, B. (2022). Consumer attitudes towards the use of autonomous vehicles: Evidence from United Kingdom taxi services. *International Journal of Data and Network Science, 6*(2), 537–550.
Herschel, G., & Idoine, C. (2016). *Combine predictive and prescriptive analytics to drive high-impact decisions*.
Holland, G. (2018). *DEWA to integrate GE's IoT platform into Dubai power plants*.
Idoine, C., & Brethenoux, E. (2018). *Combine predictive and prescriptive techniques to solve business problems*. https://www.gartner.com/en/documents/3891993.
Joghee, S., Alzoubi, H. M., & Dubey, A. R. (2020). Decisions effectiveness of FDI investment biases at real estate industry: Empirical evidence from Dubai smart city projects. *International Journal of Scientific and Technology Research, 9*(3), 3499–3503.
Kart, L. (2015). *How to get started with prescriptive analytics*.

Kashif, A. A., Bakhtawar, B., Akhtar, A., Akhtar, S., Aziz, N., & Javeid, M. S. (2021). Treatment response prediction in hepatitis c patients using machine learning techniques. *International Journal of Technology, Innovation and Management (IJTIM), 1*(2), 79–89. https://doi.org/10.54489/ijtim.v1i2.24.
Khan, M. A. (2021). Challenges facing the application of iot in medicine and healthcare. *International Journal of Computations, Information and Manufacturing (IJCIM), 1*(1), 39–55. https://doi.org/10.54489/ijcim.v1i1.32.
Lee, C., & Ahmed, G. (2021). Improving IoT privacy, data protection and security concerns. *International Journal of Technology, Innovation and Management (IJTIM), 1*(1), 18–33. https://doi.org/10.54489/ijtim.v1i1.12.
Lee, K., Azmi, N., Hanaysha, J., Alzoubi, H., & Alshurideh, M. (2022a). The effect of digital supply chain on organizational performance: An empirical study in Malaysia manufacturing industry. *Uncertain Supply Chain Management, 10*(2), 495–510.
Lee, K., Romzi, P., Hanaysha, J., Alzoubi, H., & Alshurideh, M. (2022b). Investigating the impact of benefits and challenges of IOT adoption on supply chain performance and organizational performance: An empirical study in Malaysia. *Uncertain Supply Chain Management, 10*(2), 537–550.
Leo, S., Alsharari, N. M., Abbas, J., & Alshurideh, M. T. (2021). From offline to online learning: A qualitative study of challenges and opportunities as a response to the COVID-19 pandemic in the UAE higher education context. In *Studies in systems, decision and control* (Vol. 334). https://doi.org/10.1007/978-3-030-67151-8_12.
Mahesh, K., Peter, V., Sean, C., Rachel, F., & Frank, S. (2016). *Mission analytics: Data-driven decision making in government.*
Mehmood, T. (2021). Does information technology competencies and fleet management practices lead to effective service delivery? Empirical evidence from e-commerce industry. *International Journal of Technology, Innovation and Management (IJTIM), 1*(2), 14–41.
Mehmood, T., Alzoubi, H. M., & Ahmed, G. (2019). Schumpeterian entrepreneurship theory: Evolution and relevance. *Academy of Entrepreneurship Journal, 25*(4).
Mehrez, A. A. A., Alshurideh, M., Kurdi, B. A., & Salloum, S. A. (2021). Internal factors affect knowledge management and firm performance: A systematic review. In *Advances in intelligent systems and computing, AISC* (Vol. 1261).. https://doi.org/10.1007/978-3-030-58669-0_57.
Miller, D. (2021). The best practice of teach computer science students to use paper prototyping. *International Journal of Technology, Innovation and Management (IJTIM), 1*(2), 42–63. https://doi.org/10.54489/ijtim.v1i2.17.
Mina, W. (2014). United Arab Emirates FDI outlook. *The World Economy, 37*(12), 1716–1730.
Mondol, E. P. (2021). The impact of block chain and smart inventory system on supply chain performance at retail industry. *International Journal of Computations, Information and Manufacturing (IJCIM), 1*(1), 56–76. https://doi.org/10.54489/ijcim.v1i1.30.
Nuseir, M.T., Aljumah, A., & Alshurideh, M. T. (2021a). How the business intelligence in the new startup performance in UAE during COVID-19: The mediating role of innovativeness. In *Studies in systems, decision and control* (Vol. 334). https://doi.org/10.1007/978-3-030-67151-8_4.
Nuseir, M. T, Al Kurdi, B. H., Alshurideh, M. T., & Alzoubi, H. M. (2021b). Gender discrimination at workplace: Do artificial intelligence (AI) and machine learning (ML) have opinions about it. In *The International Conference on Artificial Intelligence and Computer Vision* (pp. 301–316).
Obaid, A. J. (2021). Assessment of smart home assistants as an IoT. *International Journal of Computations, Information and Manufacturing (IJCIM), 1*(1), 18–36. https://doi.org/10.54489/ijcim.v1i1.34.
Odeh, R., Obeidat, B. Y., Jaradat, M. O., Masa'deh, R., & Alshurideh, M. T. (2021). The transformational leadership role in achieving organizational resilience through adaptive cultures: the case of Dubai service sector. *International Journal of Productivity and Performance Management.* https://doi.org/10.1108/IJPPM-02-2021-0093.
Parcero, O. J., & Ryan, J. C. (2017). Becoming a knowledge economy: The case of Qatar, UAE, and 17 benchmark countries. *Journal of the Knowledge Economy, 8*(4), 1146–1173.

Radwan, N., & Farouk, M. (2021). The growth of Internet of Things (IoT) in the management of healthcare issues and healthcare policy development. *International Journal of Technology, Innovation and Management (IJTIM), 1*(1), 69–84. https://doi.org/10.54489/ijtim.v1i1.8.

Shannak, R. O., Masa'deh, R. M. T., Al-Zu'bi, Z. M. F., Obeidat, B. Y., Alshurideh, M., & Altamony, H. (2012). A theoretical perspective on the relationship between knowledge management systems, customer knowledge management, and firm competitive advantage. *European Journal of Social Sciences, 32*(4).

Svoboda, P., Ghazal, T. M., Afifi, M. A. M., Kalra, D., Alshurideh, M. T., & Alzoubi, H. M. (2021). Information systems integration to enhance operational customer relationship management in the pharmaceutical industry. In *The International Conference on Artificial Intelligence and Computer Vision* (pp. 553–572).

Taryam, M., Alawadhi, D., Aburayya, A., Albaqa'een, A., Alfarsi, A., Makki, I., Rahmani, N., Alshurideh, M., & Salloum, S. A. (2020). Effectiveness of not quarantining passengers after having a negative COVID-19 PCR test at arrival to dubai airports. *Systematic Reviews in Pharmacy, 11*(11). https://doi.org/10.31838/srp.2020.11.197.

Yousuf, H., Zainal, A. Y., Alshurideh, M., & Salloum, S. A. (2021). Artificial intelligence models in power system analysis. In *Artificial intelligence for sustainable development: Theory, practice and future applications* (pp. 231–242). Springer.

9783031123818VOL02